D0203167

Automatic Control Systems

SEVENTH EDITION

Benjamin C. Kuo

Professor Emeritus

Department of Electrical and Computer Engineering
University of Illinois at Urbana-Champaign

JOHN WILEY & SONS, INC.

NEW YORK • CHICHESTER • WEINHEIM • BRISBANE • SINGAPORE • TORONTO

Cover photo: Courtesy of NASA.

This book was typeset and printed and bound by R. R. Donnelley. The cover was printed by Lehigh Perss.

Copyright © 1995 by John Wiley & Sons, Inc. All rights reserved.

No part of this book may be reproduced, stored in a retrieval system or transmitted in any form or by any means, electronic, mechanical, photocopying recording, scanning or otherwise, except as permitted under Sections 107 or 108 of the 1976 United States Copyright Act, without either the prior written permission of the Publisher or authorization through payment of the appropriate per-copy fee to the Copyright Clearance Center, 222 Rosewood Drive, Danvers, MA 01923, (508) 750-8400, fax (508) 750-4470. Requests to the Publisher for permission should be addressed to the Permissions Department, John Wiley & Sons, Inc. 605 Third Avenue, New York, NY 10158-0012, (212) 850-6008, E-mail: PERMREQ@WILEY.COM. To order books or for customer service call 1-800-CALL-WILEY (225-5945).

Library of Congress Cataloging-in-Publication Data

Kuo, Benjamin C.
Automatic control systems/ Benjamin C. Kuo. — 7th ed.
p. cm.
Includes bibliographical references and index.
ISBN 0-471-36608-0
1. Automatic control. I. Title.
TJ213.K8354 1995 94-6330
629.8—dc20 CIP

Printed in the United States of America.

10 9 8 7

To the government that needs control and takes most of the royalties from this book.

Laplace Transform Table (continued)

Laplace Transform $F(s)$	Time Function $f(t)$
$\frac{s}{(s+\alpha)^2}$	$(1-\alpha t)e^{-\alpha t}$
$\frac{\omega_n}{s^2+\omega_n^2}$	$\sin \omega_n t$
$\frac{s}{s^2+\omega_n^2}$	$\cos \omega_n t$
$\frac{\omega_n^2}{s(s^2+\omega_n^2)}$	$1-\cos \omega_n t$
$\frac{\omega_n^2(s+\alpha)}{s^2+\omega_n^2}$	$\omega_n \sqrt{\alpha^2+\omega_n^2} \sin(\omega_n t+\theta)$ where $\theta = \tan^{-1}(\omega_n/\alpha)$
$\frac{\omega_n}{(s+\alpha)(s^2+\omega_n^2)}$	$\frac{\omega_n}{\alpha^2+\omega_n^2} e^{-\alpha t} + \frac{1}{\sqrt{\alpha^2+\omega_n^2}} \sin(\omega_n t-\theta)$ where $\theta = \tan^{-1}(\omega_n/\alpha)$
$\frac{\omega_n^2}{s^2+2\zeta\omega_n s+\omega_n^2}$	$\frac{\omega_n}{\sqrt{1-\zeta^2}} e^{-\zeta\omega_n t} \sin \omega_n \sqrt{1-\zeta^2}\, t \quad (\zeta<1)$
$\frac{\omega_n^2}{s(s^2+2\zeta\omega_n s+\omega_n^2)}$	$1-\frac{1}{\sqrt{1-\zeta^2}} e^{-\zeta\omega_n t} \sin(\omega_n \sqrt{1-\zeta^2}\, t+\theta)$ where $\theta = \cos^{-1}\zeta \quad (\zeta<1)$
$\frac{s\omega_n^2}{s^2+2\zeta\omega_n s+\omega_n^2}$	$\frac{-\omega_n^2}{\sqrt{1-\zeta^2}} e^{-\zeta\omega_n t} \sin(\omega_n \sqrt{1-\zeta^2}\, t-\theta)$ where $\theta = \cos^{-1}\zeta \quad (\zeta<1)$
$\frac{\omega_n^2(s+\alpha)}{s^2+2\zeta\omega_n s+\omega_n^2}$	$\omega_n \sqrt{\frac{\alpha^2-2\alpha\zeta\omega_n+\omega_n^2}{1-\zeta^2}}\, e^{-\zeta\omega_n t} \sin(\omega_n \sqrt{1-\zeta^2}\, t+\theta)$ where $\theta = \tan^{-1}\frac{\omega_n\sqrt{1-\zeta^2}}{\alpha-\zeta\omega_n} \quad (\zeta<1)$
$\frac{\omega_n^2}{s^2(s^2+2\zeta\omega_n s+\omega_n^2)}$	$t-\frac{2\zeta}{\omega_n}+\frac{1}{\omega_n^2\sqrt{1-\zeta^2}} e^{-\zeta\omega_n t} \sin(\omega_n \sqrt{1-\zeta^2}\, t+\theta)$ where $\theta = \cos^{-1}(2\zeta^2-1) \quad (\zeta<1)$

Contents

Preface (Readme)

A great deal of thought, feedback, and effort have gone into the writing and preparation of the seventh edition. The two-color format that was begun in the sixth edition is now almost a standard among introductory texts on control systems. It appears that the two-color format does enhance readability and the graphical presentation of complex subjects.

A significant portion of the book has been revised and expanded. Highlights of the major revisions are as follows:

1. Computer-aided learning and design are enhanced. Computer solutions of sample problems are inserted as various points in the text.
2. Effort has been expended to make difficult subjects easier to learn.
3. Certain less important subjects that probably have not been used by most users have been deleted or reduced.
4. Problems at the end of chapters have been revised and expanded. New problems have been added. Emphasis has been placed on computer solutions.
5. Many subjects have been refined, one example being the steady-state errors of nonunity-feedback systems.
6. The material on time- and frequency-domain design is consolidated into one chapter. These two design topics are treated side by side so that the reader can make comparisons immediately.

With the assistance of Don Fowley, Executive Editor at Simon & Schuster, the book has had extensive peer review prior to revision. The author is indepted to Professor Duane Hanselman at the University of Maine for reviewing the sixth edition and for giving valuable suggestions on revision. Professor Hanselman also reviewed several chapters of the new edition and contributed to the material on op-amps. His suggestions were invaluable to the revision and certainly added new dimensions to the text.

The following paragraphs are aimed at three groups: professors who have adopted the book or who we hope will select it as their text; practicing engineers looking for answers to problems and solutions to design problems; and finally, students who are reading the book because it has been adopted for a course.

To the Professor: The material assembled in this book is an outgrowth of a senior-level control-system course taught by the author at the University of Illinois at Urbana—Champaign for over 30 years. The first six editions have been adopted by hundreds of universities in the United States and around the world and have been translated into six different languages. The seventh edition has been expanded to a total of eleven chapters and three appendices. Great emphasis has been placed on the design of control systems and computer solutions. Many systems, such as the **sun-seeker system, liquid-**

level control, dc-motor control, space-vehicle payload control, and others, are carried out as numerical examples from analysis to design. There is more material in the book than can be covered in one semester. The subject on discrete-data control systems follows the treatment of continuous-data control systems in selected chapters. Personally, I find it easier to introduce discrete-data systems as extensions to their analog counterparts, although, realistically, it is difficult to cover all the subjects on digital control in a one-semester course. Chapter 11 is devoted to the design of discrete-data control systems.

A sample section-by-section selection of a one-semester course is given in the *Instructor's Manual,* which is available from the publisher to qualified instructors. The *Manual* also contains detailed solutions to all the problems in the book, as well as additional problems for supplemental assignments and possible quiz and exam problems.

To Practicing Engineers: This book is written with the readers in mind and is very suitable for self-study. The objective is to treat subjects clearly and thoroughly. The book does not follow the theorem–proof–Q.E.D. format and is without heavy mathematics. The author has consulted extensively for the industry over his 35-year teaching career, and has participated in solving numerous control-systems problems, from aerospace systems to industrial controls, automotive controls, and control of computer peripherals. Although it is difficult to adopt all the details and realism of practical problems in a textbook at this level, some examples and problems reflect simplified versions of real-life systems.

To Students: You have had it now that you have signed up for this course and your professor has adopted this book! You had no say about the choice, although you can form and express your opinion on the book after reading it. Worse yet, one of the reasons that your professor made the selection is because he or she intends to make you work hard. Please don't misunderstand me. What I really meant is that although this is an easy book to study from (in my opinion), it is a no-nonsense book. It does not have cartoons or nice-looking photographs to amuse you. From here on it is all business and hard work. You should have had the prerequisites on subjects found in a typical linear-system course, such as how to solve linear ordinary differential equations, Laplace transform and applications, time response and frequency-domain analysis of linear systems. In this book you will not find new mathematics to which you have not been exposed before. What is interesting and challenging is that you are going to learn how to apply some of the mathematics that you have acquired during the last two or three years of study in college.

This book has a total of 188 illustrative examples. Some of these are simple, for the purpose of illustrating new ideas and subject matter. Some examples are quite elaborate, in order to bring the practical world closer to you. Furthermore, the objective of this book is to present a complex subject in a clear and thorough way. One of the important learning strategies for you as students is not to rely strictly on the textbook assigned. When studying a certain subject, go to the library and check out a few similar texts to see how other authors treat the same subject. You may gain new perspectives on the subject and discover that one author may treat the material with more care and thoroughness than the others. Do not be distracted by written-down coverage with oversimplified examples. The minute you step into the real world, you will face the design of control systems with orders that can boggle your mind.

Most advanced engineering students in college do not believe that the material they learn in the classroom is ever going to be applied directly in industry. Some of my students come back from field and interview trips totally surprised to find that the material they learned in the course on control systems is actually being used in industry today. They are surprised to find that this book is also a popular reference to practicing engineers. Unfortunately, these fact-find, eye-opening, and self-motivating trips usually occur near the end of the senior year, which is often too late for students to get motivated. I am telling you this at an early stage with the hope that the book and your professor will motivate you to learn something about control systems. Your professor will probably tell you the same thing in the lecture, but you may not believe him or her. The bottom line is that the subjects treated in this book are not difficult, but you have to do the exercises to gain full comprehension. There are many learning aids available to you: the computer programs listed in the **Preface to Computer Software for Control Systems** will save you a great deal of time in problem solving. The Review Questions and Summary at the end of each chapter, and the Answers and Hints for Selected Problems, should all be useful to you.

I hope that you will enjoy the book. It will represent another major textbook acquisition (investment) in your college career. My advise to you is not to sell it back to the bookstore at the end of the semester. If you do, not only will you be violating the *''buy low and sell high''* axiom of finance, but probably when you get into industry, your first assignment will involve the design of a control system, regardless of your specialization.

Special Acknowledgments: The author wishes to thank the professional staff at Prentice Hall: Donald Fowley, Executive Editor, Linda Ratts, Managing Editor, Susan Handy, Editorial Assistant, and many other unnamed persons who contributed in many ways to the seventh edition. Special thanks goes to Irwin Zucker, Production Editor, who handled the editing and production of the book superbly. The professional manner the staffs, from top to bottom, at Prentice Hall have supported this project made the difficult task more enjoyable.

The author also wish to express appreciation to Marcia Horton and Tim Bozik, Editors in Chief, for their vision and support in producing the sixth edition of this book in two colors, which set the standard for textbooks in control systems.

Finally, the author wishes to pay tribute to his family, especially his wife, Margaret, for her support and tolerance in many ways to this and many of his professional projects for the past 40 years. 1994 marks not only the completion of the seventh edition of *Automatic Control Systems,* but is also our fortieth wedding anniversary. Without her moral support, these projects could never have been accomplished.

B. C. Kuo

Champaign, Illinois
Year of the golden dog

Preface to Computer Software for Control Systems

In recent years, the analysis and control systems have been affected dramatically by the widespread use of computers, especially personal computers. Personal computers have become so powerful and advanced that they can be used to solve sophisticated and complex control systems problems with ease. Today, most schools and companies are equipped with personal computers that are easily accessible.

The book is written with no particular computer software in mind. Certainly, there are many fine commercial computer-software packages available, and any one of these, whether it is for the IBM PC® or compatibles or the Apple Macintosh® computer, can be used to supplement this book.

The following software is all available in both PC and Macintosh versions.

ACSP Version 2.0 by Benjamin C. Kuo

Version 2.0 in an upgraded software package for use on IBM PC, XT, AT, PS/2®, or any compatible computer. A total of eight utility programs and seven review programs are contained in two $3\frac{1}{2}$-in. diskettes. The programs offer on-screen high-resolution color graphics. The programs also support a mouse, so that data and command entries and menu selections can all be aided by the use of a mouse. Another important improvement of version 2.0 over the old version is screen-dump capability. Results of the computer analysis and design can be plotted or printed on an Epson/IBM or any compatible dot-matrix printer, as well as on the LaserJet printer, with the same resolution as on the monitor screen. The programs are described as follows:

Utility Programs

linsys linear system analysis and state-feedback design in the state-variable domain

polyroot polynomial root solving

matrixm multiplication of two matrices

pfe partial-fraction expansion of *s*-domain transfer functions

clrsp time response (impulse, step, and ramp responses) of continuous-data open- and closed-loop control systems

rootloci	root-locus diagram of *s*-domain equations
freqrp	frequency-domain analysis of linear continuous-data systems
freqcad	computer-aided design of linear continuous-data control systems in the frequency domain

Review Programs

freq	frequency-domain analysis
rootl	root-locus analysis
sfg	signal-flow graphs
state	state-variable analysis
timea	time-domain analysis
timed	time-domain design
ztform	*z*-transform analysis

ACSP is published and distributed by Prentice Hall, a division of Simon & Schuster.

MATLAB® Tools for Control System Analysis and Design, Second Edition by D. C. Hanselman and B. C. Kuo, Prentice Hall, 1995

CSAD is a menu- and mouse-driven set of **MATLAB** Tools to supplement undergraduate studies of continuous-data control systems. The following prescriptions show what the tools are and are not, and the important M-files.

- ▲ **IT IS** a set of (over 70) M-files that will run on any 3.5 version of **MATLAB**. The Control Systems Toolbox and Signals and Systems Toolbox are not needed.
- ▲ **IT IS NOT** just another text showing student how to use **MATLAB** for control system analysis and design.
- ▲ **IT IS NOT** just a collection of M-files to augment the Control System Toolbox or the Student Edition of **MATLAB**.
- ▲ **IT IS** a set of M-files and accompanying reference text that allows students to analyze and design control systems *without the burden of programming in **MATLAB** or manipulating **MATLAB** data.* The user asks for information by making menu choices and using the mouse to select points of interest on plots.

M-file Top Ten List

1. pidesign (tddesign)	design of PID-type controllers
2. lldesign (fddesign)	design of lead/lag-type controllers

3. **rlplot** comprehensive root-locus plotting
4. **tftplot** step response from a transfer-function description
5. **svplot** step response from state-variable description
6. **svdesign (svstuff)** state-variable analysis and design
7. **bplot** Bode plot
8. **mvpplot** magnitude versus phase frequency-domain plots (Nichols)
9. **plrplots** polar frequency-domain plots (Nyquist)
10. **iltplot** inverse Laplace transform (i.e., time response plots) given its Laplace transform description

B. C. Kuo

Champaign, Illinois

1 Introduction

KEYWORDS AND TOPICS

- ▲ **Control Systems**
- ▲ **Feedback**
- ▲ **Linear versus Nonlinear Systems**
- ▲ **Time-Invariant versus Time-Varying Systems**
- ▲ **Discrete-Data Systems**

1-1 Introduction

The objective of this chapter is to familiarize the reader with the following subjects:

1. What a control system is.
2. Why control systems are important.
3. What the basic components of a control system are.
4. Some examples of control system applications.
5. Why feedback is incorporated into most control systems.
6. Types of control systems.

One of the most commonly asked questions by a novice on control systems is: What is a control system? To answer the question, we can cite that in our daily life there are numerous ''objectives'' that need to be accomplished. For instance, in the domestic domain, we need to regulate the temperature and humidity of homes and buildings for comfortable living. For transportation, we need to control the automobile and airplane to go from one point to another accurately and safely. Industrially, manufacturing processes contain numerous objectives for products that will satisfy the precision and cost-effectiveness requirements. A human being is capable of performing a wide range of tasks, including decision making. Some of these tasks, such as picking up objects and walking from one point to another, are commonly carried out in a routine fashion. Under certain conditions, some of these tasks are to be performed in the best possible way. For instance, an athlete running a 100-yard dash has the objective of running that distance in the shortest possible time. A marathon runner, on the other hand, not only must run the distance as quickly as possible, but in doing so, he or she must control the consumption of energy and devise the best strategy for the race. The means of achieving these ''objectives'' usually involve the use of control systems that implement certain control strategies.

In recent years, control systems have assumed an increasingly important role in the development and advancement of modern civilization and technology. Practically every aspect of our day-to-day activities is affected by some type of control systems. Control systems are found in abundance in all sectors of industry, such as quality control of manufactured products, automatic assembly line, machine-tool control, space technology and weapon systems, computer control, transportation systems, power systems, robotics, and many others. Even the control of inventory and social and economic systems may be approached from the theory of automatic control.

▲ Control systems are in abundance in modern civilization.

1-1-1 Basic Components of a Control System

The basic ingredients of a control system can be described by:

1. Objectives of control.
2. Control system components.
3. Results or outputs.

The basic relationship between these three components is illustrated in Fig. 1-1. In more technical terms, the **objectives** can be identified with **inputs,** or **actuating signals,** ***u*,** and the results are also called **outputs,** or the **controlled variables,** ***y*.** In general, the objective of the control system is to control the outputs in some prescribed manner by the inputs through the elements of the control system.

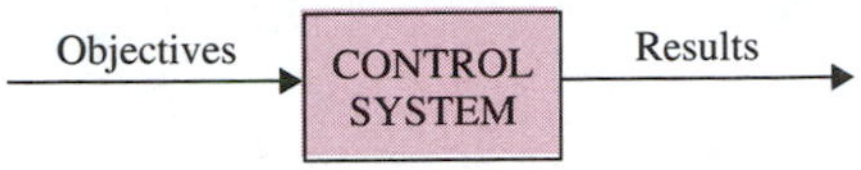

Figure 1-1 Basic components of a control system.

1-1-2 Examples of Control System Applications

Steering Control of Automobile

As a simple example of the control system fashioned in Fig. 1-1, consider the steering control of an automobile. The direction of the two front wheels can be regarded as the controlled variable, or the output, y; the direction of the steering wheel is the actuating signal, or the input, u. The control system, or process in this case, is composed of the steering mechanism and the dynamics of the entire automobile. However, if the objective is to control the speed of the automobile, then the amount of pressure exerted on the accelerator is the actuating signal, and the vehicle speed is the controlled variable. As a whole, we can regard the simplified automobile control system as one with two inputs (steering and accelerator) and two outputs (heading and speed). In this case the two controls and two outputs are independent of each other, but in general, there are systems for which the controls are coupled. Systems with more than one input and one output are called **multivariable systems.**

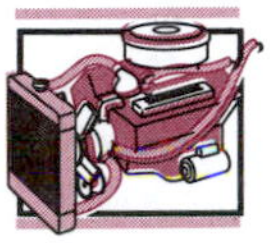

Idle-Speed Control of Automobile

As another example of a control system, we consider the idle-speed control of an automobile engine. The objective of such a control system is to maintain the engine idle speed at a relatively low value (for fuel economy) regardless of the applied engine loads (e.g., transmission, power steering, air conditioning, etc.). Without the idle-speed control, any sudden engine-load application would cause a drop in engine speed that might cause the engine to stall. Thus the main objectives of the idle-speed control system are (1) to eliminate or minimize the speed droop when engine loading is applied, and (2) to maintain the engine idle speed at a desired value. Figure 1-2 shows an idle-speed control system from the standpoint of inputs–system–outputs. In this case the throttle angle α and the load torque T_L (due to the application of air conditioning, power steering, transmission, power brake, etc.) are the inputs, and the engine speed ω is the output. The engine is the controlled process of the system.

Printwheel Control System

Figure 1-3 shows an example of the printwheel control system of a word processor or electronic typewriter. The printwheel, which typically has 96 or 100 characters, is to be rotated to position the desired character in front of the hammer for impact printing. The character selection is done in the usual manner from a keyboard. Once a certain key on the keyboard is depressed, a command microprocessor computes the direction

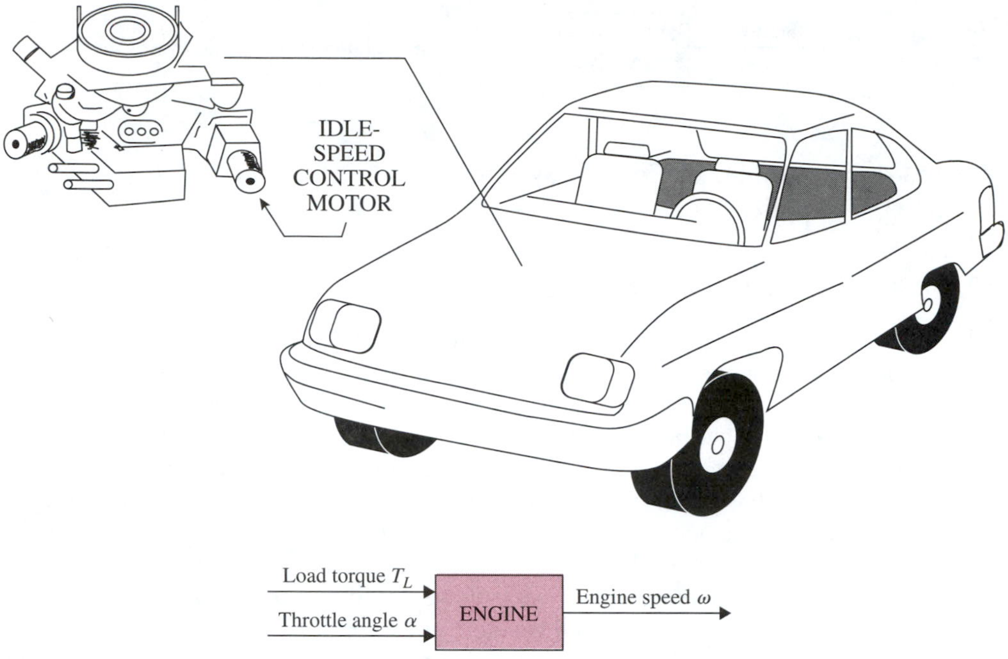

Figure 1-2 Idle-speed control system.

and the distance to be traveled and sends out a control logic signal to the power amplifier, which in turn controls the motor and drives the printwheel. In practice, the control signals generated by the microprocessor controller should be able to drive the printwheel from one position to another sufficiently fast and with high print quality, which means that the position of the printwheel should be controlled accurately. Figure 1-4 shows a typical input–output set for the system. When a reference command input is given, the signal is represented as a step function. Since the electric windings of the motor have inductance and the mechanical load has inertia, the printwheel cannot respond to the input instantaneously. Typically, it will follow the response as shown, and settle at the new position after some time t_1. Printing should not begin until the printwheel has come to a complete stop; otherwise, the character will be smeared. Figure 1-4 shows that after the printwheel has settled, the period from t_1 to t_2 is reserved for printing, so that the system is ready to receive a new command after time t_2.

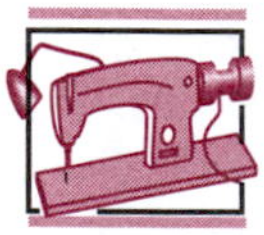

Industrial Sewing Machine

Sewing, as a basic joining operation in the garment-making process, is in principle a rather complicated and laborious operation. For low cost and high productivity, the sewing industry has to rely on sophisticated sewing machines to increase the speed and accuracy of the sewing operations. Figure 1-5 shows a typical industrial sewing machine,

(a)

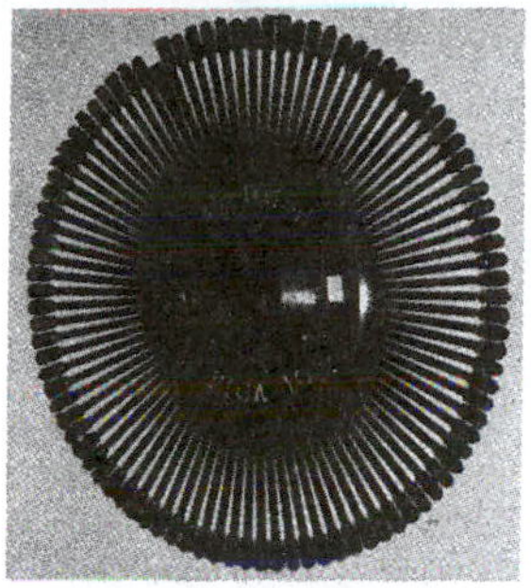

(b)

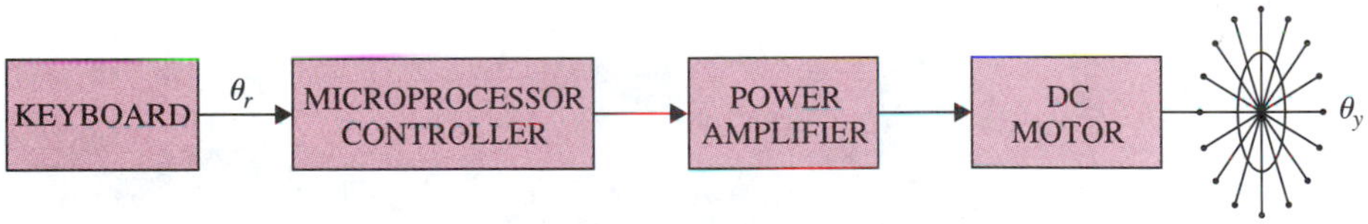

(c)

Figure 1-3 (a) Electronic word processor. (b) Printwheel. (c) Open-loop word processor control system.

which, in comparison with household machines, is strictly a single-purpose, high-precision device. It can produce only one type of stitch but is extremely fast, with a typical rate of over 100 stitches per second. One stitch corresponds to one revolution of the machine main shaft, which translates to top speeds as high as 8000 rpm. An

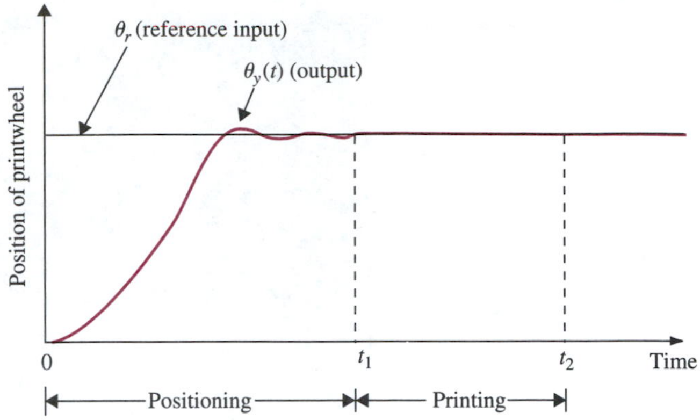

Figure 1-4 Typical input and output of the printwheel control system.

ideal velocity profile of one start–stop cycle of the machine is shown in Fig. 1-6. Typically, there should be no velocity overshoot at point A and no undershoot at point B. Acceleration time t_a, deceleration time t_b, and position search time t_{ps} should be as short as possible. When the machine reaches the stopping point, C, there should be zero or negligible oscillations. To achieve these performance objectives, the control system in the machine should be designed with stringent requirements.

Sun-Tracking Control of Solar Collectors

To achieve the goal of development of economically feasible non-fossil-fuel electrical power, the U.S. government has sponsored many organizations in research and development of solar power conversion methods, including the solar-cell conversion techniques. In most of these systems, the need for high efficiencies dictates the use of devices for sun tracking. Figure 1-7 shows a solar collector field. Figure 1-8 shows a conceptual method of efficient water extraction using solar power. During the hours of daylight

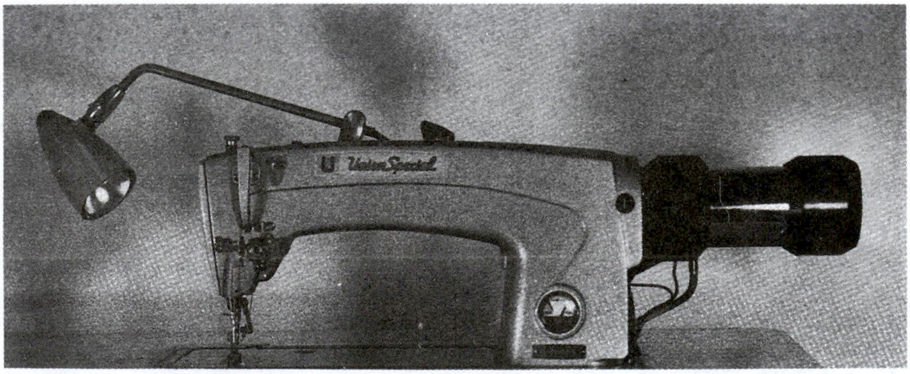

Figure 1-5 Industrial sewing machine.

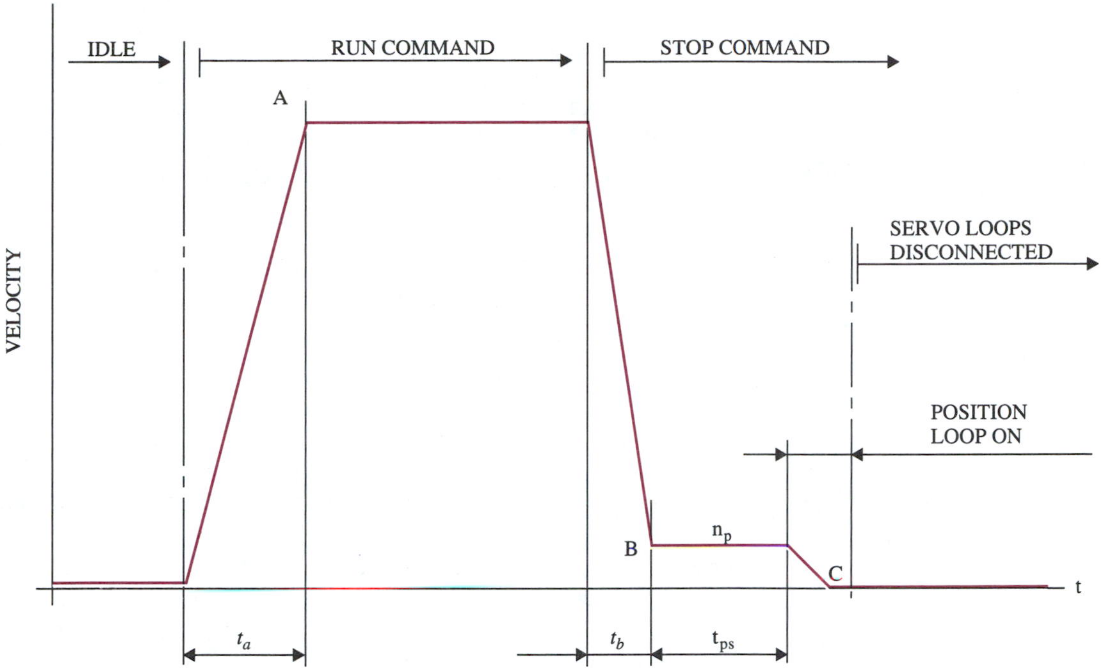

Figure 1-6 Ideal velocity profile of one start–stop cycle of an industrial sewing machine.

the solar collector would produce electricity for the purpose of pumping water from the underground water table to a reservoir (perhaps on a nearby mountain or hill), and in the early morning hours, the water would be released into the irrigation system.

One of the most important features of the solar collector is that the collector dish must track the sun accurately. Therefore, the movement of the collector dish must be

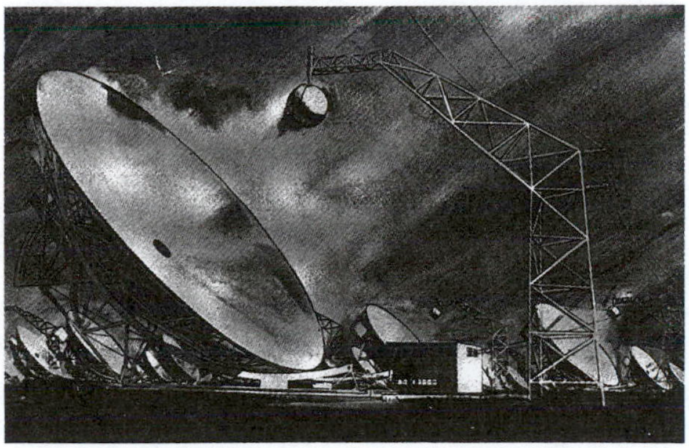

Figure 1-7 Solar collector field.

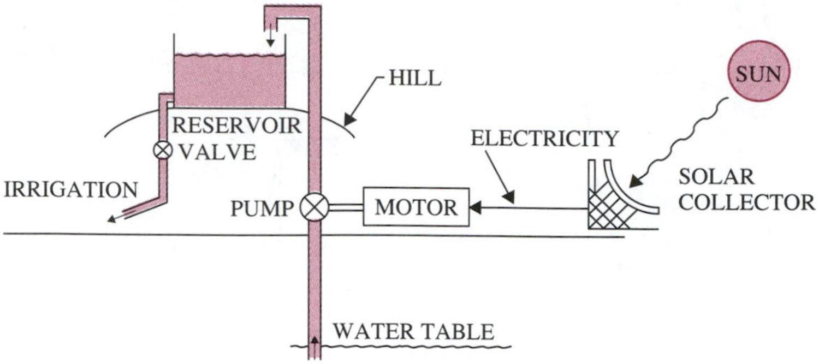

Figure 1-8 Conceptual method of efficient water extraction using solar power.

controlled by sophisticated control systems. Figure 1-9 describes the general philosophy of the sun-tracking system together with some of the most important components. The basic philosophy of the control system is that a predetermined desired rate is modified or trimmed by actual position errors determined by the sun sensor. The controller ensures that the tracking collector is pointed toward the sun in the morning and sends a "start track" command. The controller constantly calculates the sun's rate for the two axes (azimuth and elevation) of control during the day. The controller uses the sun rate and sun sensor information as inputs to generate proper motor commands to slew the collector.

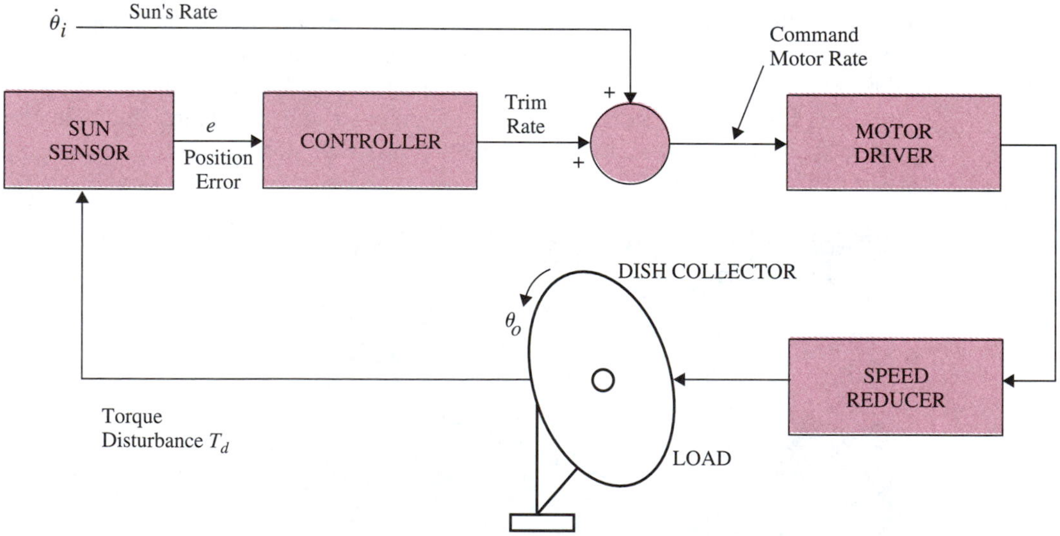

Figure 1-9 Important components of sun-tracking control system.

1-1-3 Open-Loop Control Systems (Nonfeedback Systems)

▲ Open-loop systems are economical but usually inaccurate.

The idle-speed control system and the printwheel control system illustrated in Figures 1-2 and 1-3, respectively, are rather unsophisticated and are called **open-loop control systems.** It is not difficult to see that the systems as shown would not satisfactorily fulfill critical performance requirements. For instance, if the throttle angle α is set at a certain initial value, which corresponds to a certain engine speed, when a load torque T_L is then applied, there is no way to prevent a drop in the engine speed. The only way to make the system work is to have a means of adjusting α in response to a change in the load torque in order to maintain ω at the desired level. Similarly, there is no guarantee that the printwheel will stop at the desired position once the command is given. The conventional electric washing machine is another example of an open-loop control system because, typically, the amount of machine wash time is determined entirely by the judgment and estimation of the human operator.

The elements of an open-loop control system can usually be divided into two parts: the **controller** and the **controlled process,** as shown in Fig. 1-10. An input signal or command r is applied to the controller, whose output acts as the actuating signal u; the actuating signal then controls the controlled process so that the controlled variable y will perform according to prescribed standards. In simple cases, the controller can be an amplifier, mechanical linkage, filter, or other control element, depending on the nature of the system. In more sophisticated cases, the controller can be a computer such as a microprocessor. Because of the simplicity and economy of open-loop control systems, we find this type of system in many noncritical applications.

1-1-4 Closed-Loop Control Systems (Feedback Control Systems)

What is missing in the open-loop control system for more accurate and more adaptive control is a link or feedback from the output to the input of the system. To obtain more accurate control, the controlled signal y should be fed back and compared with the reference input, and an actuating signal proportional to the difference of the input and the output must be sent through the system to correct the error. A system with one or more feedback paths such as that just described is called a **closed-loop system.**

A closed-loop idle-speed control system is shown in Fig. 1-11. The reference input ω_r sets the desired idling speed. The engine speed at idle should agree with the reference value ω_r, and any difference such as the load torque T_L is sensed by the speed transducer and the error detector. The controller will operate on the difference and provide a signal

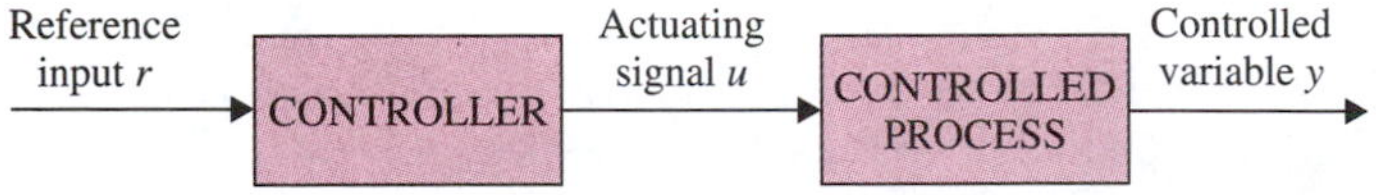

Figure 1-10 Elements of an open-loop control system.

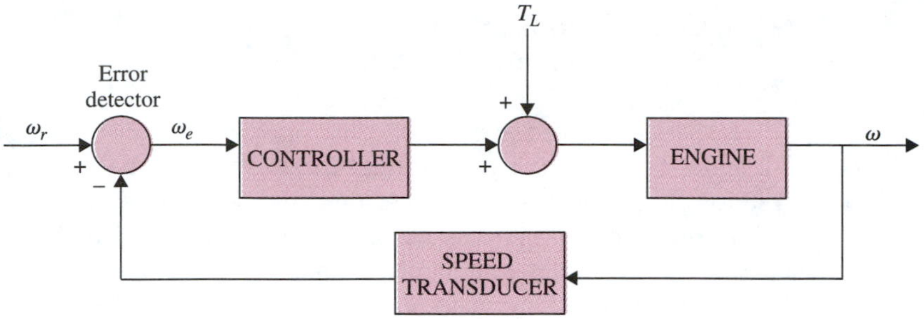

Figure 1-11 Block diagram of a closed-loop idle-speed control system.

▲ Closed-loop systems have many advantages over open-loop systems.

to adjust the throttle angle α to correct the error. Figure 1-12 compares the typical performances of open- and closed-loop idle-speed control systems. In Fig. 1-12(a), the idle speed of the open-loop system will drop and settle at a lower value after a load torque is applied. In Figure 1-12(b), the idle speed of the closed-loop system is shown to recover quickly to the preset value after the application of T_L. The objective of the idle-speed control system illustrated, also known as a **regulator system,** is to maintain the system output at a prescribed level.

Figure 1-13 shows a printwheel control system with feedback. In this case the position of the printwheel is detected by a position sensor whose output is compared with the desired position entered from the keyboard and processed in the microprocessor. The motor is thus controlled in such a way as to drive the printwheel to the desired position accurately. Information on the speed of the printwheel can also be processed in the microprocessor from the position data so that the profile of the printwheel motion can be better controlled.

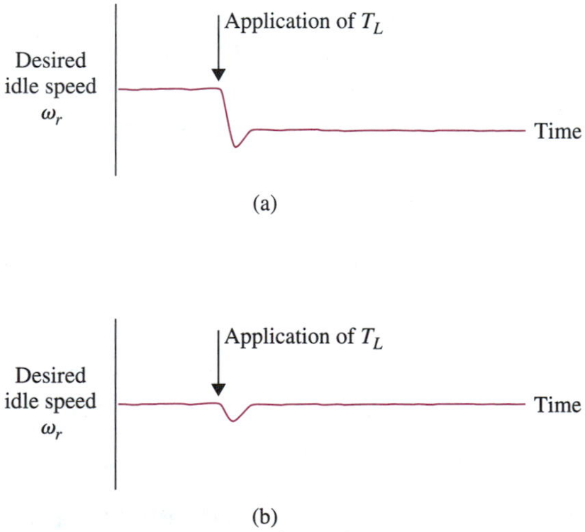

Figure 1-12 (a) Typical response of the open-loop idle-speed control system. (b) Typical response of the closed-loop idle-speed control system.

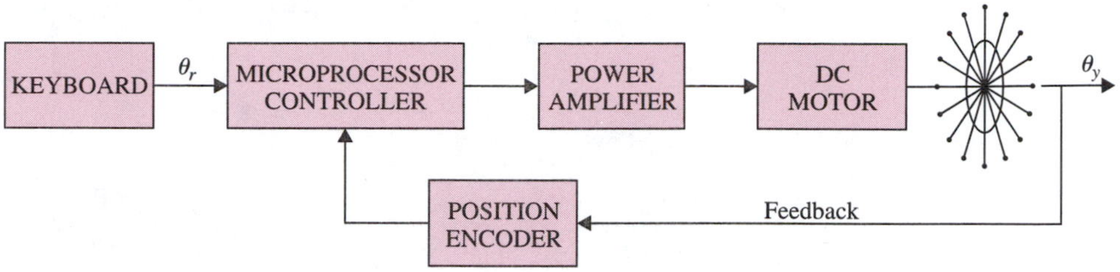

Figure 1-13 Closed-loop printwheel control system.

1-2 What Is Feedback and What Are Its Effects?

The motivation of using feedback, illustrated by the examples in Section 1-1, is somewhat oversimplified. In these examples, the use of feedback is shown to be for the purpose of reducing the error between the reference input and the system output. However, the significance of the effects of feedback in control systems is more complex than is demonstrated by these simple examples. The reduction of system error is merely one of the many important effects that feedback may have upon a system. We show in the following sections that feedback also has effects on such system performance characteristics as **stability, bandwidth, overall gain, disturbance,** and **sensitivity.**

▲ Feedback exists whenever there is a closed sequence of cause-and-effect relationships.

To understand the effects of feedback on a control system, it is essential that we examine this phenomenon in a broad sense. When feedback is deliberately introduced for the purpose of control, its existence is easily identified. However, there are numerous situations wherein a physical system that we normally recognize as an inherently nonfeedback system turn out to have feedback when it is observed in a certain manner. In general, we can state that whenever a closed sequence of **cause-and-effect relationships** exists among the variables of a system, feedback is said to exist. The viewpoint will inevitably admit feedback in a large number of systems that ordinarily would be identified as nonfeedback systems. However, with the availability of the feedback and control system theory, this general definition of feedback enables numerous systems, with or without physical feedback, to be studied in a systematic way once the existence of feedback in the sense mentioned previously is established.

We shall now investigate the effects of feedback on the various aspects of system performance. Without the necessary mathematical foundation of linear-system theory, at this point we can only rely on simple static-system notation for our discussion. Let us consider the simple feedback system configuration shown in Figure 1-14, where r is the input signal, y the output signal, e the error, and b the feedback signal. The parameters G and H may be considered as constant gains. By simple algebraic manipulations, it is simple to show that the input–output relation of the system is

$$M = \frac{y}{r} = \frac{G}{1 + GH} \tag{1-1}$$

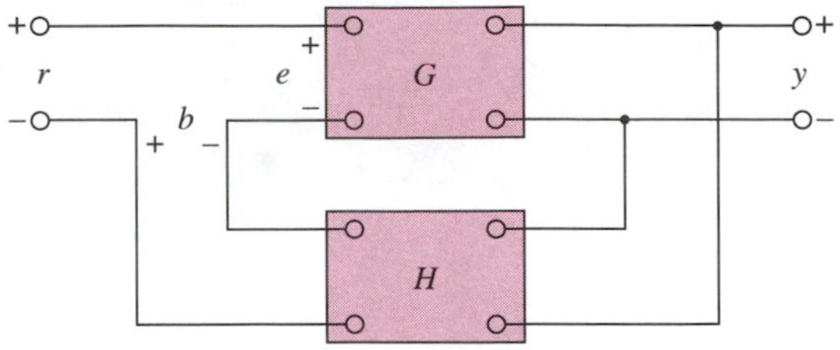

Figure 1-14 Feedback system.

Using this basic relationship of the feedback system structure, we can uncover some of the significant effects of feedback.

1-2-1 Effect of Feedback on Overall Gain

▲ Feedback may increase the gain of a system in one frequency range but decrease it in another.

As seen from Eq. (1-1), feedback affects the gain G of a nonfeedback system by a factor $1 + GH$. The system of Fig. 1-14 is said to have **negative feedback,** since a minus sign is assigned to the feedback signal. The quantity GH may itself include a minus sign, so *the general effect of feedback is that it may increase or decrease the gain G*. In a practical control system, G and H are functions of frequency, so the magnitude of $1 + GH$ may be greater than 1 in one frequency range but less than 1 in another. Therefore, *feedback could increase the system gain in one frequency range but decrease it in another.*

1-2-2 Effect of Feedback on Stability

▲ A system is unstable if its output is out of control.

Stability is a notion that describes whether the system will be able to follow the input command, or be useful in general. In a nonrigorous manner, *a system is said to be unstable if its output is out of control*. To investigate the effect of feedback on stability, we can again refer to the expression in Eq. (1-1). If $GH = -1$, the output of the system is infinite for any finite input, and the system is said to be unstable. Therefore, we may state that *feedback can cause a system that is originally stable to become unstable*. Certainly, feedback is a two-edged sword; when used improperly, it can be harmful. It should be pointed out, however, that we are only dealing with the static case here, and in general, $GH = -1$ is not the only condition for instability. The subject of system stability is treated formally in Chapter 6.

It can be demonstrated that one of the advantages of incorporating feedback is that it can stabilize an unstable system. Let us assume that the feedback system in Fig. 1-14 is unstable because $GH = -1$. If we introduce another feedback loop through a negative feedback gain of F, as shown in Fig. 1-15, the input–output relation of the overall system is

$$\frac{y}{r} = \frac{G}{1 + GH + GF} \tag{1-2}$$

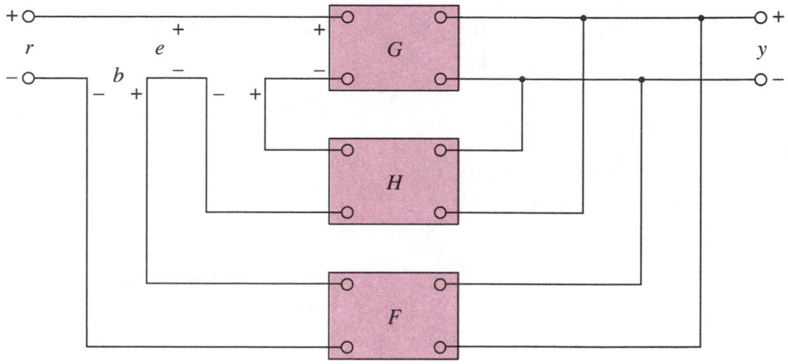

Figure 1-15 Feedback system with two feedback loops.

▲ Feedback can improve stability or be harmful to stability.

It is apparent that although the properties of G and H are such that the inner-loop feedback system is unstable, because $GH = -1$, the overall system can be stable by proper selection of the outer-loop feedback gain F. In practice, GH is a function of frequency, and the stability condition of the closed-loop system depends on the **magnitude** and **phase** of GH. The bottom line is that *feedback can improve stability or be harmful to stability if it is not applied properly.*

1-2-3 Effect of Feedback on Sensitivity

Sensitivity considerations often are important in the design of control systems. Since all physical elements have properties that change with environment and age, we cannot always consider the parameters of a control system to be completely stationary over the entire operating life of the system. For instance, the winding resistance of an electric motor changes as the temperature of the motor rises during operation. The electronic typewriter described in Section 1-1 sometimes may not operate normally when first turned on due to the still-changing system parameters during warm-up. This phenomenon is sometimes called "morning sickness." Most duplicating machines have a warm-up period during which operation is blocked out when first turned on.

In general, a good control system should be very insensitive to parameter variations but sensitive to the input commands. We shall investigate what effect feedback has on the sensitivity to parameter variations. Referring to the system shown in Fig. 1-14, we consider G to be a gain parameter that may vary. The sensitivity of the gain of the overall system, M, to the variation in G is defined as

$$S_G^M = \frac{\partial M/M}{\partial G/G} = \frac{\text{percentage change in } M}{\text{percentage change in } G} \tag{1-3}$$

▲ Feedback can increase or decrease the sensitivity of a system.

where ∂M denotes the incremental change in M due to the incremental change in G, ∂G. By using Eq. (1-1), the sensitivity function is written

$$S_G^M = \frac{\partial M}{\partial G}\frac{G}{M} = \frac{1}{1+GH} \tag{1-4}$$

This relation shows that if GH is a positive constant, the magnitude of the sensitivity function can be made arbitrarily small by increasing GH, provided that the system remains stable. It is apparent that in an open-loop system, the gain of the system will respond in a one-to-one fashion to the variation in G (i.e., $S_G^M = 1$). We again remind you that in practice, GH is a function of frequency; the magnitude of $1 + GH$ may be less than unity over some frequency ranges, so that feedback could be harmful to the sensitivity to parameter variations in certain cases. In general, the sensitivity of the system gain of a feedback system to parameter variations depends on where the parameter is located. The reader can derive the sensitivity of the system in Fig. 1-14 due to the variation of H.

1-2-4 Effect of Feedback on External Disturbance or Noise

All physical systems are subject to some types of extraneous signals or noise during operation. Examples of these signals are thermal-noise voltage in electronic circuits and brush or commutator noise in electric motors. External disturbance, such as wind gust acting on an antenna, is also quite common in control systems. Therefore, in the design of a control system, considerations should be given so that the system is insensitive to noise and disturbances and sensitive to input commands.

▲ Feedback can reduce the effect of noise.

The effect of feedback on noise and disturbance depends greatly on where these extraneous signals occur in the system. No general conclusions can be reached, but in many situations, *feedback can reduce the effect of noise and disturbance on system performance.* Let us refer to the system shown in Fig. 1-16, in which r denotes the command signal and n is the noise signal. In the absence of feedback, $H = 0$, the output y due to n acting alone is

$$y = G_2 n \tag{1-5}$$

With the presence of feedback, the system output due to n acting alone is

$$y = \frac{G_2}{1 + G_1 G_2 H} n \tag{1-6}$$

Comparing Eq. (1-6) with Eq. (1-5) shows that the noise component in the output of Eq. (1-6) is reduced by the factor $1 + G_1G_2H$ if the latter is greater than unity and the system is kept stable.

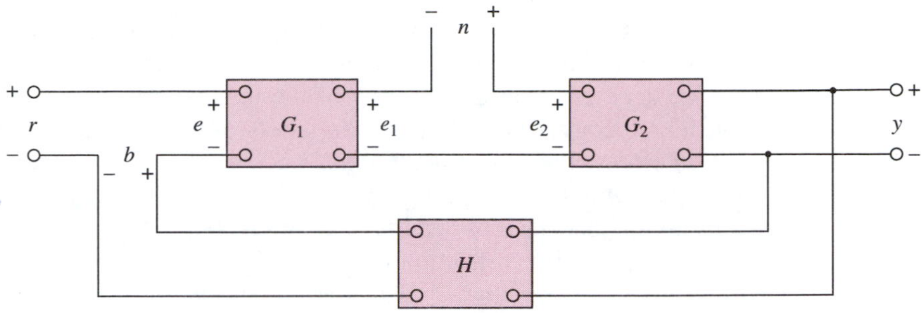

Figure 1-16 Feedback system with a noise signal.

▲ Feedback can also affect bandwidth, impedance, transient, and frequency responses.

In Chapter 10 the feedforward and forward controller configurations are used along with feedback to reduce the effects of disturbance and noise inputs. In general, feedback also has effects on such performance characteristics as bandwidth, impedance, transient response, and frequency response. These effects will become known as we continue.

1-3 Types of Feedback Control Systems

Feedback control systems may be classified in a number of ways, depending on the purpose of the classification. For instance, according to the method of analysis and design, control systems are classified as **linear** and **nonlinear, time-varying** or **time-invariant.** According to the types of signal used in the system, reference is often made to **continuous-data** and **discrete-data** systems, or **modulated** and **unmodulated** systems. Control systems are often classified according to the main purpose of the system. For instance, a **position-control system** and a **velocity-control system** control the output variables according to the way the names imply. In Chapter 7 the **type** of a control system is defined according to the form of the open-loop transfer function. In general, there are many other ways of identifying control systems according to some special features of the system. It is important that some of these more common ways of classifying control systems are known so that proper perspective is gained before embarking on the analysis and design of these systems.

1-3-1 Linear versus Nonlinear Control Systems

▲ Most real-life control systems have nonlinear characteristics to some extent.

This classification is made according to the methods of analysis and design. Strictly speaking, linear systems do not exist in practice, since all physical systems are nonlinear to some extent. Linear feedback control systems are idealized models fabricated by the analyst purely for the simplicity of analysis and design. When the magnitudes of signals in a control system are limited to ranges in which system components exhibit linear characteristics (i.e., the principle of superposition applies), the system is essentially linear. But when the magnitudes of signals are extended beyond the range of the linear operation, depending on the severity of the nonlinearity, the system should no longer be considered linear. For instance, amplifiers used in control systems often exhibit a saturation effect when their input signals become large; the magnetic field of a motor usually has saturation properties. Other common nonlinear effects found in control systems are the backlash or dead play between coupled gear members, nonlinear spring characteristics, nonlinear friction force or torque between moving members, and so on. Quite often, nonlinear characteristics are intentionally introduced in a control system to improve its performance or provide more effective control. For instance, to achieve minimum-time control, an on–off (bang-bang or relay) type of controller is used in many missile or spacecraft control systems. Typically in these systems, jets are mounted on sides of the vehicle to provide reaction torque for attitude control. These jets are often controlled in a full-on or full-off fashion, so a fixed amount of air is applied from a given jet for a certain time period to control the attitude of the space vehicle.

▲ There are no general methods for solving a wide class of nonlinear systems.

For linear systems, there exists a wealth of analytical and graphical techniques for design and analysis purposes. A majority of the material in this book is devoted to the analysis and design of linear systems. Nonlinear systems, on the other hand, are usually difficult to treat mathematically, and there are no general methods available for solving a wide class of nonlinear systems. In the design of control systems, it is practical first to design the controller based on the linear-system model by neglecting the nonlinearities of the system. The designed controller is then applied to the nonlinear system model for evaluation or redesign by computer simulation.

1-3-2 Time-Invariant versus Time-Varying Systems

When the parameters of a control system are stationary with respect to time during the operation of the system, the system is called a **time-invariant system.** In practice, most physical systems contain elements that drift or vary with time. For example, the winding resistance of an electric motor will vary when the motor is first being excited and its temperature is rising. Another example of a time-varying system is a guided-missile control system in which the mass of the missile decreases as the fuel on board is being consumed during flight. Although a time-varying system without nonlinearity is still a linear system, the analysis and design of this class of systems are usually much more complex than that of the linear **time-invariant systems.**

Continuous-Data Control Systems

A **continuous-data system** is one in which the signals at various parts of the system are all functions of the continuous time variable *t*. Among all continuous-data control systems, the signals may be further classified as ac or dc. Unlike the general definitions of ac and dc signals used in electrical engineering, **ac** and **dc control systems** carry special significance in control systems terminology. When one refers to an ac control system, it usually means that the signals in the system are *modulated* by some form of modulation scheme. On the other hand, when a dc control system is referred to, it does not mean that all the signals in the system are unidirectional; then there would be no corrective control movement. A **dc control system** simply implies that the signals are *unmodulated,* but they are still ac signals according to the conventional definition. The schematic diagram of a closed-loop dc control system is shown in Fig. 1-17. Typical waveforms of the signals in response to a step-function input are shown in the figure. Typical components of a dc control system are potentiometers, dc amplifiers, dc motors, dc tachometers, and so on.

The schematic diagram of a typical **ac control system** that performs essentially the same task as that in Fig. 1-17 is shown in Fig. 1-18. In this case, the signals in the system are modulated; that is, the information is transmitted by an ac carrier signal. Notice that the output-controlled variable still behaves similarly to that of the dc system. In this case, the modulated signals are demodulated by the low-pass characteristics of the ac motor. Ac control systems are used extensively in aircraft and missile control systems, in which noise and disturbance often create problems. By using modulated ac control systems with carrier frequencies of 400 Hz or higher, the system will be less susceptible to low-frequency noise. Typical components of an ac control system are synchros, ac amplifiers, ac motors, gyroscopes, accelerometers, and so on.

In practice, not all control systems are strictly of the ac or dc type. A system may incorporate a mixture of ac and dc components, using modulators and demodulators to match the signals at various points in the system.

Discrete-Data Control Systems

Discrete-data control systems differ from the continuous-data systems in that the signals at one or more points of the system are in the form of either a pulse train or a digital code. Usually, discrete-data control systems are subdivided into **sampled-data**

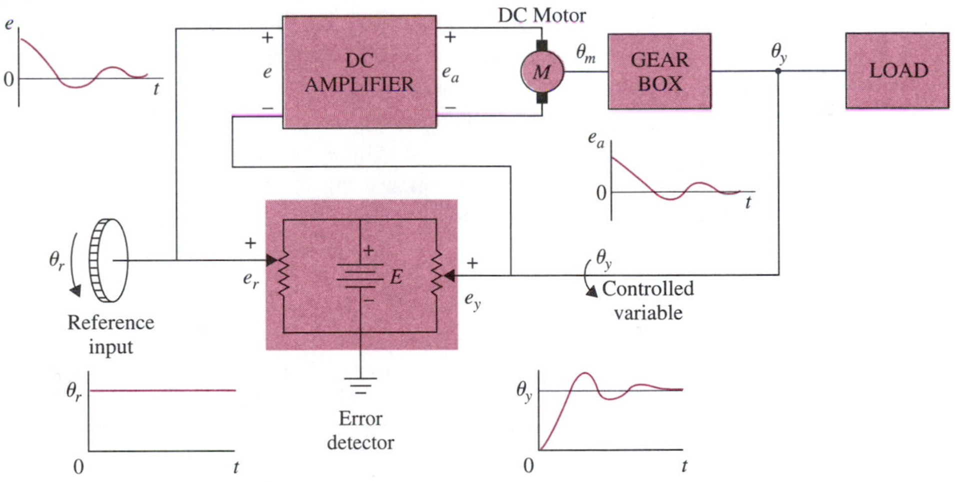

Figure 1-17 Schematic diagram of a typical dc closed-loop system.

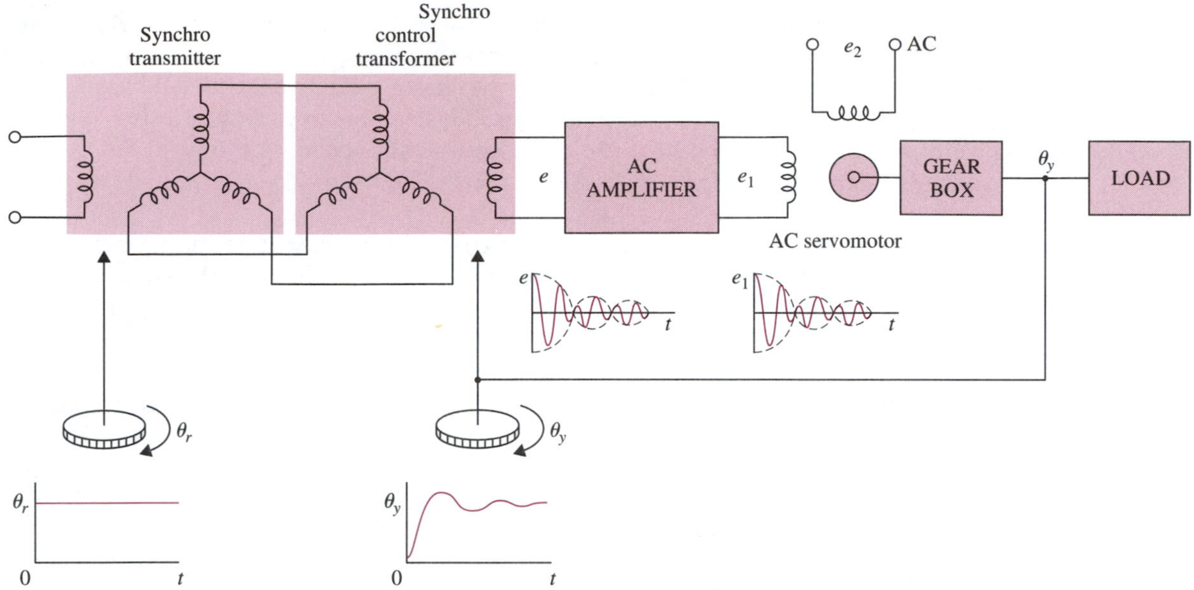

Figure 1-18 Schematic diagram of a typical ac closed-loop control system.

and **digital control systems.** Sampled-data control systems refer to a more general class of discrete-data systems in which the signals are in the form of pulse data. A digital control system refers to the use of a digital computer or controller in the system, so that the signals are digitally coded, such as in binary code. For example, the printwheel control system shown in Fig. 1-13 is a typical digital control system, since the microprocessor receives and outputs digital data.

In general, a sampled-data system receives data or information only intermittently at specific instants of time. For instance, the error signal in a control system can be supplied only in the form of pulses, in which case the control system receives no information about the error signal during the periods between two consecutive pulses.

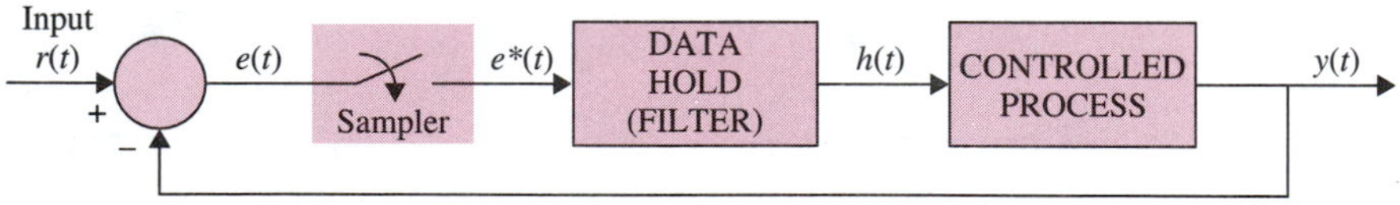

Figure 1-19 Block diagram of a sampled-data control system.

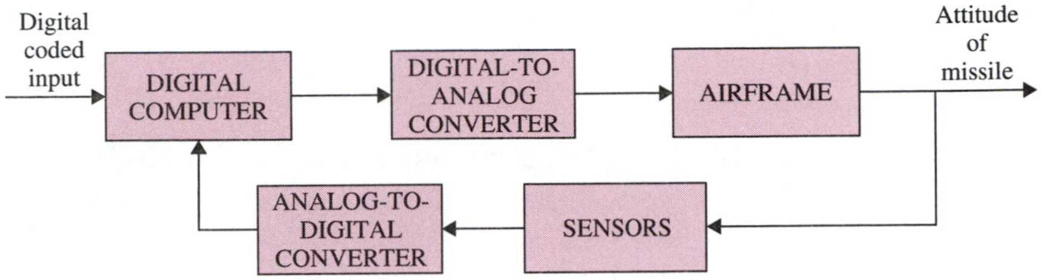

Figure 1-20 Digital autopilot system for a guided missile.

Strictly, a sampled-data system can also be classified as an ac system, since the signal of the system is pulse modulated.

Figure 1-19 illustrates how a typical sampled-data system operates. A continuous-data input signal $r(t)$ is applied to the system. The error signal $e(t)$ is sampled by a sampling device, the **sampler,** and the output of the sampler is a sequence of pulses. The sampling rate of the sampler may or may not be uniform. There are many advantages of incorporating sampling into a control system. One important advantage of the sampling operation is that expensive equipment used in the system may be time-shared among several control channels. Another advantage is that pulse data are usually less susceptible to noise.

▲ Digital control systems are usually less susceptible to noise.

Because digital computers provide many advantages in size and flexibility, computer control has become increasingly popular in recent years. Many airborne systems contain digital controllers that can pack thousands of discrete elements in a space no larger than the size of this book. Figure 1-20 shows the basic elements of a digital autopilot for guided missile control.

1-4 Summary

In this chapter we have introduced some of the basic concepts of what a control system is and what it is supposed to accomplish. The basic components of a control system are described. By demonstrating the effects of feedback in a rudimentary way, the question of why most control systems are closed-loop systems is also clarified. Most important, it is pointed out that feedback is a double-edged sword—it can benefit as well as harm a system to be controlled. This points out the challenging task in the design of a control system, which involves consideration on such performance criteria as stability, sensitivity, bandwidth, and accuracy. Finally, the various types of control systems are categorized according to the system signals, linearity, and control objectives. Several typical control system examples are given to illustrate the points of emphasis in the analysis and design of control systems. Points are also made on the fact that most systems encountered in real life are nonlinear and time varying to some extent. The concentration on the studies of linear systems is due primarily to the availability of unified and simple-to-understand analytical methods in the analysis and design of linear systems.

Review Questions

1. List the advantages and disadvantages of an open-loop system.
2. List the advantages and the disadvantages of a closed-loop system.
3. Define ac and dc control systems.
4. State the advantages of a digital control system over a continuous-data control system.
5. A closed-loop control system is usually more accurate than an open-loop system. (T) (F)
6. Feedback is sometimes used to improve the sensitivity of a control system. (T) (F)
7. If an open-loop system is unstable, applying feedback will always improve its stability. (T) (F)
8. Feedback can increase the gain of a system in one frequency range but decrease it in another. (T) (F)
9. Nonlinear elements are sometimes intentionally introduced to a control system to improve its performance. (T) (F)
10. Discrete-data control systems are more susceptible to noise, due to the nature of their signals. (T) (F)

Answers to True-and-False Questions

5. (T) 6. (T) 7. (F) 8. (T) 9. (T) 10. (F)

2 Mathematical Foundation

KEYWORDS AND TOPICS

- Complex Variables
- Laplace Transform
- Partial-Fraction Expansion
- Differential Equation
- State Equations
- Difference Equations
- Matrix Theory
- z-Transform

2-1 Introduction

The studies of control systems rely to a great extent on the use of applied mathematics. One of the major purposes of control-system studies is to develop a set of analytical tools so that the designer can arrive at reasonably predictable and reliable designs without depending completely on the drudgery of experimentation or extensive computer simulation.

For the study of classical control theory, which represents a good portion of this book, the required mathematical background includes such subjects as **complex-variable theory, differential and difference equations, Laplace transformation and z-trans-**

formation, and so on. Modern control theory, on the other hand, requires considerably more intensive mathematical background. In addition to the above-mentioned subjects, modern control theory is based on the foundation of **matrix theory, set theory, linear algebra and transformation, variational calculus, mathematical programming, probability theory,** and other advanced mathematics.

In this chapter we present the background material that is needed for the topics on control systems discussed in this book. Because of space limitations and the fact that most subjects are considered as reviews to the reader, the treatment of these mathematical subjects is not exhaustive. The reader who wishes to conduct an in-depth study of any of these subjects should refer to books that are devoted to each subject.

2-2 Complex-Variable Concept

2-2-1 Complex Variable

A complex variable s has two components: a real component σ and an imaginary component ω. Graphically, the real component of s is represented by a σ-axis in the horizontal direction, and the imaginary component is measured along the vertical $j\omega$-axis, in the complex s-plane. Figure 2-1 illustrates the complex s-plane, in which any arbitrary point $s = s_1$ is defined by the coordinates $\sigma = \sigma_1$, and $\omega = \omega_1$, or simply $s_1 = \sigma_1 + j\omega_1$.

2-2-2 Functions of a Complex Variable

The function $G(s)$ is said to be a function of the complex variable s if for every value of s, there are one or more corresponding values of $G(s)$. Since s is defined to have real and imaginary parts, the function $G(s)$ is also represented by its real and imaginary

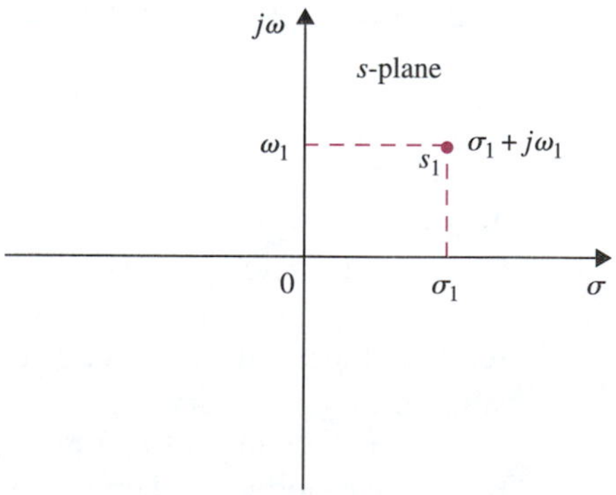

Figure 2-1 Complex s-plane.

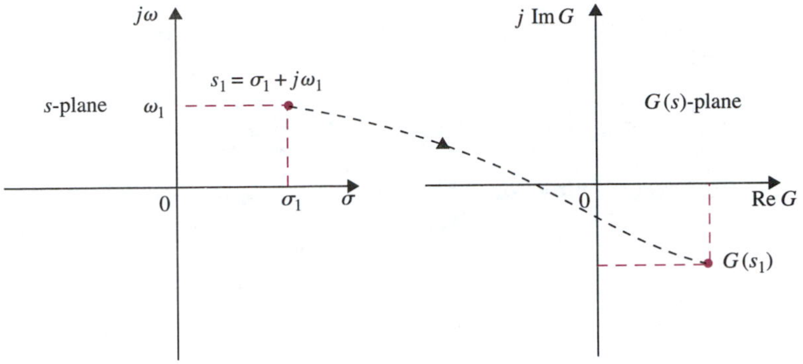

Figure 2-2 Single-valued mapping from the s-plane to the $G(s)$-plane.

parts; that is,

$$G(s) = \text{Re } G(s) + j \text{ Im } G(s) \tag{2-1}$$

where Re $G(s)$ denotes the real part of $G(s)$ and Im $G(s)$ represents the imaginary part of $G(s)$. The function $G(s)$ is also represented by the complex $G(s)$-plane, with Re $G(s)$ as the real axis and Im $G(s)$ as the imaginary axis. If for every value of s there is only one corresponding value of $G(s)$ in the $G(s)$-plane, $G(s)$ is said to be a **single-valued function,** and the mapping from points in the s-plane onto points in the $G(s)$-plane is described as **single-valued** (Fig. 2-2). If the mapping from the $G(s)$-plane to the s-plane is also single-valued, the mapping is called **one-to-one.** However, there are many functions for which the mapping from the function plane to the complex-variable plane is not single-valued. For instance, given the function

$$G(s) = \frac{1}{s(s+1)} \tag{2-2}$$

it is apparent that for each value of s, there is only one unique corresponding value for $G(s)$. However, the inverse mapping is not true; for instance, the point $G(s) = \infty$ is mapped onto two points, $s = 0$ and $s = -1$, in the s-plane.

2-2-3 Analytic Function

*A function $G(s)$ of the complex variable s is called an **analytic function** in a region of the s-plane if the function and all its derivatives exist in the region.* For instance, the function given in Eq. (2-2) is analytic at every point in the s-plane except at the point $s = 0$ and $s = -1$. At these two points, the value of the function is infinite. As another example, the function $G(s) = s + 2$ is analytic at every point in the finite s-plane.

2-2-4 Singularities and Poles of a Function

The **singularities** of a function are the points in the s-plane at which the function or its derivatives do not exist. A **pole** is the most common type of singularity and plays a very important role in studies of classical control theory.

The definition of a pole can be stated as: *If a function $G(s)$ is analytic and single-valued in the neighborhood of s_i, it is said to have a pole of order r at $s = s_i$ if the limit*

$$\lim_{s \to s_i} [(s - s_i)^r G(s)]$$

has a finite, nonzero value. In other words, the denominator of $G(s)$ must include the factor $(s - s_i)^r$, so when $s = s_i$, the function becomes infinite. If $r = 1$, the pole at $s = s_i$ is called a **simple pole.** As an example, the function

$$G(s) = \frac{10(s + 2)}{s(s + 1)(s + 3)^2} \tag{2-3}$$

has a pole of order 2 at $s = -3$ and simple poles at $s = 0$ and $s = -1$. It can also be said that the function $G(s)$ is analytic in the s-plane except at these poles.

2-2-5 Zeros of a Function

The definition of a **zero** of a function can be stated as: *If the function $G(s)$ is analytic at $s = s_i$, it is said to have a zero of order r at $s = s_i$ if the limit*

$$\lim_{s \to s_i} [(s - s_i)^{-r} G(s)]$$

▲ The total numbers of poles and zeros of a rational function are the same, counting the ones at infinity.

has a finite, nonzero value. Or, simply, $G(s)$ has a zero of order r at $s = s_i$ if $1/G(s)$ has an rth-order pole at $s = s_i$. For example, the function in Eq. (2-3) has a simple zero at $s = -2$.

If the function under consideration is a rational function of s, that is, a quotient of two polynomials of s, the total number of poles equals the total number of zeros, counting the multiple-order poles and zeros, and taking into account of the poles and zeros at infinity. The function in Eq. (2-3) has four finite poles at $s = 0, -1, -3$, and -3; there is one finite zero at $s = -2$, but there are three zeros at infinity, since

$$\lim_{s \to \infty} G(s) = \lim_{s \to \infty} \frac{10}{s^3} = 0 \tag{2-4}$$

Therefore, the function has a total of four poles and four zeros in the entire s-plane, including infinity.

2-3 Differential Equations

2-3-1 Linear Ordinary Differential Equations

A wide range of systems in engineering are modeled mathematically by differential equations. These equations generally involve derivatives and integrals of the dependent variables with respect to the independent variable. For instance, a series electric *RLC* (resistance–inductance–capacitance) network can be represented by the differential equation

$$Ri(t) + L\frac{di(t)}{dt} + \frac{1}{C}\int i(t)\,dt = e(t) \tag{2-5}$$

where R is the resistance, L the inductance, C the capacitance, $i(t)$ the current in the network, and $e(t)$ the applied voltage. In this case, $e(t)$ is the forcing function, t the independent variable, and $i(t)$ the dependent variable or the unknown that is to be determined by solving the differential equation.

Equation (2-5) is referred to as a second-order differential equation, and we refer to the system as a **second-order system.** Strictly speaking, Eq. (2-5) should be referred to as an integrodifferential equation, since an integral is involved.

In general, the differential equation of an nth-order system is written

$$\frac{d^n y(t)}{dt^n} + a_{n-1}\frac{d^{n-1}y(t)}{dt^{n-1}} + \cdots + a_1\frac{dy(t)}{dt} + a_0 y(t) = f(t) \tag{2-6}$$

which is also known as a **linear ordinary differential equation** if the coefficients a_0, $a_1, \ldots, a_{n-1}$ are not functions of $y(t)$.

In this book, since we treat only systems that contain lumped parameters, the differential equations encountered are all of the ordinary type. For systems with distributed parameters, such as in heat-transfer systems, partial differential equations are used.

2-3-2 Nonlinear Differential Equations

Many physical systems are nonlinear and must be described by nonlinear differential equations. For instance, the differential equation that describes the motion of the pendu-

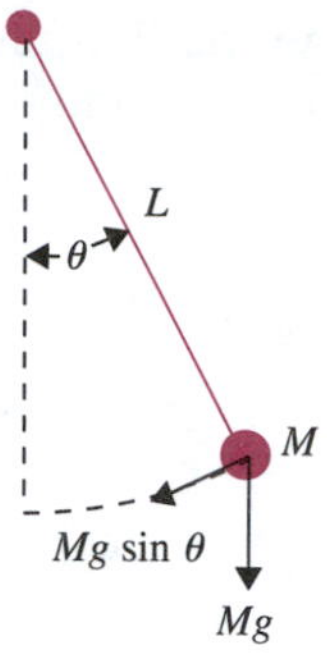

Figure 2-3 Simple pendulum.

lum shown in Fig. 2-3 is

$$ML\frac{d^2\theta(t)}{dt^2} + Mg\sin\theta(t) = 0 \tag{2-7}$$

Since $\theta(t)$ appears as a sine function, Eq. (2-7) is nonlinear, and the system is called a **nonlinear system.**

2-3-3 First-Order Differential Equations: State Equations

In general, an nth-order differential equation can be decomposed into n first-order differential equations. Since, in principle, first-order differential equations are simpler to solve than higher-order ones, there are reasons why first-order differential equations are used in the analytical studies of control systems.

For the differential equation in Eq. (2-5), if we let

$$x_1(t) = \int i(t)\,dt \tag{2-8}$$

and

$$x_2(t) = \frac{dx_1(t)}{dt} = i(t) \tag{2-9}$$

Eq. (2-5) is decomposed into the following two first-order differential equations:

$$\frac{dx_1(t)}{dt} = x_2(t) \tag{2-10}$$

$$\frac{dx_2(t)}{dt} = -\frac{1}{LC}x_1(t) - \frac{R}{L}x_2(t) + \frac{1}{L}e(t) \tag{2-11}$$

In a similar manner, for Eq. (2-6), let us define

$$
\begin{aligned}
x_1(t) &= y(t) \\
x_2(t) &= \frac{dy(t)}{dt} \\
&\ \vdots \\
x_n(t) &= \frac{d^{n-1}y(t)}{dt^{n-1}}
\end{aligned}
\tag{2-12}
$$

Then the nth-order differential equation is decomposed into n first-order differential equations:

$$
\begin{aligned}
\frac{dx_1(t)}{dt} &= x_2(t) \\
\frac{dx_2(t)}{dt} &= x_3(t) \\
&\ \vdots \\
\frac{dx_n(t)}{dt} &= -a_0x_1(t) - a_1x_2(t) - \cdots - a_{n-2}x_{n-1}(t) - a_{n-1}x_n(t) + f(t)
\end{aligned}
\tag{2-13}
$$

Notice that the last equation is obtained by equating the highest-ordered derivative term in Eq. (2-6) to the rest of the terms. In control systems theory, the set of first-order differential equations in Eq. (2-13) is called the **state equations,** and $x_1, x_2, \ldots, x_n$ are called the **state variables.**

Definition of State Variables

▲ The state of a system refers to the past, present, and future of the system.

The **state** of a system refers to the **past, present,** and **future** conditions of the system. From a mathematical sense, it is convenient to define a set of **state variables** and **state equations** to model dynamic systems. As it turns out, the variables $x_1(t), x_2(t), \ldots, x_n(t)$ defined in Eq. (2-12) are the state variables of the nth-order system described by Eq. (2-6), and the n first-order differential equations are the state equations. In general, there are some basic rules regarding the definition of a state variable and what constitutes a state equation. The state variables must satisfy the following conditions:

▲ The state variables must always be a minimal set.

1. At any initial time $t = t_0$, the state variables $x_1(t_0), x_2(t_0), \ldots, x_n(t_0)$ define the **initial states** of the system.

2. Once the inputs of the system for $t \geq t_0$ and the initial states defined above are specified, the state variables should completely define the future behavior of the system.

The state variables of a system are defined as a **minimal set** *of variables,* $x_1(t)$, $x_2(t), \ldots, x_n(t)$, *such that knowledge of these variables at any time* t_0, *and information on the input excitation subsequently applied, are sufficient to determine the state of the system at any time* $t > t_0$.

Output Equation

▲ The output of a system must always be measurable.

One should not confuse the state variables with the outputs of a system. An output of a system is a variable that can be **measured,** but a state variable does not always satisfy this requirement. For instance, in an electric motor, such state variables as the winding current, rotor velocity, and displacement can be measured physically, and these variables all qualify as output variables. On the other hand, magnetic flux can also be regarded as a state variable in an electric motor, since it represents the past, present, and future states of the motor, but it cannot be measured directly during operation, and therefore it does not ordinarily qualify as an output variable. In general, an output variable can be expressed as an algebraic combination of the state variables. For the system described by Eq. (2-6), if $y(t)$ is designated as the output, the output equation is simply $y(t) = x_1(t)$.

2-4 Laplace Transform

The Laplace transform is one of the mathematical tools used for the solution of linear ordinary differential equations. In comparison with the classical method of solving linear differential equations, the Laplace transform method has the following two attractive features:

1. The homogeneous equation and the particular integral of the solution are obtained in one operation.
2. The Laplace transform converts the differential equation into an algebraic equation in s. It is then possible to manipulate the algebraic equation by simple algebraic rules to obtain the solution in the s-domain. The final solution is obtained by taking the inverse Laplace transform.

2-4-1 Definition of the Laplace Transform

Given the real function $f(t)$ that satisfies the condition

$$\int_0^\infty |f(t)e^{-\sigma t}|\, dt < \infty \tag{2-14}$$

for some finite real σ, the Laplace transform of $f(t)$ is defined as

$$F(s) = \int_0^\infty f(t)e^{-st}\, dt \tag{2-15}$$

or

$$F(s) = \text{Laplace transform of } f(t) = \mathscr{L}[f(t)] \tag{2-16}$$

▲ The response of a causal system does not precede the input.

The variable s is referred to as the **Laplace operator,** which is a complex variable; that is, $s = \sigma + j\omega$. The defining equation in Fig. (2-15) is also known as the **one-sided Laplace transform** as the integration is evaluated from $t = 0$ to ∞. This simply means that all information contained in $f(t)$ prior to $t = 0$ is ignored or considered to be zero. This assumption does not impose any limitation on the applications of the Laplace transform to linear system problems, since in the usual time-domain studies, time reference is often chosen at $t = 0$. Furthermore, for a physical system when an input is applied at $t = 0$, the response of the system does not start sooner than $t = 0$; that is, response does not precede excitation. Such a system is also known as being **causal** or simply **physically realizable.**

Strictly, the one-sided Laplace transform should be defined from $t = 0^-$ to $t = \infty$. The symbol $t = 0^-$ implies that the limit of $t \to 0$ is taken from the left side of $t = 0$. This limiting process will take care of situations under which the function $f(t)$ has a jump discontinuity or an impulse at $t = 0$. For the subjects treated in this book, the defining equation of the Laplace transform in Eq. (2-15) is almost never used in problem solving, since the transform expressions encountered are either given or can be found from the Laplace transform table. Thus the fine point of using 0^- or 0^+ never needs to be addressed. For simplicity, we shall simply use $t = 0$ or $t = t_0$ as the initial time in all subsequent discussions.

The following examples illustrate how Eq. (2-15) is used for evaluation of the Laplace transform of $f(t)$.

Example 2-1

Let $f(t)$ be a unit-step function that is defined as

$$f(t) \begin{cases} u_s(t) = 1 & t > 0 \\ 0 & t < 0 \end{cases} \tag{2-17}$$

The Laplace transform of $f(t)$ is obtained as

$$F(s) = \mathcal{L}[u_s(t)] = \int_0^\infty u_s(t)e^{-st}\,dt = -\frac{1}{s}e^{-st}\bigg|_0^\infty = \frac{1}{s} \tag{2-18}$$

Equation (2-18) is valid if

$$\int_0^\infty |u_s(t)e^{-\sigma t}|\,dt = \int_0^\infty |e^{-\sigma t}|\,dt < \infty \tag{2-19}$$

which means that the real part of s, σ, must be greater than zero. In practice, we simply refer to the Laplace transform of the unit-step function as $1/s$, and rarely do we have to be concerned with the region in the s-plane in which the transform integral converges absolutely. ▲

Example 2-2

Consider the exponential function

$$f(t) = e^{-\alpha t} \qquad t \geq 0 \tag{2-20}$$

where α is a real constant. The Laplace transform of $f(t)$ is written

$$F(s) = \int_0^\infty e^{-\alpha t}e^{-st}\,dt = \frac{e^{-(s+\alpha)t}}{s+\alpha}\bigg|_0^\infty = \frac{1}{s+\alpha} \tag{2-21}$$

▲

2-4-2 Inverse Laplace Transformation

Given the Laplace transform $F(s)$, the operation of obtaining $f(t)$ is termed the **inverse Laplace transformation** and is denoted by

$$f(t) = \mathcal{L}^{-1}[F(s)] \tag{2-22}$$

The inverse Laplace transform integral is given as

$$f(t) = \frac{1}{2\pi j}\int_{c-j\infty}^{c+j\infty} F(s)e^{st}\,ds \tag{2-23}$$

where c is a real constant that is greater than the real parts of all the singularities of $F(s)$. Equation (2-23) represents a line integral that is to be evaluated in the s-plane. For simple functions, the inverse Laplace transform operation can be carried out simply by referring to the Laplace transform table, such as the one given in Appendix B. For complex functions, the inverse Laplace transform can be carried out by first performing a partial-fraction expansion (Section 2-5) on $F(s)$ and then using the transform table. A computer program such as the **pfe** of the **CSAD** toolbox in conjunction with **MATLAB** can also be used for partial-fraction expansion/inverse Laplace transformation. More details on these are given in Section 2-5.

2-4-3 Important Theorems of the Laplace Transform

Applications of the Laplace transform in many instances are simplified by utilization of the properties of the transform. These properties are presented by the following theorems, for which no proofs are given here.

▲ **Theorem 1.** *Multiplication by a Constant*

Let k be a constant and $F(s)$ be the Laplace transform of $f(t)$. Then

$$\mathscr{L}[kf(t)] = kF(s) \tag{2-24}$$

▲ **Theorem 2.** *Sum and Difference*

Let $F_1(s)$ and $F_2(s)$ be the Laplace transform of $f_1(t)$ and $f_2(t)$, respectively. Then

$$\mathscr{L}[f_1(t) \pm f_2(t)] = F_1(s) \pm F_2(s) \tag{2-25}$$

▲ **Theorem 3.** *Differentiation*

Let $F(s)$ be the Laplace transform of $f(t)$, and let $f(0)$ be the limit of $f(t)$ as t approaches 0. The Laplace transform of the time derivative of $f(t)$ is

$$\mathscr{L}\left[\frac{df(t)}{dt}\right] = sF(s) - \lim_{t \to 0} f(t) = sF(s) - f(0) \tag{2-26}$$

In general, for higher-order derivatives of $f(t)$,

$$\begin{aligned}\mathcal{L}\left[\frac{d^n f(t)}{dt^n}\right] &= s^n F(s) - \lim_{t\to 0}\left[s^{n-1}f(t) + s^{n-2}\frac{df(t)}{dt} + \cdots + \frac{d^{n-1}f(t)}{dt^{n-1}}\right] \\ &= s^n F(s) - s^{n-1}f(0) - s^{n-2}f^{(1)}(0) - \cdots - f^{(n-1)}(0)\end{aligned} \tag{2-27}$$

where $f^{(i)}(0)$ denotes the ith-order derivative of $f(t)$ with respect to t, evaluated at $t = 0$.

▲ **Theorem 4.** *Integration*

The Laplace transform of the first integral of $f(t)$ with respect to time is the Laplace transform of $f(t)$ divided by s; that is,

$$\mathcal{L}\left[\int_0^t f(\tau)\,d\tau\right] = \frac{F(s)}{s} \tag{2-28}$$

For nth-order integration,

$$\mathcal{L}\left[\int_0^{t_1}\int_0^{t_2}\cdots\int_0^{t_n} f(\tau)\,d\tau\,dt_1\,dt_2\cdots dt_{n-1}\right] = \frac{F(s)}{s^n} \tag{2-29}$$

▲ **Theorem 5.** *Shift in Time*

The Laplace transform of $f(t)$ delayed by time T is equal to the Laplace transform $f(t)$ multiplied by e^{-Ts}; that is,

$$\mathcal{L}[f(t - T)u_s(t - T)] = e^{-Ts}F(s) \tag{2-30}$$

where $u_s(t - T)$ denotes the unit-step function that is shifted in time to the right by T.

▲ **Theorem 6.** *Initial-Value Theorem*

If the Laplace transform of $f(t)$ is $F(s)$, then

$$\lim_{t\to 0} f(t) = \lim_{s\to\infty} sF(s) \tag{2-31}$$

if the limit exists.

▲ **Theorem 7.** *Final-Value Theorem*

▲ The final-value theorem is valid only if $sF(s)$ does not have any poles on the $j\omega$ axis and in the right half of the s-plane.

If the Laplace transform of $f(t)$ is $F(s)$, and if $sF(s)$ is analytic on the imaginary axis and in the right half of the s-plane, then

$$\lim_{t\to\infty} f(t) = \lim_{s\to 0} sF(s) \tag{2-32}$$

The final-value theorem is very useful for the analysis and design of control systems, since it gives the final value of a time function by knowing the behavior of its Laplace transform at $s = 0$. The final-value theorem is *not* valid if $sF(s)$ contains any pole whose real part is zero or positive, which is equivalent to the analytic requirement of $sF(s)$ in the right-half plane as stated in the theorem. The following examples illustrate the care that must be taken in applying the theorem.

Example 2-3

Consider the function

$$F(s) = \frac{5}{s(s^2 + s + 2)} \tag{2-33}$$

Since $sF(s)$ is analytic on the imaginary axis and in the right-half s-plane, the final-value theorem may by applied. Using Eq. (2-32), we have

$$\lim_{t\to\infty} f(t) = \lim_{s\to 0} sF(s) = \lim_{s\to 0} \frac{5}{s^2 + s + 2} = \frac{5}{2} \tag{2-34}$$

▲

Example 2-4

Consider the function

$$F(s) = \frac{\omega}{s^2 + \omega^2} \tag{2-35}$$

which is the Laplace transform of $f(t) = \sin \omega t$. Since the function $sF(s)$ has two poles on the imaginary axis of the s-plane, the final-value theorem *cannot* be applied in this case. In other words, although the final-value theorem would yield a value of zero as the final value of $f(t)$, the result is erroneous. ▲

▲ **Theorem 8.** *Complex Shifting*

The Laplace transform of $f(t)$ multiplied by $e^{\mp\alpha t}$, where α is a constant, is equal to the Laplace transform $F(s)$, with s replaced by $s \pm \alpha$; that is,

$$\mathscr{L}[e^{\mp\alpha t} f(t)] = F(s \pm \alpha) \tag{2-36}$$

▲ **Theorem 9.** *Real Convolution (Complex Multiplication)*

Let $F_1(s)$ and $F_2(s)$ be the Laplace transforms of $f_1(t)$ and $f_2(t)$, respectively, and $f_1(t) = 0$, $f_2(t) = 0$, for $t < 0$; then

$$\begin{aligned} F_1(s)F_2(s) &= \mathcal{L}[f_1(t) * f_2(t)] \\ &= \mathcal{L}\left[\int_0^t f_1(\tau)f_2(t-\tau)\,d\tau\right] = \mathcal{L}\left[\int_0^t f_2(\tau)f_1(t-\tau)\,d\tau\right] \end{aligned} \tag{2-37}$$

where the symbol "$*$" denotes **convolution** in the time domain.

Equation (2-37) shows that multiplication of two transformed functions in the complex s-domain is equivalent to the convolution of two corresponding real functions of t in the t-domain. An important fact to remember is that *the inverse Laplace transform of the product of two functions in the s-domain is* ***not*** *equal to the product of the two corresponding real functions in the t-domain*; that is, in general,

$$\mathcal{L}^{-1}[F_1(s)F_2(s)] \neq f_1(t)f_2(t) \tag{2-38}$$

There is also a dual relation to the real convolution theorem, called the **complex convolution,** or **real multiplication.** Essentially, the theorem states that multiplication in the real t-domain is equivalent to convolution in the complex s-domain; that is,

$$\mathcal{L}[f_1(t)f_2(t)] = F_1(s) * F_2(s) \tag{2-39}$$

where $*$ denotes complex convolution in this case. Details of the complex convolution formula are not given here.

Table 2-1 summarizes the theorem of the Laplace transforms represented.

2-5 Inverse Laplace Transform by Partial-Fraction Expansion

In a majority of problems in control systems, the evaluation of the inverse Laplace transform does not rely on the use of the inversion integral of Eq. (2-23). Rather, the inverse Laplace transform operation involving rational functions can be carried out using a Laplace transform table and partial-fraction expansion. The latter and the looking up of the inverse transform in the transform table can be conducted by computer programs.

Table 2-1 Theorems of Laplace Tranforms

Multiplication by a constant	$\mathscr{L}[kf(t)] = kF(s)$
Sum and difference	$\mathscr{L}[f_1(t) \pm f_2(t)] = F_1(s) \pm F_2(s)$
Differentiation	$\mathscr{L}\left[\frac{df(t)}{dt}\right] = sF(s) - f(0)$
	$\mathscr{L}\left[\frac{d^nf(t)}{dt^n}\right] = s^nF(s) - s^{n-1}f(0) - s^{n-2}f^{(1)}(0) - \cdots - sf^{(n-2)}(0) - f^{(n-1)}(0)$
	where $f^{(k)}(0) = \left.\frac{d^kf(t)}{dt^k}\right\rvert_{t=0}$
Integration	$\mathscr{L}\left[\int_0^t f(\tau)\,d\tau\right] = \frac{F(s)}{s}$
	$\mathscr{L}\left[\int_0^{t_1}\int_0^{t_2}\cdots\int_0^{t_n} f(\tau)\,d\tau\,dt_1\,dt_2\cdots dt_{n-1}\right] = \frac{F(s)}{s^n}$
Shift in time	$\mathscr{L}[f(t-T)u_s(t-T)] = e^{-Ts}F(s)$
Initial-value theorem	$\lim_{t\to 0} f(t) = \lim_{s\to\infty} sF(s)$
Final-value theorem	$\lim_{t\to\infty} f(t) = \lim_{s\to 0} sF(s)$ if $sF(s)$ does not have poles on or to the right of the imaginary axis in the s-plane
Complex shifting	$\mathscr{L}[e^{\mp\alpha t}f(t)] = F(s \pm \alpha)$
Real convolution	$F_1(s)F_2(s) = \mathscr{L}\left[\int_0^t f_1(\tau)f_2(t-\tau)\,d\tau\right] = \mathscr{L}\left[\int_0^t f_2(\tau)f_1(t-\tau)\,d\tau\right] = \mathscr{L}[f_1(t) * f_2(t)]$

2-5-1 Partial-Fraction Expansion

When the Laplace transform solution of a differential equation is a rational function in s, it can be written as

$$G(s) = \frac{Q(s)}{P(s)} \tag{2-40}$$

where $P(s)$ and $Q(s)$ are polynomials of s. It is assumed that *the order of $P(s)$ in s is greater than that of $Q(s)$*. The polynomial $P(s)$ may be written

$$P(s) = s^n + a_{n-1}s^{n-1} + \cdots + a_1 s + a_0 \tag{2-41}$$

where $a_0, a_1, \ldots, a_{n-1}$ are real coefficients. The methods of partial-fraction expansion will now be given for the cases of simple poles, multiple-order poles, and complex-conjugate poles of $G(s)$.

$G(s)$ Has Simple Poles

If all the poles of $G(s)$ are simple and real, Eq. (2-40) can be written as

$$G(s) = \frac{Q(s)}{P(s)} = \frac{Q(s)}{(s+s_1)(s+s_2)\cdots(s+s_n)} \tag{2-42}$$

where $s_1 \neq s_2 \neq \cdots \neq s_n$. Applying the partial-fraction expansion, Eq. (2-42) is written

$$G(s) = \frac{K_{s1}}{s+s_1} + \frac{K_{s2}}{s+s_2} + \cdots + \frac{K_{sn}}{s+s_n} \tag{2-43}$$

The coefficient K_{si} ($i = 1, 2, \ldots, n$) is determined by multiplying both sides of Eq. (2-42) or Eq. (2-43) by the factor $(s + s_i)$ and then setting s equal to $-s_i$. To find the coefficient K_{s1}, for instance, we multiply both sides of Eq. (2-42) by $(s + s_1)$ and let $s = -s_1$. Thus

$$K_{s1} = \left[(s+s_1)\frac{Q(s)}{P(s)}\right]\Bigg|_{s=-s_1} = \frac{Q(-s_1)}{(s_2-s_1)(s_3-s_1)\cdots(s_n-s_1)} \tag{2-44}$$

Example 2-5

Consider the function

$$G(s) = \frac{5s+3}{(s+1)(s+2)(s+3)} \tag{2-45}$$

which is written in the partial-fraction expanded form:

$$G(s) = \frac{K_{-1}}{s+1} + \frac{K_{-2}}{s+2} + \frac{K_{-3}}{s+3} \tag{2-46}$$

The coefficients K_{-1}, K_{-2}, and K_{-3} are determined as follows:

$$K_{-1} = [(s+1)X(s)]|_{s=-1} = \frac{5(-1)+3}{(2-1)(3-1)} = -1 \tag{2-47}$$

$$K_{-2} = [(s+2)X(s)]|_{s=-2} = \frac{5(-2)+3}{(1-2)(3-2)} = 7 \tag{2-48}$$

$$K_{-3} = [(s+3)X(s)]|_{s=-3} = \frac{5(-3)+3}{(1-3)(2-3)} = -6 \tag{2-49}$$

Thus Eq. (2-46) becomes

$$G(s) = \frac{-1}{s+1} + \frac{7}{s+2} - \frac{6}{s+3} \tag{2-50}$$

▲

$G(s)$ Has Multiple-Order Poles

If r of the n poles of $G(s)$ are identical, or we say that the pole at $s = -s_i$ is of multiplicity r, $G(s)$ is written

$$G(s) = \frac{Q(s)}{P(s)} = \frac{Q(s)}{(s+s_1)(s+s_2)\cdots(s+s_{n-r})(s+s_i)^r} \tag{2-51}$$

$(i \neq 1, 2, \ldots, n - r)$. Then $G(s)$ can be expanded as

$$\begin{aligned} G(s) = {} & \frac{K_{s1}}{s+s_1} + \frac{K_{s2}}{s+s_2} + \cdots + \frac{K_{s(n-r)}}{s+s_{n-r}} \\ & |\leftarrow n - r \text{ terms of simple poles} \rightarrow| \\ & + \frac{A_1}{s+s_i} + \frac{A_2}{(s+s_i)^2} + \cdots + \frac{A_r}{(s+s_i)^r} \\ & |\leftarrow r \text{ terms of repeated poles} \rightarrow| \end{aligned} \tag{2-52}$$

The $(n - r)$ coefficients, $K_{s1}, K_{s2}, \ldots, K_{s(n-r)}$, which correspond to simple poles, may be evaluated by the method described by Eq. (2-44). The determination of the coefficients that correspond to the multiple-order poles is described as follows:

$$A_r = [(s+s_i)^r G(s)]\big|_{s=-s_i} \tag{2-53}$$

$$A_{r-1} = \frac{d}{ds}[(s+s_i)^r G(s)]\big|_{s=-s_i} \tag{2-54}$$

$$A_{r-2} = \frac{1}{2!}\frac{d^2}{ds^2}[(s+s_i)^r G(s)]\big|_{s=-s_i} \tag{2-55}$$

$$\vdots$$

$$A_1 = \frac{1}{(r-1)!}\frac{d^{r-1}}{ds^{r-1}}[(s+s_i)^r G(s)]\big|_{s=-s_i} \tag{2-56}$$

Example 2-6

Consider the function

$$G(s) = \frac{1}{s(s+1)^3(s+2)} \tag{2-57}$$

By using the format of Eq. (2-52), $G(s)$ is written

$$G(s) = \frac{K_0}{s} + \frac{K_{-2}}{s+2} + \frac{A_1}{s+1} + \frac{A_2}{(s+1)^2} + \frac{A_3}{(s+1)^3} \tag{2-58}$$

The coefficients corresponding to the simple poles are

$$K_0 = [sG(s)]\big|_{s=0} = \frac{1}{2} \tag{2-59}$$

$$K_{-2} = [(s+2)G(s)]\big|_{s=-2} = \frac{1}{2} \tag{2-60}$$

and those of the third-order pole are

$$A_3 = [(s+1)^3G(s)]\big|_{s=-1} = -1 \tag{2-61}$$

$$A_2 = \frac{d}{ds}[(s+1)^3G(s)]\big|_{s=-1} = \frac{d}{ds}\left[\frac{1}{s(s+2)}\right]\Bigg|_{s=-1} = 0 \tag{2-62}$$

$$A_1 = \frac{1}{2!}\frac{d^2}{ds^2}[(s+1)^3G(s)]\big|_{s=-1} = \frac{1}{2}\frac{d^2}{ds^2}\left[\frac{1}{s(s+2)}\right]\Bigg|_{s=-1} = -1 \tag{2-63}$$

The completed partial-fraction expansion is

$$G(s) = \frac{1}{2s} + \frac{1}{2(s+2)} - \frac{1}{s+1} - \frac{1}{(s+1)^3} \tag{2-64}$$

▲

$G(s)$ Has Simple Complex-Conjugate Poles

The partial-fraction expansion of Eq. (2-43) is also valid for simple complex-conjugate poles. Since complex-conjugate poles are more difficult to handle and are of special interest in control systems studies, they deserve special treatment here.

Suppose that $G(s)$ of Eq. (2-40) contains a pair of complex poles:

$$s = -\alpha + j\omega \quad \text{and} \quad s = -\alpha - j\omega$$

The corresponding coefficients of these poles are

$$K_{-\alpha+j\omega} = (s + \alpha - j\omega)G(s)\big|_{s=-\alpha+j\omega} \tag{2-65}$$

$$K_{-\alpha-j\omega} = (s + \alpha + j\omega)G(s)\big|_{s=-\alpha-j\omega} \tag{2-66}$$

Example 2-7

Consider the function

$$G(s) = \frac{\omega_n^2}{s^2 + 2\zeta\omega_n s + \omega_n^2} \tag{2-67}$$

Let us assume that the values of ζ and ω_n are such that the poles of $G(s)$ are complex. Then $G(s)$ is expanded as follows:

$$G(s) = \frac{K_{-\alpha+j\omega}}{s + \alpha - j\omega} + \frac{K_{-\alpha-j\omega}}{s + \alpha + j\omega} \tag{2-68}$$

where

$$\alpha = \zeta\omega_n \tag{2-69}$$

and

$$\omega = \omega_n\sqrt{1 - \zeta^2} \tag{2-70}$$

The coefficients in Eq. (2-68) are determined as

$$K_{-\alpha+j\omega} = (s + \alpha - j\omega)G(s)\big|_{s=-\alpha+j\omega} = \frac{\omega_n^2}{2j\omega} \tag{2-71}$$

$$K_{-\alpha-j\omega} = (s + \alpha + j\omega)G(s)\big|_{s=-\alpha-j\omega} = -\frac{\omega_n^2}{2j\omega} \tag{2-72}$$

The complete partial-fraction expansion of Eq. (2-67) is

$$G(s) = \frac{\omega_n^2}{2j\omega}\left[\frac{1}{s + \alpha - j\omega} - \frac{1}{s + \alpha + j\omega}\right] \tag{2-73}$$

Taking the inverse Laplace transform on both sides of Eq. (2-73) gives

$$\begin{aligned} g(t) &= \frac{\omega_n^2}{2j\omega}e^{-\alpha t}(e^{j\omega t} - e^{-j\omega t}) \\ &= \frac{\omega_n}{\sqrt{1 - \zeta^2}}e^{-\zeta\omega_n t}\sin\omega_n\sqrt{1 - \zeta^2}\,t \qquad t \geq 0 \end{aligned} \tag{2-74}$$

▲

2-5-2 Computer Solution of the Partial-Fraction Expansion

The partial-fraction expansion can be carried out by a digital computer program. For example, the function **pfe** of the **CSAD** toolbox associated with **MATLAB** and **pfe** of **ACSP** can be used for performing partial-fraction expansion by entering the transfer-function expression $G(s)$. A typical run on the **CSAD** is given below for the transfer function

$$G(s) = \frac{20(s+10)}{s(s+2)^2(s^2+10s+100)} \tag{2-75}$$

pfe of CSAD

Entering **pfe** <CR> at the **MATLAB** prompt ≫, the program responds

```
Enter Numerator polynomial vector >20*[1 10] <CR>
Enter Denominator polynomial vector >([1 0],[1 2],[1 2],[1 10 100])
<CR>
```

The output of **pfe** is

```
Terms in the partial fraction expansion are:
Complex Conjugate Pole at: -5.000 +/- j8.660

        (0.02721)(S+5   )
   ----------------------------
   (S+5     )^2+(8.66      )^2

        (0.003928)(8.66)
   ----------------------------
   (S+5     )^2+(8.66      )^2

Time Domain: 0.02721*exp(-5t)*cos(8.66t)*u(t)

                + 0.003828*exp(-5t)*sin(8.66t)*u(t)

Double pole on Real Axis at: -2.000

         -0.5272     -0.9524
         ------- + -----------
           S+2      (S+2   )^2

Time Domain: -0.5272*exp(-2t)*u(t)

             -0.9524*t*exp(-2t)*u(t)

Pole at the origin:

         0.5
         ---
          S

Time Domain: 0.5*u(t)

Partial Fraction Expansion Complete
```

Notice that **CSAD** also returns the inverse Laplace transform of $G(s)$.

2-6 Application of Laplace Transform to the Solution of Linear Ordinary Differential Equations

Linear ordinary differential equations can be solved by the Laplace transform method with the aid of the theorems on Laplace transform given in Section 2-4, the partial-fraction expansion, and a table of Laplace transforms. The procedure is outlined as follows:

1. Transform the differential equation to the s-domain by Laplace transform using the Laplace transform table.
2. Manipulate the transformed algebraic equation and solve for the output variable.
3. Perform partial-fraction expansion to the transformed algebraic equation.
4. Obtain the inverse Laplace transform from the Laplace transform table.

Let us illustrate the method by several examples.

Example 2-8

Consider the differential equation

$$\frac{d^2y(t)}{dt^2} + 3\frac{dy(t)}{dt} + 2y(t) = 5u_s(t) \tag{2-76}$$

where $u_s(t)$ is the unit-step function. The initial conditions are $y(0) = -1$ and $y^{(1)}(0) = dy(t)/dt|_{t=0} = 2$. To solve the differential equation, we first take the Laplace transform on both sides of Eq. (2-76):

$$s^2Y(s) - sy(0) - y^{(1)}(0) + 3sY(s) - 3y(0) + 2Y(s) = 5/s \tag{2-77}$$

Substituting the values of the initial conditions into Eq. (2-77) and solving for $Y(s)$, we get

$$Y(s) = \frac{-s^2 - s + 5}{s(s^2 + 3s + 2)} = \frac{-s^2 - s + 5}{s(s + 1)(s + 2)} \tag{2-78}$$

Equation (2-78) is expanded by partial-fraction expansion to give

$$Y(s) = \frac{5}{2s} - \frac{5}{s + 1} + \frac{3}{2(s + 2)} \tag{2-79}$$

Taking the inverse Laplace transform of Eq. (2-79), we get the complete solution as

$$y(t) = \frac{5}{2} - 5e^{-t} + \frac{3}{2}e^{-2t} \qquad t \geq 0 \tag{2-80}$$

The first term in Eq. (2-80) is the steady-state solution or the particular integral; the last two terms represent the transient solution, or the homogeneous solution. Unlike the classical method, which requires separate steps to give the transient and steady-state solutions, the Laplace transform method gives the entire solution in one operation.

If only the magnitude of the steady-state solution of $y(t)$ is of interest, the final-value theorem of Eq. (2-32) may be applied. Thus

$$\lim_{t\to\infty} y(t) = \lim_{s\to 0} sY(s) = \lim_{s\to 0} \frac{-s^2 - s + 5}{s^2 + 3s + 2} = \frac{5}{2} \tag{2-81}$$

where we have first checked and found that the function $sY(s)$ has poles only in the left-half s-plane, so that the final-value theorem is valid. ▲

Example 2-9

Consider the linear differential equation

$$\frac{d^2y(t)}{dt^2} + 34.5\frac{dy(t)}{dt} + 1000y(t) = 1000u_s(t) \tag{2-82}$$

The initial values of $y(t)$ and $dy(t)/dt$ are zero. Taking the Laplace transform on both sides of Eq. (2-82) and solving for $Y(s)$, we have

$$Y(s) = \frac{1000}{s(s^2 + 34.5s + 1000)} = \frac{\omega_n^2}{s(s^2 + 2\zeta\omega_n s + \omega_n^2)} \tag{2-83}$$

where $\zeta = 0.5455$ and $\omega_n = 31.62$. The inverse Laplace transform of Eq. (2-83) can be executed in a number of ways. The Laplace transform table in Appendix B gives the transform pair of the pair of expressions in Eq. (2-83) directly. The result is

$$y(t) = 1 - \frac{e^{-\zeta\omega_n t}}{\sqrt{1-\zeta^2}}\sin(\omega_n\sqrt{1-\zeta^2}\,t + \theta) \qquad t \geq 0 \tag{2-84}$$

where

$$\theta = \cos^{-1}\zeta = 56.94° \tag{2-85}$$

Thus

$$y(t) = 1 - 1.193e^{-17.25t}\sin(26.5t + 56.94°) \qquad t \geq 0 \tag{2-86}$$

Equation (2-86) can be derived by performing the partial-fraction expansion of Eq. (2-83) knowing that the poles are at $s = 0$, $-\alpha + j\omega$, and $-\alpha - j\omega$, where

$$\alpha = \zeta\omega_n = 17.25 \tag{2-87}$$

$$\omega = \omega_n\sqrt{1-\zeta^2} = 26.5 \tag{2-88}$$

The partial-fraction expansion of Eq. (2-83) is written

$$Y(s) = \frac{K_0}{s} + \frac{K_{-\alpha+j\omega}}{s+\alpha-j\omega} + \frac{K_{-\alpha-j\omega}}{s+\alpha+j\omega} \tag{2-89}$$

where

$$K_0 = sY(s)|_{s=0} = 1 \tag{2-90}$$

$$K_{-\alpha+j\omega} = (s+\alpha-j\omega)Y(s)|_{s=-\alpha+j\omega} = \frac{e^{-j\phi}}{2j\sqrt{1-\zeta^2}} \tag{2-91}$$

$$K_{-\alpha-j\omega} = (s+\alpha+j\omega)Y(s)|_{s=-\alpha-j\omega} = \frac{-e^{-j\phi}}{2j\sqrt{1-\zeta^2}} \tag{2-92}$$

The angle ϕ is given by

$$\phi = 180° - \cos^{-1}\zeta \tag{2-93}$$

and is illustrated in Fig. 2-4.

The inverse Laplace transform of Eq. (2-89) is now written

$$\begin{aligned} y(t) &= 1 + \frac{1}{2j\sqrt{1-\zeta^2}} e^{-\zeta\omega_n t}[e^{j(\omega t-\phi)} - e^{-j(\omega t-\phi)}] \\ &= 1 + \frac{1}{\sqrt{1-\zeta^2}} e^{-\zeta\omega_n t}\sin(\omega_n\sqrt{1-\zeta^2}t - \phi) \qquad t \geq 0 \end{aligned} \tag{2-94}$$

Substituting Eq. (2-93) into Eq. (2-94) for ϕ, we have

$$y(t) = 1 - \frac{1}{\sqrt{1-\zeta^2}} e^{-\zeta\omega_n t}\sin(\omega_n\sqrt{1-\zeta^2}t + \cos^{-1}\zeta) \qquad t \geq 0 \tag{2-95}$$

or

$$y(t) = 1 - 1.193e^{-17.25t}\sin(26.5t + 56.94°) \qquad t \geq 0 \tag{2-96}$$

▲

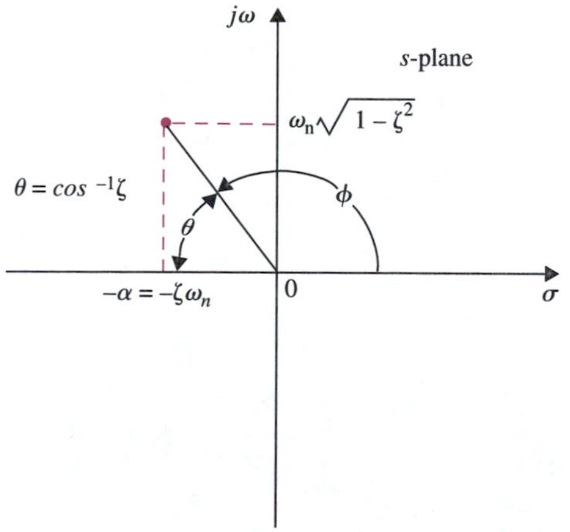

Figure 2-4 Root location in the s-plane.

2-7 Elementary Matrix Theory

In the study of modern control theory, it is often desirable to use matrix notation to simplify complex mathematical expressions. The matrix notation usually makes the equations much easier to handle and manipulate.

As a motivation for the reason to use matrix notation, let us consider the following set of n simultaneous algebraic equations:

$$\begin{aligned} a_{11}x_1 + a_{12}x_2 + \cdots + a_{1n}x_n &= y_1 \\ a_{21}x_1 + a_{22}x_2 + \cdots + a_{2n}x_n &= y_2 \\ \cdots\cdots\cdots\cdots\cdots\cdots\cdots\cdots & \\ a_{n1}x_1 + a_{n2}x_2 + \cdots + a_{nn}x_n &= y_n \end{aligned} \tag{2-97}$$

We may use the matrix equation

$$\mathbf{Ax} = \mathbf{y} \tag{2-98}$$

as a simplified representation of Eq. (2-97). The symbols **A**, **x**, and **y** are defined as **matrices,** which contain the coefficients and variables of the original equations as their elements. In terms of matrix algebra, which will be discussed shortly, Eq. (2-98) can be stated as: *The product of the matrices* **A** *and* **x** *is equal to the matrix* **y**. The three matrices involved are defined as

$$\mathbf{A} = \begin{bmatrix} a_{11} & a_{12} & \cdots & a_{1n} \\ a_{21} & a_{22} & \cdots & a_{2n} \\ \vdots & \vdots & \ddots & \vdots \\ a_{n1} & a_{n2} & \cdots & a_{nn} \end{bmatrix} \tag{2-99}$$

$$\mathbf{x} = \begin{bmatrix} x_1 \\ x_2 \\ \vdots \\ x_n \end{bmatrix} \tag{2-100}$$

$$\mathbf{y} = \begin{bmatrix} y_1 \\ y_2 \\ \vdots \\ y_n \end{bmatrix} \tag{2-101}$$

which are simply bracketed arrays of coefficients and variables. These examples of matrices prompted the definition of a matrix that follows.

2-7-1 Definition of a Matrix

▲ It is important to distinguish between matrix and determinant.

A matrix is a collection of elements arranged in a rectangular or square array. There are several ways of bracketing a matrix. In this book, the square brackets, such as those in Eqs. (2-99) through (2-101), are used to represent matrices. It is important to distinguish between a **matrix** and a **determinant.** The basic characteristics of these are listed as follows:

MATRIX	DETERMINANT
▲ An array of numbers or elements with n rows and m columns.	▲ An array of numbers or elements with n rows and n columns (always square).
▲ Does not have a value, although a square matrix ($n = m$) has a determinant.	▲ Has a value.

Some important definitions of matrices are given next.

Matrix Elements. When a matrix is written

$$\mathbf{A} = \begin{bmatrix} a_{11} & a_{12} & a_{13} \\ a_{21} & a_{22} & a_{23} \\ a_{31} & a_{32} & a_{33} \end{bmatrix} \tag{2-102}$$

a_{ij} is defined as the element in the ith **row** and the jth **column** of the matrix. As a rule, we always refer to the row first and the column last.

Order of a Matrix. *The order of a matrix refers to the total number of rows and columns of the matrix.* For example, the matrix in Eq. (2-102) has three rows and three columns and is called a 3×3 (three-by-three) matrix. A matrix with n rows and m columns is termed $n \times m$ or n by m.

Square Matrix. *A square matrix is one that has the same number of rows as columns.*

Column Matrix. *A column matrix is one that has one column and more than one row, that is, an $m \times 1$ matrix, $m > 1$.* Quite often, a column matrix is referred to as a **column vector** or simply an **m-vector** if there are m rows and one column. The matrix in Eq. (2-100) is a typical n-vector.

Row Matrix. *A row matrix is one that has one row and more than one column, that* *is, a 1 × n matrix, where n > 1.* A row matrix can also be referred to as a **row vector.**

Diagonal Matrix. *A diagonal matrix is a square matrix with $a_{ij} = 0$ for all $i \neq j$.* Examples of a diagonal matrix are

$$\begin{bmatrix} a_{11} & 0 & 0 \\ 0 & a_{22} & 0 \\ 0 & 0 & a_{33} \end{bmatrix} \begin{bmatrix} 5 & 0 \\ 0 & 3 \end{bmatrix}$$

Unity Matrix (Identity Matrix). *A unity matrix is a diagonal matrix with all the elements on the main diagonal ($i = j$) equal to 1.* A unity matrix is often designated by **I** or **U**. An example of a unity matrix is

$$\mathbf{I} = \begin{bmatrix} 1 & 0 & 0 \\ 0 & 1 & 0 \\ 0 & 0 & 1 \end{bmatrix} \tag{2-103}$$

Null Matrix. *A null matrix is one whose elements are all equal to zero.*

Symmetric Matrix. *A symmetrix matrix is a square matrix that satisfies the condition $a_{ij} = a_{ji}$ for all i and j.* A symmetric matrix has the property that if its rows are interchanged with its columns, the same matrix is obtained. Two examples of the symmetric matrix are

$$\mathbf{A} = \begin{bmatrix} 6 & 5 & 1 \\ 5 & 0 & 10 \\ 1 & 10 & -1 \end{bmatrix} \qquad \mathbf{B} = \begin{bmatrix} 1 & -4 \\ -4 & 1 \end{bmatrix} \tag{2-104}$$

Determinant of a Matrix. *With each square matrix, a determinant having the same elements and order as the matrix may be defined. The determinant of a square matrix* **A** *is designated by*

$$\det \mathbf{A} = \Delta_A = |\mathbf{A}| \tag{2-105}$$

For example, the determinant of the matrix in Eq. (2-102) is

$$|\mathbf{A}| = \begin{vmatrix} a_{11} & a_{12} & a_{13} \\ a_{21} & a_{22} & a_{23} \\ a_{31} & a_{32} & a_{33} \end{vmatrix} \tag{2-106}$$

Cofactor of a Determinant Element. *Given any* n*th-order determinant* $|\mathbf{A}|$, *the cofactor of any element* a_{ij}, A_{ij}, *is the determinant obtained by eliminating all elements of the* ith *row and* jth *column and then multiplied by* $(-1)^{i+j}$. For example, the cofactor of the element a_{11} of $|\mathbf{A}|$ in Eq. (2-106) is

$$A_{11} = (-1)^{1+1} \begin{vmatrix} a_{22} & a_{23} \\ a_{32} & a_{33} \end{vmatrix} = a_{22}a_{33} - a_{23}a_{32} \tag{2-107}$$

In general, the value of a determinant can be written in terms of the cofactors. Let $\mathbf{A}$ be an $n \times n$ matrix; then the determinant of $\mathbf{A}$ can be written in terms of the cofactor of any row or the cofactor of any column. That is,

$$\det \mathbf{A} = \sum_{j=1}^{n} a_{ij}A_{ij} \qquad (i = 1, \text{or } 2, \ldots, \text{or } n) \tag{2-108}$$

or

$$\det \mathbf{A} = \sum_{i=1}^{n} a_{ij}A_{ij} \qquad (j = 1, \text{or } 2, \ldots, \text{or } n) \tag{2-109}$$

Example 2-10

The value of the determinant in Eq. (2-106) is

$$\begin{aligned}\det \mathbf{A} = |\mathbf{A}| &= a_{11}A_{11} + a_{12}A_{12} + a_{13}A_{13} \\ &= a_{11}(a_{22}a_{33} - a_{23}a_{32}) - a_{12}(a_{21}a_{33} - a_{23}a_{31}) + a_{13}(a_{21}a_{32} - a_{22}a_{31})\end{aligned} \tag{2-110}$$

▲

Singular Matrix. *A square matrix is said to be singular if the value of its determinant is zero.* If a square matrix has a nonzero determinant, it is called a **nonsingular matrix.** When a matrix is singular, it usually means that not all the rows or not all the columns of the matrix are independent of each other. When the matrix is used to represent a set of algebraic equations, singularity of the matrix means that these equations are not independent of each other.

Example 2-11

Consider the following set of equations:

$$\begin{aligned} 2x_1 - 3x_2 + x_3 &= 0 \\ -x_1 + x_2 + x_3 &= 0 \\ x_1 - 2x_2 + 2x_3 &= 0 \end{aligned} \tag{2-111}$$

The third equation is equal to the sum of the first two. Thus these three equations are not completely independent. In matrix form these equations may be written as

$$\mathbf{Ax} = \mathbf{0} \tag{2-112}$$

where

$$\mathbf{A} = \begin{bmatrix} 2 & -3 & 1 \\ -1 & 1 & 1 \\ 1 & -2 & 2 \end{bmatrix} \qquad \mathbf{x} = \begin{bmatrix} x_1 \\ x_2 \\ x_3 \end{bmatrix} \tag{2-113}$$

and **0** is a 3 × 1 null matrix. The determinant of **A** is 0, and thus the matrix **A** is singular. In this case the rows of **A** are dependent. ▲

Transpose of a Matrix. *The transpose of a matrix* **A** *is defined as the matrix that is obtained by interchanging the corresponding rows and columns in* **A.** Let **A** be an $n \times m$ matrix that is represented by

$$\mathbf{A} = [a_{ij}]_{n,m} \tag{2-114}$$

The transpose of **A**, denoted by **A**′, is given by

$$\mathbf{A}' = \text{transpose of } \mathbf{A} = [a_{ij}]_{m,n} \tag{2-115}$$

Notice that the order of **A** is $n \times m$, but the order of **A**′ is $m \times n$.

Example 2-12

Consider the 2 × 3 matrix

$$\mathbf{A} = \begin{bmatrix} 3 & 2 & 1 \\ 0 & -1 & 5 \end{bmatrix} \tag{2-116}$$

The transpose of **A** is obtained by interchanging the rows and the columns.

$$\mathbf{A}' = \begin{bmatrix} 3 & 0 \\ 2 & -1 \\ 1 & 5 \end{bmatrix} \tag{2-117}$$

▲

SOME PROPERTIES OF MATRIX TRANSPOSE

1. $(\mathbf{A}') = \mathbf{A}$ (2-118)
2. $(k\mathbf{A})' = k\mathbf{A}'$, where k is a scalar (2-119)
3. $(\mathbf{A} + \mathbf{B})' = \mathbf{A}' + \mathbf{B}'$ (2-120)
4. $(\mathbf{AB})' = \mathbf{B}'\mathbf{A}'$ (2-121)

Adjoint of a Matrix. *Let* **A** *be a square matrix of order n. The adjoint matrix of* **A**, *denoted by adj* **A**, *is defined as*

$$\text{adj}\,\mathbf{A} = [A_{ij} \text{ of det } \mathbf{A}]_{n,n} \tag{2-122}$$

where A_{ij} denotes the cofactor of a_{ij}.

Example 2-13

Consider the 2 × 2 matrix

$$\mathbf{A} = \begin{bmatrix} a_{11} & a_{12} \\ a_{21} & a_{22} \end{bmatrix} \tag{2-123}$$

The cofactors are $A_{11} = a_{22}$, $A_{12} = -a_{21}$, $A_{21} = -a_{12}$, and $A_{22} = a_{11}$. Thus the adjoint matrix of **A** is

$$\text{adj}\,\mathbf{A} = \begin{bmatrix} A_{11} & A_{12} \\ A_{21} & A_{22} \end{bmatrix}' = \begin{bmatrix} a_{22} & -a_{21} \\ -a_{12} & a_{11} \end{bmatrix}' = \begin{bmatrix} a_{22} & -a_{12} \\ -a_{21} & a_{11} \end{bmatrix} \tag{2-124}$$

▲

Trace of a Square Matrix. *Given an n × n matrix with elements a_{ij}, the* **trace of A,** *denoted as* tr(**A**), *is defined as the sum of the elements on the main diagonal of* **A**; *that is,*

$$\text{tr}(\mathbf{A}) = \sum_{i=1}^{n} a_{ii} \tag{2-125}$$

The trace of a matrix has the following properties:

1. $\text{tr}(\mathbf{A}') = \text{tr}(\mathbf{A})$ (2-126)
2. For $n \times n$ square matrices **A** and **B**,

$$\text{tr}(\mathbf{A} + \mathbf{B}) = \text{tr}(\mathbf{A}) + \text{tr}(\mathbf{B}) \tag{2-127}$$

2-8 Matrix Algebra

When carrying out matrix operations, it is necessary to define matrix algebra in the form of addition, subtraction, multiplication, and division.

2-8-1 Equality of Matrices

Two matrices **A** and **B** are said to be equal to each other if they satisfy the following conditions:

1. They are of the same order.
2. The corresponding elements are equal; that is,

$$a_{ij} = b_{ij} \qquad \text{for every } i \text{ and } j \tag{2-128}$$

Example 2-14

$$\mathbf{A} = \begin{bmatrix} a_{11} & a_{12} \\ a_{21} & a_{22} \end{bmatrix} = \mathbf{B} = \begin{bmatrix} b_{11} & b_{12} \\ b_{21} & b_{22} \end{bmatrix} \tag{2-129}$$

implies that $a_{11} = b_{11}$, $a_{12} = b_{12}$, $a_{21} = b_{21}$, $a_{22} = b_{22}$. ▲

2-8-2 Addition and Subtraction of Matrices

Two matrices **A** and **B** can be added or subtracted to form $\mathbf{A} \pm \mathbf{B}$ if they are of the same order. That is,

$$\mathbf{A} \pm \mathbf{B} = [a_{ij}]_{n,m} \pm [b_{ij}]_{n,m} = \mathbf{C} = [c_{ij}]_{n,m} \tag{2-130}$$

where

$$c_{ij} = a_{ij} \pm b_{ij} \qquad \text{for all } i \text{ and } j \tag{2-131}$$

The order of the matrices is preserved after addition or subtraction.

Example 2-15

Consider the matrices

$$\mathbf{A} = \begin{bmatrix} 3 & 2 \\ -1 & 4 \\ 0 & -1 \end{bmatrix} \qquad \mathbf{B} = \begin{bmatrix} 0 & 3 \\ -1 & 2 \\ 1 & 0 \end{bmatrix} \tag{2-132}$$

which are of the same order. Then the sum of **A** and **B** is

$$\mathbf{C} = \mathbf{A} + \mathbf{B} \begin{bmatrix} 3+0 & 2+3 \\ -1-1 & 4+2 \\ 0+1 & -1+0 \end{bmatrix} = \begin{bmatrix} 3 & 5 \\ -2 & 6 \\ 1 & -1 \end{bmatrix} \tag{2-133}$$

▲

2-8-3 Associative Law of Matrix (Addition and Subtraction)

The associative law of scalar algebra still holds for matrix addition and subtraction. That is,

$$(\mathbf{A} + \mathbf{B}) + \mathbf{C} = \mathbf{A} + (\mathbf{B} + \mathbf{C}) \tag{2-134}$$

2-8-4 Commutative Law of Matrix (Addition and Subtraction)

The commutative law for matrix addition and subtraction states that the following matrix relationship is true:

$$\mathbf{A} + \mathbf{B} + \mathbf{C} = \mathbf{B} + \mathbf{C} + \mathbf{A} = \mathbf{A} + \mathbf{C} + \mathbf{B} \tag{2-135}$$

as well as other possible commutative combinations.

2-8-5 Matrix Multiplication

The matrices **A** and **B** may be multiplied together to form the product **AB** if they are **conformable.** This means that the number of columns of **A** must equal the number of rows of **B**. In other words, let

$$\mathbf{A} = [a_{ij}]_{n,p} \qquad \mathbf{B} = [b_{ij}]_{q,m} \tag{2-136}$$

Then **A** and **B** are conformable to form the product

$$\mathbf{C} = \mathbf{AB} = [a_{ij}]_{n,p}[b_{ij}]_{q,m} = [c_{ij}]_{n,m} \tag{2-137}$$

if and only if $p = q$. The matrix **C** will have the same number of rows as **A** and the same number of columns as **B**.

▲ Two matrices can be multiplied only if they are conformable.

It is important to note that **A** and **B** may be conformable to form **AB**, but they may not be conformable for the product **BA**, unless in Eq. (2-137), n also equals m. This points out an important fact that *the commutative law is not generally valid for matrix multiplication.* It is also noteworthy that even though **A** and **B** are conformable for both **AB** and **BA**, usually $\mathbf{AB} \neq \mathbf{BA}$. The following references are made with respect to matrix manipulation whenever they exist:

$$\mathbf{AB} = \mathbf{A} \text{ postmultiplied by } \mathbf{B} \quad \text{or} \quad \mathbf{B} \text{ premultiplied by } \mathbf{A} \tag{2-138}$$

2-8-6 Rules of Matrix Multiplication

When the matrices **A** ($n \times p$) and **B** ($p \times m$) are conformable to form the matrix **C** = **AB**, the ijth element of **C**, c_{ij}, is given by

$$c_{ij} = \sum_{k=1}^{p} a_{ik} b_{kj} \tag{2-139}$$

for $i = 1, 2, \ldots, n$ and $j = 1, 2, \ldots, m$.

Example 2-16

Given the matrices

$$\mathbf{A} = [a_{ij}]_{2,3} \qquad \mathbf{B} = [b_{ij}]_{3,1} \tag{2-140}$$

The two matrices are conformable for the product **AB** but not for **BA**. Thus

$$\mathbf{AB} = \begin{bmatrix} a_{11} & a_{12} & a_{13} \\ a_{21} & a_{22} & a_{23} \end{bmatrix} \begin{bmatrix} b_{11} \\ b_{21} \\ b_{31} \end{bmatrix} = \begin{bmatrix} a_{11}b_{11} + a_{12}b_{21} + a_{13}b_{31} \\ a_{21}b_{11} + a_{22}b_{21} + a_{23}b_{31} \end{bmatrix} \tag{2-141}$$

▲

Example 2-17

Given the matrices

$$\mathbf{A} = \begin{bmatrix} 3 & -1 \\ 0 & 1 \\ 2 & 0 \end{bmatrix} \qquad \mathbf{B} = \begin{bmatrix} 1 & 0 & -1 \\ 2 & 1 & 0 \end{bmatrix} \tag{2-142}$$

The two matrices are comformable for **AB** and **BA**.

$$\mathbf{AB} = \begin{bmatrix} 3 & -1 \\ 0 & 1 \\ 2 & 0 \end{bmatrix} \begin{bmatrix} 1 & 0 & -1 \\ 2 & 1 & 0 \end{bmatrix} = \begin{bmatrix} 1 & -1 & -3 \\ 2 & 1 & 0 \\ 2 & 0 & -2 \end{bmatrix} \tag{2-143}$$

$$\mathbf{BA} = \begin{bmatrix} 1 & 0 & -1 \\ 2 & 1 & 0 \end{bmatrix} \begin{bmatrix} 3 & -1 \\ 0 & 1 \\ 2 & 0 \end{bmatrix} = \begin{bmatrix} 1 & -1 \\ 6 & -1 \end{bmatrix} \tag{2-144}$$

Therefore, even though **AB** and **BA** both exist, they are not equal. In fact, in this case the products are not of the same order. ▲

Although the commutative law does not hold in general for matrix multiplication, the **associative** and **distributive** laws are valid. For the distributive law, we state that

$$\mathbf{A}(\mathbf{B} + \mathbf{C}) = \mathbf{AB} + \mathbf{AC} \tag{2-145}$$

if the products are conformable. For the associative law,

$$(\mathbf{AB})\mathbf{C} = \mathbf{A}(\mathbf{BC}) \tag{2-146}$$

if the product is conformable.

2-8-7 Multiplication by a Scalar k

Multiplying a matrix **A** *by any scalar k is equivalent to multiplying each element of* **A** *by k.*

2-8-8 Inverse of a Matrix (Matrix Division)

▲ Only square matrices have inverses, and only if they are nonsingular.

In the algebra of scalar quantities, when we write $y = ax$, it implies that $x = y/a$ is also true. In matrix algebra, if $\mathbf{Ax} = \mathbf{y}$, it *may be possible* to write

$$\mathbf{x} = \mathbf{A}^{-1}\mathbf{y} \tag{2-147}$$

where $\mathbf{A}^{-1}$ denotes the **matrix inverse** of **A**. The conditions that $\mathbf{A}^{-1}$ exists are:

1. **A** is a square matrix.
2. **A** must be nonsingular.
3. If $\mathbf{A}^{-1}$ exists, it is given by

$$\mathbf{A}^{-1} = \frac{\text{adj}\,\mathbf{A}}{|\mathbf{A}|} \tag{2-148}$$

Example 2-18

▲ The adjoint of a 2 × 2 matrix is obtained by interchanging the two main diagonal elements and changing the

Given the matrix

$$\mathbf{A} = \begin{bmatrix} a_{11} & a_{12} \\ a_{21} & a_{22} \end{bmatrix} \tag{2-149}$$

the inverse of **A** is given by

$$\mathbf{A}^{-1} = \frac{\text{adj}\,\mathbf{A}}{|\mathbf{A}|} = \frac{\begin{bmatrix} a_{22} & -a_{12} \\ -a_{21} & a_{11} \end{bmatrix}}{a_{11}a_{22} - a_{12}a_{21}} \tag{2-150}$$

where for **A** to be nonsingular, $|\mathbf{A}| \neq 0$, or $a_{11}a_{22} - a_{12}a_{21} \neq 0$.

signs of the elements off the diagonal.

Equation (2-150) shows that adj **A** of a 2 × 2 matrix is obtained by *interchanging the two elements on the main diagonal and changing the signs of the elements off the diagonal of* **A.** ▲

Example 2-19

Given the matrix

$$\mathbf{A} = \begin{bmatrix} a_{11} & a_{12} & a_{13} \\ a_{21} & a_{22} & a_{23} \\ a_{31} & a_{32} & a_{33} \end{bmatrix} \tag{2-151}$$

To find the inverse of **A**, the adjoint of **A** is

$$\text{adj}\,\mathbf{A} = \begin{bmatrix} a_{22}a_{33} - a_{23}a_{32} & -(a_{12}a_{33} - a_{13}a_{32}) & a_{12}a_{23} - a_{13}a_{22} \\ -(a_{21}a_{33} - a_{23}a_{31}) & a_{11}a_{33} - a_{13}a_{31} & -(a_{11}a_{23} - a_{21}a_{13}) \\ a_{21}a_{32} - a_{22}a_{31} & -(a_{11}a_{32} - a_{12}a_{31}) & a_{11}a_{22} - a_{12}a_{21} \end{bmatrix} \tag{2-152}$$

The determinant of **A** is

$$|\mathbf{A}| = a_{11}a_{22}a_{33} + a_{12}a_{23}a_{31} + a_{13}a_{32}a_{21} - a_{13}a_{22}a_{31} - a_{12}a_{21}a_{33} - a_{11}a_{23}a_{32} \tag{2-153}$$

▲

SOME PROPERTIES OF MATRIX INVERSE

1. $\mathbf{A}\mathbf{A}^{-1} = \mathbf{A}^{-1}\mathbf{A} = \mathbf{I}$ (2-154)
2. $(\mathbf{A}^{-1})^{-1} = \mathbf{A}$ (2-155)
3. In matrix algebra, in general, **AB** = **AC** does not necessarily imply that **B** = **C**. The reader can easily construct an example to illustrate this property. However, if **A** is a square matrix and is nonsingular, we can premultiply both sides of **AB** = **AC** by $\mathbf{A}^{-1}$. Then

$$\mathbf{A}^{-1}\mathbf{A}\mathbf{B} = \mathbf{A}^{-1}\mathbf{A}\mathbf{C} \tag{2-156}$$

   which leads to **B** = **C**.
4. If **A** and **B** are square matrices and are nonsingular, then

$$(\mathbf{A}\mathbf{B})^{-1} = \mathbf{B}^{-1}\mathbf{A}^{-1} \tag{2-157}$$

2-8-9 Rank of a Matrix

The rank of a matrix **A** *is the maximum number of linearly independent columns of* **A**; *or it is the order of the largest nonsingular matrix contained in* **A**.

Example 2-20

Several examples on the rank of a matrix are as follows:

$$\begin{bmatrix} 0 & 1 \\ 0 & 1 \end{bmatrix} \quad \text{rank} = 1 \qquad \begin{bmatrix} 0 & 5 & 1 & 4 \\ 3 & 0 & 3 & 2 \end{bmatrix} \quad \text{rank} = 2$$

$$\begin{bmatrix} 3 & 9 & 2 \\ 1 & 3 & 0 \\ 2 & 6 & 1 \end{bmatrix} \quad \text{rank} = 2 \qquad \begin{bmatrix} 3 & 0 & 0 \\ 1 & 2 & 0 \\ 0 & 0 & 1 \end{bmatrix} \quad \text{rank} = 3$$

▲

The following properties are useful in the determination of the rank of a matrix. Given an $n \times m$ matrix $\mathbf{A}$:

1. Rank of $\mathbf{A}'$ = rank of $\mathbf{A}$.
2. Rank of $\mathbf{A}'\mathbf{A}$ = rank of $\mathbf{A}$.
3. Rank of $\mathbf{A}\mathbf{A}'$ = rank of $\mathbf{A}$.

Properties 2 and 3 are useful in the determination of rank; since $\mathbf{A}'\mathbf{A}$ and $\mathbf{A}\mathbf{A}'$ are always square, the rank condition can be checked by evaluating the determinants of these matrices.

2-8-10 Computer-Aided Solutions of Matrices

Many commercial computer-program packages such as **MATLAB** contain routines for matrix manipulations.

2-9 Vector-Matrix Form of State Equations

The state equations defined in Section 2-3 can be expressed conveniently in vector-matrix form. Consider that a linear second-order system is described by the following state equations:

$$\frac{dx_1(t)}{dt} = a_{11}x_1(t) + a_{12}x_2(t) + b_{11}u(t) \tag{2-158}$$

$$\frac{dx_2(t)}{dt} = a_{21}x_1(t) + a_{22}x_2(t) + b_{21}u(t) \tag{2-159}$$

where $x_1(t)$ and $x_2(t)$ are the state variables; $u(t)$ is the input; a_{11}, a_{12}, a_{21}, a_{22}, b_{11}, and b_{21} are constant coefficients. Let the **state vector** be defined as the 2×1 matrix,

$$\mathbf{x}(t) = \begin{bmatrix} x_1(t) \\ x_2(t) \end{bmatrix} \tag{2-160}$$

The state equations in vector-matrix form are expressed as

$$\frac{d\mathbf{x}(t)}{dt} = \dot{\mathbf{x}}(t) = \mathbf{A}\mathbf{x}(t) + \mathbf{B}\mathbf{u}(t) \tag{2-161}$$

where

$$\mathbf{A} = \begin{bmatrix} a_{11} & a_{12} \\ a_{21} & a_{22} \end{bmatrix} \qquad \mathbf{B} = \begin{bmatrix} b_{11} \\ b_{21} \end{bmatrix} \tag{2-162}$$

Equation (2-161) can be used to represent the n state equations in Eq. (2-13), in which case $\dot{\mathbf{x}}(t)$ and $\mathbf{x}(t)$ are $n \times 1$ vectors, $\mathbf{u}(t)$ is the scalar input $f(t)$, and the coefficient matrices are

$$\mathbf{A} = \begin{bmatrix} 0 & 1 & 0 & \cdots & 0 \\ 0 & 0 & 1 & \cdots & 0 \\ \vdots & \vdots & \vdots & \ddots & \vdots \\ 0 & 0 & 0 & 0 & 1 \\ -a_0 & -a_1 & -a_2 & \cdots & -a_{n-1} \end{bmatrix} \qquad \mathbf{B} = \begin{bmatrix} 0 \\ 0 \\ \vdots \\ 0 \\ 1 \end{bmatrix} \tag{2-163}$$

For nonlinear state equations, the vector-matrix form is

$$\frac{d\mathbf{x}(t)}{dt} = \mathbf{f}[\mathbf{x}(t), \mathbf{u}(t), t] \tag{2-164}$$

where $\mathbf{f}[\cdot]$ represents an $n \times 1$ function matrix.

2-10 Difference Equations

Because digital controllers are frequently used in control systems, it is necessary to establish equations that relate digital and discrete-time signals. Just as differential equations are used to represent systems with analog signals, difference equations are used for systems with discrete or digital data. Difference equations are also used to approximate differential equations, since the former are more easily programmed on a digital computer and are easier to solve.

A linear nth-order difference equation with constant coefficients can be written as

$$y(k+n) + a_{n-1}y(k+n-1) + \cdots + a_1 y(k+1) + a_0 y(k) = f(k) \tag{2-165}$$

where $y(i)$, $i = k, k+1, \ldots, k+n$, denotes the discrete dependent variable y at the ith instant if the independent variable is time. In general, the independent variable can be any real quantity.

Similar to the case of the analog systems, it is convenient to use a set of first-order difference equations, or state equations, to represent a high-order difference equation. For the difference equation in Eq. (2-165), if we let

$$\begin{aligned} x_1(k) &= y(k) \\ x_2(k) &= x_1(k+1) = y(k+1) \\ &\vdots \\ x_{n-1}(k) &= x_{n-2}(k+1) = y(k+n-2) \\ x_n(k) &= x_{n-1}(k+1) = y(k+n-1) \end{aligned} \tag{2-166}$$

then by equating the highest-order term to the rest, the equation is written as

$$x_n(k+1) = -a_0 x_1(k) - a_1 x_2(k) - \cdots - a_{n-1}x_n(k) + f(k) \tag{2-167}$$

The first $n - 1$ state equations are taken directly from the last $n - 1$ equations in Eq. (2-166), and the last one is given by Eq. (2-167). The n state equations are written in vector-matrix form:

$$\mathbf{x}(k+1) = \mathbf{A}\mathbf{x}(k) + \mathbf{B}u(k) \tag{2-168}$$

where

$$\mathbf{x}(k) = \begin{bmatrix} x_1(k) \\ x_2(k) \\ \vdots \\ x_n(k) \end{bmatrix} \tag{2-169}$$

is the $n \times 1$ state vector, and

$$\mathbf{A} = \begin{bmatrix} 0 & 1 & 0 & \cdots & 0 \\ 0 & 0 & 1 & \cdots & 0 \\ \vdots & \vdots & \vdots & \ddots & \vdots \\ 0 & 0 & 0 & 0 & 1 \\ -a_0 & -a_1 & -a_2 & \cdots & -a_{n-1} \end{bmatrix} \qquad \mathbf{B} = \begin{bmatrix} 0 \\ 0 \\ \vdots \\ 0 \\ 1 \end{bmatrix} \tag{2-170}$$

and $u(k) = f(k)$.

2-11 The z-Transform

In Section 2-6 we have shown that the Laplace transform can be used to solve linear ordinary differential equations. The z-transform represents an operational method for solving linear difference equations and linear systems with discrete or digital data.

2-11-1 Definition of the z-Transform

Consider the sequence $y(k)$, $k = 0, 1, 2, \ldots$, where $y(k)$ could represent a sequence of numbers or events. The z-transform of $y(k)$ is defined as

$$\begin{aligned} Y(z) &= z\text{-transform of } y(k) = \mathcal{Z}[y(k)] \\ &= \sum_{k=0}^{\infty} y(k) z^{-k} \end{aligned} \tag{2-171}$$

where z is a complex variable with real and imaginary parts. The motivation of this definition will become clear later. One significance of the definition in Eq. (2-171) is that the z-transform converts the sequence of numbers in the real domain into an expression in the complex z-domain.

The following examples illustrate the derivation of the z-transforms of two simple functions.

Example 2-21

Consider the sequence

$$y(k) = e^{-\alpha k} \qquad k = 0, 1, 2, \ldots \tag{2-172}$$

where α is a real constant. Applying Eq. (2-171), the z-transform of $y(k)$ is written

$$Y(z) = \sum_{k=0}^{\infty} e^{-\alpha k} z^{-k} = 1 + e^{-\alpha} z^{-1} + e^{-2\alpha} z^{-2} + \cdots \tag{2-173}$$

which converges for $|e^{-\alpha} z^{-1}| < 1$.

Multiplying both sides of the last equation by $e^{-\alpha} z^{-1}$, subtracting the resulting equation from Eq. (2-173), and solving for $Y(z)$, the latter is expressed in a closed form:

$$Y(z) = \frac{1}{1 - e^{-\alpha} z^{-1}} = \frac{z}{z - e^{-\alpha}} \tag{2-174}$$

for $|e^{-\alpha} z^{-1}| < 1$. ▲

Example 2-22

In Example 2-21, if $\alpha = 0$, we have

$$y(k) = 1 \qquad k = 0, 1, 2, \ldots \tag{2-175}$$

which represents a sequence of ones. Then the z-transform of $y(k)$ is

$$Y(z) = 1 + z^{-1} + z^{-2} + \cdots = \frac{z}{z - 1} \tag{2-176}$$

which converges for $|z| > 1$. ▲

2-11-2 Relationship Between the Laplace Transform and the z-Transform

While the mathematicians like to talk about sequences, the engineers feel more at home dealing with signals. It may be useful to represent the sequence $y(kT)$, $k = 0, 1, 2, \ldots$ as a train of impulses separated by the time interval T. The latter is defined as the **sampling period.** The impulse at the kth time instant, $\delta(t - kT)$, carries the value of $y(kT)$. This situation occurs quite often in digital and sampled-data control systems in which a signal $y(t)$ is digitized or sampled every T seconds to form a time sequence that represents the signal at the sampling instants. Thus we can relate the sequence $y(kT)$ with a signal that can be expressed as

$$y^*(t) = \sum_{k=0}^{\infty} y(kT)\delta(t - kT) \tag{2-177}$$

Taking the Laplace transform on both sides of Eq. (2-177), we have

$$Y^*(s) = \mathscr{L}[y^*(t)] = \sum_{k=0}^{\infty} y(kT)e^{-kTs} \tag{2-178}$$

Comparing the last equation with Eq. (2-171), we see that the z-transform may be related to the Laplace transform through

$$z = e^{Ts} \tag{2-179}$$

In fact, the z-transform defined in Eq. (2-171) may be regarded as a special case with $T = 1$. The definition of the z-transform in Eq. (2-179) allows us to treat sampled systems and perform digital simulation of continuous-data systems. Thus we can summarize the definition of the z-transform as

$$\begin{aligned} Y(z) &= \mathscr{Z}[y(kT)] = \mathscr{Z}[y^*(t)] = \mathscr{Z}[Y^*(s)] \\ &= Y^*(s)\big|_{z=e^{Ts}} \end{aligned} \tag{2-180}$$

Or we can write

$$Y(z) = \mathscr{Z}[y(t)] = \mathscr{Z}[Y(s)] \tag{2-181}$$

with the understanding that the function $y(t)$ is first sampled or discretized to get $y^*(t)$ before taking the z-transform.

Example 2-23

Consider the time function

$$y(t) = e^{-\alpha t}u_s(t) \tag{2-182}$$

The z-transform of $y(t)$ is obtained by performing the following steps:

1. Represent the values of $y(t)$ at the time instants $t = kT$, $k = 0, 1, 2, \ldots,$ to form the function $y^*(t)$:

$$y^*(t) = \sum_{k=0}^{\infty} e^{-\alpha kT}\,\delta(t - kT) \tag{2-183}$$

2. Take the Laplace transform on both sides of Eq. (2-183):

$$Y^*(s) = \sum_{k=0}^{\infty} e^{-\alpha kT} e^{-kTs} = \sum_{k=0}^{\infty} e^{-(s+\alpha)kT} \tag{2-184}$$

3. Express $Y^*(s)$ into a closed form and applying Eq. (2-179), we have the z-transform,

$$Y(z) = \frac{z}{z - e^{-\alpha T}} \tag{2-185}$$

In general, the z-transforms of more complex functions may be obtained with the help of some of the z-transform theorems that follow. For engineering purposes, a z-transform table such as that given in Appendix C may be used to transform between $y(k)$ and $Y(z)$. ▲

2-11-3 Some Important Theorems of the z-Transform

Some of the commonly used theorems of the z-transform are stated in the following without proof. Just as in the case of the Laplace transform, these theorems are useful in many aspects of the z-transform analysis. For uniformity, the real sequence is expressed as $y(kT)$, and if a sampling period is not involved, T can be set to unity.

▲ **Theorem 1.** *Addition and Subtraction*

If $y_1(kT)$ and $y_2(kT)$ have z-transforms $Y_1(z)$ and $Y_2(z)$, respectively, then

$$\mathscr{Z}[y_1(kT) \pm y_2(kT)] = Y_1(z) \pm Y_2(z) \tag{2-186}$$

▲ **Theorem 2.** *Multiplication by a Constant*

$$\mathscr{Z}[\alpha y(kT)] = \alpha \mathscr{Z}[y(kT)] = \alpha Y(z) \tag{2-187}$$

where α is a constant.

▲ **Theorem 3.** *Real Translation (Time Delay and Time Advance)*

$$\mathscr{Z}[y(kT - nT)] = z^{-n} Y(z) \tag{2-188}$$

and

$$\mathcal{Z}[y(kT + nT)] = z^n \left[Y(z) - \sum_{k=0}^{n-1} y(kT)z^{-k} \right] \tag{2-189}$$

where n is a positive integer.

Equation (2-188) represents the z-transform of a time sequence that is shifted to the right by nT, and Eq. (2-189) denotes that of a time sequence shifted to the left by nT. The reason that the right-hand side of Eq. (2-189) is not just $z^n Y(z)$ is because the one-sided z-transform, similar to the Laplace transform, is defined only for $k \geq 0$. Thus the second term on the right-hand side of Eq. (2-189) simply represents the sequence that is lost after it is shifted to the left of $k = 0$.

▲ **Theorem 4.** *Complex Translation*

$$\mathcal{Z}[e^{\mp \alpha kT} y(kT)] = Y(ze^{\pm \alpha T}) \tag{2-190}$$

where α is a constant. $Y(z)$ is the z-transform of $y(kT)$.

▲ **Theorem 5.** *Initial-Value Theorem*

$$\lim_{k \to 0} y(kT) = \lim_{z \to \infty} Y(z) \tag{2-191}$$

if the limit exists.

▲ **Theorem 6.** *Final-Value Theorem*

▲ The final-value theorem is valid only if $(1 - z^{-1})Y(z)$ does not have poles on or outside the unit circle $|z| = 1$.

$$\lim_{k \to \infty} y(kT) = \lim_{z \to 1} (1 - z^{-1})Y(z) \tag{2-192}$$

if the function $(1 - z^{-1})Y(z)$ has no poles on or outside the unit circle $|z| = 1$ in the z-plane.

▲ **Theorem 7.** *Real Convolution*

$$Y_1(z)Y_2(z) = \mathcal{Z}\left[\sum_{k=0}^{N} y_1(kT)y_2(NT - kT)\right] = \mathcal{Z}\left[\sum_{k=0}^{N} y_2(kT)y_1(NT - kT)\right]$$
$$= \mathcal{Z}[y_1(kT) * y_2(kT)] \qquad (2\text{-}193)$$

where $*$ denotes real convolution in the discrete-time domain.

Thus we see that as in the Laplace transform, the z-transform of the product of two real functions $y_1(k)$ and $y_2(k)$ is **not** equal to the product of the z-transforms $Y_1(z)$ and $Y_2(z)$. One exception to this in the case of the z-transform is if one of the two functions is the delay e^{-NTs}, where N is a positive integer, then

$$\mathcal{Z}[e^{-NTs}Y(s)] = \mathcal{Z}[e^{-NTs}]\mathcal{Z}[Y(s)] = z^{-N}Y(z) \qquad (2\text{-}194)$$

Table 2-2 summarizes the theorems on the z-transform just given. The following examples illustrate the usefulness of some of these theorems.

Table 2-2 Theorems of z-Transforms

Addition and subtraction	$\mathcal{Z}[y_1(kT) \pm y_2(kT)] = Y_1(z) \pm Y_2(z)$	
Multiplication by a constant	$\mathcal{Z}[\alpha y(kT)] = \alpha\mathcal{Z}[y(kT)] = \alpha Y(z)$	
Real translation	$\mathcal{Z}[y(k - n)T] = z^{-n}Y(z)$	(time delay)
	$\mathcal{Z}[y(k + n)T] = z^{n}\left[Y(z) - \sum_{k=0}^{n-1} y(kT)z^{-k}\right]$	(time advance)
	where n = positive integer	
Complex translation	$\mathcal{Z}[e^{\mp\alpha kT}y(kT)] = Y(ze^{\pm\alpha T})$	
Initial-value theorem	$\lim_{k\to 0} y(kT) = \lim_{z\to\infty} Y(z)$	
Final-value theorem	$\lim_{k\to\infty} y(kT) = \lim_{z\to 1} (1 - z^{-1})Y(z)$	
	if $(1 - z^{-1})Y(z)$ has no poles on or outside $\lvert z\rvert = 1$.	
Real convolution	$Y_1(z)Y_2(z) = \mathcal{Z}\left[\sum_{k=0}^{n} y_1(kT)y_2(NT - kT)\right]$	
	$= \mathcal{Z}\left[\sum_{k=0}^{n} y_2(kT)y_1(NT - kT)\right]$	
	$= \mathcal{Z}[y_1(kT) * y_2(kT)]$	

Example 2-24

(*Complex Translation Theorem*) The complex translation theorem in Eq. (2-190) is used to find the z-transform of $y(t) = te^{-\alpha t}$, $t \geq 0$. Let $f(t) = t$, $t \geq 0$; then

$$F(z) = \mathcal{Z}[tu_s(t)] = \mathcal{Z}(kT) = \frac{Tz}{(z-1)^2} \tag{2-195}$$

Using the complex translation theorem in Eq. (2-190), we obtain

$$Y(z) = \mathcal{Z}[te^{-\alpha t}u_s(t)] = F(ze^{\alpha T}) = \frac{Tze^{-\alpha T}}{(z - e^{-\alpha T})^2} \tag{2-196}$$

▲

Example 2-25

(*Final-Value Theorem*) Given the function

$$Y(z) = \frac{0.792z^2}{(z-1)(z^2 - 0.416z + 0.208)} \tag{2-197}$$

determine the value of $y(kT)$ as k approaches infinity.

Since the function $(1 - z^{-1})Y(z)$ does not have any pole on or outside the unit circle $|z| = 1$ in the z-plane, the final-value theorem in Eq. (2-192) can be applied. Thus

$$\lim_{k\to\infty} y(kT) = \lim_{z\to 1} \frac{0.792z}{z^2 - 0.416z + 0.208} = 1 \tag{2-198}$$

▲

2-11-4 Inverse z-Transform

▲ The inverse z-transform of $Y(z)$ is $y(kT)$; not $y(t)$.

Just as in the Laplace transform, one of the major objectives of the z-transform is that algebraic manipulations can be made first in the z-domain, and then the final time response is determined by the inverse z-transform. *In general, the inverse z-transform of a function $Y(z)$ yields information on $y(kT)$ only, not on $y(t)$. In other words, the z-transform carries information only at the sampling instants.* With this in mind, the inverse z-transform can be carried out by one of the following three methods:

1. Partial-fraction expansion.
2. Power-series method.
3. The inverse formula.

Partial-Fraction Expansion Method

The z-transform function $Y(z)$ is expanded by partial-fraction expansion into a sum of simple recognizable terms, and the z-transform table is used to determine the corresponding $y(kT)$. In carrying out the partial-fraction expansion, there is a slight difference

▲ If $Y(z)$ has at least one zero at $z = 0$, the partial-fraction expansion of $Y(z)/z$ should first be performed.

between the z-transform and the Laplace transform procedures. With reference to the z-transform table, we note that practically all the z-transform functions have the term z in the numerator. Therefore, we should expand $Y(z)$ into the form

$$Y(z) = \frac{K_1 z}{z - e^{-\alpha T}} + \frac{K_2 z}{z - e^{-\beta T}} + \cdots \tag{2-199}$$

For this, we should first expand $Y(z)/z$ into fractions and then multiply z across to obtain the final expression. The following example will illustrate this procedure.

Example 2-26

Given the z-transform function

$$Y(z) = \frac{(1 - e^{-\alpha T})z}{(z - 1)(z - e^{-\alpha T})} \tag{2-200}$$

it is desired to find the inverse z-transform. Expanding $Y(z)/z$ by partial-fraction expansion, we have

$$\frac{Y(z)}{z} = \frac{1}{z - 1} - \frac{1}{z - e^{-\alpha T}} \tag{2-201}$$

The final expanded expression for $Y(z)$ is

$$Y(z) = \frac{z}{z - 1} - \frac{z}{z - e^{-\alpha T}} \tag{2-202}$$

From the z-transform table in Appendix C, the corresponding inverse z-transform of $Y(z)$ is found to be

$$y(kT) = 1 - e^{-\alpha kT} \qquad k = 0, 1, 2, \ldots \tag{2-203}$$

▲

▲ If $Y(z)$ does not have any zeros at $z = 0$, then perform the partial-fraction expansion of $Y(z)$ directly.

It should be pointed out that if $Y(z)$ does not contain any factors of z in the numerator, this usually means that the time sequence has a time delay, and the partial-fraction expansion of $Y(z)$ should be carried out without first dividing the function by z. The following example illustrates this situation.

Example 2-27

Consider the function

$$Y(z) = \frac{(1 - e^{-\alpha T})}{(z - 1)(z - e^{-\alpha T})} \tag{2-204}$$

which does not contain any powers of z as a factor in the numerator. In this case, the partial-fraction expansion of $Y(z)$ is carried out directly. We have

$$Y(z) = \frac{1}{z - 1} - \frac{1}{z - e^{-\alpha T}} \tag{2-205}$$

Although the z-transform table does not contain exact matches for the components in Eq. (2-205), we recognize that the inverse-z-transform of the first term on the right-hand side can be written as

$$\mathcal{Z}^{-1}\left[\frac{1}{z-1}\right] = \mathcal{Z}^{-1}\left[z^{-1}\frac{z}{z-1}\right] = \sum_{k=1}^{\infty} z^{-1}$$
$$= u_s[(k-1)T] \qquad k = 1, 2, \ldots \tag{2-206}$$

Similarly, the second term on the right-hand side of Eq. (2-205) can be identified with a time delay of T seconds. Thus the inverse z-transform of $Y(z)$ is written

$$y(kT) = (1 - e^{-\alpha(k-1)T})u_s[(k-1)T] \qquad k = 1, 2, \ldots \tag{2-207}$$

▲

2-11-5 Computer Solution of the Partial-Fraction Expansion of $Y(z)/z$

Whether the function to be expanded by partial fraction is in the form of $Y(z)/z$ or $Y(z)$, the **pfe** programs in **CSAD/MATLAB** or **ACSP** can still be applied.

Power-Series Method

The definition of the z-transform in Eq. (2-171) gives a straightforward method of carrying out the inverse z-transform. Based on Eq. (2-171) we can clearly see that in the sampled case the coefficient of z^{-k} in $Y(z)$ is simply $y(kT)$. Thus if we expand $Y(z)$ into a power series in powers of z^{-k}, we can find the values of $y(kT)$ for $k = 0, 1, 2, \ldots$.

Example 2-28

Consider the function $Y(z)$ given in Eq. (2-200), which can be expanded into a power series of z^{-1} by dividing the numerator polynomial by the denominator polynomial by long division. The result is

$$Y(z) = (1 - e^{-\alpha T})z^{-1} + (1 - e^{-2\alpha T})z^{-2} + \cdots + (1 - e^{-k\alpha T})z^{-k} + \cdots \tag{2-208}$$

Thus it is apparent that

$$y(kT) = 1 - e^{-\alpha kT} \qquad k = 0, 1, 2, \ldots \tag{2-209}$$

which is the same result as in Eq. (2-203). ▲

Inversion Formula

The time sequence $y(kT)$ can be determined from $Y(z)$ by use of the inversion formula

$$y(kT) = \frac{1}{2\pi j}\oint_{\Gamma} Y(z)z^{k-1}\,dz \tag{2-210}$$

which is a contour integration along the path Γ, which is a circle of radius $|z| = e^{cT}$ centered at the origin in the z-plane, and c is of such a value that the poles of $Y(z)z^{k-1}$ are inside the circle. The inversion formula of the z-transform is similar to that of the inverse Laplace transform integral given in Eq. (2-23). One way of evaluating the contour integration of Eq. (2-210) is by use of the residue theorem of the complex-variable theory, and the details are not covered here.

2-12 Application of the z-Transform to the Solution of Linear Difference Equations

The z-transform can be used to solve linear difference equations. As a simple example, let us consider the first-order unforced difference equation

$$y(k+1) + y(k) = 0 \tag{2-211}$$

To solve this equation, we take the z-transform on both sides of the equation. By this we mean that we multiply both sides of the equation by z^{-k} and take the sum from $k = 0$ to $k = \infty$. We have

$$\sum_{k=0}^{\infty} y(k+1)z^{-k} + \sum_{k=0}^{\infty} y(k)z^{-k} = 0 \tag{2-212}$$

By using the definition of $Y(z)$ and the real-translation theorem of Eq. (2-189) for time advance, Eq. (2-212) is written

$$z[Y(z) - y(0)] + Y(z) = 0 \tag{2-213}$$

Solving for $Y(z)$, we get

$$Y(z) = \frac{z}{z+1}y(0) \tag{2-214}$$

The inverse z-transform of Eq. (2-214) can be obtained by expanding $Y(z)$ into a power series in z^{-1} by long division. We have

$$Y(z) = (1 - z^{-1} + z^{-2} - z^{-3} + \cdots)x(0) \tag{2-215}$$

Thus $y(k)$ is written

$$y(k) = (-1)^k y(0) \qquad k = 0, 1, 2, \ldots \tag{2-216}$$

Equation (2-211) is recognized as a first-order-state equation. The z-transform solution of high-order discrete-data systems described by state equations is treated in Chapter 5. The following example shows the z-transform solution of a second-order difference equation.

Example 2-29

Consider the second-order difference equation

$$y(k+2) + 0.5y(k+1) + 0.2y(k) = u(k) \tag{2-217}$$

where

$$u(k) = u_s(k) = 1 \qquad \text{for } k = 0, 1, 2, \ldots \tag{2-218}$$

The initial conditions of $y(k)$ are $y(0) = 0$ and $y(1) = 0$.

Taking the z-transform on both sides of Eq. (2-217), we get

$$[z^2Y(z) - z^2y(0) - zy(1)] + 0.5[zY(z) - zy(0)] + 0.2Y(z) = U(z) \tag{2-219}$$

The z-transform of $u(k)$ is $U(z) = z/(z-1)$. Substituting the initial conditions of $y(k)$ and the expression of $U(z)$ into Eq. (2-219) and solving for $Y(z)$, we have

$$Y(z) = \frac{z}{(z-1)(z^2 + 0.5z + 0.2)} \tag{2-220}$$

The partial-fraction expansion of $Y(z)/z$ is

$$\frac{Y(z)}{z} = \frac{0.588}{z-1} - \frac{1.036e^{j1.283}}{z + 0.25 + j0.37} - \frac{1.036e^{-j1.283}}{z + 0.25 - j0.37} \tag{2-221}$$

where the exponents in the numerator coefficients are in radians. Taking the inverse z-transform of $Y(z)$, we get

$$\begin{aligned} y(k) &= 0.588 - 1.036(0.447)^k[e^{-j(2.165k-1.283)} + e^{j(2.165k-1.283)}] \\ &= 0.588 - 2.072(0.447)^k \cos(2.165k - 1.283) \qquad k \geq 0 \end{aligned} \tag{2-222}$$

▲

2-13 Summary

In this chapter we have presented some of the mathematical fundamentals required for the study of linear control systems. Specifically, the Laplace transform is used for the solution of linear ordinary differential equations, and the z-transform is devised for the solution of linear difference equations. Both transform methods are characterized by first transforming the real-domain equations into algebraic equations in the transform domain. The solutions are first obtained in the transform domain using the familiar methods of solving algebraic equations. The final solution in the real domain is obtained by taking the inverse transform. For engineering problems the transform tables and the partial-fraction expansion method are recommended for the inverse transformation.

Matrix algebra is introduced to simplify the modeling and manipulations of systems that involve multiple numbers of variables and equations. The definitions and properties of the matrix algebra are given. It is demonstrated that whenever computers and suitable software are available, the solutions of the transform problems and matrix manipulations can easily be obtained by use of computers.

Review Questions

1. Define the poles and zeros of a function of the complex variable s.
2. What are the advantages over the classical method of the Laplace transform method of solving linear ordinary differential equations?
3. What are state equations?
4. What is a causal system?
5. Write the defining equation of the one-sided Laplace transform.
6. Write the defining equation of the inverse Laplace transform.
7. Give the expression of the final-value theorem of the Laplace transform. What is the condition under which the theorem is valid?
8. Give the Laplace transform of the unit-step function, $u_s(t)$.
9. What is the Laplace transform of the unit-ramp function, $tu_s(t)$?
10. Give the Laplace transform of $f(t)$ shifted to the right (delayed) by T_d in terms of the Laplace transform of $f(t)$, $F(s)$.
11. If $\mathscr{L}[f_1(t)] = F_1(s)$ and $\mathscr{L}[f_2(t)] = F_2(s)$, find $\mathscr{L}[f_1(t)f_2(t)]$ in terms of $F_1(s)$ and $F_2(s)$.
12. Do you know how to handle the exponential term in performing the partial-fraction expansion of

$$F(s) = \frac{10}{(s+1)(s+2)}e^{-2s}$$

13. Do you know how to handle the partial-fraction expansion of a function whose denominator order is not greater than that of the numerator, for example,

$$F(s) = \frac{10(s^2+5s+1)}{(s+1)(s+2)}$$

14. In trying to find the inverse Laplace transform of the following function, do you have to perform the partial-fraction expansion?

$$F(s) = \frac{1}{(s+5)^3}$$

15. What is the definition of the rank of a matrix?

16. Can the rank condition be applied to a matrix that is not square?
17. Can a nonsquare matrix be symmetric?
18. Define the z-transform in terms of the Laplace operator s.
19. Give the z-transform of the unit-step function, $u_s(t)$.
20. Give the z-transform of the unit-ramp function, $tu_s(t)$.
21. Explain why the inverse z-transform of $F(z)$ is not $f(t)$.
22. List all the methods you know for performing the inverse z-transform.
23. If the function $F(z)$ has at least one zero at $z = 0$, what step must be taken before performing the partial-fraction expansion?
24. What is noticeably different between the time sequences that correspond to the following two functions?

$$F_1(z) = \frac{z}{z-1} \qquad F_2(z) = \frac{z}{z+1}$$

References

Complex Variables, Laplace Transforms, and Matrix Algebra

1. **F. B. HILDEBRAND,** *Methods of Applied Mathematics,* 2nd ed., Prentice Hall, Englewood Cliffs, NJ, 1965.
2. **R. BELLMAN,** *Introduction to Matrix Analysis.* McGraw-Hill Book Company, New York, 1960.
3. **F. AYRES, JR.,** *Theory and Problems of Matrices,* Schaum's Outline Series, McGraw-Hill Book Company, New York, 1962.
4. **B. C. KUO,** *Linear Networks and Systems,* McGraw-Hill Book Company, New York, 1967.
5. **C. R. WYLIE, JR.,** *Advanced Engineering Mathematics,* 2nd ed., McGraw-Hill Book Company, New York, 1960.

Partial-Fraction Expansion

6. **C. POTTLE,** ''On the Partial Fraction Expansion of a Rational Function with Multiple Poles by Digital Computer,'' *IEEE Trans. Circuit Theory,* Vol. CT-11, pp 161–162, Mar. 1964.
7. **B. O. WATKINS,** ''A Partial Fraction Algorithm.'' *IEEE Trans. Automatic Control,* Vol. AC-16, pp. 489–491, Oct. 1971.

z-Transforms

8. **B. C. KUO,** *Digital Control Systems,* 2nd ed., Saunders College Publishing, New York, 1992.
9. **E. I. JURY,** *Theory and Application of the z-Transform Method,* Robert E. Krieger Publishing Co., Huntington, NY, 1973.

Problems

▲ Poles and zeros

2-1. Find the poles and zeros of the following functions (including the ones at infinity, if any). Mark the finite poles with × and the finite zeros with ○ in the s-plane.

(a) $G(s) = \dfrac{10(s+2)}{s^2(s+1)(s+10)}$ (b) $G(s) = \dfrac{10s(s+1)}{(s+2)(s^2+3s+2)}$

(c) $G(s) = \dfrac{10(s+2)}{s(s^2+2s+2)}$ (d) $G(s) = \dfrac{e^{-2s}}{10s(s+1)(s+2)}$

▲ Laplace transforms

2-2. Find the Laplace transforms of the following functions. Use the theorems on Laplace transforms if applicable.

(a) $g(t) = 5te^{-5t}u_s(t)$ (b) $g(t) = (t \sin 2t + e^{-2t})u_s(t)$

(c) $g(t) = 2e^{-2t} \sin 2tu_s(t)$ (d) $g(t) = \sin 2t \cos 2tu_s(t)$

(e) $g(t) = \sum_{k=0}^{\infty} e^{-5kT}\delta(t-kT)$ where $\delta(t)$ = unit-impulse function

▲ Laplace transforms

2-3. Find the Laplace transforms of the functions shown in Fig. 2P-3. First, write a complete expression for $g(t)$ and then take the Laplace transform. Let $g_T(t)$ be the description of the function over the basic period and then delay $g_T(t)$ appropriately to get $g(t)$. Take the Laplace transform of $g(t)$ to get $G(s)$.

▲ Laplace transforms

2-4. Find the Laplace transform of the following function.

$$g(t) = \begin{cases} t+1 & 0 \le t < 1 \\ 0 & 1 \le t < 2 \\ 2-t & 2 \le t < 3 \\ 0 & t \ge 3 \end{cases}$$

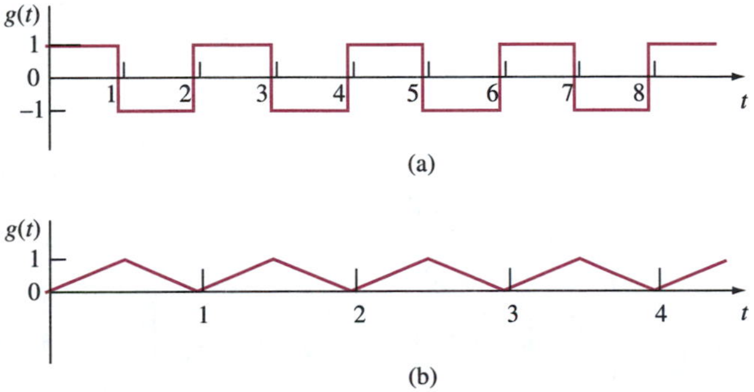

Figure 2P-3

▲ Laplace transforms

2-5. Solve the following differential equations by means of the Laplace transform.

(a) $\frac{d^2f(t)}{dt^2} + 5\frac{df(t)}{dt} + 4f(t) = e^{-2t}u_s(t)$ Assume zero initial conditions.

(b) $\frac{dx_1(t)}{dt} = x_2(t)$

$\frac{dx_2(t)}{dt} = -2x_1(t) - 3x_2(t) + u_s(t)$ $x_1(0) = 1, \quad x_2(0) = 0$

▲ Inverse Laplace transform

2-6. Find the inverse Laplace transforms of the following functions. Perform partial-fraction expansion on $G(s)$ first, then use the Laplace transform table. Use any computer program that is available for the partial fraction expansion.

(a) $G(s) = \frac{1}{s(s+2)(s+3)}$ **(b)** $G(s) = \frac{10}{(s+1)^2(s+3)}$

(c) $G(s) = \frac{100(s+2)}{s(s^2+4)(s+1)}e^{-s}$ **(d)** $G(s) = \frac{2(s+1)}{s(s^2+s+2)}$

(e) $G(s) = \frac{1}{(s+1)^3}$ **(f)** $G(s) = \frac{2(s^2+s+1)}{s(s+1.5)(s^2+5s+5)}$

▲ Matrix sums and differences

2-7. Carry out the following matrix sums and differences. Do not use the computer, as these are straightforward.

(a) $\begin{bmatrix} 5 & -6 \\ 0 & 3 \end{bmatrix} + \begin{bmatrix} 10 & -3 \\ 2 & -3 \end{bmatrix}$ **(b)** $\begin{bmatrix} 2 & -2 \\ 0 & 10 \\ 3 & 0 \end{bmatrix} - \begin{bmatrix} 10 & 0 \\ 3 & 4 \\ 0 & -4 \end{bmatrix}$

(c) $\begin{bmatrix} \frac{1}{s+1} & 3 & \frac{1}{s} \\ \frac{2}{s} & 0 & \frac{1}{s-3} \end{bmatrix} + \begin{bmatrix} 1 & 0 & -10 \\ s & \frac{1}{s} & 1 \end{bmatrix}$

▲ Conformability of matrices

2-8. Determine if the following matrices are conformable for the products **AB** and **BA**. Find the valid products.

(a) $\mathbf{A} = \begin{bmatrix} 3 \\ 2 \\ 1 \end{bmatrix}$ $\mathbf{B} = [6 \quad 1 \quad 0]$ **(b)** $\mathbf{A} = \begin{bmatrix} -1 & 2 \\ 0 & 3 \end{bmatrix}$ $\mathbf{B} = \begin{bmatrix} 0 & 10 & -10 \\ -1 & 1 & 1 \end{bmatrix}$

▲ Matrix form

2-9. Express the following sets of algebraic equations in matrix form, $\mathbf{Ax} = \mathbf{B}$.

(a)
$$\begin{aligned} x_1 + x_2 - x_3 &= 1 \\ -x_1 + 3x_2 - x_3 &= 1 \\ 3x_1 - 5x_2 - 2x_3 &= 0 \end{aligned}$$

(b)
$$\begin{aligned} x_1 + x_2 - x_3 &= 1 \\ -x_1 + 3x_2 - x_3 &= 1 \\ 2x_1 - 2x_2 \quad &= 0 \end{aligned}$$

Find the inverse of the matrix **A** to see if these equations are linearly independent. Use a computer program such as **MATLAB** to find the inverse of **A**. If the equations are not linearly independent, can you still solve for x_1, x_2, and x_3?

▲ State equations

2-10. Express the following set of first-order differential equations in the vector-matrix form

$$\frac{d\mathbf{x}(t)}{dt} = \mathbf{Ax}(t) + \mathbf{Bu}(t)$$

$$\frac{dx_1(t)}{dt} = -x_1(t) + 2x_2(t)$$

$$\frac{dx_2(t)}{dt} = -2x_2(t) + 3x_3(t) + u_1(t)$$

$$\frac{dx_3(t)}{dt} = -x_1(t) - 3x_2(t) - x_3(t) + u_2(t)$$

▲ Matrix inverses

2-11. Find the inverses of the following matrices, if they exist. Do the problems by hand and then by any available matrix computer program.

(a) $\mathbf{A} = \begin{bmatrix} 2 & 5 \\ 10 & -1 \end{bmatrix}$ **(b)** $\mathbf{A} = \begin{bmatrix} 3 & 0 & -1 \\ -2 & 1 & 2 \\ 0 & 1 & -1 \end{bmatrix}$

(c) $\mathbf{A} = \begin{bmatrix} 1 & 3 & 4 \\ -1 & 1 & 0 \\ -1 & 0 & -1 \end{bmatrix}$ **(d)** $\mathbf{A} = \begin{bmatrix} 0 & 1 & 0 \\ 2 & -2 & 3 \\ 0 & 1 & 5 \end{bmatrix}$

▲ Rank of matrix

2-12. Determine the ranks of the following matrices.

(a) $\begin{bmatrix} 3 & 2 \\ 6 & 1 \\ 3 & 0 \end{bmatrix}$ **(b)** $\begin{bmatrix} 2 & 4 & 0 & 8 \\ 1 & 2 & 6 & 3 \end{bmatrix}$

(c) $\begin{bmatrix} 1 & 0 & 0 \\ 5 & 0 & 0 \end{bmatrix}$ **(d)** $\begin{bmatrix} 1 & 0 & 0 & 1 \\ 0 & 0 & 5 & 0 \end{bmatrix}$

▲ z-Transforms

2-13. Find the z-transforms of the following functions.

(a) $f(k) = ke^{-3k}$ **(b)** $f(k) = k \sin 2k$
(c) $f(k) = e^{-2k} \sin 3k$ **(d)** $f(k) = k^2 e^{-2k}$

▲ z-Transforms

2-14. Determine the z-transforms of the following sequences.

(a) $f(kT) = kT \sin 2kT$

(b) $f(k) = \begin{cases} 1 & k = 0, 2, 4, 6, \ldots, \text{even integers} \\ -1 & k = 1, 3, 5, 7, \ldots, \text{odd integers} \end{cases}$

▲ Partial-fraction expansion, z-transform

2-15. Perform the partial-fraction expansion of the following functions, if applicable, and then find the z-transforms using the z-transform table.

(a) $F(s) = \dfrac{1}{(s+5)^3}$ **(b)** $F(s) = \dfrac{1}{s^3(s+1)}$

(c) $F(s) = \dfrac{10}{s(s+5)^2}$ **(d)** $F(s) = \dfrac{5}{s(s^2+2)}$

▲ Inverse z-transform

2-16. Find the inverse z-transforms $f(k)$ of the following functions. Apply partial-fraction expansion to $F(z)$ and then use the z-transform table.

(a) $F(z) = \dfrac{10z}{(z-1)(z-0.2)}$ **(b)** $F(z) = \dfrac{z}{(z-1)(z^2+z+1)}$

(c) $F(z) = \dfrac{z}{(z-1)(z+0.85)}$ **(d)** $F(z) = \dfrac{10}{(z-1)(z-0.5)}$

▲ z-Transform, final-value theorem

2-17. Given that $\mathscr{Z}[f(k)] = F(z)$, find the value of $f(k)$ as k approaches infinity without obtaining the inverse z-transform of $F(z)$. Use the final-value theorem of the z-transform if it is applicable.

(a) $F(z) = \dfrac{0.368z}{(z-1)(z^2-1.364z+0.732)}$ **(b)** $F(z) = \dfrac{10z}{(z-1)(z+1)}$

(c) $F(z) = \dfrac{z^2}{(z-1)(z-0.5)}$ **(d)** $F(z) = \dfrac{z}{(z-1)(z-1.5)}$

Check the answers by carrying out the long division of $F(z)$ and express it in a power series of z^{-1}.

▲ z-Transform solutions

2-18. Solve the following difference equations by means of the z-transform.

(a) $x(k+2) - x(k+1) + 0.1x(k) = u_s(k) \qquad x(0) = x(1) = 0$
(b) $x(k+2) - x(k) = 0 \qquad x(0) = 1, \quad x(1) = 0$

▲ z-Transform applications

2-19. This problem deals with application of the difference equations and the z-transform to a loan-amortization problem. Consider that a new car is purchased with a load of P_0 dollars over a period of N months, at a monthly interest rate of r percent. The principal and interest are to be paid back in N equal payments of u dollars each.

(a) Show that the difference equation that describes the loan process can be written as

$$P(k+1) = (1+r)P(k) - u$$

where

$P(k)$ = amount owed after the kth period, $k = 0, 1, 2, \ldots, N$
$P(0) = P_0$ = initial amount borrowed
$P(N) = 0$ (after N periods, owe nothing)

The last two conditions are also known as the boundary conditions.

(b) Solve the difference equation in part (a) by the recursive method, starting with $k = 0$, then $k = 1, 2, \ldots,$ and substituting successively. Show that the solution of the equation is

$$u = \frac{(1+r)^N P_0 r}{(1+r)^N - 1}$$

(c) Solve the difference equation in part (a) by means of the z-transform method.

(d) Consider that $P_0 = \$15{,}000$, $r = 0.01$ (1 percent per month), and $N = 48$ months. Find u, the monthly payment.

Additional Computer Problems

The following problems are to be solved by any computer program that is available to the reader.

2-20. Perform partial-fraction expansion of the following functions.

(a) $G(s) = \dfrac{10(s+1)}{s^2(s+4)(s+6)}$ (b) $G(s) = \dfrac{s+1}{s(s+2)(s^2+2s+2)}$

(c) $G(s) = \dfrac{5(s+2)}{s^2(s+1)(s+5)}$ (d) $G(s) = \dfrac{5e^{-2s}}{(s+1)(s^2+s+1)}$

(e) $G(s) = \dfrac{100(s^2+s+3)}{s(s^2+5s+3)}$ (f) $G(s) = \dfrac{1}{s(s^2+1)(s+0.5)^2}$

2-21. Find the inverse Laplace transforms of the functions in Problem 2-20.

2-22. Perform the partial-fraction expansion of the following z-transfer functions.

(a) $G(z) = \dfrac{5z}{(z-1)(z-0.1)}$

(b) $G(z) = \dfrac{10z(z-0.2)}{(z-1)(z-0.5)(z-0.8)}$

(c) $G(z) = \dfrac{z}{(z-1)(z-0.5)^2}$

(d) $G(z) = \dfrac{2z}{(z-1)(z^2-z+1)}$

3 Transfer Functions, Block Diagrams, Signal-Flow Graphs

KEYWORDS AND TOPICS

- ▲ Transfer Functions
- ▲ Signal Flow Graphs
- ▲ Impulse Response
- ▲ State Diagrams
- ▲ Block Diagrams

3-1 Introduction

An important first step in the analysis and design of control systems is the mathematical modeling of the controlled process. In general, given a controlled process, the set of variables that identify the dynamic characteristics of the process should first be defined. For instance, consider a motor used for control purposes. We may identify the applied voltage, current in the armature windings, developed torque on the rotor shaft, and angular displacement and velocity of the rotor, as the system variables. These variables are interrelated through established physical laws that lead to mathematical equations that describe the dynamics of the motor. Depending on the operating condition of the motor, as well as on the emphasis of the modeling, the system equations may be linear or nonlinear, time-varying or time-invariant.

The physical laws that govern the principles of operation of systems in real life can often be quite complex, and realistic characterization of the systems may require nonlinear and/or time-varying equations that are very difficult to solve. For practical reasons, to establish a class of applicable analysis and design tools for control systems, assumptions and approximations are made to the physical systems whenever possible so that these systems can be studied using linear systems theory. There are two ways of justifying the linear-systems approach. One is that the system is basically linear, or the system is operated in the linear region so that most of the conditions of linearity are satisfied. The second is that the system is basically nonlinear or operated in a nonlinear region, but to apply the linear analysis and design tools, we linearize the system about a nominal operating point, such as that described in Section 4-7. It should be kept in mind that the analysis is applicable only for the range of variables in which the linearization is valid. Therefore, the key words to the analysis and design of linear control systems are: **assumptions, identification, linearization,** and **modeling.**

3-2 Impulse Response and Transfer Functions of Linear Systems

The classical way of modeling linear systems is to use **transfer functions** to represent input–output relations between variables. One way to define the transfer function is to use the **impulse response,** which is defined in the following.

3-2-1 Impulse Response

▲ Impulse response is the output when the input is a unit impulse function.

Consider that a linear time-invariant system has the input $u(t)$ and output $y(t)$. The system can be characterized by its **impulse response** $g(t)$, which is defined as the output when the input is a unit-impulse function $\delta(t)$. Once the impulse response of a linear system is known, the output of the system, $y(t)$, with *any* input, $u(t)$, can be found by using the **transfer function.**

3-2-2 Transfer Function (Single-Input Single-Output Systems)

▲ Transfer function of a linear system is the Laplace transform of the impulse response, with zero initial conditions.

The transfer function of a linear time-invariant system is defined as the Laplace transform of the impulse response, with all the initial conditions set to zero. Let $G(s)$ denote the transfer function of a single-input single-output system, with input $u(t)$ and output $y(t)$ and impulse response $g(t)$. Then the transfer function $G(s$ is defined as

$$G(s) = \mathscr{L}[g(t)] \tag{3-1}$$

The transfer function $G(s)$ is related to the Laplace transform of the input and the output through the following relation:

$$G(s) = \frac{Y(s)}{U(s)} \tag{3-2}$$

with all the initial conditions set to zero, and $Y(s)$ and $U(s)$ are the Laplace transforms of $y(t)$ and $u(t)$, respectively.

Although the transfer function of a linear system is defined in terms of the impulse response, in practice, the input–output relation of a linear time-invariant system with continuous-data input is often described by a differential equation, so that it is more convenient to derive the transfer function directly from the differential equation. Let us consider that the input–output relation of a linear time-invariant system is described by the following nth-order differential equation with constant real coefficients:

$$\begin{aligned}\frac{d^n y(t)}{dt^n} &+ a_{n-1}\frac{d^{n-1}y(t)}{dt^{n-1}} + \cdots + a_1\frac{dy(t)}{dt} + a_0 y(t) \\ &= b_m\frac{d^m u(t)}{dt^m} + b_{m-1}\frac{d^{m-1}u(t)}{dt^{m-1}} + \cdots + b_1\frac{du(t)}{dt} + b_0 u(t)\end{aligned} \tag{3-3}$$

The coefficients $a_0, a_1, \ldots, a_{n-1}$ and $b_0, b_1, \ldots, b_m$ are real constants. Once the input $u(t)$ for $t \geq t_0$ and the initial conditions of $y(t)$ and the derivatives of $y(t)$ are specified at the initial time $t = t_0$, the output response $y(t)$ for $t \geq t_0$ is determined by solving Eq. (3-3). However, from the standpoint of linear-system analysis and design, the method of using differential equations exclusively is quite cumbersome. Thus differential equations of the form of Eq. (3-3) are seldom used in their original form for the analysis and design of control systems. It should be pointed out that although efficient subroutines are available on digital computers for the solution of high-order differential equations, *the basic philosophy of linear control theory is that of developing analysis and design tools that would avoid the exact solution of the system differential equations,* except when computer-simulation solutions are desired for final presentation or verification. In classical control theory, even computer simulation often starts with transfer functions rather than with differential equations.

To obtain the transfer function of the linear system that is represented by Eq. (3-3), we simply take the Laplace transform on both sides of the equation and assume **zero initial conditions.** The result is

$$(s^n + a_{n-1}s^{n-1} + \cdots + a_1 s + a_0)Y(s) = (b_m s^m + b_{m-1}s^{m-1} + \cdots + b_1 s + b_0)U(s) \tag{3-4}$$

The transfer function between $u(t)$ and $y(t)$ is given by

$$G(s) = \frac{Y(s)}{U(s)} = \frac{b_m s^m + b_{m-1} s^{m-1} + \cdots + b_1 s + b_0}{s^n + a_{n-1} s^{n-1} + \cdots + a_1 s + a_0} \tag{3-5}$$

The properties of the transfer function are summarized as follows:

1. The transfer function is defined only for a linear time-invariant system. It is not defined for nonlinear systems.
2. The transfer function between an input variable and an output variable of a system is defined as the Laplace transform of the impulse response. Alternatively, the transfer function between a pair of input and output variables is the ratio of the Laplace transform of the output to the Laplace transform of the input.
3. All initial conditions of the system are set to zero.
4. The transfer function is independent of the input of the system.
5. The transfer function of a continuous-data system is expressed only as a function of the complex variable s. It is not a function of the real variable, time, or any other variable that is used as the independent variable. For discrete-data systems modeled by difference equations, the transfer function is a function of z when the z-transform is used.

Proper Transfer Functions. The transfer function in Eq. (3-5) is said to be **strictly proper** if the order of the denominator polynomial is greater than that of the numerator polynomial (i.e., $n > m$). If $n = m$, the transfer function is called **proper.** The transfer function is **improper** if $m > n$.

▲ Characteristic equation of a linear system is obtained by setting the denominator polynomial of the transfer function to zero.

Characteristic Equation. *The* **characteristic equation** *of a linear system is defined as the equation obtained by setting the denominator polynomial of the transfer function to zero.* Thus, from Eq. (3-5), the characteristic equation of the system described by Eq. (3-3) is

$$s^n + a_{n-1} s^{n-1} + \cdots + a_1 s + a_0 = 0 \tag{3-6}$$

Later we shall show that stability of linear single-input single-output systems is governed completely by the roots of the characteristic equation.

3-2-3 Transfer Function (Multivariable Systems)

The definition of transfer function is easily extended to a system with multiple inputs and outputs. A system of this type is often referred to as a **multivariable system.** In a multivariable system, a differential equation of the form of Eq. (3-3) may be used to describe the relationship between an input–output pair when all other inputs are set to zero. Since the principle of superposition is valid for linear systems, the total effect on any output due to all the inputs acting simultaneously is obtained by adding up the outputs due to each input acting alone.

Examples on multivariable systems are plentiful in practice. For example, in the control of speed $\omega(t)$ of a motor subjected to an external disturbance torque, $T_d(t)$, by controlling the input voltage, $v(t)$, the system is considered to have two inputs in $v(t)$ and $T_d(t)$ and one output in $\omega(t)$. Another example is the idle-speed control system of an automobile engine. In this case the two inputs are the amounts of fuel and air intake to the engine, and the output is the idle speed of the engine. In the control of the aircraft turboprop engine shown in Fig. 3-1, the input variables are the fuel rate and the propeller blade angle. The output variables are the speed of rotation of the engine and the turbine-inlet temperature. In general, either one of the outputs is affected by the changes in both inputs. For instance, when the blade angle of the propeller is increased, the speed of rotation of the engine will decrease and the temperature usually increases. The following transfer function relations may be determined from tests performed on the system:

$$Y_1(s) = G_{11}(s)R_1(s) + G_{12}(s)R_2(s) \tag{3-7}$$
$$Y_2(s) = G_{21}(s)R_1(s) + G_{22}(s)R_2(s) \tag{3-8}$$

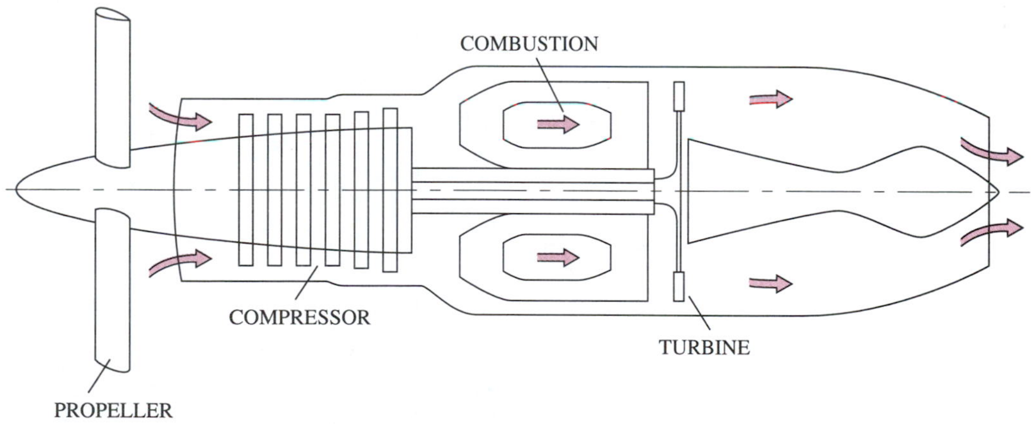

Figure 3-1 Aircraft turboprop engine.

where

$Y_1(s)$ =	speed of rotation of engine	$R_1(s)$ =	fuel rate
$Y_2(s)$ =	turbine-inlet temperature	$R_2(s)$ =	propeller blade angle

all in Laplace transform variables measured from some reference level. The transfer function $G_{11}(s)$ represents the transfer function between the fuel rate and the engine speed, with the propeller blade angle held at the reference value; that is, $R_2(s) = 0$. Similar definitions can be given to the other transfer functions, $G_{12}(s)$, $G_{21}(s)$, and $G_{22}(s)$.

In general, if a linear system has p inputs and q outputs, the transfer function between the jth input and the ith output is defined as

$$G_{ij}(s) = \frac{Y_i(s)}{R_j(s)} \tag{3-9}$$

with $R_k(s) = 0$, $k = 1, 2, \ldots, p$, $k \neq j$. Note that Eq. (3-9) is defined with only the jth input in effect, whereas the other inputs are set to zero. When all the p inputs are in action, the ith output transform is written

$$Y_i(s) = G_{i1}(s)R_1(s) + G_{i2}(s)R_2(s) + \cdots + G_{ip}(s)R_p(s) \tag{3-10}$$

It is convenient to express Eq. (3-10) in matrix-vector form:

$$\mathbf{Y}(s) = \mathbf{G}(s)\mathbf{R}(s) \tag{3-11}$$

where

$$\mathbf{Y}(s) = \begin{bmatrix} Y_1(s) \\ Y_2(s) \\ \vdots \\ Y_q(s) \end{bmatrix} \tag{3-12}$$

is the $q \times 1$ **transformed output vector;**

$$\mathbf{R}(s) = \begin{bmatrix} R_1(s) \\ R_2(s) \\ \vdots \\ R_q(s) \end{bmatrix} \tag{3-13}$$

is the $p \times 1$ **transformed input vector;** and

$$\mathbf{G}(s) = \begin{bmatrix} G_{11}(s) & G_{12}(s) & \cdots & G_{1p}(s) \\ G_{21}(s) & G_{22}(s) & \cdots & G_{2p}(s) \\ \cdot & \cdot & \cdots & \cdot \\ G_{q1}(s) & G_{q2}(s) & \cdots & G_{qp}(s) \end{bmatrix} \tag{3-14}$$

is the $q \times p$ **transfer-function matrix.**

3-3 Block Diagrams

Because of their simplicity and versatility, **block diagrams** are often used by control engineers to model all types of systems. A block diagram can be used simply to describe the composition and interconnection of a system. Or it can be used, together with transfer functions, to describe the cause-and-effect relationships throughout the system. For instance, the block diagram of Fig. 3-2(a) models an open-loop dc-motor speed-control system. The block diagram in this case simply shows how the system components are interconnected, and no mathematical details are given. If the mathematical and functional relationships of all the system elements are known, the block diagram can

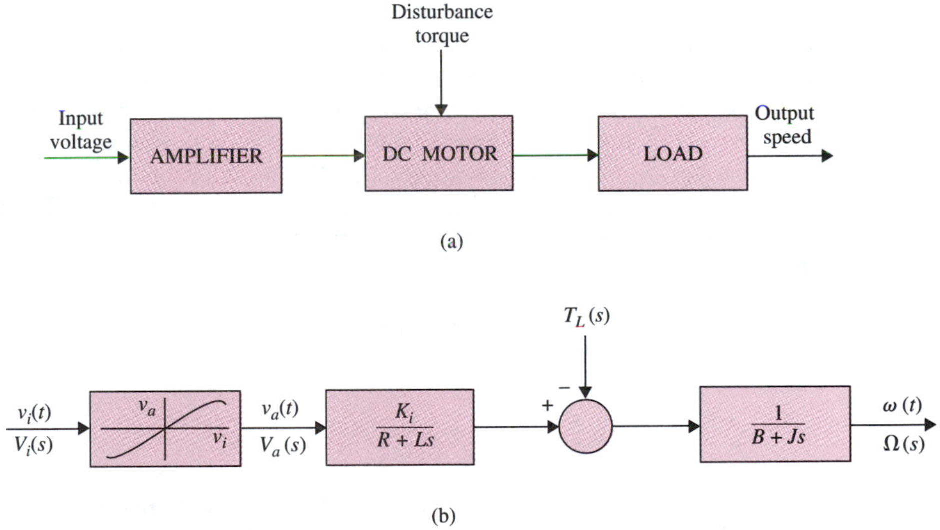

Figure 3-2 (a) Block diagram of a dc-motor control system. (b) Block diagram with transfer functions and amplifier characteristics.

be used as a tool for the analytical or computer solution of the system. In general, block diagrams can be used to model linear as well as nonlinear systems. For example, the input–output relations of the dc-motor control system may be represented by the block diagram shown in Fig. 3-2(b). In the figure, the input voltage to the motor is the output of the power amplifier, which realistically has a nonlinear characteristic. If the motor is linear, or, more appropriately, we should say that if it is operated in the linear region of its characteristics, its dynamics can be represented by transfer functions. The nonlinear amplifier gain can only be described between the time variables $v_i(t)$ and $v_a(t)$, and no transfer function exists between the Laplace transform variables $V_i(s)$ and $V_a(s)$. However, if the magnitude of $v_i(t)$ is limited to the linear range of the amplifier, the amplifier can be regarded as linear, and the amplifier can be described by the transfer function

$$\frac{V_a(s)}{V_i(s)} = K \tag{3-15}$$

where K is a constant which is the slope of the linear region of the amplifier characteristics.

3-3-1 Block Diagrams of Control Systems

We shall now define the block diagram elements used frequently in control systems and the related algebra. One of the important components of a control system is the sensing device that acts as a junction point for signal comparisons. The physical components involved are the potentiometer, syncho, resolver, differential amplifier, multiplier, and other signal-processing transducers. In general, sensing devices perform simple mathematical operations such as **addition, subtraction, multiplication** (nonlinear), and sometimes combinations of these. The block diagram representations of these operations are illustrated in Fig. 3-3. The addition and subtraction operations in Fig. 3-3(a), (b), and (c) are linear, so that the input and output variables of these block diagram elements can be time-domain variables or Laplace transform variables. Thus, in Fig. 3-3(a), the block diagram implies that

$$e(t) = r(t) - y(t) \tag{3-16}$$

or

$$E(s) = R(s) - Y(s) \tag{3-17}$$

The multiplication operation shown in Fig. 3-3(d) is nonlinear, so that the input–output relation has meaning only in the real (time) domain:

$$e(t) = r(t) \times y(t) \tag{3-18}$$

It is important to keep in mind that multiplication in the real domain *does not* carry over to the Laplace transform domain, so that $E(s) \neq R(s) \times Y(s)$. Rather, when

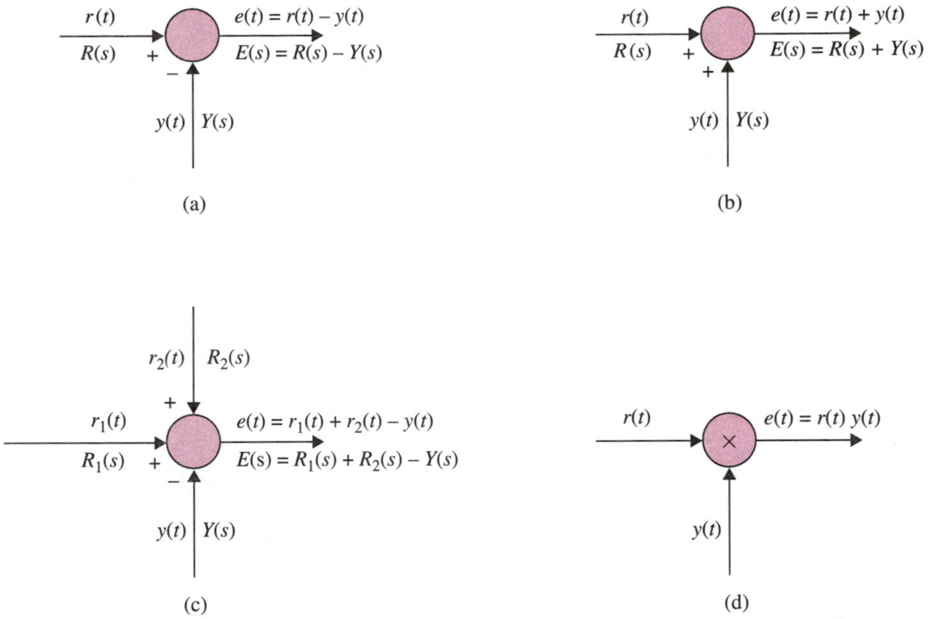

Figure 3-3 Block diagram elements of typical sensing devices of control systems. (a) Subtraction. (b) Addition. (c) Addition and subtraction. (d) Multiplication.

transformed, Eq. (3-18) should read: $E(s) = R(s) * Y(s)$, where in this case the symbol $*$ represents **complex convolution** of $R(s)$ and $Y(s)$ in the Laplace domain.

Figure 3-4 shows the block diagram of a linear feedback control system. The following terminology is defined with reference to the diagram.

$r(t), R(s)$ = reference input (command)
$y(t), Y(s)$ = output (controlled variable)
$b(t), B(s)$ = feedback signal
$u(t), U(s)$ = actuating signal = error signal $e(t), E(s)$, when $H(s) = 1$
$H(s)$ = feedback transfer function
$G(s)H(s) = L(s)$ = loop transfer function
$G(s)$ = forward-path transfer function
$M(s) = Y(s)/R(s)$ = closed-loop transfer function or system transfer function

The closed-loop transfer function $M(s)$ can be expressed as a function of $G(s)$ and $H(s)$. From Fig. 3-4 we write

$$Y(s) = G(s)U(s) \tag{3-19}$$

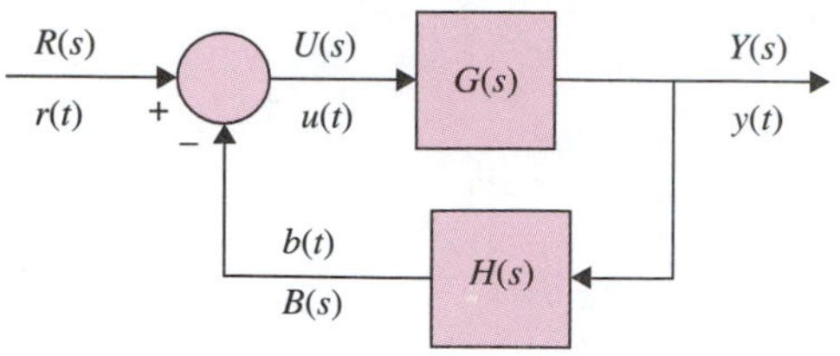

Figure 3-4 Basic block diagram of a feedback control system.

and

$$B(s) = H(s)Y(s) \tag{3-20}$$

The actuating signal is written

$$U(s) = R(s) - B(s) \tag{3-21}$$

Substituting Eq. (3-21) into Eq. (3-19) yields

$$Y(s) = G(s)R(s) - G(s)B(s) \tag{3-22}$$

Substituting Eq. (3-20) into Eq. (3-22) and then solving for $Y(s)/R(s)$ gives the closed-loop transfer function:

$$M(s) = \frac{Y(s)}{R(s)} = \frac{G(s)}{1 + G(s)H(s)} \tag{3-23}$$

In general, a control system may contain more than one feedback loop, and evaluation of the transfer function from the block diagram by means of the algebraic method just described may be tedious. Although, in principle, the block diagram of a system with one input and one output can always be reduced to the basic single-loop form of Fig. 3-4, the algebraic steps involved in the reduction process may again be quite tedious. We show in Section 3-4 that the transfer function of any linear system can be obtained directly from its block diagram by use of the signal-flow graph gain formula.

3-3-2 Block Diagrams and Transfer Functions of Multivariable Systems

In this section we illustrate the block diagram and matrix representations of multivariable systems. Two block diagram representations of a multivariable system with p inputs and q outputs are shown in Fig. 3-5. In Fig. 3-5(a), the individual input and output signals are designated, whereas in the block diagram of Fig. 3-5(b), the multiplicity of

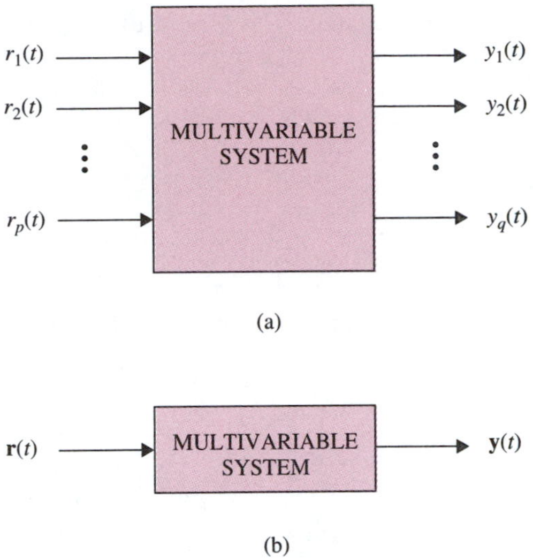

Figure 3-5 Block diagram representations of a multivariable system.

the inputs and outputs is denoted by vectors. The case of Fig. 3-5(b) is preferable in practice because of its simplicity.

Figure 3-6 shows the block diagram of a multivariable feedback control system. The transfer-function relationships of the system are expressed in vector-matrix form:

$$\mathbf{Y}(s) = \mathbf{G}(s)\mathbf{U}(s) \tag{3-24}$$
$$\mathbf{U}(s) = \mathbf{R}(s) - \mathbf{B}(s) \tag{3-25}$$
$$\mathbf{B}(s) = \mathbf{H}(s)\mathbf{Y}(s) \tag{3-26}$$

where $\mathbf{Y}(s)$ is the $q \times 1$ output vector; $\mathbf{U}(s)$, $\mathbf{R}(s)$, and $\mathbf{B}(\mathbf{s})$ are all $p \times 1$ vectors; and $\mathbf{G}(s)$ and $\mathbf{H}(s)$ are $q \times p$ and $p \times q$ transfer-function matrices, respectively. Substituting Eq. (3-26) into Eq. (3-25) and then from Eq. (3-25) to Eq. (3-24), we get

$$\mathbf{Y}(s) = \mathbf{G}(s)\mathbf{R}(s) - \mathbf{G}(s)\mathbf{H}(s)\mathbf{Y}(s) \tag{3-27}$$

Solving for $\mathbf{Y}(s)$ from Eq. (3-27) gives

$$\mathbf{Y}(s) = [\mathbf{I} + \mathbf{G}(s)\mathbf{H}(s)]^{-1}\mathbf{G}(s)\mathbf{R}(s) \tag{3-28}$$

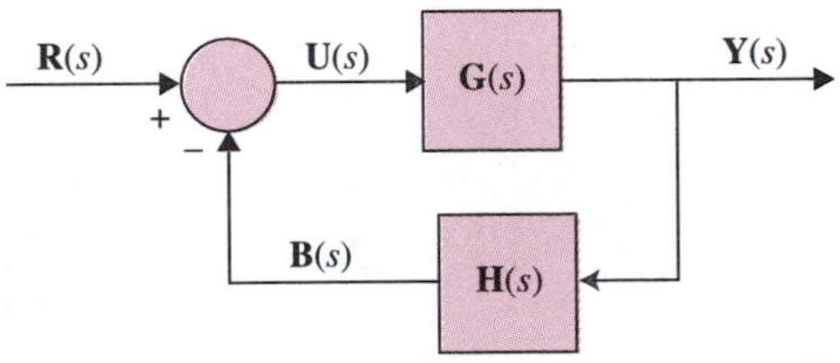

Figure 3-6 Block diagram of a multivariable feedback control system.

provided that $\mathbf{I} + \mathbf{G}(s)\mathbf{H}(s)$ is nonsingular. The closed-loop transfer matrix is defined as

$$\mathbf{M}(s) = [\mathbf{I} + \mathbf{G}(s)\mathbf{H}(s)]^{-1}\mathbf{G}(s) \tag{3-29}$$

Then Eq. (3-28) is written

$$\mathbf{Y}(s) = \mathbf{M}(s)\mathbf{R}(s) \tag{3-30}$$

Example 3-1

Consider that the forward-path transfer-function matrix and the feedback-path transfer-function matrix of the system shown in Fig. 3-6 are

$$\mathbf{G}(s) = \begin{bmatrix} \dfrac{1}{s+1} & -\dfrac{1}{s} \\ 2 & \dfrac{1}{s+2} \end{bmatrix} \qquad \mathbf{H}(s) = \begin{bmatrix} 1 & 0 \\ 0 & 1 \end{bmatrix} \tag{3-31}$$

respectively. The closed-loop transfer-function matrix of the system is given by Eq. (3-29) and is evaluated as follows:

$$\mathbf{I} + \mathbf{G}(s)\mathbf{H}(s) = \begin{bmatrix} 1 + \dfrac{1}{s+1} & -\dfrac{1}{s} \\ 2 & 1 + \dfrac{1}{s+2} \end{bmatrix} = \begin{bmatrix} \dfrac{s+2}{s+1} & -\dfrac{1}{s} \\ 2 & \dfrac{s+3}{s+2} \end{bmatrix} \tag{3-32}$$

The closed-loop transfer-function matrix is

$$\mathbf{M}(s) = [\mathbf{I} + \mathbf{G}(s)\mathbf{H}(s)]^{-1}\mathbf{G}(s) = \frac{1}{\Delta}\begin{bmatrix} \dfrac{s+3}{s+2} & \dfrac{1}{s} \\ -2 & \dfrac{s+2}{s+1} \end{bmatrix}\begin{bmatrix} \dfrac{1}{s+1} & -\dfrac{1}{s} \\ 2 & \dfrac{1}{s+2} \end{bmatrix} \tag{3-33}$$

where

$$\Delta = \frac{s+2}{s+1}\frac{s+3}{s+2} + \frac{2}{s} = \frac{s^2+5s+2}{s(s+1)} \tag{3-34}$$

Thus

$$\mathbf{M}(s) = \frac{s(s+1)}{s^2+5s+2}\begin{bmatrix} \dfrac{3s^2+9s+4}{s(s+1)(s+2)} & -\dfrac{1}{s} \\ 2 & \dfrac{3s+2}{s(s+1)} \end{bmatrix} \tag{3-35}$$

▲

3-4 Signal-Flow Graphs

A signal-flow graph (SFG) may be regarded as a simplified version of a block diagram. The SFG was introduced by S. J. Mason [2] for the cause-and-effect representation of linear systems that are modeled by algebraic equations. Besides the difference in physical appearances of the SFG and the block diagram, we may regard the signal-flow graph as constrained by more rigid mathematical rules, whereas use of the block diagram notation is more liberal. A SFG may be defined as a graphical means of portraying the input–output relationships between the variables of a set of linear algebraic equations.

Consider that a linear system is described by a set of N algebraic equations:

$$y_j = \sum_{k=1}^{N} a_{kj} y_k \qquad j = 1, 2, \ldots, N \tag{3-36}$$

It should be pointed out that these N equations are written in the form of cause-and-effect relations:

$$j\text{th effect} = \sum_{k=1}^{N} (\text{gain from } k \text{ to } j) \times (k\text{th cause}) \tag{3-37}$$

or simply

$$\text{output} = \sum \text{gain} \times \text{input} \tag{3-38}$$

This is the single most important axiom in forming the set of algebraic equations for SFGs. In the case when the system is represented by a set of integrodifferential equations, we must first transform these into Laplace transform equations and then rearrange the latter into the form of Eq. (3-36), or

$$Y_j(s) = \sum_{k=1}^{N} G_{kj}(s) Y_k(s) \qquad j = 1, 2, \ldots, N \tag{3-39}$$

3-4-1 Basic Elements of a SFG

▲ In a SFG signals can transmit through a branch only in the direction of the arrow.

When constructing a SFG, junction points, or **nodes,** are used to represent variables. The nodes are connected by line segments called **branches,** according to the cause-and-effect equations. The branches have associated branch gains and directions. *A signal can transmit through a branch only in the direction of the arrow.* In general, given a set of equations such as those of Eq. (3-36) or Eq. (3-39), the construction of the SFG is basically a matter of following through the cause-and-effect relations relating each variable in terms of itself and the others. For instance, consider that a linear system is represented by the simple algebraic equation

$$y_2 = a_{12} y_1 \tag{3-40}$$

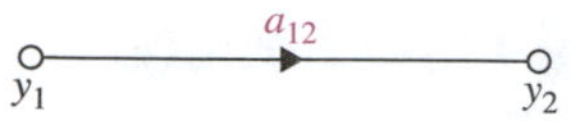

Figure 3-7 Signal-flow graph of $y_2 = a_{12}y_1$.

where y_1 is the input, y_2 the output, and a_{12} the gain, or transmittance, between the two variables. The SFG representation of Eq. (3-40) is shown in Fig. 3-7. Notice that the branch directing from node y_1 (input) to node y_2 (output) expresses the dependence of y_2 on y_1, but not the reverse. The branch between the input node and the output node should be interpreted as a unilateral amplifier with gain a_{12}, so that when a signal of one unit is applied at the input y_1, a signal of strength $a_{12}y_1$ is delivered at node y_2. Although algebraically, Eq. (3-40) can be written as

$$y_1 = \frac{1}{a_{12}} y_2 \tag{3-41}$$

the SFG of Fig. 3-7 does not imply this relationship. If Eq. (3-41) is valid as a cause-and-effect equation, a new SFG should be drawn with y_2 as the input and y_1 as the output.

Example 3-2

As an example on the construction of a SFG, consider the following set of algebraic equations:

$$\begin{aligned} y_2 &= a_{12}y_1 + a_{32}y_3 \\ y_3 &= a_{23}y_2 + a_{43}y_4 \\ y_4 &= a_{24}y_2 + a_{34}y_3 + a_{44}y_4 \\ y_5 &= a_{25}y_2 + a_{45}y_4 \end{aligned} \tag{3-42}$$

The SFG for these equations is constructed, step by step, as shown in Fig. 3-8. ▲

3-5 Summary of the Basic Properties of SFG

The important properties of the SFG that have been covered thus far are summarized as follows:

1. SFG applies only to linear systems.
2. The equations for which a SFG is drawn must be algebraic equations in the form of cause and effect.
3. Nodes are used to represent variables. Normally, the nodes are arranged from left to right, from the input to the output, following a succession of cause-and-effect relations through the system.

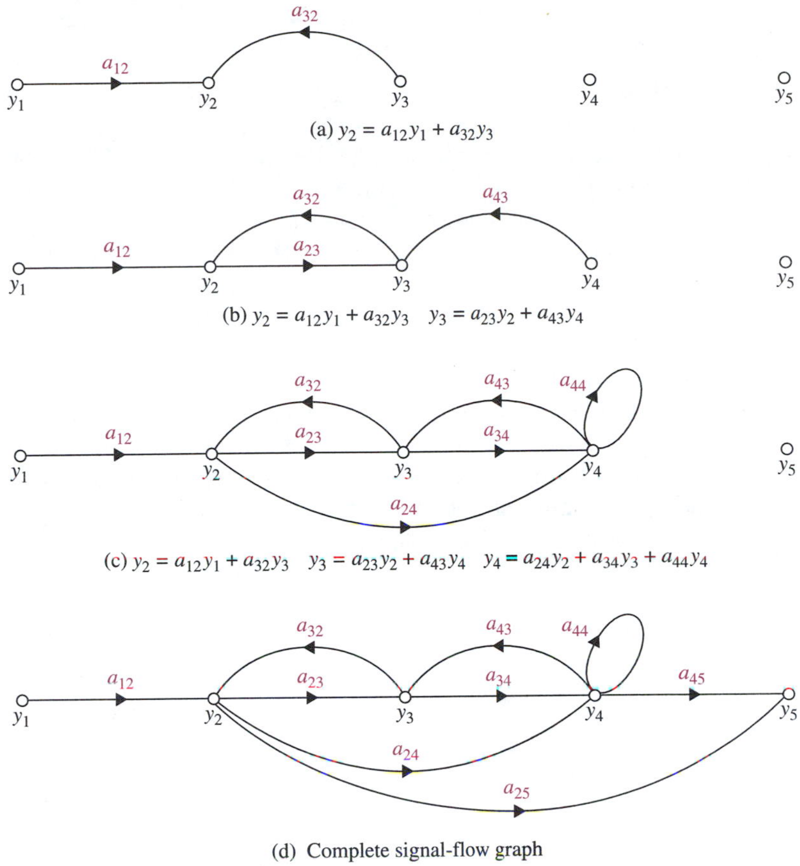

Figure 3-8 Step-by-step construction of the signal-flow graph for Eq. (3-42).

4. Signals travel along branches only in the direction described by the arrows of the branches.
5. The branch directing from node y_k to y_j represents the dependence of y_j upon y_k, but not the reverse.
6. A signal y_k traveling along a branch between y_k and y_j is multiplied by the gain of the branch, a_{kj}, so that a signal $a_{kj}y_k$ is delivered at y_j.

3-6 Definitions of SFG Terms

In addition to the branches and nodes defined earlier for the SFG, the following terms are useful for the purpose of identifying and executing the SFG algebra.

▲ An input node has only outgoing branches.

▲ An output node has only incoming branches.

Input Node (Source). *An* **input node** *is a node that has only outgoing branches.* (Example: node y_1 in Fig. 3-8.)

Output Node (Sink). *An* **output node** *is a node that has only incoming branches.* (Example: node y_5 in Fig. 3-8.) However, this condition is not always readily met by an output node. For instance, the SFG in Fig. 3-9(a) does not have any node that satisfies the condition of an output node. It may be necessary to regard y_2 and/or y_3 as output nodes, for the purpose of finding the effects at these nodes due to the input. To make y_2 an output node, we simply connect a branch with unity gain from the existing node y_2 to a new node also designated as y_2, as shown in Fig. 3-9(b). The same procedure is applied to y_3. Notice that in the modified SFG of Fig. 3-9(b), it is equivalent that the equations $y_2 = y_2$ and $y_3 = y_3$ are added to the original equations. In general, we can make any noninput node of a SFG an output by the procedure just illustrated. However, we **cannot** convert a noninput node into an input node by reversing the branch direction of the procedure described for output nodes. For instance, node y_2 of the SFG in Fig. 3-9(a) is not an input node. If we attempt to convert it into an input node by adding an incoming branch with unity gain from another identical node, y_2, the SFG of Fig. 3-10 would result. The equation that portrays the relationship at node

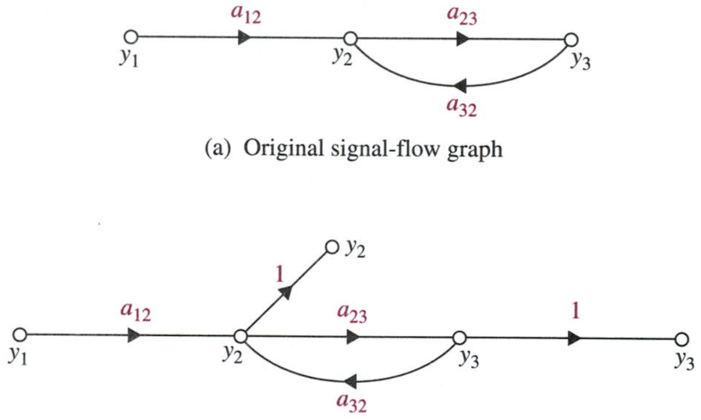

(a) Original signal-flow graph

(b) Modified signal-flow graph

Figure 3-9 Modification of a signal-flow graph so that y_2 and y_3 satisfy the requirement as output nodes.

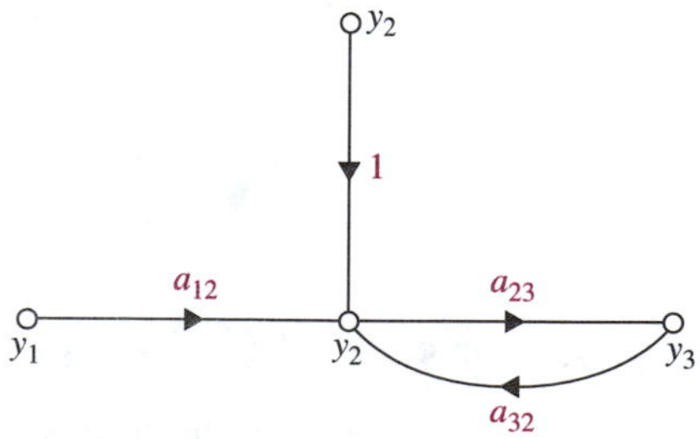

Figure 3-10 Erroneous way to make node y_2 an input node.

y_2 now reads

$$y_2 = y_2 + a_{12}y_1 + a_{32}y_3 \tag{3-43}$$

which is different from the original equation given in Fig. 3-8(a).

Path. *A* **path** *is any collection of a continuous succession of branches transversed in the same direction.* The definition of a path is entirely general, since it does not prevent any node from being traversed more than once. Therefore, as simple as the SFG of Fig. 3-9(a) is, it may have numerous paths just by traversing the branches a_{23} and a_{32} continuously.

Forward Path. *A* **forward path** *is a path that starts at an input node and ends at an output node, and along which no node is traversed more than once.* For example, in the SFG of Fig. 3-8(d), y_1 is the input node, and the rest of the nodes are all possible output nodes. The forward path between y_1 and y_2 is simply the connecting branch between the two nodes. There are two forward paths between y_1 and y_3; one contains the branches from y_1 to y_2 to y_3, and the other one contains the branches from y_1 to y_2 to y_4 (through the branch with gain a_{24}) and then back to y_3 (through the branch with gain a_{43}). The reader should try to determine the two forward paths between y_1 and y_4. Similarly, there are three forward paths between y_1 and y_5.

Loop. *A* **loop** *is a path that originates and terminates on the same node and along which no other node is encountered more than once.* For example, there are four loops in the SFG of Fig. 3-8(d). These are shown in Fig. 3-11.

Path Gain. *The product of the branch gains encountered in traversing a path is called the* **path gain.** For example, the path gain for the path y_1–y_2–y_3–y_4 in Fig. 3-8(d) is $a_{12}a_{23}a_{34}$.

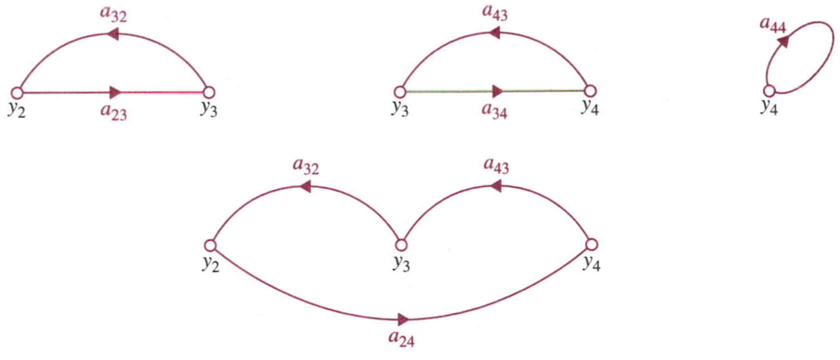

Figure 3-11 Four loops in the signal-flow graph of Fig. 3-8(d).

Forward-Path Gain. *The* **forward-path gain** *is the path gain of a forward path.*

▲ Two parts of a SFG are nontouching if they do not share a common node.

Loop Gain. *The* **loop gain** *is the path gain of a loop.* For example, the loop gain of the loop y_2–y_4–y_3–y_2 in Fig. 3-11 is $a_{24}a_{43}a_{32}$.

Nontouching Loops. *Two parts of a SFG are* **nontouching** *if they do not share a common node.* For example, the loops y_2–y_3–y_2 and y_4–y_4 of the SFG in Fig. 3-8 are nontouching loops.

3-7 SFG Algebra

Based on the properties of the SFG, we can outline the following manipulation rules and algebra of the SFG.

1. The value of the variable represented by a node is equal to the sum of all the signals entering the node. For the SFG of Fig. 3-12, the value of y_1 is equal to the sum of the signals transmitted through all the incoming branches; that is,

$$y_1 = a_{21}y_2 + a_{31}y_3 + a_{41}y_4 + a_{51}y_5 \tag{3-44}$$

2. The value of the variable represented by a node is transmitted through all branches leaving the node. In the SFG of Fig. 3-12, we have

$$\begin{aligned} y_6 &= a_{16}y_1 \\ y_7 &= a_{17}y_1 \\ y_8 &= a_{18}y_1 \end{aligned} \tag{3-45}$$

3. Parallel branches in the same direction connecting two nodes can be replaced by a single branch with gain equal to the sum of the gains of the parallel branches. An example of this case is illustrated in Fig. 3-13.
4. A series connection of unidirectional branches, as shown in Fig. 3-14, can be replaced by a single branch with gain equal to the product of the branch gains.

3-7-1 SFG of a Feedback Control System

The SFG of the single-loop feedback control system in Fig. 3-4 is drawn as shown in Fig. 3-15. Using the SFG algebra outlined above, the closed-loop transfer function in Eq. (3-23) can be obtained.

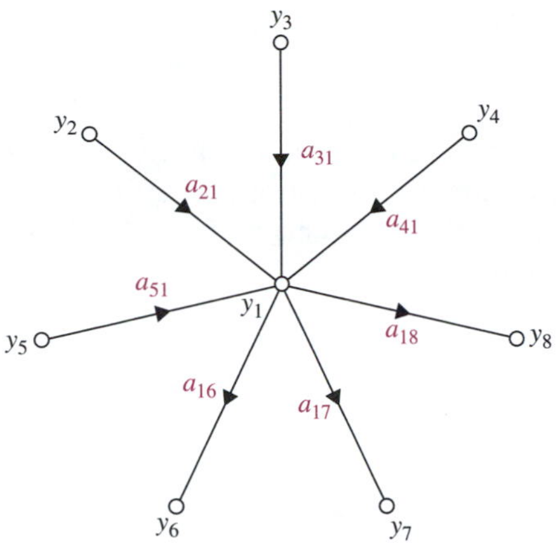

Figure 3-12 Node as a summing point and as a transmitting point.

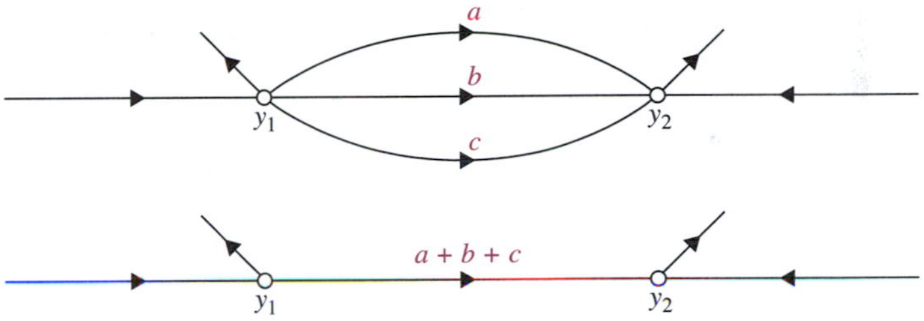

Figure 3-13 Signal-flow graph with parallel paths replaced by one with a single branch.

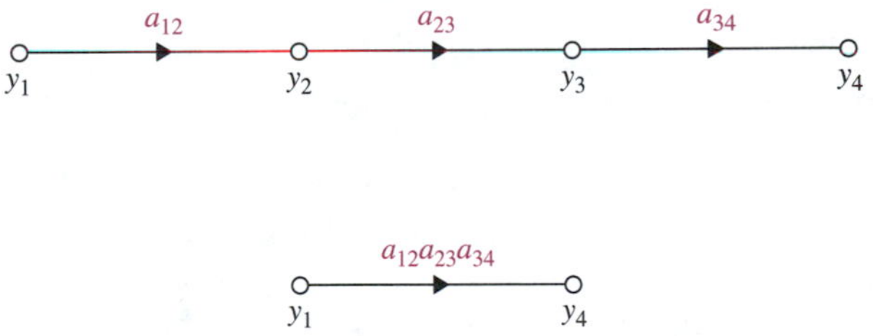

Figure 3-14 Signal-flow graph with cascaded unidirectional branches replaced by a single branch.

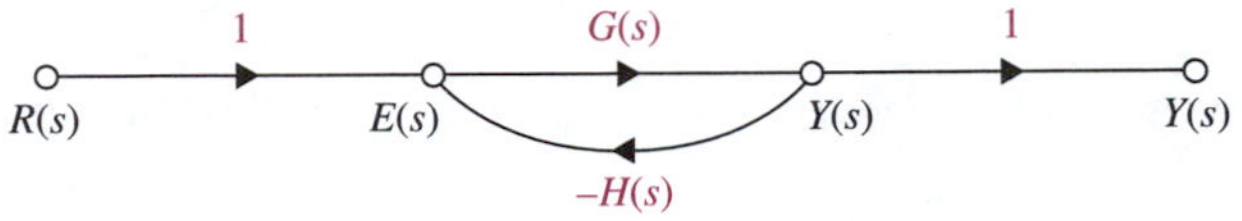

Figure 3-15 Signal-flow graph of a feedback control system.

3-8 Gain Formula for SFG

Given a SFG or block diagram, the task of solving for the input–output relations by algebraic manipulation could be quite tedious. Fortunately, there is a general gain formula available that allows the determination of the input–output relations of a SFG by inspection. Given a SFG with N forward paths and L loops, the gain between the input node y_{in} and output node y_{out} is [3].

$$M = \frac{y_{out}}{y_{in}} = \sum_{k=1}^{N} \frac{M_k \Delta_k}{\Delta} \tag{3-46}$$

where

y_{in} = input-node variable
y_{out} = output-node variable
M = gain between y_{in} and y_{out}
N = total number of forward paths between y_{in} and y_{out}
M_k = gain of the kth forward path between y_{in} and y_{out},

$$\Delta = 1 - \sum_i L_{i1} + \sum_j L_{j2} - \sum_k L_{k3} + \cdots \tag{3-47}$$

L_{mr} = gain product of the mth ($m = i, j, k, \ldots$) possible combination of r nontouching loops ($1 \leq r \leq L$)

or

Δ = 1 − (sum of the gains of **all individual** loops) + (sum of products of gains of all possible combinations of **two** nontouching loops) − (sum of products of gains of all possible combinations of **three** nontouching loops) + ··· (3-48)

Δ_k = the Δ for that part of the SFG that is nontouching with the kth forward path

▲ The SFG gain formula can only be applied between an input node and an output node.

The gain formula in Eq. (3-46) may seem formidable to use at first glance. However, Δ and Δ_k are the only terms in the formula that could be complicated if the SFG has a large number of loops and nontouching loops.

Care must be taken when applying the gain formula to ensure that it is applied between an **input node** and an **output node.**

Example 3-3

Consider that the closed-loop transfer function $Y(s)/R(s)$ of the SFG in Fig. 3-15 is to be determined by use of the gain formula, Eq. (3-46). The following results are obtained by inspection of the SFG:

1. There is only one forward path between $R(s)$ and $Y(s)$, and the forward-path gain is

$$M_1 = G(s) \tag{3-49}$$

2. There is only one loop; the loop gain is

$$L_{11} = -G(s)H(s) \tag{3-50}$$

3. There are no nontouching loops since there is only one loop. Furthermore, the forward path is in touch with the only loop. Thus $\Delta_1 = 1$, and

$$\Delta = 1 - L_{11} = 1 + G(s)H(s) \tag{3-51}$$

Using Eq. (3-46), the closed-loop transfer function is written

$$\frac{Y(s)}{R(s)} = \frac{M_1\Delta_1}{\Delta} = \frac{G(s)}{1 + G(s)H(s)} \tag{3-52}$$

which agrees with Eq. (3-23). ▲

Example 3-4

Consider the SFG shown in Fig. 3-8(d). Let us first determine the gain between y_1 and y_5 using the gain formula. The two forward paths between y_1 and y_5 and the forward-path gains are:

$M_1 = a_{12}a_{23}a_{34}a_{45}$ Forward path: $y_1 - y_2 - y_3 - y_4 - y_5$
$M_2 = a_{12}a_{25}$ Forward path: $y_1 - y_2 - y_5$
$M_3 = a_{12}a_{24}a_{45}$ Forward path: $y_1 - y_4 - y_5$

The four loops of the SFG are shown in Fig. 3-11. The loop gains are

$$L_{11} = a_{23}a_{32} \qquad L_{21} = a_{34}a_{43} \qquad L_{31} = a_{24}a_{32}a_{43} \qquad L_{41} = a_{44}$$

There is only one pair of nontouching loops; that is, the two loops are

$$y_2 - y_3 - y_2 \quad \text{and} \quad y_4 - y_4$$

Thus the product of the gains of the two nontouching loops is

$$L_{21} = a_{23}a_{32}a_{44} \tag{3-53}$$

All the loops are in touch with forward paths M_1 and M_3. Thus $\Delta_1 = \Delta_3 = 1$. Two of the loops are not in touch with forward path M_2. These are the loops $y_3 - y_4 - y_3$ and $y_4 - y_4$. Thus

$$\Delta_2 = 1 - a_{34}a_{43} - a_{44} \tag{3-54}$$

Substituting these quantities into Eq. (3-46), we have

$$\begin{aligned}\Delta &= 1 - (L_{11} + L_{12} + L_{13} + L_{14}) + L_{21}\\ &= 1 - (a_{23}a_{32} + a_{34}a_{43} + a_{24}a_{32}a_{43} + a_{44}) + a_{23}a_{32}a_{44}\end{aligned} \tag{3-55}$$

$$\frac{y_5}{y_1} = \frac{M_1\Delta_1 + M_2\Delta_2 + M_3\Delta_3}{\Delta} = \frac{(a_{12}a_{23}a_{34}a_{45}) + (a_{12}a_{25})(1 - a_{34}a_{43} - a_{44}) + a_{12}a_{24}a_{45}}{1 - (a_{23}a_{32} + a_{34}a_{43} + a_{24}a_{32}a_{43} + a_{44}) + a_{23}a_{32}a_{44}} \tag{3-56}$$

▲ Δ is the same regardless of which output node is chosen.

The reader should verify that chosing y_2 as the output,

$$\frac{y_2}{y_1} = \frac{a_{12}(1 - a_{34}a_{43} - a_{44})}{\Delta} \tag{3-57}$$

where Δ is given in Eq. (3-55). ▲

Example 3-5

Consider the SFG in Fig. 3-16. The following input–output relations are obtained by use of the gain formula:

$$\frac{y_2}{y_1} = \frac{1 + G_3H_2 + H_4 + G_3H_2H_4}{\Delta} \tag{3-58}$$

$$\frac{y_4}{y_1} = \frac{G_1G_2(1 + H_4)}{\Delta} \tag{3-59}$$

$$\frac{y_6}{y_1} = \frac{y_7}{y_1} = \frac{G_1G_2G_3G_4 + G_1G_5(1 + G_3H_2)}{\Delta} \tag{3-60}$$

where

$$\begin{aligned}\Delta = 1 &+ G_1H_1 + G_3H_2 + G_1G_2G_3H_3 + H_4 + G_1G_3H_1H_2 \\ &+ G_1H_1H_4 + G_3H_2H_4 + G_1G_2G_3H_3H_4 + G_1G_3H_1H_2H_4\end{aligned} \tag{3-61}$$

▲

3-8-1 Application of the Gain Formula between Output Nodes and Noninput Nodes

It was pointed out earlier that the gain formula can be applied only between a pair of input and output nodes. Often, it is of interest to find the relation between an output-node variable and a noninput-node variable. For example, in the SFG of Fig. 3-16, it may be of interest to find the relation y_7/y_2, which represents the dependence of y_7 on y_2; the latter is not an input.

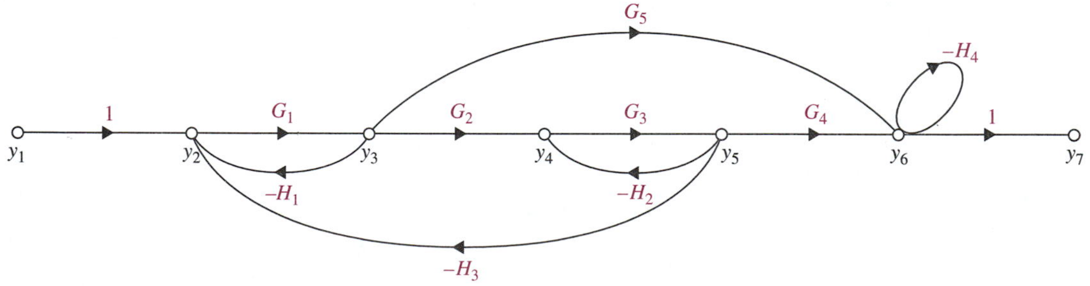

Figure 3-16 Signal-flow graph for Example 3-5.

We can show that by including an input node the gain formula can still be applied to find the gain between a noninput node and an output node. Let y_{in} be an input and y_{out} be an output node of a SFG. The gain, y_{out}/y_2, where y_2 is not an input, may be written as

$$\frac{y_{out}}{y_2} = \frac{y_{out}/y_{in}}{y_2/y_{in}} = \frac{\sum M_k\Delta_k \big|_{\text{from } y_{in} \text{ to } y_{out}} / \Delta}{\sum M_k\Delta_k \big|_{\text{from } y_{in} \text{ to } y_2} / \Delta} \tag{3-62}$$

Since Δ is independent of the inputs and the outputs, Eq. (3-62) is written

$$\frac{y_{out}}{y_2} = \frac{\sum M_k\Delta_k \big|_{\text{from } y_{in} \text{ to } y_{out}}}{\sum M_k\Delta_k \big|_{\text{from } y_{in} \text{ to } y_2}} \tag{3-63}$$

Notice that Δ does not appear in the last equation.

Example 3-6

From the SFG in Fig. 3-16, the gain between y_2 and y_7 is written

$$\frac{y_7}{y_2} = \frac{y_7/y_1}{y_2/y_1} = \frac{G_1G_2G_3G_4 + G_1G_5(1 + G_3H_2)}{1 + G_3H_2 + H_4 + G_3H_2H_4} \tag{3-64}$$

▲

3-9 Application of the Gain Formula to Block Diagrams

Because of the similarity between the block diagram and the SFG, the gain formula in Eq. (3-46) can be applied to determine the input–output gain of either. In general, the gain formula can be applied directly to a block diagram. However, in complex systems, to be able to identify all the loops and nontouching parts clearly, it may be helpful if an equivalent SFG is drawn for the block diagram before applying the gain formula.

Example 3-7

To illustrate how an equivalent SFG of a block diagram is constructed and how the gain formula is applied to a block diagram, consider the block diagram shown in Fig. 3-17(a). The equivalent SFG of the system is shown in Fig. 3-17(b). Notice that since a node on the SFG is interpreted as a summing point of all incoming signals to the node, the negative feedbacks on the block diagram are represented by assigning negative gains to the feedback paths on the SFG.

The closed-loop transfer function of the system is obtained by applying Eq. (3-46) to either the block diagram or the SFG in Fig. 3-17:

$$\frac{Y(s)}{R(s)} = \frac{G_1G_2G_3 + G_1G_4}{\Delta} \tag{3-65}$$

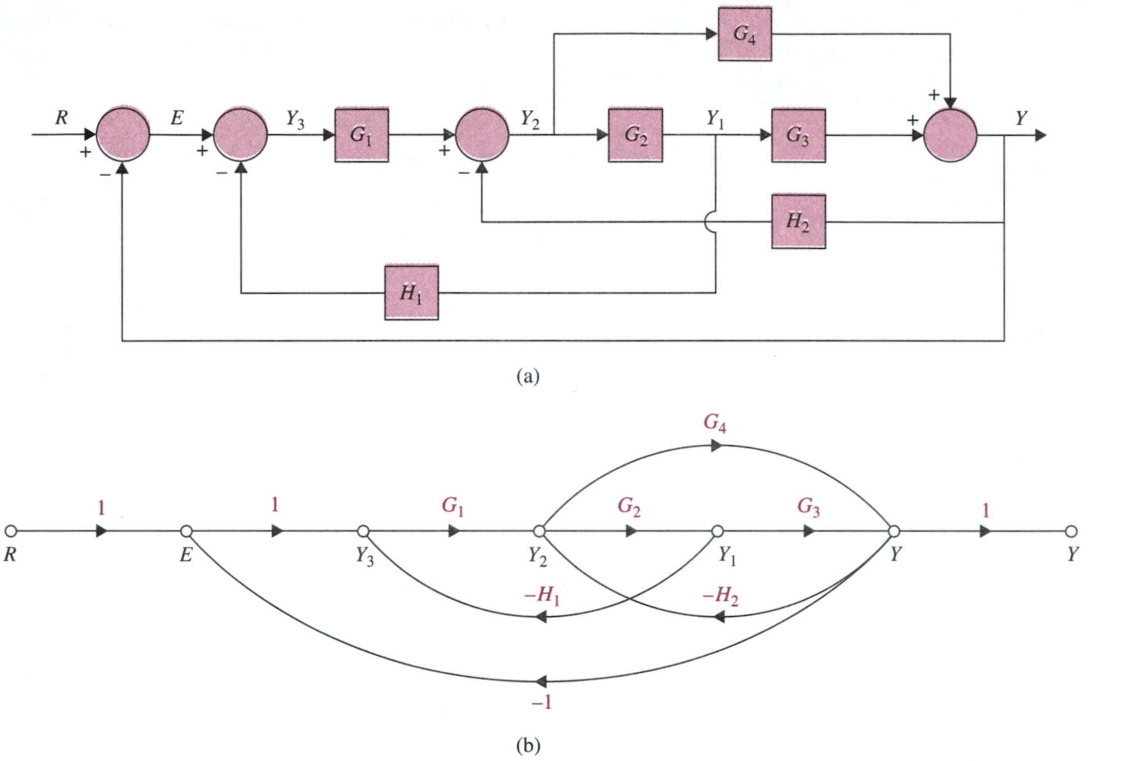

Figure 3-17 (a) Block diagram of a control system. (b) Equivalent signal-flow graph.

where

$$\Delta = 1 + G_1G_2H_1 + G_2G_3H_2 + G_1G_2G_3 + G_4H_2 + G_1G_4 \tag{3-66}$$

Similarly,

$$\frac{E(s)}{R(s)} = \frac{1 + G_1G_2H_1 + G_2G_3H_2 + G_4H_2}{\Delta} \tag{3-67}$$

$$\frac{Y(s)}{E(s)} = \frac{G_1G_2G_3 + G_1G_4}{1 + G_1G_2H_1 + G_2G_3H_2 + G_4H_2} \tag{3-68}$$

The last expression is obtained using Eq. (3-62). ▲

3-10 State Diagram

In this section we introduce the **state diagram,** which is an extension of the SFG to portray state equations and differential equations. The significance of the state diagram is that it forms a close relationship among the state equations, computer simulation, and transfer functions. A state diagram is constructed following all the rules of the SFG using the Laplace-transformed state equations.

The basic elements of a state diagram are similar to the conventional SFG, except for the **integration** operation. Let the variables $x_1(t)$ and $x_2(t)$ be related by the first-order differentiation

$$\frac{dx_1(t)}{dt} = x_2(t) \tag{3-69}$$

Integrating both sides of Eq. (3-69) with respect to t from the initial time t_0, we get

$$x_1(t) = \int_{t_0}^{t} x_2(\tau)\, d\tau + x_1(t_0) \tag{3-70}$$

Since the SFG algebra does not handle integration in the time domain, we must take the Laplace transform on both sides of Eq. (3-70). We have

$$\begin{aligned} X_1(s) &= \mathcal{L}\left[\int_{t_0}^{t} x_2(\tau)\, d\tau\right] + \frac{x_1(t_0)}{s} \\ &= \mathcal{L}\left[\int_{0}^{t} x_2(\tau)\, d\tau - \int_{0}^{t_0} x_2(\tau)\, d\tau\right] + \frac{x_1(t_0)}{s} \\ &= \frac{X_2(s)}{s} - \mathcal{L}\left[\int_{0}^{t_0} x_2(\tau)\, d\tau\right] + \frac{x_1(t_0)}{s} \end{aligned} \tag{3-71}$$

Since the past history of the integrator is represented by $x_1(t_0)$ and the state transition is assumed to start at $\tau = t_0$, $x_2(\tau) = 0$ for $0 < \tau < t_0$. Thus Eq. (3-71) becomes

$$X_1(s) = \frac{X_2(s)}{s} + \frac{x_1(t_0)}{s} \qquad \tau \geq t_0 \tag{3-72}$$

Equation (3-72) is now algebraic and can be represented by a SFG, as shown in Fig. 3-18. An alternative SFG with fewer elements for Eq. (3-72) is shown in Fig. 3-19.

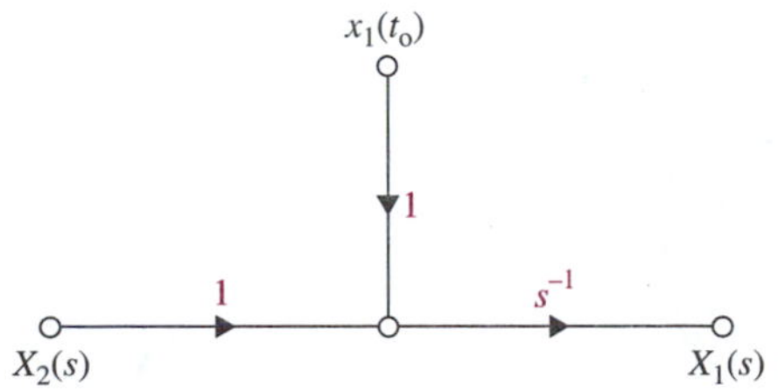

Figure 3-18 Signal-flow-graph representation of $X_1(s) = [X_2(s)/s] + [x_1(t_0)/s]$.

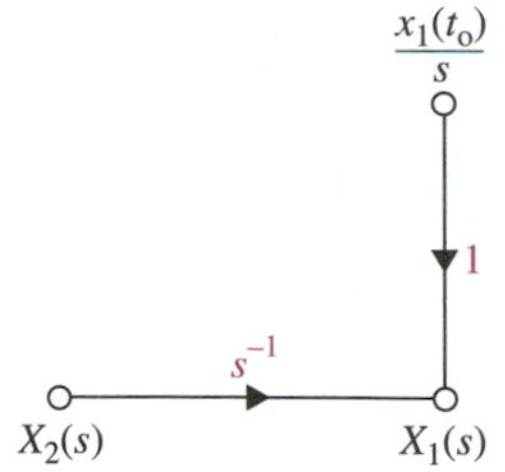

Figure 3-19 Signal-flow-graph representation of $X_1(s) = [X_2(s)/s] + [x_1(t_0)/s]$.

Figure 3-19 shows that *the output of the integrator is equal to s^{-1} times the input, plus the initial condition $x_1(t_0)/s$.*

Before embarking on several illustrative examples on the construction of state diagrams, let us point out the important usages of the state diagram.

1. A state diagram can be constructed directly from the system's differential equation. This allows the determination of the state variables and the state equations.
2. A state diagram can be constructed from the system's transfer function. This step is defined as the **decomposition** of transfer functions (Section 5-11).
3. The state diagram can be used for the programming of the system on an analog computer or simulation on a digital computer.
4. The state-transition equation in the Laplace transform domain may be obtained from the state diagram by means of the SFG gain formula.
5. The transfer functions of a system can be determined from the state diagram.
6. The state equations and the output equations can be determined from the state diagram.

The details of these techniques follow.

3-10-1 From Differential Equations to State Diagram

▲ The outputs of the integrators on the state diagram are usually defined as the state variable.

When a linear system is described by a high-order differential equation, a state diagram can be constructed from these equations, although a direct approach is not always the most convenient. Consider the following differential equation:

$$\frac{d^n y(t)}{dt^n} + a_n \frac{d^{n-1} y(t)}{dt^{n-1}} + \cdots + a_2 \frac{dy(t)}{dt} + a_1 y(t) = r(t) \tag{3-73}$$

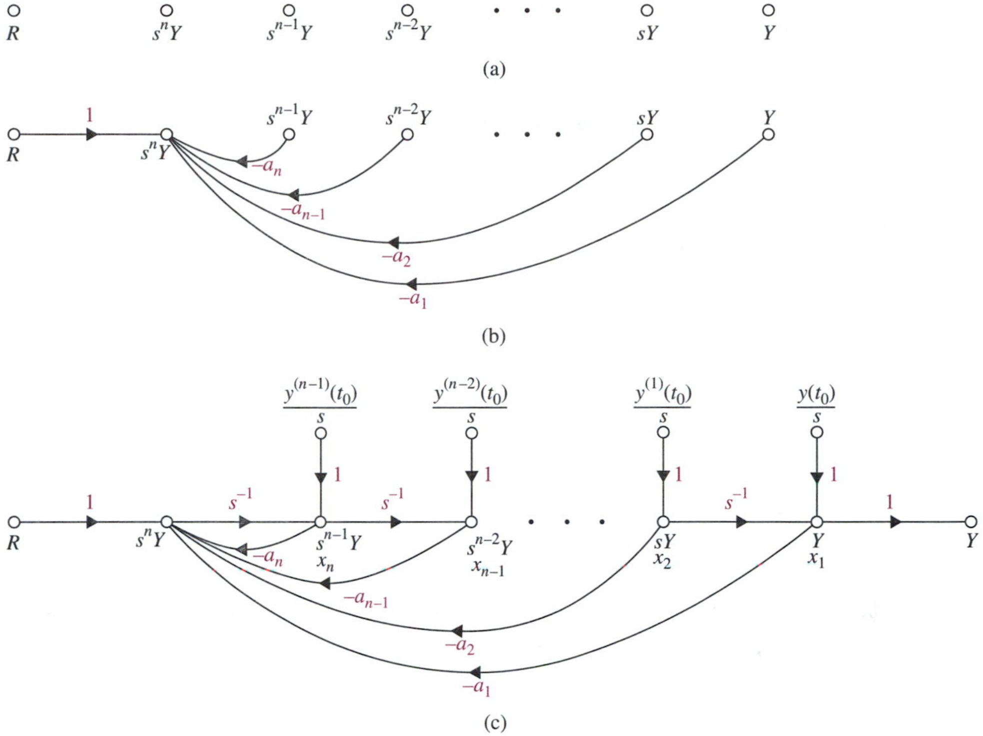

Figure 3-20 State-diagram representation of the differential equation of Eq. (3-73).

To construct a state diagram using this equation, we rearrange the equation as

$$\frac{d^n y(t)}{dt^n} = -a_n \frac{d^{n-1}y(t)}{dt^{n-1}} - \cdots - a_2 \frac{dy(t)}{dt} - a_1 y(t) + r(t) \tag{3-74}$$

As a first step, the nodes representing $R(s)$, $s^nY(s)$, $s^{n-1}Y(s)$, . . . , $sY(s)$, and $Y(s)$ are arranged from left to right, as shown in Fig. 3-20. Since $s^iY(s)$ corresponds to $d^iy(t)/dt^i$, $i = 0, 1, 2, \ldots, n$, in the Laplace domain, as the next step, the nodes in Fig. 3-20(a) are connected by branches to portray Eq. (3-74), resulting in Fig. 3-20(b). Finally, the integrator branches with gains of s^{-1} are inserted and the initial conditions are added to the outputs of the integrators according to the basic scheme in Fig. 3-19. The complete state diagram is drawn as shown in Fig. 3-20(c). *The outputs of the integrators are defined as the state variables,* $x_1, x_2, \ldots, x_n$. This is usually the natural choice of state variables once the state diagram is drawn.

When the differential equation has derivatives of the input on the right side, the problem of drawing the state diagram directly is not as straightforward as just illustrated. We will show that, in general, it is more convenient to obtain the transfer function from the differential equation first and then arrive at the state diagram through decomposition (Section 5-9).

Example 3-8

Consider the differential equation

$$\frac{d^2y(t)}{dt^2} + 3\frac{dy(t)}{dt} + 2y(t) = r(t) \tag{3-75}$$

Equating the highest-ordered term of the last equation to the rest of the terms, we have

$$\frac{d^2y(t)}{dt^2} = -3\frac{dy(t)}{dt} - 2y(t) + r(t) \tag{3-76}$$

Following the procedure just outlined, the state diagram of the system is drawn as shown in Fig. 3-21. The state variables x_1 and x_2 are assigned as shown. ▲

3-10-2 From State Diagram to Transfer Function

The transfer function between an input and an output is obtained from the state diagram by use of the gain formula and by setting all other inputs and initial states to zero. The following example shows how the transfer function is obtained directly from a state diagram.

Example 3-9

Consider the state diagram of Fig. 3-21. The transfer function between $R(s)$ and $Y(s)$ is obtained by applying the gain formula between these two nodes and setting the initial states to zero. We have

$$\frac{Y(s)}{R(s)} = \frac{1}{s^2 + 3s + 2} \tag{3-77}$$

▲

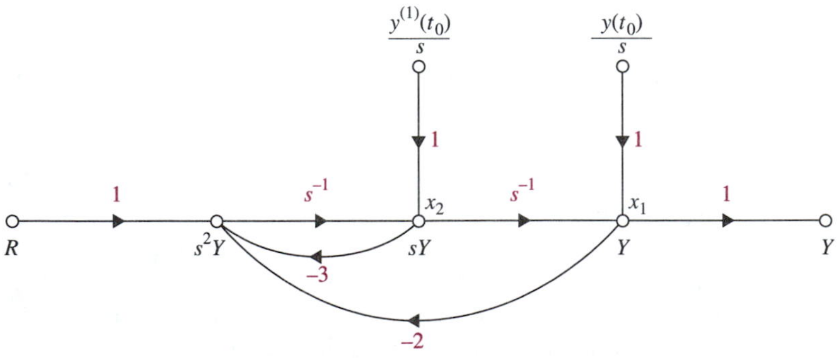

Figure 3-21 State diagram for Eq. (3-75).

3-10-3 From State Diagram to State and Output Equations

The state equations and the output equations can be obtained directly from the state diagram by use of the SFG gain formula. The general form of a state equation and the output equation for a linear system is described in Section 2-3 and presented again below.

State equation: $$\frac{dx(t)}{dt} = ax(t) + br(t) \tag{3-78}$$

Output equation: $$y(t) = cx(t) + dr(t) \tag{3-79}$$

where $x(t)$ is the state variable, $r(t)$ the input, $y(t)$ the output, and a, b, c, and d are constant coefficients. Based on the general form of the state and output equations, the following procedure of deriving the state and output equations from the state diagram is outlined:

1. Delete the initial states and the integrator branches with gains s^{-1} from the state diagram, since the state and output equations do not contain the Laplace operator s or the initial states.
2. For the state equations, regard the nodes that represent the derivatives of the state variables as output nodes, since these variables appear on the left-hand side of the state equations. The output $y(t)$ in the output equation is naturally an output node variable.
3. Regard the state variables and the inputs as input variables on the state diagram, since these variables are found on the right-hand side of the state and output equations.
4. Apply the SFG gain formula to the state diagram.

Example 3-10

Figure 3-22 shows the state diagram of Fig. 3-21 with the integrator branches and the initial states eliminated. Using $dx_1(t)/dt$ and $dx_2(t)/dt$ as the output nodes and $x_1(t)$, $x_2(t)$, and $r(t)$ as the input nodes, and applying the gain formula between these nodes, the state equations are obtained as

$$\begin{aligned} \frac{dx_1(t)}{dt} &= x_2(t) \\ \frac{dx_2(t)}{dt} &= -2x_1(t) - 3x_2(t) + r(t) \end{aligned} \tag{3-80}$$

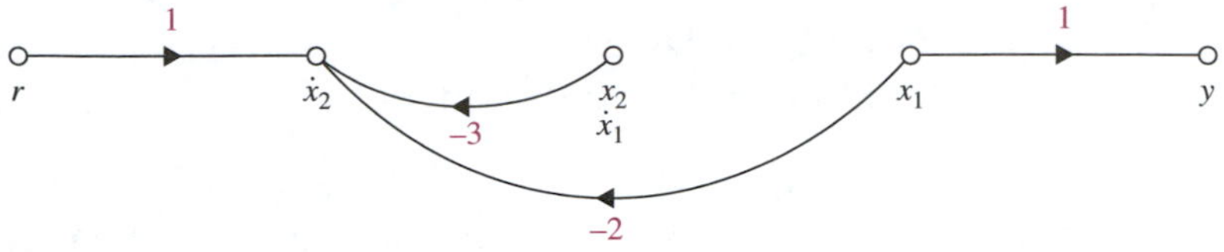

Figure 3-22 State diagram of Fig. 3-21 with the initial states and the integrator branches eliminated.

Applying the gain formula with $x_1(t)$, $x_2(t)$, and $r(t)$ as input nodes and $y(t)$ as the output node, the output equation is written

$$y(t) = x_1(t) \tag{3-81}$$

▲

Example 3-11

As another example on the determination of the state equations from the state diagram, consider the state diagram shown in Fig. 3-23(a). This example will also emphasize the importance of applying the gain formula. Figure 3-23(b) shows the state diagram with the initial states and the integrator branches deleted. Notice that in this case the state diagram in Fig. 3-23(b) still contains a loop. By applying the gain formula to the state diagram in Fig. 3-23(b) with $\dot{x}_1(t)$, $\dot{x}_2(t)$, and $\dot{x}_3(t)$ as output-node variables and $r(t)$, $x_1(t)$, $x_2(t)$, and $x_3(t)$ as input nodes, the state equations are obtained as follows in vector-matrix form:

$$\begin{bmatrix} \dfrac{dx_1(t)}{dt} \\ \dfrac{dx_2(t)}{dt} \\ \dfrac{dx_3(t)}{dt} \end{bmatrix} = \begin{bmatrix} 0 & 1 & 0 \\ \dfrac{-(a_2 + a_3)}{1 + a_0 a_3} & -a_1 & \dfrac{1 - a_0 a_2}{1 + a_0 a_3} \\ 0 & 0 & 0 \end{bmatrix} \begin{bmatrix} x_1(t) \\ x_2(t) \\ x_3(t) \end{bmatrix} + \begin{bmatrix} 0 \\ 0 \\ 1 \end{bmatrix} r(t) \tag{3-82}$$

The output equation is

$$y(t) = \frac{1}{1 + a_0 a_3} x_1(t) + \frac{a_0}{1 + a_0 a_3} x_3(t) \tag{3-83}$$

▲

3-11 Transfer Functions of Discrete-Data Systems

Discrete-data control systems have unique features in that the signals in these systems are either in the form of pulse trains or are digitally coded, and the controlled processes often contain analog components. For instance, a dc motor, which is an analog device, can be controlled either by a controller that puts out analog signals or by a digital

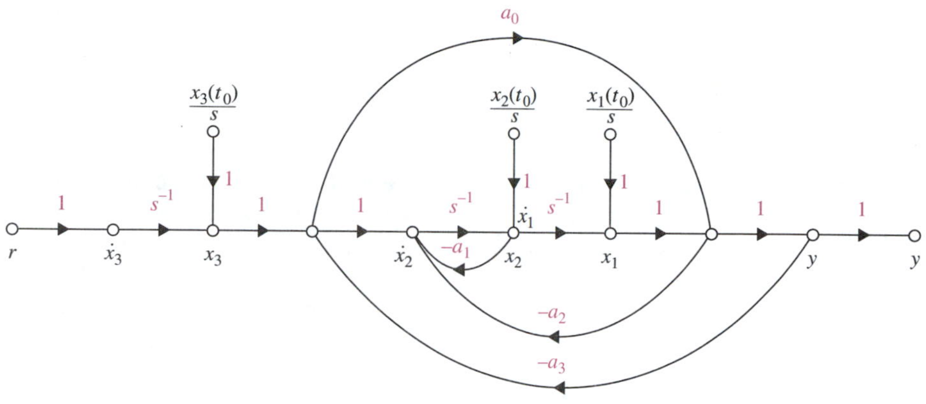

(a)

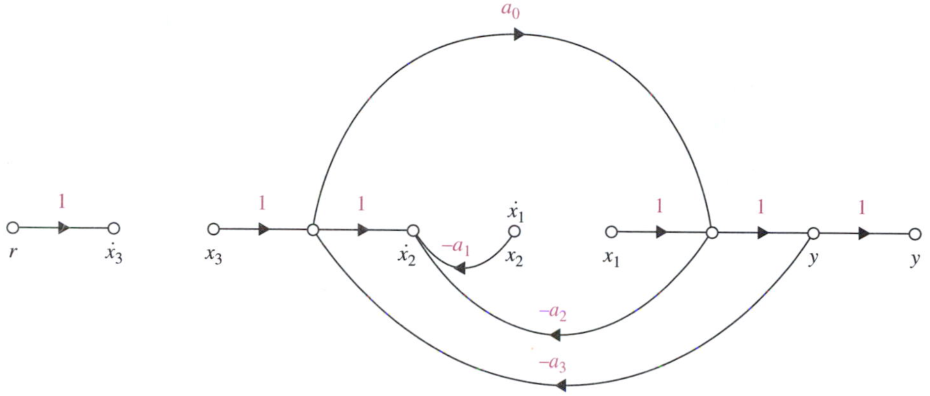

(b)

Figure 3-23 (a) State diagram. (b) State diagram in part (a) with all initial states and integrators eliminated.

controller that outputs digital data. In the latter case, an interface such as a digital-to-analog (D/A) converter is necessary to couple the digital component to the analog devices. Figure 3-24 shows the block diagram of a typical discrete-data system. The input and output of the digital controller can be represented by number sequences with numbers separated by the sampling period T. For linear operation, the D/A converter

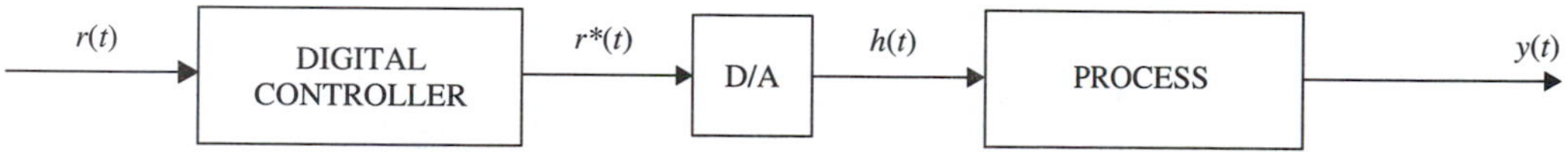

Figure 3-24 Block diagram of a typical discrete-data control system.

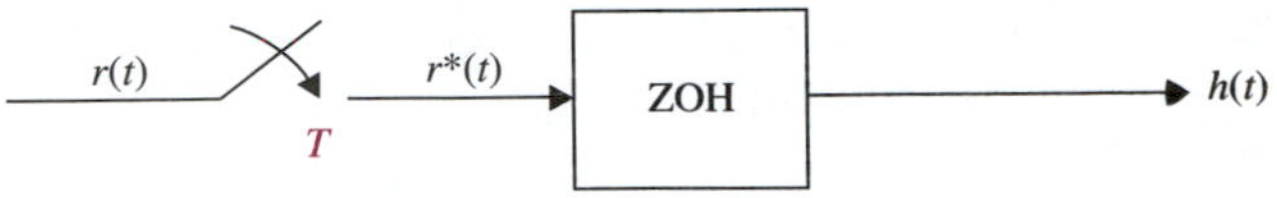

Figure 3-25 Sample-and-hold.

▲ The ideal sampler is not a physical entity. It is used only for the representation of discrete data.

can be represented by a **sample-and-hold (S/H)** device, which consists of a sampler and a data-hold device. The S/H that is most often used for the analysis of discrete-data systems consists of an **ideal sampler** and the **zero-order hold (ZOH).** Thus the system shown in Fig. 3-24 can be functionally represented by the block diagram in Fig. 3-25. Figure 3-26 shows the typical operation of the ideal sampler and the ZOH. The continuous data $r(t)$ are sampled with a sampling period T by the ideal sampler. The output of the ideal sampler, $r^*(t)$, is a train of impulses with the magnitudes of $r(t)$ at T carried by the strengths of the impulses. It should be pointed out that the ideal sampler is not a physical entity. It is devised simply to represent the discrete-time signal mathematically. In Fig. 3-26 the arrows at the sampling instants represent impulses. Since by definition, an impulse has zero pulse width and infinite height, the lengths of the arrows simply represent the areas under the impulses and are the magnitudes of the input signal $r(t)$ at the sampling instants. The ZOH simply holds the magnitude of the signal carried by the incoming impulse at a given time instant, say, kT, for the entire sampling period t, until the next impulse arrives at $t = (k + 1)T$. The output of the

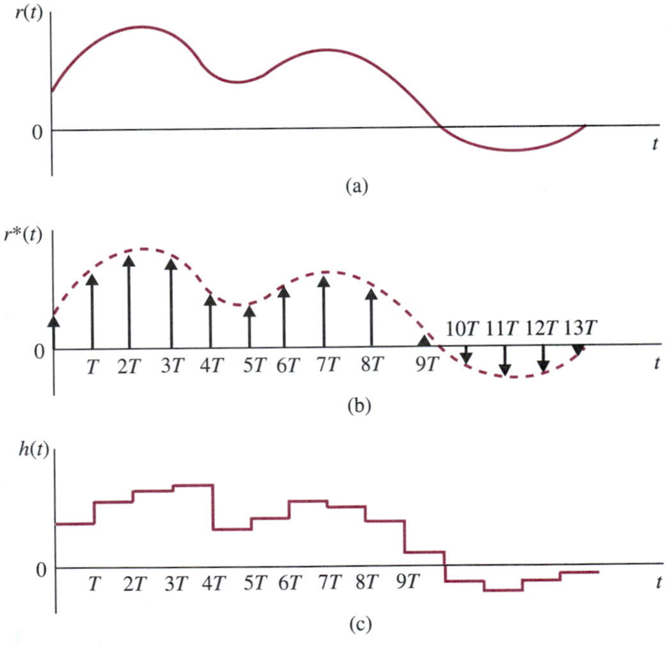

Figure 3-26 (a) Input signal to ideal sampler. (b) Output signal of ideal sampler. (c) Output signal of ZOH.

ZOH is a staircase approximation of the input to the ideal sampler, $r(t)$. As the sampling period T approaches zero, the output of the ZOH, $h(t)$, approaches $r(t)$; that is,

$$\lim_{T \to 0} h(t) = r(t) \tag{3-84}$$

However, since the output of the sampler, $r^*(t)$, is an impulse train, its limit as T approaches zero *does not* have any physical meaning. Based on the discussions above, a typical open-loop discrete-data system is modeled as shown in Fig. 3-27.

There are several ways of deriving the transfer-function representation of the system in Fig. 3-27. The following derivation is based on the Fourier series representation of the signal $r^*(t)$. We begin by writing

$$r^*(t) = r(t)\delta_T(t) \tag{3-85}$$

where $\delta_T(t)$ is the unit-impulse train,

$$\delta_T(t) = \sum_{k=-\infty}^{\infty} \delta(t - kT) \tag{3-86}$$

Since $\delta_T(t)$ is a periodic function with period T, it can be expressed as a Fourier series:

$$\delta_T(t) = \sum_{n=-\infty}^{\infty} C_n e^{j2\pi nt/T} \tag{3-87}$$

where C_n is the Fourier coefficient and is given by

$$C_n = \frac{1}{T}\int_0^T \delta_T(t) e^{-jn\omega_s t}\, dt \tag{3-88}$$

where $\omega_s = 2\pi/T$ is the **sampling frequency** in rad/sec.

Since the unit impulse is defined as a pulse with a width of δ and a height of $1/\delta$, and $\delta \to 0$, C_n is written

$$C_n = \lim_{\delta \to 0} \frac{1}{T\delta}\int_0^\delta e^{-jn\omega_s t}\, dt = \lim_{\delta \to 0} \frac{1 - e^{-jn\omega_s \delta}}{jn\omega_s T\delta} = \frac{1}{T} \tag{3-89}$$

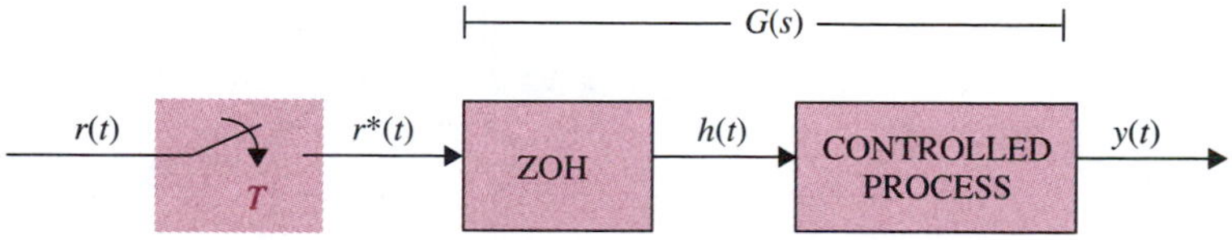

Figure 3-27 Block diagram of a discrete-data system.

Substituting Eq. (3-89) in Eq. (3-87) and then the latter in Eq. (3-85), we get

$$r^*(t) = \frac{1}{T}\sum_{n=-\infty}^{\infty} r(t)e^{-jn\omega_s t} \tag{3-90}$$

Taking the Laplace transform on both sides of Eq. (3-90) and using the complex shifting property of Eq. (2-36), we get

$$R^*(s) = \frac{1}{T}\sum_{n=-\infty}^{\infty} R(s - jn\omega_s) = \frac{1}{T}\sum_{n=-\infty}^{\infty} R(s + jn\omega_s) \tag{3-91}$$

Equation (3-91) represents the Laplace transform of the sampled signal $r^*(t)$. It is an alternative expression to Eq. (2-178). Using the method introduced in Chapter 2, $R^*(s)$ can be written as

$$R^*(s) = \sum_{k=0}^{\infty} r(kT)e^{-kTs} \tag{3-92}$$

Since the summing limits of $R^*(s)$ are from $-\infty$ to ∞, if s is replaced by $s + jm\omega_s$ in Eq. (3-91), where m is any integer, we have

$$R^*(s + jm\omega_s) = R^*(s) \tag{3-93}$$

Pulse-Transfer Function

We are now ready to derive the transfer function of the discrete-data system shown in Fig. 3-27. The Laplace transform of the system output y(t) is written

$$Y(s) = G(s)R^*(s) \tag{3-94}$$

Although the output $y(t)$ is obtained from $Y(s)$ by taking the inverse Laplace transform on both sides of Eq. (3-94), this step is difficult to execute because $G(s)$ and $R^*(s)$ represent different types of signals. To overcome this problem, we apply a fictitious sampler at the output of the system, as shown in Fig. 3-28. The fictititious sample S_2 has the same sampling period T and is synchronized to the original sampler S_1. The sampled form of $y(t)$ is $y^*(t)$. Applying Eq. (3-91) to $y^*(t)$ and using Eq. (3-94), we have

$$Y^*(s) = \frac{1}{T}\sum_{n=-\infty}^{\infty} G(s + jn\omega_s)R^*(s + jn\omega_s) \tag{3-95}$$

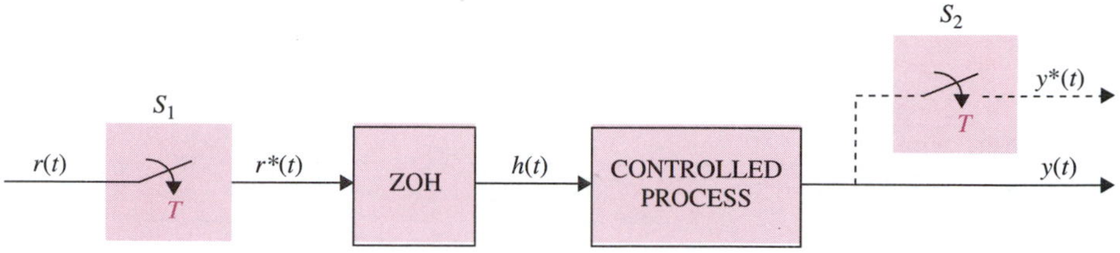

Figure 3-28 Discrete-data system with fictitious sampler.

In view of the relationship in Eq. (3-93), Eq. (3-95) is written

$$Y^*(s) = R^*(s)\frac{1}{T}\sum_{n=-\infty}^{\infty} G(s + jn\omega_s) = R^*(s)G^*(s) \tag{3-96}$$

where $G^*(s)$ is defined the same way as $R^*(s)$ in Eq. (3-91) and is called the **pulse-transfer function** of $G(s)$.

z-Transfer Function

Now that all the functions in Eq. (3-96) are in sampled form, where $R^*(s)$, $G^*(s)$, and $Y^*(s)$ all have the form of Eq. (3-92), we can take the *z*-transform on both sides of the equation by substituting $z = e^{Ts}$. We have

$$Y(z) = G(z)R(z) \tag{3-97}$$

where $G(z)$ is defined as the ***z*-transfer function** of $G(s)$ and is given by

$$G(z) = \sum_{k=0}^{\infty} g(kT)z^{-k} \tag{3-98}$$

Thus for the discrete-data system shown in Figs. 3-27 and 3-28, the *z*-transform of the output is equal to the *z*-transfer function of the process and the *z*-transform of the input.

3-11-1 Transfer Functions of Discrete-Data Systems with Cascade Elements

The transfer-function representation of discrete-data systems with elements connected in cascade is slightly more involved than that for continuous-data systems, because of the variation of having or not having samplers between the elements. Figure 3-29 shows

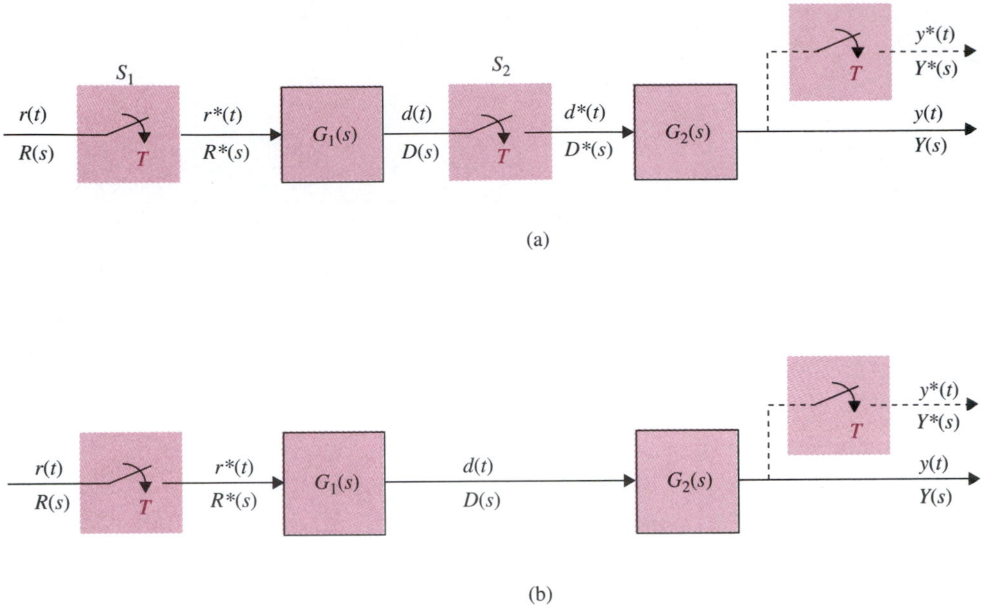

Figure 3-29 (a) Discrete-data system with cascaded elements and sampler separates the two elements. (b) Discrete-data system with cascaded elements and no sampler in between.

two different situations of a discrete-data system that contains two elements connected in cascade. In the system of Fig. 3-29(a), the two elements are separated by the sampler S_2, which is synchronized to, and has the same period as, the sampler S_1. The two elements in the system of Fig. 3-29(b) are connected together directly. It is important to distinguish these two cases when deriving the pulse-transfer function and the z-transfer function. For the system in Fig. 3-29(a), the output of $G_1(s)$ is written

$$D(s) = G_1(s)R^*(s) \tag{3-99}$$

and the system output is

$$Y(s) = G_2(s)D^*(s) \tag{3-100}$$

Taking the pulse tranform on both sides of Eq. (3-99) and using Eq. (3-93), we have

$$D^*(s) = G_1^*(s)R^*(s) \tag{3-101}$$

Now substituting Eq. (3-101) in Eq. (3-100) and taking the pulse transform, we get

$$Y^*(s) = G_1^*(s)G_2^*(s)R^*(s) \tag{3-102}$$

The corresponding z-transform expression of Eq. (3-102) is

$$Y(z) = G_1(z)G_2(s)R(z) \tag{3-103}$$

▲ The z-transform of two systems separated by a sampler is equal to the product of the z-transforms of the two systems.

We conclude that *the z-transform of two systems separated by a sampler is equal to the product of the z-transforms of the two systems.*

The Laplace transform of the output of the system in Fig. 3-29(b) is

$$Y(s) = G_1(s)G_2(s)R^*(s) \tag{3-104}$$

Taking the pulse transform on both sides of the last equation, we get

$$Y^*(s) = [G_1(s)G_2(s)]^*R^*(s) \tag{3-105}$$

where

$$[G_1(s)G_2(s)]^* = \frac{1}{T}\sum_{n=-\infty}^{\infty} G_1(s + jn\omega_s)G_2(s + jn\omega_s) \tag{3-106}$$

Notice that since $G_1(s)$ and $G_2(s)$ are not separated by a sampler, they have to be treated as one system when taking the pulse transform.

Taking the z-transform on both sides of Eq. (3-105) gives

$$Y(z) = \mathscr{Z}\{[G_1(s)G_2(s)]^*\}R(z) \tag{3-107}$$

Let

$$\mathscr{Z}\{[G_1(s)G_2(s)]^*\} = G_1G_2(z) \tag{3-108}$$

Then Eq. (3-107) is written

$$Y(z) = G_1G_2(z)R(z) \tag{3-109}$$

3-11-2 Transfer Function of the Zero-Order Hold

Based on the description of the ZOH given earlier, its impulse response is shown in Fig. 3-30. The transfer function of the ZOH is written

$$G_h(s) = \mathscr{L}[g_h(t)] = \frac{1 - e^{-Ts}}{s} \tag{3-110}$$

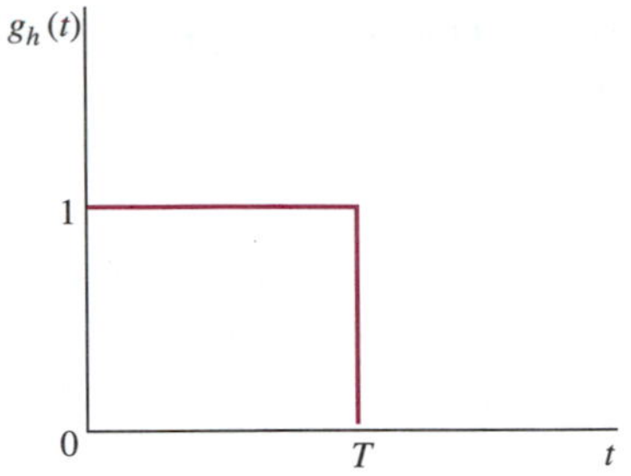

Figure 3-30 Impulse response of the ZOH.

Thus if the ZOH is connected in cascade with a linear process with transfer function $G_p(s)$, as shown in Fig. 3-27, the z-transform of the combination is written

$$G(z) = \mathscr{Z}[G_h(s)G_p(s)] = \mathscr{Z}\left(\frac{1 - e^{-Ts}}{s}G_p(s)\right) \tag{3-111}$$

By using the time-delay property of z-transforms, Eq. (2-188), Eq. (3-111) is simplified to

$$G(z) = (1 - z^{-1})\mathscr{Z}\left(\frac{G_p(s)}{s}\right) \tag{3-112}$$

Example 3-12

Consider that for the system shown in Fig. 3-27,

$$G_p(s) = \frac{1}{s(s + 0.5)} \tag{3-113}$$

The sampling period is 1 sec. The z-transfer function of the system between the input and the output is determined using Eq. (3-112).

$$\begin{aligned} G(z) &= (1 - z^{-1})\mathscr{Z}\left(\frac{1}{s^2(s + 0.5)}\right) \\ &= (1 - z^{-1})\mathscr{Z}\left(\frac{2}{s^2} - \frac{4}{s} + \frac{4}{s + 0.5}\right) = \frac{0.426z + 0.361}{z^2 - 1.606z + 0.606} \end{aligned} \tag{3-114}$$

▲

3-11-3 Transfer Functions of Closed-Loop Discrete-Data Systems

The transfer functions of closed-loop discrete-data systems are derived using the following procedures:

1. Regard the outputs of samplers as "inputs" to the system.
2. All other noninputs of the system are treated as outputs.
3. Write cause-and-effect equations between the inputs and the outputs of the system using the SFG gain formula.
4. Take the pulsed transform or the z-transform of the equations obtained in step 3, and manipulate these equations to get the pulse-transfer function or the z-transfer function.

Reference [6] describes the sampled signal-flow graph that can be used to implement step 4 using the SFG gain formula.

The following examples illustrate the algebraic procedure of finding the transfer functions of closed-loop discrete-data systems.

Example 3-13

Consider the closed-loop discrete-data system shown in Fig. 3-31. The output of the sampler is regarded as an input to the system. Thus the system has inputs $R(s)$ and $E^*(s)$. The signals $E(s)$ and $Y(s)$ are regarded as the outputs of the system.

Writing the cause-and-effect equations for $E(s)$ and $Y(s)$ using the gain formula, we get

$$E(s) = R(s) - G(s)H(s)E^*(s) \tag{3-115}$$
$$Y(s) = G(s)E^*(s) \tag{3-116}$$

Notice that the right-hand side of the last two equations contains only the inputs $R(s)$ and $E^*(s)$ and the transfer functions. Taking the pulse transform on both sides of Eq. (3-115) and solving for $E^*(s)$, we get

$$E^*(s) = \frac{R^*(s)}{1 + [G(s)H(s)]^*} \tag{3-117}$$

Substituting $E^*(s)$ from Eq. (3-117) into Eq. (3-116), we get

$$Y(s) = \frac{G(s)}{1 + [G(s)H(s)]^*}R^*(s) \tag{3-118}$$

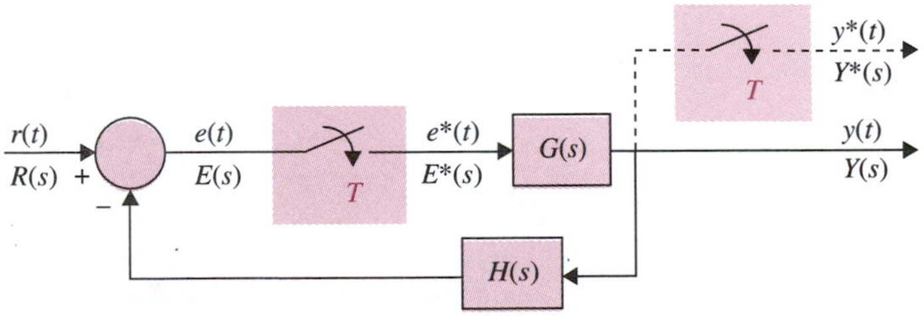

Figure 3-31 Closed-loop discrete-data system.

Taking the pulse transform on both sides of Eq. (3-118) and using Eq. (3-93), we arrive at the pulse-transfer function of the closed-loop system,

$$\frac{Y^*(s)}{R^*(s)} = \frac{G^*(s)}{1 + [G(s)H(s)]^*} \tag{3-119}$$

Taking the z-transform on both sides of Eq. (3-119), we have

$$\frac{Y(z)}{R(z)} = \frac{G(z)}{1 + GH(z)} \tag{3-120}$$

▲

Example 3-14

We show in this example that although it is possible to define an input–output transfer function for the system in Fig. 3-31, in general, this may not be possible for all discrete-data systems. Let us consider the system shown in Fig. 3-32, which has a sampler in the feedback path. In this case, the output of the sampler, $Y^*(s)$, and $R(s)$, are the inputs of the system; $Y(s)$ and $E(s)$ are regarded as the outputs. Writing $E(s)$ and $Y(s)$ in terms of the inputs using the gain formula, we get

$$Y(s) = G(s)E(s) \tag{3-121}$$
$$E(s) = R(s) - H(s)Y^*(s) \tag{3-122}$$

Taking the pulse transform on both sides of Eqs. (3-121) and (3-122), and after simple algebraic manipulations, the pulse transform of the output is written

$$Y^*(s) = \frac{[G(s)R(s)]^*}{1 + [G(s)H(s)]^*} \tag{3-123}$$

Note that the input $R(s)$ and the transfer function $G(s)$ are now combined as one function, $[G(s)R(s)]^*$, and we cannot define a transfer function in the form of $Y^*(s)/R^*(s)$. The z-transform of the output is written

$$Y(z) = \frac{GR(z)}{1 + GH(z)} \tag{3-124}$$

▲

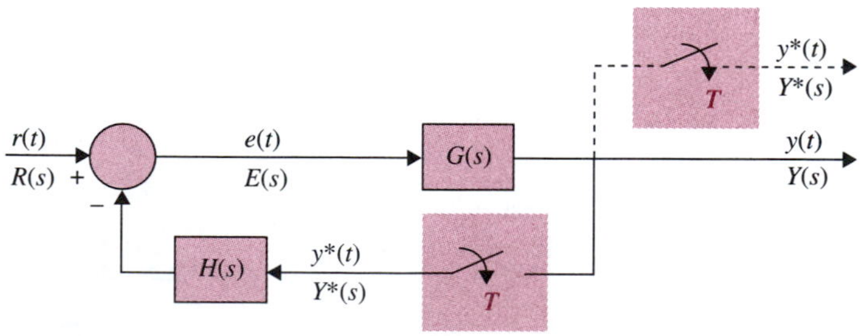

Figure 3-32 Closed-loop discrete-data system.

Although we have been able to arrive at the input–output transfer function and transfer relation of the systems in Figs. 3-31 and 3-32 by algebraic means without difficulty, for more complex system configurations, the algebraic method may become tedious. The signal-flow graph method may be extended to the analysis of discrete-data systems; the reader may refer to reference [6] for details.

3-12 Summary

In this chapter we have introduced the mathematical modeling of linear systems using transfer functions, block diagrams, and signal-flow graphs. The transfer function of a linear system is defined in terms of the impulse response as well as the differential equation. Multivariable and single-variable systems are treated.

The block diagram representation is shown to be a versatile method of portraying linear and nonlinear systems. A block diagram may or may not have mathematical implications.

A powerful method of representing the interrelationships between signals of a linear system is the signal-flow graph. When applied properly, a SFG allows the derivation of the transfer functions between input and output variables of a linear system using the gain formula. The state diagram is a SFG that is applied to dynamic systems that are represented by differential equations.

Finally, the transfer functions of discrete-data systems are defined, and the rules of governing the manipulations of transfer functions between interconnected discrete-data systems are described.

Review Questions

1. Define the transfer function of a linear time-invariant system in terms of its impulse response.
2. When defining the transfer function, what happens to the initial conditions of the system?
3. Define the characteristic equation of a linear system in terms of the transfer function.
4. What is referred to as a multivariable system?
5. Can signal-flow graphs be applied to nonlinear systems?
6. How can signal-flow graphs be applied to systems that are described by differential equations?
7. Define the input node of a SFG.
8. Define the output node of a SFG.
9. State the form to which the equations must be conditioned before drawing the SFG.
10. What does the arrow on the branch of a SFG represent?
11. Explain how a noninput node of a SFG can be made into an output node.
12. Can the gain formula be applied between any two nodes of a SFG?
13. Explain the nontouching loops of a SFG.
14. Does the Δ of a SFG depend on which input–output pair is selected?
15. List the advantages and utilities of the state diagram.

16. Given the state diagram of a linear dynamic system, how do you define the state variables?
17. Given the state diagram of a linear dynamic system, how do you find the transfer function between a pair of input and output variables?
18. Given the state diagram of a linear dynamic system, how do you write the state equations of the system from the state diagram?
19. What is an ideal sampler, and what are its properties?
20. What is a zero-order hold, and what are its properties?
21. Is the ZOH a linear device?
22. Can you always express the input–output relationship as a transfer function for a discrete-data system?

References

Block Diagrams and Signal-Flow Graphs

1. **T. D. GRAYBEAL,** ''Block Diagram Network Transformation,'' *Electr. Eng.,* Vol. 70, pp. 985–990, 1951.
2. **S. J. MASON,** ''Feedback Theory—Some Properties of Signal Flow Graphs,'' *Proc. IRE,* Vol. 41, No. 9, pp. 1144–1156, Sept. 1953.
3. **S. J. MASON,** ''Feedback Theory—Further Properties of Signal Flow Graphs,'' *Proc. IRE,* Vol. 44, No. 7, pp. 920–926, July 1956.
4. **L. P. A. ROBICHAUD, M. BOISVERT, and J. ROBERT,** *Signal Flow Graphs and Applications,* Prentice Hall, Englewood Cliffs, NJ, 1962.
5. **B. C. KUO,** *Linear Networks and Systems,* McGraw-Hill Book Company, New York, 1967.

Discrete-Data Control Systems

6. **B. C. KUO,** *Digital Control Systems,* 2nd ed., Saunders College Publishing, New York, 1992.
7. **G. F. FRANKLIN, J. D. POWELL, and M. L. WORKMAN,** *Digital Control of Dynamic Systems,* 2nd ed., Addison-Wesley Publishing Co., Reading, MA, 1990.
8. **C. L. PHILLIPS and H. T. NAGLE, JR.,** *Digital Control Systems Analysis and Design,* Prentice Hall, Englewood Cliffs, NJ, 1984.
9. **K. OGATA,** *Discrete-Time Control Systems,* Prentice Hall, Englewood Cliffs, NJ, 1987.

Problems

▲ Transfer function

3-1. The following differential equations represent linear time-invariant systems, where $r(t)$ denotes the input, and $y(t)$ the output. Find the transfer function $Y(s)/R(s)$ for each system.

(a) $$\frac{d^3y(t)}{dt^3} + 2\frac{d^2y(t)}{dt^2} + 5\frac{dy(t)}{dt} + 6y(t) = 3\frac{dr(t)}{dt} + r(t)$$

(b) $$\frac{d^4y(t)}{dt^4} + 10\frac{d^2y(t)}{dt^2} + \frac{dy(t)}{dt} + 5y(t) = 5r(t)$$

(c) $$\frac{d^3y(t)}{dt^3} + 10\frac{d^2y(t)}{dt^2} + 2\frac{dy(t)}{dt} + y(t) + 2\int_0^t y(\tau)\,d\tau = \frac{dr(t)}{dt} + 2r(t)$$

(d) $$2\frac{d^2y(t)}{dt^2} + \frac{dy(t)}{dt} + 5y(t) = r(t) + 2r(t-1)$$

▲ Transfer functions, final value

3-2. The block diagram of an electric train control is shown in Fig. 3P-2. The system parameters and variables are

$e_r(t)$ = voltage representing the desired train speed, V
$v(t)$ = speed of train, ft/sec
M = mass of train = 30,000 lb/sec^2
K = amplifier gain
K_t = gain of speed indicator = 0.15 V/ft/sec

To determine the transfer function of the controller, we apply a step function of 1 V to the input of the controller [i.e., $e_c(t) = u_s(t)$]. The output of the controller is measured and is described by the following equation:

$$f(t) = 100(1 - 0.3e^{-6t} - 0.7e^{-10t})u_s(t)$$

(a) Find the transfer function $G_c(s)$ of the controller.

(b) Derive the forward-path transfer function $V(s)/E(s)$ of the system. The feedback path is opened in this case.

(c) Derive the closed-loop transfer function $V(s)/E_r(s)$ of the system.

(d) Assuming that K is set at a value so that the train will not run away (unstable), find the steady-state speed of the train in ft/sec when the input is $e_r(t) = u_s(t)$.

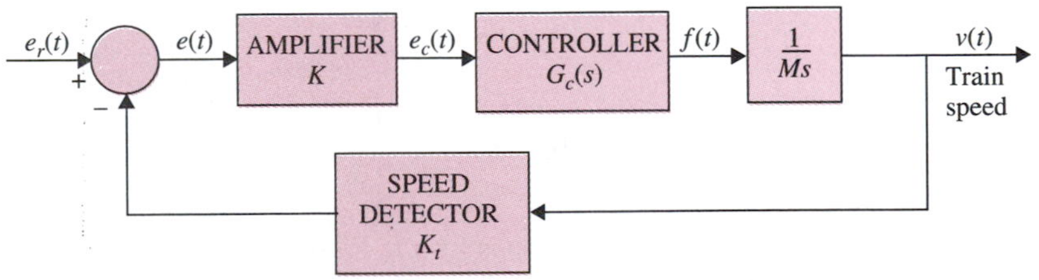

Figure 3P-2

▲ Transfer function, final value

3-3. Repeat Problem 3-2 when the output of the controller is measured and is described by the following expression:

$$f(t) = 100(1 - 0.3e^{-6(t-0.5)})u_s(t - 0.5)$$

when a step input of 1 V is applied to the controller.

▲ Multivariable system

3-4. A linear time-invariant multivariable system with inputs $r_1(t)$ and $r_2(t)$ and outputs $y_1(t)$ and $y_2(t)$ is described by the following set of differential equations:

$$\frac{d^2y_1(t)}{dt^2} + 2\frac{dy_1(t)}{dt} + 3y_2(t) = r_1(t) + r_2(t)$$

$$\frac{d^2y_2(t)}{dt^2} + 3\frac{dy_1(t)}{dt} + y_1(t) - y_2(t) = r_2(t) + \frac{dr_1(t)}{dt}$$

Find the following transfer functions:

$$\left.\frac{Y_1(s)}{R_1(s)}\right|_{R_2=0} \quad \left.\frac{Y_2(s)}{R_1(s)}\right|_{R_2=0} \quad \left.\frac{Y_1(s)}{R_2(s)}\right|_{R_1=0} \quad \left.\frac{Y_2(s)}{R_2(s)}\right|_{R_1=0}$$

▲ Multivariable system

3-5. The aircraft turboprop engine shown in Fig. 3-1 is controlled by the closed-loop system with block diagram shown in Fig. 3P-5. The engine is modeled as a multivariable system with input vector $\mathbf{E}(s)$ that contains the fuel rate and propeller blade angle, and output vector $\mathbf{Y}(s)$, consisting of the engine speed and turbine-inlet temperature. The transfer-function matrices are given as

$$\mathbf{G}(s) = \begin{bmatrix} \dfrac{2}{s(s+2)} & 10 \\ \dfrac{5}{s} & \dfrac{1}{s+1} \end{bmatrix} \qquad \mathbf{H}(s) = \begin{bmatrix} 1 & 0 \\ 0 & 1 \end{bmatrix}$$

Find the closed-loop transfer function matrix $[\mathbf{I} + \mathbf{G}(s)\mathbf{H}(s)]^{-1}\mathbf{G}(s)$.

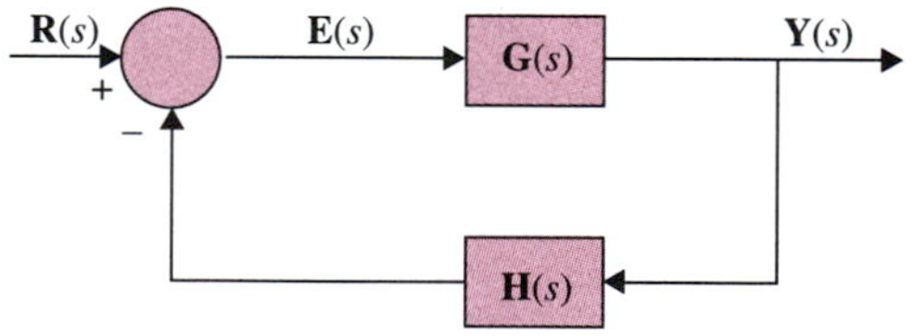

Figure 3P-5

▲ Signal-flow graphs

3-6. Draw signal-flow graphs for the following sets of algebraic equations. These equations should be arranged in the form of cause-and-effect relations before signal-flow graphs can be drawn. Show that there are many possible signal-flow graphs for each set of equations.

(a) $x_1 = \quad - x_2 - 3x_3 + 3$ (b) $2x_1 + 3x_2 + x_3 = -1$
$x_2 = 5x_1 - 2x_2 + x_3$ $x_1 - 2x_2 - x_3 = 1$
$x_3 = 4x_1 + x_2 - 5x_3 + 5$ $3x_2 + x_3 = 0$

▲ Equivalent SFG of block diagrams

3-7. The block diagram of a control system is shown in Fig. 3P-7. Draw an equivalent SFG for the system. Find the following transfer functions by applying the gain formula of the SFG directly to the block diagram. Compare the answers by applying the gain formula to the equivalent SFG.

$$\left.\frac{Y(s)}{R(s)}\right|_{N=0} \quad \left.\frac{Y(s)}{N(s)}\right|_{R=0} \quad \left.\frac{E(s)}{R(s)}\right|_{N=0} \quad \left.\frac{E(s)}{N(s)}\right|_{R=0}$$

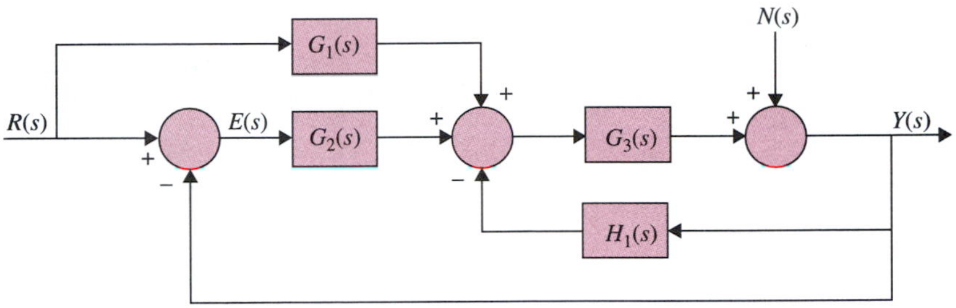

Figure 3P-7

▲ Transfer functions of SFG

3-8. Apply the gain formula to the SFGs shown in Fig. 3P-8 to find the following transfer functions:

$$\frac{Y_5}{Y_1} \quad \frac{Y_4}{Y_1} \quad \frac{Y_2}{Y_1} \quad \frac{Y_5}{Y_2}$$

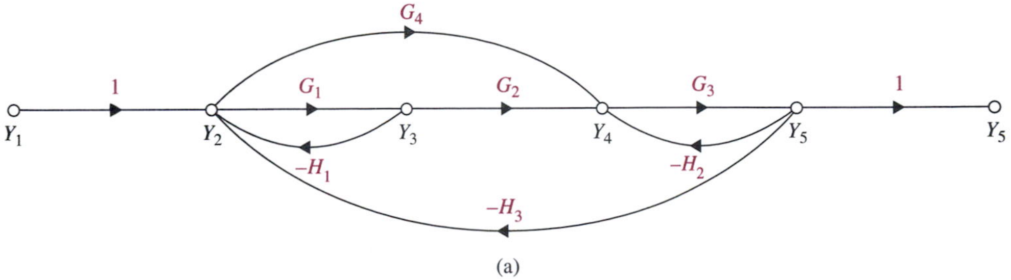

Figure 3P-8

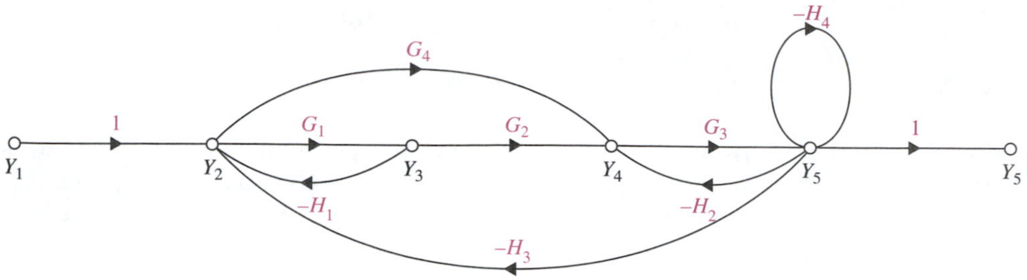

(b)

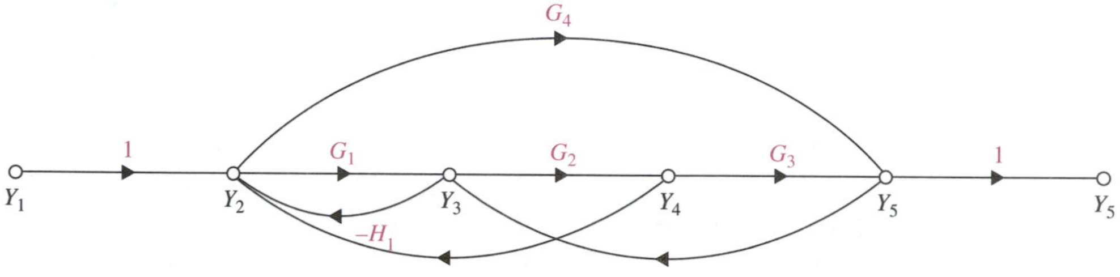

(c)

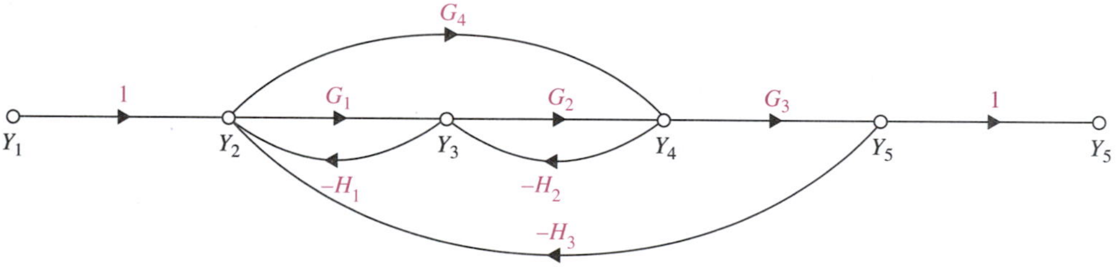

(d)

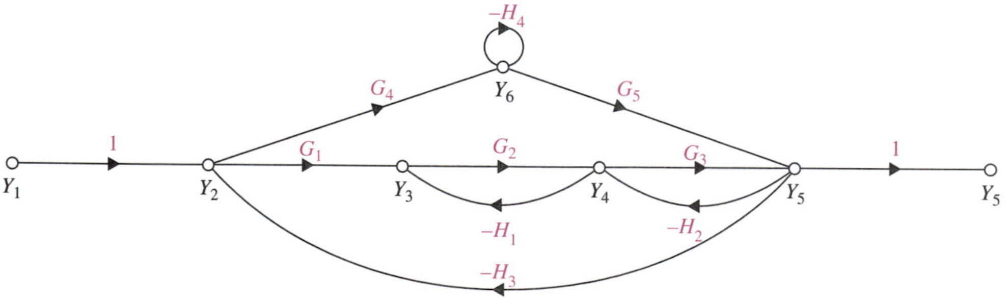

(e)

Figure 3P-8 Continued

▲ Transfer functions of SFG

3-9. Find the transfer functions Y_7/Y_1 and Y_2/Y_1 of the SFGs shown in Fig. 3P-9.

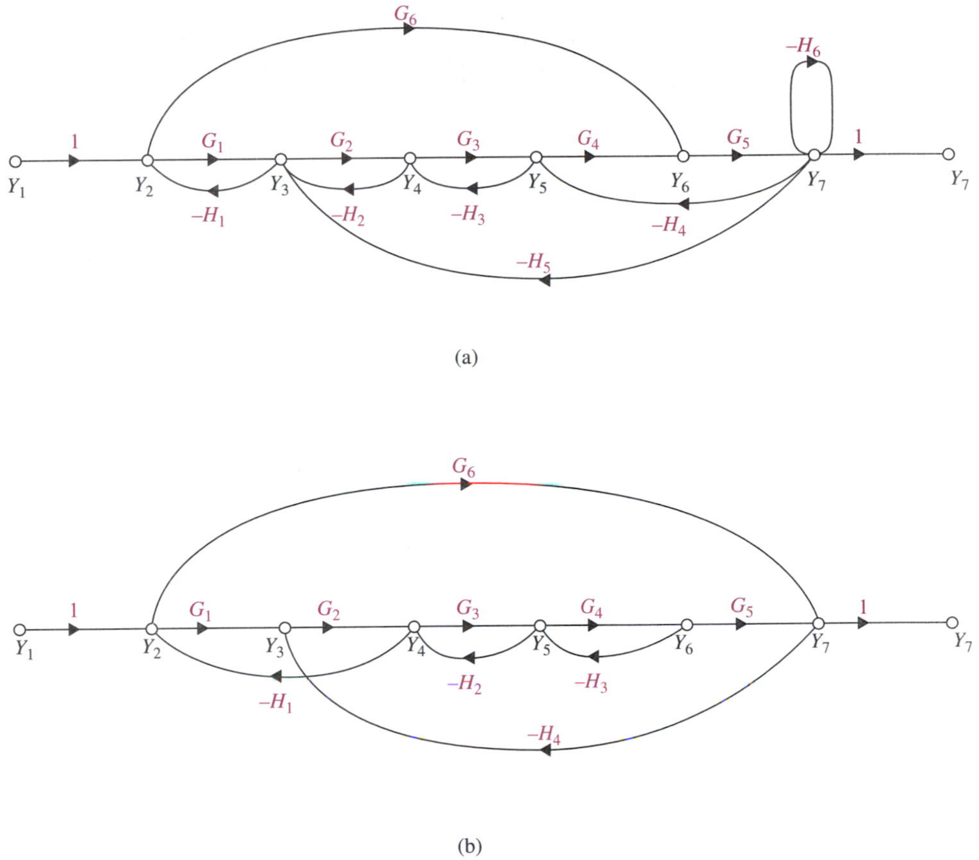

Figure 3P-9

▲ Application of SFG

3-10. Signal-flow graphs may be used to solve a variety of electric network problems. Shown in Fig. 3P-10 is the equivalent circuit of an electronic circuit. The voltage source $e_d(t)$ represents a disturbance voltage. The objective is to find the value of the constant k so that the output voltage $e_o(t)$ is not affected by $e_d(t)$. To solve the problem, it is best first to write a set of cause-and-effect equations for the network. This involves a combination of node and loop equations. Then construct a SFG using these equations. Find the gain e_o/e_d with all other inputs set to zero. For e_d not to effect e_o, set e_o/e_d to zero.

▲ SFG

3-11. Show that the two systems shown in Fig. 3P-11(a) and (b) are equivalent.

▲ SFG

3-12. Show that the two systems shown in Fig. 3P-12(a) and (b) are not equivalent.

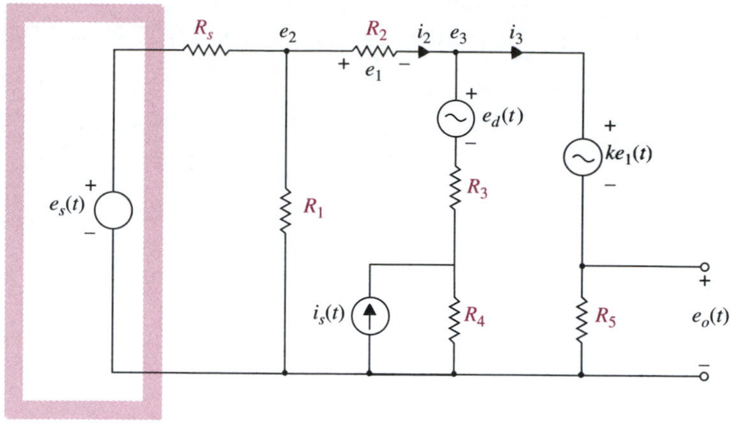

Figure 3P-10

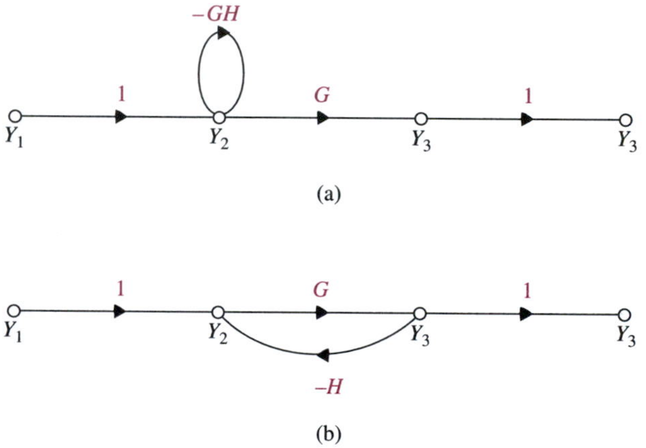

Figure 3P-11

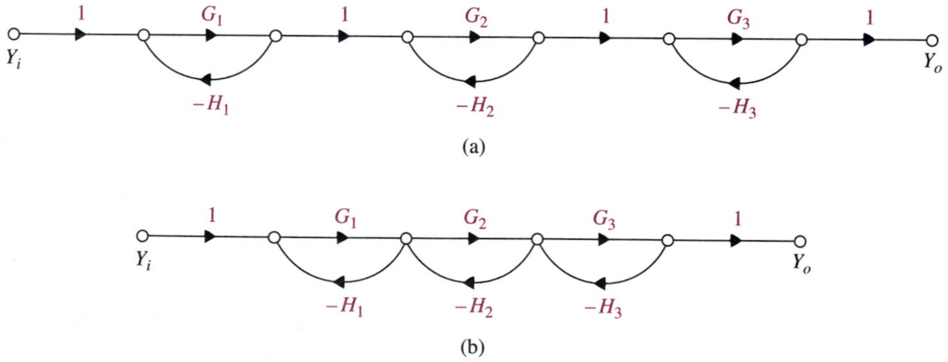

Figure 3P-12

▲ Transfer functions of SFG

3-13. Find the following transfer functions for the SFG shown in Fig. 3P-13.

$$\left.\frac{Y_6}{Y_1}\right|_{Y_7=0} \quad \text{and} \quad \left.\frac{Y_6}{Y_7}\right|_{Y_1=0}$$

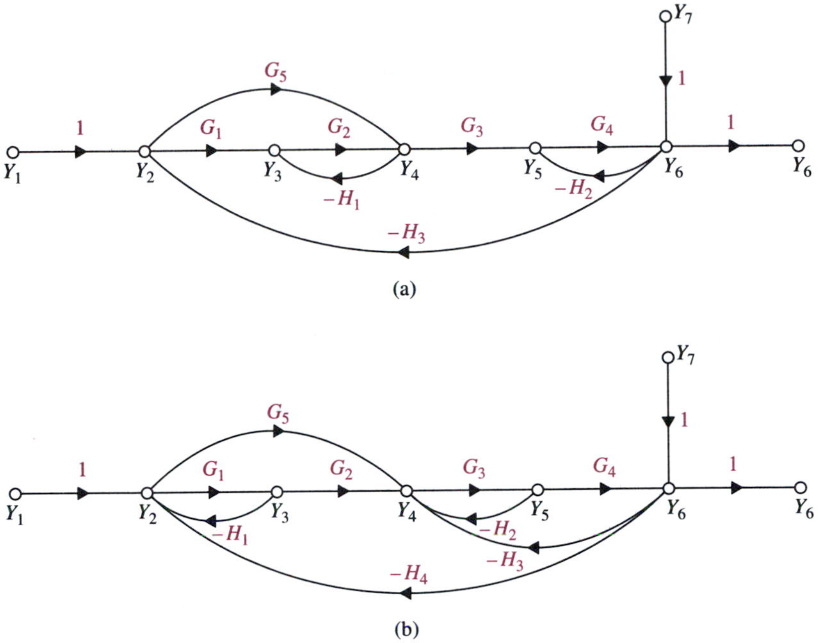

Figure 3P-13

▲ Transfer functions of SFG

3-14. Find the following transfer functions for the SFG shown in Fig. 3P-14. Comment on why the results for parts (c) and (d) are not the same.

(a) $\left.\frac{Y_7}{Y_1}\right|_{Y_8=0}$ **(b)** $\left.\frac{Y_7}{Y_8}\right|_{Y_1=0}$ **(c)** $\left.\frac{Y_7}{Y_4}\right|_{Y_8=0}$ **(d)** $\left.\frac{Y_7}{Y_4}\right|_{y_1=0}$

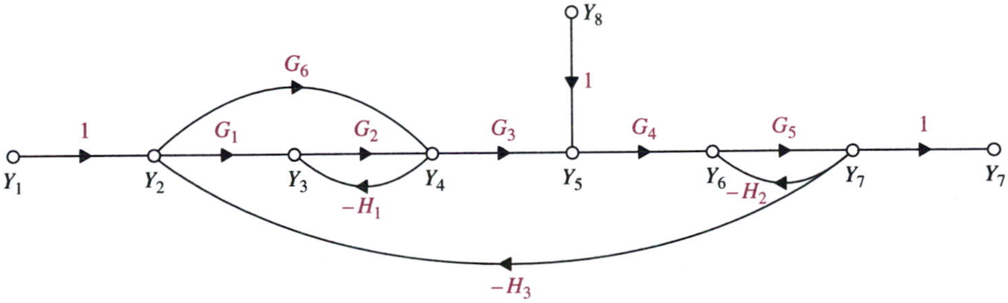

Figure 3P-14

▲ Transfer functions of block diagrams

3-15. The block diagram of a feedback control system is shown in Fig. 3P-15. Find the following transfer functions.

(a) $\left.\frac{Y(s)}{R(s)}\right|_{N=0}$ (b) $\left.\frac{Y(s)}{E(s)}\right|_{N=0}$ (c) $\left.\frac{Y(s)}{N(s)}\right|_{R=0}$

(d) Find the output $Y(s)$ when $R(s)$ and $N(s)$ are applied simultaneously.

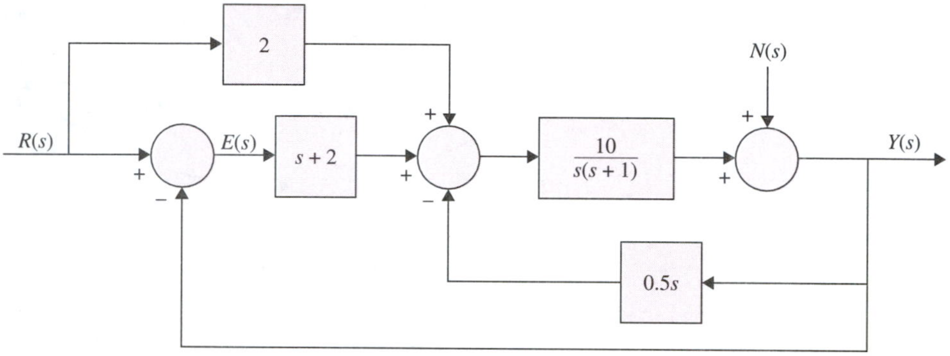

Figure 3P-15

▲ Transfer functions of block diagrams

3-16. The block diagram of a feedback control system is shown in Fig. 3P-16.

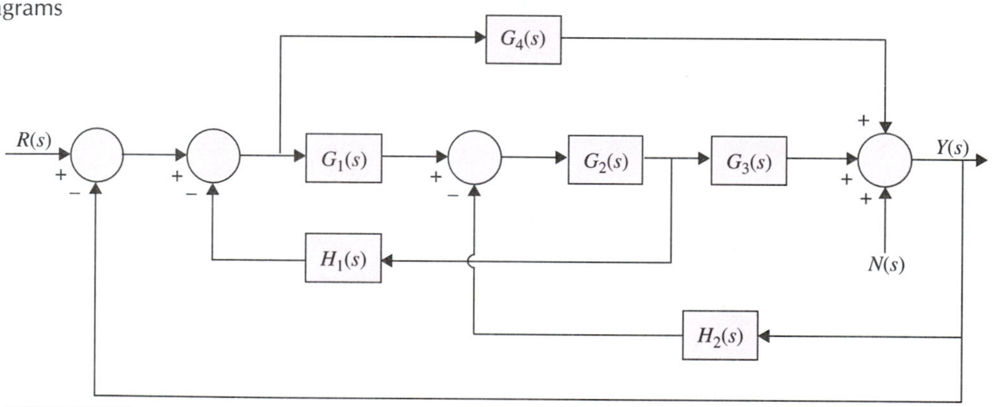

Figure 3P-16

(a) Applying the SFG gain formula directly to the block diagram to find the transfer functions

$$\left.\frac{Y(s)}{R(s)}\right|_{N=0} \qquad \left.\frac{Y(s)}{N(s)}\right|_{R=0}$$

Express $Y(s)$ in terms of $R(s)$ and $N(s)$ when both inputs are applied simultaneously.

(b) Find the desired relation among the transfer functions $G_1(s)$, $G_2(s)$, $G_3(s)$, $G_4(s)$, $H_1(s)$, and $H_2(s)$ so that the output $Y(s)$ is not affected by the disturbance signal $N(s)$ at all.

▲ Antenna control system

3-17. Figure 3P-17 shows the block diagram of the antenna control system of the solar-collector field shown in Fig. 1-7. The signal $N(s)$ denotes the wind gust disturbance that acted on the antenna. The feedforward transfer function $G_d(s)$ is used to eliminate the effect of $N(s)$ on the output $Y(s)$. Find the transfer function $Y(s)/N(s)|_{R=0}$. Determine the expression of $G_d(s)$ so that the effect of $N(s)$ is eliminated entirely.

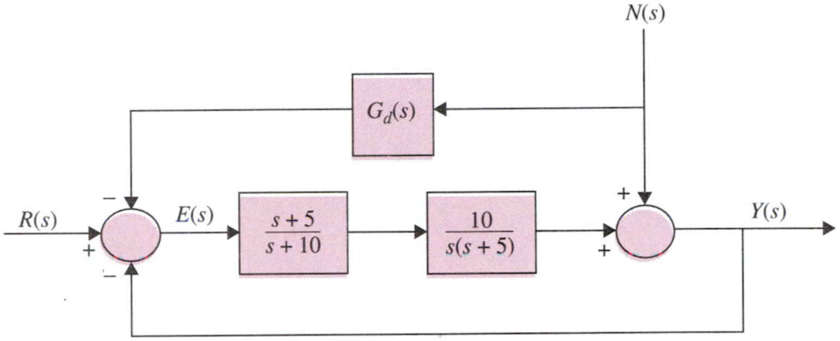

Figure 3P-17

▲ Transfer function of dc-motor control system

3-18. Figure 3P-18 shows the block diagram of a dc-motor control system. The signal $N(s)$ denotes the frictional torque at the motor shaft.

(a) Find the transfer function $H(s)$ so that the output $Y(s)$ is not affected by the disturbance torque $N(s)$.

(b) With $H(s)$ as determined in part (a), find the value of K so that the steady-state value of $e(t)$ is equal to 0.1 when the input is a unit-ramp function, $r(t) = tu_s(t)$, $R(s) = 1/s^2$, and $N(s) = 0$. Apply the final-value theorem.

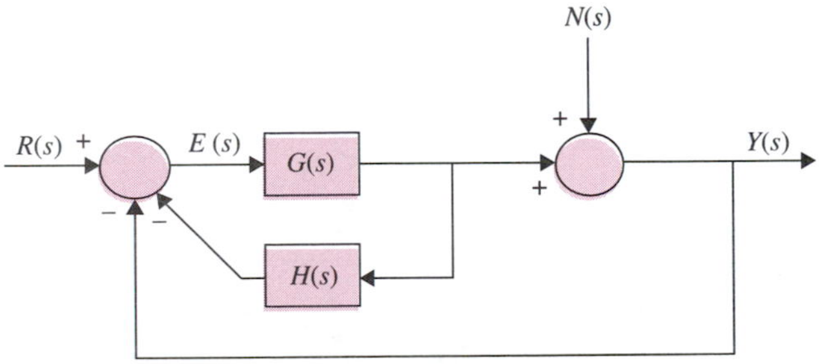

$$G(s) = \frac{K(s+3)}{s(s+1)(s+2)}$$

Figure 3P-18

▲ Transfer function of position-control system

3-19. The block diagram of the position-control system of the electronic word processor shown in Fig. 1-3 is shown in Fig. 3P-19.

(a) Find the loop transfer function $\Theta_o(s)/\Theta_e(s)$ (the outer feedback path is open).

(b) Find the closed-loop transfer function $\Theta_o(s)/\Theta_r(s)$.

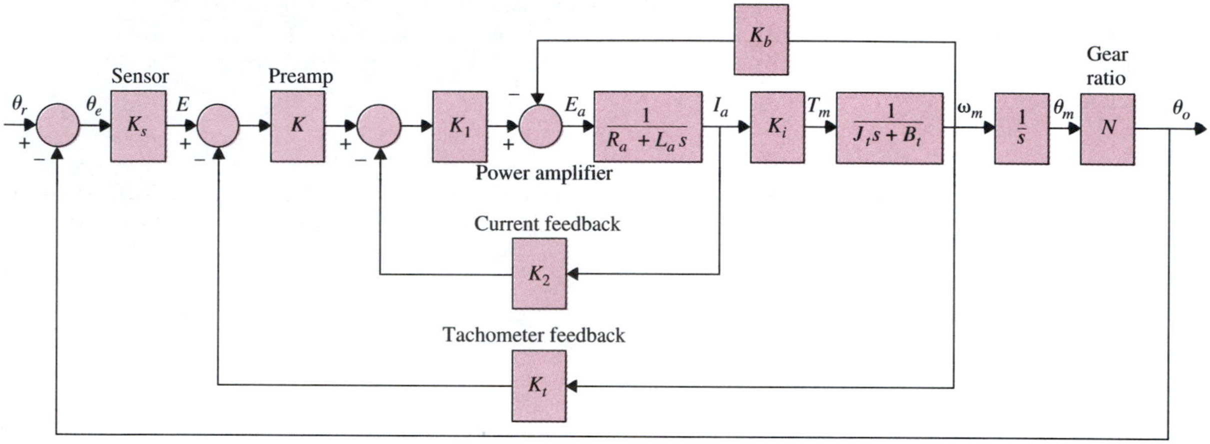

Figure 3P-19

▲ Turboprop engine

3-20. The coupling between the signals of the turboprop engine shown in Fig. 3-1 is shown in Fig. 3P-20. The signals are defined as: $R_1(s)$ = fuel rate, $R_2(s)$ = propeller blade angle, $Y_1(s)$ = engine speed, and $Y_2(s)$ = turbine-inlet temperature.

(a) Draw an equivalent SFG for the system.

(b) Find the Δ of the system using the SFG gain formula.

(c) Find the following transfer functions:

$$\left.\frac{Y_1(s)}{R_1(s)}\right|_{R_2=0} \qquad \left.\frac{Y_1(s)}{R_2(s)}\right|_{R_1=0} \qquad \left.\frac{Y_2(s)}{R_1(s)}\right|_{R_2=0} \qquad \left.\frac{Y_2(s)}{R_2(s)}\right|_{R_1=0}$$

(d) Express the transfer functions in matrix form, $\mathbf{Y}(s) = \mathbf{G}(s)\mathbf{R}(s)$.

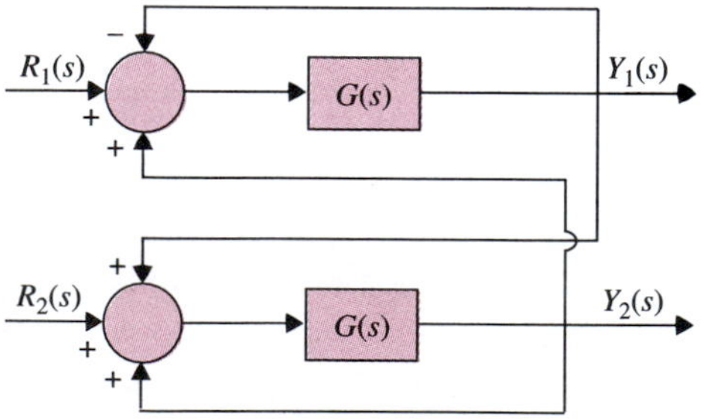

Figure 3P-20

▲ Transfer function of system with conditional feedback

3-21. Figure 3P-21 shows a control system with conditional feedback. The transfer function, $G_p(s)$, denotes the controlled process, and $G_c(s)$ and $H(s)$ are the controller transfer functions.

(a) Derive the transfer functions $Y(s)/R(s)|_{N=0}$ and $Y(s)/N(s)|_{R=0}$. Find $Y(s)/R(s)|_{N=0}$ when $G_c(s) = G_p(s)$.

(b) Let

$$G_p(s) = G_c(s) = \frac{100}{(s+1)(s+5)}$$

Find the output response $y(t)$ when $N(s) = 0$ and $r(t) = u_s(t)$.

(c) With $G_p(s)$ and $G_c(s)$ as given in part (b), select $H(s)$ among the following choices such that when $n(t) = u_s(t)$ and $r(t) = 0$, the steady-state value of $y(t)$ is equal to zero. (There may be more than one answer.)

$$H(s) = \frac{10}{s(s+1)} \qquad H(s) = \frac{10}{(s+1)(s+2)}$$

$$H(s) = \frac{10(s+1)}{s+2} \qquad H(s) = \frac{K}{s^n} \quad (n = \text{positive integer; select } n)$$

Keep in mind that the poles of the closed-loop transfer function must all be in the left-half s-plane for the final-value theorem to be valid.

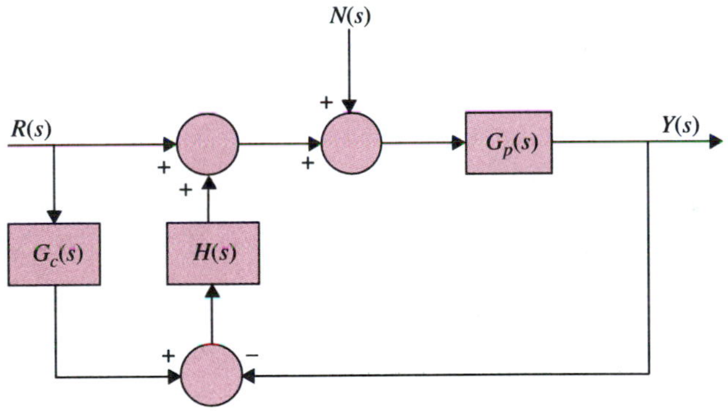

Figure 3P-21

▲ State diagram, characteristic equation, transfer function

3-22. **(a)** Draw a state diagram for the following state equations:

$$\frac{dx_1(t)}{dt} = -2x_1(t) + 3x_2(t)$$

$$\frac{dx_2(t)}{dt} = -5x_1(t) - 5x_2(t) + 2r(t)$$

(b) Find the characteristic equation of the system.

(c) Find the transfer functions $X_1(s)/R(s)$ and $X_2(s)/R(s)$.

▲ State diagram, state equations, characteristic equation, transfer function, final value

3-23. The differential equation of a linear system is

$$\frac{d^3y(t)}{dt^3} + 5\frac{d^2y(t)}{dt^2} + 6\frac{dy(t)}{dt} + 10y(t) = r(t)$$

where $y(t)$ is the output and $r(t)$ is the input.

(a) Draw a state diagram for the system.

(b) Write the state equation from the state diagram. Define the state variables from right to left in ascending order.

(c) Find the characteristic equation and its roots. Use any computer program to find the roots.

(d) Find the transfer function $Y(s)/R(s)$.

(e) Perform a partial-fraction expansion of $Y(s)/R(s)$ and find the output $y(t)$ for $t \geq 0$ when $r(t) = u_s(t)$. Find the final value of $y(t)$ by use of the final-value theorem. Use a computer program such as **pfe** of **CSAD.**

▲ State diagram, state equations, characteristic equation, transfer function, final value

3-24. Repeat Problem 3-23 for the following differential equation:

$$\frac{d^4y(t)}{dt^4} + 4\frac{d^3y(t)}{dt^3} + 3\frac{d^2y(t)}{dt^2} + 5\frac{dy(t)}{dt} + y(t) = r(t)$$

▲ Transfer function, characteristic equation, steady state response

3-25. The block diagram of a feedback control system is shown in Fig. 3P-25.

(a) Derive the following transfer functions:

$$\left.\frac{Y(s)}{R(s)}\right|_{N=0} \qquad \left.\frac{Y(s)}{N(s)}\right|_{R=0} \qquad \left.\frac{E(s)}{R(s)}\right|_{N=0}$$

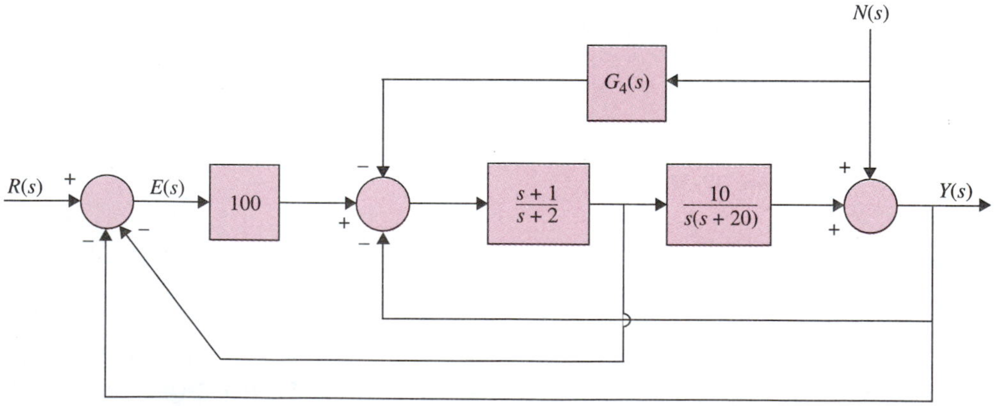

Figure 3P-25

(b) The controller with the transfer function $G_4(s)$ is for the reduction of the effect of the noise $N(s)$. Find $G_4(s)$ so that the output $Y(s)$ is totally independent of $N(s)$.

(c) Find the characteristic equation and its roots when $G_4(s)$ is as determined in part (b).

(d) Find the steady-state value of $e(t)$ when the input is a unit-step function. Set $N(s) = 0$.

(e) Find $y(t)$ for $t \geq 0$ when the input is a unit-step function. Use $G_4(s)$ as determined in part (b).

▲ Time response, discrete-data system

3-26. A linear time-invariant discrete-data system has an output that is described by the time sequence

$$y(kT) = 1 - e^{-2kT} \qquad k = 0, 1, 2, \ldots$$

when the system is subject to an input sequence described by $r(kT) = 1$ for all $k \geq 0$. Find the transfer function $G(z) = Y(z)/R(s)$.

▲ Transfer function, discrete-data system

3-27. Find the transfer functions $Y(z)/R(z)$ of the discrete-data systems shown in Fig. 3P-27. The sampling period is 0.5 sec.

▲ Numerical integration

3-28. It is well known that the transfer function of an analog integrator is

$$G(s) = \frac{Y(s)}{X(s)} = \frac{1}{s}$$

where $X(s)$ and $Y(s)$ are the Laplace transforms of the input and the output of the integrator, respectively. There are many ways of implementing integration digitally. In a basic computer course, the rectangular integration is described by the schemes shown in Fig. 3P-28. The continuous signal $x(t)$ is approximated by a staircase signal; T is the sampling period. The integral of $x(t)$, which is the area under $x(t)$, is approximated by the area under the rectangular approximation signal.

(a) Let $y(kT)$ denote the digital approximation of the integral of $x(t)$ from $t = 0$ to $t = kT$. Then $y(kT)$ can be written as

$$y(kT) = y[(k - 1)T] + Tx(kT) \qquad (1)$$

where $y[(k - 1)T]$ denotes the area under $x(t)$ from $t = 0$ to $t = (k - 1)T$. Take the z-transform on both sides of Eq. (1) and show that the transfer function of the digital integrator is

$$G(z) = \frac{Y(z)}{X(z)} = \frac{Tz}{z - 1}$$

(b) The rectangular integration described in Fig. 3P-28(a) can be interpreted as a sample-and-hold operation, as shown in Fig. 3P-28(b). The signal $x(t)$ is first sent

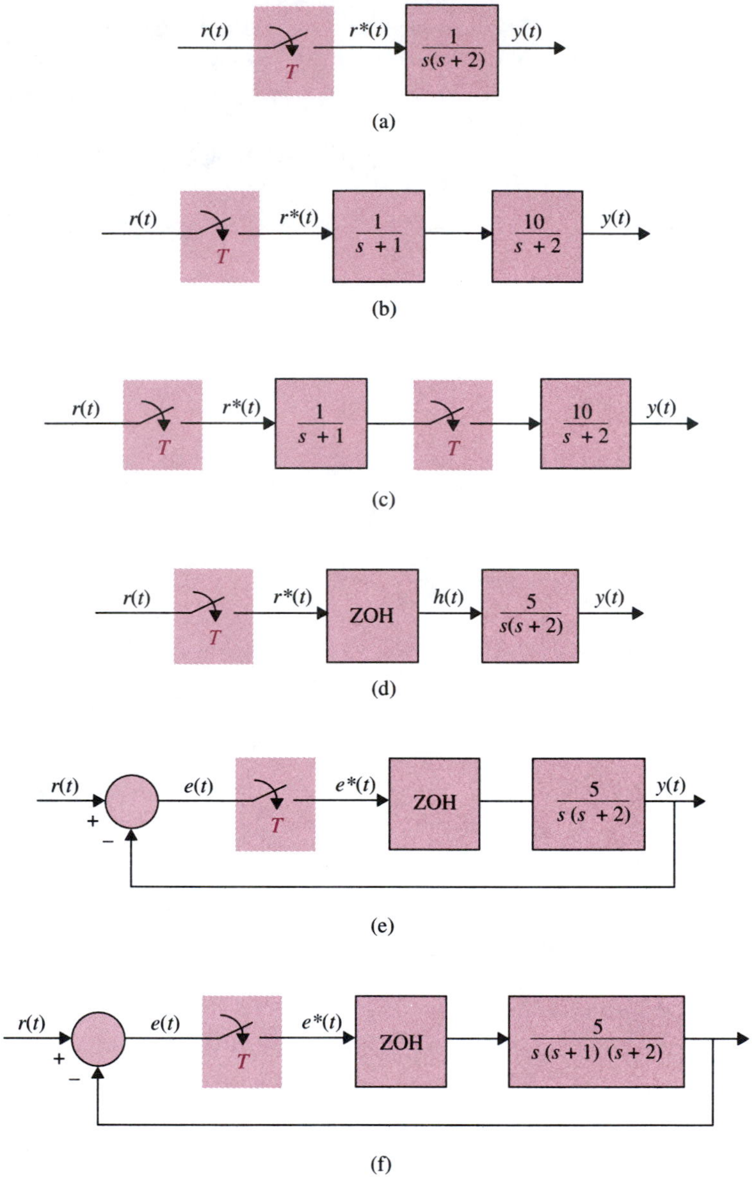
r(t)
T
r*(t)
1/s(s + 2)
y(t)
(a)
r(t)
T
r*(t)
1/(s + 1)
10/(s + 2)
y(t)
(b)
r(t)
T
r*(t)
1/(s + 1)
T
10/(s + 2)
y(t)
(c)
r(t)
T
r*(t)
ZOH
h(t)
5/s(s + 2)
y(t)
(d)
r(t)
+
−
e(t)
T
e*(t)
ZOH
5/s (s + 2)
y(t)
(e)
r(t)
+
−
e(t)
T
e*(t)
ZOH
5/s (s + 1) (s + 2)
(f)

Figure 3P-27

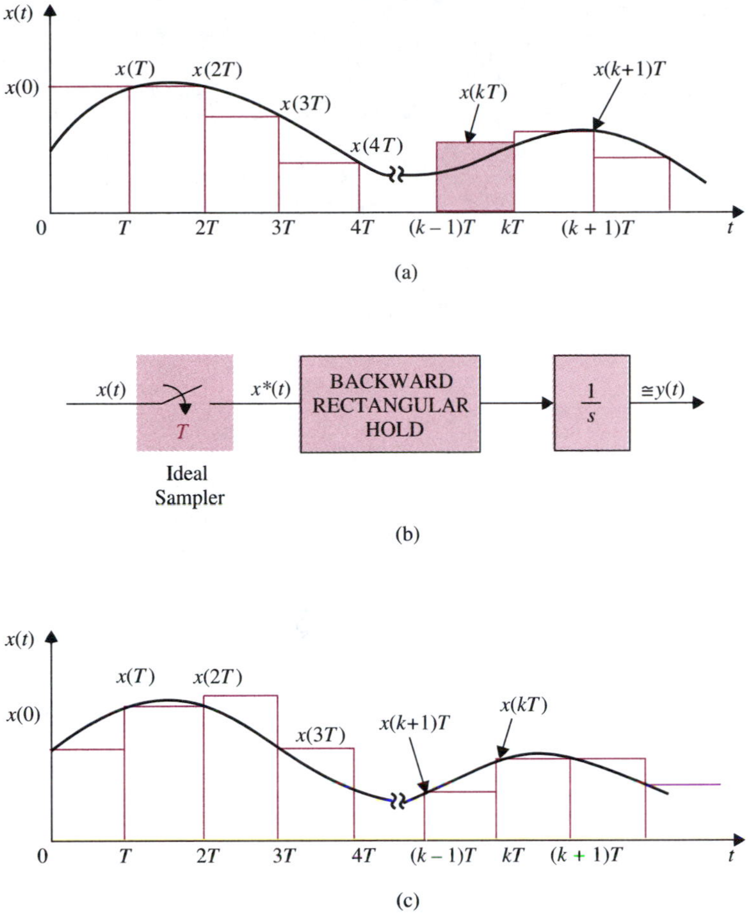

Figure 3P-28

through an ideal sampler with sampling period T. The output of the sampler is the sequence $x(0)$, $x(T)$, . . . , $x(kT)$, ... These numbers are then sent through a ''backward'' hold device to give the rectangle of height $x(kT)$ during the time interval from $(k - 1)T$ to kT. Verify the result obtained in part (a) for $G(z)$ using the ''backward'' sample-and-hold interpretation.

(c) As an alternative, we can use a ''forward'' rectangular hold, as shown in Fig. 3P-28(c). Find the transfer function $G(z)$ for such a rectangular integrator.

4 Mathematical Modeling of Physical Systems

KEYWORDS AND TOPICS

- ▲ System Modeling
- ▲ Mechanical Systems
- ▲ Potentiometers
- ▲ Tachometers
- ▲ Encoders
- ▲ DC Motors
- ▲ Op-amps
- ▲ Linearization of Nonlinear Systems
- ▲ Systems with Time Delays

4-1 Introduction

One of the most important tasks in the analysis and design of control systems is mathematical modeling of the systems. In the preceding chapter, we have introduced two most common methods of modeling linear systems namely the transfer function method and the state-variable method. The transfer function is valid only for linear time-invariant systems, whereas the state equations can be applied to linear as well as nonlinear systems.

Although the analysis and design of linear control systems have been well developed, their counterparts for nonlinear systems are usually quite complex. Therefore, the control systems engineer often has the task of determining not only how to accurately describe a system mathematically, but, more important, how to make proper assumptions and approximations, whenever necessary, so that the system may be realistically characterized by a linear mathematical model. It is not difficult to understand that the analytical and computer simulations of any system are only as good as the model used to describe it. It should also be emphasized that the modern control engineer should place special emphasis on the mathematical modeling of systems so that analysis and design problems can conveniently be solved by computers. Therefore, the main objectives of this chapter are:

1. To introduce some commonly used components in control systems.
2. To demonstrate the mathematical modeling of control systems and components.
3. To demonstrate how the modeling will lead to computer solutions.
4. To linearize nonlinear systems.

The intended purpose is to introduce the method of modeling. Since numerous types of control system components are available, the coverage here is by no means exhaustive.

4-2 Equations of Electric Networks

The classical way of writing equations of electric networks is based on the loop method or the node method, which are formulated from the two laws of Kirchhoff. A modern method of writing network equations is the state-variable method, which is illustrated briefly in Section 2-3. Since the electrical networks found in most control systems are rather simple, the subject is treated here only at the introductory level. The state-variable modeling of electric networks can be illustrated by the following two examples.

Example 4-1

Let us consider the *RLC* network shown in Fig. 4-1(a). A practical way is to assign the current in the inductor *L*, $i(t)$, and the voltage across the capacitor *C*, $e_c(t)$, as the state variables. The reason for this choice is because the state variables are directly related to the energy-storage elements of a system. The inductor is a storage for kinetic energy, and the capacitor is a storage of electric potential energy. By assigning $i(t)$ and $e_c(t)$ as state variables, we have a complete description of the past history (via the initial states) and the present and future states of the network.

The state equations for the network in Fig. 4-1(a) are written by first equating the current in *C* and the voltage across *L* in terms of the state variables and the applied voltage $e(t)$. We have

Current in *C*:
$$C\frac{de_c(t)}{dt} = i(t) \tag{4-1}$$

Voltage in *L*:
$$L\frac{di(t)}{dt} = -e_c(t) - Ri(t) + e(t) \tag{4-2}$$

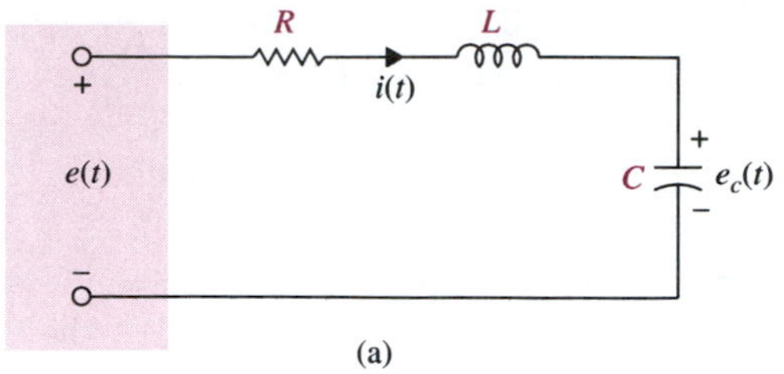

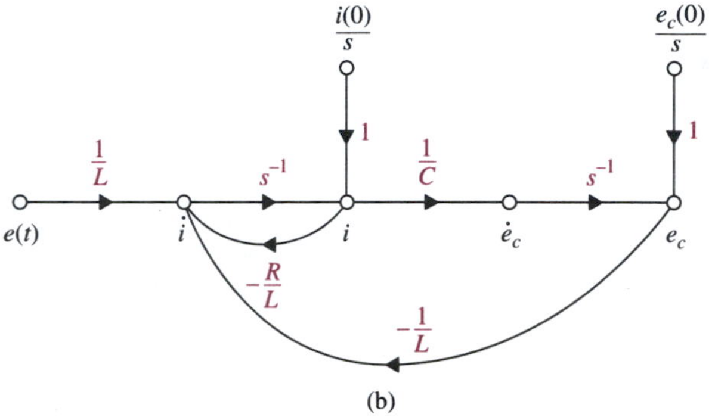

Figure 4-1 (a) *RLC* network. (b) State diagram.

In vector-matrix form, the state equations are expressed as

$$\begin{bmatrix} \dfrac{de_c(t)}{dt} \\ \dfrac{di(t)}{dt} \end{bmatrix} = \begin{bmatrix} 0 & \dfrac{1}{C} \\ -\dfrac{1}{L} & -\dfrac{R}{L} \end{bmatrix} \begin{bmatrix} e_c(t) \\ i(t) \end{bmatrix} + \begin{bmatrix} 0 \\ \dfrac{1}{L} \end{bmatrix} e(t) \tag{4-3}$$

The state diagram of the network is shown in Fig. 4-1(b). Notice that the outputs of the integrators are defined as the state variables. The transfer functions of the system are obtained by applying the SFG gain formula to the state diagram when all the initial states are set to zero.

$$\frac{E_c(s)}{E(s)} = \frac{(1/LC)s^{-2}}{1 + (R/L)s^{-1} + (1/LC)s^{-2}} = \frac{1}{1 + RCs + LCs^2} \tag{4-4}$$

$$\frac{I(s)}{E(s)} = \frac{(1/L)s^{-1}}{1 + (R/L)s^{-1} + (1/LC)s^{-2}} = \frac{Cs}{1 + RCs + LCs^2} \tag{4-5}$$

▲

Example 4-2

As another example of writing the state equations of an electric network, consider the network shown in Fig. 4-2(a). According to the foregoing discussion, the voltage across the capacitor, $e_c(t)$, and the currents in the inductors, $i_1(t)$ and $i_2(t)$, are assigned as state variables, as shown in Fig. 4-2(a). The state equations of the network are obtained by writing the voltages across the inductors and the currents in the capacitor in terms of the three state variables. The state

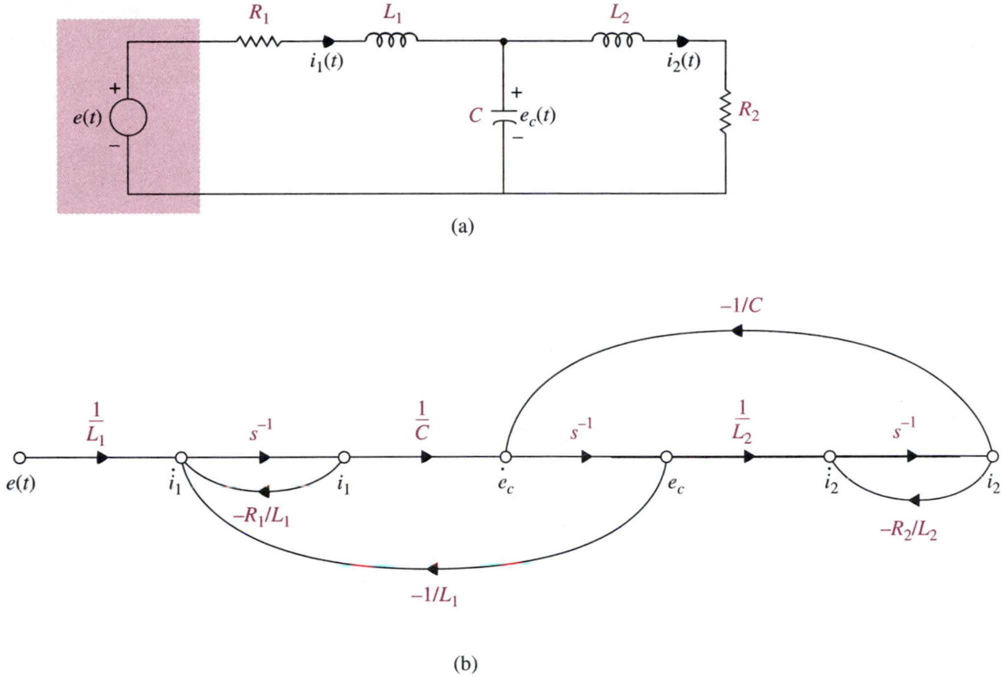

Figure 4-2 (a) Network of Example 4-2. (b) State diagram.

equations are

$$L_1 \frac{di_1(t)}{dt} = R_1 i_1(t) - e_c(t) + e(t) \tag{4-6}$$

$$L_2 \frac{di_2(t)}{dt} = -R_2 i_2(t) + e_c(t) \tag{4-7}$$

$$C \frac{de_c(t)}{dt} = i_1(t) - i_2(t) \tag{4-8}$$

In vector-matrix form, the state equations are written as

$$\begin{bmatrix} \dfrac{di_1(t)}{dt} \\ \dfrac{di_2(t)}{dt} \\ \dfrac{de_c(t)}{dt} \end{bmatrix} = \begin{bmatrix} -\dfrac{R_1}{L_1} & 0 & -\dfrac{1}{L_1} \\ 0 & -\dfrac{R_2}{L_2} & \dfrac{1}{L_2} \\ \dfrac{1}{C} & -\dfrac{1}{C} & 0 \end{bmatrix} \begin{bmatrix} i_1(t) \\ i_2(t) \\ e_c(t) \end{bmatrix} + \begin{bmatrix} \dfrac{1}{L_1} \\ 0 \\ 0 \end{bmatrix} e(t) \tag{4-9}$$

The state diagram of the network, without the initial states, is shown in Fig. 4-2(b). The transfer functions between $I_1(s)$ and $I_2(s)$, and $E_c(s)$ and $E(s)$, respectively, are written from the state diagram.

$$\frac{I_1(s)}{E(s)} = \frac{L_2Cs^2 + R_2Cs + 1}{\Delta} \tag{4-10}$$

$$\frac{I_2(s)}{E(s)} = \frac{1}{\Delta} \tag{4-11}$$

$$\frac{E_c(s)}{E(s)} = \frac{L_2s + R_2}{\Delta} \tag{4-12}$$

where

$$\Delta = L_1L_2Cs^3 + (R_1L_2 + R_2L_1)Cs^2 + (L_1 + L_2 + R_1R_2C)s + R_1 + R_2 \tag{4-13}$$

▲

4-3 Modeling of Mechanical System Elements

Most control systems contain mechanical as well as electrical components, although some systems even have hydraulic and pneumatic elements. From a mathematical viewpoint the descriptions of electrical and mechanical elements are analogous. In fact, we can show that given an electrical device, there is usually an analogous mechanical counterpart mathematically, and vice versa.

The motion of mechanical elements can be described in various dimensions as **translational, rotational,** or combinations. The equations governing the motion of mechanical systems are often formulated directly or indirectly from Newton's law of motion.

4-3-1 Translational Motion

The motion of translation is defined as a motion that takes place along a straight line. The variables that are used to describe translational motion are **acceleration, velocity,** and **displacement.** Newton's law of motion states that the *algebraic sum of forces acting on a rigid body in a given direction is equal to the product of the mass of the body and its acceleration in the same direction.* The law can be expressed as

$$\sum \text{forces} = Ma \tag{4-14}$$

where M denotes the mass and a is the acceleration in the direction considered. For translational motion, the following system elements are usually involved.

1. Mass. *Mass is considered as a property of an element that stores the kinetic energy of translational motion.* Mass is analogous to inductance of electric networks. If W denotes the weight of a body, then M is given by

$$M = \frac{W}{g} \tag{4-15}$$

where g is the acceleration of free fall of the body due to gravity. ($g = 32.174$ ft/sec^2 in British units, and $g = 9.8066$ m/sec^2 in SI units.)

The consistent sets of basic units in British and SI units are as follows:

Units	Mass M	Acceleration	Force
SI	kilogram (kg)	m/sec^2	Newton (N)
British	slug	ft/sec^2	pound (lb force)

Conversion factors between these and other secondary units are as follows:

Force

$$1 \text{ N} = 0.2248 \text{ lb (force)} = 3.5969 \text{ oz (force)}$$

Mass

$$\begin{aligned} 1 \text{ kg} = 1000 \text{ g} &= 2.2046 \text{ lb (mass)} \\ &= 35.274 \text{ oz (mass)} \\ &= 0.06852 \text{ slug} \end{aligned}$$

Distance

$$\begin{aligned} 1 \text{ m} &= 3.2808 \text{ ft} = 39.37 \text{ in.} \\ 1 \text{ in.} &= 25.4 \text{ mm} \\ 1 \text{ ft} &= 0.3048 \text{ m} \end{aligned}$$

Figure 4-3 illustrates the situation where a force is acting on a body with mass M. The force equation is written

$$f(t) = Ma(t) = M\frac{d^2y(t)}{dt^2} = M\frac{dv(t)}{dt} \tag{4-16}$$

where $v(t)$ denotes velocity.

2. Linear Spring. In practice, a linear spring may be a model of an actual spring or a compliance of a cable or a belt. In general, *a spring is considered to be an*

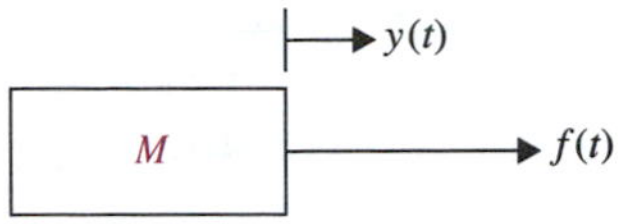

Figure 4-3 Force–mass system.

element that stores potential energy. It is analogous to a capacitor in electric networks. All springs in real life are nonlinear to some extent. However, if the deformation of the spring is small, its behavior can be approximated by a linear relationship:

$$f(t) = Ky(t) \tag{4-17}$$

where K is the **spring constant,** or simply **stiffness.**

The two basic unit systems for the spring constant are as follows:

Units	Spring Constant K
SI	N/m
British	lb/ft

Equation (4-17) implies that the force acting on the spring is directly proportional to the displacement (deformation) of the spring. A model of a linear spring element is shown in Fig. 4-4. If the spring is preloaded with a preload tension of T, Eq. (4-17) should be modified to

$$f(t) - T = Ky(t) \tag{4-18}$$

Friction for Translation Motion

Whenever there is motion or tendency of motion between two physical elements, frictional forces exist. The frictional forces encountered in physical systems are usually of a nonlinear nature. The characteristics of the frictional forces between two contacting surfaces often depend on such factors as the composition of the surfaces, the pressure between the surfaces, their relative velocity, and others, so that an exact mathematical description of the frictional force is difficult. Three different types of friction are

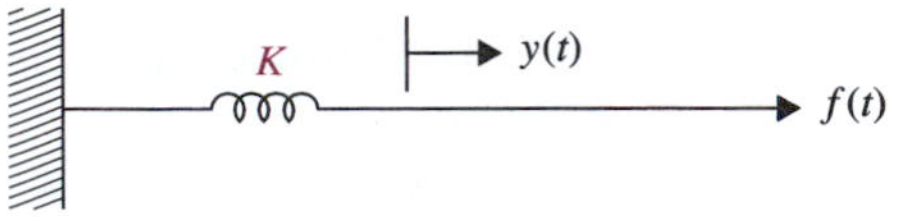

Figure 4-4 Force–spring system.

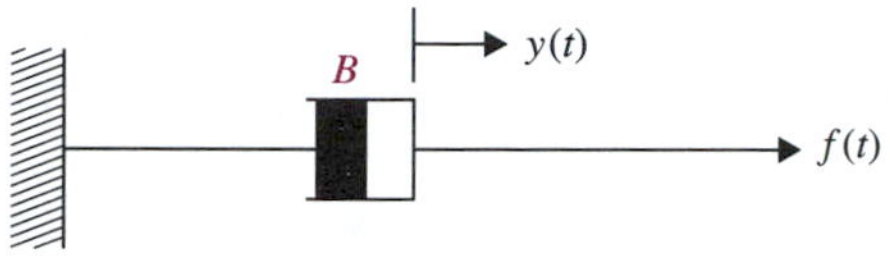

Figure 4-5 Dashpot for viscous friction.

commonly used in practical systems: **viscous friction, static friction,** and **Coulomb friction.** These are discussed below.

1. Viscous Friction. *Viscous friction represents a retarding force that is a linear relationship between the applied force and velocity.* The schematic diagram element for viscous friction is often represented by a dashpot, such as that shown in Fig. 4-5. The mathematical expression of viscous friction is

$$f(t) = B\frac{dy(t)}{dt} \tag{4-19}$$

where B is the **viscous frictional coefficient.**
The units of B are as follows:

Units	Viscous Frictional Coefficient B
SI	N/m/sec
British	lb/ft/sec

Figure 4-6(a) shows the functional relation between the viscous frictional force and velocity.

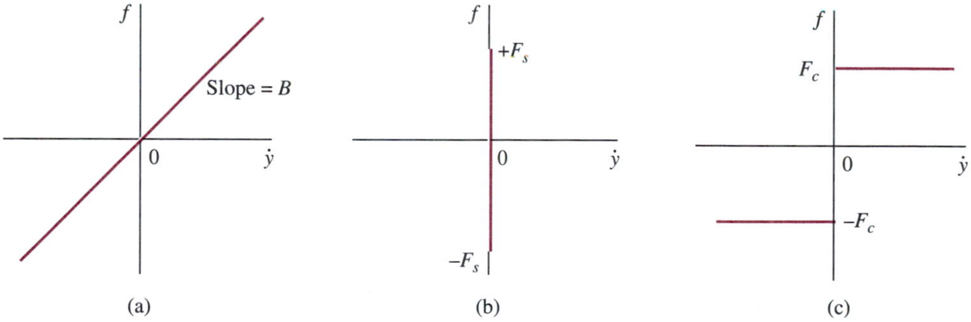

Figure 4-6 Functional relationships of linear and nonlinear frictional forces. (a) Viscous friction. (b) Static friction. (c) Coulomb friction.

2. Static Friction. *Static friction represents a retarding force that tends to prevent motion from beginning.* The static frictional force can be represented by the expression

$$f(t) = \pm(F_s)\big|_{\dot{y}=0} \tag{4-20}$$

which is defined as a frictional force that exists only when the body is stationary but has a tendency to move. The sign of the friction depends on the direction of motion or the initial direction of velocity. The force-to-velocity relation of static friction is illustrated in Fig. 4-6(b). Notice that once motion begins, the static frictional force vanishes and other frictions take over.

3. Coulomb Friction. *Coulomb friction is a retarding force that has a constant amplitude with respect to the change of velocity, but the sign of the frictional force changes with the reversal of the direction of velocity.* The mathematical relation for the Coulomb friction is given by

$$f(t) = F_c \frac{dy(t)/dt}{|dy(t)/dt|} \tag{4-21}$$

where F_c is the **Coulomb friction coefficient.** The functional description of the friction-to-velocity relation is shown in Fig. 4-6(c).

It should be pointed out that the three types of frictions cited here are merely practical models that have been devised to portray frictional phenomena found in physical systems. They are by no means exhaustive or guaranteed to be accurate. In many unusual situations, we have to use other frictional models to represent the actual phenomenon accurately. One such example is rolling dry friction [4,5], which is used to model friction in high-precision ball bearings used in spacecraft systems. It turns out that rolling dry friction has nonlinear hysteresis properties which make it impossible for use in linear system modeling.

4-3-2 Rotational Motion

The rotational motion of a body can be defined as motion about a fixed axis. The extension of Newton's law of motion for rotational motion states that the *algebraic sum of moments or torque about a fixed axis is equal to the product of the inertia and the angular acceleration about the axis;* or,

$$\sum \text{torques} = J\alpha \tag{4-22}$$

where J denotes the inertia and α is the angular acceleration. The other variables generally used to describe the motion of rotation are **torque** T, **angular velocity** ω, and **angular displacement** θ. The elements involved with the rotational motion are as follows:

1. Inertia. *Inertia, J, is considered to be the property of an element that stores the kinetic energy of rotational motion.* The inertia of a given element depends on the geometric composition about the axis of rotation and its density. For instance, the inertia of a circular disk or shaft about its geometric axis is given by

$$J = \tfrac{1}{2} Mr^2 \tag{4-23}$$

When a torque is applied to a body with inertia J, as shown in Fig. 4-7, the torque equation is written

$$T(t) = J\alpha(t) = J\frac{d\omega(t)}{dt} = J\frac{d^2\theta(t)}{dt^2} \tag{4-24}$$

where $\theta(t)$ is the angular displacement, $\omega(t)$ the angular velocity, and $\alpha(t)$ the angular acceleration.

The SI and British units for inertia and the variables in Eq. (4-24) are tabulated as follows:

Units	Inertia	Torque	Angular Displacement
SI	kg-m^2	N-m	rad
		dyn-cm	rad
British	slug-ft^2	lb-ft	
	lb-ft-sec^2		rad
	oz-in.-sec^2	oz-in.	

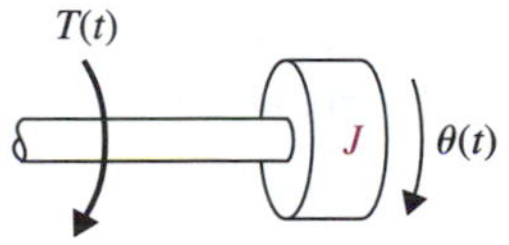

Figure 4-7 Torque–inertia system.

The following conversion factors are often found useful:

Angular Displacement

$$1 \text{ rad} = \frac{180}{\pi} = 57.3 \text{ deg}$$

Angular Velocity

$$1 \text{ rpm} = \frac{2\pi}{60} = 0.1047 \text{ rad/sec}$$

$$1 \text{ rpm} = 6 \text{ deg/sec}$$

Torque

$$1 \text{ g-cm} = 0.0139 \text{ oz-in.}$$
$$1 \text{ lb-ft} = 192 \text{ oz-in.}$$
$$1 \text{ oz-in.} = 0.00521 \text{ lb-ft}$$

Inertia

$$1 \text{ g-cm} = 1.417 \times 10^{-5} \text{ oz-in.-sec}^2$$
$$1 \text{ lb-ft-sec}^2 = 192 \text{ oz-in.-sec}^2 = 32.2 \text{ lb-ft}^2$$
$$1 \text{ oz-in.-sec}^2 = 386 \text{ oz-in}^2$$
$$1 \text{ g-cm-sec}^2 = 980 \text{ g-cm}^2$$

2. Torsional Spring. As with the linear spring for translational motion, a **torsional spring constant *K*,** in torque per unit angular displacement, can be devised to represent the compliance of a rod or a shaft when it is subject to an applied torque. Figure 4-8 illustrates a simple torque–spring system that can be represented by the equation

$$T(t) = K\theta(t) \tag{4-25}$$

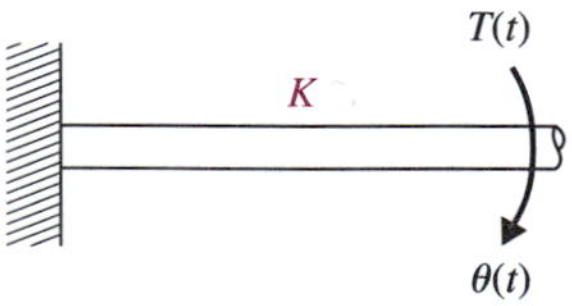

Figure 4-8 Torque–torsional spring system.

The units of the spring constant K in the SI and British systems are as follows:

Units	Spring Constant K
SI	N-m/rad
British	ft-lb/rad

If the torsional spring is preloaded by a preload torque of TP, Eq. (4-25) is modified to

$$T(t) - TP = K\theta(t) \tag{4-26}$$

3. Friction for Rotational Motion. The three types of friction described for translational motion can be carried over to the motion of rotation. Therefore, Eqs. (4-19), (4-20), and (4-21) can be replaced, respectively, by their counterparts:

Viscous friction: $$T(t) = B\frac{d\theta(t)}{dt} \tag{4-27}$$

Static friction: $$T(t) = \pm(F_s)|_{\dot{\theta}=0} \tag{4-28}$$

Coulomb friction: $$T(t) = F_c\frac{d\theta(t)/dt}{|d\theta(t)/dt|} \tag{4-29}$$

4-3-3 Conversion between Translational and Rotational Motions

In motion-control systems it is often necessary to convert rotational motion into translation. For instance, a load may be controlled to move along a straight line through a rotary motor-and-screw assembly, such as that shown in Fig. 4-9. Figure 4-10 shows

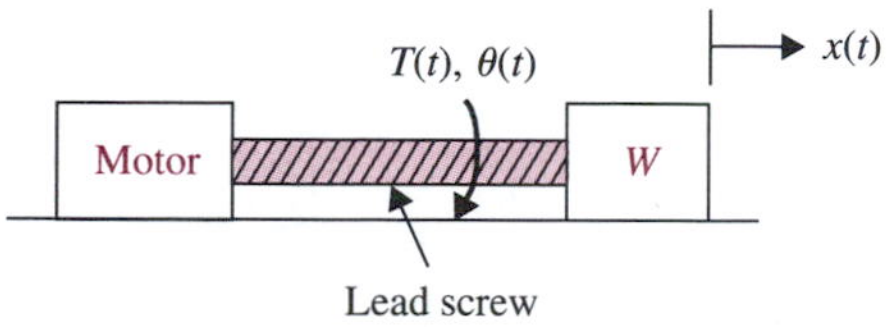

Figure 4-9 Rotary-to-linear motion-control system (lead screw).

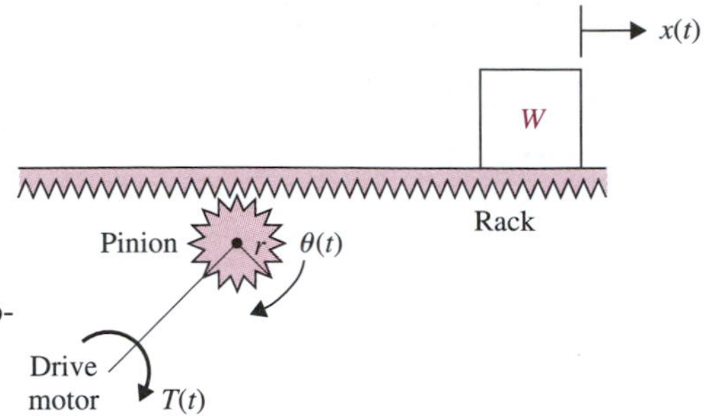

Figure 4-10 Rotary-to-linear motion-control system (rack and pinion)

a similar situation in which a rack-and-pinion is used as a mechanical linkage. Another familiar system in motion control is the control of a mass through a pulley by a rotary motor, such as the control of a printwheel in an electric typewriter (Fig. 4-11). The systems shown in Figs. 4-9, 4-10, and 4-11 can all be represented by a simple system with an equivalent inertia connected directly to the drive motor. For instance, the mass in Fig. 4-11 can be regarded as a point mass that moves about the pulley, which has a radius r. By disregarding the inertia of the pulley, the equivalent inertia that the motor sees is

$$J = Mr^2 = \frac{W}{g} r^2 \tag{4-30}$$

If the radius of the pinion in Fig. 4-10 is r, the equivalent inertia that the motor sees is also given by Eq. (4-30).

Now consider the system of Fig. 4-9. The lead of the screw, L, is defined as the linear distance that the mass travels per revolution of the screw. In principle, the two systems in Figs. 4-10 and 4-11 are equivalent. In Fig. 4-10 the distance traveled by the mass per revolution of the pinion is $2\pi r$. By using Eq. (4-30) as the equivalent inertia for the system of Fig. 4-9, we have

$$J = \frac{W}{g}\left(\frac{L}{2\pi}\right)^2 \tag{4-31}$$

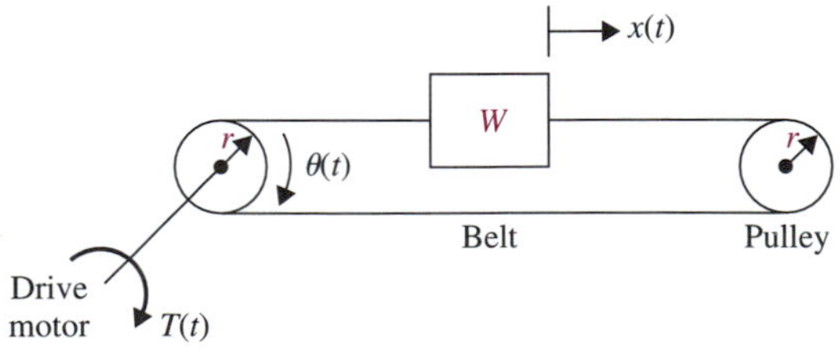

Figure 4-11 Rotary-to-linear motion-control system (belt and pulley).

where, in British units,

J = inertia (oz-in.-sec^2) L = screw lead (in.)
W = weight (oz) g = gravitational force (386.4 in./sec^2)

4-3-4 Gear Trains, Levers, and Timing Belts

A gear train, lever, or timing belt over a pulley is a mechanical device that transmits energy from one part of the system to another in such a way that force, torque, speed, and displacement may be altered. These devices can also be regarded as matching devices used to attain maximum power transfer. Two gears are shown coupled together in Fig. 4-12. The inertia and friction of the gears are neglected in the ideal case considered.

The relationships between the torques, T_1 and T_2, angular displacement θ_1 and θ_2, and the teeth numbers N_1 and N_2 of the gear train are derived from the following facts:

1. *The number of teeth on the surface of the gears is proportional to the radii r_1 and r_2 of the gears; that is,*

$$r_1 N_2 = r_2 N_1 \tag{4-32}$$

2. *The distance traveled along the surface of each gear is the same.* Thus

$$\theta_1 r_1 = \theta_2 r_2 \tag{4-33}$$

3. *The work done by one gear is equal to that of the other since there are assumed to be no losses.* Thus

$$T_1 \theta_1 = T_2 \theta_2 \tag{4-34}$$

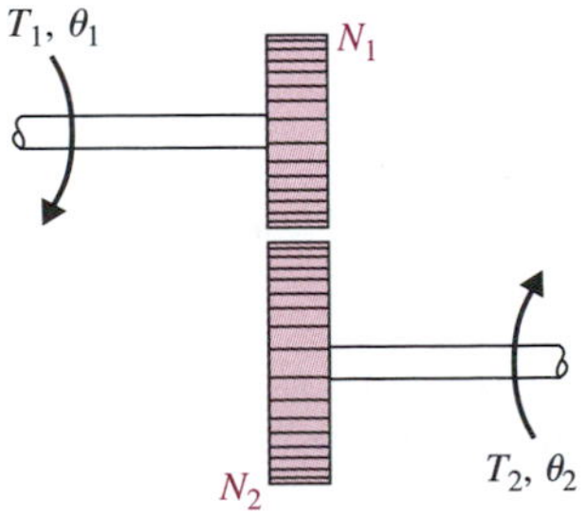

Figure 4-12 Gear train.

If the angular velocities of the two gears, ω_1 and ω_2, are brought into the picture, Eqs. (4-32) through (4-34) lead to

$$\frac{T_1}{T_2} = \frac{\theta_2}{\theta_1} = \frac{N_1}{N_2} = \frac{\omega_2}{\omega_1} = \frac{r_1}{r_2} \tag{4-35}$$

In practice, gears do have inertia and friction between the coupled gear teeth that often cannot be neglected. An equivalent representation of a gear train with viscous friction, Coulomb friction, and inertia, considered as lumped parameters, is shown in Fig. 4-13, where T denotes the applied torque, T_1 and T_2 are the transmitted torques, F_{c1}, and F_{c2}, are the Coulomb friction coefficients, and B_1 and B_2 are the viscous friction coefficients. The torque equation for gear 2 is

$$T_2(t) = J_2 \frac{d^2\theta_2(t)}{dt^2} + B_2 \frac{d\theta_2(t)}{dt} + F_{c2} \frac{\omega_2}{|\omega_2|} \tag{4-36}$$

The torque equation on the side of gear 1 is

$$T(t) = J_1 \frac{d^2\theta_1(t)}{dt^2} + B_1 \frac{d\theta_1(t)}{dt} + F_{c1} \frac{\omega_1}{|\omega_1|} + T_1(t) \tag{4-37}$$

By use of Eq. (4-35), Eq. (4-36) is converted to

$$T_1(t) = \frac{N_1}{N_2} T_2(t) = \left(\frac{N_1}{N_2}\right)^2 J_2 \frac{d^2\theta_1(t)}{dt^2} + \left(\frac{N_1}{N_2}\right)^2 B_2 \frac{d\theta_1(t)}{dt} + \frac{N_1}{N_2} F_{c2} \frac{\omega_2}{|\omega_2|} \tag{4-38}$$

Equation (4-38) indicates that *it is possible to reflect inertia, friction, compliance, torque, speed, and displacement from one side of a gear train to the other.* The following

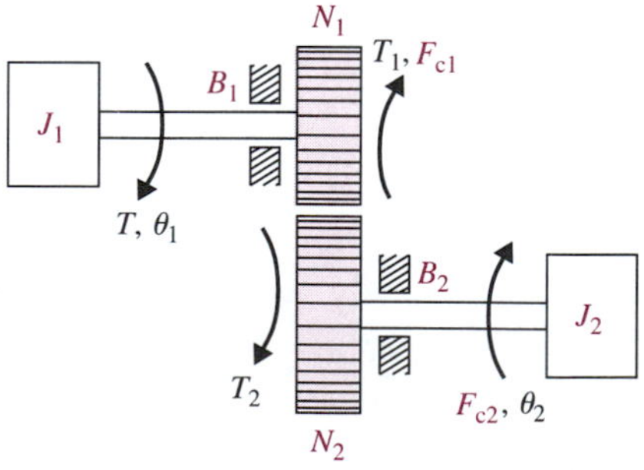

Figure 4-13 Gear train with friction and inertia.

quantities are obtained when reflecting from gear 2 to gear 1:

Inertia: $\left(\frac{N_1}{N_2}\right)^2 J_2$ Angular displacement: $\frac{N_1}{N_2}\theta_2$

Viscous friction coefficient: $\left(\frac{N_1}{N_2}\right)^2 B_2$ Angular velocity: $\frac{N_1}{N_2}\omega_2$

Torque: $\frac{N_1}{N_2}T_2$ Coulomb friction torque: $\frac{N_1}{N_2}F_{c2}\frac{\omega_2}{|\omega_2|}$

Similarly, gear parameters and variables can be reflected from gear 1 to gear 2 simply by interchanging the subscripts in the expressions above.

If a torsional spring effect is present, the spring constant is also multiplied by $(N_1/N_2)^2$ in reflecting from gear 2 to gear 1. Now substituting Eq. (4-38) into Eq. (4-37), we get

$$T(t) = J_{1e}\frac{d^2\theta_1(t)}{dt^2} + B_{1e}\frac{d\theta_1(t)}{dt} + T_F \tag{4-39}$$

where

$$J_{1e} = J_1 + \left(\frac{N_1}{N_2}\right)^2 J_2 \tag{4-40}$$

$$B_{1e} = B_1 + \left(\frac{N_1}{N_2}\right)^2 B_2 \tag{4-41}$$

$$T_F = F_{c1}\frac{\omega_1}{|\omega_1|} + \frac{N_1}{N_2}F_{c2}\frac{\omega_2}{|\omega_2|} \tag{4-42}$$

Example 4-3

Given a load that has inertia 0.05 oz-in.-sec^2 and Coulomb friction torque 2 oz-in., find the inertia and frictional torque reflected through a 1 : 5 gear train ($N_1/N_2 = 1/5$, with N_2 on the load side). The reflected inertia on the side of N_1 is $(\frac{1}{5})^2 \times 0.05 = 0.002$ oz-in.-sec^2. The reflected Coulomb friction is $\frac{1}{5} \times 2 = 0.4$ oz-in. ▲

Timing belts and chain drives serve the same purpose as the gear train except that they allow the transfer of energy over a longer distance without using an excessive number of gears. Figure 4-14 shows the diagram of a belt or chain drive between two pulleys. Assuming that there is no slippage between the belt and the pulleys, it is easy

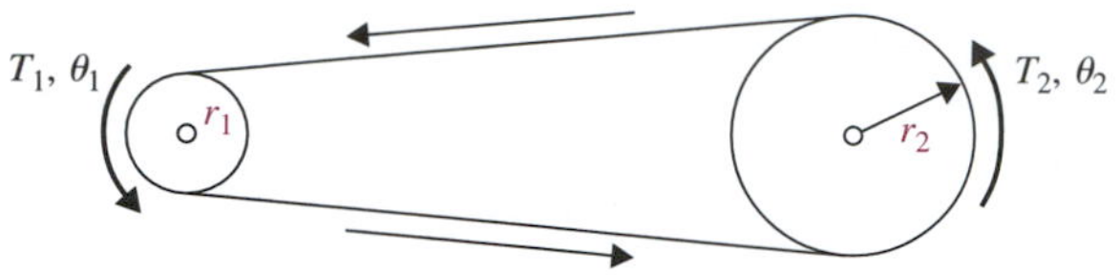

Figure 4-14 Belt or chain drive.

to see that Eq. (4-35) still applies to this case. In fact, the reflection or transmittance of torque, inertia, friction, and so on, is similar to that of a gear train.

The lever system shown in Fig. 4-15 transmits translational motion and forces in the same way that gear trains transmit rotational motion. The relation between the forces and distances is

$$\frac{f_1}{f_2} = \frac{l_2}{l_1} = \frac{x_2}{x_1} \tag{4-43}$$

4-3-5 Backlash and Dead Zone

Backlash and dead zone are commonly found in gear trains and similar mechanical linkages where the coupling is not perfect. In a majority of situations, backlash may give rise to undesirable inaccuracy, oscillations, and instability in control systems. In addition, it has a tendency to wear out the mechanical elements. Regardless of the actual mechanical elements, a physical model of backlash or dead zone between an input and an output member is shown in Fig. 4-16. The model can be used for a rotational system as well as for a translational system. The amount of backlash is $b/2$ on either size of the reference position.

In general, the dynamics of the mechanical linkage with backlash depend on the relative inertia-to-friction ratio of the output member. If the inertia of the output member is very small compared with that of the input member, the motion is controlled, predominantly by friction. This means that the output member will not coast whenever there

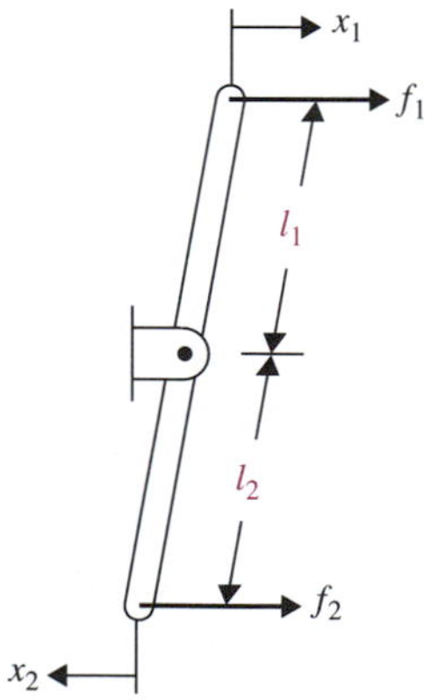

Figure 4-15 Lever system.

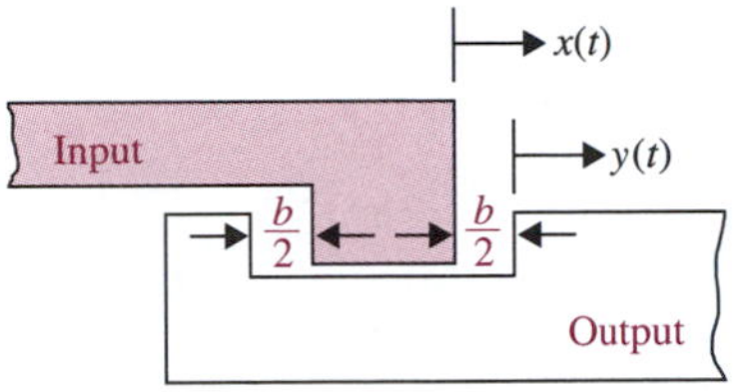

Figure 4-16 Physical model of backlash between two mechanical elements.

is no contact between the two members. When the output is driven by the input, the two members will travel together until the input member reverses its direction; then the output member will be at a standstill until the backlash is taken up on the other side, at which time it is assumed that the output member instantaneously takes on the velocity of the input member. The transfer characteristic between the input and output displacements of a system with backlash with negligible output inertia is shown in Fig. 4-17.

4-4 Equations of Mechanical Systems

The equations of a linear mechanical system are written by first constructing a model of the system containing interconnected linear elements and then by applying Newton's law of motion to the **free-body diagram.** For translational motion, the equation of motion is Eq. (4-14), and for rotational motion, Eq. (4-22) is used. The following examples illustrate how equations of mechanical systems are written.

Example 4-4

Consider the mass–spring–friction system shown in Figure 4-18(a). The linear motion concerned is in the horizontal direction. The free-body diagram of the system is shown in Fig. 4-18(b). The force equation of the system is

$$f(t) = M\frac{d^2y(t)}{dt^2} + B\frac{dy(t)}{dt} + Ky(t) \qquad (4\text{-}44)$$

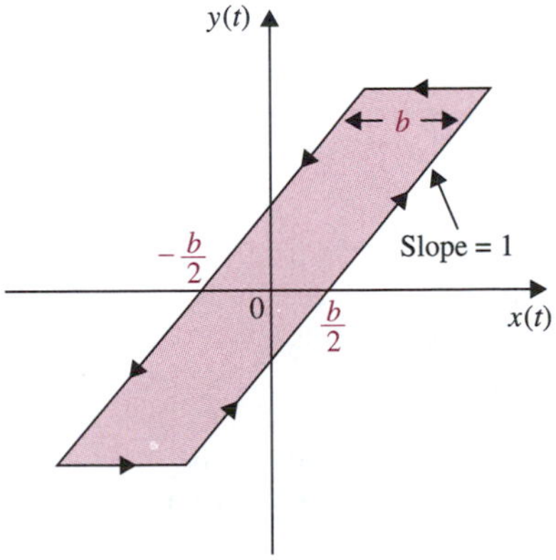

Figure 4-17 Input–output characteristic of backlash.

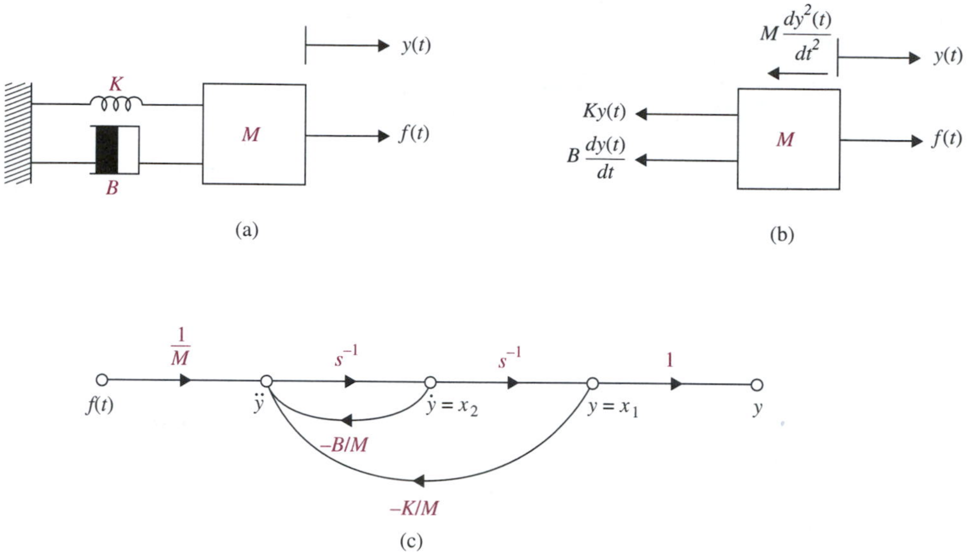

Figure 4-18 (a) Mass–spring–friction system. (b) Free-body diagram. (c) State diagram.

Equation (4-44) is rearranged by equating the highest-order derivative term to the rest of the terms:

$$\frac{d^2y(t)}{dt^2} = -\frac{B}{M}\frac{dy(t)}{dt} - \frac{K}{M}y(t) + \frac{1}{M}f(t) \tag{4-45}$$

The state diagram of the system is constructed as shown in Fig. 4-18(c). By defining the outputs of the integrators on the state diagram as state variables x_1 and x_2, the state equations are

$$\frac{dx_1(t)}{dt} = x_2(t) \tag{4-46}$$

$$\frac{dx_2(t)}{dt} = -\frac{K}{M}x_1(t) - \frac{B}{M}x_2(t) + \frac{1}{M}f(t) \tag{4-47}$$

It is not difficult to see that this mechanical system is analogous to a series *RLC* electric network. With this analogy, the state equations can be written directly using a different set of state variables. Consider that mass M is analogous to inductance L, the spring constant K is analogous to the inverse of capacitance $1/C$, and the viscous-friction coefficient B is analogous to resistance R. It is logical to assign $v(t)$, the velocity, and $f_k(t)$, the force acting on the spring, as state variables, since the former is analogous to the current in L and the latter is analogous to the voltage across C. Then writing the force on M and the velocity of the spring as functions of the state variables and the input force $f(t)$, we have:

$$\text{Force on mass:} \quad M\frac{dv(t)}{dt} = -Bv(t) - f_k(t) + f(t) \tag{4-48}$$

$$\text{Velocity of spring:} \quad \frac{1}{K}\frac{df_k(t)}{dt} = v(t) \tag{4-49}$$

The state equations are obtained by dividing both sides of Eq. (4-48) by M and multiplying Eq. (4-49) by K.

▲ The state variables and state equations of a dynamic system are not unique.

This simple example illustrates that the state equation and state variables of a dynamic system are not unique. The transfer function between $Y(s)$ and $F(s)$ is obtained by taking the Laplace transform on both sides of Eq. (4-44) with zero initial conditions:

$$\frac{Y(s)}{F(s)} = \frac{1}{Ms^2 + Bs + K} \tag{4-50}$$

The same result is obtained by applying the gain formula to Fig. 4-18(c). ▲

Example 4-5

As another example of writing the dynamic equations of a mechanical system with translational motion, consider the system shown in Fig. 4-19(a). Since the spring is deformed when it is subject to a force $f(t)$, two displacements, y_1 and y_2, must be assigned to the endpoints of the spring. The free-body diagrams of the system are shown in Fig. 4-19(b). The force equations are

$$f(t) = K[y_1(\text{t}) - y_2(t)] \tag{4-51}$$

$$K[y_1(t) - y_2(t)] = M\frac{d^2y_2(t)}{dt^2} + B\frac{dy_2(t)}{dt} \tag{4-52}$$

These equations are rearranged as

$$y_1(\text{t}) = y_2(t) + \frac{1}{K}f(t) \tag{4-53}$$

$$\frac{d^2y_2(t)}{dt^2} = -\frac{B}{M}\frac{dy_2(t)}{dt} + \frac{K}{M}[y_1(t) - y_2(t)] \tag{4-54}$$

By using the last two equations, the state diagram of the system is drawn in Fig. 4-19(c). The state variables are defined as $x_1(t) = y_2(t)$ and $x_2(t) = dy_2(t)/dt$. The state equations are written directly from the state diagram:

$$\frac{dx_1(t)}{dt} = x_2(t) \tag{4-55}$$

$$\frac{dx_2(t)}{dt} = -\frac{B}{M}x_2(t) + \frac{1}{M}f(t) \tag{4-56}$$

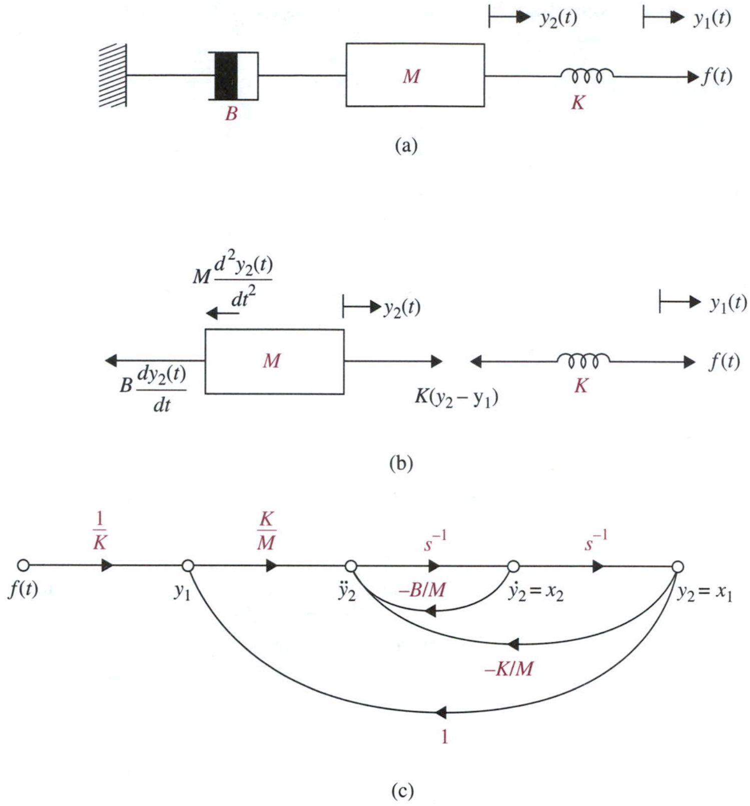

Figure 4-19 Mechanical system for Example 4-5. (a) Mass–spring–friction system. (b) Free-body diagram. (c) State diagram.

As an alternative, we can assign the velocity $v(t)$ of the mass M as one state variable and the force $f_k(t)$ on the spring as the other state variable. We have

$$\frac{dv(t)}{dt} = -\frac{B}{M}v(t) + \frac{1}{M}f_k(t) \tag{4-57}$$

$$f_k(t) = f(t) \tag{4-58}$$

One may wonder why there is only one state equation in Eq. (4-48), whereas there are two state variables in $v(t)$ and $f_k(t)$. The two state equations of Eqs. (4-55) and (4-56) clearly show that the system is of the second order. The situation is better explained by referring to the analogous electric network of the system shown in Fig. 4-20. Although the network has two energy-storage elements in L and C, and thus there should be two state variables, the voltage across the capacitance, e_c(t), in this case is redundant, since it is equal to the applied voltage $e(t)$. Equations (4-57) and (4-56) can provide only the solutions to the velocity of M, $v(t)$, which

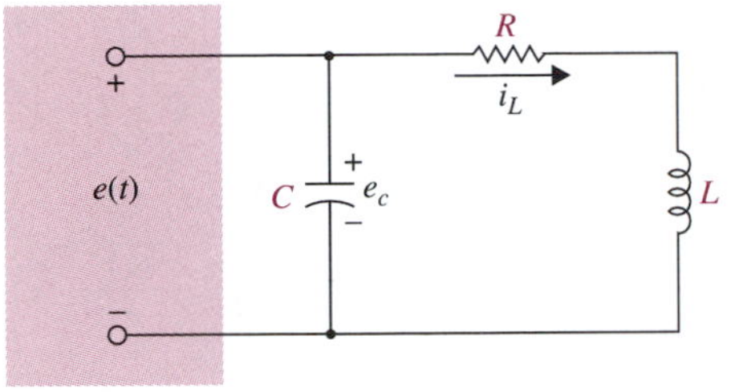

Figure 4-20 Electric network analogous to the mechanical system in Fig. 4-19.

is the same as $dy_2(t)/dt$, once $f(t)$ is specified. Then $y_2(t)$ is determined by integrating $v(t)$ with respect to t. The displacement $y_1(t)$ is then found using Eq. (4-51). On the other hand, Eqs. (4-55) and (4-56) give the solutions to $y_2(t)$ and $dy_2(t)/dt$ directly, and $y_1(t)$ is obtained from Eq. (4-51).

The transfer functions of the system are obtained by applying the gain formula to the state diagram in Fig. 4-19(c).

$$\frac{Y_2(s)}{F(s)} = \frac{1}{s(Ms + B)} \tag{4-59}$$

$$\frac{Y_1(s)}{F(s)} = \frac{Ms^2 + Bs + K}{Ks(Ms + B)} \tag{4-60}$$

▲

Example 4-6

In this example the equations for the mechanical system in Fig. 4-21(a) are to be written. Then we are to draw the state diagram and derive the transfer functions for the system. The free-body diagrams for the two masses are shown in Fig. 4-21(b), with the reference directions of the displacements $y_1(t)$ and $y_2(t)$ as indicated. The Newton's force equations for the system are written directly from the free-body diagrams:

$$f(t) + M_1 g = M_1 \frac{d^2 y_1(t)}{dt^2} + B_1 \left[\frac{dy_1(t)}{dt} - \frac{dy_2(t)}{dt}\right] + K_1[y_1(t) - y_2(t)] \tag{4-61}$$

$$M_2 g = -B_1 \left[\frac{dy_1(t)}{dt} - \frac{dy_2(t)}{dt}\right] - K_1[y_1(t) - y_2(t)] + M_2 \frac{d^2 y_2(t)}{dt^2} + B_2 \frac{dy_2(t)}{dt} + K_2 y_2(t) \tag{4-62}$$

We may now decompose these two second-order simultaneous differential equations into four state equations by defining the following state variables:

$$x_1(t) = y_1(t) \tag{4-63}$$

$$x_2(t) = \frac{dy_1(t)}{dt} = \frac{dx_1(t)}{dt} \tag{4-64}$$

$$x_3(t) = y_2(t) \tag{4-65}$$

$$x_4(t) = \frac{dy_2(t)}{dt} = \frac{dx_3(t)}{dt} \tag{4-66}$$

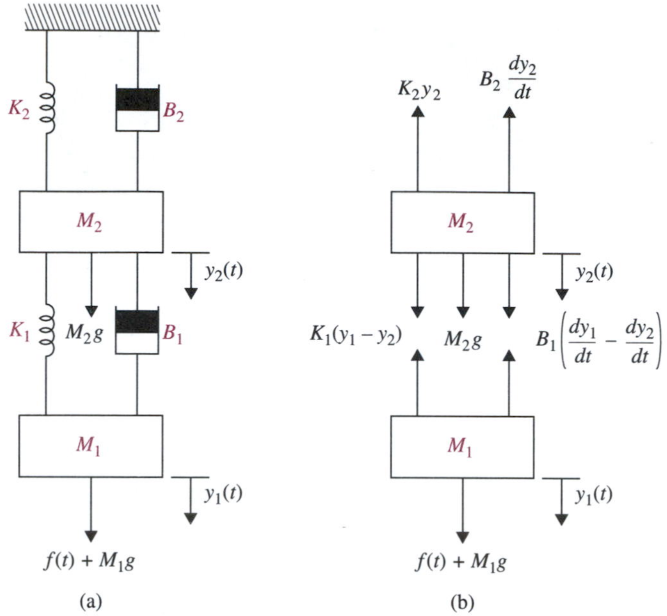

Figure 4-21 Mechanical system for Example 4-6.

Equations (4-64) and (4-66) form two state equations naturally; the other two are obtained by substituting Eqs. (4-63) through (4-66) into Eqs. (4-61) and (4-62). Rearranging, we have the following state equations:

$$\frac{dx_1(t)}{dt} = x_2(t) \tag{4-67}$$

$$\frac{dx_2(t)}{dt} = -\frac{K_1}{M_1}[x_1(t) - x_3(t)] - \frac{B_1}{M_1}[x_2(t) - x_4(t)] + \frac{1}{M_1}f(t) + g \tag{4-68}$$

$$\frac{dx_3(t)}{dt} = x_4(t) \tag{4-69}$$

$$\frac{dx_4(t)}{dt} = \frac{K_1}{M_2}x_1(t) + \frac{B_1}{M_2}x_2(t) - \frac{K_1 + K_2}{M_2}x_3(t) - \frac{1}{M_2}(B_1 + B_2)x_4(t) + g \tag{4-70}$$

If we are interested in the displacements $y_1(t)$ and $y_2(t)$, the output equations are

$$y_1(t) = x_1(t) \tag{4-71}$$
$$y_2(t) = x_3(t) \tag{4-72}$$

The state diagram of the system, according to the equations written above, is as shown in Fig. 4-22. Alternatively, the state diagram can be drawn by first rearranging Eqs. (4-61) and (4-62). The output transforms $Y_1(s)$ and $Y_2(s)$ are written in terms of the input $F(s)$ and g by

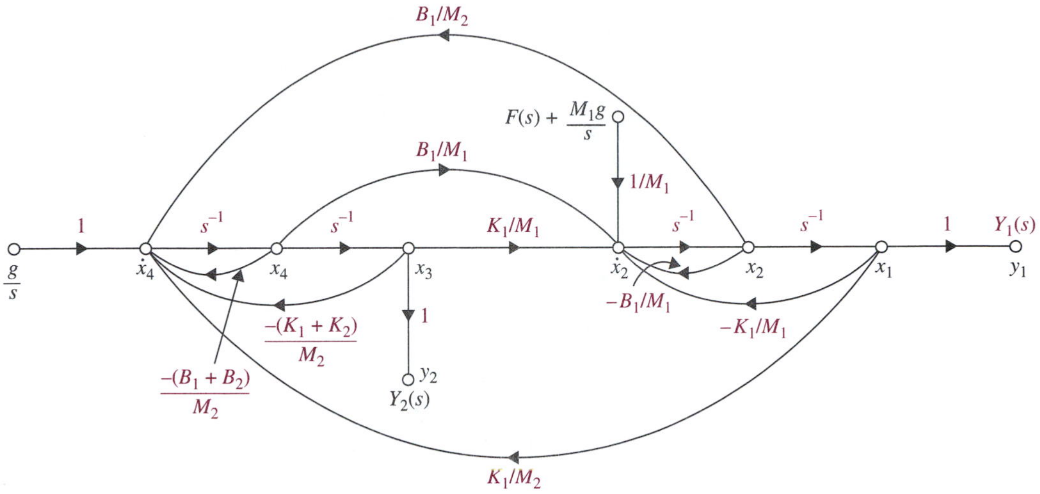

Figure 4-22 State diagram for the mechanical system of Fig. 4-21.

applying the gain formula to the state diagram:

$$Y_1(s) = \frac{M_2s^2 + (B_1 + B_2)s + (K_1 + K_2)}{\Delta}F(s) + \frac{M_1M_2s^2 + (M_1B_1 + M_1B_2 + M_2B_1)s + M_1K_1 + M_1K_2 + M_2K_1}{\Delta}\frac{g}{s} \tag{4-73}$$

$$Y_2(s) = \frac{B_1s + K_1}{\Delta}F(s) + \frac{M_1M_2s^2 + B_1(M_1 + M_2)s + K_1(M_1 + M_2)}{\Delta}\frac{g}{s} \tag{4-74}$$

where

$$\Delta = M_1M_2s^4 + [M_1(B_1 + B_2) + B_1M_2]s^3 + [M_1(K_1 + K_2) + K_1M_2 + B_1B_2]s^2 + (K_1B_2 + B_1K_2)s + K_1K_2 \tag{4-75}$$

The transfer functions $Y_1(s)/F(s)$ and $Y_2(s)/F(s)$ can be obtained from Eqs. (4-73) and (4-74), respectively, by setting the gravitational force g to zero. This is equivalent to assuming that the system is initially at the equilibrium state and that the gravitational force is balanced by preloading of the springs.

The state equations can also be written directly from the mechanical system by assigning the state variables as $v_1(t) = dy_1(t)/dt$ and $v_2(t) = dy_2(t)/dt$, and the forces on the two springs as $f_{k1}(\mathrm{t})$ and $f_{k2}(t)$. Then if we write the forces acting on the masses and the velocities of the springs as functions of the four state variables and the external forces, the state equations are:

$$\text{Force on } M_1: \quad M_1\frac{dv_1(t)}{dt} = -B_1v_1(t) + B_2v_2(t) - f_{k1}(t) + f(t) + M_1g \tag{4-76}$$

Force on M_2: $$M_2 \frac{dv_2(t)}{dt} = B_1 v_1(t) - (B_1 + B_2) v_2(t) + f_{k1}(t) - f_{k2}(t) + M_2 g \tag{4-77}$$

Velocity of K_1: $$\frac{df_{k1}(t)}{dt} = K_1[v_1(t) - v_2(t)] \tag{4-78}$$

Velocity of K_2: $$\frac{df_{k2}(t)}{dt} = K_2 v_2(t) \tag{4-79}$$

▲

Example 4-7

The rotational system shown in Fig. 4-23(a) consists of a disk mounted on a shaft that is fixed at one end. The moment of inertia of the disk about the axis of rotation is J. The edge of the disk is riding on the surface, and the viscous friction coefficient between the two surfaces is B. The inertia of the shaft is negligible, but the torsional spring constant is K.

Assume that a torque is applied to the disk, as shown; then the torque or moment equation about the axis of the shaft is written from the free-body diagram of Fig. 4-23(b):

$$T(t) = J\frac{d^2\theta(t)}{dt^2} + B\frac{d\theta(t)}{dt} + K\theta(t) \tag{4-80}$$

Notice that this system is analogous to the translational system in Fig. 4-18. The state equations may be written by defining the state variables as $x_1(t) = \theta(t)$ and $x_2(t) = dx_1(t)$. ▲

Example 4-8

Figure 4-24(a) shows the diagram of a motor coupled to an inertial load through a shaft with a spring constant K. A nonrigid coupling between two mechanical components in a control system often causes torsional resonances that can be transmitted to all parts of the system. The system variables and parameters are defined as follows:

- ▲ $T_m(t)$ = motor torque
- ▲ B_m = motor viscous friction coefficient
- ▲ K = spring constant of the shaft
- ▲ $\theta_L(t)$ = load displacement
- ▲ $\omega_L(t)$ = load velocity
- ▲ J_m = motor inertia
- ▲ J_L = load inertia
- ▲ $\theta_m(t)$ = motor displacement
- ▲ $\omega_m(t)$ = motor velocity

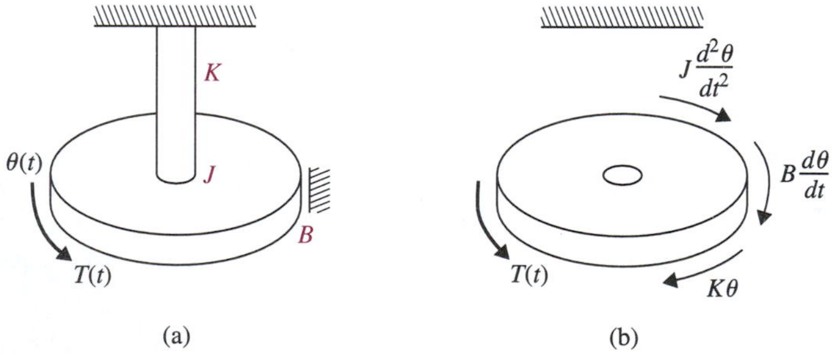

Figure 4-23 Rotational system for Example 4-7.

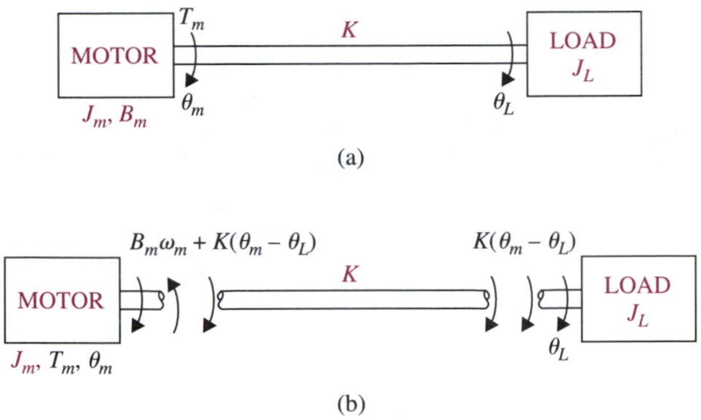

Figure 4-24 (a) Motor–load system. (b) Free-body diagram.

The free-body diagrams of the system are shown in Fig. 4-24(b). The torque equations of the system are

$$\frac{d^2\theta_m(t)}{dt^2} = -\frac{B_m}{J_m}\frac{d\theta_m(t)}{dt} - \frac{K}{J_m}[\theta_m(t) - \theta_L(t)] + \frac{1}{J_m}T_m(t) \tag{4-81}$$

$$K[\theta_m(t) - \theta_L(t)] = J_L\frac{d^2\theta_L(t)}{dt^2} \tag{4-82}$$

In this case the system contains three energy-storage elements in J_m, J_L, and K. Thus there should be three state variables. Care should be taken in constructing the state diagram and assigning the state variables so that a minimum number of the latter are incorporated. Equations (4-81) and (4-82) are rearranged as

$$\frac{d^2\theta_m(t)}{dt^2} = -\frac{B_m}{J_m}\frac{d\theta_m(t)}{dt} - \frac{K}{J_m}[\theta_m(t) - \theta_L(t)] + \frac{1}{J_m}T_m(t) \tag{4-83}$$

$$\frac{d^2\theta_L(t)}{dt^2} = \frac{K}{J_L}[\theta_m(t) - \theta_L(t)] \tag{4-84}$$

The state diagram with three integrators is shown in Fig. 4-25. The clue given by Eqs. (4-83) and (4-85) is that $\theta_m(t)$ and $\theta_L(t)$ appear only as the difference, $\theta_m(t)-\theta_L(t)$, in these equations. From the state diagram in Fig. 4-25, the state variables are defined as $x_1(t) = \theta_m(t) - \theta_L(t)$, $x_2(t) = d\theta_L(t)/dt$, and $x_3(t) = d\theta_m(t)/dt$. The state equations are

$$\frac{dx_1(t)}{dt} = x_3(t) - x_2(t) \tag{4-85}$$

$$\frac{dx_2(t)}{dt} = \frac{K}{J_L}x_1(t) \tag{4-86}$$

$$\frac{dx_3(t)}{dt} = -\frac{K}{J_m}x_1(t) - \frac{B_m}{J_m}x_3(t) + \frac{1}{J_m}T_m(t) \tag{4-87}$$

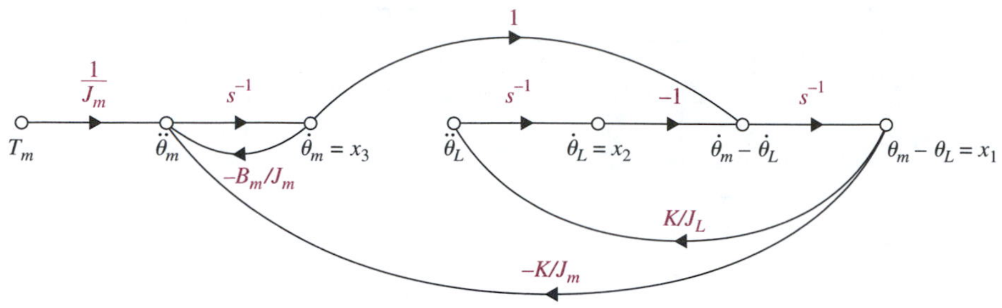

Figure 4-25 State diagram of the system of Fig. 4-24.

The transfer functions between $\Theta_m(s)$ and $T_m(s)$ and $\Theta_L(s)$ and $T_m(s)$ are written by applying the gain formula to the state diagram in Fig. 4-25:

$$\frac{\Theta_m(s)}{T_m(s)} = \frac{X_3(s)}{sT_m(s)} = \frac{J_L s^2 + K}{s[J_m J_L s^3 + B_m J_L s^2 + K(J_m + J_L)s + B_m K]} \tag{4-88}$$

$$\frac{\Theta_L(s)}{T_m(s)} = \frac{X_2(s)}{sT_m(s)} = \frac{K}{s[J_m J_L s^3 + B_m J_L s^2 + K(J_m + J_L)s + B_m K]} \tag{4-89}$$

▲

4-5 Sensors and Encoders in Control Systems

Sensors and encoders are important components used to monitor the performance and for feedback in control systems. In this section the principle of operation and applications of some of the sensors and encoders that are commonly used in control systems are described.

4-5-1 Potentiometer

A potentiometer is an electromechanical transducer that converts mechanical energy into electrical energy. The input to the device is in the form of a mechanical displacement, either linear or rotational. When a voltage is applied across the fixed terminals of the potentiometer, the output voltage, which is measured across the variable terminal and ground, is proportional to the input displacement, either linearly or according to some nonlinear relation.

Rotary potentiometers are available commercially in single-revolution or multirevolution form, with limited or unlimited rotational motion. The potentiometers commonly are made with wirewound or conductive plastic resistance material. Figure 4-26 shows a rotary potentiometer, and Fig. 4-27 shows linear potentiometer that contains a built-in operational amplifier. For precision control, the conductive plastic potentiometer is preferable, since it has infinite resolution, long rotational life, good output smoothness, and low static noise.

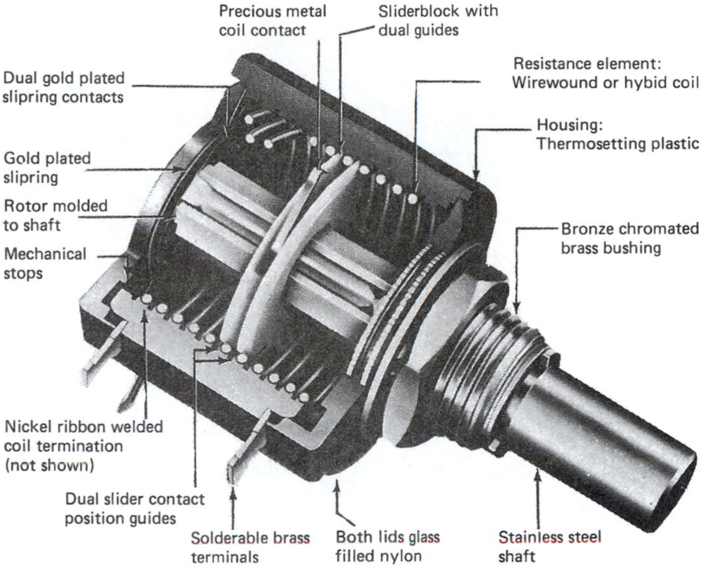

Figure 4-26 Ten-turn rotary potentiometer. (Courtesy of Helipot Divison of Beckman Instruments, Inc.)

Figure 4-28 shows the equivalent-circuit representation of a potentiometer, linear or rotary. Since the voltage across the variable terminal and reference is proportional to the shaft displacement of the potentiometer, when a voltage is applied across the fixed terminals, the device can be used to indicate the absolute position of a system or the relative position of two mechanical outputs. Figure 4-29(a) shows the arrangement

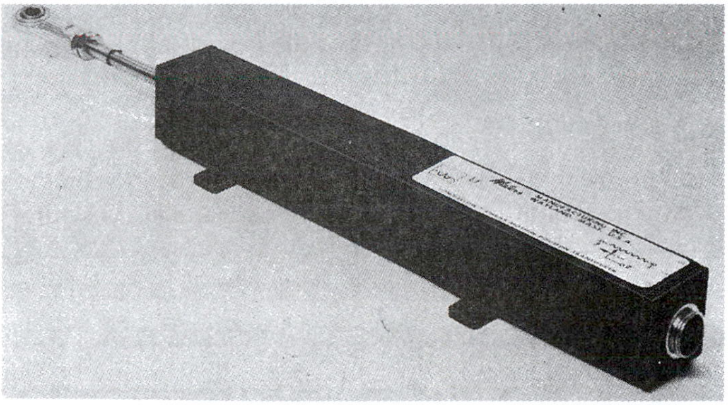

Figure 4-27 Linear motion potentiometer with built-in operational amplifier. (Courtesy of Waters Manufacturing, Inc.)

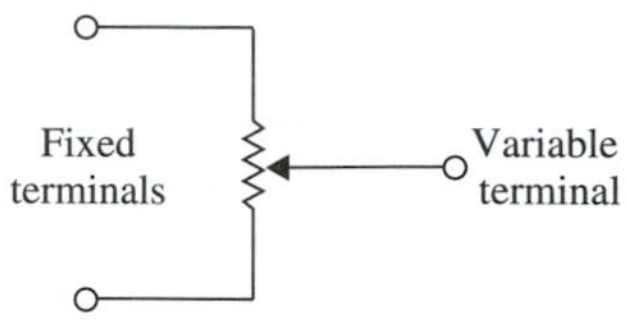

Figure 4-28 Circuit representation of a potentiometer.

when the housing of the potentiometer is fixed at reference; the output voltage $e(t)$ will be proportional to the shaft position $\theta_c(t)$ in the case of a rotary motion. Then

$$e(t) = K_s \theta_c(t) \tag{4-90}$$

where K_s is the proportional constant. For an N-turn potentiometer, the total displacement of the variable arm is $2\pi N$ radians. The proportional constant K_s is given by

$$K_s = \frac{E}{2\pi N} \quad \text{V/rad} \tag{4-91}$$

where E is the magnitude of the reference voltage applied to the fixed terminals. A more flexible arrangement is obtained by using two potentiometers connected in parallel, as shown in Fig. 4-29(b). This arrangement allows the comparison of two remotely located shaft positions. The output voltage is taken across the variable terminals of the two potentiometers and is given by

$$e(t) = K_s[\theta_1(t) - \theta_2(t)] \tag{4-92}$$

Figure 4-30 illustrates the block diagram representation of the setups in Fig. 4-29.

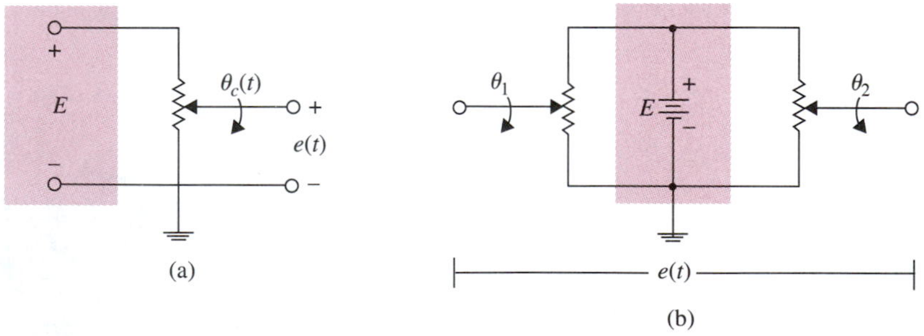

Figure 4-29 (a) Potentiometer used as a position indicator. (b) Two potentiometers used to sense the positions of two shafts.

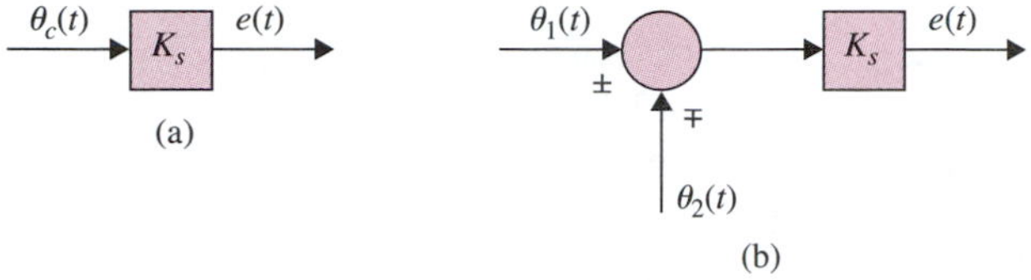

Figure 4-30 Block diagram representations of potentiometer arrangements in Fig. 4-29.

In dc-motor control systems, potentiometers are often used for position feedback. Figure 4-31(a) shows the schematic diagram of a typical dc-motor position-control system. The potentiometers are used in the feedback path to compare the actual load position with the desired reference position. If there is a discrepancy between the load position and the reference input, an error signal is generated by the potentiometers that will drive the motor in such a way that this error is minimized quickly. As shown in Fig. 4-31(a), the error signal is amplified by a dc amplifier whose output drives the armature of a permanent-magnet dc motor. Typical waveforms of the signals in the system when the input $\theta_r(t)$ is a step function are shown in Fig. 4-31(b). Note that the electric signals are all unmodulated. *In control systems terminology, a dc signal usually refers to an unmodulated signal. On the other hand, an ac signal refers to signals that are modulated by a modulation process.* These definitions are different from those commonly used in electrical engineering, where dc refers simply to unidirectional signals and ac indicates alternating signals.

▲ In control-system terminology, a dc signal usually refers to an unmodulated signal.

Figure 4-32(a) illustrates a control system that serves essentially the same purpose as that of the system in Fig. 4-31(a), except that ac signals prevail. In this case the voltage applied to the error detector is sinusoidal. The frequency of this signal is usually much higher than that of the signal that is being transmitted through the system. Control systems with ac signals are usually found in aerospace systems that are more susceptible to noise.

Typical signals of the ac control system are shown in Fig. 4-32(b). The signal $v(t)$ is referred to as the carrier whose frequency is ω_c, or

$$v(\text{t}) = E \sin \omega_c t \tag{4-93}$$

Analytically, the output of the error signal is given by

$$e(\text{t}) = K_s \theta_e(t) v(t) \tag{4-94}$$

where $\theta_e(t)$ is the difference between the input displacement and the load displacement, or

$$\theta_e(t) = \theta_r(t) - \theta_r(t) \tag{4-95}$$

For the $\theta_e(t)$ shown in Fig. 4-32(b), $e(t)$ becomes a **suppressed-carrier-modulated** signal. A reversal in phase of $e(\text{t})$ occurs whenever the signal crosses the zero-magnitude axis. This reversal in phase causes the ac motor to reverse in direction according to the desired sense of correction of the error signal $\theta_e(t)$. The name *suppressed-carrier*

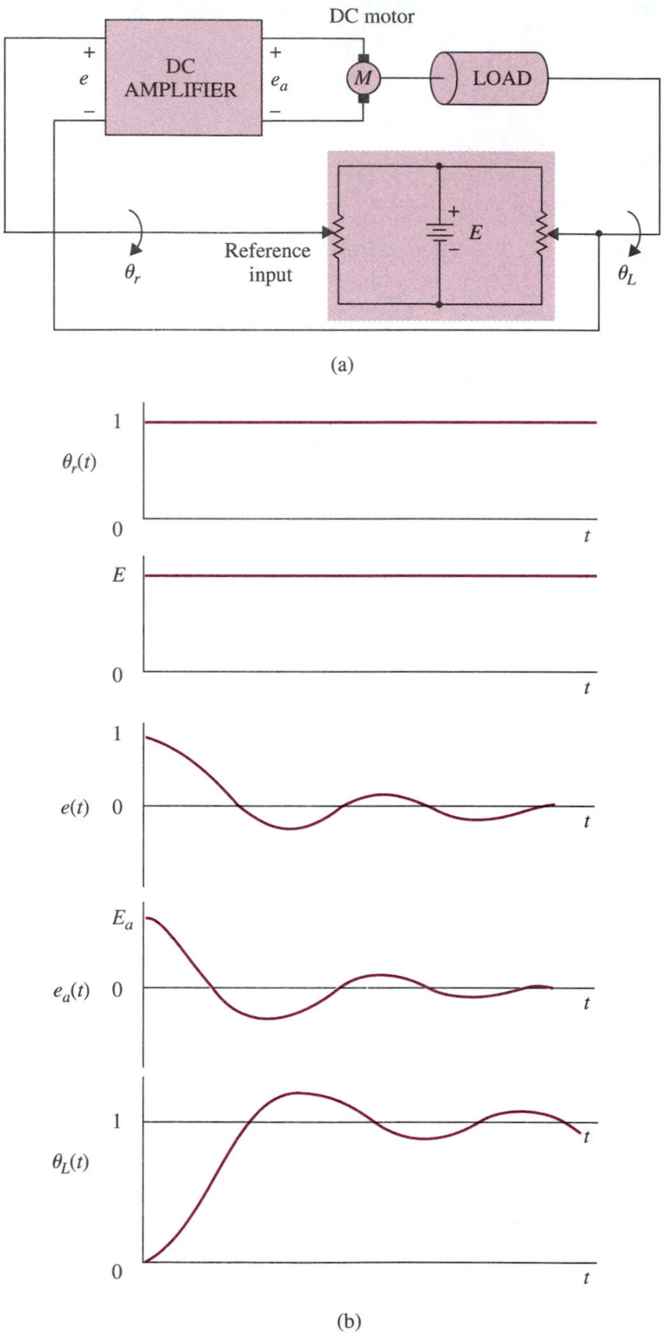

Figure 4-31 (a) Dc-motor position-control system with potentiometers as error sensors. (b) Typical waveforms of signals in the control system of part (a).

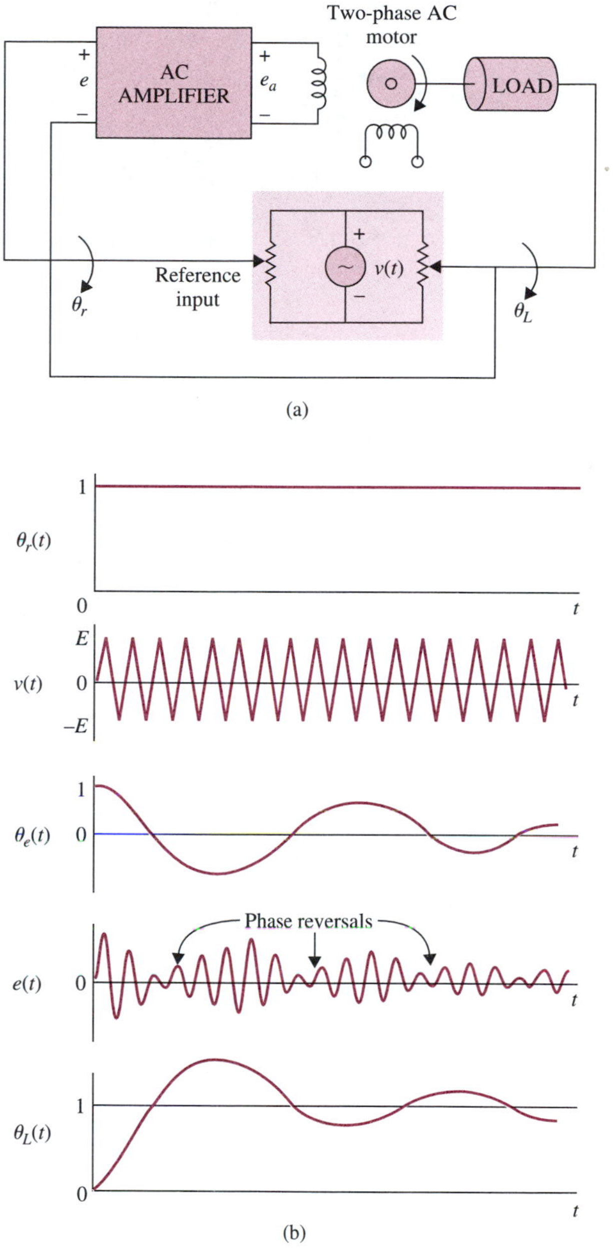

Figure 4-32 (a) Ac control system with potentiometers as error detectors. (b) Typical waveforms of signals in the control system of part (a).

modulation stems from the fact that when a signal $\theta_e(t)$ is modulated by a carrier signal $v(t)$ according to Eq. (4-94), the resultant signal $e(t)$ no longer contains the original carrier frequency ω_c. To illustrate this, let us assume that $\theta_e(t)$ is also a sinusoid given by

$$\theta_e(t) = \sin \omega_s t \tag{4-96}$$

where, normally, $\omega_s \ll \omega_c$. By use of familiar trigonometric relations, substituting Eqs. (4-93) and (4-96) into Eq. (4-94), we get

$$e(t) = \tfrac{1}{2}K_s E[\cos(\omega_c - \omega_s)t - \cos(\omega_c + \omega_s)t] \tag{4-97}$$

Therefore, $e(t)$ no longer contains the carrier frequency ω_c or the signal frequency ω_s, but has only the two sidebands $\omega_c + \omega_s$ and $\omega_c - \omega_s$.

When the modulated signal is transmitted through the system, the motor acts as a demodulator, so that the displacement of the load will be of the same form as the dc signal before modulation. This is seen clearly from the waveforms of Fig. 4-32(b). It should be pointed out that a control system need not contain all-dc or all-ac components. It is quite common to couple dc component to an ac component through a modulator or an ac device to a dc device through a demodulator. For instance, the dc amplifier of the system in Fig. 4-31(a) may be replaced by an ac amplifier that is preceded by a modulator and followed by a demodulator.

4-5-2 Tachometers

Tachometers are electromechanical devices that convert mechanical energy into electrical energy. The device works essentially as a voltage generator, with the output voltage proportional to the magnitude of the angular velocity of the input shaft. In control systems, most of the tachometers used are of the dc variety (i.e., the output voltage is

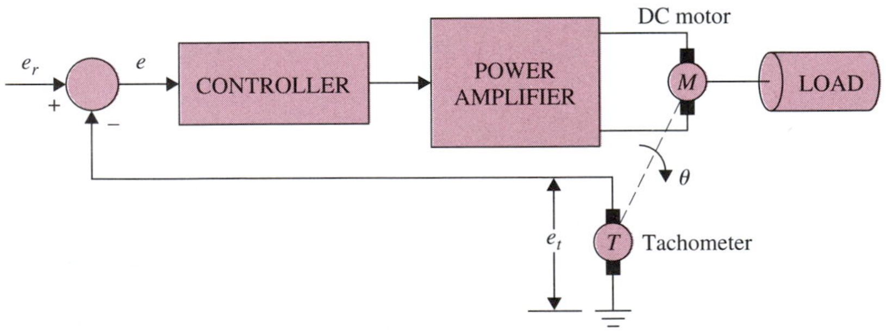

Figure 4-33 Velocity-control system with tachometer feedback.

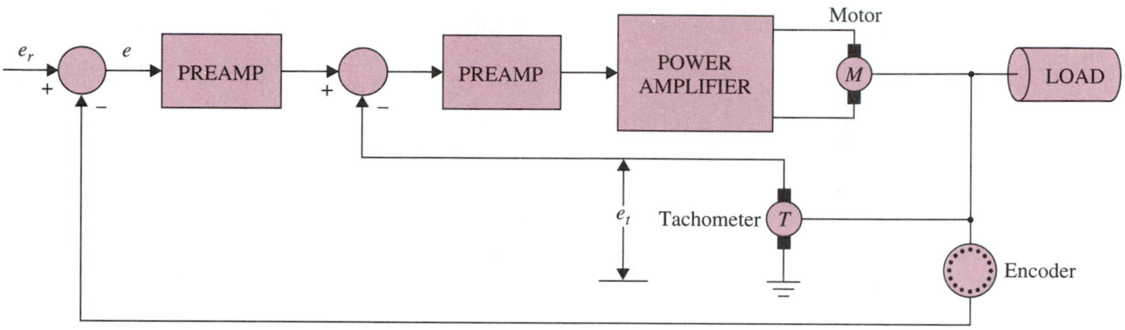

Figure 4-34 Position-control system with tachometer feedback.

a dc signal). Dc tachometers are used in control systems in many ways; they can be used as velocity indicators to provide shaft-speed readout or to provide velocity feedback or speed control or stabilization. Figure 4-33 is a block diagram of a typical velocity-control system in which the tachometer output is compared with the reference voltage, which represents the desired velocity to be achieved. The difference between the two signals, or the error, is amplified and used to drive the motor, so that the velocity will eventually reach the desired value. In this type of application, the accuracy of the tachometer is highly critical, as the accuracy of the speed control depends on it.

▲ *Note:* Tachometers are often used in control systems to improve stability.

In a position-control system, velocity feedback is often used to improve the stability or the damping of the closed-loop system. Figure 4-34 shows the block diagram of such an application. In this case the tachometer feedback forms an inner loop to improve the damping characteristics of the system, and the accuracy of the tachometer is not so critical.

The third and most traditional use of a dc tachometer is in providing the visual speed readout of a rotating shaft. Tachometers used in this capacity are generally connected directly to a voltmeter calibrated in revolutions per minute (rpm).

Mathematical Modeling of Tachometers

The dynamics of the tachometer can be represented by the equation

$$e_t(t) = K_t \frac{d\theta(t)}{dt} = K_t \omega(t) \tag{4-98}$$

where $e_t(t)$ is the output voltage, $\theta(t)$ the rotor displacement in radians, $\omega(t)$ the rotor velocity in rad/sec, and K_t the **tachometer constant** in V/rad/sec. The value of K_t is usually given as a catalog parameter in **volts per 1000 rpm** (V/krpm).

The transfer function of a tachometer is obtained by taking the Laplace transform on both sides of Eq. (4-98). The result is

$$\frac{E_t(s)}{\Theta(s)} = K_t s \tag{4-99}$$

where $E_t(s)$ and $\Theta(s)$ are the Laplace transforms of $e_t(t)$ and $\theta(t)$, respectively.

4-5-3 Incremental Encoder

Incremental encoders are frequently found in modern control systems for converting linear or rotary displacement into digitally coded or pulse signals. The encoders that output a digital signal are known as **absolute encoders.** In the simplest terms, absolute encoders provide as output a distinct digital code indicative of each particular least significant increment of resolution. **Incremental encoders,** on the other hand, provide a pulse for each increment of resolution but do not make distinctions between the increments. In practice, the choice of which type of encoder to use depends on economics and control objectives. For the most part, the need for absolute encoders has much to do with the concern for data loss during power failure or the applications involving periods of mechanical motion without the readout under power. However, the incremental encoder's simplicity in construction, low cost, ease of application, and versatility have made it by far one of the most popular encoders in control systems.

Incremental encoders are available in rotary and linear forms. Figures 4-35 and 4-36 show typical rotary and linear incremental encoders, respectively. A typical rotary incremental encoder has four basic parts: a light source, a rotary disk, a stationary mask, and a sensor, as shown in Fig. 4-37. The disk has alternate opaque and transparent sectors. Any pair of these sectors represents an incremental period. The mask is used to pass or block the light beam between the light source and the photosensor located behind the mask. For encoders with relatively low resolution, the mask is not necessary. For fine-resolution encoders (up to thousands of increments per revolution), a multiple-slit mask is often used to maximize reception of the shutter light.

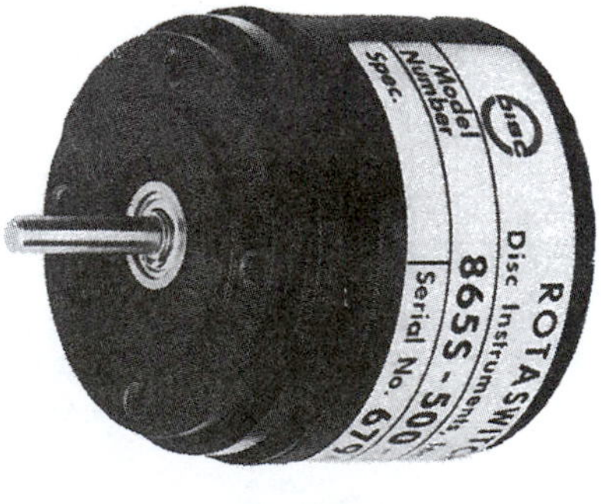

Figure 4-35 Rotary incremental encoder. (Courtesy of DISC Instruments, Inc.)

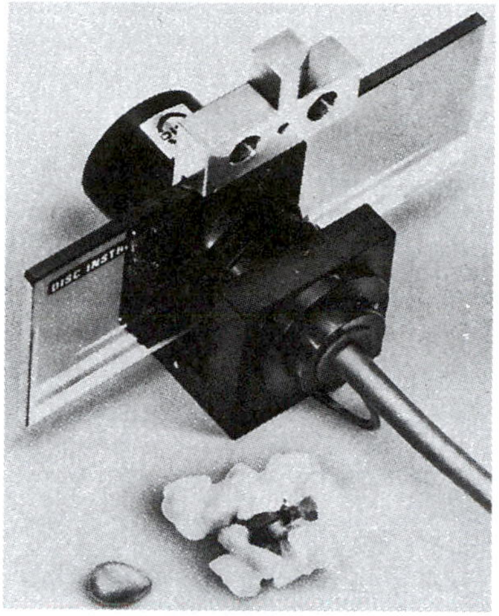

Figure 4-36 Linear incremental encoder. (Courtesy of DISC Instruments, Inc.)

The waveforms of the sensor outputs are generally triangular or sinusoidal, depending on the resolution required. Square-wave signals compatible with digital logic are derived by using a linear amplifier followed by a comparator. Figure 4-38(a) shows a typical rectangular output waveform of a single-channel incremental encoder. In this case pulses are produced for both directions of shaft rotation. A dual-channel encoder with two sets of output pulses is necessary for direction sensing and other control functions. When the phase of the two-output pulse train is 90° apart electrically, the two signals are said to be in quadrature, as shown in Fig. 4-38(b); the signals uniquely define 0-to-1 and 1-to-0 logic transitions with respect to the direction of rotation of the encoder disk, so that a direction-sending logic circuit can be constructed to decode the signals. Figure 4-39 shows the single-channel output and the quadrature outputs with sinusoidal waveforms. The sinusoidal signals from the incremental encoder can be used for fine-position control in feedback control systems. The following example illustrates some applications of the incremental encoder in control systems.

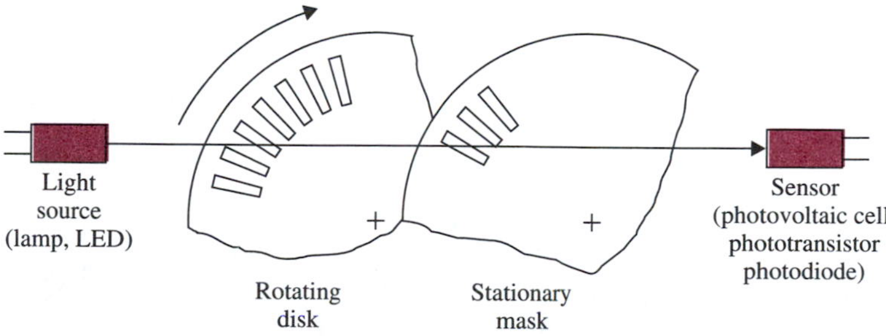

Figure 4-37 Typical increment optomechanics.

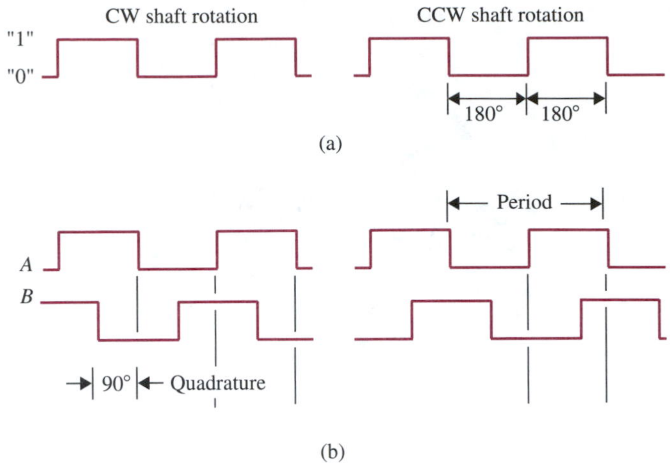

Figure 4-38 (a) Typical rectangular output waveform of a single-channel encoder device (bidirectional). (b) Typical dual-channel encoder signals in quadrature (bidirectional).

Example 4-9

Consider an incremental encoder that generates two sinusoidal signals in quadrature as the encoder disk rotates. The output signals of the two channels are shown in Fig. 4-40 over one cycle. Note that the two encoder signals generate four zero crossings per cycle. These zero crossings can be used for position indication, position control, or speed measurements in control systems. Let us assume that the encoder shaft is coupled directly to the rotor shaft of a motor that directly drives the printwheel of an electronic typewriter or word processor. The printwheel has 96 character positions on its periphery, and the encoder has 480 cycles. Thus there are $480 \times 4 = 1920$ zero crossings per revolution. For the 96-character printwheel, this corresponds

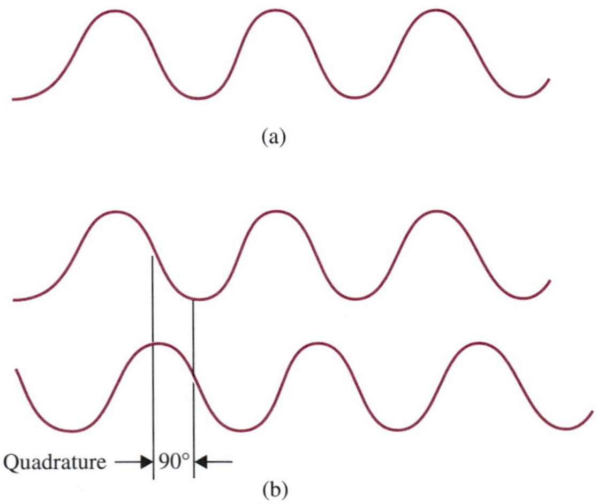

Figure 4-39 (a) Typical sinusoidal output waveform of a single-channel encoder device. (b) Typical dual-channel encoder signals in quadrature.

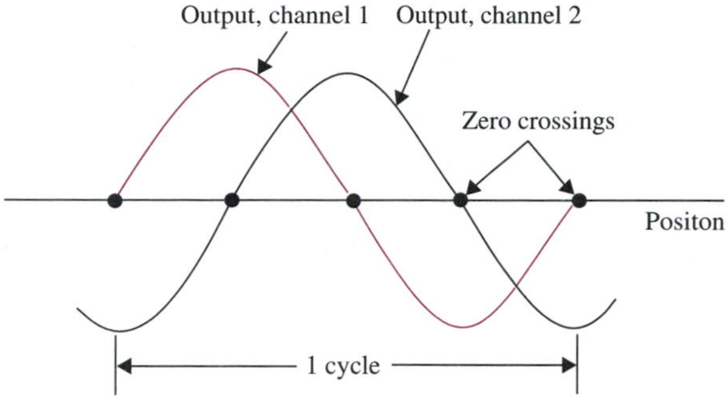

Figure 4-40 One cycle of the output signals of a dual-channel incremental encoder.

to 1920/96 = 20 zero crossings per character; that is, there are 20 zero crossings between two adjacent characters.

One way of measuring the velocity of the printwheel is to count the number of pulses generated by an electronic clock that occur between consecutive zero crossings of the encoder outputs. Let us assume that a 500-kHz clock is used (i.e., the clock generates 500,000 pulses/sec). If the counter counts, say 500 clock pulses while the encoder rotates from the zero crossing to the next, the shaft speed is

$$\begin{aligned}\frac{500{,}000 \text{ pulses/sec}}{500 \text{ pulses/zero crossing}} &= 1000 \text{ zero crossings/sec} \\ &= \frac{1000 \text{ zero crossings/sec}}{1920 \text{ zero crossings/rev}} = 0.52083 \text{ rev/sec} \\ &= 31.25 \text{ rpm} \qquad (4\text{-}100)\end{aligned}$$

The encoder arrangement described can be used for fine-position control of the printwheel. Let the zero crossing A of the waveforms in Fig. 4-40 correspond to a character position on the printwheel (the next character position is 20 zero crossings away), and the point corresponds to a stable equilibrium point. The coarse-position control of the system must first drive the printwheel position to within one zero crossing on either side of position A; then by using the slope of the sine wave at position A, the control system should null the error quickly. ▲

4-6 DC Motors in Control Systems

Direct-current (dc) motors are one of the most widely used prime movers in industry today. Years ago a majority of the small servomotors used for control purposes were of the ac variety. In reality, ac motors are more difficult to control, especially for position control, and their characteristics are quite nonlinear, which makes the analytical task more difficult. Dc motors, on the other hand, are more expensive, because of the brushes and commutators, and variable-flux dc motors are suitable only for certain types of control applications. Before permanent-magnet technology was fully developed, the torque per unit volume or weight of a dc motor with a permanent-magnet (PM) field was far from desirable. Today, with the development of the rare-earth magnet, it is possible to achieve very high torque-to-volume PM dc motors at reasonable cost. Furthermore, the advances made in brush-and-commutator technology have made these wearable parts practically maintenance-free. The advancements made in power electronics have made brushless dc motors quite popular in high-performance control systems. Advanced manufacturing techniques have also produced dc motors with ironless rotors that have very low inertia, thus achieving a very high torque-to-inertia ratio, and low-time-constant properties have opened new applications for dc motors in computer peripheral equipment such as tape drives, printers, disk drives, and word processors, as well as in the automation and machine-tool industries.

4-6-1 Basic Operational Principles of DC Motors

The dc motor is basically a torque transducer that converts electric energy into mechanical energy. The torque developed on the motor shaft is directly proportional to the field flux and the armature current. As shown in Fig. 4-41, a current-carrying conductor is established in a magnetic field with flux ϕ, and the conductor is located at a distance r from the center of rotation. The relationship among the developed torque, flux ϕ, and current i_a is

$$T_m = K_m \phi i_a \tag{4-101}$$

where T_m is the motor torque (N-m, lb-ft, or oz-in.), ϕ the magnetic flux (webers), i_a the armature current (amperes), and K_m is a proportional constant.

In addition to the torque developed by the arrangement shown in Fig. 4-41, when the conductor moves in the magnetic field, a voltage is generated across its terminals. This voltage, the **back emf,** which is proportional to the shaft velocity, tends to oppose the current flow. The relationship between the back emf and the shaft velocity is

$$e_b = K_m \phi \omega_m \tag{4-102}$$

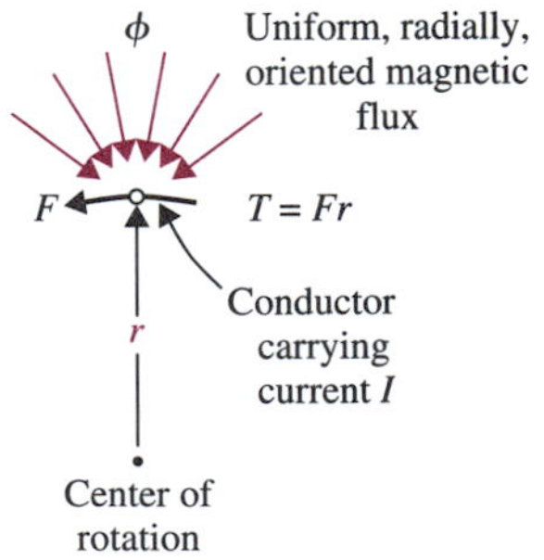

Figure 4-41 Torque production in a dc motor.

where e_b denotes the back emf (volts), and ω_m is the shaft velocity (rad/sec) of the motor. Equations (4-101) and (4-102) form the basis of the dc-motor operation.

4-6-2 Basic Classifications of PM DC Motors

In general, the magnetic field of a dc motor can be produced by field windings or permanent magnets. Due to the popularity of PM dc motors in control system applications, we concentrate on this type of motor. PM dc motors can be classified according to commutation scheme and armature design. Conventional dc motors have mechanical brushes and commutators. However, in one important class of dc motors the commutation is done electronically; this type of motor is called a **brushless dc motor.**

According to the armature construction, the PM dc motor can be broken down into three types of armature design: **iron-core, surface-wound,** and **moving-coil** motors.

Iron-Core PM DC Motors

The rotor and stator configuration of an iron-core PM dc motor is shown in Fig. 4-42. The permanent-magnet material can be barium–ferrite, Alnico, or rare-earth compound. The magnetic flux produced by the magnet passes through a laminated rotor structure that contains slots. The armature conductors are placed in the rotor slots. This type of dc motor is characterized by relatively high rotor inertia (since the rotating part consists of the armature windings), high inductance, low cost, and high reliability.

Surface-Wound DC Motors

Figure 4-43 shows the rotor construction of a surface-wound PM dc motor. The armature conductors are bonded to the surface of a cylindrical rotor structure, which is made of laminated disks fastened to the motor shaft. Since no slots are used on the rotor in this

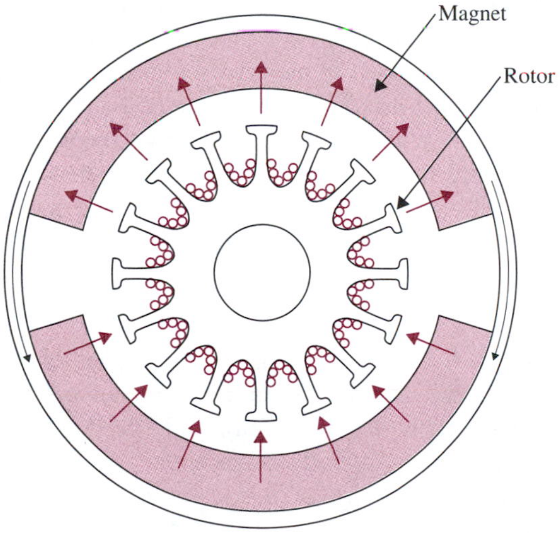

Figure 4-42 Cross-sectional view of a permanent-magnet iron-core dc motor.

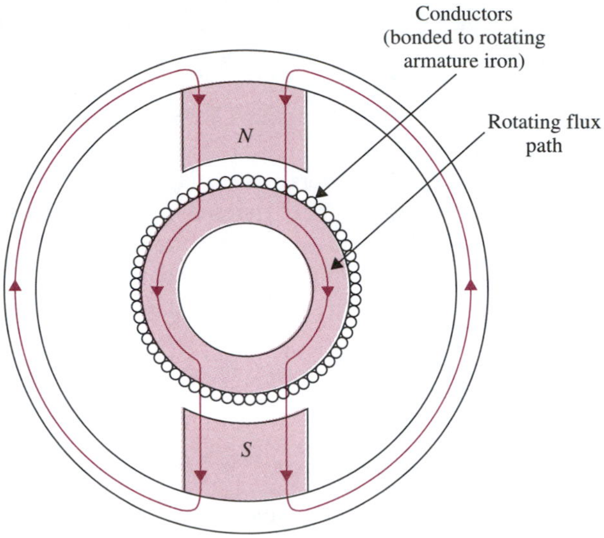

Figure 4-43 Cross-sectional view of a surface-wound permanent-magnet dc motor.

design, the armature has no "cogging" effect. Since the conductors are laid out in the air gap between the rotor and the permanent-magnet field, this type of motor has lower inductance than that of the iron-core structure.

Moving-Coil DC Motors

Moving-coil motors are designed to have very low moments of inertia and very low armature inductance. This is achieved by placing the armature conductors in the air gap between a stationary flux return path and the permanent-magnet structure, as shown in Fig. 4-44. In this case, the conductor structure is supported by nonmagnetic material—usually epoxy resins or fiber glass—to form a hollow cylinder. One end of the cylinder forms a hub, which is attached to the motor shaft. A cross-sectional view of such a motor is shown in Fig. 4-45. Since all unnecessary elements have been removed from the armature of the moving-coil motor, its moment of inertia is very low. Since the conductors in the moving-coil armature are not in direct contact with iron, the motor inductance is very low; values of less than 100 μH are common in this type of motor. The low-inertia and low-inductance properties make the moving-coil motor one of the best actuator choices for high-performance control systems.

Brushless DC Motors

Brushless dc motors differ from the previously mentioned dc motors in that they employ electrical (rather than mechanical) commutation of the armature current. The configuration of the brushless dc motor most commonly used—especially for incremental-motion applications—is one in which the rotor consists of magnets and "back iron" support, and whose commutated windings are located external to the rotating parts, as shown in Fig. 4-46. Compared to the conventional dc motors, such as the one shown

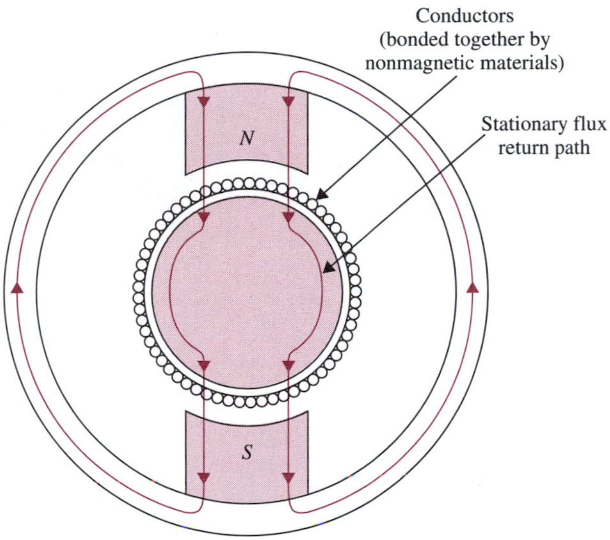

Figure 4-44 Cross-sectional view of a moving-coil permanent-magnet dc motor.

in Fig. 4-42, it is an ‘‘inside-out’’ configuration. Depending on the specific application, brushless dc motors can be used when a low moment of inertia is called for, such as the spindle drive in high-performance disk drives used in computers.

4-6-3 Mathematical Modeling of PM DC Motors

Since dc motors are used extensively in control systems, for analytical purposes, it is necessary to establish mathematical models for dc motors for control applications. We use the equivalent circuit diagram in Fig. 4-47 to represent a PM dc motor. The armature

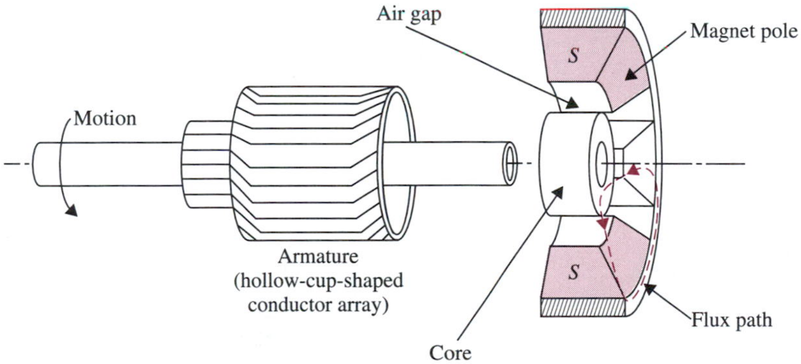

Figure 4-45 Cross-sectional side view of a moving-coil dc motor.

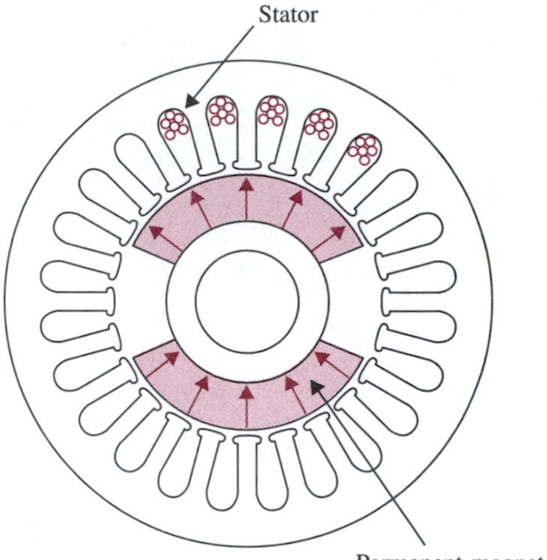

Figure 4-46 Cross-sectional view of a permanent-magnet brushless dc motor.

is modeled as a circuit with resistance R_a connected in series with an inductance L_a, and a voltage source e_b representing the back emf in the armature when the rotor rotates. The motor variables and parameters are defined as follows:

- ▲ $i_a(t)$ = armature current
- ▲ R_a = armature resistance
- ▲ $e_b(t)$ = back emf
- ▲ $T_L(t)$ = load torque
- ▲ $T_m(t)$ = motor torque
- ▲ $\theta_m(t)$ = rotor displacement
- ▲ K_i = torque constant
- ▲ L_a = armature inductance
- ▲ $e_a(t)$ = applied voltage
- ▲ K_b = back-emf constant
- ▲ ϕ = magnetic flux in the air gap
- ▲ $\omega_m(t)$ = rotor angular velocity
- ▲ J_m = rotor inertia
- ▲ B_m = viscous-friction coefficient

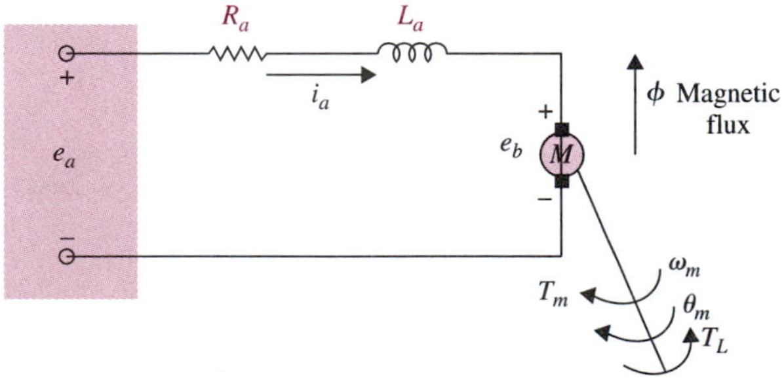

Figure 4-47 Model of a separately excited dc motor.

With reference to the circuit diagram of Fig. 4-47, the control of the dc motor is applied at the armature terminals in the form of the applied voltage $e_a(t)$. For linear analysis, we assume that the torque developed by the motor is proportional to the air-gap flux and the armature current. Thus

$$T_m(t) = K_m(t)\phi i_a(t) \tag{4-103}$$

Since ϕ is constant, Eq. (4-103) is written

$$T_m(t) = K_i i_a(t) \tag{4-104}$$

where K_i is the **torque constant** in N-m/A, lb-ft/A, or oz-in./A.

Starting with the control input voltage $e_a(t)$, the cause-and-effect equations for the motor circuit in Fig. 4-47 are

$$\frac{di_a(t)}{dt} = \frac{1}{L_a}e_a(t) - \frac{R_a}{L_a}i_a(t) - \frac{1}{L_a}e_b(t) \tag{4-105}$$

$$T_m(t) = K_i i_a(t) \tag{4-106}$$

$$e_b(t) = K_b\frac{d\theta_m(t)}{dt} = K_b\omega_m(t) \tag{4-107}$$

$$\frac{d^2\theta_m(t)}{dt^2} = \frac{1}{J_m}T_m(t) - \frac{1}{J_m}T_L(t) - \frac{B_m}{J_m}\frac{d\theta_m(t)}{dt} \tag{4-108}$$

where $T_L(t)$ represents a load frictional torque such as Coulomb friction.

Equations (4-105) through (4-108) consider that $e_a(t)$ is the cause of all causes; Eq. (4-105) considers that $di_a(t)/dt$ is the immediate effect due to the applied voltage $e_a(t)$, then in Eq. (4-106), $i_a(t)$ causes the torque $T_m(t)$, Eq. (4-107) defines the back emf, and finally, in Eq. (4-108), the torque $T_m(t)$ causes the angular velocity $\omega_m(t)$ and displacement $\theta_m(t)$.

The state variables of the system can be defined as $i_a(t)$, $\omega_m(t)$, and $\theta_m(t)$. By direct substitution and eliminating all the nonstate variables from Eqs. (4-105) through (4-108), the state equations of the dc-motor system are written in vector-matrix form:

$$\begin{bmatrix} \dfrac{di_a(t)}{dt} \\ \dfrac{d\omega_m(t)}{dt} \\ \dfrac{d\theta_m(t)}{dt} \end{bmatrix} = \begin{bmatrix} -\dfrac{R_a}{L_a} & -\dfrac{K_b}{L_a} & 0 \\ \dfrac{K_i}{J_m} & -\dfrac{B_m}{J_m} & 0 \\ 0 & 1 & 0 \end{bmatrix} \begin{bmatrix} \dfrac{1}{L_a} \\ 0 \\ 0 \end{bmatrix} e_a(t) - \begin{bmatrix} 0 \\ \dfrac{1}{J_m} \\ 0 \end{bmatrix} T_L(t) \tag{4-109}$$

Notice that in this case, $T_L(t)$ is treated as a second input in the state equations.

The state diagram of the system is drawn as shown in Fig. 4-48, using Eq. (4-109). The transfer function between the motor displacement and the input voltage is obtained from the state diagram as

$$\frac{\Theta_m(s)}{E_a(s)} = \frac{K_i}{L_a J_m s^3 + (R_a J_m + B_m L_a)s^2 + (K_b K_i + R_a B_m)s} \tag{4-110}$$

where $T_L(t)$ has been set to zero.

▲ Dc motor is essentially an integrating device.

Figure 4-49 shows a block-diagram representation of the dc-motor system. The advantage of using the block diagram is that it gives a clear picture of the transfer-function relation between each block of the system. Since an s can be factored out from the denominator of Eq. (4-110), *the significance of the transfer function $\Theta_m(s)/E_a(s)$ is that the dc motor is essentially an integrating device between these two variables.* This is expected, since if $e_a(t)$ is a constant input, the output motor displacement will behave as the output of an integrator; that is, it will increase linearly with time.

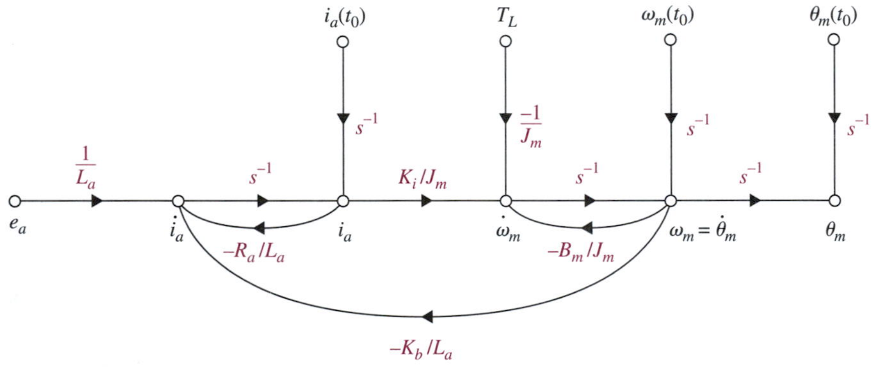

Figure 4-48 State diagram of a dc motor.

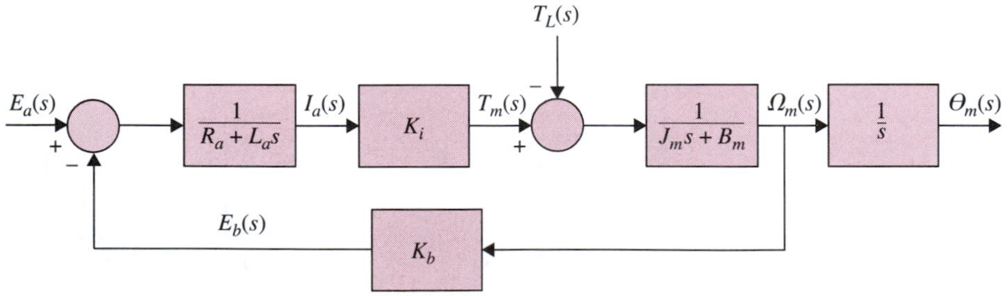

Figure 4-49 Black diagram of a dc motor system.

Although a dc motor by itself is basically an open-loop system, the state diagram of Fig. 4-48 and the block diagram of Fig. 4-49 show that the motor has a "built-in" feedback loop caused by the back emf. Physically, the back emf represents the feedback of a signal that is proportional to the negative of the speed of the motor. As seen from Eq. (4-110), the back-emf constant K_b represents an added term to the resistance R_a and the viscous-friction coefficient B_m. Therefore, *the back-emf is equivalent to an "electric friction," which tends to improve the stability of the motor, and in general, the stability of the system.*

▲ Back emf generally improves system stability.

Relation between K_i and K_b

Although functionally the torque constant K_i and back-emf constant K_b are two separate parameters, for a given motor, their values are closely related. To show the relationship, we write the mechanical power developed in the armature as

$$P = e_b(t)i_a(t) \tag{4-111}$$

The mechanical power is also expressed as

$$P = T_m(t)\omega_m(t) \tag{4-112}$$

where, in SI units, $T_m(t)$ is in N-m and $\omega_m(t)$ is in rad/sec. Now substituting Eqs. (4-106) and (4-107) in Eq. (4-111), we get

$$P = T_m(t)\omega_m(t) = K_b\omega_m(t)\frac{T_m(t)}{K_i} \tag{4-113}$$

from which we get

$$K_b \text{ (V/rad/sec)} = K_i \text{ (N-m/A)} \tag{4-114}$$

Thus we see that in the SI unit system, the values of K_b and K_i are identical if K_b is represented in V/rad/sec and K_i is in N-m/A.

In the British unit system, we convert Eq. (4-111) into horsepower (hp); that is,

$$P = \frac{e_b(t)i_a(t)}{746} \quad \text{hp} \tag{4-115}$$

In terms of torque and angular velocity, P is

$$P = \frac{T_m(t)\omega_m(t)}{550} \quad \text{hp} \tag{4-116}$$

where $T_m(t)$ is in ft-lb and $\omega_m(t)$ is in rad/sec. Using Eq. (4-106) and (4-107) and equating the last two equations, we get

$$\frac{K_b \omega_m(t) T_m(t)}{746 K_i} = \frac{T_m(t)\omega_m(t)}{550} \tag{4-117}$$

Thus

$$K_b = \frac{746}{550} K_i = 1.356 K_i \tag{4-118}$$

where K_b is in V/rad/sec and K_i is in ft-lb/A.

4-6-4 Torque–Speed Curves of a DC Motor

The torque–speed curves of a dc motor describe the static-torque-producing capability of the motor with respect to the applied voltage and motor speed. With reference to Fig. 4-50, in the steady state, the effect of the inductance is zero, and the torque equation of the motor is

$$T_m = K_i I_a = \frac{K_i(E_a - K_b \Omega_m)}{R_a} \tag{4-119}$$

where T_m, I_a, E_a, and Ω_m represent the steady-state values of the motor torque, current, applied voltage, and speed, respectively.

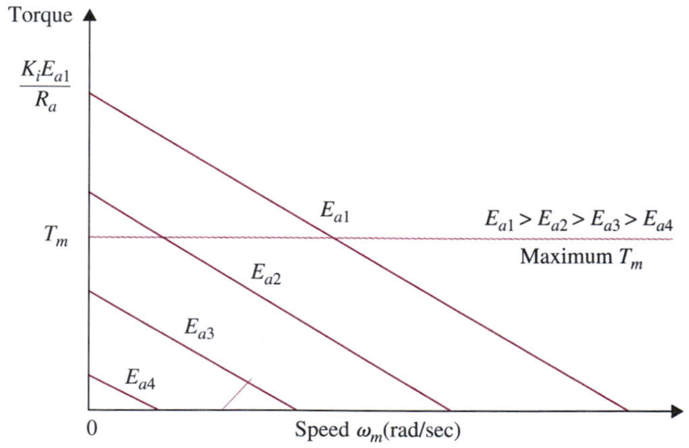

Figure 4-50 Typical torque–speed curves of a dc motor.

For a given applied voltage E_a, Eq. (4-119) describes the straight-line relation of the torque–speed characteristics of the dc motor. In reality, the motor may be subject to two types of saturation or limitations. One limitation is that as the armature current increases from the increase in E_a, the magnetic circuit will saturate, so that the motor torque cannot exceed a certain maximum value. The second limitation is due to the maximum current that the motor can handle due to the heat-dissipation rating of the motor.

Figure 4-50 shows a typical set of torque–speed curves for various applied voltages. The slope of these curves is determined from Eq. (4-119) and is expressed as

$$k = \frac{dT_m}{d\Omega_m} = -\frac{K_i K_b}{R_a} \tag{4-120}$$

The limit on the torque due to magnetic saturation is shown by the dashed line in the figure. In practice, the torque–speed curves of a dc motor can be determined experimentally by a dynamometer.

4-6-5 Torque–Speed Curves of an Amplifier/DC-Motor System

Since the dc motor is always driven by a power amplifier that acts as a source of energy, it is more practical to present the torque–speed curves of the amplifier–motor combination, especially if the amplifier gain is subject to saturation. Figure 4-51 shows the equivalent block diagram of an amplifier–motor system. The amplifier has a gain of K_1. The saturation level of the output voltage of the amplifier is E_L. Current feedback with gain K_2 is also introduced for stabilization of the system. The steady-state torque equation of the system is written from Fig. 4-51 by treating E_{in} and Ω_m as inputs, and T_m as the output. When the amplifier is not saturated, the motor torque is

$$T_m = \frac{K_i(K_1 E_{\text{in}} - K_b \Omega_m)}{R_a + K_1 K_2} \tag{4-121}$$

Without amplifier saturation, the torque–speed curves described by Eq. (4-121) are again a family of straight lines, as illustrated in Fig. 4-50. The slope of the torque–speed curves is

$$k = \frac{dT_m}{d\Omega_m} = -\frac{K_1 K_b}{R_a + K_1 K_2} \tag{4-122}$$

Comparing Eq. (4-122) with Eq. (4-120), we see that the effect of the current feedback of the amplifier is that the gain K_2 reduces the slope of the torque–speed curves of the motor.

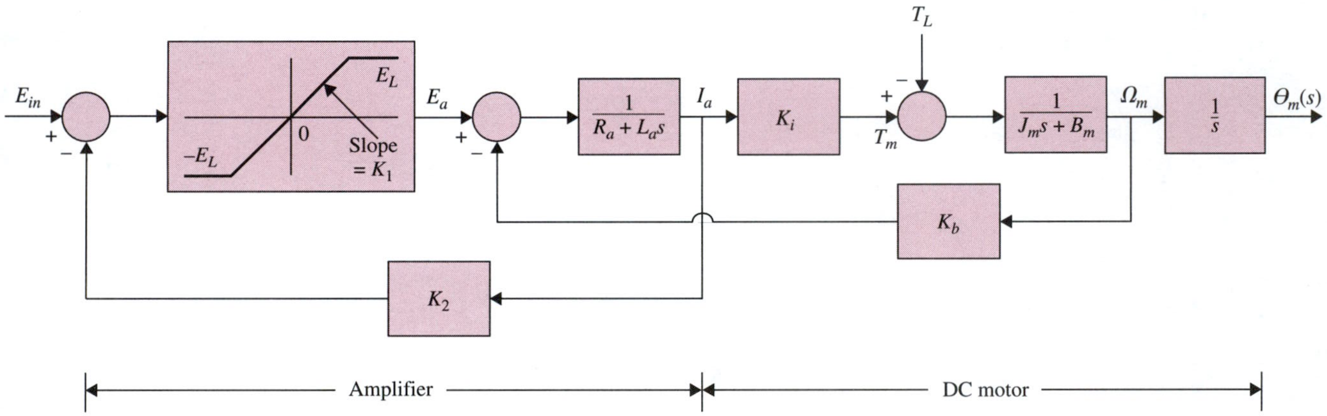

Figure 4-51 Equivalent block diagram of an amplifier/dc-motor system.

When the amplifier is subject to saturation, $|E_a| \leq E_L$. When $|E_a| = E_L$, the torque equation in the steady state is

$$T_m = \frac{K_i}{R_a}(E_L - K_b\Omega_m) \tag{4-123}$$

This gives the maximum blocked-rotor ($\Omega_m = 0$) torque as $T_m = K_iE_L/R_a$. The slope of the torque–speed curve under amplifier saturation is $-K_iK_b/R_a$. Figure 4-52 shows the torque–speed curves of the amplifier–motor combination when the amplifier is subject to saturation.

4-7 Linearization of Nonlinear Systems

From the discussions given in the preceding sections, we should realize that most components and actuators found in physical systems have nonlinear characteristics. In practice, we may find that some devices have moderate nonlinear characteristics, or the nonlinear properties would occur if they are driven into certain operating regions. For these devices, modeling by linear system models may give quite accurate analytical results over a relatively wide range of operating conditions. However, there are numerous physical devices that possess strong nonlinear characteristics. For these devices, strictly, a linearized model is valid only for a limited range of operation, and often only at the

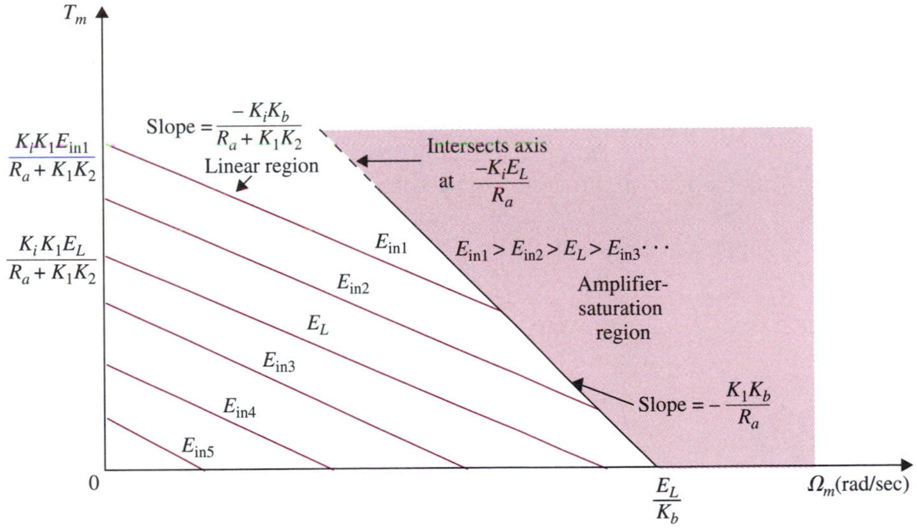

Figure 4-52 Torque–speed curve of amplifier/dc-motor system with amplifier saturation and current feedback.

operating point at which the linearization is carried out. More important, when a nonlinear system is linearized at an operating point, the linear model may contain time-varying elements.

Let us represent a nonlinear system by the following vector-matrix state equations:

$$\frac{d\mathbf{x}(t)}{dt} = \mathbf{f}[\mathbf{x}(t), \mathbf{r}(t)] \tag{4-124}$$

where $\mathbf{x}(t)$ represents the $n \times 1$ state vector, $\mathbf{r}(t)$ the $p \times 1$ input vector, and $\mathbf{f}[\mathbf{x}(t), \mathbf{r}(t)]$ denotes an $n \times 1$ function vector. In general, $\mathbf{f}$ is a function of the state vector and the input vector.

Being able to represent a nonlinear and/or time-varying system by state equations is a distinct advantage of the state-variable approach over the transfer-function method, since the latter is defined strictly only for linear time-invariant systems. As a simple illustrative example, the following nonlinear state equations are given:

$$\frac{dx_1(t)}{dt} = x_1(t) + x_2^2(t) \tag{4-125}$$

$$\frac{dx_2(t)}{dt} = x_1(t) + r(t) \tag{4-126}$$

Since nonlinear systems are usually difficult to analyze and design, it would be desirable to perform a linearization whenever the situation justifies.

A linearization process that depends on expanding the nonlinear state equations into a Taylor series about a nominal operating point or trajectory is now described. All the terms of the Taylor series or order higher than the first are discarded, and the linear approximation of the nonlinear state equations at the nominal point results.

Let the nominal operating trajectory be denoted by $\mathbf{x}_0(t)$, which corresponds to the nominal input $\mathbf{r}_0(t)$ and some fixed initial states. Expanding the nonlinear state equation of Eq. (4-124) into a Taylor series about $\mathbf{x}(t) = \mathbf{x}_0(t)$ and neglecting all the higher-order terms yields

$$\dot{x}_i(t) = f_i(\mathbf{x}_0, \mathbf{r}_0) + \sum_{j=1}^{n} \left.\frac{\partial f_i(\mathbf{x}, \mathbf{r})}{\partial x_j}\right|_{\mathbf{x}_0, \mathbf{r}_0} (x_j - x_{0j}) + \sum_{j=1}^{p} \left.\frac{\partial f_i(\mathbf{x}, \mathbf{r})}{\partial r_j}\right|_{\mathbf{x}_0, \mathbf{r}_0} (r_j - r_{0j}) \tag{4-127}$$

where $i = 1, 2, \ldots, n$. Let

$$\Delta x_i = x_i - x_{0i} \tag{4-128}$$

and

$$\Delta r_j = r_j - r_{0j} \tag{4-129}$$

Then

$$\Delta \dot{x}_i = \dot{x}_i - \dot{x}_{0i} \tag{4-130}$$

Since

$$\dot{x}_{0i} = f_i(\mathbf{x}_0, \mathbf{r}_0) \tag{4-131}$$

Equation (4-127) is written

$$\Delta \dot{x}_i = \sum_{j=1}^{n} \left. \frac{\partial f_i(\mathbf{x}, \mathbf{r})}{\partial x_j} \right|_{\mathbf{x}_0, \mathbf{r}_0} \Delta \dot{x}_j + \sum_{j=1}^{p} \left. \frac{\partial f_i(\mathbf{x}, \mathbf{r})}{\partial r_j} \right|_{\mathbf{x}_0, \mathbf{r}_0} \Delta r_j \tag{4-132}$$

Equation (4-132) may be written in vector-matrix form:

$$\Delta \dot{\mathbf{x}} = \mathbf{A}^* \Delta \mathbf{x} + \mathbf{B}^* \Delta r \tag{4-133}$$

where

$$\mathbf{A}^* = \begin{bmatrix} \dfrac{\partial f_1}{\partial x_1} & \dfrac{\partial f_1}{\partial x_2} & \cdots & \dfrac{\partial f_1}{\partial x_n} \\ \dfrac{\partial f_2}{\partial x_1} & \dfrac{\partial f_2}{\partial x_2} & \cdots & \dfrac{\partial f_2}{\partial x_n} \\ \cdot & \cdot & \cdots & \cdot \\ \dfrac{\partial f_n}{\partial x_1} & \dfrac{\partial f_n}{\partial x_2} & \cdots & \dfrac{\partial f_n}{\partial x_n} \end{bmatrix} \tag{4-134}$$

$$\mathbf{B}^* = \begin{bmatrix} \dfrac{\partial f_1}{\partial r_1} & \dfrac{\partial f_1}{\partial r_2} & \cdots & \dfrac{\partial f_1}{\partial r_p} \\ \dfrac{\partial f_2}{\partial r_1} & \dfrac{\partial f_2}{\partial r_2} & \cdots & \dfrac{\partial f_2}{\partial r_p} \\ \cdot & \cdot & \cdots & \cdot \\ \dfrac{\partial f_n}{\partial r_1} & \dfrac{\partial f_n}{\partial r_2} & \cdots & \dfrac{\partial f_n}{\partial r_n} \end{bmatrix} \tag{4-135}$$

The following examples serve to illustrate the linearization procedure just described.

Example 4-10

Figure 4-53 shows the block diagram of a control system with a saturation nonlinearity. The state equations of the system are

$$\dot{x}_1(t) = f_1(t) = x_2(t) \tag{4-136}$$
$$\dot{x}_2(t) = f_2(t) = u(t) \tag{4-137}$$

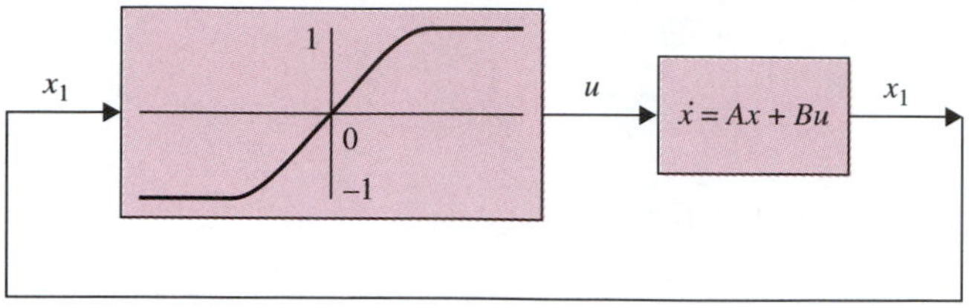

Figure 4-53 Nonlinear control system.

where the input–output relation of the saturation nonlinearity is represented by

$$u(t) = (1 - e^{-K|x_1(t)|})\ \text{SGN}\ x_1(t) \tag{4-138}$$

where

$$\text{SGN}\ x_1(t) = \begin{cases} +1 & x_1(t) > 0 \\ -1 & x_1(t) < 0 \end{cases} \tag{4-139}$$

Substituting Eq. (4-138) into Eq. (4-137) and using Eq. (4-132), we have the linearized state equations

$$\Delta\dot{x}_1(t) = \frac{\partial f_1(t)}{\partial x_2}\Delta x_2(t) = \Delta x_2(t) \tag{4-140}$$

$$\Delta\dot{x}_2(t) = \frac{\partial f_2(t)}{\partial x_1(t)}\Delta x_1(t) = Ke^{-K|x_{01}|}\,\Delta x_1 \tag{4-141}$$

where x_{01} denotes a nominal value of $x_1(t)$. Notice that the last two equations are linear and are valid only for small signals. In vector-matrix form, these linearized state equations are written as

$$\begin{bmatrix} \Delta\dot{x}_1(t) \\ \Delta\dot{x}_2(t) \end{bmatrix} = \begin{bmatrix} 0 & 1 \\ a & 0 \end{bmatrix} \begin{bmatrix} \Delta x_1(t) \\ \Delta x_2(t) \end{bmatrix} \tag{4-142}$$

where

$$a = Ke^{-K|x_{01}|} = \text{constant} \tag{4-143}$$

It is of interest to check the significance of the linearization. If x_{01} is chosen to be at the origin of the nonlinearity, $x_{01} = 0$, then $a = K$; Eq. (4-141) becomes

$$\Delta\dot{x}_2(t) = K\,\Delta x_1(t) \tag{4-144}$$

Thus the linearized model is equivalent to having a linear amplifier with a constant gain K. On the other hand, if x_{01} is a large number, the nominal operating point will lie on the saturated portion of the nonlinearity, and $a = 0$. This means that any small variation in $x_1(t)$ [small $\Delta x_1(t)$] will give rise to practically no change in $\Delta\dot{x}_2(t)$. ▲

Example 4-11

In Example 4-10 the linearized system turns out to be time-invariant. As mentioned earlier, linearization of a nonlinear system often results in a linear time-varying system. Consider the following nonlinear system:

$$\dot{x}_1(t) = \frac{-1}{x_2^2(t)} \tag{4-145}$$

$$\dot{x}_2(t) = u(t)x_1(t) \tag{4-146}$$

These equations are to be linearized about the nominal trajectory $[x_{01}(t), x_{02}(t)]$, which is the solution to the equations with the initial conditions $x_1(0) = x_2(0) = 1$ and the input $u(t) = 0$.

Integrating both sides of Eq. (4-137) with respect to t under the conditions above, we have

$$x_2(t) = x_2(0) = 1 \tag{4-147}$$

Then Eq. (4-145) gives

$$x_1(t) = -t + 1 \tag{4-148}$$

Therefore, the nominal trajectory about which Eqs. (4-145) and (4-146) are to be linearized is described by

$$x_{01}(t) = -t + 1 \tag{4-149}$$

$$x_{02}(t) = 1 \tag{4-150}$$

Now evaluating the coefficients of Eq. (4-132), we get

$$\frac{\partial f_1(t)}{\partial x_1(t)} = 0 \qquad \frac{\partial f_1(t)}{\partial x_2(t)} = \frac{2}{x_2^3(t)} \qquad \frac{\partial f_2(t)}{\partial x_1(t)} = u(t) \qquad \frac{\partial f_2(t)}{\partial u(t)} = x_1(t) \tag{4-151}$$

Equation (4-132) gives

$$\Delta\dot{x}_1(t) = \frac{2}{x_{02}^3(t)}\Delta x_2(t) \tag{4-152}$$

$$\Delta\dot{x}_2(t) = u_0(t)\,\Delta x_1(t) + x_{01}(t)\,\Delta u(t) \tag{4-153}$$

By substituting Eqs. (4-149) and (4-150) into Eqs. (4-152) and (4-153), the linearized equations are

$$\begin{bmatrix} \Delta\dot{x}_1(t) \\ \Delta\dot{x}(t) \end{bmatrix} = \begin{bmatrix} 0 & 2 \\ 0 & 0 \end{bmatrix} \begin{bmatrix} \Delta x_1(t) \\ \Delta x_2(t) \end{bmatrix} + \begin{bmatrix} 0 \\ 1-t \end{bmatrix} \Delta u(t) \tag{4-154}$$

which is a set of linear state equations with time-varying coefficients. ▲

Example 4-12

Figure 4-54 shows the diagram of a magnetic-ball-suspension system. The objective of the system is to control the position of the steel ball by adjusting the current in the electromagnet through the input voltage $e(t)$. The differential equations of the system are

$$M\frac{d^2y(t)}{dt^2} = Mg - \frac{i^2(t)}{y(t)} \tag{4-155}$$

$$e(t) = Ri(t) + L\frac{di(t)}{dt} \tag{4-156}$$

where

- $e(t)$ = input voltage
- $i(t)$ = winding current
- L = winding inductance
- g = gravitational acceleration
- $y(t)$ = ball position
- R = winding resistance
- M = mass of ball

Let us define the state variables as $x_1(t) = y(t)$, $x_2(t) = dy(t)/dt$, and $x_3(t) = i(t)$. The state equations of the system are

$$\frac{dx_1(t)}{dt} = x_2(t) \tag{4-157}$$

$$\frac{dx_2(t)}{dt} = g - \frac{1}{M}\frac{x_3^2(t)}{x_1(t)} \tag{4-158}$$

$$\frac{dx_3(t)}{dt} = -\frac{R}{L}x_3(t) + \frac{1}{L}e(t) \tag{4-159}$$

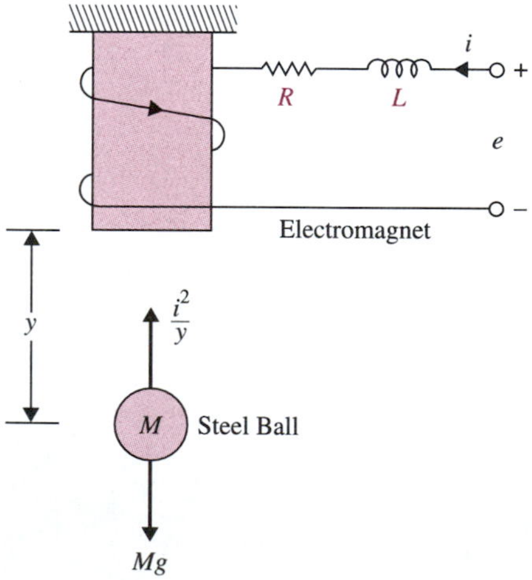

Figure 4-54 Magnetic-ball-suspension system.

Let us linearize the system about the equilibrium point $y_0(t) = x_{01} =$ constant. Then

$$x_{02}(t) = \frac{dx_{01}(t)}{dt} = 0 \tag{4-160}$$

$$\frac{d^2y_0(t)}{dt^2} = 0 \tag{4-161}$$

The nominal value of $i(t)$ is determined by substituting Eq. (4-161) into Eq. (4-155). Thus

$$i_0(t) = x_{03}(t) = \sqrt{Mgx_{01}} \tag{4-162}$$

The linearized state equation is expressed in the form of Eq. (4-133), with the coefficient matrices $\mathbf{A}^*$ and $\mathbf{B}^*$ evaluated as

$$\mathbf{A}^* = \begin{bmatrix} 0 & 1 & 0 \\ \dfrac{x_{03}^2}{Mx_{01}^2} & 0 & \dfrac{-2x_{03}}{Mx_{01}} \\ 0 & 0 & -\dfrac{R}{L} \end{bmatrix} = \begin{bmatrix} 0 & 1 & 0 \\ \dfrac{g}{x_{01}} & 0 & -2\left(\dfrac{g}{Mx_{01}}\right)^{1/2} \\ 0 & 0 & -\dfrac{R}{L} \end{bmatrix} \tag{4-163}$$

$$\mathbf{B}^* = \begin{bmatrix} 0 \\ 0 \\ \dfrac{1}{L} \end{bmatrix} \tag{4-164}$$

▲

4-8 Systems with Transportation Lags

Thus far the systems considered all have transfer functions that are quotients of polynomials. In practice, pure time delays may be encountered in various types of systems, especially systems with hydraulic, pneumatic, or mechanical transmissions. Systems with computer control also have time delays, since it takes time for the computer to execute numerical operations. In these systems the output will not begin to respond to an input until after a given time interval. Figure 4-55 illustrates systems in which transportation lags or pure time delays are observed. Figure 4-55(a) outlines an arrangement in which two different fluids are to be mixed in appropriate proportions. To assure that a homogeneous solution is measured, the monitoring point is located some distance from the mixing point. A time delay therefore exists between the mixing point and the place where the change in concentration is detected. If the rate of flow of the mixed solution is v inches per second and d is the distance between the mixing and the metering points, the time lag is given by

$$T_d = \frac{d}{v} \qquad \text{seconds} \tag{4-165}$$

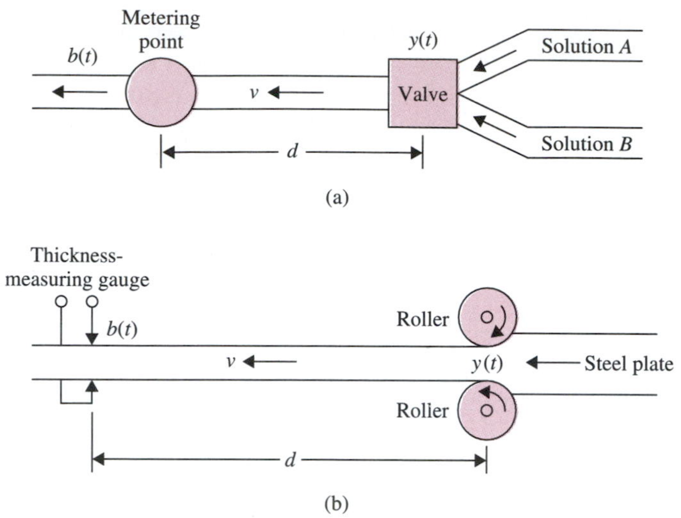

Figure 4-55 Physical systems with transportation lags.

If it is assumed that the concentration of the mixing point is $y(t)$ and that it is reproduced without change T_d seconds later at the monitoring point, the quantity measured is

$$b(t) = y(t - T_d) \tag{4-166}$$

The Laplace transform of Eq. (4-166) is

$$B(s) = e^{-T_d s}Y(s) \tag{4-167}$$

where $Y(s)$ is the Laplace transform of $y(t)$. The transfer function between $b(t)$ and $y(t)$ is

$$\frac{B(s)}{Y(s)} = e^{-T_d s} \tag{4-168}$$

Figure 4-55(b) illustrates the control of thickness of rolled steel plates. The transfer function between the thickness at the rollers and the measuring point is again given by Eq. (4-168).

4-8-1 Approximation of the Time-Delay Function by Rational Functions

Systems that are described inherently by transcendental transfer functions are more difficult to handle. Many analytical tools, such as the Routh–Hurwitz criterion (Chapter

6), are restricted to rational transfer functions. The root-locus technique (Chapter 8) is also more easily applied only to systems with rational transfer functions.

There are many ways approximating $e^{-T_d s}$ by a rational function. One way is to approximate the exponential function by a Maclaurin series; that is,

$$e^{-T_d s} \cong 1 - T_d s + \frac{T_d^2 s^2}{2} \tag{4-169}$$

or

$$e^{-T_d s} \cong \frac{1}{1 + T_d s + T_d^2 s^2/2} \tag{4-170}$$

where only three terms of the series are used. Apparently, the approximations are not valid when the magnitude of $T_d s$ is large.

A better approximation is to use the Padé approximation [6,7], which is given in the following for a two-term approximation:

$$e^{-T_d s} \cong \frac{1 - T_d s/2}{1 + T_d s/2} \tag{4-171}$$

The characteristic of the approximation in Eq. (4-171) is that the transfer function contains a zero in the right-half s-plane so that the step response of the approximating system may exhibit a small negative undershoot near $t = 0$.

4-9 Operational Amplifiers

Operational amplifiers, or simply **op-amps,** offer a convenient way to build, implement, or realize continuous-data or s-domain transfer functions. In control systems, op-amps are often used to implement the controllers or compensators that evolve from the control system design process. As a result, it is beneficial to illustrate common op-amp configurations. An in-depth presentation of op-amps is beyond the scope of this book. For those interested, many texts are available that are devoted to all aspects of op-amp circuit design and applications [8,9].

Our primary goal here is to show how to implement first-order transfer functions with op-amps while keeping in mind that higher-order transfer functions are also im-

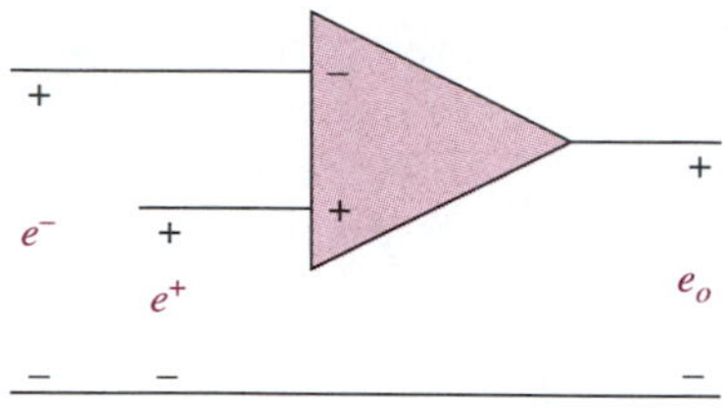

Figure 4-56 Schematic diagram of an op-amp.

portant. In fact, simple high-order transfer functions can be implemented by connecting first-order op-amp configurations. Only a representative sampling of the multitude of op-amp configurations is discussed.

4-9-1 Ideal Op-Amp

When good engineering practice is used, an op-amp circuit can be analyzed accurately by considering the op-amp to be ideal. The ideal op-amp circuit is shown in Fig. 4-56 and has the following properties:

1. The voltage between the + and − terminals is zero; that is, $e^+ = e^-$. This property is commonly called the *virtual ground* or *virtual short.*
2. The currents into the + and − input terminals are zero. Thus the input impedance is infinite.
3. The impendance seen looking into the output terminal is zero. Thus the output is an ideal voltage source.
4. The input–output relationship is $e_o = A(e^+ - e^-)$, where the gain A approaches infinity.

Application of these ideal properties allows one to determine the input–output relationship for many op-amp configurations. An op-amp cannot be used as shown in Fig. 4-56. Rather, linear operation requires the addition of feedback of the output signal to the − input terminal.

4-9-2 Sums and Differences

As illustrated in Chapter 3, one of the most fundamental elements in a block diagram or a SFG is the addition or subtraction of signals. When these signals are voltages, op-amps provide a simple way to add or subtract signals as shown in Fig. 4-57, where all the resistors have the same value. Using superposition and the ideal properties given

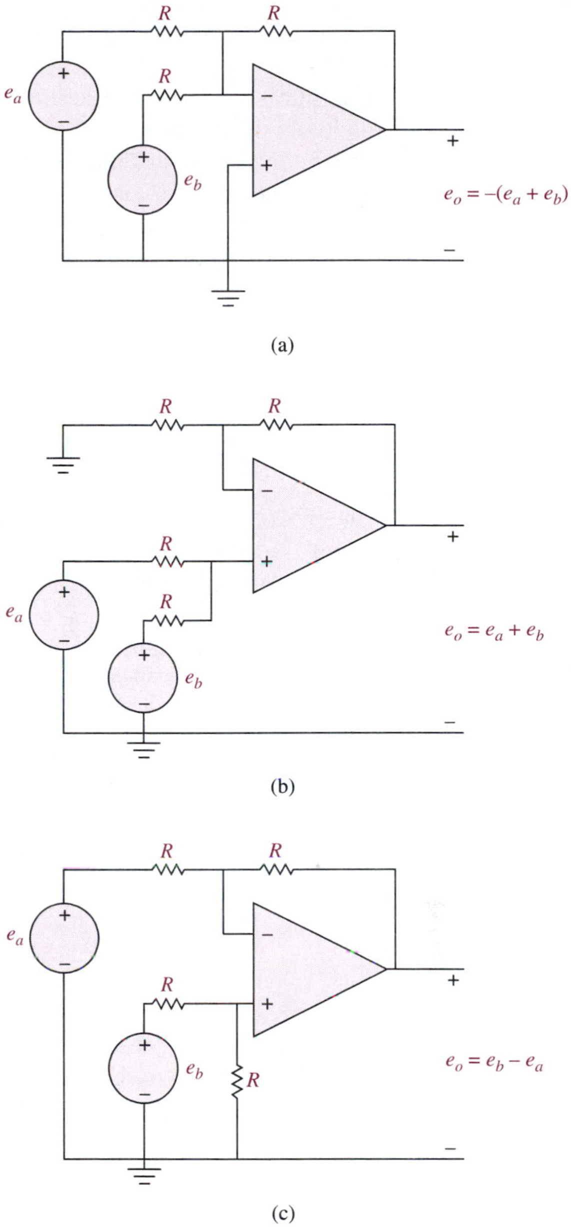

Figure 4-57 Op-amps used to add and subtract signals.

above, the input–output relationship in Fig. 4-57(a) is $v_o = -(v_a - v_b)$. Thus the output is the negative sum of the input voltages. When a positive sum is desired, the circuit shown in Fig. 4-57(b) can be used. Here the output is given by $e_o = e_a + e_b$. Modifying Fig. 4-57(b) slightly gives the differencing circuit shown in Fig. 4-57(c), which has an input–output relationship of $e_o = e_b - e_a$.

4-9-3 First-Order Op-Amp Configurations

In addition to adding and subtracting signals, op-amps can be used to implement transfer functions of continuous-data systems. While many alternatives are available, we will explore only those that use the inverting op-amp configuration shown in Fig. 4-58. In the figure, $Z_1(s)$ and $Z_2(s)$ are impedances commonly composed of resistors and capacitors. Inductors are not commonly used because they tend to be bulkier and more expensive. Using ideal op-amp properties, the input–output relationship, or transfer function, of the circuit shown in Fig. 4-58 can be written in a number of ways as

$$G(s) = \frac{E_o(s)}{E_i(s)} = -\frac{Z_2(s)}{Z_1(s)} = \frac{-1}{Z_1(s)Y_2(s)}$$
$$= -Z_2(s)Y_1(s) = -\frac{Y_1(s)}{Y_2(s)} \tag{4-172}$$

where $Y_1(s) = 1/Z_1(s)$ and $Y_2(s) = 1/Z(s)$ are the admittances associated with the circuit impedances. The different transfer function forms given in Eq. (4-172) apply conveniently to the different compositions of the circuit impedances.

Using the inverting op-amp configuration shown in Fig. 4-58 and using resistors and capacitors as elements to compose $Z_1(s)$, and $Z_2(s)$, Table 4-1 illustrates a number of common transfer-function implementations. As shown in the table, it is possible to implement poles and zeros along the negative real axis as well as at the origin in the s-plane. Because the inverting op-amp configuration was used, all the transfer functions have negative gains. The negative gain is usually not an issue, since it is simple to add a gain of -1 to the input and output signal to make the net gain positive.

Example 4-13

As an example on op-amp realization of transfer functions, consider the transfer function

$$G(s) = K_P + \frac{K_I}{s} + K_D s \tag{4-173}$$

where K_P, K_D, and K_I are real constants. In Chapter 10 this transfer function will be called the **PID controller,** since the first term is a **P**roportional gain, the second an **I**ntegral term, and the third a **D**erivative term. Using Table 4-1, the proportional gain can be implemented using line (a), the integral term can be implemented using line (b), and the derivative term can be imple-

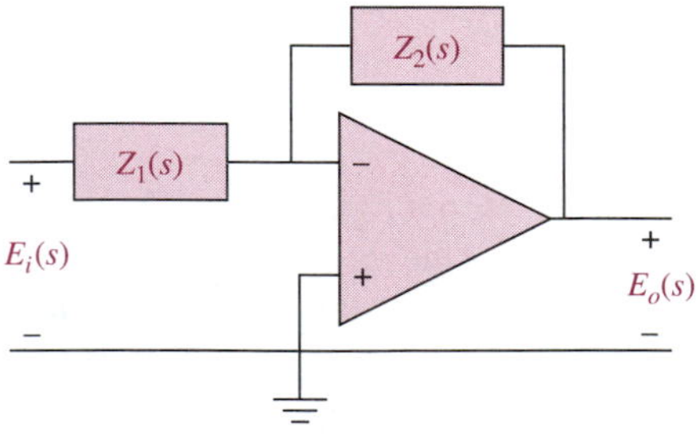

Figure 4-58 Inverting op-amp configuration.

Table 4-1 Inverting Op-Amp Transfer Functions

	Input Element	Feedback Element	Transfer Function	Comments
(a)	R_1 $Z_1 = R_1$	R_2 $Z_2 = R_2$	$-\dfrac{R_2}{R_1}$	Inverting Gain, e.g., if $R_1 = R_2$, $e_o = -e_i$
(b)	R_1 $Z_1 = R_1$	C_2 $Y_2 = sC_2$	$\left(\dfrac{-1}{R_1C_2}\right)\dfrac{1}{s}$	Pole at the Origin, i.e., an integrator.
(c)	C_1 $Y_1 = sC_1$	R_2 $Z_2 = R_2$	$(-R_2C_1)s$	Zero at the Origin, i.e., a differentiator.
(d)	R_1 $Z_1 = R_1$	R_2, C_2 $Y_2 = \dfrac{1}{R_2} + sC_2$	$\dfrac{-\dfrac{1}{R_1C_2}}{s + \dfrac{1}{R_2C_2}}$	Pole at $\dfrac{-1}{R_2C_2}$ with a DC gain of $-R_2/R_1$.
(e)	R_1 $Z_1 = R_1$	R_2 C_2 $Z_2 = R_2 + \dfrac{1}{sC_2}$	$\dfrac{-R_2}{R_1}\left(\dfrac{s + 1/R_2C_2}{s}\right)$	Pole at the origin and a Zero at $-1/R_2C_2$, i.e., a PI Controller.
(f)	R_1, C_1 $Y_1 = \dfrac{1}{R_1} + sC_1$	R_2 $Z_2 = R_2$	$-R_2C_1\left(s + \dfrac{1}{R_1C_1}\right)$	Zero at $s = \dfrac{-1}{R_1C_1}$, i.e., a PD Controller.
(g)	R_1, C_1 $Y_1 = \dfrac{1}{R_1} + sC_1$	R_2, C_2 $Y_2 = \dfrac{1}{R_2} + sC_2$	$\dfrac{\dfrac{-C_1}{C_2}\left(s + \dfrac{1}{R_1C_1}\right)}{s + \dfrac{1}{R_2C_2}}$	Pole at $s = \dfrac{-1}{R_2C_2}$ and a Zero at $s = \dfrac{-1}{R_1C_1}$, i.e., A Lead or Lag Controller

mented using line (c). By superposition, the output of $G(s)$ is the sum of the responses due to each term in $G(s)$. This sum can be implemented by adding an additional input resistance to the circuit shown in Fig. 4-57(a). By making the sum negative, the negative gains of the proportional, integral, and derivative term implementations are canceled, giving the desired result shown in

Fig. 4-59. The transfer functions of the components of the op-amp circuit in Fig. 4-59 are:

Proportional: $$\frac{E_P(s)}{E(s)} = -\frac{R_2}{R_1} \tag{4-174}$$

Integral: $$\frac{E_I(s)}{E(s)} = -\frac{1}{R_iC_is} \tag{4-175}$$

Derivative: $$\frac{E_D(s)}{E(s)} = -R_dC_ds \tag{4-176}$$

The output voltage is

$$E_o(s) = -[E_P(s) + E_I(s) + E_D(s)] \tag{4-177}$$

Thus the transfer function of the PID op-amp circuit is

$$G(s) = \frac{E_o(s)}{E(s)} = \frac{R_2}{R_1} + \frac{1}{R_iC_is} + R_dC_ds \tag{4-178}$$

By equating Eqs. (4-173) and (4-178), the design is completed by choosing the values of the resistors and the capacitors of the op-amp circuit so that the desired values of K_P, K_I, and K_D

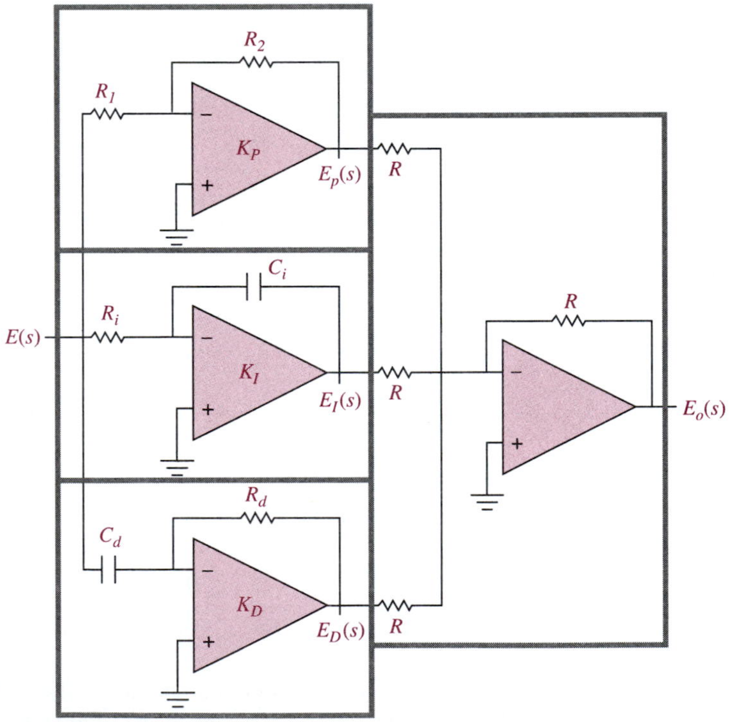

Figure 4-59 Implementation of a PID controller.

are matched. The design of the controller should be guided by the availability of standard capacitors and resistors.

It is important to note that Fig. 4-59 is just one of many possible implementations of Eq. (4-173). For example, it is possible to implement the PID controller with just three op-amps. Also, it is common to add components to limit the high-frequency gain of the differentiator and to limit the integrator output magnitude, which is often referred to as *antiwindup* protection. One advantage of the implementation shown in Fig. 4-59 is that each of the three constants, K_P, K_I, and K_D, can be adjusted or tuned individually by varying resistor values in their respective op-amp circuits. Op-amps are also used in control systems for A/D and D/A converters, sampling devices, and for realization of nonlinear elements for system compensation. ▲

4-10 Sun-Seeker System

In this section we model a sun-seeker control system whose purpose is to control the attitude of a space vehicle so that it will track the sun with high accuracy. In the system described, tracking the sun in only one plane is accomplished. A schematic diagram of the system is shown in Fig. 4-60. The principal elements of the error discriminator are two small rectangular silicon photovoltaic cells mounted behind a rectangular slit in an enclosure. The cells are mounted in such a way that when the sensor is pointed at the sun, the beam of light from the slit overlaps both cells. The silicon cells are used as

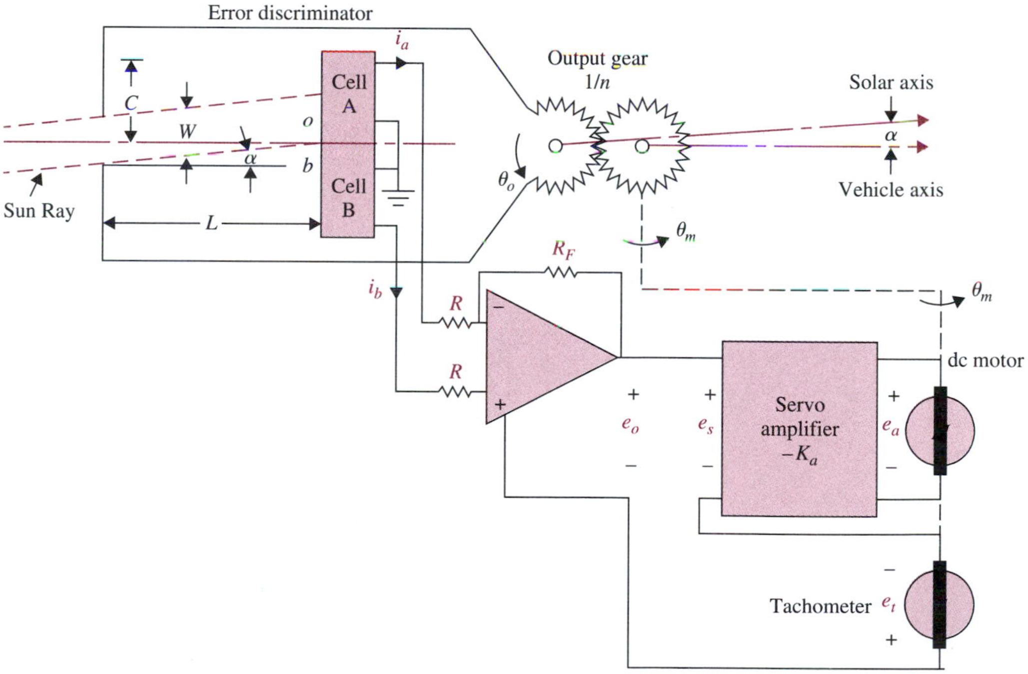

Figure 4-60 Schematic diagram of a sun-seeker system.

current sources and connected in opposite polarity to the input of an op-amp. Any difference in the short-circuit current of the two cells is sensed and amplified by the op-amp. Since the current of each cell is proportional to the illumination on the cell, an error signal will be present at the output of the amplifier when the light from the slit is not centered precisely on the cells. This error voltage, when fed to the servoamplifier, will cause the motor to drive the system back into alignment. The description of each part of the system is given as follows.

4-10-1 Coordinate System

The center of the coordinate system is considered to be at the output gear of the system. The reference axis is taken to be the fixed frame of the dc motor, and all rotations are measured with respect to this axis. The solar axis or the line from the output gear to the sun makes an angle $\theta_r(t)$ with respect to the reference axis, and $\theta_o(t)$ denotes the vehicle axis with respect to the reference axis. The objective of the control system is to maintain the error between $\theta_r(t)$ and $\theta_o(t)$, $\alpha(t)$, near zero:

$$\alpha(t) = \theta_r(t) - \theta_o(t) \tag{4-179}$$

The coordinate system described is illustrated in Fig. 4-61.

4-10-2 Error Discriminator

When the vehicle is aligned perfectly with the sun, $\alpha(t) = 0$, and $i_a(t) = i_b(t) = I$, or $i_a(t) = i_b(t) = 0$. From the geometry of the sun ray and the photovoltaic cells shown in Fig. 4-60, we have

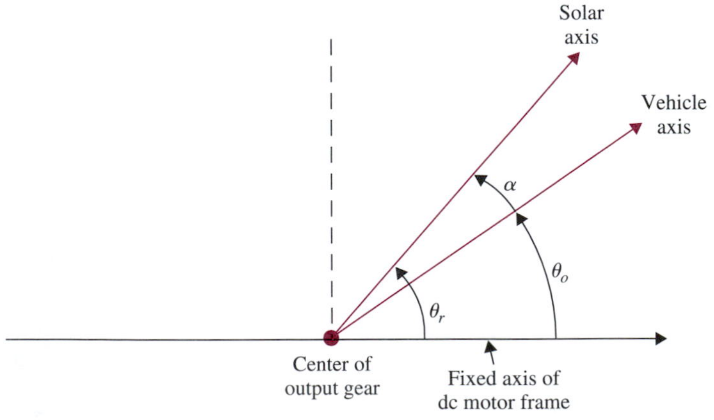

Figure 4-61 Coordinate system of the sun-seeker system.

$$oa = \frac{W}{2} + L\tan\alpha(t) \tag{4-180}$$

$$ob = \frac{W}{2} - L\tan\alpha(t) \tag{4-181}$$

where oa denotes the width of the sun ray that shines on cell A, and ob is the same on cell B, for a given $\alpha(t)$. Since the current $i_a(t)$ is proportional to oa, and $i_b(t)$ to ob, we have

$$i_a(t) = I + \frac{2LI}{W}\tan\alpha(t) \tag{4-182}$$

$$i_b(t) = I - \frac{2LI}{W}\tan\alpha(t) \tag{4-183}$$

for $0 \le \tan\alpha(t) \le W/2L$. For $W/2L \le \tan\alpha(t) \le (C - W/2)/L$, the sun ray is completely on cell A, and $i_a(t) = 2I$, $i_b(t) = 0$. For $(C - W/2)L \le \tan\alpha(t) \le (C + W/2)/L$, $i_a(t)$ decreases linearly from $2I$ to zero. $i_a(t) = i_b(t) = 0$ for $\tan\alpha(t) \ge (C + W/2)/L$. Therefore, the error discriminator may be represented by the nonlinear characteristic of Fig. 4-62, where for small angle $\alpha(t)$, $\tan\alpha(t)$ has been approximated by $\alpha(t)$ on the abscissa.

4-10-3 Op-Amp

The relationship between the output of the op-amp and the currents $i_a(t)$ and $i_b(t)$ is

$$e_o(t) = -R_F[i_a(t) - i_b(t)] \tag{4-184}$$

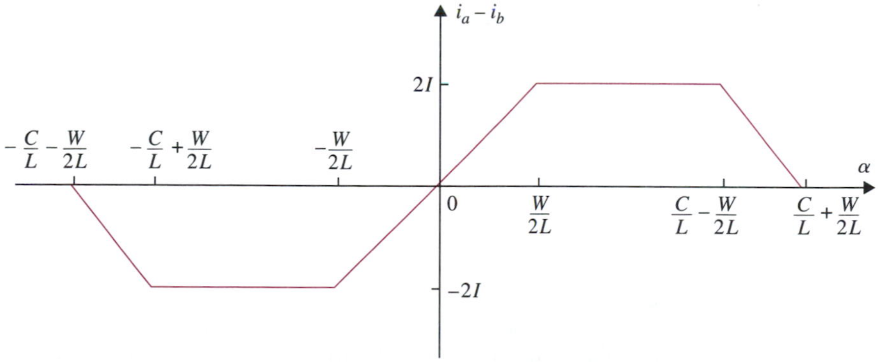

Figure 4-62 Nonlinear characteristic of the error discriminator. The abscissa is $\tan\alpha$ but is approximated by α for small values of α.

4-10-4 Servoamplifier

The gain of the servoamplifier is $-K$. With reference to Fig. 4-60, the output of the servoamplifier is expressed as

$$e_a(t) = -K[e_o(t) + e_t(t)] = -Ke_s(t) \tag{4-185}$$

4-10-5 Tachometer

The output voltage of the tachometer, e_t, is related to the angular velocity of the motor through the tachometer constant K_t:

$$e_t(t) = K_t\omega_m(t) \tag{4-186}$$

The angular position of the output gear is related to the motor position through the gear ratio, $1/n$. Thus

$$\theta_o = \frac{1}{n}\theta_m \tag{4-187}$$

4-10-6 DC Motor

The dc motor has been modeled in Section 4-6. The equations are

$$\begin{aligned} e_a(t) &= R_a i_a(t) + e_b(t) \\ e_b(t) &= K_b\omega_m(t) \\ T_m(t) &= K_i i_a(t) \\ T_m(t) &= J\frac{d\omega_m(t)}{dt} + B\omega_m(t) \end{aligned} \tag{4-188}$$

where J and B are the inertia and viscous friction coefficients seen at the motor shaft. The inductance of the motor is neglected in Eq. (4-188). A block diagram that characterizes all the functional relations of the system is shown in Fig. 4-63.

4-11 Summary

This chapter is devoted to the mathematical modeling of physical systems. The basic mathematical relations of linear electrical and mechanical systems are described. For linear systems, differential equations, state equations, and transfer functions are the fundamental tools of modeling. The operations and mathematical descriptions of some of the commonly used components in control systems, such as error detectors, tachometers, and dc motors, are presented in this chapter.

Due to space limitations and the intended scope of this book, only some of the physical devices used in practice are described. The main purpose of this chapter is to illustrate the methods of system modeling, and the coverage is not intended to be exhaustive.

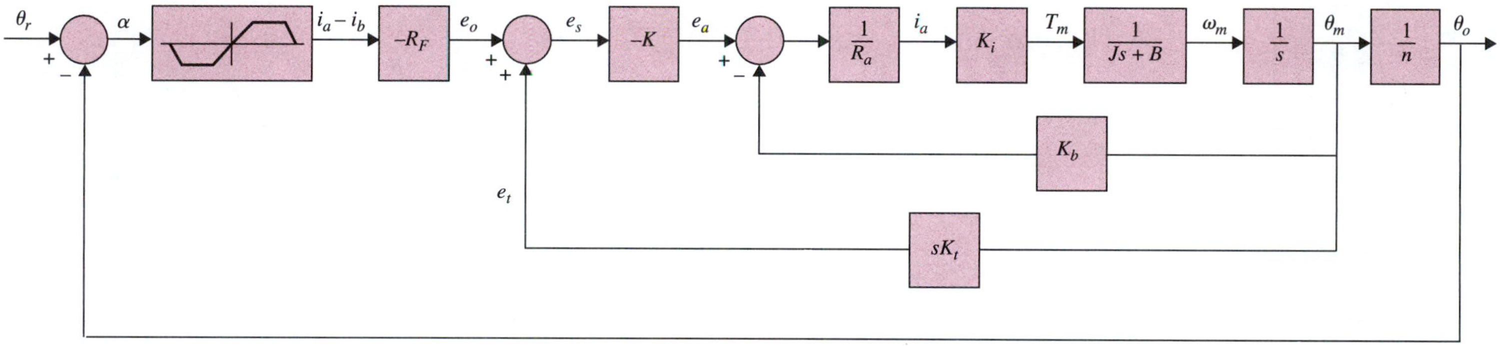

Figure 4-63 Block diagram of the sun-seeker system.

Since nonlinear systems cannot be ignored in the real world, and this book is not devoted to the subject, Section 4-7 introduced the linearization of nonlinear systems at a nominal operating point. Once the linearized model is determined, the performance of the nonlinear system can be investigated under the small-signal conditions at the designated operating point.

Systems with pure time delays are modeled, and methods of approximating the transfer functions by rational ones are described. Finally, operational amplifiers and their applications to the implementation of transfer functions are described.

Review Questions

1. Describe how state variables should be defined for a physical system.
2. The state variables of a dynamic system are not equal to the number of energy-storage elements under what condition?
3. Among the three types of friction described, which type is governed by a linear mathematical relation?
4. Given a two-gear system with angular displacement θ_1 and θ_2, numbers of teeth N_1 and N_2, and torques T_1 and T_2, write the mathematical relations between these variables and parameters.
5. How are potentiometers used in control systems?
6. Digital encoders are used in control systems for position and speed detection. Consider that an encoder is set up to output 3600 zero crossings per revolution. What is the angular rotation of the encoder shaft in degrees if 16 zero crossings are detected?
7. The encoder described in Review Question 6 and an electronic clock with a frequency of 1 MHz are used for speed measurement. What is the average speed of the encoder shaft in rpm if 500 clock pulses are detected between two consecutive zero crossings of the encoder?
8. State the advantages of dc motors for control systems applications.
9. What are the sources of nonlinearities in a dc motor?
10. What are the effects of inductance and inertia in a dc motor?
11. What is back emf in a dc motor, and how does it affect the performance of a control system?
12. What are the electrical and mechanical time constants of an electric motor?
13. Under what condition is the torque constant K_i of a dc motor valid, and how is it related to the back-emf constant K_b?
14. For an amplifier/dc-motor system if the amplifier has saturation characteristics, what performance parameters does the saturation limit? (Refer to Fig. 4-52.)
15. An inertial and a frictional load are driven by a dc motor with torque T_m. The dynamic equation of the system is

$$T_m(t) = J_m \frac{d\omega_m(t)}{dt} + B_m\omega_m$$

If the inertia is doubled, how will it affect the steady-state speed of the motor? How will the steady-state speed be affected if, instead, the fricitonal coefficient B_m is doubled? What is the mechanical constant of the system?

16. What is a tachometer, and how is it used in control systems?
17. Does the linearization technique described in this chapter always result in a linear time-invariant system?
18. Give the transfer function of a pure time delay T_d.

References

State-Variable Analysis of Electric Networks

1. **B. C. KUO,** *Linear Circuits and Systems,* McGraw-Hill Book Company, New York, 1967.

Mechanical Systems

2. **R. CANNON,** *Dynamics of Physical Systems,* McGraw-Hill Book Company, New York, 1967.

DC Motors

3. **B. C. KUO and J. TAL, eds.,** *Incremental Motion Control,* Vol. 1, *DC Motors and Control Systems,* SRL Publishing Co., Champaign, IL, 1979.

Rolling Dry Friction

4. **P. B. DAHL,** *A Solid Friction Model,* Report TOR-0158 (3107-18)-1, Aerospace Corporation, El Segundo, CA, May 1968.
5. **N. A. OSBORN and D. L. RITTENHOUSE,** ''The Modeling of Friction and Its Effects on Fine Pointing Control,'' *AIAA Mechanics and Control of Flight Conference,* Paper 74-875, Aug. 1974.

Padé Approximation

6. **J. G. TRUXAL,** *Automatic Feedback Control System Synthesis,* McGraw-Hill Book Company, New York, 1955.
7. **H. S. WALL,** *Continued Fractions,* Chapter 20, D. Van Nostrand Co., New York, 1948.

Operational Amplifiers

8. **E. J. KENNEDY,** *Operational Amplifier Circuits,* Holt, Rinehart and Winston, Fort Worth, TX, 1988.
9. **J. V. WAIT, L. P. HUELSMAN, and G. A. KORN,** *Introduction to Operational Amplifier Theory and Applications,* 2nd ed., McGraw-Hill Book Company, New York, 1992.

Problems

▲ State diagrams, state equations, transfer functions

4-1. Write the force equations of the linear translational systems shown in Fig. 4P-1.

(a) Draw state diagrams using a minimum number of integrators. Write the state equations from the state diagrams.

(b) Define the state variables as follows:

(i) $x_1 = y_2$, $x_2 = dy_2/dt$, $x_3 = y_1$, and $x_4 = dy_1/dt$

(ii) $x_1 = y_2$, $x_2 = y_1$, and $x_3 = dy_1/dt$

(iii) $x_1 = y_1$, $x_2 = y_2$, and $x_3 = dy_2/dt$

Write the state equations and draw the state diagram with these state variables. Find the transfer functions $Y_1(s)/F(s)$ and $Y_2(s)/F(s)$.

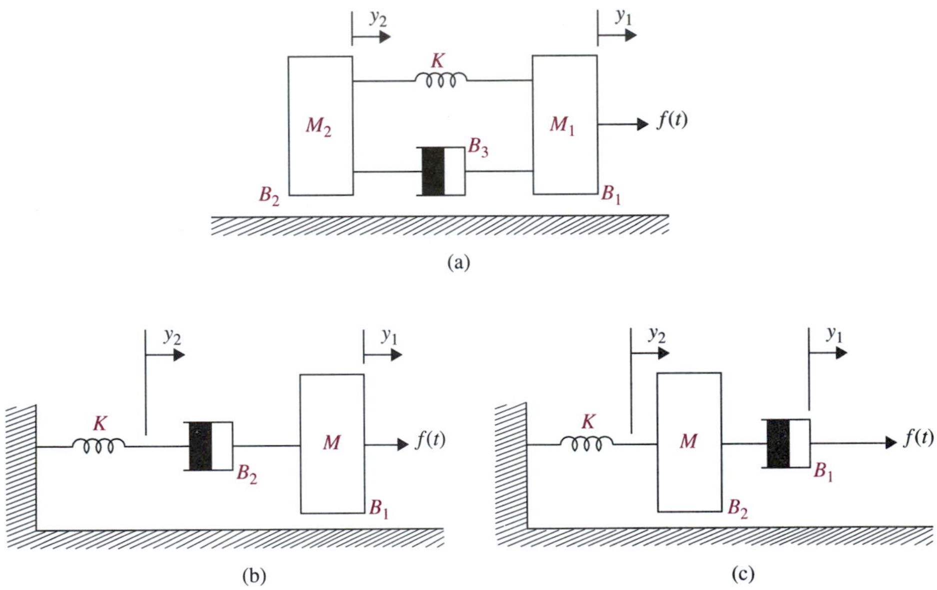

Figure 4P-1

▲ Force equations, state diagram, state equations, transfer functions

4-2. Write the force equations of the linear translational system shown in Fig. 4P-2. Draw the state diagram using a minimum number of integrators. Write the state equations from the state diagram. Find the transfer functions $Y_1(s)/F(s)$ and $Y_2(s)/F(s)$. Set $Mg = 0$ for the transfer functions.

▲ Rotational system, torque equations, state equations, transfer functions

4-3. Write the torque equations of the rotational systems shown in Fig. 4P-3. Draw state diagrams using a minimum number of integrators. Write the state equations from the state diagrams. Find the transfer function $\Theta(s)/T(s)$ for the system in Fig. 4P-3(a). Find

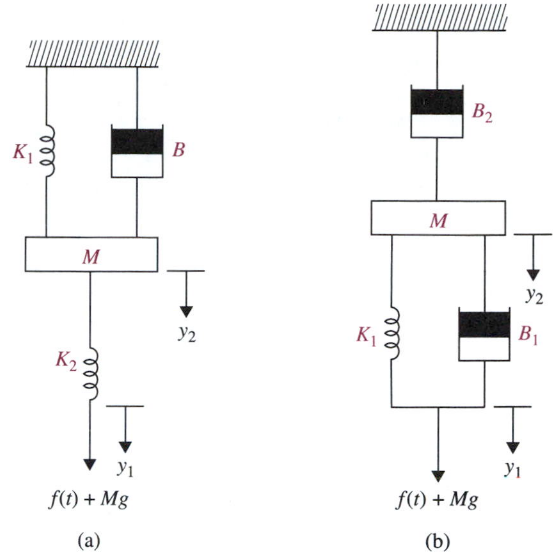

Figure 4P-2

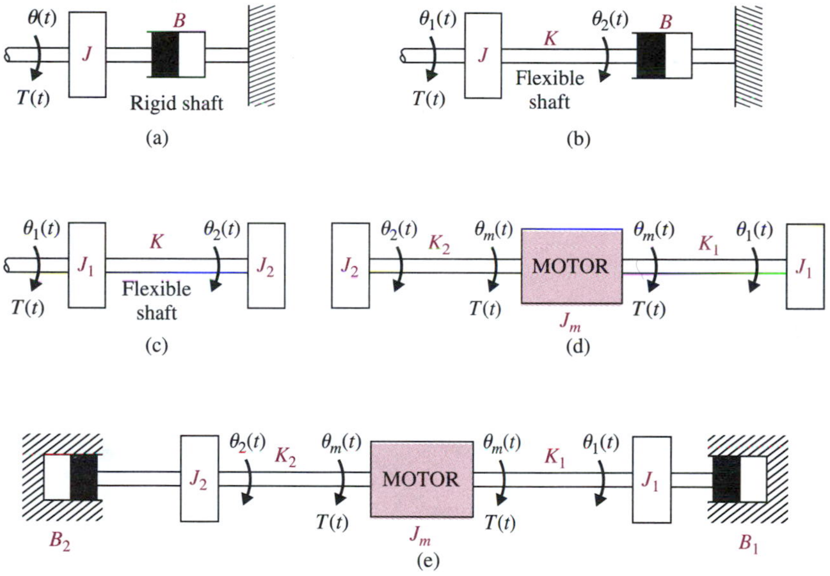

Figure 4P-3

the transfer functions $\Theta_1(s)/T(s)$ and $\Theta_2(s)/T(s)$ for the systems in parts (b), (c), (d), and (e).

▲ Motor control system

4-4. An open-loop motor control system is shown in Fig. 4P-4. The potentiometer has a maximum range of 10 turns (20π rad). Find the transfer functions $E_o(s)/T_m(s)$. The

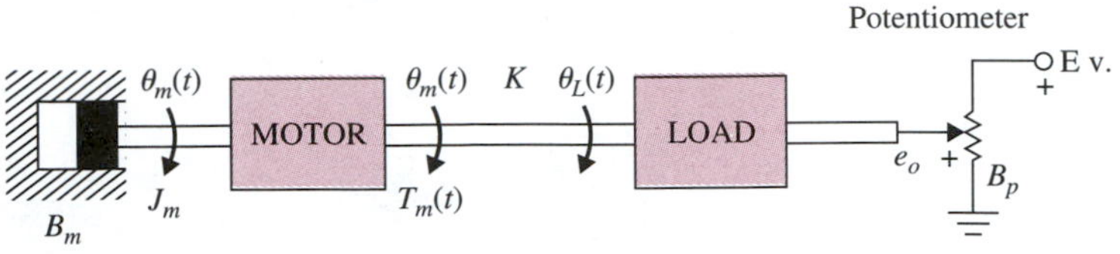

Figure 4P-4

following parameters and variables are defined: $\theta_m(t)$ is the motor displacement, $\theta_L(t)$ the load displacement, $T_m(t)$ the motor torque, J_m the motor inertia, B_m the motor viscous-friction coefficient, B_p the potentiometer viscous-friction coefficient, $e_o(t)$ the output voltage, and K the torsional spring constant.

▲ Gear-train system

4-5. Write the torque equations of the gear-train system shown in Fig. 4P-5. The moments of inertia of gears are lumped as J_1, J_2, and J_3. $T_m(t)$ is the applied torque; N_1, N_2, N_3, and N_4 are the number of gear teeth. Assume rigid shafts.

(a) Assume that J_1, J_2, and J_3 are negligible. Write the torque equations of the system. Find the total inertia the motor sees.

(b) Repeat part (a) with moments of inertia J_1, J_2, and J_3.

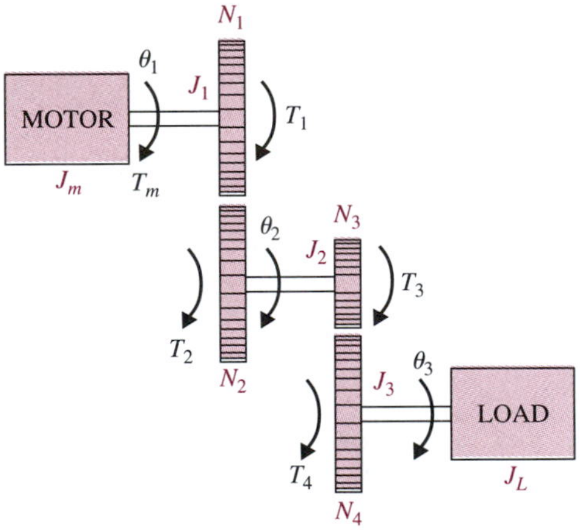

Figure 4P-5

▲ Trailer towing system

4-6. A vehicle towing a trailer through a spring–damper coupling hitch is shown in Fig. 4P-6. The following parameters and variables are defined: M is the mass of the trailer, K_h the spring constant of the hitch, B_h the viscous-damping coefficient of the hitch, B_t the viscous-friction coefficient of the trailer, $y_1(t)$ the displacement of the towing vehicle, $y_2(t)$ the displacement of the trailer, and $f(t)$ the force of the towing vehicle.

(a) Write the differential equation of the system.

(b) Write the state equations by defining the following state variables: $x_1(t) = y_1(t) - y_2(t)$ and $x_2(t) = dy_2(t)\ dt$.

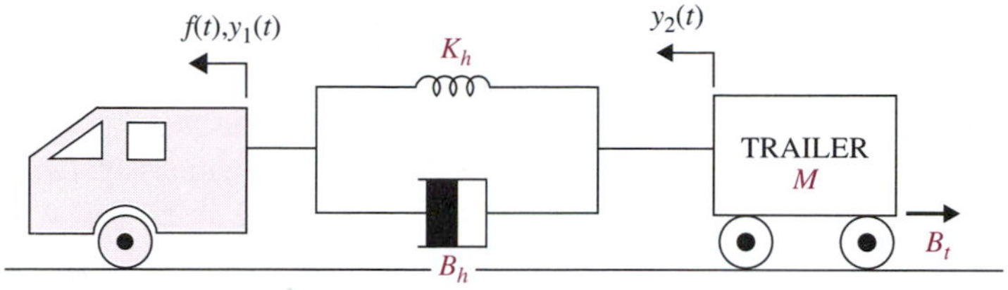

Figure 4P-6

▲ Motor-load system

4-7. Figure 4P-7 shows a motor-load system coupled through a gear train with gear ratio $n = N_1/N_2$. The motor torque is $T_m(t)$, and $T_L(t)$ represents a load torque.

(a) Find the optimum gear ratio n^* such that the load acceleration $\alpha_L = d^2\theta_L/dt^2$ is maximized.

(b) Repeat part (a) when the load torque is zero.

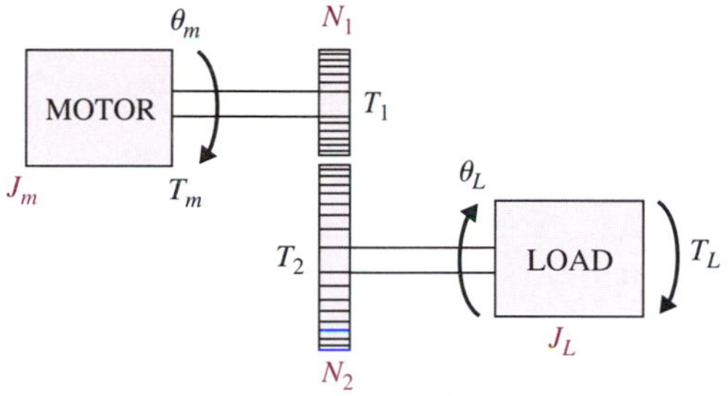

Figure 4P-7

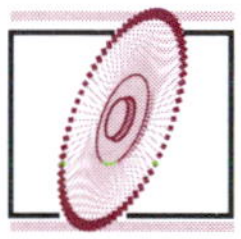

▲ Printwheel control system

4-8. Figure 4P-8 shows the simplified diagram of the printwheel control system of a word processor. The printwheel is controlled by a dc motor through belts and pulleys. Assume that the belts are rigid. The following parameters and variables are defined: $T_m(t)$ is the motor torque, $\theta_m(t)$ the motor displacement, $y(t)$ the linear displacement of the printwheel, J_m the motor inertia, B_m the motor viscous-friction coefficient, r the pulley radius, and M the mass of the printwheel.

(a) Write the differential equation of the system.

(b) Find the transfer function $Y(s)/T_m(s)$.

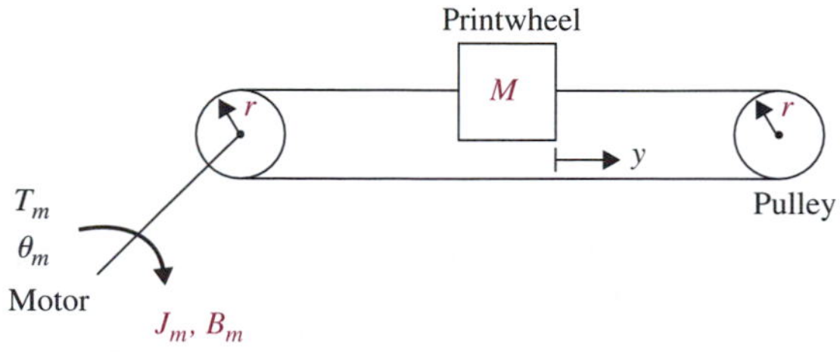

Figure 4P-8

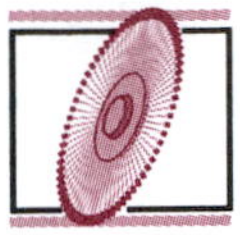

▲ Printwheel control system

4-9. Figure 4P-9 shows the diagram of a printwheel system with belts and pulleys. The belts are modeled as linear springs with spring constants K_1 and K_2.

(a) Write the differential equations of the system using θ_m and y as the dependent variables.

(b) Write the state equations using $x_1 = r\theta_m - y$, $x_2 = dy/dt$, and $x_3 = \omega_m = d\theta_m/dt$ as the state variables.

(c) Draw a state diagram for the system.

(d) Find the transfer function $Y(s)/T_m(s)$.

(e) Find the characteristic equation of the system.

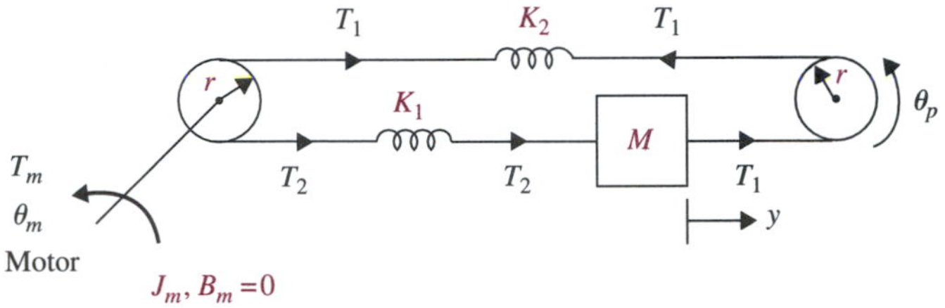

Figure 4P-9

▲ Motor-load system

4-10. The schematic diagram of a motor-load system is shown in Fig. 4P-10. The following parameters and variables are defined: $T_m(t)$ is the motor torque, $\omega_m(t)$ the motor velocity, $\theta_m(t)$ the motor displacement, $\omega_L(t)$ the load velocity, $\theta_L(t)$ the load displacement, K the torsional spring constant, J_m the motor inertia, B_m the motor viscous-friction coefficient, and B_L the load viscous-friction coefficient.

(a) Write the torque equations of the system.

(b) Find the transfer functions $\Theta_L(s)/T_m(s)$ and $\Theta_m(s)/T_m(s)$.

(c) Find the characteristic equation of the system.

(d) Let $T_m(t) = T_m$ be a constant applied torque; show that $\omega_m = \omega_L =$ constant in the steady state. Find the steady-state speeds ω_m and ω_L.

(e) Repeat part (d) when the value of J_L is doubled but J_m stays the same.

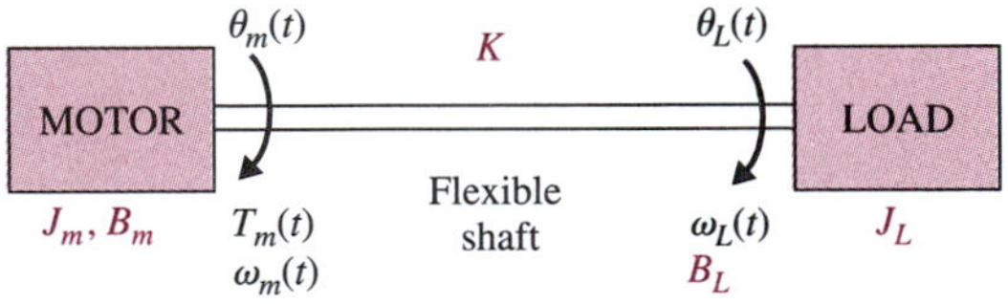

Figure 4P-10

▲ Motor-tachometer system

4-11. The schematic diagram of a control system containing a motor coupled to a tachometer and an inertial load is shown in Fig. 4P-11. The following parameters and variables are defined: T_m is the motor torque, J_m the motor inertia, J_t the tachometer inertia, J_L the

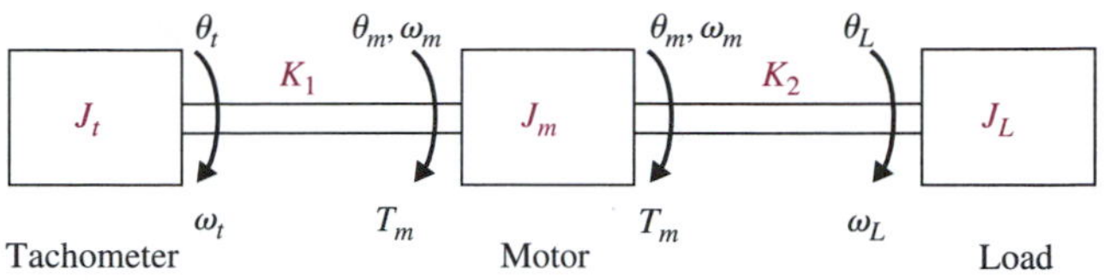

Figure 4P-11

load inertia, K_1 and K_2 the spring constants of the shafts, θ_t the tachometer displacement, θ_m the motor velocity, θ_L the load displacement, ω_t the tachometer velocity, ω_L the load velocity, and B_m the motor viscous-friction coefficient.

(a) Write the state equations of the system using θ_L, ω_L, θ_t, ω_t, θ_m, and ω_m as the state variables (in the listed order). The motor torque T_m is the input.

(b) Draw a state diagram with T_m at the left and ending with θ_L on the far right. The state diagram should have a total of 10 nodes. Leave out the initial states.

(c) Find the following transfer functions:

$$\frac{\Theta_L(s)}{T_m(s)} \qquad \frac{\Theta_t(s)}{T_m(s)} \qquad \frac{\Theta_m(s)}{T_m(s)}$$

(d) Find the characteristic equation of the system.

▲ Dc-motor system

4-12. The voltage equation of a dc motor is written as

$$e_a(t) = R_a i_a(t) + L_a \frac{di_a(t)}{dt} + K_b \omega_m(t)$$

where $e_a(t)$ is the applied voltage, $i_a(t)$ the armature current, R_a the armature resistance, L_a the armature inductance, K_b the back-emf constant, $\omega_m(t)$ the motor velocity, and $\omega_r(t)$ the reference input voltage. Taking the Laplace transform on both sides of the

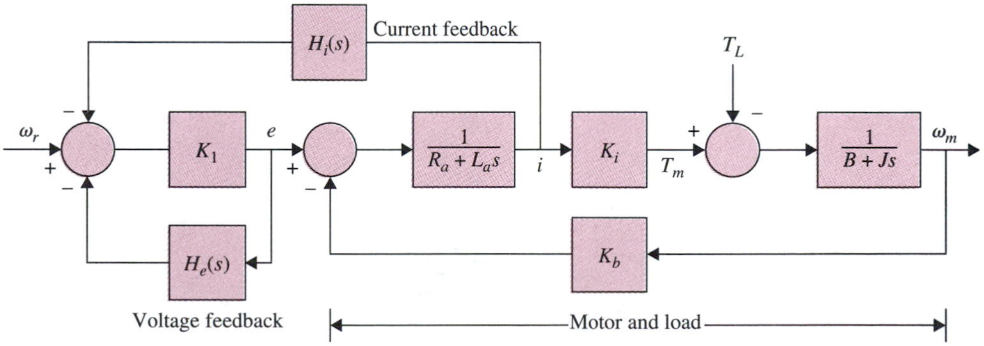

Figure 4P-12

voltage equation with zero initial conditions and solving for $\Omega_m(s)$, we get

$$\Omega_m(s) = \frac{E_a(s) - (R_a + L_a s)I_a(s)}{K_b}$$

which shows that the velocity information can be generated by feeding back the armature voltage and current. The block diagram in Fig. 4P-12 shows a dc-motor system, with voltage and current feedbacks, for speed control.

(a) Let K_I be a very high gain amplifier. Show that when $H_i(s)/H_e(s) = -(R_a + L_a s)$, the motor velocity $\omega_m(t)$ is totally independent of the load-disturbance torque T_L.

(b) Find the transfer function between $\Omega_m(s)$ and $\Omega_r(s)$ $(T_L = 0)$ when $H_i(s)$ and $H_e(s)$ are selected as in part (a).

▲ Attitude control of guided missile

4-13. This problem deals with the attitude control of a guided missile. When traveling through the atmosphere, a missile encounters aerodynamic forces that tend to cause instability in the attitude of the missile. The basic concern from the flight-control standpoint is the lateral force of the air, which tends to rotate the missile about its center of gravity. If the missile centerline is not aligned with the direction in which the center of gravity C is traveling, as shown in Fig. 4P-13, with angle θ, which is also called the angle of attack, a side force is produced by the drag of the air through which the missile travels. The total force F_a may be considered to be applied at the center of pressure P. As shown in Fig. 4P-13, this side force has a tendency to cause the missile to tumble end over end, especially if the point P is in front of the center of gravity C. Let the angular acceleration of the missile about the point C, due to the side force, be denoted by α_F. Normally, α_F is directly proportional to the angle of attack θ and is given by

$$\alpha_F = \frac{K_F d_1}{J}\theta$$

where K_F is a constant that depends on such parameters as dynamic pressure, velocity of the missile, air density, and so on, and

J = missile moment of inertia about C
d_1 = distance between C and P

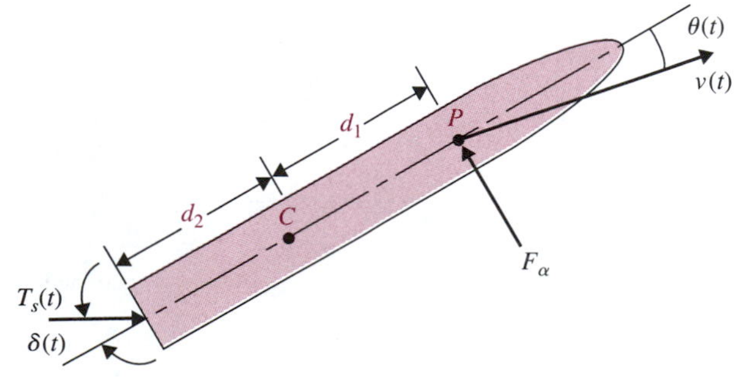

Figure 4P-13

The main objective of the flight-control system is to provide the stabilizing action to counter the effect of the side force. One of the standard control means is to use gas injection at the tail of the missile to deflect the direction of the rocket engine thrust T_s, as shown in the figure.

(a) Write a torque differential equation to relate among T_s, δ, θ, and the system parameters given. Assume that δ is very small, so that sin $\delta(t)$ is approximated by $\delta(t)$.

(b) Assume that T_s is a constant torque. Find the transfer function $\Theta(s)/\Delta(s)$, where $\Theta(s)$ and $\Delta(s)$ are the Laplace transforms of $\theta(t)$ and $\delta(t)$, respectively. Assume that $\delta(t)$ is very small.

(c) Repeat parts (a) and (b) with points C and P interchanged. The d_1 in the expression of α_F should be changed to d_2.

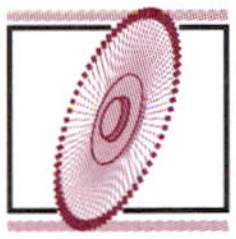

▲ Printwheel control system

4-14. Figure 4P-14(a) shows the schematic diagram of a dc-motor control system for control of the printwheel described in Fig. 1-3. The load in this case is the printwheel, which is coupled directly to the motor shaft. The following parameters and variables are defined: K_s is the error-detector gain (V/rad), K_i the torque constant (oz-in./A), K the amplifier gain (V/V), K_b the back-emf constant (V/rad/sec), n the gear train ratio $= \theta_2/\theta_m = T_m/T_2$, B_m the motor viscous-friction coefficient (oz-in.-sec), J_m the motor inertia (oz-in.-sec^2), K_L the torsional spring constant of the motor shaft (oz-in./rad), and J_L the load inertia (oz-in.-sec^2).

(a) Write the cause-and-effect equations of the system. Rearrange these equations into the form of state equations with $x = \theta_o$, $x = \omega_o$, $x = \theta_m$, $x = \omega_m$, and $x = i_a$.

(b) Draw a state diagram using the nodes shown in Fig. 4P-14(b).

(c) Derive the forward-path transfer function (with the outer feedback path open): $G(s) = \Theta_o(s)/\Theta_e(s)$. Find the closed-loop transfer function $M(s) = \Theta_o(s)/\Theta_r(s)$.

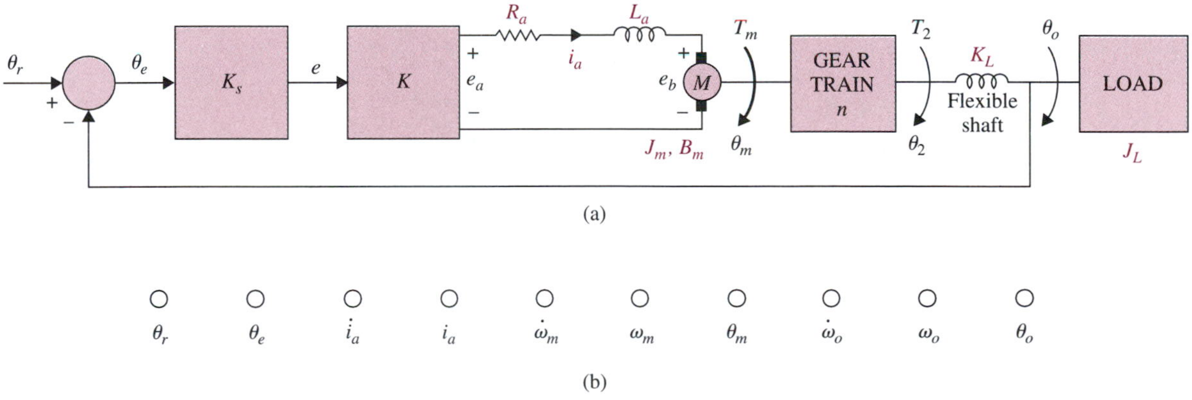

Figure 4P-14

(d) Repeat part (c) when the motor shaft is rigid (i.e., $K_L = \infty$). Show that you can obtain the solutions by taking the limit of K_L approaches infinity in the results in part (c).

▲ Amplifier-dc-motor control system

4-15. Figure 4P-15 shows the schematic diagram of an amplifier/dc-motor system with current feedback. The system is used for the control of the control surfaces of modern aircrafts. The amplifier has the voltage-saturation characteristic shown. The torque–speed curves of the system are determined experimentally and are shown in the figure. The output voltage of the amplifier saturates at 100 V ($E_L = 100$ V).

(a) Determine the torque–speed curve equation (without saturation) from the block diagram, under steady state, by use of the SFG gain formula, with T_m as output and E_{in} and Ω_m as inputs. In Fig. 4P-15, Ω_m is shown to be an input to K_b, not as a feedback from the motor output. This reflects how the torque–speed curve

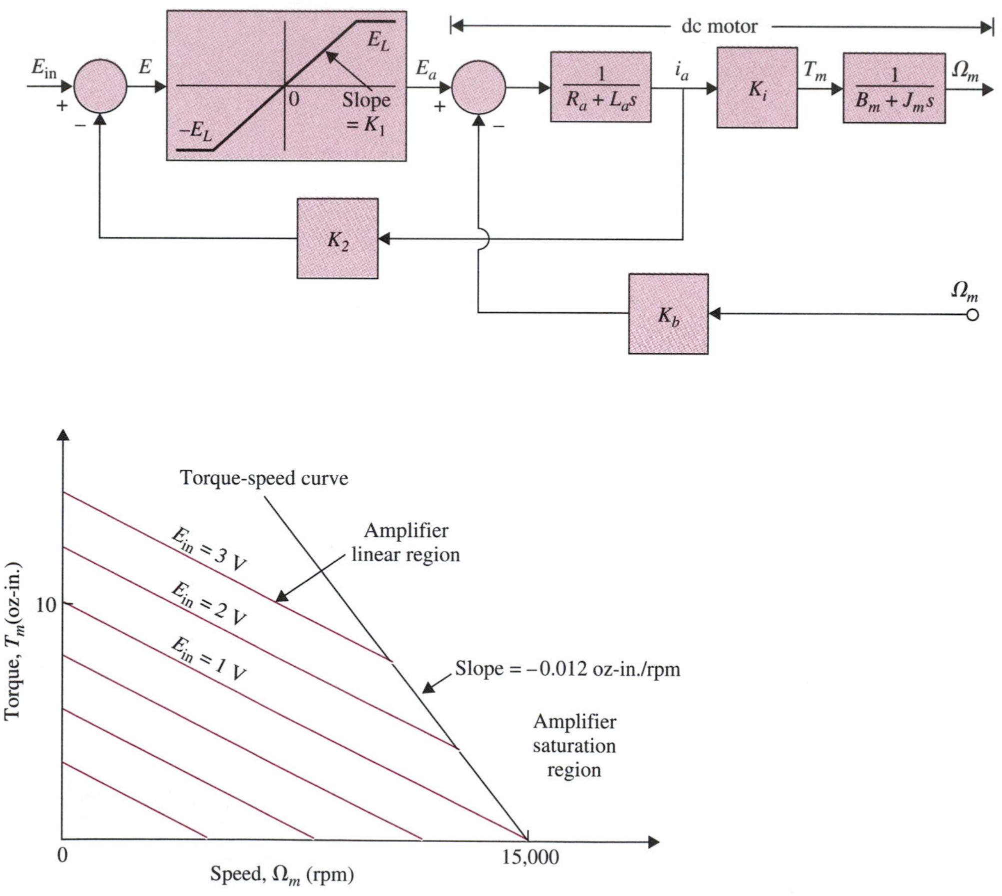

Figure 4P-15

experimentation is conducted. Determine the slope of the torque–speed curves when the amplifier output is saturated at $E_L = 100$ V. What is the effect of the current feedback gain K_2 on the torque–speed curve? Note that at steady state, all the terms that are multiplied by s in the block diagram are zero. When the amplifier is saturated, its output voltage is E_L volts, the input voltage E_{in} no longer matters, and the current-feedback gain K_2 is also ineffective. The effective input to the system is E_L.

(b) From the torque–speed curve data given, find the following.

(i) For the motor:

Torque constant K_i (oz-in./A)
Back-emf constant K_b (V/rpm)
Armature resistance R_a (Ω)

(ii) For the amplifier:

Forward amplifier gain K_1 (V/V)
Current-feedback gain K_2 (V/V)

▲ Voice-coil motor

4-16. The schematic diagram of a voice-coil motor (VCM), used as a linear actuator in a disk memory-storage system, is shown in Fig. 4P-16(a). The VCM consists of a cylindrical permanent magnet (PM) and a voice coil. When current is sent through the coil, the magnetic field of the PM interacts with the current-carrying conductor, causing the coil to move linearly. The voice coil of the VCM in Fig. 4P-16(a) consists of a primary coil and a shorted-turn coil. The later is installed for the purpose of effectively reducing the

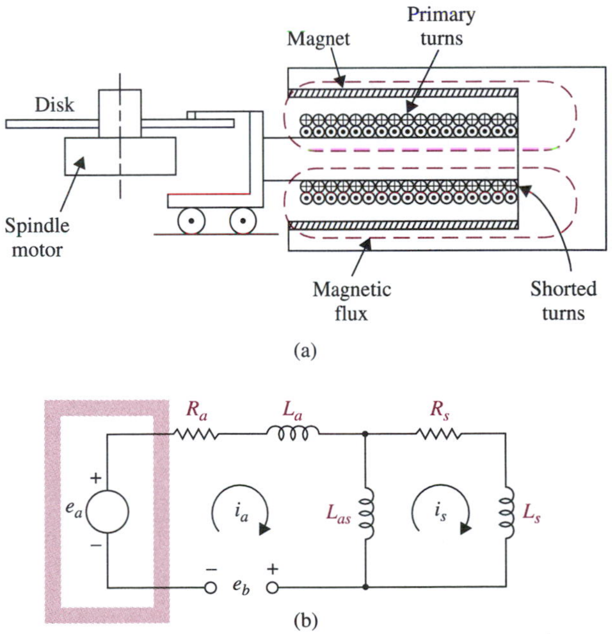

Figure 4P-16

electric constant of the device. Figure 4P-16(b) shows the equivalent circuit of the coils. The following parameters and variables are defined: $e_a(t)$ is the applied coil voltage, $i_a(t)$ the primary-coil current, $i_s(t)$ the shorted-turn coil current, R_a the primary-coil resistance, L_a the primary-coil inductance, L_{as} the mutual inductance between the primary and shorted-turn coils, $v(t)$ the velocity of the voice coil, $y(t)$ the displacement of the voice coil, $f(t) = K_i v(t)$ the force of the voice coil, K_i the force constant, K_b the back-emf constant, $e_b(t) = K_b v(t)$ the back emf, M_T the total mass of the voice coil and load, and B_T the total viscous-friction coefficient of the voice coil and load.

(a) Write the differential equations of the system.

(b) Draw a block diagram of the system with $E_a(s)$, $I_a(s)$, $I_s(s)$, $V(s)$, and $Y(s)$ as variables.

(c) Derive the transfer function $Y(s)/E_a(s)$.

▲ Dc-motor control system

4-17. A dc-motor position-control system is shown in Fig. 4P-17(a). The following parameters and variables are defined: e is the error voltage, e_r the reference input, θ_L the load position, K_A the amplifier gain, e_a the motor input voltage, e_b the back emf, i_a the motor current, T_m the motor torque, J_m the motor inertia = 0.03 oz-in.-s², B_m the motor viscous-friction coefficient = 10 oz-in.-s, K_L the torsional spring constant = 50,000 oz-in./rad, J_L the load inertia = 0.05 oz-in.-s², K_i the motor torque constant = 21 oz-in./A, K_b

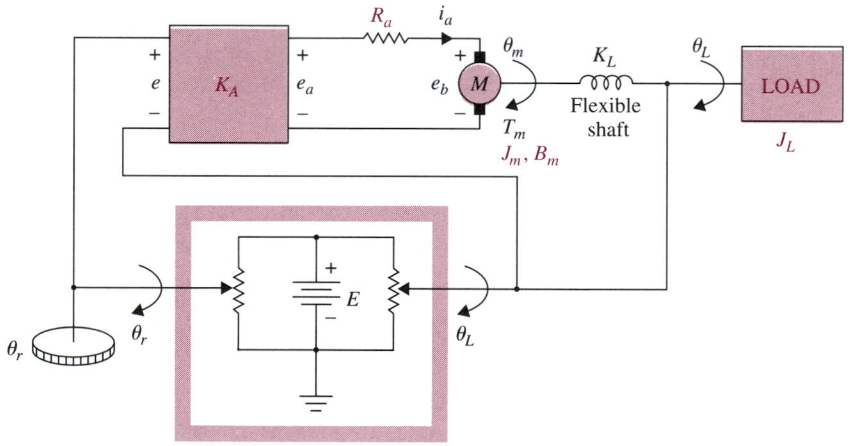

(a)

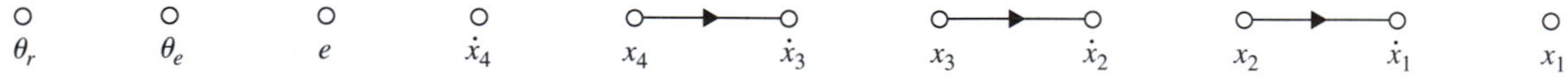

(b)

Figure 4P-17

the back-emf constant = 15.5 V/1000 rpm, K_s the error-detector gain = $E/2\pi$, E the error-detector applied voltage = 2π volts, R_a the motor resistance = 1.15 Ω, and $\theta_e = \theta_r - \theta_L$.

(a) Write the state equations of the system using the following state variables: $x = \theta_L$, $x = d\theta_L/dt = \omega_L$, $x = \theta$, and $x = d\theta_m/dt = \omega_m$.

(b) Draw a state diagram using the nodes shown in Fig. 4P-17(b).

(c) Derive the forward-path transfer function $G(s) = \Theta_L(s)/\Theta_e(s)$ when the outer feedback path from θ_L is opened. Find the poles of $G(s)$.

(d) Derive the closed-loop transfer function $M(s) = \Theta_L(s)\Theta_e(s)$. Find the poles of $M(s)$ when K_A = 1,2738, and 5476. Locate these poles in the s-plane, and comment on the significance of these values of K_A.

▲ Temperature control of air-flow system

4-18. Figure 4P-18(a) shows the setup of the temperature control of an airflow system. The hot-water reservoir supplies the water that flows into the heat exchanger for heating the air. The temperature sensor senses the air temperature T_{AO} and sends it to be compared with the reference temperature T_r. The temperature error T_e is sent to the controller, which has the transfer function $G_c(s)$. The output of the controller, $u(t)$, which is an electric signal, is converted to a pneumatic signal by a transducer. The output of the actuator controls the water-flow rate through the three-way valve. Figure 4P-18(b) shows the block diagram of the system.

The following parameters and variables are defined: dM_w is the flow rate of the heating fluid = $K_M u$, K_M = 0.054 kg/sec/V, T_W the water temperature = $K_R dM_W$, K_R = 65°C/kg/sec, and T_{AO} the output air temperature.

Heat-transfer equation between water and air:

$$\tau_c \frac{dT_{AO}}{dt} = T_w - T_{AO} \qquad \tau_c = 10 \text{ sec}$$

Temperature sensor equation:

$$\tau_s \frac{dT_s}{dt} = T_{AO} - T_s \qquad \tau_s = 2 \text{ sec}$$

(a) Draw a functional block diagram that includes all the transfer functions of the system.

(b) Derive the transfer function $T_{AO}(s)/T_r(s)$ when $G_c(s) = 1$.

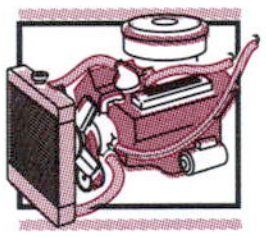

▲ Idle-speed control system

4-19. The objective of this problem is to develop a linear analytical model of the automobile engine for the idle-speed control system shown in Fig. 1-2. The input of the system is the throttle position that controls the rate of airflow into the manifold (see Fig. 4P-19). Engine torque is developed from the buildup of manifold pressure due to air intake and the intake of the air–gas mixture into the cylinder. The engine variations are as follows:

$q_i(t)$ = amount of airflow across throttle into manifold
$dq_i(t)/dt$ = rate of airflow across throttle into manifold
$q_m(t)$ = average air mass in manifold

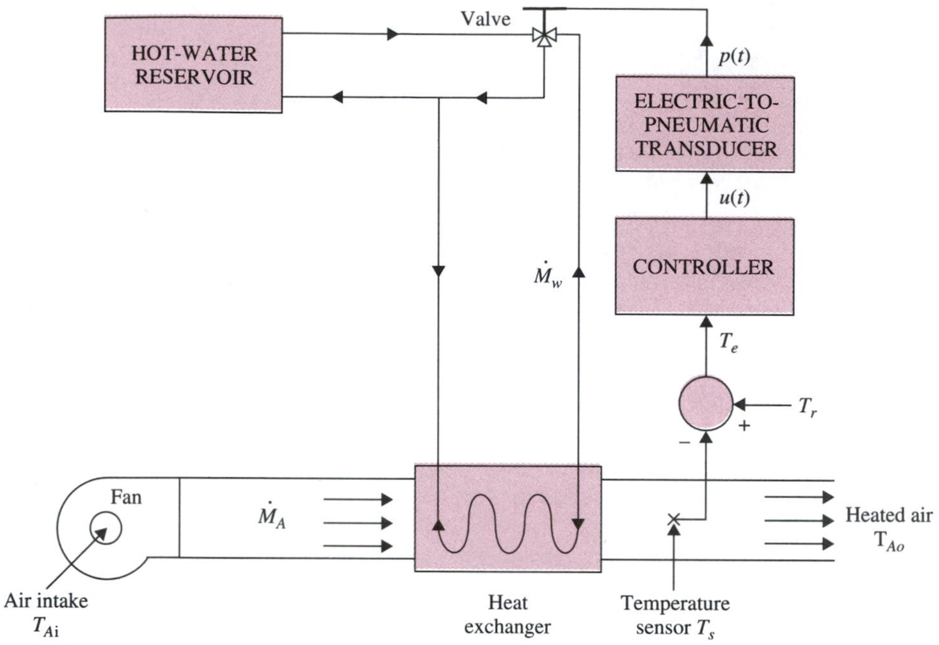

(a)

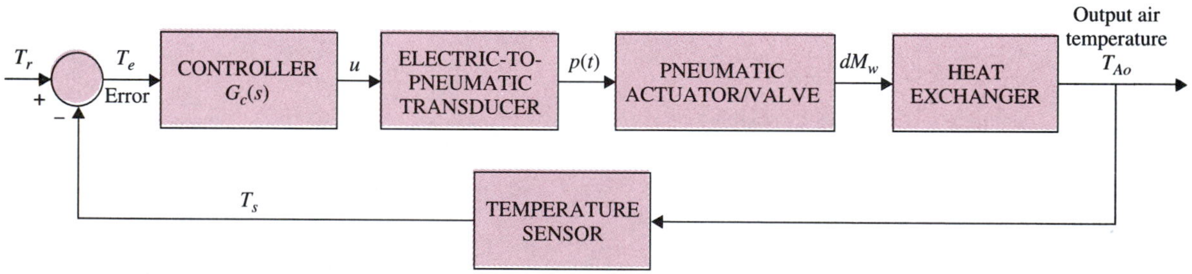

(b)

Figure 4P-18

$q_o(t)$ = amount of air leaving intake manifold through intake valves
$dq_o(t)/dt$ = rate of air leaving intake manifold through intake valves
$T(t)$ = engine torque
T_d = disturbance torque due to application of auto accessories = constant
$\omega(t)$ = engine speed

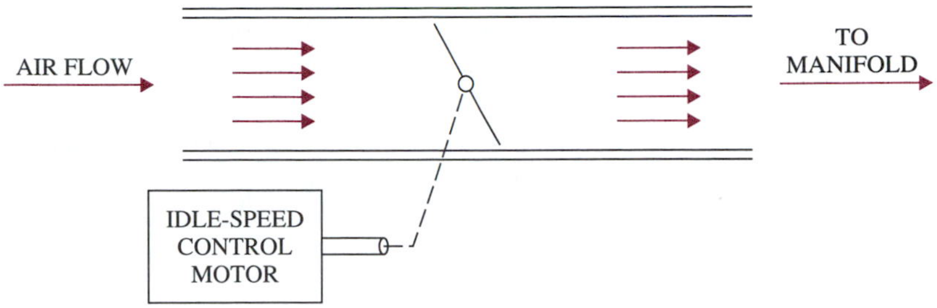

Figure 4P-19

$\alpha(t)$ = throttle position
τ_D = time delay in engine
J_e = inertia of engine

The following assumptions and mathematical relations between the engine variables are given:

1. The rate of airflow into the manifold is linearly dependent on the throttle position:

$$\frac{dq_i(t)}{dt} = K_1\alpha(t) \qquad K_1 = \text{proportional constant}$$

2. The rate of airflow leaving the manifold depends linearly on the air mass in the manifold and the engine speed:

$$\frac{dq_o(t)}{dt} = K_2 q_m(t) + K_3\omega(t) \qquad K_2, K_3 = \text{constants}$$

3. A pure time delay of τ_D seconds exist between the change in the manifold air mass and the engine torque:

$$T(t) = K_4 q_m(t - \tau_D) \qquad K_4 = \text{constant}$$

4. The engine drag is modeled by a viscous-friction torque $B\omega(t)$, where B is the viscous-friction coefficient.
5. The average air mass $q_m(t)$ is determined from

$$q_m(t) = \int \left[\frac{dq_i(t)}{dt} - \frac{dq_o(t)}{dt}\right] dt$$

6. The equation describing the mechanical components is

$$T(t) = J\frac{d\omega(t)}{dt} + B\omega(t) + T_d$$

(a) Draw a functional block diagram of the system with $\alpha(t)$ as input, $\omega(t)$ as output, and T_d as the disturbance input. Show the transfer function of each block.

(b) Find the transfer function $\Omega(s)/\alpha(s)$ of the system.

(c) Find the characteristic equation and show that it is not rational with constant coefficients.

(d) Approximate the engine time delay by

$$e^{-\tau_D s} \cong \frac{1 - \tau_D s/2}{1 + \tau_D s/2}$$

and repeat parts (b) and (c).

▲ Phase-locked loop control

4-20. Phase-locked loops are control systems used for precision motor-speed control. The basic elements of a phase-locked-loop system incorporating a dc motor are shown in Fig. 4P-20(a). An input pulse train represents the reference frequency or desired output speed. The digital encoder produces digital pulses that represent motor speed. The phase detector compares the motor speed and the reference frequency and sends an error voltage to the filter (controller) that governs the dynamic response of the system. Phase detector gain $= K_p$, encoder gain $= K_e$, counter gain $= 1/N$, and dc-motor torque constant $= K_i$. Assume zero inductance and zero friction for the motor.

(a) Derive the transfer function $E_c(s)/E(s)$ of the filter shown in Fig. 4P-20(b). Assume that the filter sees infinite impedance at the output and zero impedance at the input.

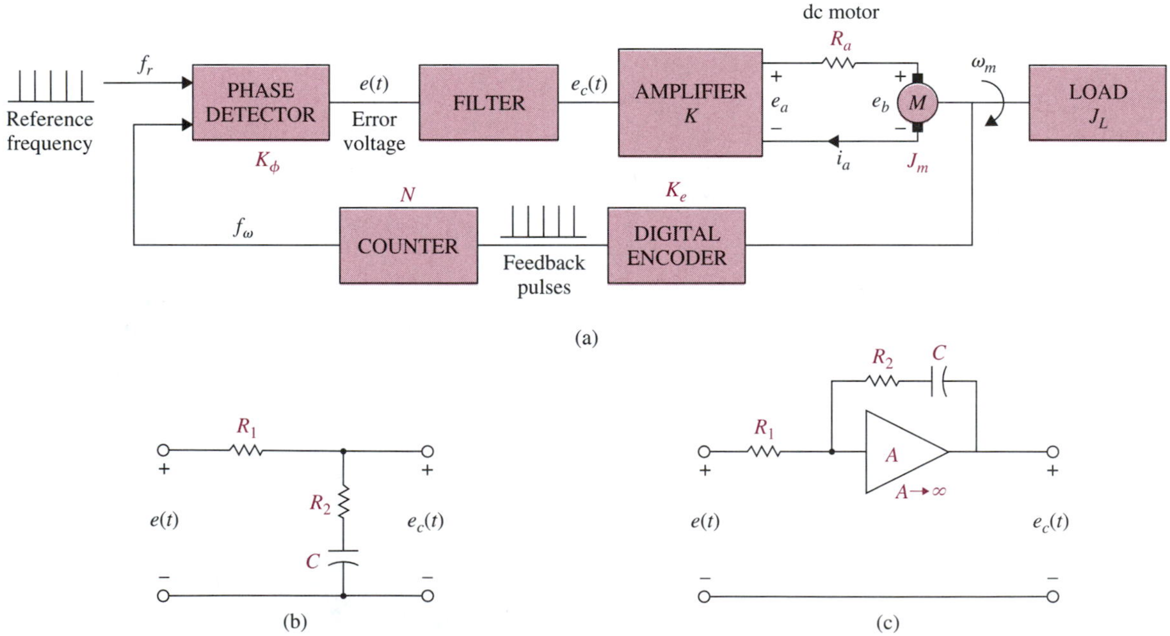

Figure 4P-20

(b) Draw a functional block diagram of the system with gains or transfer functions in the blocks.

(c) Derive the forward-path transfer function $\Omega_m(s)/E(s)$ when the feedback path is open.

(d) Find the closed-loop transfer function $\Omega_m(s)/F_r(s)$.

(e) Repeat parts (a), (c), and (d) for the filter shown in Fig. 4P-20(c).

(f) The digital encoder has an output of 36 pulses per revolution. The reference frequency f_r is fixed at 120 pulses/sec. Find K_e in pulses/rad. The idea of using the counter N is that with f_r fixed, various desired output speeds can be attained by changing the value of N. Find N if the desired output speed is 200 rpm. Find N if the desired output speed is 1800 rpm.

▲ Robot arm control

4-21. The linearized model of a robot arm system driven by a dc motor is shown in Fig. 4P-21. The system parameters and variables are given as follows:

DC MOTOR

- ▲ T_m = motor torque = $K_i i_a$
- ▲ K_i = torque constant
- ▲ i_a = armature current of motor
- ▲ J_m = motor inertia
- ▲ B_m = motor viscous-friction coefficient
- ▲ θ_m = motor-shaft displacement

ROBOT ARM

- ▲ J_L = inertia of arm
- ▲ T_L = disturbance torque on arm
- ▲ θ_L = arm displacement
- ▲ K = torsional spring constant
- ▲ B = viscous-friction coefficient of shaft between the motor and arm
- ▲ B_L = viscous-friction coefficient of the robot arm shaft

(a) Write the differential equations for the system with $i_a(t)$ and $T_L(t)$ as input and $\theta_m(t)$ and $\theta_L(t)$ as outputs.

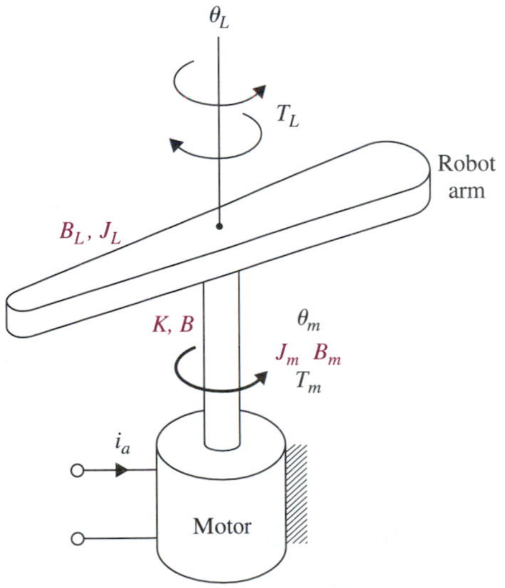

Figure 4P-21

(b) Draw a SFG using $i_a(s)$, $T_L(s)$, $\Theta_m(s)$, and $\Theta_L(s)$ as node variables.

(c) Express the transfer-function relations as

$$\begin{bmatrix} \Theta_m(s) \\ \Theta_L(s) \end{bmatrix} = \mathbf{G}(s) \begin{bmatrix} I_a(s) \\ -T_L(s) \end{bmatrix}$$

Find $\mathbf{G}(s)$.

▲ Electric train control system

4-22. The following differential equations describe the motion of an electric train in a traction system:

$$\frac{dx(t)}{dt} = v(t)$$

$$\frac{dv(t)}{dt} = -k(v) - g(x) + f(t)$$

where

$x(t)$ = linear displacement of train
$v(t)$ = linear velocity of train
$k(v)$ = resistance force on train [odd function of v, with the properties $k(0) = 0$ and $dk(v)/dv = 0$
$g(x)$ = gravitational force for a nonlevel track or due to curvature of track
$f(t)$ = tractive force

The electric motor that provides the tractive force is described by the following equations:

$$e(t) = K_b\phi(t)v(t) + R_a i_a(t)$$
$$f(t) = K_i\phi(t)i_a(t)$$

where $e(t)$ is the applied voltage, $i_a(t)$ the armature current, i_f(t) the field current, R_a the armature resistance, $\phi(t)$ the magnetic flux from a separately excited field $= K_f i_f(t)$, and K_i the force constant.

(a) Consider that the motor is a dc series motor with the armature and field windings connected in series, so that $i_a(t) = i_f(t)$, $g(x) = 0$, $k(v) = Bv(t)$, and $R_a = 0$. Show that the system is described by the following nonlinear state equations:

$$\frac{dx(t)}{dt} = v(t)$$

$$\frac{dv(t)}{dt} = -Bv(t) + \frac{K_i}{K_b K_f v^2(t)} e^2(t)$$

(b) Consider that for the conditions stated in part (a), $i_a(t)$ is the input of the system [instead of $e(t)$]. Derive the state equations of the system.

(c) Consider the same conditions as in part (a), but with $\phi(t)$ as the input. Derive the state equations.

▲ Broom-balancing system

4-23. Figure 4P-23(a) shows a well-known "broom-balancing" system in control systems. The objective of the control system is to maintain the broom in the upright position by means of the force $u(t)$ applied to the car as shown. In practical applications, the system is analogous to a one-dimensional control problem of the balancing of a unicycle or a missile immediately after launching. The free-body diagram of the system is shown in Fig. 4P-23(b), where

f_x = force at broom base in horizontal direction
f_y = force at broom base in vertical direction
M_b = mass of broom
g = gravitational acceleration
M_c = mass of car
J_b = moment of inertia of broom about center of gravity CG = $M_b L^2/3$

(a) Write the force equations in the x and the y directions at the pivot point of the broom. Write the torque equation about the center of gravity (CG) of the broom. Write the force equation of the car in the horizontal direction.

(b) Express the equations obtained in part (a) as state equations by assigning the state variables as $x_1 = \theta$, $x_2 = d\theta/dt$, $x_3 = x$, and $x_4 = dx/dt$. Simplify these equations for small θ by making the approximations: $\sin\theta \cong \theta$ and $\cos\theta \cong 1$.

(c) Obtain a small-signal linearized state-equation model for the system in the form

$$\frac{d\,\Delta\mathbf{x}(t)}{dt} = \mathbf{A}^*\,\Delta\mathbf{x}(t) + \mathbf{B}^*\,\Delta\mathbf{r}(t)$$

at the equilibrium point $x_{01}(t) = 1$, $x_{02}(t) = 0$, $x_{03}(t) = 0$, and $x_{04}(t) = 0$.

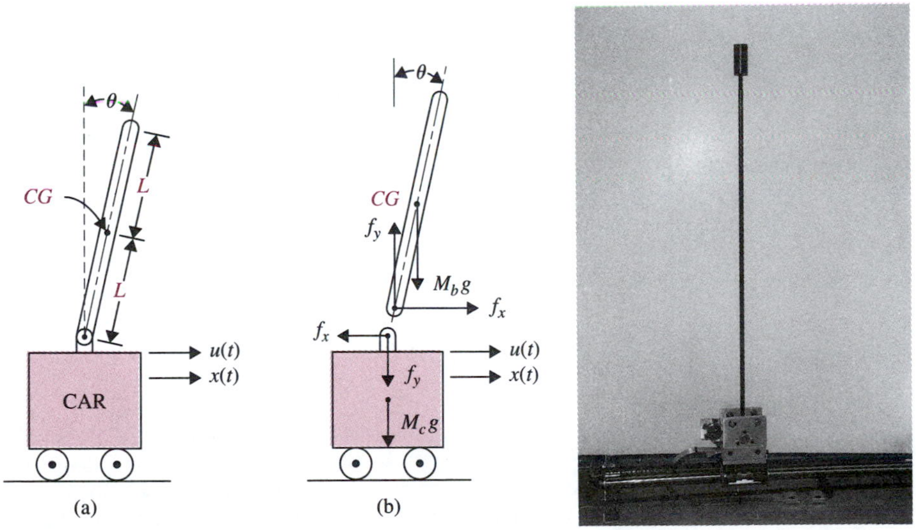

Figure 4P-23

▲ Ball-suspension control system

4-24. Figure 4P-24 shows the schematic diagram of a ball-suspension control system. The steel ball is suspended in the air by the electromagnetic force generated by the electromagnet. The objective of the control is to keep the metal ball suspended at the nominal equilibrium position by controlling the current in the magnet with the voltage $e(t)$. The practical application of this system is the magnetic levitation of trains or magnetic bearings in high-precision control systems.

The resistance of the coil is R, and the inductance is $L(y) = L/y(t)$, where L is a constant. The applied voltage $e(t)$ is a constant with amplitude E.

(a) Let E_{eq} be a nominal value of E. Find the nominal values of $y(t)$ and $dy(t)/dt$ at equilibrium.

(b) Define the state variables at $x_1(t) = i(t)$, $x_2(t) = y(t)$, and $x_3(t) = dy(t)/dt$. Find the nonlinear state equations in the form of

$$\frac{d\mathbf{x}(t)}{dt} = \mathbf{f}(\mathbf{x}, e)$$

(c) Linearize the state equations about the equilibrium point and express the linearized state equations as

$$\frac{d\,\Delta\mathbf{x}(t)}{dt} = \mathbf{A}\,\Delta\mathbf{x}(t) + \mathbf{B}\,\Delta e(t)$$

The force generated by the electromagnet is $Ki^2(t)/y(t)$, where K is a proportional constant and the gravitational force on the steel ball is Mg.

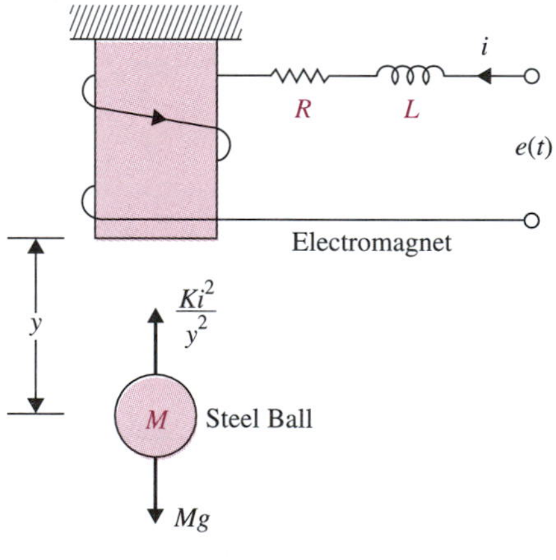

Figure 4P-24

▲ Ball-suspension control system

4-25. Figure 4P-25(a) shows the schematic diagram of a ball-suspension system. The steel ball is suspended in the air by the electromagnetic force generated by the electromagnet. The objective of the control is to keep the metal ball suspended at the nominal position by controlling the current in the electromagnet. When the system is at the stable equilibrium point, any small perturbation of the ball position from its floating equilibrium position will cause control to return the ball to the equilibrium position. The free-body

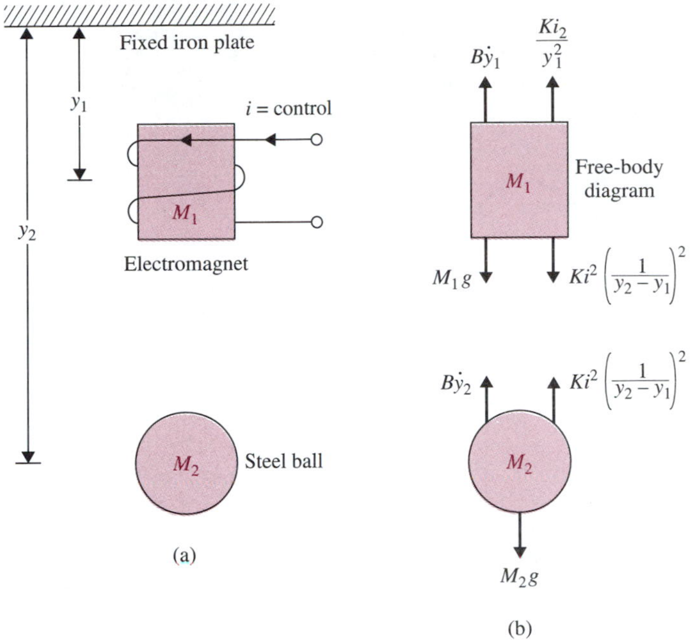

Figure 4P-25

diagram of the system is shown in Fig. 4P-25(b), where

M_1 = mass of electromagnet = 2.0
M_2 = mass of steel ball = 1.0
B = viscous-friction coefficient of air = 0.1
K = proportional constant of electromagnet = 1.0
g = gravitational acceleration = 32.2

Assume that all units are consistent. Let the stable equilibrium values of the variable, $i(t)$, $y_1(t)$, and $y_2(t)$ be I, Y_1, and Y_2, respectively. The state variables are defined as $x_1(t) = y_1(t)$, $x_2(t) = dy_1(t)/dt$, $x_3(t) = y_2(t)$, and $x_4(t) = dy_2(t)/dt$.

(a) Given $Y_1 = 1$, find I and Y_2.

(b) Write the nonlinear state equations of the system in the form of $d\mathbf{x}(t)/dt = \mathbf{f}(\mathbf{x}, i)$.

(c) Find the state equations of the linearized system about the equilibrium state I, Y_1, and Y_2 in the form

$$\frac{d\mathbf{x}(t)}{dt} = \mathbf{A}^* \Delta\mathbf{x}(t) + \mathbf{B}^* \Delta i(t)$$

▲ Steel-rolling process control

4-26. The schematic diagram of a steel-rolling process is shown in Fig. 4P-26. The steel plate is fed through the rollers at a constant speed of V ft/sec. The distance between the rollers and the point where the thickness is measured is d feet. The rotary displacement of the motor, $\theta_m(t)$, is converted to the linear displacement $y(t)$ by the gearbox and linear-

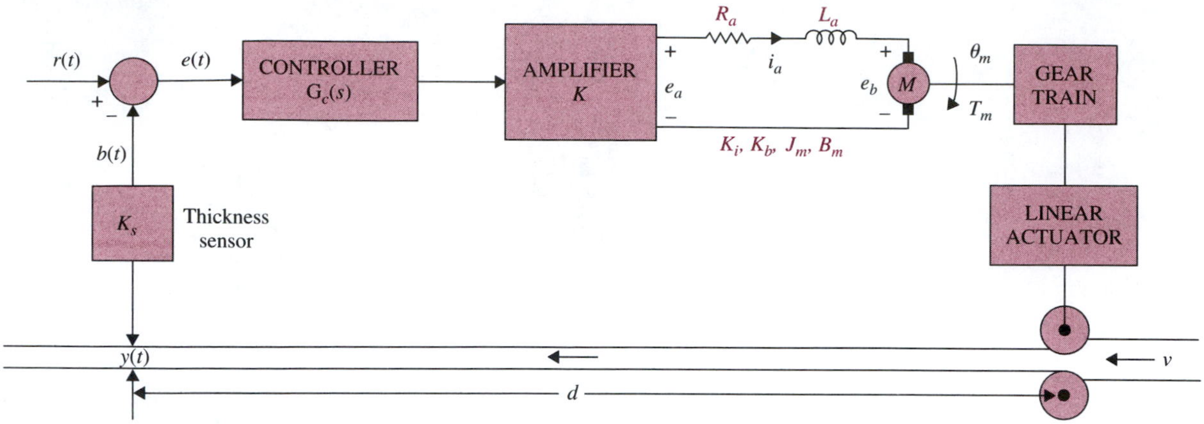

Figure 4P-26

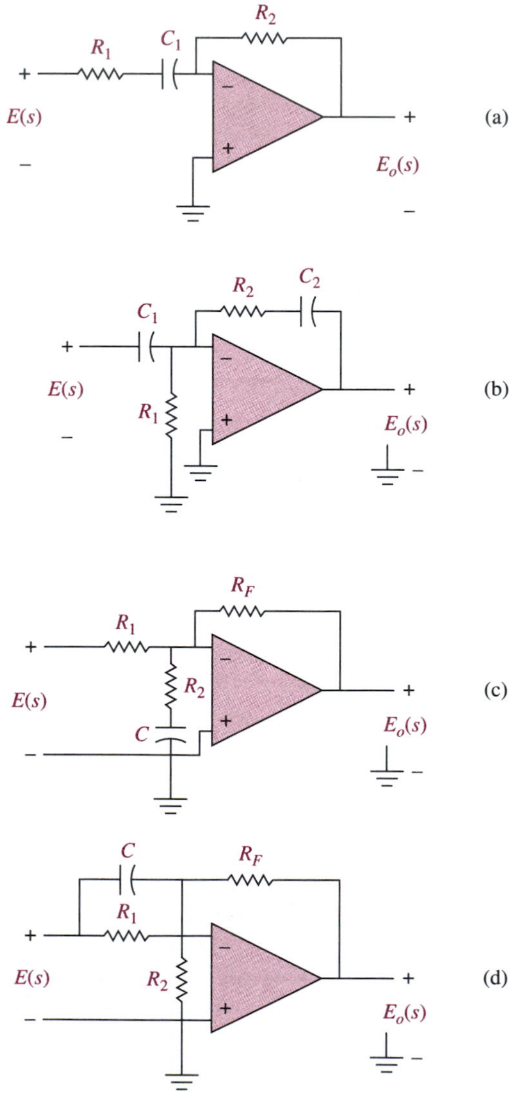

Figure 4P-27

actuator combination; $y(t) = n\theta_m(t)$, where n is a positive constant in ft/rad. The equivalent inertia of the load that is reflected to the motor shaft is J_L.

(a) Draw a functional block diagram for the system.

(b) Derive the forward-path transfer function $Y(s)/E(s)$ and the closed-loop transfer function $Y(s)/R(s)$.

▲ Op-amp circuits

4-27. Find the transfer functions $E_o(s)/E(s)$ for the circuits shown in Fig. 4P-27 on page 224.

5 State-Variable Analysis

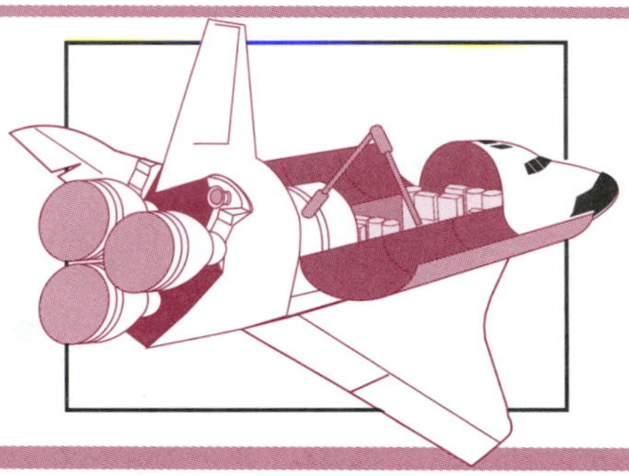

KEYWORDS AND TOPICS

- State Variables
- State Diagram
- State-Transition Matrix
- State-Transition Equation
- Characteristic Equation
- Eigenvalues
- Eigenvectors
- Similarity Transformations
- Decomposition
- Controllability
- Observability

5-1 Introduction

In Chapter 2 the concept and definition of state variables and state equations are presented, together with several illustrative examples on how the state variables are chosen and how state equations are written for linear and nonlinear dynamic systems. In Chapter 3, the signal-flow-graph method is extended to the modeling of the state equations and the result is the **state diagram.** In contrast to the transfer-function approach to the analysis and design of linear control systems, the state-variable method is regarded as modern, since it is the underlying force for optimal control. The basic characteristic of the state-variable formulation is that linear and nonlinear systems, time-invariant and time-varying systems, and single-variable and multivariable systems can all be modeled in a unified manner. Transfer functions, on the other hand, are defined only for linear time-invariant systems.

The objective of this chapter is to introduce the basic methods of state variables and state equations so that the reader can gain a working knowledge of the subject for further studies when the state-space approach is used for modern and optimal control design. Specifically, the closed-form solutions of linear time-invariant state equations are presented. Various transformations that may be used to facilitate the analysis and design of linear control systems in the state-variable domain are introduced. The relationship between the conventional transfer-function approach and the state-variable approach is established, so that the analyst will be able to investigate a system problem with various alternative methods. Finally, the controllability and observability of linear systems are defined and their applications investigated.

5-2 Vector-Matrix Representation of State Equations

Let the n state equations of an nth-order dynamic system be represented as

$$\frac{dx_i(t)}{dt} = f_i[x_1(t), x_2(t), \ldots, x_n(t), u_1(t), u_2(t), \ldots, u_p(t), w_1(t), w_2(t), \ldots, w_v(t)] \tag{5-1}$$

where $i = 1, 2, \ldots, n$. The ith state variable is represented by $x_i(t)$; $r_j(t)$ denotes the jth input for $j = 1, 2, \ldots, p$; and $w_k(t)$ denotes the kth disturbance input, with $k = 1, 2, \ldots, v$.

Let the variables $y_1(t), y_2(t), \ldots, y_q(t)$ be the q output variables of the system. In general, the output variables are functions of the state variables and the input variables. The **output equations** can be expressed as

$$y_j(t) = g_j[x_1(t), x_2(t), \ldots, x_n(t), u_1(t), u_2(t), \ldots, u_p(t), w_1(t), w_2(t), \ldots, w_v(t)] \tag{5-2}$$

where $j = 1, 2, \ldots, q$.

The set of n state equations in Eq. (5-1) and q output equations in Eq. (5-2) together form the **dynamic equations.**

For ease of expression and manipulation, it is convenient to represent the dynamic equations in vector-matrix form. Let us define the following vectors:

State Vector

$$\mathbf{x}(t) = \begin{bmatrix} x_1(t) \\ x_2(t) \\ \vdots \\ x_n(t) \end{bmatrix} \quad (n \times 1) \tag{5-3}$$

Input Vector

$$\mathbf{u}(t) = \begin{bmatrix} u_1(t) \\ u_2(t) \\ \vdots \\ u_p(t) \end{bmatrix} \quad (p \times 1) \tag{5-4}$$

Output Vector

$$\mathbf{y}(t) = \begin{bmatrix} y_1(t) \\ y_2(t) \\ \vdots \\ y_q(t) \end{bmatrix} \quad (q \times 1) \tag{5-5}$$

Disturbance Vector

$$\mathbf{w}(t) = \begin{bmatrix} w_1(t) \\ w_2(t) \\ \vdots \\ w_v(t) \end{bmatrix} \quad (v \times 1) \tag{5-6}$$

By using these vectors, the n state equations of Eq. (5-1) can be written

$$\frac{d\mathbf{x}(t)}{dt} = \mathbf{f}[\mathbf{x}(t), \mathbf{u}(t), \mathbf{w}(t)] \tag{5-7}$$

where $\mathbf{f}$ denotes an $n \times 1$ column matrix that contains the functions $f_1, f_2, \ldots, f_n$ as elements. Similarly, the q output equations in Eq. (5-2) become

$$\mathbf{y}(t) = \mathbf{g}[\mathbf{x}(t), \mathbf{u}(t), \mathbf{w}(t)] \tag{5-8}$$

where $\mathbf{g}$ denotes a $q \times 1$ column matrix that contains the functions $g_1, g_2, \ldots, g_q$ as elements.

For a linear time-invariant system, the dynamic equations are written as

State Equations

$$\frac{d\mathbf{x}(t)}{dt} = \mathbf{A}\mathbf{x}(t) + \mathbf{B}\mathbf{u}(t) + \mathbf{E}\mathbf{w}(t) \tag{5-9}$$

Output Equations

$$\mathbf{y}(t) = \mathbf{C}\mathbf{x}(t) + \mathbf{D}\mathbf{u}(t) + \mathbf{H}\mathbf{w}(t) \tag{5-10}$$

where

$$\mathbf{A} = \begin{bmatrix} a_{11} & a_{12} & \cdots & a_{1n} \\ a_{21} & a_{22} & \cdots & a_{2n} \\ \vdots & \vdots & \ddots & \vdots \\ a_{n1} & a_{n2} & \cdots & a_{nn} \end{bmatrix} \quad (n \times n) \tag{5-11}$$

$$\mathbf{B} = \begin{bmatrix} b_{11} & b_{12} & \cdots & b_{1p} \\ b_{21} & b_{22} & \cdots & b_{2p} \\ \vdots & \vdots & \ddots & \vdots \\ b_{n1} & b_{n2} & \cdots & b_{np} \end{bmatrix} \quad (n \times p) \tag{5-12}$$

$$\mathbf{C} = \begin{bmatrix} c_{11} & c_{12} & \cdots & c_{1n} \\ c_{21} & c_{22} & \cdots & c_{2n} \\ \vdots & \vdots & \ddots & \vdots \\ c_{q1} & c_{q2} & \cdots & c_{qn} \end{bmatrix} \quad (q \times n) \tag{5-13}$$

$$\mathbf{D} = \begin{bmatrix} d_{11} & d_{12} & \cdots & d_{1p} \\ d_{21} & d_{22} & \cdots & d_{2p} \\ \vdots & \vdots & \ddots & \vdots \\ d_{q1} & d_{q2} & \cdots & d_{qp} \end{bmatrix} \quad (q \times p) \tag{5-14}$$

$$\mathbf{E} = \begin{bmatrix} e_{11} & e_{12} & \cdots & e_{1v} \\ e_{21} & e_{22} & \cdots & e_{2v} \\ \vdots & \vdots & \ddots & \vdots \\ e_{n1} & e_{n2} & \cdots & e_{nv} \end{bmatrix} \quad (n \times v) \tag{5-15}$$

$$\mathbf{H} = \begin{bmatrix} h_{11} & h_{12} & \cdots & h_{1v} \\ h_{21} & h_{22} & \cdots & h_{2v} \\ \vdots & \vdots & \ddots & \vdots \\ h_{q1} & h_{q2} & \cdots & h_{qn} \end{bmatrix} \quad (q \times v) \tag{5-16}$$

5-3 State-Transition Matrix

Once the state equations of a linear time-invariant system are expressed in the form of Eq. (5-9), the next step often involves the solutions of these equations given the initial state vector $\mathbf{x}(t_0)$, the input vector $\mathbf{u}$(t), and the disturbance vector $\mathbf{w}$(t), for $t \geq t_0$. The first term on the right-hand side of Eq. (5-9) is known as the homogeneous part of the state equation, and the last two terms represent the forcing functions $\mathbf{u}(t)$ and $\mathbf{w}(t)$.

The **state-transition matrix** is defined as a matrix that satisfies the linear homogeneous state equation:

$$\frac{d\mathbf{x}(t)}{dt} = \mathbf{A}\mathbf{x}(t) \tag{5-17}$$

Let $\boldsymbol{\phi}(t)$ be the $n \times n$ matrix that represents the state-transition matrix; then it must satisfy the equation

$$\frac{d\boldsymbol{\phi}(t)}{dt} = \mathbf{A}\boldsymbol{\phi}(t) \tag{5-18}$$

Furthermore, let $\mathbf{x}(0)$ denote the initial state at $t = 0$; then $\boldsymbol{\phi}(t)$ is also defined by the matrix equation

$$\mathbf{x}(t) = \boldsymbol{\phi}(t)\mathbf{x}(0) \tag{5-19}$$

which is the solution of the homogeneous state equation for $t \geq 0$.

One way of determining $\boldsymbol{\phi}(t)$ is by taking the Laplace transform on both sides of Eq. (5-17); we have

$$s\mathbf{X}(s) - \mathbf{x}(0) = \mathbf{A}\mathbf{X}(s) \tag{5-20}$$

Solving for $\mathbf{X}(s)$ from Eq. (5-20), we get

$$\mathbf{X}(s) = (s\mathbf{I} - \mathbf{A})^{-1}\mathbf{x}(0) \tag{5-21}$$

where it is assumed that the matrix $(s\mathbf{I} - \mathbf{A})$ is nonsingular. Taking the inverse Laplace transform on both sides of Eq. (5-21) yields

$$\mathbf{x}(t) = \mathcal{L}^{-1}[(s\mathbf{I} - \mathbf{A})^{-1}]\mathbf{x}(0) \qquad t \geq 0 \tag{5-22}$$

By comparing Eq. (5-19) with Eq. (5-22), the state-transition matrix is identified to be

$$\boldsymbol{\phi}(t) = \mathcal{L}^{-1}[(s\mathbf{I} - \mathbf{A})^{-1}] \tag{5-23}$$

An alternative way of solving the homogeneous state equation is to assume a solution, as in the classic method of solving linear differential equations. We let the solution to Eq. (5-17) be

$$\mathbf{x}(t) = e^{\mathbf{A}t}\mathbf{x}(0) \tag{5-24}$$

for $t \geq 0$, where $e^{\mathbf{A}t}$ represents the following power series of the matrix $\mathbf{A}t$, and

$$e^{\mathbf{A}t} = \mathbf{I} + \mathbf{A}t + \frac{1}{2!}\mathbf{A}^2t^2 + \frac{1}{3!}\mathbf{A}^3t^3 + \cdots \tag{5-25}$$

It is easy to show that Eq. (5-24) is a solution of the homogeneous state equation, since, from Eq. (5-25),

$$\frac{de^{\mathbf{A}t}}{dt} = \mathbf{A}e^{\mathbf{A}t} \tag{5-26}$$

Therefore, in addition to Eq. (5-23), we have obtained another expression for the state-transition matrix:

$$\boldsymbol{\phi}(t) = e^{\mathbf{A}t} = \mathbf{I} + \mathbf{A}t + \frac{1}{2!}\mathbf{A}^2t^2 + \frac{1}{3!}\mathbf{A}^3t^3 + \cdots \tag{5-27}$$

Equation (5-27) can also be obtained directly from Eq. (5-23). This is left as an exercise for the reader (Problem 5-2).

5-3-1 Significance of the State-Transition Matrix

Since the state transition matrix satisfies the homogeneous state equation, it represents the **free response** of the system. In other words, it governs the response that is excited by the initial conditions only. In view of Eqs. (5-23) and (5-27), the state-transition matrix is dependent only on the matrix **A**, and therefore is sometimes referred to as the **state-transition matrix of A.** As the name implies, the state-transition matrix $\boldsymbol{\phi}(t)$ completely defines the transition of the states from the initial time $t = 0$ to any time t when the inputs are zero.

5-3-2 Properties of the State-Transition Matrix

The state-transition matrix $\boldsymbol{\phi}(t)$ possesses the following properties:

1. $\boldsymbol{\phi}(0) = \mathbf{I}$ (the identity matrix) (5-28)

▲ ***Proof*** Equation (5-28) follows directly from Eq. (5-27) by setting $t = 0$.

2. $\boldsymbol{\phi}^{-1}(t) = \boldsymbol{\phi}(-t)$ (5-29)

▲ ***Proof*** Postmultiplying both sides of Eq. (5-27) by $e^{-\mathbf{A}t}$, we get

$$\boldsymbol{\phi}(t)e^{-\mathbf{A}t} = e^{\mathbf{A}t}e^{-\mathbf{A}t} = \mathbf{I} \tag{5-30}$$

Then, premultiplying both sides of Eq. (5-30) by $\boldsymbol{\phi}^{-1}(t)$, we get

$$e^{-\mathbf{A}t} = \boldsymbol{\phi}^{-1}(t) \tag{5-31}$$

Thus

$$\boldsymbol{\phi}(-t) = \boldsymbol{\phi}^{-1}(t) = e^{-\mathbf{A}t} \tag{5-32}$$

An interesting result from this property of $\boldsymbol{\phi}(t)$ is that Eq. (5-24) can be rearranged to read

$$\mathbf{x}(0) = \boldsymbol{\phi}(-t)\mathbf{x}(t) \tag{5-33}$$

which means that the state-transition process can be considered as bilateral in time. That is, the transition in time can take place in either direction.

3. $$\boldsymbol{\phi}(t_2 - t_1)\boldsymbol{\phi}(t_1 - t_0) = \boldsymbol{\phi}(t_2 - t_0) \qquad \text{for any } t_0,\ t_1,\ t_2 \tag{5-34}$$

▲ ***Proof***

$$\begin{aligned}\boldsymbol{\phi}(t_2 - t_1)\boldsymbol{\phi}(t_1 - t_0) &= e^{\mathbf{A}(t_2-t_1)}e^{\mathbf{A}(t_1-t_0)} \\ &= e^{\mathbf{A}(t_2-t_0)} = \boldsymbol{\phi}(t_2 - t_0)\end{aligned} \tag{5-35}$$

This property of the state-transition matrix is important since it implies that a state-transition process can be divided into a number of sequential transitions. Figure 5-1 illustrates that the transition from $t = t_0$ to $t = t_2$ is equal to the transition from t_0 to t_1, and then from t_1 to t_2. In general, of course, the state-transition process can be divided into any number of parts.

4. $$[\boldsymbol{\phi}(t)]^k = \boldsymbol{\phi}(kt) \qquad \text{for } k = \text{positive integer} \tag{5-36}$$

▲ ***Proof***

$$\begin{aligned}[\boldsymbol{\phi}(t)]^k &= e^{\mathbf{A}t}e^{\mathbf{A}t}\cdots e^{\mathbf{A}t} \qquad (k \text{ terms}) \\ &= e^{k\mathbf{A}t} = \boldsymbol{\phi}(kt)\end{aligned} \tag{5-37}$$

5-4 State-Transition Equation

The **state-transition equation** *is defined as the solution of the linear homogeneous state equation.* The linear time-invariant state equation

$$\frac{d\mathbf{x}(t)}{dt} = \mathbf{A}\mathbf{x}(t) + \mathbf{B}\mathbf{u}(t) + \mathbf{E}\mathbf{w}(t) \tag{5-38}$$

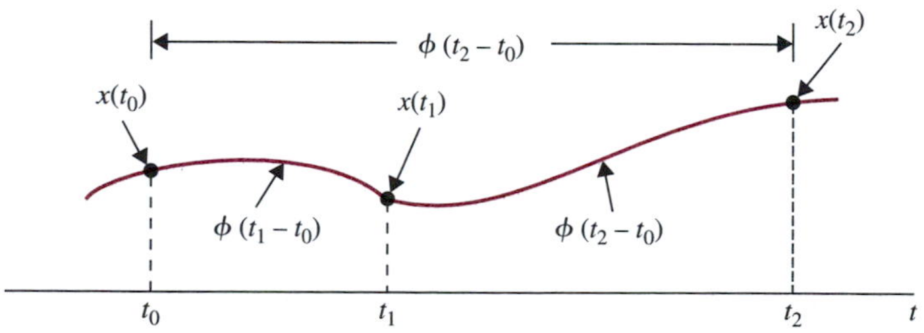

Figure 5-1 Property of the state-transition matrix.

can be solved using either the classical method of solving linear differential equations or the Laplace transform method. The Laplace transform solution is presented in the following.

Taking the Laplace transform on both sides of Eq. (5-38), we have

$$s\mathbf{X}(s) - \mathbf{x}(0) = \mathbf{AX}(s) + \mathbf{BU}(s) + \mathbf{EW}(s) \tag{5-39}$$

where $\mathbf{x}(0)$ denotes the initial-state vector evaluated at $t = 0$. Solving for $\mathbf{X}(s)$ in Eq. (5-39) yields

$$\mathbf{X}(s) = (s\mathbf{I} - \mathbf{A})^{-1}\mathbf{x}(0) + (s\mathbf{I} - \mathbf{A})^{-1}[\mathbf{BU}(s) + \mathbf{EW}(s)] \tag{5-40}$$

The state-transition equation of Eq. (5-38) is obtained by taking the inverse Laplace transform on both sides of Eq. (5-40):

$$\begin{aligned}\mathbf{x}(t) &= \mathscr{L}^{-1}[(s\mathbf{I} - \mathbf{A})^{-1}]\mathbf{x}(0) + \mathscr{L}^{-1}\{(s\mathbf{I} - \mathbf{A})^{-1}[\mathbf{BU}(s) + \mathbf{EW}(s)]\} \\ &= \boldsymbol{\phi}(t)\mathbf{x}(0) + \int_0^t \boldsymbol{\phi}(t-\tau)[\mathbf{Bu}(\tau) + \mathbf{Ew}(\tau)]\,d\tau \qquad t \geq 0\end{aligned} \tag{5-41}$$

The state-transition equation in Eq. (5-41) is useful only when the initial time is defined to be at $t = 0$. In the study of control systems, especially discrete-data control systems, it is often desirable to break up a state-transition process into a sequence of transitions, so a more flexible initial time must be chosen. Let the initial time be represented by t_0 and the corresponding initial state by $\mathbf{x}(t_0)$, and assume that the input $\mathbf{u}(t)$ and the disturbance $\mathbf{w}(t)$ are applied at $t \geq 0$. We start with Eq. (5-41) by setting $t = t_0$, and solving for $\mathbf{x}(0)$, we get

$$\mathbf{x}(0) = \boldsymbol{\phi}(-t_0)\mathbf{x}(t_0) - \boldsymbol{\phi}(-t_0)\int_0^{t_0} \boldsymbol{\phi}(t_0-\tau)[\mathbf{Bu}(\tau) + \mathbf{Ew}(\tau)]\,d\tau \tag{5-42}$$

where the property on $\boldsymbol{\phi}(t)$ of Eq. (5-29) has been applied.

Substituting Eq. (5-42) into Eq. (5-41) yields

$$\begin{aligned}\mathbf{x}(t) = \boldsymbol{\phi}(t)\boldsymbol{\phi}(-t_0)\mathbf{x}(t_0) &- \boldsymbol{\phi}(t)\boldsymbol{\phi}(-t_0)\int_0^{t_0} \boldsymbol{\phi}(t_0-\tau)[\mathbf{Bu}(\tau) + \mathbf{Ew}(\tau)]\,d\tau \\ &+ \int_0^t \boldsymbol{\phi}(t-\tau)[\mathbf{Bu}(\tau) + \mathbf{Ew}(\tau)]\,d\tau\end{aligned} \tag{5-43}$$

Now by using the property of Eq. (5-34) and combining the last two integrals, Eq. (5-43) becomes

$$\mathbf{x}(t) = \boldsymbol{\phi}(t-t_0)\mathbf{x}(t_0) + \int_{t_0}^t \boldsymbol{\phi}(t-\tau)[\mathbf{Bu}(\tau) + \mathbf{Ew}(\tau)]\,d\tau \qquad t \geq t_0 \tag{5-44}$$

It is apparent that Eq. (5-44) reverts to Eq. (5-41) when $t_0 = 0$.

Once the state-transition equation is determined, the output vector can be expressed as a function of the initial state and the input vector simply by substituting $\mathbf{x}(t)$ from Eq. (5-44) into Eq. (5-10). Thus the output vector is

$$\mathbf{y}(t) = \mathbf{C}\boldsymbol{\phi}(t - t_0)\mathbf{x}(t_0) + \int_{t_0}^{t} \mathbf{C}\boldsymbol{\phi}(t - \tau)[\mathbf{B}\mathbf{u}(\tau) + \mathbf{E}\mathbf{w}(\tau)]\, d\tau + \mathbf{D}\mathbf{u}(t) + \mathbf{H}\mathbf{w}(t) \qquad t \geq t_0 \tag{5-45}$$

The following example illustrates the determination of the state-transition matrix and equation.

Example 5-1

Consider the state equation

$$\begin{bmatrix} \dfrac{dx_1(t)}{dt} \\ \dfrac{dx_2(t)}{dt} \end{bmatrix} = \begin{bmatrix} 0 & 1 \\ -2 & -3 \end{bmatrix} \begin{bmatrix} x_1(t) \\ x_2(t) \end{bmatrix} + \begin{bmatrix} 0 \\ 1 \end{bmatrix} u(t) \tag{5-46}$$

The problem is to determine the state-transition matrix $\boldsymbol{\phi}(t)$ and the state vector $\mathbf{x}(t)$ for $t \geq 0$ when the input is $u(t) = 1$ for $t \geq 0$. The coefficient matrices are identified to be

$$\mathbf{A} = \begin{bmatrix} 0 & 1 \\ -2 & -3 \end{bmatrix} \qquad \mathbf{B} = \begin{bmatrix} 0 \\ 1 \end{bmatrix} \qquad \mathbf{E} = 0 \tag{5-47}$$

Therefore,

$$s\mathbf{I} - \mathbf{A} = \begin{bmatrix} s & 0 \\ 0 & s \end{bmatrix} - \begin{bmatrix} 0 & 1 \\ -2 & -3 \end{bmatrix} = \begin{bmatrix} s & -1 \\ 2 & s+3 \end{bmatrix} \tag{5-48}$$

The inverse matrix of $(s\mathbf{I} - \mathbf{A})$ is

$$(s\mathbf{I} - \mathbf{A})^{-1} = \frac{1}{s^2 + 3s + 2} \begin{bmatrix} s+3 & 1 \\ -2 & s \end{bmatrix} \tag{5-49}$$

The state transition matrix of $\mathbf{A}$ is found by taking the inverse Laplace transform of Eq. (5-49). Thus

$$\boldsymbol{\phi}(t) = \mathscr{L}^{-1}[(s\mathbf{I} - \mathbf{A})^{-1}] = \begin{bmatrix} 2e^{-t} - e^{-2t} & e^{-t} - e^{-2t} \\ -2e^{-t} + 2e^{-2t} & -e^{-t} + 2e^{-2t} \end{bmatrix} \tag{5-50}$$

The state-transition equation for $t \geq 0$ is obtained by substituting Eq. (5-50), **B**, and $u(t)$ into Eq.(5-41). We have

$$\mathbf{x}(t) = \begin{bmatrix} 2e^{-t} - e^{-2t} & e^{-t} - e^{-2t} \\ -2e^{-t} + 2e^{-2t} & -e^{-t} + 2e^{-2t} \end{bmatrix} \mathbf{x}(0)$$
$$= \int_0^t \begin{bmatrix} 2e^{-(t-\tau)} - e^{-2(t-\tau)} & e^{-(t-\tau)} - e^{-2(t-\tau)} \\ -2e^{-(t-\tau)} + e^{-2(t-\tau)} & -e^{-(t-\tau)} + 2e^{-2(t-\tau)} \end{bmatrix} \begin{bmatrix} 0 \\ 1 \end{bmatrix} d\tau \tag{5-51}$$

or

$$\mathbf{x}(t) = \begin{bmatrix} 2e^{-t} - e^{-2t} & e^{-t} - e^{-2t} \\ -2e^{-t} + 2e^{-2t} & -e^{-t} + 2e^{-2t} \end{bmatrix} \mathbf{x}(0) + \begin{bmatrix} 0.5 - e^{-t} + 0.5e^{-2t} \\ e^{-t} - e^{-2t} \end{bmatrix} \quad t \geq 0 \tag{5-52}$$

As an alternative, the second term of the state-transition equation can be obtained by taking the inverse Laplace transform of $(s\mathbf{I} - \mathbf{A})^{-1}\mathbf{B}U(s)$. Thus we have

$$\mathcal{L}^{-1}[(s\mathbf{I} - \mathbf{A})^{-1}]\mathbf{B}U(s) = \mathcal{L}^{-1}\left(\frac{1}{s^2 + 3s + 2}\begin{bmatrix} s+3 & 1 \\ -2 & s \end{bmatrix}\begin{bmatrix} 0 \\ 1 \end{bmatrix}\frac{1}{s}\right)$$
$$= \mathcal{L}^{-1}\left(\frac{1}{s^2 + 3s + 2}\begin{bmatrix} \frac{1}{s} \\ 1 \end{bmatrix}\right) = \begin{bmatrix} 0.5 - e^{-t} + 0.5e^{-2t} \\ e^{-t} - e^{-2t} \end{bmatrix} \quad t \geq 0 \tag{5-53}$$

▲

5-4-1 State-Transition Equation Determined from the State Diagram

Equations (5-40) and (5-41) show that the Laplace transform method of solving the state equations requires the carrying out of the matrix inverse of $(s\mathbf{I} - \mathbf{A})$. We shall now show that the state diagram described in Chapter 3 and the SFG gain formula can be used to solve for the state-transition equation in the Laplace domain of Eq. (5-40). Let the initial time be t_0; then Eq. (5-40) is written

$$\mathbf{X}(s) = (s\mathbf{I} - \mathbf{A})^{-1}\mathbf{x}(t_0) + (s\mathbf{I} - \mathbf{A})^{-1}[\mathbf{B}\mathbf{U}(s) + \mathbf{E}\mathbf{W}(s)] \qquad t \geq t_0 \tag{5-54}$$

The last equation can be written directly from the state diagram using the gain formula, with $X_i(s)$, $i = 1, 2, \ldots, n$, as the input nodes. The following example illustrates the state-diagram method of finding the state-transition equations for the system described in Example 5-1.

Example 5-2

The state diagram for the system described by Eq. (5-46) is shown in Fig. 5-2 with t_0 as the initial time. The outputs of the integrators are assigned as state variables. Applying the gain formula to the state diagram in Fig. 5-2, with $X_1(s)$ and $X_2(s)$ as output nodes, and $x_1(t_0)$, $x_2(t_0)$, and $U(s)$ as input nodes, we have

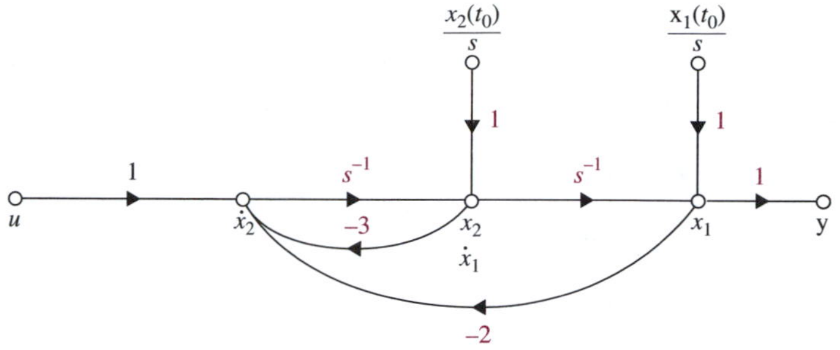

Figure 5-2 State diagram for Eq. (5-46).

$$X_1(s) = \frac{s^{-1}(1 + 3s^{-1})}{\Delta} x_1(t_0) + \frac{s^{-2}}{\Delta} x_2(t_0) + \frac{s^{-2}}{\Delta} U(s) \tag{5-55}$$

$$X_2(s) = \frac{-2s^{-2}}{\Delta} x_1(t_0) + \frac{s^{-1}}{\Delta} x_2(t_0) + \frac{s^{-1}}{\Delta} U(s) \tag{5-56}$$

where

$$\Delta = 1 + 3s^{-1} + 2s^{-2} \tag{5-57}$$

After simplification, Eqs. (5-55) and (5-56) are presented in vector-matrix form:

$$\begin{bmatrix} X_1(s) \\ X_2(s) \end{bmatrix} = \frac{1}{(s+1)(s+2)} \begin{bmatrix} s+3 & 1 \\ -2 & s \end{bmatrix} \begin{bmatrix} x_1(t_0) \\ x_2(t_0) \end{bmatrix} + \frac{1}{(s+1)(s+2)} \begin{bmatrix} 1 \\ s \end{bmatrix} U(s) \tag{5-58}$$

The state-transition equation for $t \geq t_0$ is obtained by taking the inverse Laplace transform on both sides of Eq. (5-58).

Consider that the input $u(t)$ is a unit-step function applied at $t = t_0$. Then the following inverse Laplace transform relationships are identified:

$$\mathcal{L}^{-1}\left(\frac{1}{s}\right) = u_s(t - t_0) \qquad t \geq t_0 \tag{5-59}$$

$$\mathcal{L}^{-1}\left(\frac{1}{s+a}\right) = e^{-a(t-t_0)} u_s(t - t_0) \qquad t \geq t_0 \tag{5-60}$$

Since the initial time is defined to be t_0, the Laplace transform expressions here do not have the delay factor $e^{-t_0 s}$. The inverse Laplace transform of Eq. (5-58) is

$$\begin{bmatrix} x_1(t) \\ x_2(t) \end{bmatrix} = \begin{bmatrix} 2e^{-(t-t_0)} - e^{-2(t-t_0)} & e^{-(t-t_0)} - e^{-2(t-t_0)} \\ -2e^{-(t-t_0)} + 2e^{-2(t-t_0)} & -e^{(t-t_0)} + 2e^{-2(t-t_0)} \end{bmatrix} \begin{bmatrix} x_1(t_0) \\ x_2(t_0) \end{bmatrix} + \begin{bmatrix} 0.5u_s(t-t_0) - e^{-(t-t_0)} + 0.5e^{-2(t-t_0)} \\ e^{-(t-t_0)} - e^{-2(t-t_0)} \end{bmatrix} \qquad t \geq t_0 \tag{5-61}$$

The reader should compare this result with that in Eq. (5-52), which is obtained for $t \geq 0$. ▲

Example 5-3

In this example we illustrate the utilization of the state-transition method to a system with input discontinuity. An *RL* network is shown in Fig. 5-3. The history of the network is completely specified by the initial current of the inductance, $i(0)$ at $t = 0$. Consider that at $t = 0$, the voltage shown in Fig. 5-4 is applied to the network. The state equation of the network for $t > 0$ is

$$\frac{di(t)}{dt} = -\frac{R}{L}i(t) + \frac{1}{L}e_{\text{in}}(t) \tag{5-62}$$

Comparing the last equation with Eq. (5-9), the scalar coefficients of the state equation are identified to be

$$A = -\frac{R}{L} \qquad B = \frac{1}{L} \qquad E = 0 \tag{5-63}$$

The state-transition matrix is

$$\boldsymbol{\phi}(t) = e^{-At} = e^{-Rt/L} \tag{5-64}$$

The conventional approach of solving for $i(t)$ for $t \geq 0$ is to express the input voltage as

$$e(t) = E_{\text{in}}u_s(t) + E_{\text{in}}u_s(t - t_1) \tag{5-65}$$

where $u_s(t)$ is the unit-step function. The Laplace transform of $e(t)$ is

$$E_{\text{in}}(s) = \frac{E_{\text{in}}}{s}(1 + e^{-t_1 s}) \tag{5-66}$$

Then

$$(s\mathbf{I} - \mathbf{A})^{-1}\mathbf{B}\mathbf{U}(s) = \frac{E_{\text{in}}}{Ls(s + R/L)}(1 + e^{-t_1 s}) \tag{5-67}$$

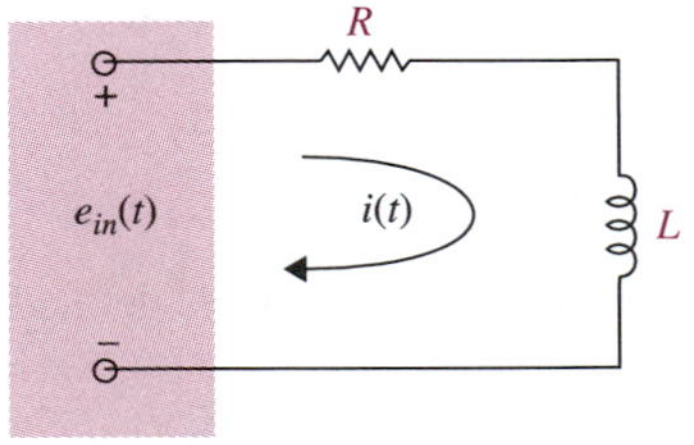

Figure 5-3 *RL* network.

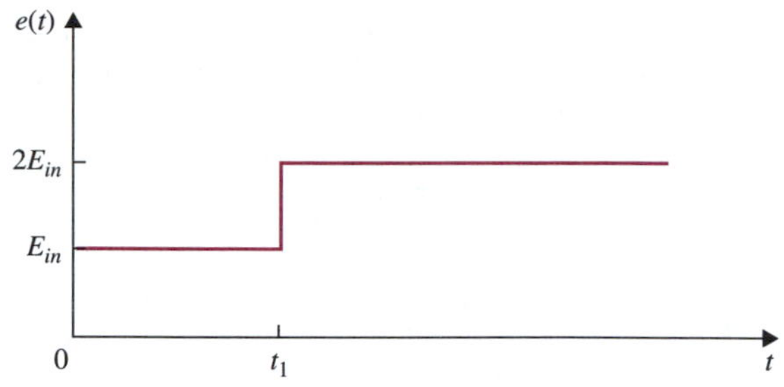

Figure 5-4 Input voltage waveform for the network in Fig. 5-3.

By substituting Eq. (5-67) into Eq. (5-41), the state-transition equation, the current for $t \geq 0$ is obtained:

$$i(t) = e^{-Rt/L} i(0) u_s(t) + \frac{E_{\text{in}}}{R}(1 - e^{-Rt/L}) u_s(t) + \frac{E_{\text{in}}}{R}(1 - e^{-R(t-t_1)/L}) u_s(t - t_1) \qquad (5\text{-}68)$$

Using the state-transition approach, we can divide the transition period into two parts: $t = 0$ to $t = t_1$, and $t = t_1$ to $t = \infty$. First, for the time interval $0 \leq t \leq t_1$, the input is

$$e(t) = E_{\text{in}} u_s(t) \qquad 0 \leq t < t_1 \qquad (5\text{-}69)$$

Then

$$(s\mathbf{I} - \mathbf{A})^{-1}\mathbf{B}\mathbf{U}(s) = \frac{E_{\text{in}}}{Ls(s + R/L)} = \frac{E_{\text{in}}}{Rs[1 + (L/R)s]} \qquad (5\text{-}70)$$

Thus the state-transition equation for the time interval $0 \leq t \leq t_1$ is

$$i(t) = \left[e^{-Rt/L} i(0) + \frac{E_{\text{in}}}{R}(1 - e^{-Rt/L}) \right] u_s(t) \qquad (5\text{-}71)$$

Substituting $t = t_1$ into Eq. (5-71), we get

$$i(t_1) = e^{-Rt_1/L} i(0) + \frac{E_{\text{in}}}{R}(1 - e^{-Rt_1/L}) \qquad (5\text{-}72)$$

The value of $i(t)$ at $t = t_1$ is now used as the initial state for the next transition period of $t_1 \leq t < \infty$. The amplitude of the input for the interval is $2E_{\text{in}}$. The state-transition equation for the second transition period is

$$i(t) = e^{-R(t-t_1)/L} i(t_1) + \frac{2E_{\text{in}}}{R}(1 - e^{-R(t-t_1)/L}) \qquad t \geq t_1 \qquad (5\text{-}73)$$

where $i(t_1)$ is given by Eq. (5-72).

This example illustrates two possible ways of solving a state-transition problem. In the first approach, the transition is treated as one continuous process, whereas in the second, the transition period is divided into parts over which the input can be more easily presented. Although the first approach requires only one operation, the second method yields relatively simple results to the state-transition equation, and it often presents computational advantages. Notice that in the second method, the state at $t = t_1$ is used as the initial state for the next transition period, which begins at t_1. ▲

5-5 Relationship between State Equations and High-Order Differential Equations

In the preceding sections we defined the state equations and their solutions for linear time-invariant systems. Although it is usually possible to write the state equations directly from the schematic diagram of a system, in practice, the system may have been described by a high-order differential equation or transfer function. It becomes necessary to investigate how state equations can be written directly from the high-order differential equation or the transfer function. In Chapter 2 we have illustrated how the state variables of an nth-order differential equation in Eq. (2-6) are intuitively defined in Eq. (2-12). The results are the n state equations in Eq. (2-13).

The state equations are written in vector-matrix form:

$$\frac{d\mathbf{x}(t)}{dt} = \mathbf{A}\mathbf{x}(t) + \mathbf{B}u(t) \tag{5-74}$$

where

$$\mathbf{A} = \begin{bmatrix} 0 & 1 & 0 & \cdots & 0 \\ 0 & 0 & 1 & \cdots & 0 \\ \vdots & \vdots & \vdots & \ddots & \vdots \\ 0 & 0 & 0 & \cdots & 1 \\ -a_0 & -a_1 & -a_2 & \cdots & -a_{n-1} \end{bmatrix} \quad (n \times n) \tag{5-75}$$

$$\mathbf{B} = \begin{bmatrix} 0 \\ 0 \\ \vdots \\ 0 \\ 1 \end{bmatrix} \quad (n \times 1) \tag{5-76}$$

Notice that the last row of **A** contains the negative values of the coefficients of the homogeneous part of the differential equation in ascending order, except for the coefficient of the highest-order term, which is unity. **B** is a column matrix with the last row equal to 1, and the rest of the elements are all zeros. The state equations in Eq. (5-74) with **A** and **B** given in Eqs. (5-75) and (5-76) are known as the **phase-variable canonical form (PVCF),** or the **controllability canonical form (CCF).**

The output equation of the system is written

$$y(t) = \mathbf{C}\mathbf{x}(t) = x_1(t) \tag{5-77}$$

where

$$\mathbf{C} = [1 \quad 0 \quad 0 \quad \cdots \quad 0] \tag{5-78}$$

We have shown earlier that the state variables of a given system are not unique. In general, we seek the most convenient way of assigning the state variables as long as the definition of state variables is satisfied. Unfortunately, there is no intuitive way of defining the state variables for the general nth-order differential equation given in Eq. (3-3), which differs from Eq. (2-6) by having derivatives of the input on the right-hand side of the equation. However, in Section 5-9 we show that by first writing the transfer function and then drawing the state diagram of the system by decomposition of the transfer function, the state variables and state equations of any system can be found very easily. The following example serves the purpose of reviewing the method described in Eqs. (2-12) and (2-13) and above.

Example 5-4

Consider the differential equation

$$\frac{d^3y(t)}{dt^3} + 5\frac{d^2y(t)}{dt^2} + \frac{dy(t)}{dt} + 2y(t) = u(t) \tag{5-79}$$

Rearranging the last equation so that the highest-order derivative term is equated to the rest of the terms, we have

$$\frac{d^3y(t)}{dt^3} = -5\frac{d^2y(t)}{dt^2} - \frac{dy(t)}{dt} - 2y(t) + u(t) \tag{5-80}$$

The state variables are defined as

$$\begin{aligned} x_1(t) &= y(t) \\ x_2(t) &= \frac{dy(t)}{dt} \\ x_3(t) &= \frac{d^2y(t)}{dt^2} \end{aligned} \tag{5-81}$$

Then the state equations are represented by the vector-matrix equation

$$\frac{d\mathbf{x}(t)}{dt} = \mathbf{A}\mathbf{x}(t) + \mathbf{B}u(t) \tag{5-82}$$

where $\mathbf{x}(t)$ is the 2×1 state vector, $u(t)$ the scalar input, and

$$\mathbf{A} = \begin{bmatrix} 0 & 1 & 0 \\ 0 & 0 & 1 \\ -2 & -1 & -5 \end{bmatrix} \qquad \mathbf{B} = \begin{bmatrix} 0 \\ 0 \\ 1 \end{bmatrix} \tag{5-83}$$

The output equation is

$$y(t) = x_1(t) = [1 \quad 0 \quad 0]\mathbf{x}(t) \tag{5-84}$$

▲

5-6 Relationship between State Equations and Transfer Functions

We have presented the methods of modeling a linear time-invariant system by transfer functions and by dynamic equations. It is of interest to investigate the relationship between these two representations.

In Eq. (3-5) the transfer function of a linear single-input single-output system is defined in terms of the coefficients of the system's differential equation. Similarly, Eq. (3-11) gives the transfer-function matrix relation for a multivariable system that has p inputs and q outputs. Now we investigate the transfer-function matrix relation using the dynamic equation notation.

Consider that a linear time-invariant system is described by the dynamic equations

$$\frac{d\mathbf{x}(t)}{dt} = \mathbf{A}\mathbf{x}(t) + \mathbf{B}\mathbf{u}(t) + \mathbf{E}\mathbf{w}(t) \tag{5-85}$$

$$\mathbf{y}(t) = \mathbf{C}\mathbf{x}(t) + \mathbf{D}\mathbf{u}(t) + \mathbf{H}\mathbf{w}(t) \tag{5-86}$$

where

$\mathbf{x}(t) = n \times 1$ state vector $\quad \mathbf{y}(t) = q \times 1$ output vector
$\mathbf{u}(t) = p \times 1$ input vector $\quad \mathbf{w}(t) = v \times 1$ disturbance vector

and **A**, **B**, **C**, **D**, **E**, and **H** are coefficient matrices of appropriate dimensions.

Taking the Laplace transform on both sides of Eq. (5-85) and solving for $\mathbf{X}(s)$, we have

$$\mathbf{X}(s) = (s\mathbf{I} - \mathbf{A})^{-1}\mathbf{x}(0) + (s\mathbf{I} - \mathbf{A})^{-1}[\mathbf{B}\mathbf{U}(s) + \mathbf{E}\mathbf{W}(s)] \tag{5-87}$$

The Laplace transform of Eq. (5-86) is

$$\mathbf{Y}(s) = \mathbf{CX}(s) + \mathbf{DU}(s) + \mathbf{HW}(s) \tag{5-88}$$

Substituting Eq. (5-87) into Eq. (5-88), we have

$$\mathbf{Y}(s) = \mathbf{C}(s\mathbf{I} - \mathbf{A})^{-1}\mathbf{x}(0) + \mathbf{C}(s\mathbf{I} - \mathbf{A})[\mathbf{BU}(s) + \mathbf{EW}(s)] + \mathbf{DU}(s) + \mathbf{HW}(s) \tag{5-89}$$

Since the definition of a transfer function requires that the initial conditions be set to zero, $\mathbf{x}(0) = \mathbf{0}$; thus Eq. (5-89) becomes

$$\mathbf{Y}(s) = [\mathbf{C}(s\mathbf{I} - \mathbf{A})^{-1}\mathbf{B} + \mathbf{D}]\mathbf{U}(s) + [\mathbf{C}(s\mathbf{I} - \mathbf{A})\mathbf{E} + \mathbf{H}]\mathbf{W}(s) \tag{5-90}$$

Let us define

$$\mathbf{G}_r(s) = \mathbf{C}(s\mathbf{I} - \mathbf{A})^{-1}\mathbf{B} + \mathbf{D} \tag{5-91}$$
$$\mathbf{G}_w(s) = \mathbf{C}(s\mathbf{I} - \mathbf{A})^{-1}\mathbf{E} + \mathbf{H} \tag{5-92}$$

where $\mathbf{G}_r(s)$ is a $q \times p$ transfer-function matrix between $\mathbf{u}(t)$ and $\mathbf{y}(t)$ when $\mathbf{w}(t) = \mathbf{0}$, and $\mathbf{G}_v(s)$ is a $q \times v$ transfer-function matrix between $\mathbf{w}(t)$ and $\mathbf{y}(t)$ when $\mathbf{u}(t) = \mathbf{0}$. Then Eq. (5-90) becomes

$$\mathbf{Y}(s) = \mathbf{G}_r(s)\mathbf{U}(s) + \mathbf{G}_w(s)\mathbf{W}(s) \tag{5-93}$$

Example 5-5

Consider that a multivariable system is described by the differential equations

$$\frac{d^2y_1(t)}{dt^2} + 4\frac{dy_1(t)}{dt} - 3y_2(t) = u_1(t) + 2w(t) \tag{5-94}$$

$$\frac{dy_1(t)}{dt} + \frac{dy_2(t)}{dt} + y_1(t) + 2y_2(t) = u_2(t) \tag{5-95}$$

The state variables of the system are assigned as

$$x_1(t) = y_1(t)$$
$$x_2(t) = \frac{dy_1(t)}{dt} \tag{5-96}$$
$$x_3(t) = y_2(t)$$

These state variables are defined by mere inspection of the two differential equations, as no particular reasons for the definitions are given other than that these are the most convenient.

Now equating the first term of each of Eqs. (5-94) and (5-95) to the rest of the terms and using the state-variable relations of Eq. (5-96), we arrive at the following state equations and output equations in vector-matrix form:

$$\begin{bmatrix} \dfrac{dx_1(t)}{dt} \\ \dfrac{dx_2(t)}{dt} \\ \dfrac{dx_3(t)}{dt} \end{bmatrix} = \begin{bmatrix} 0 & 1 & 0 \\ 0 & -4 & 3 \\ -1 & -1 & -2 \end{bmatrix} \begin{bmatrix} x_1(t) \\ x_2(t) \\ x_3(t) \end{bmatrix} + \begin{bmatrix} 0 & 0 \\ 1 & 0 \\ 0 & 1 \end{bmatrix} \begin{bmatrix} u_1(t) \\ u_2(t) \end{bmatrix} + \begin{bmatrix} 0 \\ 2 \\ 0 \end{bmatrix} w(t) \tag{5-97}$$

$$\begin{bmatrix} y_1(t) \\ y_2(t) \end{bmatrix} = \begin{bmatrix} 1 & 0 & 0 \\ 0 & 0 & 1 \end{bmatrix} \begin{bmatrix} x_1(t) \\ x_2(t) \\ x_3(t) \end{bmatrix} = \mathbf{C}\mathbf{x}(t) \tag{5-98}$$

To determine the transfer-function matrix of the system using the state-variable formulation, we substitute the **A**, **B**, **C**, **D**, and **E** matrices into Eq. (5-90). First, we form the matrix $(s\mathbf{I} - \mathbf{A})$:

$$(s\mathbf{I} - \mathbf{A}) = \begin{bmatrix} s & -1 & 0 \\ 0 & s+4 & -3 \\ 1 & 1 & s+2 \end{bmatrix} \tag{5-99}$$

The determinant of $(s\mathbf{I} - \mathbf{A})$ is

$$|s\mathbf{I} - \mathbf{A}| = s^3 + 6s^2 + 11s + 3 \tag{5-100}$$

Thus

$$(s\mathbf{I} - \mathbf{A})^{-1} = \frac{1}{|s\mathbf{I} - \mathbf{A}|} \begin{bmatrix} s^2 + 6s + 11 & s+2 & 3 \\ -3 & s(s+2) & 3s \\ -(s+4) & -(s+1) & s(s+4) \end{bmatrix} \tag{5-101}$$

The transfer-function matrix between $\mathbf{u}(t)$ and $\mathbf{y}(t)$ is

$$\mathbf{G}_r(s) = \mathbf{C}(s\mathbf{I} - \mathbf{A})^{-1}\mathbf{B} = \frac{1}{s^3 + 6s^2 + 11s + 3} \begin{bmatrix} s+2 & 3 \\ -(s+1) & s(s+4) \end{bmatrix} \tag{5-102}$$

and that between $\mathbf{w}(t)$ and $\mathbf{y}(t)$ is

$$\mathbf{G}_w(s) = \mathbf{C}(s\mathbf{I} - \mathbf{A})^{-1}\mathbf{E} = \frac{1}{s^3 + 6s + 11s + 3} \begin{bmatrix} 2(s+2) \\ -2(s+1) \end{bmatrix} \tag{5-103}$$

Using the conventional approach, we take the Laplace transform on both sides of Eqs. (5-94) and (5-95) and assume zero initial conditions. The resulting transformed equations are written in vector-matrix form as

$$\begin{bmatrix} s(s+4) & -3 \\ s+1 & s+2 \end{bmatrix} \begin{bmatrix} Y_1(s) \\ Y_2(s) \end{bmatrix} = \begin{bmatrix} U_1(s) \\ U_2(s) \end{bmatrix} + \begin{bmatrix} 2 \\ 0 \end{bmatrix} W(s) \tag{5-104}$$

Solving for $\mathbf{Y}(s)$ from Eq. (5-104), we obtain

$$\mathbf{Y}(s) = \mathbf{G}_r(s)\mathbf{U}(s) + \mathbf{G}_w(s)W(s) \tag{5-105}$$

where

$$\mathbf{G}_r(s) = \begin{bmatrix} s(s+4) & -3 \\ s+1 & s+2 \end{bmatrix}^{-1} \tag{5-106}$$

$$\mathbf{G}_w(s) = \begin{bmatrix} s(s+4) & -3 \\ s+1 & s+2 \end{bmatrix}^{-1} \begin{bmatrix} 2 \\ 0 \end{bmatrix} \tag{5-107}$$

which will give the same results as in Eqs. (5-102) and (5-103), respectively, when the matrix inverses are carried out. ▲

5-7 Characteristic Equation, Eigenvalues, and Eigenvectors

The characteristic equation plays an important role in the study of linear systems. It can be defined with respect to the differential equation, the transfer function, or the state equations.

Characteristic Equation from Differential Equation

Consider that a linear time-invariant system is described by the differential equation

$$\frac{d^n y(t)}{dt^n} + a_{n-1}\frac{d^{n-1}y(t)}{dt^{n-1}} + \cdots + a_1\frac{dy(t)}{dt} + a_0 y(t) \\ + b_m\frac{d^m u(t)}{dt^m} + b_{m-1}\frac{d^{m-1}u(t)}{dt^{m-1}} + \cdots + b_1\frac{du(t)}{dt} + b_0 u(t) \tag{5-108}$$

where $n > m$. By defining the operator s as

$$s^k = \frac{d^k}{dt^k} \qquad k = 1, 2, \ldots, n \tag{5-109}$$

Eq. (5-108) is written

$$(s^n + a_{n-1}s^{n-1} + \cdots + a_1 s + a_0)y(t) = (b_m s^m + b_{m-1}s^{m-1} + \cdots + b_1 s + b_0)u(t) \tag{5-110}$$

The **characteristic equation** of the system is defined as

$$s^n + a_{n-1}s^{n-1} + \cdots + a_1 s + a_0 = 0 \tag{5-111}$$

which is obtained by setting the homogeneous part of Eq. (5-110) to zero.

Example 5-6

Consider the differential equation in Eq. (5-79). The characteristic equation is obtained by inspection,

$$s^3 + 5s^2 + s + 2 = 0 \tag{5-112}$$

▲

▲ The characteristic equation is obtained by setting the denominator polynomial of the transfer function to zero.

Characteristic Equation from Transfer Function

The transfer function of the system described by Eq. (5-108) is

$$G(s) = \frac{b_m s^m + b_{m-1}s^{m-1} + \cdots + b_1 s + b_0}{s^n + a_{n-1}s^{n-1} + \cdots + a_1 s + a_0} \tag{5-113}$$

The characteristic equation is obtained by equating the denominator polynomial of the transfer function to zero.

Example 5-7

The transfer function of the system described by the differential equation in Eq. (5-79) is

$$\frac{Y(s)}{U(s)} = \frac{1}{s^3 + 5s^2 + s + 2} \tag{5-114}$$

The same characteristic equation as in Eq. (5-112) is obtained by setting the denominator polynomial of Eq. (5-114) to zero. ▲

Characteristic Equation from State Equations

From the state-variable approach, we can write Eq. (5-91) as

$$\begin{aligned}\mathbf{G}_r(s) &= \mathbf{C}\frac{\text{adj}(s\mathbf{I} - \mathbf{A})}{|s\mathbf{I} - \mathbf{A}|}\mathbf{B} + \mathbf{D} \\ &= \frac{\mathbf{C}[\text{adj}(s\mathbf{I} - \mathbf{A})]\mathbf{B} + |s\mathbf{I} - \mathbf{A}|\mathbf{D}}{|s\mathbf{I} - \mathbf{A}|}\end{aligned} \tag{5-115}$$

Setting the denominator of the transfer-function matrix $\mathbf{G}_r(s)$ to zero, we get the characteristic equation as

$$|s\mathbf{I} - \mathbf{A}| = 0 \tag{5-116}$$

which is an alternative form of the characteristic equation and should lead to the same equation as in Eq. (5-111). *An important property of the characteristic equation is that if the coefficients of* **A** *are real, the coefficients of* $|s\mathbf{I} - \mathbf{A}|$ *are also real.*

Example 5-8

The matrix **A** for the state equations of the differential equation in Eq. (5-79) is given in Eq. (5-83). The characteristic equation of **A** is

$$|s\mathbf{I} - \mathbf{A}| = \begin{vmatrix} s & -1 & 0 \\ 0 & s & -1 \\ 2 & 1 & s+5 \end{vmatrix} = s^3 + 5s^2 + s + 2 = 0 \tag{5-117}$$

▲

▲ Eigenvalues of **A** are also roots of the characteristic equation.

5-7-1 Eigenvalues

The roots of the characteristic equation are often referred to as the eigenvalues of the matrix **A**. Some of the important properties of eigenvalues are given as follows.

1. If the coefficients of **A** are all real, its eigenvalues are either real or in complex conjugate pairs.
2. If $\lambda_1, \lambda_2, \ldots, \lambda_n$ are the eigenvalues of **A**, then

$$\text{tr}(\mathbf{A}) = \sum_{i=1}^{n} \lambda_i \tag{5-118}$$

   That is, the trace of **A** is the sum of all the eigenvalues of **A**.
3. If λ_i, $i = 1, 2, \ldots, n$, is an eigenvalue of **A**, it is an eigenvalue of $\mathbf{A}'$.
4. If **A** is nonsingular, with eigenvalues λ_i, $i = 1, 2, \ldots, n$, then $1/\lambda_i$, $i = 1, 2, \ldots, n$, are the eigenvalues of $\mathbf{A}^{-1}$.

Example 5-9

The eigenvalues or the roots of the characteristic equation of the **A** matrix in Eq. (5-83) are obtained by solving for the roots of Eq. (5-117). The results are

$$s = -0.06047 + j0.63738 \qquad s = -0.06047 - j0.63738 \qquad s = -4.87906 \tag{5-119}$$

▲

5-7-2 Eigenvectors

Eigenvectors are useful under many topics in modern control, one of which is in the similarity transformation that will be discussed in a forthcoming section. *Any nonzero vector* $\mathbf{p}_i$ *that satisfies the matrix equation*

$$(\lambda_i \mathbf{I} - \mathbf{A})\mathbf{p}_i = \mathbf{0} \tag{5-120}$$

where λ_i, $i = 1, 2, \ldots, n$, *denotes the* ith *eigenvalue of* $\mathbf{A}$, *is called the* **eigenvector** *of* $\mathbf{A}$ *associated with the eigenvalue* λ_i. If $\mathbf{A}$ has distinct eigenvalues, the eigenvectors can be solved directly from Eq. (5-120).

Example 5-10

Consider that the state equation of Eq. (5-38) has the coefficient matrices

$$\mathbf{A} = \begin{bmatrix} 1 & -1 \\ 0 & -1 \end{bmatrix} \qquad \mathbf{B} = \begin{bmatrix} 1 \\ 1 \end{bmatrix} \qquad \mathbf{E} = \mathbf{0} \tag{5-121}$$

The characteristic equation of $\mathbf{A}$ is

$$|\lambda \mathbf{I} - \mathbf{A}| = \lambda^2 - 1 \tag{5-122}$$

The eigenvalues are $\lambda_1 = 1$ and $\lambda_2 = -1$. Let the eigenvectors be written as

$$\mathbf{p}_1 = \begin{bmatrix} p_{11} \\ p_{21} \end{bmatrix} \qquad \mathbf{p}_2 = \begin{bmatrix} p_{12} \\ p_{22} \end{bmatrix} \tag{5-123}$$

Substituting $\lambda_1 = 1$ and $\mathbf{p}_1$ into Eq. (5-120), we get

$$\begin{bmatrix} 0 & 1 \\ 0 & 2 \end{bmatrix} \begin{bmatrix} p_{11} \\ p_{21} \end{bmatrix} = \begin{bmatrix} 0 \\ 0 \end{bmatrix} \tag{5-124}$$

Thus $p_{21} = 0$, and p_{11} is arbitrary, which in this case can be set equal to 1.

Similarly, for $\lambda_2 = -1$, Eq. (5-120) becomes

$$\begin{bmatrix} -2 & 1 \\ 0 & 0 \end{bmatrix} \begin{bmatrix} p_{12} \\ p_{22} \end{bmatrix} = \begin{bmatrix} 0 \\ 0 \end{bmatrix} \tag{5-125}$$

which leads to

$$-2p_{12} + p_{22} = 0 \tag{5-126}$$

Equation (5-126) has two unknowns, which means that one can be set arbitrarily. Let $p_{12} = 1$; then $p_{22} = 2$. The eigenvectors are

$$\mathbf{p}_1 = \begin{bmatrix} 1 \\ 0 \end{bmatrix} \qquad \mathbf{p}_2 = \begin{bmatrix} 1 \\ 2 \end{bmatrix} \tag{5-127}$$

▲

Generalized Eigenvectors

It should be pointed out that if **A** has multiple-order eigenvalues and is nonsymmetric, not all the **eigenvectors** can be found using Eq. (5-120). Let us assume that among the n eigenvalues of **A**, $q(<n)$ are distinct. The eigenvectors that correspond to the q distinct eigenvalues can be determined in the usual manner from

$$(\lambda_i \mathbf{I} - \mathbf{A})\mathbf{p}_i = \mathbf{0} \tag{5-128}$$

where λ_i denotes the ith distinct eigenvalue, $i = 1, 2, \ldots, q$. Among the remaining high-order eigenvalues, let λ_j be of the mth order ($m \leq n - q$); the corresponding eigenvectors, called the **generalized eigenvectors,** can be determined from the following m vector equations:

$$\begin{aligned}
(\lambda_j \mathbf{I} - \mathbf{A})\mathbf{p}_{n-q+1} &= \mathbf{0} \\
(\lambda_j \mathbf{I} - \mathbf{A})\mathbf{p}_{n-q+2} &= -\mathbf{p}_{n-q+1} \\
(\lambda_j \mathbf{I} - \mathbf{A})\mathbf{p}_{n-q+3} &= -\mathbf{p}_{n-q+2} \\
&\vdots \\
(\lambda_j \mathbf{I} - \mathbf{A})\mathbf{p}_{n-q+m} &= -\mathbf{p}_{n-q+m-1}
\end{aligned} \tag{5-129}$$

Example 5-11

Given the matrix

$$\mathbf{A} = \begin{bmatrix} 0 & 6 & -5 \\ 1 & 0 & -2 \\ 3 & 2 & 4 \end{bmatrix} \tag{5-130}$$

The eigenvalues of **A** are $\lambda_1 = 2$, $\lambda_2 = \lambda_3 = 1$. Thus **A** has a second-order eigenvalue at 1. The eigenvector that is associated with $\lambda_1 = 2$ is determined using Eq. (5-128). Thus

$$(\lambda_1 \mathbf{I} - \mathbf{A})\mathbf{p}_1 = \begin{bmatrix} 2 & -6 & 5 \\ -1 & 2 & -2 \\ -3 & -2 & -2 \end{bmatrix} \begin{bmatrix} p_{11} \\ p_{21} \\ p_{31} \end{bmatrix} = \mathbf{0} \tag{5-131}$$

Since there are only two independent equations in Eq. (5-131), we arbitrarily set $p_{11} = 2$, and we have $p_{21} = -1$ and $p_{31} = -2$. Thus

$$\mathbf{p}_1 = \begin{bmatrix} 2 \\ -1 \\ -2 \end{bmatrix} \tag{5-132}$$

For the generalized eigenvectors that are associated with the second-order eigenvalue, we substitute $\lambda_2 = 1$ into the first equation of Eq. (5-129). We have

$$(\lambda_2\mathbf{I} - \mathbf{A})\mathbf{p}_2 = \begin{bmatrix} 1 & -6 & 5 \\ -1 & 1 & -2 \\ -3 & -2 & -3 \end{bmatrix} \begin{bmatrix} p_{12} \\ p_{22} \\ p_{32} \end{bmatrix} = \mathbf{0} \tag{5-133}$$

Setting $p_{12} = 1$ arbitrarily, we have $p_{22} = -\frac{3}{7}$ and $p_{32} = -\frac{5}{7}$. Thus

$$\mathbf{p}_2 = \begin{bmatrix} 1 \\ -\frac{3}{7} \\ -\frac{5}{7} \end{bmatrix} \tag{5-134}$$

Substituting $\lambda_3 = 1$ into the second equation of Eq. (5-129), we have

$$(\lambda_3\mathbf{I} - \mathbf{A})\mathbf{p}_3 = \begin{bmatrix} 1 & -6 & -5 \\ -1 & 1 & -2 \\ -3 & -2 & -3 \end{bmatrix} \begin{bmatrix} p_{13} \\ p_{23} \\ p_{33} \end{bmatrix} = -\mathbf{p}_2 = \begin{bmatrix} -1 \\ \frac{3}{7} \\ \frac{5}{7} \end{bmatrix} \tag{5-135}$$

Setting p_{13} arbitrarily to 1, we have the generalized eigenvector

$$\mathbf{p}_3 = \begin{bmatrix} 1 \\ -\frac{22}{49} \\ -\frac{46}{49} \end{bmatrix} \tag{5-136}$$

▲

5-8 Similarity Transformations

Given the dynamic equations of a SISO system in the form

$$\frac{d\mathbf{x}(t)}{dt} = \mathbf{A}\mathbf{x}(t) + \mathbf{B}u(t) \tag{5-137}$$

$$y(t) = \mathbf{C}\mathbf{x}(t) + \mathbf{D}u(t) \tag{5-138}$$

where $\mathbf{x}(t)$ is the $n \times 1$ state vector, $u(t)$ and $y(t)$ are the scalar input and output, respectively. When carrying out analysis and design in the state domain, it is often advantageous to transform these equations into particular forms. For example, as we will show later, the controllable canonical form (CCF) has many interesting properties, which makes it convenient for controllability tests and state-feedback design.

Let us consider that the dynamic equations of Eq. (5-137) and (5-138) are transformed into another set of equations of the same dimension by the following transformation:

$$\mathbf{x}(t) = \mathbf{P}\overline{\mathbf{x}}(t) \tag{5-139}$$

where $\mathbf{P}$ is an $n \times n$ nonsingular matrix, so that

$$\overline{\mathbf{x}}(t) = \mathbf{P}^{-1}\mathbf{x}(t) \tag{5-140}$$

The transformed dynamic equations are written

$$\frac{d\overline{\mathbf{x}}(t)}{dt} = \overline{\mathbf{A}}\overline{\mathbf{x}}(t) + \overline{\mathbf{B}}u(t) \tag{5-141}$$

$$\overline{y}(t) = \overline{\mathbf{C}}\overline{\mathbf{x}}(t) + \overline{\mathbf{D}}u(t) \tag{5-142}$$

Taking the derivative on both sides of Eq. (140) with respect to t, we have

$$\begin{aligned}\frac{d\overline{\mathbf{x}}(t)}{dt} &= \mathbf{P}^{-1}\frac{d\mathbf{x}(t)}{dt} = \mathbf{P}^{-1}\mathbf{A}\mathbf{x}(t) + \mathbf{P}^{-1}\mathbf{B}u(t) \\ &= \mathbf{P}^{-1}\mathbf{A}\mathbf{P}\overline{\mathbf{x}}(t) + \mathbf{P}^{-1}\mathbf{B}u(t)\end{aligned} \tag{5-143}$$

Comparing Eq. (5-143) with Eq. (5-141), we get

$$\overline{\mathbf{A}} = \mathbf{P}^{-1}\mathbf{A}\mathbf{P} \tag{5-144}$$

and

$$\overline{\mathbf{B}} = \mathbf{P}^{-1}\mathbf{B} \tag{5-145}$$

Using Eq. (5-139), Eq. (5-142) is written

$$\overline{y}(t) = \mathbf{C}\mathbf{P}\mathbf{x}(t) + \overline{\mathbf{D}}u(t) \tag{5-146}$$

Comparing Eq. (5-146) with (5-138), we see that

$$\overline{\mathbf{C}} = \mathbf{CP} \qquad \overline{\mathbf{D}} = \mathbf{D} \tag{5-147}$$

The transformation described above is called a **similarity transformation,** since in the transformed system such properties as the characteristic equation, eigenvectors, eigenvalues, and transfer function are all preserved by the transformation. We shall describe the controllability canonical form (CCF), the observability canonical form (OCF), and the diagonal canonical form (DCF) transformations in the following. The transformation equations are given without proofs.

5-8-1 Invariance Properties of the Similarity Transformations

One of the important properties of the similarity transformations is that the characteristic equation, eigenvalues, eigenvectors, and transfer functions are invariant under the transformations.

Characteristic Equation, Eigenvalues, Eigenvectors

The characteristic equation of the system described by Eq. (5-141) is $|s\mathbf{I} - \overline{\mathbf{A}}| = 0$ and is written

$$|s\mathbf{I} - \overline{\mathbf{A}}| = |s\mathbf{I} - \mathbf{P}^{-1}\mathbf{AP}| = |s\mathbf{P}^{-1}\mathbf{P} - \mathbf{P}^{-1}\mathbf{AP}| \tag{5-148}$$

Since the determinant of a product matrix is equal to the product of the determinants of the matrices, Eq. (5-148) becomes

$$|s\mathbf{I} - \overline{\mathbf{A}}| = |\mathbf{P}^{-1}|\,|s\mathbf{I} - \mathbf{A}|\,|\mathbf{P}| = |s\mathbf{I} - \mathbf{A}| \tag{5-149}$$

Thus the characteristic equation is preserved, which leads naturally to the same eigenvalues and eigenvectors.

Transfer-Function Matrix

From Eq. (5-91), the transfer-function matrix of the system of Eqs. (5-141) and (5-142) is

$$\begin{aligned}\overline{\mathbf{G}}(s) &= \overline{\mathbf{C}}(s\mathbf{I} - \overline{\mathbf{A}})^{-1}\overline{\mathbf{B}} + \overline{\mathbf{D}} \\ &= \mathbf{CP}(s\mathbf{I} - \mathbf{P}^{-1}\mathbf{AP})\mathbf{P}^{-1}\mathbf{B} + \mathbf{D}\end{aligned} \tag{5-150}$$

which is simplified to

$$\overline{\mathbf{G}}(s) = \mathbf{C}(s\mathbf{I} - \mathbf{A})^{-1}\mathbf{B} + \mathbf{D} = \mathbf{G}(s) \tag{5-151}$$

5-8-2 Controllability Canonical Form

Consider the dynamic equations given in Eqs. (5-137) and (5-138). The characteristic equation of **A** is

$$|s\mathbf{I} - \mathbf{A}| = s^n + a_{n-1}s^{n-1} + \cdots + a_1 s + a_0 = 0 \tag{5-152}$$

The dynamic equations in Eqs. (5-137) and (5-138) are transformed into the controllability canonical form (CCF) of the form of Eqs. (5-141) and (5-142) by the transformation of Eq. (5-139), with

▲ To transform **A** and **B** into **CCF**, the matrix **S** must be nonsingular.

$$\mathbf{P} = \mathbf{SM} \tag{5-153}$$

where

$$\mathbf{S} = [\mathbf{B} \quad \mathbf{AB} \quad \mathbf{A}^2\mathbf{B} \quad \cdots \quad \mathbf{A}^{n-1}\mathbf{B}] \tag{5-154}$$

and

$$\mathbf{M} = \begin{bmatrix} a_1 & a_2 & \cdots & a_{n-1} & 1 \\ a_2 & a_3 & \cdots & 1 & 0 \\ \vdots & \vdots & \ddots & \vdots & \vdots \\ a_{n-1} & 1 & \cdots & 0 & 0 \\ 1 & 0 & \cdots & 0 & 0 \end{bmatrix} \tag{5-155}$$

Then

$$\overline{\mathbf{A}} = \mathbf{P}^{-1}\mathbf{A}\mathbf{P} = \begin{bmatrix} 0 & 1 & 0 & \cdots & 0 \\ 0 & 0 & 1 & \cdots & 0 \\ \vdots & \vdots & \vdots & \ddots & \vdots \\ 0 & 0 & 0 & \cdots & 1 \\ -a_0 & -a_1 & -a_2 & \cdots & -a_{n-1} \end{bmatrix} \tag{5-156}$$

$$\overline{\mathbf{B}} = \mathbf{P}^{-1}\mathbf{B} = \begin{bmatrix} 0 \\ 0 \\ \vdots \\ 0 \\ 1 \end{bmatrix} \tag{5-157}$$

The matrices $\overline{\mathbf{C}}$ and $\overline{\mathbf{D}}$ are given by Eq. (5-147) and do not follow any particular pattern. The CCF transformation requires that $\mathbf{P}^{-1}$ exists, which implies that the matrix $\mathbf{S}$ must have an inverse, since the inverse of $\mathbf{M}$ always exists on account of its determinant is $(-1)^{n-1}$, which is nonzero. The $n \times n$ matrix $\mathbf{S}$ in Eq. (5-154) is later defined as the **controllability matrix.**

Example 5-12

Consider that the coefficient matrices of the state equations in Eq. (5-137) are

$$\mathbf{A} = \begin{bmatrix} 1 & 2 & 1 \\ 0 & 1 & 3 \\ 1 & 1 & 1 \end{bmatrix} \qquad \mathbf{B} = \begin{bmatrix} 1 \\ 0 \\ 1 \end{bmatrix} \tag{5-158}$$

The state equations are to be transformed to the CCF.

The characteristic equation of $\mathbf{A}$ is

$$|s\mathbf{I} - \mathbf{A}| = \begin{vmatrix} s-1 & -2 & -1 \\ 0 & s-1 & -3 \\ -1 & -1 & s-1 \end{vmatrix} = s^3 - 3s^2 - s - 3 = 0 \tag{5-159}$$

Thus the coefficients of the characteristic equation are identified as $a_0 = -3$, $a_1 = -1$, and $a_2 = -3$. From Eq. (5-155),

$$\mathbf{M} = \begin{bmatrix} a_1 & a_2 & 1 \\ a_2 & 1 & 0 \\ 1 & 0 & 0 \end{bmatrix} = \begin{bmatrix} -1 & -3 & 1 \\ -3 & 1 & 0 \\ 1 & 0 & 0 \end{bmatrix} \tag{5-160}$$

The controllability matrix is

$$\mathbf{S} = [\mathbf{B} \quad \mathbf{AB} \quad \mathbf{A}^2\mathbf{B}] = \begin{bmatrix} 1 & 2 & 10 \\ 0 & 3 & 9 \\ 1 & 2 & 7 \end{bmatrix} \tag{5-161}$$

We can show that **S** is nonsingular, so that the system can be transformed into the CCF. Substituting **S** and **M** into Eq. (5-153), we get

$$\mathbf{P} = \mathbf{SM} = \begin{bmatrix} 3 & -1 & 1 \\ 0 & 3 & 0 \\ 0 & -1 & 1 \end{bmatrix} \tag{5-162}$$

Thus, from Eqs. (5-156) and (5-157), the CCF model is given by

$$\overline{\mathbf{A}} = \mathbf{P}^{-1}\mathbf{AP} = \begin{bmatrix} 0 & 1 & 0 \\ 0 & 0 & 1 \\ 3 & 1 & 3 \end{bmatrix} \qquad \overline{\mathbf{B}} = \mathbf{P}^{-1}\mathbf{B} = \begin{bmatrix} 0 \\ 0 \\ 1 \end{bmatrix} \tag{5-163}$$

which could have been determined once the coefficients of the characteristic equation are known. However, the exercise above is to show how the CCF transformation matrix **P** is obtained. ▲

Computer Solutions

Doing matrix manipulations manually is probably the most tedious of all the mathematical endeavors. The M-files in **MATLAB** and **CSAD** can readily be used for the matrix manipulations conducted above. In fact, the function **tr2ccf(A,B)** returns the transformation matrix **P** once the state equation matrices **A** and **B** are entered. The following example gives the results for the system in Example 5-12 using **tr2ccf** of **CSAD** with **MATLAB.**

```
≫A=[1 2 1; 0 1 3; 1 1 1]

A=
        1  2  1

        0  1  3

        1  1  1
```

```
>B=[1; 1; 1]

B=
        1

        0

        1

>P=tr2ccf(A, B) <CR>

P=
        3.0000  -1.0000  1.0000

        0.0000   3.0000  0.0000

        0.0000  -1.0000  1.0000
```

5-8-3 Observability Canonical Form

A dual form of transformation of the CCF is the **observability canonical form** (OCF). The system described by Eqs. (5-137) and (5-138) is transformed to the OCF by the transformation

$$\mathbf{x}(t) = \mathbf{Q}\overline{\mathbf{x}}(t) \tag{5-164}$$

The transformed equations are as given in Eqs. (5-141) and (5-142). Thus

$$\overline{\mathbf{A}} = \mathbf{Q}^{-1}\mathbf{A}\mathbf{Q} \qquad \overline{\mathbf{B}} = \mathbf{Q}^{-1}\mathbf{B} \qquad \overline{\mathbf{C}} = \mathbf{C}\mathbf{Q} \qquad \overline{\mathbf{D}} = \mathbf{D} \tag{5-165}$$

where

$$\overline{\mathbf{A}} = \mathbf{Q}^{-1}\mathbf{A}\mathbf{Q} = \begin{bmatrix} 0 & 0 & \cdots & 0 & -a_0 \\ 1 & 0 & \cdots & 0 & -a_1 \\ 0 & 1 & \cdots & 0 & -a_2 \\ \vdots & \vdots & \ddots & \vdots & \vdots \\ 0 & 0 & \cdots & 0 & -a_{n-1} \end{bmatrix} \tag{5-166}$$

$$\overline{\mathbf{C}} = \mathbf{C}\mathbf{Q} = [0 \quad 0 \quad \cdots \quad 0 \quad 1] \tag{5-167}$$

The elements of the matrices $\overline{\mathbf{B}}$ and $\overline{\mathbf{D}}$ are not restricted to any form. Notice that $\overline{\mathbf{A}}$ and $\overline{\mathbf{C}}$ are the transpose of the $\overline{\mathbf{A}}$ and $\overline{\mathbf{B}}$ in Eqs. (5-156) and (5-157), respectively.

The OCF transformation matrix $\mathbf{Q}$ is given by

$$\mathbf{Q} = (\mathbf{MV})^{-1} \tag{5-168}$$

where $\mathbf{M}$ is as given in Eq. (5-155), and

$$\mathbf{V} = \begin{bmatrix} \mathbf{C} \\ \mathbf{CA} \\ \mathbf{CA}^2 \\ \vdots \\ \mathbf{CA}^{n-1} \end{bmatrix} \quad (n \times n) \tag{5-169}$$

The matrix $\mathbf{V}$ is often defined as the **observability matrix,** and $\mathbf{V}^{-1}$ must exist in order for the OCF transformation to be possible.

Example 5-13

Consider that the coefficient matrices of the system described by Eqs. (5-137) and (5-138) are

$$\mathbf{A} = \begin{bmatrix} 1 & 2 & 1 \\ 0 & 1 & 3 \\ 1 & 1 & 1 \end{bmatrix} \quad \mathbf{B} = \begin{bmatrix} 1 \\ 0 \\ 1 \end{bmatrix} \quad \mathbf{C} = [1 \quad 1 \quad 0] \quad \mathbf{D} = \mathbf{0} \tag{5-170}$$

Since the matrix $\mathbf{A}$ is identical to that of the system in Example 5-12, the matrix $\mathbf{M}$ the same as that in Eq. (5-160). The observability matrix is

$$\mathbf{V} = \begin{bmatrix} \mathbf{C} \\ \mathbf{CA} \\ \mathbf{CA}^2 \end{bmatrix} = \begin{bmatrix} 1 & 1 & 0 \\ 1 & 3 & 4 \\ 5 & 6 & 10 \end{bmatrix} \tag{5-171}$$

We can show that $\mathbf{V}$ is nonsingular, so that the system can be transformed into the OCF. Substituting $\mathbf{V}$ and $\mathbf{M}$ into Eq. (5-168), we have the OCF transformation matrix,

$$\mathbf{Q} = (\mathbf{MV})^{-1} = \begin{bmatrix} 0.3333 & -0.1667 & 0.3333 \\ -0.3333 & 0.1667 & 0.6667 \\ 0.1667 & 0.1667 & 0.1667 \end{bmatrix} \tag{5-172}$$

From Eq. (5-165), the OCF model of the system is described by

$$\overline{\mathbf{A}} = \mathbf{Q}^{-1}\mathbf{AQ} = \begin{bmatrix} 0 & 0 & 3 \\ 1 & 0 & 1 \\ 0 & 1 & 3 \end{bmatrix} \qquad \overline{\mathbf{C}} = \mathbf{CQ} = [0 \quad 0 \quad 1] \qquad \overline{\mathbf{B}} = \mathbf{Q}^{-1}\mathbf{B} = \begin{bmatrix} 3 \\ 2 \\ 1 \end{bmatrix} \tag{5-173}$$

Thus $\overline{\mathbf{A}}$ and $\overline{\mathbf{C}}$ are of the OCF form given in Eqs. (5-166) and (5-167), respectively, and $\overline{\mathbf{B}}$ does not conform to any particular form. ▲

Computer Solutions

The function **tr2ocf** of **CSAD** finds the transformation matrix **Q** for the OCF transformation given the matrices **A** and **C**. The following results are obtained by using **CSAD/MATLAB.**

```
≫A=[1 2 1; 0 1 3; 1 1 1]

A=
        1  2  1

        0  1  3

        1  1  1

≫C=[1 1 0]

C=
        1  1  0

≫Q=tr2ocf(A, C)

Q=
       0.3333  -0.1667  0.3333

      -0.3333   0.1667  0.6667

       0.1667   0.1667  0.1667
```

5-8-4 Diagonal Canonical Form

Given the dynamic equations in Eqs. (5-137) and (5-138), if **A** has *distinct* eigenvalues, there is a nonsingular transformation

$$\mathbf{x}(t) = \mathbf{T}\overline{\mathbf{x}}(t) \tag{5-174}$$

which transforms these equations to the dynamic equations of Eqs. (5-141) and (5-142), where

$$\overline{\mathbf{A}} = \mathbf{T}^{-1}\mathbf{A}\mathbf{T} \qquad \overline{\mathbf{B}} = \mathbf{T}^{-1}\mathbf{B} \qquad \overline{\mathbf{C}} = \mathbf{C}\mathbf{T} \qquad \overline{\mathbf{D}} = \mathbf{D} \tag{5-175}$$

The matrix $\overline{\mathbf{A}}$ is a diagonal matrix,

$$\overline{\mathbf{A}} = \begin{bmatrix} \lambda_1 & 0 & 0 & \cdots & 0 \\ 0 & \lambda_2 & 0 & \cdots & 0 \\ 0 & 0 & \lambda_3 & \cdots & 0 \\ \vdots & \vdots & \vdots & \ddots & \vdots \\ 0 & 0 & 0 & \cdots & \lambda_n \end{bmatrix} \quad (n \times n) \tag{5-176}$$

where $\lambda_1, \lambda_2, \ldots, \lambda_n$ are the n distinct eigenvalues of $\mathbf{A}$. The coefficient matrices $\overline{\mathbf{B}}$, $\overline{\mathbf{C}}$, and $\overline{\mathbf{D}}$ are given in Eq. (5-175) and do not follow any particular form.

It is apparent that one of the advantages of the diagonal canonical form (DCF) is that the transformed state equations are *decoupled* from each other and therefore can be solved individually.

We show in the following that the DCF transformation matrix $\mathbf{T}$ can be formed by use of the eigenvectors of $\mathbf{A}$ as its columns; that is,

$$\mathbf{T} = [\mathbf{p}_1 \quad \mathbf{p}_2 \quad \mathbf{p}_3 \quad \cdots \quad \mathbf{p}_n] \tag{5-177}$$

where $\mathbf{p}_i$, $i = 1, 2, \ldots, n$, denotes the eigenvector associated with the eigenvalue λ_i. This is proved by use of Eq. (5-120), which is written as

$$\lambda_i \mathbf{p}_i = \mathbf{A}\mathbf{p}_i \qquad i = 1, 2, \ldots, n \tag{5-178}$$

Now forming the $n \times n$ matrix,

$$\begin{aligned} [\lambda_1\mathbf{p}_1 \quad \lambda_2\mathbf{p}_2 \quad \cdots \quad \lambda_n\mathbf{p}_n] &= [\mathbf{A}\mathbf{p}_1 \quad \mathbf{A}\mathbf{p}_2 \quad \cdots \quad \mathbf{A}\mathbf{p}_n] \\ &= \mathbf{A}[\mathbf{p}_1 \quad \mathbf{p}_2 \quad \cdots \quad \mathbf{p}_n] \end{aligned} \tag{5-179}$$

Equation (5-179) is written

$$[\mathbf{p}_1 \quad \mathbf{p}_2 \quad \cdots \quad \mathbf{p}_n]\overline{\mathbf{A}} = \mathbf{A}[\mathbf{p}_1 \quad \mathbf{p}_2 \quad \cdots \quad \mathbf{p}_n] \tag{5-180}$$

where $\overline{\mathbf{A}}$ is as given in Eq. (5-176). Thus if we let

$$\mathbf{T} = [\mathbf{p}_1 \quad \mathbf{p}_2 \quad \cdots \quad \mathbf{p}_n] \tag{5-181}$$

Eq. (5-180) is written

$$\overline{\mathbf{A}} = \mathbf{T}^{-1}\mathbf{A}\mathbf{T} \tag{5-182}$$

If the matrix **A** is of the CCF and **A** has distinct eigenvalues, the DCF transformation matrix is the Vandermonde matrix,

$$\mathbf{T} = \begin{bmatrix} 1 & 1 & 1 & \cdots & 1 \\ \lambda_1 & \lambda_2 & \lambda_3 & \cdots & \lambda_n \\ \lambda_1^2 & \lambda_2^2 & \lambda_3^2 & \cdots & \lambda_n^2 \\ \vdots & \vdots & \vdots & \ddots & \vdots \\ \lambda_1^{n-1} & \lambda_2^{n-1} & \lambda_3^{n-1} & \cdots & \lambda_n^{n-1} \end{bmatrix} \tag{5-183}$$

where $\lambda_1, \lambda_2, \ldots, \lambda_n$ are the eigenvalues of **A**. This can be proven by substituting the CCF of **A** in Eq. (5-75) into Eq. (5-120). The result is that the ith eigenvector $\mathbf{p}_i$ is equal to the ith column of **T** in Eq. (5-183).

Example 5-14

Consider the matrix

$$\mathbf{A} = \begin{bmatrix} 0 & 1 & 0 \\ 0 & 0 & 1 \\ -6 & -11 & -6 \end{bmatrix} \tag{5-184}$$

which has eigenvalues $\lambda_1 = -1$, $\lambda_2 = -2$, and $\lambda_3 = -3$. Since **A** is CCF, to transform it into DCF the transformation matrix can be the Vandermonde matrix in Eq. (5-183). Thus

$$\mathbf{T} = \begin{bmatrix} 1 & 1 & 1 \\ \lambda_1 & \lambda_2 & \lambda_3 \\ \lambda_1^2 & \lambda_2^2 & \lambda_3^2 \end{bmatrix} = \begin{bmatrix} 1 & 1 & 1 \\ -1 & -2 & -3 \\ 1 & 4 & 9 \end{bmatrix} \tag{5-185}$$

Thus the DCF of **A** is written

$$\overline{\mathbf{A}} = \mathbf{T}^{-1}\mathbf{A}\mathbf{T} = \begin{bmatrix} -1 & 0 & 0 \\ 0 & -2 & 0 \\ 0 & 0 & -3 \end{bmatrix} \tag{5-186}$$

▲

Computer Solutions

The function **tr2df** of **CSAD** finds the transformation matrix **T** given the system matrix **A**. To demonstrate this, we enter the matrix in Eq. (5-184) in **CSAD**.

```
>>A=[0 1 0; 0 0 1; -6 -11 -6]

A =
      0    1    0

      0    0    1

     -6  -11   -6

>>T=tr2df(A) <CR>

T =
     -0.5774    0.2182   -0.1048

      0.5774   -0.4364    0.3145

     -0.5774    0.8729   -0.9435
```

Notice that the transformation matrix **T** computed by the computer is different than the one given in Eq. (5-185). However, we can show that the corresponding columns of the two matrices are linearly dependent. This further shows that the eigenvectors are not unique.

5-8-5 Jordan Canonical Form

In general, when the **A** matrix has multiple-order eigenvalues, unless the matrix is symmetric with real elements, it cannot be transformed into a diagonal matrix. However, there exists a similarity transformation in the form of Eq. (5-182) such that the $\overline{\mathbf{A}}$ is almost diagonal. The matrix $\overline{\mathbf{A}}$ is called the **Jordan canonical form (JCF)**. A typical JCF is shown below.

$$\overline{\mathbf{A}} = \left[\begin{array}{ccc|c|c} \lambda_1 & 1 & 0 & 0 & 0 \\ 0 & \lambda_1 & 1 & 0 & 0 \\ 0 & 0 & \lambda_1 & 0 & 0 \\ \hline 0 & 0 & 0 & \lambda_2 & 0 \\ \hline 0 & 0 & 0 & 0 & \lambda_3 \end{array}\right] \tag{5-187}$$

where it is assumed that **A** has a third-order eigenvalue λ_1 and distinct eigenvalues λ_2 and λ_3.

The JCF generally has the following properties:

1. The elements on the main diagonal are the eigenvalues.
2. All the elements below the main diagonal are zero.
3. Some of the elements immediately above the multiple-order eigenvalues on the main diagonal are 1s, as illustrated in Eq. (5-187).
4. The 1s, together with the eigenvalues, form typical blocks called **Jordan blocks.** As shown in Eq. (5-187), the Jordan blocks are enclosed by dashed lines.
5. When the nonsymmetrical **A** matrix has multiple-order eigenvalues, its eigenvectors are not linearly independent. For an $n \times n$ **A**, there are only $r(r < n)$ linearly independent eigenvectors.
6. The number of Jordan blocks is equal to the number of independent eigenvectors, r. There is one and only one linearly independent eigenvector associated with each Jordan block.
7. The number of 1s above the main diagonal is equal to $n - r$.

To perform the JCF transformation, the transformation matrix **T** *is again formed by using the eigenvectors and generalized eigenvectors as its columns.*

Example 5-15

Consider the matrix given in Eq.(5-130). We have shown that the matrix has eigenvalues 2, 1, and 1. Thus the DCF transformation matrix can be formed by using the eigenvector and generalized eigenvector given in Eqs. (5-132), (5-134), and (5-136), respectively. That is,

$$\mathbf{T} = [\mathbf{p}_1 \quad \mathbf{p}_2 \quad \mathbf{p}_3] = \begin{bmatrix} 2 & 1 & 1 \\ -1 & -\frac{3}{7} & -\frac{22}{49} \\ -2 & -\frac{5}{7} & -\frac{46}{49} \end{bmatrix} \tag{5-188}$$

Thus the DCF is

$$\overline{\mathbf{A}} = \mathbf{T}^{-1}\mathbf{A}\mathbf{T} = \left[\begin{array}{c:cc} 2 & 0 & 0 \\ \hdashline 0 & 1 & 1 \\ 0 & 0 & 1 \end{array}\right] \tag{5-189}$$

Note that in this case there are two Jordan blocks and there is one element of 1 above the main diagonal. ▲

5-9 Decomposition of Transfer Functions

Up to this point, various methods of characterizing a linear system have been presented. It will be useful to summarize briefly and gather thoughts at this point before proceeding to the main topics of decomposition of transfer functions covered in this section. It has been shown that the starting point of modeling of a linear system may be the system's differential equation, transfer function, or dynamic equations. It is demonstrated that all these methods are closely related. Further, the state diagram defined in Chapter 3 is shown to be a useful tool that not only can lead to the solutions of state equations, but also serves as a vehicle of transformation from one form of description to the others. The block diagram of Fig. 5-5 shows the relationships between the various ways of describing a linear system. The block diagram shows that starting, for instance, with the differential equation of a system, one can get to the solution by use of the transfer-function method or state-equation method. The block diagram also shows that the majority of the relationships are bilateral, so a great deal of flexibility exists between the methods.

One subject remains to be discussed, which involves the construction of the state diagram from the transfer function between the input and the output. The process of going from the transfer function to the state diagram is called **decomposition.** In general, there are three basic methods of decomposition of transfer functions: **direct decomposition, cascade decomposition,** and **parallel decomposition.** Each of these three schemes of decomposition has its own merits and is best suited for a particular purpose.

5-9-1 Direct Decomposition

The direct decomposition is applied to an input–output transfer function that is not in factored form. Consider that the transfer function of an nth-order SISO system between

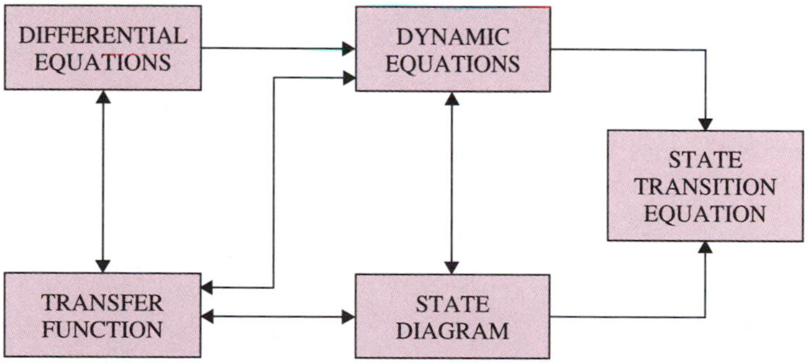

Figure 5-5 Block diagram showing the relationships among various methods of describing linear systems.

the input $U(s)$ and output $Y(s)$ is

$$\frac{Y(s)}{U(s)} = \frac{b_{n-1}s^{n-1} + b_{b-2}s^{n-2} + \cdots + b_1 s + b_0}{s^n + a_{n-1}s^{n-1} + \cdots + a_1 s + a_0} \tag{5-190}$$

where we have assumed that the order of the denominator is at least one higher than that of the numerator. We shall show that the direct decomposition can be conducted in at least two ways, one leading to a state diagram that corresponds to the CCF, the other to the OCF.

Direct Decomposition to CCF

To objective is to construct a state diagram from the transfer function of Eq. (5-190). The following steps are outlined:

1. Express the transfer function in negative powers of s. This is done by multiplying the numerator and the denominator of the transfer function by s^{-n}.
2. Multiply the numerator and the denominator of the transfer function by a dummy variable $X(s)$. By implementing the last two steps, Eq. (5-190) becomes

$$\frac{Y(s)}{U(s)} = \frac{b_{n-1}s^{-1} + b_{n-2}s^{-2} + \cdots + b_1 s^{-n+1} + b_0 s^{-n}}{1 + a_{n-1}s^{-1} + \cdots + a_1 s^{-n+1} + a_0 s^{-n}} \frac{X(s)}{X(s)} \tag{5-191}$$

3. The numerators and the denominators on both sides of Eq. (5-191) are equated to each other, respectively. The results are

$$Y(s) = (b_{n-1}s^{n-1} + b_{n-2}s^{n-2} + \cdots + b_1 s^{-n+1} + b_0 s^{-n})X(s) \tag{5-192}$$
$$U(s) = (1 + a_{n-1}s^{-1} + \cdots + a_1 s^{-n+1} + a_0 s^{-n})X(s) \tag{5-193}$$

4. To construct a state diagram using the two equations in Eqs. (5-192) and (5-193), they must first be in the proper cause-and-effect relation. It is apparent that Eq. (5-192) already satisfies this prerequisite. However, Eq. (5-193) has the input on the left-hand side of the equation and must be rearranged. Equation (5-193) is rearranged as

$$X(s) = U(s) - (a_{n-1}s^{-1} + a_{n-2}s^{-2} + \cdots + a_1 s^{-n+1} + a_0 s^{-n})X(s) \tag{5-194}$$

The state diagram is drawn as shown in Fig. 5-6 using Eqs. (5-192) and (5-194). For simplicity, the initial states are not drawn on the diagram. The state variables $x_1(t)$,

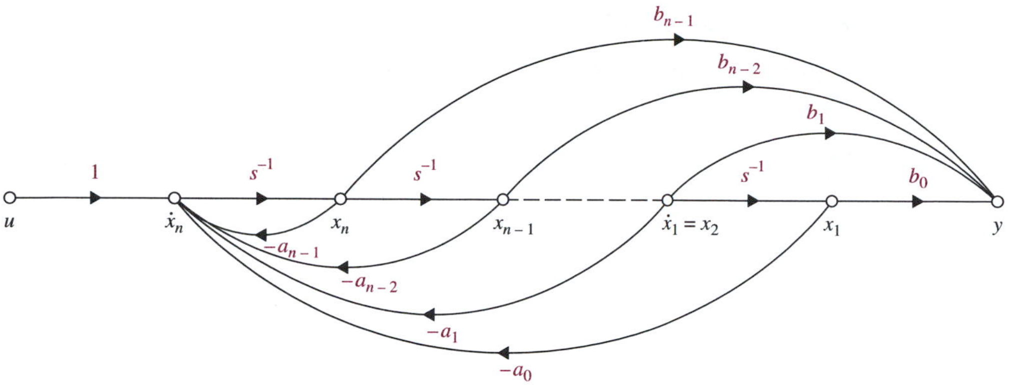

Figure 5-6 CCF state diagram of the transfer function in Eq. (5-190) by direct decomposition.

$x_2(t), \ldots, x_n(t)$ are defined as the outputs of the integrators and arranged in order from the right to the left on the state diagram. The state equations are obtained by applying the SFG gain formula to Fig. 5-6 with the derivatives of the state variables as the outputs and the state variables and $u(t)$ as the inputs, and overlooking the integrator branches. The output equation is determined by applying the gain formula between the state variables, the input, and the output $y(t)$. The dynamic equations are written

$$\frac{d\mathbf{x}(t)}{dt} = \mathbf{A}\mathbf{x}(t) + \mathbf{B}u(t) \tag{5-195}$$

$$y(t) = \mathbf{C}\mathbf{x}(t) + Du(t) \tag{5-196}$$

where

$$\mathbf{A} = \begin{bmatrix} 0 & 1 & 0 & \cdots & 0 \\ 0 & 0 & 1 & \cdots & 0 \\ \vdots & \vdots & \vdots & \ddots & \vdots \\ 0 & 0 & 0 & 0 & 1 \\ -a_0 & -a_1 & -a_2 & \cdots & -a_{n-1} \end{bmatrix} \qquad \mathbf{B} = \begin{bmatrix} 0 \\ 0 \\ \vdots \\ 0 \\ 1 \end{bmatrix} \tag{5-197}$$

$$\mathbf{C} = [b_0 \quad b_1 \quad \cdots \quad b_{n-2} \quad b_{n-1}] \qquad D = 0 \tag{5-198}$$

Apparently, **A** and **B** in Eq. (5-197) are of the CCF.

Direct Decomposition to OCF

Multiplying the numerator and the denominator of Eq. (5-190) by s^{-n}, the equation is expanded as

$$\begin{aligned}(1 + a_{n-1}s^{-1} + \cdots + a_1 s^{-n+1} + a_0 s^{-n})Y(s)\\ = (b_{n-1}s^{-1} + b_{n-2}s^{-2} + \cdots + b_1 s^{-n+1} + b_0 s^{-n})U(s)\end{aligned} \tag{5-199}$$

or

$$\begin{aligned}Y(s) = -(a_{n-1}s^{-1} + \cdots + a_1 s^{-n+1} + a_0 s^{-n})Y(s)\\ + (b_{n-1}s^{-1} + b_{n-2}s^{-2} + \cdots + b_1 s^{-n+1} + b_0 s^{-n})U(s)\end{aligned} \tag{5-200}$$

Figure 5-7 shows the state-diagram form using Eq. (5-200). The outputs of the integrators are designated as the state variables. However, unlike the usual convention, the state variables are assigned in *descending order* from right to left. Applying the SFG gain formula to the state diagram, the dynamic equations are written as in Eqs. (5-195) and (5-196), with

$$\mathbf{A} = \begin{bmatrix} 0 & 0 & \cdots & 0 & -a_0 \\ 1 & 0 & \cdots & 0 & -a_1 \\ 0 & 1 & \cdots & 0 & -a_2 \\ \vdots & \vdots & \ddots & \vdots & \vdots \\ 0 & 0 & \cdots & 0 & -a_{n-1} \end{bmatrix} \qquad \mathbf{B} = \begin{bmatrix} b_0 \\ b_1 \\ b_2 \\ \vdots \\ b_{n-1} \end{bmatrix} \tag{5-201}$$

and

$$\mathbf{C} = [0 \quad 0 \quad \cdots \quad 0 \quad 1] \qquad D = 0 \tag{5-202}$$

The matrices **A** and **C** are of OCF.

It should be pointed out that given the dynamic equations of a system, the input–output transfer function is unique. However, given the transfer function, the state model is not unique, as shown by the CCF, the OCF, and DCF, and many other possibilities. In fact, even for any one of these canonical forms, for example, the CCF, while the matrices **A** and **B** are defined, the elements of **C** and **D** could still be different, depending on how the state diagram is drawn, better described as how the transfer function is decomposed. In other words, referring to Fig. 5-6, while the feedback branches

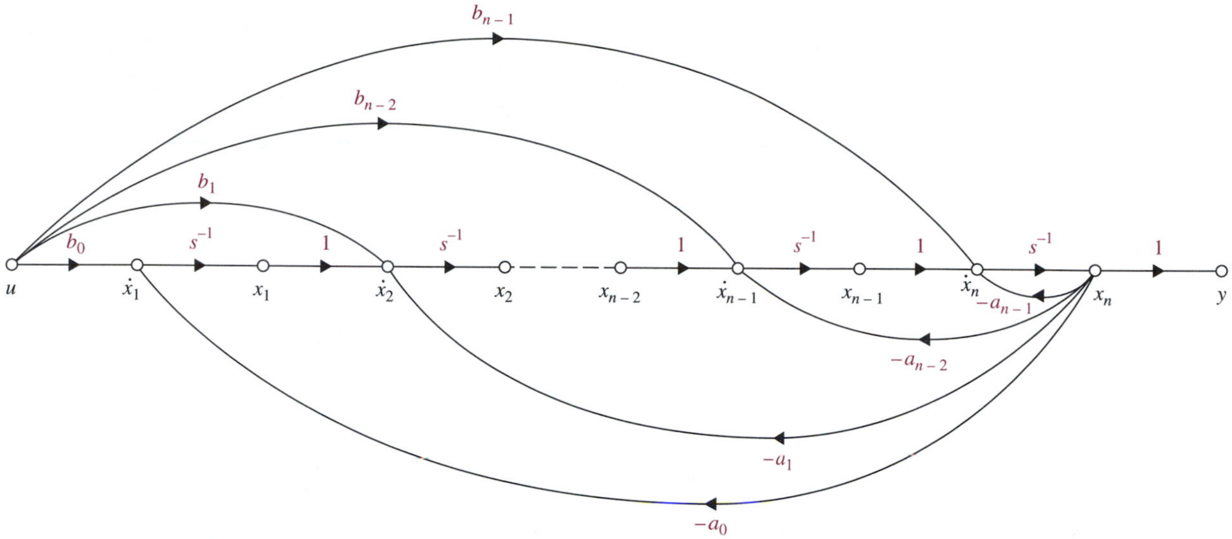

Figure 5-7 OCF state diagram of the transfer function in Eq. (5-190) by direct decomposition.

are fixed, the feedforward branches that contain the coefficients of the numerator of the transfer function could still be manipulated to change the contents of **C**.

Example 5-16

Consider the following input–output transfer function:

$$\frac{Y(s)}{U(s)} = \frac{2s^2 + s + 5}{s^3 + 6s^2 + 11s + 4} \tag{5-203}$$

The CCF state diagram of the system is shown in Fig. 5-8, which is drawn from the following equations:

$$Y(s) = (2s^{-1} + s^{-2} + 5s^{-3})X(s) \tag{5-204}$$

$$X(s) = U(s) - (6s^{-1} + 11s^{-2} + 4s^{-3})X(s) \tag{5-205}$$

The dynamic equations of the system in CCF are

$$\begin{bmatrix} \dfrac{dx_1(t)}{dt} \\ \dfrac{dx_2(t)}{dt} \\ \dfrac{dx_3(t)}{dt} \end{bmatrix} = \begin{bmatrix} 0 & 1 & 0 \\ 0 & 0 & 1 \\ -4 & -11 & -6 \end{bmatrix} \begin{bmatrix} x_1(t) \\ x_2(t) \\ x_3(t) \end{bmatrix} + \begin{bmatrix} 0 \\ 0 \\ 1 \end{bmatrix} u(t) \tag{5-206}$$

$$y(t) = [5 \quad 1 \quad 2]\mathbf{x}(t) \tag{5-207}$$

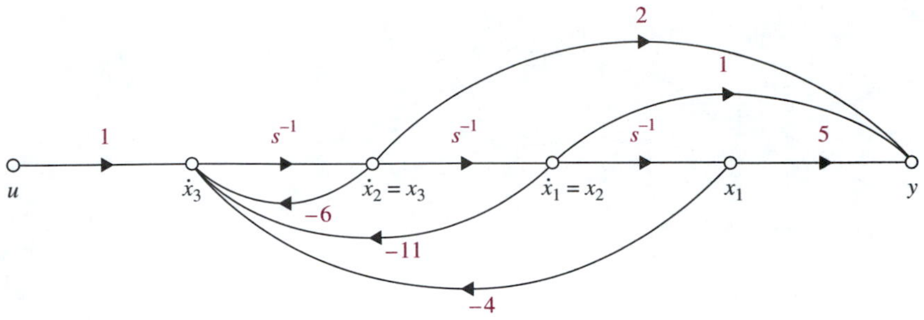

Figure 5-8 CCF state diagram of the transfer function in Eq. (5-203).

For the OCF, Eq. (5-203) is expanded to

$$Y(s) = (2s^{-1} + s^{-2} + 5s^{-3})U(s) - (6s^{-1} + 11s^{-2} + 4s^{-3})Y(s) \tag{5-208}$$

which leads to the OCF state diagram shown in Fig. 5-9. The OCF dynamic equations are written

$$\begin{bmatrix} \dfrac{dx_1(t)}{dt} \\ \dfrac{dx_2(t)}{dt} \\ \dfrac{dx_3(t)}{dt} \end{bmatrix} = \begin{bmatrix} 0 & 0 & -4 \\ 1 & 0 & -11 \\ 0 & 1 & -6 \end{bmatrix} \begin{bmatrix} x_1(t) \\ x_2(t) \\ x_3(t) \end{bmatrix} + \begin{bmatrix} 2 \\ 1 \\ 5 \end{bmatrix} u(t) \tag{5-209}$$

$$y(t) = [0 \quad 0 \quad 1]\mathbf{x}(t) \tag{5-210}$$

▲

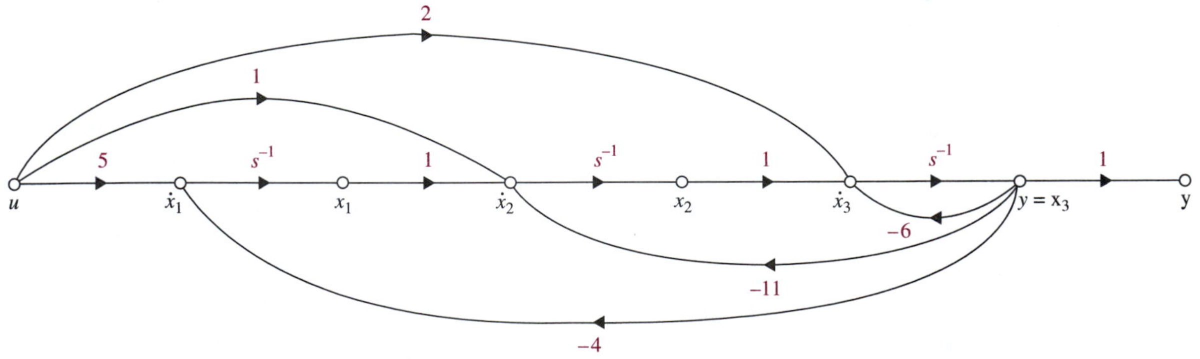

Figure 5-9 OCF state diagram of the transfer function in Eq. (5-203).

Computer Solutions

The **CSAD/MATLAB** toolbox has functions that can be used to compute the CCF and OCF from transfer functions. The **CSAD** command **[A, B, C, D]=tf2ccf(N,D)** gives the matrices **A, B, C,** and **D** in CCF once the numerator polynomial N and the denominator polynomial D of the transfer function are entered. Similarly, the command **[A, B, C, D]=tf2ocf(N,D)** gives the OCF. The **CSAD** function **svstuff** has the **Transformation** option, which conducts the CCF, OCF, or DCF transformation given the system dynamics in either the state-variable form or the transfer function form.

5-9-2 Cascade Decomposition

Cascade compensation refers to transfer functions that are written as products of simple first- or second-order components. Consider the following transfer function, which is the product of two first-order transfer functions.

$$\frac{Y(s)}{U(s)} = K\left(\frac{s+b_1}{s+a_1}\right)\left(\frac{s+b_2}{s+a_2}\right) \tag{5-211}$$

where a_1, a_2, b_1, and b_2 are real constants. Each of the first-order transfer functions is decomposed by the direct decomposition, and the two state diagrams are connected in cascade, as shown in Fig. 5-10. The state equations are obtained by regarding the derivatives of the state variables as outputs, the state variables and $u(t)$ as inputs, and apply the SFG gain formula to the state diagram in Fig. 5-10. The integrator branches are neglected when applying the gain formula. The results are

$$\begin{bmatrix} \dfrac{dx_1(t)}{dt} \\ \dfrac{dx_2(t)}{dt} \end{bmatrix} = \begin{bmatrix} -a_1 & b_2 - a_2 \\ 0 & -a_2 \end{bmatrix} \begin{bmatrix} x_1(t) \\ x_2(t) \end{bmatrix} + \begin{bmatrix} K \\ K \end{bmatrix} u(t) \tag{5-212}$$

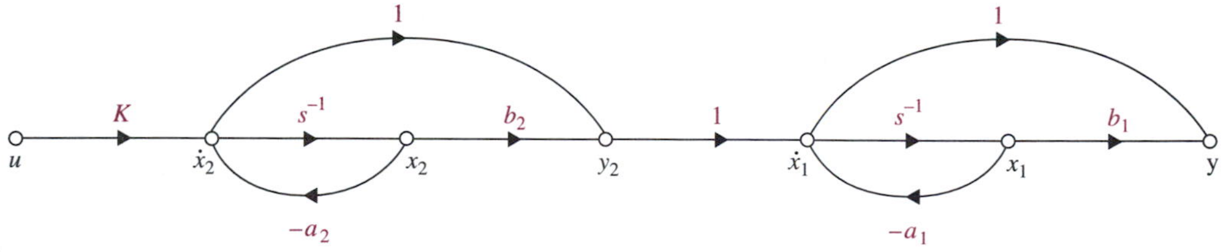

Figure 5-10 State diagram of the transfer function in Eq. (5-211) by cascade decomposition.

The output equation is obtained by regarding the state variables and $u(t)$ as inputs and $y(t)$ as the output, and applying the gain formula to Fig. 5-10. Thus

$$y(t) = [b_1 - a_1 \quad b_2 - a_2]\mathbf{x}(t) + Ku(t) \tag{5-213}$$

When the overall transfer function has complex poles or zeros, the individual factors related to these poles or zeros should be in second-order form. As an example, consider the transfer function

$$\frac{Y(s)}{U(s)} = \left(\frac{s+5}{s+2}\right)\left(\frac{s+1.5}{s^2+3s+4}\right) \tag{5-214}$$

where the poles of the second term are complex. The state diagram of the system with the two subsystems connected in cascade is shown in Fig. 5-11. The dynamic equations of the system are

$$\begin{bmatrix} \dfrac{dx_1(t)}{dt} \\ \dfrac{dx_2(t)}{dt} \\ \dfrac{dx_3(t)}{dt} \end{bmatrix} = \begin{bmatrix} 0 & 1 & 0 \\ -4 & -3 & 3 \\ 0 & 0 & -2 \end{bmatrix} \begin{bmatrix} x_1(t) \\ x_2(t) \\ x_3(t) \end{bmatrix} + \begin{bmatrix} 0 \\ 1 \\ 1 \end{bmatrix} u(t) \tag{5-215}$$

$$y(t) = [1.5 \quad 1 \quad 0]\mathbf{x}(t) \tag{5-216}$$

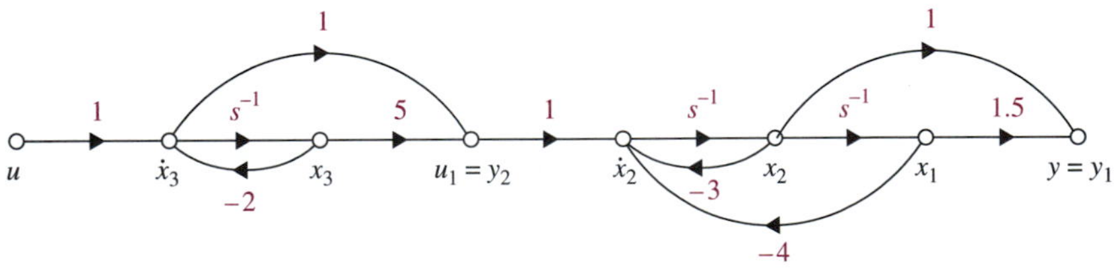

Figure 5-11 Stage diagram of the transfer function in Eq. (5-214) by cascade decomposition.

Computer Solutions

The function **svseries** of the **CSAD** toolbox can be used to find the dynamic equations of two systems connected in cascade. For more than two systems connected in cascade, simply execute **svseries** repeatedly.

Let us apply **svseries** to find the dynamic equations of the system in Eq. (5-214) as two systems connected in cascade. Let us designate the second-order portion of the transfer function as that of system 1 and that of the first-order portion as system 2. From Fig. 5-11 the dynamic equations of the two individual systems are written, and the coefficient matrices are:

$$\text{System 1:}\quad \mathbf{A}_1 = \begin{bmatrix} 0 & 1 \\ -4 & -3 \end{bmatrix} \qquad \mathbf{B}_1 = \begin{bmatrix} 0 \\ 1 \end{bmatrix} \qquad \mathbf{C}_1 = [1.5 \quad 1] \qquad \mathbf{D}_1 = 0 \tag{5-217}$$

$$\text{System 2:}\quad a_2 = -2 \qquad b_2 = 1 \qquad c_2 = 3 \qquad d_2 = 1 \tag{5-218}$$

To execute the function **svseries,** we enter the following data at the **CSAD** prompt:

```
>>a1=[0 1; -4 -3] <CR>
>>b1=[0; 1] <CR>
>>c1=[1.5 1] <CR>
>>d1=0 <CR>
>>a2=-2 <CR>
>>b2=1 <CR>
>>c2=3 <CR>
>>d2=1 <CR>
>>[a,b,c,d]=svseries(a1,b1,c1,d1,a2,b2,c2,d2) <CR>
```

The **svseries** will return the same results as given in Eqs. (5-215) and (5-216).

5-9-3 Parallel Decomposition

When the denominator of the transfer function is in factored form, the transfer function may be expanded by partial-fraction expansion. The resulting state diagram will consist of simple first- or second-order systems connected in parallel, which leads to state equations in DCF or JCF, the latter for the case with multiple-order eigenvalues.

Consider that a second-order system is represented by the transfer function

$$\frac{Y(s)}{U(s)} = \frac{Q(s)}{(s + a_1)(s + a_2)} \tag{5-219}$$

where $Q(s)$ is a polynomial of order less than 2, and a_1 and a_2 are real and distinct. Although analytically a_1 and a_2 may be complex, realistically, complex numbers are difficult to implement on a computer. Equation (5-219) is expanded by partial fraction as

$$\frac{Y(s)}{U(s)} = \frac{K_1}{s + a_1} + \frac{K_2}{s + a_2} \tag{5-220}$$

where K_1 and K_2 are real constants.

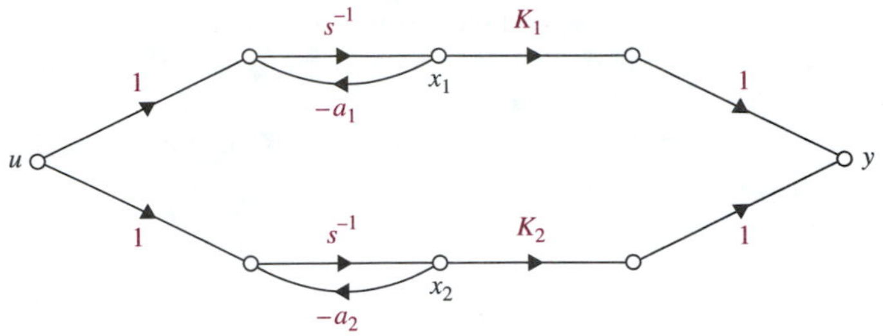

Figure 5-12 State diagram of the transfer function of Eq. (5-219) by parallel decomposition.

The state diagram of the system is drawn by the parallel combination of the state diagrams of each of the first-order terms in Eq. (5-220), as shown in Fig. 5-12. The dynamic equations of the system are

$$\begin{bmatrix} \dfrac{dx_1(t)}{dt} \\ \dfrac{dx_2(t)}{dt} \end{bmatrix} = \begin{bmatrix} -a_1 & 0 \\ 0 & -a_2 \end{bmatrix} \begin{bmatrix} x_1(t) \\ x_2(t) \end{bmatrix} + \begin{bmatrix} 1 \\ 1 \end{bmatrix} u(t) \tag{5-221}$$

$$y(t) = [K_1 \quad K_2]\mathbf{x}(t) \tag{5-222}$$

Thus the state equations are of the DCF.

The conclusion is that for transfer functions with distinct poles, parallel decomposition will lead to the DCF for the state equations. For transfer functions with multiple-order eigenvalues, parallel decomposition to a state diagram with a minimum number of integrators will lead to the JCF state equations. The following example will clarify this point.

Example 5-17

Consider the following transfer function and its partial-fraction expansion:

$$\frac{Y(s)}{U(s)} = \frac{2s^2 + 6s + 5}{(s+1)^2(s+2)} = \frac{1}{(s+1)^2} + \frac{1}{s+1} + \frac{1}{s+2} \tag{5-223}$$

Note that the transfer function is of the third order, and although the total order of the terms on the right-hand side of Fig. (5-223) is four, only three integrators should be used in the state diagram, which is drawn as shown in Fig. 5-13. The minimum number of three integrators is used, with one integrator being shared by two channels. The state equations of the system are written directly from Fig. 5-13.

$$\begin{bmatrix} \dfrac{dx_1(t)}{dt} \\ \dfrac{dx_2(t)}{dt} \\ \dfrac{dx_3(t)}{dt} \end{bmatrix} = \begin{bmatrix} -1 & 1 & 0 \\ 0 & -1 & 0 \\ 0 & 0 & -2 \end{bmatrix} \begin{bmatrix} x_1(t) \\ x_2(t) \\ x_3(t) \end{bmatrix} + \begin{bmatrix} 0 \\ 1 \\ 1 \end{bmatrix} u(t) \tag{5-224}$$

which is recogized to be the JCF. ▲

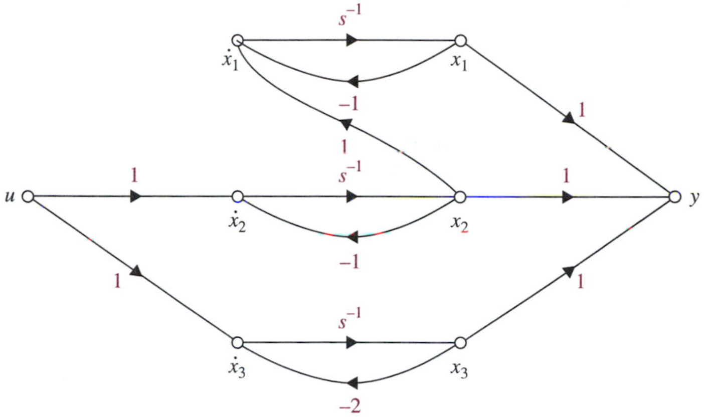

Figure 5-13 State diagram of the transfer function of Eq. (5-223) by parallel decomposition.

Computer Solutions

The function **svparall** of the **CSAD** toolbox can be used to find the dynamic equations of two systems with distinct eigenvalues connected in parallel. The procedure is similar to that of **svseries** and will not be elaborated here.

5-10 Controllability of Linear Systems

The concepts of **controllability** and **observability** introduced first by Kalman [3] play an important role in both theoretical and practical aspects of modern control. The conditions on controllability and observability essentially govern the existence of a solution to an optimal control problem. This seems to be the basic difference between optimal control theory and classical control theory. In the classical control theory, the design techniques are dominated by trial-and-error methods, so that given a set of design specifications the designer at the outset does not know if any solution exists. Optimal control theory, on the other hand, for the most part, has criteria for determining at the outset if the design solution exists or not for the system parameters and design objectives.

We shall show that the condition of controllability of a system is closely related to the existence of solutions of state feedback for the purpose of placing the eigenvalues of the system arbitrarily. The concept of observability relates to the condition of observing or estimating the state variables from the output variables, which are generally measurable.

One way of illustrating the motivation of investigating controllability and observability can be made by referring to the block diagram shown in Fig. 5-14. Figure 5-14(a) shows a system with the process dynamics described by

$$\frac{d\mathbf{x}(t)}{dt} = \mathbf{A}\mathbf{x}(t) + \mathbf{B}\mathbf{u}(\mathbf{t}) \tag{5-225}$$

The closed-loop system is formed by feeding back the state variables through a constant-feedback gain matrix **K**. Thus, from Fig. 5-14,

$$\mathbf{u}(t) = -\mathbf{K}\mathbf{x}(\mathrm{t}) + \mathbf{r}(t) \tag{5-226}$$

where **K** is a $p \times n$ feedback matrix with constant elements. The closed-loop system is thus described by

$$\frac{d\mathbf{x}(t)}{dt} = (\mathbf{A} - \mathbf{B}\mathbf{K})\mathbf{x}(t) + \mathbf{B}\mathbf{r}(t) \tag{5-227}$$

This problem is also known as the **pole-placement design** through state feedback. The design objective in this case is to find the feedback matrix **K** such that the eigenvalues

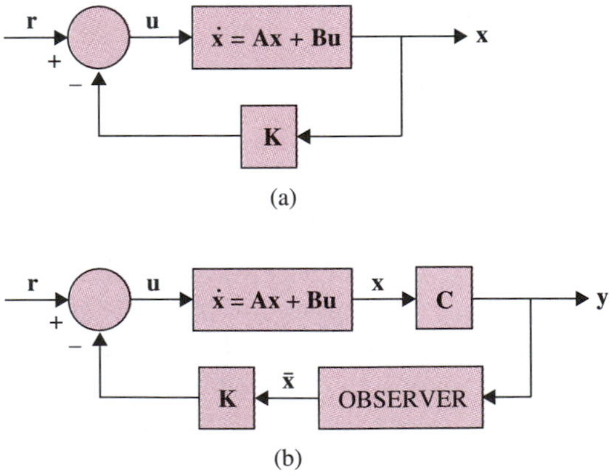

Figure 5-14 (a) Control system with state feedback. (b) Control system with observer and state feedback.

of (**A** − **BG**), or of the closed-loop system, are of certain prescribed values. The word "pole" refers here to the poles of the closed-loop transfer function, which are the same as the eigenvalues of (**A** − **BG**).

We shall show later that the existence of the solution to the pole-placement design with arbitrarily assigned pole values through state feedback is based directly on the controllability of the states of the system. The result is that *if the system of Eq. (5-225) is controllable, there exists a constant-feedback matrix* **K** *that allows the eigenvalues of* (**A** − **BG**) *to be assigned arbitrarily.*

Once the closed-loop system is designed, one has to deal with the practical problem of implementing the feedback of state variables. There are two practical problems with implementing state feedback control. One is that the number of state variables may be excessive, so that the cost of sensing each of these state variables for feedback may be prohibitive. The other problem is that not all the state variables are physically accessible. Therefore, it may be necessary to design and construct an **observer** that will estimate the state vector from the output vector $\mathbf{y}(t)$. Figure 5-14(b) shows the block diagram of a closed-loop system with an observer. The observed state vector $\bar{\mathbf{x}}(t)$ is used to generate the control $\mathbf{u}(t)$ through the feedback matrix **K**. *The condition that such an observer can be designed for the system is called the observability of the system.*

5-10-1 General Concept of Controllability

The concept of controllability can be stated with reference to the block diagram of Fig. 5-15. *The process is said to be* ***completely controllable*** *if every state variable of the process can be controlled to reach a certain objective in finite time by some unconstrained control* $\mathbf{u}(t)$. Intuitively, it is simple to understand that if any one of the state variables is independent of the control $\mathbf{u}(t)$, there would be no way of driving this particular state variable to a desired state in finite time by means of a control effort. Therefore, this particular state is said to be uncontrollable, and as long as there is at least one uncontrollable state, the system is said not to be completely controllable, or simply uncontrollable.

As a simple example of an uncontrollable system, Fig. 5-16 illustrates the state diagram of a linear system with two state variables. Since the control $u(t)$ affects only the state $x_1(t)$, the state $x_2(t)$ is uncontrollable. In other words, it would be impossible to drive $x_2(t)$ from an initial state $x_2(t_0)$ to a desired state $x_2(t_f)$ in a finite time interval $t_f - t_0$ by the control $u(t)$. Therefore, the entire system is said to be uncontrollable.

The concept of controllability given above refers to the states and is sometimes referred to as **state controllability.** Controllability can also be defined for the outputs of the system, so there is a difference between state controllability and output controllability.

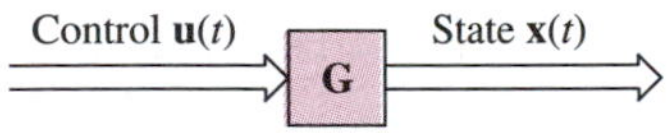

Figure 5-15 Linear time-invariant system.

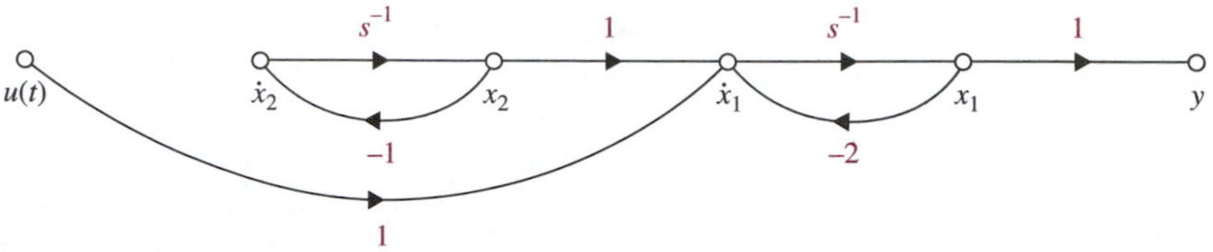

Figure 5-16 State diagram of the system that is not state controllable.

5-10-2 Definition of State Controllability

Consider that a linear time-invariant system is described by the following dynamic equations:

$$\frac{d\mathbf{x}(t)}{dt} = \mathbf{A}\mathbf{x}(t) + \mathbf{B}\mathbf{u}(t) \tag{5-228}$$

$$\mathbf{y}(t) = \mathbf{C}\mathbf{x}(t) + \mathbf{D}\mathbf{u}(t) \tag{5-229}$$

where $\mathbf{x}(t)$ is the $n \times 1$ state vector, $\mathbf{u}(t)$ is the $\mathbf{r} \times 1$ input vector, and $\mathbf{y}(t)$ is the $\mathbf{p} \times 1$ output vector. **A, B, C,** and **D** are coefficients of appropriate dimensions.

The state $\mathbf{x}(t)$ *is said to be controllable at* $t = t_0$ *if there exists a piecewise-continuous input* $\mathbf{u}(t)$ *that will drive the state to any final state* $\mathbf{x}(t_f)$ *for a finite time* $(t_f - t_0) \geq 0$. *If every state* $\mathbf{x}(t_0)$ *of the system is controllable in a finite time interval, the system is said to be completely state controllable or simply controllable.*

The following theorem shows that the condition of controllability depends on the coefficient matrices **A** and **B** of the system. The theorem also gives one method of testing for state controllability.

▲ Theorem 5-1

For the system described by the state equation of Eq. (5-228) to be **completely state controllable,** it is necessary and sufficient that the following $n \times nr$ **controllability matrix** has a rank of n:

$$\mathbf{S} = [\mathbf{B} \quad \mathbf{AB} \quad \mathbf{A}^2\mathbf{B} \quad \cdots \quad \mathbf{A}^{n-1}\mathbf{B}] \tag{5-230}$$

Since the matrices **A** and **B** are involved, sometimes we say that the pair [**A, B**] is controllable which implies that **S** is of rank n.

The proof of this theorem is given in any standard textbook on optimal control systems [3]. The idea is to start with the state-transition equation of Eq. (5-44) and then proceed to show that Eq. (5-230) must be satisfied in order that all the states are accessible by the input.

Although the criterion of state controllability given in Theorem 5-1 is quite straightforward, manually, it is not very easy to test for high-order and/or systems with many inputs. If **S** is nonsquare, we can form the matrix **SS′**, which is $n \times n$; then if **SS′** is nonsingular, **S** has rank n.

5-10-3 Alternative Tests on Controllability

There are several other alternative methods of testing controllability, and some of these may be more convenient to apply than the condition in Eq. (5-230).

▲ Theorem 5-2

For a SISO system described by the state equation of Eq. (5-228), the pair [**A, B**] is completely controllable if **A** and **B** are in CCF or transformable into CCF by a similarity transformation.

The proof of this theorem is straightforward, since it was established in Section 5-8 that the CCF transformation requires that the controllability matrix **S** be nonsingular. Since the CCF transformation in Section 5-8 was defined only for SISO systems, the theorem applies only to this type of systems.

▲ Theorem 5-3

For a system described by the state equation of Eq. (5-228), if **A** is in DCF or JCF, the pair [**A, B**] is completely controllable if all the elements in the rows of **B** that correspond to the last row of each Jordan block are nonzero.

The proof of this theorem comes directly from the definition of controllability. Let us assume that **A** is diagonal and that it has distinct eigenvalues. Then the pair [**A, B**] is controllable if **B** does not have any row with all zeros. The reason is that if **A** is diagonal, all the states are decoupled from each other, and if any row of **B** contains all zero elements, the corresponding state would not be accessed from any of the inputs, and that state would be uncontrollable.

For a system in JCF, such as the **A** and **B** matrices illustrated below, for controllability, only the elements in the row of **B** that corresponds to the last row of the Jordan block are not all zeros. The elements in the other rows of **B** need not all be nonzero, since the corresponding states are still coupled through the 1s in the Jordan blocks of **A**.

$$\mathbf{A} = \left[\begin{array}{ccc|c} \lambda_1 & 1 & 0 & 0 \\ 0 & \lambda_1 & 1 & 0 \\ 0 & 0 & \lambda_1 & 0 \\ \hline 0 & 0 & 0 & \lambda_2 \end{array}\right] \qquad \mathbf{B} = \begin{bmatrix} b_{11} & b_{12} \\ b_{21} & b_{22} \\ b_{31} & b_{32} \\ b_{41} & b_{42} \end{bmatrix} \tag{5-231}$$

Thus the condition of controllability for the **A** and **B** in Eq. (5-231) is $b_{31} \neq 0$, $b_{32} \neq 0$, $b_{41} \neq 0$, and $b_{42} \neq 0$.

Example 5-18

The following matrices are for a system with two identical eigenvalues, but the matrix **A** is diagonal.

$$\mathbf{A} = \begin{bmatrix} \lambda_1 & 0 \\ 0 & \lambda_1 \end{bmatrix} \qquad \mathbf{B} = \begin{bmatrix} b_{11} \\ b_{21} \end{bmatrix} \tag{5-232}$$

The system is uncontrollable, since the two state equations are dependent; that is, it would not be possible to control the states independently by the input. We can easily show that in this case $\mathbf{S} = [\mathbf{B} \quad \mathbf{AB}]$ is singular. ▲

Example 5-19

Consider the system shown in Fig. 5-16, which was reasoned earlier to be uncontrollable. Let us investigate the same system using the condition of Eq. (5-230). The state equations of the system are written in the form of Eq. (5-228) with

$$\mathbf{A} = \begin{bmatrix} -2 & 1 \\ 0 & -1 \end{bmatrix} \qquad \mathbf{B} = \begin{bmatrix} 1 \\ 0 \end{bmatrix} \tag{5-233}$$

Thus, from Eq. (5-230), the controllability matrix is

$$\mathbf{S} = [\mathbf{B} \quad \mathbf{AB}] = \begin{bmatrix} 1 & -2 \\ 0 & 0 \end{bmatrix} \tag{5-234}$$

which is singular, and the system is uncontrollable. ▲

Example 5-20

Consider that a third-order system has the coefficient matrices

$$\mathbf{A} = \begin{bmatrix} 1 & 2 & -1 \\ 0 & 1 & 0 \\ 1 & -4 & 3 \end{bmatrix} \qquad \mathbf{B} = \begin{bmatrix} 0 \\ 0 \\ 1 \end{bmatrix} \tag{5-235}$$

The controllability matrix is

$$\mathbf{S} = [\mathbf{B} \quad \mathbf{AB} \quad \mathbf{A}^2\mathbf{B}] = \begin{bmatrix} 0 & -1 & -4 \\ 0 & 0 & 0 \\ 1 & 3 & 8 \end{bmatrix} \tag{5-236}$$

which is singular. Thus the system is not controllable.

The eigenvalues of **A** are $\lambda_1 = 2$, $\lambda_2 = 2$, and $\lambda_3 = 1$. The JCF of **A** and **B** are obtained with the transformation $\mathbf{x}(t) = \mathbf{T}\overline{\mathbf{x}}(t)$, where

$$\mathbf{T} = \begin{bmatrix} 1 & 0 & 0 \\ 0 & 0 & 1 \\ -1 & 1 & 2 \end{bmatrix} \tag{5-237}$$

Then

$$\overline{\mathbf{A}} = \mathbf{T}^{-1}\mathbf{A}\mathbf{T} = \begin{bmatrix} 2 & 1 & 0 \\ 0 & 2 & 0 \\ 0 & 0 & 1 \end{bmatrix} \qquad \overline{\mathbf{B}} = \mathbf{T}^{-1}\mathbf{B} = \begin{bmatrix} 0 \\ -1 \\ 0 \end{bmatrix} \tag{5-238}$$

Since the last row of $\overline{\mathbf{B}}$, which corresponds to the Jordan block for the eigenvalue λ_3, is zero, the transformed state variable $\bar{x}_3(t)$ is uncontrollable. From the transformation matrix **T** in Eq. (5-237), $x_2 = \bar{x}_3$, which means that x_2 is uncontrollable in the original system. ▲

Computer Solutions

The function **svdesign** in the **CSAD** toolbox has an option that returns the controllability condition of a SISO system once the parameters of the dynamic equations or the transfer function are entered.

5-11 Observability of Linear Systems

The concept of observability was covered in Section 5-10. Essentially, *a system is completely observable if every state variable of the system affects some of the outputs.* In other words, it is often desirable to obtain information on the state variables from the measurements of the outputs and the inputs. If any one of the states cannot be observed from the measurements of the outputs, the state is said to be unobservable, and the system is not completely observable, or simply unobservable. Figure 5-17 shows the state diagram of a linear system in which the state x_2 is not connected to the output $y(t)$ in any way. Once we have measured $y(\mathrm{t})$, we can observe the state $x_1(t)$, since

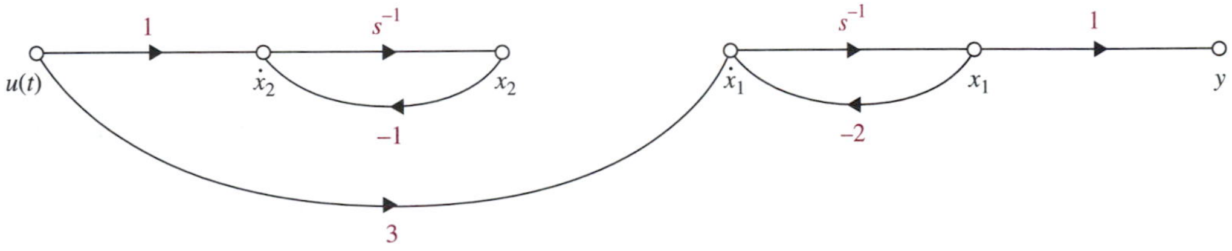

Figure 5-17 State diagram of a system that is not observable.

$x_1(t) = y(t)$. However, the state x_2 cannot be observed from the information on $y(t)$. Thus the system is unobservable.

5-11-1 Definition of Observability

Given a linear time-invariant system that is described by the dynamic equations of Eqs. (5-228) and (5-229), the state $\mathbf{x}(t_0)$ *is said to be observable if given any input* $u(t)$, *there exists a finite time* $t_f \geq t_0$ *such that the knowledge of* $\mathbf{u}(t)$ *for* $t_0 \leq t < t_f$; *matrices* **A, B, C,** *and* **D;** *and the output* $\mathbf{y}(t)$ *for* $t_0 \leq t < t_f$ *are sufficient to determine* $\mathbf{x}(t_0)$. *If every state of the system is observable for a finite* t_f, *we say that the system is completely observable, or simply observable.*

The following theorem shows that the condition of observability depends on the matrices **A** and **C** of the system. The theorem also gives one method of testing observability.

▲ **Theorem 5-4**

For the system described by Eqs. (5-228) and (5-229) to be completely observable, it is necessary and sufficient that the following $n \times np$ **observability matrix** has a rank of n:

$$\mathbf{V} = \begin{bmatrix} \mathbf{C} \\ \mathbf{CA} \\ \mathbf{CA}^2 \\ \vdots \\ \mathbf{CA}^{n-1} \end{bmatrix} \tag{5-239}$$

The condition is also referred to as the pair [**A, C**] being observable. In particular, if the system has only one output, **C** is a $1 \times n$ row matrix; **V** is an $n \times n$ square matrix. Then the system is completely observable if **V** is nonsingular.

The proof of this theorem is not given here. It is based on the principle that Eq. (5-239) must be satisfied so that $\mathbf{x}(t_0)$ can be uniquely determined from the output $\mathbf{y}(t)$.

5-11-2 Alternative Tests on Observability

Just as with controllability, there are several other alternative methods of testing observability. These are described in the following.

▲ **Theorem 5-5**

For a SISO system described by the dynamic equations of Eqs. (5-228) and (5-229), the pair [**A, C**] is completely observable if **A** and **C** are in OCF or transformable into OCF by a similarity transformation.

The proof of this theorem is straightforward, since it was established in Section 5-8 that the OCF transformation requires that the observability matrix **V** be nonsingular.

▲ Theorem 5-6

For a system described by the dynamic equations of Eqs. (5-228) and (5-229), if **A** is in DCF or JCF, the pair [**A**, **C**] is completely observable if all the elements in the columns of **C** that correspond to the first row of each Jordan block are nonzero.

Note that this theorem is a dual of the test of controllability given in Theorem 5-3. If the system has distinct eigenvalues, **A** is diagonal; then the condition on observability is that none of the columns of **C** can contain all zeros.

Example 5-21

Consider the system shown in Fig. 5-17, which was earlier defined to be unobservable. The dynamic equations of the system are expressed in the form of Eqs. (5-228) and (5-229) with

$$\mathbf{A} = \begin{bmatrix} -2 & 0 \\ 0 & -1 \end{bmatrix} \qquad \mathbf{B} = \begin{bmatrix} 3 \\ 1 \end{bmatrix} \qquad \mathbf{C} = [1 \quad 0] \tag{5-240}$$

Thus the observability matrix is

$$\mathbf{V} = \begin{bmatrix} \mathbf{C} \\ \mathbf{CA} \end{bmatrix} = \begin{bmatrix} 1 & 0 \\ -2 & 0 \end{bmatrix} \tag{5-241}$$

which is singular. Thus the pair [**A**, **C**] is uncontrollable. In fact, since **A** is of DCF and the second column of **C** is zero, this means that the state $x_2(t)$ is unobservable, as conjectured from Fig. 5-17. ▲

Computer Solutions

The function **svdesign** in the **CSAD** toolbox has an option that returns the observability condition of a SISO system once the parameters of the dynamic equations or the transfer function are entered.

5-12 Relationship among Controllability, Observability, and Transfer Functions

In the classical analysis of control systems, transfer functions are used for modeling of linear time-invariant systems. Although controllability and observability are concepts and tools of modern control theory, we shall show that they are closely related to the properties of transfer functions.

▲ Theorem 5-7

If the input–output transfer function of a linear system has pole–zero cancellation, the system will be either uncontrollable or unobservable, or both, depending on how the state variables are defined. On the other hand, if the input–output transfer function does not have pole–zero cancellation, the system can always be represented by dynamic equations as a completely controllable and observable system.

The proof of this theorem is not given here. The importance of this theorem is that if a linear system is modeled by a transfer function with no pole–zero cancellation, we are assured that it is a controllable and observable system, no matter how the state-variable model is derived. Let us amplify this point further by referring to the following SISO system.

$$\mathbf{A} = \begin{bmatrix} -1 & 0 & 0 & 0 \\ 0 & -2 & 0 & 0 \\ 0 & 0 & -3 & 0 \\ 0 & 0 & 0 & -4 \end{bmatrix} \qquad \mathbf{B} = \begin{bmatrix} 1 \\ 1 \\ 0 \\ 0 \end{bmatrix} \qquad \mathbf{C} = [1 \quad 0 \quad 1 \quad 0] \qquad D = 0 \tag{5-242}$$

Since **A** is a diagonal matrix, the controllability and observability conditions of the four states are determined by inspection. They are:

x_1: controllable and observable (C and O)
x_2: controllable but unobservable (C but UO)
x_3: uncontrollable but observable (UC but O)
x_4: uncontrollable and uncontrollable (UC and UO)

The block diagram of the system in Fig. 5-18 shows the DCF decomposition of the system. Clearly, the transfer function of the controllable and observable system should be

$$\frac{Y(s)}{U(s)} = \frac{1}{s+1} \tag{5-243}$$

whereas the transfer function that corresponds to the dynamics described in Eq. (5-242) is

$$\frac{Y(s)}{U(s)} = \mathbf{C}(s\mathbf{I} - \mathbf{A})^{-1}\mathbf{B} = \frac{(s+2)(s+3)(s+4)}{(s+1)(s+2)(s+3)(s+4)} \tag{5-244}$$

which has three pole–zero cancellations. This simpleminded example illustrates that ‘‘minimum-order’’ transfer function without pole–zero cancellation is the only component that corresponds to a system that is controllable and observable.

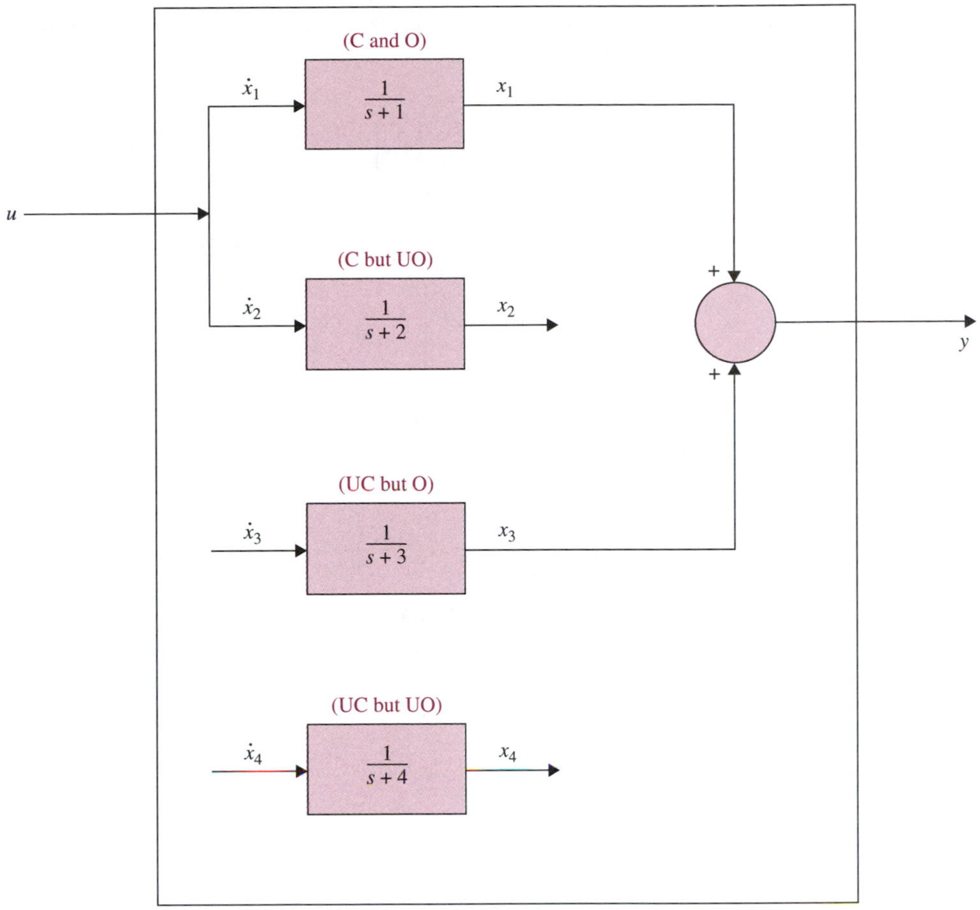

Figure 5-18 Block diagram showing the controllable, uncontrollable, observable, and unobservable components of the system described in Eq. (5-242).

Example 5-22

Let us consider the transfer function

$$\frac{Y(s)}{U(s)} = \frac{s+2}{(s+1)(s+2)} \tag{5-245}$$

which is a reduced form of Eq. (5-244). Equation (5-245) is decomposed into CCF and OCF as follows (the function **svdesign** of the **CSAD** toolbox can be used for these operations):

$$CCF: \quad \mathbf{A} = \begin{bmatrix} 0 & 1 \\ -2 & -3 \end{bmatrix} \quad \mathbf{B} = \begin{bmatrix} 0 \\ 1 \end{bmatrix} \quad \mathbf{C} = [1 \quad 1] \tag{5-246}$$

Since the CCF transformation can be made, the pair [**A**, **B**] of the CCF is controllable. The observability matrix is

$$\mathbf{V} = \begin{bmatrix} \mathbf{C} \\ \mathbf{CA} \end{bmatrix} = \begin{bmatrix} 1 & 1 \\ -2 & -2 \end{bmatrix} \tag{5-247}$$

which is singular, and the pair [**A**, **C**] of the CCF is unobservable.

$$OCF: \quad \mathbf{A} = \begin{bmatrix} 0 & -2 \\ 1 & -3 \end{bmatrix} \qquad \mathbf{B} = \begin{bmatrix} 1 \\ 1 \end{bmatrix} \qquad \mathbf{C} = [0 \quad 1] \tag{5-248}$$

Since the OCF transformation can be made, the pair [**A**, **C**] of the OCF is observable. However, the controllability matrix is

$$\mathbf{S} = [\mathbf{B} \quad \mathbf{AB}] = \begin{bmatrix} 1 & -2 \\ 1 & -2 \end{bmatrix} \tag{5-249}$$

which is singular, and the pair [**A**, **B**] of the OCF is uncontrollable.

The conclusion that can be drawn from this example is that given a system that is modeled by a transfer function, the controllability and observability conditions of the system depend on how the state variables are defined. ▲

5-13 Invariant Theorems on Controllability and Observability

We now investigate the effects of the similarity transformations on controllability and observability. The effects of controllability and observability due to state feedback will be investigated.

▲ **Theorem 5-8.** *Invariant Theorem on Similarity Transformations*

Consider that the system described by the dynamic equations of Eqs. (5-228) and (5-229). The similarity transformation $\mathbf{x}(t) = \mathbf{P}\overline{\mathbf{x}}(t)$, where **P** is nonsingular, transforms the dynamic equations to

$$\frac{d\overline{\mathbf{x}}(t)}{dt} = \overline{\mathbf{A}}\overline{\mathbf{x}}(t) + \overline{\mathbf{B}}\mathbf{u}(t) \tag{5-250}$$

$$\overline{\mathbf{y}}(t) = \overline{\mathbf{C}}\mathbf{x}(t) + \overline{\mathbf{D}}\mathbf{u}(t) \tag{5-251}$$

where

$$\overline{\mathbf{A}} = \mathbf{P}^{-1}\mathbf{AP} \qquad \overline{\mathbf{B}} = \mathbf{P}^{-1}\mathbf{B} \tag{5-252}$$

The controllability of $[\overline{\mathbf{A}}, \overline{\mathbf{B}}]$ and the observability of $[\overline{\mathbf{A}}, \overline{\mathbf{C}}]$ are not affected by the transformation.

In other words, controllability and observability are preserved through similar transformations. The theorem is easily proven by showing that the ranks of $\overline{\mathbf{S}}$ and $\mathbf{S}$ and the ranks of $\overline{\mathbf{V}}$ and $\mathbf{V}$ are identical, where $\overline{\mathbf{S}}$ and $\overline{\mathbf{V}}$ are the controllability and observability matrices, respectively, of the transformed system.

▲ **Theorem 5-9.** *Theorem on Controllability of Closed-Loop Systems with State Feedback*

If the open-loop system

$$\frac{d\mathbf{x}(t)}{dt} = \mathbf{A}\mathbf{x}(t) + \mathbf{B}\mathbf{u}(t) \tag{5-253}$$

is completely controllable, the closed-loop system obtained through state feedback,

$$\mathbf{u}(t) = \mathbf{r}(t) - \mathbf{K}\mathbf{x}(t) \tag{5-254}$$

so that the state equation becomes

$$\frac{d\mathbf{x}(t)}{dt} = (\mathbf{A} - \mathbf{B}\mathbf{K})\mathbf{x}(t) + \mathbf{B}\mathbf{r}(t) \tag{5-255}$$

is also completely controllable. On the other hand, if [**A**, **B**] is uncontrollable, there is no **K** that will make the pair [**A** − **BK**, **B**] controllable. In other words, if an open-loop system is uncontrollable, it cannot be made controllable through state feedback.

▲ ***Proof*** The controllability of [**A**, **B**] implies that there exists a control $\mathbf{u}(t)$ over the time interval $[t_0, t_f]$ such that the initial state $\mathbf{x}(t_0)$ is driven to the final state $\mathbf{x}(t_f)$ over the finite time interval $t_f - t_0$. We can write Eq. (5-254) as

$$\mathbf{r}(t) = \mathbf{u}(t) + \mathbf{K}\mathbf{x}(t) \tag{5-256}$$

which is the control of the closed-loop system. Thus if $\mathbf{u}(t)$ exists which can drive $\mathbf{x}(t_0)$ to any $\mathbf{x}(t_f)$ in finite time, we cannot find an input $\mathbf{r}(t)$ that will do the same to $\mathbf{x}(t)$, since otherwise we can set $\mathbf{u}(t)$ as in Eq. (5-254) to control the open-loop system.

▲ **Theorem 5-10.** *Theorem on Observability of Closed-Loop Systems with State Feedback*

If an open-loop system is controllable and observable, state feedback of the form of Eq. (5-254) could destroy observability. In other words, the observability of open-loop and closed-loop systems due to state feedback are unrelated.

The following example illustrates the relation between observability and state feedback.

Example 5-23

Let the coefficient matrices of a linear system be

$$\mathbf{A} = \begin{bmatrix} 0 & 1 \\ -2 & -3 \end{bmatrix} \quad \mathbf{B} = \begin{bmatrix} 1 \\ 1 \end{bmatrix} \quad \mathbf{C} = [1 \quad 2] \tag{5-257}$$

We can show that the pair [**A**, **B**] is controllable and [**A**, **D**] is observable. Let the state feedback be defined as

$$u(t) = r(t) - \mathbf{K}\mathbf{x}(t) \tag{5-258}$$

where

$$\mathbf{K} = [k_1 \quad k_2] \tag{5-259}$$

Then the closed-loop system is described by the state equation

$$\frac{d\mathbf{x}(t)}{dt} = (\mathbf{A} - \mathbf{BK})\mathbf{x}(t) + \mathbf{B}r(t) \tag{5-260}$$

$$\mathbf{A} - \mathbf{BK} = \begin{bmatrix} -k_1 & 1 - k_2 \\ -2 - k_1 & -3 - k_2 \end{bmatrix} \tag{5-261}$$

The observability matrix of the closed-loop system is

$$\mathbf{V} = \begin{bmatrix} \mathbf{C} \\ \mathbf{C}(\mathbf{A} - \mathbf{BK}) \end{bmatrix} = \begin{bmatrix} 1 & 2 \\ -k_1 - 4 & -3k_2 - 5 \end{bmatrix} \tag{5-262}$$

The determinant of **V** is

$$|\mathbf{V}| = 6k_1 - 3k_2 + 3 \tag{5-263}$$

Thus if k_1 and k_2 are chosen so that $|\mathbf{V}| = 0$, the closed-loop system would be uncontrollable. ▲

5-14 State Equations of Linear Discrete-Data Systems

Just as for continuous-data systems, the modern way of modeling a discrete-data system is by discrete state equations. As described earlier, when dealing with discrete-data systems, we often encounter two different situations. The first one is that the system contains continuous-data components, but the signals at certain points of the system are discrete with respect to time, due to sample-and-hold operations. In this case the components of the system are still described by differential equations, but because of the discrete-time data, the differential equations are discretized to yield a set of difference equations. The second situation involves systems that are completely discrete with respect to time, and the system dynamics should be difference equations from the outset.

5-14-1 Discrete State Equations

Let us consider the discrete-data control system with a sample-and-hold device, as shown in Fig. 5-19. Typical signals that appear at various points in the system are shown in the figure. The output signal, $y(t)$, ordinarily is a continuous-data signal. The output of the sample-and-hold, $h(t)$, is a sequence of steps. Therefore, we can

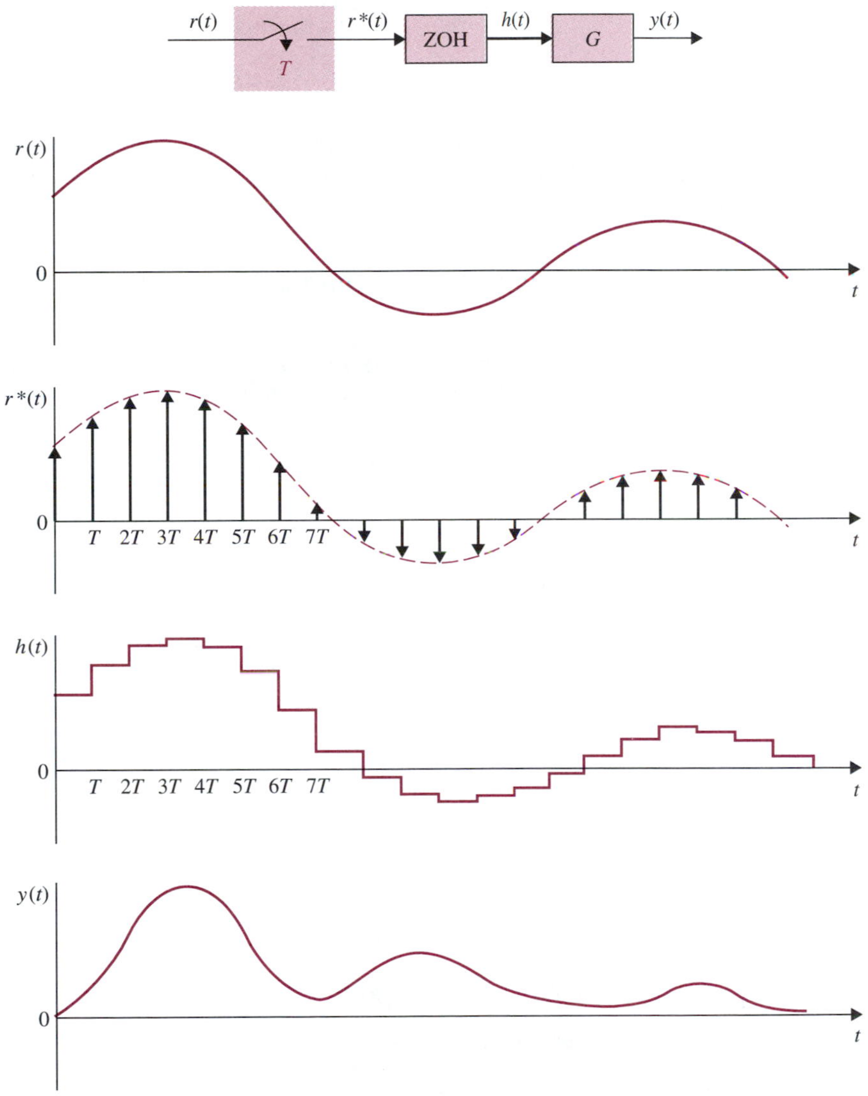

Figure 5-19 Discrete-data system with sample-and-hold.

write

$$h(t) = h(kT) = r(kT) \tag{5-264}$$

for $kT \le t < (k+1)T$, $k = 0, 1, 2, \ldots$.

Now we let the linear process G be described by the state equation and output equation:

$$\frac{d\mathbf{x}(t)}{dt} = \mathbf{A}\mathbf{x}(t) + \mathbf{B}h(t) \tag{5-265}$$

$$y(t) = \mathbf{C}\mathbf{x}(t) + \mathbf{D}h(t) \tag{5-266}$$

where $\mathbf{x}(t)$ is the $n \times n$ state vector, and $h(t)$ and $y(t)$ are the scalar input and output, respectively. The matrices **A**, **B**, **C**, and **D** are coefficient matrices. By using Eq. (5-44), the state-transition equation is

$$\mathbf{x}(t) = \boldsymbol{\phi}(t - t_0)\mathbf{x}(t_0) + \int_{t_0}^{t} \boldsymbol{\phi}(t - \tau)\mathbf{B}h(\tau)\, d\tau \tag{5-267}$$

for $t \ge t_0$. If we are interested only in the responses at the sampling instants, we let $t = (k+1)T$ and $t_0 = kT$. Then Eq. (5-267) becomes

$$\mathbf{x}[(k+1)T] = \boldsymbol{\phi}(T)\mathbf{x}(kT) + \int_{kT}^{(k+1)T} \boldsymbol{\phi}[(k+1)T - \tau]\mathbf{B}h(\tau)\, d\tau \tag{5-268}$$

Since $h(t)$ is piecewise constant, as defined in Eq. (5-264), the input $h(\tau)$ in Eq. (5-268) can be taken outside of the integral sign. Equation (5-268) is written

$$\mathbf{x}[(k+1)T] = \boldsymbol{\phi}(T)\mathbf{x}(kT) + \int_{kT}^{(k+1)T} \boldsymbol{\phi}[(k+1)T - \tau]\mathbf{B}\, d\tau\, h(kT) \tag{5-269}$$

or

$$\mathbf{x}[(k+1)T] = \boldsymbol{\phi}(T)\mathbf{x}(kT) + \boldsymbol{\theta}(T)h(kT) \tag{5-270}$$

where

$$\boldsymbol{\theta}(T) = \int_{kT}^{(k+1)T} \boldsymbol{\phi}[(k+1)T - \tau]\mathbf{B}\, d\tau = \int_{0}^{T} \boldsymbol{\phi}(T - \tau)\mathbf{B}\, d\tau \tag{5-271}$$

Equation (5-270) is of the form of a set of linear first-order difference equations in vector-matrix form, and is referred to as the **vector-matrix discrete state equation.**

5-14-2 Solutions of the Discrete State Equations: Discrete State-Transition Equations

The discrete state equations represented by Eq. (5-270) can be solved by means of a simple recursion procedure. By setting $k = 0, 1, 2, \ldots$ successively in Eq. (5-270), the following equations result:

$$k = 0: \quad \mathbf{x}(T) = \boldsymbol{\phi}(T)\mathbf{x}(0) + \boldsymbol{\theta}(T)h(0) \tag{5-272}$$

$$k = 1: \quad \mathbf{x}(2T) = \boldsymbol{\phi}(2T) = \boldsymbol{\phi}(T)\mathbf{x}(T) + \boldsymbol{\theta}(T)h(T) \tag{5-273}$$

$$k = 2: \quad \mathbf{x}(3T) = \boldsymbol{\phi}(3T) = \boldsymbol{\phi}(T)\mathbf{x}(2T) + \boldsymbol{\theta}(T)h(2T) \tag{5-274}$$

$$\vdots \qquad \vdots$$

$$k = n - 1: \quad \mathbf{x}(nT) = \boldsymbol{\phi}(T)\mathbf{x}[(n-1)T] + \theta(T)h[(n-1)T] \tag{5-275}$$

Substituting Eq. (5-272) into Eq. (5-273), and then Eq. (5-273) into Eq. (5-274), and so on, we obtain the following solution for Eq. (5-270):

$$\mathbf{x}(nT) = \boldsymbol{\phi}^n(T)\mathbf{x}(0) + \sum_{i=0}^{n-1} \boldsymbol{\phi}^{n-i-1}(T)\boldsymbol{\theta}(T)h(iT) \tag{5-276}$$

where, from Eq. (5-36), $\boldsymbol{\phi}^n(T) = [\boldsymbol{\phi}(T)]^n = \boldsymbol{\phi}(nT)$.

Equation (5-276) is defined as the **discrete state-transition equation** of the discrete-data system. It is interesting to note that Eq. (5-276) is analogous to its continuous-data counterpart in Eq. (5-41). The state-transition equation of Eq. (5-267) describes the state of the system of Fig. 5-19 for all values of t, whereas the discrete state-transition equation in Eq. (5-276) describes the states only at the sampling instants $t = nT$, $n = 0, 1, 2, \ldots$.

With nT considered as the initial time, where n is any positive integer, the state-transition equation is

$$\mathbf{x}[(n+N)T] = \boldsymbol{\phi}^N(T)\mathbf{x}(nT) + \sum_{i=0}^{N-1} \boldsymbol{\phi}^{N-i-1}(T)\boldsymbol{\theta}(T)h[(n+i)T] \tag{5-277}$$

where N is a positive integer.

The output of the system at the sampling instants is obtained by substituting $t = nT$ and Eq. (5-276) into Eq. (5-266), yielding

$$\begin{aligned} y(nT) &= \mathbf{C}\mathbf{x}(nT) + \mathbf{D}h(nT) \\ &= \mathbf{C}\boldsymbol{\phi}(nT)\mathbf{x}(0) + \mathbf{C}\sum_{i=0}^{n-1} \boldsymbol{\phi}[(n-i-1)T]\boldsymbol{\theta}(T)h(iT) + \mathbf{D}h(nT) \end{aligned} \tag{5-278}$$

An important advantage of the state-variable method over the z-transform method is that it can be modified easily to describe the states and the output between sampling instants. In Eqs. (5-267), if we let $t = (n + \Delta)T$, where $0 < \Delta \leq 1$ and $t_0 = nT$, we get

$$\begin{aligned} \mathbf{x}[(n+\Delta)T] &= \boldsymbol{\phi}(\Delta T)\mathbf{x}(nT) + \int_{nT}^{(n+\Delta)T} \boldsymbol{\phi}[(n+\Delta)T - \tau]\mathbf{B}\, d\tau\, h(nT) \\ &= \boldsymbol{\phi}(\Delta T)\mathbf{x}(nT) + \theta(\Delta T)h(nT) \end{aligned} \tag{5-279}$$

By varing the value of Δ between 0 and 1, the information on the state variables between the sampling instants is described completely by Eq. (5-279).

When a linear system has only discrete data through the system, its dynamics can be described by a set of discrete state equations:

$$\mathbf{x}[(k+1)T] = \mathbf{A}\mathbf{x}(kT) + \mathbf{B}\mathbf{r}(kT) \tag{5-280}$$

and output equations:

$$\mathbf{y}(kT) = \mathbf{C}\mathbf{x}(kT) + \mathbf{D}\mathbf{r}(kT) \tag{5-281}$$

where **A**, **B**, **C**, and **D** are coefficient matrices of the appropriate dimensions. Notice that Eq. (5-280) is basically of the same form as Eq. (5-270). The only difference in the two situations is the starting point of system representation. In the case of Eq. (5-270), the starting point is the continuous-data state equations of Eq. (5-265); $\boldsymbol{\phi}(T)$ and $\boldsymbol{\theta}(T)$ are determined from the **A** and **B** matrices and must satisfy the conditions and properties of the state-transition matrix. In the case of Eq. (5-280), the equation itself represents an outright description of the discrete-data system, and there are no restrictions on the matrices **A** and **B**.

The solution of Eq. (5-280) follows directly from that of Eq. (5-270), and is

$$\mathbf{x}(nT) = \mathbf{A}^n\mathbf{x}(0) + \sum_{i=0}^{n-1} \mathbf{A}^{n-i-1}\mathbf{B}\mathbf{r}(iT) \tag{5-282}$$

where

$$\mathbf{A}^n = \underbrace{\mathbf{AAAA} \cdots \mathbf{A}}_{|\leftarrow\ n\ \rightarrow|} \tag{5-283}$$

5-15 z-Transform Solution of Discrete State Equations

In Section 2-12 we illustrated the solution of a simple discrete state equation by the z-transform method. In this section the discrete state equations in vector-matrix form of an nth-order system are solved by z-transformation. Consider the discrete state equations

$$\mathbf{x}[(k+1)T] = \mathbf{A}\mathbf{x}(kT) + \mathbf{B}\mathbf{r}(kT) \tag{5-284}$$

Taking the z-transform on both sides of the last equation, we get

$$z\mathbf{X}(z) - z\mathbf{x}(0) = \mathbf{A}\mathbf{X}(z) + \mathbf{B}\mathbf{R}(z) \tag{5-285}$$

Solving for $\mathbf{X}(z)$ from the last equation, we get

$$\mathbf{X}(z) = (z\mathbf{I} - \mathbf{A})^{-1}z\mathbf{x}(0) + (z\mathbf{I} - \mathbf{A})^{-1}\mathbf{B}\mathbf{R}(z) \tag{5-286}$$

Taking the inverse z-transform on both sides of Eq. (5-286), we have

$$\mathbf{x}(nT) = \mathcal{Z}^{-1}[(z\mathbf{I} - \mathbf{A})^{-1}z]\mathbf{x}(0) + \mathcal{Z}^{-1}[(z\mathbf{I} - \mathbf{A})^{-1}\mathbf{B}\mathbf{R}(z)] \tag{5-287}$$

To carry out the inverse z-transform operation of Eq. (5-287), we write the z-transform of $\mathbf{A}^n$ as

$$\mathcal{Z}(\mathbf{A}^n) = \sum_{n=0}^{\infty} \mathbf{A}^n z^{-n} = \mathbf{I} + \mathbf{A}z^{-1} + \mathbf{A}^2 z^{-2} + \cdots \tag{5-288}$$

Premultiplying both sides of Eq. (5-288) by $\mathbf{A}z^{-1}$ and subtracting the result from Eq. (5-288), we get

$$(\mathbf{I} - \mathbf{A}z^{-1})\mathcal{Z}(\mathbf{A}^n) = \mathbf{I} \tag{5-289}$$

Therefore, solving for $\mathcal{Z}(\mathbf{A}^n)$ from Eq. (5-289) yields

$$\mathcal{Z}(\mathbf{A}^n) = (\mathbf{I} - \mathbf{A}z^{-1})^{-1} = (z\mathbf{I} - \mathbf{A})^{-1}z \tag{5-290}$$

or

$$\mathbf{A}^n = \mathcal{Z}^{-1}[(z\mathbf{I} - \mathbf{A})^{-1}z] \tag{5-291}$$

Equation (5-291) represents a way of finding $\mathbf{A}^n$ by using the z-transform method. Similarly, we can prove that

$$\mathcal{Z}^{-1}[z\mathbf{I} - \mathbf{A})^{-1}\mathbf{B}\mathbf{R}(z)] = \sum_{i=0}^{n-1} \mathbf{A}^{n-i-1}\mathbf{B}\mathbf{r}(iT) \tag{5-292}$$

Now we substitute Eqs. (5-291) and (5-292) into Eq. (5-287), and $\mathbf{x}(kT)$ becomes

$$\mathbf{x}(nT) = \mathbf{A}^n\mathbf{x}(0) + \sum_{i=0}^{n-1} \mathbf{A}^{n-i-1}\mathbf{B}\mathbf{r}(iT) \tag{5-293}$$

which is identical to Eq. (5-281).

5-15-1 Transfer-Function Matrix and the Characteristic Equation

Once a discrete-data system is modeled by the dynamic equations of Eqs. (5-280) and (5-281), the transfer-function relation of the system can be expressed in terms of the coefficient matrices. By setting the initial state $\mathbf{x}(0)$ to zero, Eq. (5-286) becomes

$$\mathbf{X}(z) = (z\mathbf{I} - \mathbf{A})^{-1}\mathbf{B}\mathbf{R}(z) \tag{5-294}$$

Substituting Eq. (5-294) into the z-transformed version of Eq. (5-281), we have

$$\mathbf{Y}(z) = [\mathbf{C}(z\mathbf{I} - \mathbf{A})^{-1}\mathbf{B} + \mathbf{D}]\mathbf{R}(z) = \mathbf{G}(z)\mathbf{R}(z) \tag{5-295}$$

where the transfer-function matrix of the system is defined as

$$\mathbf{G}(z) = \mathbf{C}(z\mathbf{I} - \mathbf{A})^{-1}\mathbf{B} + \mathbf{D} \tag{5-296}$$

or

$$\mathbf{G}(z) = \frac{\mathbf{C}[\text{adj}(z\mathbf{I} - \mathbf{A})\mathbf{B} + |z\mathbf{I} - \mathbf{A}|\mathbf{D}]}{|z\mathbf{I} - \mathbf{A}|} \tag{5-297}$$

The characteristic equation of the system is defined as

$$|z\mathbf{I} - \mathbf{A}| = 0 \tag{5-298}$$

In general, a linear time-invariant discrete-data system with one input and one output can be described by the following difference equation with constant coefficients:

$$\begin{aligned} &y[(k+n)T] + a_{n-1}y[(k+n-1)T] + a_{n-2}y[(k+n-2)T] \\ &\quad + \cdots + a_1 y[(k+1))T] + a_0 y(kT) \\ &\quad = b_m r[(k+m)T] + b_{m-1} r[(k+m-1)T] \\ &\quad + \cdots + b_1 r[(k+1)T] + b_0 r(kT) \qquad n \geq m \end{aligned} \tag{5-299}$$

Taking the z-transform on both sides of Eq. (5-299) and setting zero initial conditions, the transfer function of the system is written

$$\frac{Y(z)}{R(s)} = \frac{b_m z^m + b_{m-1} z^{m-1} + \cdots + b_1 z + b_0}{z^n + a_{n-1} z^{n-1} + \cdots + a_1 z + a_0} \qquad n \geq m \tag{5-300}$$

The characteristic equation is obtained by equating the denominator polynomial of the transfer function to zero.

$$z^n + a_{n-1} z^{n-1} + \cdots + a_1 z + a_0 = 0 \tag{5-301}$$

Example 5-24

Consider that a discrete-data system is described by the difference equation

$$y(k+2) + 5y(k+1) + 3y(k) = r(k+1) + 2r(k) \tag{5-302}$$

The transfer function of the system is simply

$$\frac{Y(z)}{R(z)} = \frac{z+2}{z^2 + 5z + 3} \tag{5-303}$$

The characteristic equation is

$$z^2 + 5z + 3 = 0 \tag{5-304}$$

The state variables of the system may be defined as

$$x_1(k) = y(k) \tag{5-305}$$

$$x_2(k) = x_1(k+1) - r(k) \tag{5-306}$$

Substituting the last two equations into Eq. (5-302) gives the two state equations as

$$x_1(k+1) = x_2(k) + r(k) \tag{5-307}$$

$$x_2(k+1) = -3x_1(k) - 5x_2(k) - 3r(k) \tag{5-308}$$

from which we have the matrices **A** and **B**:

$$\mathbf{A} = \begin{bmatrix} 0 & 1 \\ -3 & -5 \end{bmatrix} \qquad \mathbf{B} = \begin{bmatrix} 1 \\ -3 \end{bmatrix} \tag{5-309}$$

The same characteristic equation as in Eq. (5-304) is obtained by using $|z\mathbf{I} - \mathbf{A}| = 0$. ▲

5-16 State Diagrams of Discrete-Data Systems

When a discrete-data system is described by difference or discrete state equations, a **discrete state diagram** may be constructed for the system. Similar to the relations between the analog-computer block diagram and the state diagram for continuous-data systems, the elements of a discrete state diagram resemble the computing elements of a digital computer. Some of the operations of a digital computer are **multiplication by a constant, addition of several variables,** and **time delay** or **shifting.** The discrete state diagram can be used to determine the transfer functions as well as for digital implementation of the system. The mathematical description of these basic digital computations and their corresponding z-transform expressions are as follows:

1. *Multiplication by a constant*:

$$x_2(kT) = ax_1(kT) \tag{5-310}$$
$$X_2(z) = aX_1(z) \tag{5-311}$$

2. *Summing*:

$$x_2(kT) = x_1(kT) + x_3(kT) \tag{5-312}$$
$$X_2(z) = X_1(z) + X_3(z) \tag{5-313}$$

3. *Shifting or time delay*:

$$x_2(kT) = x_1[(k+1)T] \tag{5-314}$$
$$X_2(z) = zX_1(z) - zx_1(0) \tag{5-315}$$

or

$$X_1(z) = z^{-1}X_2(z) + x_1(0) \tag{5-316}$$

The state diagram representation of these operations are illustrated in Fig. 5-20. The initial time $t = 0$ in Eqs. (5-315) and (5-316) can be generalized to $t = t_0$. Then the equations represent the discrete-time state transition from $t \geq t_0$.

Example 5-25

Consider again the difference equation in Eq. (5-302), which is

$$y(k+2) + 5y(k+1) + 3y(k) = r(k+1) + 2r(k) \tag{5-317}$$

▲ The state variables are defined as the outputs of the delay units on the state diagram.

One way of constructing the discrete state diagram for the system is to use the state equations. In this case, the state equations are already defined in Eqs. (5-307) and (5-308). By using essentially the same principle as the state diagram for continuous-data systems, the state diagram for Eqs. (5-307) and (5-308) is shown in Fig. 5-21. The time-delay unit z^{-1} is used to relate $x_1(k+1)$ to $x_2(k)$. *The state variables are defined as the outputs of the delay units on the state diagram.*

As an alternative, the discrete state diagram can be drawn directly from the difference equation via the transfer function, using the decomposition schemes (Fig. 5-22). The decomposition of a discrete-data transfer function follows basically the same procedure as that of an analog transfer function covered in Section 5-9, and the details are not repeated.

The state-transition equations of the system can be obtained directly from the state diagram using the SFG gain formula. By referring to $X_1(z)$ and $X_2(z)$ as the output nodes and $x_1(0)$, $x_2(0)$,

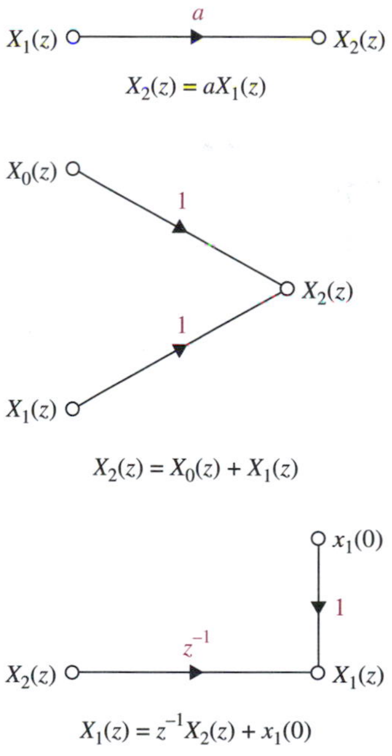

Figure 5-20 Basic elements of a discrete state diagram.

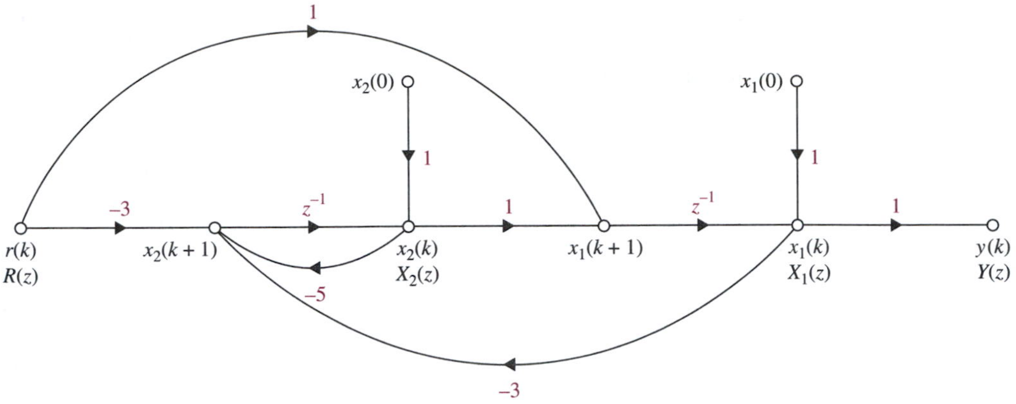

Figure 5-21 Discrete state diagram of the system described by the difference equation of Eq. (5-302) or by the state equations of Eqs. (5-307) and (5-308).

and $R(z)$ as input nodes in Fig. 5-21, the state-transition equations are written as

$$\begin{bmatrix} X_1(z) \\ X_2(z) \end{bmatrix} = \frac{1}{\Delta}\begin{bmatrix} 1+5z^{-1} & z^{-1} \\ -3z^{-1} & 1 \end{bmatrix}\begin{bmatrix} x_1(0) \\ x_2(0) \end{bmatrix} + \frac{1}{\Delta}\begin{bmatrix} z^{-1}(1+5z^{-1}) - 3z^{-2} \\ -3z^{-1} - 3z^{-2} \end{bmatrix} R(z) \tag{5-318}$$

where

$$\Delta = 1 + 5z^{-1} + 3z^{-2} \tag{5-319}$$

The same transfer function between $R(z)$ and $Y(z)$ as in Eq. (5-303) can be obtained directly from the state diagram in Fig. 5-21 by applying the SFG gain formula between these two nodes. ▲

5-16-1 State Diagrams for Sampled-Data Systems

When a discrete-data system has continuous-data as well as discrete-data elements, with the two types of elements separated by sample-and-hold devices, a state-diagram model for the sample-and-hold (zero-order hold) must be established.

Consider that the input of the zero-order hold is denoted by $e^*(t)$, which is a train of impulses, and the output by $h(t)$. Since the zero-order hold simply holds the strength of the input impulse at the sampling instant until the next input comes, the signal $h(t)$ is a sequence of steps. The input–output relation in the Laplace domain is

$$H(s) = \frac{1 - e^{-Ts}}{s} E^*(s) \tag{5-320}$$

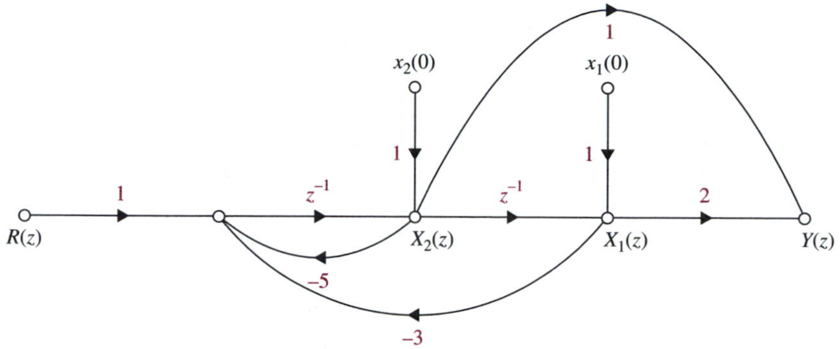

(a) Direct decomposition

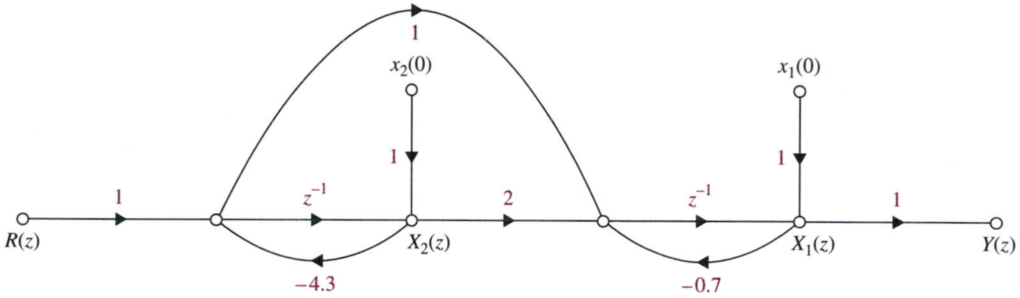

(b) Cascade decomposition

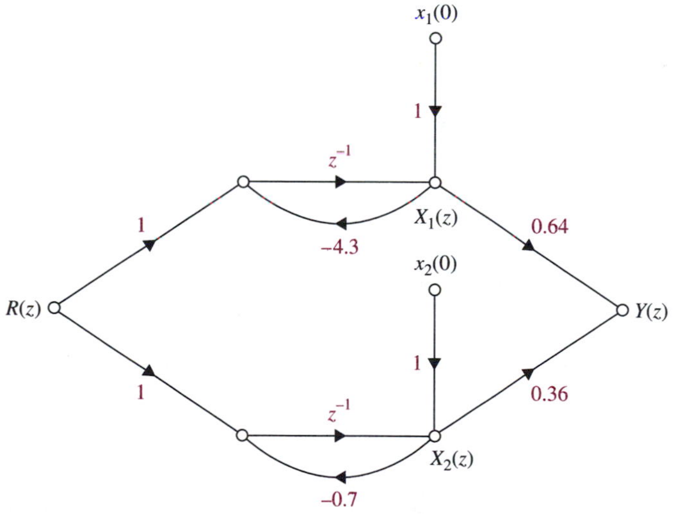

(c) Parallel decomposition

Figure 5-22 State diagrams of the transfer function $y(z)/R(z) = (z + 2)/(z^2 + 5z + 3)$ by the three methods of decomposition.

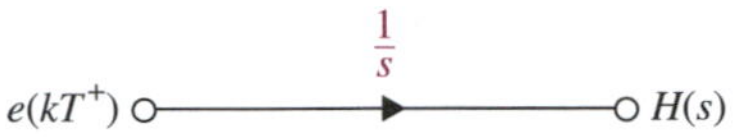

Figure 5-23 State-diagram representation of the zero-order hold.

In the time domain, the relation is simply

$$h(t) = e(kT^+) \tag{5-321}$$

for $kT \leq t < (k+1)T$.

In the state-diagram notation, we need the relation between $H(s)$ and $e(kT^+)$. For this purpose, we take the Laplace transform on both sides of Eq. (5-321) to give

$$H(s) = \frac{e(kT^+)}{s} \tag{5-322}$$

for $kT \leq t < (k+1)T$. The state-diagram representation of the zero-order hold is shown in Fig. 5-23.

Example 5-26

As an illustrative example on how the state diagram of a sampled-data system is constructed, let us consider the system in Fig. 5-24. We shall demonstrate the various ways of modeling the input–output relations of the system. First, the Laplace transform of the output of the system is written in terms of the input to the zero-order hold.

$$Y(s) = \frac{1 - e^{-Ts}}{s} \frac{1}{s+1} E^*(s) \tag{5-323}$$

Taking the z-transform on both sides of Eq. (5-323), we get

$$Y(z) = \frac{1 - e^{-T}}{z - e^{-T}} E(z) \tag{5-324}$$

Figure 5-25 shows the state diagram for Eq. (5-324). The discrete dynamic equations of the system are written directly from the state diagram.

$$x_1[(k+1)T] = -e^{-T}x_1(kT) + (1 - e^{-T})e(kT^+) \tag{5-325}$$
$$y(kT) = x_1(kT) \tag{5-326}$$

▲

Figure 5-24 Sampled-data system.

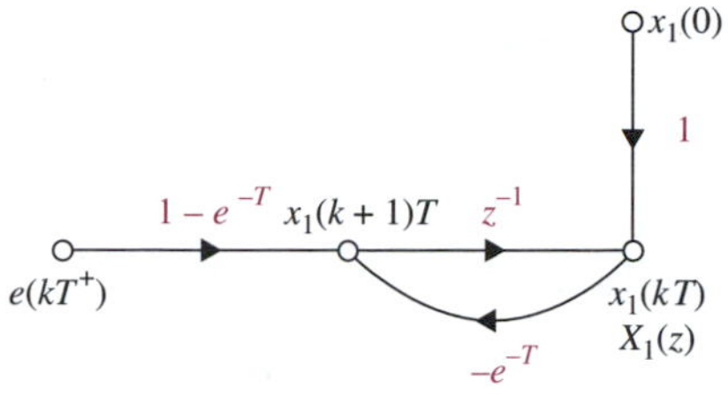

Figure 5-25 Discrete state diagram of the system in Fig. 5-24.

5-17 Final Illustrative Example: Magnetic-Ball-Suspension System

As a final example to illustrate some of the material presented in this chapter, let us consider the magnetic-ball-suspension system shown in Fig. 5-26. The objective of the system is to regulate the current of the electromagnet so that the ball will be suspended at a fixed distance from the end of the magnet. The dynamic equations of the system are

$$M\frac{d^2x(t)}{dt^2} = Mg - \frac{ki^2(t)}{x(t)} \tag{5-327}$$

$$v(t) = Ri(t) + L\frac{di(t)}{dt} \tag{5-328}$$

where Eq. (5-327) is nonlinear. The system variables and parameters are as follows:

- ▲ $v(t)$ = input voltage (V)
- ▲ $i(t)$ = winding current (A)
- ▲ k = proportional constant = 1.0
- ▲ L = winding inductance = 0.01 H

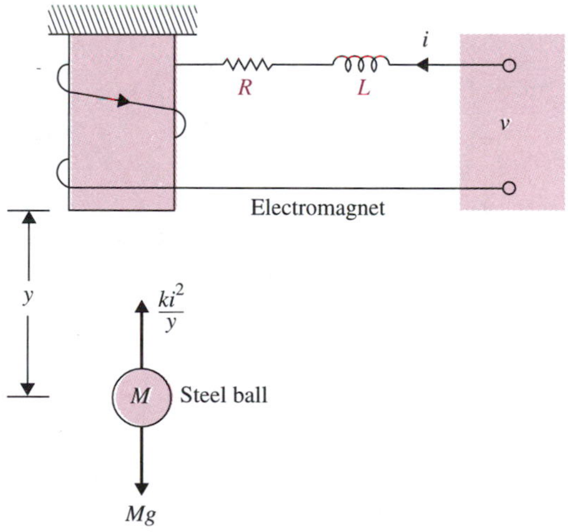

Figure 5-26 Ball-suspension system.

- R = winding resistance = 1 Ω
- M = ball mass = 1.0 kg
- $x(t)$ = ball position (m)
- g = gravitational acceleration = 32.2 m/sec^2

The state variables are defined as

$$x_1(t) = x(t)$$
$$x_2(t) = \frac{dx(t)}{dt} \tag{5-329}$$
$$x_3(t) = i(t)$$

The state equations are

$$\frac{dx_1(t)}{dt} = x_2(t) \tag{5-330}$$
$$\frac{dx_2(t)}{dt} = g - \frac{k}{M}\frac{x_3^2(t)}{x_1(t)} \tag{5-331}$$
$$\frac{dx_3(t)}{dt} = -\frac{R}{L}x_3(t) + \frac{v(t)}{L} \tag{5-332}$$

These nonlinear state equations are linearized about the equilibrium point, $x_1(t) = x(t) = 0.5$ m, using the method described in Chapter 4. After substituting the parameter values, the linear equations are written

$$\Delta\dot{\mathbf{x}}(t) = \mathbf{A}^*\Delta\mathbf{x}(t) + \mathbf{B}^*\Delta v(t) \tag{5-333}$$

where $\Delta\mathbf{x}(t)$ denotes the state vector and $\Delta v(t)$ the input voltage of the linearized system. The coefficient matrices are

$$\mathbf{A}^* = \begin{bmatrix} 0 & 1 & 0 \\ 64.4 & 0 & -16 \\ 0 & 0 & -100 \end{bmatrix} \qquad \mathbf{B} = \begin{bmatrix} 0 \\ 0 \\ 100 \end{bmatrix} \tag{5-334}$$

All the analysis performed in the following can be carried out using the **LYNSYS** program of the **ACSP** software package or the **svdesign** function from the **CSAD/MATLAB** toolbox. To show the analytical method, we carry out the steps of the derivations as follows.

Characteristic Equation

$$|s\mathbf{I} - \mathbf{A}^*| = \begin{vmatrix} s & -1 & 0 \\ -64.4 & s & 16 \\ 0 & 0 & s+100 \end{vmatrix} = s^3 + 100s^2 - 64.4s - 6440 = 0 \tag{5-335}$$

Eigenvalues. The eigenvalues of **A***, or the roots of the characteristic equation, are

$$s = -100 \qquad s = -8.025 \qquad s = 8.025$$

State-Transition Matrix. The state-transition matrix of **A*** is

$$\boldsymbol{\phi}(t) = \mathscr{L}^{-1}[(s\mathbf{I} - \mathbf{A}^*)^{-1}] = \mathscr{L}^{-1}\left(\begin{bmatrix} s & -1 & 0 \\ -64.4 & s & 16 \\ 0 & 0 & s+100 \end{bmatrix}^{-1}\right) \tag{5-336}$$

or

$$\boldsymbol{\phi}(t) = \mathscr{L}^{-1}\left(\frac{1}{(s+100)(s+8.025)(s-8.025)} \begin{bmatrix} s(s+100) & s+100 & -16 \\ 64.4(s+100) & s(s+100) & -16s \\ 0 & 0 & s^2 - 64.4 \end{bmatrix}\right) \tag{5-337}$$

By performing the partial-fraction expansion and carrying out the inverse Laplace transform, the state-transition matrix is

$$\boldsymbol{\phi}(t) = \begin{bmatrix} 0 & 0 & -0.0016 \\ 0 & 0 & 0.16 \\ 0 & 0 & 1 \end{bmatrix} e^{-100t} + \begin{bmatrix} 0.5 & -0.062 & 0.0108 \\ -4.012 & 0.5 & -0.087 \\ 0 & 0 & 0 \end{bmatrix} e^{-8.025t} + \begin{bmatrix} 0.5 & 0.062 & -0.0092 \\ 4.012 & 0.5 & -0.074 \\ 0 & 0 & 0 \end{bmatrix} e^{8.025t} \tag{5-338}$$

Since the last term in Eq. (5-338) has a positive exponent, the response of $\boldsymbol{\phi}(t)$ increases with time, and the system is unstable. This is expected, since without control, the steel ball would be attracted by the magnet until it hits the bottom of the magnet.

Transfer Function. Let us define the ball position $x(t)$ as the output $y(t)$, and the input as $v(t)$. Then the input–output transfer function of the system is

$$\begin{aligned} \frac{Y(s)}{V(s)} &= \mathbf{C}^*(s\mathbf{I} - \mathbf{A}^*)^{-1}\mathbf{B}^* = [1 \quad 0 \quad 0](s\mathbf{I} - \mathbf{A}^*)^{-1}\mathbf{B}^* \\ &= \frac{-1600}{(s+100)(s+8.025)(s-8.025)} \end{aligned} \tag{5-339}$$

Controllability. The controllability matrix is

$$\mathbf{S} = [\mathbf{B}^* \quad \mathbf{A}^*\mathbf{B}^* \quad \mathbf{A}^{*2}\mathbf{B}^*] = \begin{bmatrix} 0 & 0 & -1{,}600 \\ 0 & -1{,}600 & 160{,}000 \\ 100 & -10{,}000 & 1{,}000{,}000 \end{bmatrix} \tag{5-340}$$

Since the rank of **S** is three, the system is completely controllable.

Observability. The observability of the system depends on which variable is defined at the output. For state-feedback control (discussed in Chapter 10) the full controller requires feeding back all three state variables, x_1, x_2, and x_3. However, for reasons of economy, we may want to feed back only one of the three state variables. To make the problem more general, we may want to investigate which state if chosen as the output would render the system unobservable.

1. $y(t)$ = ball position = $x(t)$: $\mathbf{C}^* = [1 \quad 0 \quad 0]$
   The observability matrix is

$$\mathbf{V} = \begin{bmatrix} \mathbf{C}^* \\ \mathbf{C}^*\mathbf{A}^* \\ \mathbf{C}^*\mathbf{A}^{*2} \end{bmatrix} = \begin{bmatrix} 1 & 0 & 0 \\ 0 & 1 & 0 \\ 64.4 & 0 & -16 \end{bmatrix} \tag{5-341}$$

   which has a rank of three. Thus the system is completely observable.

2. $y(t)$ = ball velocity = $dx(t)/dt$: $\mathbf{C}^* = [0 \quad 1 \quad 0]$
   The observability matrix is

$$\mathbf{V} = \begin{bmatrix} \mathbf{C}^* \\ \mathbf{C}^*\mathbf{A}^* \\ \mathbf{C}^*\mathbf{A}^{*2} \end{bmatrix} = \begin{bmatrix} 0 & 1 & 0 \\ 64.4 & 0 & -16 \\ 0 & 64.4 & 1600 \end{bmatrix} \tag{5-342}$$

   which has a rank of three. Thus the system is completely observable.

3. $y(t)$ = winding current = $i(t)$: $\mathbf{C}^* = [0 \quad 0 \quad 1]$
   The observability matrix is

$$\mathbf{V} = \begin{bmatrix} \mathbf{C}^* \\ \mathbf{C}^*\mathbf{A}^* \\ \mathbf{C}^*\mathbf{A}^{*2} \end{bmatrix} = \begin{bmatrix} 0 & 0 & 1 \\ 0 & 0 & -100 \\ 0 & 0 & -10{,}000 \end{bmatrix} \tag{5-343}$$

   which has a rank of one. Thus the system is unobservable. The physical interpretation of this result is that if we choose the current $i(t)$ as the measurable output, we would not be able to reconstruct the state variables from the measured information.

The interested reader should enter the data of this system to any available computer program and verify the results obtained.

5-18 Summary

This chapter is devoted to the state-variable analysis of linear systems. The fundamentals on state variables and state equations are introduced in Chapter 2 and 4. Formal discussions on these subjects are covered in this chapter. Specifically, the state-transition matrix and state-transition equations are introduced. The relationship between state equations and transfer functions is established. Given the transfer function of a linear system, the state equations of the system can be obtained by means of decomposition of the transfer function. Given the state equations and the output equations, the transfer function can be determined either analytically or directly from the state diagram.

Characteristic equations and eigenvalues are defined in terms of the state equations and the transfer function. Eigenvectors of **A** are also defined for distinct and multiple-order eigenvalues.

Similarity transformations to controllability canonical form (CCF), observability canonical form (OCF), diagonal canonical form (DCF), and Jordan canonical form (JCF) are discussed.

State controllability and observability of linear time-invariant systems are defined and illustrated.

The state formulation of a discrete-data system is covered in this chapter. The analysis closely parallels the state-variable formulation of continuous-data systems. A final example on the magnetic-ball-suspension system summarizes the important elements of the state-variable analysis of linear systems.

Review Questions

1. What are the components of the dynamic equations of a linear system?
2. Given the state equations of a linear system as

$$\frac{d\mathbf{x}(t)}{dt} = \mathbf{A}\mathbf{x}(t) + \mathbf{B}\mathbf{u}(t)$$

state two expressions of the state-transition matrix $\boldsymbol{\phi}(t)$ in terms of **A**.

3. List the properties of the state-transition matrix $\boldsymbol{\phi}(t)$.
4. Given the state equations as in Review Question 2, write the state-transition equation.
5. List the advantages of expressing a linear system in the controllability canonical form (CCF). Given an example of **A** and **B** in CCF.
6. Given the state equations in the form in Review Question 2, give the conditions for **A** and **B** to be transformable into CCF.
7. Express the characteristic equation in terms of the matrix **A**.
8. List the three methods of decomposition of a transfer function.

9. What special forms will the state equations be in if the transfer function is decomposed by direct decomposition?
10. What special form will the state equations be in if the transfer function is decomposed by parallel decomposition?
11. What is the advantage of using cascade decomposition?
12. State the relationship between the CCF and controllability.
13. For controllability, does the magnitude of the inputs have to be finite?
14. State the condition of controllability in terms of the matrices **A** and **B**.
15. What is the motivation behind the concept of observability?
16. State the condition of observability in terms of the matrices **A** and **C**.
17. What can be said about the controllability and observability conditions if the transfer function has pole–zero cancellation?
18. State the relationship between OCF and observability.

References

State Variables and State Equations

1. **B. C. KUO,** *Linear Networks and Systems,* McGraw-Hill Book Company, New York, 1967.
2. **R. A. GABEL** and **R. A. ROBERTS,** *Signals and Linear Systems,* 3rd ed., John Wiley & Sons, New York, 1987.

Controllability and Observability

3. **R. E. KALMAN,** "On the General Theory of Control Systems," *Proc. IFAC,* Vol. 1, pp. 481–492, Butterworth, London, 1961.
4. **W. L. BROGAN,** *Modern Control Theory,* 2nd ed., Prentice Hall, Englewood Cliffs, NJ, 1985.

Problems

▲ Dynamic equations

5-1. The following differential equations represent linear time-invariant systems. Write the dynamic equations (state equations and output equations) in vector-matrix form.

(a) $$\frac{d^2y(t)}{dt^2} + 4\frac{dy(t)}{dt} + y(t) = 5r(t)$$

(b) $$2\frac{d^3y(t)}{dt^3} + 3\frac{d^2y(s)}{dt^2} + 5\frac{dy(t)}{dt} + 2y(t) = r(t)$$

(c) $$\frac{d^3y(t)}{dt^3} + 5\frac{d^2y(t)}{dt^2} + 3\frac{dy(t)}{dt} + y(t) + \int_0^t y(\tau)\,d\tau = r(t)$$

(d) $$\frac{d^4y(t)}{dt^4} + 1.5\frac{d^3y(t)}{dt^3} + 2.5\frac{dy(t)}{dt} + y(t) = 2r(t)$$

▲ State-transition matrix

5-2. By use of Eq. (5-23), show that

$$\boldsymbol{\phi}(t) = \mathbf{I} + \mathbf{A}t + \frac{1}{2!}\mathbf{A}^2t^2 + \frac{1}{3!}\mathbf{A}^3t^3 + \cdots$$

▲ State-transition matrix

5-3. The state equations of a linear time-invariant system are represented by

$$\frac{d\mathbf{x}(t)}{dt} = \mathbf{A}\mathbf{x}(t) + \mathbf{B}\mathbf{u}(t)$$

Find the state transition matrix $\boldsymbol{\phi}(t)$, the characteristic equation, and the eigenvalues of **A** for the following cases.

(a) $\mathbf{A} = \begin{bmatrix} 0 & 1 \\ -2 & -1 \end{bmatrix} \quad \mathbf{B} = \begin{bmatrix} 0 & 1 \\ 1 & 0 \end{bmatrix}$

(b) $\mathbf{A} = \begin{bmatrix} 0 & 1 \\ -4 & -5 \end{bmatrix} \quad \mathbf{B} = \begin{bmatrix} 1 \\ 1 \end{bmatrix}$

(c) $\mathbf{A} = \begin{bmatrix} -3 & 0 \\ 0 & -3 \end{bmatrix} \quad \mathbf{B} = \begin{bmatrix} 0 \\ 1 \end{bmatrix}$

(d) $\mathbf{A} = \begin{bmatrix} 3 & 0 \\ 0 & -3 \end{bmatrix} \quad \mathbf{B} = \begin{bmatrix} 0 \\ 1 \end{bmatrix}$

(e) $\mathbf{A} = \begin{bmatrix} 0 & 2 \\ -2 & 0 \end{bmatrix} \quad \mathbf{B} = \begin{bmatrix} 0 \\ 1 \end{bmatrix}$

(f) $\mathbf{A} = \begin{bmatrix} -1 & 0 & 0 \\ 0 & -2 & 1 \\ 0 & 0 & -2 \end{bmatrix} \quad \mathbf{B} = \begin{bmatrix} 0 \\ 1 \\ 0 \end{bmatrix}$

(g) $\mathbf{A} = \begin{bmatrix} -5 & 1 & 0 \\ 0 & -5 & 1 \\ 0 & 0 & -5 \end{bmatrix} \quad \mathbf{B} = \begin{bmatrix} 0 \\ 0 \\ 1 \end{bmatrix}$

Find $\boldsymbol{\phi}(t)$ and the characteristic equation using a computer program if available.

▲ State-transition equation

5-4. Find the state-transition equation of each of the systems described in Problem 5-3 for $t \geq 0$. Assume that $\mathbf{x}(0)$ is the initial state vector and that the components of the input vector $\mathbf{u}(t)$ are all unit-step functions.

▲ State-transition matrix

5-5. Find out if the matrices given in the following can be state-transition matrices. [*Hint:* Check the properties of $\boldsymbol{\phi}(t)$.]

(a) $\begin{bmatrix} -e^{-t} & 0 \\ 0 & 1-e^{-t} \end{bmatrix}$ (b) $\begin{bmatrix} 1-e^{-t} & 0 \\ 1 & e^{-t} \end{bmatrix}$

(c) $\begin{bmatrix} 1 & 0 \\ 1-e^{-t} & e^{-t} \end{bmatrix}$ (d) $\begin{bmatrix} e^{-2t} & te^{-2t} & t^2e^{-2t}/2 \\ 0 & e^{-2t} & te^{-2t} \\ 0 & 0 & e^{-2t} \end{bmatrix}$

▲ Eigenvalues, transfer functions

5-6. Given a system described by the dynamic equations

$$\frac{d\mathbf{x}(t)}{dt} = \mathbf{A}\mathbf{x}(t) + \mathbf{B}u(t) \qquad y(t) = \mathbf{C}\mathbf{x}(t)$$

where

(a) $\mathbf{A} = \begin{bmatrix} 0 & 1 & 0 \\ 0 & 0 & 1 \\ -1 & -2 & -3 \end{bmatrix} \quad \mathbf{B} = \begin{bmatrix} 0 \\ 0 \\ 1 \end{bmatrix} \quad \mathbf{C} = [1 \quad 0 \quad 0]$

(b) $\mathbf{A} = \begin{bmatrix} -1 & 1 \\ 0 & -1 \end{bmatrix} \quad \mathbf{B} = \begin{bmatrix} 0 \\ 1 \end{bmatrix} \quad \mathbf{C} = [1 \quad 1]$

(c) $\mathbf{A} = \begin{bmatrix} 0 & 1 & 0 \\ 0 & 0 & 1 \\ 0 & -1 & -2 \end{bmatrix} \quad \mathbf{B} = \begin{bmatrix} 0 \\ 0 \\ 1 \end{bmatrix} \quad \mathbf{C} = [1 \quad 1 \quad 0]$

(i) Find the eigenvalues of **A**. Use a computer program such as **LINSYS** of **ACSP** or **svstuff** of **CSAD/MATLAB** if any of these is available to check the answers. You may get the characteristic equation and solve for the roots using a computer program.

(ii) Find the transfer-function relation between $\mathbf{X}(s)$ and $U(s)$.

(iii) Find the transfer function $Y(s)/U(s)$.

▲ State equations

5-7. Given the dynamic equations of a time-invariant system,

$$\frac{d\mathbf{x}(t)}{dt} = \mathbf{A}\mathbf{x}(t) + \mathbf{B}u(t) \qquad y(t) = \mathbf{C}\mathbf{x}(t)$$

where

$$\mathbf{A} = \begin{bmatrix} 0 & 1 & 0 \\ 0 & 0 & 1 \\ -1 & -2 & -3 \end{bmatrix} \quad \mathbf{B} = \begin{bmatrix} 0 \\ 0 \\ 1 \end{bmatrix} \quad \mathbf{C} = [1 \quad 1 \quad 0]$$

Find the matrices $\mathbf{A}_1$ and $\mathbf{B}_1$ so that the state equations are written as

$$\frac{d\overline{\mathbf{x}}(t)}{dt} = \mathbf{A}_1\overline{\mathbf{x}}(\mathrm{t}) + \mathbf{B}_1 u(t)$$

where

$$\overline{\mathbf{x}}(t) = \begin{bmatrix} x_1(t) \\ y(t) \\ \dfrac{dy(t)}{dt} \end{bmatrix}$$

▲ CCF

5-8. Given the dynamic equations

$$\frac{d\mathbf{x}(t)}{dt} = \mathbf{A}\mathbf{x}(t) + \mathbf{B}u(\mathrm{t}) \qquad y(t) = \mathbf{C}\mathbf{x}(t)$$

where

(a) $\mathbf{A} = \begin{bmatrix} 0 & 2 & 0 \\ 1 & 2 & 0 \\ -1 & 0 & 1 \end{bmatrix} \quad \mathbf{B} = \begin{bmatrix} 0 \\ 1 \\ 1 \end{bmatrix} \quad \mathbf{C} = [1 \quad 0 \quad 1]$

(b) $\mathbf{A} = \begin{bmatrix} 0 & 2 & 0 \\ 1 & 2 & 0 \\ -1 & 1 & 1 \end{bmatrix} \quad \mathbf{B} = \begin{bmatrix} 1 \\ 1 \\ 0 \end{bmatrix} \quad \mathbf{C} = [1 \quad 0 \quad 1]$

(c) $\mathbf{A} = \begin{bmatrix} -2 & 1 & 0 \\ 0 & -2 & 0 \\ -1 & -2 & -3 \end{bmatrix} \quad \mathbf{B} = \begin{bmatrix} 1 \\ 1 \\ 1 \end{bmatrix} \quad \mathbf{C} = [1 \quad 0 \quad 0]$

(d) $\mathbf{A} = \begin{bmatrix} -1 & 1 & 0 \\ 0 & -1 & 1 \\ 0 & 0 & -1 \end{bmatrix} \quad \mathbf{B} = \begin{bmatrix} 0 \\ 1 \\ 1 \end{bmatrix} \quad \mathbf{C} = [1 \quad 0 \quad 1]$

(e) $\mathbf{A} = \begin{bmatrix} 1 & 1 \\ -2 & -3 \end{bmatrix} \quad \mathbf{B} = \begin{bmatrix} 0 \\ 1 \end{bmatrix} \quad \mathbf{C} = [1 \quad 0]$

Find the transformation $\mathbf{x}(t) = \mathbf{P}\overline{\mathbf{x}}(t)$ that transforms the state equations into the controllability canonical form (CCF), if possible. If the **CSAD** software is available, check the answers using the computer program.

▲ OCF

5-9. For the systems described in Problem 5-8, find the transformation $\mathbf{x}(t) = \mathbf{Q}\overline{\mathbf{x}}(t)$ so that the state equations are transformed into the observability canonical form (OCF), if possible. If the **CSAD** software is available, check the answers using the computer program.

▲ DCF, JCF

5-10. For the systems described in Problem 5-8, find the transformation $\mathbf{x}(t)\ \mathbf{T}\overline{\mathbf{x}}(t)$ so that the state equations are transformed into the diagonal canonical form (DCF) if **A** has distinct eigenvalues, and Jordan canonical form (JCF) if **A** has at least one multiple-order eigenvalue.

▲ CCF

5-11. The state equation of a linear system is described by

$$\frac{d\mathbf{x}(t)}{dt} = \mathbf{A}\mathbf{x}(t) + \mathbf{B}u(t)$$

The coefficient matrices are given as follows. Explain why the state equations cannot be transformed into the controllability canonical form (CCF).

(a) $\mathbf{A} = \begin{bmatrix} -2 & 0 \\ 0 & -1 \end{bmatrix} \qquad \mathbf{B} = \begin{bmatrix} 0 \\ 1 \end{bmatrix}$

(b) $\mathbf{A} = \begin{bmatrix} -1 & 0 & 0 \\ 0 & -1 & 0 \\ 0 & 0 & -1 \end{bmatrix} \qquad \mathbf{B} = \begin{bmatrix} 1 \\ 2 \\ 3 \end{bmatrix}$

(c) $\mathbf{A} = \begin{bmatrix} 1 & 2 \\ 1 & 1 \end{bmatrix} \qquad \mathbf{B} = \begin{bmatrix} 2 \\ \sqrt{2} \end{bmatrix}$

(d) $\mathbf{A} = \begin{bmatrix} -2 & 1 & 0 \\ 0 & -2 & 0 \\ -1 & -2 & -3 \end{bmatrix} \qquad \mathbf{B} = \begin{bmatrix} 1 \\ 0 \\ 1 \end{bmatrix}$

▲ State equations, transfer functions

5-12. The equations that describe the dynamics of a motor control system are

$$e_a(t) = R_a i_a(t) + L_a \frac{di_a(t)}{dt} + K_b \frac{d\theta_m(t)}{dt}$$

$$T_m(t) = K_i i_a(t)$$

$$T_m(t) = J \frac{d^2\theta_m(t)}{dt^2} + B \frac{d\theta_m(t)}{dt} + K\theta_m(t)$$

$$e_a(t) = K_a e(t)$$

$$e(t) = K_s[\theta_r(t) - \theta_m(t)]$$

(a) Assign the state variables as $x_1(t) = \theta_m(t)$, $x_2(t) = d\theta_m(t)/dt$, and $x_3(t) = i_a(t)$. Write the state equations in the form

$$\frac{d\mathbf{x}(t)}{dt} = \mathbf{A}\mathbf{x}(t) + \mathbf{B}\theta_r(t)$$

Write the output equation in the form $y(t) = \mathbf{C}\mathbf{x}(t)$, where $y(t) = \theta_m(t)$.

(b) Find the transfer function $G(s) = \Theta_m(s)/E(s)$ when the feedback path from $\Theta_m(s)$ to $E(s)$ is broken. Find the closed-loop transfer function $M(s) = \Theta_m(s)/\Theta_r(s)$.

▲ State transition matrix

5-13. Given the matrix **A** of a linear system described by the state equation

$$\frac{d\mathbf{x}(t)}{dt} = \mathbf{A}\mathbf{x}(t) + \mathbf{B}u(t)$$

where

(a) $\mathbf{A} = \begin{bmatrix} 0 & 1 \\ -1 & 0 \end{bmatrix}$ **(b)** $\mathbf{A} = \begin{bmatrix} -1 & 0 \\ 0 & -2 \end{bmatrix}$ **(c)** $\mathbf{A} = \begin{bmatrix} 0 & 1 \\ 1 & 0 \end{bmatrix}$

Find the state-transition matrix $\boldsymbol{\phi}(t)$ using the following methods.

(i) Infinite-series expansion of $e^{\mathbf{A}t}$ and express it in a closed form.

(ii) The inverse Laplace transform of $(s\mathbf{I} - \mathbf{A})^{-1}$.

▲ State equations, inverse Laplace transform, characteristic equation

5-14. The schematic diagram of a feedback control system using a dc motor is shown in Fig. 5P-14. The torque developed by the motor is $T_m(t) = K_i i_a(t)$, where K_i is the torque constant. The constants of the system are

▲ $K_s = 2$	▲ $R = 2\ \Omega$	▲ $R_s = 0.1\ \Omega$
▲ $K_b = 5$ V/rad/sec	▲ $K_i = 5$ N-m/A	▲ $L_a \cong 0$ H
▲ $J_m + J_L = 0.1$ N-m-sec^2	▲ $B_m \cong 0$ N-m-sec	▲ $K = 10$

Assume that all the units are consistent so that no conversion is necessary.

(a) Let the state variables be assigned as $x_1 = \theta_y$ and $x_2 = d\theta_y/dt$. Let the output be $y = \theta_y$. Write the state equations in vector-matrix form. Show that the matrices **A** and **B** are in CCF.

(b) Let $\theta_r(t)$ be a unit-step function. Find $\mathbf{x}(t)$ in terms of $\mathbf{x}(0)$, the initial state. Use the Laplace transform table.

(c) Find the characteristic equation and the eigenvalues of **A**.

(d) Comment on the purpose of the feedback resistor R_s.

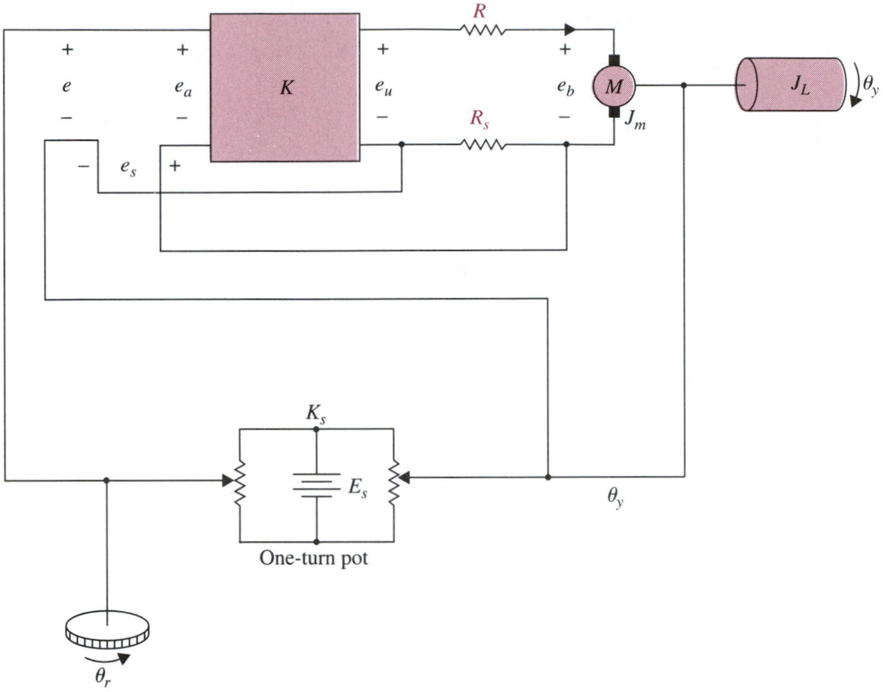

Figure 5P-14

▲ State equations, inverse Laplace tansform, characteristic equation

5-15. Repeat Problem 5-14 with the following system parameters:

▲ $K_s = 1$	▲ $K = 9$	▲ $R_a = 0.1\ \Omega$
▲ $R_s = 0.1\ \Omega$	▲ $K_b = 1$ V/rad/sec	▲ $K_i = 1$ N-m/A
▲ $L_a \cong 0$ H	▲ $J_m + J_L = 0.01$ N-m-sec^2	▲ $B_m \cong 0$ N-m-sec

▲ Transfer functions, dynamic equations, final value

5-16. The block diagram of a feedback control system is shown in Fig. 5P-16.

(a) Find the forward-path transfer function $Y(s)/E(s)$ and the closed-loop transfer function $Y(s)/R(s)$.

(b) Write the dynamic equations in the form of

$$\frac{d\mathbf{x}(t)}{dt} = \mathbf{A}\mathbf{x}(t) + \mathbf{B}r(t) \qquad y(t) = \mathbf{C}\mathbf{x}(t) + \mathbf{D}r(\mathrm{t})$$

Find **A, B. C.** and **D** in terms of the system parameters.

(c) Apply the final-value theorem to find the steady-state value of the output $y(t)$ when the input $r(t)$ is a unit-step function. Assume that the closed-loop system is stable.

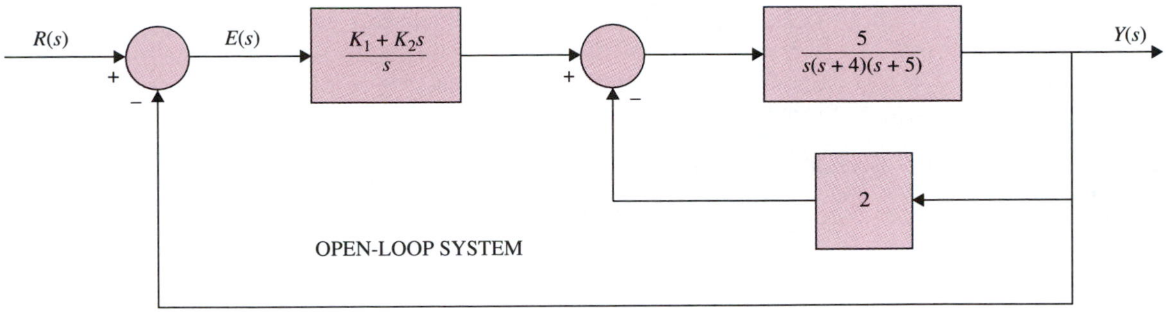

Figure 5P-16

▲ Adjoint matrix, characteristic equation

5-17. For the linear time-invariant system whose state equations have the coefficient matrices given by Eqs. (5-156) and (5-157) (CCF), show that

$$\operatorname{adj}(s\mathbf{I} - \mathbf{A})\mathbf{B} = \begin{bmatrix} 1 \\ s \\ s^2 \\ \vdots \\ s^{n-1} \end{bmatrix}$$

and the characteristic equation of **A** is

$$s^n + a_{n-1}s^{n-1} + \cdots + a_1 s + a_0 = 0$$

▲ State equations, state-transition equation, state-transition matrix, characteristic equation

5-18. A linear time-invariant system is described by the differential equation

$$\frac{d^3y(t)}{dt^3} + 3\frac{d^2y(t)}{dt^2} + 3\frac{dy(t)}{dt} + y(t) = r(t)$$

(a) Let the state variables be defined as $x_1 = y$, $x_2 = dy/dt$ and $x_3 = d^2y/dt^2$. Write the state equations of the system in vector-matrix form.

(b) Find the state-transition matrix $\boldsymbol{\phi}(t)$ of **A**.

(c) Let $y(0) = 1$, $dy(0)/dt = 0$, $d^2y(0)/dt^2 = 0$, and $r(t) = u_s(t)$. Find the state-transition equation of the system.

(d) Find the characteristic equation and the eigenvalues of **A**.

▲ State equations, state-transition matrix, characteristic equation

5-19. A spring–mass–friction system is described by the following differential equation:

$$\frac{d^2y(t)}{dt^2} + 2\frac{dy(t)}{dt} + y(t) = r(t)$$

(a) Define the state variables as $x_1(t) = y(t)$ and $x_2(t) = dy(t)/dt$. Write the state equations in vector-matrix form. Find the state-transition matrix $\boldsymbol{\phi}(t)$ of **A**.

(b) Define the state variables as $x_1(t) = y(t)$ and $x_2(t) = y(t) + dy(t)/dt$. Write the state equations in vector-matrix form. Find the state-transition matrix $\boldsymbol{\phi}(t)$ of $\mathbf{A}$.

(c) Show that the characteristic equations, $|s\mathbf{I} - \mathbf{A}| = 0$, for parts (a) and (b) are identical.

▲ State-transition matrix, eigenvalues

5-20. Given the state equations $d\mathbf{x}(t)/dt = \mathbf{Ax}(t)$, where

$$\mathbf{A} = \begin{bmatrix} \sigma & -\omega \\ \omega & \sigma \end{bmatrix} \qquad \sigma \text{ and } \omega \text{ are real numbers}$$

(a) Find the state-transition matrix of $\mathbf{A}$.

(b) Find the eigenvalues of $\mathbf{A}$.

▲ Dynamic equations

5-21. (a) Show that the input–output transfer functions of the two systems shown in Fig. 5P-21 are the same.

(b) Write the dynamic equations of the system in Fig. 5P-21 as

$$\frac{d\mathbf{x}(t)}{dt} = \mathbf{A}_1\mathbf{x}(t) + \mathbf{B}_1 u_1(t) \qquad y_1(t) = \mathbf{C}_1\mathbf{x}(t)$$

and those of the system in Fig. 5P-21(b) as

$$\frac{d\mathbf{x}(t)}{dt} = \mathbf{A}_2\mathbf{x}(t) + \mathbf{B}_2 u_2(t) \qquad y_2(t) = \mathbf{C}_2\mathbf{x}(t)$$

Find $\mathbf{A}_1$, $\mathbf{B}_1$, $\mathbf{C}_1$, $\mathbf{A}_2$, $\mathbf{B}_2$, and $\mathbf{C}_2$. Show that $\mathbf{A}_2 = \mathbf{A}_1'$.

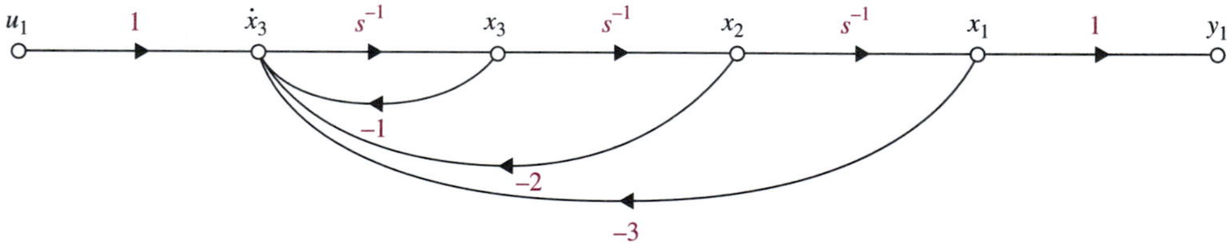

(a)

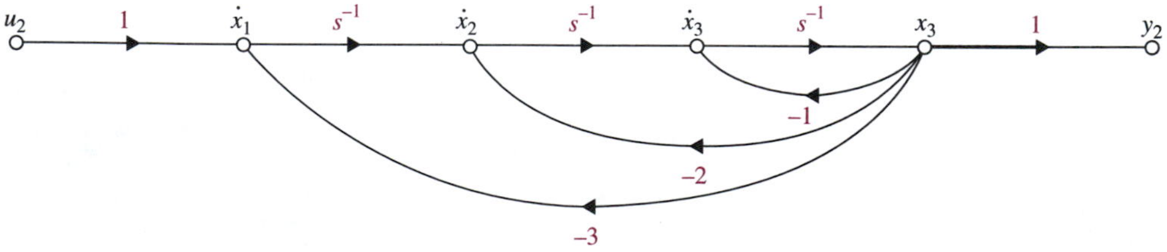

(b)

Figure 5P-21

▲ State diagrams

5-22. Draw the state diagrams for the following systems.

$$\frac{d\mathbf{x}(t)}{dt} = \mathbf{A}\mathbf{x}(t) + \mathbf{B}\mathbf{u}(t)$$

(a) $\mathbf{A} = \begin{bmatrix} -3 & 2 & 0 \\ -1 & 0 & 1 \\ -2 & -3 & -4 \end{bmatrix} \quad \mathbf{B} = \begin{bmatrix} 0 \\ 0 \\ 1 \end{bmatrix}$

(b) Same **A** as in part (a), but with

$$\mathbf{B} = \begin{bmatrix} 0 & 1 \\ 1 & 0 \\ 1 & 0 \end{bmatrix}$$

▲ State diagrams, state equations

5-23. Draw state diagrams for the following transfer functions by direct decomposition. Assign the state variables from right to left for $x_1, x_2, \ldots$. Write the state equations from the state diagram and show that the equations are in CCF.

(a) $G(s) = \dfrac{10}{s^3 + 8.5s^2 + 20.5s + 15}$ **(b)** $G(s) = \dfrac{10(s+2)}{s^2(s+1)(s+3.5)}$

(c) $G(s) = \dfrac{5(s+1)}{s(s+2)(s+10)}$ **(d)** $G(s) = \dfrac{1}{s(s+5)(s^2+2s+2)}$

▲ State diagrams, state equations

5-24. Draw state diagrams for the systems described in Problem 5-23 by parallel decomposition. Make certain that the state diagrams contain a minimum number of integrators. The constant branch gains must be real. Write the state equations from the state diagram.

▲ State diagrams, state equations

5-25. Draw the state diagrams for the systems described in Problem 5-23 by means of cascade decomposition. Assign the state variables in ascending order from right to left. Write the state equations from the state diagram.

▲ State diagrams, dynamic equations, state-transition equations

5-26. The block diagram of a feedback control system is shown in Fig. 5P-26.

(a) Draw a state diagram for the system by first decomposing $G(s)$ by direct decomposition. Assign the state variables in ascending order, $x_1, x_2, \ldots$, from right to left. In addition to the state-variable-related nodes, the state diagram should contain nodes for $R(s)$, $E(s)$, and $Y(s)$.

(b) Write the dynamic equations of the system in vector-matrix form.

(c) Find the state-transition equations of the system using the state equations found in part (b). The initial state vector is $\mathbf{x}(0)$, and $r(t) = u_s(t)$.

(d) Find the output $y(t)$ for $t \geq 0$ with the initial state $\mathbf{x}(0)$, and $r(t) = u_s(t)$.

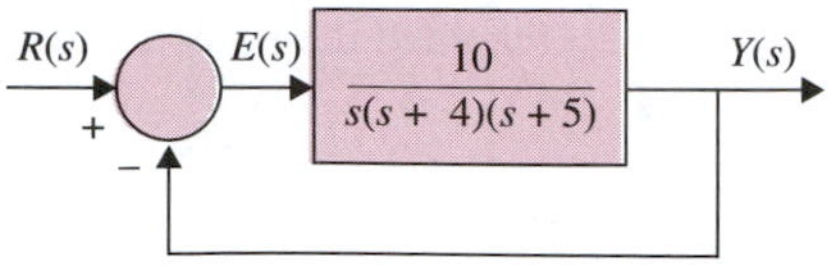

Figure 5P-26

▲ Transfer function, state diagram, state-transition equations

5-27. (a) Find the closed-loop transfer function $Y(s)/R(s)$, and draw the state diagram for the system shown in Fig. 5P-26.

(b) Perform a direct decomposition to $Y(s)/R(s)$, and draw the state diagram.

(c) Assign the state variables from right to left in ascending order, and write the state equations in vector-matrix form.

(d) Find the state-transition equations of the system using the state equations found in part (c). The initial state vector is $\mathbf{x}(0)$, and $r(t) = u_s(t)$.

(e) Find the output $y(t)$ for $t \geq 0$ with the initial state $\mathbf{x}(0)$, and $r(t) = u_s(t)$.

▲ State diagram, state equations, transfer function

5-28. The block diagram of a linearized idle-speed engine-control system of an automobile is shown in Fig. 5P-28. The system is linearized about a nominal operating point, so that all the variables represent linear-perturbed quantities. The following variables are defined: $T_m(t)$ is the engine torque, T_D the constant load-disturbance torque, $\omega(t)$ the engine speed, $u(t)$ the input voltage to the throttle actuator, and α the throttle angle. The time delay in the engine model can be approximated by

$$e^{-0.2s} \cong \frac{1 - 0.1s}{1 + 0.1s}$$

(a) Draw a state diagram for the system by decomposing each block individually. Assign the state variables from right to left in ascending order.

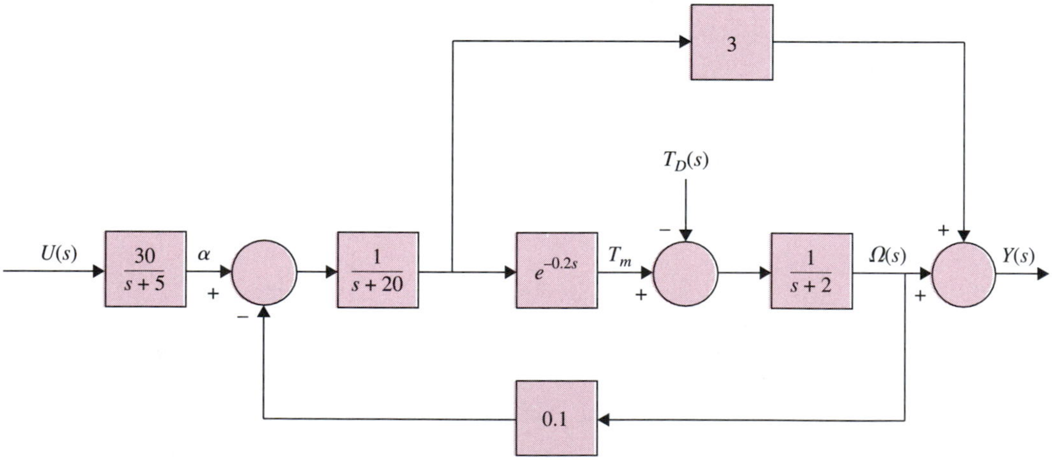

Figure 5P-28

(b) Write the state equations from the state diagram obtained in part (a), in the form of

$$\frac{d\mathbf{x}(t)}{dt} = \mathbf{A}\mathbf{x}(t) + \mathbf{B}\begin{bmatrix} u(t) \\ T_D(t) \end{bmatrix}$$

(c) Write $Y(s)$ as a function of $U(s)$ and $T_D(s)$. Write $\Omega(s)$ as a function of $U(s)$ and $T_D(s)$.

▲ Dynamic equations

5-29. The state diagram of a linear system is shown in Fig. 5P-29.

(a) Assign state variables on the state diagram from right to left in ascending order. Create additional artificial nodes if necessary so that the state-variable nodes satisfy as ''input nodes'' after the integrator branches are deleted.

(b) Write the dynamic equations of the system from the state diagram in part (a).

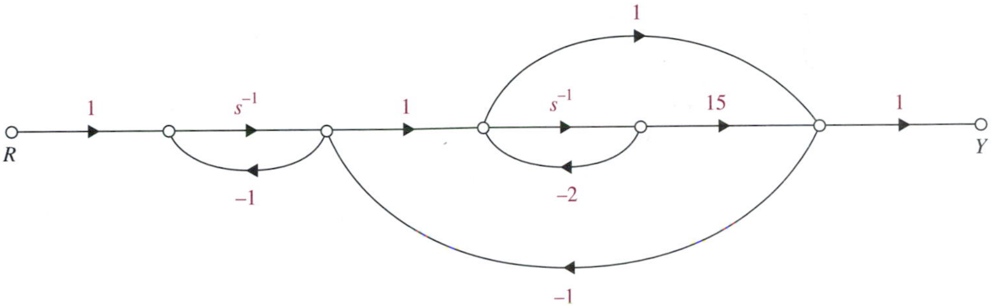

Figure 5P-29

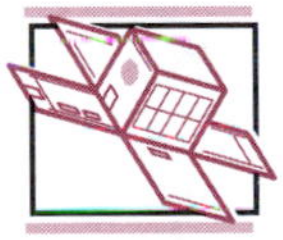

5-30. The block diagram of a linear spacecraft control system is shown in Fig. 5P-30.

(a) Determine the transfer function $Y(s)/R(s)$.

(b) Find the characteristic equation and its roots of the system. Show that the roots of the characteristic equation are not dependent on K.

▲ Transfer function, characteristic equation, state diagram, controllability, observability

(c) When $K = 1$, draw a state diagram for the system by decomposing $Y(s)/R(s)$, using a minimum number of integrators.

(d) Repeat part (c) when $K = 4$.

(e) Determine the values of K that must be avoided if the system is to be both state controllable and observable.

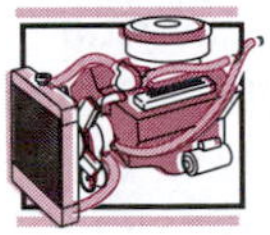

5-31. A considerable amount of effort is being spent by automobile manufacturers to meet the exhaust-emission-performance standards set by the government. Modern automobile-power-plant systems consist of an internal combustion engine that has an internal cleanup device called a catalytic converter. Such a system requires control of such variables as the engine air/fuel (A/F) ratio, ignition-spark timing, exhaust-gas recirculation, and

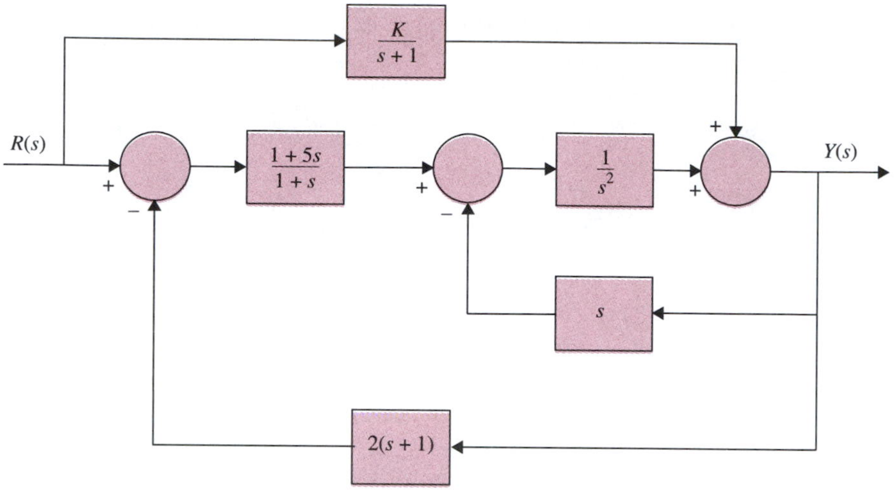

Figure 5P-30

▲ State diagram, characteristic equation

injection air. The control system problem considered in this problem deals with control of the A/F ratio. In general, depending on fuel composition and other factors, a typical stoichiometric A/F is 14.7 : 1, that is, 14.7 grams of air to each gram of fuel. An A/F greater or less than stoichiometry will cause high hydrocarbons, carbon monoxide, and nitrous oxides in the tailpipe emission. The control system shown in Fig. 5P-31 is devised to control the air/fuel ratio so that a desired output is achieved for a given input command. The sensor senses the composition of the exhaust-gas mixture entering the catalytic converter. The electronic controller detects the difference or the error between the command and the error and computes the control signal necessary to achieve the desired exhaust-gas composition. The output y(t) denotes the effective air/fuel ratio. The transfer function of the engine is given by

$$G_p(s) = \frac{Y(s)}{U(s)} = \frac{e^{-T_d s}}{1 + 0.5s}$$

where $T_d = 0.2$ sec is the time delay and is approximated by

$$e^{-T_d s} = \frac{1}{e^{T_d s}} = \frac{1}{1 + T_d s + T_d^2 s^2/2! + \cdots} \cong \frac{1}{1 + T_d s + T_d^2 s^2/2!}$$

The gain of the sensor is 1.0.

(a) Using the approximation for $e^{-T_d s}$ given above, find the expression for $G_p(s)$. Decompose $G_p(s)$ by direct decomposition, and draw the state diagram with u(t) as the input and $y(t)$ as the output. Assign state variables from right to left in ascending order, and write the state equations in vector-matrix form.

(b) Assuming that the controller is a simple amplifier with a gain of 1 [i.e., $u(t) = e$(t)], find the characteristic equation and its roots of the closed-loop system.

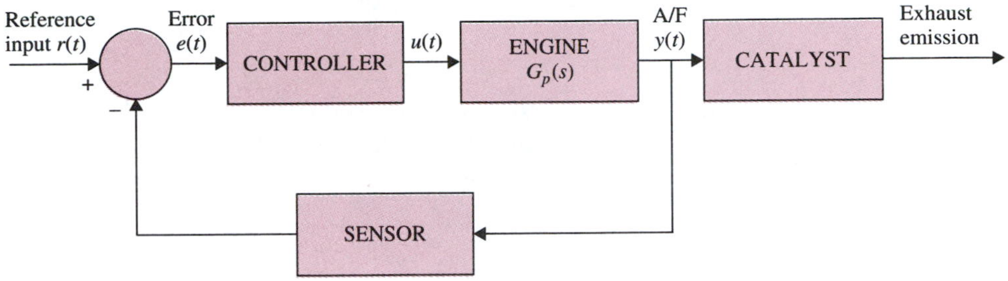

Figure 5P-31

▲ State diagram, characteristic equation

5-32. Repeat Problem 5-31 when the time delay of the automobile engine is approximated as

$$e^{-T_d s} \cong \frac{1 - T_d s/3}{1 + \frac{2}{3}T_d s + \frac{1}{6}T_d^2 s^2} \qquad T_d = 0.2 \text{ sec}$$

▲ State equation, state diagram, transfer functions

5-33. The schematic diagram in Fig. 5P-33 shows a permanent-magnet dc-motor-control system with a viscous-inertia damper. The system can be used for the control of the printwheel of an electronic word processor. A mechanical damper such as the viscous-inertia type is sometimes used in practice as a simple and economical way of stabilizing a control system. The damping effect is achieved by a rotor suspended in a viscous fluid. The differential and algebraic equations that describe the dynamics of the system are as

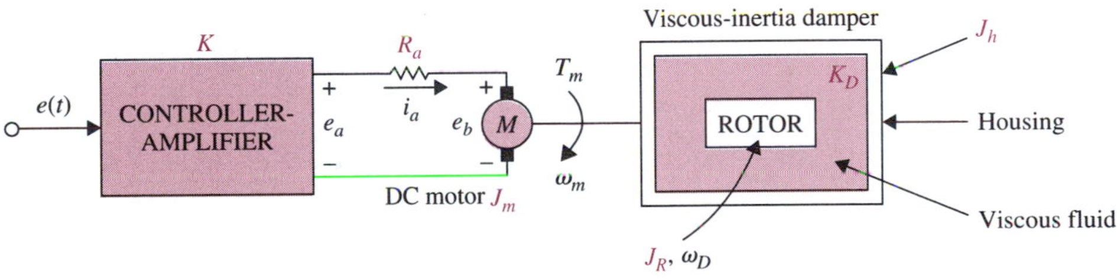

(a)

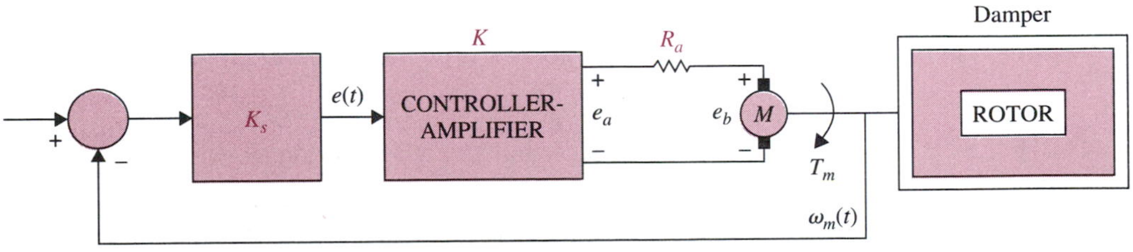

(b)

Figure 5P-33

follows:

▲ $e(t) = K_s[\omega_r(t) - \omega_m(t)]$	▲ $K_s = 1$ V/rad/sec
▲ $e_a(t) = Ke(t) = R_a i_a(t) + e_b(t)$	▲ $K = 10$
▲ $e_b(t) = K_b \omega_m(t)$	▲ $K_b = 0.0706$ V/rad/sec
▲ $T_m(t) = J\dfrac{d\omega_m(t)}{dt} + K_D[\omega_m(t) - \omega_D(t)]$	▲ $J = J_h + J_m = 0.1$ oz-in.-sec^2
▲ $T_m(t) = K_i i_a(t)$	▲ $K_i = 10$ oz-in./A
▲ $K_D[\omega_m(t) - \omega_D(t)] = J_R\dfrac{d\omega_D(t)}{dt}$	▲ $J_R = 0.05$ oz-in.-sec^2
▲ $R_a = 1\ \Omega$	▲ $K_D = 1$ oz-in.-sec

(a) Let the state variables be defined as $x_1(t) = \omega_m(t)$ and $x_2(t) = \omega_D(t)$. Write the state equations for the open-loop system with $e(t)$ as the input. (Open-loop refers to the feedback path from ω_m to e being open.)

(b) Draw the state diagram for the overall system using the state equations found in part (a) and $e(t) = K_s[\omega_r(t) - \omega_m(t)]$.

(c) Derive the open-loop transfer function $\Omega_m(s)/E(s)$ and the closed-loop transfer function $\Omega_m(s)/\Omega_r(s)$.

(d) Find the characteristic equation and its roots of the closed-loop system.

▲ Controllability

5-34. Determine the state controllability of the system shown in Fig. 5P-34.

(a) $a = 1, b = 2, c = 2,$ and $d = 1$.

(b) Are there any nonzero values for $a, b, c,$ and d such that the system is uncontrollable?

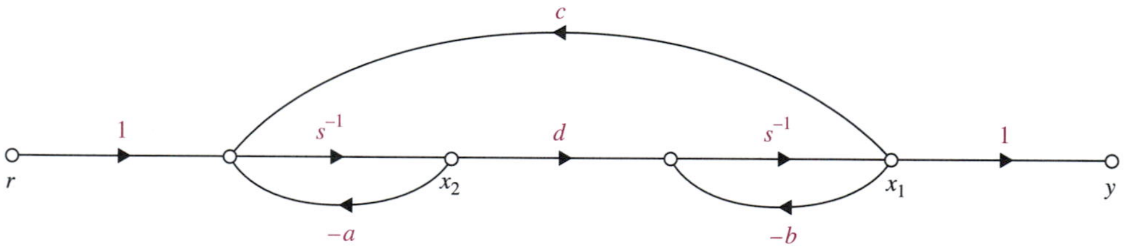

Figure 5P-34

▲ Controllability

5-35. Determine the controllability of the following systems.

(a) $\mathbf{A} = \begin{bmatrix} -1 & 0 & 0 \\ 0 & -1 & 0 \\ 0 & 0 & -1 \end{bmatrix} \quad \mathbf{B} = \begin{bmatrix} 1 \\ 1 \\ 1 \end{bmatrix}$

(b) $\mathbf{A} = \begin{bmatrix} -1 & 0 & 0 \\ 0 & -2 & 0 \\ 0 & 0 & -3 \end{bmatrix} \quad \mathbf{B} = \begin{bmatrix} 1 \\ 1 \\ 1 \end{bmatrix}$

▲ Controllability, observability

5-36. Determine the controllability and observability of the system shown in Fig. 5P-36 by the following methods.

(a) Conditions on the **A, B, C,** and **D** matrices.

(b) Conditions on the pole–zero cancellation of the transfer functions.

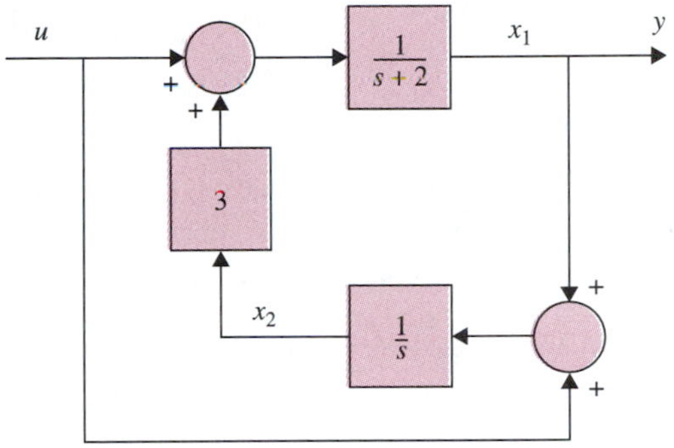

Figure 5P-36

▲ Controllability, observability

5-37. The transfer function of a linear control system is

$$\frac{Y(s)}{R(s)} = \frac{s + \alpha}{s^3 + 7s^2 + 14s + 8}$$

(a) Determine the value(s) of α so that the system is either uncontrollable or unobservable.

(b) With the value(s) of α found in part (a), define the state variables so that one of them is uncontrollable.

(c) With the value(s) of α found in part (a), define the state variables so that one of them is unobservable.

▲ Controllability

5-38. Consider the system described by the state equation

$$\frac{d\mathbf{x}(t)}{dt} = \mathbf{A}\mathbf{x}(t) + \mathbf{B}u(t)$$

where

$$\mathbf{A} = \begin{bmatrix} 0 & 1 \\ -1 & a \end{bmatrix} \qquad \mathbf{B} = \begin{bmatrix} 1 \\ b \end{bmatrix}$$

Find the region in the a–b plane such that the system is completely controllable.

▲ Controllability, observability

5-39. Determine the condition on b_1, b_2, c_1, and c_2 so that the following system is completely controllable and observable.

$$\frac{d\mathbf{x}(t)}{dt} = \mathbf{A}\mathbf{x}(t) + \mathbf{B}u(t) \qquad y(t) = \mathbf{C}\mathbf{x}(t)$$

$$\mathbf{A} = \begin{bmatrix} 1 & 1 \\ 0 & 1 \end{bmatrix} \qquad \mathbf{B} = \begin{bmatrix} b_1 \\ b_2 \end{bmatrix} \qquad \mathbf{C} = [d_1 \quad d_2]$$

▲ State diagram, characteristic equation, eigenvalues, controllability, observability

5-40. The schematic diagram of Fig. 5P-40 represents a control system whose purpose is to hold the level of the liquid in the tank at a desired level. The liquid level is controlled by a float whose position $h(t)$ is monitored. The input signal of the open-loop system is $e(t)$. The system parameters and equations are as follows.

- ▲ Motor resistance $R_a = 10\ \Omega$
- ▲ Torque constant $K_i = 10$ oz-in./A
- ▲ Back-emf constant $K_b = 0.0706$ V/rad/sec
- ▲ Load inertia $J_L = 10$ oz-in.-sec^2
- ▲ Amplifier gain $K_a = 50$
- ▲ Motor inductance $L_a = 0$ H
- ▲ Rotor inertia $J_m = 0.005$ oz-in.-sec^2
- ▲ Gear ratio $n = N_1/N_2 = 1/100$
- ▲ Load and motor friction = negligible
- ▲ Area of tank $A = 50$ ft^2

- ▲ $e_a(t) = R_a i_a(t) + K_b \omega_m(t)$
- ▲ $T_m(t) = K_i i_a(t) = (J_m + n^2 J_L)\dfrac{d\omega_m(t)}{dt}$
- ▲ $\omega_m(t) = \dfrac{d\theta_m(t)}{dt}$
- ▲ $\theta_y(t) = n\theta_m(t)$

The number of valves connected to the tank from the reservoir is $N = 10$. All the valves have the same characteristics and are controlled simultaneously by θ_y. The equations that govern the volume of flow are as follows:

- ▲ $q_i(t) = K_I N \theta_y(t)$
- ▲ $q_o(t) = K_o h(t)$
- ▲ $h(t) = \dfrac{\text{volume of tank}}{\text{area of tank}} = \dfrac{1}{A}\displaystyle\int [q_i(t) - q_o(t)]\, dt$
- ▲ $K_I = 10$ ft^3/sec-rad
- ▲ $K_o = 50$ ft^2/sec

(a) Define the state variables as $x_1(t) = h(t)$, $x_2(t) = \theta_m(t)$, and $x_3(t) = d\theta_m(t)/dt$. Write the state equations of the system in the form of $d\mathbf{x}(t)/dt = \mathbf{A}\mathbf{x}(t) + \mathbf{B}e_i(t)$. Draw a state diagram for the system.

(b) Find the characteristic equation and the eigenvalues of the **A** matrix found in part (a).

(c) Show that the open-loop system is completely controllable; that is, the pair [**A**, **B**] is controllable.

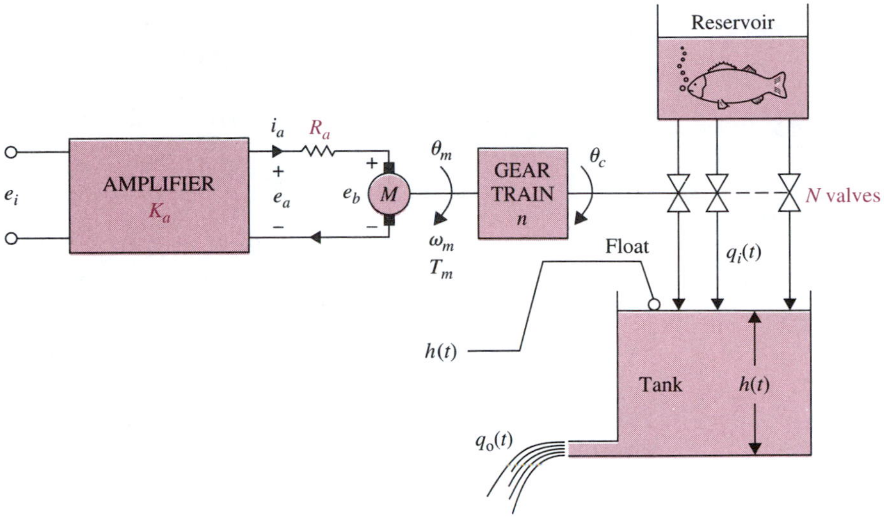

Figure 5P-40

(d) For reasons of economy, only one of the three state variables is measured and fed back for control purposes. The output equation is $y = \mathbf{Cx}$, where $\mathbf{C}$ can be one of the following forms:

(1) $\mathbf{C} = [1 \quad 0 \quad 0]$ **(2)** $\mathbf{C} = [0 \quad 1 \quad 0]$ **(3)** $\mathbf{C} = [0 \quad 0 \quad 1]$

Determine which case (or cases) corresponds to a completely observable system.

▲ Characteristic equation, controllability, observability

5-41. The "broom-balancing" control system described in Problem 4-23 has the following parameters:

$$M_b = 1 \text{ kg} \qquad M_c = 10 \text{ kg} \qquad L = 1 \text{ m} \qquad g = 32.2 \text{ ft/sec}^2$$

The small-signal linearized state equation model of the system is

$$\Delta\dot{\mathbf{x}}(t) = \mathbf{A}^*\Delta\mathbf{x}(t) + \mathbf{B}^*\Delta r(t)$$

where

$$\mathbf{A}^* = \begin{bmatrix} 0 & 1 & 0 & 0 \\ 25.92 & 0 & 0 & 0 \\ 0 & 0 & 0 & 1 \\ -2.36 & 0 & 0 & 0 \end{bmatrix} \qquad \mathbf{B}^* = \begin{bmatrix} 0 \\ -0.0732 \\ 0 \\ 0.0976 \end{bmatrix}$$

(a) Find the characteristic equation of $\mathbf{A}^*$ and its roots.

(b) Determine the controllability of $[\mathbf{A}^*, \mathbf{B}^*]$.

(c) For reason of economy, only one of the state variables are to be measured for feedback. The output equation is written

$$\Delta y(t) = \mathbf{C}^* \Delta \mathbf{x}(t)$$

where

(1) $\mathbf{C}^* = [1 \quad 0 \quad 0 \quad 0]$ (2) $\mathbf{C}^* = [0 \quad 1 \quad 0 \quad 0]$

(3) $\mathbf{C}^* = [0 \quad 0 \quad 1 \quad 0]$ (4) $\mathbf{C}^* = [0 \quad 0 \quad 0 \quad 1]$

Determine which **C*** corresponds to an observable system.

▲ Controllability

5-42. The double-inverted pendulum shown in Fig. 5P-42 is modeled approximately by the following linear state equation:

$$\frac{d\mathbf{x}(t)}{dt} = \mathbf{A}\mathbf{x}(t) + \mathbf{B}u(\mathrm{t})$$

where

$$\mathbf{x}(t) = \begin{bmatrix} \theta_1(t) \\ \dot{\theta}_1(t) \\ \theta_2(t) \\ \dot{\theta}_2(t) \\ x(t) \\ \dot{x}(t) \end{bmatrix}$$

$$\mathbf{A} = \begin{bmatrix} 0 & 1 & 0 & 0 & 0 & 0 \\ 16 & 0 & -8 & 0 & 0 & 0 \\ 0 & 0 & 0 & 1 & 0 & 0 \\ -16 & 0 & 16 & 0 & 0 & 0 \\ 0 & 0 & 0 & 0 & 0 & 1 \\ 0 & 0 & 0 & 0 & 0 & 0 \end{bmatrix} \qquad \mathbf{B} = \begin{bmatrix} 0 \\ -1 \\ 0 \\ 0 \\ 0 \\ 1 \end{bmatrix}$$

Determine the controllability of the states.

5-43. The block diagram of a simplified control system for the Large Space Telescope (LST) is shown in Fig. 5P-43. For simulation and control purposes, it would be desirable to model the system by state equations and by a state diagram.

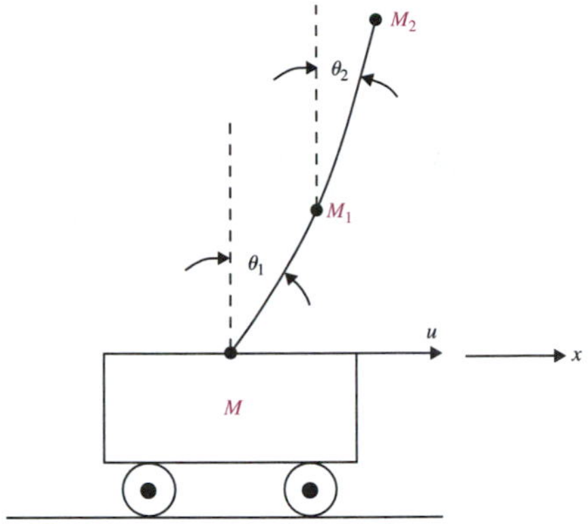

Figure 5P-42

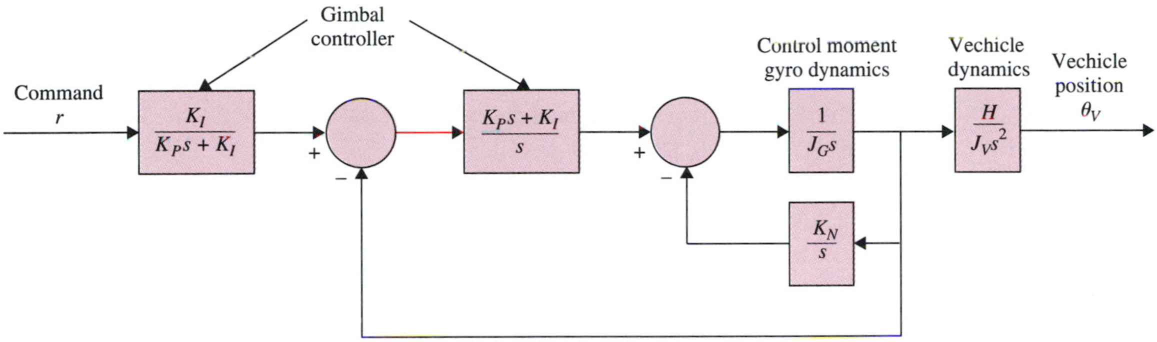

Figure 5P-43

▲ State equations, state diagram, transfer function, characteristic equation

(a) Draw a state diagram for the system and write the state equations in vector-matrix form. The state diagram should contain a minimum number of state variables, so it would be helpful if the transfer function of the system is written first.

(b) Find the characteristic equation of the system.

▲ Controllability observability, output feedback

5-44. The state diagram shown in Fig. 5P-44 represents two subsystems connected in cascade.

(a) Determine the controllability and observability of the system.

(b) Consider that output feedback is applied by feeding back y_2 to u_2, that is, $u_2 =$

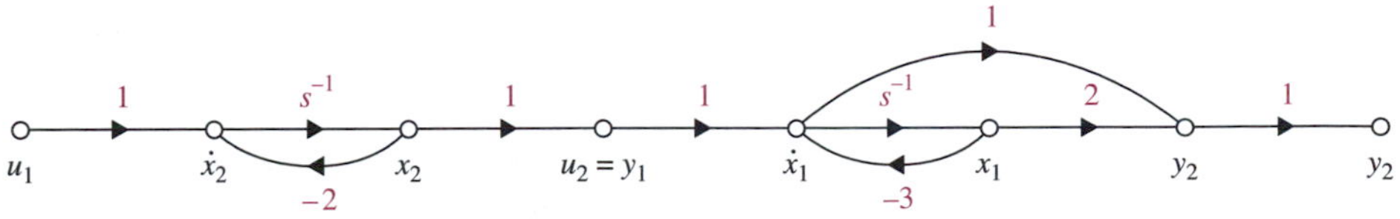

Figure 5P-44

$-ky_2$, where k is a real constant. Determine how the value of k affects the controllability and observability of the system.

▲ Controllability, state feedback

5-45. Given the system

$$\frac{d\mathbf{x}(t)}{dt} = \mathbf{A}\mathbf{x}(t) + \mathbf{B}u(t) \qquad y(t) = \mathbf{C}\mathbf{x}(t)$$

where

$$\mathbf{A} = \begin{bmatrix} 0 & 1 \\ -1 & -3 \end{bmatrix} \qquad \mathbf{B} = \begin{bmatrix} 1 \\ 2 \end{bmatrix} \qquad \mathbf{C} = [1 \quad 1]$$

(a) Determine the state controllability and observability of the system.

(b) Let $u(t) = -\mathbf{K}\mathbf{x}(t)$, where $\mathbf{K} = [k_1 \quad k_2]$, where k_1 and k_2 are real constants. Determine if and how controllability and observability of the closed-loop system are affected by the elements of $\mathbf{K}$.

▲ Vector-matrix discrete state equations

5-46. The block diagram of a sampled-data system is shown in Fig. 5P-46. The state equations of the controlled process are

$$\frac{dx_1(t)}{dt} = x_2(t) \qquad \frac{dx_2(t)}{dt} = -2x_1(t) - 3x_2(t) + h(t)$$

where $h(t)$ is the output of the sample-and-hold [i.e., $u(t)$ is constant during the sampling period T]. The sampling period T is 1 sec.

(a) Find the vector-matrix discrete state equations in the form of

$$\mathbf{x}[(k+1)T] = \boldsymbol{\phi}(T)\mathbf{x}(kT) + \theta(T)u(kT)$$

(b) Find $\mathbf{x}(NT)$ as functions of $\mathbf{x}(0)$ and $u(kT)$ for $k = 0, 1, 2, \ldots, N$.

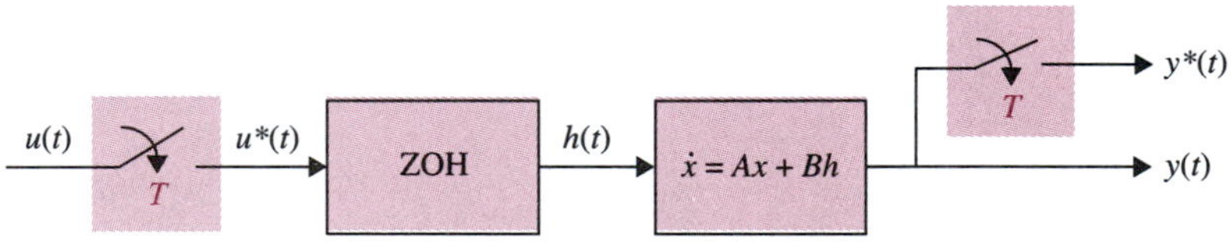

Figure 5P-46

▲ Vector-matrix discrete state equations

5-47. Repeat Problem 5-46 for the linear sampled-data system with the following state equations:

$$\frac{dx_1(t)}{dt} = x_1(t) \qquad \frac{dx_2(t)}{dt} = u(t)$$

The sampling period is 0.001 sec.

▲ Discrete-data system, transfer function

5-48. **(a)** Find the transfer function $\mathbf{X}(z)/U(z)$ for the system described in Problem 5-46.
(b) Find the characteristic equation of the system described in Problem 5-46.

▲ Discrete-data system, transfer function

5-49. **(a)** Find the transfer function $\mathbf{X}(z)/U(z)$ for the system described in Problem 5-47.
(b) Find the characteristic equation and its roots of the system described in Problem 5-47.

▲ Digital control, state diagram characteristic equation

5-50. Draw a state diagram for the digital control system represented by the following dynamic equations:

$$\mathbf{x}(k+1) = \mathbf{A}\mathbf{x}(k) + \mathbf{B}u(k) \qquad y(k) = x_1(k)$$

$$\mathbf{A} = \begin{bmatrix} 0 & 1 & -1 \\ 0 & 1 & 2 \\ 5 & 3 & -1 \end{bmatrix} \qquad \mathbf{B} = \begin{bmatrix} 0 \\ 0 \\ 1 \end{bmatrix}$$

Find the characteristic equation of the system.

▲ Digital control, state diagram, transfer function

5-51. The state diagram of a digital control system is shown in Fig. 5P-51. Write the dynamic equations. Find the transfer function $Y(z)/R(z)$.

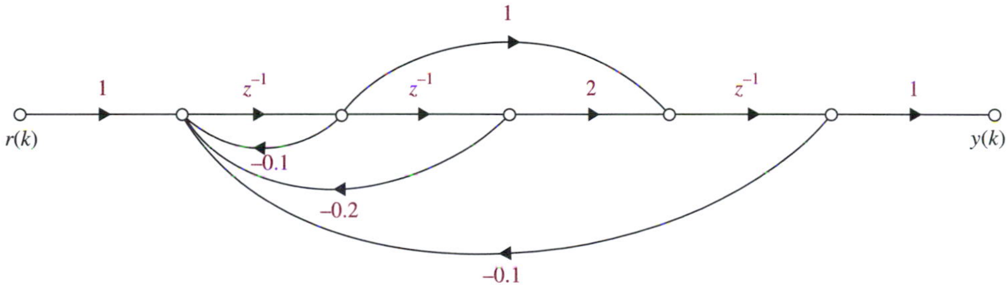

Figure 5P-51

▲ Sampled-data system, state equations, state diagram

5-52. The block diagram of a sampled-data system is shown in Fig. 5P-52. Write the discrete state equations of the system. Draw a state diagram for the system.

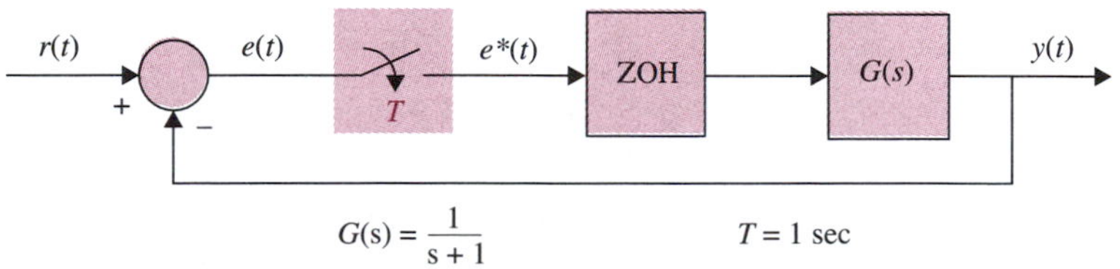

Figure 5P-52

Additional Computer Problems

5-53. The torque equation for part (a) of Problem 4-13 is

$$J\frac{d^2\theta(t)}{dt^2} = K_F d_1 \theta(t) + T_s d_2 \delta(t)$$

where $K_F d_1 = 1$ and $J = 1$. Define the state variables as $x_1 = \theta$ and $x_2 = d\theta/dt$. Find the state-transition matrix $\boldsymbol{\phi}(t)$ using **LINSYS** of **ACSP** or any other available computer program.

5-54. Starting with the state equation $d\mathbf{x}(t)/dt = \mathbf{A}\mathbf{x}(t) + \mathbf{B}\theta_r$ obtained in Problem 5-14, use **LINSYS** or **MATLAB** or any other available computer program to find the following.

(a) The state-transition matrix of $\mathbf{A}$, $\boldsymbol{\phi}(t)$.
(b) The characteristic equation of $\mathbf{A}$.
(c) The eigenvalues of $\mathbf{A}$.
(d) Compute and plot the unit-step response of $y(t) = \theta_y(t)$ for 3 sec. Set all the initial conditions to zero.

6 Stability of Linear Control Systems

KEYWORDS AND TOPICS

- ▲ **Stability Definitions**
- ▲ **Absolute Stability**
- ▲ **Relative Stability**
- ▲ **Routh–Hurwitz Criterion**
- ▲ **Stability of Discrete-Data Systems**

6-1 Introduction

From the studies of linear differential equations with constant coefficients of SISO systems, we learned that the homogeneous solution that corresponds to the transient response of the system is governed by the roots of the characteristic equation. Basically, the design of linear control systems may be regarded as a problem of arranging the location of the poles and zeros of the system transfer function such that the system will perform according to the prescribed specifications.

Among the many forms of performance specifications used in design, the most important requirement is that the system be stable. An unstable system is generally considered to be useless. When all types of systems are considered—linear, nonlinear, time-invariant, and time-varying—the definition of stability can be given in many different forms. We deal only with the stability of linear SISO time-invariant systems in the following discussions.

For analysis and design purposes, we can classify stability as **absolute stability** and **relative stability.** Absolute stability refers to the condition of whether the system is stable or unstable; it is a *yes* or *no* answer. Once the system is found to be stable, it is of interest to determine how stable it is, and this degree of stability is a measure of relative stability.

In preparation for the definition of stability, we define the two following types of responses for linear-time-invariant systems:

1. **Zero-state response.** The zero-state response is due to the input only; all the initial conditions of the system are zero.
2. **Zero-input response.** The zero-input response is due to the initial conditions only; all the inputs are zero.

From the principle of superposition, when a system is subject to both inputs and initial conditions, the total response is written

total response = zero-state response + zero-input response

The definitions above apply to continuous- as well as discrete-data systems.

6-2 Bounded-Input Bounded-Output Stability: Continuous-Data Systems

Let $u(t)$, $y(t)$, and $g(t)$ be the input, output, and impulse response of a linear time-invariant system, respectively. *With zero initial conditions, the system is said to be bounded-input bounded-output (BIBO) stable, or simply stable, if its output $y(t)$ is bounded to a bounded input $u(t)$.*

The convolution integral relating $u(t)$, $y(t)$, and $g(t)$ is

$$y(t) = \int_0^\infty u(t-\tau)g(\tau)\,d\tau \tag{6-1}$$

Taking the absolute value of both sides of the equation, we get

$$|y(t)| = \left|\int_0^\infty u(t-\tau)g(\tau)\,d\tau\right| \tag{6-2}$$

or

$$|y(t)| \le \int_0^\infty |u(t-\tau)||g(\tau)|\,d\tau \tag{6-3}$$

If $u(t)$ is bounded,

$$|u(t)| \le M \tag{6-4}$$

where M is a finite positive number. Then

$$|y(t)| \le M\int_0^\infty |g(\tau)|\,d\tau \tag{6-5}$$

Thus, if $y(t)$ is to be bounded, or

$$|y(t)| \le N < \infty \tag{6-6}$$

where N is a finite positive number, the following condition must hold:

$$M\int_0^\infty |g(\tau)|\,d\tau \le N < \infty \tag{6-7}$$

Or, for some finite positive Q,

$$\int_0^\infty |g(\tau)|\,d\tau \le Q < \infty \tag{6-8}$$

The condition given in Eq. (6-8) implies that the area under the $|g(\tau)|$ versus τ curve must be finite.

6-2-1 Relationship between Characteristic Equation Roots and Stability

To show the relation between the roots of the characteristic equation and the condition in Eq. (6-8), we write the transfer function $G(s)$, according to the Laplace transform definition, as

$$G(s) = \mathscr{L}[g(\tau)] = \int_0^\infty g(t)e^{-st}\,dt \tag{6-9}$$

Taking the absolute value on both sides of the last equation, we have

$$|G(s)| = \left|\int_0^\infty g(t)e^{-st}\,dt\right| \le \int_0^\infty |g(t)||e^{-st}|\,dt \tag{6-10}$$

since $|e^{-st}| = |e^{-\sigma t}|$, where σ is the real part of s. When s assumes a value of a pole of $G(s)$, $G(s) = \infty$, Eq. (6-10) becomes

$$\infty \leq \int_0^\infty |g(t)||e^{-\sigma t}|\, dt \tag{6-11}$$

▲ For BIBO stability, the roots of the characteristic equation must all lie in the left-half s-plane.

▲ A system that is BIBO stable is simply called stable; otherwise, it is unstable.

If one or more roots of the characteristic equation are in the right-half s-plane or on the $j\omega$-axis, $\sigma \geq 0$; then

$$|e^{-\sigma t}| \leq M = 1 \tag{6-12}$$

Equation (6-11) becomes

$$\infty \leq \int_0^\infty M|g(t)|\, dt = \int_0^\infty |g(t)|\, dt \tag{6-13}$$

which violates the BIBO stability requirement. Thus, *for BIBO stability, the roots of the characteristic equation, or the poles of $G(s)$, cannot be located in the right-half s-plane or on the $j\omega$-axis, or they must all lie in the left-half s-plane. A system is said to be unstable if it is not BIBO stable.* When a system has roots on the $j\omega$-axis, say, at $s = j\omega_0$ and $s = -j\omega_0$, if the input is a sinusoid, $\sin \omega_0 t$, the output will be of the form of $t \sin \omega_0 t$, which is unbounded, and the system is unstable.

6-3 Zero-Input and Asymptotic Stability of Continuous-Data Systems

In this section we define **asymptotic stability** and **zero-input stability** and establish their relations with BIBO stability.

Zero-input stability refers to the stability condition when the input is zero, and the system is driven only by its initial conditions. We shall show that the zero-input stability also depends on the roots of the characteristic equation. Let the input of an nth-order system be zero, and the output due to the initial conditions be $y(t)$. Then $y(t)$ can be expressed as

$$y(t) = \sum_{k=0}^{n-1} g_k(t) y^{(k)}(t_0) \tag{6-14}$$

where

$$y^{(k)}(t_0) = \left.\frac{d^k y(t)}{dt^k}\right|_{t=t_0} \tag{6-15}$$

and $g_k(t)$ denotes the zero-input response due to $y^{(k)}(t_0)$. The zero-input stability is defined as follows: *If the zero-input response $y(t)$, subject to the finite initial conditions, $y^{(k)}(t_0)$, reaches zero as t approaches infinity, the system is said to be zero-input stable, or stable; otherwise, the system is unstable.*

Mathematically, the foregoing definition can be stated: *A linear time-invariant system is zero-input stable if for any set of finite $y^{(k)}(t_0)$, there exists a positive number M, which depends on $y^{(k)}(t_0)$, such that*

$$\text{(1)} \quad |y(t)| \le M < \infty \qquad \text{for all } t \ge t_0 \tag{6-16}$$

and

$$\text{(2)} \quad \lim_{t\to\infty} |y(t)| = 0 \tag{6-17}$$

Because the condition in the last equation requires that the magnitude of $y(t)$ reaches zero as time approaches infinity, the zero-input stability is also known at the **asymptotic stability.**

Taking the absolute value on both sides of Eq. (6-14), we get

$$|y(t)| = \left| \sum_{k=0}^{n-1} g_k(t) y^{(k)}(t_0) \right| \le \sum_{k=0}^{n-1} |g_k(t)||y^{(k)}(t_0)| \tag{6-18}$$

Since all the initial conditions are assumed to be finite, the condition in Eq. (6-16) requires that the following condition be true:

$$\sum_{k=0}^{n-1} |g_k(t)| < \infty \qquad \text{for all } t \ge 0 \tag{6-19}$$

Let the n characteristic equation roots be expressed as $s_i = \sigma_i + j\omega_i$, $i = 1, 2, \ldots, n$. Then, if m of the n roots are simple and the rest are of multiple order, $y(t)$ will be of the form

$$y(t) = \sum_{i=1}^{m} K_i e^{s_i t} + \sum_{i=0}^{n-m-1} L_i t^i e^{s_i t} \tag{6-20}$$

where K_i and L_i are constant coefficients. Since the exponential terms $e^{s_i t}$ in Eq. (6-20) control the response $y(t)$ as $t \to \infty$, to satisfy the two conditions in Eqs. (6-16) and

(6-17), the real parts of s_i must be negative. In other words, the roots of the characteristic equation must all be in the left-half s-plane.

From the preceding discussions, we see that *for linear time-invariant systems, BIBO and zero-input stability and asymptotic stability all have the same requirement that the roots of the characteristic equation all be located in the left-half s-plane. Thus if a system is BIBO stable, it must also be zero-input or asymptotically stable.* For this reason, we shall simply refer to the stability condition of a linear system as **stable** or **unstable.** The latter condition refers to the condition that at least one of the characteristic equation roots is not in the left-half s-plane. For practical reasons, we often refer to the situation when the characteristic equation has simple roots on the $j\omega$-axis and none in the right-half plane as **marginally stable** or **marginally unstable.** *An exception to this is if the system is intended to be an integrator, or in the case of control systems, a velocity control system, the system would have root(s) at $s = 0$ and would be considered as stable.* Similarly, if the system is designed to be an oscillator, the characteristic equation would have simple roots on the $j\omega$-axis and the system would be regarded as stable.

▲ Systems with characteristic equation roots intentionally placed on the $j\omega$-axis are defined to be stable.

Since the roots of the characteristic equation are the same as the eigenvalues of **A** of the state equations, the stability condition places the same restrictions on the eigenvalues. Let the characteristic equation roots or eigenvalues of **A** of a linear continuous-data time-invariant SISO system be $s_i = \sigma_i + j\omega_i$, $i = 1, 2, \ldots, n$. If any of the roots is complex, it is in complex-conjugate pairs. The possible stability conditions of the system are summarized in Table 6-1 with respect to the roots of the characteristic equation.

The following example illustrates the stability conditions of systems with reference to the poles of the system transfer functions that are also the roots of the characteristic equation.

Table 6-1 Stability Conditions of Linear Continuous-Data Time-Invariant SISO Systems

Stability Condition	Root Values
Asymptotically stable or simply stable	$\sigma_i < 0$ for all i, $i = 1, 2, \ldots, n$. (All the roots are in the left-half s-plane.)
Marginally stable or marginally unstable	$\sigma_i = 0$ for any i for simple roots, and no $\sigma_i > 0$ for $i = 1, 2, \ldots, n$. (At least one simple root, and no multiple-order roots on the $j\omega$-axis; and no roots in the right-half s-plane.) (Note exceptions.)
Unstable	$\sigma_i > 0$ for any i, or $\sigma_i = 0$ for any multiple-order root. $i = 1, 2, \ldots, n$. (At least one simple root in the right-half s-plane or at least one multiple-order root on the $j\omega$-axis.)

Example 6-1

$$M(s) = \frac{20}{(s+1)(s+2)(s+3)}$$ BIBO or asymptotically stable, or simply stable.

$$M(s) = \frac{20(s+1)}{(s-1)(s^2+2s+2)}$$ Unstable due to the pole at $s = 1$.

$$M(s) = \frac{20(s-1)}{(s+2)(s^2+4)}$$ Marginally stable or marinally unstable due to $s = \pm j2$.

$$M(s) = \frac{10}{(s^2+4)^2(s+10)}$$ Unstable due to the multiple-order pole at $s = \pm j2$.

$$M(s) = \frac{10}{s^4+30s^3+s^2+10s}$$ Stable if the pole at $s = 0$ is placed intentionally.

▲

6-4 Methods of Determining Stability

The discussions in the preceding sections lead to the conclusion that the stability of linear time-invariant SISO systems can be determined by checking on the location of the roots of the characteristic equation of the system. For all practical purposes, there is no need to compute the complete system response to determine stability. The regions of stability and instability in the s-plane are illustrated in Fig. 6-1. When the system parameters are all known, the roots of the characteristic equation can be solved by means of a root-finding computer program which is readily available in many commercially prepared packages. For example, the M-file **roots(c)** of **MATLAB** solves for the roots of a polynomial, and **eig(A)** of **MATLAB** finds the eigenvalues of **A**. Several

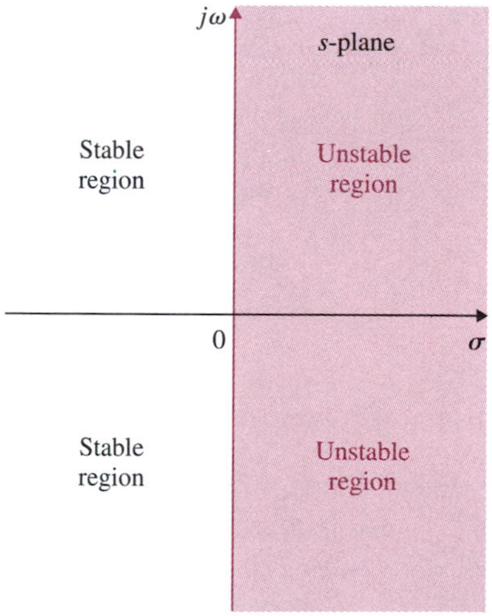

Figure 6-1 Stable and unstable regions in the s-plane.

functions in the **CSAD** toolbox contain features that allow computation of the poles of the system transfer function. The best one to use is **svdesign,** which has the option of finding the eigenvalues with the system described by either the system transfer function or dynamic equations. In **Program CC, zeros** can be used to find the zeros of a polynomial, and **stability** returns the poles of the closed-loop transfer function and the stability condition once the open-loop transfer function is entered. These are mere examples of what computer programs can do in finding the characteristic equation roots.

For design purposes, there will be unknown or variable parameters embedded in the characteristic equation, and it would not be feasible to use the root-finding programs. The methods outlined below are well known for the determination of stability of linear continuous-data systems, without involving root solving.

1. **Routh–Hurwitz criterion.** This criterion is an algebraic method that provides information on the absolute stability of a linear time-invariant system that has a characteristic equation with constant coefficients. The criterion tests whether any of the roots of the characteristic equation lie in the right-half s-plane. The number of roots that lie on the $j\omega$-axis and in the right-half plane is also indicated.
2. **Nyquist criterion.** This criterion is a semigraphical method that gives information on the difference between the number of poles and zeros of the closed-loop transfer function that are in the right-half s-plane by observing the behavior of the Nyquist plot of the loop transfer function.
3. **Bode diagram.** This diagram is a plot of the magnitude of the loop transfer function $G(j\omega)H(j\omega)$ in decibels and the phase of $G(j\omega)H(j\omega)$ in degrees, all versus frequency ω. The stability of the closed-loop system can be determined by observing the behavior of these plots.

Thus, as will be evident throughout the book, most of the analysis and design techniques on control systems represent alternative methods of solving the same problem. The designer simply has to make the proper choice of the best analytical tool, depending on the particular situation. Details of the Routh–Hurwitz stability criterion are presented in the following section.

6-5 Routh–Hurwitz Criterion

The Routh–Hurwitz criterion represents a method of determining the location of zeros of a polynomial with constant real coefficients with respect to the left half and right half of the s-plane, without actually solving for the zeros. *Since root-finding computer programs can solve for the zeros of a polynomial with ease, the value of the Routh–Hurwitz criterion is at best limited to equations with at least one unknown parameter.*

Consider that the characteristic equation of a linear time-variant SISO system is of the form

$$F(s) = a_n s^n + a_{n-1} s^{n-1} + \cdots + a_1 s + a_0 = 0 \tag{6-21}$$

where all the coefficients are real. In order that Eq. (6-21) not have roots with positive real parts, it is *necessary and sufficient* that the following conditions hold:

1. All the coefficients of the equation have the same sign.
2. None of the coefficients vanishes.

The conditions above are based on the laws of algebra, which related the coefficients of Eq. (6-21) as follows:

$$\frac{a_{n-1}}{a_n} = -\sum \text{all roots} \tag{6-22}$$

$$\frac{a_{n-2}}{a_n} = \sum \text{products of the roots taken two at a time} \tag{6-23}$$

$$\frac{a_{n-3}}{a_n} = -\sum \text{products of the roots taken three at a time} \tag{6-24}$$

$$\vdots$$

$$\frac{a_0}{a_n} = (-1)^n \text{products of all the roots} \tag{6-25}$$

Thus all these ratios must be positive and nonzero unless at least one of the roots has a positive real part.

The two necessary conditions for Eq. (6-21) to have no roots in the right-half s-plane can easily be checked by inspection of the equation. However, these conditions are not sufficient, for it is quite possible that an equation with all its coefficients nonzero and of the same sign still may not have all the roots in the left half of the s-plane.

6-5-1 Hurwitz Criterion [1]

The Routh–Hurwitz criterion is based on the Hurwitz criterion, which is stated as follows. *The necessary and sufficient condition that all roots of Eq. (6-21) lie in the left half of the s-plane is that the equation's* ***Hurwitz determinant,*** D_k, $k = 1, 2, \ldots, n$, *must all be positive.*

The Hurwitz determinants of Eq. (6-21) are given by

$$D_1 = a_{n-1} \qquad D_2 = \begin{vmatrix} a_{n-1} & a_{n-3} \\ a_n & a_{n-2} \end{vmatrix} \qquad D_3 = \begin{vmatrix} a_{n-1} & a_{n-3} & a_{n-5} \\ a_n & a_{n-2} & a_{n-4} \\ 0 & a_{n-1} & a_{n-3} \end{vmatrix} \qquad \cdots$$

$$D_n = \begin{vmatrix} a_{n-1} & a_{n-3} & a_{n-5} & \cdots & 0 \\ a_n & a_{n-2} & a_{n-4} & \cdots & 0 \\ 0 & a_{n-1} & a_{n-3} & \cdots & 0 \\ \vdots & \vdots & \vdots & \ddots & \vdots \\ 0 & 0 & 0 & \cdots & a_0 \end{vmatrix} \tag{6-26}$$

where the coefficients with indices larger than n or negative should be replaced with zeros. At first glance, the application of the Hurwitz determinants may seem to be formidable for high-order equations, because of the amount of work involved in evaluating the determinants in Eq. (6-26). Fortunately, Routh simplified the process by introducing a tabulation method in place of the Hurwitz determinants.

6-5-2 Routh's Tabulation

The first step in the simplification of the Hurwitz criterion, now called the Routh–Hurwitz criterion, is to arrange the coefficients of the equation in Eq. (6-21) into two rows. The first row consists of the first, third, fifth, . . . , coefficients, and the second row consists of the second, fourth, sixth, . . . , coefficients, all counting from the highest-order term, as shown in the following tabulation:

$$\begin{array}{ccccc} a_n & a_{n-2} & a_{n-4} & a_{n-6} & \cdots \\ a_{n-1} & a_{n-3} & a_{n-5} & a_{n-7} & \cdots \end{array}$$

The next step is to form the following array of numbers by the indicated operations, illustrated here for a sixth-order equation:

$$a_6 s^6 + a_5 s^5 + a_4 s^4 + a_3 s^3 + a_2 s^2 + a_1 s + a_0 = 0 \tag{6-27}$$

s^6	a_6	a_4	a_2	a_0
s^5	a_5	a_3	a_1	0
s^4	$\dfrac{a_5a_4 - a_6a_3}{a_5} = A$	$\dfrac{a_5a_2 - a_6a_1}{a_5} = B$	$\dfrac{a_5a_0 - a_6 \times 0}{a_5} = a_0$	0
s^3	$\dfrac{Aa_3 - a_5B}{A} = C$	$\dfrac{Aa_1 - a_5a_0}{A} = D$	$\dfrac{A \times 0 - a_5 \times 0}{A} = 0$	0
s^2	$\dfrac{BC - AD}{C} = E$	$\dfrac{Ca_0 - A \times 0}{C} = a_0$	$\dfrac{C \times 0 - A \times 0}{C} = 0$	0
s^1	$\dfrac{ED - Ca_0}{E} = F$	0	0	0
s^0	$\dfrac{Fa_0 - E \times 0}{F} = a_0$	0	0	0

The array above is called the **Routh's tabulation** or **Routh's array.** The column of s's on the left side is used for identification purposes. The reference column keeps track of the calculations, and the last row of the Routh's tabulation should always be the s^0 row.

Once the Routh's tabulation has been completed, the last step in the application of the criterion is to investigate the *signs* of the coefficients in the *first column* of the tabulation, which contains information on the roots of the equation. The following conclusions are made: *The roots of the equation are all in the left half of the s-plane if all the elements of the first column of the Routh's tabulation are of the same sign. The number of changes of signs in the elements of the first column equals the number of roots with positive real parts or in the right-half s-plane.*

The reason for the foregoing conclusion is simple, based on the requirements of the Hurwitz determinants. The relations between the elements in the first column of the Routh's tabulation and the Hurwitz determinants are:

s^6	$a_5 = a_6$
s^5	$a_5 = D_1$
s^4	$A = \dfrac{D_2}{D_1}$
s^3	$C = \dfrac{D_3}{D_2}$
s^2	$E = \dfrac{D_4}{D_3}$
s^1	$F = \dfrac{D_5}{D_4}$
s^0	$a_0 = \dfrac{D_6}{D_5}$

Therefore, if all the Hurwitz determinants are positive, the elements in the first column of the Routh's tabulation would be of the same sign.

The following examples illustrate the applications of the Routh–Hurwitz criterion when the tabulation terminates without complications.

Example 6-2

Consider the equation

$$(s-2)(s+1)(s-3) = s^3 - 4s^2 + s + 6 = 0 \tag{6-28}$$

which has one negative coefficient. Thus, from the necessary condition, we know without applying Routh's test that not all the roots of the equation are in the left-half s-plane. In fact, from the factored form of the equation, we know that there are two roots in the right-half s-plane, at $s = 2$ and $s = 3$. For the purpose of illustrating the Routh–Hurwitz criterion, Routh's tabulation is made as follows:

	s^3	1	1
Sign change	s^2	-4	6
Sign change	s^1	$\dfrac{(-4)(1)-(6)(1)}{-4} = 2.5$	0
	s^0	$\dfrac{(2.5)(6)-(-4)(0)}{2.5} = 6$	0

Since there are two sign changes in the first column of the tabulation, the equation has two roots located in the right-half s-plane. This agrees with the known result. ▲

Example 6-3

Consider the equation

$$2s^4 + s^3 + 3s^2 + 5s + 10 = 0 \tag{6-29}$$

Since the equation has no missing terms and the coefficients are all of the same sign, it satisfies the necessary condition for not having roots in the right half or on the imaginary axis of the s-plane. However, the sufficient condition must still be checked. Routh's tabulation is made as follows:

	s^4	2	3	10
	s^3	1	5	0
Sign change	s^2	$\dfrac{(1)(3)-(2)(5)}{1} = -7$	10	0
Sign change	s^1	$\dfrac{(-7)(5)-(1)(10)}{-7} = 6.43$	0	0
	s^0	10	0	0

Since there are two changes in sign in the first column of the tabulation, the equation has two roots in the right half of the s-plane. Solving for the roots of Eq. (6-29), we have the four roots at $s = -1.0055 \pm j0.93311$ and $s = 0.7555 \pm j1.4444$. Clearly, the last two roots are in the right-half s-plane, which cause the system to be unstable. ▲

6-5-3 Special Cases When Routh's Tabulation Terminates Prematurely

The equations considered in the two preceding examples are designed so that Routh's tabulation can be carried out without any complications. Depending on the coefficients of the equation, the following difficulties may occur that prevent Routh's tabulation from completing properly:

1. The first element in any one row of Routh's tabulation is zero, but the others are not.
2. The elements in one row of Routh's tabulation are all zero.

In the first case, if a zero appears in the first element of a row, the elements in the next row will all become infinite, and Routh's tabulation cannot continue. To remedy the situation, we *replace the zero element in the first column by an arbitrary small positive number ε, and then proceed with Routh's tabulation.* This is illustrated by the following example.

Example 6-4

Consider the characteristic equation of a linear system:

$$s^4 + s^3 + 2s^2 + 2s + 3 = 0 \tag{6-30}$$

Since all the coefficients are nonzero and of the same sign, we need to apply the Routh–Hurwitz criterion. Routh's tabulation is carried out as follows:

s^4	1	2	3
s^3	1	2	0
s^2	0	3	

Since the first element of the s^2 row is zero, the element in the s^1 row would all be infinite. To overcome this difficulty, we replace the zero in the s^2 row by a small positive number ε and then proceed with the tabulation. Starting with the s^2 row, the results are as follows:

	s^2	ε	3
Sign change			
	s^1	$\dfrac{2\varepsilon - 3}{\varepsilon} \cong -\dfrac{3}{\varepsilon}$	0
Sign change			
	s^0	3	0

Since there are two sign changes in the first column of Routh's tabulation, the equation in Eq. (6-30) has two roots in the right-half s-plane. Solving for the roots of Eq. (6-30), we get $s = -0.09057 \pm j0.902$ and $s = 0.4057 \pm j1.2928$; the last two roots are clearly in the right-half s-plane. ▲

It should be noted that the ε-method described may not give correct results if the equation has pure imaginary roots [2,3].

In the second special case, when all the elements in one row of Routh's tabulation are zeros before the tabulation is properly terminated, it indicates that one or more of the following conditions may exist:

1. The equation has at least one pair of real roots with equal magnitude but opposite signs.
2. The equation has one or more pairs of imaginary roots.
3. The equation has pairs of complex-conjugate roots forming symmetry about the origin of the s-plane (e.g., $s = -1 \pm j1$, $s = 1 \pm j1$).

▲ The coefficients of the auxiliary equation are those of the row just above the row of zeros in Routh's tabulation.

▲ The roots of the auxiliary equation must also satisfy the original equation.

The situation with the entire row of zeros can be remedied by using the **auxiliary equation** $A(s) = 0$, which is formed from the coefficients of the row just above the row of zeros in Routh's tabulation. The auxiliary equation is always an even polynomial; that is, only even powers of s appear. *The roots of the auxiliary equation also satisfy the original equation.* Thus, by solving the auxiliary equation, we also get some of the roots of the original equation. To continue with Routh's tabulation when a row of zeros appears, we conduct the following steps:

1. Form the auxiliary equation $A(s) = 0$ by use of the coefficients from the row just preceding the row of zeros.
2. Take the derivative of the auxiliary equation with respect to s; this gives $dA(s)/ds = 0$.
3. Replace the row of zeros with the coefficients of $dA(s)/ds = 0$.

4. Continue with Routh's tabulation in the usual manner with the newly formed row of coefficients replacing the row of zeros.
5. Interpret the change of signs, if any, of the coefficients in the first column of Routh's tabulation in the usual manner.

Example 6-5

Consider the following equation, which may be the characteristic equation of a linear control system:

$$s^5 + 4s^4 + 8s^3 + 8s^2 + 7s + 4 = 0 \tag{6-31}$$

Routh's tabulation is

s^5	1	8	7
s^4	4	8	4
s^3	6	6	0
s^2	4	4	
s^1	0	0	

Since a row of zeros appear prematurely, we form the auxiliary equation using the coefficients of the s^2 row:

$$A(s) = 4s^2 + 4 = 0 \tag{6-32}$$

The derivative of $A(s)$ with respect to s is

$$\frac{dA(s)}{ds} = 8s = 0 \tag{6-33}$$

from which the coefficients 8 and 0 replace the zeros in the s^1 row of the original tabulation. The remaining portion of the Routh's tabulation is

s^1	8	0	coefficients of $dA(s)/ds$
s^0	4		

Since there are no sign changes in the first column of the entire Routh's tabulation, the equation in Eq. (6-31) does not have any root in the right-half s-plane. Solving the auxiliary equation in Eq. (6-32), we get the two roots at $s = j$ and $s = -j$, which are also two of the roots of Eq. (6-31). Thus the equation has two roots on the $j\omega$-axis, and the system is marginally stable. These imaginary roots caused the initial Routh's tabulation to have the entire row of zeros in the s^1 row. ▲

Since all zeros occurring in a row that corresponds to an odd power of s creates an auxiliary equation that has only even powers of s, the roots of the auxiliary equation may all lie on the $j\omega$-axis. For design purposes, we can use the all-zero-row condition to solve for the marginal value of a system parameter for system stability. The following example illustrates the realistic value of the Routh–Hurwitz criterion in a simple design problem.

Example 6-6

Consider that a third-order control system has the characteristic equation

$$s^3 + 3408.3s^2 + 1{,}204{,}000s + 1.5 \times 10^7 K = 0 \tag{6-34}$$

The Routh–Hurwitz criterion is best suited to determine the critical value of K for stability, that is, the value of K for which at least one root will lie on the $j\omega$-axis and none in the right-half s-plane. Routh's tabulation of Eq. (6-34) is made as follows:

s^3	1	$1{,}204{,}000$
s^2	3408.3	$1.5 \times 10^7 K$
s^1	$\dfrac{410.36 \times 10^7 - 1.5 \times 10^7 K}{3408.3}$	0
s^0	$1.5 \times 10^7 K$	

For the system to be stable, all the roots of Eq. (6-34) must be in the left-half s-plane, and thus all the coefficients in the first column of Routh's tabulation must have the same sign. This leads to the following conditions:

$$\frac{410.36 \times 10^7 - 1.5 \times 10^7 K}{3408.3} > 0 \tag{6-35}$$

and

$$1.5 \times 10^7 K > 0 \tag{6-36}$$

From the inequality of Eq. (6-35), we have $K < 273.57$, and the condition in Eq. (6-36) gives $K > 0$. Therefore, the condition of K for the system to be stable is

$$0 < K < 273.57 \tag{6-37}$$

If we let $K = 273.57$, the characteristic equation in Eq. (6-34) will have two roots on the $j\omega$-axis. To find these roots, we substitute $K = 273.57$ in the auxiliary equation, which is obtained from Routh's tabulation by using the coefficients of the s^2 row. Thus

$$A(s) = 3408.3s^2 + 4.1036 \times 10^9 = 0 \tag{6-38}$$

which has roots at $s = j1097.27$ and $s = -j1097.27$, and the corresponding value of K at these roots is 273.57. Also, if the system is operated with $K = 273.57$, the zero-input response of the system will be an undamped sinusoid with a frequency of 1097.27 rad/sec. ▲

Example 6-7

As another example of using the Routh–Hurwitz criterion for simple design problems, consider that the characteristic equation of a closed-loop control system is

$$s^3 + 3Ks^2 + (K + 2)s + 4 = 0 \tag{6-39}$$

It is desired to find the range of K so that the system is stable. Routh's tabulation of Eq. (6-39) is

s^3	1	$K + 2$
s^2	$3K$	4
s^1	$\dfrac{3K(K+2) - 4}{3K}$	0
s^0	4	

From the s^2 row, the condition of stability is $K > 0$, and from the s^1 row, the condition of stability is

$$3K^2 + 6K - 4 > 0 \tag{6-40}$$

or

$$K < -2.528 \quad \text{or} \quad K > 0.528$$

When the conditions of $K > 0$ and $K > 0.528$ are compared, it is apparent that the latter requirement is more stringent. Thus for the closed-loop system to be stable, K must satisfy

$$K > 0.528$$

The requirement of $K < -2.528$ is disregarded since K cannot be negative. ▲

It should be reiterated that the Routh–Hurwitz criterion is valid only if the characteristic equation is algebraic with real coefficients. If any one of the coefficients is complex, or if the equation is not algebraic, such as containing exponential functions or sinusoidal functions of s, the Routh–Hurwitz criterion simply cannot be applied.

Another limitation of the Routh–Hurwitz criterion is that it is valid only for the determination of roots of the characteristic equation with respect to the left half or the right half on the s-plane. The stability boundary is the $j\omega$-axis of the s-plane. The criterion *cannot* be applied to any other stability boundaries in a complex plane, such as the unit circle in the z-plane, which is the stability boundary of discrete-data systems.

6-6 Stability of Discrete-Data Systems

The definitions of BIBO and zero-input stability can readily be extended to linear time-invariant SISO discrete-data control systems.

6-6-1 BIBO Stability

Let $u(kT)$, $y(kT)$, and $g(kT)$ be the input, output, and impulse sequence of a linear time-invariant SISO discrete-data system, respectively. With *zero initial conditions, the system is said to be BIBO stable, or simply stable, if its output sequence, $y(kT)$ is bounded to a bounded input $u(kT)$.* As with the treatment in Section 6-1, we can show that for the system to be BIBO stable, the following condition must be met:

$$\sum_{k=0}^{\infty} |g(kT)| < \infty \tag{6-41}$$

6-6-2 Zero-Input Stability

For zero-input stability, the output sequence of the system must satisfy the following conditions:

1. $|y(kT)| \leq M < \infty$ (6-42)

2. $\lim_{k\to\infty} |y(kT)| = 0$ (6-43)

Thus zero-input stability can also be referred to as **asymptotic stability.** We can show that *both the BIBO stability and the zero-input stability of discrete-data systems require that the roots of the characteristic equation lie inside the unit circle $|z| = 1$ in the z-plane.* This is not surprising, since the $j\omega$-axis of the s-plane is mapped onto the unit circle in the z-plane. The regions of stability and instability for discrete-data systems in the z-plane are shown in Fig. 6-2. Let the characteristic equation roots of a linear

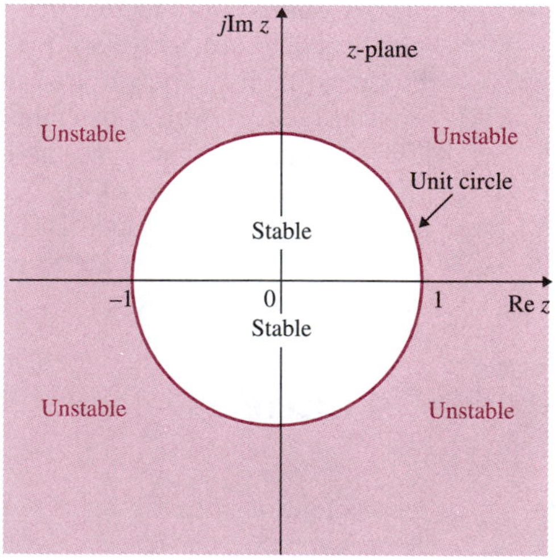

Figure 6-2 Stable and unstable regions for discrete-data systems in the z-plane.

Table 6-2 Stability Conditions of Linear Time-Invariant Discrete-Data SISO Systems

Stability Condition	Root Values
Asymptotically stable or simply stable	$\lvert z_i \rvert < 1$ for all i, $i = 1, 2, \ldots, n$. (All the roots are inside the unit circle.)
Marginally stable or marginally unstable	$\lvert z_i \rvert = 1$ for any i for simple roots, and no $\lvert z_i \rvert > 1$ for $i = 1, 2, \ldots, n$. (At least one simple root, and no multiple-order roots on the unit circle and no roots outside the unit circle.)
Unstable	$\lvert z_i \rvert > 1$ for any i, or $\lvert z_i \rvert = 1$ for any multiple-order root. $i = 1, 2, \ldots, n$. (At least one simple root outside the unit circle and at least one multiple-order root on the unit circle.)

discrete-data time-invariant SISO system be z_i, $i = 1, 2, \ldots, n$. The possible stability conditions of the system are summarized in Table 6-2 with respect to the roots of the characteristic equation.

The following example illustrates the relationship between the closed-loop transfer function poles, which are the characteristic equation roots, and the stability condition of the system.

Example 6-8

$$M(z) = \frac{5z}{(z-0.2)(z-0.8)} \quad \text{Stable system}$$

$$M(z) = \frac{5z}{(z+1.2)(z-0.8)} \quad \text{Unstable system due to the pole at } z = -1.2$$

$$M(z) = \frac{5(z+1)}{z(z-1)(z-0.8)} \quad \text{Marginally stable due to } z = 1$$

$$M(z) = \frac{5(z+1.2)}{z^2(z+1)^2(z+0.1)} \quad \text{Unstable due to second-order pole at } z = -1$$

▲

6-7 Stability Tests of Discrete-Data Systems

We pointed out in Section 6-6 that the stability test of a linear discrete-data system is essentially a problem of investigating whether all the roots of the characteristic equation are inside the unit circle $\lvert z \rvert = 1$ in the z-plane. The Nyquist criterion, root-locus diagram, and the Bode diagram, originally devised for continuous-data systems, can all be extended to the stability studies of discrete-data systems. One except is the Routh–Hurwitz criterion, which in its original form is restricted to only the imaginary axis of the s-plane as the stability boundary, and thus can only be applied to continuous-data systems.

6-7-1 Bilinear Transformation Method [9]

We can still apply the Routh–Hurwitz criterion to discrete-data systems if we could find a transformation that transforms the unit circle in the z-plane onto the imaginary axis of another complex plane. We cannot use the z-transform relation $z = \exp(Ts)$ or $s = (\ln z)/T$, since it would transform an algebraic equation in z into a nonalgebraic equation in s, and the Routh test still cannot be applied. However, there are many bilinear transformations of the form

$$z = \frac{ar + b}{cr + d} \tag{6-44}$$

where a, b, c, d are real constants, *and* r is a complex variable, that will transform circles in the z-plane onto straight lines in the r-plane. One such transformation that transforms the interior of the unit circle of the z-plane onto the left half of the r-plane is

$$z = \frac{1 + r}{1 - r} \tag{6-45}$$

which is referred to as the ***r*-transformation.** Once the characteristic equation in z is transformed into the r domain using Eq. (6-45), the Routh–Hurwitz criterion can again be applied to the equation in r.

The r-transformation given in Eq. (6-45) is probably the simplest form that can be used for manual manipulation to transform an equation $F(z)$ to an equation in r. Another transformation that is often used in discrete-data control system design in the frequency domain is

$$z = \frac{(2/T) + w}{(2/T) - w} \tag{6-46}$$

or

$$w = \frac{2}{T}\frac{z - 1}{z + 1} \tag{6-47}$$

which is called the ***w*-transformation.** Note that the w-transformation becomes the r-transformation when $T = 2$. The advantage of the w-transformation over the r-

transformation is that the imaginary axis of the w-plane resembles that of the s-plane. To show this, we substitute

$$z = e^{j\omega T} = \cos \omega T + j \sin \omega T \tag{6-48}$$

into Eq. (6-47); we get

$$w = \frac{2}{T}\frac{\cos wT + j \sin \omega T - 1}{\cos \omega T + j \sin \omega T + 1} \tag{6-49}$$

Rationalizing Eq. (6-49), and simplifying, we get

$$w = j\omega_w = j\frac{2}{T}\tan\frac{\omega T}{2} \tag{6-50}$$

Thus the unit circle in the z-plane is mapped onto the imaginary axis $w = j\omega_w$ in the w-plane. The relationship between ω_w and ω, the real frequency, is

$$\omega_w = \frac{2}{T}\tan\frac{\omega T}{2} = \frac{\omega_s}{\pi}\tan\frac{\pi\omega}{\omega_s} \tag{6-51}$$

where ω_s is the sampling frequency in rad/sec. The correlation between ω and ω_w is that they both go to 0 and ∞ at the same time. For Routh–Hurwitz criterion, of course, the w-transformation is more difficult to use, especially since the sampling period T appears in Eq. (6-47). However, if computer programs are available for the transformations, the difference is insignificant.

The program **rwt** of the **DCSP** software package that accompanies the text [9] can be used to perform either the r- or w-transformation. In **Program CC, WPLANE, G1, G2** converts and transfers function $G_1(z)$ into $G_2(w)$ using Eq. (6-46). The program must first be in the digital (DIG) mode, and the program will prompt for the sampling period T.

The following examples illustrate the application of the r-transformation to a characteristic equation in z so that the equation can be tested by Routh–Hurwitz criterion in the r-domain.

Example 6-9

Consider that the characteristic equation of a discrete-data control system is

$$z^3 + 5.94z^2 + 7.7z - 0.368 = 0 \tag{6-52}$$

Substituting Eq. (6-45) into Eq. (6-52) and simplifying, we get

$$3.128r^3 - 11.74r^2 + 2.344r + 14.27 = 0 \tag{6-53}$$

Routh's tabulation of the last equation is

	r^3	3.128	2.344
Sign change			
	r^2	−11.74	14.27
Sign change			
	r^1	6.146	0
	r^0	14.27	

Since there are two sign changes in the first column of the tabulation, Eq. (6-53) has two roots in the right half of the r-plane. This corresponds to Eq. (6-52) having two roots outside the unit circle in the z-plane. This result can be checked by solving the two equations in z and r. For Eq. (6-52), the roots are: $z = -2.0$, $z = -3.984$, and $z = 0.0461$. The three corresponding roots in the r-plane are: $r = 3.0$, $r = 1.67$, and $r = -0.9117$, respectively. ▲

Example 6-10

Let us consider a design problem using the bilinear transformation and Routh–Hurwitz criterion. The characteristic equation of a linear discrete-data control system is given as

$$F(z) = z^3 + z^2 + z + K = 0 \tag{6-54}$$

where K is a real constant. The problem is to find the range of values of K so that the system is stable. We first transform $F(z)$ into an equation in r using the bilinear transformation of Eq. (6-45). The result is

$$(1 - K)r^3 + (1 + 3K)r^2 + 3(1 - K)r + 3 + K = 0 \tag{6-55}$$

Routh's tabulation of the last equation is

r^3	$1 - K$	$3(1 - K)$
r^2	$1 + 3K$	$3 + K$
r^1	$\dfrac{8K(1-K)}{1+3K}$	0
r^0	$3 + K$	

For a stable system, the numbers in the first column of the tabulation must be of the same sign. We can show that these numbers cannot be all negative, since the conditions contradict each other. Next, for all the numbers to be positive, we have the following conditions:

$$1 - K > 0 \qquad 1 + 3K > 0 \qquad K > 0 \qquad 3 + K > 0$$

which lead to the condition for stability:

$$0 < K < 1 \tag{6-56}$$

▲

6-7-2 Direct Stability Tests

There are stability tests that can be applied directly to the characteristic equation in z with reference to the unit circle in the z-plane. One of the first methods that gives the necessary and sufficient conditions for the characteristic equation roots to lie inside the unit circle is the Schur–Cohn criterion [5]. A simpler tabulation method was devised by Jury and Blanchard [6,7] and is called **Jury's stability criterion** [10]. R. H. Raible [8] devised an alternative tabular form of Jury's stability test. Unfortunately, these analytical tests all become very tedious for equations higher than the second order, especially when the equation has unknown parameter(s) in it. Then there is no reason to use any of these tests if all the coefficients of the equation are known, since we can always use the root-finding program and a computer. Weighing all the pros and cons, this author believes that when the characteristic equation has at least one unknown parameter, the bilinear transformation method is still the best manual method for determining stability of linear discrete-data systems. However, it is useful to introduce the necessary condition of stability that can be checked by inspection.

Consider that the characteristic equation of a linear time-invariant discrete-data system is

$$F(z) = a_n z^n + a_{n-1} z^{n-1} + \cdots + a_1 z + a_0 = 0 \qquad (6\text{-}57)$$

where all the coefficients are real. Among all the conditions provided in Jury's test, the following *necessary* conditions must be satisfied for $F(z)$ to have no roots on or outside the unit circle.

$$\begin{aligned} F(1) &> 0 & & \\ F(-1) &> 0 & &\text{if } n = \text{even integer} \\ F(-1) &< 0 & &\text{if } n = \text{odd integer} \\ |a_0| &< a_n & & \end{aligned} \qquad (6\text{-}58)$$

If an equation of the form of Eq. (6-57) violates any of the conditions above, not all of the roots are inside the unit circle, and the system would not be stable. Apparently, these necessary conditions can be checked easily by inspection.

Example 6-11

Consider the equation

$$F(z) = z^3 + z^2 + 0.5z + 0.25 = 0 \qquad (6\text{-}59)$$

Applying the conditions in Eq. (6-58), we have

$$F(1) = 2.75 > 0 \quad \text{and} \quad F(-1) = -0.25 < 0 \qquad \text{for } n = 3\text{, which is odd}$$
$$|a_0| = 0.25 < a_3 = 1$$

Thus the conditions in Eq. (6-58) are all satisfied, but nothing can be said about the stability of the system. ▲

Example 6-12

Consider the equation

$$F(z) = z^3 + z^2 + 0.5z + 1.25 = 0 \tag{6-60}$$

The conditions in Eq. (6-58) are

$F(-1) = 0.75 > 0$ for $n = 3$, which is odd
$|a_0| = 1.25$ which is not less than a_3, which equals 1

Since for odd n $F(-1)$ must be negative, the equation in Eq. (6-60) has at least one root outside the unit circle. The condition on the absolute value of a_0 is also not met. ▲

Second-Order Systems

The conditions in Eq. (6-58) become *necessary and sufficient* when the system is of the second order. That is, the necessary and sufficient conditions for the second-order equation

$$F(z) = a_2 z^2 + a_1 z + a_0 = 0 \tag{6-61}$$

to have no roots on or outside the unit circle are

$$\begin{aligned} F(1) &> 0 \\ F(-1) &> 0 \\ |a_0| &< a_2 \end{aligned} \tag{6-62}$$

Example 6-13

Consider the equation

$$F(z) = z^2 + z + 0.25 = 0 \tag{6-63}$$

Applying the conditions in Eq. (6-62), we have

$F(1) = 2.25 > 0$ $F(-1) = 0.25 > 0$ for $n = 2$, which is even
$|a_0| = 0.25 < a_2 = 1$

Thus the conditions in Eq. (6-62) are all satisfied. The two roots in Eq. (6-63) are all inside the unit circle, and the system is stable. ▲

6-8 Summary

In this chapter the definitions of BIBO stability, zero-input, or asymptotic stability of linear time-invariant continuous-data and discrete-data systems are given. It is shown that the condition of these types of stability is related directly to the roots of the characteristic equation. For a continuous-data system to be stable, the roots of the characteristic equation must all be located in the left half of the s-plane. For a discrete-data system to be stable, the roots of the characteristic equation must all be inside the unit circle $|z| = 1$ in the z-plane.

The necessary condition for a polynomial $F(s)$ to have no zeros on the $j\omega$-axis and in the right half of the s-plane is that all its coefficients must be of the same sign and none can vanish. The necessary and sufficient conditions of $F(s)$ to have zeros only in the left-half s-plane are checked with the Routh–Hurwitz criterion. The value of the Routh–Hurwitz criterion is diminished if the characteristic equation can be solved using a root-finding routine on the computer.

For discrete-data systems, the characteristic equation $F(z)$ must be checked for roots on and outside the unit circle in the z-plane. The Routh–Hurwitz criterion **cannot** be applied directly to this situation. Although there are direct methods similar to the Routh–Hurwitz criterion that can be used to check the roots in the z-plane with respect to the stability boundary, these turn out to be quite cumbersome, except for systems of the second order. A reliable method is to use a bilinear transformation, which transforms the unit circle in the z-plane onto the imaginary axis of another complex-variable plane, so that the Routh–Hurwitz criterion can be applied to the transformed equation.

Review Questions

1. Can the Routh–Hurwitz criterion be directly applied to the stability analysis of the following systems?

(a) Continuous-data system with the characteristic equation

$$s^4 + 5s^3 + 2s^2 + 3s + 2e^{-2s} = 0$$

(b) Continuous-data system with the characteristic equation

$$s^4 - 5s^3 + 3s^2 + Ks + K^2 = 0$$

(c) Discrete-data system with the characteristic equation

$$z^3 + 2z^2 + 0.5z + 0.1 = 0$$

2. The first two rows of Routh's tabulation of a third-order system are

$$\begin{array}{lll} s^3 & 2 & 2 \\ s^2 & 4 & 4 \end{array}$$

Select the correct answer from the following choices

(a) The equation has one root in the right-half s-plane.

(b) The equation has two roots on the $j\omega$-axis at $s = j$ and $-j$. The third root is in the left-half plane.

(c) The equation has two roots on the $j\omega$-axis at $s = 2j$ and $s = -2j$. The third root is in the left-half plane.

(d) The equation has two roots on the $j\omega$-axis at $s = 2j$ and $s = -2j$. The third root is in the right-half plane.

3. If the numbers in the first column of Routh's tabulation turn out to be all negative, the equation for which the tabulation is made has at least one root not in the left half of the s-plane. **(T)** **(F)**

4. The roots of the auxiliary equation, $A(s) = 0$, of Routh's tabulation of a characteristic equation must also be the roots of the latter. **(T)** **(F)**

5. The following characteristic equation of the discrete-data system represents an unstable system since it contains a negative coefficient.

$$z^3 - z^2 + 0.1z + 0.5 = 0$$ **(T)** **(F)**

6. The following characteristic equation of a continuous-data system represents an unstable system since it contains a negative coefficient.

$$s^3 - s^2 + 5s + 10 = 0$$ **(T)** **(F)**

7. The following characteristic equation of a discrete-data system represents an unstable system since the coefficient of the z term is zero.

$$z^2 - 0.5 = 0$$ **(T)** **(F)**

8. The following characteristic equation of a continuous-data system represents an unstable system since there is a zero coefficient.

$$s^3 + 5s^2 + 4 = 0$$ **(T)** **(F)**

9. When a row of Routh's tabulation contains all zeros before the tabulation ends, this means that the equation has roots on the imaginary axis of the s-plane. **(T)** **(F)**

10. Without conducting detailed tests, it is easy to see that the following z-transform characteristic equation must have at least one root outside the unit circle in the z-plane.

$$z^4 + 5z^3 + z^2 + z + 2 = 0$$ **(T)** **(F)**

Answers to the true-and-false questions are given following the Problems section.

References

1. **F. R. GANTMACHER,** *Matrix Theory,* Vol. II, Chelsea Publishing Company, New York, 1964.

2. **K. J. KHATWANI,** "On Routh–Hurwitz Criterion," *IEEE Trans. Automatic Control,* Vol. AC-26, p. 583, Apr. 1981.

3. **S. K. PILLAI,** "The ε Method of the Routh–Hurwitz Criterion," *IEEE Trans. Automatic Control,* Vol. AC-26, p. 584, Apr. 1981.
4. **E. I. JURY,** "The Number of Roots of a Real Polynomial Inside (or Outside) the Unit Circle Using the Determinant Method," *IEEE Trans. Automatic Control,* Vol. AC-10, pp. 371–372, July 1965.
5. **M. L. COHEN,** "A Set of Stability Constraints on the Denominator of a Sampled-Data Filter," *IEEE Trans. Automatic Control,* Vol. AC-11, pp. 327–328, Apr. 1966.
6. **E. I. JURY and J. BLANCHARD,** "A Stability Test for Linear Discrete Systems in Table Form," *IRE Proc.,* Vol. 49, No. 12, pp. 1947–1948, Dec. 1961.
7. **E. I. JURY and B. D. O. ANDERSON,** "A Simplified Schur–Cohen Test," *IEEE Trans. Automatic Control,* Vol. AC-18, pp. 157–163, Apr. 1973.
8. **R. H. RAIBLE,** "A Simplification of Jury's Tabular Form," *IEEE Trans. Automatic Control,* Vol. AC-19, pp. 248–250, June 1974.
9. **B. C. KUO,** *Digital Control Systems,* 2nd ed., Saunders College Publishing, Philadelphia, 1992.
10. **E. I. JURY,** *Theory and Application of the z-Transform Method,* John Wiley & Sons, New York, 1964.

Problems

▲ System stability

6-1. Without using the Routh–Hurwitz criterion, determine if the following systems are asymptotically stable, marginally stable, or unstable. In each case, the closed-loop system transfer function is given.

(a) $M(s) = \dfrac{10(s+2)}{s^3 + 3s^2 + 5s}$ **(b)** $M(s) = \dfrac{s-1}{(s+5)(s^2+2)}$

(c) $M(s) = \dfrac{K}{s^3 + 5s + 5}$ **(d)** $M(s) = \dfrac{100(s-1)}{(s+5)(s^2+2s+2)}$

(e) $M(s) = \dfrac{100}{s^3 - 2s^2 + 3s + 10}$ **(f)** $M(s) = \dfrac{10(s+12.5)}{s^4 + 3s^3 + 50s^2 + s + 10^6}$

▲ Routh–Hurwitz criterion

6-2. Using the Routh–Hurwitz criterion, determine the stability of the closed-loop system that has the following characteristic equations. Determine the number of roots of each equation that are in the right-half s-plane and on the $j\omega$-axis.

(a) $s^3 + 25s^2 + 10s + 450 = 0$
(b) $s^3 + 25s^2 + 10s + 50 = 0$
(c) $s^3 + 25s^2 + 250s + 10 = 0$
(d) $2s^4 + 10s^3 + 5.5s^2 + 5.5s + 10 = 0$
(e) $s^6 + 2s^5 + 8s^4 + 15s^3 + 20s^2 + 16s + 16 = 0$
(f) $s^4 + 2s^3 + 10s^2 + 20s + 5 = 0$

▲ Routh–Hurwitz criterion on margin of stability

6-3. For each of the characteristic equations of feedback control systems given, determine the range of K so that the system is asymptotically stable. Determine the value of K so that the system is marginally stable and the frequency of sustained oscillation if applicable.

(a) $s^4 + 25s^3 + 15s^2 + 20s + K = 0$
(b) $s^4 + Ks^3 + 2s^2 + (K + 1)s + 10 = 0$
(c) $s^3 + (K + 2)s^2 + 2Ks + 10 = 0$
(d) $s^3 + 20s^2 + 5s + 10K = 0$
(e) $s^4 + Ks^3 + 5s^2 + 10s + 10K = 0$
(f) $s^4 + 125s^3 + s^2 + 5s + K = 0$

▲ Region of stability in the parameter plane

6-4. The loop transfer function of a single-loop feedback control system is given as

$$G(s)H(s) = \frac{K(s + 5)}{s(s + 2)(1 + Ts)}$$

The parameters K and T may be represented in a plane with K as the horizontal axis and T as the vertical axis. Determine the regions in the T versus K parameter plane in which the closed-loop system is asymptotically stable, and unstable. Indicate the boundary on which the system is marginally stable.

▲ Margin of stability, sustained oscillation

6-5. Given the forward-path transfer function of unity-feedback control systems,

(a) $G(s) = \dfrac{K(s + 4)(s + 20)}{s^3(s + 100)(s + 500)}$ (b) $G(s) = \dfrac{K(s + 10)(s + 20)}{s^2(s + 2)}$

(c) $G(s) = \dfrac{K}{s(s + 10)(s + 20)}$ (d) $G(s) = \dfrac{K(s + 1)}{s^3 + 2s^2 + 3s + 1}$

Apply the Routh–Hurwitz criterion to determine the stability of the closed-loop system as a function of K. Determine the value of K that will cause sustained constant-amplitude oscillations in the system. Determine the frequency of oscillation.

▲ Stability of system in state equation form, state feedback control

6-6. A controlled process is modeled by the following state equations.

$$\frac{dx_1(t)}{dt} = x_1(t) - 2x_2(t) \qquad \frac{dx_2(t)}{dt} = 10x_1(t) + u(t)$$

The control $u(t)$ is obtained from state feedback, such that

$$u(t) = -k_1x_1(t) - k_2x_2(t)$$

where k_1 and k_2 are real constants. Determine the region in the k_1 versus k_2 parameter plane in which the closed-loop system is asymptotically stable.

▲ Stability of system in state equation form, state feedback control

6-7. A linear time-invariant system is described by the following state equations:

$$\frac{d\mathbf{x}(t)}{dt} = \mathbf{A}\mathbf{x}(t) + \mathbf{B}u(t)$$

where

$$\mathbf{A} = \begin{bmatrix} 0 & 1 & 0 \\ 0 & 0 & 1 \\ 0 & -4 & -3 \end{bmatrix} \qquad \mathbf{B} = \begin{bmatrix} 0 \\ 0 \\ 1 \end{bmatrix}$$

The closed-loop system is implemented by state feedback, so that $u(t) = -\mathbf{K}\mathbf{x}(t)$, where $\mathbf{K} = [k_1 \quad k_2 \quad k_3]$; and k_1, k_2, and k_3 are real constants. Determine the constraints on the elements of $\mathbf{K}$ so that the closed-loop system is asymptotically stable.

▲ Stablizability with state feedback

6-8. Given the system in state equation form,

$$\frac{d\mathbf{x}(t)}{dt} = \mathbf{A}\mathbf{x}(t) + \mathbf{B}u(t)$$

where

(a) $$\mathbf{A} = \begin{bmatrix} 1 & 0 & 0 \\ 0 & -3 & 0 \\ 0 & 0 & -2 \end{bmatrix} \qquad \mathbf{B} = \begin{bmatrix} 1 \\ 0 \\ 1 \end{bmatrix}$$

(b) $$\mathbf{A} = \begin{bmatrix} 1 & 0 & 0 \\ 0 & -2 & 0 \\ 0 & 0 & 3 \end{bmatrix} \qquad \mathbf{B} = \begin{bmatrix} 0 \\ 1 \\ 1 \end{bmatrix}$$

Can the system be stabilized by state feedback $u(t) = -\mathbf{K}\mathbf{x}(t)$, where $\mathbf{K} = [k_1 \quad k_2 \quad k_3]$?

▲ System with tachometer feedback

6-9. The block diagram of a motor-control system with tachometer feedback is shown in Fig. 6P-9. Find the range of the tachometer constant K_t so that the system is asymptotically stable.

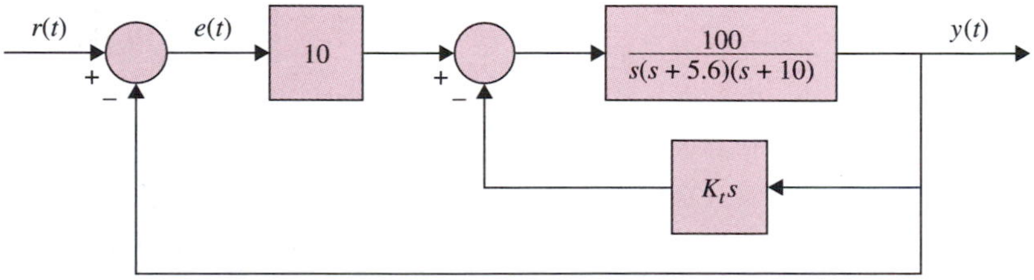

Figure 6P-9

▲ Stability in parameter plane

6-10. The block diagram of a control system is shown in Fig. 6P-10. Find the region in the K versus α plane for the system to be asymptotically stable. (Use K as the vertical and α as the horizontal axis.)

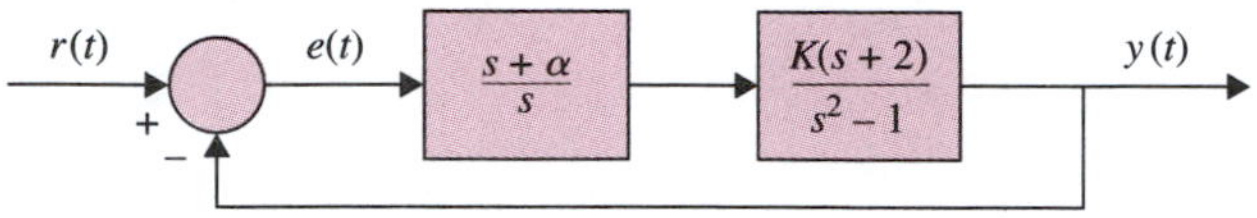

Figure 6P-10

▲ Stability of attitude control of a missile

6-11. The attitude control of a missile shown in Fig. 4P-13 is accomplished by thrust vectoring. The transfer function between the thrust angle $\Delta(s)$ and the angle of attack $\Theta(s)$ is represented by

$$G_p(s) = \frac{\Theta(s)}{\Delta(s)} = \frac{K}{s^2 - a}$$

where K and α are positive real constants. The block diagram of the control system is shown in Fig. 6P-11.

(a) In Fig. 6P-11, consider that only the attitude-sensor loop is in operation, but $K_t = 0$. Determine the relationship between K, K_s, and α so that the missile will oscillate back and forth (marginally stable). Find the condition when the missile will tumble end over end (unstable).

(b) Consider that both loops are in operation. Determine the relationships between K, K_s, K_t, and α so that the missile will oscillate back and forth. Repeat for the missile tumbling end over end.

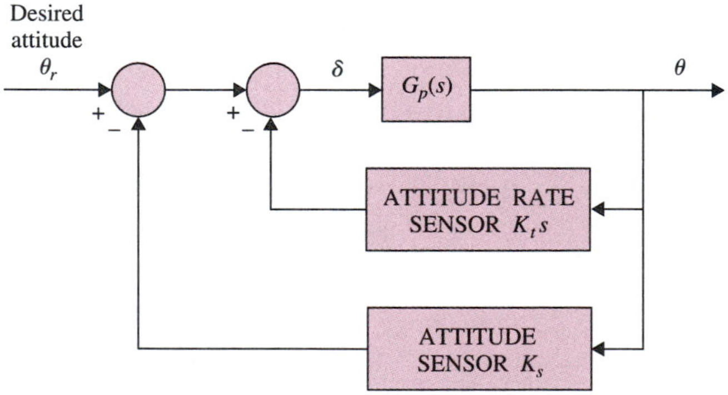

Figure 6P-11

▲ Application of Routh–Hurwitz criterion to relative stability

6-12. The conventional Routh–Hurwitz criterion gives information only on the location of the zeros of a polynomial $F(s)$ with respect to the left and right halves of the s-plane. Devise a linear transformation $s = f(p, \alpha)$, where p is a complex variable, so that the Routh–Hurwitz criterion can be applied to determine whether $F(s)$ has zeros to the right of the line $s = -\alpha$, where α is a positive real number. Apply the transformation to the following characteristic equations to determine how many roots are to the right of the line $s = -1$ in the s-plane.

(a) $F(s) = s^2 + 5s + 3 = 0$ **(b)** $s^3 + 3s^2 + 3s + 1 = 0$

(c) $F(s) = s^3 + 4s^2 + 3s + 10 = 0$ **(d)** $s^3 + 4s^2 + 4s + 4 = 0$

▲ Stability of liquid-level control system

6-13. Figure 6P-13 shows the closed-loop version of the liquid-level control system described in Problem 5-40. All the system parameters and equations are given in Problem 5-40. The error detector is modeled as

$$K_s = 1 \text{ V/ft} \qquad e(t) = K_s[r(t) - h(t)]$$

(a) Draw a functional block diagram for the overall system, showing the functional relationship between the transfer functions.

(b) Let the transfer function of the controller, $G_c(s)$, be unity. Find the open-loop transfer function $G(s) = H(s)/E(s)$ and the closed-loop transfer function $M(s) = H(s)/R(s)$. Find the characteristic equation of the closed-loop system. Note that all the system parameters are specified except for N, the number of inlets.

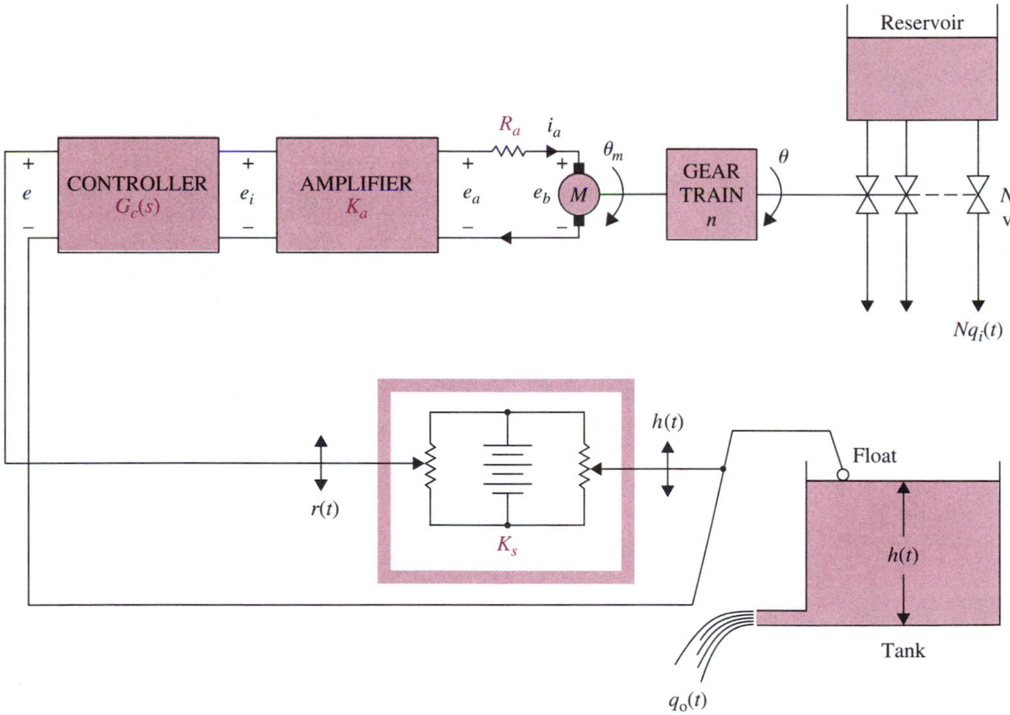

Figure 6P-13

(c) Apply the Routh–Hurwitz criterion to the characteristic equation and determine the maximum number of inlets (positive integer) so that the system is asymptotically stable. Set $G_c(s) = 1$.

▲ Stability of liquid-level control system

6-14. The block diagram shown in Fig. 6P-14 represents the liquid-level control system described in Problem 6-13. The following data are given: $K_a = 50$, $K_I = 50$, $K_b = 0.0706$, $J = 0.006$, $R_a = 10$, $K_i = 10$, and $n = 1/100$. The values of A, N, and K_o will be assigned in the following.

(a) Let the number of inlets, N, and the tank area, A, be the variable parameters, and $K_o = 40$. Find the ranges of N and A so that the closed-loop system is asymptotically stable. Indicate the region in the N versus A plane in which the system is stable. Use the fact that N is an integer. If the cross-sectional area of the tank, A, is infinitely large, what is the maximum value of N (integer) for stability?

(b) It is useful to investigate the relationship between the gear ratio n and K_o, which control the liquid inflow and outflow, respectively. Find the stable region in the N versus K_o plane for K_o up to 100. Let $A = 50$.

(c) Let $N = 10$ and $A = 50$. K_I and K_o are variable parameters. Find the stable region in the K_I versus K_o plane.

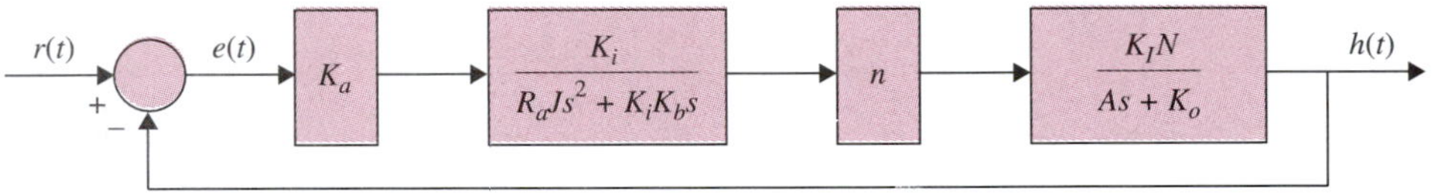

Figure 6P-14

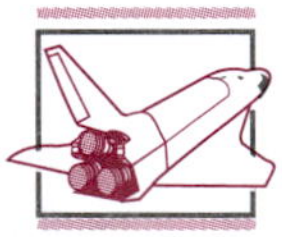

▲ Stability of space-shuttle pointing control system

6-15. The payload of a space-shuttle pointing control system is modeled as a pure mass M. The payload is suspended by magnetic bearings so that no friction is encountered in the control. The attitude of the payload in the y-direction is controlled by magnetic actuators located at the base. The total force produced by the magnetic actuators is $f(t)$. The controls of the other degrees of motion are independent and are not considered here. Since there are experiments located on the payload, electric power must be brought to the payload through cables. The linear spring with spring constant K_s is used to model the cable attachment. The dynamic system model for control of the y-axis motion is shown in Fig. 6P-15. The force equation of motion in the y-direction is

$$f(t) = K_s y(t) + M\frac{d^2y(t)}{dt^2}$$

where $K_s = 0.5$ N-m/m and $M = 500$ kg. The magnetic actuators are controlled through state feedback, so that

$$f(t) = -K_P y(t) - K_D\frac{dy(t)}{dt}$$

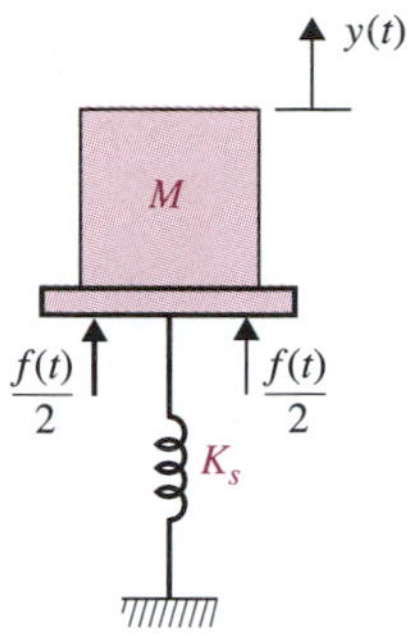

Figure 6P-15

(a) Draw a functional block diagram for the system.

(b) Find the characteristic equation of the closed-loop system.

(c) Find the region in the K_D versus K_p plane in which the system is asymptotically stable.

▲ Stability of inventory control system

6-16. An inventory-control system is modeled by the following differential equations:

$$\frac{dx_1(t)}{dt} = -x_2(t) + u(t)$$

$$\frac{dx_2(t)}{dt} = -Ku(t)$$

where $x_1(t)$ is the level of inventory, $x_2(t)$ the rate of sales of product, $u(t)$ the production rate, and K a real constant. Let the output of the system by $y(\mathrm{t}) = x_1(t)$, and $r(t)$ be the reference set point for the desired inventory level. Let $u(t) = r(t) - y(t)$. Determine the constraint on K so that the closed-loop system is asymptotically stable.

▲ Stability of discrete-data systems

6-17. Apply the w-transform to the following characteristic equations of discrete-data control systems, and determine the conditions of stability (asymptotically stable, and marginally stable, or unstable) by means of the Routh–Hurwitz criterion.

(a) $z^2 + 1.5z - 1 = 0$ (b) $z^3 + z^2 + 3z + 0.2 = 0$

(c) $z^3 - 1.2z^2 - 2z + 3 = 0$ (d) $z^3 - z^2 - 2z + 0.5 = 0$

Check the answers by solving for the roots of the equations using a root-finding computer program.

▲ Stability of digital control system

6-18. A digital control system is described by the state equation

$$x(k+1) = (0.368 - 0.632K)x(k) + Kr(k)$$

where $r(k)$ is the input and $x(k)$ is the state variable. Determine the values of K for the system to be asymptotically stable.

▲ Stability of digital control system

6-19. The characteristic equation of a linear digital control system is

$$z^3 + z^2 + 1.5Kz - (K + 0.5) = 0$$

Determine the values of K for the system to be asymptotically stable.

▲ Stability of discrete-data control system

6-20. The block diagram of a discrete-data control system is shown in Fig. 6P-20.

(a) For $T = 0.1$ sec, find the values of K so that the system is asymptotically stable at the sampling instants.
(b) Repeat part (a) when the sampling period is 0.5 sec.
(c) Repeat part (a) when the sampling period is 1.0 sec.

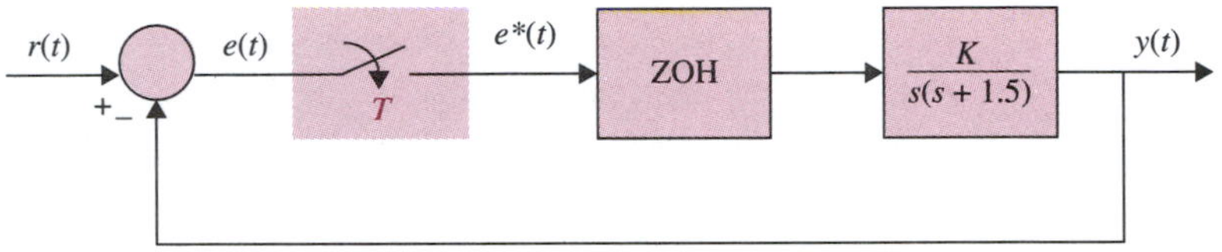

Figure 6P-20

Additional Computer Problems

6-21. Use a root-finding computer program to find the roots of the following characteristic equations of linear continuous-data systems, and determine the stability condition of the systems.

(a) $s^3 + 10s^2 + 10s + 130 = 0$ **(b)** $s^4 + 12s^3 + s^2 + 2s + 10 = 0$
(c) $s^4 + 12s^3 + 10s^2 + 10s + 10 = 0$ **(d)** $s^4 + 12s^3 + s^2 + 10s + 1 = 0$
(e) $s^6 + 6s^5 + 125s^4 + 100s^3 + 100s^2 + 20s + 10 = 0$
(f) $s^5 + 125s^4 + 100s^3 + 100s^2 + 20s + 10 = 0$

6-22. Use a root-finding computer program to find the roots of the following characteristic equations of linear discrete-data control systems, and determine the stability condition of the systems.

(a) $z^3 + 2z^2 + 1.2z + 0.5 = 0$ **(b)** $z^3 + z^2 + z - 0.5 = 0$
(c) $0.5z^3 + z^2 + 1.5z + 0.5 = 0$
(d) $z^4 + 0.5z^3 + 0.25z^2 + 0.1z - 0.25 = 0$

Answers to True-and-False Review Questions

3. (F) **4.** (T) **5.** (F) **6.** (T) **7.** (F) **8.** (T) **9.** (F)
10. (T)

7 Time-Domain Analysis of Control Systems

KEYWORDS AND TOPICS

- Test Signals
- Steady-State Error
- Error Constants
- Transient Response
- Damping Ratio
- Natural Undamped Frequency
- Critical Damping
- Unit-Step Response
- Effects of Adding Poles and Zeros to Transfer Functions
- Approximation of High-Order Systems by Low-Order Systems

7-1 Time Response of Continuous-Data Systems: Introduction

Since time is used as an independent variable in most control systems, it is usually of interest to evaluate the state and output responses with respect to time, or simply, the **time response.** In the analysis problem, a reference input signal is applied to a system, and the performance of the system is evaluated by studying the system response in the time domain. For instance, if the objective of the control system is to have the output variable track the input signal, starting at some initial time and initial condition, it is necessary to compare the input and output responses as functions of time. Therefore, in most control system problems, the final evaluation of the performance of the system is based on the time responses.

The time response of a control system is usually divided into two parts: the **transient response** and the **steady-state response.** Let $y(t)$ denote the time response of a continuous-data system; then, in general, it can be written as

$$y(t) = y_t(t) + y_{ss}(t) \tag{7-1}$$

where $y_t(t)$ denotes the transient response and $y_{ss}(t)$ denotes the steady-state response.

▲ Transient response is defined as the part of the time response that goes to zero as time goes to infinity.

▲ Steady-state response is the part of the total response that remains after the transient has died out.

In control systems, *transient response is defined as the part of the time response that goes to zero as time becomes very large.* Thus $y_t(t)$ has the property

$$\lim_{t\to\infty} y_t(t) = 0 \tag{7-2}$$

The steady-state response is simply the part of the total response that remains after the transient has died out. Thus the steady-state response can still vary in a fixed pattern, such as a sine wave, or a ramp function that increases with time.

All real stable control systems exhibit transient phenomena to some extent before the steady state is reached. Since inertia, mass, and inductance are unavoidable in physical systems, the responses of a typical control system cannot follow sudden changes in the input instantaneously, and transients are usually observed. Therefore, the control of the transient response is necessarily important, as it is a significant part of the dynamic behavior of the system; and the deviation between the output response and the input or the desired response, before the steady state is reached, must be closely controlled.

The steady-state response of a control system is also very important, since it indicates where the system output ends up at when time becomes large. For a position-control system, the steady-state response when compared with the desired reference position gives an indication of the final accuracy of the system. In general, if the steady-state response of the output does not agree with the desired reference exactly, the system is said to have a **steady-state error.**

The study of a control system in the time domain essentially involves the evaluation of the transient and the steady-state responses of the system. In the design problem, specifications are usually given in terms of the transient and the steady-state performances, and controllers are designed so that the specifications are all met by the designed system.

7-2 Typical Test Signals for the Time Response of Control Systems

Unlike electric networks and communication systems, the inputs to many practical control systems are not exactly known ahead of time. In many cases, the actual inputs of a control system may vary in random fashion with respect to time. For instance, in a radar tracking system for antiaircraft missiles, the position and speed of the target to be tracked may vary in an unpredictable manner, so that they cannot be predetermined. This poses a problem for the designer, since it is difficult to design a control system so that it will perform satisfactorily for all possible forms of input signals. For the purpose of analysis and design, it is necessary to assume some basic types of test inputs so that the performance of a system can be evaluated. By selecting these basic test signals properly, not only is the mathematical treatment of the problem systematized, but the response due to these inputs allows prediction of the system's performance to other more complex inputs. In the design problem, performance criteria may be specified with respect to these test signals so that the system may be designed to meet the criteria. This approach is particularly useful for linear systems, since the response to complex signals can be determined by superposing those due to simple test signals.

When the response of a linear time-invariant system is analyzed in the frequency domain, a sinusoidal input with variable frequency is used. When the input frequency is swept from zero to beyond the significant range of the system characteristics, curves in terms of the amplitude ratio and phase between the input and the output are drawn as functions of frequency. It is possible to predict the time-domain behavior of the system from its frequency-domain characteristics.

To facilitate the time-domain analysis, the following deterministic test signals are used.

Step-Function Input. The step-function input represents an instantaneous change in the reference input. For example, if the input is an angular position of a mechanical shaft, a step input represents the sudden rotation of the shaft. The mathematical representation of a step function of magnitude R is

$$r(t) = \begin{cases} R & t \geq 0 \\ 0 & t < 0 \end{cases} \tag{7-3}$$

where R is a real constant. Or,

$$r(t) = Ru_s(t) \tag{7-4}$$

where $u_s(t)$ is the unit-step function. The step function as a function of time is shown in Fig. 7-1(a). The step function is very useful as a test signal since its initial instantaneous

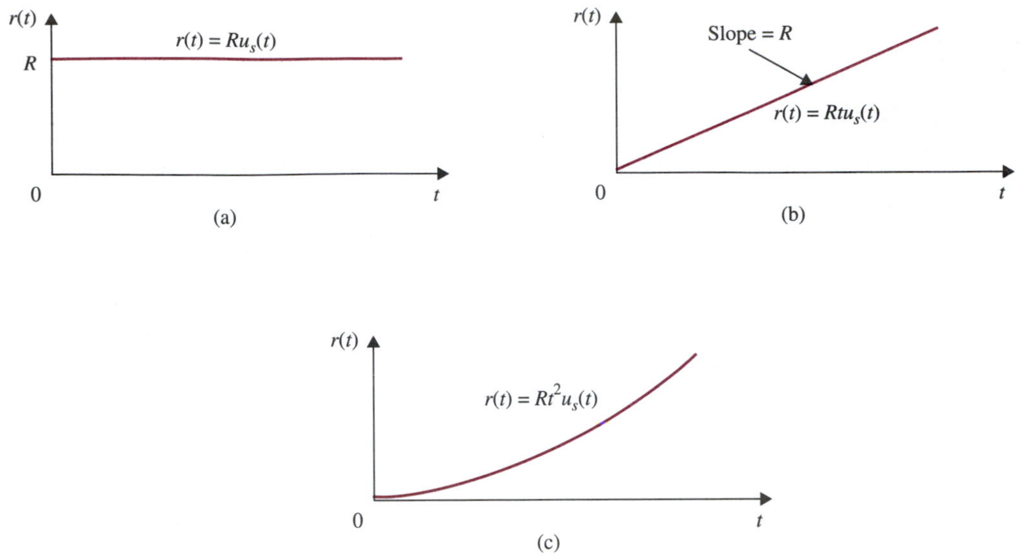

Figure 7-1 Basic time-domain test signals for control systems. (a) Step function. (b) Ramp function. (c) Parabolic function.

jump in amplitude reveals a great deal about a system's quickness in responding to inputs with abrupt changes. Also, since the step function contains, in principle, a wide band of frequencies in its spectrum, as a result of the jump discontinuity, it is equivalent to the application of numerous sinusoidal signals with a wide range of frequencies.

Ramp-Function Input. The ramp function is a signal that changes constantly with time. Mathematically, a ramp function is represented by

$$r(t) = Rtu_s(t) \tag{7-5}$$

where R is a real constant. The ramp function is shown in Fig. 7-1(b). If the input variable represents the angular displacement of a shaft, the ramp input denotes the constant-speed rotation of the shaft. The ramp function has the ability to test how the system would respond to a signal that changes linearly with time.

Parabolic-Function Input. The parabolic function represents a signal that is one order faster than the ramp function. Mathematically, it is represented as

$$r(t) = \frac{Rt^2}{2}u_s(t) \tag{7-6}$$

where R is a real constant, and the factor $\frac{1}{2}$ is added for mathematical convenience since the Laplace transform of $r(t)$ is simply R/s^3. The graphical representation of the parabolic function is shown in Fig. 7-1(c).

These signals all have the common feature that they are simple to describe mathematically. From the step function to the parabolic function, the signals become progressively faster with respect to time. In theory we can define signals with still higher rates, such as t^3, which is called the *jerk function,* and so on. However, in reality, we seldom find it necessary or feasible to use a test signal faster than a parabolic function. This is because, as we shall see later, in order to track a high-order input accurately, the system must have high-order integrations in the loop, which usually leads to serious stability problems.

7-3 Steady-State Error

One of the objectives of most control systems is that the system output response follows a specific reference signal accurately in the steady state. The difference between the output and the reference in the steady state was defined earlier as the **steady-state error.** In the real world, because of friction and other imperfections, and the natural composition of the system, the steady state of the output response seldom agrees exactly with the reference. Therefore, steady-state errors in control systems are almost unavoidable. In a design problem, one of the objectives is to keep the steady-state error to a minimum, or below a certain tolerable value, and at the same time the transient response must satisfy a certain set of specifications.

The accuracy requirement on control systems depends to a great extent on the control objectives of the system. For instance, the final position accuracy of an elevator would be far less stringent than the pointing accuracy on the control of the Large Space Telescope (LST), which is a telescope mounted on board a space shuttle. The accuracy of position control of such a system is often measured in microradians.

7-3-1 Steady-State Error Caused by Nonlinear Systems Elements

In many instances steady-state errors of control systems are attributed to some nonlinear system characteristics, such as nonlinear friction or dead zone. For instance, if an amplifier used in a control system has the input–output characteristics shown in Fig. 7-2, then, when the amplitude of the amplifier input signal falls within the dead zone, the output of the amplifier would be zero, and the control would not be able to correct the error if any exists. Dead-zone nonlinearity characteristics shown in Fig. 7-2 are not limited to amplifiers. The flux-to-current relation of the magnetic field of an electric motor may exhibit a similar characteristic. As the current of the motor falls below the dead zone D, no magnetic flux, and thus no torque, will be produced by the motor to move the load.

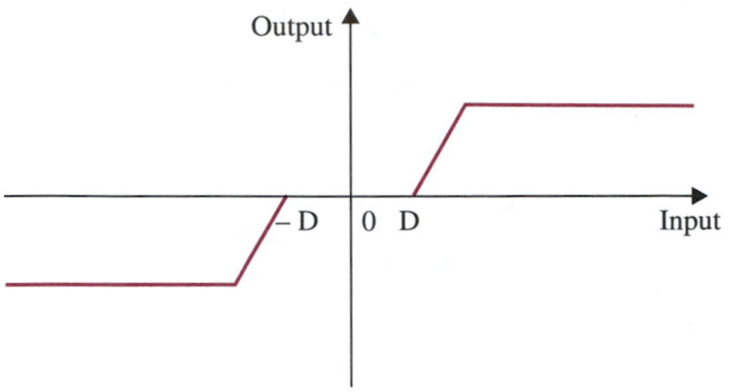

Figure 7-2 Typical input–output characteristics of an amplifier with dead zone and saturation.

The output signals of digital components used in control systems, such as a microprocessor, can take on only discrete or quantized levels. This property is illustrated by the quantization characteristics shown in Fig. 7-3. When the input to the quantizer is within $\pm q/2$, the output is zero, and the system may generate an error in the output

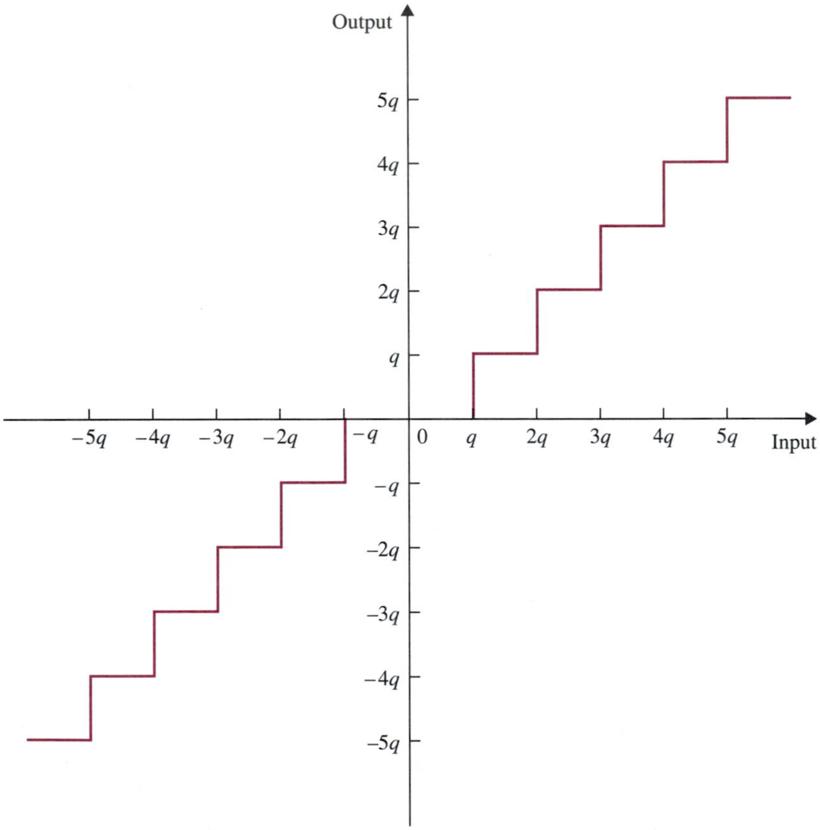

Figure 7-3 Typical input–output characteristics of a quantizer.

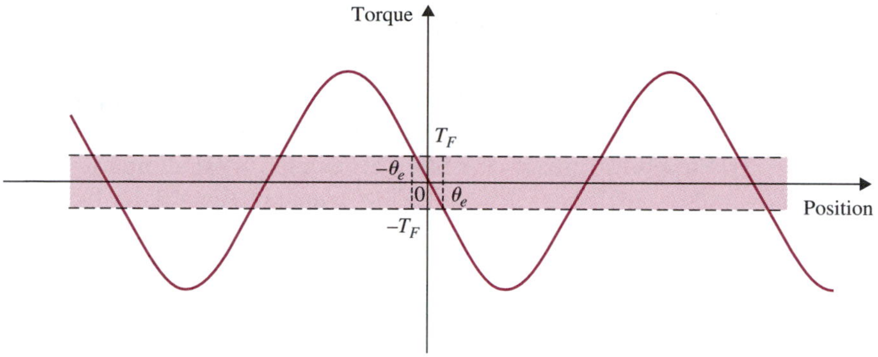

Figure 7-4 Torque–angle curve of a motor or closed-loop system with Coulomb friction.

whose magnitude is related to $\pm q/2$. This type of error is also known as the **quantization error** in digital control systems.

When the control of physical objects is involved, friction is almost always present. Coulomb friction is a common cause of steady-state position errors in control systems. Figure 7-4 shows a restoring torque–position curve of a control system. The torque curve typically could be generated by a step motor or a switched-reluctance motor, or from a closed-loop system with a position encoder. Point 0 designates a stable equilibrium point on the torque curve, as well as the other periodic interesting points along the axis where the slope on the torque curve is negative. The torque on either side of point 0 represents a restoring torque that tends to return the output to the equilibrium point when some angular-displacement disturbance takes place. When there is no friction, the position error should be zero, since there is always a restoring torque as long as the position is not at the stable equilibrium point. If the rotor of the motor sees a Coulomb friction torque T_F, the motor torque must first overcome this frictional torque before producing any motion. Thus, as the motor torque falls below T_F as the rotor position approaches the stable equilibrium point, it may stop at any position inside the error band bounded by $\pm\theta_e$, as shown in Fig. 7-4.

Although it is relatively simple to comprehend the effects of nonlinearities on errors, and to establish maximum upper bounds on the error magnitudes, it is difficult to establish general and closed-form solutions for nonlinear systems. Usually, exact and detailed analysis of errors in nonlinear control systems can be carried out only by computer simulations.

7-3-2 Steady-State Error of Linear Continuous-Data Control Systems

Linear control systems are subject to steady-state errors for somewhat different causes than nonlinear systems, although the reason is still that the system no longer "sees" the error, and no corrective effort is exerted. In general, the steady-state errors of linear control systems depend on the type of the reference signal and the type of the system.

Definition of the Steady-State Error with Respect to System Configuration

Before embarking on the steady-state error analysis, we must clarify what is meant by system error. In general, we can regard the error as a signal that should quickly be reduced to zero, if possible. Let us refer to the closed-loop system shown in Fig. 7-5, where $r(t)$ is the input, $u(t)$ the actuating signal, $b(t)$ the feedback signal, and $y(t)$ is the output. The error of the system may be defined as

$$e(t) = \text{reference signal} - y(t) \tag{7-7}$$

where the *reference signal* is the signal that the output $y(t)$ is to track. When the system has unity feedback [i.e., $H(s) = 1$], the input $r(t)$ is the reference signal, and the error is simply

$$e(t) = r(t) - y(t) \tag{7-8}$$

The steady-state error is defined as

$$e_{ss} = \lim_{t \to \infty} e(t) \tag{7-9}$$

When $H(s)$ is not unity, the actuating signal $u(t)$ in Fig. 7-5 may or may not be the error, depending on the form and the purpose of $H(s)$. Let us assume that the objective of the system in Fig. 7-5 is to have the output $y(t)$ track the input $r(t)$ as closely as possible, and the system transfer functions are

$$G(s) = \frac{1}{s^2(s+12)} \qquad H(s) = \frac{5(s+1)}{s+5} \tag{7-10}$$

We can show that if $H(s) = 1$, the characteristic equation is

$$s^3 + 12s^2 + 1 = 0 \tag{7-11}$$

which has roots in the right-half s-plane, and the closed-loop system is unstable. We can show that the $H(s)$ given in Eq. (7-10) stabilizes the system and the characteristic

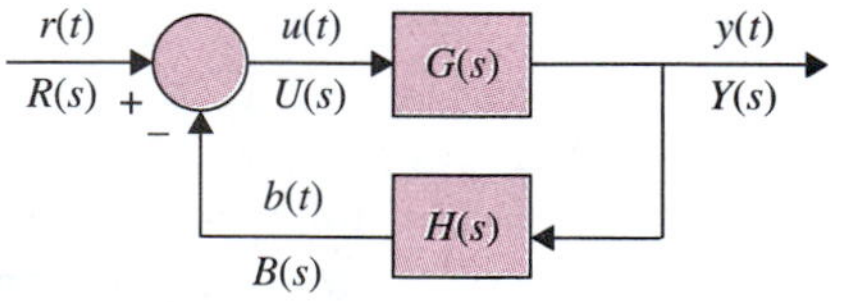

Figure 7-5 Nonunity-feedback control system.

equation becomes

$$s^4 + 17s^3 + 60s^2 + 5s + 5 = 0 \tag{7-12}$$

In this case the system error may still be defined as in Eq. (7-8).

However, consider a velocity control system, in which case a step input is used to control the system output, which contains a ramp in the steady state. The system transfer functions may be of the form

$$G(s) = \frac{1}{s^2(s + 12)} \qquad H(s) = K_t s \tag{7-13}$$

where $H(s)$ is the transfer function of an electromechanical or electronic tachometer, and K_t is the tachometer constant. The system error should be defined as in Eq. (7-7), where the *reference signal* is the *desired velocity,* and not $r(t)$. In this case, since $r(t)$ and $y(t)$ are not of the same dimension, it would be meaningless to define the error as in Eq. (7-8). To illustrate the system further, let $K_t = 10$ V/rad/sec. This means that for a unit-step input of 1 V, the desired velocity in the steady state is 1/10 or 0.1 rad/sec, since when this is achieved, the output voltage of the tachometer would be 1 V, and the steady-state error would be zero. The closed-loop transfer function of the system is

$$M(s) = \frac{Y(s)}{R(s)} = \frac{G(s)}{1 + G(s)H(s)} = \frac{1}{s(s^2 + 12s + 10)} \tag{7-14}$$

For a unit-step function input, $R(s) = 1/s$. The output time response is

$$y(t) = 0.1t - 0.12 - 0.000796e^{-11.1t} + 0.1208e^{-0.901t} \qquad t \geq 0 \tag{7-15}$$

Since the exponential terms of $y(t)$ in Eq. (7-15) all diminish as $t \to \infty$, the steady-state part of $y(t)$ is $0.1t - 0.12$. Thus the steady-state error of the system is

$$e_{ss} = \lim_{t\to\infty} [0.1t - y(t)] = 0.12 \tag{7-16}$$

More explanations will be given of how to define the reference signal when $H(s) \neq 1$ later when the general discussion on the steady-state error of nonunity feedback systems is given.

Not all system errors are defined with respect to the response due to the input. Figure 7-6 shows a system with a disturbance $d(t)$, in addition to the input $r(t)$. The output due to $d(t)$ acting alone may also be considered as an error. For these reasons, the definition of system error has not been unified in the literature.

To establish a systematic study of the steady-state error for linear systems, we shall classify three types of systems and treat these separately.

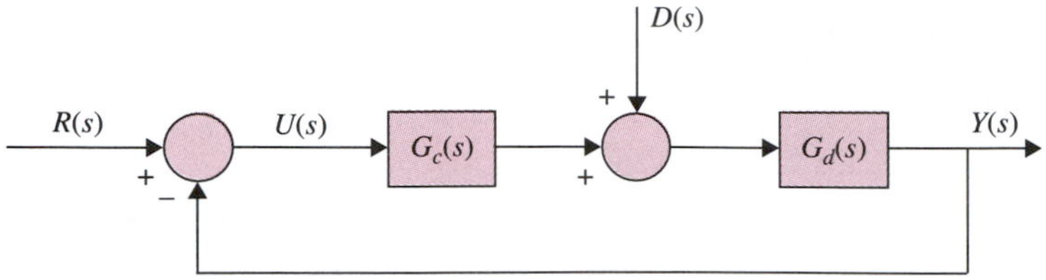

Figure 7-6 System with disturbance input.

1. Systems with unity feedback; $H(s) = 1$.
2. Systems with nonunity feedback, but $H(0) = K_H = \text{constant}$.
3. Systems with nonunity feedback, and $H(s)$ has zeros at $s = 0$ of order N.

The objective here is to establish a definition of the error with respect to one basic system configuration so that some fundamental relationships can be determined between the steady-state error and the system parameters.

Type of Control Systems: Unity-Feedback Systems

Consider that a control system with unity feedback can be represented by or simplified to the block diagram in Fig. 7-7. The steady-state error of the system is written

$$e_{ss} = \lim_{t\to\infty} e(t) = \lim_{s\to 0} sE(s) = \lim_{s\to 0} \frac{sR(s)}{1 + G(s)} \tag{7-17}$$

▲ System **type** is defined here only for unity-feedback systems.

Clearly, e_{ss} depends on the characteristics of $G(s)$. More specifically, we can show that e_{ss} depends on the number of poles that $G(s)$ has at $s = 0$. This number is known at the **type** of the control system, or simply, system **type.** We can show that the steady-state error e_{ss} depends on the **type** of the control system. Let us formalize the system **type** by referring to the form of the forward-path transfer function $G(s)$. In general,

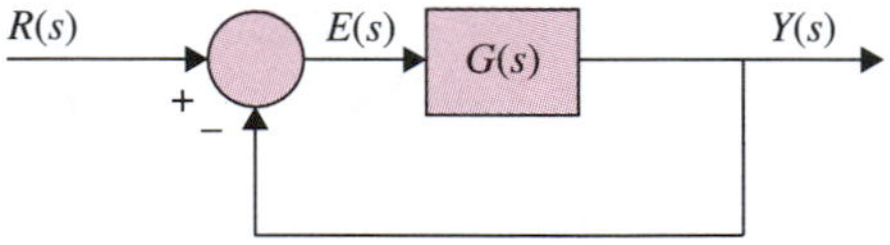

Figure 7-7 Unity-feedback system.

$G(s)$ can be expressed for convenience as

$$G(s) = \frac{K(1 + T_1 s)(1 + T_2 s) \cdots (1 + T_{m1} s + T_{m2} s^2)}{s^j (1 + T_a s)(1 + T_b s) \cdots (1 + T_{n1} s + T_{n2} s^2)} e^{-T_d s} \tag{7-18}$$

where K and all the T's are real constants. The system **type** refers to the **order** of the pole of $G(s)$ at $s = 0$. Thus the closed-loop system having the forward-path transfer function of Eq. (7-18) is type j, where $j = 0, 1, 2, \ldots$. The total number of terms in the numerator and the denominator and the values of the coefficients are not important to the system type, as system type refers only to the number of poles $G(s)$ has at $s = 0$. The following example illustrates the system type with reference to the form of $G(s)$.

Example 7-1

$$G(s) = \frac{K(1 + 0.5s)}{s(1 + s)(1 + 2s)(1 + s + s^2)} \qquad \text{type 1} \tag{7-19}$$

$$G(s) = \frac{K(1 + 2s)}{s^3} \qquad \text{type 3} \tag{7-20}$$

▲

Now let us investigate the effects of the types of inputs on the steady-state error. We shall consider only the step, ramp, and parabolic inputs.

Steady-State Error of System with a Step-Function Input

When the input $r(t)$ to the control system of Fig. 7-7 is a step function with magnitude R, $R(s) = R/s$, the steady-state error is written from Eq. (7-17),

$$e_{ss} = \lim_{s \to 0} \frac{sR(s)}{1 + G(s)} = \lim_{s \to 0} \frac{R}{1 + G(s)} = \frac{R}{1 + \lim_{s \to 0} G(s)} \tag{7-21}$$

For convenience, we define

$$K_p = \lim_{s \to 0} G(s) \tag{7-22}$$

▲ The **step-error constant** K_p is defined only when the system is subject to a step input.

as the **step-error constant.** Then Eq. (7-21) becomes

$$e_{ss} = \frac{R}{1 + K_p} \tag{7-23}$$

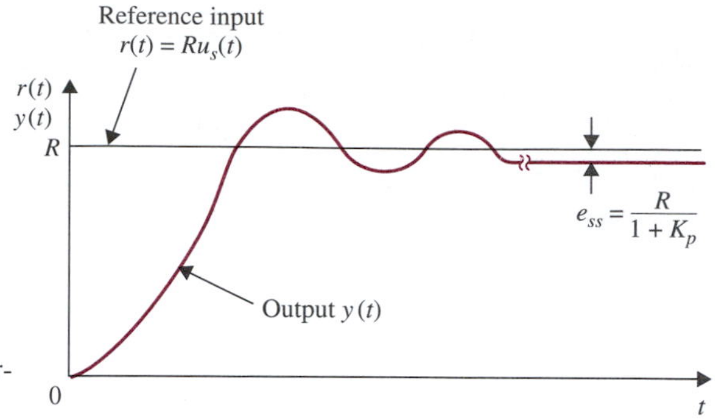

Figure 7-8 Typical steady-state error due to a step input.

A typical e_{ss} due to a step input when K_p is finite and nonzero is shown in Fig. 7-8. We see from Eq. (7-23) that for e_{ss} to be zero, when the input in a step function, K_p must be infinite. If $G(s)$ is described by Eq. (7-18), we see that for K_p to be infinite, j must be at least equal to unity; that is, $G(s)$ must have at least one pole at $s = 0$. Therefore, we can summarize the steady-state error due to a step-function input as follows:

▲ Type 0 system: $e_{ss} = \dfrac{R}{1 + K_p} = \text{constant}$

▲ Type 1 or higher system: $e_{ss} = 0$

Steady-State Error of System with a Ramp-Function Input

When the input to the control system of Fig. 7-7 is a ramp function with magnitude R,

$$r(t) = Rtu_s(t) \tag{7-24}$$

where R is a real constant, the Laplace transform of $r(t)$ is

$$R(s) = \frac{R}{s^2} \tag{7-25}$$

The steady-state error is written using Eq. (7-17),

$$e_{ss} = \lim_{s \to 0} \frac{R}{s + sG(s)} = \frac{R}{\lim\limits_{s \to 0} sG(s)} \tag{7-26}$$

▲ The **ramp-error constant** K_v is defined only when the system is subject to a ramp input.

We define the **ramp-error constant** as

$$K_v = \lim_{s \to 0} sG(s) \tag{7-27}$$

Then Eq. (7-26) becomes

$$e_{ss} = \frac{R}{K_v} \tag{7-28}$$

which is the steady-state error when the input is a ramp function. A typical e_{ss} due to a ramp input when K_v is finite and nonzero is illustrated in Fig. 7-9.

Equation (7-28) shows that for e_{ss} to be zero when the input is a ramp function, K_v must be infinite. Using Eqs. (7-18) and (7-27), we obtain

$$K_v = \lim_{s \to 0} sG(s) = \lim_{s \to 0} \frac{K}{s^{j-1}} \qquad j = 0, 1, 2, \ldots \tag{7-29}$$

Thus, for K_v to be infinite, j must be at least equal to 2, or the system must be of type 2 or higher. The following conclusions may be stated with regard to the steady-state error of a system with ramp input:

▲ Type 0 system: $e_{ss} = \infty$

▲ Type 1 system: $e_{ss} = \dfrac{R}{K_v} =$ constant

▲ Type 2 or higher system: $e_{ss} = 0$

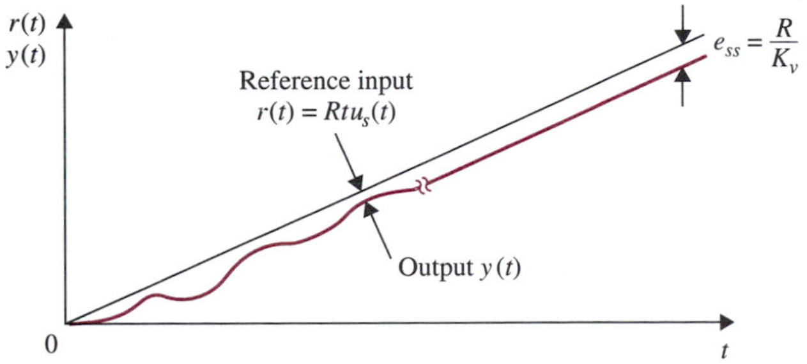

Figure 7-9 Typical steady-state error due to a ramp input.

Steady-State Error of System with a Parabolic Input

When the input is described by the standard parabolic form,

$$r(t) = \frac{Rt^2}{2} u_s(t) \tag{7-30}$$

the Laplace transform of $r(t)$ is

$$R(s) = \frac{R}{s^3} \tag{7-31}$$

The steady-state error of the system in Fig. 7-7 is

$$e_{ss} = \frac{R}{\lim_{s \to 0} s^2 G(s)} \tag{7-32}$$

▲ The **parabolic error constant** K_a is defined only if the system is subject to a parabolic input.

A typical e_{ss} of a system with a nonzero and finite K_a due to a parabolic input is shown in Fig. 7-10.

Defining the **parabolic-error constant** as

$$K_a = \lim_{s \to 0} s^2 G(s) \tag{7-33}$$

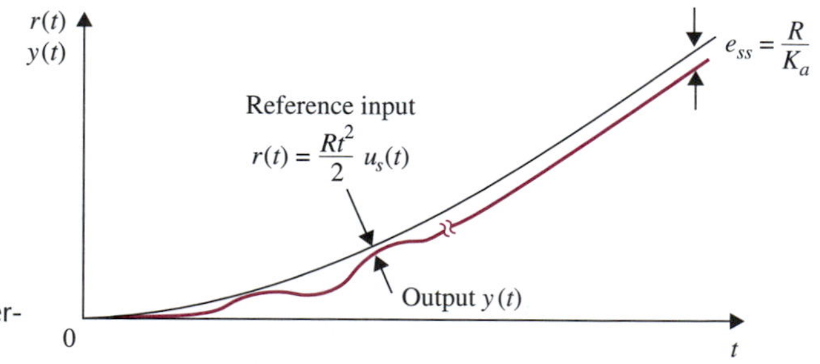

Figure 7-10 Typical steady-state error due to a parabolic input.

the steady-state error becomes

$$e_{ss} = \frac{R}{K_a} \tag{7-34}$$

Following the pattern set with the step and ramp inputs, the steady-state error due to the parabolic input is zero if the system is of type 3 or greater. The following conclusions are made with regard to the steady-state error of a system with parabolic input:

- ▲ Type 0 system: $e_{ss} = \infty$
- ▲ Type 1 system: $e_{ss} = \infty$
- ▲ Type 2 system: $e_{ss} = \dfrac{R}{K_a} = \text{constant}$
- ▲ Type 3 or higher system: $e_{ss} = 0$

We cannot emphasize often enough that in order for these results to be valid, the closed-loop system must be stable.

By using the method described, the steady-state error of any linear closed-loop system subject to an input with order higher than the parabolic function can also be derived if necessary. As a summary of the error analysis, Table 7-1 shows the relations

Table 7-1 Summary of the Steady-State Errors Due to Step, Ramp, and Parabolic Inputs for Unity-Feedback Systems

Type of System j	Error Constants K_p	K_v	K_a	Steady-State Error e_{ss}: Step Input $\frac{R}{1+K_p}$	Ramp Input $\frac{R}{K_v}$	Parabolic Input $\frac{R}{K_a}$
0	K	0	0	$\frac{R}{1+K}$	∞	∞
1	∞	K	0	0	$\frac{R}{K}$	∞
2	∞	∞	K	0	0	$\frac{R}{K}$
3	∞	∞	∞	0	0	0

among the error constants, the types of systems with reference to Eq. (7-18), and the input types.

As a summary, the following points should be noted when applying the error-constant analysis presented above.

1. The step-, ramp-, and parabolic-error constants are significant for the error analysis only when the input signal is a step function, ramp function, and parabolic function, respectively.
2. Since the error constants are defined with respect to the forward-path transfer function $G(s)$, the method is applicable to only the system configuration shown in Fig. 7-7. Since the error analysis relies on the use of the final-value theorem of the Laplace transform, it is important first to check to see if $sE(s)$ has any poles on the $j\omega$-axis or in the right-half s-plane.
3. The steady-state error properties summarized in Table 7-1 are for systems with unity feedback only.
4. The steady-state error of a system with an input that is a linear combination of the three basic types of inputs can be determined by superimposing the errors due to each input component.
5. When the system configuration differs from that of Fig. 7-7, we can either simplify the system to the form of Fig. 7-7, or establish the error signal and apply the final-value theorem. The error constants defined here may or may not apply, depending on the individual situation.

When the steady-state error is infinite, that is, when the error increases continuously with time, the error-constant method does not indicate how the error varies with time. This is one of the disadvantages of the error-constant method. The error-constant method also does not apply to systems with inputs that are sinusoidal, since the final-value theorem cannot be applied. The following examples illustrate the utility of the error constants and their values in the determination of the steady-state errors of linear control systems with unit feedback.

Example 7-2

Consider that the system shown in Fig. 7-7 has the following transfer functions. The error constants and steady-state errors are calculated for the three basic types of inputs using the error constants.

(a) $$G(s) = \frac{K(s+3.15)}{s(s+1.5)(s+0.5)} \qquad H(s) = 1 \qquad \text{Type 1 system}$$

▲ Step input:	Step-error constant $K_p = \infty$	$e_{ss} = \dfrac{R}{1+K_p} = 0$
▲ Ramp input:	Ramp-error constant $K_v = 4.2K$	$e_{ss} = \dfrac{R}{K_v} = \dfrac{R}{4.2K}$
▲ Parabolic input	Parabolic-error constant $K_a = 0$	$e_{ss} = \dfrac{R}{K_a} = \infty$

These results are valid only if the value of K stays within the range that corresponds to a stable closed-loop system, which is $0 < K < 1.3043$.

(b) $G(s) = \dfrac{K}{s^2(s+12)}$ $\quad H(s) = 1$ — Type 2 system

The closed-loop system is unstable for all values of K, and error analysis is meaningless.

(c) $G(s) = \dfrac{5(s+1)}{s^2(s+12)(s+5)}$ — Type 2 system

We can show that the closed-loop system is stable. The steady-state errors are calculated for the three basic types of inputs.

▲ Step input:	Step-error constant: $K_p = \infty$	$e_{ss} = \dfrac{R}{1+K_p} = 0$
▲ Ramp input:	Ramp-error constant: $K_v = \infty$	$e_{ss} = \dfrac{R}{K_v} = 0$
▲ Parabolic input:	Parabolic-error constant: $K_a = \frac{1}{12}$	$e_{ss} = \dfrac{R}{K_a} = 12R$

▲

Relationship between Steady-State Error and the Closed-Loop Transfer Function: Unity and Nonunity Feedback Systems

In the preceding section, the steady-state error of a closed-loop system is related to the forward-path transfer function of the system, which is usually known. Often, the closed-loop transfer function is derived in the analysis process, and it would be of interest to establish the relationships between the steady-state error and the coefficients of the closed-loop transfer function. As it turns out, the closed-loop transfer function can be used to find the steady-state error of systems with unity as well as nonunity feedback. For the present discussion, let us impose the following condition:

$$\lim_{s \to 0} H(s) = H(0) = K_H = \text{constant} \tag{7-35}$$

Since the signal that is fed back to be compared with the input in the steady state is K_H times the steady-state output, when this feedback signal equals the input, the steady-state error would be zero. Thus we can define the reference signal as $r(t)/K_H$ and the error signal as

$$e(t) = \frac{1}{K_H} r(t) - y(t) \tag{7-36}$$

or in the transform domain,

$$E(s) = \frac{1}{K_H} R(s) - Y(s) = \frac{1}{K_H}[1 - K_H M(s)]R(s) \tag{7-37}$$

where $M(s)$ is the closed-loop transfer function. Notice that the development above includes the unity-feedback case for which $K_H = 1$. Let us assume that $M(s)$ does not have any poles at $s = 0$, and is of the form

$$M(s) = \frac{Y(s)}{R(s)} = \frac{b_m s^m + b_{m-1}s^{m-1} + \cdots + b_1 s + b_0}{s^n + a_{n-1}s^{n-1} + \cdots + a_1 s + a_0} \tag{7-38}$$

where $n > m$. We further require that all the poles of $M(s)$ are in the left-half s-plane. The steady-state error of the system is written

$$e_{ss} = \lim_{s \to 0} sE(s) = \lim_{s \to 0} \frac{1}{K_H}[1 - K_H M(s)]sR(s) \tag{7-39}$$

Substituting Eq. (7-38) into Eq. (7-39) and simplifying, we get

$$e_{ss} = \frac{1}{K_H} \lim_{s \to 0} \frac{s^n + \cdots + (a_1 - b_1 K_H)s + (a_0 - b_0 K_H)}{s^n + a_{n-1}s^{n-1} + \cdots + a_1 s + a_0} sR(s) \tag{7-40}$$

We consider the three basic types of inputs for $r(t)$.

1. **Step input.** $R(s) = R/s$.
For a step-function input, the steady-state error in Eq. (7-40) becomes

$$e_{ss} = \frac{1}{K_H}\left(1 - \frac{b_0 K_H}{a_0}\right) R \tag{7-41}$$

Thus the steady-state error due to a step input can be zero only if

$$a_0 - b_0 K_H = 0 \tag{7-42}$$

or

$$M(0) = \frac{b_0}{a_0} = \frac{1}{K_H} \tag{7-43}$$

▲ For a system with unity feedback, the steady-state error due to a step input is zero only if the constant terms in the numerator and denominator of the closed-loop transfer function are equal.

This means that *for a unity-feedback system,* $K_H = 1$, *the constant terms of the numerator and denominator of* $M(s)$ *must be equal in order for the steady-state error to be zero.*

2. **Ramp input.** $R(s) = R/s^2$.
For a ramp function input, the steady-state error in Eq. (7-40) becomes

$$e_{ss} = \frac{1}{K_H}\lim_{s\to 0}\frac{s^n + \cdots + (a_1 - b_1K_H)s + (a_0 - b_0K_H)}{s(s^n + a_{n-1}s^{n-1} + \cdots + a_1 s + a_0)} R \tag{7-44}$$

The following values of e_{ss} are possible:

$$e_{ss} = 0 \qquad \text{if } a_0 - b_0K_H = 0 \text{ and } a_1 - b_1K_H = 0 \tag{7-45}$$

$$e_{ss} = \frac{a_1 - b_1K_H}{a_0K_H} R = \text{constant} \qquad \text{if } a_0 - b_0K_H = 0 \text{ and } a_1 - b_1K_H \neq 0 \tag{7-46}$$

$$e_{ss} = \infty \qquad \text{if } a_0 - b_0K_H \neq 0 \tag{7-47}$$

3. Parabolic input. $R(s) = R/s^3$.
For a parabolic input, the steady-state error in Eq. (7-40) becomes

$$e_{ss} = \frac{1}{K_H}\lim_{s\to 0}\frac{s^n + \cdots + (a_2 - b_2K_H)s^2 + (a_1 - b_1K_H)s + (a_0 - b_0K_H)}{s^2(s^n + a_{n-1}s^{n-1} + \cdots + a_1 s + a_0)}R \tag{7-48}$$

The following values of e_{ss} are possible:

$$e_{ss} = 0 \qquad \text{if } a_i - b_iK_H = 0 \quad \text{for} \quad i = 0, 1, \text{ and } 2 \tag{7-49}$$

$$e_{ss} = \frac{a_2 - b_2K_H}{a_0K_H}R = \text{constant} \qquad \text{if } a_i - b_iK_H = 0 \quad \text{for} \quad i = 0 \text{ and } 1 \tag{7-50}$$

$$e_{ss} = \infty \qquad \text{if } a_i - b_iK_H \neq 0 \quad \text{for} \quad i = 0 \text{ or } 1 \tag{7-51}$$

Example 7-3

Consider the system described in Example 7-2(a). The closed-loop transfer function of the system is given in the following. The steady-state errors for step, ramp, and parabolic inputs are found directly from the closed-loop transfer functions.

$$M(s) = \frac{Ks + 3.15K}{s^3 + 2s^2 + (0.75 + K)s + 3.15K} \tag{7-52}$$

Since the system has unity feedback, $K_H = 1$. Comparing the coefficients of $M(s)$ with those of Eq. (7-38), we have $a_0 = 3.15K$, $a_1 = 0.75 + K$, $a_2 = 2$, $b_0 = 3.15K$, and $b_1 = K$.

- ▲ Step input: $e_{ss} = 0$ since the constant terms in the numerator and the denominator of $M(s)$ are the same
- ▲ Ramp input: $e_{ss} = \dfrac{a_1 - b_1K_H}{a_0K_H}R = \dfrac{0.75 + K - K}{3.15K}R = \dfrac{R}{4.2K}$
- ▲ Parabolic input: $e_{ss} = \infty$ since $a_1 - b_1K_H = 0.75 + K - K \neq 0$

These results agree with those obtained in Example 7-2(a). ▲

Example 7-4

The forward-path and closed-loop transfer functions of the system shown in Fig. 7-7 are given below. The system is assumed to have unity feedback, so that $H(s) = 1$, and thus $K_H = 1$.

$$G(s) = \frac{5(s+1)}{s^2(s+12)(s+5)} \qquad M(s) = \frac{5(s+1)}{s^4 + 17s^3 + 60s^2 + 5s + 5} \tag{7-53}$$

Comparing the coefficients of $M(s)$ with those of Eq. (7-38), we have $a_0 = 5$, $a_1 = 5$, $a_2 = 60$, $a_1 = 17$, $b_0 = 5$, $b_1 = 5$, and $b_2 = 0$.

The poles of $M(s)$ are all in the left-half s-plane. Thus the system is stable. The steady-state errors due to the three basic types of inputs are evaluated as follows:

▲ Step input: $e_{ss} = 0$ since $a_0 = b_0 \ (= 5)$

▲ Ramp input: $e_{ss} = 0$ since $a_0 = b_0 \ (= 5)$ and $a_1 = b_1 \ (= 5)$

▲ Parabolic input: $e_{ss} = \dfrac{a_2 - b_2 K_H}{a_0 K_H} R = \dfrac{60}{5} R = 12R$

Since this is a type 2 system with unity feedback, the same results are obtained with the error-constant method as in Example 7-2(c). ▲

Example 7-5

Consider that the system shown in Fig. 7-5 has the following transfer functions:

$$G(s) = \frac{1}{s^2(s+12)} \qquad H(s) = \frac{5(s+1)}{s+5} \tag{7-54}$$

Then $K_H = 1$. The closed-loop transfer function is

$$M(s) = \frac{Y(s)}{R(s)} = \frac{G(s)}{1 + G(s)H(s)} = \frac{s+5}{s^4 + 17s^3 + 60s^2 + 5s + 5} \tag{7-55}$$

Comparing the last equation with Eq. (7-38), we have $a_0 = 5$, $a_1 = 5$, $a_2 = 60$, $b_0 = 5$, $b_1 = 1$, $b_2 = 0$. The steady-state errors of the system are calculated for the three basic types of inputs.

▲ Unit-step input, $r(t) = u_s(t)$: $e_{ss} = \dfrac{a_0 - b_0 K_H}{a_0} = 0$

▲ Unit-ramp input, $r(t) = tu_s(t)$: $e_{ss} = \dfrac{a_1 - b_1 K_H}{a_0 K_H} = \dfrac{5-1}{5} = 0.8$

▲ Unit-parabolic input, $r(t) = tu_s(t)/2$: $e_{ss} = \infty$ since $a_1 - b_1 K_H \neq 0$

It would be illuminating to calculate the steady-state errors of the system from the difference between the input and the output, and compare them with the results obtained above.

Applying the unit-step, unit-ramp, and unit-parabolic inputs to the system described by Eq. (7-55) and taking the inverse Laplace transform of $Y(s)$, the outputs are:

$$\text{Unit-step input:}\quad y(t) = 1 - 0.00056e^{-12.05t} - 0.0001381e^{-4.886t} - 0.9993e^{-0.0324t}\cos 0.2898t - 0.1301e^{-0.03024t}\sin 0.2898t \qquad t \geq 0 \tag{7-56}$$

Thus the steady-state value of $y(t)$ is unity, and the steady-state error is zero.

$$\text{Unit-ramp input:}\quad y(t) = t - 0.8 + 4.682 \times 10^{-5}e^{-12.05t} + 2.826 \times 10^{-5}e^{-4.886t} + 0.8e^{-0.03024t}\cos 0.2898t - 3.365e^{-0.0302t}\sin 0.2898t \qquad t \geq 0 \tag{7-57}$$

Thus the steady-state portion of $y(t)$ is $t - 0.8$, and the steady-state error to a unit ramp is 0.8.

$$\text{Unit-parabolic input:}\quad y(t) = 0.5t^2 - 0.8t - 11.2 - 3.8842 \times 10^{-6}e^{-12.05t} - 5.784 \times 10^{-6}e^{-4.886t} + 11.2e^{-0.03024t}\cos 0.2898t + 3.9289e^{-0.03024t}\sin 0.2898t \qquad t \geq 0 \tag{7-58}$$

The steady-state portion of $y(t)$ is $0.5t^2 - 0.8t - 11.2$. Thus the steady-state error is $0.8t + 11.2$, which becomes infinite as time goes to infinity.

It is easy to see that these results agree with those obtained using the error analysis with the closed-loop transfer function. ▲

Example 7-6

Consider that the system shown in Fig. 7-5 has the following transfer functions:

$$G(s) = \frac{1}{s^2(s+12)} \qquad H(s) = \frac{10(s+1)}{s+5} \tag{7-59}$$

Thus

$$K_H = \lim_{s \to 0} H(s) = 2 \tag{7-60}$$

The closed-loop transfer function is

$$M(s) = \frac{Y(s)}{R(s)} = \frac{G(s)}{1 + G(s)H(s)} = \frac{s+5}{s^4 + 17s^3 + 60s^2 + 10s + 10} \tag{7-61}$$

Comparing the coefficients of $M(s)$ with those of Eq. (7-38), we have $a_0 = 10$, $a_1 = 10$, $a_2 = 60$, $a_3 = 17$, $b_0 = 5$, $b_1 = 1$, and $b_2 = 0$. The steady-state errors of the system due to the three basic types of inputs are calculated as follows:

$$\text{Unit-step input, } r(t) = u_s(t):\quad e_{ss} = \frac{1}{K_H}\left(\frac{a_0 - b_0K_H}{a_0}\right) = \frac{1}{2}\left(\frac{10 - 5 \times 2}{10}\right) = 0 \tag{7-62}$$

Solving for the unit-step response using the $M(s)$ in Eq. (7-61), we get

$$y(t) = 0.5u_s(t) + \text{transient terms} \tag{7-63}$$

Thus the steady-state value of $y(t)$ is 0.5, and since $K_H = 2$, the steady-state error due to a unit-step input is zero.

Unit-ramp input, $r(t) = tu_s(t)$: $$e_{ss} = \frac{1}{K_H}\left(\frac{a_1 - b_1K_H}{a_0}\right) = \frac{1}{2}\left(\frac{10 - 1 \times 2}{10}\right) = 0.4 \tag{7-64}$$

The unit-ramp response of the system is written

$$y(t) = [0.5t - 0.4]u_s(t) + \text{transient terms} \tag{7-65}$$

Thus, using Eq. (7-36), the steady-state error is calculated as

$$e(t) = \frac{1}{K_H}r(t) - y(t) = 0.4u_s(t) - \text{transient terms} \tag{7-66}$$

The steady-state error is 0.4 as calculated in Eq. (7-64).

Unit-parabolic input, $r(t) = t^2u_s(t)/2$: $e_{ss} = \infty$ since $a_1 - b_1K_H \neq 0$

The unit-parabolic input is

$$y(t) = [0.25t^2 - 0.4t - 2.6]u_s(t) + \text{transient terms} \tag{7-67}$$

Thus the steady-state error is $0.4t + 2.6$, which increases with time. ▲

Steady-State Error of Nonunity Feedback: $H(s)$ Has Nth Order Zero at $s = 0$

When $H(s)$ has an Nth-order zero at $s = 0$, this corresponds to the desired output being proportional to the Nth-order derivative of the input in the steady state. Thus the reference signal can be defined as $R(s)/K_Hs^N$, and the error signal in the transform domain may be defined as

$$E(s) = \frac{1}{K_Hs^N}R(s) - Y(s) \tag{7-68}$$

where

$$K_H = \lim_{s \to 0} \frac{H(s)}{s^N} \tag{7-69}$$

Let $N = 1$; then the transfer function of $M(s)$ in Eq. (7-38) will have a pole at $s = 0$, or $a_0 = 0$. The steady-state error is written, from Eq. (7-68).

$$e_{ss} = \frac{1}{K_H} \lim_{s \to 0} \left[\frac{s^{n-1} + \cdots + (a_2 - b_1 K_H)s + (a_1 - b_0 K_H)}{s^n + a_{n-1}s^{n-1} + \cdots + a_1 s} \right] sR(s) \tag{7-70}$$

For a step input of magnitude R, Eq. (7-70) is written

$$e_{ss} = \frac{1}{K_H} \lim_{s \to 0} \left[\frac{s^{n-1} + \cdots + (a_2 - b_1 K_H)s + (a_1 - b_0 K_H)}{s^n + a_{n-1}s^{n-1} + \cdots + a_1 s} \right] R \tag{7-71}$$

Thus the steady-state error is

$$e_{ss} = 0 \qquad \text{if } a_2 - b_1 K_H = 0 \quad \text{and} \quad a_1 - b_0 K_H = 0 \tag{7-72}$$

$$e_{ss} = \frac{a_2 - b_1 K_H}{a_1 K_H} R = \text{constant} \qquad \text{if } a_1 - b_0 K_H = 0 \quad \text{but} \quad a_2 - b_1 K_H \neq 0 \tag{7-73}$$

$$e_{ss} = \infty \qquad \text{if } a_1 - b_0 K_H \neq 0 \tag{7-74}$$

We shall use the following example to illustrate these results.

Example 7-7

Consider that the transfer functions of the system shown in Fig. 7-5 has the following transfer functions:

$$G(s) = \frac{1}{s^2(s + 12)} \qquad H(s) = \frac{10s}{s + 5} \tag{7-75}$$

Thus

$$K_H = \lim_{s \to 0} \frac{H(s)}{s} = 2 \tag{7-76}$$

The closed-loop transfer function is

$$M(s) = \frac{Y(s)}{R(s)} = \frac{s + 5}{s^4 + 17s^3 + 60s^2 + 10s} \tag{7-77}$$

The velocity control system is considered to be stable, as defined in Chapter 5, although $M(s)$ has a pole at $s = 0$, since the objective is to control velocity with a step input. The coefficients are identified to be: $a_0 = 0$, $a_1 = 10$, $a_2 = 60$, $b_0 = 5$, and $b_1 = 1$.

For a unit-step input, the steady-state error, from Eq. (7-73), is

$$e_{ss} = \frac{1}{K_H}\left(\frac{a_2 - b_1 K_H}{a_1}\right) = \frac{1}{2}\left(\frac{60 - 1 \times 2}{10}\right) = 2.9 \tag{7-78}$$

To verify this result, we find the unit-step response using the closed-loop transfer function in Eq. (7-77). The result is

$$y(t) = (0.5t - 2.9)u_s(t) + \text{transient terms} \tag{7-79}$$

From the discussion that leads to Eq. (7-68), the reference signal is $tu_s(t)/K_H = 0.5tu_s(t)$ in the steady state; thus the steady-state error is 2.9. ▲

7-4 Unit-Step Response and Time-Domain Specifications

As defined earlier, the transient portion of the time response is that part which goes to zero as time becomes large. Nevertheless, the transient response of a control system is necessarily important, since both the amplitude and time duration of the transient response must be kept within tolerable or prescribed limits. For example, in the automobile idle-speed control system described in Chapter 1, in addition to striving for a desirable idle speed in the steady state, the transient droop in engine speed must not be excessive, and the recovery in speed should be made as quickly as possible.

For linear control systems, the characterization of the transient response is often done by use of the unit-step function $u_s(t)$ as the input. The response of a control system when the input is a unit-step function is called the **unit-step response.** Figure 7-11 illustrates a typical unit-step response of a linear control system. With reference to the unit-step response, performance criteria commonly used for the characterization of linear control systems in the time domain are defined as follows:

1. **Maximum overshoot.** Let $y(t)$ be the unit-step response. Let y_{max} denote the maximum value of $y(t)$, and y_{ss} be the steady-state value of $y(t)$, and $y_{max} \geq y_{ss}$. The maximum overshoot of $y(t)$ is defined as

$$\text{maximum overshoot} = y_{max} - y_{ss} \tag{7-80}$$

The maximum overshoot is often represented as a percentage of the final value of the step response; that is,

$$\text{percent maximum overshoot} = \frac{\text{maximum overshoot}}{y_{ss}} \times 100\% \tag{7-81}$$

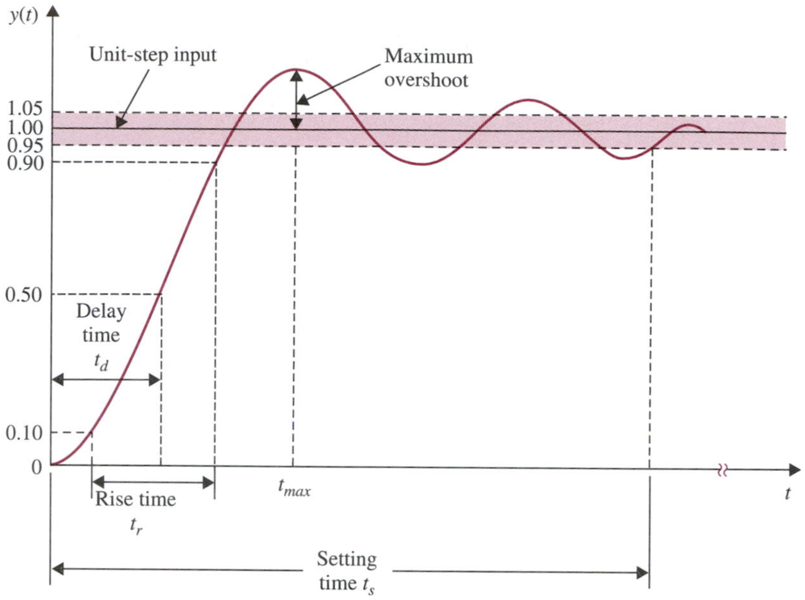

Figure 7-11 Typical unit-step response of a control system.

The maximum overshoot is often used to measure the relative stability of a control system. A system with a large overshoot is usually undesirable. For design purposes, the maximum overshoot is often given as a time-domain specification. The unit-step illustrated in Fig. 7-11 shows that the maximum overshoot occurs at the first overshoot. For some systems, the maximum overshoot may occur at a later peak, and if the system transfer function has an odd number of zeros in the right-half *s*-plane, a negative undershoot may even occur [3,4] (Problem 7-23).

2. **Delay Time.** The delay time t_d is defined as the time required for the step response to reach 50 percent of its final value. This is shown in Fig. 7-11.
3. **Rise time.**The rise time t_r is defined as the time required for the step response to rise from 10 to 90 percent of its final value, as shown in Fig. 7-11. An alternative measure is to represent the rise time as the reciprocal of the slope of the step response at the instant that the response is equal to 50 percent of its final value.
4. **Settling time.** The settling time t_s is defined as the time required for the step response to decrease and stay within a specified percentage of its final value. A frequently used figure is 5 percent.

The four quantities just defined give a direct measure of the transient characteristics of a control system in terms of the unit-step response. These time-domain specifications are relatively easy to measure when the step response is well defined, as shown in Fig. 7-11. Analytically, these quantities are difficult to establish, except for simple systems lower than the third order.

7-5 Transient Response of a Prototype Second-Order System

Although true second-order control systems are rare in practice, their analysis generally helps to form a basis for the understanding of analysis and design of higher-order systems, especially the ones that can be approximated by second-order systems. Consider that a second-order control system with unity feedback is represented by the block diagram shown in Fig. 7-12. The open-loop transfer function of the system is

$$G(s) = \frac{Y(s)}{E(s)} = \frac{\omega_n^2}{s(s + 2\zeta\omega_n)} \tag{7-82}$$

where ζ and ω_n are real constants. The closed-loop transfer function of the system is

$$\frac{Y(s)}{R(s)} = \frac{\omega_n^2}{s^2 + 2\zeta\omega_n s + \omega_n^2} \tag{7-83}$$

The system in Fig. 7-12 with the transfer functions given by Eqs. (7-82) and (7-83) is defined as the **prototype second-order system.**

The characteristic equation of the prototype second-order system is obtained by setting the denominator of Eq. (7-83) to zero:

$$\Delta(s) = s^2 + 2\zeta\omega_n s + \omega_n^2 = 0 \tag{7-84}$$

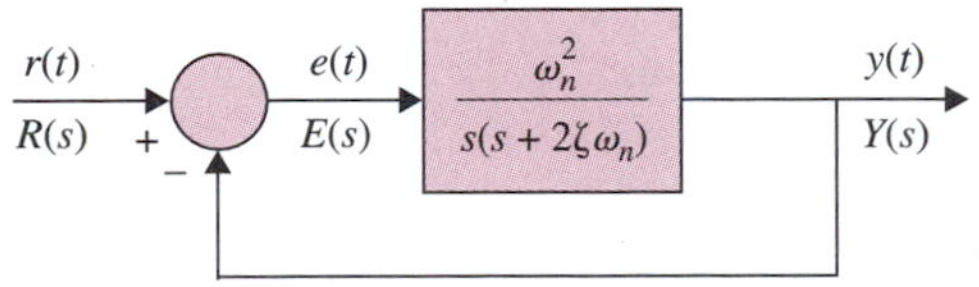

Figure 7-12 Prototype second-order control system.

For a unit-step function input, $R(s) = 1/s$, the output response of the system is obtained by taking the inverse Laplace transform of the output transform

$$Y(s) = \frac{\omega_n^2}{s(s^2 + 2\zeta\omega_n s + \omega_n^2)} \tag{7-85}$$

This can be done by referring to the Laplace transform table in Appendix B. The result is

$$y(t) = 1 - \frac{e^{-\zeta\omega_n t}}{\sqrt{1-\zeta^2}} \sin(\omega_n \sqrt{1-\zeta^2}\, t + \cos^{-1}\zeta) \qquad t \geq 0 \tag{7-86}$$

Figure 7-13 shows the unit-step responses of Eq. (7-86) plotted as functions of the normalized time $\omega_n t$ for various values of ζ. As seen, the response becomes more oscillatory with larger overshoot as ζ decreases. When $\zeta \geq 1$, the step response does not exhibit any overshoot; that is, $y(t)$ never exceeds its final value during the transient. The responses also show that ω_n has a direct effect on the rise time, delay time, and settling time, but does not affect the overshoot. These will be studied in more detail in the following sections.

7-5-1 Damping Ratio and Damping Factor

The effects of the system parameters ζ and ω_n on the step response $y(t)$ of the prototype second-order system can be studied by referring to the roots of the characteristic equation in Eq. (7-84). The two roots can be expressed as

$$\begin{aligned} s_1, s_2 &= -\zeta\omega_n \pm j\omega_n \sqrt{1-\zeta^2} \\ &= -\alpha \pm j\omega \end{aligned} \tag{7-87}$$

where

$$\alpha = \zeta\omega_n \tag{7-88}$$

and

$$\omega = \omega_n \sqrt{1-\zeta^2} \tag{7-89}$$

The physical significance of ζ and α is now investigated. As seen from Eqs. (7-88) and (7-86), α appears as the constant that is multiplied to t in the exponential term of $y(t)$. Therefore, α controls the rate of rise or decay of the unit-step response $y(t)$. In other words, α controls the "damping" of the system and is called the **damping factor,** or the **damping constant.** The inverse of α, $1/\alpha$, is proportional to the time constant of the system.

▲ When the damping is critical, $\zeta = 1$.

When the two roots of the characteristic equation are real and equal, we called the system **critically damped.** From Eq. (7-87) we see that critical damping occurs

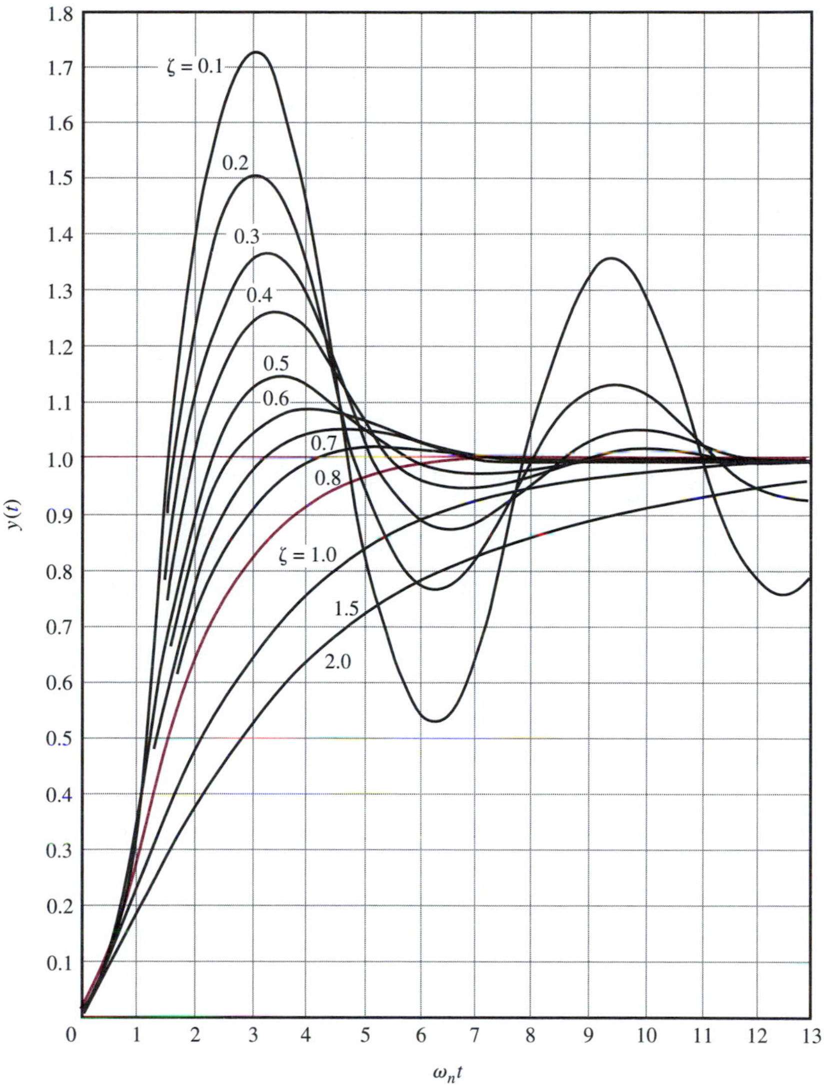

Figure 7-13 Unit-step responses of the prototype second-order system with various damping ratios.

when $\zeta = 1$. Under this condition, the damping factor is simply $\alpha = \omega_n$. Thus we can regard ζ as the damping ratio, that is,

$$\zeta = \text{damping ratio} = \frac{\alpha}{\omega_n} = \frac{\text{actual damping factor}}{\text{damping factor at critical damping}} \tag{7-90}$$

7-5-2 Natural Undamped Frequency

The parameter ω_n is defined as the **natural undamped frequency.** As seen from Eq. (7-87), when $\zeta = 0$, the damping is zero, the roots of the characteristic equation are imaginary, and Eq. (7-86) shows that the unit-step response is purely sinusoidal. Therefore, ω_n corresponds to the frequency of the undamped sinusoidal response. Equation (7-87) shows that when $0 < \zeta < 1$, the imaginary parts of the roots have the magnitude of ω. Since when $\zeta \neq 0$ the response of $y(t)$ is not a periodic function, ω defined in Eq. (7-89) is not a frequency. For the purpose of reference, ω is sometimes defined as the **conditional frequency,** or the **damped frequency.**

Figure 7-14 illustrates the relationships between the location of the characteristic equation roots and α, ζ, ω_n, and ω. For the complex-conjugate roots shown:

- ω_n is the radial distance from the roots to the origin of the s-plane.
- α is the real part of the roots.
- ω is the imaginary part of the roots.
- ζ is the cosine of the angle between the radial line to the roots and the negative axis when the roots are in the left-half s-plane, or

$$\zeta = \cos\theta \tag{7-91}$$

Figure 7-15 shows in the s-plane, (a) the constant-ω_n loci, (b) the constant-ζ loci, (c) the constant-α loci, and (d) the constant-ω loci. Regions in the s-plane are identified

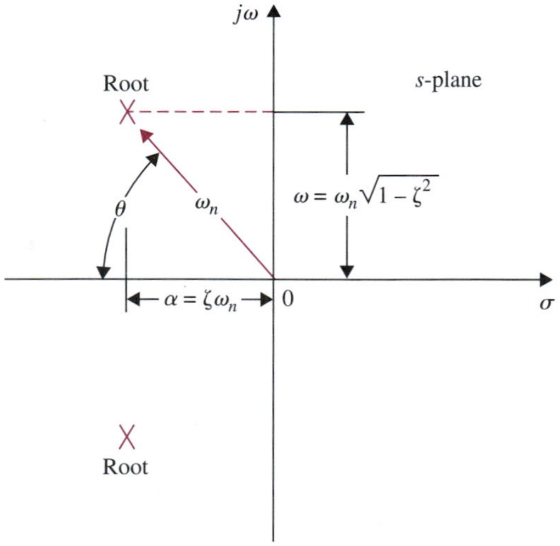

Figure 7-14 Relationship between the characteristic-equation roots of the prototype second-order system and α, ζ, ω_n, and ω.

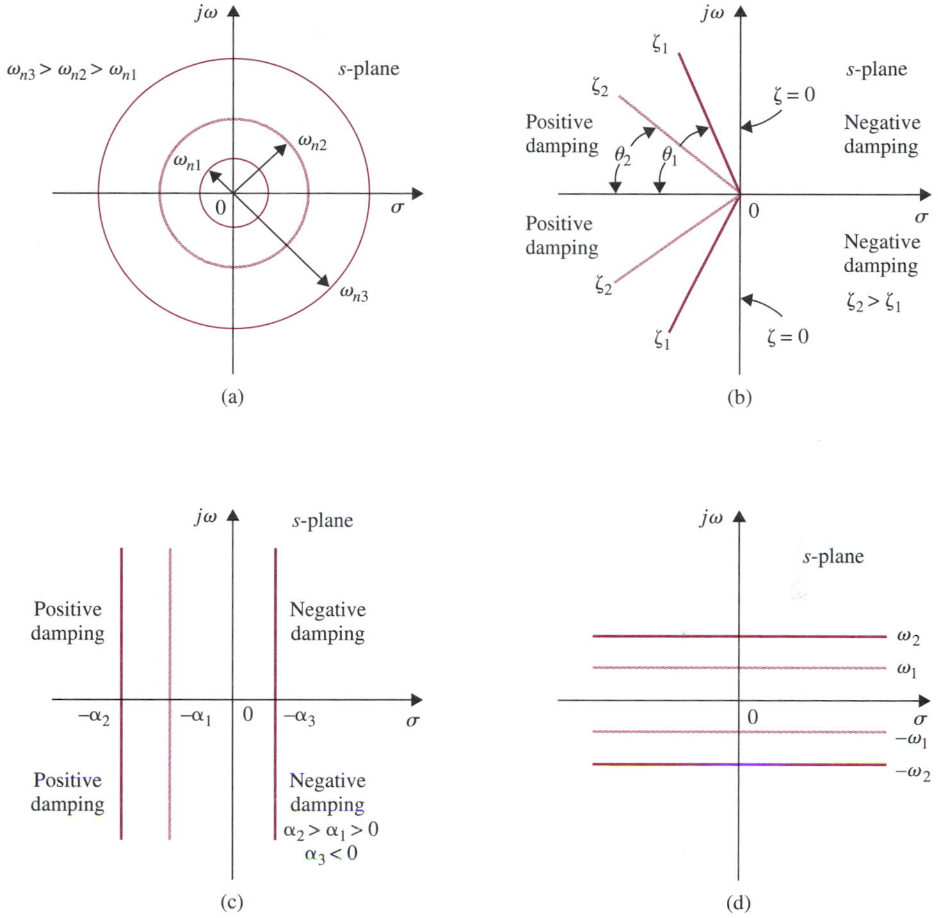

Figure 7-15 (a) Constant-natural-undamped-frequency loci. (b) Constant-damping-ratio loci. (c) Constant-damping-factor loci. (d) Constant-conditional-frequency loci.

with the system damping as follows:

- The left-half s-plane corresponds to positive damping (i.e., the damping factor or damping ratio is positive). Positive damping causes the unit-step response to settle to a constant final value in the steady state due to the negative exponent of $\exp(-\zeta\omega_n t)$. The system is stable.
- The right-half s-plane corresponds to negative damping. Negative damping gives a response that grows in magntiude without bound with time, and the system is unstable.

▲ The imaginary axis corresponds to zero damping ($\alpha = 0$ or $\zeta = 0$). Zero damping results in a sustained oscillation response, and the system is marginally stable or marginally unstable.

Thus we have demonstrated with the help of the simple prototype second-order system that the location of the characteristic equation roots plays an important role in the transient response of the system.

The effect of the characteristic equation roots on the damping of the second-order system is further illustrated by Figs. 7-16 and 7-17. In Fig. 7-16, ω_n is held constant while the damping ratio ζ is varied from $-\infty$ to $+\infty$. The following classification of the system dynamics with respect to the value of ζ is made:

▲ $0 < \zeta < 1$: $s_1, s_2 = -\zeta\omega_n \pm j\omega_n\sqrt{1-\zeta^2}$ $(-\zeta\omega_n < 0)$ *underdamped*
▲ $\zeta = 1$: $s_1, s_2 = -\omega_n$ *critically damped*
▲ $\zeta > 1$: $s_1, s_2 = -\zeta\omega_n \pm \omega_n\sqrt{\zeta^2-1}$ *overdamped*
▲ $\zeta = 0$: $s_1, s_2 = \pm j\omega_n$ *undamped*
▲ $\zeta < 0$: $s_1, s_2 = -\zeta\omega_n \pm j\omega_n\sqrt{1-\zeta^2}$ $(-\zeta\omega_n > 0)$ *negatively damped*

Figure 7-17 illustrates typical unit-step responses that correspond to the various root locations shown above.

In practical applications, only stable systems that correspond to $\zeta > 0$ are of interest. Figure 7-13 gives the unit-step responses of Eq. (7-86) plotted as functions of the normalized time $\omega_n t$ for various values of the damping ratio ζ. As seen, the response becomes more oscillatory as ζ decreases in value. When $\zeta \geq 1$, the step response does not exhibit any overshoot; that is, $y(t)$ never exceeds its final value during the transient.

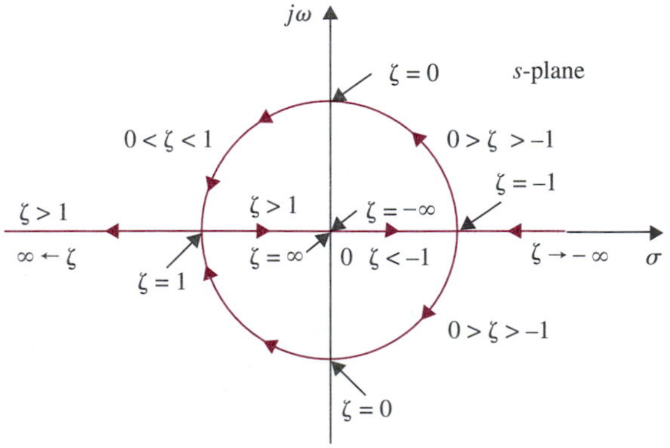

Figure 7-16 Locus of roots of the characteristic equation of the prototype second-order system, Eq. (7-84), when ω_n is held constant while the damping ratio is varied from $-\infty$ to ∞.

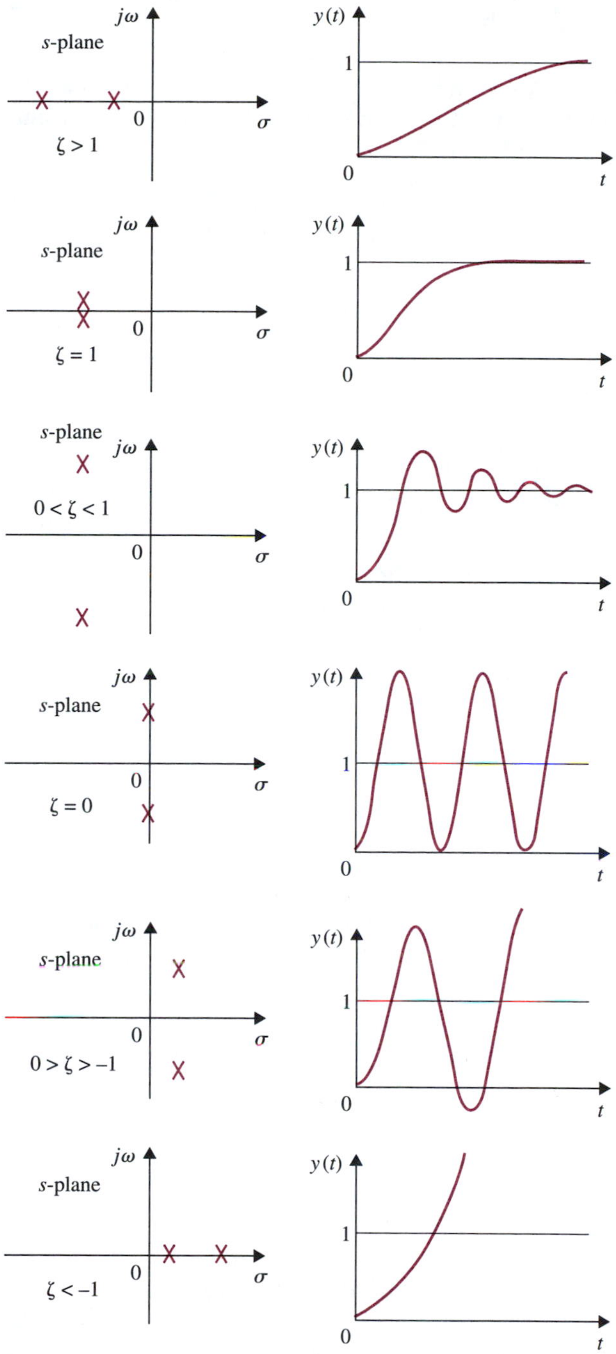

Figure 7-17 Step-response comparison for various characteristic-equation-root locations in the s-plane.

7-5-3 Maximum Overshoot

The exact relation between the damping ratio and the amount of overshoot can be obtained by taking the derivative of Eq. (7-86) with respect to t and setting the result to zero. Thus

$$\frac{dy(t)}{dt} = \frac{\omega_n e^{-\zeta\omega_n t}}{\sqrt{1-\zeta^2}}\left[\zeta \sin(\omega t + \theta) - \sqrt{1-\zeta^2}\cos(\omega t + \theta)\right] \qquad t \geq 0 \tag{7-92}$$

▲ Maximum overshoot is defined only for a step-function input.

where ω and θ are defined in Eqs. (7-89) and (7-91), respectively. We can show that the quantity inside the square brackets in Eq. (7-92) can be reduced to $\sin \omega t$. Thus Eq. (7-92) is simplified to

$$\frac{dy(t)}{dt} = \frac{\omega_n}{\sqrt{1-\zeta^2}} e^{-\zeta\omega_n t} \sin \omega_n \sqrt{1-\zeta^2}\,t \qquad t \geq 0 \tag{7-93}$$

Setting $dy(t)/dt$ to zero, we have the solutions $t = \infty$ and

$$\omega_n \sqrt{1-\zeta^2}\,t = n\pi \qquad n = 0, 1, 2, \ldots \tag{7-94}$$

from which we get

$$t = \frac{n\pi}{\omega_n \sqrt{1-\zeta^2}} \qquad n = 0, 1, 2, \ldots \tag{7-95}$$

The solution at $t = \infty$ is the maximum of $y(t)$ only when $\zeta \geq 1$. For the unit-step responses shown in Fig. 7-13, the first overshoot is the maximum overshoot. This corresponds to $n = 1$ in Eq. (7-95). Thus the time at which the maximum overshoot occurs is

$$t_{\max} = \frac{\pi}{\omega_n \sqrt{1-\zeta^2}} \tag{7-96}$$

With reference to Fig. 7-13, the overshoots occur at odd values of n, that is, n = 1, 3, 5, . . . , and the undershoots occur at even values of n. Whether the extremum is an overshoot or an undershoot, the time at which it occurs is given by Eq. (7-95). It should be noted that *although the unit-step response for $\zeta \neq 0$ is not periodic, the overshoots and undershoots of the response do occur at periodic intervals, as shown in Fig. 7-18.*

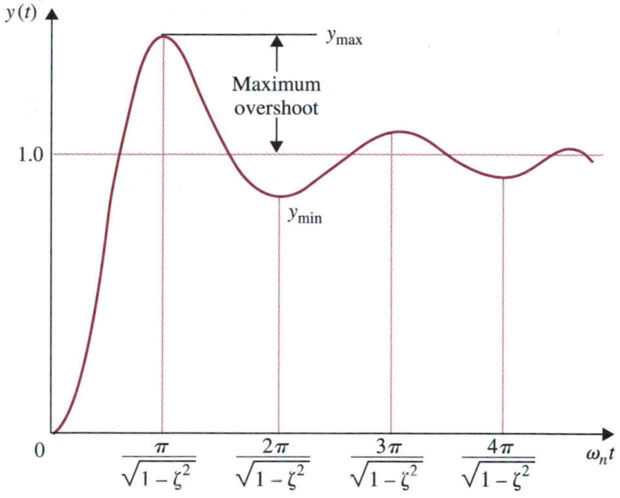

Figure 7-18 Unit-step responses illustrating that the maxima and minima occur at periodic intervals.

The magnitudes of the overshoots and the undershoots can be determined by substituting Eq. (7-95) into Eq. (7-86). The result is

$$y(t)\big|_{\max\ or\ \min} = 1 - \frac{e^{-n\pi\zeta/\sqrt{1-\zeta^2}}}{\sqrt{1-\zeta^2}}\sin(n\pi + \theta) \qquad n = 1, 2, \ldots \tag{7-97}$$

or

$$y(t)\big|_{\max\ or\ \min} = 1 + (-1)^{n-1}e^{-n\pi\zeta/\sqrt{1-\zeta^2}} \qquad n = 1, 2, \ldots \tag{7-98}$$

The maximum overshoot is obtained by letting $n = 1$ in Eq. (7-98). Therefore,

$$\text{maximum overshoot} = y_{\max} - 1 = e^{-\pi\zeta/\sqrt{1-\zeta^2}} \tag{7-99}$$

and the percent maximum overshoot is

$$\text{percent maximum overshoot} = 100e^{-\pi\zeta/\sqrt{1-\zeta^2}} \tag{7-100}$$

Equation (7-99) shows that the maximum overshoot of the step response of the prototype second-order system is a function only of the damping ratio ζ. The relationship between the percent maximum overshoot and the damping ratio given in Eq. (7-100) is plotted in Fig. 7-19. The time $t_{\max}$ in Eq. (7-96) is a function of both ζ and ω_n.

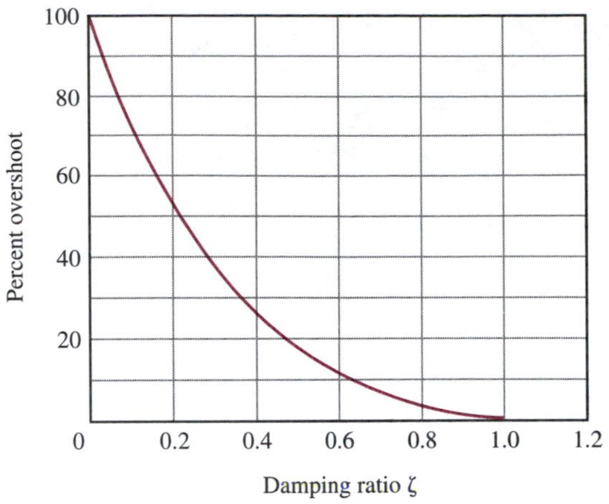

Figure 7-19 Percent overshoot as a function of damping ratio for the step response of the prototype second-order system.

7-5-4 Delay Time and Rise Time

It is more difficult to determine the exact analytical expressions of the delay time t_d, rise time t_r, and settling time t_s, even for just the simple prototype second-order system. For instance, for the delay time, we would have to set $y(t) = 0.5$ in Eq. (7-86) and solve for t. An easier way would be to plot $\omega_n t_d$ versus ζ, as shown in Fig. 7-20, and

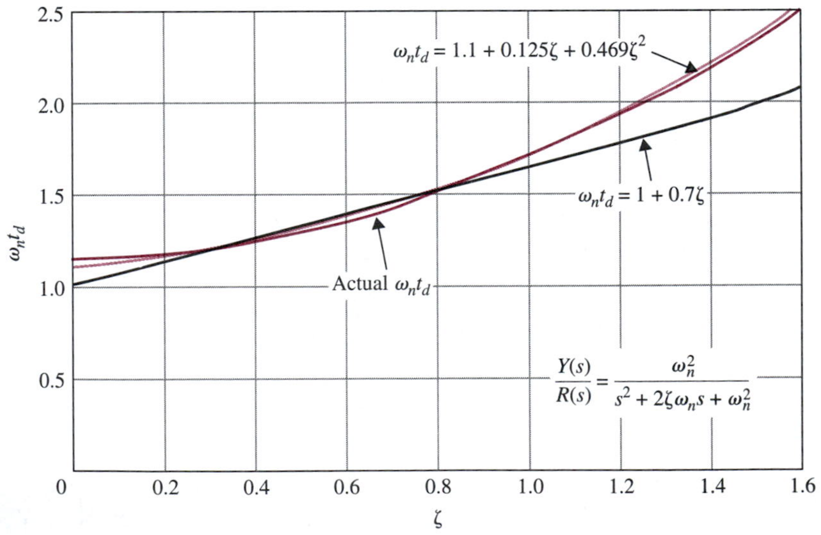

Figure 7-20 Normalized delay time versus ζ for the prototype second-order system.

then approximate the curve by a straight line or a curve over the range of $0 < \zeta < 1$. From Fig. 7-20, the delay time for the prototype second-order system is approximated as

$$t_d \cong \frac{1 + 0.7\zeta}{\omega_n} \qquad 0 < \zeta < 1.0 \tag{7-101}$$

We can obtain a better approximation by using a second-order equation for t_d:

$$t_d \cong \frac{1.1 + 0.125\zeta + 0.469\zeta^2}{\omega_n} \qquad 0 < \zeta < 1.0 \tag{7-102}$$

For the rise time t_r, which is the time for the step response to reach from 10 to 90 percent of its final value, the exact value can be determined directly from the responses of Fig. 7-13. The plot of $\omega_n t_r$ versus ζ is shown in Fig. 7-21. In this case, the relation can again be approximated by a straight line over a limited range of ζ:

$$t_r \cong \frac{0.8 + 2.5\zeta}{\omega_n} \qquad 0 < \zeta < 1 \tag{7-103}$$

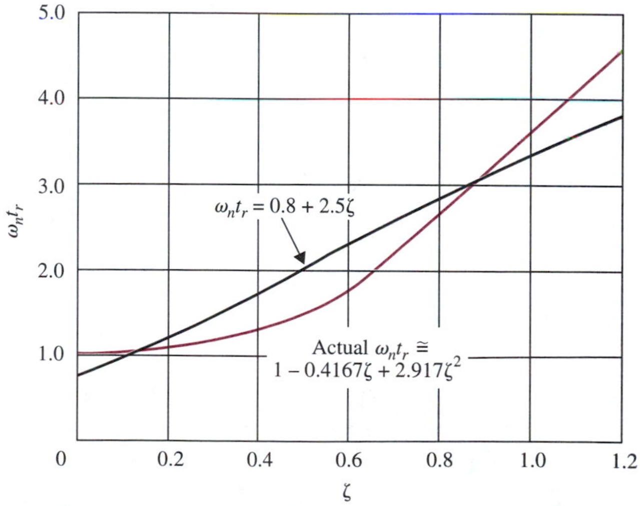

Figure 7-21 Normalized rise time versus ζ for the prototype second-order system.

A better approximation can be obtained by using a second-order equation:

$$t_r \cong \frac{1 - 0.4167\zeta + 2.917\zeta^2}{\omega_n} \qquad 0 < \zeta < 1 \tag{7-104}$$

From this discussion, the following conclusions can be made on the rise time and delay time of the prototype second-order system:

- t_r and t_d are proportional to ζ and inversely proportional to ω_n.
- Increasing (decreasing) the natural undamped frequency ω_n will reduce (increase) t_r and t_d.

7-5-5 Settling Time

From Fig. 7-13 we see that when $0 < \zeta < 0.69$ the unit-step response has a maximum overshoot greater than 5 percent, and the response can enter the band between 0.95 and 1.05 for the last time from either the top or the bottom. When ζ is greater than 0.69, the overshoot is less than 5 percent, and the response can enter the band between 0.95 and 1.05 only from the bottom. Figure 7-22(a) and (b) show the two different situations. Thus the settling time has a discontinuity at $\zeta = 0.69$.

The exact analytical description of the settling time t_s is difficult to obtain. We can obtain an approximation for t_s for $0 < \zeta < 0.69$ by using the envelope of the damped sinusoid of $y(t)$, as shown in Fig. 7-22(a). When the settling time corresponds to an intersect with the upper envelope of $y(t)$, the following relation is obtained:

$$1 + \frac{1}{\sqrt{1-\zeta^2}} e^{-\zeta\omega_n t_s} = 1.05 \tag{7-105}$$

When the settling time corresponds to an intersect with the bottom envelope of $y(t)$, t_s must satisfy the following condition:

$$1 - \frac{1}{\sqrt{1-\zeta^2}} e^{-\zeta\omega_n t_s} = 0.95 \tag{7-106}$$

It is apparent that the same result is obtained using either Eq. (7-105) or Eq. (7-106).

Solving Eq. (7-105) for $\omega_n t_s$, we have

$$\omega_n t_s = -\frac{1}{\zeta}\ln(0.05\sqrt{1-\zeta^2}) \tag{7-107}$$

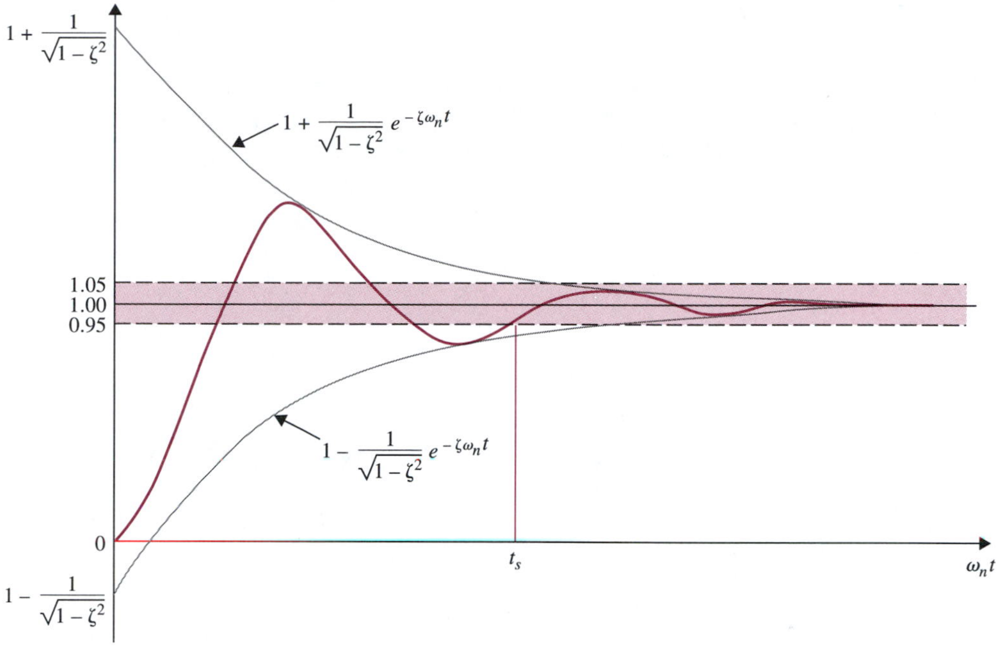

(a) $0 < \zeta < 0.69$

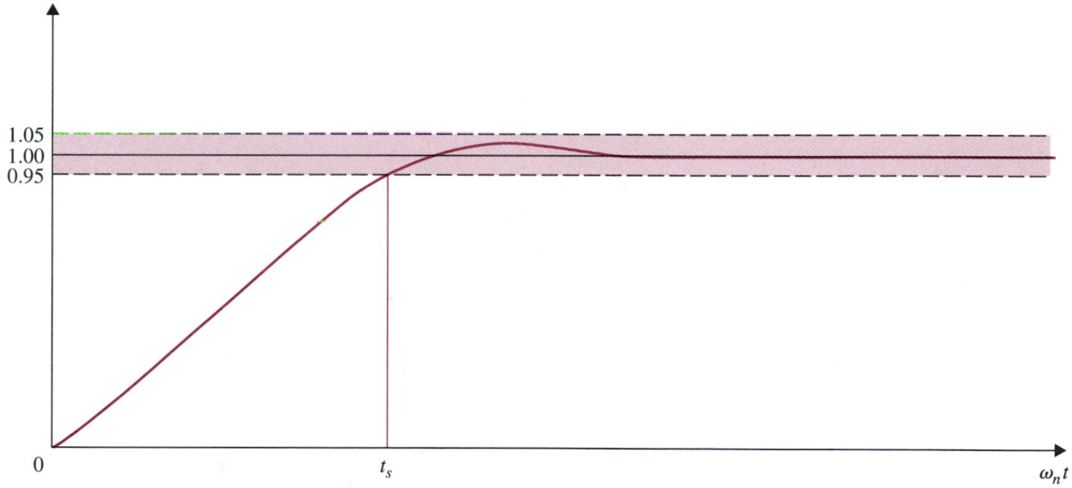

(b) $\zeta > 0.69$

Figure 7-22 Settling time of the unit-step response.

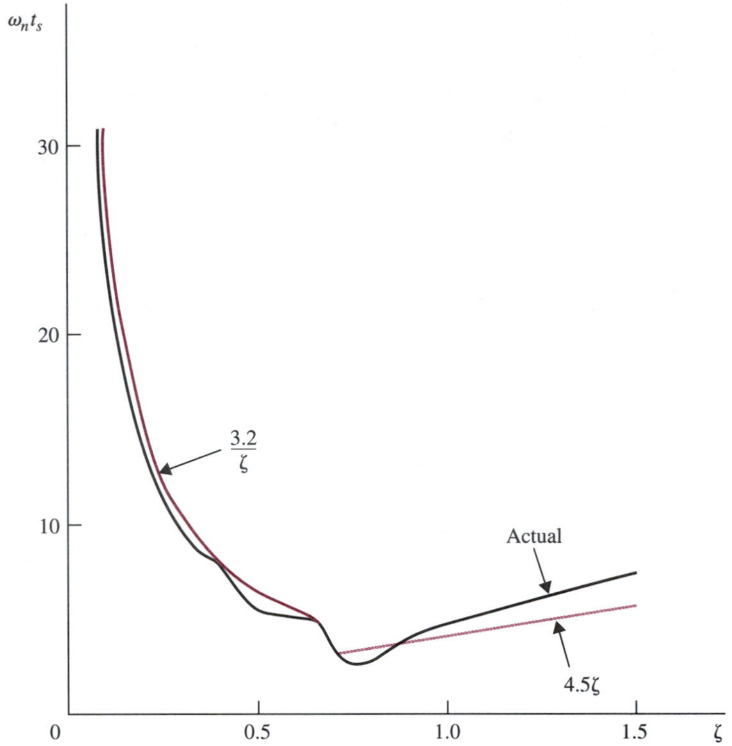

(c)

Figure 7-22 Continued.

The right-hand side of Eq. (7-107) varies between 3.0 and 3.32 as ζ varies from 0 to 0.69. We can approximate the settling time for the prototype second-order system as

$$t_s \cong \frac{3.2}{\zeta\omega_n} \qquad 0 < \zeta < 0.69 \tag{7-108}$$

The approximation will be poor for small values of ζ (<0.3).

When the damping ratio ζ is greater than 0.69, the unit-step response will always enter the band between 0.95 and 1.05 from below. We can show by observing the responses in Fig. 7-13 that the value of $\omega_n t_s$ is almost directly proportional to ζ. The following approximation is used for t_s for $\zeta > 0.69$.

$$t_s = \frac{4.5\zeta}{\omega_n} \qquad \zeta > 0.69 \tag{7-109}$$

Figure 7-22(c) shows the actual values of $\omega_n t_s$ versus ζ for the prototype second-order system described by Eq. (7-83), along with the approximations using Eqs. (7-108) and (7-109) for their respective effective ranges. The numerical values are tabulated in Table 7-2. The actual values of the normalized settling time, $\omega_n t_s$, is computed using the function **svdesign** in the **CSAD** toolbox. The option **Attributes** computes the rise time, delay time, settling time, maximum overshoot, and t_{max} from the step response.

We can summarize the relationships between t_s and the system parameters as follows:

- ▲ For $\zeta < 0.69$, the settling time is inversely proportional to ζ and ω_n. A practical way of reducing the settling time is to increase ω_n while holding ζ constant. Although the response will be more oscillatory, the maximum overshoot depends only on ζ and can be controlled independently.
- ▲ For $\zeta > 0.69$, the settling time is proportional to ζ and inversely porportional to ω_n. Again, t_s can be reduced by increasing ω_n.

Table 7-2 Comparison of Settling Times of Prototype Second-Order System, $\omega_n t_s$

ζ	Actual	$\frac{3.2}{\zeta}$	4.5ζ
0.10	28.7	30.2	
0.20	13.7	16.0	
0.30	10.0	10.7	
0.40	7.5	8.0	
0.50	5.2	6.4	
0.60	5.2	5.3	
0.62	5.16	5.16	
0.64	5.00	5.00	
0.65	5.03	4.92	
0.68	4.71	4.71	
0.69	4.35	4.64	
0.70	2.86		3.15
0.80	3.33		3.60
0.90	4.00		4.05
1.00	4.73		4.50
1.10	5.50		4.95
1.20	6.21		5.40
1.50	8.20		6.75

It should be commented that the settling time for $\zeta > 0.69$ is truly a measure of how fast the step response rises to its final value. It seems that for this case, the rise and delay times should be adequate to describe the response behavior. As the name implies, settling time should be used to measure how fast the step response settles to its final value.

Keep in mind that while the definitions on $y_{\max}$, $t_{\max}$, t_d, t_r, and t_s, apply to a system of any order, the damping ratio ζ and the natural undamped frequency ω_n apply strictly only to a second-order system whose closed-loop transfer function is given in Eq. (7-83). Naturally, the relationships between t_d, t_r, and t_s and ζ and ω_n are valid only for the same second-order system model. However, these relationships can be used to measure the performance of higher-order systems that can be approximated by second-order ones, under situations that some of the higher-order poles can be neglected.

7-6 Time-Domain Analysis of a Position-Control System

▲ The maximum overshoot of the prototype second-order system depends only on ζ.

In this section we analyze the performance of a system using the time-domain criteria established in Section 7-5. The purpose of the control system considered is to control the positions of the control fins of a modern airship. Due to the requirements of improved response and reliability, the control surfaces of modern aircraft are controlled by electric actuators with electronics controls. Gone are the days when the ailerons, rudder, and elevators of the aircraft are all linked to the cockpit control through mechanical linkages. The so-called "fly-by-wire" control system used in modern aircraft control implies that the attitude control of aircraft is no longer controlled entirely by mechanical linkages. Figure 7-23 illustrates the controlled surfaces and the block diagram of one axis of such a position-control system. Figure 7-24 shows the analytical block diagram of the system using the amplifier/dc-motor model given in Fig. 4-51. The system is simplified to the extent that saturation of the amplifier gain and motor torque, gear backlash, and shaft compliances have all been neglected. (As students, you guys and dolls have really got it made! When you get into the real world, some of these nonlinearities have to be considered to make things work.)

The objective of the system is to have the output of the system, $\theta_y(t)$, follow the input $\theta_r(t)$. The following system parameters are given initially:

▲ Gain or encoder	$K_s = 1$ V/rad
▲ Gain or preamplifier	$K =$ variable
▲ Gain of power amplifier	$K_1 = 10$ V/V
▲ Gain of current feedback	$K_2 = 0.5$ V/A
▲ Gain of tachometer feedback	$K_t = 0$ V/rad/sec
▲ Armature resistance of motor	$R_a = 5.0\ \Omega$
▲ Armature inductance of motor	$L_a = 0.003$ H

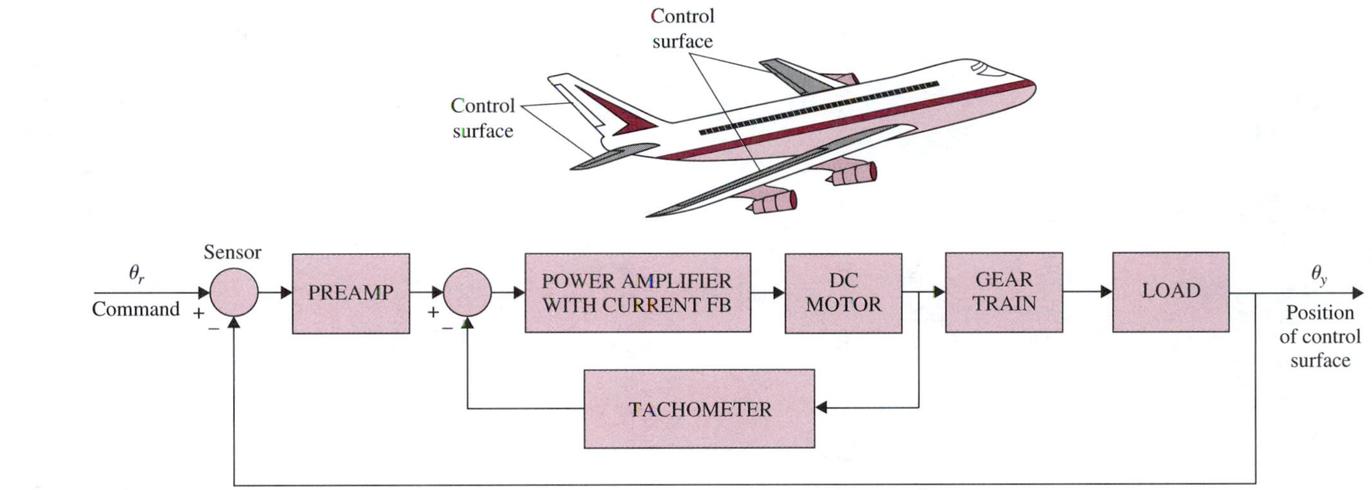

Figure 7-23 Block diagram of an attitude-control system of an aircraft.

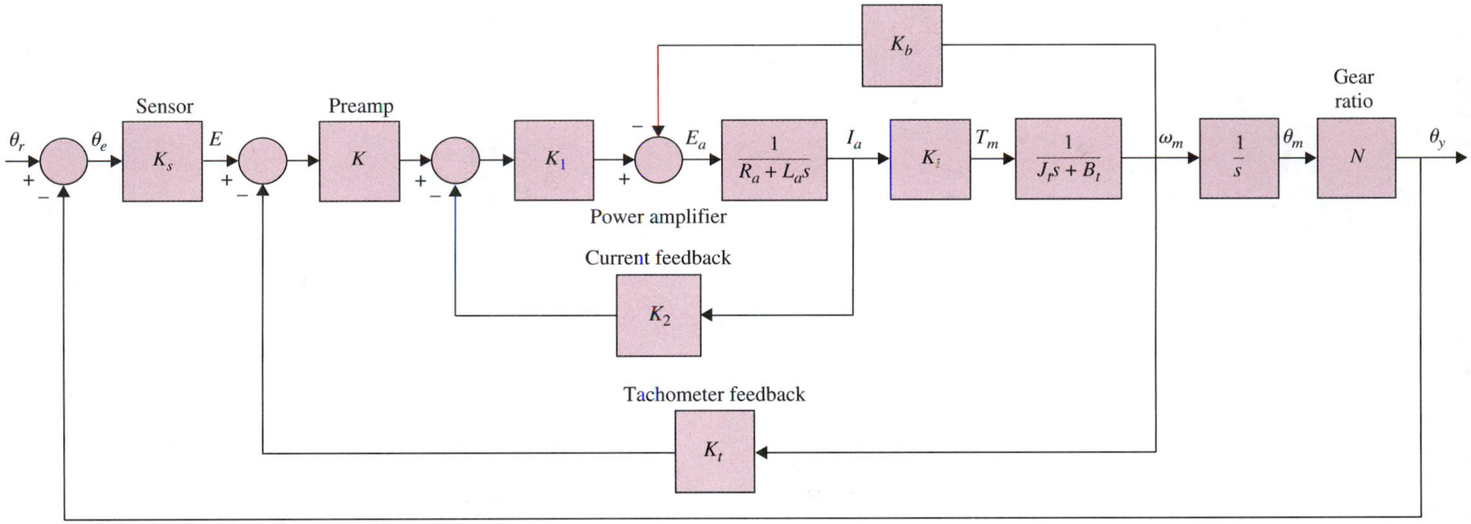

Figure 7-24 Transfer-function block diagram of the system shown in Fig. 7-23.

▲ Torque constant of motor	$K_i = 9.0$ oz-in./A
▲ Back-emf constant of motor	$K_b = 0.0636$ V/rad/sec
▲ Inertia of motor rotor	$J_m = 0.0001$ oz-in.-sec^2
▲ Inertia of load	$J_L = 0.01$ oz-in.-sec^2
▲ Viscous-friction coefficient of motor	$B_m = 0.005$ oz-in.-sec
▲ Viscous-friction coefficient of load	$B_L = 1.0$ oz-in.-sec
▲ Gear-train ratio between motor and load	$N = \theta_y/\theta_m = 1/10$

Since the motor shaft is coupled to the load through a gear train with a gear ratio of N, $\theta_y = N\theta_m$, the total inertia and viscous-friction coefficient seen by the motor are

$$J_t = J_m + N^2 J_L = 0.0001 + 0.01/100 = 0.0002 \text{ oz-in.-sec}^2 \tag{7-110}$$

$$B_t = B_m + N^2 B_L = 0.005 + 1/100 = 0.015 \text{ oz-in.-sec} \tag{7-111}$$

respectively. The forward-path transfer function of the unity-feedback system is written from Fig. 7-24 by applying the SFG gain formula:

$$G(s) = \frac{\Theta_y(s)}{\Theta_e(s)}$$

$$= \frac{K_s K_1 K_i K N}{s[L_a J_t s^2 + (R_a J_t + L_a B_t + K_1 K_2 J_t)s + R_a B_t + K_1 K_2 B_t + K_i K_b + K K_1 K_t K_i]} \tag{7-112}$$

The system is of the third order, since the highest-order term in $G(s)$ is s^3. The electrical time constant of the amplifier–motor system is

$$\tau_a = \frac{L_a}{R_a + K_1 K_2} = \frac{0.003}{5 + 5} = 0.0003 \text{ sec} \tag{7-113}$$

The mechanical time constant of the motor-load system is

$$\tau_t = \frac{J_t}{B_t} = \frac{0.0002}{0.015} = 0.01333 \text{ sec} \tag{7-114}$$

Since the electrical time constant is much smaller than the mechanical time constant because of the low inductance of the motor, we can perform an initial approximation by neglecting the armature inductance L_a. The result is a second-order approximation of the third-order system. Later we will show that this is not the best way of approximating a high-order system by a low-order system. The forward-path transfer function is now

$$G(s) = \frac{K_s K_1 K_i K N}{s[(R_a J_t + K_1 K_2 J_t)s + R_a B_t + K_1 K_2 B_t + K_i K_b + K K_1 K_i K_t]}$$
$$= \frac{K_s K_1 K_i K N/(R_a J_t + K_1 K_2 J_t)}{s[s + (R_a B_t + K_1 K_2 B_t + K_i K_b + K K_1 K_i K_t)/(R_a J_t + K_1 K_2 J_t)]} \qquad (7\text{-}115)$$

Substituting the system parameters in Eq. (7-115), we get

$$G(s) = \frac{4500K}{s(s + 361.2)} \qquad (7\text{-}116)$$

Comparing Eq. (7-115) and (7-116) with the prototype second-order transfer function of Eq. (7-82), we have

$$\text{natural undamped frequency } \omega_n = \pm\sqrt{\frac{K_s K_1 K_i K N}{R_a J_t + K_1 K_2 J_t}} = \pm\sqrt{4500K} \quad \text{rad/sec} \qquad (7\text{-}117)$$

$$\text{damping ratio } \zeta = \frac{R_a B_t + K_1 K_2 B_t + K_i K_b + K K_1 K_i K_t}{2\sqrt{K_s K_1 K_i K N (R_a J_t + K_1 K_2 J_t)}} = \frac{2.692}{\sqrt{K}} \qquad (7\text{-}118)$$

Thus we see that the natural undamped frequency ω_n is proportional to the square root of the amplifier gain K, whereas the damping ratio ζ is inversely proportional to $\sqrt{K}$.

The characteristic equation of the unity-feedback control system is

$$s^2 + 361.2s + 4500K = 0 \qquad (7\text{-}119)$$

7-6-1 Unit-Step Transient Response

For time-domain analysis, it is informative to analyze the system performance by applying the unit-step input with zero initial conditions. In this way it is possible to characterize the system performance in terms of the maximum overshoot and some of the other measures, such as rise time, delay time, and settling time, if necessary.

Let the reference input be a unit-step function $\theta_r(t) = u_s(t)$ rad; then $\Theta(s) = 1/s$. The output of the system, with zero initial conditions, is

$$\theta_y(t) = \mathscr{L}^{-1}\left[\frac{4500K}{s(s^2 + 361.2s + 4500K)}\right] \qquad (7\text{-}120)$$

The inverse Laplace transform of the right-hand side of Eq. (7-120) is carried out using the Laplace transform table in Appendix B, or using Eq. (7-86) directly. When the value of K is known, the function **pfe** of the **CSAD** toolbox can be used to carry out the inverse Laplace transform. The following results are obtained for the three values of K indicated using **pfe** of **CSAD.**

$$K = 7.248\ (\zeta = 1.0):\quad \theta_y(t) = (1 - 151e^{-180t} + 150e^{-180.2t})u_s(t) \tag{7-121}$$

$$K = 14.5\ (\zeta = 0.707):\quad \theta_y(t) = (1 - e^{-180.6t}\cos 180.6t - 0.9997e^{-180.6t}\sin 180.6t)u_s(t) \tag{7-122}$$

$$K = 181.17\ (\zeta = 0.2):\quad \theta_y(t) = (1 - e^{-180.6t}\cos 884.7t - 0.2041e^{-180.6t}\sin 884.7t)u_s(t) \tag{7-123}$$

The three responses are plotted as shown in Fig. 7-25. Table 7-3 gives the comparison of the characteristics of the three unit-step responses for the three values of K used. When $K = 181.17$, $\zeta = 0.2$, the system is lightly damped, and the maximum overshoot is 52.7 percent, which is excessive. When the value of K is set at 7.248, $\zeta = 1.0$, and the system is critically damped. The unit-step response does not have any overshoot or oscillation. When K is set at 14.5, the damping ratio is 0.707 and the overshoot is 4.3 percent.

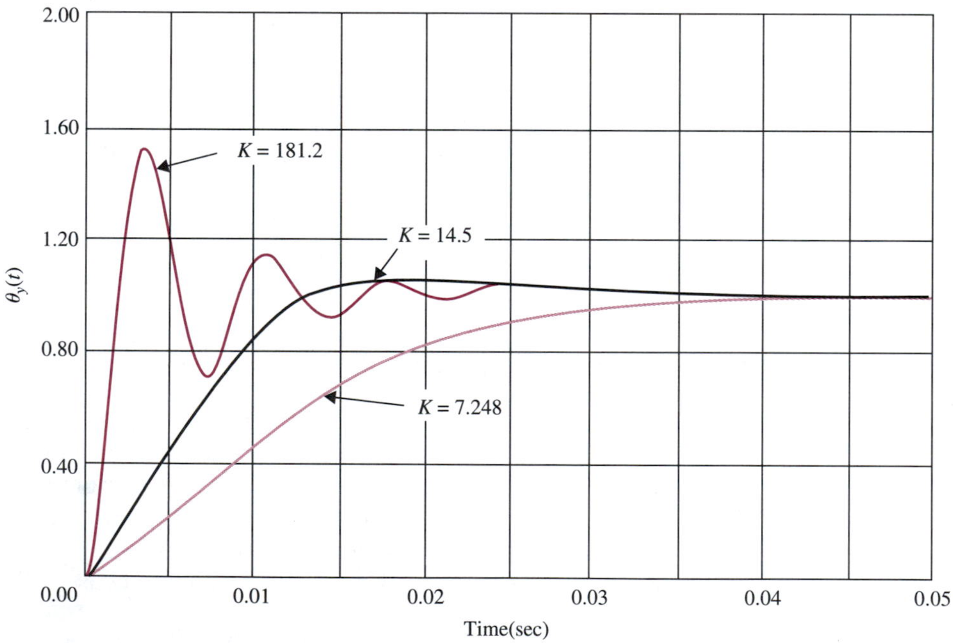

Figure 7-25 Unit-step responses of the attitude-control system in Fig. 7-24; $L_a = 0$.

Table 7-3 Comparison of the Performance of the Second-Order Position-Control System with the Gain K Varies

Gain K	ζ	ω_n (rad/sec)	Maximum Overshoot (%)	t_d (sec)	t_r (sec)	t_s (sec)	t_{max} (sec)
7.25	1.000	180.62	0	0.00929	0.0186	0.0259	—
14.50	0.707	255.44	4.3	0.00560	0.0084	0.0114	0.01735
181.20	0.200	903.00	52.2	0.00125	0.00136	0.0150	0.00369

It should be pointed out that in practice, it would be time consuming, even with the aid of a computer, to compute the time response for each change of a system parameter for either analysis or design purposes. Indeed, one of the main objectives of studying control systems theory using either the conventional or the modern approach is to establish methods so that the total reliance on computer simulation can be reduced. The motivation behind this discussion is to show that the performance of some control systems can be predicted by investigating the roots of the characteristic equation of the system. For the characteristic equation of Eq. (7-119), the roots are

$$s_1 = -180.6 + \sqrt{32{,}616 - 4500K} \tag{7-124}$$
$$s_2 = -180.6 - \sqrt{32{,}616 - 4500K} \tag{7-125}$$

For $K = 7.248$, 14.5, and 181.2, the roots of the characteristic equation are tabulated as follows:

$K = 7.248$:	$s_1 = s_2 = -180.6$	
$K = 14.5$:	$s_1 = -180.6 + j180.6$	$s_2 = -180.6 - j180.6$
$K = 181.2$:	$s_1 = -180.6 + j884.7$	$s_2 = -180.6 + j884.7$

These roots are marked as shown in Fig. 7-26. The trajectories of the two characteristic equation roots when K varies continuously from $-\infty$ to ∞ are also shown in Fig. 7-26. These root trajectories are called the **root loci** (see Chapter 8) of Eq. (7-119) and are used extensively for the analysis and design of linear control systems.

From Eqs. (7-124) and (7-125) we see that the two roots are real and negative for values of K between 0 and 7.248. This means that the system is overdamped, and the step response will have no overshoot for this range of K. For values of K greater than 7.248, the natural undamped frequency will increase with $\sqrt{K}$. When K is negative, one of the roots is positive, which corresponds to a time response that increases monotonically with time, and the system is unstable. The dynamic characteristics of the transient step response as determined from the root loci of Fig. 7-26 are summarized as follows:

Amplifier Gain	Characteristic Equation Roots	System Dynamics
$0 < K < 7.248$	Two negative distinct real roots	Overdamped ($\zeta > 1$)
$K = 7.248$	Two negative equal real roots	Critically damped ($\zeta = 1$)
$7.248 < K < \infty$	Two complex-conjugate roots with negative real parts	Underdamped ($\zeta < 1$)
$-\infty < K < 0$	Two distinct real roots, one positive and one negative	Unstable system ($\zeta < 0$)

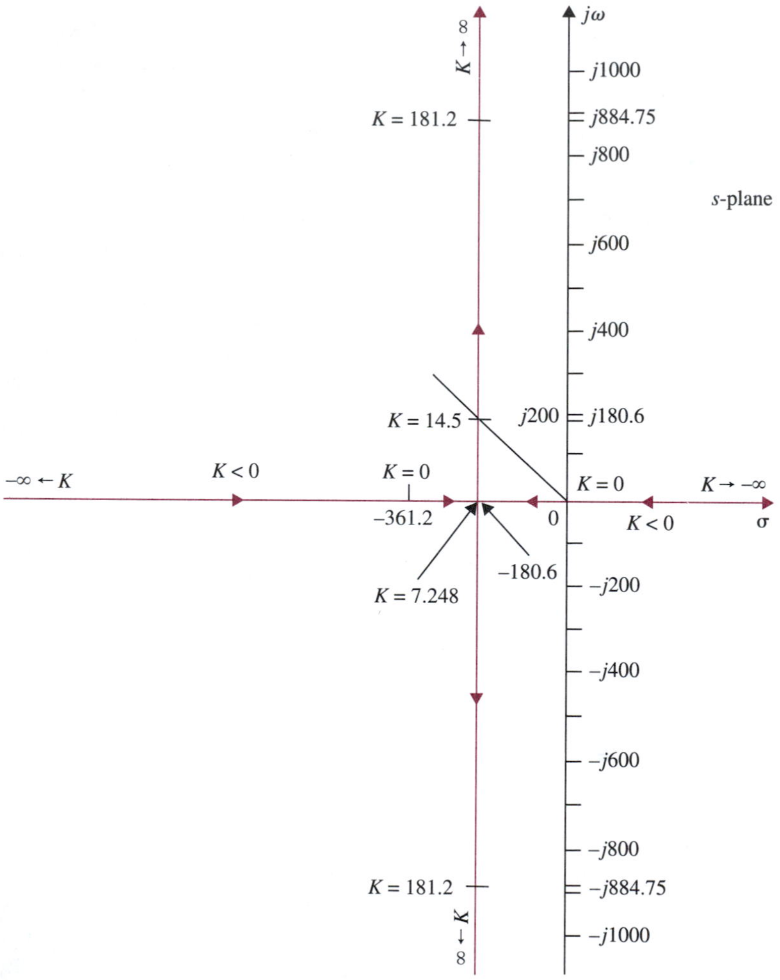

Figure 7-26 Root loci of the characteristic equation in Eq. (7-119) as K varies.

7-6-2 Steady-State Response

Since the forward-path transfer function in Eq. (7-116) has a simple pole at $s = 0$, the system is of type 1. This means that the steady-state error of the system is zero for all positive values of K when the input is a step function. Substituting Eq. (7-116) into Eq. (7-22), the step-error constant is obtained as

$$K_p = \lim_{s \to 0} \frac{4500K}{s(s + 361.2)} = \infty \tag{7-126}$$

Thus the steady-state error of the system due to a step input, as given by Eq. (7-23), is zero. The unit-step responses in Fig. 7-25 verify this result. The zero-steady-state condition is achieved because only viscous friction is considered in the simplified system model. In the practical case, Coulomb friction is almost always present, so that the steady-state positioning accuracy of the system can never be perfect.

7-6-3 Time Response to a Unit-Ramp Input

The control of position may often be effected by control of the profile of the output rather than just applying a step input. In other words, the system may be designed to follow a reference profile that represents the desired trajectory. It may be necessary to investigate the ability of the position-control system to follow a ramp-function input.

For a unit-ramp input, $\theta_r(t) = tu_s(t)$. The output response of the system in Fig. 7-24 is

$$\theta_y(t) = \mathscr{L}^{-1}\left[\frac{4500K}{s^2(s^2 + 361.2s + 4500K)}\right] \tag{7-127}$$

which can be solved by using the Laplace transform table in Appendix B. The result is

$$\theta_y(t) = \left[t - \frac{2\zeta}{\omega_n} + \frac{e^{-\zeta\omega_n t}}{\omega_n\sqrt{1-\zeta^2}}\sin(\omega_n\sqrt{1-\zeta^2}t + \theta)\right]u_s(t) \tag{7-128}$$

where

$$\theta = \cos^{-1}(2\zeta^2 - 1) \qquad (\zeta < 1) \tag{7-129}$$

The values of ζ and ω_n are given in Eqs. (7-118) and (7-117), respectively. The output $\theta_y(t)$ can be obtained applying the function **pfe** of the **CSAD** toolbox to Eq. (7-127) if all the system parameters are known. The ramp responses of the system for the three values of K are obtained with **pfe** of **CSAD** and are presented below.

$$K = 7.248: \quad \theta_y(t) = (t - 0.01107 - 0.8278e^{-181.2t} + 0.8389e^{-180t})u_s(t) \qquad (7\text{-}130)$$

$$K = 14.5: \quad \theta_y(t) = (t - 0.005536 + 0.005536e^{-180.6t}\cos 180.6t - 5.467 \times 10^{-7}e^{-180.6t}\sin 180.6t)u_s(t) \qquad (7\text{-}131)$$

$$K = 181.2: \quad \theta_y(t) = (t - 0.000443 + 0.000443e^{-180.6t}\cos 884.7t - 0.00104e^{-180.6t}\sin 884.7t)u_s(t) \qquad (7\text{-}132)$$

These ramp responses are plotted as shown in Fig. 7-27. Notice that the steady-state error of the ramp response is not zero. The last term in Eq. (7-128) is the transient response. The steady-state portion of the unit-ramp response is

$$\lim_{t\to\infty} \theta_y(t) = \lim_{t\to\infty} \left[\left(t - \frac{2\zeta}{\omega_n} \right) u_s(t) \right] \qquad (7\text{-}133)$$

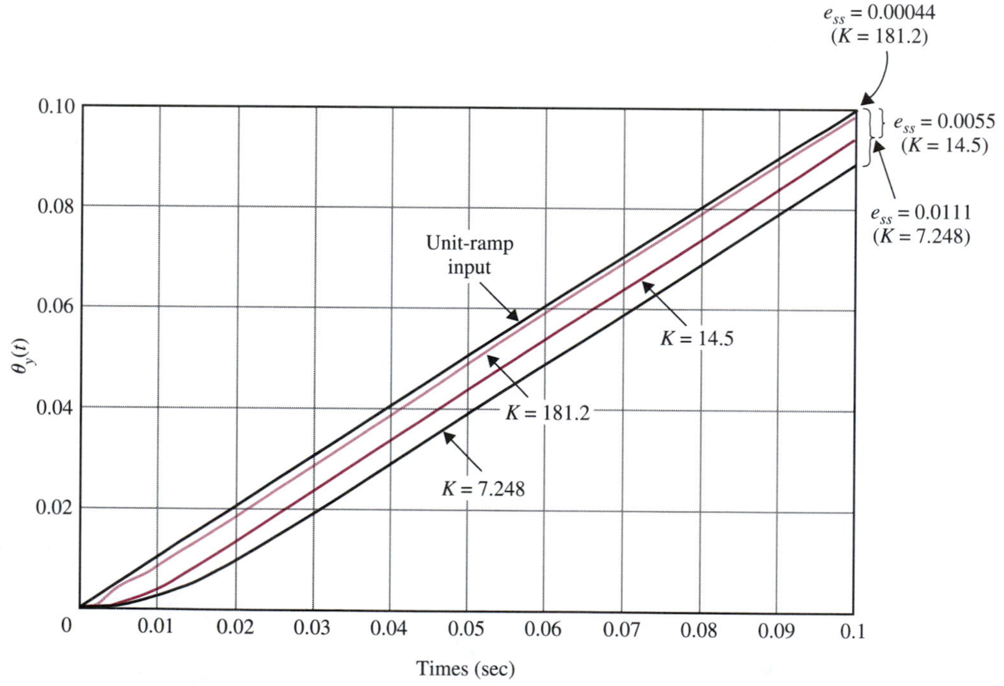

Figure 7-27 Unit-ramp responses of the attitude-control system in Fig. 7-24; $L_a = 0$.

Thus the steady-state error of the system due to a unit-ramp input is

$$e_{ss} = \frac{2\zeta}{\omega_n} = \frac{0.0803}{K} \tag{7-134}$$

which is a constant.

A more direct method of determining the steady-state error due to a ramp input is to use the ramp-error constant K_v. From Eq. (7-27),

$$K_v = \lim_{s \to 0} sG(s) = \lim_{s \to 0} \frac{4500K}{s + 361.2} = 12.46K \tag{7-135}$$

Thus the steady-state error is

$$e_{ss} = \frac{1}{K_v} = \frac{0.0803}{K} \tag{7-136}$$

which agrees with the result in Eq. (7-134).

The result in Eq. (7-136) shows that the steady-state error is inversely proportional to K. For $K = 14.5$, which corresponds to a damping ratio of 0.707, the steady-state error is 0.0055 rad, or more appropriately, 0.55 percent of the ramp-input magnitude. Apparently, if we attempt to improve the steady-state accuracy of the system due to ramp inputs by increasing the value of K, the transient step response will become more oscillatory and with a higher overshoot. This phenomenon is rather typical in all control systems. For higher-order systems, if the loop gain of the system is too high, the system can become unstable. This lays the foundation for using controller in the system loop so that the transient and the steady-state error can be improved simultaneously.

7-6-4 Time Response of a Third-Order System

▲ All second-order systems with positive coefficients in the characteristic equations are stable.

In the preceding section we have shown that the prototype second-order system, obtained by neglecting the armature inductance, is always stable for all positive values of K. It is not difficult to prove that, in general, all second-order systems with positive coefficients in the characteristic equations are stable.

Let us investigate the performance of the position-control system with the armature inductance $L_a = 0.003$ H. The forward-path transfer function of Eq. (7-112) becomes

$$G(s) = \frac{1.5 \times 10^7 K}{s(s^2 + 3408.3s + 1{,}204{,}000)} = \frac{1.5 \times 10^7 K}{s(s + 400.26)(s + 3008)} \tag{7-137}$$

The closed-loop transfer function is

$$\frac{\Theta_y(s)}{\Theta_r(s)} = \frac{1.5 \times 10^7 K}{s^3 + 3408.3s^2 + 1{,}204{,}000s + 1.5 \times 10^7 K} \tag{7-138}$$

The system is now of the third order, and the characteristic equation is

$$s^3 + 3408.3s^2 + 1{,}204{,}000s + 1.5 \times 10^7 K = 0 \tag{7-139}$$

Transient Response

The roots of the characteristic equation are tabulated for the three values of K used earlier for the second-order system:

▲ $K = 7.248$:	$s_1 = -156.21$	$s_2 = -230.33$	$s_3 = -3021.8$
▲ $K = 14.5$:	$s_1 = -186.53 + j192$	$s_2 = -186.53 - j192$	$s_3 = -3035.2$
▲ $K = 181.2$:	$s_1 = -57.49 + j906.6$	$s_2 = -57.49 - j906.6$	$s_3 = -3293.3$

Comparing these results with those of the approximating second-order system, we see that when $K = 7.428$, the second-order system is critically damped, whereas the third-order system has three distinct real roots, and the system is slightly overdamped. The root at -3021.8 corresponds to a time constant of 0.33 millisecond, which is over 13 times faster than the next-fastest time constant due to the pole at -230.33. Thus the transient response due to the pole at -3021.8 decays rapidly, and the pole can be neglected from the transient standpoint. The output transient response is dominated by the two roots at -156.21 and -230.33. This analysis is verified by writing the transformed output response as

$$\Theta_y(s) = \frac{10.87 \times 10^7}{s(s + 156.21)(s + 230.33)(s + 3021.8)} \tag{7-140}$$

Taking the inverse Laplace transform of Eq. (7-140), we get

$$\theta_y(t) = (1 - 3.28e^{-156.21t} + 2.28e^{-230.33t} - 0.0045e^{-3021.8t})u_s(t) \tag{7-141}$$

▲ The dominant roots of a system are the roots that are closest to the $j\omega$-axis in the left-half s-plane.

The last term in Eq. (7-141), which is due to the root at -3021.8, decays to zero very rapidly. Furthermore, the magnitude of the term at $t = 0$ is very small compared to the other two transient terms. This simply demonstrates that, in general, the contribution of roots that lie relatively far to the left in the s-plane to the transient response will be small. The roots that are closer to the imaginary axis will dominate the transient response, and these are defined as the **dominant roots** of the characteristic equation, or of the system.

When $K = 14.5$, the second-order system has a damping ratio of 0.707, since the real and imaginary parts of the two characteristic equation roots are identical. For the third-order system, it should be remembered that damping ratio is strictly not defined. However, since the effect on transient of the root at -3021.8 is negligible, the two roots that dominate the transient response correspond to a damping ratio of 0.697. Thus for $K = 14.5$, the second-order approximation by setting L_a to zero is not a bad one. It should be noted, however, that *just because the second-order approximation is justified for $K = 14.5$ does not mean that the approximation is valid for all values of K.*

When $K = 181.2$, the two complex-conjugate roots of the third-order system again dominate the transient response, and the equivalent damping ratio due to the two roots is only 0.0633, which is much smaller than the value of 0.2 for the second-order system. Thus we see that the justification and accuracy of the second-order approximation diminish as the value of K is increased. Figure 7-28 illustrates the root loci of the third-order characteristic equation of Eq. (7-139) as K varies. When $K = 181.2$, the real root at -3293.3 still contributes little to the transient response, but the two complex-conjugate roots at $-57.49 \pm j906.6$ are much closer to the $j\omega$-axis than those of the second-order system for the same K, which are at $-180.6 \pm j884.75$. This explains why the third-order system is a great deal less stable than the second-order system when $K = 181.2$.

By using the Routh–Hurwitz criterion, the marginal value of K for stability is found to be 273.57. With this critical value of K, the closed-loop transfer function becomes

$$\frac{\Theta_y(s)}{\Theta_r(s)} = \frac{1.0872 \times 10^8}{(s + 3408.3)(s^2 + 1.204 \times 10^6)} \tag{7-142}$$

The roots of the characteristic equation are at $s = -3408.3$, $-j1097.3$, and $j1097.3$. These points are shown on the root loci in Fig. 7-28.

The unit-step response of the system when $K = 273.57$ is

$$\theta_y(t) = [1 - 0.094e^{-3408.3t} - 0.952 \sin(1097.3t + 72.16^0)]u_s(t) \tag{7-143}$$

The steady-state response is an undamped sinusoid with a frequency of 1097.3 rad/sec, and the system is said to be marginally stable. When K is greater than 273.57, the two complex-conjugate roots will have positive real parts, the sinusoidal component of the time response will increase with time, and the system is unstable. Thus we see that *the third-order system is capable of being unstable, whereas the second-order system obtained with $L_a = 0$ is stable for all finite positive values of K.*

Figure 7-29 shows the unit-step responses of the third-order system for the three values of K used. The responses for $K = 7.248$ and $K = 14.5$ are very close to those of the second-order system with the same values of K, which are shown in Fig. 7-25. However, the two responses for $K = 181.2$ are quite different.

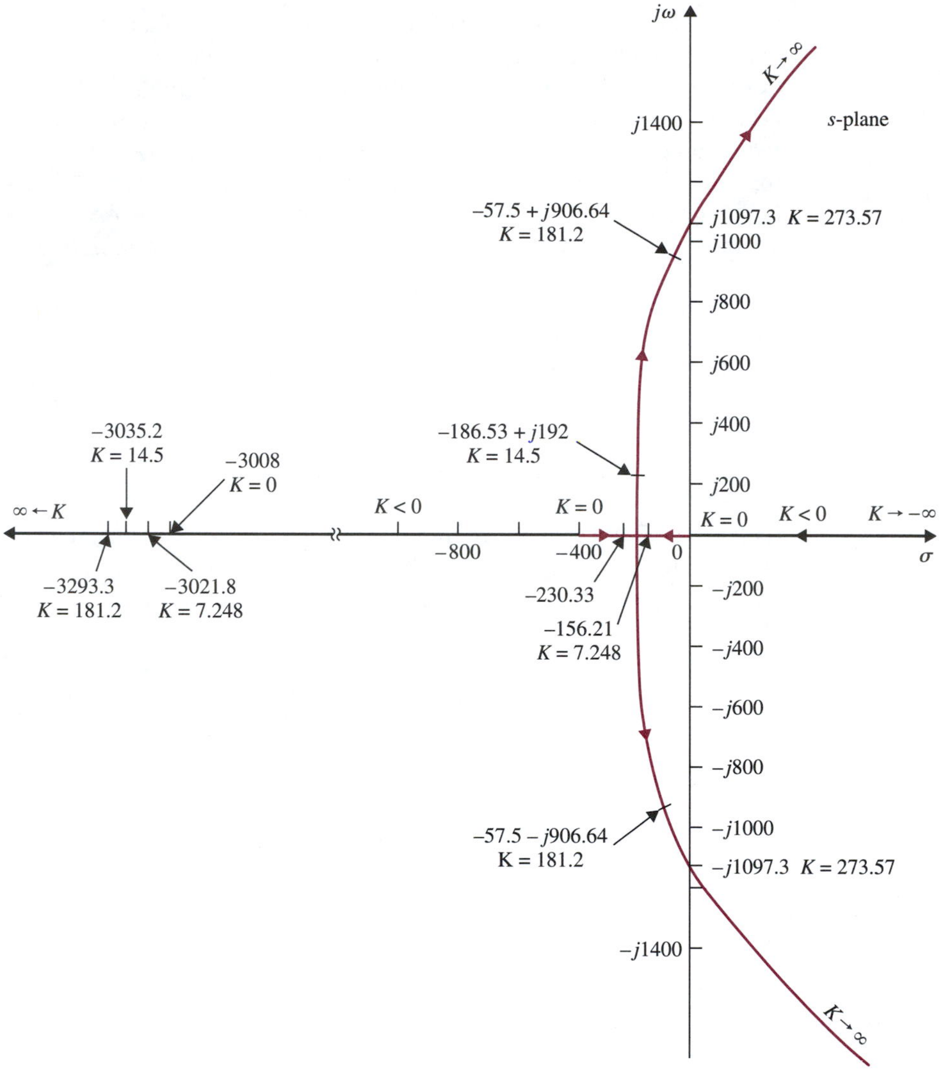

Figure 7-28 Root loci of the third-order attitude-control system.

Steady-State Response

From Eq. (7-137) we see that when the inductance is restored, the third-order system is still of type 1. The value of K_v is still the same as that given in Eq. (7-135). Thus the inductance of the motor does not affect the steady-state performance of the system, provided that the system is stable. This is expected, since L_a affects only the rate of

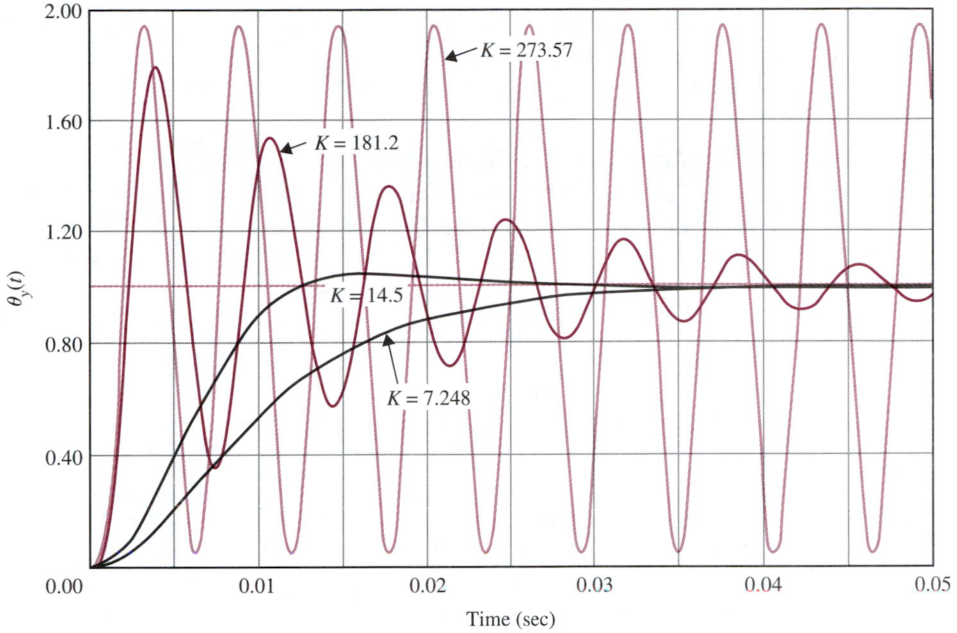

Figure 7-29 Unit-step responses of the third-order attitude-control system.

change and not the final value of the motor current. *A good engineer should always try to reason and interpret the analytical results with the physical system.*

7-7 Effects of Adding Poles and Zeros to Transfer Functions

The position-control system discussed in Section 7-6 reveals important properties of the time responses of typical second- and third-order closed-loop systems. Specifically, the effects on the transient response relative to the location of the roots of the characteristic equation are demonstrated. However, in practice, successful design of a control system cannot depend only on choosing values of the system parameters so that the characteristic equation roots are properly placed. We shall show that *although the roots of the characteristic equation, which are the poles of the closed-loop transfer function, affect the transient response of linear time-invariant control systems, particularly the stability, the zeros of the transfer function, if there are any, are also important.* Thus the addition of poles and zeros and/or cancellation of undesirable poles and zeros of the transfer function often are necessary in achieving satisfactory time-domain performance of control systems. In this section we show that the addition of poles and zeros to loop and closed-loop transfer functions has varying effects on the transient reponse of the transient response of the closed-loop system.

7-7-1 Addition of a Pole to the Forward-Path Transfer Function: Unity-Feedback Systems

For the position-control system described in Section 7-6, when the motor inductance is neglected, the system is of second order, and the forward-path transfer function is of the prototype given in Eq. (7-116). When the motor inductance is restored, the system is of third order, and the forward-path transfer function is given in Eq. (7-137). Comparing the two transfer functions of Eqs. (7-116) and (7-137), we see that the effect of the motor inductance is equivalent to adding a pole at $s = -3008$ to the forward-path transfer function of Eq. (7-116), while shifting the pole at -361.2 to -400.26, and the proportional constant is also increased. The apparent effect of adding a pole to the forward-path transfer function is that the third-order system can now become unstable if the value of the amplifier gain K exceeds 273.57. As shown by the root-locus diagrams of Figs. 7-26 and 7-28, the new pole of $G(s)$ at $s = -3008$ essentially ''pushes'' and ''bends'' the complex-conjugate portion of the root loci of the second-order system toward the right-half s-plane. Actually, owing to the specific value of the inductance chosen, the additional pole of the third-order system is far to the left of the pole at -400.26, so that its effect is small except when the value of K is relatively large.

To study the general effect of the addition of a pole, and its relative location, to a forward-path transfer function of a unity-feedback system, consider the transfer function

$$G(s) = \frac{\omega_n^2}{s(s + 2\zeta\omega_n)(1 + T_p s)} \tag{7-144}$$

The pole at $s = -1/T_p$ is considered to be added to the prototype second-order transfer function. The closed-loop transfer function is written

$$M(s) = \frac{Y(s)}{R(s)} = \frac{G(s)}{1 + G(s)} = \frac{\omega_n^2}{T_p s^3 + (1 + 2\zeta\omega_n T_p)s^2 + 2\zeta\omega_n s + \omega_n^2} \tag{7-145}$$

Figure 7-30 illustrates the unit-step responses of the closed-loop system when $\omega_n = 1$, $\zeta = 1$, and $T_p = 0$, 1, 2, and 5. These responses again show that

the addition of a pole to the foreward-path transfer function generally has the effect of increasing the maximum overshoot of the closed-loop system.

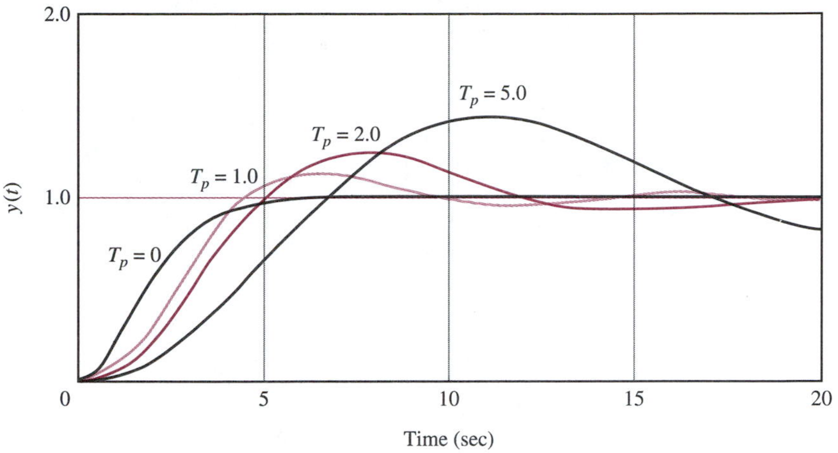

Figure 7-30 Unit-step responses of the system with the closed-loop transfer function in Eq. (7-145): $\zeta = 1$, $\omega_n = 1$, and $T_p = 0, 1, 2,$ and 5.

As the value of T_p increases, the pole at $-1/T_p$ moves closer to the origin in the s-plane, and the maximum overshoot increases. These responses also show that the *added pole increases the rise time of the step response.* This is not surprising, since the *additional pole has the effect of reducing the bandwidth* (*see Chapter* 9) *of the system,* thus cutting out the high-frequency components of the signal transmitted through the system.

The same conclusion can be drawn from the unit-step responses of Fig. 7-31, which are obtained with $\omega_n = 1$, $\zeta = 0.25$, and $T_p = 0, 0.2, 0.667,$ and 1.0. In this case, when T_p is greater than 0.667, the amplitude of the unit-step response increases with time, and the system is unstable.

7-7-2 Addition of a Pole to the Closed-Loop Transfer Function

Since the poles of the closed-loop transfer function are roots of the characteristic equation, they control the transient response of the system directly. Consider the closed-loop transfer function

$$M(s) = \frac{Y(s)}{R(s)} = \frac{\omega_n^2}{(s^2 + 2\zeta\omega_n s + \omega_n^2)(1 + T_p s)} \tag{7-146}$$

where the term $(1 + T_p s)$ is added to a prototype second-order transfer function. Figure 7-32 illustrates the unit-step response of the system with $\omega_n = 1$, $\zeta = 0.5$, and $T_p =$

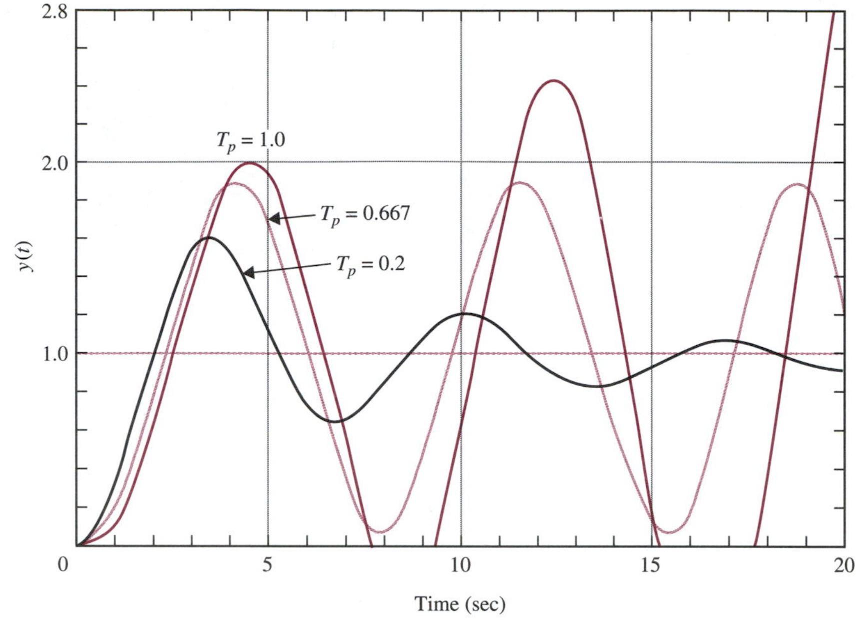

Figure 7-31 Unit-step responses of the system with the closed-loop transfer function in Eq. (7-145): $\zeta = 0.25$, $\omega_n = 1$, and $T_p = 0$, 0.2, 0.667, and 1.0.

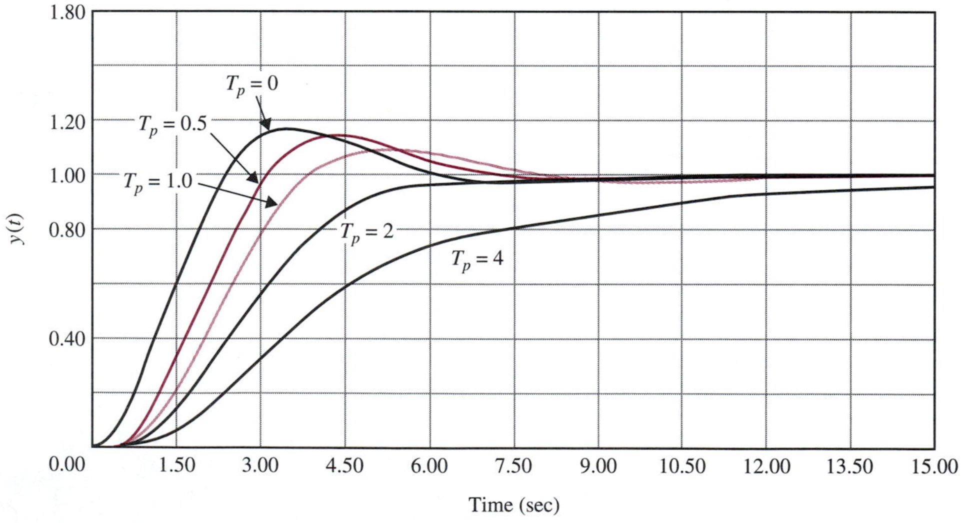

Figure 7-32 Unit-step responses of the system with the closed-loop transfer function in Eq. (7-146); $\zeta = 0.5$, $\omega_n = 1$, and $T_p = 0$, 0.5, 1.0, 2.0, and 4.0.

0, 0.5, 1, 2, and 4. As the pole at $s = -1/T_p$ is moved toward the origin in the s-plane, the rise time increases and the maximum overshoot decreases. Thus

as far as the overshoot is concerned, adding a pole to the closed-loop transfer function has just the opposite effect to that of adding a pole to the forward-path transfer function.

7-7-3 Addition of a Zero to the Closed-Loop Transfer Function

Figure 7-33 shows the unit-step responses of the closed-loop system with the transfer function

$$M(s) = \frac{Y(s)}{R(s)} = \frac{\omega_n^2(1 + T_z s)}{(s^2 + 2\zeta\omega_n s + \omega_n^2)} \tag{7-147}$$

where $\omega_n = 1$, $\zeta = 0.5$, and $T_z = 0$, 1, 3, 6, and 10. In this case we see that

adding a zero to the closed-loop transfer function decreases the rise time and increases the maximum overshoot of the step response.

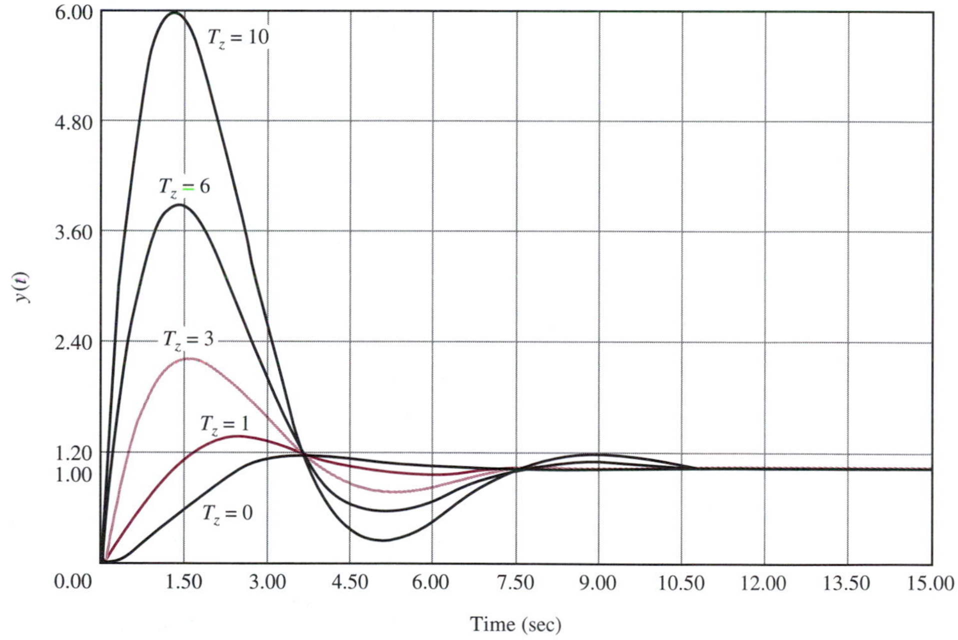

Figure 7-33 Unit-step responses of the system with the closed-loop transfer function in Eq. (7-147): $T_z = 0$, 1, 3, 6, and 10.

We can analyze the general case by writing Eq. (7-147) as

$$M(s) = \frac{Y(s)}{R(s)} = \frac{\omega_n^2}{s^2 + 2\zeta\omega_n s + \omega_n^2} + \frac{T_z \omega_n^2 s}{s^2 + 2\zeta\omega_n s + \omega_n^2} \tag{7-148}$$

For a unit-step input, let the output response that corresponds to the first term of the right-hand side of Eq. (7-148) be $y_1(t)$. Then the total unit-step response is

$$y(t) = y_1(t) + T_z \frac{dy_1(t)}{dt} \tag{7-149}$$

Figure 7-34 shows why the addition of the zero at $s = -1/T_z$ reduces the rise time and increases the maximum overshoot, according to Eq. (7-149). In fact, as T_z approaches infinity, the maximum overshoot also approaches infinity, yet the system is still stable as long as the overshoot is finite and ζ is positive.

7-7-4 Addition of a Zero to the Forward-Path Transfer Function: Unity-Feedback Systems

Let us consider that a zero at $-1/T_z$ is added to the forward-path transfer function of a third-order system, so that

$$G(s) = \frac{6(1 + T_z s)}{s(s+1)(s+2)} \tag{7-150}$$

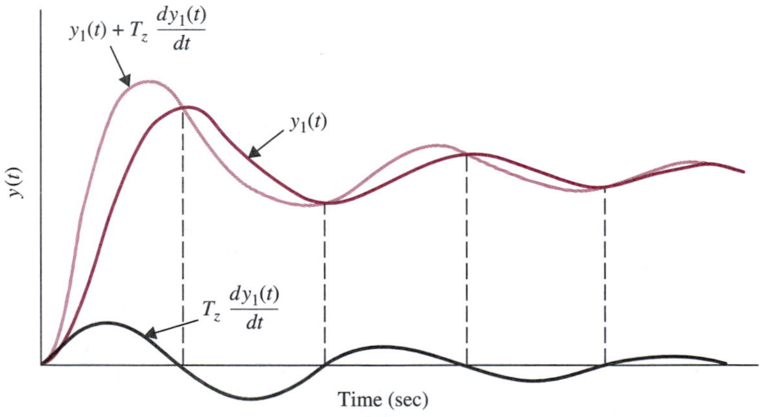

Figure 7-34 Unit-step responses showing the effect of adding a zero to the closed-loop transfer function.

The closed-loop transfer function is

$$M(s) = \frac{Y(s)}{R(s)} = \frac{6(1 + T_z s)}{s^3 + 3s^2 + (2 + 6T_z)s + 6} \tag{7-151}$$

The difference between this case and that of adding a zero to the closed-loop transfer function is that in the present case, not only does the term $(1 + T_z s)$ appear in the numerator of $M(s)$ but the denominator of $M(s)$ contains T_z. The term $(1 + T_z s)$ in the numerator of $M(s)$ increases the maximum overshoot, but T_z appears in the coefficient of the s term in the denominator, which has the effect of improving damping or reducing the maximum overshoot. Figure 7-35 illustrates the unit-step responses when $T_z = 0$, 0.2, 0.5, 2, 5, and 10. Notice that when $T_z = 0$, the closed-loop system is on the verge of becoming unstable. When $T_z = 0.2$ and 0.5, the maximum overshoots are reduced, owing mainly to the improved damping. As T_z increases beyond 2, although the damping is still further improved, the $(1 + T_z s)$ term in the numerator becomes more dominant, so that the maximum overshoot actually becomes greater as T_z is increased further.

An important finding from these discussions is that

although the characteristic equation roots are generally used to study the relative damping and relative stability of linear control systems, the zeros of the transfer function should not be overlooked in their effects on the transient performance of the system.

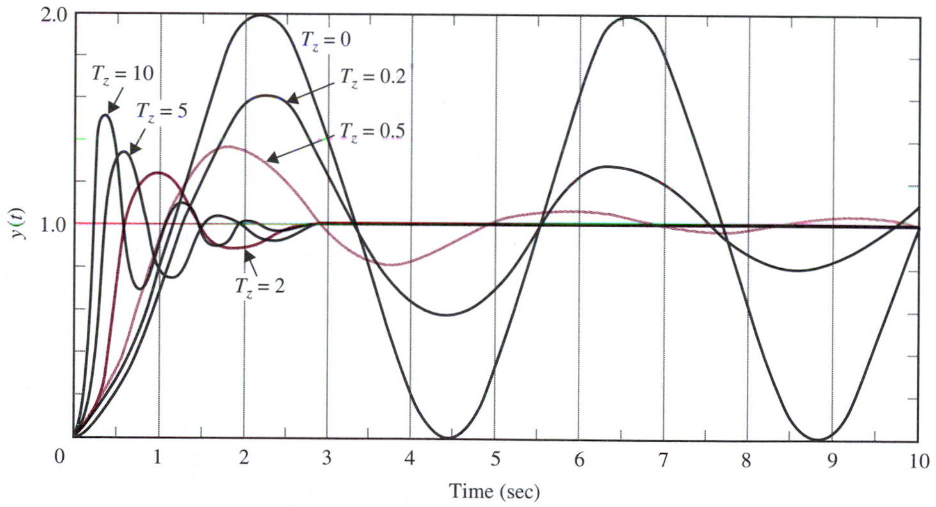

Figure 7-35 Unit-step responses of the system with the closed-loop transfer function in Eq. (7-151): $T_z = 0$, 0.2, 0.5, 2, 5, and 10.

7-8 Dominant Poles of Transfer Functions

From the discussions given in the preceding sections, it becomes apparent that the location of the poles of a transfer function in the s-plane affects greatly the transient response of the system. For analysis and design purposes, it is important to sort out the poles that have a dominant effect on the transient response and call these the **dominant poles.**

Since most control systems found in practice are of orders higher than two, it would be useful to establish guidelines on the approximation of high-order systems by lower-order ones insofar as the transient response is concerned. In design, we can use the dominant poles to control the dynamic performance of the system, whereas the **insignificant poles** are used for the purpose of ensuring that the controller transfer function can be realized by physical components.

For all practical purposes, we can qualitatively sectionalize the s-plane into regions in which the dominant and insignificant poles can lie, as shown in Fig. 7-36. We intentionally do not assign specific values to the coordinates, since these are all relative to a given system.

The poles that are close to the imaginary axis in the left-half s-plane give rise to transient responses that will decay relatively slowly, whereas the poles that are far away from the axis (relative to the dominant poles) correspond to fast-decaying time responses. The distance D between the dominant region and the least significant region shown in Fig. 7-36 will be subject to discussion. The question is: How large a pole is considered to be really large?'' It has been recognized in practice and in the literature that if the magnitude of the real part of a pole is at least 5 to 10 times that of a dominant pole or a pair of complex dominant poles, the pole may be regarded as insignificant insofar as the transient response is concerned.

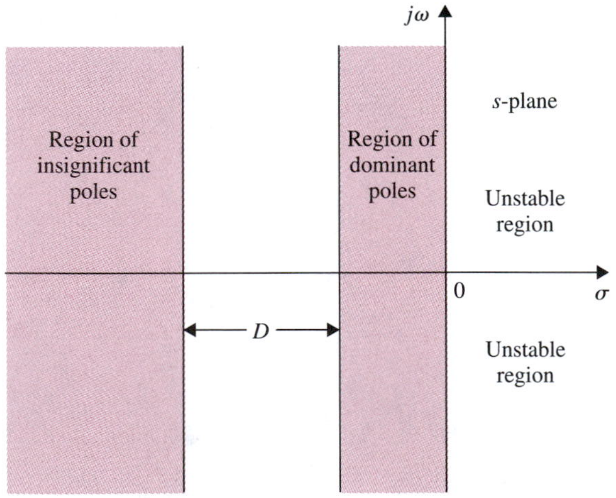

Figure 7-36 Regions of dominant and insignificant poles in the s-plane.

We must point out that the regions shown in Fig. 7-36 are selected merely for the definitions of dominant and insignificant poles. For design purposes, such as in the pole-placement design, the dominant poles and the insignificant poles, as selected by the designer, should most likely be located in the tinted regions in Fig. 7-37. Again, we do not show any absolute coordinates, except that the desired region of the dominant poles is centered around the line that corresponds to $\zeta = 0.707$. It should also be noted that in design, we cannot place the insignificant poles arbitrarily far to the left in the s-plane, or these may require unrealistic system parameter values when the pencil-and-paper design is to be implemented by physical components.

7-8-1 Relative Damping Ratio

When a system is higher than the second order, strictly, we can no longer use the damping ratio ζ and the natural undamped frequency ω_n, which are defined for the prototype second-order systems. However, if the system dynamics can be accurated represented by a pair of complex-conjugate dominant poles, we can still use ζ and ω_n to indicate the dynamics of the transient response, and the damping ratio in this case is referred to as the **relative damping ratio** of the system. For example, consider the closed-loop transfer function

$$M(s) = \frac{Y(s)}{R(s)} = \frac{20}{(s+10)(s^2+2s+2)} \tag{7-152}$$

The pole at $s = -10$ is 10 times the real part of the complex-conjugate poles, which are at $-1 \pm j1$. We can refer to the relative damping ratio of the system as 0.707.

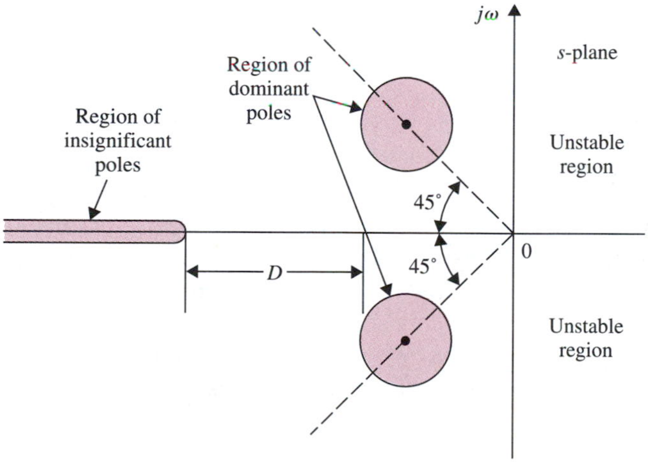

Figure 7-37 Regions of dominant and insignificant poles in the s-plane.

7-8-2 Proper Way of Neglecting the Insignificant Poles with Consideration of the Steady-State Response

Thus far we have provided guidelines of neglecting insignificant poles of a transfer function from the standpoint of the transient response. However, in going through with the mechanics, the steady-state performance must also be considered. Let us consider the transfer function in Eq. (7-152); the pole at -10 can be neglected from the transient standpoint. To do this, we should first express Eq. (7-152) as

$$M(s) = \frac{20}{10(s/10 + 1)(s^2 + 2s + 2)} \tag{7-153}$$

Then we reason that $|s/10| << 1$ when the absolute value of s is much smaller than 10, owing to the dominant nature of the complex poles. The term $s/10$ can be neglected when compared with 1. Then Eq. (7-153) is approximated by

$$M(s) \cong \frac{20}{10(s^2 + 2s + 2)} \tag{7-154}$$

This way, the steady-state performance of the third-order system will not be affected by the approximation. In other words, the third-order system described by Eq. (7-152) and the second-order system approximated by Eq. (7-154) all have a final value of unity when a unit-step input is applied. On the other hand, if we simply "throw away" the term $(s + 10)$ in Eq. (7-152), the approximating second-order system would have a steady-state value of 5 when a unit-step input is applied.

7-9 Approximation of High-Order Systems by Low-Order Systems: The Formal Approach

In Section 7-8 we have shown that a high-order control system often contains less important poles that have little effect on the system response. Thus, given a high-order system, it is desirable to find a low-order approximating system, if possible, so that the analysis and design effort is reduced. This means that given a high-order transfer function $M_H(s)$, how can we find a low-order transfer function $M_L(s)$ as an approximation, in the sense that the responses of the two systems are similar according to some prescribed criterion?

In Section 7-8 we set up practical but nonrigorous guidelines for neglecting the poles that are far to the left in the s-plane relative to the dominant poles. However, in general, the transfer function may not have the so-called dominant poles, or the latter may not be obviously defined, so a more scientific method may be necessary to arrive at a low-order equivalent.

In this section we introduce a method [2] of approximating a high-order system by a low-order one in the sense that the frequency responses of the two systems are similar.

Let the high-order system transfer function be represented by

$$M_H(s) = K\frac{1 + b_1 s + b_2 s^2 + \cdots + b_m s^m}{1 + a_1 s + a_2 s^2 + \cdots + a_n s^n} \tag{7-155}$$

where $n \geq m$. Let the transfer function of the approximating low-order system be

$$M_L(s) = K\frac{1 + c_1 s + c_2 s^2 + \cdots + c_q s^q}{1 + d_1 + d_2 s^2 + \cdots + d_p s^p} \tag{7-156}$$

where $n \geq p \geq q$. Notice that the zero-frequency ($s = 0$) gain K of the two transfer functions is the same. This will ensure that the steady-state behavior of the high-order system is preserved in the low-order system. Furthermore, we assume that the poles of $M_H(s)$ and $M_L(s)$ are all in the left-half s-plane, since we are not interested in unstable systems. The transfer functions $M_L(s)$ and $M_L(s)$ generally refer to the closed-loop transfer function, but if necessary, they can be treated as the loop transfer functions.

7-9-1 Approximation Criterion

The criterion of finding the low-order $M_L(s)$, given $M_H(s)$, is that the following relation should be satisfied as closely as possible:

$$\frac{|M_H(j\omega)|^2}{|M_L(j\omega)|^2} = 1 \qquad \text{for } 0 \leq \omega \leq \infty \tag{7-157}$$

The last condition implies that the amplitude characteristics of the two systems in the frequency domain ($s = j\omega$) are similar. It is hoped that this will lead to similar time responses for the two systems.

The approximation procedure involves the following two steps:

1. Choose the appropriate orders of the numerator polynomial, q, and the denominator polynomial, p, of $M_L(s)$.
2. Determine the coefficients c_i, $i = 1, 2, \ldots, q$, and d_j, $j = 1, 2, \ldots, p$, so that the condition in Eq. (7-157) is approached.

By using Eqs. (7-155) and (7-156), the ratio of $M_H(s)$ to $M_L(s)$ is

$$\frac{M_H(s)}{M_L(s)} = \frac{(1 + b_1 s + b_2 s^2 + \cdots + b_m s^m)(1 + d_1 s + d_2 s^2 + \cdots + d_p s^p)}{(1 + a_1 s + a_2 s^2 + \cdots + a_n s^n)(1 + c_1 s + c_2 s^2 + \cdots + c_q s^q)}$$
$$= \frac{1 + m_1 s + m_2 s^2 + \cdots + m_u s^u}{1 + l_1 s + l_2 + \cdots + l_v s^v} \tag{7-158}$$

where $u = m + p$ and $v = n + q$.

Equation (7-157) is written as

$$\frac{|M_H(j\omega)|^2}{|M_L(j\omega)|^2} = \left. \frac{M_H(s)M_H(-s)}{M_L(s)M_L(s)} \right|_{s=j\omega} \tag{7-159}$$

where $M_H(s)M_H(-s)$ and $M_L(s)M_L(-s)$ are even polynomials of s (i.e., they contain only even powers of s). Thus Eq. (7-159) can be written as

$$\frac{|M_H(j\omega)|^2}{|M_L(j\omega)|^2} = \left. \frac{1 + e_2 s^2 + e_4 s^4 + \cdots + e_{2u} s^{2u}}{1 + f_2 s^2 + f_4 s^4 + \cdots + f_{2v} s^{2v}} \right|_{s=j\omega} \tag{7-160}$$

Dividing the numerator by the denominator once on the right-hand side of Eq. (7-160), we have

$$\frac{|M_H(j\omega)|^2}{|M_L(j\omega)|^2} = 1 + \left. \frac{(e_2 - f_2)s^2 + (e_4 - f_4)s^4 + \cdots}{1 + f_2 s^2 + f_4 s^4 + \cdots + f_{2v} s^{2v}} \right|_{s=j\omega} \tag{7-161}$$

If $u = v$, the last term in the numerator of Eq. (7-161) will be $(e_{2u} - f_{2u})s^{2u}$. However, if $u < v$, as in most practical cases, then beyond the term $(e_{2u} - f_{2u})s^{2u}$, in addition, there will be

$$-f_{2(u+1)} s^{2(u+1)} - \cdots - f_{2v} s^{2v}$$

in the numerator of Eq. (7-161).

We see that to satisfy the condition of Eq. (7-157), one possible set of approximating solutions is obtained from Eq. (7-161):

$$\begin{aligned} e_2 &= f_2 \\ e_4 &= f_4 \\ e_6 &= f_6 \\ &\vdots \\ e_{2u} &= f_{2u} \end{aligned} \tag{7-162}$$

if $u = v$. If $u < v$, the error generated by the low-order model is

$$|\varepsilon| = \frac{|M_H(j\omega)|^2}{|M_L(j\omega)|^2} - 1 = \left. \frac{-f_{2(u+1)}s^{2(u+1)} - \cdots - f_{2v}s^{2v}}{1 + f_2 s^2 + f_4 s^4 + \cdots + f_{2v}s^{2v}} \right|_{s=j\omega} \tag{7-163}$$

The conditions in Eq. (7-162) are used to solve for the unknown coefficients in $M_L(s)$ once $M_H(s)$ is given. This is done by writing

$$\begin{aligned} \frac{M_H(s)M_H(-s)}{M_L(s)M_L(-s)} &= \frac{(1 + m_1 s + m_2 s^2 + \cdots + m_u s^u)[1 - m_1 s + m_2 s^2 - \cdots + (-1)^u m_u s^u]}{(1 + l_1 s + l_2 s^2 + \cdots + l_v s^v)[1 - l_1 s + l_2 s^2 + \cdots + (-1)^v l_v s^v]} \\ &= \frac{1 + e_2 s^2 + e_4 s^4 + \cdots + e_{2u}s^{2u}}{1 + f_2 s^2 + f_4 s^4 + \cdots + f_{2v}s^{2v}} \end{aligned} \tag{7-164}$$

Equating both sides of Eq. (7-164), we can express $e_2, e_4, \ldots, e_{2u}$ in terms of $m_1, m_2, \ldots, m_u$. Similar relationships can be obtained for $f_2, f_4, \ldots, f_{2v}$ in terms of $l_1, l_2, \ldots, l_v$.

As an illustrative example, for $u = 8$, Eq. (7-164) gives

$$\begin{aligned} e_2 &= 2m_2 - m_1^2 \\ e_4 &= 2m_4 - 2m_1 m_3 + m_2^2 \\ e_6 &= 2m_6 - 2m_1 m_5 + 2m_2 m_4 - m_3^2 \\ e_8 &= 2m_8 - 2m_1 m_7 + 2m_2 m_6 - 2m_3 m_5 + m_4^2 \\ e_{10} &= 2m_2 m_8 - 2m_3 m_7 + 2m_4 m_6 - m_5^2 \\ e_{12} &= 2m_4 m_8 - 2m_3 m_7 + m_6^2 \\ e_{14} &= 2m_6 m_8 - m_7^2 \\ e_{16} &= m_8^2 \end{aligned} \tag{7-165}$$

In general,

$$e_{2x} = \sum_{i=0}^{x-1} (-1)^i 2m_i m_{2x-i} + (-1)^x m_x^2 \tag{7-166}$$

for $x = 1, 2, \ldots, u$, and $m_0 = 1$. Similarly,

$$f_{2y} = \sum_{i=0}^{y-1} (-1)^i 2 l_i l_{2y-i} + (-1)^y l_y^2 \tag{7-167}$$

for $y = 1, 2, \ldots, v$, and $l_0 = 1$.

The following examples illustrate the method of simplification of linear systems outlined in this section.

Example 7-8

Consider that the forward-path transfer function of a unity-feedback control system is given as

$$G(s) = \frac{8}{s(s^2 + 6s + 12)} \tag{7-168}$$

The closed-loop transfer function is

$$M_H(s) = \frac{Y(s)}{R(s)} = \frac{8}{s^3 + 6s^2 + 12s + 8}$$
$$= \frac{1}{1 + 1.5s + 0.75s^2 + 0.125s^3} \tag{7-169}$$

The poles of the closed-loop transfer function are at $s = -2, -2$, and -2. The low-order approximating system is considered for the following two cases.

Case 1

Consider that the simplified low-order system is of the second order, and the transfer function is of the form

$$M_L(s) = \frac{1}{1 + d_1 s + d_2 s^2} \tag{7-170}$$

Equation (7-158) gives

$$\frac{M_H(s)}{M_L(s)} = \frac{1 + d_1 s + d_2 s^2}{1 + 1.5s + 0.75s^2 + 0.125s^3}$$
$$= \frac{1 + m_1 s + m_2 s^2}{1 + l_1 s + l_2 s^2 + l_3 s^3} \tag{7-171}$$

Thus

▲ $l_1 = 1.5 \qquad l_2 = 0.75 \qquad l_3 = 0.125 \qquad d_1 = m_1 \qquad d_2 = m_2$

Using Eq. (7-164), we have

$$\frac{M_H(s)M_H(-s)}{M_L(s)M_L(-s)} = \frac{1 + e_2 s^2 + e_4 s^4}{1 + f_2 s^2 + f_4 s^4 + f_6 s^6} \tag{7-172}$$

Now using Eqs. (7-162) and (7-165), the following nonlinear equations are obtained:

$$e_2 = f_2 = 2m_2 - m_1^2 = 2d_2 - d_1^2$$
$$e_4 = f_4 = 2m_4 - 2m_4 - 2m_1 m_3 = m_2^2 = d_2^2$$

Similarly,

$$f_2 = 2l_2 - l_1^2 = 1.5 - (1.5)^2 = -0.75$$
$$f_4 = 2l_4 - 2l_1 l_3 + l_2^2 = 0.1875$$
$$f_6 = 2l_6 - 2l_1 l_5 + 2l_2 l_4 - l_3^2 = -l_3^2 = -0.156$$

Solving the equations above, we have

$$d_1^2 = 1.616 \qquad d_2^2 = 0.1875$$

Thus

$$d_1 = 1.271 \qquad d_2 = 0.433$$

where we have taken the positive values so that $M_L(s)$ will represent a stable system. Thus the second-order simplified system transfer function is

$$M_L(s) = \frac{1}{1 + 1.271s + 0.433s^2} = \frac{2.31}{s^2 + 2.936s + 2.31} \tag{7-173}$$

The poles of $M_L(s)$ are at $s = -1.468 + j0.384$ and $-1.468 - j0.384$. Thus the third-order system with the transfer function M_H which has real poles at -2, -2, and -2 is approximated by the second-order transfer function with two complex-conjugate poles. The forward-path transfer function of the second-order system is

$$G_L(s) = \frac{2.31}{s(s + 2.936)} \tag{7-174}$$

Viewing from the open-loop standpoint, the third-order system with open-loop poles at $s = 0$, $-3 + j1.732$ and $-3 - j1.732$ and a gain of 8 is approximated by a second-order system with the open-loop poles at $s = 0$ and $s = -2.936$ and a gain of 2.31.

Figure 7-38 shows the unit-step responses of the systems with the closed-loop transfer functions $M_H(s)$ and $M_L(s)$. Notice that the second-order approximating system has a faster rise time. This is due to the two complex-conjugate poles of $M_L(s)$.

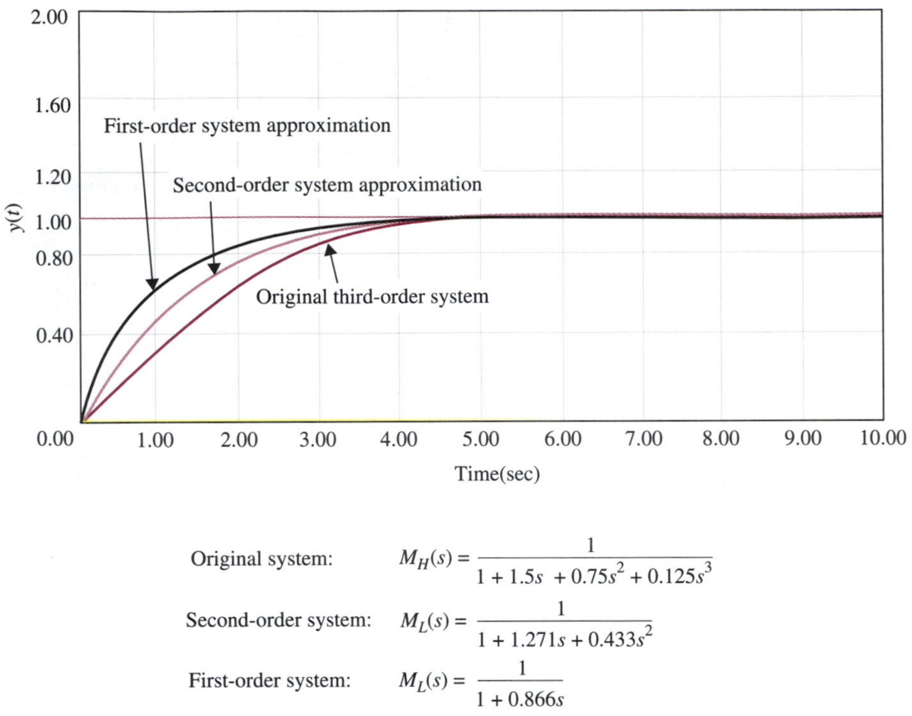

Figure 7-38 Approximation of a third-order system by a first-order system and a second-order system.

Case 2

Now let us use a first-order system to approximate the closed-loop transfer function of Eq. (7-169). We choose

$$M_L(s) = \frac{1}{1 + d_1 s} \tag{7-175}$$

Then

$$\frac{M_H(s)}{M_L(s)} = \frac{1 + d_1 s}{1 + 1.5s + 0.75s^2 + 0.125s^3} = \frac{1 + m_1 s}{1 + l_1 s + l_2 s^2 + l_3 s^3} \tag{7-176}$$

Thus

$$l_1 = 1.5 \qquad l_2 = 0.75 \qquad l_3 = 0.125 \qquad d_1 = m_1$$

and

$$\frac{M_H(s)M_H(-s)}{M_L(s)M_L(-s)} = \frac{1 + e_2 s^2}{1 + f_2 s^2 + f_4 s^4 + f_6 s^6} \tag{7-177}$$

Thus

$$e_2 = f_2 = 2d_2 - d_1^2 = -d_1^2 = -0.75$$
$$f_4 = 0.1875$$
$$f_6 = -0.156$$

This gives $d_1 = 0.866$. The transfer function of $M_L(s)$ is

$$M_L(s) = \frac{1}{1 + 0.866s} = \frac{1.1547}{s + 1.1547} \tag{7-178}$$

The forward-path transfer function of the first-order system is

$$G_L(s) = \frac{1.1546}{s} \tag{7-179}$$

The unit-step response of the first-order system is shown in Fig. 7-38. As expected, the first-order system gives a more inferior approximation to the third-order system than the second-order system. ▲

Example 7-9

Consider the following closed-loop transfer function:

$$M_H(s) = \frac{1}{(1 + s + 0.5s^2)(1 + Ts)} = \frac{1}{1 + (1 + T)s + (0.5 + T)s^2 + 0.5s^3} \tag{7-180}$$

where T is a positive variable parameter. We shall investigate the approximation of $M_H(s)$ by a second-order model when T takes on various values.

Let the second-order system be modeled by the transfer function

$$M_L(s) = \frac{1}{1 + d_1 s + d_2 s^2} \tag{7-181}$$

Substituting Eqs. (7-180) and (7-181) into Eq. (7-158), we get

$$\frac{M_H(s)}{M_L(s)} = \frac{1 + d_1 s + d_2 s^2}{1 + (1 + T)s + (0.5 + T)s^2 + 0.5Ts^3} = \frac{1 + m_1 s + m_2 s^2}{1 + l_1 s + l_2 s^2 + l_3 s^3} \tag{7-182}$$

Then

$$l_1 = 1 + T \qquad l_2 = 0.5 + T \qquad l_3 = 0.5T \tag{7-183}$$

From Eq. (7-166) we get

$$e_2 = f_2 = 2d_2 - d_1^2 \qquad e_4 = f_4 = m_2^2 = d_2^2 \tag{7-184}$$

From Eq. (7-167) we get

$$\begin{aligned} f_2 &= 2l_2 - l_1^2 = 2(0.5 + T) - (1 + T)^2 = -T^2 \\ f_4 &= 2l_4 - 2l_1l_3 + l_2^2 = -2(1 + T)(0.5T) + (0.5 + T)^2 = 0.25 \\ f_6 &= -l_3^2 = -(0.5T)^2 = -0.25T^2 \end{aligned} \tag{7-185}$$

From the equations in Eqs. (7-183) through (7-185), we solve for the values of d_1 and d_2, and the results are

$$d_1 = \sqrt{1 + T^2} \qquad d_2 = 0.5$$

Thus the transfer function of the second-order approximating system is

$$M_L(s) = \frac{1}{1 + \sqrt{1 + T^2}\, s + 0.5s^2} \tag{7-186}$$

The two poles of $M_L(s)$ are at

$$s = -\sqrt{1 + T^2} + j\sqrt{1 - T^2} \quad \text{and} \quad s = -\sqrt{1 + T^2} - j\sqrt{1 - T^2}$$

The poles of $M_H(s)$ are at $s = -1 + j$, $-1 - j$, and $-1/T$. The two poles of $M_L(s)$ are tabulated for $T = 0$ to $T = 6$:

▲ $T = 0$	$s = -1.000 + j$	$s = -1.000 - j$
▲ $T = 0.1$	$s = -1.005 + j0.995$	$s = -1.005 - j0.995$
▲ $T = 0.25$	$s = -1.031 + j0.968$	$s = -1.031 - j0.968$
▲ $T = 0.5$	$s = -1.118 + j0.866$	$s = -1.118 - j0.866$
▲ $T = 0.75$	$s = -1.250 + j0.661$	$s = -1.250 - j0.661$
▲ $T = 1.0$	$s = -1.414$	$s = -1.414$
▲ $T = 2.0$	$s = -3.968$	$s = -0.504$
▲ $T = 4.0$	$s = -7.996$	$s = -0.250$
▲ $T = 6.0$	$s = -11.999$	$s = -0.167$

Notice that as the value of T decreases, the original system approaches a second-order system, as the pole at $-1/T$ moves toward $-\infty$. Thus when $T = 0$, $M_L(s)$ is identical to $M_H(s)$, and the approximate solution is exact. As the value of T increases, the pole at $-1/T$ moves toward the origin and becomes more dominant; the poles of $M_L(s)$ move toward the real axis, and eventually when $T \geq 1$, the poles become real. As T increases, one of the real poles of the second-order system moves toward $-\infty$, and the other approached $-1/T$. Thus the second-order approximation is good for small and large values of T.

Figures 7-39 through 7-42 illustrate the unit-step responses of the third-order system and the second-order approximating system when $T = 0.1$, 0.5, 1.0, and 6.0, respectively. When the value of T is very small, the unit-step responses of the third-order and the second-order systems are very close. The error between the two responses increases as T is increased to 0.5 and 1.0. Figure 7-42 shows that the response of the two systems again moves closer to each other when $T = 6$. As the value of T increases, the real pole at $-1/T$ becomes dominant relative to the two complex-conjugate poles, and the second-order model for $M_L(s)$ in Eq. (7-181) would give a better approximation to the third-order system. ▲

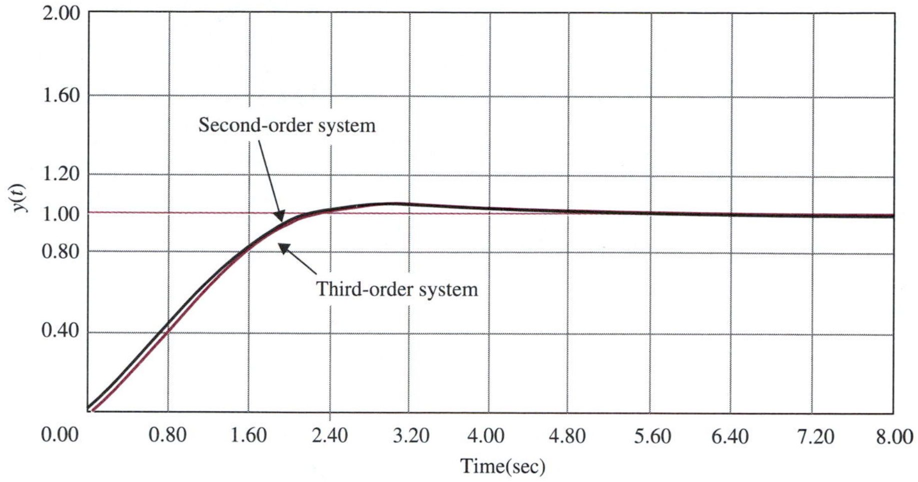

Original system: $M_H(s) = \dfrac{1}{(1 + s + 0.5s^2)(1 + 0.1s)}$

Second-order system: $M_H(s) = \dfrac{1}{s^2 + 2.01s + 2}$

Figure 7-39 Approximation of a third-order system by a second-order system.

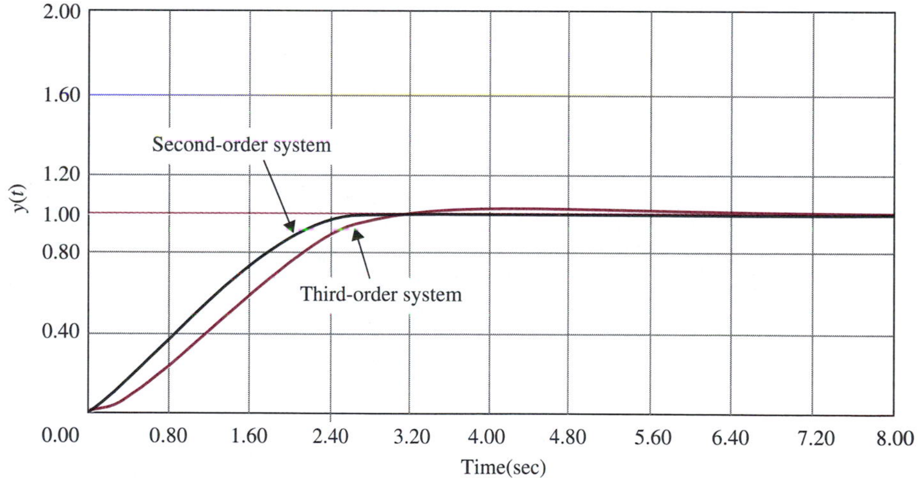

Original system: $M_H(s) = \dfrac{1}{(1 + s + 0.5s^2)(1 + 0.5s)}$

Second-order system: $M_H(s) = \dfrac{2}{s^2 + 2.2361s + 2}$

Figure 7-40 Approximation of a third-order system by a second-order system.

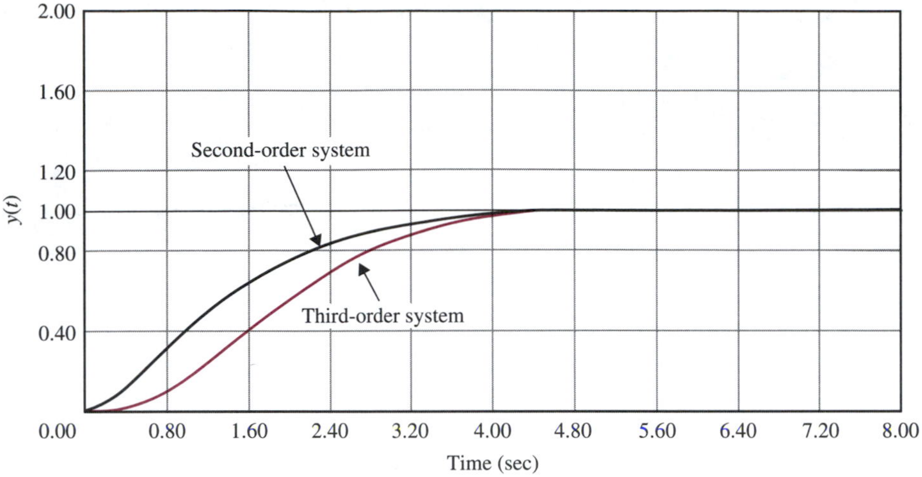

Original system: $M_H(s) = \dfrac{1}{(1 + s + 0.5s^2)(1 + s)}$

Second-order system: $M_L(s) = \dfrac{2}{s^2 + 2.828s + 2}$

Figure 7-41 Approximation of a third-order system by a second-order system.

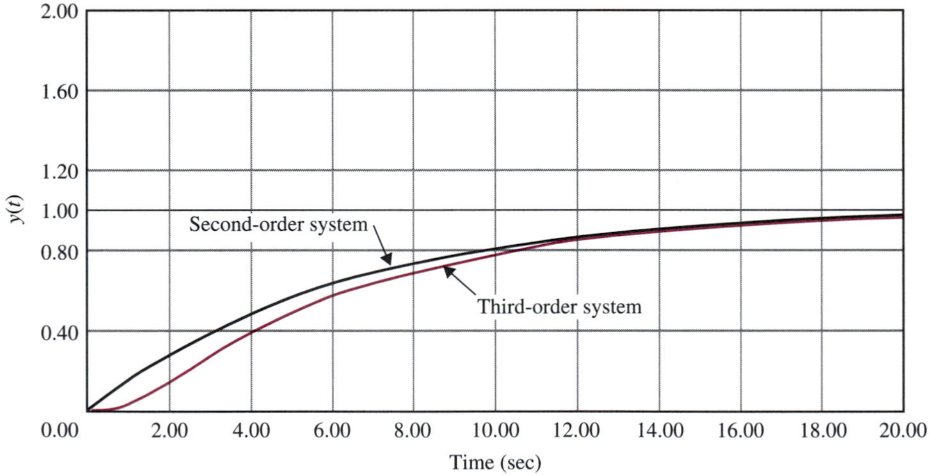

Original system: $M_H(s) = \dfrac{1}{(1 + s + 0.5s^2)(1 + 6s)}$

Second-order system: $M_L(s) = \dfrac{2}{s^2 + 12.17s + 2}$

Figure 7-42 Approximation of a third-order system by a second-order system.

Example 7-10

The third-order position-control system described in Section 7-6 was approximated by a second-order system by neglecting the small motor inductance. The second-order model arrived at in Eq. (7-116) was intended for all values of K. It was shown that the approximation was relatively good for small values of K and inferior for large values of K. Now we shall use the approximating method presented in this section in hope of arriving at a better second-order approximation for a wider range of K.

The closed-loop transfer function of the system is given in Eq. (7-138), and is repeated:

$$M_H(s) = \frac{1.5 \times 10^7 K}{s^3 + 3408.3s^2 + 1{,}204{,}000s + 1.5 \times 10^7 K} \tag{7-187}$$

For a given value of K, this third-order transfer function is to be approximated by a second-order transfer function of the form

$$M_L(s) = \frac{1}{1 + d_1 s + d_2 s^2} \tag{7-188}$$

Once $M_L(s)$ is determined, the forward-path transfer function of the approximating second-order system is written

$$G_L(s) = \frac{d_1/d_2}{s(s + d_1/d_2)} \tag{7-189}$$

The coefficients d_1 and d_2 are determined for $K = 7.248$, 14.5, and 100, using the method described in this section. The results are tabulated as follows, where the first column of $G(s)$ denotes the approximation with $L_a = 0$, and the second column gives the corresponding $G_L(s)$ with the values of d_1 and d_2 indicated.

K	$G(s)$	$G_L(s)$
7.248	$\dfrac{326{,}166}{s(s + 361.2)}$	$\dfrac{35{,}829.5}{s(s + 385.5)}$
14.5	$\dfrac{65{,}250}{s(s + 361.2)}$	$\dfrac{71{,}684.6}{s(s + 378.2)}$
100.0	$\dfrac{450{,}000}{s(s + 361.2)}$	$\dfrac{494{,}300}{s(s + 188.84)}$

Comparisons of the unit-step responses of the third-order system, and the second-order systems with $L_a = 0$ and with the approximation method are shown in Fig. 7-43 for the three values of K indicated. When the value of K is relatively small, both second-order approximations are quite good. For $K = 100$, the second-order system with $L_a = 0$ has a lower maximum overshoot and is less oscillatory than those of the true third-order system. The second-order system from the approximation method has a step response that is closer to the actual response, although the maximum overshoot is slightly larger. ▲

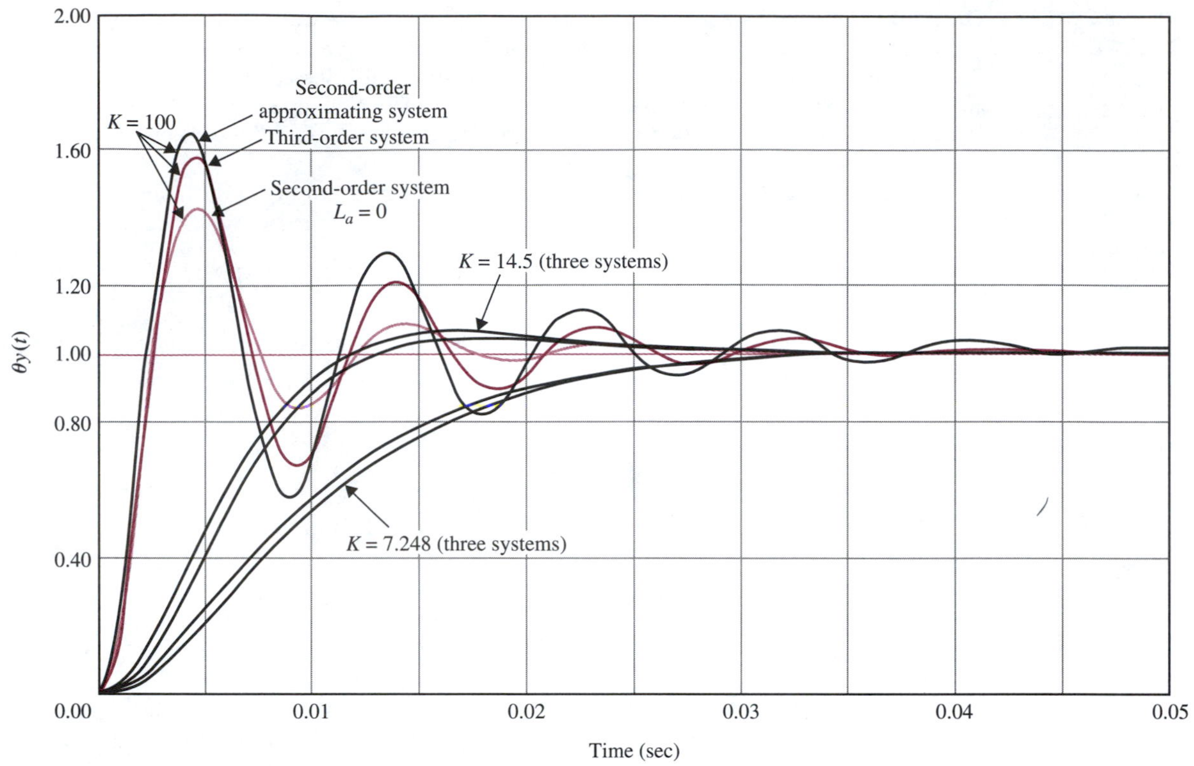

Figure 7-43 Comparison of the unit-step responses of the systems in Example 7-10.

One of the problems that surfaces from this investigation is that the low-order approximation method described in this section provides a system that is less damped than the high-order system. This means that when the original system is lightly damped, it is possible that the dominant roots of the low-order system may end up in the right-half s-plane. For the system treated in Example 7-10, we run into this difficulty when $K = 181.2$, one of the values chosen in Section 7-6. Since d_1 and d_2 are solved from $d_1^2 = 2d_2 - f_2$, which comes from Eqs. (7-162) and (7-165), if f_2 is greater than $2d_2$, no real solution could be obtained for d_1. For the system in Example 7-10, we can show that when $K = 129.3$, the two poles of the second-order approximating system will be on the $j\omega$-axis. To approximate the third-order system when K is large would require that the second-order transfer function $M_L(s)$ contain a zero.

In conclusion, we would like to point out that the problem of reduced-order approximation of a high-order system is a complex subject. Most of the existing methods have flaws of one type or another. The serious reader should refer to the literature [1,2] for more in-depth treatment of the subject.

7-10 Time-Domain Properties of Discrete-Data Systems

7-10-1 Time Response of Discrete-Data Control Systems

To carry out the design of discrete-data control systems in the time domain or z-domain, we must first study the time- and z-domain properties of these systems. We learned in Chapter 3 that the output responses of most discrete-data control systems are functions of the continuous-time variable t. Thus time-domain specifications such as the maximum overshoot, rise time, damping ratio, and so on, can still be applied to discrete-data systems. The only difference is that to make use of analytical tools such as z-transforms, the continuous data found in a discrete-data system are sampled so that the independent time variable is kT, where T is the sampling period in seconds. Also, instead of working in the s-plane, the transient performance of a discrete-data system is characterized by poles and zeros of the transfer function in the z-plane.

The objectives of the following sections are as follows:

1. To present methods of finding the discretized time responses of discrete-data control systems.
2. To describe the important characteristics of the discretized time response $y(kT)$.
3. To establish the significance of pole and zero locations in the z-plane.
4. To provide comparison between time responses of continuous-data and discrete-data control systems.

Let us refer to the block diagram of the discrete-data control system shown in Fig. 7-44. The transfer function of the system is

$$\frac{Y(z)}{R(z)} = \frac{G(z)}{1 + GH(z)} \tag{7-190}$$

where $GH(z)$ denotes the z-transform of $G(s)H(s)$. Once the input $R(z)$ is given, the output sequence $y(kT)$ can be determined using one of the following two methods:

1. Take the inverse z-transform of $Y(z)$ using the z-transform table.
2. Expand $Y(z)$ into a power series of z^{-k}.

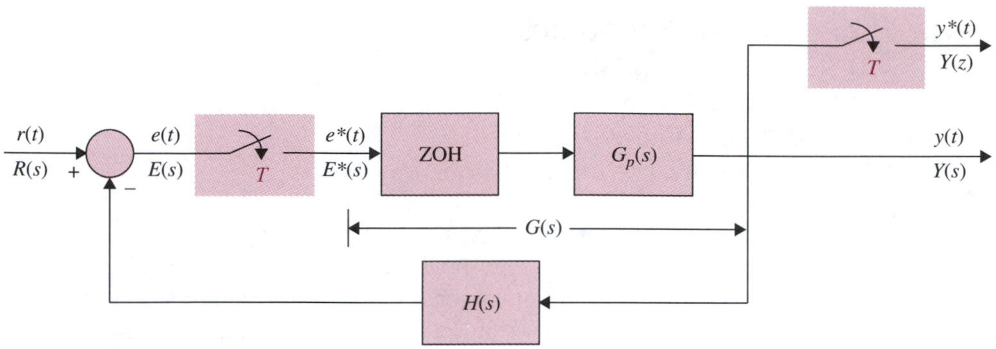

Figure 7-44 Block diagram of the discrete-data control system.

The z-transform of the output is defined as

$$Y(z) = \sum_{k=0}^{\infty} y(kT)z^{-k} \tag{7-191}$$

The discrete-time response $y(kT)$ can be determined by referring to the coefficient of z^{-k} for $k = 0, 1, 2, \ldots$. Remember that $y(kT)$, $k = 0, 1, 2, \ldots$ contains only the sampled information on $y(t)$ at the sampling instants. If the sampling period is large relative to the most significant time constant of the system, $y(kT)$ may not be an accurate representation of $y(t)$.

Example 7-11

Consider that the position-control system described in Section 7-6 has discrete data in the forward path, so that the system is now described by the block diagram of Fig. 7-44. For $K = 14.5$, the transfer function of the controlled process is

$$G_p(s) = \frac{65{,}250}{s(s + 361.2)} \tag{7-192}$$

The forward-path transfer function of the discrete-data system is

$$G_{h0}G_p(z) = \mathcal{Z}[G_{h0}G_p(s)] = (1 - z^{-1})\mathcal{Z}\left[\frac{G_p(s)}{s}\right] \tag{7-193}$$

For a sampling period of $T = 0.001$ sec, the z-transfer function in Eq. (7-193) is evaluated as

$$G_{h0}G_p(z) = \frac{0.029z + 0.0257}{z^2 - 1.697z + 0.697} \tag{7-194}$$

The closed-loop transfer function of the system is

$$\frac{Y(z)}{R(z)} = \frac{G_{h0}G_p(z)}{1 + G_{h0}G_p(z)} = \frac{0.029z + 0.0257}{z^2 - 1.668z + 0.7226} \tag{7-195}$$

where $R(z)$ and $Y(z)$ represent the z-transforms of the input and the output, respectively. For a unit-step input, $R(z) = z/(z - 1)$. The output transform $Y(z)$ becomes

$$Y(z) = \frac{z(0.029z + 0.0257)}{(z - 1)(z^2 - 1.668z + 0.7226)} \tag{7-196}$$

The output sequence $y(kT)$ can be determined by dividing the numerator polynomial of $Y(z)$ by its denominator polynomial to yield a power series in z^{-1}. Figure 7-45 shows the plot of $y(kT)$, as dots, versus kT when $T = 0.001$ sec. For comparison, the unit-step response of the continuous-data system in Section 7-6 with $K = 14.5$ is shown in the same figure. As seen from Fig. 7-45, when the sampling period is small, the output responses of the discrete-data and the continuous-data systems are very similar. The maximum value of $y(kT)$ is 1.0731, or a 7.31 percent maximum overshoot, as against the 4.3 percent maximum overshoot for the continuous-data system.

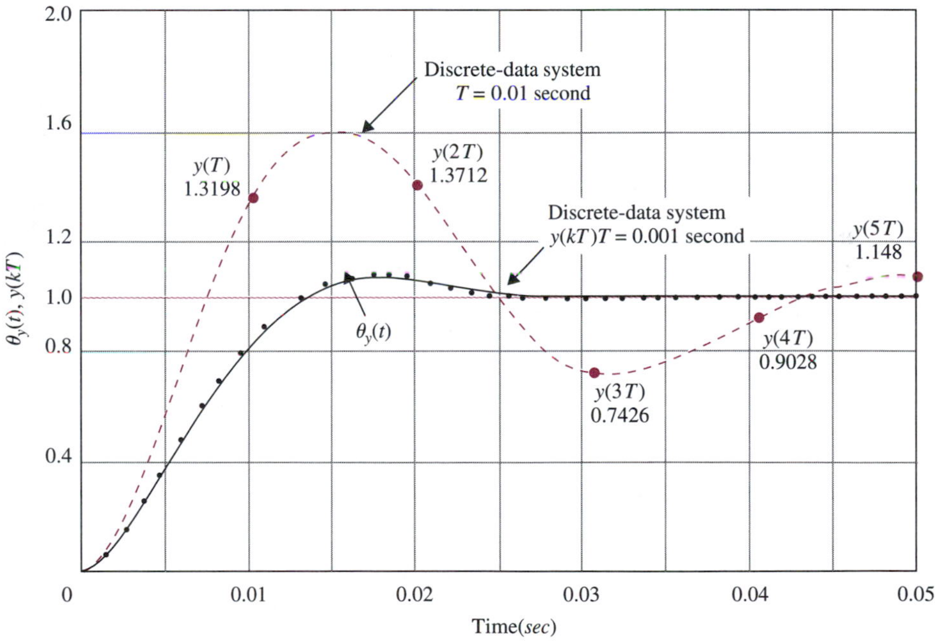

Figure 7-45 Comparison of unit-step responses of discrete-data and continuous-data systems.

When the sampling period is increased to 0.01 sec, the forward-path transfer function of the discrete-data system is

$$G_{h0}G_p(z) = \frac{1.3198z + 0.4379}{z^2 - 1.027z + 0.027} \tag{7-197}$$

and the closed-loop transfer function is

$$\frac{Y(z)}{R(z)} = \frac{1.3198z + 0.4379}{z^2 + 0.2929z + 0.4649} \tag{7-198}$$

The output sequence $y(kT)$ with $T = 0.01$ sec is shown in Fig. 7-45 with $k = 0, 1, 2, 3, 4,$ and 5. The true continuous-time output of the discrete-data system is shown as the dotted curve. Notice that the maximum value of $y(kT)$ is 1.3712, but the true maximum overshoot is considerably higher than that. Thus the larger sampling period only makes the system less stable, but the sampled output no longer gives an accurate measure of the true output.

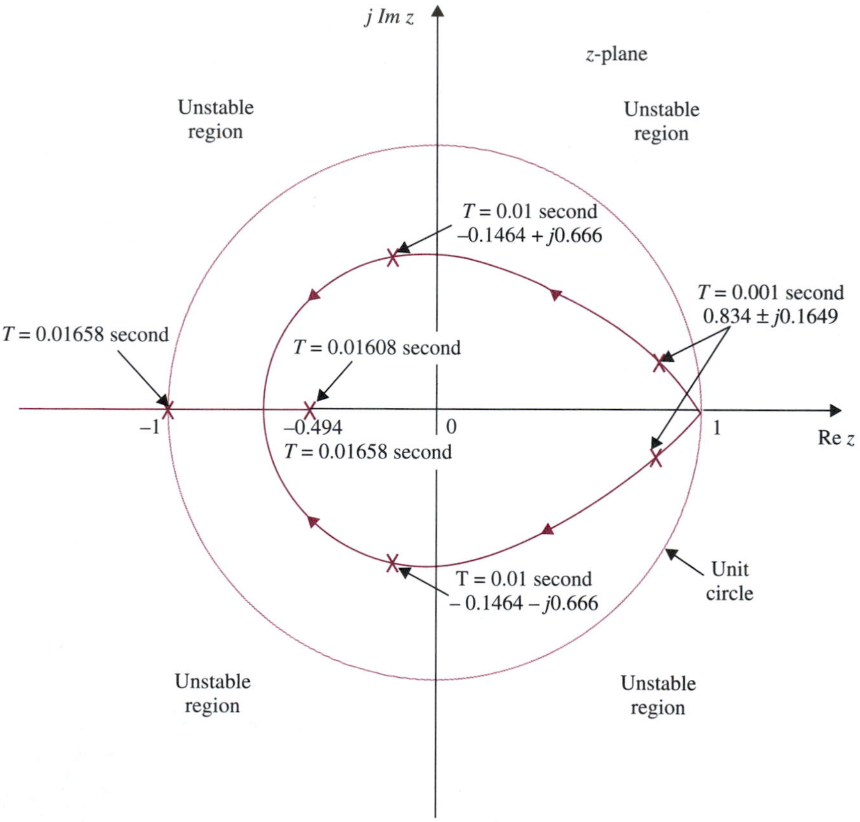

Figure 7-46 Trajectories of roots of a second-order discrete-data control system as the sampling period T varies.

▲ Two equal characteristic equation roots on the negative real axis in the z-plane do not correspond to critical damping.

▲ When one or more characteristic equation roots lie on the negative axis in the z-plane, response will oscillate with positive and negative peaks.

When the sampling period is increased to 0.01658 sec, the characteristic equation of the discrete-data system is

$$z^2 + 1.4938z + 0.4939 = 0 \tag{7-199}$$

which has roots at $z = -0.494$ and $z = -1.000$. The root at -1.000 causes the step response of the system to oscillate with a constant amplitude, and the system is marginally stable. Thus for all sampling periods greater than 0.01658 sec, the discrete-data system will be unstable. From Section 7-6 we learned that the second-order continuous-data system is always stable for finite positive values of K. For the discrete-data system, the sample-and-hold has the effect of making the system less stable, and if the value of T is too large, the second-order system can become unstable. Figure 7-46 shows the trajectories of the two characteristic-equation roots of the discrete-data system as the sampling period T varies. Notice that when the sampling period is very small, the two characteristic-equation roots are very close to the $z = 1$ point and are complex. When $T = 0.01608$ sec, the two roots become equal and real and are negative. Unlike the continuous-data system, the case of two identical roots on the *negative* real axis in the z-plane *does not* correspond to critical damping. For discrete-data systems, when one or more characteristic-equation roots lie on the negative real axis of the z-plane, the system response will oscillate with positive and negative peaks. Figure 7-47 shows the oscillatory response of $y(kT)$ when $T = 0.01658$ sec, which is the critical value for stability. Beyond this value of T, one root will move outside the unit circle, and the system becomes unstable. ▲

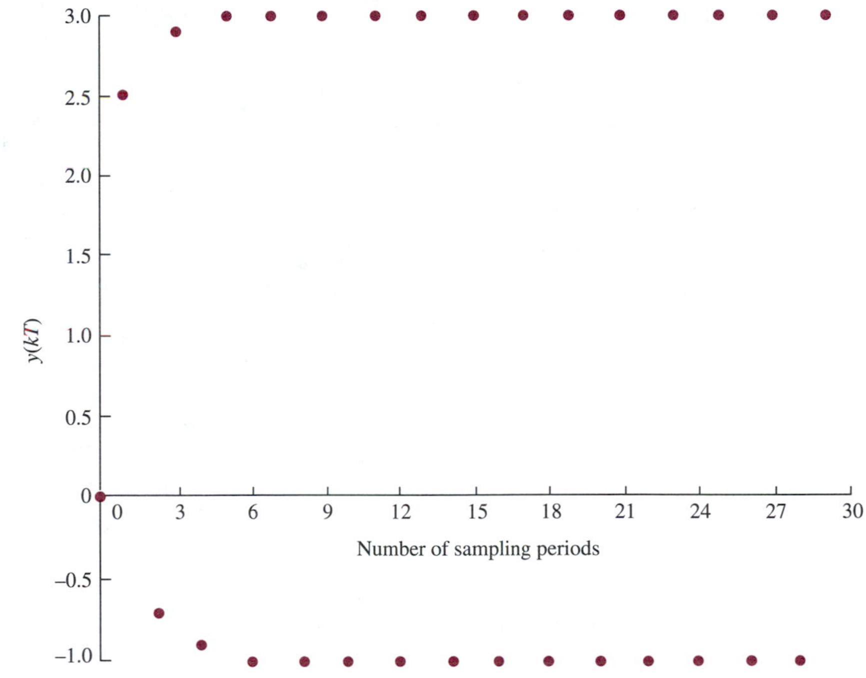

Figure 7-47 Oscillatory response of a discrete-data system with the sampling period $T = 0.01658$ sec.

7-10-2 Mapping between *s*-Plane and *z*-Plane Trajectories

For analysis and design purposes, it is important to study the relation between the location of the characteristic-equation roots in the z-plane and the time response of the discrete-data system. In Chapter 3 the periodic property of the Laplace transform of the Laplace transform of the sampled signal $R^*(s)$ is established by Eq. (3-93) [i.e., $R^*(s + jm\omega_s) = R^*(s)$, where m is an integer]. In other words, given any point s_1 in the s-plane, the function $R^*(s)$ has the same value at all periodic points $s = s_1 + jm\omega_s$. Thus the s-plane is divided into an infinite number of periodic strips, as shown in Fig. 7-48(a). The strip between $\omega = \omega_s/2$ is called the **primary strip** and all others at higher frequencies are called the **complementary strips.** Figure 7-48(b) shows the mapping of the periodic strips from the s-plane to the z-plane, and the details are explained as follows.

1. The $j\omega$-axis in the s-plane is mapped onto the unit circle $|z| = 1$ in the z-plane.
2. The boundaries of the period strips, $s = jm\omega_s/2$, $m = \pm1, \pm3, \pm5, \ldots$, are mapped onto the negative real axis of the z-plane. The portion inside the unit circle corresponds to $\sigma < 0$, and the portion outside the unit circle corresponds to $\sigma > 0$.
3. The centerlines of the periodic strips, $s = jm\omega_s$, $m = 0, \pm2, \pm4, \ldots$, are mapped onto the positive real axis of the z-plane. The portion inside the unit circle corresponds to $\sigma < 0$, and the portion outside the unit circle corresponds to $\sigma > 0$.
4. Regions shown in the periodic strips in the left-half s-plane are mapped onto the interior of the unit circle in the z-plane.
5. The point $z = 1$ in the z-plane corresponds to the origin, $s = 0$, in the x-plane.
6. The origin, $z = 0$, in the z-plane corresponds to $s = -\infty$ in the s-plane.

In the time-domain analysis of continuous-data systems, we devise the damping factor α, the damping ratio ζ, and the natural undamped frequency ω_n to characterize the system dynamics. The same parameters can be defined for discrete-data systems with respect to the characteristic-equation roots in the z-plane. The loci of the constant-α, constant-ζ, constant-ω, and constant-ω_n in the z-plane are described in the following.

Constant-Damping Loci. For a constant-damping factor $\sigma = \alpha$ in the s-plane, the corresponding trajectory in the z-plane is described by

$$z = e^{\alpha T} \tag{7-200}$$

which is a circle centered at the origin with a radius of $e^{\alpha T}$, as shown in Fig. 7-49.

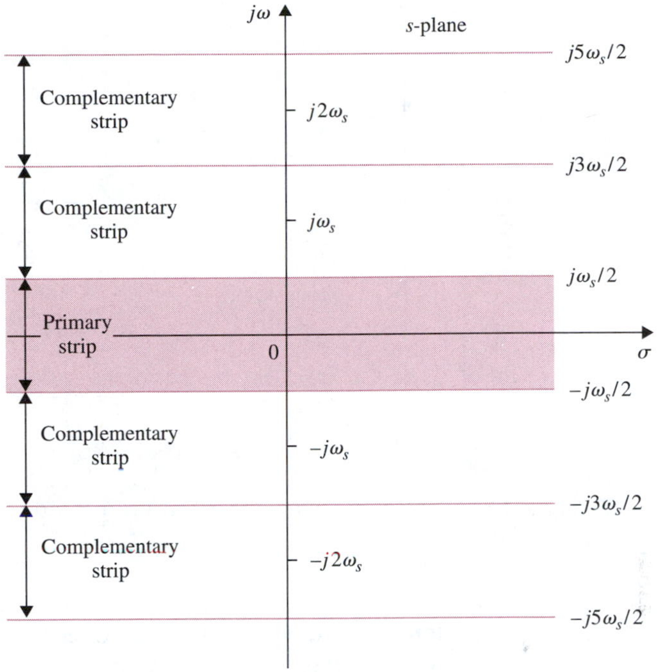

(a)

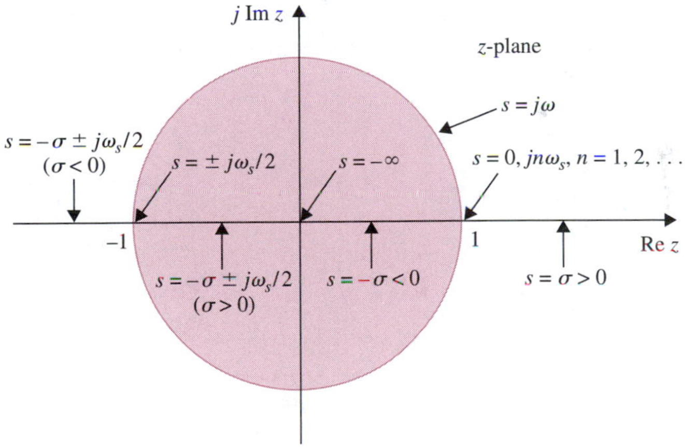

(b)

Figure 7-48 Periodic strips in the s-plane and the corresponding points and lines between the s-plane and the z-plane.

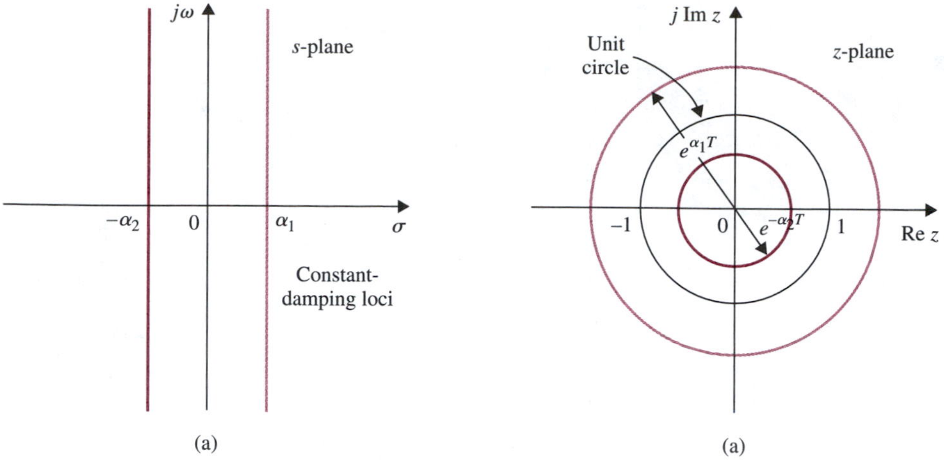

Figure 7-49 Constant-damping loci in the s-plane and the z-plane.

Constant-Frequency Loci. The constant-frequency $\omega = \omega_1$ locus in the s-plane is a horizontal line parallel to the σ-axis. The corresponding z-plane locus is a straight line emanating from the origin at an angle of $\theta = \omega_1 T$ radians, measured from the real axis, as shown in Fig. 7-50.

Constant-Natural-Undamped-Frequency Loci. The constant-ω_n loci in the s-plane are concentric circles with the center at the origin, and the radius is ω_n. The corresponding

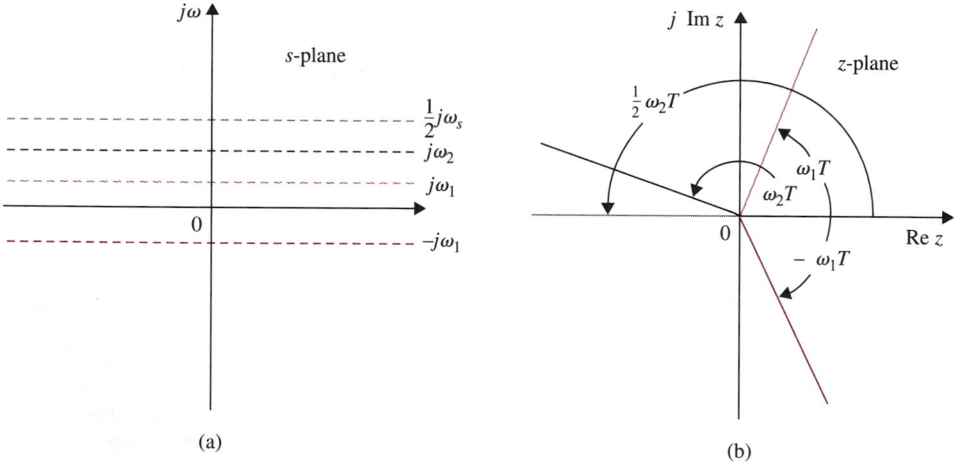

Figure 7-50 Constant-frequency loci in the s-plane and the z-plane.

constant-ω_n loci in the z-plane are shown in Fig. 7-51 for $\omega_n = \omega_s/16$ to $\omega_s/2$. Only the loci inside the unit circle are shown.

Constant-Damping Ratio Loci. For a constant-damping ratio ζ, the s-plane loci are described by

$$s = -\omega \tan \beta + j\omega \tag{7-201}$$

The constant-ζ loci in the z-plane are described by

$$z = e^{Ts} = e^{-2\pi(\tan\beta)/\omega_s} \angle 2\pi\omega/\omega_s \tag{7-202}$$

where

$$\beta = \sin^{-1}\zeta = \text{constant} \tag{7-203}$$

For a given value of β, the constant-ζ locus in the z-plane, described by Eq. (7-203), is a logarithmic spiral for $0° < \beta < 90°$. Figure 7-52 shows several typical constant-ζ loci in the top half of the z-plane.

7-10-3 Relation between Characteristic-Equation Roots and Transient Response

Based on the discussions given in the preceding section, we can establish the basic relation between the characteristic-equation roots and the transient response of a discrete-data system, keeping in mind that, in general, the zeros of the closed-loop transfer

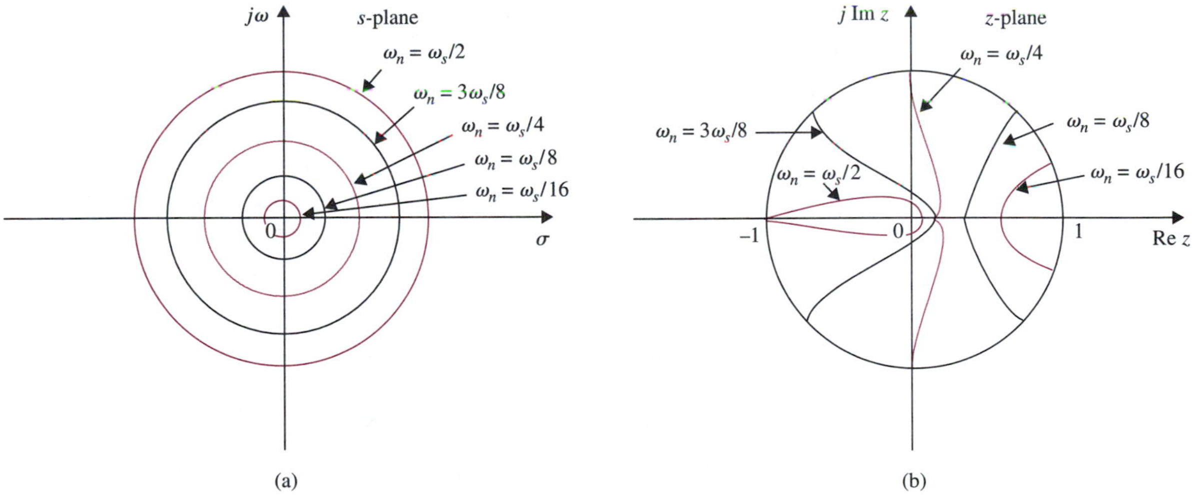

Figure 7-51 Constant-natural-undamped-frequency loci in the s-plane and the z-plane.

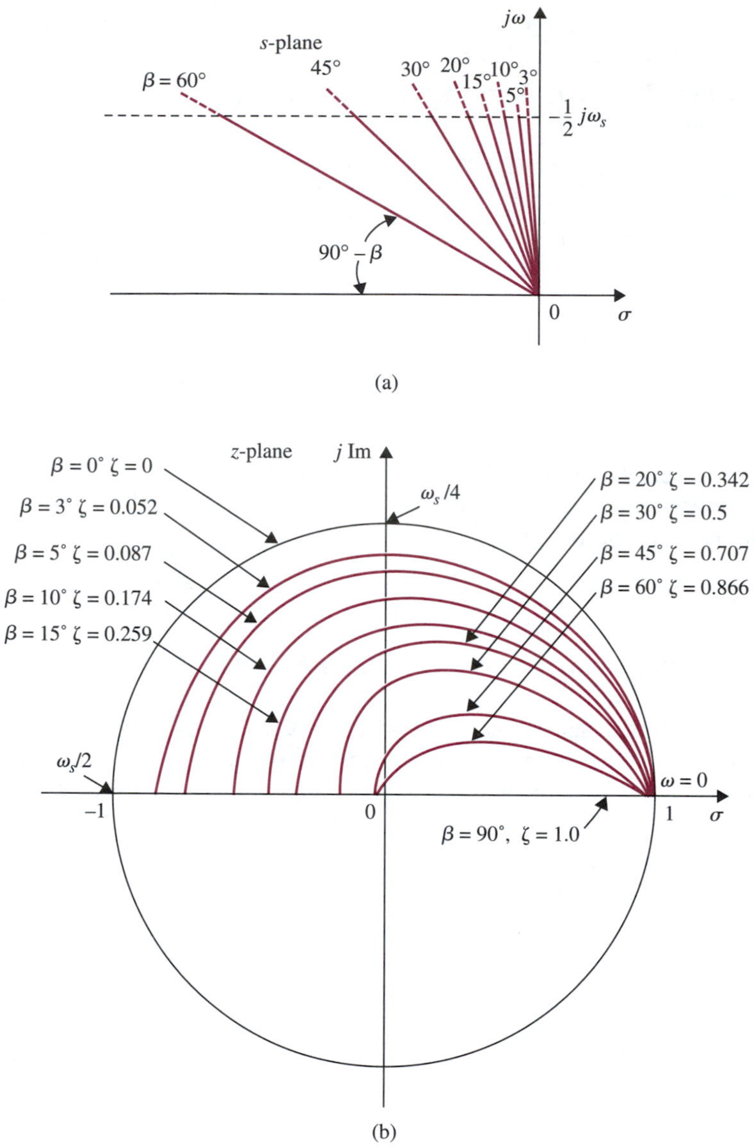

Figure 7-52 Constant-damping-ratio loci in the s-plane and the z-plane.

function will also play an important role on the response, but not on the stability, of the system.

Roots on the Positive Real Axis in the z-Plane. Roots on the positive real axis inside the unit circle of the z-plane give rise to responses that decay exponentially with an increase of kT. Typical responses relative to the root locations are shown in Fig. 7-53. The roots closer to the unit circle will decay slower. When the root is at $z = 1$, the response has a constant amplitude. Roots outside the unit circle correspond to unstable systems, and the responses will increase with kT.

Roots on the Negative Real Axis in the *z*-Plane. The negative real axis of the z-plane corresponds to the boundaries of the periodic strips in the s-plane. For example, when $s = -\sigma_1 \pm j\omega_s/2$, the complex-conjugate points are on the boundaries of the primary strip in the s-plane. The corresponding z-plane points are

$$z = e^{-\sigma_1 T} e^{\pm j\omega_s T/2} = -e^{-\sigma_1 T} \qquad (7\text{-}204)$$

which are on the negative real axis of the z-plane. For the frequency of $\omega_s/2$, the output sequence will have exactly one sample in each one-half period of the envelope. Thus the output sequence will occur in alternating positive and negative pulses, as shown in Fig. 7-53(b).

Complex-Conjugate Roots in the *z*-Plane. Complex-conjugate roots inside the unit circle in the z-plane correspond to oscillatory responses that decay with an increase in kT. Roots that are closer to the unit circle will decay slower. As the roots move toward the second and third quadrants, the frequency of oscillation of the responses increases. Refer to Fig. 7-53 for typical examples.

7-10-4 Steady-State-Error Analysis of Discrete-Data Control Systems

Since the input and output signals of a typical discrete-data control system are continuous-time functions, as shown in the block diagram of Fig. 7-44, the error signal should still be defined as

$$e(t) = r(t) - y(t) \qquad (7\text{-}205)$$

where $r(t)$ is the input and $y(t)$ is the output. The error analysis conducted here is only for unity-feedback systems with $H(s) = 1$. Due to the discrete data that appear inside the system, z-transform or difference equations are often used, so that the input and output are represented in sampled form, $r(kT)$ and $y(kT)$, respectively. Thus the error signal is more appropriately represented by $e^*(t)$ or $e(kT)$. That is,

$$e^*(t) = r^*(t) - y^*(t) \qquad (7\text{-}206)$$

or

$$e(kT) = r(kT) - y(kT) \qquad (7\text{-}207)$$

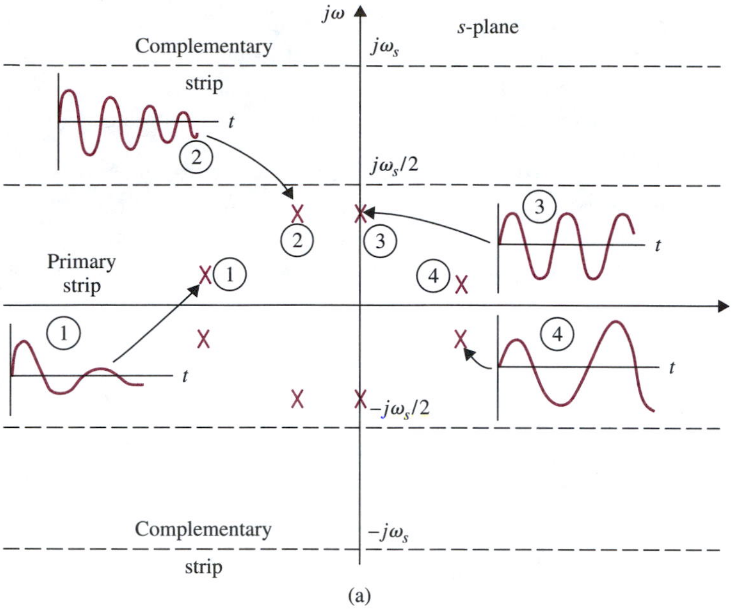

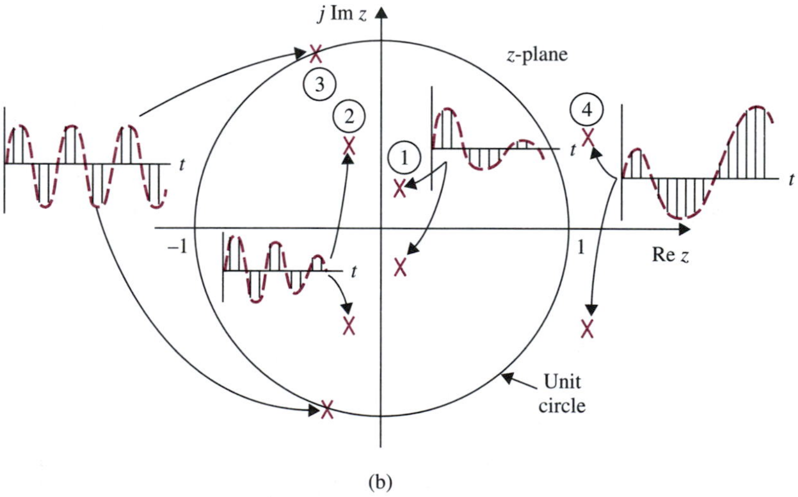

Figure 7-53 (a) Transient responses corresponding to various pole locations of $Y^*(s)$ in the s-plane (complex-conjugate poles only). (b) Transient-response sequence corresponding to various pole locations of $Y(z)$ in the z-plane.

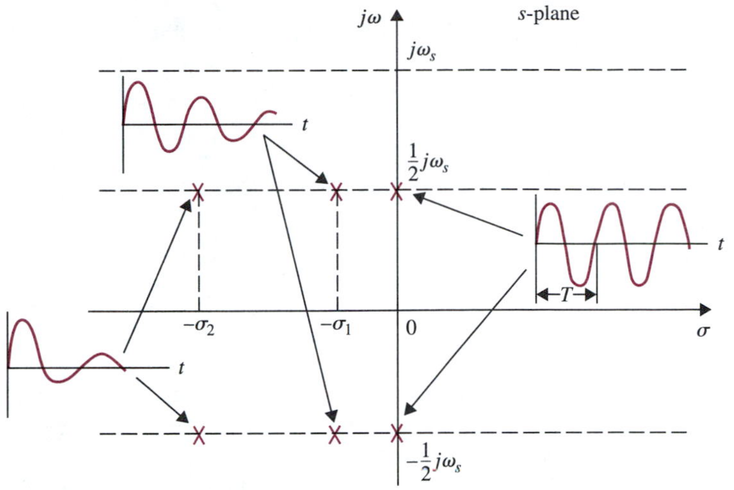

(c)

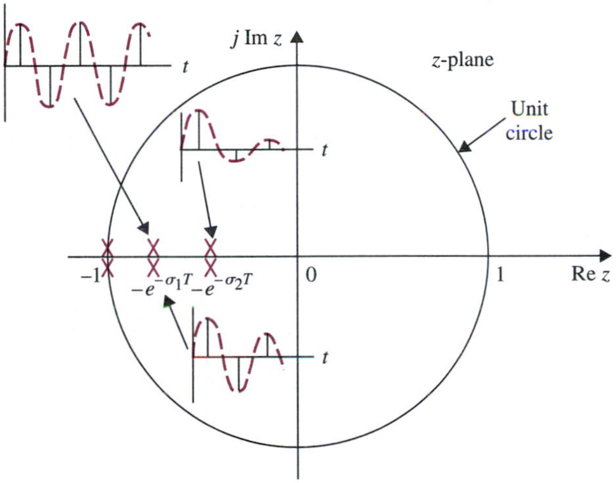

(d)

Figure 7-53 (cont.) (c) Transient responses corresponding to various pole locations of $Y^*(s)$ in the s-plane (complex-conjugate poles on the boundaries between periodic strips). (d) Transient-response sequence corresponding to various pole locations of $Y(z)$ in the z-plane.

The steady-state error at the sampling instants is defined as

$$e_{ss}^* = \lim_{t\to\infty} e(t) = \lim_{k\to\infty} e(kT) \qquad \text{(7-208)}$$

By using the final-value theorem of the z-transform, the steady-state error is

$$e_{ss}^* = \lim_{k\to\infty} e(kT) = \lim_{z\to 1} (1 - z^{-1})E(z) \qquad \text{(7-209)}$$

provided that the function $(1 - z^{-1})E(z)$ does not have any pole on or outside the unit circle in the z-plane. It should be pointed out that since the true error of the system is $e(t)$; e_{ss}^* predicts only the steady-state error of the system at the sampling instants.

By expressing $E(z)$ in terms of $R(z)$ and $G_{h0}G_p(z)$, Eq. (11-50) is written

$$e_{ss}^* = \lim_{k\to\infty} e(kT) = \lim_{z\to 1} (1 - z^{-1}) \frac{R(z)}{1 + G_{h0}G_p(z)} \qquad \text{(7-210)}$$

This expression shows that the steady-state error depends on the reference input $R(z)$ as well as the forward-path transfer function $G_{h0}G_p(z)$. Just as in the continuous-data systems, we shall consider only the three basic types of input signals and the associated error constants and relate e_{ss}^* to these and the type of the system.

Let the transfer function of the controlled process in the system of Fig. 7-44 be of the form

$$G_p(s) = \frac{K(1 + T_a s)(1 + T_b s)\cdots(1 + T_m s)}{s^j(1 + T_1 s)(1 + T_2 s)\cdots(1 + T_n s)} \qquad \text{(7-211)}$$

where $j = 0, 1, 2, \ldots$. The transfer function $G_{h0}G_p(z)$ is

$$G_{h0}G_p(s) = (1 - z^{-1})\mathscr{Z}\left[\frac{K(1 + T_a s)(1 + T_b s)\cdots(1 + T_m s)}{s^{j+1}(1 + T_1 s)(1 + T_2 s)\cdots(1 + T_n s)}\right] \qquad \text{(7-212)}$$

Steady-State Error Due to a Step-Function Input

When the input to the system, $r(t)$, in Fig. 7-44 is a step function with magnitude R, the z-transform of $r(t)$ is

$$R(z) = \frac{Rz}{z - 1} \qquad \text{(7-213)}$$

Substituting $R(z)$ into Eq. (7-210), we get

$$e_{ss}^* = \lim_{z \to 1} \frac{R}{1 + G_{h0}G_p(z)} = \frac{R}{1 + \lim_{z \to 1} G_{h0}G_p(z)} \tag{7-214}$$

Let the **step-error constant** be defined as

$$K_p^* = \lim_{z \to 1} G_{h0}G_p(z) \tag{7-215}$$

Equation (7-214) becomes

$$e_{ss}^* = \frac{R}{1 + K_p^*} \tag{7-216}$$

Thus we see that the steady-state error of the discrete-data control system in Fig. 7-44 is related to the step-error constant K_p^* in the same way as in the continuous-data case, except that K_p^* is given by Eq. (7-215).

We can relate K_p^* to the system type as follows. For a type-0 system, $j = 0$ in Eq. (7-212), and the equation becomes

$$G_{h0}G_p(z) = (1 - z^{-1})\mathscr{Z}\left[\frac{K(1 + T_a s)(1 + T_b s) \cdots (1 + T_m s)}{s(1 + T_1 s)(1 + T_2 s) \cdots (1 + T_n s)}\right] \tag{7-217}$$

Performing partial-fraction expansion to the function inside the square brackets of Eq. (7-217), we get

$$\begin{aligned} G_{h0}G_p(z) &= (1 - z^{-1})\mathscr{Z}\left[\frac{K}{s} + \text{terms due to the nonzero poles}\right] \\ &= (1 - z^{-1})\left[\frac{Kz}{z - 1} + \text{terms due to the nonzero poles}\right] \end{aligned} \tag{7-218}$$

Since the terms due to the nonzero poles do not contain the term $(z - 1)$ in the denominator, the step-error constant is written

$$K_p^* = \lim_{z \to 1} G_{h0}G_p(z) = \lim_{z \to 1} (1 - z^{-1}) \frac{Kz}{z - 1} = K \tag{7-219}$$

Similarly, for a type 1 system, $G_{h0}G_p(z)$ will have an s^2 term in the denominator that corresponds to a term $(z-1)^2$. This causes the step-error constant K_p^* to be infinite. The same is true for any system type greater than 1. The summary of the error constants and the steady-state error due to a step input is as follows:

System Type	K_p^*	e_{ss}^*
0	K	$R/(1+K)$
1	∞	0
2	∞	0

Steady-State Error Due to a Ramp-Function Input

When the reference input to the system in Fig. 7-44 is a ramp function of magnitude R, $r(t) = Rtu_s(t)$. The steady-state error in Eq. (7-210) becomes

$$\begin{aligned} e_{ss}^* &= \lim_{z\to 1} \frac{RT}{(z-1)[1+G_{h0}G_p(z)]} \\ &= \frac{R}{\lim\limits_{z\to 1} [(z-1)/T]G_{h0}G_p(z)} \end{aligned} \tag{7-220}$$

Let the **ramp-error constant** be defined as

$$K_v^* = \frac{1}{T}\lim_{z\to 1} [(z-1)G_{h0}G_p(z)] \tag{7-221}$$

Then Eq. (7-220) becomes

$$e_{ss}^* = \frac{R}{K_v^*} \tag{7-222}$$

The ramp-error constant K_v^* is meaningful only when the input $r(t)$ is a ramp function and if the function $(z-1)G_{h0}G_p(z)$ in Eq. (7-221) does not have any poles on or outside the unit circle $|z| = 1$. The relations between the steady-state error e_{ss}^*, K_v^*, and the system type when the input is a ramp function with magnitude R are summarized as follows.

System Type	K_v^*	e_{ss}^*
0	0	∞
1	K	R/K
2	∞	0

Steady-Error Due to a Parabolic-Function Input

When the input is a parabolic function, $r(t) = Rtu_s(t)/2$; the z-transform of $r(t)$ is

$$R(z) = \frac{RT^2 z(z+1)}{2(z-1)^3} \tag{7-223}$$

From Eq. (7-210), the steady-state error at the sampling instants is

$$\begin{aligned} e_{ss}^* &= \frac{T^2}{2} \lim_{z \to 1} \frac{R(z+1)}{(z-1)^2[1 + G_{h0}G_p(z)]} \\ &= \frac{R}{(1/T^2) \lim_{z \to 1} (z-1)^2 G_{h0}G_p(z)} \end{aligned} \tag{7-224}$$

By defining the **parabolic-error constant** as

$$K_a^* = \frac{1}{T^2} \lim_{z \to 1} [(z-1)^2 G_{h0}G_p(z)] \tag{7-225}$$

the steady-state error due to a parabolic-function input is

$$e_{ss}^* = \frac{R}{K_a^*} \tag{7-226}$$

The relations between the steady-state error e_{ss}^*, K_a^*, and the system type when the input is a parabolic function with its z-transform described by Eq. (7-223) are summarized as follows.

System Type	K_a^*	e_{ss}^*
0	0	∞
1	0	∞
2	K	R/K
3	∞	0

7-11 Summary

This chapter is devoted to the time-domain analysis of linear continuous-data and discrete-data control systems. The time response of control system is divided into the transient and the steady-state responses. The steady-state error is a measure of the accuracy of the system when time approaches infinity. When the system has unity feedback, for the step, ramp, and parabolic inputs, the steady-state error is characterized by the error constants K_p, K_v, and K_a, respectively, as well as the system **type.** When applying the steady-state-error analysis, the final-value theorem of the Laplace transform is the basis; it should be ascertained that the closed-loop system is stable or the error analysis is invalid. The error constants are not defined for systems with nonunity feedback. For nonunity-feedback systems, a method of determining the steady-state error is introduced by using the closed-loop transfer function.

The transient response is characterized by such criteria as the **maximum overshoot, rise time, delay time,** and **settling time,** and such parameters as **damping ratio, natural undamped frequency,** and **time constant.** The analytical expressions of these parameters can all be related to the system parameters simply if the transfer function is of the second-order prototype. For second-order systems that are not of the prototype, and higher-order systems, the analytical relationships between the transient parameters and the system constants are difficult to determine. Computer simulations are recommended for these systems.

Time-domain analysis of a position-control system is conducted. The transient and steady-state analyses are carried out first by approximating the system as a second-order system. The effect of varying the amplifier gain K on the transient and steady-state performance is demonstrated. The concept of the root-locus technique is introduced. The system is then analyzed as a third-order system, and it is shown that the second-order approximation is accurate only for low values of K.

The effects of adding poles and zeros to the forward-path and closed-loop transfer functions are demonstrated. The dominant poles of transfer functions are also discussed. This established the significance of the location of the poles of transfer function in the s-plane, and under what condition the insignificant poles (and zeros) can be neglected as far as the transient response is concerned.

In Section 7-9 a method of approximating a high-order system by an equivalent low-order system is introduced. The low-order system tracks the high-order one in the sense that the frequency responses of the two systems are similar.

Finally, the last portion of this chapter is devoted to the time-domain analysis of discrete-data control systems. It is shown that practically all the steady-state and transient analysis of the continuous-data systems can be extended to discrete-data systems. Remember that when the z-transform method is used, the time response addressed is the sampled signal $y(kT)$. The interpretation of the pole–zero configuration in the z-plane should be made with respect to the unit circle $|z| = 1$.

Review Questions

1. Give the definitions of the error constants, K_p, K_v, and K_a.
2. Specify the type of input to which the error constant K_p is dedicated.
3. Specify the type of input to which the error constant K_v is dedicated.
4. Specify the type of input to which the error constant K_a is dedicated.
5. Define an error constant if the input to a unity-feedback control system is described by $r(t) = t^3 u_s(t)/6$.
6. Give the definition of the system **type** of a linear time-invariant system.
7. If a unity-feedback control system type is 2, it is certain that the steady-state error of the system to a step input or a ramp input will be zero. **(T)** **(F)**
8. Linear and nonlinear frictions will generally degrade the steady-state error of a control system. **(T)** **(F)**
9. The maximum overshoot of a unit-step response of the second-order prototype system will never exceed 100 percent when the damping ratio ζ and the natural undamped frequency ω_n are all positive. **(T)** **(F)**
10. For the second-order prototype system, when the undamped natural frequency ω_n increases, the maximum overshoot of the output stays the same. **(T)** **(F)**
11. The maximum overshoot of the following system will never exceed 100 percent when ζ, ω_n, and T are all positive.

$$\frac{Y(s)}{R(s)} = \frac{\omega_n^2(1 + Ts)}{s^2 + 2\zeta\omega_n s + \omega_n^2} \qquad \textbf{(T)} \qquad \textbf{(F)}$$

12. Increasing the undamped natural frequency will generally reduce the rise time of the step response. **(T)** **(F)**
13. Increasing the undamped natural frequency will generally reduce the settling time of the step response. **(T)** **(F)**
14. Adding a zero to the forward-path transfer function will generally improve the system damping, and thus will always reduce the maximum overshoot of the system. **(T)** **(F)**
15. Given the following characteristic equation of a linear control system, increasing the value of K will increase the frequency of oscillation of the system:

$$s^3 + 3s^2 + 5s + K = 0 \qquad \textbf{(T)} \qquad \textbf{(F)}$$

16. For the characteristic equation given in Review Question 15, increasing the coefficient of the s^2 term will generally improve the damping of the system. **(T)** **(F)**

17. The location of the roots of the characteristic equation in the s-plane will give a definite indication on the maximum overshoot of the transient response of the system. (T) (F)

18. The following transfer function $G(s)$ can be approximated by $G_L(s)$ since the pole at -20 is much larger than the dominant pole at $s = -1$.

$$G(s) = \frac{10}{s(s+1)(s+20)} \qquad G_L(s) = \frac{10}{s(s+1)}$$ (T) (F)

19. Give definitions of the error constants K_p^*, K_v^*, and K_a^* of discrete-data control systems with unity feedback.

20. Point out the difference between the time responses of a system with a pole at $z = 1$ and a system with a pole at $z = -1$.

21. When the characteristic-equation roots of a discrete-data system are found in the second and third quadrants inside the unit circle in the z-plane, the frequency of oscillation of the system will generally be higher than if the roots are in the first and fourth quadrants. (T) (F)

Answers to the true-and-false questions are given following the Problems section.

References

Simplification of Linear Systems

1. **E. J. DAVISON,** "A Method for Simplifying Linear Dynamic Systems," *IEEE Trans. Automatic Control,* Vol. AC-11, pp. 93–101, Jan. 1966.
2. **T. C. HSIA,** "On the Simplification of Linear Systems," *IEEE Trans. Automatic Control,* Vol. AC-17, pp. 372–374, June 1972.

Undershoot in Step Response

3. **M. VIDYASAGAR,** "On Undershoot and Nonminimum Phase Zeros," *IEEE Trans. Automatic Control,* Vol. AC-31, p. 440, May 1986.
4. **T. NORIMATSU and M. ITO,** "On the Zero Non-regular Control System," *J. Inst. Electr. Eng. Japan,* Vol. 81, pp. 566–575, 1961.

Discrete-Data Control Systems

See the references at the end of Chapter 3.

Problems

▲ *s*-Domain specifications

7-1. A pair of complex-conjugate poles in the s-plane is required to meet the various specifications that follow. For each specification, sketch the region in the s-plane in which the poles should be located.

(a) $\zeta \geq 0.707$ $\omega_n \geq 2$ rad/sec (positive damping)
(b) $0 \leq \zeta \leq 0.707$ $\omega_n \leq 2$ rad/sec (positive damping)
(c) $\zeta \leq 0.5$ $1 \leq \omega_n \leq 5$ rad/sec (positive damping)
(d) $0.5 \leq \zeta \leq 0.707$ $\omega_n \leq 5$ rad/sec (positive and negative damping)

▲ System type

7-2. Determine the type of the following unity-feedback systems for which the forward-path transfer functions are given.

(a) $G(s) = \dfrac{K}{(1+s)(1+10s)(1+20s)}$ (b) $G(s) = \dfrac{10e^{-0.2s}}{(1+s)(1+10s)(1+20s)}$

(c) $G(s) = \dfrac{10(s+1)}{s(s+5)(s+6)}$ (d) $G(s) = \dfrac{100(s-1)}{s^2(s+5)(s+6)^2}$

(e) $G(s) = \dfrac{10(s+1)}{s^3(s^2+5s+5)}$ (f) $G(s) = \dfrac{100}{s^3(s+2)^2}$

▲ Error constants

7-3. Determine the step, ramp, and parabolic error constants of the following unity-feedback control systems. The forward-path transfer functions are given.

(a) $G(s) = \dfrac{1000}{(1+0.1s)(1+10s)}$ (b) $G(s) = \dfrac{100}{s(s^2+10s+100)}$

(c) $G(s) = \dfrac{K}{s(1+0.1s)(1+0.5s)}$ (d) $G(s) = \dfrac{100}{s^2(s^2+10s+100)}$

(e) $G(s) = \dfrac{1000}{s(s+10)(s+100)}$ (f) $G(s) = \dfrac{K(1+2s)(1+4s)}{s^2(s^2+s+1)}$

▲ Steady-state error of unity-feedback systems

7-4. For the unity-feedback control systems described in Problem 7-3, determine the steady-state error for a unit-step input, a unit-ramp input, and a parabolic input, $(t^2/2)u_s(t)$. Check the stability of the system before applying the final-value theorem.

▲ Steady-state error of nonunity-feedback systems

7-5. The following transfer functions are given for a single-loop nonunity-feedback control system. Find the steady-state errors due to a unit-step input, a unit-ramp input, and a parabolic input, $(t^2/2)u_s(t)$.

(a) $G(s) = \dfrac{1}{s^2+s+2}$ $H(s) = \dfrac{1}{s+1}$

(b) $G(s) = \dfrac{1}{s(s+5)}$ $H(s) = 5$

(c) $G(s) = \dfrac{1}{s^2(s+10)}$ $H(s) = \dfrac{s+1}{s+5}$

(d) $G(s) = \dfrac{1}{s^2(s+12)}$ $H(s) = 5(s+2)$

▲ Steady-state error of single-loop systems

7-6. Find the steady-state errors of the following single-loop control systems for a unit-step input, a unit-ramp input, and a parabolic input, $(t^2/2)u_s(t)$. For systems that include a parameter K, find its value so that the answers are valid.

(a) $M(s) = \dfrac{s+4}{s^4 + 16s^3 + 48s^2 + 4s + 4}, K_H = 1$

(b) $M(s) = \dfrac{K(s+3)}{s^3 + 3s^2 + (K+2)s + 3K}, K_H = 1$

(c) $M(s) = \dfrac{s+5}{s^4 + 15s^3 + 50s^2 + 10s}, H(s) = \dfrac{10s}{s+5}$

(d) $M(s) = \dfrac{K(s+5)}{s^4 + 17s^3 + 60s^2 + 5Ks + 5K}, K_H = 1$

▲ Error constants and steady-state errors

7-7. The block diagram of a control system is shown in Fig. 7P-7. Find the step-, ramp-, and parabolic-error constants. The error signal is defined to be $e(t)$. Find the steady-state errors in terms of K and K_t when the following inputs are applied. Assume that the system is stable.

(a) $r(t) = u_s(t)$ (b) $r(t) = tu_s(t)$ (c) $r(t) = (t^2/2)u_s(t)$

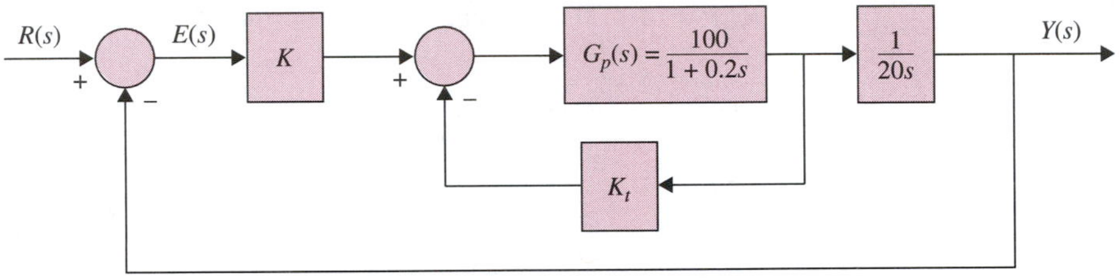

Figure 7P-7

▲ Error constants and steady-state errors

7-8. Repeat Problem 7-7 when the transfer function of the process is, instead,

$$G_p(s) = \frac{100}{(1+0.1s)(1+0.5s)}$$

What constraints must be made, if any, on the values of K and K_t so that the answers are valid? Determine the minimum steady-state error that can be achieved with a unit-ramp input by varying the values of K and K_t.

▲ Steady-state error

7-9. For the position-control system shown in Fig. 3P-19, determine the following.

(a) Find the steady-state value of the error signal $\theta_e(t)$ in terms of the system parameters when the input is a unit-step function.

(b) Repeat part (a) when the input is a unit-ramp function. Assume that the system is stable.

▲ Steady-state error

7-10. The block diagram of a feedback control system is shown in Fig. 7P-10. The error signal is defined to be $e(t)$.

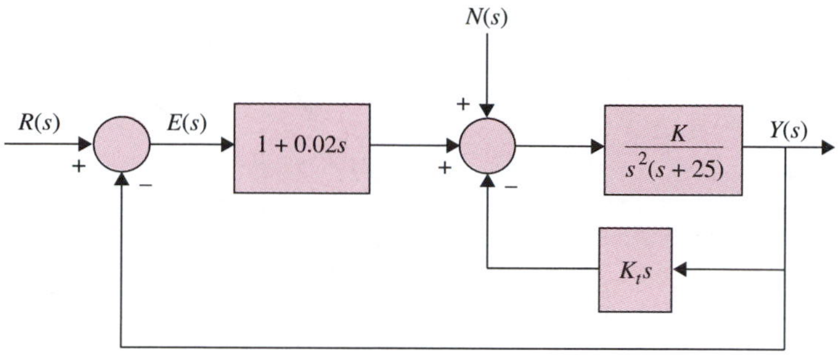

Figure 7P-10

(a) Find the steady-state error of the system in terms of K and K_t when the input is a unit-ramp function. Give the constraints on the values of K and K_t so that the answer is valid. Let $n(t) = 0$ for this part.

(b) Find the steady-state value of $y(t)$ when $n(t)$ is a unit-step function. Let $r(t) = 0$. Assume that the system is stable.

▲ Steady-state error due to input and disturbance

7-11. The block diagram of a linear control system is shown in Fig. 7P-11, where $r(t)$ is the reference input and $n(t)$ is the disturbance.

(a) Find the steady-state value of $e(t)$ when $n(t) = 0$ and $r(t) = tu_s(t)$. Find the conditions on the values of α and K so that the solution is valid.

(b) Find the steady-state value of $y(t)$ when $r(t) = 0$ and $n(t) = u_s(t)$.

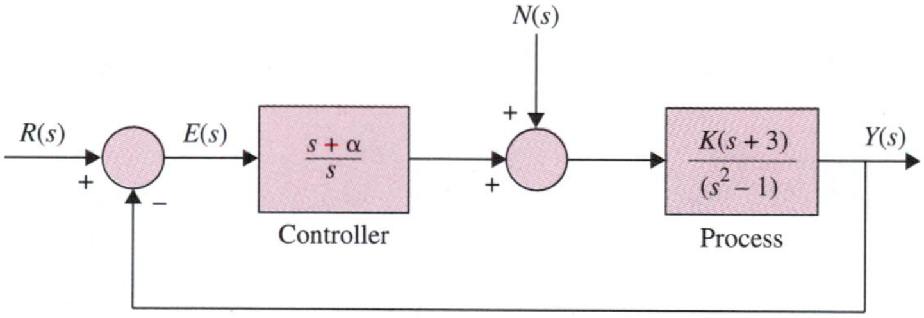

Figure 7P-11

▲ Transfer function of second-order prototype system

7-12. The unit-step response of a linear control system is shown in Fig. 7P-12. Find the transfer function of a second-order prototype system to model the system.

▲ Maximum overshoot and rise time

7-13. For the control system shown in Fig. 7P-7, find the values of K and K_t so that the maximum overshoot of the output is approximately 4.3 percent and the rise time t_r is

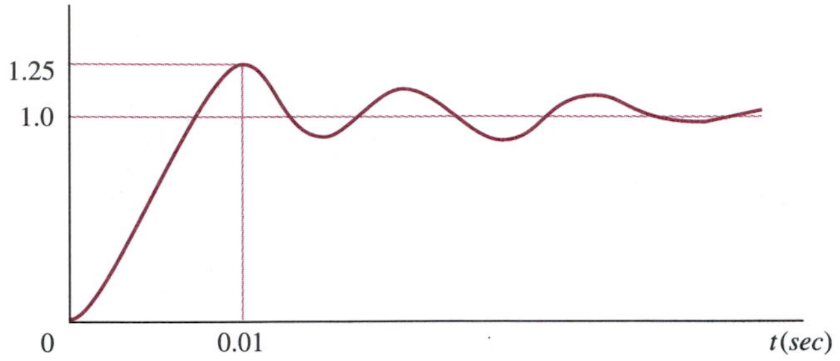

Figure 7P-12

approximately 0.2 sec. Use Eq. (7-104) for the rise-time relationship. Simulate the system with any time-response simulation program to check the accuracy of your solutions.

▲ Maximum overshoot and rise time

7-14. Repeat Problem 7-13 with a maximum overshoot of 10 percent and a rise time of 0.1 sec.

7-15. Repeat Problem 7-13 with a maximum overshoot of 20 percent and a rise time of 0.05 sec.

▲ Maximum overshoot and delay time

7-16. For the control system shown in Fig. 7P-7, find the values of K and K_t so that the maximum overshoot of the output is approximately 4.3 percent and the delay time t_d is approximately 0.1 sec. Use Eq. (7-102) for the delay-time relationship. Simulate the system with a computer program to check the accuracy of your solutions.

▲ Maximum overshoot and delay time

7-17. Repeat Problem 7-16 with a maximum overshoot of 10 percent and a delay time of 0.05 sec.

7-18. Repeat Problem 7-16 with a maximum overshoot of 20 percent and a delay time of 0.01 sec.

▲ Damping ratio and settling time

7-19. For the control system shown in Fig. 7P-7, find the values of K and K_t so that the damping ratio of the system is 0.6 and the settling time of the unit-step response is 0.1 sec. Use Eq. (7-108) for the settling-time relationship. Simulate the system with a computer program to check on the accuracy of your results.

▲ Maximum overshoot and settling time

7-20. **(a)** Repeat Problem 7-19 with a maximum overshoot of 10 percent and a settling time of 0.05 sec.

(b) Repeat Problem 7-19 with a maximum overshoot of 20 percent and a settling time of 0.01 sec.

▲ Damping ratio and settling time

7-21. Repeat Problem 7-19 with a damping ratio of 0.707 and a settling time of 0.1 sec. Use Eq. (7-109) for the settling-time relationship.

▲ Damping ratio, rise time, steady-state error

7-22. The forward-path transfer function of a control system with unity feedback is

$$G(s) = \frac{K}{s(s+a)(s+30)}$$

where a and K are real constants.

(a) Find the values of a and K so that the relative damping ratio of the complex roots of the characteristic equation is 0.5 and the rise time of the unit-step response is approximately 1 sec. Use Eq. (7-104) as an approximation of the rise time. With the values of a and K found, determine the actual rise time using computer simulation.

(b) With the values of a and K found in part (a), find the steady-state errors of the system when the reference input is (i) a unit-step function, and (ii) a unit-ramp function.

▲ Time response

7-23. The block diagram of a linear control system is shown in Fig. 7P-23.

(a) By means of trial and error, find the value of K so that the characteristic equation has two equal real roots and the system is stable. You may use any root-finding computer program to solve this problem.

(b) Find the unit-step response of the system when K has the value found in part (a). Use any computer simulation program for this. Set all the initial conditions to zero.

(c) Repeat part (b) when $K = -1$. What is peculiar about the step response for small t, and what may have caused it?

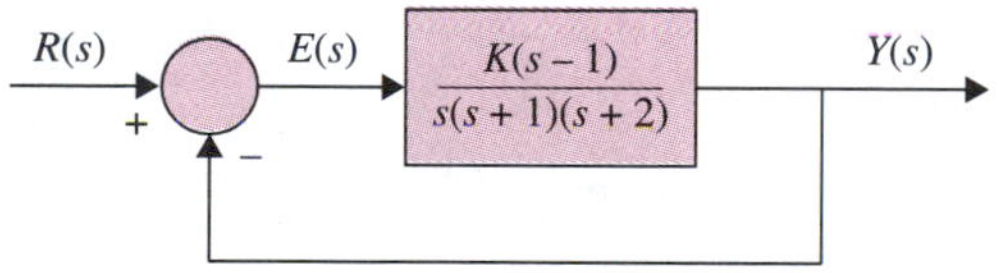

Figure 7P-23

▲ State feedback and parameter plane

7-24. A controlled process is represented by the following dynamic equations:

$$\frac{dx_1(t)}{dt} = -x_1(t) + 5x_2(t)$$

$$\frac{dx_2(t)}{dt} = -6x_1(t) + u(t)$$

$$y(t) = x_1(t)$$

The control is obtained through state feedback with

$$u(t) = -k_1x_1(t) - k_2x_2(t) + r(t)$$

where k_1 and k_2 are real constants and $r(t)$ is the reference input.

(a) Find the locus in the k_1 versus k_2 plane (k_1 = vertical axis) on which the overall system has a natural undamped frequency of 10 rad/sec.

(b) Find the locus in the k_1 versus k_2 plane on which the overall system has a damping ratio of 0.707.

(c) Find the values of k_1 and k_2 such that $\zeta = 0.707$ and $\omega_n = 10$ rad/sec.

(d) Let the error signal be defined as $e(t) = r(t) - y(t)$. Find the steady-state error when $r(t) = u_s(t)$ and k_1 and k_2 are at the values found in part (c).

(e) Find the locus in the k_1 versus k_2 plane on which the steady-state error due to a unit-step input is zero.

▲ Parameter plane and special trajectories

7-25 The block diagram of a linear control system is shown in Fig. 7P-25. Construct a parameter plane of K_P versus K_D (K_P is the vertical axis) and show the following trajectories or regions in the plane.

(a) Unstable and stable regions.

(b) Trajectories on which the damping is critical ($\zeta = 1$).

(c) Region in which the system is overdamped ($\zeta > 1$).

(d) Region in which the system is underdamped ($\zeta < 1$).

(e) Trajectory on which the parabolic-error constant K_a is 1000 $\sec^{-2}$.

(f) Trajectory on which the natural undamped frequency ω_n is 50 rad/sec.

(g) Trajectory on which the system is either uncontrollable or unobservable. (*Hint:* Look for pole–zero cancellation.)

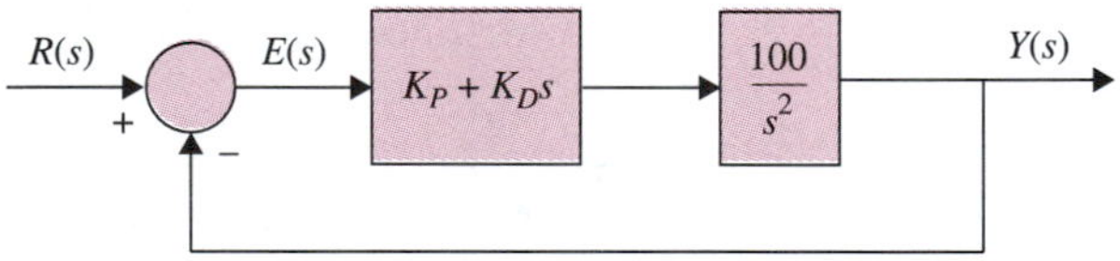

Figure 7P-25

▲ Steady-state performances

7-26. The block diagram of a linear control system is shown in Fig. 7P-26. The fixed parameters of the system are given as $T = 0.1$, $J = 0.01$, and $K_i = 10$.

(a) When $r(t) = tu_s(t)$ and $T_d(t) = 0$, determine how the values of K and K_t affect the steady-state value of $e(t)$. Find the restrictions on K and K_t so that the system is stable.

(b) Let $r(t) = 0$. Determine how the values of K and K_t affect the steady-state value of $y(t)$ when the disturbance input $T_d(t) = u_s(t)$.

(c) Let $K_t = 0.01$ and $r(t) = 0$. Find the minimum steady-state value of $y(t)$ that can be obtained by varying K when $T_d(t)$ is a unit-step function. Find the value of this K. From the transient standpoint, would you operate the system at this value of K? Explain.

(d) Assume that it is desired to operate the system with the value of K as selected in part (c). Find the value of K_t so that the complex roots of the characteristic equation will have a real part of -2.5. Find all the three roots of the characteristic equation.

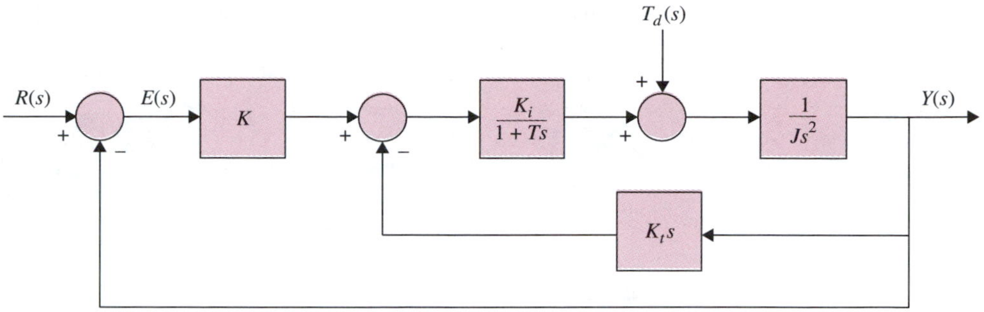

Figure 7P-26

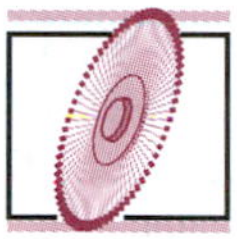

▲ Printwheel control system

7-27. The dc-motor control system for controlloing a printwheel described in Problem 4-14 has the forward-path transfer function

$$G(s) = \frac{\Theta_o(s)}{\Theta_e(s)} = \frac{nK_sK_iK_LK}{\Delta(s)}$$

where

$$\Delta(s) = s[L_aJ_mJ_Ls^4 + J_L(R_aJ_m + B_mL_a)s^3 + (n^2K_LL_aJ_L + K_LL_aJ_m + K_iK_bJ_L + R_aB_mJ_L)s^2 + (n^2R_aK_LJ_m + R_aK_LJ_m + B_mK_LL_a)s + R_aB_mK_L + K_iK_bK_L]$$

and where $K_i = 9$ oz-in./A, $K_b = 0.636$ V/rad/sec, $R_a = 5\ \Omega$, $L_a = 1$ mH, $K_s = 1$ V/rad, $n = 1/10$, $J_m = J_L = 0.001$ oz-in.-sec², and $B_m \cong 0$. The characteristic equation of the closed-loop system is

$$\Delta(s) + nK_sK_iK_LK = 0$$

(a) Let $K_L = 10{,}000$ oz-in./rad. Write the forward-path transfer function $G(s)$ and find the poles of $G(s)$. Find the critical value of K for the closed-loop system to be stable. Find the roots of the characteristic equation of the closed-loop system when K is at marginal stability.

(b) Repeat part (a) when $K_L = 1000$ oz-in./rad.

(c) Repeat part (a) when $K_L = \infty$ (i.e., the motor shaft is rigid).

(d) Compare the results of parts (a), (b), and (c), and comment on the effects of the values of K_L on the poles of $G(s)$ and the roots of the characteristic equation.

▲ Guided-missile control system

7-28. The block diagram of the guided-missile attitude-control system described in Problem 4-13 is shown in Fig. 7P-28. The command input is $r(t)$, and $d(t)$ represents disturbance input. The objective of this problem is to study the effect of the controller $G_c(s)$ on the steady-state and transient responses of the system.

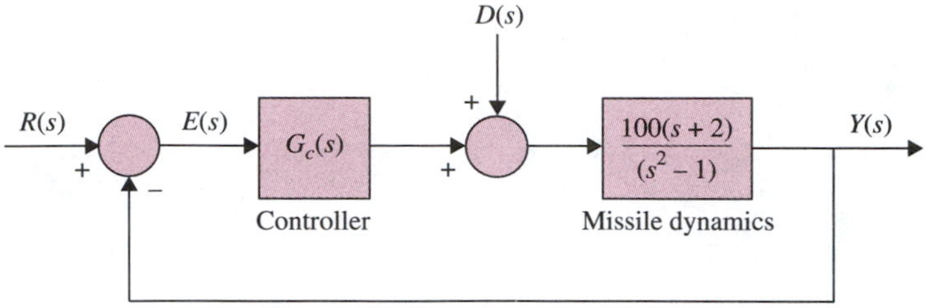

Figure 7P-28

(a) Let $G_c(s) = 1$. Find the steady-state error of the system when $r(t)$ is a unit-step function. Set $d(t) = 0$.

(b) Let $G_c(s) = (s + \alpha)/s$. Find the steady-state error when $r(t)$ is a unit-step function.

(c) Obtain the unit-step response of the system for $0 \leq t \leq 0.5$ sec with $G_c(s)$ as given in part (b), and $\alpha = 5, 50,$ and 500. Assume zero initial conditions. Record the maximum overshoot of $y(t)$ for each case. Use any available computer simulation program. Comment on the effect of varying the value of α of the controller on the transient response.

(d) Set $r(t) = 0$ and $G_c(s) = 1$. Find the steady-state value of $y(t)$ when $d(t) = u_s(t)$.

(e) Let $G_c(s) = (s + \alpha)/s$. Find the steady-state value of $y(t)$ when $d(t) = u_s(t)$.

(f) Obtain the output response for $0 \leq t \leq 0.5$ sec, with $G_c(s)$ as given in part (e) when $r(t) = 0$ and $d(t) = u_s(t)$; $\alpha = 5, 50,$ and 500. Use zero initial conditions.

(g) Comment on the effect of varying the value of α of the controller on the transient responses of $y(t)$ due to $r(t)$ and $d(t)$.

▲ Reduced-order liquid-level control system

7-29. The block diagram shown in Fig. 7P-29 represents the liquid-level control system described in Problem 6-13. The liquid level is represented by $h(t)$, and N denotes the number of inlets.

(a) Since one of the poles of the open-loop transfer function is relatively far to the left on the real axis of the s-plane at $s = -10$, it is suggested that this pole can be neglected. Approximate the system by a second-order system by neglecting the pole of $G(s)$ at $s = -10$. The approximation should be valid for both the transient and the steady-state responses. Apply the formulas for the maximum overshoot and the peak time t_{max} to the second-order model for $N = 1$ and $N = 10$.

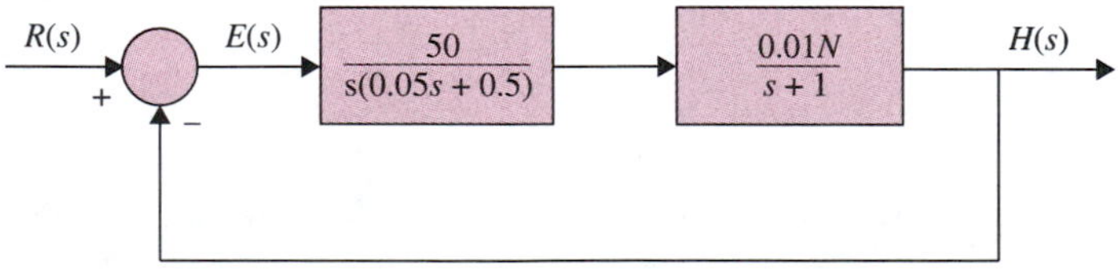

Figure 7P-29

(b) Obtain the unit-step response (with zero initial conditions) of the original third-order system with $N = 1$ and then with $N = 10$. Compare the responses of the original system with those of the second-order approximating system. Comment on the accuracy of the approximation as a function of N.

▲ Effects of moving open-loop zeros

7-30. The forward-path transfer function of a unity-feedback control system is

$$G(s) = \frac{1 + T_z s}{s(s + 1)^2}$$

Compute and plot the unit-step responses of the closed-loop system for $T_z = 0, 0.5, 1, 10$, and 50. Assume zero initial conditions. Use any computer simulation program that is available. Comment on the effects of the various values of T_z on the step response.

▲ Effect of moving open-loop pole

7-31. The forward-path transfer function of a unity-feedback control system is

$$G(s) = \frac{1}{s(s + 1)^2(1 + T_p s)}$$

Compute and plot the unit-step responses of the closed-loop system for $T_p = 0, 0.5$, and 0.707. Assume zero initial conditions. Use any computer simulation program. Find the critical value of T_p so that the closed-loop system is marginally stable. Comment on the effects of the pole at $s = -1/T_p$ in $G(s)$.

▲ Reduced order of liquid-level control system

7-32. This problem is devoted to the approximation of a third-order system by a second-order one, using the method described in Section 7-9. The liquid-level control system described in Problems 6-13 and 7-29 has the forward-path transfer function

$$G(s) = \frac{10N}{s(s + 1)(s + 10)}$$

(a) For $N = 1$, find the transfer function $G_L(s)$ of a second-order system that approximates the third-order system. Compute the unit-step responses of the systems, and observe the closeness of the approximation. Compare the roots of the characteristic equations of the two closed-loop systems.

(b) Repeat part (a) with $N = 2$.

(c) Repeat part (a) with $N = 3$.

(d) Repeat part (a) with $N = 4$.

(e) Repeat part (a) with $N = 5$.

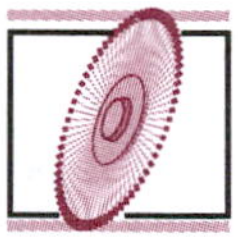

▲ Reduced-order printwheel control system

7-33. The dc-motor control system for controlling a printwheel described in Problem 4-14 has the following forward-path transfer function when $K_L = \infty$.

$$G(s) = \frac{891{,}100K}{s(s^2 + 5000s + 566{,}700)}$$

The system is to be approximated by a second-order system using the method described in Section 7-9.

(a) For $K = 1$, find the second-order transfer function $G_L(s)$ that approximates $G(s)$. Compute and plot the unit-step responses of the systems, and evaluate the closeness of the approximation. Compare the roots of the characteristic equations of the two closed-loop systems.
(b) Repeat part (a) with $K = 100$.
(c) Repeat part (a) with $K = 1000$.

▲ Reduced-order approximation

7-34. Approximate the control system shown in Fig. 7P-23 when $K = -1$ by a second-order system with the closed-loop transfer function

$$M(s) = \frac{Y(s)}{R(s)} = \frac{1 + c_1 s}{1 + d_1 s + d_2 s^2}$$

where c_1, d_1, and d_2 are real constants. Use the method outlined in Section 7-9. Compute and plot the unit-step responses of the original third-order system ($K = -1$) and the second-order approximating system. Compare the roots of the characteristic equations of the two systems.

7-35. A unity-feedback control system has the forward-path transfer function

$$G(s) = \frac{K}{s(s+1)(s+2)(s+20)}$$

(a) For $K = 10$, the fourth-order system is to be approximated by a second-order system with the closed-loop transfer function

$$L(s) = \frac{G_L(s)}{1 + G_L(s)} = \frac{1}{1 + d_1 s + d_2 s^2}$$

Find d_1 and d_2 using the method outlined in Section 7-9. Compute and plot the unit-step responses of the fourth- and second-order systems and compare. Find the roots of the characteristic equations of the two systems.

(b) For $K = 10$, approximate the system by a third-order model with

$$L(s) = \frac{1}{1 + d_1 s + d_2 s^2 + d_3 s^3}$$

Carry out the tasks specified in part (a).

(c) Repeat parts (a) and (b) with $K = 40$.

▲ Sampled-data system

7-36. The block diagram of a sampled-data control system is shown in Fig. 7P-36.

(a) Derive the forward-path and the closed-loop transfer functions of the system in z-transforms. The sampling period is 0.1 sec.

(b) Compute the unit-step response $y(kT)$ for $k = 0$ to 100.

(c) Repeat parts (a) and (b) for $T = 0.05$ sec.

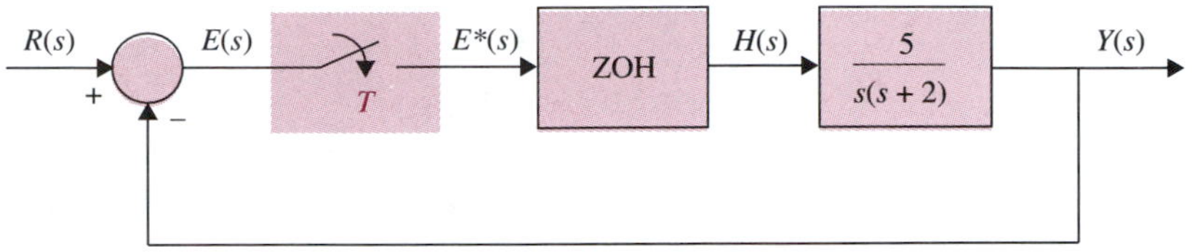

Figure 7P-36

▲ Sampled-data system, error constants

7-37. The block diagram of a sampled-data control system is shown in Fig. 7P-37.

(a) Find the error constants K_p^*, K_v^*, and K_a^*.

(b) Derive the transfer functions $Y(z)/E(z)$ and $Y(z)/R(z)$.

(c) For $T = 0.1$ sec, find the critical value of K for system stability.

(d) Compute the unit-step response $y(kT)$ for $k = 0$ to 50 for $T = 0.1$ sec and $K_t = 5$.

(e) Repeat part (d) for $T = 0.1$ sec and $K_t = 1$.

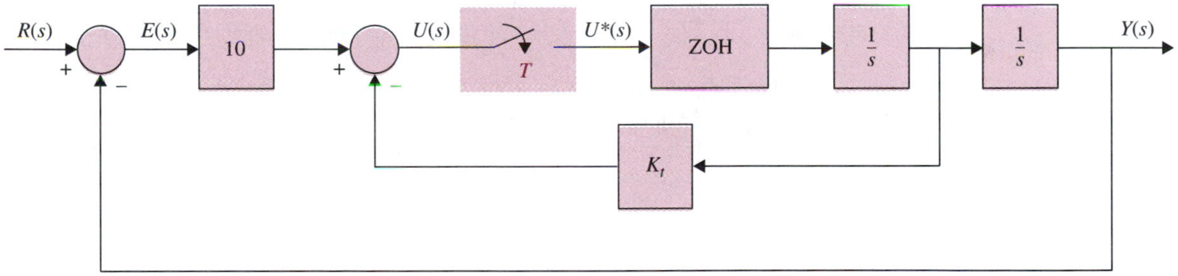

Figure 7P-37

▲ Digital dc-motor control systems

7-38. The forward-path dc-motor control system described in Problem 5-33 is now incorporated in a digital control system, as shown in Fig. 7P-38(a). The microprocessor takes the information from the encoder and computes the velocity information. This generates the sequence of numbers $\omega(kT)$, $k = 0, 1, 2, \ldots$. The microprocessor then generates

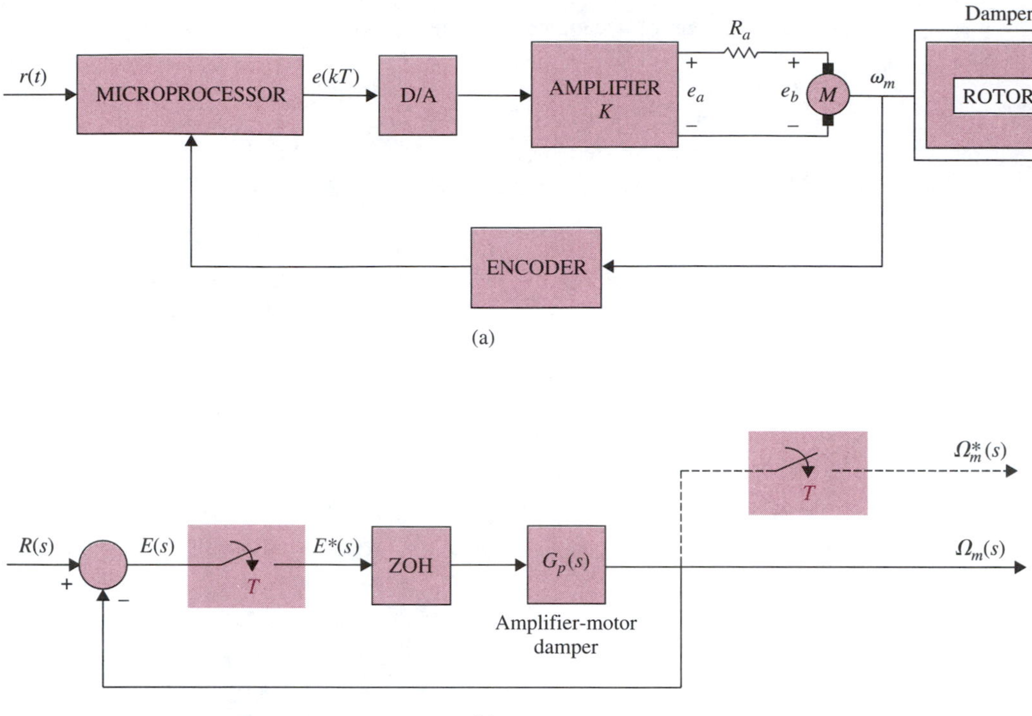

Figure 7P-38

the error signal $e(kT) = r(kT) - \omega(kT)$. The digital control is modeled by the block diagram shown in Fig. 7P-38(b). Use the parameter values given in Problem 5-33.

(a) Find the transfer function $\Omega(z)/E(z)$ with the sampling period $T = 0.1$ sec.

(b) Find the closed-loop transfer function $\Omega(z)/R(z)$. Find the characteristic equation and its roots. Locate these roots in the z-plane. Show that the closed-loop system is unstable when $T = 0.1$ sec.

(c) Repeat parts (a) and (b) for $T = 0.01$ and 0.001 sec. Use any computer simulation program.

(d) Find the error constants K_p^*, K_v^*, and K_a^*. Find the steady-state error $e(kT)$ as $k \to \infty$ when the input $r(t)$ is a unit-step function, a unit-ramp function, and parabolic function $t^2u_s(t)/2$.

Additional Computer Problems

Compare and plot the unit-step responses of the unity-feedback closed-loop systems with the forward-path transfer functions given. Assume zero initial conditions. Use any computer simulation program.

7-39. **(a)** For $T_z = 0, 1, 5, 20,$

$$G(s) = \frac{1 + T_z s}{s(s + 0.55)(s + 1.5)}$$

(b) For $T_z = 0, 1, 5, 20,$

$$G(s) = \frac{1 + T_z s}{s^2 + 2s + 2}$$

7-40. **(a)** For $T_p = 0, 0.5, 1.0,$

$$G(s) = \frac{2}{(s^2 + 2s + 2)(1 + T_p s)}$$

(b) For $T_p = 0, 0.5, 1.0,$

$$G(s) = \frac{10}{s(s + 5)(1 + T_p s)}$$

7-41. $$G(s) = \frac{K}{s(s + 1.25)(s^2 + 2.5s + 10)}$$

(a) For $K = 5$. **(b)** For $K = 10$. **(c)** For $K = 30$.

7-42. $$G(s) = \frac{K(s + 2.5)}{s(s + 1.25)(s^2 + 2.5s + 10)}$$

(a) For $K = 5$. **(b)** For $K = 10$. **(c)** For $K = 30$.

Answers to True-and-False Review Questions

7. **(F)** **8.** **(T)** **9.** **(T)** **10.** **(T)** **11.** **(F)** **12.** **(T)** **13.** **(T)**
14. **(F)** **15.** **(T)** **16.** **(T)** **17.** **(F)** **18.** **(F)** **21.** **(T)**

8 Root-Locus Technique

KEYWORDS AND TOPICS

- ▲ Root Loci
- ▲ Properties of Root Loci
- ▲ Root Sensitivity
- ▲ Root Contours
- ▲ Root Loci of Discrete-Data Systems

8-1 Introduction

▲ Root loci are trajectories of roots of characteristic equation when a system parameter varies.

In the preceding chapter, we have demonstrated the importance of the poles and zeros of the closed-loop transfer function of a linear control system on the dynamic performance of the system. The roots of the characteristic equation, which are the poles of the closed-loop transfer function, determine the absolute and the relative stability of linear SISO systems. Keep in mind that the transient properties of the system also depend on the zeros of the closed-loop transfer function.

An important study in linear control systems is the investigation of the trajectories of the roots of the characteristic equation—or, simply, the **root loci**—when a certain system parameter varies. In Chapter 7, several examples illustrated the usefulness of the root loci of the characteristic equation in the study of linear control systems.

The basic properties and the systematic construction of the root loci are first due to W. R. Evans [1,3]. In this chapter we show how to construct these loci by following some simple rules. For plotting the root loci accurately, one can always use a root-locus program and a digital computer. For example, the programs **ROOTLOCI** in the **ACSP** package, **rlplot** in the **CSAD** toolbox, and **ROOT LOCUS** of the **Program CC,** just to name a few, can all be used to generate the data and plots of the root loci given the loop transfer function. If we were learning to be a technician, all we have to do is get familiar with the application of one of these programs. However, it is far more important to learn the basics of the root loci, their properties, as well as how to interpret the data provided by the root loci for analysis and design purposes. As an intelligent engineer, we must know if the data provided by the root loci are indeed correct, and be able to derive vital information from the root loci. The material in this chapter is prepared with these objectives in mind.

The root-locus technique is not confined to the study of control systems. In general, the method can be applied to study the behavior of roots of any algebraic equation with variable parameters. The general root-locus problem can be formulated by referring to the following algebraic equation of the complex variable, say, s:

$$F(s) = P(s) + KQ(s) = 0 \tag{8-1}$$

where $P(s)$ is an nth-order polynomial of s,

$$P(s) = s^n + a_{n-1}s^{n-1} + \cdots + a_1 s + a_0 \tag{8-2}$$

and $Q(s)$ is an mth-order polynomial of s; n and m are positive integers.

$$Q(s) = s^m + b_{m-1}s^{m-1} + \cdots + b_1 s + b_0 \tag{8-3}$$

For the present, we do not place any limitations on the relative magnitudes between n and m. K in Eq. (8-1) is a real constant that can vary from $-\infty$ to $+\infty$.

The coefficients $a_0, a_2, \ldots, a_{n-1}, b_1, b_2, \ldots, b_{m-1}$ are considered to be real and fixed.

▲ Root loci of multiple-variable parameters are called root contours.

Root loci of multiple-variable parameters can be treated by varying one parameter at a time. The resultant loci are called the **root contours,** and the subject is treated in Section 8-6. By replacing s with z in Eqs. (8-1) through (8-3), we can construct the root loci of the characteristic equation of a linear discrete-data system in a similar fashion.

For the purpose of identification, in this book we define the following categories of root loci based on the sign of K:

▲ RL refers to root loci with positive K.

▲ CRL refers to root loci with negative K.

1. **RL:** the portion of the root loci when K is positive; $0 \le K < \infty$.
2. **CRL (complementary root loci):** the portion of the root loci when K is negative; $-\infty < K \le 0$.
3. **RC (root contours):** contour of roots when more than one parameter varies.
4. **Root loci:** refers to the total root loci for $-\infty < K < \infty$.

8-2 Basic Properties of the Root Loci

Since our main interest is control systems, let us consider the closed-loop transfer function of a single-loop control system:

$$\frac{Y(s)}{R(s)} = \frac{G(s)}{1 + G(s)H(s)} \tag{8-4}$$

keeping in mind that the transfer function of multiple-loop SISO systems can also be expressed in a similar form. The characteristic equation of the closed-loop system is obtained by setting the denominator polynomial of $Y(s)/R(s)$ to zero. Thus the roots of the characteristic equation must satisfy

$$1 + G(s)H(s) = 0 \tag{8-5}$$

Suppose that $G(s)H(s)$ contains a variable parameter K as a multiplying factor, such that the rational function can be written as

$$G(s)H(s) = \frac{KQ(s)}{P(s)} \tag{8-6}$$

where $P(s)$ and $Q(s)$ are polynomials as defined in Eqs. (8-2) and (8-3), respectively. Equation (8-5) is written

$$1 + \frac{KQ(s)}{P(s)} = \frac{P(s) + KQ(s)}{P(s)} = 0 \tag{8-7}$$

The numerator polynomial of Eq. (8-7) is identical to Eq. (8-1). Thus by considering that the loop transfer function $G(s)H(s)$ can be written in the form of Eq. (8-6), we have identified the root loci of a control system with the general root-locus problem.

When the variable parameter K does not appear as a multiplying factor of $G(s)H(s)$, we can always condition the functions in the form of Eq. (8-1). As an illustrative example, consider that the characteristic equation of a control system is

$$s(s + 1)(s + 2) + s^2 + (3 + 2K)s + 5 = 0 \tag{8-8}$$

To express the last equation in the form of Eq. (8-7), *we divide both sides of the equation by the terms that do not contain K,* and we get

$$1 + \frac{2Ks}{s(s + 1)(s + 2) + s^2 + 3s + 5} = 0 \tag{8-9}$$

Comparing the last equation with Eq. (8-7), we get

$$\frac{Q(s)}{P(s)} = \frac{2s}{s^3 + 4s^2 + 5s + 5} \tag{8-10}$$

Now K is isolated as a multiplying factor to the function $Q(s)/P(s)$.

We shall show that the root loci of Eq. (8-5) can be constructed based on the properties of $Q(s)/P(s)$. In case where $G(s)H(s) = KQ(s)/P(s)$, the root-locus problem is another example in which the characteristics of the closed-loop system, in this case represented by the roots of the characteristic equation, are determined from the knowledge of the loop transfer function $G(s)H(s)$.

Now we are ready to investigate the conditions under which Eq. (8-5) or Eq. (8-7) is satisfied. Let us express $G(s)H(s)$ as

$$G(s)H(s) = KG_1(s)H_1(s) \tag{8-11}$$

where $G_1(s)H_1(s)$ does not contain the variable parameter K. Then Eq. (8-5) is written

$$G_1(s)H_1(s) = -\frac{1}{K} \tag{8-12}$$

To satisfy Eq. (8-12), the following conditions must be satisfied simultaneously:

Condition on Magnitude

$$|G_1(s)H_1(s)| = \frac{1}{|K|} \qquad -\infty < K < \infty \tag{8-13}$$

Condition on Angles

$$\angle G_1(s)H_1(s) = (2i + 1)\pi \qquad K \geq 0$$
$$= \text{odd multiples of } \pi \text{ radians or } 180° \tag{8-14}$$
$$\angle G_1(s)H_1(s) = 2i\pi \qquad K \leq 0$$
$$= \text{even multiples of } \pi \text{ radians or } 180° \tag{8-15}$$

where $i = 0, \pm 1, \pm 2, \ldots$ (any integer).

In practice, the conditions stated in Eqs. (8-13) through (8-15) play different roles in the construction of the root loci.

- ▲ The conditions on angles in Eq. (8-14) or Eq. (8-15) are used to determine the trajectories of the root loci in the s-plane.
- ▲ Once the root loci are drawn, the values of K on the loci are determined by using the condition on magnitude in Eq. (8-13).

The construction of the root loci is basically a graphical problem, although some of the properties are derived analytically. The graphical construction of the root loci is based on knowledge of the poles and zeros of the function $G(s)H(s)$. In other words, $G(s)H(s)$ must first be written as

$$G(s)H(s) = KG_1(s)H_1(s) = \frac{K(s + z_1)(s + z_2)\cdots(s + z_m)}{(s + p_1)(s + p_2)\cdots(s + p_n)} \tag{8-16}$$

where the zeros and poles of $G(s)H(s)$ are real or in complex-conjugate pairs.

Applying the conditions in Eqs. (8-13), (8-14), and (8-15) to Eq. (8-16), we have

$$|G_1(s)H_1(s)| = \frac{\prod_{k=1}^{m} |s + z_k|}{\prod_{j=1}^{n} |s + p_j|} = \frac{1}{|K|} \qquad -\infty < K < \infty \tag{8-17}$$

For $0 \leq K < \infty$ (RL):

$$\angle G_1(s)H_1(s) = \sum_{k=1}^{m} \angle(s + z_k) - \sum_{j=1}^{n} \angle(s + p_j) = (2i + 1) \times 180° \tag{8-18}$$

For $-\infty < K \le 0$ (CRL):

$$\angle G_1(s)H_1(s) = \sum_{k=1}^{m} \angle(s + z_k) - \sum_{j=1}^{n} \angle(s + p_j) = 2i \times 180^\circ \tag{8-19}$$

where $i = 0, \pm 1, \pm 2, \ldots$.

The graphical interpretation of Eq. (8-18) is that any point s_1 on the RL that corresponds to a positive value of K must satisfy the condition:

> ***The difference between the sums of the angles of the vectors drawn from the zeros and those from the poles of $G(s)H(s)$ to s_1 is an odd multiple of 180°.***

For negative values of K, any point s_1 on the CRL must satisfy the condition:

> ***The difference between the sums of the angles of the vectors drawn from the zeros and those from the poles of $G(s)H(s)$ to s_1 is an even multiple of 180°, including zero degrees.***

Once the root loci are constructed, the values of K along the loci can be determined by writing Eq. (8-17) as

$$|K| = \frac{\prod_{j=1}^{n} |s + p_j|}{\prod_{k=1}^{m} |s + z_k|} \tag{8-20}$$

The value of K at any point s_1 on the root loci is obtained from Eq. (8-20) by substituting the value of s_1 into the equation. Graphically, the numerator of Eq. (8-20) represents the product of the lengths of the vectors drawn from the poles of $G(s)H(s)$ to s_1, and the denominator represents the product of lengths of the vectors drawn from the zeros of $G(s)H(s)$ to s_1. If the point s_1 is on the RL, K is positive; if s_1 is on the CRL, then K is negative.

To illustrate the use of Eqs. (8-18) to (8-20) for the construction of the root loci, let us consider the function

$$G(s)H(s) = \frac{K(s + z_1)}{s(s + p_2)(s + p_3)} \tag{8-21}$$

The location of the poles and zero of $G(s)H(s)$ are arbitrarily assigned as shown in Fig. 8-1. Let us select an arbitrary trial point s_1 in the s-plane and draw vectors directed

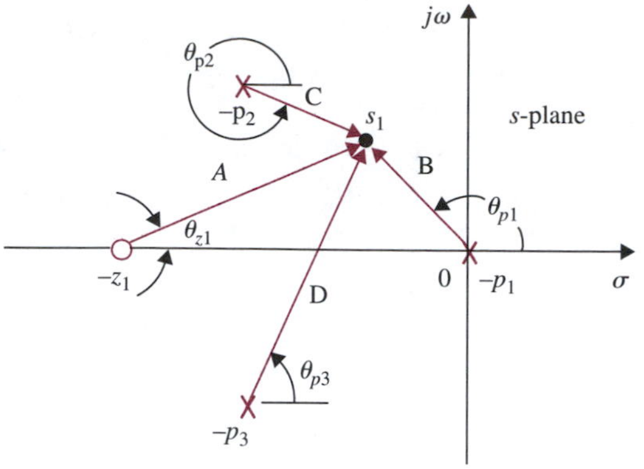

Figure 8-1 Pole–zero configuration of $G(s)H(s) = K(s + z_1)/[s(s + p_2)(s + p_3)]$.

from the poles and zeros of $G(s)H(s)$ to the point. If s_1 is indeed a point on the RL (K is positive), it must satisfy Eq. (8-18); that is, the angles of the vectors shown in Fig. 8-1 must satisfy

$$\begin{aligned} &\angle(s_1 + z_1) - \angle s_1 - \angle(s_1 + p_2) - \angle(s_1 + p_3) \\ &= \theta_{z1} - \theta_{p1} - \theta_{p2} - \theta_{p3} = (2i + 1) \times 180^\circ \end{aligned} \tag{8-22}$$

where $i = 0, \pm 1, \pm 2, \ldots$. As shown in Fig. 8-1, the angles of the vectors are measured with the positive real axis as reference. Similarly, if s_1 is a point on the CRL (K is negative), it must satisfy Eq. (8-19); that is,

$$\begin{aligned} &\angle(s_1 + z_1) - \angle s_1 - \angle(s_1 + p_2) - \angle(s_1 + p_3) \\ &= \theta_{z1} - \theta_{p1} - \theta_{p2} - \theta_{p3} = 2i \times 180^\circ \end{aligned} \tag{8-23}$$

where $i = 0, \pm 1, \pm 2, \ldots$.

If s_1 is found to satisfy either Eq. (8-22) or Eq. (8-23), Eq. (8-20) is used to find the magnitude of K at the point. As shown in Fig. 8-1, the lengths of the vectors are represented by A, B, C, and D. The magnitude of K is

$$|K| = \frac{|s_1||s_1 + p_2||s_1 + p_3|}{|s_1 + z_1|} = \frac{BCD}{A} \tag{8-24}$$

The sign of K depends on whether s_1 is on the RL or the CRL. Thus, given the function $G(s)H(s)$ with K as a multiplying factor and the poles and zeros are known, the construction of the root loci of the zeros of $1 + G(s)H(s)$ involves the following two steps:

1. A search for all the s_1 points in the s-plane that satisfy Eq. (8-18) for positive K. If the root loci for negative values of K are required, Eq. (8-19) must be satisfied.
2. Use Eq. (8-20) to find the magnitude of K on the root loci.

We have established the basic conditions on the construction of the root-locus diagram. However, if we were to use the trial-and-error method just described, the search for all the root-locus points in the s-plane that satisfy Eq. (8-18) or Eq. (8-19) and Eq. (8-20) would be a very tedious task. Years ago, when Evans first invented the root-locus technique, digital computer technology was still at its infancy; he had to devise a special tool, called the **Spirule,** which can be used to assist in adding and subtracting angles of vectors quickly, according to Eq. (8-18) or Eq. (8-19). Even with the Spirule, for the device to be effective, the user still has to know the general proximity of the roots in the s-plane.

With the availability of digital computers and efficient root-finding subroutines, the Spirule and the trial-and-error method have long become obsolete. Nevertheless, even with a high-speed computer and an effective root-locus program, the analyst should still have an in-depth understanding of the properties of the root loci in order to be able to manually sketch the root loci of simple and moderately complex systems and to interpret the computer results correctly, when applying the root loci for analysis and design of control systems.

8-3 Properties and Construction of the Root Loci

The following properties of the root loci are useful for the purpose of constructing the root loci manually and for the understanding of the root loci. The properties are developed based on the relation between the poles and zeros of $G(s)H(s)$ and the zeros of $1 + G(s)H(s)$, which are the roots of the characteristic equation.

8-3-1 $K = 0$ and $K = \pm\infty$ Points

The $K = 0$ points on the root loci are at the poles of $G(s)H(s)$.
The $K = \pm\infty$ points on the root loci are at the zeros of $G(s)H(s)$.

The poles and zeros referred to here include those at infinity, if any.

The reason for this is seen from the condition of the root loci given by Eq. (8-12),

$$G_1(s)H_1(s) = -\frac{1}{K} \tag{8-25}$$

As the magnitude of K approaches zero, $G_1(s)H_1(s)$ approaches infinity, so s must approach the poles of $G_1(s)H_1(s)$ or of $G(s)H(s)$. Similarly, as the magnitude of K approached infinity, s must approach the zeros of $G(s)H(s)$.

Example 8-1

Consider the equation

$$s(s + 2)(s + 3) + K(s + 1) = 0 \tag{8-26}$$

When $K = 0$, the three roots of the equation are at $s = 0$, -2, and -3. When the magnitude of K is infinite, the three roots of the equation are at $s = -1$, ∞, and ∞. It is useful to consider that infinity in the s-plane is a point concept. We can visualize that the finite s-plane is only a small portion of a sphere with an infinite radius. Then infinity in the s-plane is a point on the opposite side of the sphere that we face.

Dividing both sides of Eq. (8-26) by the terms that do not contain K, we get

$$1 + G(s)H(s) = 1 + \frac{K(s+1)}{s(s+2)(s+3)} = 0 \tag{8-27}$$

which gives

$$G(s)H(s) = \frac{K(s+1)}{s(s+2)(s+3)} \tag{8-28}$$

Thus the three roots of Eq. (8-26) when $K = 0$ are the same as the poles of the function $G(s)H(s)$. The three roots of Eq. (8-26) when $K = \pm\infty$ are at the three zeros of $G(s)H(s)$, including those at infinity. The three points on the root loci at which $K = 0$ and those at which $K = \pm\infty$ are shown in Fig. 8-2. ▲

8-3-2 Number of Branches on the Root Loci

A branch of the root loci is the locus of one root when K varies between $-\infty$ and ∞. The following property of the root loci results, since the number of branches of the root loci must equal the number of roots of the equation.

The number of branches of the root loci of Eq. (8-1) or Eq. (8-5) is equal to the order of the polynomial.

▲ It is important to keep track of the total number of branches of the root loci.

Keeping track of the individual branches and the total number of branches of the root-locus diagram is important in making certain that the plot is done correctly. This is particularly true when the root-locus plot is done by a computer, since unless each root-locus branch is coded by a different color, it is up to the user to make the distinctions.

Example 8-2

The number of branches of the root loci of

$$s(s+2)(s+3) + K(s+1) = 0 \tag{8-29}$$

is three, since the equation is of the third order. In other words, the equation has three roots, and thus there should be three root loci. ▲

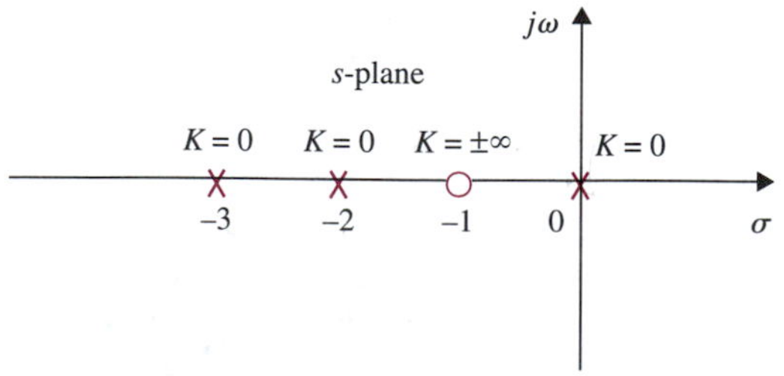

Figure 8-2 Points at which $K = 0$ on the root loci of $s(s+2)(s+3) + K(s+1) = 0$.

8-3-3 Symmetry of the Root Loci

▲ It is important to pay attention to the symmetry of the root loci.

The root loci are symmetrical with respect to the real axis of the s-plane. In general, the root loci are symmetrical with respect to the axes of symmetry of the pole–zero configuration of $G(s)H(s)$.

The reason behind this property is because for real coefficients in Eq. (8-1), the roots must be real or in complex-conjugate pairs. In general, if the poles and zeros of $G(s)H(s)$ is symmetrical to an axis in addition to the real axis in the s-plane, we can regard this axis of symmetry as if it were the real axis of a new complex plane obtained through a linear transformation.

Example 8-3

Consider the equation

$$s(s + 1)(s + 2) + K = 0 \tag{8-30}$$

Dividing both sides of the equation by the terms that do not contain K, we get

$$G(s)H(s) = \frac{K}{s(s + 1)(s + 2)} \tag{8-31}$$

The root loci of Eq. (8-30) are shown in Fig. 8-3 for $K = -\infty$ to $K = \infty$. Since the pole–zero configuration of $G(s)H(s)$ is symmetrical with respect to the real axis as well as the $s = -1$ axis, the root-locus plot is symmetrical to the two axes.

As a review of all the properties of the root loci presented thus far, we conduct the following exercise with regard to the root loci in Fig. 8-3. The points at which $K = 0$ are at the poles of $G(s)H(s)$, $s = 0$, -1, and -2. The function $G(s)H(s)$ has three zeros at $s = \infty$ at which $K = \pm\infty$. The reader should try to trace out the three separate branches of the root loci by starting from one of the $K = -\infty$ points, through the $K = 0$ point on the same branch, and ending at $K = \infty$ at $s = \infty$. ▲

Example 8-4

When the pole–zero configuration of $G(s)H(s)$ is symmetrical with respect to a point in the s-plane, the root loci will also be symmetrical to that point. This is illustrated by the root-locus plot of

$$s(s + 2)(s + 1 + j)(s + 1 - j) + K = 0 \tag{8-32}$$

shown in Fig. 8-4. ▲

8-3-4 Angles of Asymptotes of the Root Loci: Behavior of the Root Loci at $|s| = \infty$

▲ Asymptotes of root loci refers to behavior of root loci at $s \to \infty$.

As shown by the root loci in Figs. 8-3 and 8-4, when n, the order of $P(s)$, is not equal to m, the order of $Q(s)$, some of the loci will approach infinity in the s-plane. The properties of the root loci near infinity in the s-plane are described by the **asymptotes** of the loci when $|s| \to \infty$. In general when $n \neq m$, there will be $2|n - m|$ asymptotes

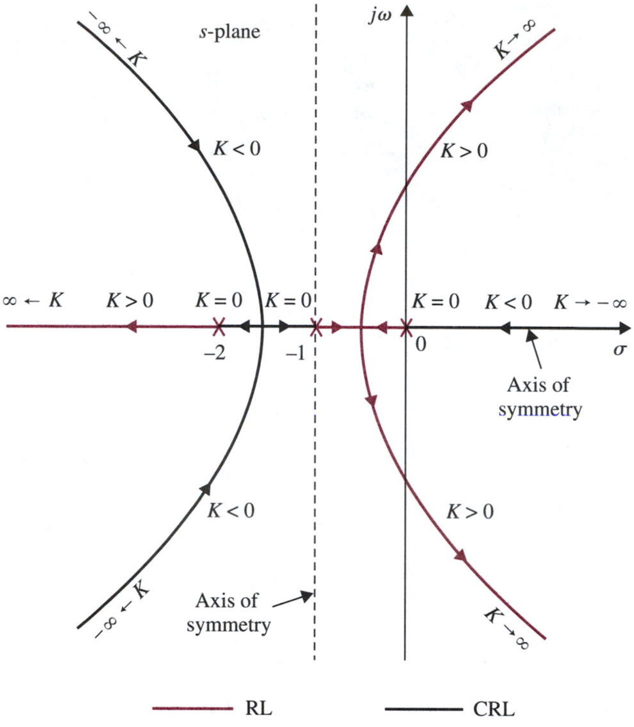

Figure 8-3 Root loci of $s(s+1)(s+2)+K=0$, showing the properties of symmetry.

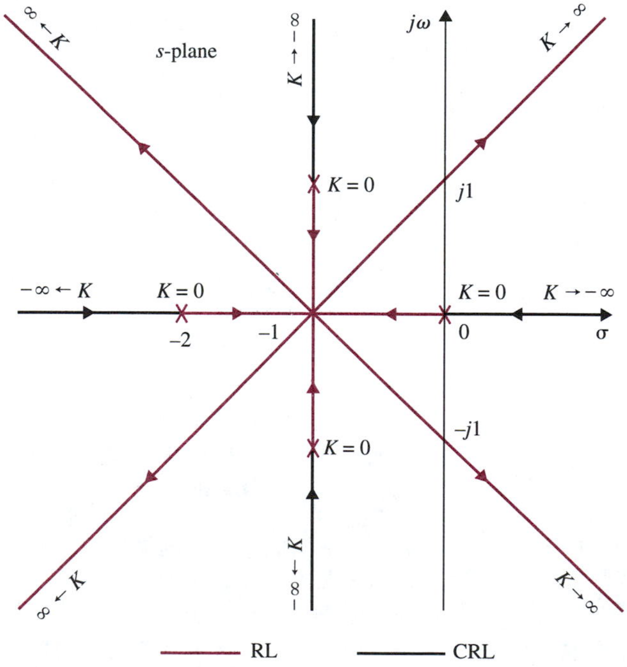

Figure 8-4 Root loci of $s(s+2)(s^2+2s+2)+K=0$, showing the properties of symmetry.

that describe the behavior of the root loci at $|s| = \infty$. The angles of the asymptotes and their intersect with the real axis of the s-plane are described as follows.

For large values of s, the root loci for $K \geq 0$ (RL) are asymptotic to asymptotes with angles given by

$$\theta_i = \frac{2i+1}{|n-m|} \times 180^\circ \qquad n \neq m \tag{8-33}$$

where $i = 0, 1, 2, \ldots, |n-m| - 1$; n and m are the number of finite poles and zeros of $G(s)H(s)$, respectively. For $K \leq 0$ (CRL), the angles of the asymptotes are

$$\theta_i = \frac{2i}{|n-m|} \times 180^\circ \qquad n \neq m \tag{8-34}$$

where $i = 0, 1, 2, \ldots, |n-m| - 1$.

8-3-5 Intersect of the Asymptotes (Centroid)

The intersect of the $2|n-m|$ asymptotes of the root loci lies on the real axis of the s-plane, at

$$\sigma_1 = \frac{\Sigma \text{ finite poles of } G(s)H(s) - \Sigma \text{ finite zeros of } G(s)H(s)}{n-m} \tag{8-35}$$

where n is the number of finite poles and m is the number of finite zeros of $G(s)H(s)$, respectively. The intersect of the asymptotes σ_1 represents the center of gravity of the root loci, and is always a real number.

Since the poles and zeros of $G(s)H(s)$ are either real or in complex-conjugate pairs, the imaginary parts in the numerator of Eq. (8-35) always cancel each other. Thus, in Eq. (8-35), the terms in the summations may be replaced by the real parts of the poles and zeros of $G(s)H(s)$, respectively. That is,

$$\sigma_1 = \frac{\Sigma \text{ real parts of poles of } G(s)H(s) - \Sigma \text{ real parts of zeros of } G(s)H(s)}{n-m} \tag{8-36}$$

Example 8-5

Consider the transfer function

$$G(s)H(s) = \frac{K(s+1)}{s(s+4)(s^2+2s+2)} \tag{8-37}$$

which corresponds to the characteristic equation

$$s(s+4)(s^2+2s+2) + K(s+1) = 0 \tag{8-38}$$

$s(s^3+2s^2+2s+4s^2+8s+8) = s^4+6s^3+10s^2+8s$

The pole–zero configuration of $G(s)H(s)$ is shown in Fig. 8-5. From the six properties of the root loci discussed so far, the following information concerning the root loci of Eq. (8-38) when K varies from $-\infty$ to ∞ is obtained:

1. $K = 0$: The points at which $K = 0$ on the root loci are at the poles of $G(s)H(s)$: $s = 0, -4, -1 + j$, and $-1 - j$.
2. $K = \pm\infty$; The points at which $K = \pm\infty$ on the root loci are at the zeros of $G(s)H(s)$: $s = -1, \infty, \infty$, and ∞.
3. There are four root-loci branches, since Eqs. (8-37) and (8-38) are of the fourth order.
4. The root loci are symmetrical to the real axis.
5. Since the number of finite poles of $G(s)H(s)$ exceeds the number of finite zeros of $G(s)H(s)$ by three ($n - m = 4 - 1 = 3$), when $K = \pm\infty$, the root loci approach $s = \infty$ along six asymptotes. The angles of the asymptotes of the RL ($K \geq 0$) are

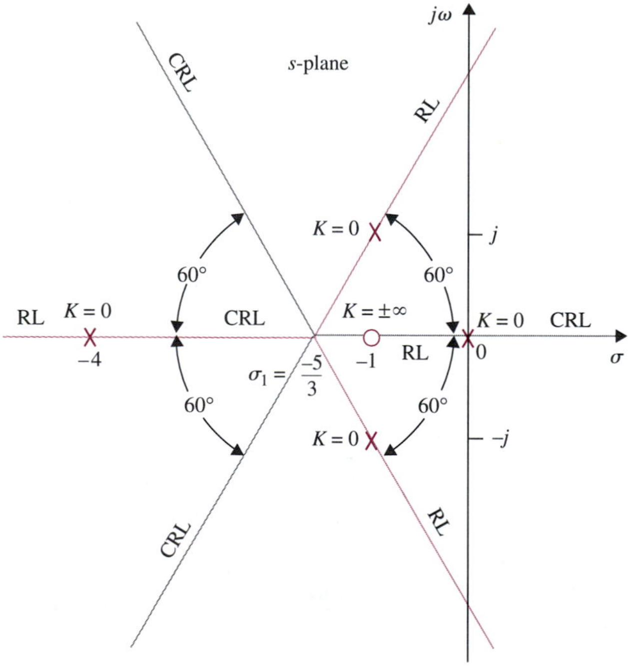

Figure 8-5 Asymptotes of the root loci of $s(s+4)(s^2+2s+2) + K(s+1) = 0$.

given by Eq. (8-33):

$$i = 0: \quad \theta_0 = \frac{180°}{3} = 60°$$

$$i = 1: \quad \theta_1 = \frac{540°}{3} = 180°$$

$$i = 2: \quad \theta_2 = \frac{900°}{3} = 300°$$

The angles of the asymptotes of the CRL ($K \leq 0$) are given by Eq. (8-34) and are calculated to be 0°, 120°, and 240°.

6. The intersect of the asymptotes is given by Eq. (8-36):

$$\sigma_1 = \frac{(-4 - 1 - 1) - (-1)}{4 - 1} = -\frac{5}{3} \tag{8-39}$$

The asymptotes of the root loci are shown in Fig. 8-5. ▲

Example 8-6

As further illustrations, the asymptotes of the root loci of several equations are shown in Fig. 8-6. ▲

8-3-6 Root Loci on the Real Axis

▲ The entire real axis of the *s*-plane is occupied by root loci.

The entire real axis of the s-plane is occupied by the root loci (i.e., either the RL or the CRL).

1. ***RL: On a given section of the real axis, RL are found in the section only if the total number of poles and zeros of G(s)H(s) to the right of the section is odd.***
2. ***CRL: On a given section of the real axis, CRL are found in the section only if the total number of real poles and zeros of G(s)H(s) to the right of the section is even. Complex poles and zeros of G(s)H(s) do not affect the type of root loci found on the real axis.***

These properties are arrived at based on the following observations:

1. At any point s_1 on the real axis, the angles of the vectors drawn from the complex-conjugate poles and zeros of $G(s)H(s)$ add up to zero. Thus only the real zeros and poles of $G(s)H(s)$ contribute to the angular relations in Eqs. (8-18) and (8-19).

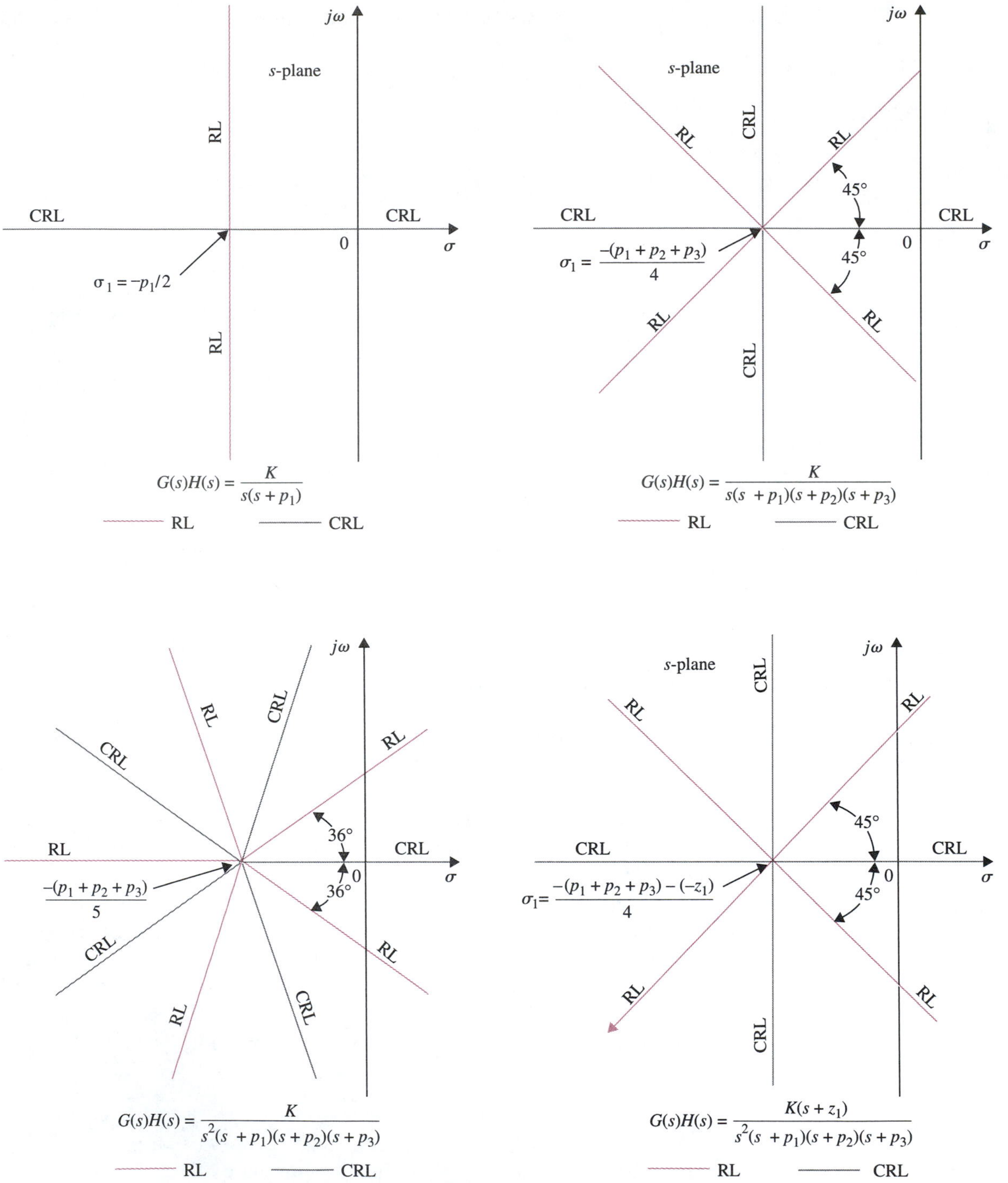

Figure 8-6 Examples of the asymptotes of the root loci (RL and CRL).

2. Only the real poles and zeros of $G(s)H(s)$ that lie to the right of the point s_1 contribute to Eqs. (8-18) and (8-19), since real poles and zeros that lie to the left of the point contribute zero degree to Eqs. (8-18) and (8-19).
3. Each real pole of $G(s)H(s)$ to the right of s_1 contributes $-180°$, and each real zero of $G(s)H(s)$ to the right of s_1 contributes $+180°$ to Eqs. (8-18) and (8-19).

The last observation shows that for s_1 to be a point on the RL, there must be an **odd** number of poles and zeros of $G(s)H(s)$ to the right of the point. For s_1 to be a point on the CRL, the total number of poles and zeros of $G(s)H(s)$ to the right of the point must be **even.** The following example illustrates the determination of the properties of the root loci on the real axis of the s-plane.

Example 8-7

The root loci on the real axis for two pole–zero configurations of $G(s)H(s)$ are shown in Fig. 8-7. Notice that the entire real axis is occupied by either the RL or the CRL. ▲

8-3-7 Angles of Departure and Angles of Arrival of the Root Loci

The angle of departure or arrival of a root locus at a pole or zero, respectively, of G(s)H(s) denotes the angle of the tangent to the locus near the point.

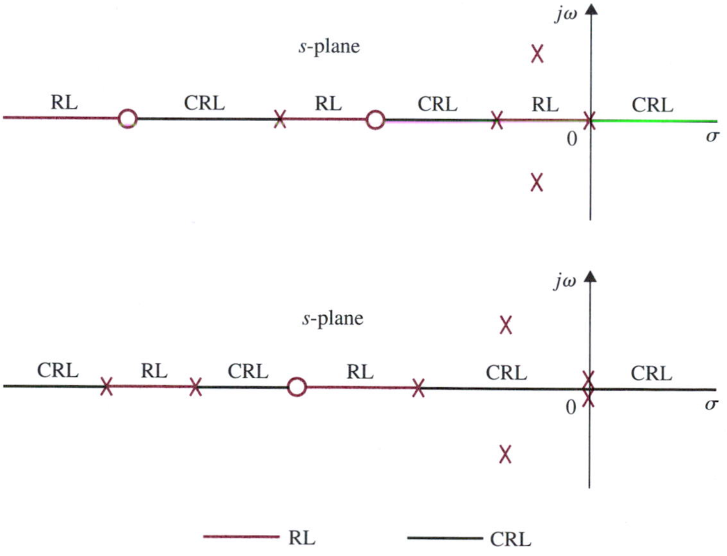

Figure 8-7 Properties of root loci on the real axis.

The angles of departure and arrival are determined using Eq. (8-18) for RL and Eq. (8-19) for CRL. The details are illustrated by the following example.

Example 8-8

For the root-locus diagram shown in Fig. 8-8, the root locus near the pole $s = -1 + j$ may be more accurately sketched by knowing the angle at which the root locus leaves the pole. As shown in Fig. 8-8, the angle of departure of the root locus at $s = -1 + j$ is represented by θ_2, measured with respect to the real axis. Let us assign s_1 to be a point on the RL leaving the pole at $-1 + j$ and is very close to the pole. Then s_1 must satisfy Eq. (8-18). Thus

$$\angle G(s_1)H(s_1) = -(\theta_1 + \theta_2 + \theta_3 + \theta_4) = (2i + 1)180° \tag{8-40}$$

where i is any integer. Since s_1 is assumed to be very close to the pole at $-1 + j$, the angles

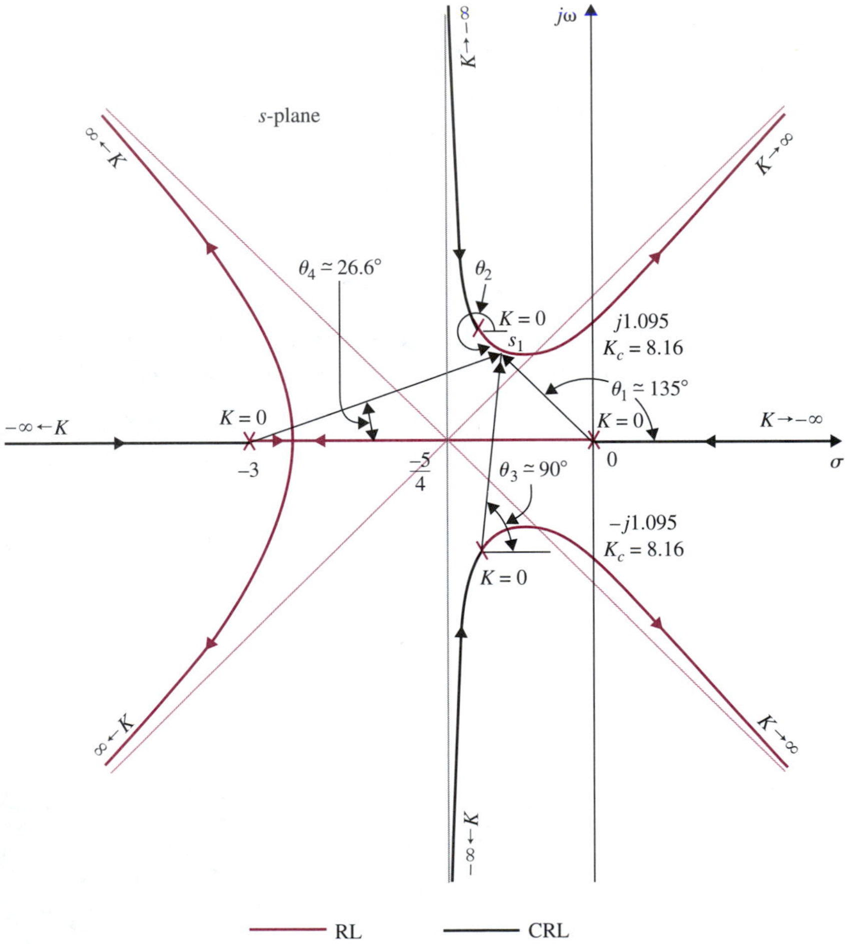

Figure 8-8 Root loci of $s(s + 3)(s^2 + 2s + 2) + K = 0$ to illustrate the angles of departure or arrival.

of the vectors drawn from the other three poles are approximated by considering that s_1 is at $-1 + j$. From Fig. 8-8, Eq. (8-40) is written

$$-(135° + \theta_2 + 90° + 26.6°) = (2i + 1)180° \tag{8-41}$$

where θ_2 is the only unknown angle. In this case, we can set i to be -1, and the result for θ_2 is $-71.6°$.

When the angle of departure or arrival of a RL at a simple pole or zero of $G(s)H(s)$ is determined, the angle of arrival or departure of the CRL at the same point differs from this angle by 180°, Eq. (8-19) is now used. Figure 8-8 shows that the angle of arrival of the CRL at $-1 + j$ is 108.4°, which is $180° - 71.6°$. Similarly, for the root-locus diagram in Fig. 8-8, we can show that the CRL arrives at the pole $s = -3$ with an angle of 180°, and the RL leaves the same pole at 0°. For the pole at $s = 0$, the angle of arrival of the CRL is 0°, whereas the angle of departure of the RL is 180°. These angles are also determined from knowledge of the type of root loci on sections of the real axis separated by the poles and zeros of $G(s)H(s)$. Since the total angles of the vectors drawn from complex-conjugate poles and zeros to any point on the real axis add up to be zero, the angles of arrival and departure of root loci on the real axis are not affected by the complex poles and zeros of $G(s)H(s)$. ▲

Example 8-9

In this example we examine the angles of departure and arrival of the root loci at a multiple-order pole or zero of $G(s)H(s)$. Consider that a $G(s)H(s)$ has a multiple-order (third-order) pole on the real axis, as shown in Fig. 8-9. Only the real poles and zeros of $G(s)H(s)$ are shown, since the complex ones do not affect the type or the angles of arrival and departure of the root loci on the real axis. For the third-order pole at $s = -2$, there are three RL leaving and three CRL arriving at the point. To find the angles of departure of the RL, we assign a point s_1 on one of the RL near $s = -2$, and apply Eq. (8-18). The result is

$$-\theta_1 - 3\theta_2 + \theta_3 = (2i + 1)180° \tag{8-42}$$

where θ_1 and θ_3 denote the angles of the vectors drawn from the pole at 0 and the zero at -3, respectively, to s_1. The angle θ_2 is multiplied by 3, since there are three poles at $s = -2$, so

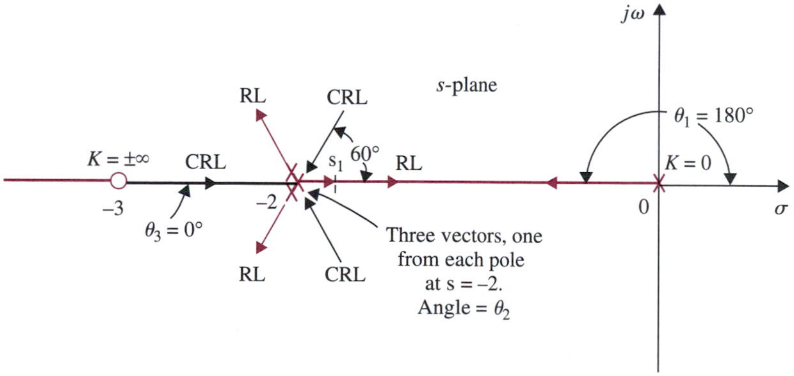

Figure 8-9 Angles of departure and arrival at a third-order pole.

that there are three vectors drawn from -2 to s_1. Setting i to zero in Eq. (8-42), and since $\theta_1 = 180°$, $\theta_3 = 0°$, we have $\theta_2 = 0°$, which is the angle of departure of the RL that lies between $s = 0$ and $s = -2$. For the angles of departure of the other two RLs we set $i = 1$ and $i = 2$ successively in Eq. (8-42), and we have $\theta_2 = 120°$ and $-120°$. Similarly, for the three CRL that arrive at $s = -2$, Eq. (8-19) is used, and the angles of arrivals are found to be 60°, 180°, and $-60°$. ▲

8-3-8 Intersection of the Root Loci with the Imaginary Axis

▲ Routh–Hurwitz criterion may be used to find the intersects of the root loci on the imaginary axis.

The points where the root loci intersect the imaginary axis of the s-plane, if any, and the corresponding values of K, may be determined by means of the Routh–Hurwitz criterion. For complex situations, when the root loci have multiple numbers of intersections on the imaginary axis, the intersects and the critical values of K can be determined with the help of the root-locus computer program. The Bode diagram method in Chapter 9, associated with the frequency response, can also be used for this purpose.

Example 8-10

The root loci shown in Fig. 8-8 is for the equation

$$s(s + 3)(s^2 + 2s + 2) + K = 0 \tag{8-43}$$

Figure 8-8 shows that the RL intersect the $j\omega$-axis at two points. Applying the Routh–Hurwitz criterion to Eq. (8-43), and by solving the auxiliary equation, we have the critical value of K for stability at $K = 8.16$, and the corresponding crossover points on the $j\omega$-axis are at $\pm j1.095$. ▲

8-3-9 Breakaway Points (Saddle Points) on the Root Loci

Breakaway points on the root loci of an equation correspond to multiple-order roots of the equation.

▲ A root locus plot may have more than one breakaway point.

▲ Breakaway points may be complex conjugates in the s-plane.

Figure 8-10(a) illustrates a case in which two branches of the root loci meet at the breakaway point on the real axis and then depart from the axis in opposite directions. In this case the breakaway point represents a double root of the equation when the value of K is assigned the value corresponding to the point. Figure 8-10(b) shows another common situation when two complex-conjugate root loci approach the real axis, meet at the breakaway point, and then depart in opposite directions along the real axis. In general, a breakaway point may involve more than two root loci. Figure 8-10(c) illustrates a situation when the breakaway point represents a fourth-order root.

A root-locus diagram can have, of course, more than one breakaway point. Moreover, the breakaway points need not always be on the real axis. Because of the conjugate symmetry of the root loci, the breakaway points not on the real axis must be in complex-

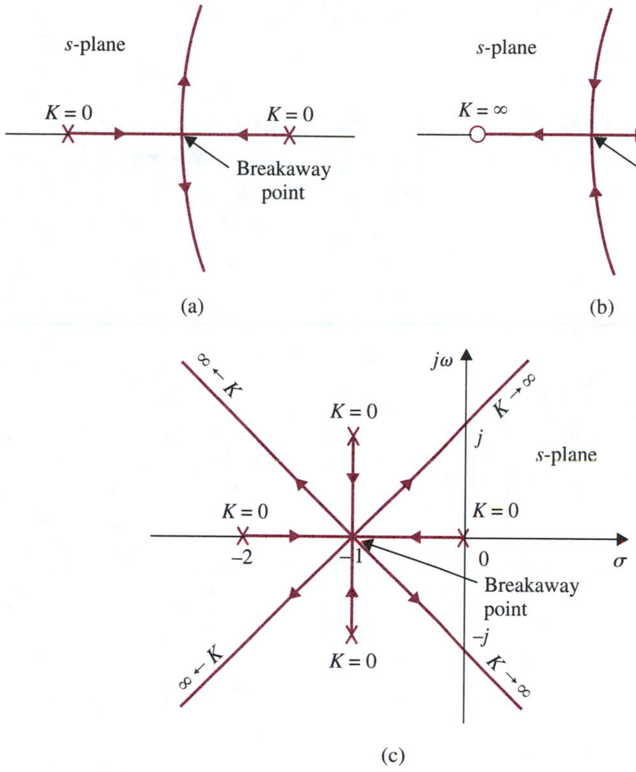

Figure 8-10 Examples of breakaway points on the real axis in the *s*-plane.

conjugate pairs. Refer to Fig. 8-13 for an example of root loci with complex breakaway points. The properties of the breakaway points of root loci are given as follows.

The breakaway points on the root loci of* $1 + KG_1(s)H_1(s) = 0$ *must satisfy

$$\frac{dG_1(s)H_1(s)}{ds} = 0 \qquad (8\text{-}44)$$

It is important to point out that the condition for the breakaway point given in Eq. (8-44) is *necessary* but *not sufficient.* In other words, all breakaway points on the RL and the CRL must satisfy Eq. (8-44), but not all solutions of Eq. (8-44) are breakaway points. To be a breakaway point, the solution of Eq. (8-44) must also satisfy the equation $1 + KG_1(s)H_1(s) = 0$, that is, must also be a point on the root loci for some real K. In general, the following conclusions may be made with regard to the solutions of Eq. (8-44):

1. All *real* solutions of Eq. (8-44) are breakaway points on the root loci (RL and CRL), since the entire real axis of the s-plane is occupied by the root loci.
2. The complex-conjugate solutions of Eq. (8-44) are breakaway points only if they satisfy the characteristic equation or are points on the root loci.
3. Since the condition of the root loci is

$$K = -\frac{1}{G_1(s)H_1(s)} \tag{8-45}$$

taking the derivative on both sides of the equation with respect to s, we have

$$\frac{dK}{ds} = \frac{dG_1(s)H_1(s)/ds}{[G_1(s)H_1(s)]^2} \tag{8-46}$$

Thus the breakaway point condition can also be written as

$$\frac{dK}{ds} = 0 \tag{8-47}$$

where K is expressed as in Eq. (8-45).

Angle of Arrival and Departure of Root Loci at the Breakaway Points

The angles at which the root loci arrive or depart from a breakaway point depend on the number of loci that are involved at the point. For example, the root loci shown in Fig. 8-10(a) and (b) all arrive and break away at 90° apart, whereas in Fig. 8-10(c), the four root loci arrive and depart with angles 45° apart. In general,

n root loci (RL or CRL) arrive or depart a breakaway point at 180/n degrees apart.

Many root-locus computer programs have features that will obtain the breakaway points, which is one of the properties of the root loci that is tedious to do manually.

Example 8-11

Consider the second-order equation

$$s(s + 2) + K(s + 4) = 0 \tag{8-48}$$

Based on some of the properties of the root loci described thus far, the root loci of Eq. (8-48) are sketched as shown in Fig. 8-11 for $-\infty < K < \infty$. It can be proven that the complex portion of the root loci is a circle. The two breakaway points are all on the real axis, one between 0 and -2 and the other between -4 and $-\infty$. From Eq. (8-48), we have

$$G_1(s)H_1(s) = \frac{s + 4}{s(s + 2)} \tag{8-49}$$

Applying Eq. (8-44), the breakaway points on the root loci must satisfy

$$\frac{dG_1(s)H_1(s)}{ds} = \frac{s(s + 2) - 2(s + 1)(s + 4)}{s^2(s + 2)^2} = 0 \tag{8-50}$$

or

$$s^2 + 8s + 8 = 0 \tag{8-51}$$

Solving Eq. (8-51), we find the two breakaway points of the root loci at $s = -1.172$ and -6.828. Figure 8-11 shows that the two breakaway points are all on the RL. ▲

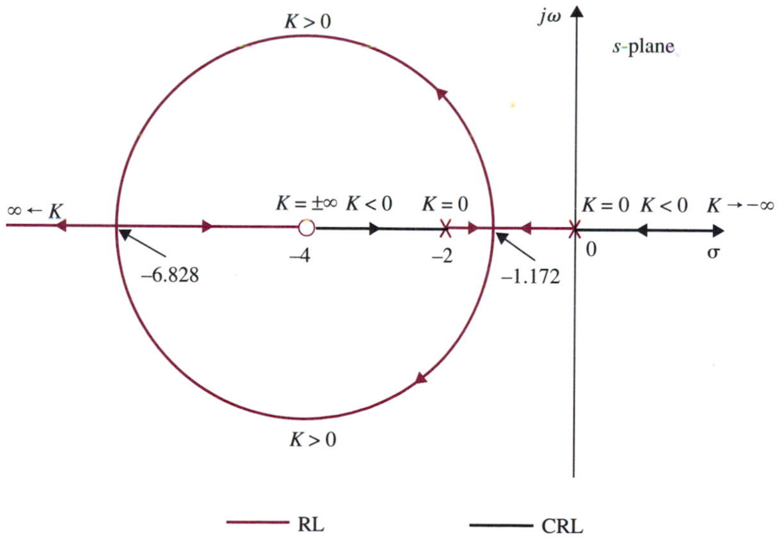

Figure 8-11 Root loci of $s(s + 2) + K(s + 4) = 0$.

Example 8-12

Consider the equation

$$s^2 + 2s + 2 + K(s + 2) = 0 \tag{8-52}$$

The equivalent $G(s)H(s)$ is obtained by dividing both sides of Eq. (8-52) by the terms that do not contain K. We have

$$G(s)H(s) = \frac{K(s + 2)}{s^2 + 2s + 2} \tag{8-53}$$

Based on the poles and zeros of $G(s)H(s)$, the root loci of Eq. (8-52) are plotted as shown in Fig. 8-12. The plot shows that the RL and the CRL each has a breakaway point. These breakaway points are determined from

$$\frac{dG_1(s)H_1(s)}{ds} = \frac{d}{ds}\left(\frac{s+2}{s^2 + 2s + 2}\right) = \frac{s^2 + 2s + 2 - 2(s + 1)(s + 2)}{(s^2 + 2s + 2)^2} = 0 \tag{8-54}$$

or

$$s^2 + 4s + 2 = 0 \tag{8-55}$$

The solution of this equation gives the breakaway point at $s = -0.586$, which is on the CRL, and $s = -3.414$, which is a breakaway point of the RL. ▲

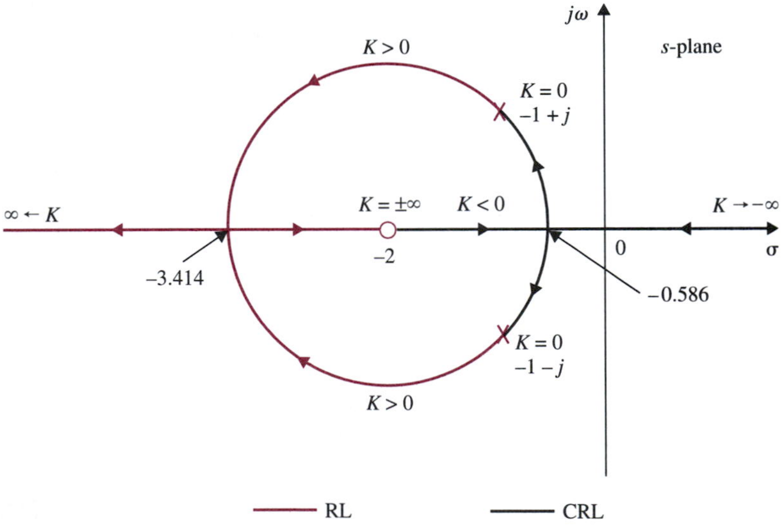

Figure 8-12 Root loci of $s^2 + 2s + 2 + K(s + 2) = 0$.

Example 8-13

Figure 8-13 shows the root loci of the equation

$$s(s + 4)(s^2 + 4s + 20) + K = 0 \tag{8-56}$$

Dividing both sides of the last equation by the terms that do not contain K, we have

$$1 + KG_1(s)H_1(s) = 1 + \frac{K}{s(s + 4)(s^2 + 4s + 20)} = 0 \tag{8-57}$$

Since the poles of $G_1(s)H_1(s)$ are symmetrical about the axes $\sigma = -2$ and $\omega = 0$ in the s-plane, the root loci of the equation are also symmetrical with respect to these two axes. Taking the

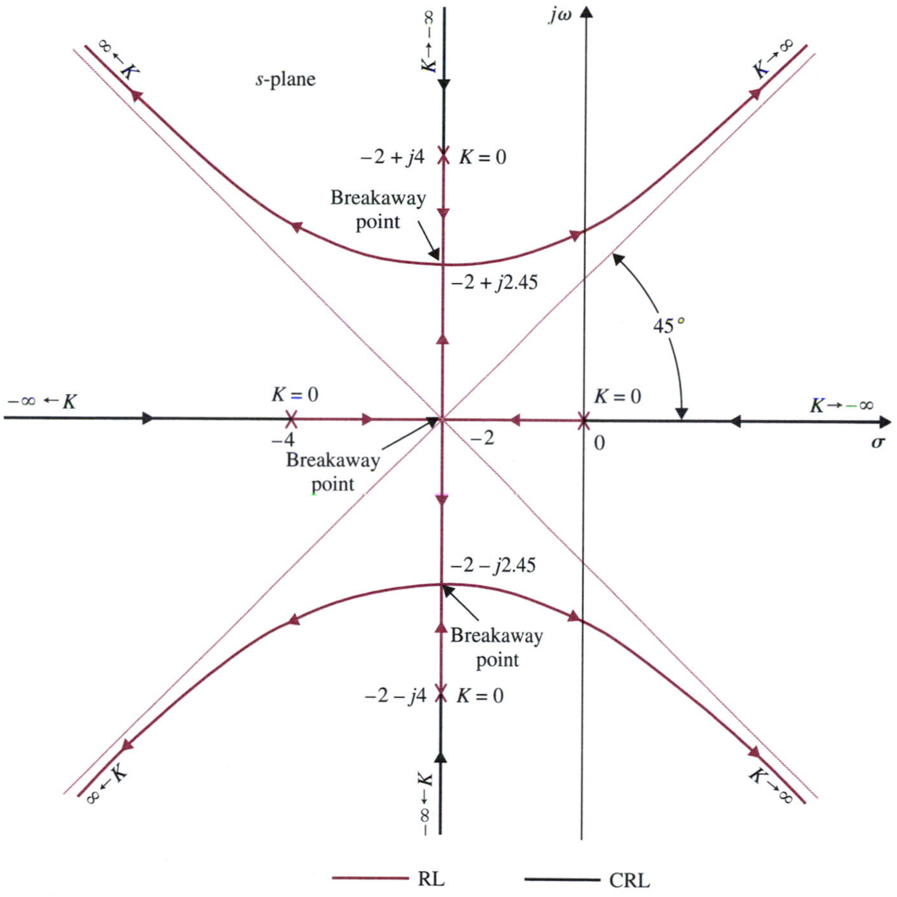

Figure 8-13 Root loci of $s(s + 4)(s^2 + 4s + 20) + K = 0$.

derivative of $G_1(s)H_1(s)$ with respect to s, we get

$$\frac{dG_1(s)H_1(s)}{ds} = -\frac{4s^3 + 24s^2 + 72s + 80}{[s(s+4)(s^2+4s+20)]^2} = 0 \tag{8-58}$$

or

$$s^3 + 6s^2 + 18s + 20 = 0 \tag{8-59}$$

The solutions of the last equation are $s = -2$, $-2 + j2.45$, and $-2 - j2.45$. In this case Fig. 8-13 shows that all the solutions of Eq. (8-59) are breakaway points on the root loci, and two of these points are complex. ▲

Example 8-14

In this example we show that not all the solutions of Eq. (8-44) are breakaway points on the root loci. The root loci of the equation

$$s(s^2 + 2s + 2) + K = 0 \tag{8-60}$$

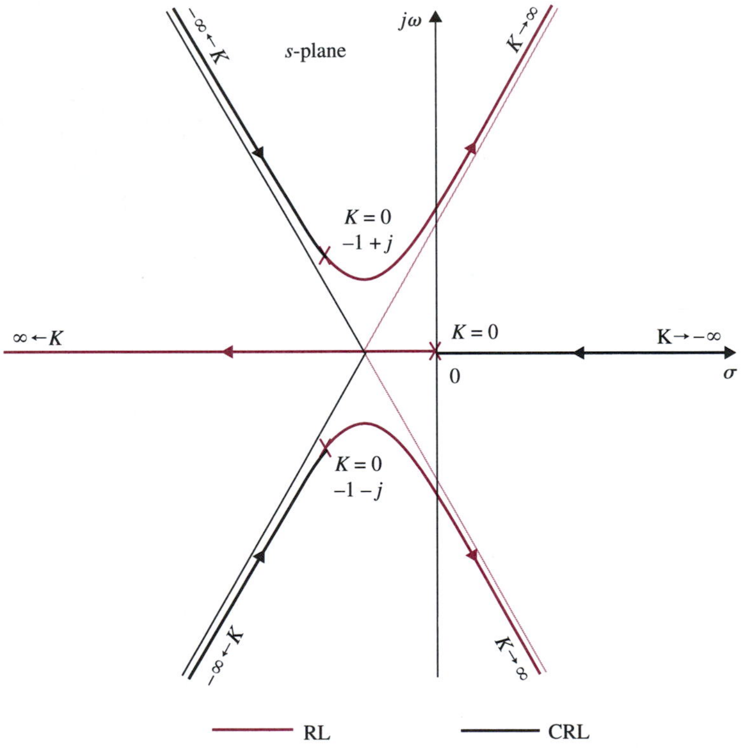

Figure 8-14 Root loci of $s(s^2 + 2s + 2) + K = 0$.

are shown in Fig. 8-14. The root loci show that neither the RL nor the CRL has any breakaway point in this case. However, writing Eq. (8-60) as

$$1 + KG_1(s)H_1(s) = 1 + \frac{K}{s(s^2 + 2s + 2)} = 0 \tag{8-61}$$

and applying Eq. (8-44), we have the equation for the breakaway points:

$$3s^2 + 4s + 2 = 0 \tag{8-62}$$

The roots of Eq. (8-62) are $s = -0.667 + j0.471$ and $-0.667 - j0.471$. These two roots are not breakaway points on the root loci, since they do not satisfy Eq. (8-60) for any real value of K. ▲

8-3-10 Root Sensitivity [20–22]

▲ The root sensitivity at the breakaway points is infinite.

The condition on the breakaway points on the root loci in Eq. (8-47) leads to the **root sensitivity** of the characteristic equation. The sensitivity of the roots of the characteristic equation when K varies is defined as the **root sensitivity,** and is given by

$$S_K = \frac{ds/s}{dK/K} = \frac{K}{s}\frac{ds}{dK} \tag{8-63}$$

▲ From the root sensitivity standpoint a system should not be operated at the breakaway points.

Thus Eq. (8-63) shows that *the root sensitivity at the breakaway points is infinite.* From the root sensitivity standpoint, we should avoid selecting the value of K to operate at the breakaway points, which correspond to multiple-order roots of the characteristic equation. In the design of control systems, not only is it important to arrive at a system that has the desired characteristics, but just as important, the system should be insensitive to parameter variations. For instance, a system may perform satisfactorily at a certain K, but if it is very sensitive to the variation of K, it may get into the undesirable performance region or become unstable if K varies by only a small amount. In formal control system terminology, a system that is insensitive to parameter variations is called a **robust system.** Thus the root-locus study of control systems must involve not only the shape of the root loci with respect to the variable parameter K, but also how the roots along the loci vary with the variation of K.

Example 8-15

Figure 8-15 shows the root-locus diagram of

$$s(s + 1) + K = 0 \tag{8-64}$$

with K incremented uniformly over 100 values from -20 to 20. The root loci were computed with the **ROOTLOCI** program of **ACSP** and plotted digitally. Each dot on the root-locus plot

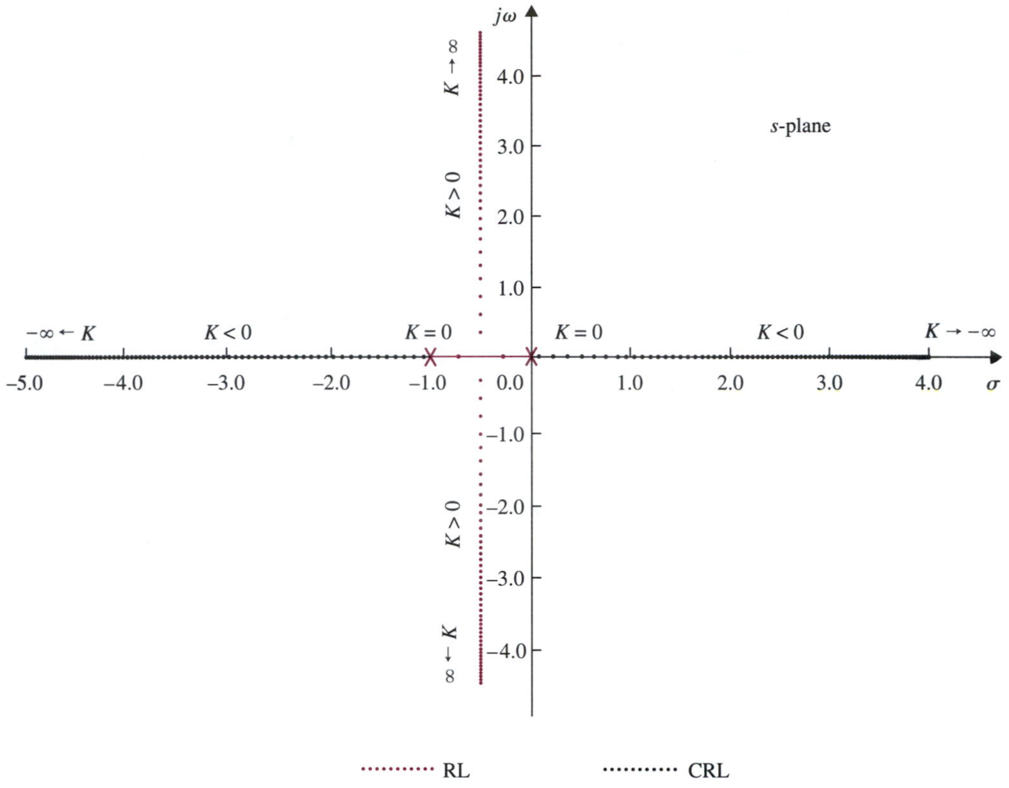

Figure 8-15 Root loci of $s(s + 1) + K = 0$, showing the root sensitivity with respect to K.

represents one root for a distinct value of K. Thus we see that the root sensitivity is low when the magnitude of K is large. As the magnitude of K decreases, the movements of the roots become larger for the same incremental change in K. At the breakaway point, $s = -0.5$, the root sensitivity is infinite.

Figure 8-16 shows the root loci of

$$s^2(s + 1)^2 + K(s + 2) = 0 \tag{8-65}$$

with K incremented uniformly over 200 values from -40 to 50. Again, the loci show that the root sensitivity increases as the roots approach the breakaway points at $s = 0, -0.543, -1.0$, and -2.457.

We can investigate the root sensitivity further by using the expression in Eq. (8-47). For the second-order equation in Eq. (8-64),

$$\frac{dK}{ds} = -2s - 1 \tag{8-66}$$

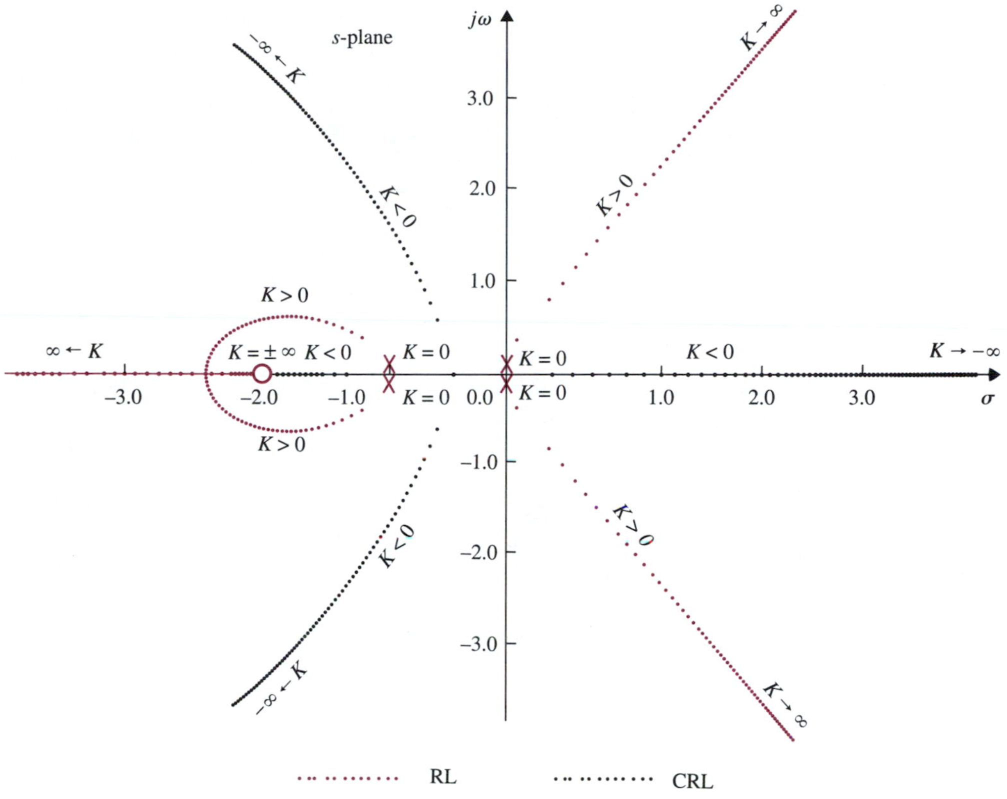

Figure 8-16 Root loci of $s^2(s + 1) + K(s + 2) = 0$, showing the root sensitivity with respect to K.

From Eq. (8-64), $K = -s(s + 1)$; the root sensitivity becomes

$$S_K = \frac{ds}{dK}\frac{K}{s} = \frac{s + 1}{2s + 1} \tag{8-67}$$

where $s = \sigma + j\omega$, and s must take on the values of the roots of Eq. (8-64). For the roots on the real axis, $\omega = 0$. Thus Eq. (8-67) leads to

$$|S_K|_{\omega=0} = \left|\frac{\sigma + 1}{2\sigma + 1}\right| \tag{8-68}$$

When the two roots are complex, $\sigma = -0.5$ for all values of ω, Eq. (8-67) gives

$$|S_K|_{\sigma=-0.5} = \left(\frac{0.25 + \omega^2}{4\omega^2}\right)^{1/2} \tag{8-69}$$

From Eq. (8-69) it is apparent that the sensitivities of the pair of complex-conjugate roots are the same, since ω appears only as ω^2 in the equation. Equation (8-68) indicates that the sensitivities of the two real roots are different for a given value of K. Table 8-1 gives the magnitudes of the sensitivities of the two roots of Eq. (8-64) for several values of K, where $|S_{K1}|$ denotes the root sensitivity of the first root and $|S_{K2}|$ denotes that of the second root. These values indicate that although the two real roots reach $\sigma = -0.5$ for the same value of $K = 0.25$, and each root travels the same distance from $s = 0$ and $s = -1$, respectively, the sensitivities of the two real roots are not the same. ▲

8-3-11 Calculation of K on the Root Loci

Once the root loci are constructed, the values of K at any point s_1 on the loci can be determined by use of the defining equation of Eq. (8-20). Graphically, the magnitude of K can be written as

$$|K| = \frac{\Pi \text{ lengths of vectors drawn from the poles of } G_1(s)H_1(s) \text{ to } s_1}{\Pi \text{ lengths of vectors drawn from the zeros of } G_1(s)H_1(s) \text{ to } s_1} \tag{8-70}$$

Example 8-16

As an illustration on the determination of the value of K on the root loci, the root loci of the equation

$$s^2 + 2s + 2 + K(s + 2) = 0 \tag{8-71}$$

are shown in Fig. 8-17. The value of K at the point s_1 is given by

$$K = \frac{A \times B}{C} \tag{8-72}$$

Table 8-1 Root Sensitivity

K	Root 1	$\lvert S_{K1}\rvert$	Root 2	$\lvert S_{K2}\rvert$
0	0	1.000	-1.000	0
0.04	-0.042	1.045	-0.958	0.454
0.16	-0.200	1.333	-0.800	0.333
0.24	-0.400	3.000	-0.600	2.000
0.25	-0.500	∞	-0.500	∞
0.28	$-0.5 + j0.173$	1.527	$-0.5 - j0.173$	1.527
0.40	$-0.5 + j0.387$	0.817	$-0.5 - j0.387$	0.817
1.20	$-0.5 + j0.975$	0.562	$-0.5 - j0.975$	0.562
4.00	$-0.5 + j1.937$	0.516	$-0.5 - j1.937$	0.516
∞	$-0.5 + j\infty$	0.500	$-0.5 - j\infty$	0.500

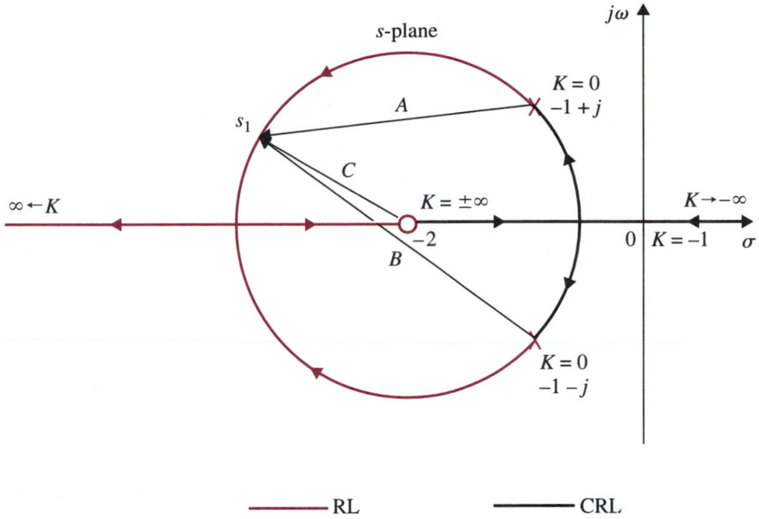

Figure 8-17 Graphical method of finding the values of K on the root loci.

where A and B are the lengths of the vectors drawn from the poles of $G(s)H(s) = K(s + 2)/(s^2 + 2s + 2)$ to the point s_1, and C is the length of the vector drawn from the zero of $G(s)H(s)$ to s_1. In this case, s_1 is on the RL, so K is positive. In general, the value of K at the point where the root loci intersect the imaginary axis can also be found by the method just described. Figure 8-17 shows that the value of K at $s = 0$ is -1. The computer method and the Routh–Hurwitz criterion are other convenient alternatives of finding the critical value of K for stability. ▲

In summary, except for extremely complex cases, the properties on the root loci presented above should be adequate for making a reasonably accurate sketch of the root-locus diagram short of plotting it point by point. The computer program can be used to solve for the exact root locations, the breakaway points, and some of the other specific details of the root loci, including the plotting of the final loci. However, one cannot rely on the computer solution completely, since the users still have to decide on the range and resolution of K so that the root-locus plot has a reasonable appearance. For quick reference, the important properties described are summarized in Table 8-2.

The following example illustrates the construction of a root-locus diagram manually, step by step, using the root-locus properties given in Table 8-2.

Example 8-17

Consider the equation

$$s(s + 5)(s + 6)(s^2 + 2s + 2) + K(s + 3) = 0 \tag{8-73}$$

Dividing both sides of the last equation by the terms that do not contain K, we have

$$G(s)H(s) = \frac{K(s+3)}{s(s+5)(s+6)(s^2+2s+2)} \tag{8-74}$$

The following properties of the root loci are determined:

1. The $K = 0$ points are at the poles of $G(s)H(s)$: $s = 0, -5, -6, -1 + j$, and $-1 - j$.
2. The $K = \pm\infty$ points are at the zeros of $G(s)H(s)$: $s = -3, \infty, \infty, \infty, \infty$.
3. There are five separate branches on the root loci.
4. The root loci are symmetrical with respect to the real axis of the s-plane.

Table 8-2 Properties of the Root Loci of $1 + KG_1(s)H_1(s) = 0$

1. $K = 0$ points	The $K = 0$ points are at the poles of $G(s)H(s)$, including those at $s = \infty$.
2. $K = \pm\infty$ points	The $K = \pm\infty$ points are at the zeros of $G(s)H(s)$, including those at $s = \infty$.
3. Number of separate root loci	The total number of root loci is equal to the order of the equation $F(s) = 0$.
4. Symmetry of root loci	The root loci are symmetrical about the axes of symmetry of the pole–zero configuration of $G(s)H(s)$.
5. Asymptotes of root loci as $s \to \infty$	For large values of s, the RL ($K > 0$) are asymptotic to asymptotes with angles given by $\theta_i = \frac{2i+1}{\lvert n-m \rvert} \times 180°$. For CRL ($K < 0$), $\theta_i = \frac{2i}{\lvert n-m \rvert} \times 180°$, where $i = 0, 1, 2, \ldots, \lvert n-m \rvert - 1$; n = number of finite poles of $G(s)H(s)$; m = number of finite zeros of $G(s)H(s)$
6. Intersection of the asymptotes	(a) The intersection of the asymptotes lies only on the real axis in the s-plane. (b) The point of intersection of the asymptotes is given by $\sigma_1 = \frac{\Sigma \text{ real parts of poles of } G(s)H(s) - \Sigma \text{ real parts of zeros of } G(s)H(s)}{n-m}$

Table 8-2 Continued

7. Root loci on the real axis	RL ($K > 0$) are found in a section of the real axis only if the total number of real poles and zeros of $G(s)H(s)$ to the **right** of the section is **odd.** If the total number of real poles and zeros to the right of a given section is **even,** CRL ($K < 0$) are found.
8. Angles of departure	The angle of departure or arrival of the root loci from a pole or a zero of $G(s)H(s)$ can be determined by assuming a point s_1 that is very close to the pole, or zero, and applying the equation $$\begin{aligned}\angle G(s_1)H(s_1) &= \sum_{k=1}^{m} \angle(s_1 + z_k) - \sum_{j=1}^{n} \angle(s_1 + p_j) \\ &= 2(i+1)180^\circ \qquad RL \\ &= 2i \times 180^\circ \qquad CRL\end{aligned}$$ where $i = 0, \pm1, \pm2, \ldots$.
9. Intersection of the root loci with the imaginary axis	The crossing points of the root loci on the imaginary axis and the corresponding values of K may be found by use of the Routh–Hurwitz criterion.
10. Breakaway points	The breakaway points on the root loci are determined by finding the roots of $dK/ds = 0$, or $dG(s)H(s)/ds = 0$. These are necessary conditions only.
11. Calculation of the values of K on the root loci	The absolute value of K at any point s_1 on the root loci is determined from the equation $$\lvert K \rvert = \frac{1}{\lvert G_1(s_1)H_1(s_1) \rvert}$$

5. Since $G(s)H(s)$ has five poles and one finite zero, four RL and CRL should approach infinity along the asymptotes. The angles of the asymptotes of the RL are given by [Eq. (8-33)]

$$\theta_i = \frac{2i+1}{|n-m|}180^\circ = \frac{2i+1}{|5-1|}180^\circ \qquad 0 \le K < \infty \tag{8-75}$$

for $i = 0, 1, 2, 3$. Thus the four root loci that approach infinity as K approaches infinity should approach asymptotes with angles of 45°, −45°, 135°, and −135°, respectively. The angles of the asymptotes of the CRL at infinity are given by [Eq. (8-34)]

$$\theta_i = \frac{2i}{|n-m|}180^\circ = \frac{2i}{|5-1|}180^\circ \qquad -\infty < K \le 0 \tag{8-76}$$

for $i = 0, 1, 2, 3$. Thus as K approaches $-\infty$, four CRL should approach infinity along asymptotes with angles of 0°, 90°, 180°, and 270°.

6. The intersection of the asymptotes is given by [Eq. (8-36)]

$$\sigma_1 = \frac{\Sigma(-5-6-1-1)-(-3)}{4} = -2.5 \tag{8-77}$$

The results from these six steps are illustrated in Fig. 8-18. It should be pointed out that in general the properties of the asymptotes do not indicate on which side of the asymptotes the root loci lie. The asymptotes indicate nothing more than the behavior of the root loci as $s \to \infty$. In fact, the root locus can even cross an asymptote in the finite s domain. The segments of the RL and the CRL shown in Fig. 8-18 can be plotted accurately only if additional information is obtained.

7. Root loci on the real axis: There are RL on the real axis between $s = 0$ and -3, $s = -5$ and -6. There are CRL on the remaining portions of the real axis, that is, between $s = -3$ and -5, and $s = -6$ and $-\infty$, as shown in Fig. 8-19.

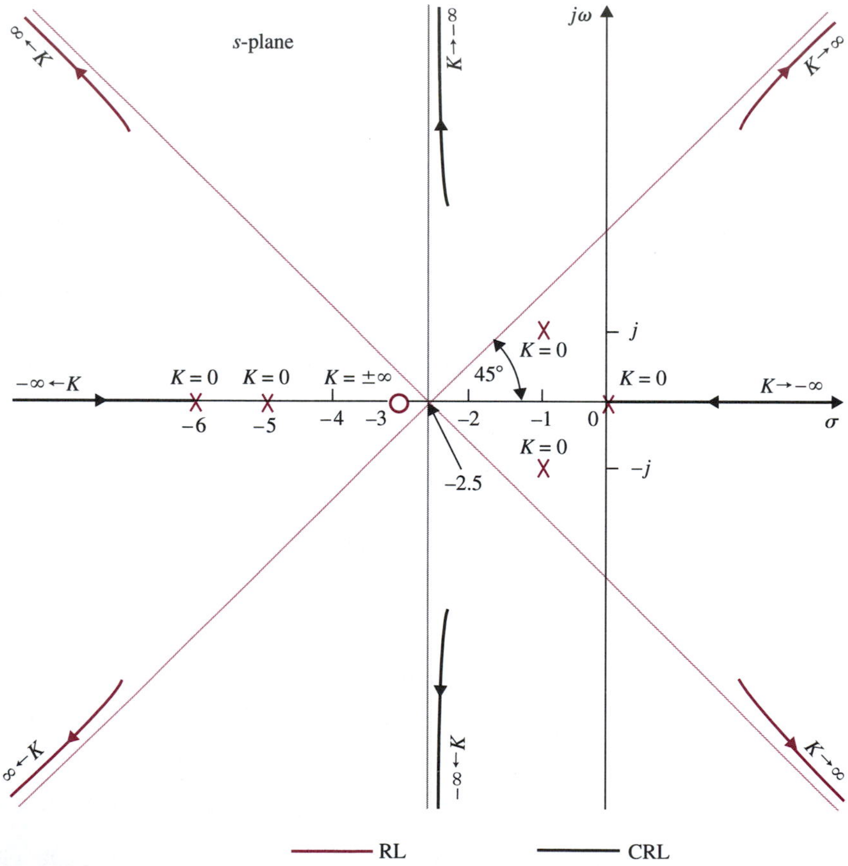

Figure 8-18 Preliminary calculations of the root loci of $s(s+5)(s+6)(s^2+2s+2)+K(s+3)=0$.

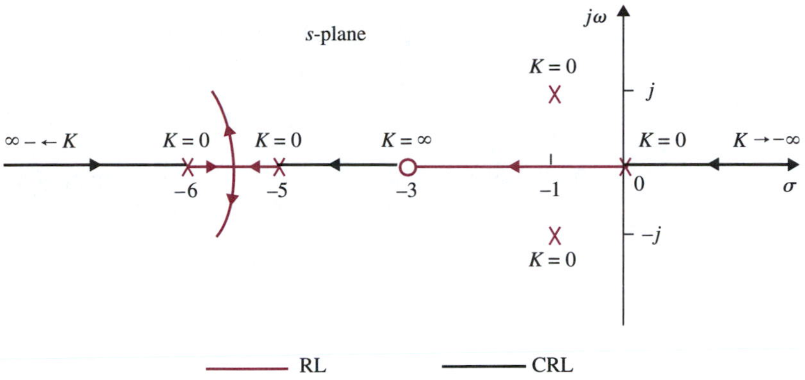

Figure 8-19 Root loci of $s(s + 5)(s + 6)(s^2 + 2s + 2) + K(s + 3) = 0$ on the real axis.

8. Angles of departure: The angle of departure θ of the RL leaving the pole at $-1 + j$ is determined using Eq. (8-18). If s_1 is a point on the RL leaving the pole at $-1 + j$, and s_1 is very close to the pole, as shown in Fig. 8-20, Eq. (8-18) gives

$$\angle(s_1 + 3) - \angle s_1 - \angle(s_1 + 1 + j) - \angle(s_1 + 5) - \angle(s_1 + 1 - j) \\ = (2i + 1)180° \tag{8-78}$$

or

$$26.6° - 135° - 90° - 14° - 11.4° - \theta \cong (2i + 1)180° \tag{8-79}$$

for $i = 0, \pm1, \pm2, \ldots$ Therefore, selecting $i = 2$ gives

$$\theta \cong -43.8° \tag{8-80}$$

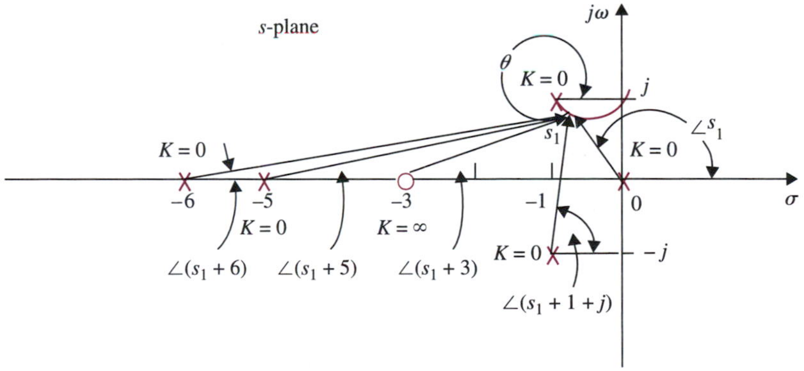

Figure 8-20 Computation of angle of departure of the root loci of $s(s + 5)(s + 6)(s^2 + 2s + 2) + K(s + 3) = 0$.

Similarly, Eq. (8-19) is used to determine the angle of arrival θ of the CRL arriving at the pole $-1 + j$. It is easy to see that θ' differs from θ by 180°; thus

$$\theta' = 180° - 43.8° = 136.2° \tag{8-81}$$

9. The intersection of the root loci on the imaginary axis is determined using Routh's tabulation. Equation (8-73) is written

$$s^5 + 13s^4 + 54s^3 + 82s^2 + (60 + K)s + 3K = 0 \tag{8-82}$$

Routh's tabulation is

s^5	1	54	$60 + K$
s^4	13	82	$3K$
s^3	47.7	$0.769K$	0
s^2	$65.6 - 0.212K$	$3K$	0
s^1	$\dfrac{3940 - 105K - 0.163K^2}{65.6 - 0.212K}$	0	0
s^0	$3K$	0	0

For Eq. (8-82) to have no roots on the imaginary axis or in the right half of the s-plane, the elements in the first column of Routh's tabulation must all be of the same sign. Thus the following inequalities must be satisfied:

$$65.6 - 0.212K > 0 \quad \text{or} \quad K < 309 \tag{8-83}$$
$$3940 - 105K - 0.163K^2 > 0 \quad \text{or} \quad K < 35 \tag{8-84}$$
$$K > 0 \tag{8-85}$$

Thus all the roots of Eq. (8-82) will stay in the left-half s-plane if K lies between 0 and 35, which means that the root loci of Eq. (8-82) cross the imaginary axis when $K = 35$ and $K = 0$. The coordinates at the crossover points on the imaginary axis that correspond to $K = 35$ are determined from the auxiliary equation

$$A(s) = (65.6 - 0.212K)s^2 + 3K = 0 \tag{8-86}$$

which is obtained by using the coefficients from the row just above the row of zeros in the s^1 row that would have happened when K is set to 35. Substituting $K = 35$ in Eq. (8-86), we get

$$58.2s^2 + 105 = 0 \tag{8-87}$$

The roots of Eq. (8-87) are $s = j1.34$ and $-j1.34$, which are the points at which the root loci cross the $j\omega$-axis.

10. Breakaway points: Based on the information gathered from the preceding nine steps, a trial sketch of the root loci indicates that there can be only one breakaway point on the entire root loci, and the point should lie between the two poles of

$G(s)H(s)$ at $s = -5$ and -6. To find the breakaway point, we take the derivative on both sides of Eq. (8-74) with respect to s and set it to zero; the resulting equation is

$$s^5 + 13.5s^4 + 66s^3 + 142s^2 + 123s + 45 = 0 \tag{8-88}$$

Since only one breakaway point is expected, only one root of Eq. (8-88) is the correct solution of the breakaway point. The five roots of Eq. (8-88) are

$$s = -3.33 + j1.204 \qquad s = -3.33 - j1.204$$
$$s = -0.656 + j0.468 \qquad s = -0.656 - j0.468$$
$$s = -5.53$$

Clearly, the breakaway point is at -5.53. The other four solutions do not satisfy Eq. (8-82) and are not breakaway points. Based on the information obtained in the last 10 steps, the root loci of Eq. (8-82) are sketched as shown in Fig. 8-21. ▲

8-4 Computer Solutions

In this section we demonstrate the computer solutions of the root loci using **rlplot** from the **CSAD** toolbox and **root** from **Program CC.**

8-4-1 rlplot from CSAD

The transfer function $G(s)H(s)$ in Eq. (8-74) is entered to **rlplot** in $N(s)/D(s)$ form.

```
>>n=[1 3] <CR>
>>d=pmult([1 0],[1 5],[1 6],[1 2 2]) <CR>
>>rlplot <CR>
```

The following menu will appear:

```
    - RLPLOT OPTIONS -

K values       Zeta line
Single K       Wn circle
Find K         Rule info
Grid           New/clear
Title          Display TF
               Quit

Option? >k <CR>      (Select K values to plot root loci)
Enter k values to plot >linespace(-3000,3000) <CR>      (K = -3000 to 3000)
```

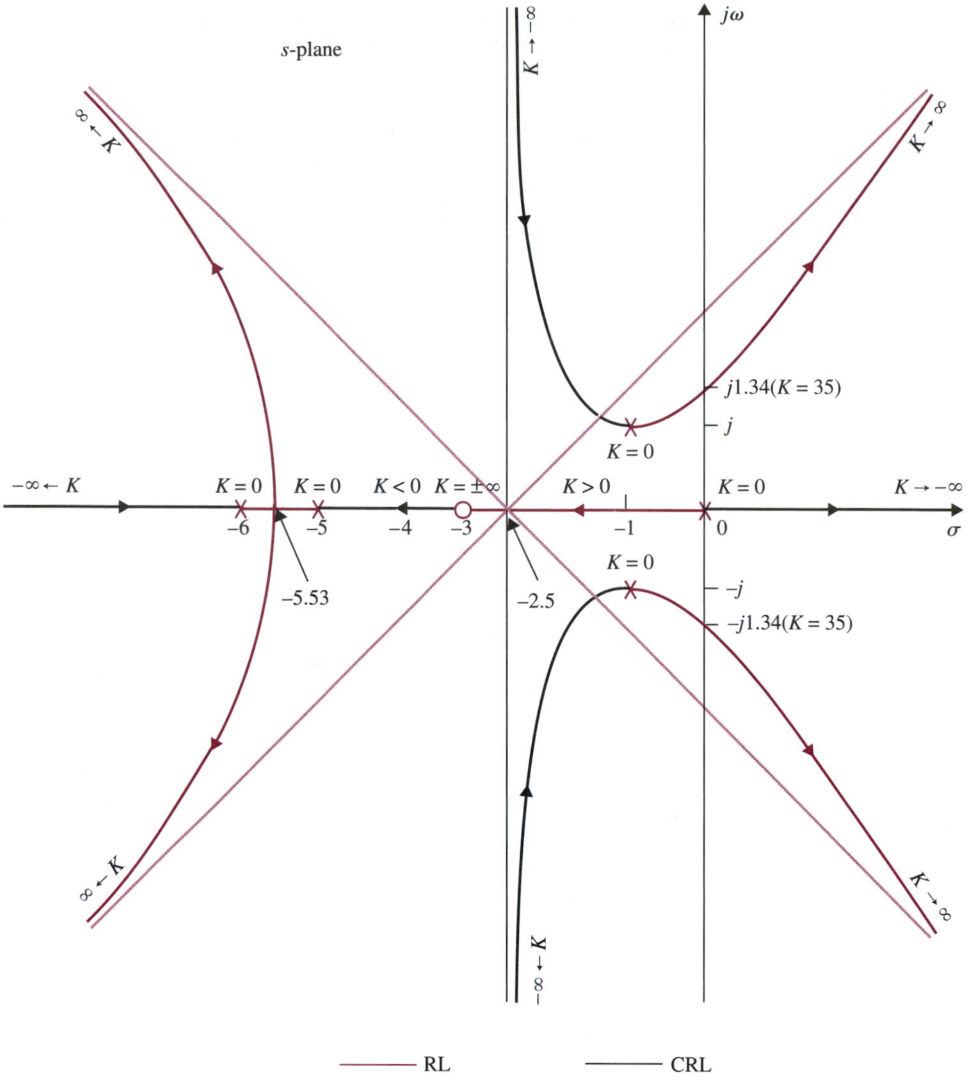

Figure 8-21 Root loci of $s(s + 5)(s + 6)(s^2 + 2s + 2) + K(s + 3) = 0$.

This command will produce a root-locus plot of 100 (default) linearly spaced points starting at $K = -3000$ to $K = 3000$. The root-locus plot is shown in Fig. 8-22. These plots are obtained with the Shift-PrtSc key for print-screen with **MATLAB.** With a color monitor, the RL are plotted in red and the CRL are in green.

Selecting the **Rule info** option from the menu, we get the following information on the root loci:

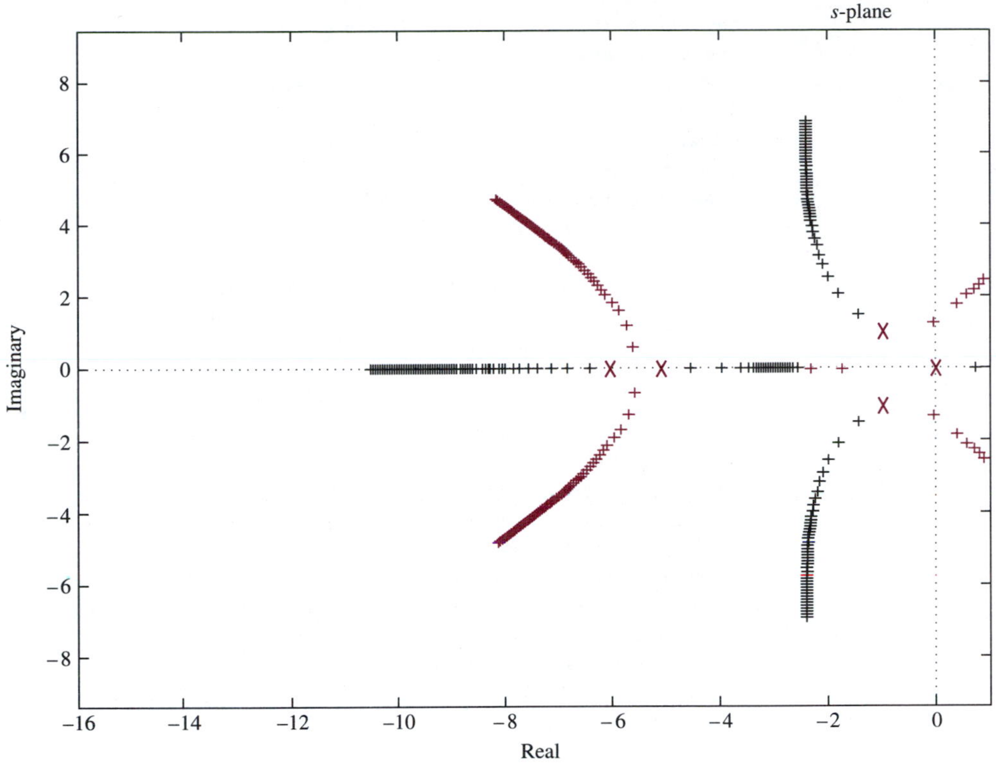

Figure 8-22 Root loci of $G(s)H(s) = \dfrac{K(s+3)}{s(s+5)(s+6)(s^2+2s+2)}$ for $-3000 < K < 3000$ plotted with **rlplot** of **CSAD.**

```
K=0 points                              K=inf points

           0                            -3.0000e+00
-6.0000e+000                                 ∞
-5.0000e+000                                 ∞
-1.0000e+000+ 1.0000e+000i                   ∞
-1.0000e+000- 1.0000e+000i                   ∞

Asymptote angles in degrees for
K>0          K<0
 45            0
135           90
225          180
315          270
```

```
Asymptote Intersection is: -2.5
Press any key to continue . . .
K>0 root loci appears on the real axis between:
0 and -3
-5 and -6
K<0 root loci appears along all
other sections of the real axis.
Angles of departure for K>0 are
Pole                             Angle (deg)
-1.0000+000- 1.0000e+000i        4.3781e+001
-1.0000+000+ 1.0000e+000i        -4.3781e+001
Possible Breakaway Points, and associatd |K| values are:
    Point                         |K|
-5.5257e+000                     1.1718e+001       (This is the only breakaway point)
-3.3311e+000+ 1.2040e+001i       1.2830e+002
-3.3311e+000- 1.2040e+001i       1.2830e+002
-6.5604e-000+ 4.6768e-001i       7.5478e+000
-6.5604e-000- 4.6768e-001i       7.5478e+000
Find a jw-axis crossing? (y/n) [y] ><CR>
Enter a value of K that gives stability [0.001]><CR>
Enter a value of K that gives Instability [1000]><CR>
A jw-axis crossing occurs at K=35.519
Associated system poles are
-5.5803 + 0.7077i
-5.5803 - 0.0707i
-1.8393
-0.0000 + 1.3531i
-0.0000 - 1.3531i
```

The program has other features, such as finding the value of the roots for a given K, finding the root values and K at a point on the root loci indicated by the mouse cursor, and so on.

8-4-2 Root of Program CC

Program CC has a command, **root, g,** that computes and plots the root loci of the characteristic equation given the transfer function $G(s)$. The following input command is given for the transfer function in Eq. (8-74):

```
>g=(s+3)/(s*(s+5)*(s+6)*(s^2+2*s+2)) <CR>
>root,g,auto <CR>      (auto implies that automatic scales are selected for both axes)
```

The root-locus plot of Fig. 8-23 is shown on the monitor screen for RL and CRL by using the N option for a negative sign of gain. However, screen dump on the Student's version of **CC** takes a long time to execute. The **I** option displays information on open-loop poles, zeros, breakpoints, angles of departure and arrival, and asymptotic behavior. Other details of the program are not elaborated here.

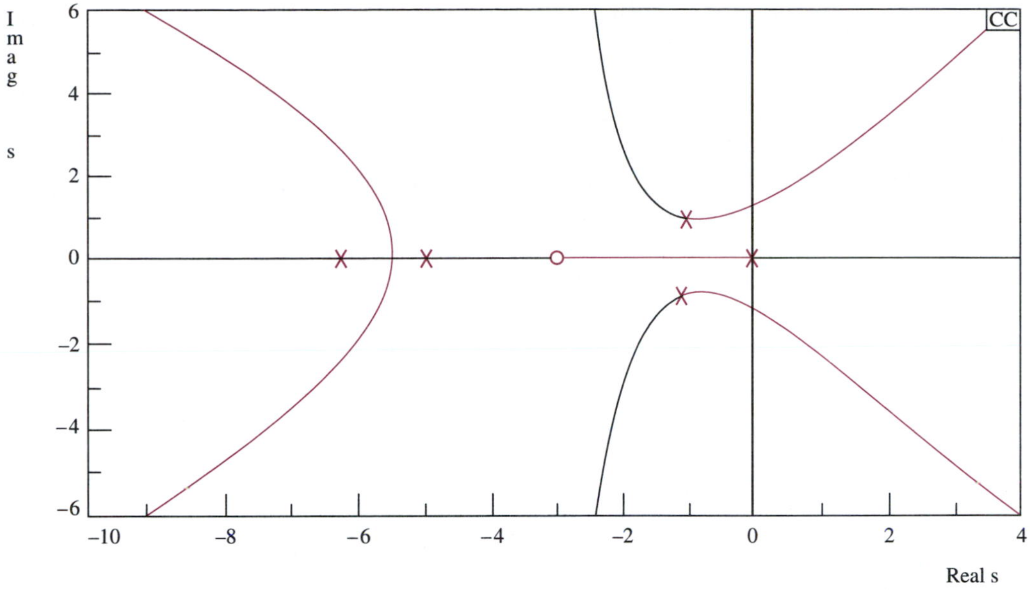

Figure 8-23 Root loci of $G(s)H(s) = \dfrac{K(s+3)}{s(s+5)(s+6)(s^2+2s+2)}$ plotted with **root** of **Program CC.**

8-5 Some Important Aspects of the Construction of the Root Loci

One of the important aspects of the root-locus technique is that for most control systems with moderate complexity, the analyst or designer can obtain vital information on the performance of the system by making a quick sketch of the root loci using some or all of the properties of the root loci. It is of importance to understand all the properties of the root loci even when the diagram is to be plotted with the help of a digital computer program. From the design standpoint, it is useful to learn the effects on the root loci when poles and zeros of $G(s)H(s)$ are added or moved around in the s-plane. Some of these properties are helpful in construction of the root-locus diagram.

8-5-1 Effects of Adding Poles and Zeros to $G(s)H(s)$

The general problem of controller design in control systems may be treated as an investigation of the effects to the root loci when poles and zeros are added to the loop transfer function $G(s)H(s)$.

▲ Adding a pole to $G(s)H(s)$ has the effect of pushing the root loci to the right.

Addition of Poles to $G(s)H(s)$

Adding a pole to $G(s)H(s)$ has the effect of pushing the root loci toward the right-half plane.

The effect of adding a pole to $G(s)H(s)$ can be illustrated by several examples.

Example 8-18

Consider the function

$$G(s)H(s) = \frac{K}{s(s+a)} \qquad a > 0 \tag{8-89}$$

The RL of $1 + G(s)H(s) = 0$ are shown in Fig. 8-24(a). These root loci are constructed based on the poles of $G(s)H(s)$, which are at $s = 0$ and $-a$. Now let us introduce a pole at $s = -b$, with $b > a$. The function $G(s)H(s)$ now becomes

$$G(s)H(s) = \frac{K}{s(s+a)(s+b)} \tag{8-90}$$

Figure 8-24(b) shows that the pole at $s = -b$ causes the complex part of the root loci to bend toward the right-half s-plane. The angles of the asymptotes for the complex roots are changed from $\pm 90°$ to $\pm 60°$. The intersect of the asymptotes is also moved from $-a/2$ to $-(a+b)/2$ on the real axis. If $G(s)H(s)$ represents the loop transfer function of a control system, the system with the root loci in Fig. 8-24(b) may become unstable if the value of K exceeds the critical value for stability, whereas the system represented by the root loci in Fig. 8-24(a) is always stable for $K > 0$. Figure 8-24(c) shows the root loci when another pole is added to $G(s)H(s)$ at $s = -c$, $c > b$. The system is now of the fourth order, and the two complex root loci are bent farther to the right. The angles of asymptotes of these two complex loci are now $\pm 45°$. The stability condition of the fourth-order system is even more acute than that of the third-order system. Figure 8-24(d) illustrates that the addition of a pair of complex-conjugate poles to the transfer function of Eq. (8-89) will result in a similar effect. Therefore, we may draw a general conclusion that the addition of poles to $G(s)H(s)$ has the effect of moving the dominant portion of the root loci toward the right-half s-plane. ▲

Addition of Zeros to $G(s)H(s)$

▲ Adding zeros to $G(s)H(s)$ tends to push the root loci to the left.

Adding left-half plane zeros to the function $G(s)H(s)$ generally has the effect of moving and bending the root loci toward the left-half s-plane.

The following example illustrates the effect of adding a zero and zeros to $G(s)H(s)$ on the root loci.

Example 8-19

Figure 8-25(a) shows the RL of the $G(s)H(s)$ in Eq. (8-89) with a zero added at $s = -b$ ($b > a$). The complex-conjugate part of the RL of the original system is bent toward the left and forms a circle. Thus if $G(s)H(s)$ is the loop transfer function of a control system, the relative stability of the system is improved by the addition of the zero. Figure 8-25(b) shows that a similar effect will result if a pair of complex-cojugate zeros is added to the function of Eq.

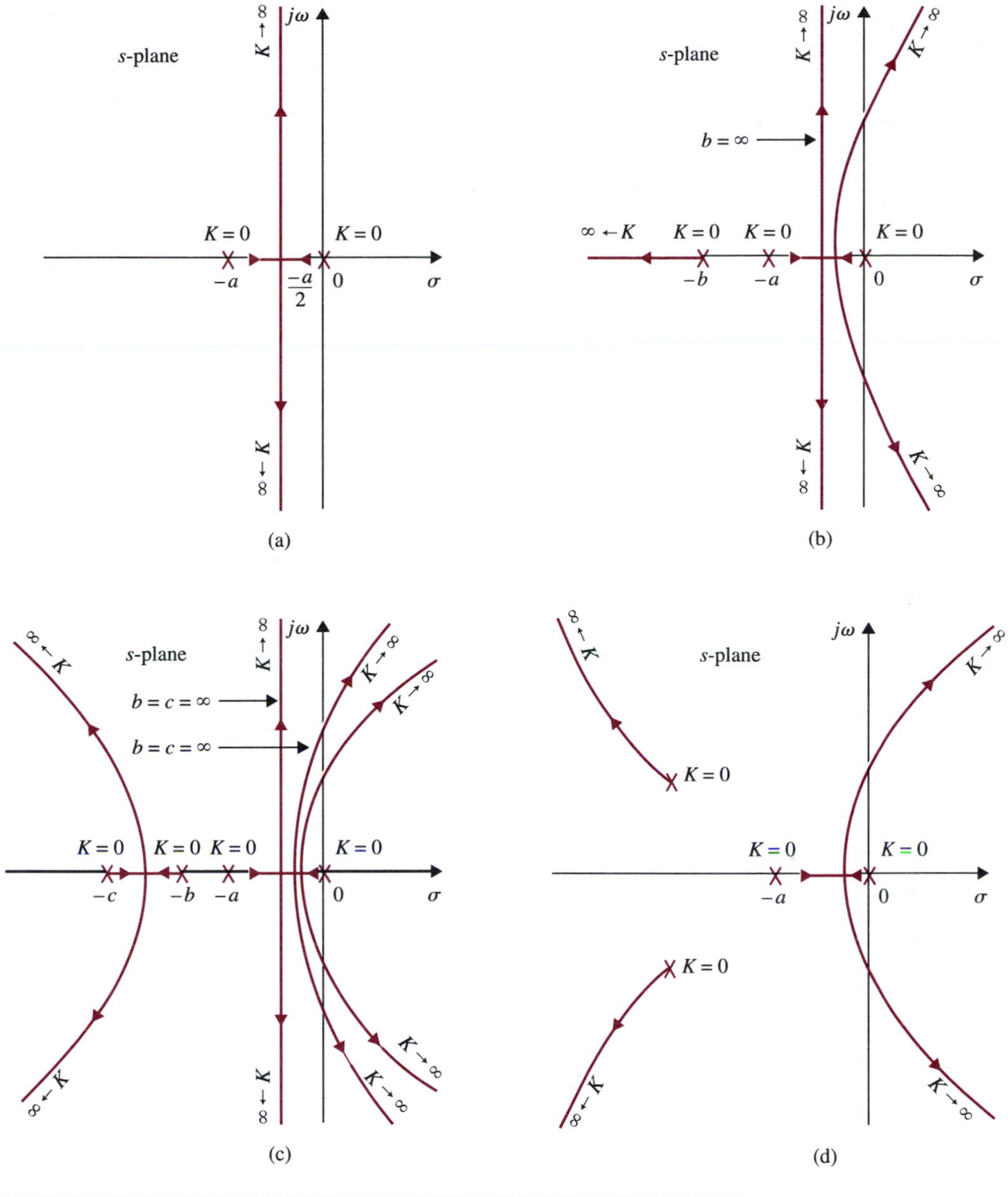

Figure 8-24 Root-locus diagrams that show the effects of adding poles to $G(s)H(s)$.

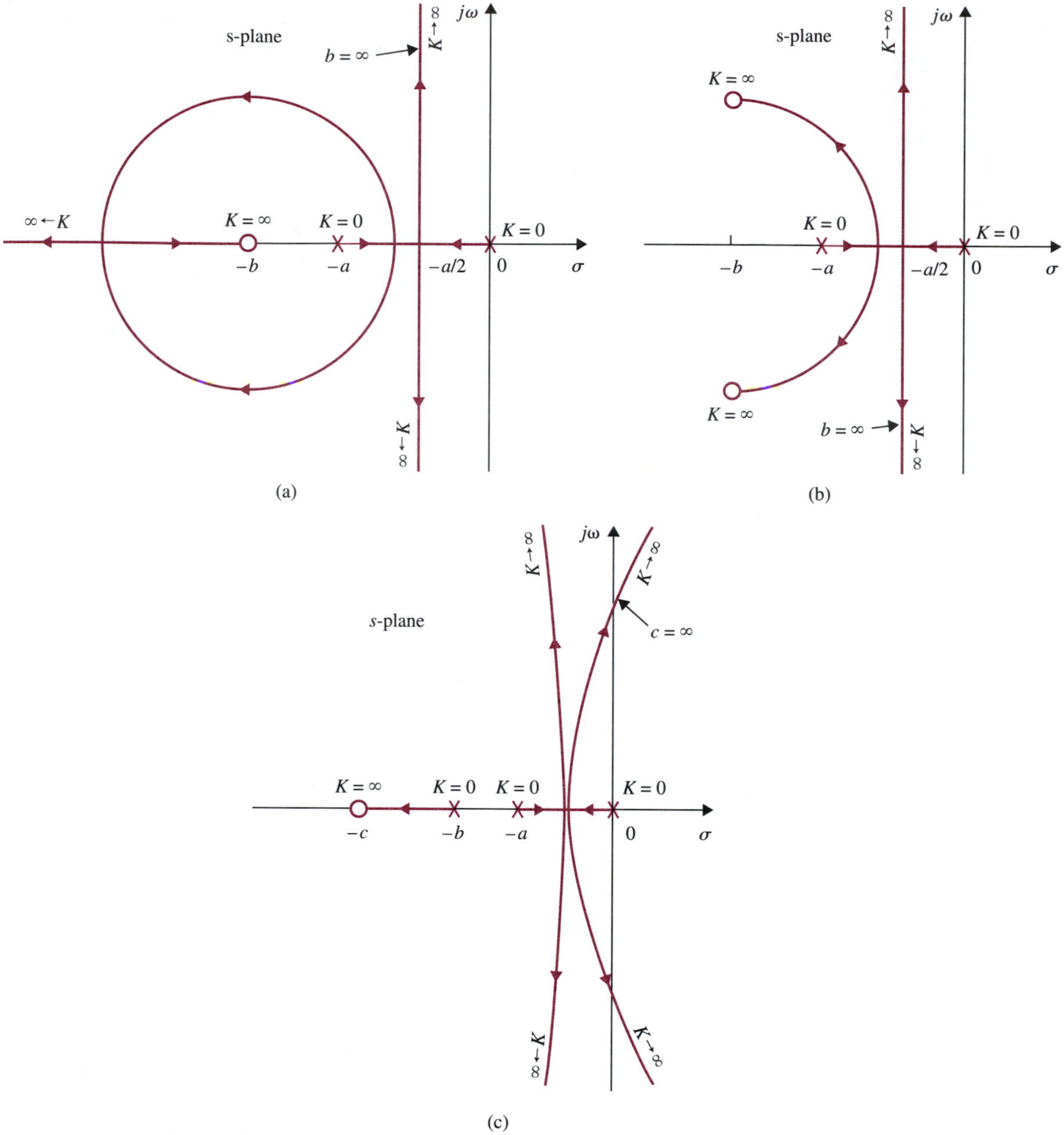

Figure 8-25 Root-locus diagrams that show the effects of adding a zero to $G(s)H(s)$.

(8-89). Figure 8-25(c) shows the RL when a zero at $s = -c$ is added to the transfer function of Eq. (8-90). ▲

Example 8-20

Consider the equation

$$s^2(s + a) + K(s + b) = 0 \tag{8-91}$$

Dividing both sides of Eq. (8-91) by the terms that do not contain K, we have the loop transfer function,

$$G(s)H(s) = \frac{K(s + b)}{s^2(s + a)} \tag{8-92}$$

It can be shown that the nonzero breakaway points depend on the value of a, and are

$$s = -\frac{a + 3}{4} \pm \frac{1}{4}\sqrt{a^2 - 10a + 9} \tag{8-93}$$

Figure 8-26 shows the RL of Eq. (8-91) with $b = 1$ and several values of a. The results are summarized as follows:

- ▲ Figure 8-26(a): $a = 10$. Breakaway points: $s = -2.5$ and -4.0.
- ▲ Figure 8-26(b): $a = 9$. The two breakaway points given by Eq. (8-93) converge to one point at $s = -3$. Note the change in the RL when the pole at $-a$ is moved from -10 to -9.

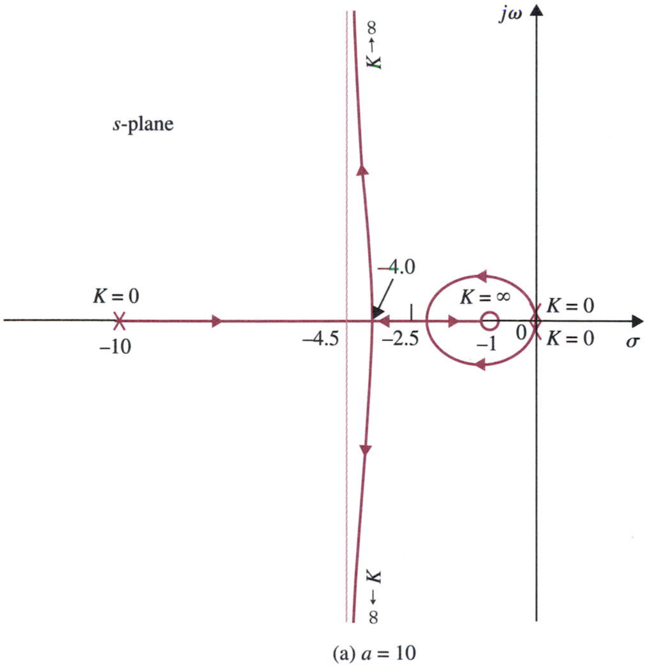

Figure 8-26 Root-locus diagrams that show the effects of moving a pole of $G(s)H(s)$. $G(s)H(s) = K(s + 1)/[s^2(s + a)]$.

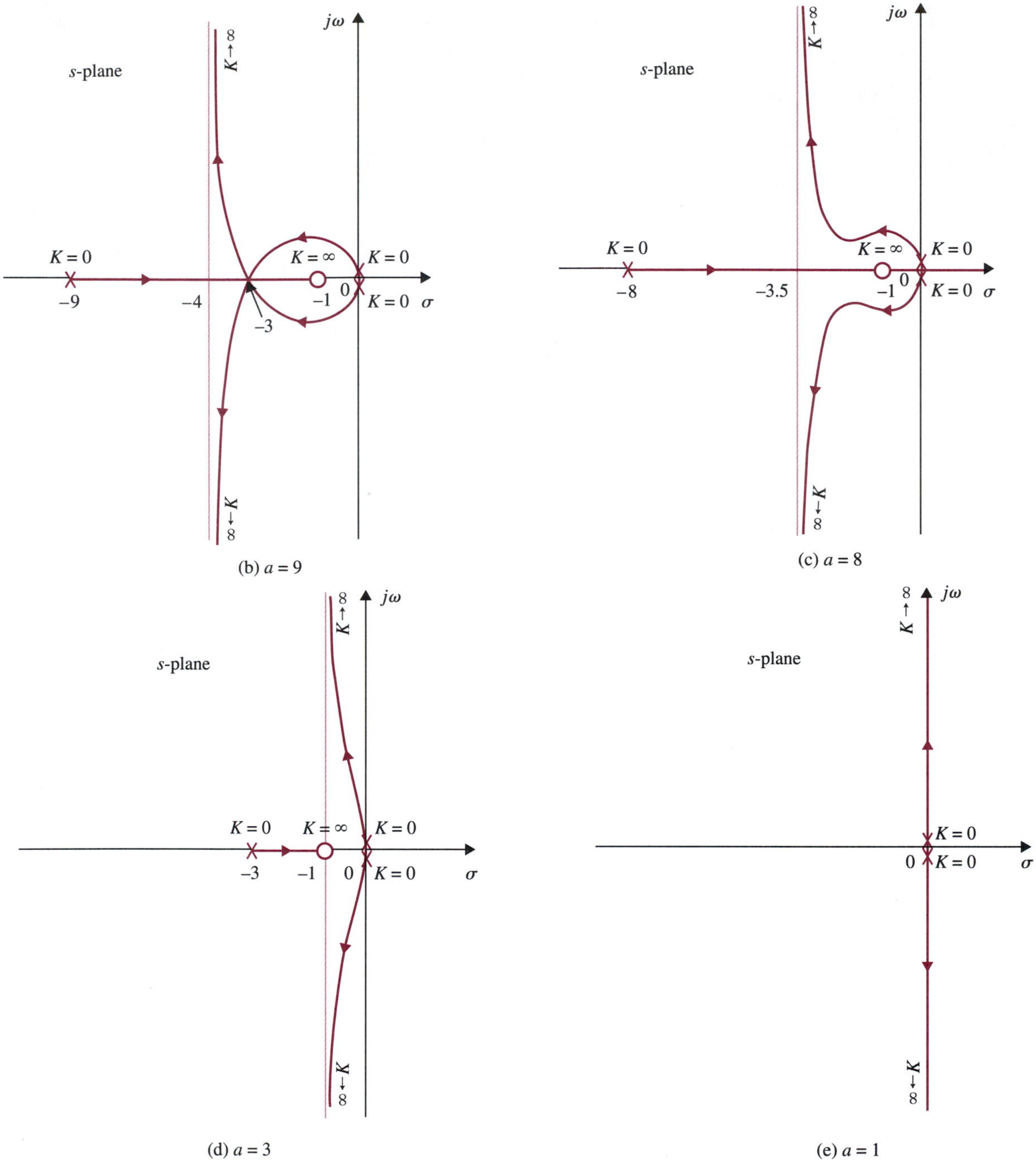

Figure 8-26 Continued.

For values of a less than 9, the values of s as given by Eq. (8-93) no longer satisfy Eq. (8-91), which means that there are no finite breakaway points, other than the one at $s = 0$.

▲ Figure 8-26(c): $a = 8$. No breakaway point on RL, except at $s = 0$.

As the pole at $s = -a$ is moved farther to the right along the real axis, the complex portion of the RL is pushed farther toward the right-half plane.

▲ Figure 8-26(d): $a = 3$.

▲ Figure 8-26(e): $a = b = 1$. The pole at $s = -a$ and the zero at $-b$ cancel each other, and the RL degenerate into a second-order case and lie entirely on the $j\omega$-axis. ▲

Example 8-21

Consider the equation

$$s(s^2 + 2s + a) + K(s + 2) = 0 \tag{8-94}$$

which leads to the equivalent $G(s)H(s)$ as

$$G(s)H(s) = \frac{K(s+2)}{s(s^2 + 2s + a)} \tag{8-95}$$

The objective is to study the RL for various values of a (>0). The breakaway point equation of the RL is determined as

$$s^3 + 4s^2 + 4s + a = 0 \tag{8-96}$$

Figure 8-27 shows the root loci of Eq. (8-94) under the following conditions.

▲ Figure 8-27(a): $a = 1$. Breakaway points: $s = -0.38$, -1.0, and -2.618, with the last point being on the CRL. As the value of a is increased from unity, the two double poles of $G(s)H(s)$ at $s = -1$ will move vertically up and down with the real parts equal to -1. The breakaway points at $s = -0.38$ and $s = -2.618$ will move to the left, whereas the breakaway point at $s = -1$ will move to the right.

▲ Figure 8-27(b): $a = 1.12$. Breakaway points: $s = -0.493$, -0.857, and -2.65. Since the real parts of the poles and zeros of $G(s)H(s)$ are not affected by the value of a, the intersect of the asymptotes is always at $s = 0$.

▲ Figure 8-27(c): $a = 1.185$. Breakaway points: $s = -0.667$, -0.667, and -2.667. The two breakaway points of the RL that lie between $s = 0$ and -1 converge to a point.

▲ Figure 8-27(d): $a = 3$. Breakaway point: $s = -3$. When a is greater than 1.185, Eq. (8-96) yields only one solution for the breakaway point.

The reader may investigate the difference between the root loci in Fig. 8-27(c) and (d), and fill in the evolution of the loci when the value of a is gradually changed from 1.185 to 3 and beyond. ▲

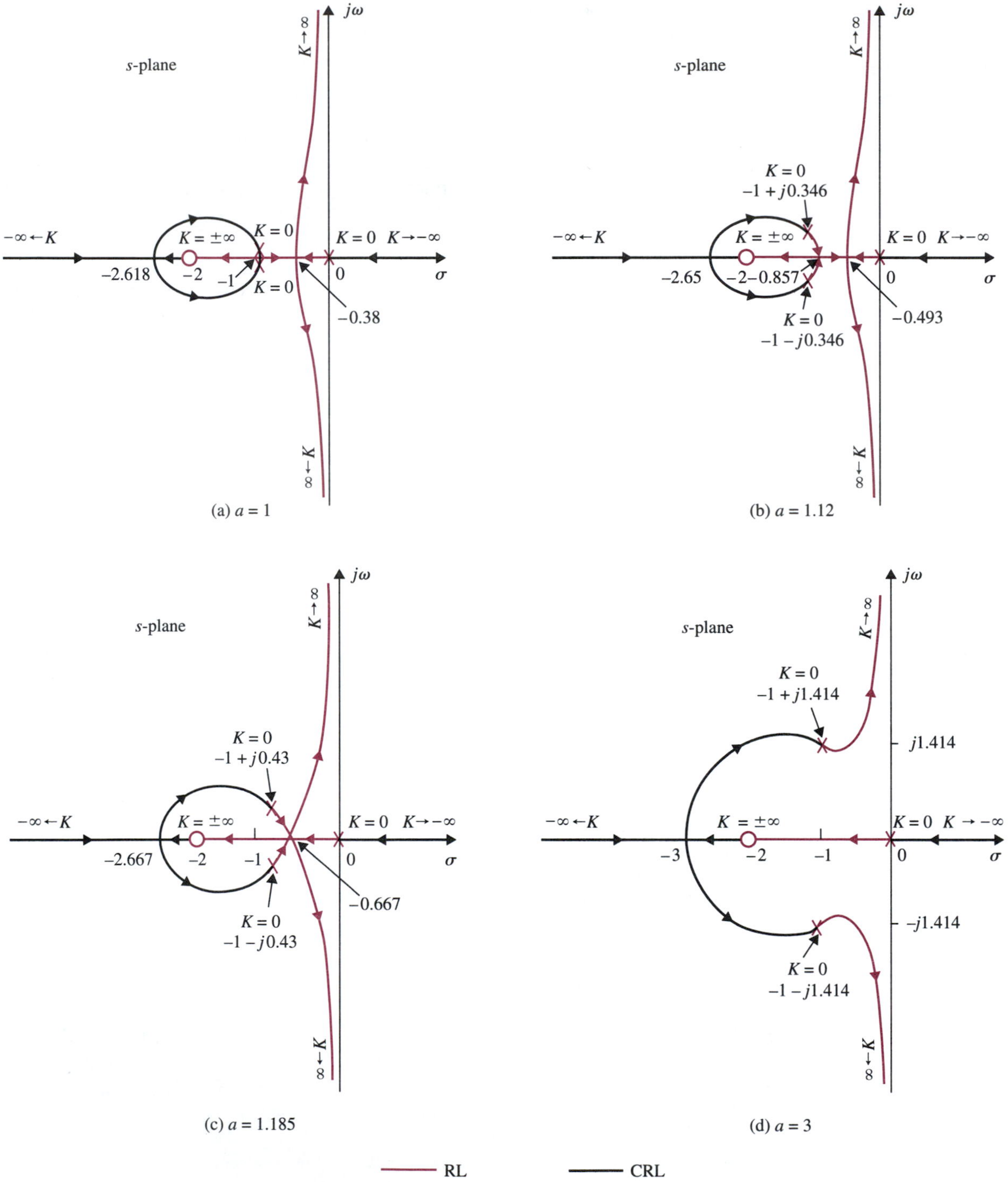

Figure 8-27 Root-locus diagrams that show the effects of moving a pole of $G(s)H(s) = K(s + 2)/[s(s^2 + 2s + 2)]$.

8-6 Root Contours: Multiple-Parameter Variation

The root-locus technique discussed thus far is limited to only one variable parameter in K. In many control systems problems, often the effects of varying several parameters should be investigated. For example, when designing a controller that is represented by a transfer function with poles and zeros, it would be useful to investigate the effects on the characteristic equation roots when these poles and zeros take on various values. In Section 8-5 the root loci of equations with two variable parameters are studied by fixing one parameter and assigning different values to the other. In this section the multiparameter problem is investigated through a more systematic method of embedding. When more than one parameter varies continuously from $-\infty$ to ∞, the root loci are referred to as the **root contours** (RCs). It will be shown that the root contours still possess the same properties as the single-parameter root loci, so that the methods of construction discussed thus far are all applicable.

The principle of root contour can be described by considering the equation

$$P(s) + K_1Q_1(s) + K_2Q_2(s) = 0 \tag{8-97}$$

where K_1 and K_2 are the variable parameters, and $P(s)$, $Q_1(s)$, and $Q_2(s)$ are polynomials of s. The first step involves setting the value of one of the parameters to zero. Let us set K_2 to zero. Then Eq. (8-97) becomes

$$P(s) + K_1Q_1(s) = 0 \tag{8-98}$$

which now has only one variable parameter in K_1. The root loci of Eq. (8-98) may be determined by dividing both sides of the equation by $P(s)$. Thus

$$1 + \frac{K_1Q_1(s)}{P(s)} = 0 \tag{8-99}$$

The last equation is of the form of $1 + K_1G_1(s)H_1(s) = 0$, so that we can construct the root loci of Eq. (8-98) based on the pole–zero configuration of $G_1(s)H_1(s)$. Next we restore the value of K_2, while considering that the value of K_1 is fixed, and divide both sides of Eq. (8-97) by the terms that do not contain K_2. We have

$$1 + \frac{K_2Q_2(s)}{P(s) + K_1Q_1(s)} = 0 \tag{8-100}$$

which is of the form of $1 + K_2G_2(s)H_2(s) = 0$. The root contours of Eq. (8-97) when K_2 varies (while K_1 is fixed) are constructed based on the pole–zero configuration of

$$G_2(s)H_2(s) = \frac{Q_2(s)}{P(s) + K_1Q_1(s)} \tag{8-101}$$

It is important to note that the poles of $G_2(s)H_2(s)$ are identical to the roots of Eq. (8-98). Thus the root contours of Eq. (8-97) when K_2 varies must all start ($K_2 = 0$) at the points that lie on the root loci of Eq. (8-98). This is the reason why one root-contour problem is considered to be embedded in another. The same procedure may be extended to more than two variable parameters. The following examples illustrate the construction of RCs when multiparameter-variation situations exist.

Example 8-22

Consider the equation

$$s^3 + K_2 s^2 + K_1 s + K_1 = 0 \tag{8-102}$$

where K_1 and K_2 are the variable parameters, which vary from 0 to ∞.

As the first step, we let $K_2 = 0$, and Eq. (8-102) becomes

$$s^3 + K_1 s + K_1 = 0 \tag{8-103}$$

Dividing both sides of the last equation by s^3, which is the term that does not contain K_1, we have

$$1 + \frac{K_1(s+1)}{s^3} = 0 \tag{8-104}$$

The RC of Eq. (8-103) are drawn based on the pole–zero configuration of

$$G_1(s)H_1(s) = \frac{s+1}{s^3} \tag{8-105}$$

as shown in Fig. 8-28(a). Next, we let K_2 vary between 0 and ∞ while holding K_1 at a constant nonzero value. Dividing both sides of Eq. (8-102) by the terms that do not contain K_2, we have

$$1 + \frac{K_2 s^2}{s^3 + K_1 s + K_1} = 0 \tag{8-106}$$

Thus the root contours of Eq. (8-102) when K_2 varies may be drawn from the pole–zero configuration of

$$G_2(s)H_2(s) = \frac{s^2}{s^3 + K_1 s + K_1} \tag{8-107}$$

The zeros of $G_2(s)H_2(s)$ are at $s = 0, 0$; but the poles are at the zeros of $1 + K_1G_1(s)H_1(s)$, which are found on the RL of Fig. 8-28(a). Thus, for fixed K_1, the RC when K_2 varies must all emanate from the RC of Fig. 8-28(a). Figure 8-28(b) shows the RC of Eq. (8-102) when K_2 varies from 0 to ∞, for $K_1 = 0.0184$, 0.25, and 2.56. ▲

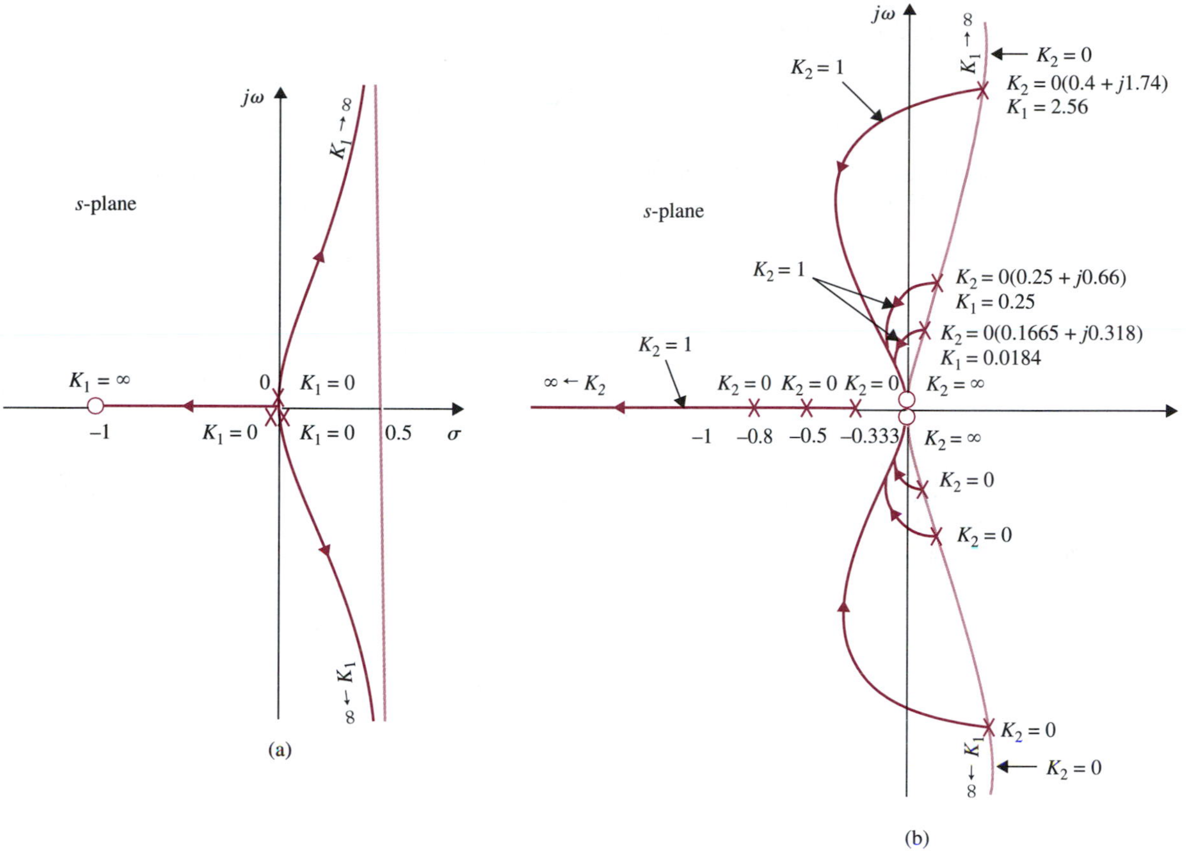

Figure 8-28 Root contours of $s^3 + K_2 s^2 + K_1 s + K_1 = 0$. (a) $K_2 = 0$. (b) K_2 varies and K_1 is constant.

Example 8-23

Consider the loop transfer function

$$G(s)H(s) = \frac{K}{s(1 + Ts)(s^2 + 2s + 2)} \tag{8-108}$$

of a closed-loop control system. It is desired to construct the RC of the characteristic equation with K and T as variable parameters. The characteristic equation of the system is

$$s(1 + Ts)(s^2 + 2s + 2) + K = 0 \tag{8-109}$$

First, we set the value of T to zero. The characteristic equation becomes

$$s(s^2 + 2s + 2) + K = 0 \tag{8-110}$$

The RC of this equation when K varies are drawn based on the pole–zero configuration of

$$G_1(s)H_1(s) = \frac{1}{s(s^2 + 2s + 2)} \tag{8-111}$$

as shown in Fig. 8-29(a). Next, we let K be fixed and consider that T is the variable parameter.

Dividing both sides of Eq. (8-109) by the terms that do not contain T, we get

$$1 + TG_2(s)H_2(s) = 1 + \frac{Ts^2(s^2 + 2s + 2)}{s(s^2 + 2s + 2) + K} = 0 \tag{8-112}$$

The RCs when T varies are constructed based on the pole–zero configuration of $G_2(s)H_2(s)$. When $T = 0$, the points on the RCs are at the poles of $G_2(s)H_2(s)$, which are on the RC of Eq. (8-110). When $T = \infty$, the roots of Eq. (8-109) are at the zeros of $G_2(s)H_2(s)$, which are at $s = 0, 0, -1 + j$, and $-1 - j$. Figure 8-29(b) shows the pole–zero configuration of $G_2(s)H_2(s)$ for $K = 10$. Notice that $G_2(s)H_2(s)$ has three finite poles and four finite zeros. The RCs for Eq. (8-109) when T varies are shown in Figs. 8-30, 8-31, and 8-32 for three different values of K. The RCs in Fig. 8-31 show that when $K = 0.5$ and $T = 0.5$, the characteristic equation in Eq. (8-109) has a quadruple root at $s = -1$. ▲

Example 8-24

As an example to illustrate the effect of the variation of a zero of $G(s)H(s)$, consider the function

$$G(s)H(s) = \frac{K(1 + Ts)}{s(s + 1)(s + 2)} \tag{8-113}$$

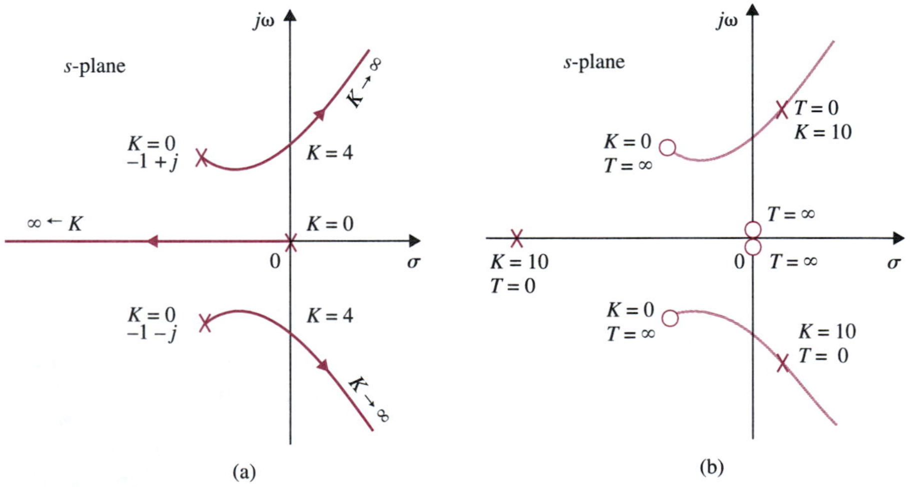

Figure 8-29 (a) Root loci for $s(s^2 + 2s + 2) + K = 0$. (b) Pole–zero configuration of $G_2(s)H_2(s) = Ts^2(s^2 + 2s + 2)/[s(s^2 + 2s + 2) + K]$.

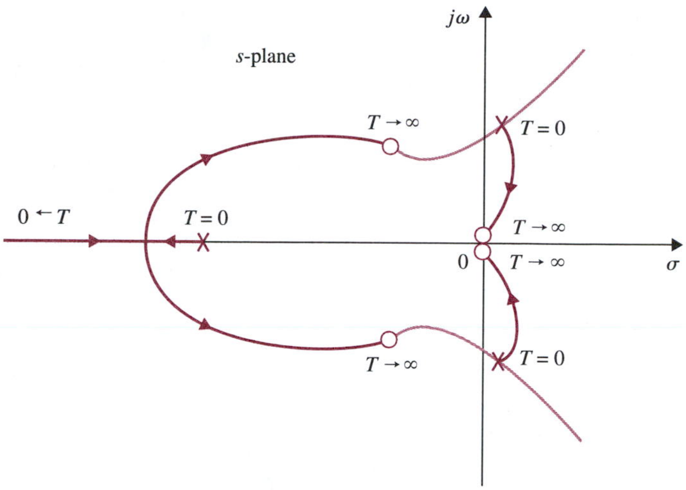

Figure 8-30 Root contours for $s(1 + Ts)(s^2 + 2s + 2) + K = 0$. $K > 4$.

The characteristic equation is

$$s(s + 1)(s + 2) + K(1 + Ts) = 0 \tag{8-114}$$

Let us first set T to zero and consider the effect of varying K. Equation (8-114) becomes

$$s(s + 1)(s + 2) + K = 0 \tag{8-115}$$

This leads to

$$G_1(s)H_1(s) = \frac{1}{s(s + 1)(s + 2)} \tag{8-116}$$

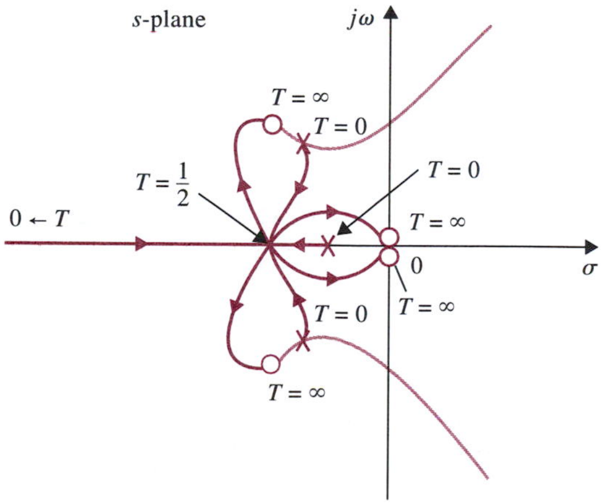

Figure 8-31 Root contours for $s(1 + Ts)(s^2 + 2s + 2) + K = 0$. $K = 0.5$.

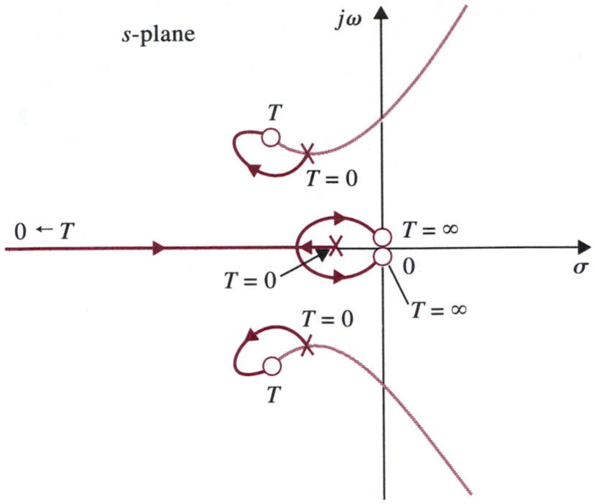

Figure 8-32 Root contours for $s(1 + Ts)(s^2 + 2s + 2) + K = 0$. $K < 0.5$

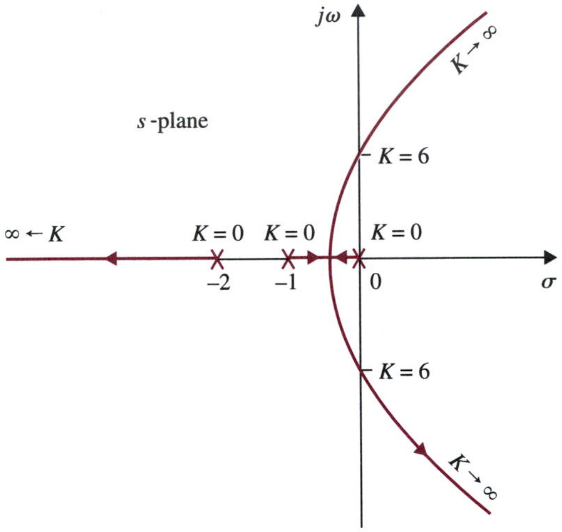

Figure 8-33 RC for $s(s + 1)(s + 2) + K = 0$.

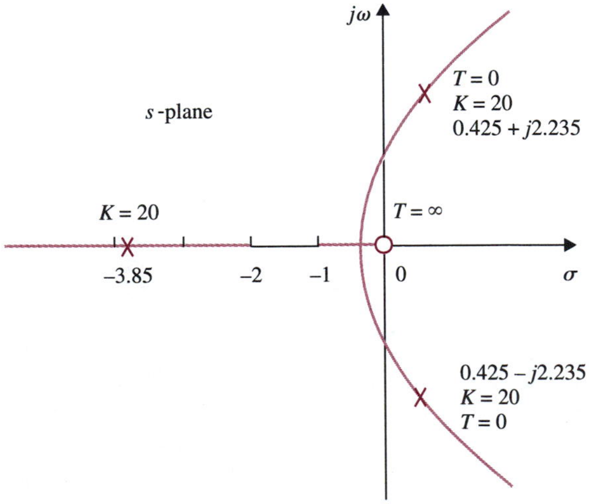

Figure 8-34 Pole–zero configuration of $G_2(s)H_2(s) = TKs/[s(s + 1)(s + 2) + K]$. $K = 20$.

The RL of Eq. (8-115) are drawn based on the pole–zero configuration of Eq. (8-116) and are shown in Fig. 8-33.

When the K is fixed and nonzero, we divide both sides of Eq. (8-114) by the terms that do not contain T, and we get

$$1 + TG_2(s)H_2(s) = 1 + \frac{TKs}{s(s+1)(s+2)+K} = 0 \tag{8-117}$$

The points that correspond to $T = 0$ on the RCs are at the poles of $G_2(s)H_2(s)$ or the zeros of $s(s+1)(s+2)+K$, whose RC are sketched as shown in Fig. 8-33 when K varies. If we choose $K = 20$ just as an illustration, the pole–zero configuration of $G_2(s)H_2(s)$ is shown in Fig. 8-34. The RCs of Eq. (8-114) for $0 \le T < \infty$ are shown in Fig. 8-35 for three different values of K.

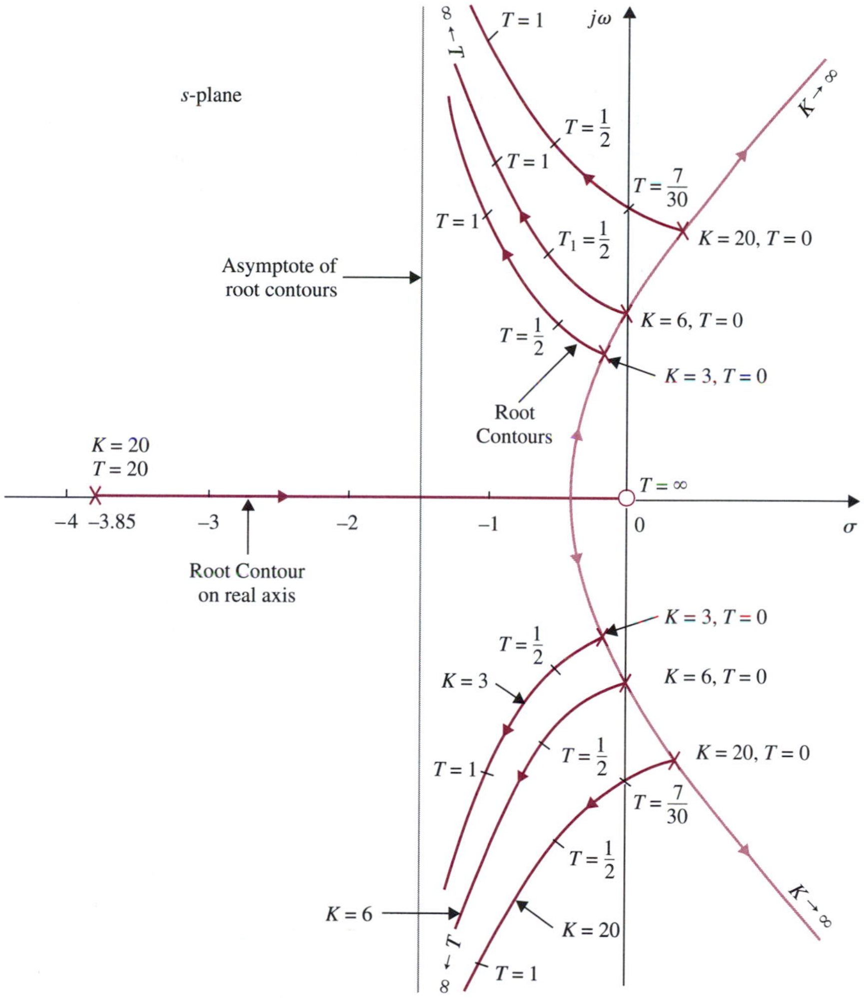

Figure 8-35 Root contours of $s(s+1)(s+2) + K + KTs = 0$.

Since $G_2(s)H_2(s)$ has three poles and one zero, the angles of the asymptotes of the RCs when T varies are at 90° and −90°. We can show that the intersection of the asymptotes is always at $s = 1.5$. This is because the sum of the poles of $G_2(s)H_2(s)$, which is given by the negative of the coefficient of the s^2 term in the denominator polynomial of Eq. (8-117), is 3, the sum of the zeros of $G_2(s)H_2(s)$ is 0, and $n - m$ in Eq. (8-36) is 2.

The RCs in Fig. 8-35 show that adding a zero to the loop transfer function generally improves the relative stability of the closed-loop system by moving the characteristic equation roots toward the left in the s-plane. As shown in Fig. 8-35, for $K = 20$, the system is stabilized for all values of T greater than 0.2333. However, the largest relative damping ratio that the system can have by increasing T is only approximately 30 percent. ▲

8-7 Root Loci of Discrete-Data Systems [23,24]

The root-locus technique can be applied to discrete-data systems without any complications. With the z-transformed transfer function, the root loci for discrete-data systems are plotted in the z-plane rather than in the s-plane. Let us consider the discrete-data control system shown in Fig. 8-36. The characteristic equation roots of the system satisfy the following equation:

$$1 + GH^*(s) = 0 \tag{8-118}$$

in the s-plane, or

$$1 + GH(z) = 0 \tag{8-119}$$

in the z-plane. From Chapter 3, $GH^*(s)$ is written

$$GH^*(s) = \frac{1}{T} \sum_{n=-\infty}^{\infty} G(s + jn\omega_s)H(s + jn\omega_s) \tag{8-120}$$

which is an infinite series. Thus the poles and zeros of $GH^*(s)$ in the s-plane will be infinite in number. This evidently makes construction of the root loci of Eq. (8-118)

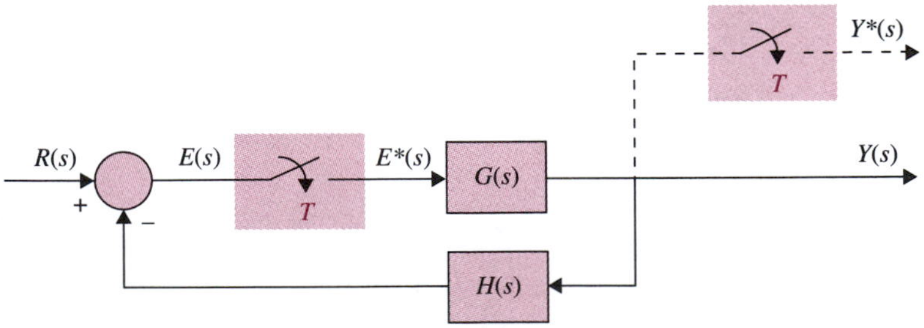

Figure 8-36 Discrete-data control system.

in the s-plane quite complex. As an illustration, consider that for the system of Fig. 8-36,

$$G(s)H(s) = \frac{K}{s(s+1)} \tag{8-121}$$

Substituting Eq. (8-121) into Eq. (8-120), we get

$$GH^*(s) = \frac{1}{T}\sum_{n=-\infty}^{\infty} \frac{K}{(s + jn\omega_s)(s + jn\omega_s + 1)} \tag{8-122}$$

which has poles at $s = -jn\omega_s$ and $s = -1 - jn\omega_s$, where n takes on all integers between $-\infty$ and ∞. The pole configuration of $GH^*(s)$ is shown in Fig. 8-37(a). By

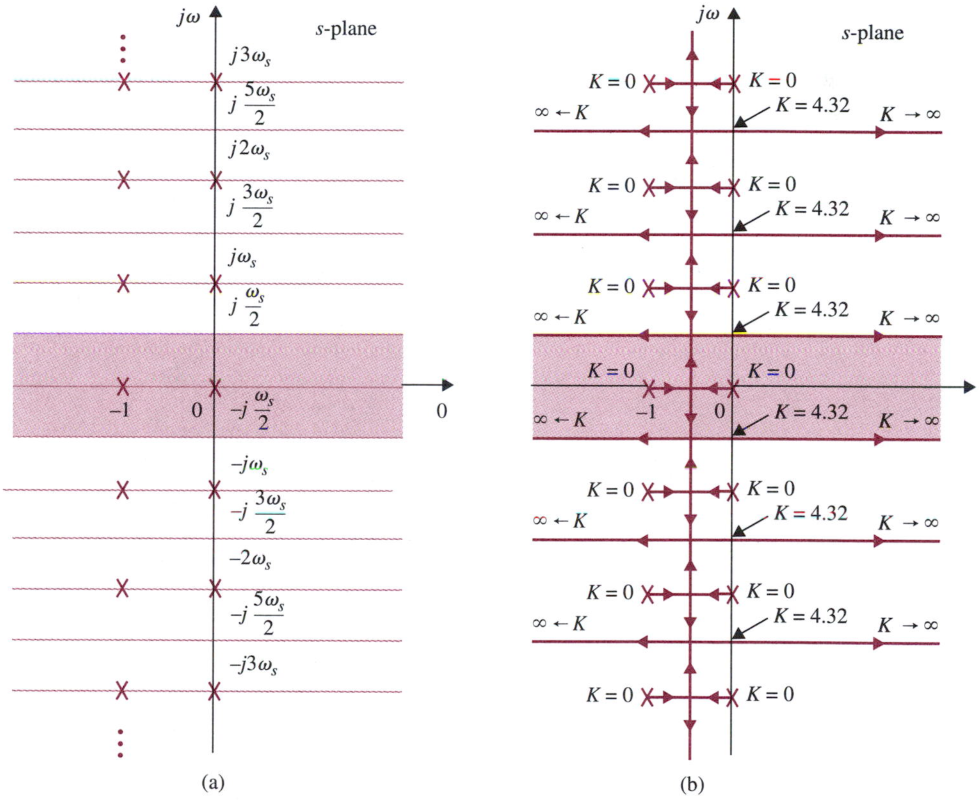

Figure 8-37 Pole configuration of $GH^*(s)$ and the root-locus diagram in the s-plane for the discrete-data system in Fig. 8-36 with $G(s)H(s) = K/[s(s+1)]$. $T = 1$ sec.

▲ The same procedures of construction of root loci of continuous-data systems can be applied to root loci of discrete-data systems in the z-plane.

using the properties of the RL in the s-plane, RL of $1 + GH^*(s) = 0$ are drawn as shown in Fig. 8-37(b) for the sampling period $T = 1$ sec. The RL contain an infinite number of branches, and these clearly indicate that the closed-loop system is unstable for all values of K greater than 4.32. In contrast, it is well known that the same system without sampling is stable for all positive values of K.

The root-locus problem for discrete-data systems is simplified if the root loci are constructed in the z-plane using Eq. (8-119). Since Eq. (8-119) is, in general, a rational function in z with constant coefficients, its poles and zeros are finite in number, and the number of root loci is finite in the z-plane. The same procedures of construction for continuous-data systems are directly applicable in the z-plane for discrete-data systems. The following examples illustrate the constructions of root loci for discrete-data systems in the z-plane.

Example 8-25

Consider that for the discrete-data system shown in Fig. 8-36 the loop transfer function in the z domain is

$$GH(z) = \frac{0.632\,Kz}{(z-1)(z-0.368)} \tag{8-123}$$

The RL of the closed-loop characteristic equation are constructed based on the pole–zero configuration of $GH(z)$, as shown in Fig. 8-38. Notice that when the value of K exceeds 4.32, one of the two roots moves outside the unit circle and the system becomes unstable. The constant-damping-ratio locus may be superimposed on the RL to determine the required value of K for a specified damping ratio. In Fig. 8-38 the constant-damping-ratio locus for $\zeta = 0.5$ is drawn, and the intersection with the RL gives the desired value of $K = 1$. For the same system, if the

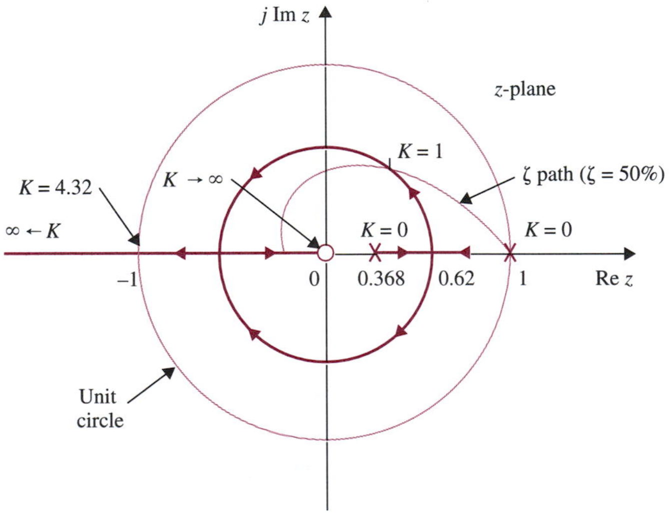

Figure 8-38 Root-locus diagram of a discrete-data control system without zero-order hold. $G(s)H(s) = K/[s(s+1)]$. $T = 1$ sec.

sampling period T is increased to 2 sec, the z-transform loop transfer function becomes

$$GH(z) = \frac{0.865\,Kz}{(z-1)(z-0.135)} \tag{8-124}$$

The RL for this case are shown in Fig. 8-39. Note that although the complex part of the RL for $T = 2$ sec takes the form of a smaller circle than that when $T = 1$ sec, the system is actually less stable, since the marginal value of K for stability is 2.624, as compared with the marginal K of 4.32 for $T = 1$ sec.

Next, let us consider that a zero-order hold is inserted between the sampler and the controlled process $G(s)$ in the system of Fig. 8-36. The loop transfer function of the system with the zero-order hold is

$$G_{ho}GH(z) = \frac{K[(T-1+e^{-T})z - Te^{-T} + 1 - e^{-T}]}{(z-1)(z-e^{-T})} \tag{8-125}$$

The RL of the system with ZOH for $T = 1$ and 2 sec are shown in Fig. 8-40(a) and (b), respectively. In this case, the marginal value of stability for K is 2.3 for $T = 1$ sec and 1.46 for $T = 2$ sec. Comparing the root loci of the system with and without the ZOH, we see that the ZOH reduces the stability margin of the discrete-data system.

In conclusion, the root loci of discrete-data systems can be constructed in the z-plane using essentially the same properties as those of the continuous-data systems in the s-plane. However, the absolute and relative stability conditions of the discrete-data system must be investigated with respect to the unit circle and the other interpretation of performance with respect to the regions in the z-plane. ▲

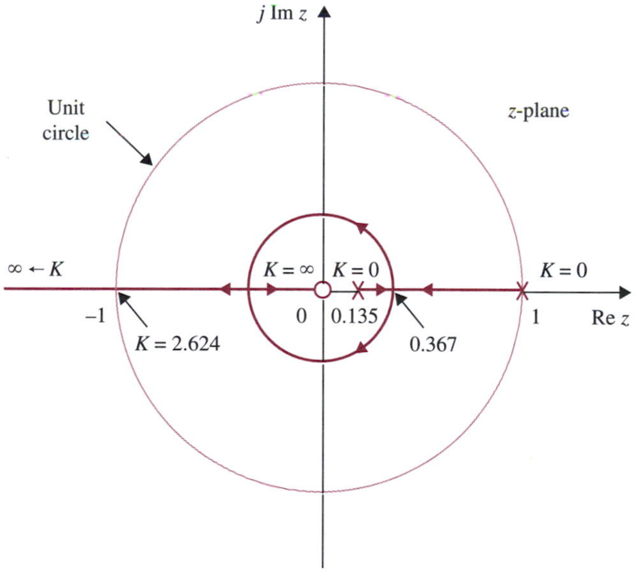

Figure 8-39 Root-locus diagram of a discrete-data control system without zero-order hold. $G(s)H(s) = K/[s(s+1)]$. $T = 2$ sec.

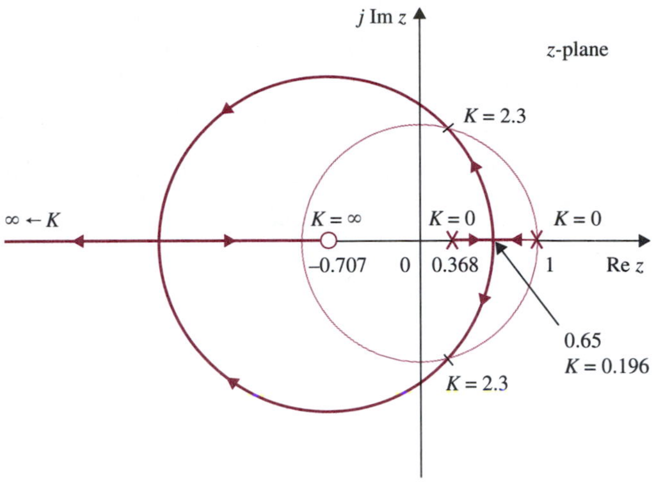

(a) Root loci for $T = 1$ seconds

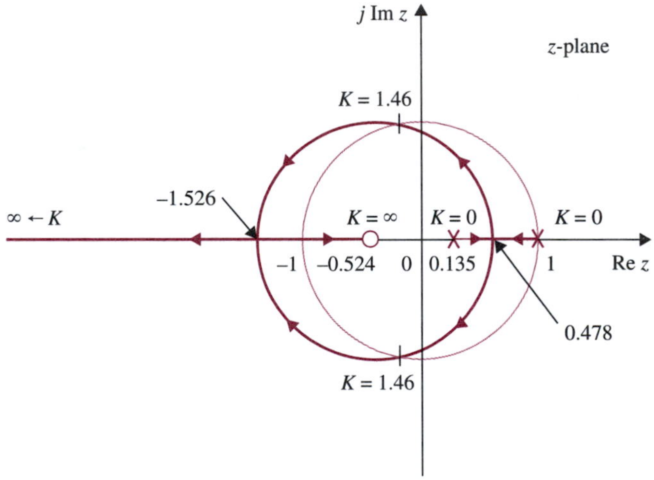

(b) Root loci for $T = 2$ seconds

Figure 8-40 Root-locus diagrams of discrete-data control systems with sample-and-hold. $G(s)H(s) = K/[s(s + 1)]$.

8-8 Summary

In this chapter we have introduced the root-locus technique for linear control systems. The technique represents a graphical method of investigating the roots of the characteristic equation of a linear time-invariant system when one or more parameters vary. In Chapter 10 the root-locus method will be used heavily for the design of control systems. However, keep in mind that the characteristic equation roots give exact indication on the absolute stability of linear SISO systems but give only qualitative information on the relative stability, since the zeros of the closed-loop transfer function, if any, also play an important role on the dynamic performance of the system.

The root-locus technique is also applied to discrete-data systems with the characteristic equation expressed in the z-transform. The properties and construction of the root loci in the z-plane are essentially the same as those of the continuous-data systems in the s-plane, except that the interpretation of the root locations to system performance must be made with respect to the unit circle $|z| = 1$ and the significance of the regions in the z-plane.

The material in this chapter provides a guide to the basics of root loci and the manual sketching of the root loci. Computer programs such as **rootloci** in **ACSP, rlplot** in **CSAD/ MATLAB,** and **root** in **Program CC,** just to name a few, can all be used to plot the root loci and give detail information of the plot. However, this author believes that computer programs can be used only as a tool, and the intelligent investigator cannot be without understanding of the fundamental of the subject.

The root-locus technique can also be applied to linear system with a pure time delay in the system loop [25,26]. The subject is not treated here, since systems with pure time delays are more easily treated with the frequency-domain methods discussed in Chapter 9.

Review Questions

The following questions and true-and-false problems all refer to the equation $P(s) + KQ(s) = 0$, where $P(s)$ and $Q(s)$ are polynomials of s with constant coefficients.

1. Give the condition from which the root loci are constructed.
2. Determine the points on the complete root loci at which $K = 0$, with reference to the poles and zeros of $Q(s)/P(s)$.
3. Determine the points on the root loci at which $K = \pm\infty$, with reference to the poles and zeros of $Q(s)/P(s)$.
4. Give the significance of the breakaway points with respect to the roots of $P(s) + KQ(s) = 0$.
5. Give the equation of intersect of the asymptotes.
6. The asymptotes of the root loci refer to the angles of the root loci when $K = \pm\infty$. **(T) (F)**
7. There is only one intersect of the asymptotes of the complete root loci. **(T) (F)**
8. The intersect of the asymptotes must always be on the real axis. **(T) (F)**
9. The breakaway points of the root loci must always be on the real axis. **(T) (F)**
10. Given the equation $1 + KG_1(s)H_1(s) = 0$, where $G_1(s)H_1(s)$ is a rational function of s and does not contain K, the roots of $dG_1(s)H_1(s)/ds$ are all breakaway points on the root loci $(-\infty < K < \infty)$. **(T) (F)**
11. At the breakaway points on the root loci, the root sensitivity is infinite. **(T) (F)**
12. Without modification, all the rules and properties for the construction of root loci in the s-plane can be applied to the construction of root loci of discrete-data systems in the z-plane. **(T) (F)**

13. The determination of the intersections of the root loci in the s-plane with the $j\omega$-axis can be made by solving the auxiliary equation of Routh's tabulation of the equation. **(T)** **(F)**
14. Adding a pole to $Q(s)/P(s)$ has the general effect of pushing the root loci to the right, whereas adding a zero pushes the loci to the left. **(T)** **(F)**

Answers to the true-and-false questions are given following the Problems section.

References

General Subjects

1. **W. R. EVANS,** "Graphical Analysis of Control Systems," *Trans. AIEE,* Vol. 67, pp. 547–551, 1948.
2. **W. R. EVANS,** "Control System Synthesis by Root Locus Method," *Trans. AIEE,* Vol. 69, pp. 66–69, 1950.
3. **W. R. EVANS,** *Control System Dynamics,* McGraw-Hill Book Company, New York, 1954.

Construction and Properties of Root Loci

4. **C. C. MACDUFF,** *Theory of Equations,* pp. 29–104, John Wiley & Sons, New York, 1954.
5. **C. S. LORENS and R. C. TITSWORTH,** "Properties of Root Locus Asymptotes," *IRE Trans. Automatic Control,* Vol. AC-5, pp. 71–72, Jan. 1960.
6. **C. A. STAPLETON,** "On Root Locus Breakaway Points," *IRE Trans. Automatic Control,* Vol. AC-7, pp. 88–89, Apr. 1962.
7. **M. J. REMEC,** "Saddle-Points of a Complete Root Locus and an Algorithm for Their Easy Location in the Complex Frequency Plane," *Proc. Natl. Electronics Conf.,* Vol. 21, pp. 605–608, 1965.
8. **C. F. CHEN,** "A New Rule for Finding Breaking Points of Root Loci Involving Complex Roots," *IEEE Trans. Automatic Control,* Vol. AC-10, pp. 373–374, July 1965.
9. **V. KRISHRAN,** "Semi-analytic Approach to Root Locus," *IEEE Trans. Automatic Control,* Vol. AC-11, pp. 102–108, Jan. 1966.
10. **R. H. LABOUNTY and C. H. HOUPIS,** "Root Locus Analysis of a High-Grain Linear System with Variable Coefficients; Application of Horowitz's Method," *IEEE Trans. Automatic Control,* Vol. AC-11, pp. 255–263, Apr. 1966.
11. **A. FREGOSI and J. FEINSTEIN,** "Some Exclusive Properties of the Negative Root Locus," *IEEE Trans. Automatic Control,* Vol. AC-14, pp. 304–305, June 1969.

Analytical Representation of Root Loci

12. **G. A. BENDRIKOV and K. F. TEODORCHIK,** "The Analytic Theory of Constructing Root Loci," *Automatic and Remote Control,* pp. 340–344, Mar. 1959.
13. **K. STEIGLITZ,** "Analytical Approach to Root Loci," *IRE Trans. Automatic Control,* Vol. AC-6, pp. 326–332, Sept. 1961.

14. **C. WOJCIK,** "Analytical Representation of Root Locus," *Trans. ASME, J. Basic Eng.,* Ser. D, Vol. 86, Mar. 1964.

15. **C. S. CHANG,** "An Analytical Method for Obtaining the Root Locus with Positive and Negative Gain," *IEEE Trans. Automatic Control,* Vol. AC-10, pp. 92–94, Jan. 1965.

16. **B. P. BHATTACHARYYA,** "Root Locus Equations of the Fourth Degree," *Internat. J. Control,* Vol. 1, No. 6, pp. 533–556, 1965.

Computer-Aided Plotting of Root Loci

17. **D. J. DODA,** "The Digital Computer Makes Root Locus Easy," *Control Eng.,* May 1958.

18. **Z. KLAGSBRUNN and Y. WALLACH,** "On Computer Implementation of Analytic Root-Locus Plotting," *IEEE Trans. Automatic Control,* Vol. AC-13, pp. 744–745, Dec. 1968.

19. **R. H. ASH and G. R. ASH,** "Numerical Computation of Root Loci Using the Newton--Raphson Technique," *IEEE Trans. Automatic Control,* Vol. AC-13, pp. 576–582, Oct. 1968.

Root Sensitivity

20. **J. G. TRUXAL and M. HOROWITZ,** "Sensitivity Consideration in Active Network Synthesis," *Proc. 2nd Midwest Symposium on Circuit Theory,* East Lansing, MI, 1956.

21. **R. Y. HUANG,** "The Sensitivity of the Poles of Linear Closed-Loop Systems," *Trans. AIEE, Appl. Ind.,* Vol. 77, Part 2, pp. 182–187, Sept. 1958.

22. **H. UR,** "Root Locus Properties and Sensitivity Relations in Control Systems," *IRE Trans. Automatic Control,* Vol. AC-5, pp. 57–65, Jan. 1960.

Root Loci for Discrete-Data Systems

23. **M. MORI,** "Root Locus Method of Pulse Transfer Function for Sampled-Data Control Systems," *IRE Trans. Automatic Control,* Vol. AC-3, pp. 13–20, Nov. 1963.

24. **B. C. KUO,** *Digital Control Systems,* 2nd ed., Saunders College Publishing, Philadelphia, 1992.

Root Loci for Systems with Time Delays

25. **Y. CHU,** "Feedback Control System with Dead-Time Lag or Distributed Lag by Root-Locus Method," *Trans. AIEE,* Vol. 70, Part 2, p. 291, 1951.

26. **B. C. KUO,** *Automatic Control Systems,* 5th ed., Prentice Hall, Englewood Cliffs, NJ, 1987.

Problems

Any computer software for root locus can be used to solve the problems in this chapter. Do the problems without relying on the computer first, and then carry out the computer solutions if a root-locus program is available. Determine all the vital information and mark on the root loci, including the points at which the variable parameter (e.g., K) equals zero and $\pm\infty$, asymptotic properties, intersections on the $j\omega$-axis and the corresponding parameter value, and the breakaway points. Indicate the direction of increase of the variable parameter on the loci with arrows.

▲ Asymptotes

8-1. Find the angles of the asymptotes and the intersect of the asymptotes of the root loci of the following equations when K varies from $-\infty$ to ∞.

(a) $s^4 + 4s^3 + 4s^2 + (K + 8)s + K = 0$
(b) $s^3 + 5s^2 + (K + 1)s + K = 0$
(c) $s^2 + K(s^3 + 3s^2 + 2s + 8) = 0$
(d) $s^3 + 2s^2 + 3s + K(s^2 - 1)(s + 3) = 0$
(e) $s^5 + 2s^4 + 3s^3 + K(s^2 + 3s + 5) = 0$
(f) $s^4 + 2s^2 + 10 + K(s + 5) = 0$

▲ Angles of departure and arrival

8-2. For the loop transfer functions that follow, find the angle or departure or arrival of the root loci at the designated pole or zero.

(a) $G(s)H(s) = \dfrac{Ks}{(s + 1)(s^2 + 1)}$
Angle of arrival ($K < 0$) and angle of departure ($K > 0$) at $s = j$.

(b) $G(s)H(s) = \dfrac{Ks}{(s - 1)(s^2 + 1)}$
Angle of arrival ($K < 0$) and angle of departure ($K > 0$) at $s = j$.

(c) $G(s)H(s) = \dfrac{K}{s(s + 2)(s^2 + 2s + 2)}$
Angle of departure ($K > 0$) at $s = -1 + j$.

(d) $G(s)H(s) = \dfrac{K}{s^2(s^2 + 2s + 2)}$
Angle of departure ($K > 0$) at $s = -1 + j$.

(e) $G(s)H(s) = \dfrac{K(s^2 + 2s + 2)}{s^2(s + 2)(s + 3)}$
Angle of arrival ($K > 0$) at $s = -1 + j$.

▲ $K = 0$ and $K = \pm\infty$ points and RL and CRL on real axis

8-3. Mark the $K = 0$ and $K = \pm\infty$ points and the RL and CRL on the real axis for the pole–zero configurations shown in Fig. 8P-3. Add arrows on the root loci on the real axis in the direction of increasing K.

▲ Breakaway points

8-4. Find all the breakaway points of the root loci of the systems described by the pole–zero configurations shown in Fig. 8P-3.

▲ Root-loci construction

8-5. Construct the root-locus diagram for each of the following control systems for which the poles and zeros of $G(s)H(s)$ are given. The characteristic equation is obtained by equating the numerator of $1 + G(s)H(s)$ to zero.

(a) Poles at $0, -5, -6$; zero at -8.
(b) Poles at $0, -1, -3, -4$; no finite zeros.

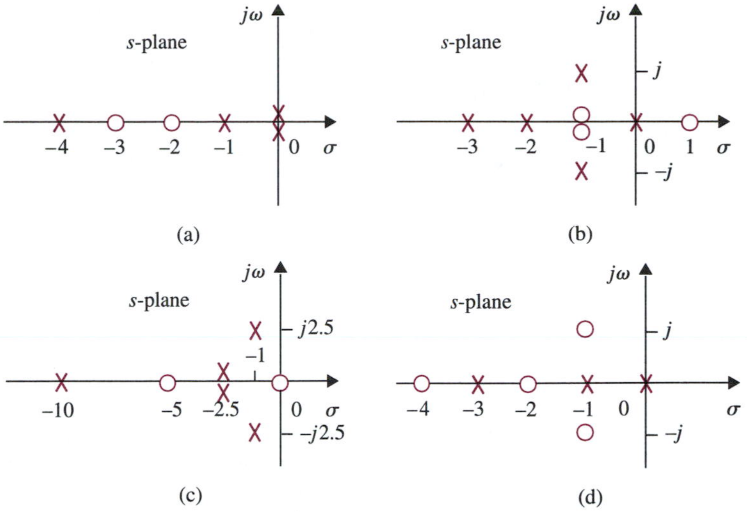

Figure 8P-3

(c) Poles at 0, 0, −2, −2; zero at −4.
(d) Poles at 0, $-1 + j$, $-1 - j$; zero at −2.
(e) Poles at 0, $-1 + j$, $-1 - j$; zero at −5.
(f) Poles at 0, $-1 + j$, $-1 - j$, −4.
(g) Poles at 0, 0, −8, −8; zeros at −4, −4.
(h) Poles at 0, 0, −8, −8; no finite zeros.
(i) Poles at 0, 0, −8, −8; zeros at $-4 + j2$, $-4 - j2$.
(j) Poles at −2, 2; zeros at 0, 0.
(k) Poles at j, $-j$, $j2$, $-j2$; zeros at −2, 2.
(l) Poles at j, $-j$, $j2$, $-j2$; zeros at −1, 1.
(m) Poles at 0, 0, 0, 1; zeros at −1, −2, −3.
(n) Poles at 0, 0, 0, −100, −200; zeros at −5, −40.
(o) Poles at 0, −1, −2; zero at 1.

▲ Root-loci construction

8-6. The characteristic equation of linear control systems are given as follows. Construct the root loci for $K \geq 0$.

(a) $s^3 + 3s^2 + (K + 2)s + 5K = 0$
(b) $s^3 + s^2 + (K + 2)s + 3K = 0$
(c) $s^3 + 5Ks^2 + 10 = 0$
(d) $s^4 + (K + 3)s^3 + (K + 1)s^2 + (2K + 5) + 10 = 0$
(e) $s^3 + 2s^2 + 2s + K(s^2 - 1)(s + 2) = 0$
(f) $s^3 - 2s + K(s + 4)(s + 1) = 0$
(g) $s^4 + 6s^3 + 9s^2 + K(s^2 + 4s + 5) = 0$

(h) $s^3 + 2s^2 + 2s + K(s^2 - 2)(s + 4) = 0$

(i) $s(s^2 - 1) + K(s + 2)(s + 0.5) = 0$

(j) $s^4 + 2s^3 + 2s^2 + 2Ks + 5K = 0$

▲ Root-loci construction

8-7. The forward-path transfer functions of a unity-feedback control systems are given in the following.

(a) $G(s) = \dfrac{K(s+4)}{s(s^2+4s+4)(s+5)(s+6)}$ (b) $G(s) = \dfrac{K}{s(s+2)(s+5)(s+10)}$

(c) $G(s) = \dfrac{K(s^2+2s+10)}{s(s+5)(s+10)}$ (d) $G(s) = \dfrac{K(s^2+4)}{(s+2)^2(s+5)(s+6)}$

Construct the root loci for $K \geq 0$. Find the value of K that makes the relative damping ratio of the closed-loop system (measured by the dominant complex characteristic equation roots) equal to 0.707 if such solution exists.

▲ Root loci, breakaway points

8-8. A unit-feedback control system has the forward-path transfer function given in the following. Construct the root-locus diagram for $K \geq 0$. Find the values of K at all the breakaway points.

(a) $G(s) = \dfrac{K}{s(s+10)(s+20)}$ (b) $G(s) = \dfrac{K}{s(s+1)(s+3)(s+5)}$

▲ Root-loci construction

8-9. The forward-path transfer function of a unity-feedback control system is

$$G(s) = \frac{K}{(s+4)^n}$$

Construct the root loci of the characteristic equation of the closed-loop system for $K \geq \infty$, with **(a)** $n = 1$, **(b)** $n = 2$, **(c)** $n = 3$, **(d)** $n = 4$, and **(e)** $n = 5$.

▲ Root loci of system with tachometer feedback

8-10. The characteristic equation of the control system shown in Fig. 7P-10 when $K = 100$ is

$$s^3 + 25s^2 + (100K_t + 2)s + 100 = 0$$

Construct the root loci of the equation for $K_t \geq 0$.

▲ Root loci of system with tachometer feedback

8-11. The block diagram of a control system with tachometer feedback is shown in Fig. 8P-11.

(a) Construct the root loci of the characteristic equation for $K \geq 0$ when $K_t = 0$.

(b) Set $K = 10$. Construct the root loci of the characteristic equation for $K_t \geq 0$.

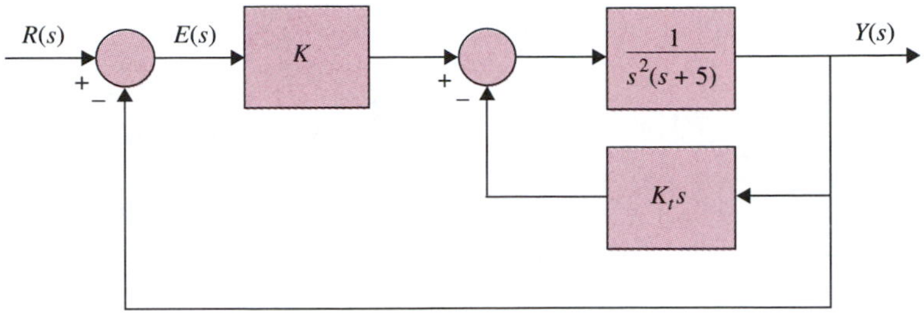

Figure 8P-11

▲ Root loci of dc-motor control system

8-12. The characteristic equation of the dc-motor control system described in Problem 7-27 can be approximated as

$$2.05J_L s^3 + (1 + 10.25J_L)s^2 + 116.84s + 1843 = 0$$

when $K_L = \infty$, and the load inertia J_L is considered as a variable parameter. Construct the root loci of the characteristic equation for $J_L \geq 0$.

▲ Root loci of control system

8-13. The forward-path transfer function of the control system shown in Fig. 7P-11 is

$$G(s) = \frac{K(s + \alpha)(s + 3)}{s(s^2 - 1)}$$

(a) Construct the root loci for $K \geq 0$ with $\alpha = 5$.

(b) Construct the root loci for $\alpha \geq 0$ with $K = 10$.

▲ Root loci of liquid-level control system

8-14. The characteristic equation of the liquid-level control system described in Problem 6-13 is written

$$0.06s(s + 12.5)(As + K_o) + 250N = 0$$

(a) For $A = K_o = 50$, construct the root loci of the characteristic equation as N varies from 0 to ∞.

(b) For $N = 10$ and $K_o = 50$, construct the root loci of the characteristic equation for $A \geq 0$.

(c) For $A = 50$ and $N = 20$, construct the root loci for $K_o \geq 0$.

▲ Root loci of liquid-level control system

8-15. Repeat Problem 8-14 for the following cases.

(a) $A = K_o = 100$.

(b) $N = 20$ and $K_o = 50$.

(c) $A = 100$ and $N = 20$.

▲ Root loci of control system

8-16. The transfer functions of a single-feedback-loop control system are

$$G(s) = \frac{K}{s^2(s+1)(s+5)} \qquad H(s) = 1$$

(a) Construct the loci of the zeros of $1 + G(s)$ for $K \geq 0$.

(b) Repeat part (a) when $H(s) = 1 + 5s$.

▲ Root loci of control system

8-17. The transfer functions of a single-feedback-loop control system are

$$G(s) = \frac{10}{s^2(s+1)(s+5)} \qquad H(s) = 1 + T_d s$$

Construct the root loci of the characteristic equation for $T_d \geq 0$.

▲ Root loci of dc-motor control system

8-18. For the dc-motor control system described in Problem 7-27, it is of interest to study the effects of the motor-shaft compliance K_L on the system performance.

(a) Let $K = 1$, with the other system parameters as given in Problem 7-27. Find an equivalent $G(s)H(s)$ with K_L as the gain factor. Construct the root loci of the characteristic equation for $K_L \geq 0$. The system can be approximated as a fourth-order system by canceling the large negative pole and zero of $G(s)H(s)$ that are very close to each other.

(b) Repeat part (a) with $K = 1000$.

▲ Root loci of dc-motor control system; J_L varies

8-19. The characteristic equation of the dc-motor control system described in Problem 7-27 is given in the following when the motor shaft is considered to be rigid ($K_L = \infty$). Let $K = 1$, $J_m = 0.001$, $L_a = 0.001$, $n = 0.1$, $R_a = 5$, $K_i = 9$, $K_b = 0.0636$, $B_m = 0$, and $K_s = 1$.

$$L_a(J_m + n^2 J_L)s^3 + (R_a J_m + n^2 R_a J_L + B_m L_a)s^2 + (R_a B_m + K_i K_b)s + nK_s K_i K = 0$$

Construct the root loci for $J_L \geq 0$ to show the effects of variation of the load inertia on system performance.

▲ Root loci with two variable parameters

8-20. Given the equation

$$s^3 + \alpha s^2 + Ks + K = 0$$

it is desired to investigate the root loci of this equation for $-\infty < K < \infty$ and for various values of α.

(a) Construct the root loci for $-\infty < K < \infty$ when $\alpha = 12$.

(b) Repeat part (a) when $\alpha = 4$.

(c) Determine the value of α so that there is only one nonzero breakaway point on the entire root loci for $-\infty < K < \infty$. Construct the root loci.

▲ Breakaway points of root loci

8-21. The forward-path transfer function of a unity-feedback control system is

$$G(s) = \frac{K(s+\alpha)}{s^2(s+3)}$$

Determine the values of α so that the root loci ($-\infty < K < \infty$) will have zero, one, and two breakaway points, respectively, not including the one at $s = 0$. Construct the root loci for $-\infty < K < \infty$ for all three cases.

▲ Exercise on angle of departure and arrival

8-22. The pole–zero configuration of $G(s)H(s)$ of a single-feedback-loop control system is shown in Fig. 8P-22(a). Without actually plotting, apply the angle-of-departure (and angle-of-arrival) property of the root loci to determine which root-locus diagram shown is the correct one.

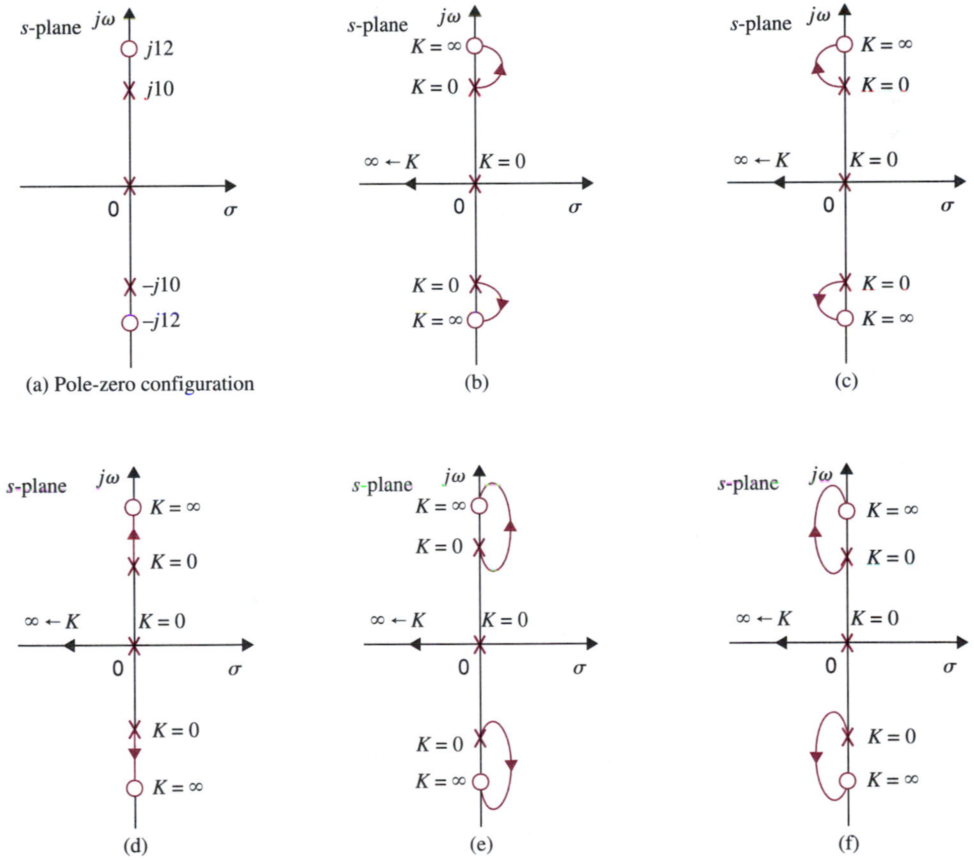

Figure 8P-22

▲ Root loci of sampled-data system

8-23. The block diagram of a sampled-data control system is shown in Fig. 8P-23.

(a) Construct the root loci in the z-plane for the system for $K \geq 0$, without the zero-order hold, when $T = 0.5$ sec and then with $T = 0.1$ sec. Find the marginal values of K for stability.

$$G(s) = \frac{K}{s(s+5)}$$

(b) Repeat part (a) when the system has a zero-order hold, as shown in Fig. 8P-23.

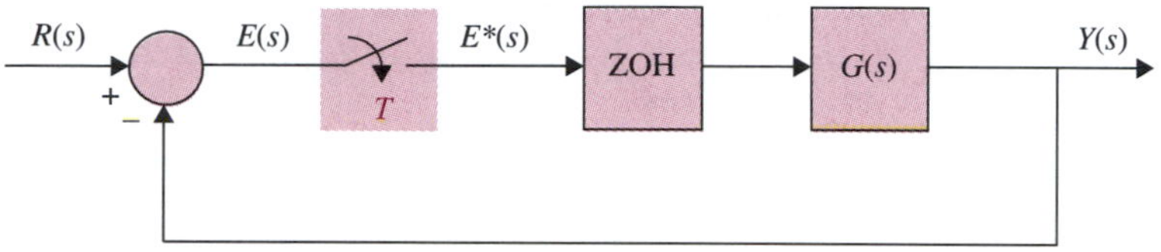

Figure 8P-23

▲ Root loci of sampled-data system

8-24. The system shown in Fig. 8P-23 has the following transfer function:

$$G(s) = \frac{Ke^{-0.1s}}{s(s+1)(s+2)}$$

Construct the root loci in the z-plane for $K \geq 0$, with $T = 0.1$ sec.

▲ Root loci of discrete-data systems

8-25. The characteristic equations of linear discrete-data control systems are given in the following. Construct the root loci for $K \geq 0$. Determine the marginal value of K for stability.

(a) $z^3 + Kz^2 + 1.5Kz - (K+1) = 0$

(b) $z^2 + (0.15K - 1.5)z + 1 = 0$

(c) $z^2 + (0.1K - 1)z + 0.5 = 0$

(d) $z^2 + (0.4 + 0.14K)z + (0.5 + 0.5K) = 0$

(e) $(z-1)(z^2 - z + 0.4) + 4 \times 10^{-5}K(z+1)(z+0.7) = 0$

Answers to True-and-False Review Questions

6. (F) **7.** (T) **8.** (T) **9.** (F) **10.** (F) **11.** (T) **12.** (T)
13. (T) **14.** (T)

9 Frequency-Domain Analysis

KEYWORDS AND TOPICS

- Frequency Response
- Frequency-Domain Specifications
- Peak Resonance
- Resonant Frequency
- Bandwidth
- Nyquist Criterion
- Gain Margin
- Phase Margin
- Bode Plot
- Nichols Chart
- Sensitivity
- Frequency Response of Discrete-Data Systems

9-1 Introduction

In practice the performance of a control system is measured more realistically by its time-domain characteristics. The reason is that the performance of most control systems is judged based on the time responses due to certain test signals. This is in contrast to the analysis and design of communication systems for which the frequency response is of more importance, since most of the signals to be processed are either sinusoidal or composed of sinusoidal components. We learned in Chapter 7 that the time response of a control system is usually more difficult to determine analytically, especially for high-order systems. In design problems there are no unified methods of arriving at a designed system that meets the time-domain performance specifications, such as maximum overshoot, rise time, delay time, settling time, and so on. On the other hand, in the frequency domain a wealth of graphical methods are available that are not limited to low-order systems. It is important to realize that there are correlating relations between the frequency- and time-domain performances in a linear system, so that the time-domain properties of the system can be predicted based on the frequency-domain characteristics. The frequency domain is also more convenient for measurements of system sensitivity to noise and parameter variations. With these in mind, we may consider the primary motivation of conducting control systems analysis and design in the frequency domain to be convenience and the availability of the existing analytical tools. Another reason is that it presents an alternative point of view to control system problems, which often provides valuable or crucial information in the complex analysis and design of control systems. Therefore, to conduct frequency-domain analysis of a linear control system does not imply that the system will be subject only to a sine-wave input. It may never be. Rather, from the frequency-response studies, we will be able to project the time-domain performance of the system.

The starting point for frequency-domain analysis of a linear system is its transfer function. It is well known from linear system theory that when the input to a linear time-invariant system is sinusoidal with amplitude R and frequency ω_0.

$$r(t) = R \sin \omega_0 t \tag{9-1}$$

the steady-state output of the system, $y(t)$, will be a sinusoid with the same frequency ω_0, but possibly with different amplitude and phase; that is,

$$y(t) = Y \sin(\omega_0 t + \phi) \tag{9-2}$$

where Y is the amplitude of the output sine wave and ϕ is the phase shift in degrees or radians.

Let the transfer function of a linear SISO system be $M(s)$; then the Laplace transforms of the input and the output are related through

$$Y(s) = M(s)R(s) \tag{9-3}$$

For sinusoidal steady-state analysis, we replace s by $j\omega$, and Eq. (9-3) becomes

$$Y(j\omega) = M(j\omega)R(j\omega) \tag{9-4}$$

By writing the function $Y(j\omega)$ as

$$Y(j\omega) = |Y(j\omega)|\angle Y(j\omega) \tag{9-5}$$

with similar definitions for $M(j\omega)$ and $R(j\omega)$, Eq. (9-4) leads to the magnitude relation between the input and the output:

$$|Y(j\omega)| = |M(j\omega)||R(j\omega)| \tag{9-6}$$

and the phase relation:

$$\angle Y(j\omega) = \angle M(j\omega) + \angle R(j\omega) \tag{9-7}$$

Thus for the input and output signals described by Eqs. (9-1) and (9-2), respectively, the amplitude of the output sinusoid is

$$Y = R|M(j\omega_0)| \tag{9-8}$$

and the phase of the output is

$$\phi = \angle M(j\omega_0) \tag{9-10}$$

Thus, by knowing the transfer function $M(s)$ of a linear system, the magnitude characteristic, $|M(j\omega)|$, and the phase characteristic, $\angle M(j\omega)$, completely describe the steady-state performance when the input is a sinusoid. The crux of frequency-domain analysis is that the amplitude and phase characteristics of a closed-loop system can be used to predict both time-domain transient and steady-state system performances.

9-1-1 Frequency Response of Closed-Loop Systems

For the single-loop control system configuration studied in the preceding chapters, the closed-loop transfer function is

$$M(s) = \frac{Y(s)}{R(s)} = \frac{G(s)}{1 + G(s)H(s)} \tag{9-11}$$

Under the sinusoidal steady-state, $s = j\omega$, Eq. (9-11) becomes

$$M(j\omega) = \frac{Y(j\omega)}{R(j\omega)} = \frac{G(j\omega)}{1 + G(j\omega)H(j\omega)} \tag{9-12}$$

The sinusoidal steady-state transfer function $M(j\omega)$ may be expressed in terms of its magnitude and phase; that is,

$$M(j\omega) = |M(j\omega)|\angle M(j\omega) \tag{9-13}$$

Or $M(j\omega)$ can be expressed in terms of its real and imaginary parts:

$$M(j\omega) = \text{Re}[M(j\omega)] + j\text{Im}[M(j\omega)] \tag{9-14}$$

The magnitude of $M(j\omega)$ is

$$|M(j\omega)| = \left|\frac{G(j\omega)}{1 + G(j\omega)H(j\omega)}\right| = \frac{|G(j\omega)|}{|1 + G(j\omega)H(j\omega)|} \tag{9-15}$$

and the phase of $M(j\omega)$ is

$$\angle M(j\omega) = \phi_M(j\omega) = \angle G(j\omega) - \angle[1 + G(j\omega)H(j\omega)] \tag{9-16}$$

If $M(s)$ represents the input–output transfer function of an electric filter, the magnitude and phase of $M(j\omega)$ indicate the filter characteristics on the input signal. Figure 9-1 shows the gain and phase characteristics of an ideal low-pass filter that has a sharp cutoff frequency at ω_c. It is well known that an ideal filter characteristic is physically unrealizable. In many ways the design of control systems is quite similar to filter design, and the control system is regarded as a signal processor. In fact, if the ideal low-pass-filter characteristics shown in Fig. 9-1 were physically realizable, they would be highly desirable for a control system, since all signals would be passed without distortion below the frequency ω_c, and eliminated completely at frequencies above ω_c, where noise may lie.

If ω_c is increased indefinitely, the output $Y(j\omega)$ would be identical to the input $R(j\omega)$ for all frequencies. Such a system would follow a step-function input in the time

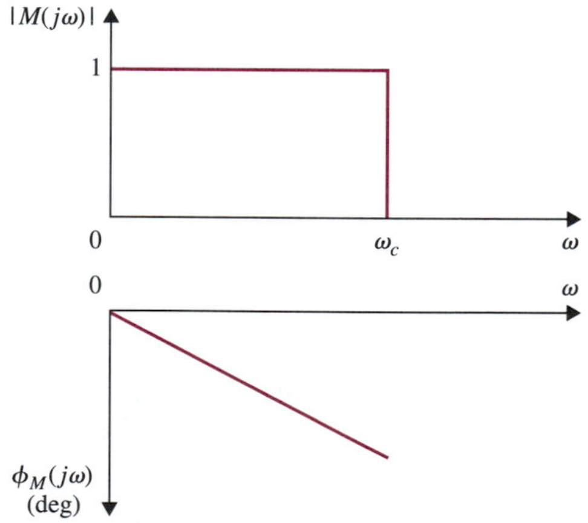

Figure 9-1 Gain–phase characteristics of an ideal low-pass filter.

domain exactly. From Eq. (9-15) we see that for $|M(j\omega)|$ to be unity at all frequencies, the magnitude of $G(j\omega)$ must be infinite. An infinite magnitude of $G(j\omega)$ is, of course, impossible to achieve in practice, nor would it be desirable, since most control systems become unstable when their loop gains become very high. Furthermore, all control systems are subject to noise during operation. Thus, in addition to responding to the input signal, the system should be able to reject and suppress noise and unwanted signals. For control systems with high-frequency noise, such as air-frame vibration of an aircraft, the frequency response should have a finite cutoff frequency ω_c. The phase characteristics of the frequency response of a control system are also of importance, as we shall see that they affect the stability of the system.

Figure 9-2 illustrates typical gain and phase characteristics of a control system. As shown by Eqs. (9-15) and (9-16), the gain and phase of a closed-loop system can be determined from the foreward-path and loop transfer functions. In practice, the frequency responses of $G(s)$ and $H(s)$ can often be determined by applying sine-wave inputs to the system and sweeping the frequency from 0 to a value beyond the frequency range of the system.

9-1-2 Frequency-Domain Specifications

In the design of linear control systems using frequency-domain methods, it is necessary to define a set of specifications so that the performance of the system can be identified. Specifications such as the maximum overshoot, damping ratio, and so on, used in the time domain, can no longer be used directly in the frequency domain. The following frequency-domain specifications are often used in practice.

Resonant Peak M_r

The resonant peak M_r is the maximum value of $|M(j\omega)|$.

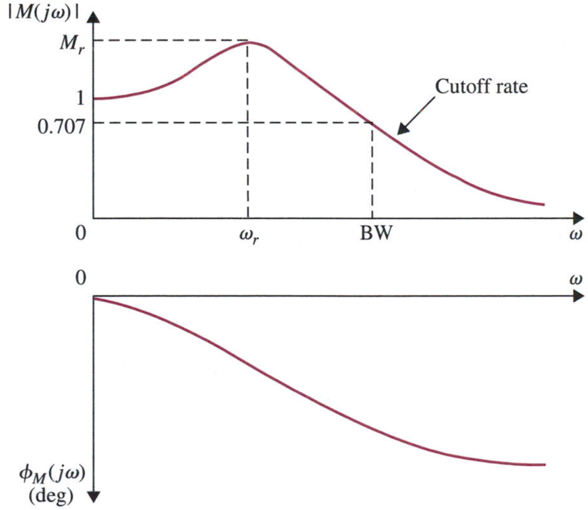

Figure 9-2 Typical gain–phase characteristics of a feedback control system.

▲ M_r indicates the relative stability of a stable closed-loop system.

In general, the magnitude of M_r gives indication of the relative stability of a stable closed-loop system. Normally, a large M_r corresponds to a large maximum overshoot of the step response. For most control systems, it is generally accepted in practice that the desirable value of M_r should be between 1.1 and 1.5.

Resonant Frequency ω_r

The resonant frequency ω_r is the frequency at which the peak resonance M_r occurs.

Bandwidth BW

The bandwidth BW is the frequency at which $|M(j\omega)|$ drops to 70.7 percent of, or 3 dB down from, its zero-frequency value.

▲ BW gives an indication of the transient response properties of a control system.

▲ BW gives an indication of the noise-filtering characteristics and robustness of the system.

In general, the bandwidth of a control system gives an indication of the transient-response properties in the time domain. A large bandwidth corresponds to a faster rise time, since higher-frequency signals are more easily passed through the system. Conversely, if the bandwidth is small, only signals of relatively low frequencies are passed, and the time response will be slow and sluggish. Bandwidth also indicates the noise-filtering characteristics and the robustness of the system. The robustness represents a measure of the sensitivity of a system to parameter variations. A robust system is one that is insensitive to parameter variations.

Cutoff Rate. Often, bandwidth alone is inadequate to indicate the ability of a system in distinguishing signals from noise. Sometimes it may be necessary to look at the slope of $|M(j\omega)|$, which is called the cutoff rate of the frequency response, at high frequencies. Apparently, two systems can have the same bandwidth, but the cutoff rates may be different. The performance criteria for the frequency domain defined above are illustrated in Fig. 9-2. Other important criteria for the frequency domain are defined in later sections of this chapter.

9-2 M_r, ω_r, and Bandwidth of the Prototype Second-Order System

9-2-1 Resonant Peak and Resonant Frequency

For the prototype second-order system defined in Section 7-5, the resonant peak M_r, the resonant frequency ω_r, and the bandwidth BW are all uniquely related to the damping ratio ζ and the natural undamped frequency ω_n of the system.

Consider the closed-loop transfer function of the prototype second-order system,

$$M(s) = \frac{Y(s)}{R(s)} = \frac{\omega_n^2}{s^2 + 2\zeta\omega_n s + \omega_n^2} \tag{9-17}$$

At sinusoidal steady state, $s = j\omega$, Eq. (9-17) becomes

$$M(j\omega) = \frac{Y(j\omega)}{R(j\omega)} = \frac{\omega_n^2}{(j\omega)^2 + 2\zeta\omega_n(j\omega) + \omega_n^2} = \frac{1}{1 + j2(\omega/\omega_n)\zeta - (\omega/\omega_n)^2} \tag{9-18}$$

We can simplify Eq. (9-18) by letting $u = \omega/\omega_n$. Then Eq. (9-18) becomes

$$M(ju) = \frac{1}{1 + j2u\zeta - u^2} \tag{9-19}$$

The magnitude and phase of $M(ju)$ are

$$|M(ju)| = \frac{1}{[(1 - u^2)^2 + (2\zeta u)^2]^{1/2}} \tag{9-20}$$

and

$$\angle M(ju) = \phi_M(ju) = -\tan^{-1}\frac{2\zeta u}{1 - u^2} \tag{9-21}$$

respectively. The resonant frequency is determined by setting the derivative of $|M(ju)|$ with respect to u to zero. Thus

$$\frac{d|M(ju)|}{du} = -\frac{1}{2}[(1 - u^2)^2 + (2\zeta u)^2]^{-3/2}(4u^3 - 4u + 8u\zeta^2) = 0 \tag{9-22}$$

from which we get

$$4u^3 - 4u + 8u\zeta^2 = 4u(u^2 - 1 + 2\zeta^2) = 0 \tag{9-23}$$

In normalized frequency, the roots of Eq. (9-23) are $u_r = 0$ and

$$u_r = \sqrt{1 - 2\zeta^2} \tag{9-24}$$

The solution of $u_r = 0$ merely indicates that the slope of the $|M(ju)|$ versus ω curve is zero at $\omega = 0$; it is not a true maximum if ζ is less than 0.707. Equation (9-24) gives the resonant frequency

$$\omega_r = \omega_n\sqrt{1 - 2\zeta^2} \tag{9-25}$$

Since frequency is a real quantity, Eq. (9-25) is meaningful only for $2\zeta^2 \leq 1$, or $\zeta \leq 0.707$. This means simply that for all values of ζ greater than 0.707, the resonant frequency is $\omega_r = 0$, and $M_r = 1$.

Substituting Eq. (9-24) in Eq. (9-20) for u and simplifying, we get

$$M_r = \frac{1}{2\zeta\sqrt{1-\zeta^2}} \qquad \zeta \leq 0.707 \tag{9-26}$$

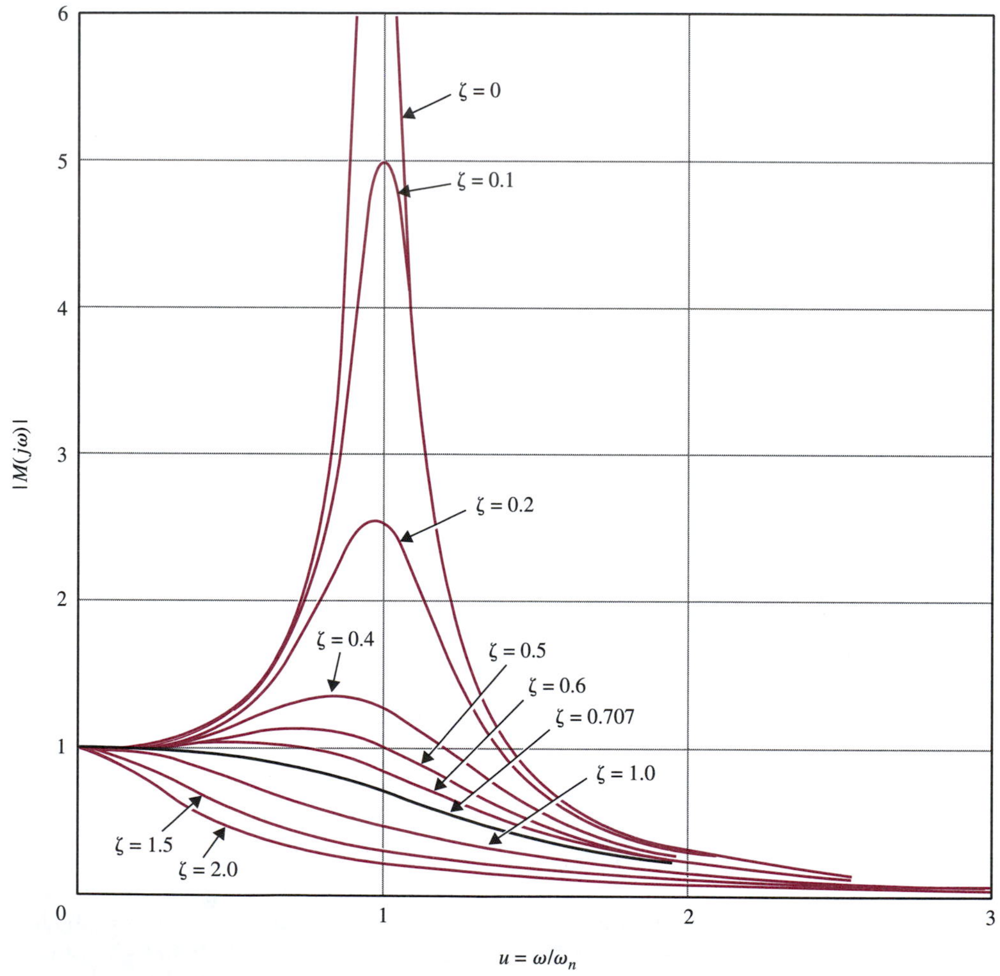

Figure 9-3 Magnification versus normalized frequency of a second-order closed-loop control system.

▲ For the prototype second-order system, $M_r = 1$ and $\omega_r = 0$ when $\zeta \geq 0.707$.

▲ For the prototype second-order system, M_r is a function of ζ only.

It is important to note that for the prototype second-order system, M_r is a function of the damping ratio ζ only, and ω_r is a function of both ζ and ω_n. Furthermore, although taking the derivative of $|M(ju)|$ with respect to u is a valid method of determining M_r and ω_r for higher-order systems, this analytical method is quite tedious and is not recommended. Graphical methods to be discussed and computer methods are much more efficient for general systems.

Figure 9-3 illustrates the plots of $|M(ju)|$ of Eq. (9-20) versus u for various values of ζ. Notice that if the frequency scale were unnormalized, the value of $\omega_r = u_r\omega_n$ would increase when ζ decreases, as indicated by Eq. (9-25). When $\zeta = 0$, $\omega_r = \omega_n$. Figures 9-4 and 9-5 illustrate the relationship between M_r and ζ, and u_r $(=\omega_r/\omega_n)$ and ζ, respectively.

9-2-2 Bandwidth

According with the definition of the bandwidth, we set the value of $|M(ju)|$ to $1/\sqrt{2} \cong 0.707$.

$$|M(ju)| = \frac{1}{[(1-u^2)^2 + (2\zeta u)^2]^{1/2}} = \frac{1}{\sqrt{2}} \tag{9-27}$$

Thus

$$[(1-u^2)^2 + (2\zeta u)^2]^{1/2} = \sqrt{2} \tag{9-28}$$

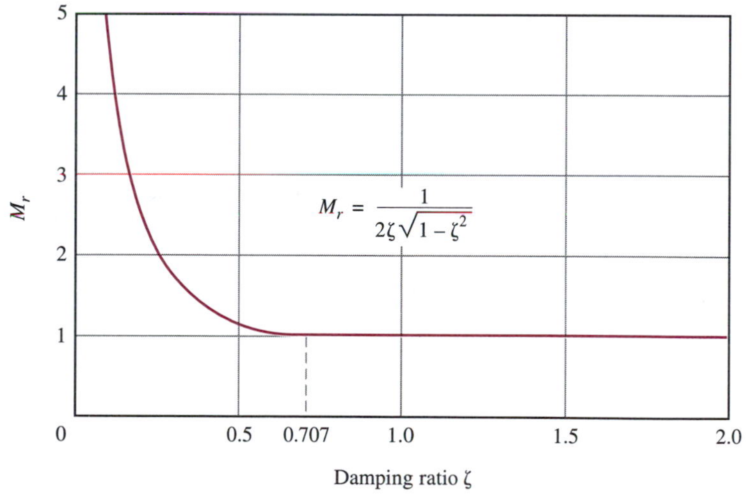

Figure 9-4 M_r versus damping ratio for a second-order system.

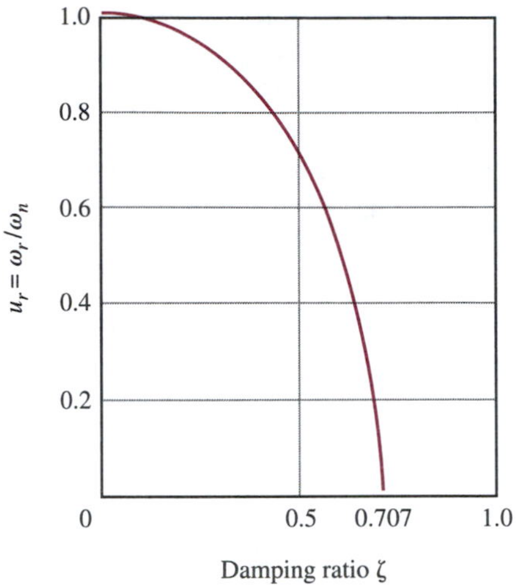

Figure 9-5 Normalized resonant frequency versus the damping ratio for a second-order system, $u_r = \sqrt{1 - 2\zeta^2}$.

which leads to

$$u^2 = (1 - 2\zeta^2) \pm \sqrt{4\zeta^4 - 4\zeta^2 + 2} \qquad (9\text{-}29)$$

▲ BW/ω_n decreases monotonically as the damping ratio ζ decreases.

The plus sign should be chosen in Eq. (9-29), since u must be a positive real quantity for any ζ. Therefore, the bandwidth of the prototype second-order system is determined from Eq. (9-29) as

$$\mathrm{BW} = \omega_n \left[(1 - 2\zeta^2) + \sqrt{\zeta^4 - 4\zeta^2 + 2}\right]^{1/2} \qquad (9\text{-}30)$$

▲ BW is directly proportional to ω_n.

Figure 9-6 shows a plot of BW/ω_n as a function of ζ. Notice that as ζ increases, BW/ω_n decreases monotonically. Even more important, Eq. (9-30) shows that BW is directly proportional to ω_n.

We have established some simple relationships between the time-domain response and the frequency-domain characteristics of the prototype second-order system. The summary of these relationships are as follows.

▲ When a system is unstable, the M_r no longer have any meaning.

1. The resonant peak M_r of the closed-loop frequency response depends on ζ only [Eq. (9-26)]. When ζ is zero, M_r is infinite. When ζ is negative, the system is unstable and the value of M_r ceases to have any meaning. As ζ

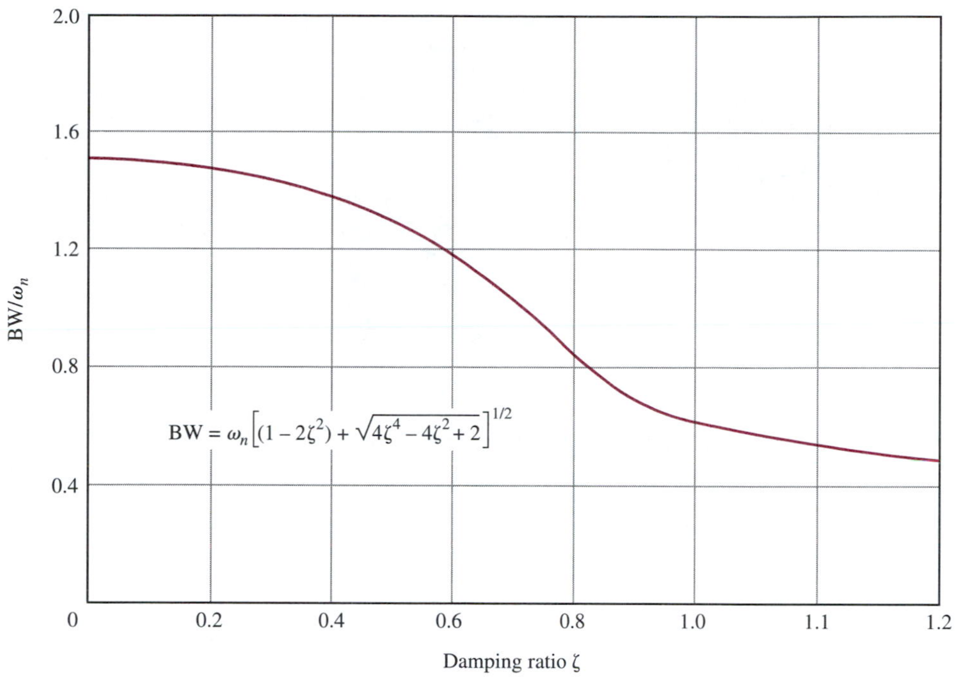

Figure 9-6 Bandwidth/ω_n versus the damping ratio for a second-order system.

increases, M_r decreases. For $\zeta \geq 0.707$, $M_r = 1$ (see Fig. 9-4) and $\omega_r = 0$ (see Fig. 9-5). Comparing with the unit-step time response, the maximum overshoot in Eq. (7-99) also depends only on ζ. However, the maximum overshoot = 0 when $\zeta \geq 1$.

▲ Bandwidth and rise time are inversely proportional to each other.

2. Bandwidth is directly proportional to ω_n [Eq. (9-30)]; that is, BW increases and decreases linearly with ω_n. BW also decreases with increase in ζ for a fixed ω_n (see Fig. 9-6). For the unit-step response, rise time increases as ω_n decreases [Eq. (7-104) and Fig. 7-21]. Therefore, BW and rise time are inversely proportional to each other.
3. Bandwidth and M_r are proportional to each other for $0 \leq \zeta \leq 0.707$.

The correlation between pole locations, unit-step response, and the magnitude of the frequency response for the prototype second-order system are summarized in Fig. 9-7.

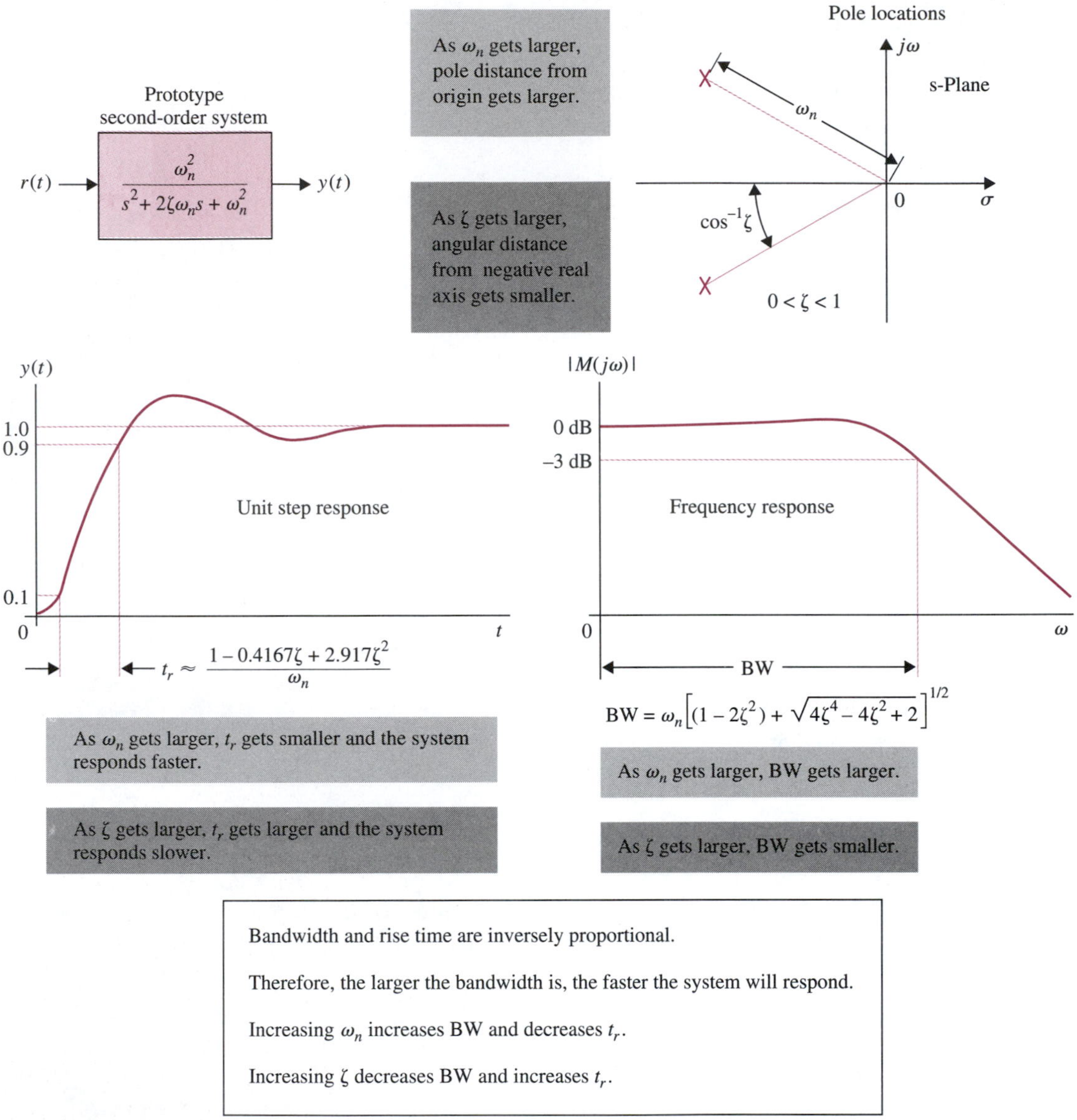

Figure 9-7 Correlation between pole locations, unit-step response, and the magnitude of frequency response of the prototype second-order system.

9-3 Effects of Adding a Zero to the Forward-Path Transfer Function

The relationships between the time-domain and frequency-domain responses arrived at in Section 9-2 apply only to the prototype second-order system described by Eq. (9-17). When other second-order or higher-order systems are involved, the relationships are different and may be more complex. It is of interest to consider the effects on the frequency-domain response when poles and zeros are added to the prototype second-order transfer function. It is simpler to study the effects of adding poles and zeros to the closed-loop transfer function; however, it is more realistic from a design standpoint to modify the forward-path transfer function.

The closed-loop transfer function of Eq. (9-17) may be considered as that of a unity-feedback control system with the prototype second-order forward-path transfer function

$$G(s) = \frac{\omega_n^2}{s(s + 2\zeta\omega_n)} \tag{9-31}$$

Let us add a zero at $s = -1/T$ to the transfer function so that Eq. (9-31) becomes

$$G(s) = \frac{\omega_n^2(1 + Ts)}{s(s + 2\zeta\omega_n)} \tag{9-32}$$

The closed-loop transfer function is

$$M(s) = \frac{\omega_n^2(1 + Ts)}{s^2 + (2\zeta\omega_n + T\omega_n^2)s + \omega_n^2} \tag{9-33}$$

▲ The general effect of adding a zero to the forward-path transfer function is to increase the BW of the closed-loop system.

In principle, M_r, ω_r, and BW of the system can all be derived using the same steps used in Section 9-2. However, since there are now three parameters in ζ, ω_n, and T, the exact expression for M_r, ω_r, and BW are difficult to obtain analytically even though the system is still second order. After a lengthy derivation, the bandwidth of the system is found to be

$$\text{BW} = \left(-b + \tfrac{1}{2}\sqrt{b^2 + 4\omega_n^4}\right)^{1/2} \tag{9-34}$$

where

$$b = 4\zeta^2\omega_n^2 + 4\zeta\omega_n^3 T - 2\omega_n^2 - \omega_n^4 T^2 \tag{9-35}$$

While is is difficult to see how each of the parameters in Eq. (9-34) affects the bandwidth, Fig. 9-8 shows the relationship between BW and T for $\zeta = 0.707$ and $\omega_n = 1$. Notice that

the general effect of adding a zero to the forward-path transfer function is to increase the bandwidth of the closed-loop system.

However, as shown in Fig. 9-8, over a range of small values of T, the bandwidth is actually decreased. Figure 9-9(a) and (b) give the plots of $|M(j\omega)|$ of the closed-loop system that has the $G(s)$ of Eq. (9-32) as its forward-path transfer function; $\omega_n = 1$, $\zeta = 0.707$ and 0.2, respectively, and T takes on various values. These curves verify that the bandwidth generally increases with the increase of T by the addition of a zero to $G(s)$ but for a range of small values of T for which BW is actually decreased. Figures 9-10 and 9-11 show the corresponding unit-step responses of the closed-loop system. These curves show that a high bandwidth corresponds to a faster rise time. However, as T become very large, the zero of the closed-loop transfer function, which is at $s = -1/T$, moves very close to the origin, causing the system to have a large time constant. Thus Fig. 9-10 illustrates the situation that the rise time is fast, but the large time constant of the zero near the origin of the s-plane causes the time response to drag out in reaching the final steady state (i.e., the settling time will be longer).

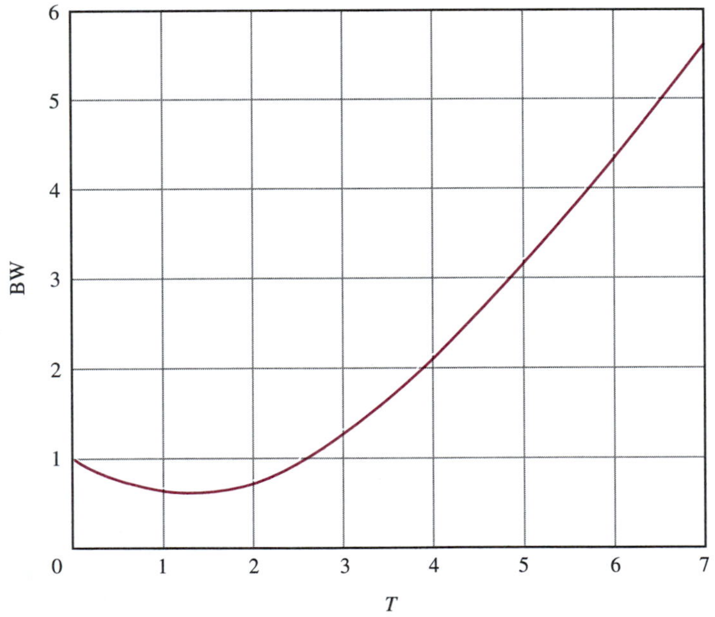

Figure 9-8 Bandwidth of a second-order system with open-loop transfer function $G(s) = (1 + Ts)/[s(s + 1.414)]$.

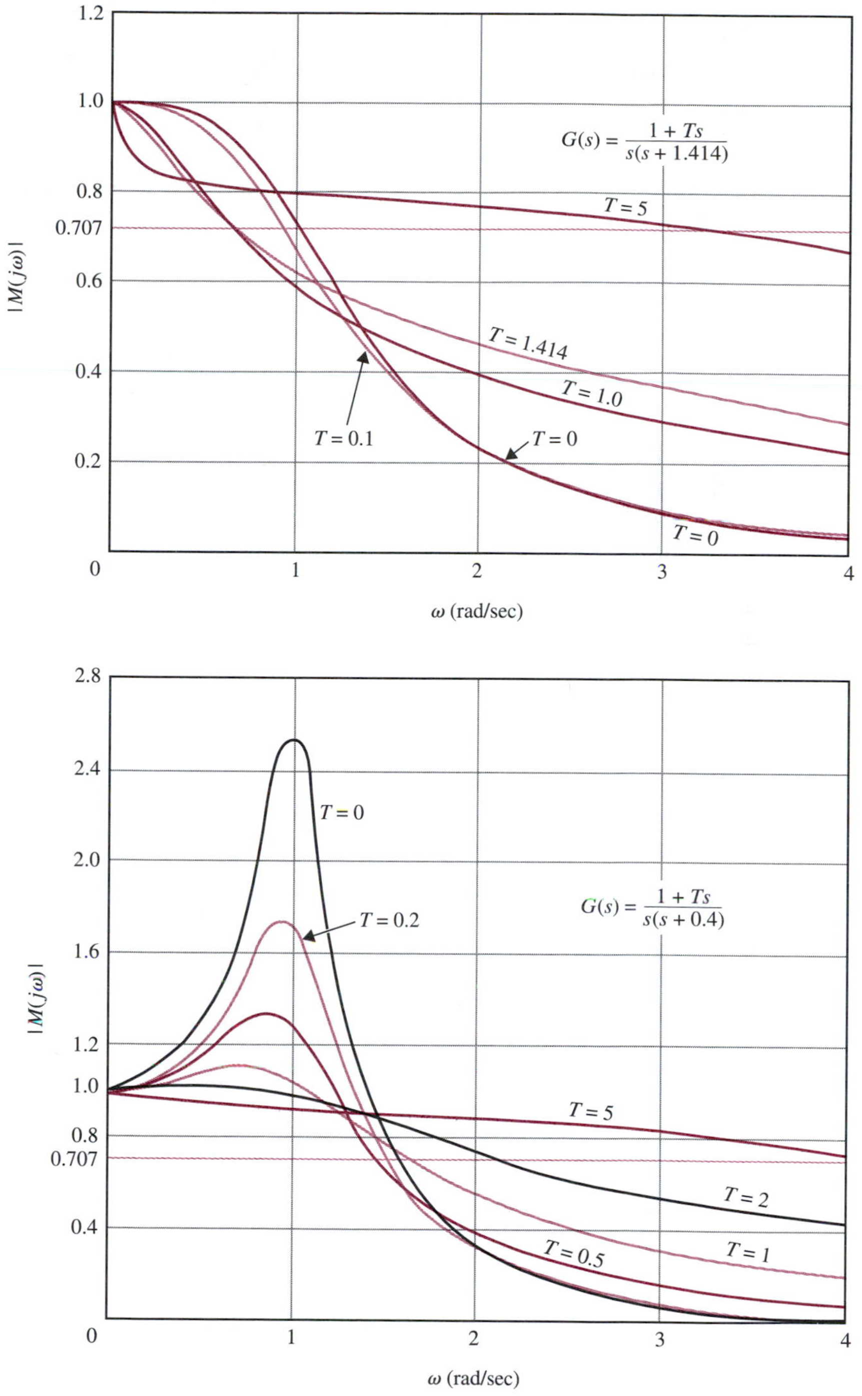

Figure 9-9 Magnification curves for a second-order system with a forward-path transfer function $G(s)$.

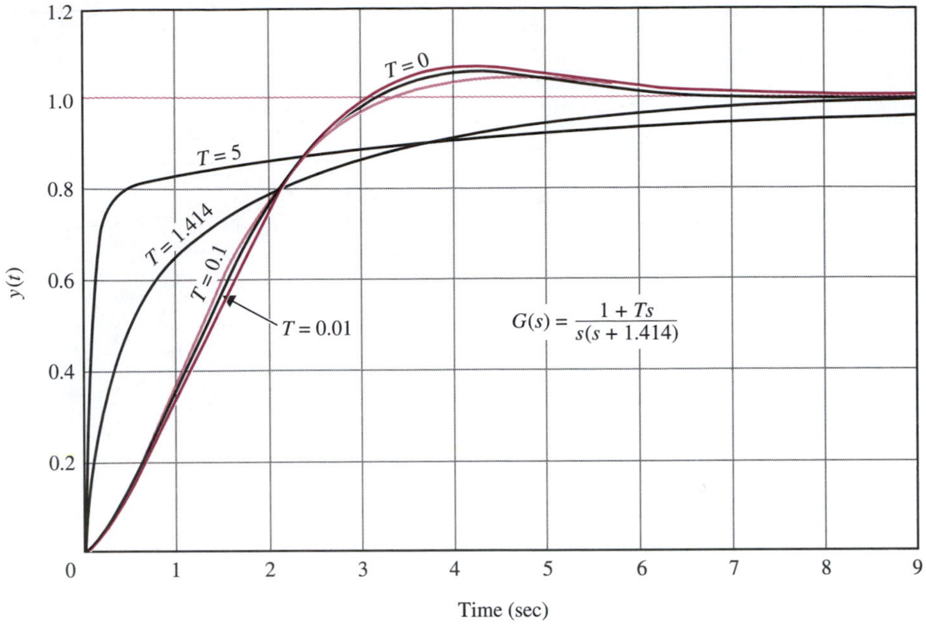

Figure 9-10 Unit-step responses of a third-order system with a forward-path transfer function $G(s)$.

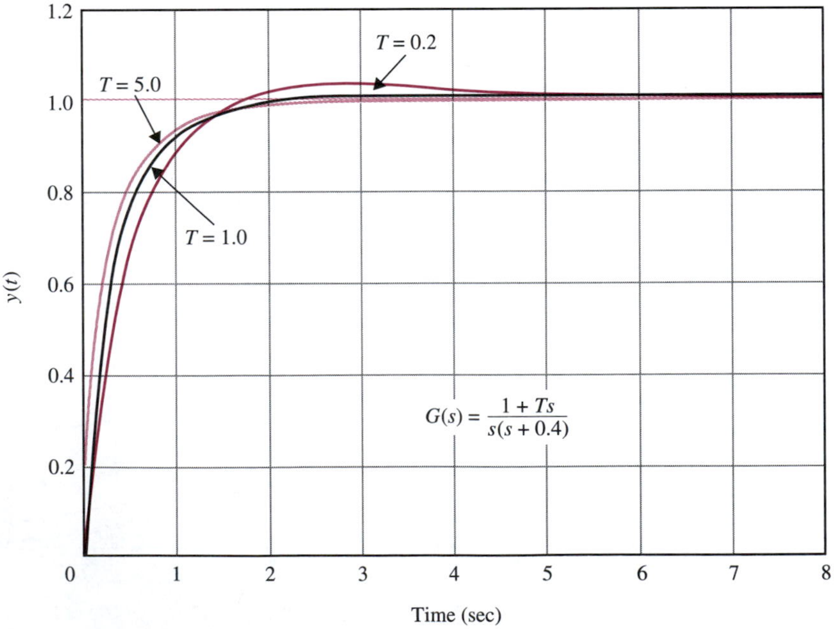

Figure 9-11 Unit-step responses of a second-order system with a forward-path transfer function $G(s)$.

9-4 Effects of Adding a Pole to the Forward-Path Transfer Function

Adding a pole at $s = -1/T$ to the forward-path transfer function of Eq. (9-31) leads to

$$G(s) = \frac{\omega_n^2}{s(s + 2\zeta\omega_n)(1 + Ts)} \tag{9-36}$$

▲ Adding a pole to the forward-path transfer function makes the closed-loop system less stable and decreases the bandwidth.

The derivation of the bandwidth of the closed-loop system with $G(s)$ given in Eq. (9-36) is quite tedious. We can obtain a qualitative indication on the bandwidth properties by referring to Fig. 9-12, which shows the plots of $|M(j\omega)|$ versus ω for $\omega_n = 1$, $\zeta = 0.707$, and various values of T. Since the system is now of the third order, it can be unstable for a certain set of system parameters. It can be shown that for $\omega_n = 1$ and $\zeta = 0.707$, the system is stable for all positive values of T. The $|M(j\omega)|$ versus ω curves of Fig. 9-12 show that for small values of T, the bandwidth of the system is slightly increased by the addition of the pole, but M_r is also increased. When T becomes large, the pole added to $G(s)$ has the effect of decreasing the bandwidth but increasing

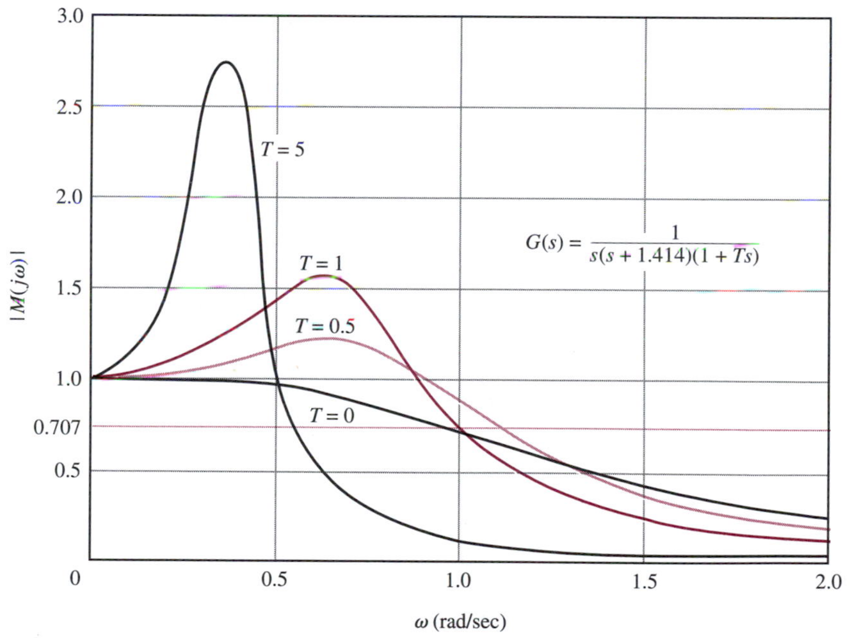

Figure 9-12 Magnification curves for a third-order system with a forward-path transfer function $G(s)$.

M_r. Thus we can conclude that, in general,

the effect of additing a pole to the forward-path transfer function is to make the closed-loop system less stable, while decreasing the bandwidth.

The unit-step responses of Fig. 9-13 show that for larger values of T, $T = 1$ and $T = 5$, the following relations are observed:

1. The rise time increases with the decrease of bandwidth.
2. The larger values of M_r also correspond to a larger maximum overshoot in the unit-step responses.

▲ When $M_r = \infty$ the closed-loop system is marginally stable. When the system is unstable, M_r no longer has any meaning.

The correlation between M_r and the maximum overshoot of the step response is meaningful only when the system is stable. When $G(j\omega) = -1$, $|M(j\omega)|$ is infinite, and the closed-loop system is marginally stable. On the other hand, when the system is unstable, the value of $|M(j\omega)|$ is analytically finite, but it no longer has any significance.

The objective of these last two sections is to demonstrate the simple relationships between BW, M_r, and the time-domain response. Typical effects on BW of adding a pole and a zero to the foreward-path transfer function are investigated. No attempt is made to include all general cases.

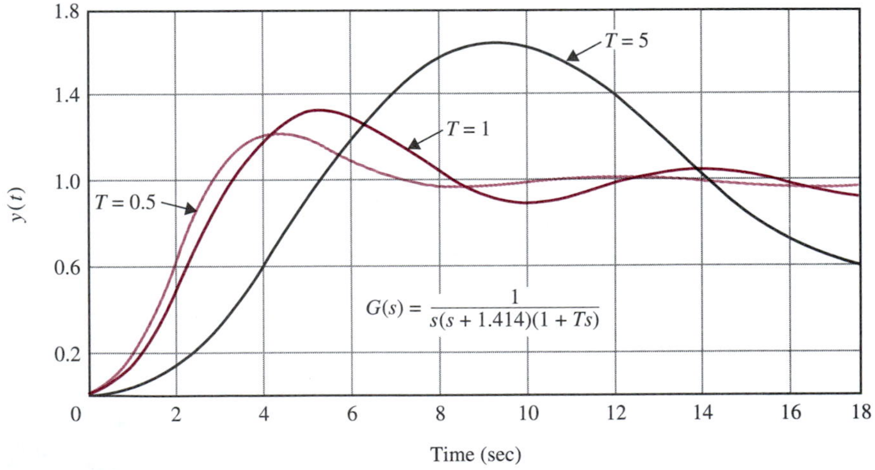

Figure 9-13 Unit-step responses of a third-order system with a forward-path transfer function $G(s)$.

9-5 Nyquist Stability Criterion: Fundamentals

▲ The Nyquist plot of $L(j\omega)$ is done in polar coordinates as ω varies from 0 to ∞.

Thus far we have presented two methods of determining the stability of linear SISO systems: the Routh–Hurwitz criterion and the root-locus method of determining stability by locating the roots of the characteristic equation in the s-plane. Of course, if the coefficients of the characteristic equation are all known, we can solve for the roots of the equation by use of a computer program.

The Nyquist criterion is a semigraphical method that determines the stability of a closed-loop system by investigating the properties of the frequency-domain plot, the **Nyquist plot,** of the loop transfer function $G(s)H(s)$, or $L(s)$. Specifically, the Nyquist plot of $L(s)$ is a plot of $L(j\omega)$ in the polar coordinates of $\text{Im}[L(j\omega)]$ versus $\text{Re}[L(j\omega)]$ as ω varies from 0 to ∞. This is another example of using the properties of the loop transfer function to find the performance of the closed-loop system. The Nyquist criterion has the following features that make it an alternative method that is attractive for the analysis and design of control systems.

▲ Nyquist criterion also gives indication on relative stability.

1. In addition to providing the absolute stability, like the Routh–Hurwitz criterion, the Nyquist criterion also gives information on the relative stability of a stable system and the degree of instability of an unstable system. It also gives indication on how the system stability may be improved, if needed.
2. The Nyquist plot of $G(s)H(s)$ or of $L(s)$ is very easy to obtain, especially with the aid of a computer.
3. The Nyquist plot of $G(s)H(s)$ gives information on the frequency-domain characteristics, such as M_r, ω_r, BW, and others, with ease.
4. The Nyquist plot is useful for systems with pure time delay which cannot be treated with the Routh–Hurwitz criterion, and are difficult to analyze with the root-locus method.

9-5-1 Stability Problem

The Nyquist criterion represents a method of determining the location of the characteristic equation roots with respect to the left half and the right half of the s-plane. Unlike the root-locus method, the Nyquist criterion does not give the exact location of the characteristic equation roots.

Let us consider that the closed-loop transfer function of a SISO system is

$$M(s) = \frac{G(s)}{1 + G(s)H(s)} \tag{9-37}$$

where $G(s)H(s)$ can assume the following form:

$$G(s)H(s) = \frac{K(1 + T_1 s)(1 + T_2 s) \cdots (1 + T_m s)}{s^p(1 + T_a s)(1 + T_b s) \cdots (1 + T_n s)} e^{-T_d s} \qquad (9\text{-}38)$$

where the T's are real or complex-conjugate coefficients and T_d is a real time delay.

Since the characteristic equation is obtained by setting the denominator polynomial of $M(s)$ to zero, the roots of the characteristic equation are also the zeros of $1 + G(s)H(s)$. Or, the characteristic equation roots must satisfy

$$\Delta(s) = 1 + G(s)H(s) = 0 \qquad (9\text{-}39)$$

In general, for a system with multiple number of loops, the denominator of $M(s)$ can be written as

$$\Delta(s) = 1 + L(s) = 0 \qquad (9\text{-}40)$$

where $L(s)$ is the loop transfer function and is of the form of Eq. (9-38).

Before embarking on the details of the Nyquist criterion, it is useful to summarize the pole–zero relationships of the various system transfer functions.

Identification of Poles and Zeros

loop transfer function zeros: zeros of $L(s)$
loop transfer function poles: poles of $L(s)$
closed-loop transfer function poles: zeros of $1 + L(s)$
$= $ roots of the characteristic equation
poles of $1 + L(s)$ = poles of $L(s)$

Stability Conditions. We define two types of stability with respect to the system configuration.

1. **Open-loop stability.** A system is said to be **open-loop stable** if the poles of the loop transfer function $L(s)$ are all in the left-half s-plane. For a single-loop system this is equivalent to the system being stable when the loop is opened at any point.
2. **Closed-loop stability.** A system is said to be **closed-loop stable,** or simply stable, if the poles of the closed-loop transfer function or the zeros of $1 + L(s)$ are all in the left-half s-plane. Exceptions to the definitions above are systems with poles or zeros intentionally placed at $s = 0$.

9-5-2 Definition of Encircled and Enclosed

Since the Nyquist criterion is a graphical method, we need to establish the concepts of **encircled** and **enclosed,** which are used for the interpretation of the Nyquist plots for stability.

Encircled

> ***A point or region in a complex function plane is said to be encircled by a closed path if it is found inside the path.***

For example, point A in Fig. 9-14 is encircled by the closed path Γ, since A is *inside* the closed path. Point B is not encircled by the closed path Γ, since it is *outside* the path. Furthermore, when the closed path Γ has a direction assigned to it, the encirclement, if made, can be in the clockwise (CW) or the counterclockwise (CCW) direction. As shown in Fig. 9-14, point A is encircled by Γ in the CCW direction. We can say that the region *inside* the path is encircled in the prescribed direction, and the region *outside* the path is not encircled.

Enclosed

> ***A point or region is said to be enclosed by a closed path if it is encircled in the CCW direction, or the point or region lies to the left of the path when the path is traversed in the prescribed direction.***

The concept of enclosure is particularly useful if only a portion of the closed path is shown. For example, the shaded regions in Figs. 9-15 are considered to be *enclosed* by the close path Γ. In other words, point A in Fig. 9-15(a) is *enclosed* by Γ, but point A in Fig. 9-15(b) is not. However, point B and all the points in the shaded region outside Γ in Fig. 9-15(b) are *enclosed.*

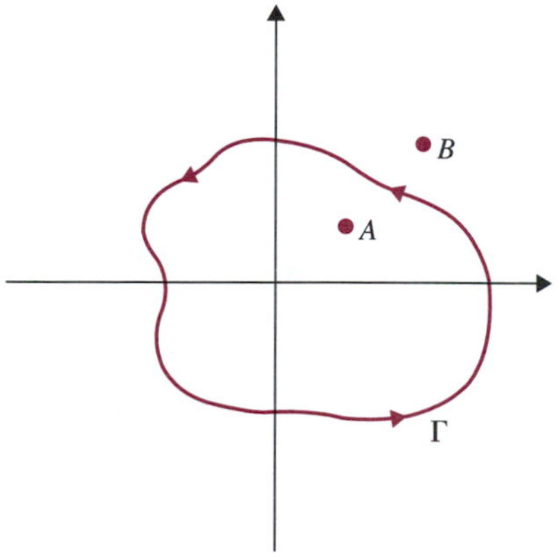

Figure 9-14 Definition of encirclement.

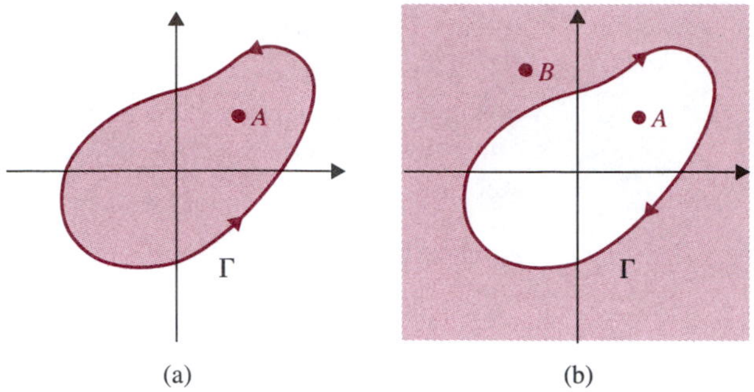

Figure 9-15 Definition of enclosed points and regions. (a) Point A is enclosed by Γ. (b) Point A is not enclosed, but B is enclosed by the locus Γ.

9-5-3 Number of Encirclements and Enclosures

When a point is encircled by a closed path Γ, a number N can be assigned to the number of times it is encircled. The magnitude of N can be determined by drawing an arrow from the point to any arbitrary point s_1 on the closed path Γ and then let s_1 follow the path in the prescribed direction until it returns to the starting point. The total *net* number of revolutions traversed by this arrow is N, or the net angle is $2\pi N$ radians. For example, point A in Fig. 9-16(a) is *encircled once* or 2π radians by Γ, and point B is *encircled twice* or 4π radians, all in the CW direction. In Fig. 9-16(b), point A is *enclosed once,* and point B is *enclosed twice,* by Γ. By definition, N is positive for CCW encirclement and negative for CW encirclement.

9-5-4 Principle of the Argument

The Nyquist criterion was originated as an engineering application of the well-known ''principle of the argument'' concept in complex-variable theory. The principle is stated in the following in a heuristic manner.

Let $\Delta(s)$ be a single-valued function of the form of the right-hand side of Eq. (9-38), which has a finite number of poles in the s-plane. *Single-valued* means that for each point in the s-plane, there is one and only one corresponding point, including infinity, in the complex $\Delta(s)$-plane. As defined in Chapter 8, infinity in the complex plane is interpreted as a point.

Suppose that a continuous closed path Γ_s is arbitrarily chosen in the s-plane, as shown in Fig. 9-17(a). If Γ_s does not go through any poles of $\Delta(s)$, then the trajectory Γ_Δ mapped by $\Delta(s)$ into the $\Delta(s)$-plane is also a closed one, as shown in Fig. 9-17(b). Starting from a point s_1, the Γ_s locus is traversed in the arbitrarily chosen direction (CW in the illustrated case), through the points s_2 and s_3, and then returning to s_1 after going through all the points on the Γ_s locus, as shown in Fig. 9-17(a). The corresponding Γ_Δ locus will start from the point $\Delta(s_1)$ and go through points $\Delta(s_2)$ and $\Delta(s_3)$, corresponding to s_1, s_2, and s_3, respectively, and finally return to the starting point, $\Delta(s_1)$. The direction of traverse of Γ_Δ can be either CW or CCW, that is, in the same direction or

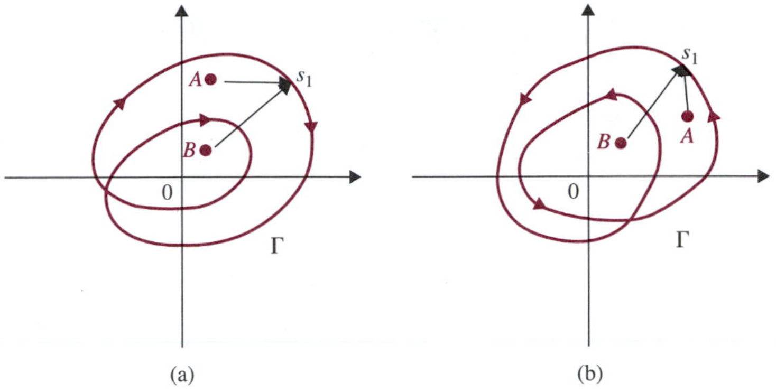

Figure 9-16 Definition of the number of encirclements and enclosures.

the opposite direction as that of Γ_s, depending on the function $\Delta(s)$. In Fig. 9-17(b) the direction of Γ_Δ is arbitrarily assigned, for illustration purposes, to be CCW.

▲ Do not attempt to relate $\Delta(s)$ with $L(s)$. They are not the same.

Although the mapping from the s-plane to the $\Delta(s)$-plane is single-valued, the reverse process is not a single-valued mapping. For example, consider the function

$$\Delta(s) = \frac{K}{s(s+1)(s+2)} \tag{9-41}$$

which has poles $s = 0, -1$, and -2 in the s-plane. For each point in the s-plane, there is only one corresponding point in the $\Delta(s)$-plane. However, for each point in the $\Delta(s)$-

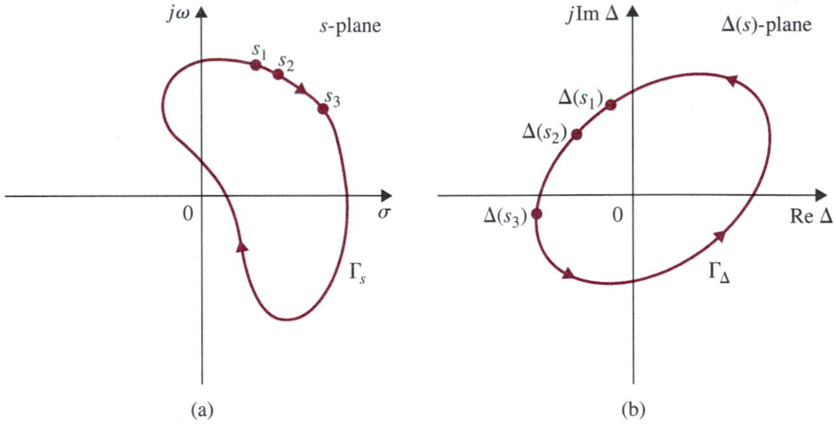

Figure 9-17 (a) Arbitrarily chosen closed path in the s-plane. (b) Corresponding locus Γ_Δ in the $\Delta(s)$-plane.

plane, the function maps into three corresponding points in the s-plane. The simplest way to illustrate this is to write Eq. (9-41) as

$$s(s+1)(s+2) - \frac{K}{\Delta(s)} = 0 \tag{9-42}$$

If $\Delta(s)$ is a real constant, which represents a point on the real axis in the $\Delta(s)$-plane, the third-order equation in Eq. (9-42) gives three roots in the s-plane. The reader should recognize the parallel of this situation to the root-locus diagram that essentially represents the mapping of $\Delta(s) = -1 + j0$ onto the loci of roots of the characteristic equation in the s-plane, for a given value of K. Thus the root loci of Eq. (9-41) have three individual branches in the s-plane.

The principle of the argument can be stated:

Let $\Delta(s)$ be a single-valued function that has a finite number of poles in the s-plane. Suppose that an arbitrary closed path Γ_s is chosen in the s-plane so that the path does not go through any one of the poles or zeros of $\Delta(s)$; the corresponding Γ_Δ locus mapped in the $\Delta(s)$-plane will encircle the origin as many times as the difference between the number of zeros and poles of $\Delta(s)$ that are encircled by the s-plane locus Γ_s.

In equation form, the principle of the argument is stated as

$$N = Z - P \tag{9-43}$$

where

N = number of encirclements of the origin made by the $\Delta(s)$-plane locus Γ_Δ
Z = number of zeros of $\Delta(s)$ encircled by the s-plane locus Γ_s in the s-plane
P = number of poles of $\Delta(s)$ encircled by the s-plane locus Γ_s in the s-plane

In general, N can be positive ($Z > P$), zero ($Z = P$), or negative ($Z < P$). These three situations are described in more detail as follows.

1. $N > 0$ ($Z > P$). If the s-plane locus encircles more zeros than poles of $\Delta(s)$ in a certain prescribed direction (CW or CCW), N is a positive integer. In this case the $\Delta(s)$-plane locus Γ_Δ will encircle the origin of the $\Delta(s)$-plane N times in the same direction as that of Γ_s.

2. $N = 0$ $(Z = P)$. If the s-plane locus encircles as many poles as zeros, or no poles and zeros, of $\Delta(s)$, the $\Delta(s)$-plane locus Γ_Δ will not encircle the origin of the $\Delta(s)$-plane.
3. $N < 0$ $(Z < P)$. If the s-plane locus encircles more poles than zeros of $\Delta(s)$ in a certain direction, N is a negative integer. In this case the $\Delta(s)$-plane locus Γ_Δ will encircle the origin N times in the *opposite* direction as that of Γ_s.

A convenient way of determining N with respect to the origin (or any point) of the Δ-plane is to draw a line from the point in any direction to a point as far as necessary; the number of *net* intersections of this line with the $\Delta(s)$ locus gives the magnitude of N. Figure 9-18 gives several examples of this method of determining N. In these illustrated cases it is assumed that the Γ_s locus has a CCW sense.

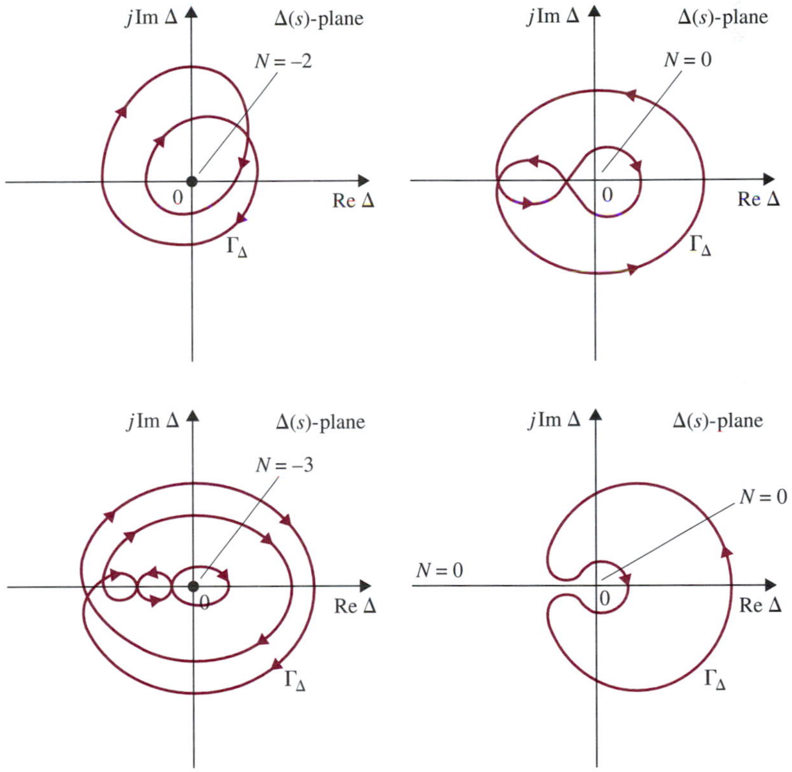

Figure 9-18 Examples of the determination of N in the $\Delta(s)$-plane.

Critical Point

For convenience, we shall designate the origin of the Δ-plane as the **critical point** from which the value of N is determined. Later, we shall designate other points in the complex-function plane as critical points, dependent on the way the Nyquist criterion is applied.

A rigorous proof of the principle of the argument is not given here. The following illustrative example may be considered as a heuristic explanation of the principle. Let us consider that the function $\Delta(s)$ is of the form

$$\Delta(s) = \frac{K(s + z_1)}{(s + p_1)(s + p_2)} \tag{9-44}$$

where K is a positive real number. The poles and zeros of $\Delta(s)$ are assumed to be as shown in Fig. 9-19(a). The function $\Delta(s)$ can be written as

$$\begin{aligned}\Delta(s) &= |\Delta(s)|\angle\Delta(s) \\ &= \frac{K|s + z_1|}{|s + p_1||s + p_2|}[\angle(s + z_1) - \angle(s + p_1) - \angle(s + p_2)]\end{aligned} \tag{9-45}$$

Figure 9-19(a) shows an arbitrarily chosen trajectory Γ_s in the s-plane, with the arbitrary point s_1 on the path, and Γ_s does not pass through any of the poles and the

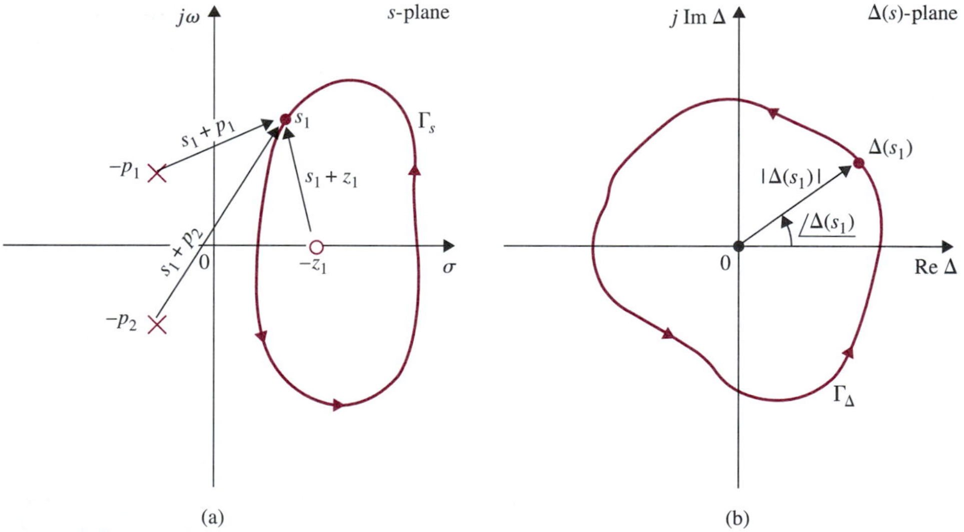

Figure 9-19 (a) Pole–zero configuration of $\Delta(s)$ in Eq. (9-45) and the s-plane trajectory Γ_s. (b) $\Delta(s)$-plane locus Γ_Δ, which corresponds to the Γ_s locus of (a) through the mapping of Eq. (9-45).

zeros of $\Delta(s)$. The function $\Delta(s)$ evaluated at $s = s_1$ is

$$\Delta(s_1) = \frac{K(s_1 + z_1)}{(s_1 + p_1)(s_1 + p_2)} \tag{9-46}$$

The term $(s_1 + z_1)$ can be represented graphically by the vector drawn from $-z_1$ to s_1. Similar vectors can be drawn for $(s_1 + p_1)$ and $(s_1 + p_2)$. Thus $\Delta(s_1)$ is represented by the vectors drawn from the finite poles and zeros of $\Delta(s)$ to the point s_1, as shown in Fig. 9-19(a). Now, if the point s_1 is moved along the locus Γ_s in the prescribed CCW direction until it returns to the starting point, the angles generated by the vectors drawn from the two poles that are not encircled by Γ_s when s_1 completes one round trip are zero, whereas the vector $(s_1 + z_1)$ drawn from the zero at $-z_1$, which is encircled by Γ_s, generates a positive angle (CCW) of 2π radians, which means that the corresponding $\Delta(s)$ plot must go around the origin 2π radians, or one revolution, in the CCW direction, as shown in Fig. 9-19(b). This is why only the poles and zeros of $\Delta(s)$ that are inside the Γ_s trajectory in the s-plane will contribute to the value of N of Eq. (9-43). Since the poles of $\Delta(s)$ contribute to a negative phase, and zeros contribute to positive phase, the value of N depends only on the difference between Z and P. For the case illustrated in Fig. 9-19(a), $Z = 1$ and $P = 0$. Thus

▲ Z and P refer only to the zeros and poles, respectively of $\Delta(s)$ that are encircled by Γ_s.

$$N = Z - P = 1 \tag{9-47}$$

which means that the $\Delta(s)$-plane locus Γ_Δ should encircle the origin once in the same direction as that of the s-plane locus Γ_s. It should be kept in mind that Z and P refer only to the zeros and poles, respectively, of $\Delta(s)$ that are encircled by Γ_s, and not the total number of zeros and poles of $\Delta(s)$.

In general, the net angle traversed by the $\Delta(s)$-plane locus, as the s-plane locus is traversed once in any direction, is equal to

$$2\pi(Z - P) = 2\pi N \qquad \text{radians} \tag{9-48}$$

This equation implies that if there are N more zeros than poles of $\Delta(s)$, which are encircled by the s-plane locus Γ_s, in a prescribed direction, the $\Delta(s)$-plane locus will encircle the origin N times in the *same direction* as that of Γ_s. Conversely, if N more poles than zeros are encircled by Γ_s in a given direction, N in Eq. (9-48) will be negative, and the $\Delta(s)$-plane locus must encircle the origin N times in the *opposite direction* to that of Γ_s.

A summary of all the possible outcomes of the principle of the argument is given in Table 9-1.

9-5-5 Nyquist Path

Years ago when Nyquist was faced with the problem of solving the stability problem, which involves determining if the function $\Delta(s) = 1 + L(s)$ has zeros in the right-half s-plane, he apparently discovered that the principle of the argument could be applied

Table 9-1 Summary of All Possible Outcomes of the Principle of the Argument

		$\Delta(s)$-Plane Locus	
$N = Z - P$	Sense of the s-Plane Locus	Number of Encirclements of the Origin	Direction of Encirclement
$N > 0$	CW	N	CW
	CCW	N	CW
$N < 0$	CW	N	CCW
	CCW	N	CW
$N = 0$	CW	0	No encirclement
	CCW	0	No encirclement

▲ The Nyquist path is defined to encircle the entire right-half s-plane.

to solve the stability problem if the s-plane locus Γ_s is taken to be one that encircles the entire right-half of the s-plane. Of course, as an alternative, Γ_s can be chosen to encircle the entire left-half s-plane, as the solution is a relative one. Figure 9-20 illustrates a Γ_s locus with a CCW sense, which encircles the entire right half of the s-plane. This path is chosen to be the s-plane trajectory Γ_s for the Nyquist criterion, since in mathematics, CCW is traditionally defined to be the positive sense. The path Γ_s shown in Fig. 9-20 is defined to be the **Nyquist path.** Since the Nyquist path must not pass through any poles and zeros of $\Delta(s)$, the small semicircles shown along the $j\omega$-axis in Fig. 9-20 are used to indicate that the path should go around these poles and zeros if

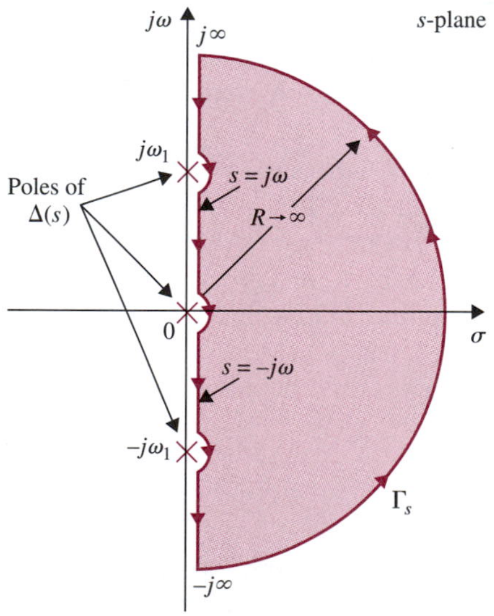

Figure 9-20 Nyquist path.

they fall on the $j\omega$-axis. It is apparent that if any pole or zero of $\Delta(s)$ lies inside the right-half s-plane, it will be encircled by the Nyquist path Γ_s.

9-5-6 Nyquist Criterion and the $L(s)$ or the $G(s)H(s)$ Plot

The Nyquist criterion is a direct application of the principle of the argument when the s-plane locus is the Nyquist path of Fig. 9-20. In principle, once the Nyquist path is specified, the stability of a closed-loop system can be determined by plotting the $\Delta(s) = 1 + L(s)$ locus when s takes on values along the Nyquist path, and investigating the behavior of the $\Delta(s)$ plot with respect to the **critical point,** which in this case is the origin of the $\Delta(s)$-plane.

Since the function $L(s)$ is generally known,

> ***it would be simpler to construct the L(s) plot that corresponds to the Nyquist path, and the same conclusion on the stability of the closed-loop system can be obtained by observing the behavior of the L(s) plot with respect to the (−1, j0) point in the L(s)-plane.***

This is because the origin of the $\Delta(s) = 1 + L(s)$ plane corresponds to the $(-1, j0)$ point in the $L(s)$-plane. Thus the $(-1, j0)$ point in the $L(s)$-plane becomes the critical point for the determination of closed-loop stability.

For single-loop systems, $L(s) = G(s)H(s)$, the previous development leads to the determination of the closed-loop stability by investigating the behavior of the $G(s)H(s)$ plot with respect to the $(-1, j0)$ point of the $G(s)H(s)$-plane. Thus the Nyquist stability criterion is another example of using the loop transfer function properties to find the behavior of closed-loop systems.

Thus, given a control system that has the characteristic equation given by equating the numerator polynomial of $1 + L(s)$ to zero, where $L(s)$ is the loop transfer function, the application of the Nyquist criterion to the stability problem involves the following steps.

1. The Nyquist path Γ_s is defined in the s-plane, as shown in Fig. 9-20.
2. The $L(s)$ plot corresponding to the Nyquist path is constructed in the $L(s)$-plane.
3. The value of N, the number of encirclements of the $(-1, j0)$ point made by the $L(s)$ plot, is observed.
4. The Nyquist criterion follows from Eq. (9-43),

$$N = Z - P \tag{9-49}$$

where

N = number of encirclements of the $(-1, j0)$ point made by the $L(s)$ plot
Z = number of zeros of $1 + L(s)$ that are inside the Nyquist path (i.e., the right-half s-plane)
P = number of poles of $1 + L(s)$ that are inside the Nyquist path (i.e., the right-half s-plane); notice that the poles of $1 + L(s)$ are the same as those of $L(s)$

The stability requirements for the two types of stability defined earlier are interpreted in terms of Z and P.

For closed-loop stability, Z must equal zero.
For open-loop stability, P must equal zero.

Thus the condition of stability according to the Nyquist criterion is stated as

$$N = -P \tag{9-50}$$

That is,

for a closed-loop system to be stable, the L(s) plot must encircle the (−1, j0) point as many times as the number of poles of L(s) that are in the right-half s-plane, and the encirclement, if any, must be made in the clockwise direction (if Γ_s is defined in the CCW sense).

9-6 Nyquist Criterion for Systems with Minimum-Phase Transfer Functions

We shall first apply the Nyquist criterion to systems with $L(s)$ that are **minimum-phase transfer functions.** The properties of the minimum-phase transfer functions are described in Appendix A, and are summarized as follows:

1. A minimum-phase transfer function does not have poles or zeros in the right-half s-plane or on the $j\omega$-axis, excluding the origin.
2. For a minimum-phase transfer function $L(s)$ with m zeros and n poles, excluding the poles at $s = 0$, when $s = j\omega$ and as ω varies from ∞ to 0, the total phase variation of $L(j\omega)$ is $(n - m)\pi/2$ radians.

3. The value of a minimum-phase transfer function cannot become zero or infinity at any finite nonzero frequency.
4. A nonminimum-phase transfer function will always have a more positive phase shift as ω varies from ∞ to 0. Or equally true, it will always have a more negative phase shift as ω varies from 0 to ∞.

▲ A minimum-phase transfer function does not have poles or zeros in the right-half s-plane or on the $j\omega$-axis, except at $s = 0$.

Since a majority of the loop transfer functions encountered in the real world satisfy condition 1 above and are of the minimum-phase type, it would be prudent to investigate application of the Nyquist criterion to this class of systems. As it turns out, this is quite simple.

Since a minimum-phase $L(s)$ does not have any poles or zeros in the right-half s-plane or on the $j\omega$-axis, except at $s = 0$, $P = 0$, and the poles of $\Delta(s) = 1 + L(s)$ also have the same properties. Thus the Nyquist criterion for a system with $L(s)$ being a minimum-phase transfer function is simplified to

$$N = 0 \tag{9-51}$$

Thus the Nyquist criterion can be stated:

For a closed-loop system with loop transfer function L(s) that is of minimum-phase type, the system is closed-loop stable if the plot of L(s) that corresponds to the Nyquist path does not encircle the critical point (−1, j0) in the L(s)-plane.

Furthermore, if the system is unstable, $Z \neq 0$; N in Eq. (9-51) would be a positive integer, which means that the critical point $(-1, j0)$ is **enclosed** N times (corresponding to the direction of the Nyquist path defined here). Thus the Nyquist criterion of stability for systems with minimum-phase loop transfer functions can be further simplified to:

For a closed-loop system with loop transfer function L(s) that is of minimum-phase type, the system is closed-loop stable if the L(s) plot that corresponds to the Nyquist path does not enclose the (−1, j0) point. If the (−1, j0) is enclosed by the Nyquist plot, the system is unstable.

▲ For $L(s)$ that is minimum-phase type, Nyquist criterion can be checked by plotting the segment of $L(j\omega)$ from $\omega = \infty$ to 0.

Since the region that is enclosed by a trajectory is defined as the region that lies to the left when the trajectory is traversed in the prescribed direction, the *Nyquist criterion can be checked simply by plotting the segment of* $L(j\omega)$ *from* $\omega = \infty$ *to 0, or, points on the positive* $j\omega$*-axis.* This simplifies the procedure considerably, since the plot can be made easily on a computer. The only drawback to this method is that the Nyquist plot that corresponds to the $j\omega$ axis tells only whether the critical point is enclosed or not; and if it does, not how many times. Thus if the system is found to be unstable, the enclosure property does not give information on how many roots of the characteristic equation are in the right-half s-plane. However, in practice, this information may not be vital. From this point on we shall define *the $L(j\omega)$ plot that corresponds to the positive $j\omega$-axis of the s-plane as the Nyquist plot of $L(s)$.*

9-6-1 Application of the Nyquist Criterion to Minimum-Phase Transfer Functions That Are Not Strictly Proper

Just as in the case of the root locus, it is often necessary in design to create an equivalent loop transfer function $L_{eq}(s)$ so that a variable parameter K will appear as a multiplying factor in $L_{eq}(s)$ [i.e., $L(s) = KL_{eq}(s)$]. Since the equivalent loop transfer function does not correspond to any physical entity, it may not have more poles than zeros, and the transfer function is not strictly proper as defined in Chapter 3. In principle, there is no difficulty in constructing the Nyquist plot of a transfer function that is not strictly proper, and the Nyquist criterion can be applied for stability studies without any complications. However, some computer programs may not be prepared for handling improper transfer functions, and it may be necessary to reformulate the equation for compatibility with the computer program. To consider this case, consider that the characteristic equation of a system with a variable parameter K is conditioned to

$$1 + KL_{eq}(s) = 0 \tag{9-52}$$

If $L_{eq}(s)$ does not have more poles than zeros, we can rewrite the last equation as

$$1 + \frac{1}{KL_{eq}(s)} = 0 \tag{9-53}$$

by dividing both sides of the equation by $KL_{eq}(s)$. Now we can plot the Nyquist plot of $1/L_{eq}(s)$, and the critical point is still $(-1, j0)$ for $K > 0$. The variable parameter on the Nyquist plot is now $1/K$. Thus, with this minor adjustment, the Nyquist criterion can still be applied.

9-7 Relation Between the Root Loci and the Nyquist Plot

Since both the root-locus analysis and the Nyquist criterion deal with the location of the roots of the characteristic equation of a linear SISO system, the two analyses are closely related. Exploring the relationship between the two methods will enhance the understanding of both methods. Given the characteristic equation

$$1 + L(s) = 1 + KG_1(s)H_1(s) = 0 \qquad (9\text{-}54)$$

the Nyquist plot of $L(s)$ in the $L(s)$-plane is the mapping of the Nyquist path in the s-plane. Since the root loci of Eq. (9-54) must satisfy the conditions

$$\angle KG_1(s)H_1(s) = (2i + 1)\pi \qquad K \geq 0 \qquad (9\text{-}55)$$

$$\angle KG_1(s)H_1(s) = 2i\pi \qquad K \leq 0 \qquad (9\text{-}56)$$

for $i = 0, \pm 1, \pm 2, \ldots$, the root loci simply represent a mapping of the real axis of the $L(s)$-plane or the $G(s)H(s)$-plane onto the s-plane. In fact, for the RL, $K \geq 0$, the mapping points are on the negative real axis of the $L(s)$-plane; and for the CRL, $K \leq 0$, the mapping points are on the positive real axis of the $L(s)$-plane. It was pointed out earlier that the mapping from the s-plane to the function plane for a rational function is single valued, but the reverse process is multivalued. As a simple illustration, the Nyquist plot of a type 1 third-order transfer function $G(s)H(s)$ that corresponds to points on the $j\omega$-axis of the s-plane is shown in Fig. 9-21. The root loci for the same system are shown in Fig. 9-22 as a mapping of the real axis of the $G(s)H(s)$-plane onto the s-plane. Note that in this case, each point of the $G(s)H(s)$-plane corresponds to three points in the s-plane. The $(-1, j0)$ point of the $G(s)H(s)$-plane corresponds to the two points where the root loci intersect the $j\omega$-axis and a point on the real axis.

The Nyquist plot and the root loci each represents the mapping of only a very limited portion of one domain to the other. In general, it would be useful to consider

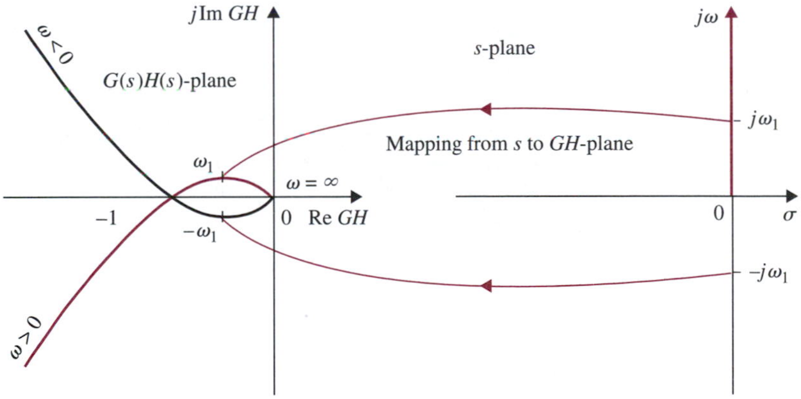

Figure 9-21 Polar plot of $G(s)H(s) = K/[s(s + a)(s + b)]$ interpreted as a mapping of the $j\omega$-axis of the s-plane onto the $G(s)H(s)$-plane.

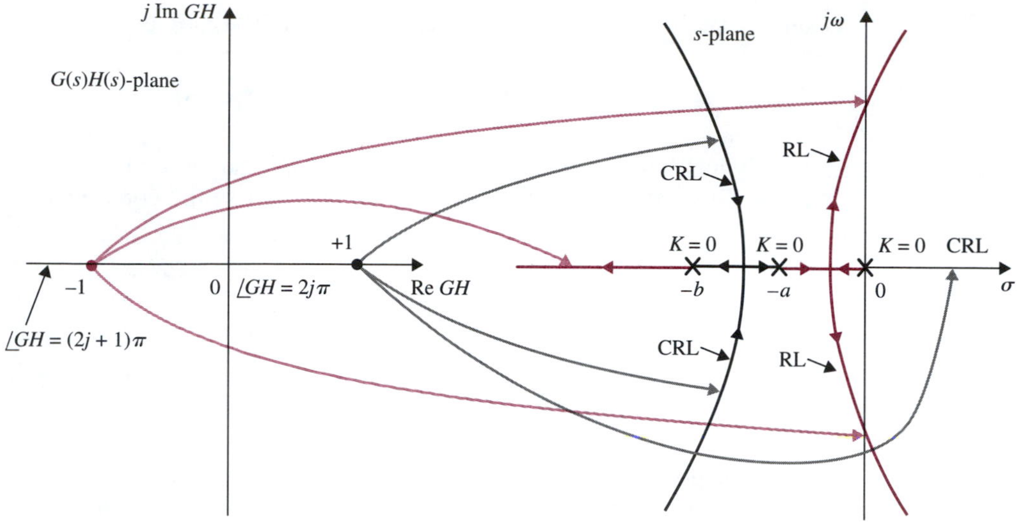

Figure 9-22 Root-locus diagram of $G(s)H(s) = K/[s(s + a)(s + b)]$ interpreted as a mapping of the real axis of the $G(s)H(s)$-plane onto the s-plane.

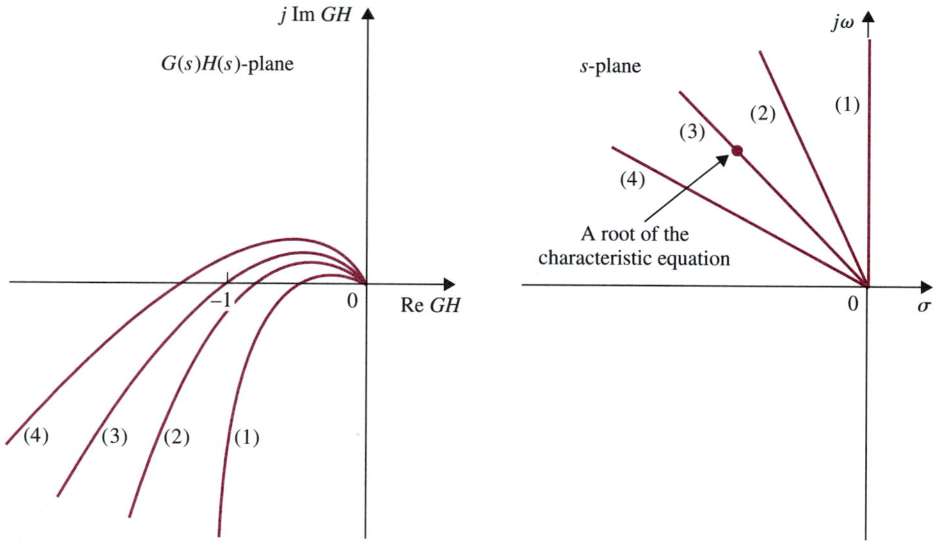

Figure 9-23 $G(s)H(s)$ plots that correspond to constant-damping-ratio lines in the s-plane.

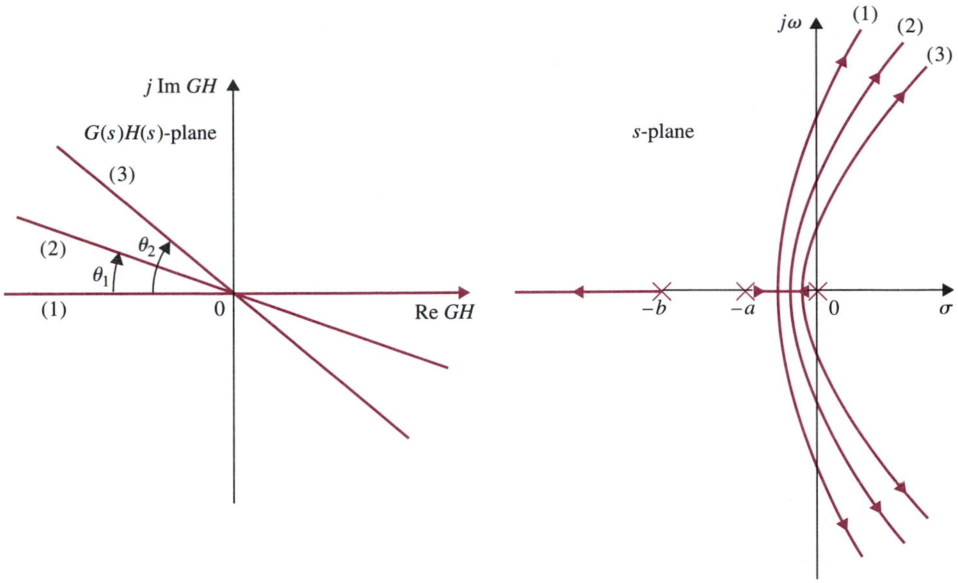

Figure 9-24 Root loci that correspond to different phase-angle loci in the $G(s)\ H(s)$-plane. (The CRL are not shown.)

the mapping of points other than those on the $j\omega$-axis of the s-plane and on the real axis of the $G(s)H(s)$-plane. For instance, we may use the mapping of the constant-damping-ratio lines in the s-plane onto the $G(s)H(s)$-plane for the purpose of determining relative stability of the closed-loop system. Figure 9-23 illustrates the $G(s)H(s)$ plots that correspond to different constant-damping-ratio lines in the s-plane. As shown by curve (3) in Fig. 9-23, when the $G(s)H(s)$ curve passes through the $(-1, j0)$ point, it means that Eq. (9-54) is satisfied, and the corresponding trajectory in the s-plane passes through the root of the characteristic equation. Similarly, we can construct the root loci that correspond to the straight lines rotated at various angles from the real axis in the $G(s)H(s)$-plane, as shown in Fig. 9-24. Notice that these root loci now satisfy the condition of

$$\angle KG_1(s)H_1(s) = (2j + 1)\pi - \theta \qquad K \geq 0 \tag{9-57}$$

or the root loci of Fig. 9-24 must satisfy the equation

$$1 + G(s)H(s)e^{j\theta} = 0 \tag{9-58}$$

for the various values of θ indicated.

9-8 Illustrative Examples: Nyquist Criterion Minimum-Phase Transfer Functions

The following examples serve to illustrate the application of the Nyquist criterion to systems with minimum-phase loop transfer functions.

Example 9-1

Consider that a single-loop feedback control system has the loop transfer function

$$L(s) = G(s)H(s) = \frac{K}{s(s+2)(s+10)} \tag{9-59}$$

which is of minimum-phase type. The stability of the closed-loop system can be conducted by investigating whether the Nyquist plot of $L(j\omega)$ for $\omega = \infty$ to 0 encloses the $(-1, j0)$ point. The Nyquist plot of $L(j\omega)$ may be plotted using a computer program such as **plrplot** of **CSAD, freqrp** of **ACSP,** or **Nyquist** of **Program CC.** Figure 9-25 shows the Nyquist plot of $L(j\omega)$ for $\omega = \infty$ to 0. However, since we are interested only in whether the critical point is enclosed or not, in general, it is not necessary to produce an accurate Nyquist plot. Since the area that is enclosed by the Nyquist plot is to the left of the curve, traversed in the direction that corresponds to $\omega = \infty$ to 0 on the Nyquist path, all that is necessary to determine stability is to find the point or points at which the Nyquist plot crosses the real axis in the $L(j\omega)$-plane. In many cases, information on the intersection on the real axis and the properties of $L(j\omega)$ at $\omega = \infty$ and $\omega = 0$ would allow the sketching of the Nyquist plot without actual plotting. We can use the following steps to obtain a sketch of the Nyquist plot of $L(j\omega)$.

1. Substitute $s = j\omega$ in $L(s)$. Setting $s = j\omega$ in Eq. (9-59), we get

$$L(j\omega) = \frac{K}{j\omega(j\omega+2)(j\omega+10)} \tag{9-60}$$

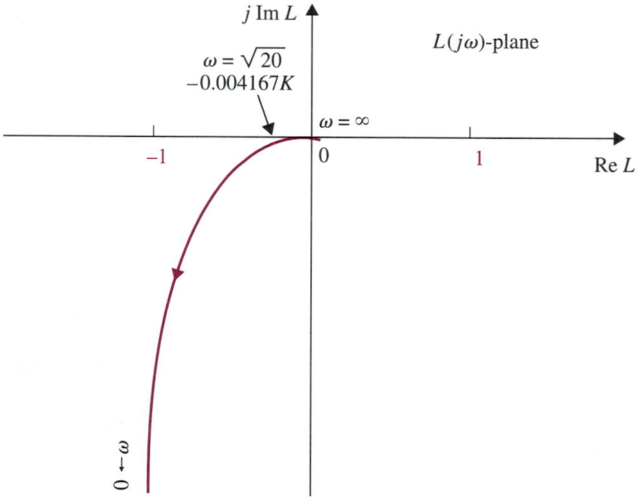

Figure 9-25 Nyquist plot of $L(s) = \dfrac{K}{s(s+2)(s+10)}$ for $\omega = \infty$ to $\omega = 0$.

2. Substituting $\omega = 0$ in Eq. (9-60), we get the zero-frequency property of $L(j\omega)$,

$$L(j0) = \infty\angle -90^\circ \tag{9-61}$$

3. Substituting $\omega = \infty$ in Eq. (9-60), the property of the Nyquist plot at infinite frequency is established.

$$L(j\infty) = 0\angle -270^\circ \tag{9-62}$$

Apparently, these results are verified by the plot shown in Fig. 9-25.

4. To find the intersect(s) of the Nyquist plot with the real axis, if any, we rationalize $L(j\omega)$ by multiplying the numerator and the denominator of the equation by the complex conjugate of the denominator. Thus Eq. (9-60) becomes

$$\begin{aligned} L(j\omega) &= \frac{K[-12\omega^2 - j\omega(20-\omega^2)]}{[-12\omega^2 + j\omega(20-\omega^2)][-12\omega^2 - j\omega(20-\omega^2)]} \\ &= \frac{K[-12\omega^2 - j\omega(20-\omega^2)]}{144\omega^4 + \omega^2(20-\omega^2)^2} \\ &= \frac{K[-12\omega - j(20-\omega^2)]}{\omega[144\omega^2 + (20-\omega^2)]} \end{aligned} \tag{9-63}$$

5. To find the possible intersects on the real axis, we set the imaginary part of $L(j\omega)$ to zero. The result is

$$\text{Im}[L(j\omega)] = \frac{-K(20-\omega^2)}{\omega[144\omega^2 + (20-\omega^2)]} = 0 \tag{9-64}$$

The solutions of the last equation are $\omega = \infty$, which is known to be a solution at $L(j\omega) = 0$, and

$$\omega = \pm\sqrt{20} \qquad \text{rad/sec} \tag{9-65}$$

Since ω is positive, the correct answer is $\omega = \sqrt{20}$ rad/sec. Substituting this frequency into Eq. (9-63), we have the intersect on the real axis of the $L(j\omega)$-plane at

$$L\left(j\sqrt{20}\right) = -\frac{12K}{2880} = -0.004167K \tag{9-66}$$

The last five steps should lead to an adequate sketch of the Nyquist plot of $L(j\omega)$ short of plotting it. Thus we see that if K is less than 240, the intersect of the $L(j\omega)$ locus on the real axis would be to the right of the critical point $(-1, j0)$; the latter is not enclosed and the system is stable. If $K = 240$, the Nyquist plot of $L(j\omega)$ would intersect the real axis at the -1 point, and the system would be marginally stable. In this case the characteristic equation would have two roots on the $j\omega$-axis in the s-plane at $s = \pm j\sqrt{20}$. If the gain is increased to a value beyond 240, the intersect would be to the left of the -1 point on the real axis, and the system would be unstable. When K is negative, we can use the $(+1, j0)$ point in the $L(j\omega)$-plane as the critical point. Figure 9-25 shows that under this condition, the +1 point on the real axis would be enclosed for all negative values of K, and the system would always be unstable. Thus the Nyquist criterion leads to the conclusion that the system is stable in the range of $0 < K < 240$. Note that application of the Routh–Hurwitz stability criterion leads to the same result. Figure 9-26 shows the root loci of the characteristic equation of the system described by the loop transfer function in Eq. (9-59). The correlation between the Nyquist criterion and the root loci is easily observed. ▲

Example 9-2

Consider the characteristic equation

$$Ks^3 + (2K + 1)s^2 + (2K + 5)s + 1 = 0 \tag{9-67}$$

Dividing both sides of Eq. (9-67) by the terms that do not contain K, we have

$$1 + KL_{eq}(s) = 1 + \frac{Ks(s^2 + 2s + 2)}{s^2 + 5s + 1} = 0 \tag{9-68}$$

Thus

$$L_{eq}(s) = \frac{s(s^2 + 2s + 2)}{s^2 + 5s + 1} \tag{9-69}$$

which is an improper function. We can obtain the information to manually sketch the Nyquist plot of $L_{eq}(s)$ to determine the stability of the system. Setting $s = j\omega$ in Eq. (9-69), we get

$$L_{eq}(j\omega) = \frac{K\omega[-2\omega + j(2 - \omega^2)]}{(1 - \omega^2) + 5j\omega} \tag{9-70}$$

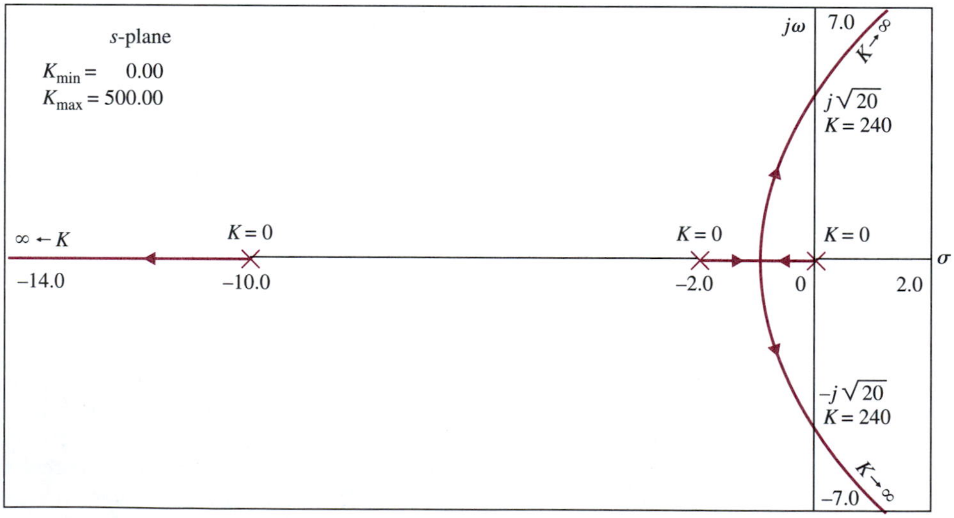

Figure 9-26 RL of $L(s) = \dfrac{K}{s(s+2)(s+10)}$.

From Eq. (9-70) we obtain the two endpoints of the Nyquist plot:

$$L_{eq}(j0) = 0\angle 90° \quad \text{and} \quad L_{eq}(j\infty) = \infty\angle 90° \tag{9-71}$$

Rationalizing Eq. (9-70) by multiplying its numerator and denominator by the complex conjugate of the denominator, we get

$$L_{eq}(j\omega) = K\frac{\omega^2[5(2-\omega^2) - 2(1-\omega^2)] + j\omega[10\omega^2 + (2-\omega^2)(1-\omega^2)]}{(1-\omega^2)^2 + 25\omega^2} \tag{9-72}$$

To find the possible intersects of the $L_{eq}(j\omega)$ plot on the real axis, we set the imaginary part of Eq. (9-72) to zero. We get $\omega = 0$ and

$$\omega^4 + 7\omega^2 + 2 = 0 \tag{9-73}$$

We can show that all the four roots of Eq. (9-73) are imaginary, which indicates that the $L_{eq}(j\omega)$ locus intersects the real axis only at $\omega = 0$. Using the information given by Eq. (9-71) and the fact that there are no other intersections on the real axis than at $\omega = 0$, the Nyquist plot of $L_{eq}(j\omega)$ is sketched as shown in Fig. 9-27. Notice that this plot is sketched without any detailed data computed on $L_{eq}(j\omega)$, and in fact, could be grossly inaccurate. However, the sketch is adequate to determine the stability of the system. Since the Nyquist plot in Fig. 9-27 does not enclose the $(-1, j0)$ point as ω varies from ∞ to 0, the system is stable for all finite positive values of K.

Figure 9-28 shows the Nyquist plot of $1/L_{eq}(s)$, where

$$\frac{1}{L_{eq}(j\omega)} = \frac{(1-\omega^2) + 5j\omega}{K[-2\omega^2 + j\omega(2-\omega^2)]} \tag{9-74}$$

for $\omega = \infty$ to 0. The plot again does not enclose the $(-1, j0)$ point, and the system is again stable for all positive values of K by interpreting the Nyquist plot of $1/L_{eq}(j\omega)$.

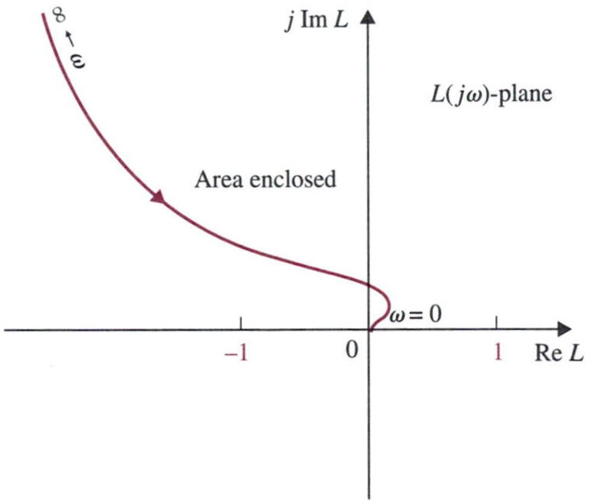

Figure 9-27 Nyquist plot of $L(j\omega)$ for

$$L(s) = \frac{K(s^3 + 2s^2 + 2s)}{s^2 + 5s + 1} \qquad K = 1$$

for $\omega = \infty$ to $\omega = 0$.

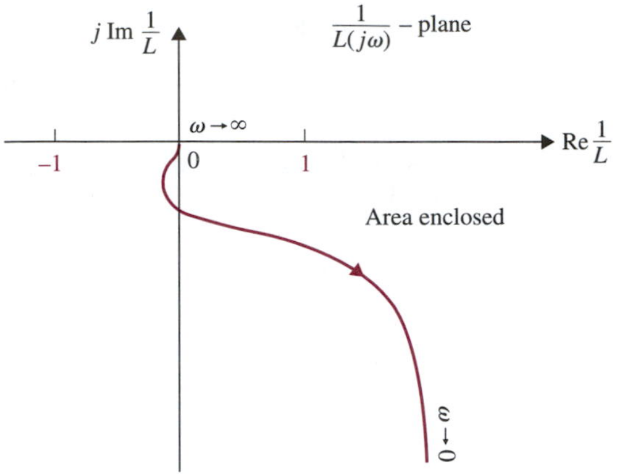

Figure 9-28 Nyquist plot of $1/L(j\omega)$ for

$$L(s) = \frac{K(s^3 + 2s^2 + 2s)}{s^2 + 5s + 1} \qquad K = 1$$

for $\omega = \infty$ to $\omega = 0$.

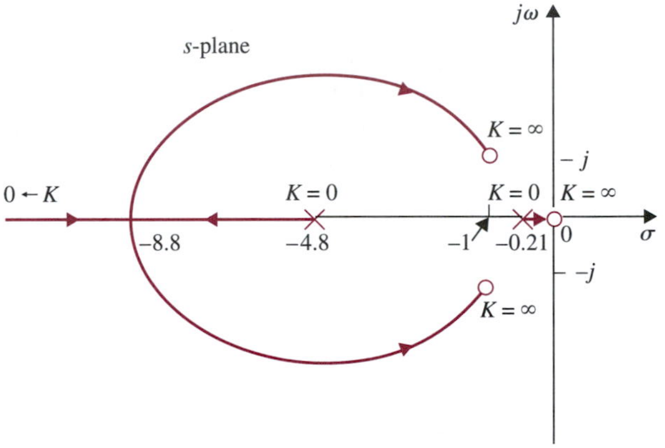

Figure 9-29 RL of $L(s) = \dfrac{Ks(s^2 + 2s + 2)}{s^2 + 5s + 1}$.

Figure 9-29 shows the RL of Eq. (9-67) for $K > 0$, using the pole–zero configuration of $L_{eq}(s)$ of Eq. (9-69). Since the RL stays in the left-half s-plane for all positive values of K, the system is stable for $0 < K < \infty$, which agrees with the conclusion obtained with the Nyquist criterion. ▲

9-9 General Nyquist Criterion for Minimum- and Nonminimum-Phase Transfer Functions

The original Nyquist criterion presented in Section 9-8 is cumbersome to apply when the loop transfer function is of the nonminimum-phase type. We shall show that if the loop transfer function is of the nonminimum-phase type, then plotting the Nyquist plot of $L(s)$ only for $s = j\omega$ to $s = j0$ and not enclosing the $(-1, j0)$ point in the $L(s)$-plane is only a necessary but not sufficient condition for closed-loop stability. For a system with a nonminimum-phase loop transfer function $L(s)$, the original Nyquist criterion requires that the $L(s)$ plot that corresponds to the entire Nyquist path in Fig. 9-20 be made. If the loop transfer function $L(s)$ has poles or zeros on the $j\omega$-axis, the Nyquist path of Fig. 9-20 must have small indentations around them on the $j\omega$-axis. This adds even more complexity to construction of the $L(s)$ plot. Computer programs such as the **freqrp** of **ACSP, plrplot** of **CSAD,** and **Nyquist** of **Program CC** can all be used to construct the plots of only functions that correspond to the positive $j\omega$-axis of the s-plane. The rest of the Nyquist plot that corresponds to the small indentations and the large semicircle on the Nyquist path have to be done manually. With modern computer facilities and software, the analyst should not be burdened with the chores of plotting manually. Therefore, we are introducing *a simplified Nyquist criterion that can be applied to minimum as well as nonminimum-phase transfer functions by using only the positive $j\omega$-axis of the Nyquist path and then observe the behavior of the Nyquist plot with reference to the* $(-1, j0)$ *point.*

Yeung and Lai [6,7] introduced a general, yet simplified version of the Nyquist criterion that allows the determination of stability of closed-loop systems of minimum- as well as nonminimum-phase loop transfer functions by using only the positive part of the $j\omega$-axis of the Nyquist path. However, if the system is of the minimum-phase type, the test of whether the $(-1, j0)$ point is enclosed is still simpler to apply. We shall show that for nonminimum-phase systems, if the $(-1, j0)$ point is enclosed, the system is still unstable. However, if the $(-1, j0)$ point is not enclosed, an additional angle condition is all that must be satisfied by the Nyquist plot of $L(s)$ for the system to be stable.

Let us consider the two Nyquist paths shown in Fig. 9-30. Apparently, the Nyquist path Γ_{s1} in Fig. 9-30(a) is the original one shown in Fig. 9-20, whereas the path Γ_{s2} in Fig. 9-30(b) encircles not only the entire right-half s-plane, but also all the poles and zeros of $L(s)$ on the $j\omega$-axis, if there are any. Let us define the following quantities.

Z = number of zeros of $1 + L(s)$ that are in the right-half s-plane

P = number of poles of $L(s)$, or of $1 + L(s)$, that are in the right-half s-plane

P_ω = number of poles of $L(s)$, or of $1 + L(s)$, that are on the $j\omega$-axis, including the origin

N_1 = number of times the $(-1, j0)$ point of the $L(s)$-plane that is encircled by the Nyquist plot of $L(s)$ corresponding to Γ_{s1}

N_2 = number of times the $(-1, j0)$ point of the $L(s)$-plane that is encircled by the Nyquist plot of $L(s)$ corresponding to Γ_{s2}

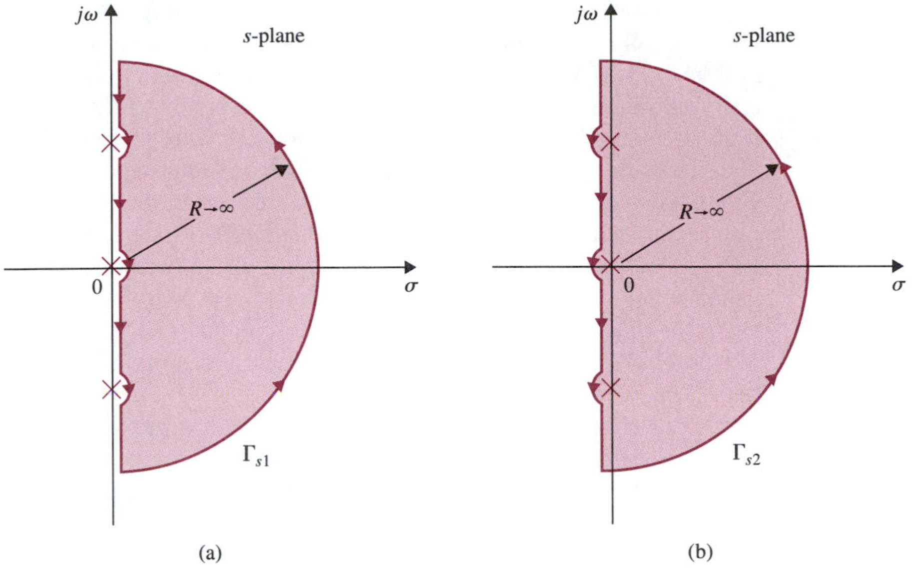

Figure 9-30 (a) Nyquist path. (b) An alternative Nyquist path. (*Source:* K. S. Yeung, "A Reformulation of Nyquist's Criterion," *IEEE Trans, Educ.*, Vol. E-28 (© 1985, IEEE.)

Then, with reference to the two Nyquist paths in Fig. 9-30, and according to the Nyquist criterion,

$$N_1 = Z - P \tag{9-75}$$

and

$$N_2 = Z - P - P_\omega \tag{9-76}$$

Let Φ_1 and Φ_2 represent the net angles traversed by the Nyquist plot of $L(s)$ with respect to the $(-1, j0)$ point, corresponding to Γ_{s1} and Γ_{s2}, respectively. Then

$$\Phi_1 = N_1 \times 360^\circ = (Z - P)360^\circ \tag{9-77}$$
$$\Phi_2 = N_2 \times 360^\circ = (Z - P - P_\omega)360^\circ \tag{9-78}$$

Let us consider that each of the Nyquist paths Γ_{s1} and Γ_{s2} is composed of three portions:

1. The portion from $s = -j\infty$ to $+j\infty$ along the semicircle with infinite radius.

2. The portion along the $j\omega$-axis, excluding all the small indentations.
3. All the small indentations on the $j\omega$-axis.

Since the Nyquist paths in Fig. 9-30 are symmetrical about the real axis in the s-plane, the angles traversed by the Nyquist plots are identical for positive and negative values of ω. Thus Φ_1 and Φ_2 are written

$$\Phi_1 = 2\Phi_{11} + \Phi_{12} + \Phi_{13} \tag{9-79}$$
$$\Phi_2 = 2\Phi_{11} - \Phi_{12} + \Phi_{13} \tag{9-80}$$

where

Φ_{11} = angle traversed by the Nyquist plot of $L(s)$ with respect to the $(-1, j0)$ point, corresponding to the positive $j\omega$-axis or the $-j\omega$-axis of the s-plane, excluding the small indentations

Φ_{12} = angle traversed by the Nyquist plot of $L(s)$ with respect to the $(-1, j0)$ point, corresponding to the small indentations on the $j\omega$-axis of Γ_{s1} [since on Γ_{s2} the directions of the small indentations are opposite to that of Γ_{s1}, the sign of Φ_{12} in Eq. (9-79) is negative]

Φ_{13} = angle traversed by the Nyquist plot of $L(s)$ with respect to the $(-1, j0)$ point, corresponding to the semicircle with infinite radius on the Nyquist paths

For a transfer function $L(s)$ that does not have more zeros than poles, the Nyquist plot of $L(s)$ that corresponds to the infinite semicircle must either be a point on the real axis or a trajectory around the origin of the $L(s)$-plane. Thus the angle Φ_{13} traversed by the phasor drawn from the $(-1, j0)$ point to the Nyquist plot along the semicircle with infinite radius is always zero.

Now adding Eq. (9-79) to Eq. (9-80) and using Eqs. (9-77) and (9-78), we get

$$\begin{aligned}\Phi_1 + \Phi_2 &= 4\Phi_{11} \\ &= (2Z - 2P - P_\omega)360^\circ\end{aligned} \tag{9-81}$$

Solving for Φ_{11}, we get

$$\Phi_{11} = (Z - P - 0.5P_\omega)180^\circ \tag{9-82}$$

The equation states:

The total angle traversed by the phasor drawn from the $(-1, j0)$ point to the $L(s)$ Nyquist plot that corresponds to the portion on the positive $j\omega$-axis of the s-plane, excluding the small indentations, if any,

= [the number of zeros of $1 + L(s)$ in the right-half s-plane
− the number of poles of $L(s)$ in the right-half s-plane
− 0.5 (the number of poles of $L(s)$ on the $j\omega$-axis)]180° (9-83)

▲ In general, the Nyquist criterion can be carried out by drawing the Nyquist plot of $L(j\omega)$ that corresponds to $\omega = \infty$ to $\omega = 0$.

▲ The angle Φ_{11} is an **angle variation,** so that $\Phi_{11} = -270°$ is not the same as $\Phi_{11} = 90°$.

Thus *the Nyquist stability criterion can be carried out by constructing only the Nyquist plot that corresponds to the $s = j\infty$ to $s = 0$ portion on the Nyquist path.* Furthermore, if the closed-loop system is unstable, by knowing the values of Φ_{11}, P_ω, and P, Eq. (9-82) gives the number of roots of the characteristic equation that are in the right-half s-plane.

For the closed-loop system to be stable, Z must equal zero. Thus the Nyquist criterion for stability of the closed-loop system is

$$\Phi_{11} = -(0.5P_\omega + P)180° \tag{9-84}$$

Since P_ω and P cannot be negative, Eq. (9-84) indicates:

> ***If the phase traversed by the Nyquist plot of $L(j\omega)$ as ω varies from ∞ to 0, Φ_{11}, is positive with respect to the $(-1, j0)$ point, the closed-loop system is unstable.***

However, if Φ_{11} is negative, it still has to satisfy Eq. (9-83) for the system to be stable.

With reference to the Nyquist plot of $L(j\omega)$ and the $(-1, j0)$ point, we see that when the angle variation Φ_{11} is positive, it corresponds to the $(-1, j0)$ point being enclosed—thus the condition that the Nyquist plot of $L(j\omega)$ not enclosing the $(-1, j0)$ point is a necessary condition for closed-loop stability for nonminimum-phase systems. However, if the $(-1, j0)$ point is not enclosed by the Nyquist plot of $L(j\omega)$, for the nonminimum-phase system to be closed-loop stable, the angle variation Φ_{11} still has to satisfy Eq. (9-83).

9-9-1 System with Minimum-Phase Loop Transfer Functions

If $L(s)$ is of the minimum-phase type, $P = 0$ and P_ω denotes the number of poles of $L(s)$ that are at the origin; Eq. (9-82) becomes

$$\Phi_{11} = (Z - 0.5P_\omega)180° \tag{9-85}$$

For closed-loop stability, $Z = 0$; Eq. (9-85) becomes

$$\Phi_{11} = -P_\omega \times 90° \tag{9-86}$$

Since P_ω denotes the number of poles of $L(s)$ that are at the origin, it is easy to see that if the $(-1, j0)$ point is not enclosed by the Nyquist plot of $L(s)$, Φ_{11} will always be given by Eq. (9-86). Thus when $L(s)$ is of the minimum-phase type, the condition that the $(-1, j0)$ point not be enclosed by the Nyquist plot is a necessary and sufficient condition for closed-loop stability.

9-9-2 Systems with Improper Loop Transfer Functions

Equation (9-82) is derived based on the condition that $\Phi_{13} = 0$, which is true only if $L(s)$ is strictly proper; that is, it has more poles than zeros. For improper transfer functions, we can again use the method discussed in Section 9-6 by plotting the Nyquist plot of $1/L(s)$.

9-10 Illustrative Examples: General Nyquist Criterion for Minimum- and Nonminimum-Phase Transfer Functions

In the following example we show that the Nyquist plot of a nonminimum-phase transfer function does not enclose the $(-1, j0)$ point, yet the system is unstable.

Example 9-3

Consider that the loop transfer function of a control system is given by

$$L(s) = \frac{s^2 - s + 1}{s(s^2 - 6s + 5)} \tag{9-87}$$

Since $L(s)$ has one pole at the origin and two poles in the right-half s-plane, $P_\omega = 1$ and $P = 2$. From Eq. (9-84), the closed-loop system is stable if the following condition is satisfied:

$$\Phi_{11} = -(0.5P_\omega + P)180° = -450° \tag{9-88}$$

The Nyquist plot of $L(j\omega)$ for $\omega = \infty$ to $\omega = 0$ is plotted with **plrplot** of **CSAD,** as shown in Fig. 9-31. Apparently, in this case the $(-1, j0)$ point is not enclosed by the Nyquist plot. However, since Φ_{11} is $-90°$, and not $-450°$, the system is unstable. Substituting $\Phi_{11} = 90°$ into Eq. (9-82) and solving for Z, we have $Z = 2$, which means that there are two closed-loop poles in the right-half s-plane. ▲

Example 9-4

Consider the system described in Example 9-1. The loop transfer function of the system, $L(s)$, is given in Eq. (9-59). The Nyquist plot of $L(j\omega)$ is shown in Fig. 9-25. It is shown in Example 9-1 that for closed-loop stability, the $(-1, j0)$ point in the $L(j\omega)$-plane must be to the left of the intersect of the Nyquist plot with the real axis.

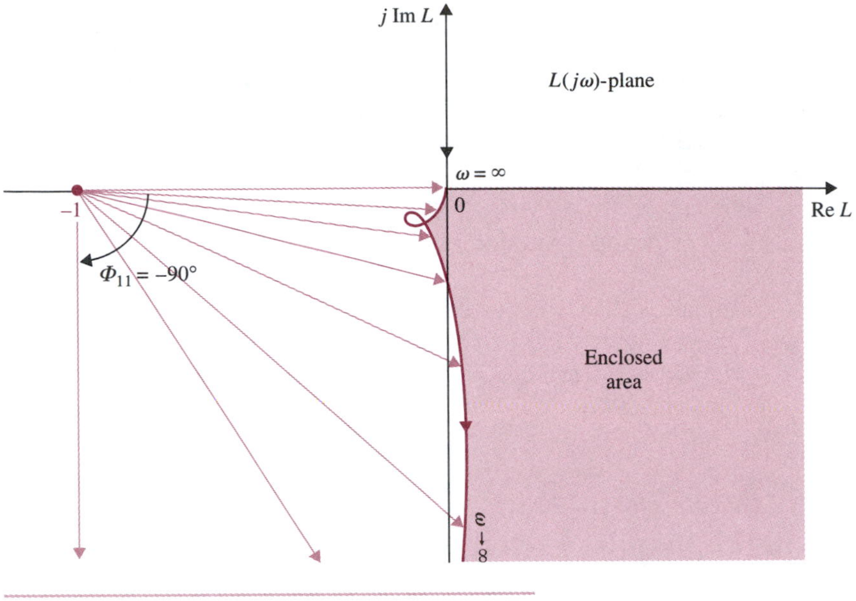

Figure 9-31 Nyquist plot of $L(s) = \dfrac{s^2 - s + 1}{s(s^2 - 6s + 5)}$.

Since $L(s)$ is of the minimum-phase type, $P = 0$, and it has one pole at the origin; thus $P_\omega = 1$. Substituting these quantities into Eq. (9-85), we have

$$\Phi_{11} = (Z - 0.5)180° \tag{9-89}$$

For closed-loop stability, Z must equal zero; thus the last equation gives $\Phi_{11} = -90°$. This means that the phasor drawn from the $(-1, j0)$ to the Nyquist plot, from $\omega = \infty$ to $\omega = 0$, must equal $-90°$, or $90°$ in the CW direction. Figure 9-32(a) shows that if the $(-1, j0)$ point is to the left of the intersect of $L(j\omega)$ with the real axis, Φ_{11} is indeed $-90°$. On the other hand, if the $(-1, j0)$ point is to the right of the intersect, then, Φ_{11} is $+270°$, or, more easily observed that the critical point is enclosed, and the system would be unstable. Substituting, $\Phi_{11} = 270°$ into Eq. (9-89), we get $Z = 2$, which means that the characteristic equation has two roots in the right-half s-plane. Thus for systems with minimum-phase loop transfer functions, although the Nyquist criterion with the "enclosure" test is easier to observe, when the system is unstable, it does not tell how many characteristic equation roots are in the right-half plane; the general Nyquist criterion does. ▲

Example 9-5

Consider that a control system has the loop transfer function

$$L(s) = \frac{K(s-1)}{s(s+1)} \tag{9-90}$$

We observed from Eq. (9-90) that $P_\omega = 1$, and $P = 0$. The function $L(s)$ is of the nonminimum-phase type, since it has a zero at $s = 1$. Thus the Nyquist criterion on enclosure cannot be used

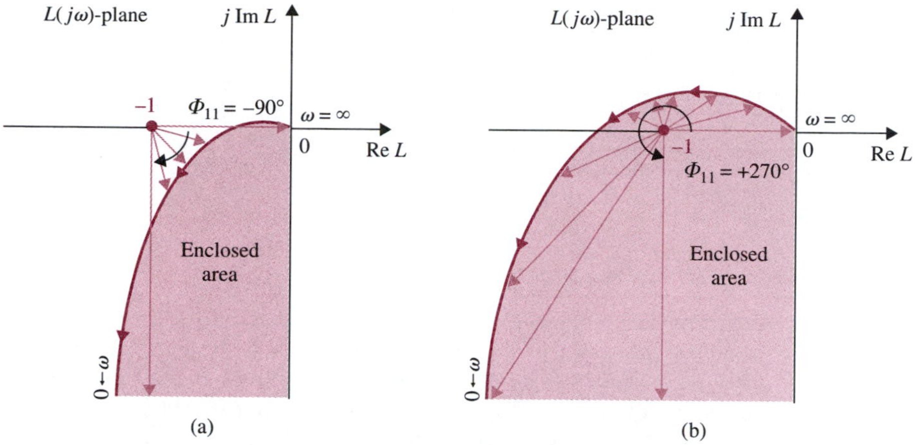

Figure 9-32 Nyquist plots of $L(s) = \dfrac{K}{s(s+2)(s+10)}$.
(a) $K < 240$ (b) $K > 240$.

entirely in this case. From Eq. (9-84), the requirement for closed-loop stability is

$$\Phi_{11} = -(0.5P_\omega + P)180° = -90° \tag{9-91}$$

Thus the stability criterion requires that the phasor drawn from the $(-1, j0)$ point to the Nyquist plot of $L(j\omega)$ should traverse $-90°$ as ω varies from ∞ to 0.

To sketch the Nyquist plot of $L(s)$ that corresponds to the positive portion of the $j\omega$-axis of the s-plane, we set $s = j\omega$ in Eq. (9-90). We get

$$L(j\omega) = \frac{K(j\omega - 1)}{j\omega(j\omega + 1)} = \frac{K(j\omega - 1)}{-\omega^2 + j\omega} \tag{9-92}$$

When $\omega = \infty$,

$$L(j\infty) = \left.\frac{K}{j\omega}\right|_{\omega=\infty} = 0\angle -90° \tag{9-93}$$

When $\omega = 0$,

$$L(j0) = \left.\frac{K}{j\omega}\right|_{\omega=0} = \infty\angle 90° \tag{9-94}$$

To find the intersect of the $L(j\omega)$ plot on the real axis, we rationalize the function by multiplying the numerator and the denominator of Eq. (9-92) by $-\omega^2 - j\omega$. We have

$$L(j\omega) = \frac{K(j\omega - 1)(-\omega^2 - j\omega)}{\omega^4 + \omega^2} = \frac{K[2\omega + j(1 - \omega^2)]}{\omega(\omega^2 + 1)} \tag{9-95}$$

Setting the imaginary part of $L(j\omega)$ to zero and solving for ω^2, we have

$$\omega^2 = \pm 1 \text{ rad/sec} \tag{9-96}$$

For $\omega = 1$,

$$L(j1) = K \tag{9-97}$$

Nyquist Plot for $K > 0$

Based on the information acquired above, the Nyquist plot of $L(j\omega)$ that corresponds to the positive portion of the $j\omega$-axis is sketched as shown in Fig. 9-33 for $K > 0$. Figure 9-33 shows that as ω varies from ∞ to 0 along the Nyquist plot, the net angle Φ_{11} traversed by the phasor drawn from the $(-1, j0)$ point to the Nyquist plot is $+90°$. Thus the system is unstable since Φ_{11} is positive. We can also see readily that the $(-1, j0)$ is enclosed by the Nyquist plot, and the same conclusion on closed-loop stability can be drawn.

From Eq. (9-82),

$$\Phi_{11} = (Z - 0.5P_\omega - P)180° = (Z - 0.5)180° = 90° \tag{9-98}$$

Thus $Z = 1$, which means that the characteristic equation of the closed-loop system has one root in the right-half s-plane. The characteristic equation of the system is

$$s^2 + (1 + K)s - K = 0 \tag{9-99}$$

We can easily verify that stability requires that

$$0 > K > -1 \tag{9-100}$$

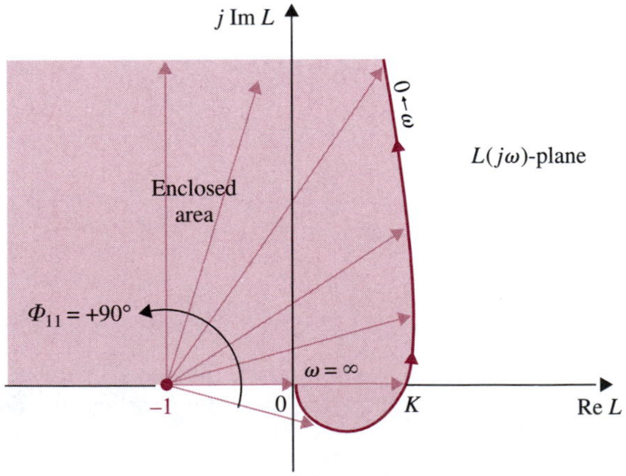

Figure 9-33 Nyquist plot of the system in Example 9-5.
$L(s) = K(s - 1)/[s(s + 1)]$

Nyquist Plot for $K < 0$

Figure 9-34(a) shows the Nyquist plot of $L(j\omega)$ when K lies between 0 and -1. Notice that the plot is obtained by *rotating* the $L(j\omega)$ plot of Fig. 9-33 by 180° about the origin. As ω is varied from ∞ to 0, the angle Φ_{11} in Fig. 9-34(a) has a net rotation of $-90°$, which agrees with the stability requirement in Eq. (9-91), and the closed-loop system is stable. It should be remembered that since $L(j\omega)$ is of the nonminimum-phase type, the fact that the Nyquist plot of $L(j\omega)$ of Fig. 9-34(a) does not enclose the $(-1, j0)$ is *not* the reason that the system is stable.

Figure 9-34(b) shows the Nyquist plot when $K < -1$. Now we see that the $(-1, j0)$ point is enclosed. Checking the value of Φ_{11}, we have $\Phi_{11} = 270°$, which differs from the required $-90°$, so that system is unstable. Using Eq. (9-82) yields

$$\Phi_{11} = (Z - 0.5P_\omega - P)180° = (Z - 0.5)180° = 270° \tag{9-101}$$

Thus $Z = 2$, which means that the characteristic equation has two roots in the right-half s-plane.

In general, when K changes sign, it is not necessary to redraw the Nyquist plot as shown in Fig. 9-34. Equation (9-40) can be written as

$$1 + L(s) = 1 + KL_1(s) = 0 \tag{9-102}$$

where K is positive. For negative K, Eq. (9-102) can be written as

$$1 - KL_1(s) = 0 \tag{9-103}$$

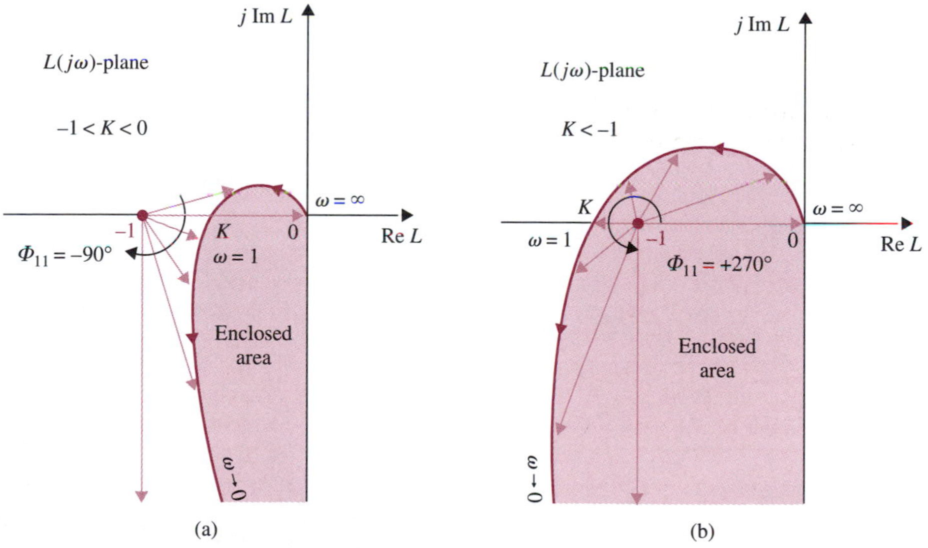

Figure 9-34 Nyquist plots of the system in Example 9-5. $L(s) = K(s - 1)/[s(s + 1)]$. (a) $-1 < K < 0$. (b) $K < -1$.

or

$$KL_1(s) = 1 \tag{9-104}$$

where K is now positive. Thus Eq. (9-104) shows that when K is negative, we can still use the $L(j\omega)$ plot for positive K, but designate the $(+1, j0)$ point as the critical point for stability analysis.

Figure 9-35 shows the Nyquist plot of Eq. (9-90) for $K > 0$. When K is negative, the $(+1, j0)$ point is regarded as the critical point. As shown in Fig. 9-35, for $-1 < K < 0$, Φ_{11} is $-90°$, which is the required value, and the system is stable. When $K < -1$, the $(+1, j0)$ point is enclosed by the Nyquist plot, and the system is unstable. These results agree with those obtained from Fig. 9-34 when the Nyquist plots for $K < 0$ were actually constructed.

It is of interest to compare the Nyquist stability analysis with the root-locus analysis. Figure 9-36 shows the root loci of the characteristic equation of the system with the loop transfer function given in Eq. (9-90). The stability condition of the system as a function of K is clearly indicated by the root-locus diagram. The RL between 0 and +1 on the real axis indicates that the system is unstable for $0 < K < \infty$. The CRL indicates that for negative values of K, the system is unstable for $-\infty < K < -1$, and the system is stable only for the range of $-1 < K < 0$. The CRL cross the $j\omega$-axis at $\omega = \pm 1$ rad/sec, which are the values of ω at which the Nyquist plot of $L(j\omega)$ intersects the negative real axis. ▲

Example 9-6

Consider the control system shown in Fig. 9-37. It is desired to determine the range of K for which the system is stable. The loop transfer function of the system is

$$L(s) = \frac{Y(s)}{E(s)} = \frac{10K(s+2)}{s^3 + 3s^2 + 10} \tag{9-105}$$

The poles of $L(s)$ are found to be at $s = -3.72$ $0.361 + j1.6$, and $0.361 - j1.6$. Or, we can use the Routh–Hurwitz criterion to verify that $L(s)$ has two poles in the right-half s-plane. Thus $P = 2$ and $P_\omega = 0$. The transfer function $L(s)$ is of the nonminimum-phase type. From Eq.

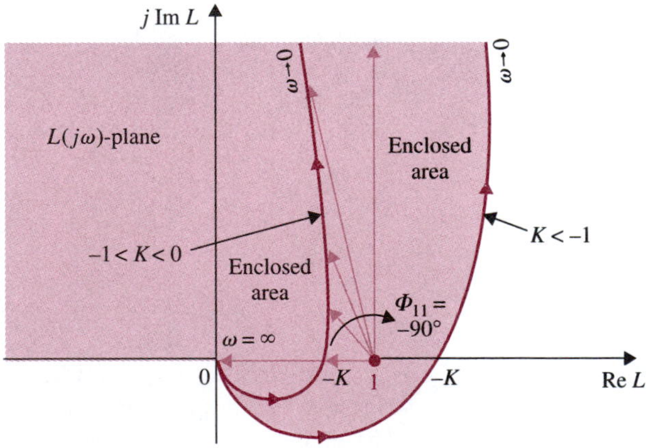

Figure 9-35 Nyquist plot of $L(s) = K(s-1)/[s(s+1)]$ with $K < 0$. The $(+1, j0)$ point is the critical point.

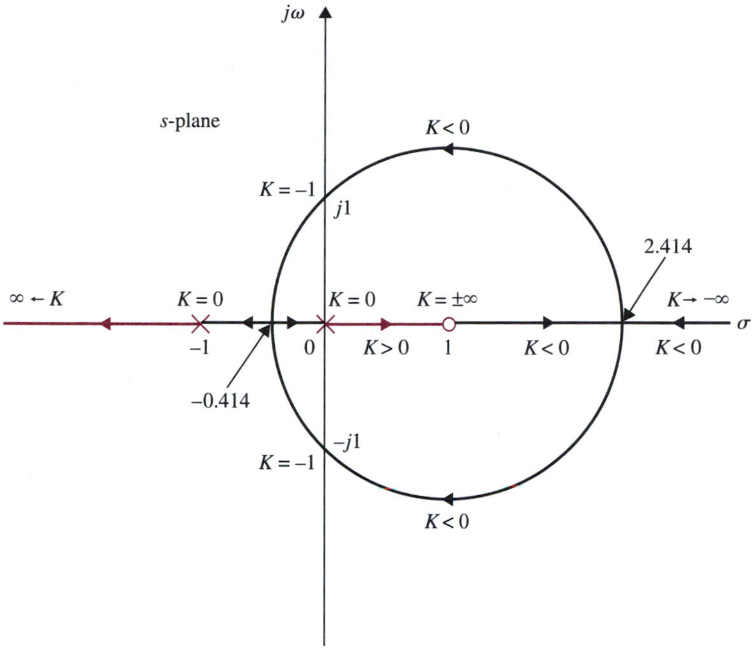

Figure 9-36 Complete root loci of system in Example 9-5.

(9-84), the requirement for the closed-loop system to be stable is

$$\Phi_{11} = -(0.5P_\omega + P)180^\circ = -360^\circ \tag{9-106}$$

Setting $s = j\omega$, Eq. (9-105) becomes

$$L(j\omega) = \frac{10K(j\omega + 2)}{(10 - 3\omega^2) - j\omega^3} \tag{9-107}$$

At $\omega = \infty$,

$$L(j\infty) = 0\angle 180^\circ \tag{9-108}$$

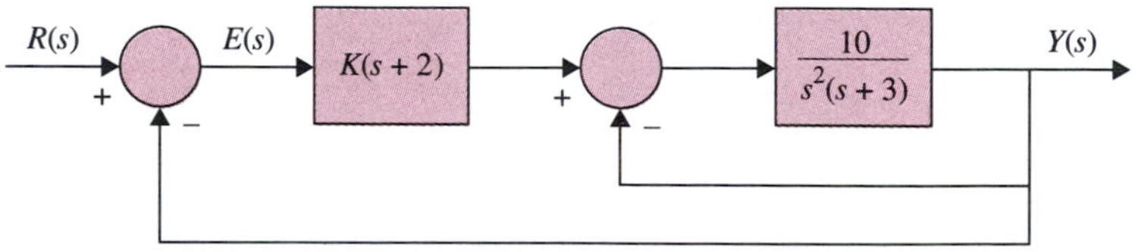

Figure 9-37 Block diagram of the control system in Example 9-6.

At $\omega = 0$,

$$L(j0) = 2K \tag{9-109}$$

To find the intersect on the real axis of the $L(j\omega)$-plane, we rationalize $L(j\omega)$ as

$$L(j\omega) = \frac{10K\{2(10 - 3\omega^2) - \omega^4 + j[\omega(10 - 3\omega^2) + 2\omega^3]\}}{(10 - 3\omega^2)^2 + \omega^6} \tag{9-110}$$

Setting the imaginary part of $L(j\omega)$ to zero, we have

$$\omega(10 - 3\omega^2) + 2\omega^3 = 0 \tag{9-111}$$

The solutions of Eq. (9-111) are $\omega = 0$ and $\omega = \pm\sqrt{10} = 3.16$ rad/sec, which are the frequencies at which the $L(j\omega)$ plot intersects the real axis of the $L(j\omega)$-plane. When $\omega = 0$, we already have $L(j0) = 2K$ in Eq. (9-109). When $\omega = 3.16$ rad/sec,

$$L(j3.16) = -K \tag{9-112}$$

Figure 9-38(a) shows the Nyquist plot of $L(j\omega)$ for $0 < K < 1$. Since the $(-1, j0)$ point is enclosed by the Nyquist plot, the closed-loop system is unstable. We can also show that the

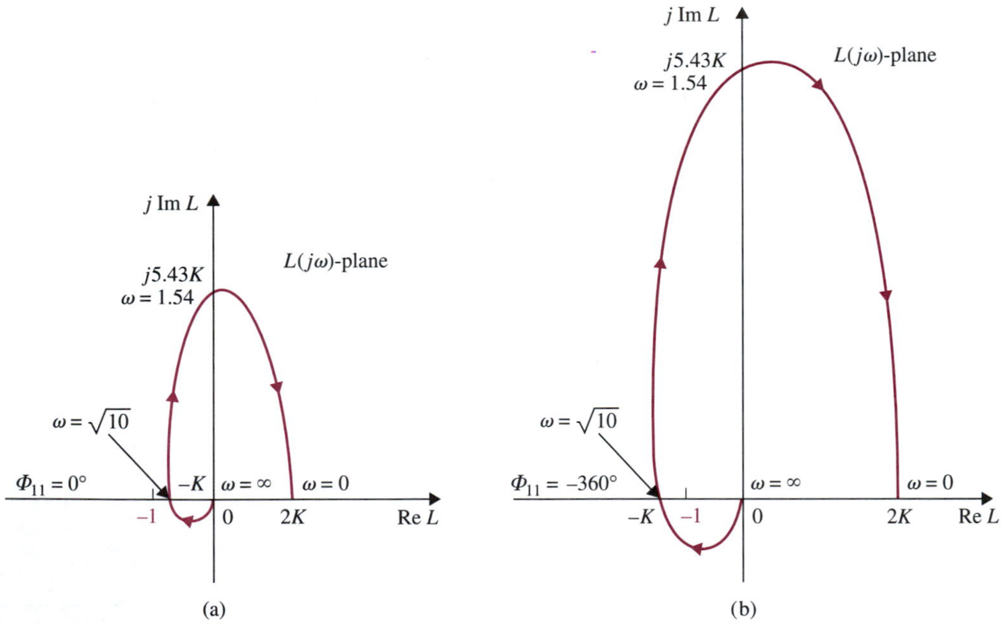

Figure 9-38 Nyquist plots of the system in Example 9-6. (a) $1 > K > 0$. (b) $K > 1$.

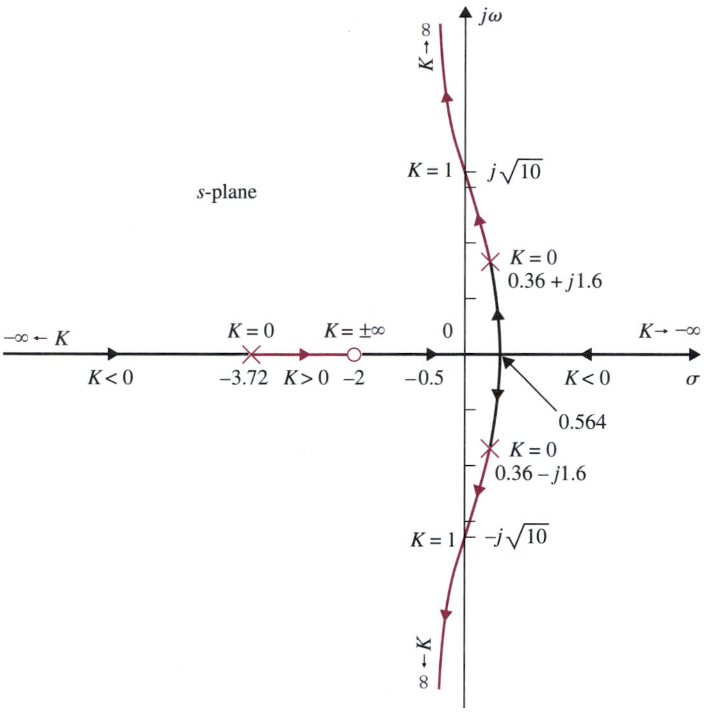

Figure 9-39 Complete root loci of the system in Example 9-6.

angle traversed by Φ_{11} is 0°, not −360°, as required in Eq. (9-106). Figure 9-38(b) shows the Nyquist plot of $L(j\omega)$ when K is greater than unity. In this case, the angle Φ_{11} rotates a total of −360°; thus the system is stable.

When K is negative, we can use the plots in Fig. 9-38, and regard the $(+1, j0)$ point as the critical point. The following stability conditions are observed:

▲ $2K < -1$: The $(+1, j0)$ point lies between 0 and $2K$, and is not enclosed by $L(j\omega)$, but $\Phi_{11} = -180°$; the system is unstable. For stability, Φ_{11} must equal −360°.

▲ $-1 < 2K < 0$: The $(+1, j0)$ point is to the right of the point $2K$, and is enclosed by the Nyquist plot; the system is unstable. In this case, $\Phi_{11} = 0°$.

The conclusion is that the system is stable for $K > 1$. The root loci of the system are shown in Fig. 9-39. Clearly, when K is negative, one branch of the CRL will always stay in the right-half plane, and the system is unstable. The system is stable only for $K > 1$, and the RL cross the $j\omega$-axis at $\omega = \pm 3.16$ rad/sec, which corresponds to the frequency at which the $L(j\omega)$ plot intersects the negative real axis. The value of K at the crossing point on the $j\omega$-axis is 1. ▲

Example 9-7

Consider a control system with the loop transfer function

$$L(s) = \frac{K}{(s+2)(s^2+4)} \tag{9-113}$$

which has a pair of imaginary poles at $s = j2$ and $-j2$. Thus $P_\omega = 2$ and $P = 0$. To apply the Nyquist criterion in the original form, we would have to define the Nyquist path with small indentation around the poles on the $j\omega$-axis.

Instead of constructing the entire Nyquist plot, the portion that corresponds to $s = j\infty$ to $j0$ is plotted as shown in Fig. 9-40. The data for this Nyquist plot are easily obtained using any of the frequency-domain programs mentioned earlier.

From Eq. (9-84), the value of Φ_{11} required for stability is

$$\Phi_{11} = -(0.5P_\omega + P)180° = -180° \tag{9-114}$$

As seen from Fig. 9-40, the magnitude of $L(j\omega)$ goes to infinity when $\omega = 2$ rad/sec. When K is positive, the critical point $(-1, j0)$ is not enclosed by the Nyquist plot, and the

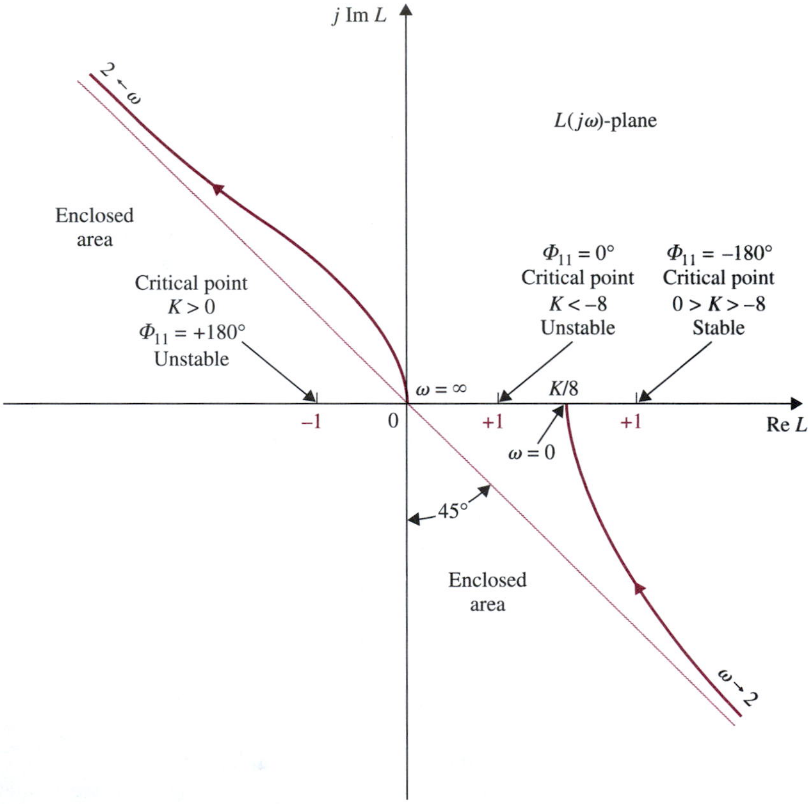

Figure 9-40 Nyquist plot of the control system in Example 9-7.

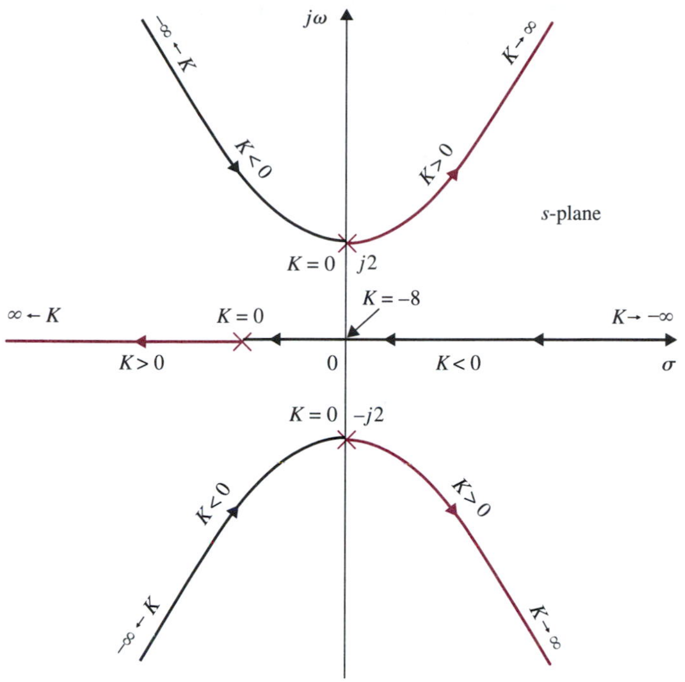

Figure 9-41 Root loci of the system in Example 9-7.

system is unstable. For the angle check, when ω varies from ∞ to 2, the angle Φ_{11} is $+135°$, and for the portion of $\omega = 2$ to 0, Φ_{11} is $+45°$. Thus the total Φ_{11} is $+180°$, not $-180°$. The system is unstable for all positive values of K.

When K is negative, the critical point in Fig. 9-40 is at $(+1, j0)$. Figure 9-40 shows that if the $(+1, j0)$ point lies between 0 and $K/8$, it is enclosed by the Nyquist path, and the system is unstable. Thus the system is unstable for $K < -8$. When the $(+1, j0)$ point is to the right of the $K/8$ point, Φ_{11} from $\omega = \infty$ to $\omega = 2$ is $-45°$, and from $\omega = 2$ to $\omega = 0$ is $-135°$. Thus the total Φ_{11} as ω varies from ∞ to 0 is $-180°$, which agrees with the value required in Eq. (9-114). The system is stable for $0 > K > -8$. The summary of the Nyquist criterion application to this system is as follows:

Range of K	Φ_{11} (deg) for $\omega = \infty$ to 2	Φ_{11} (deg) for $\omega = 2$ to 0	Total Φ_{11} (deg)	Critical Point	Stability Condition
$K > 0$	+135	+45	+180	−1 point enclosed	Unstable
$K < -8$	−45	+45	0	+1 point enclosed	Unstable
$-8 < K < 0$	−45	−135	−180		Stable

The root loci of the characteristic equation of the system are constructed in Fig. 9-41 using the pole–zero configuration of Eq. (9-113). The stability condition of $-8 < K < 0$ is easily arrived at from the root loci. ▲

9-11 Effects of Addition of Poles and Zeros to $L(s)$ on the Shape of the Nyquist Locus

Since the performance of a control system is often affected by adding and moving poles and zeros of the loop transfer function, it is important to investigate how the Nyquist plot is affected when poles and zeros are added to $L(s)$.

Let us begin with a first-order transfer function

$$L(s) = \frac{K}{1 + T_1 s} \tag{9-115}$$

where T_1 is a positive real constant. The Nyquist plot of $L(j\omega)$ for $0 \leq \omega \leq \infty$ is a semicircle, as shown in Fig. 9-42. The figure also shows the interpretation of the closed-loop stability with respect to the critical point for all values of K between $-\infty$ and ∞.

Addition of Poles at $s = 0$

Consider that a pole at $s = 0$ is added to the transfer function of Eq. (9-115); then

$$L(s) = \frac{K}{s(1 + T_1 s)} \tag{9-116}$$

Since adding a pole at $s = 0$ is equivalent to dividing $L(s)$ by $j\omega$, the phase of $L(j\omega)$ is reduced by 90° at both zero and infinite frequencies. In addition, the magnitude of $L(j\omega)$ at $\omega = 0$ becomes infinite. Figure 9-43 illustrates the Nyquist plot of $L(j\omega)$ in Eq. (9-116) and the closed-loop stability interpretations with respect to the critical points

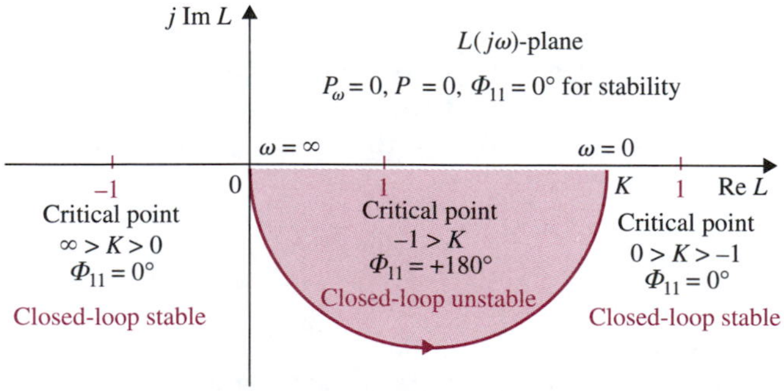

Figure 9-42 Nyquist plot of $L(s) = K/(1 + T_1 s)$.

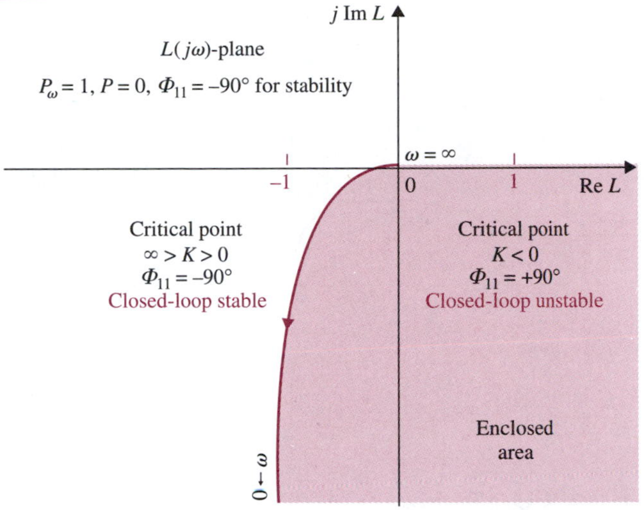

Figure 9-43 Nyquist plot of $L(s) = K/[s(1 + T_1 s)]$.

for $-\infty < K < \infty$. In general, adding a pole of multiplicity p at $s = 0$ to the transfer function of Eq. (9-115) will give the following properties to the Nyquist plot of $L(j\omega)$:

$$\lim_{\omega \to \infty} \angle L(j\omega) = -(p+1)90^\circ \tag{9-117}$$

$$\lim_{\omega \to 0} \angle L(j\omega) = -p \times 90^\circ \tag{9-118}$$

$$\lim_{\omega \to \infty} |L(j\omega)| = 0 \tag{9-119}$$

$$\lim_{\omega \to 0} |L(j\omega)| = \infty \tag{9-120}$$

The following example illustrates the effects of adding multiple-order poles to $L(s)$.

Example 9-8

Figure 9-44 shows the Nyquist plot of

$$L(s) = \frac{K}{s^2(1 + T_1 s)} \tag{9-121}$$

and the critical points, with stability interpretations. Figure 9-45 illustrates the same for

$$L(s) = \frac{K}{s^3(1 + T_1 s)} \tag{9-122}$$

▲

▲ Adding poles at $s = 0$ to a loop transfer function will reduce stability of the closed-loop system.

The conclusion from these illustrations is that the addition of poles at $s = 0$ to a loop transfer function will affect the stability of the closed-loop system adversely. A system that has a loop transfer function with more than one pole at $s = 0$ (type 2 or higher) is likely to be unstable or difficult to stabilize.

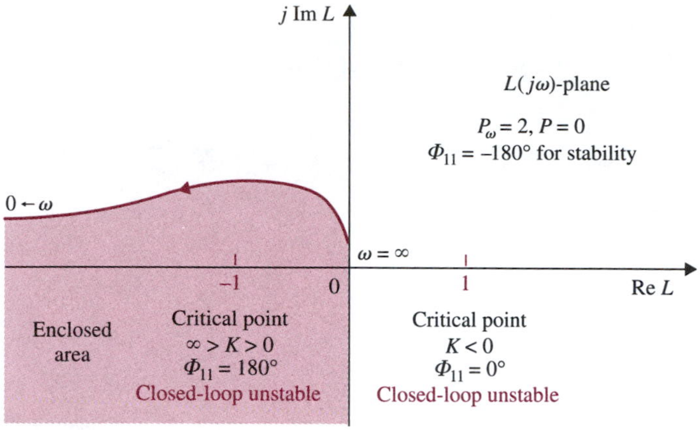

Figure 9-44 Nyquist plot of $L(s) = K/[s^2(1 + T_1 s)]$.

Addition of Finite Nonzero Poles

When a pole at $s = -1/T_2$ $(T_2 > 0)$ is added to the function $L(s)$ of Eq. (9-115), we have

$$L(s) = \frac{K}{(1 + T_1 s)(1 + T_2 s)} \tag{9-123}$$

The Nyquist plot of $L(j\omega)$ at $\omega = 0$ is not affected by the addition of the pole, since

$$\lim_{\omega \to 0} L(j\omega) = K \tag{9-124}$$

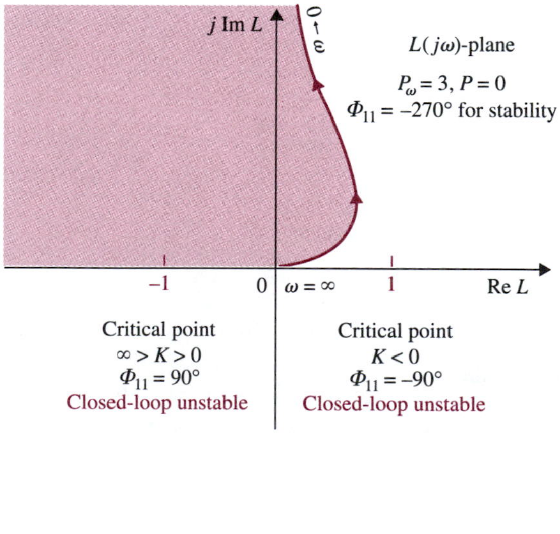

Figure 9-45 Nyquist plot of $L(s) = K/[s^3(1 + T_1 s)]$.

The value of $L(j\omega)$ at $\omega = \infty$ is

$$\lim_{\omega\to\infty} L(j\omega) = \lim_{\omega\to\infty} \frac{-K}{T_1 T_2 \omega^2} = 0 \angle -180° \tag{9-125}$$

Thus the effect of adding a pole at $s = -1/T_2$ to the transfer function of Eq. (9-115) is to shift the phase of the Nyquist plot by $-90°$ at $\omega = \infty$, as shown in Fig. 9-46. The figure also shows the Nyquist plot of

$$L(s) = \frac{K}{(1 + T_1 s)(1 + T_2 s)(1 + T_3 s)} \tag{9-126}$$

▲ Adding nonzero poles to the loop transfer function also reduce stability of the closed-loop system.

where two nonzero poles have been added to the transfer function of Eq. (9-115) (T_1, T_2, T_3, > 0). In this case, the Nyquist plot at $\omega = \infty$ is rotated clockwise by another 90° from that of Eq. (9-123). *These examples show the adverse effects on closed-loop stability when poles are added to the loop transfer function.* The closed-loop systems with the loop transfer functions of Eqs. (9-115) and (9-123) are all stable as long as K is positive. The system represented by Eq. (9-126) is unstable if the intersect of the Nyquist plot on the negative real axis is to the left of the $(-1, j0)$ point when K is positive.

Addition of Zeros

It was demonstrated in Chapter 6 that adding zeros to the loop transfer function has the effect of reducing the overshoot and the general effect of stabilization. In terms of the Nyquist criterion, this stabilization effect is easily demonstrated, since the multiplication of the term $(1 + T_d s)$ to the loop transfer function increases the phase of $L(s)$ by 90° at $\omega = \infty$. The following example illustrates the effect on stability of adding a zero at $-1/T_d$ to a loop transfer function.

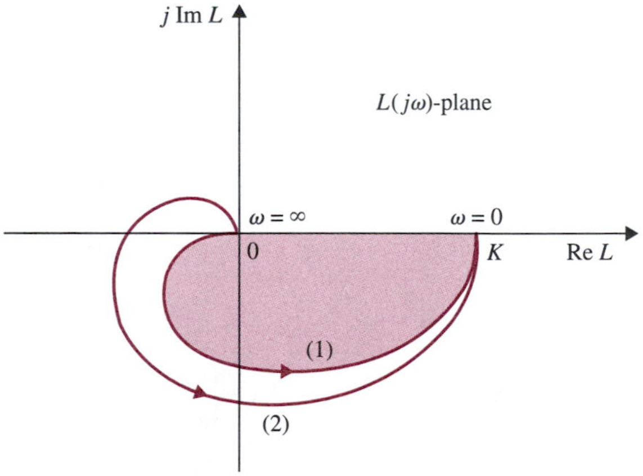

Figure 9-46 Nyquist plots. Curve (1): $L(s) = K/(1 + T_1 s)(1 + T_2 s)$. Curve (2): $L(s) = K/(1 + T_1 s)(1 + T_2 s)(1 + T_3 s)$.

Example 9-9

Consider that the loop transfer function of a closed-loop control system is

$$L(s) = \frac{K}{s(1 + T_1 s)(1 + T_2 s)} \tag{9-127}$$

It can be shown that the closed-loop system is stable for

$$0 < K < \frac{T_1 + T_2}{T_1 T_2} \tag{9-128}$$

Suppose that a zero at $s = -1/T_d$ ($T_d > 0$) is added to the transfer function of Eq. (9-127); then

$$L(s) = \frac{K(1 + T_d s)}{s(1 + T_1 s)(1 + T_2 s)} \tag{9-129}$$

▲ Adding zeros to the loop transfer function has the effect of stabilizing the closed-loop system.

The Nyquist plots of the two transfer functions of Eqs. (9-127) and (9-129) are shown in Fig. 9-47. The effect of the zero in Eq. (9-129) is to add 90° to the phase of the $L(j\omega)$ in Eq. (9-127) at $\omega = \infty$ while not affecting the value at $\omega = 0$. The intersect on the negative real axis of the $L(j\omega)$-plane is moved from $-KT_1T_2/(T_1 + T_2)$ to $-K(T_1T_2 - T_dT_1 - T_dT_2)/(T_1 + T_2)$. Thus the system with the loop transfer function in Eq. (9-129) is stable for

$$0 < K < \frac{T_1 + T_2}{T_1T_2 - T_d(T_1 + T_2)} \tag{9-130}$$

which for positive T_d and K has a higher upper bound than that of Eq. (9-128). ▲

9-12 Stability Analysis of Multiloop Systems

The Nyquist stability analyses conducted in the preceding sections are all directed toward the loop transfer function $L(s)$. It does not matter whether the system is with

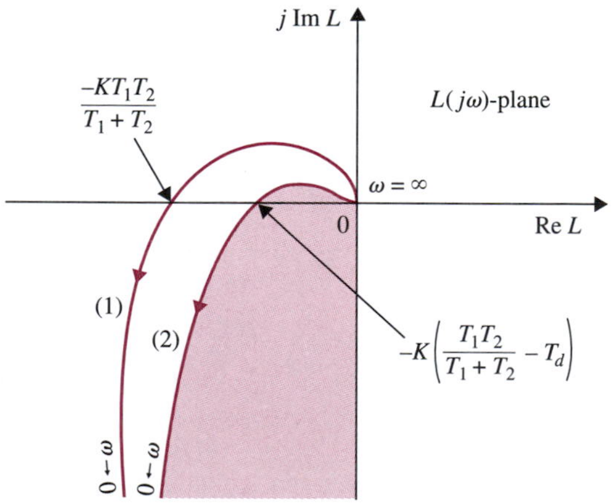

Figure 9-47 Nyquist plots. Curve (1): $L(s) = K/[s(1 + T_1 s)(1 + T_2 s)]$. Curve (2): $L(s) = K(1 + T_d s)/[s(1 + T_1 s)(1 + T_2 s)]$; $T_d < T_1, T_2$.

single loop or multiple loops, since once the loop transfer function is obtained, stability analysis can be conducted using either the Routh–Hurwitz criterion, root loci, or the Nyquist criterion.

For multiloop feedback systems, it may be advantageous to analyze the stability of the system by working from the inner loop toward the outer loop, one at a time. This way, more insight may be gained on the stability of the individual loops of the system. The following example will illustrate this approach.

Example 9-10

Figure 9-48 shows the block diagram of a system that controls the gun turret of a tank. During servicing the turret control system, the mechanic accidentally opened the outer loop of the system. With the power turned on, the gun turret went out of control, and finally flew apart. The purpose of this example is to show that it is inadequate to investigate just the stability of the overall system. In general, for a multiloop control system, one should conduct a systematic stability analysis of all the inner loops of the system. It is admissible to have unstable inner loops, as long as the overall system is stable. However, if such a situation exists, it is important to forewarn or take precautionary measure to prevent opening the loops during operation.

The loop transfer function of the inner loop is

$$G(s) = \frac{6}{s(s+1)(s+2)} \tag{9-131}$$

Figure 9-49 shows the Nyquist plot of $G(s)$. Since the plot intersects the real axis at the -1 point at $\omega = 1.414$ rad/sec, the inner loop is marginally stable. Therefore, if the outer loop of

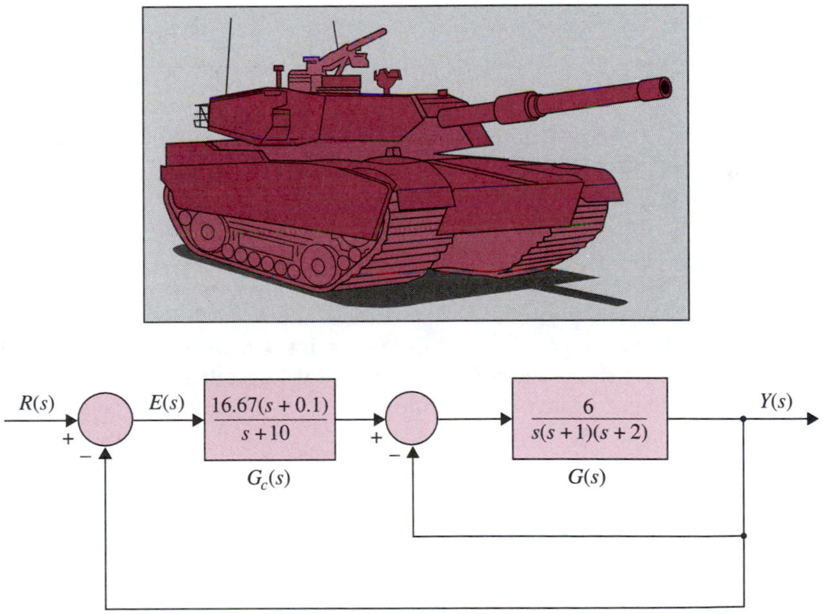

Figure 9-48 Multiloop feedback control system for tank turret control.

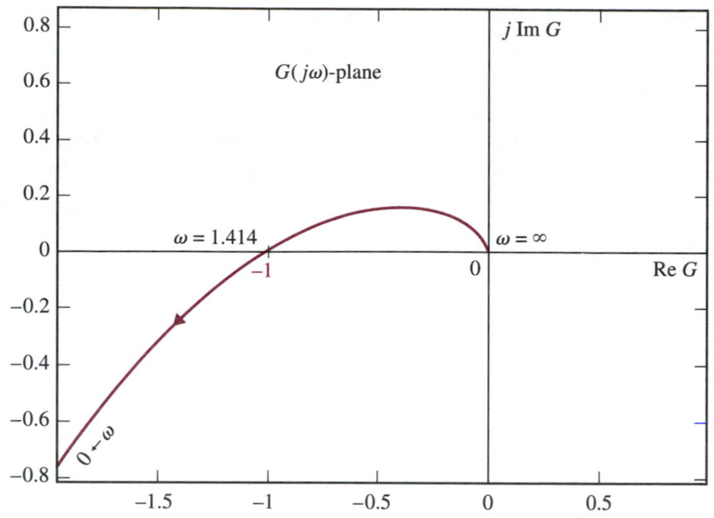

Figure 9-49 Nyquist plot of $G(s) = \dfrac{6}{s(s+1)(s+2)}$.

the system is opened, the system will oscillate continuously with a frequency of 1.414 rad/sec. The loop transfer function of the overall system is

$$L(s) = \frac{G_c(s)G(s)}{1 + G(s)} = \frac{100(s + 0.1)}{(s + 10)(s^3 + 3s^2 + 2s + 6)} \tag{9-132}$$

Since $L(s)$ has two poles on the $j\omega$-axis and the rest are in the left-half s-plane, $P_\omega = 2$ and $P = 0$. The function is also of the nonminimum-phase type, so we must use Eq. (9-84) for the stability test of the overall system. Thus

$$\Phi_{11} = -(0.5P_\omega + P)180° = -180° \tag{9-133}$$

The Nyquist plot of $L(s)$ is plotted as shown in Fig. 9-50. The angle Φ_{11} for $\omega = \infty$ to $\omega =$ 1.414 rad/sec is $-90°$, and from $\omega = 1.414$ rad/sec to $\omega = 0$ is $-90°$. Thus the total value of Φ_{11} for $\omega = \infty$ to $\omega = 0$ is $-180°$, and the overall system is stable. In general, when more than two loops are involved, the proper way is to start with the stability of the innermost loop by opening all the outer loops and then add one loop at a time until the outermost loop is closed. ▲

9-13 Stability of Linear Control Systems with Pure Time Delays

Systems with time delays and their modeling were discussed in Section 4-8. Closed-loop systems with time delays in the loops generally will be subject to more stability

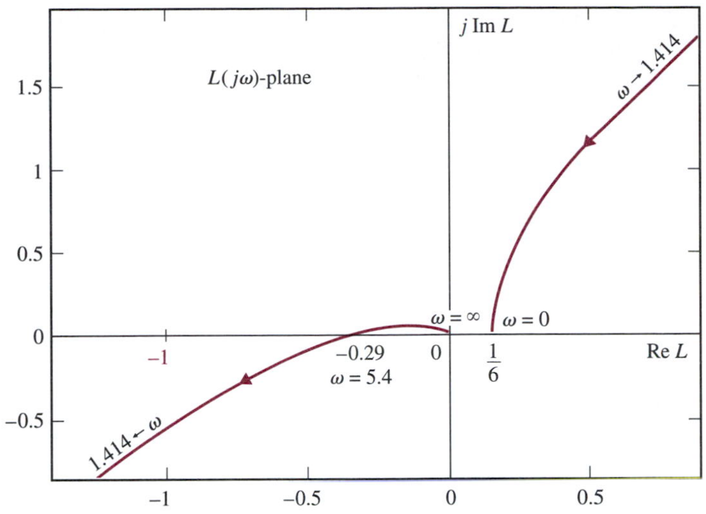

Figure 9-50 Nyquist plot of $L(s) = \dfrac{100(s+0.1)}{(s+10)(s^3+3s^2+2s+6)}$.

▲ Routh-Hurwitz criterion is not applicable to systems with time delays.

problems than systems without delays. Since a pure time delay T_d is modeled by the transfer function e^{-sT_d}, the characteristic equation of the system will no longer have constant coefficients. Therefore, the Routh–Hurwitz criterion is *not* applicable. The root-locus method can be applied to systems with pure time delays, but the construction of the root loci is quite complex. In this section we show that the Nyquist criterion can be readily applied to a system with a pure time delay.

Let us consider that the loop transfer function of a control system with pure time delay is expressed in a form similar to Eq. (9-38), or

$$L(s) = L_1(s)e^{-T_d s} \tag{9-134}$$

where $L_1(s)$ is a rational function of s with constant coefficients, and T_d is the pure time delay in seconds.

The stability of the system can be investigated by constructing the Nyquist plot of $L(s)$ and observe its behavior with respect to the $(-1, j0)$ point. The effect of the exponential term in Eq. (9-134) is that it rotates the phasor $L_1(j\omega)$ at each ω by an angle ωT_d radians in the clockwise direction. The amplitude of $L_1(j\omega)$ is not affected by the time delay, since the amplitude of $e^{-j\omega T_d}$ is unity for all frequencies.

Most control systems are of type 1 or higher, the magnitude of $L(j\omega)$ usually approaches zero as ω approaches infinity. Thus the Nyquist plot of the transfer function of Eq. (9-134) will usually spiral toward the origin in the clockwise direction as ω approaches infinity, and there is an infinite number of intersects on the real axis of the $L(j\omega)$-plane. Once the Nyquist plot of the $L(j\omega)$ is constructed, the stability of the system is determined in the usual manner by investigating the angle Φ_{11}. The following example illustrates the stability analysis of a closed-loop system with pure time delay using the Nyquist criterion.

Example 9-11

Figure 9-51 shows the Nyquist plots of

$$L(s) = L_1(s)e^{-T_d s} = \frac{e^{-T_d s}}{s(s+1)(s+2)} \tag{9-135}$$

for several values of T_d. These Nyquist plots show that when T_d is zero, the system is stable. The stability condition deteriorates as T_d increases, since the intersect on the real axis by the $L(j\omega)$ moves closer to the $(-1, j0)$ point. When $T_d = 2.09$ sec, the $L(j\omega)$ plot passes through

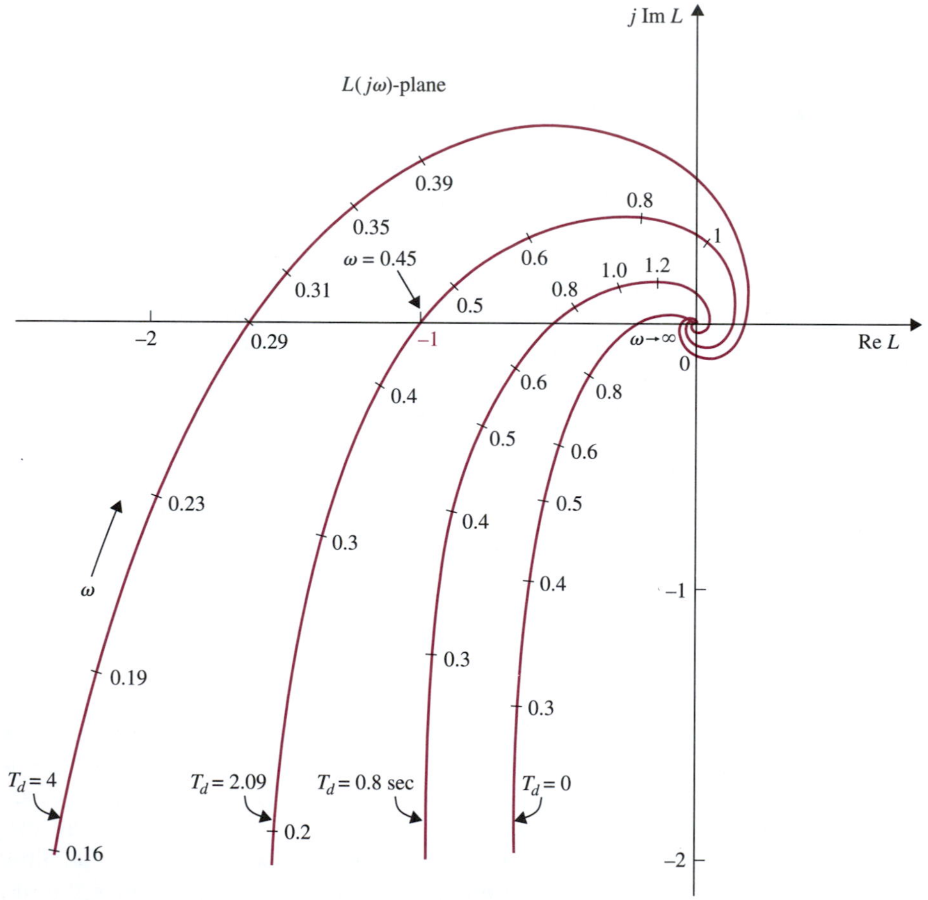

Figure 9-51 Nyquist plot of $L(s) = e^{-T_d s}/[s(s+1)(s+2)]$.

▲ Pure time delay usually deteriorates the stability of closed-loop systems.

the $(-1, j0)$ point, and the system is marginally stable. The system is unstable when T_d is greater than 2.09, since the $(-1, j0)$ point would be enclosed, or Φ_{11} would be positive.

Unlike the rational-function case, the analytical solution of the intersects on the real axis of the $L(j\omega)$-plane is not trivial, since the equations that govern the intersects are no longer algebraic. To demonstrate this, the transfer function of Eq. (9-135) is rationalized in the usual manner by multiplying its numerator and denominator by the complex conjugate of the denominator. The result is

$$L(j\omega) = \frac{(\cos \omega T_d - j \sin \omega T_d)[-3\omega^2 - j\omega(2-\omega^2)]}{9\omega^4 + \omega^2(2-\omega^2)^2} \tag{9-136}$$

The condition of $L(j\omega)$ intersecting the real axis is obtained by setting the imaginary part of Eq. (9-136) to zero.

$$3\omega^2 \sin \omega T_d - \omega(2-\omega^2)\cos \omega T_d = 0 \tag{9-137}$$

which is not easily solved for ω, given T_d. ▲

9-13-1 Critical Trajectory

Thus far we have used either the $(-1, j0)$ point or the $(+1, j0)$ point of the $L(j\omega)$-plane as the critical point for stability analysis when the gain K is either positive or negative, respectively. Under certain conditions, we can relax the critical point idea by extending it into a trajectory.

Referring to Eq. (9-134), we recognize that the roots of the characteristic equation satisfy

$$L_1(s)e^{-T_d s} = -1 \tag{9-138}$$

The right-hand side of the last equation points to the fact that the $(-1, j0)$ is the critical point for stability analysis of the closed-loop system. Equation (9-138) can be written as

$$L_1(s) = -e^{T_d s} \tag{9-139}$$

When $s = j\omega$, the left-hand side of the last equation gives the Nyquist plot of the loop transfer function when there is no time delay. The exponential term in Eq. (9-139) has a magnitude of one for all values of ω, and its phase is $-\omega T_d$ radians. Thus the right-hand side of Eq. (9-139) describes a **critical trajectory,** which is a circle with unit radius, and centered at the origin of the $L_1(j\omega)$-plane. When $\omega = 0$, the trajectory starts at the point $(-1, j0)$, as ω increases, the critical point traces out the unit circle in the CCW direction. The following example illustrates the application of the Nyquist criterion to the stability study of a closed-loop system with a pure time delay.

Example 9-12

Given the loop transfer function of a closed-loop control system with a pure time delay,

$$L(s) = L_1(s)e^{-T_d s} = \frac{e^{-T_d s}}{s(s+1)(s+2)} \tag{9-140}$$

The problem is to find the marginal value of the time delay T_d so that the system is stable.

Figure 9-52 shows the Nyquist plot of $L_1(j\omega)$ together with the critical trajectory $-e^{j\omega T_d}$. The frequency at which the $L_1(j\omega)$ plot intersects the critical trajectory is found by setting the magnitude of $L_1(j\omega)$ to unity; that is,

$$|L_1(j\omega)| = \frac{1}{|-3\omega^2 + j\omega(2-\omega^2)|} = 1 \tag{9-141}$$

Solving Eq. (9-141) yields the positive solution for ω to be 0.446 rad/sec, which is the frequency at which the Nyquist plot of $L(j\omega)$, such as those shown in Fig. 9-51, intersects the $(-1, j0)$ point in the $L(j\omega)$-plane. Since the angle in radians measured from the $(-1, j0)$ point to the point of intersect of the $L_1(j\omega)$ plot and the critical trajectory is ωT_d, where $\omega = 0.446$ rad/sec, we can find the critical value of T_d for stability by equating

$$\angle L_1(j0.446) = 0.446T_d + \pi \quad \text{rad} \tag{9-142}$$

or

$$-\tan^{-1}\left(\frac{2-\omega^2}{-3\omega}\right)\Bigg|_{\omega=0.446} = 0.446T_d + \pi \quad \text{rad} \tag{9-143}$$

which leads to

$$0.932 \text{ rad} = 0.446T_d \quad \text{rad} \tag{9-144}$$

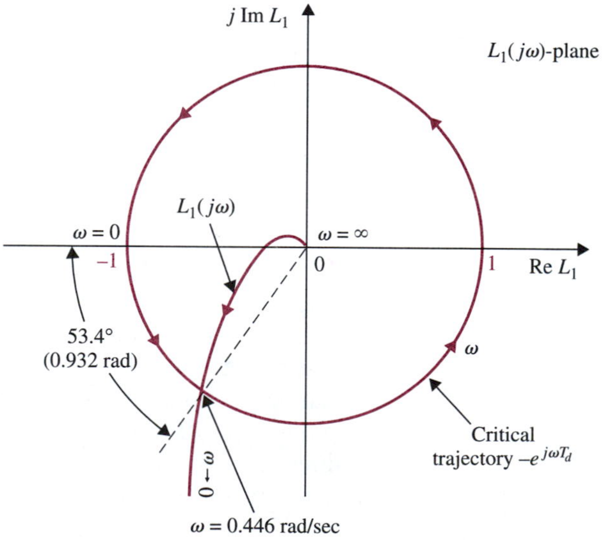

Figure 9-52 Nyquist plot of $L_1(s) = 1/[s(s+1)(s+2)]$ and the critical trajectory $-e^{j\omega T_d}$

Solving for T_d from Eq. (9-144), we get the critical value of the time delay to be 2.09 sec, which agrees with the result shown in Fig. 9-51 using the Nyquist plot of $L(j\omega)$. ▲

Although the method just presented still involves the calculation of the properties of the intersection between the Nyquist plot of $L_1(j\omega)$ and the critical trajectory for the time delay, it does provide another perspective on the application of the Nyquist criterion to systems with pure time delays. The valuable lesson learned is that the critical point for stability analysis can be generalized to a trajectory whenever the particular situation requires. Computer programs such as **freqrp** and **plrplot** all can handle the frequency-domain plots of systems with pure time delays with ease.

9-13-2 Approximation of $e^{-T_d s}$

One of the advantages of the frequency-domain analysis is that systems with pure time delays in the loop can be conducted by plotting the frequency response of the loop transfer function directly with the aid of a computer. However, for analysis and design techniques that require a rational function, the series Padé approximation of the time delay described in Section 4-8 can be applied. We can use the frequency-domain plot to evaluate the accuracy of these approximations. An illustrative example follows.

Example 9-13

By using the series approximation of $e^{-T_d s}$ given in Eq. (4-170), Eq. (9-135) becomes

$$L(s) = \frac{1}{s(s+1)(s+2)(1 + T_d s + T_d^2 s^2/2)} \tag{9-145}$$

Or by using the Padé approximation in Eq. (4-171), Eq. (9-135) becomes

$$L(s) = \frac{1 - T_d s/2}{s(s+1)(s+2)(1 + T_d s/2)} \tag{9-146}$$

Figure 9-53 shows the polar plots of Eqs. (9-135), (9-145), and (9-146) for $T_d = 0.8$. The intersects of the polar plots of Eqs. (9-135), (9-145), and (9-146) are at -0.5164, -0.525, and -0.5076, respectively. ▲

9-14 Relative Stability: Gain Margin and Phase Margin

▲ Relative stability is used to indicate how stable a system is.

We have demonstrated in Sections 9-2 through 9-4 the general relationship between the resonance peak M_r of the frequency response and the maximum overshoot of the time response. Comparisons and correlations between frequency-domain and time-domain parameters such as these are useful in prediction of the performance of control systems. In general, we are interested not only in the absolute stability of a system, but also how stable it is. The latter is often called **relative stability.** In the time domain, relative stability is measured by parameters such as maximum overshoot and damping

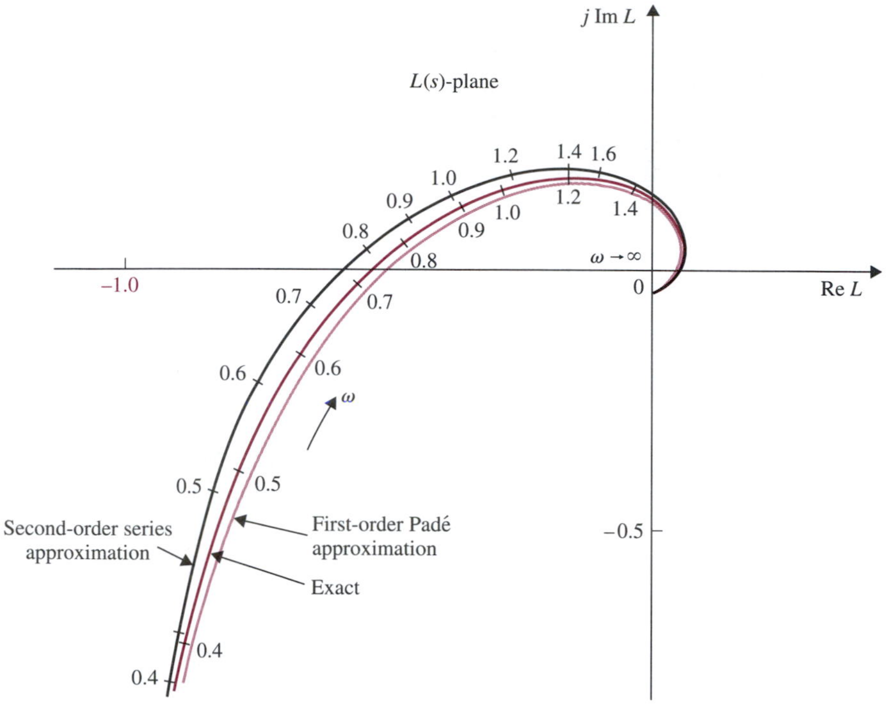

Figure 9-53 Approximation of Nyquist plot of Figure 9-51 by a two-term power-series approximation; $T_d = 0.8$ sec.

ratio. In the frequency domain, the resonance peak M_p can be used to indicate relative stability. Another way of measuring relative stability in the frequency domain is by how close the Nyquist plot of $L(j\omega)$ is to the $(-1, j0)$ point.

To demonstrate the concept of relative stability in the frequency domain, the Nyquist plots and the corresponding step responses and frequency responses of a typical third-order system are shown in Fig. 9-54 for four different values of loop gain K. It is assumed that the function $L(j\omega)$ is of minimum-phase type, so that the enclosure of the $(-1, j0)$ point is sufficient for stability analysis. The four cases are evaluated as follows.

1. **Figure 9-54(a); the loop gain K is low.** The Nyquist plot of $L(j\omega)$ intersects the negative real axis at a point that is quite far to the right of the $(-1, j0)$ point. The corresponding step response is quite well damped, and the value of M_r of the frequency response is low.
2. **Figure 9-54(b); K is increased.** The intersect is moved closer to the $(-1, j0)$ point; the system is still stable, since the critical point is not enclosed, but the step response has a larger maximum overshoot, and M_r is also larger.

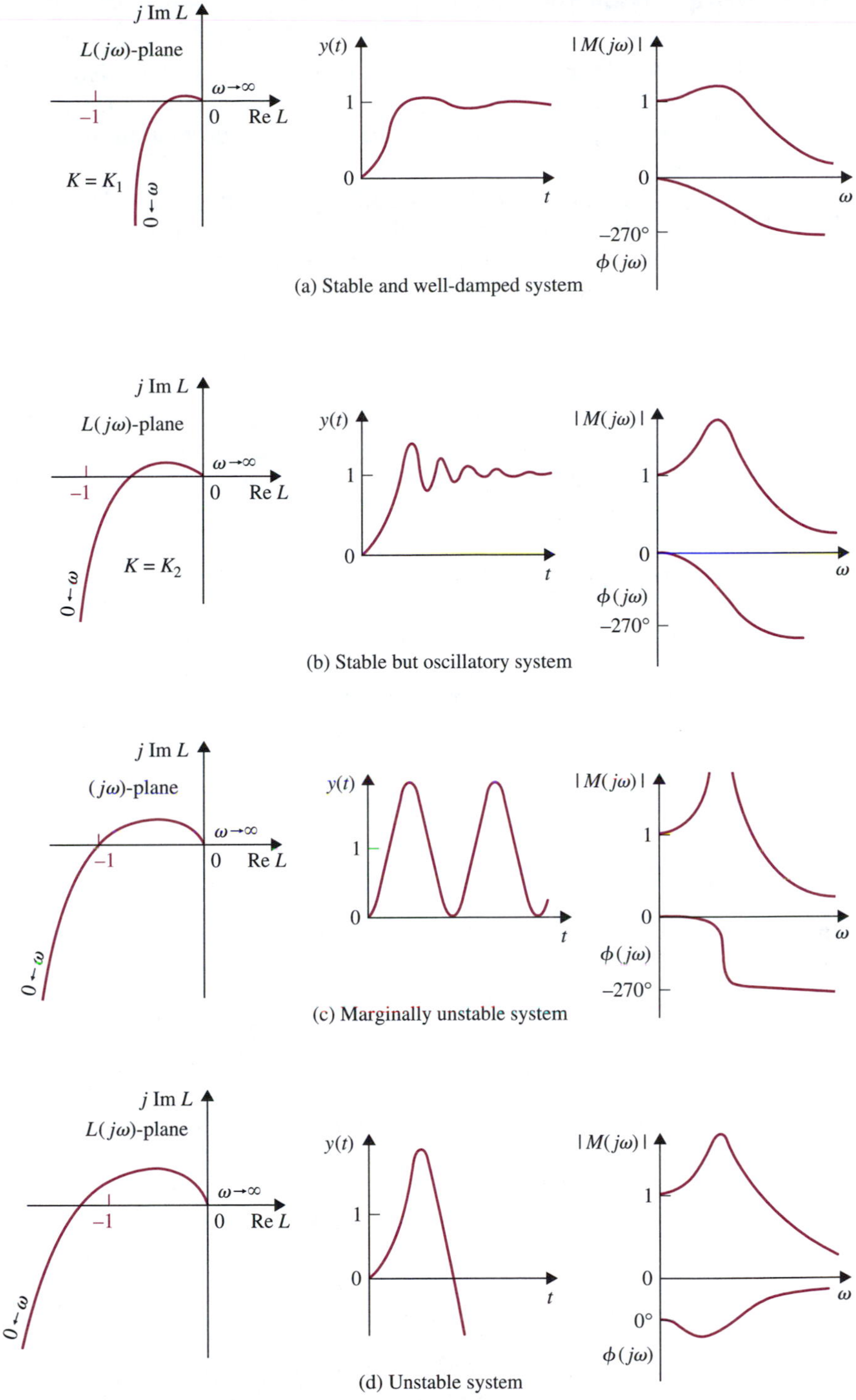

Figure 9-54 Correlation among Nyquist plots, step responses, and frequency responses.

3. **Figure 9-54(c); *K* is increased further.** The Nyquist plot now passes through the $(-1, j0)$ point, and the system is marginally stable. The step response becomes oscillatory with constant amplitude, and M_r becomes infinite.

▲ M_r ceases to have any meaning when the closed-loop system is unstable.

4. **Figure 9-54(d); *K* relatively very large.** The Nyquist plot now encloses the $(-1, j0)$ point, and the system is unstable. The step response becomes unbounded. The magnitude curve of $|M(j\omega)|$ versus ω ceases to have any significance. In fact, for the unstable system, the value of M_r is still finite!

In all the analysis above, the phase curve $\phi(j\omega)$ of the closed-loop frequency response also gives qualitative information about stability. Notice that the negative slope of the phase curve becomes steeper as the relative stability decreases. When the system is unstable, the slope beyond the resonant frequency becomes positive. In practice, the phase characteristics of the closed-loop system are seldom used for analysis and design purposes.

9-14-1 Gain Margin

Gain margin (GM) is one of the most frequently used criteria for measuring the relative stability of control systems. In the frequency domain, gain margin is used to indicate the closeness of the intersection of the negative real axis made by the Nyquist plot of $L(j\omega)$ to the $(-1, j0)$ point. Before giving the definition of gain margin, let us first define the **phase crossover** on the Nyquist plot and the **phase-crossover frequency.**

▲ The definition of gain margin given here is for minimum-phase loop transfer functions.

Phase Crossover. A phase crossover on the $L(j\omega)$ plot is a point at which the plot intersects the negative real axis.

Phase-Crossover Frequency. The **phase-crossover frequency** ω_p is the frequency at the phase crossover, or where

$$\angle L(j\omega_p) = 180° \tag{9-147}$$

▲ Gain margin is measured at the phase crossover.

The Nyquist plot of a loop transfer function $L(j\omega)$ that is of minimum-phase type is shown in Fig. 9-55. The phase-crossover frequency is denoted as ω_p, and the magnitude of $L(j\omega)$ at $\omega = \omega_p$ is designated as $|L(j\omega_p)|$. Then the gain margin of the closed-loop system that has $L(s)$ as its loop transfer function is defined as

$$\begin{aligned} \text{gain margin} = \text{GM} &= 20\log_{10}\frac{1}{|L(j\omega_p)|} \\ &= -20\log_{10}|L(j\omega_p)| \quad \text{dB} \end{aligned} \tag{9-148}$$

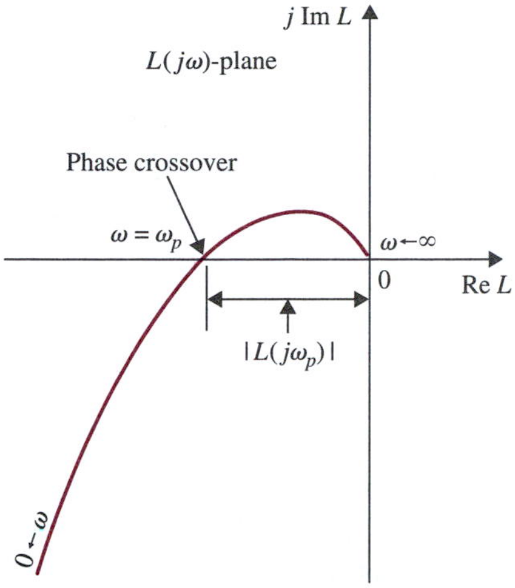

Figure 9-55 Definition of the gain margin in the polar coordinates.

On the basis of this definition, we can draw the following conclusions about the gain margin of the system shown in Fig. 9-55, depending on the properties of the Nyquist plot.

1. The $L(j\omega)$ plot does not intersect the negative real axis (no finite nonzero phase crossover).

$$|L(j\omega_p)| = 0 \qquad \text{GM} = \infty \text{ dB} \tag{9-149}$$

2. The $L(j\omega)$ plot intersects the negative real axis between (phase crossover lies between) 0 and the -1 point.

$$0 < |L(j\omega_p)| < 1 \qquad \text{GM} > 0 \text{ dB} \tag{9-150}$$

3. The $L(j\omega)$ plot passes through (phase crossover is at) the $(-1, j0)$ point.

$$|L(j\omega_p)| = 1 \qquad \text{GM} = 0 \text{ dB} \tag{9-151}$$

4. The $L(j\omega)$ plot encloses (phase crossover is to the left of) the $(-1, j0)$ point.

$$|L(j\omega_p)| > 1 \qquad \text{GM} < 0 \text{ dB} \tag{9-152}$$

Based on the foregoing discussions, the physical significance of gain margin can be summarized as:

▲ Gain margin is the amount of gain in dB that can be added to the loop before the closed-loop system becomes unstable.

Gain margin is the amount of gain in decibels (dB) that can be added to the loop before the closed-loop system becomes unstable.

- ▲ When the Nyquist plot does not intersect the negative real axis at any finite nonzero frequency, the gain margin is infinite in dB; this means that, theoretically, the value of the loop gain can be increased to infinity before instability occurs.
- ▲ When the Nyquist plot of $L(j\omega)$ passes through the $(-1, j0)$ point, the gain margin is 0 dB, which implies that the loop gain can no longer be increased, as the system is at the margin of instability.
- ▲ When the phase crossover is to the left of the $(-1, j0)$ point, the phase margin is negative in dB, and the loop gain must be reduced by the gain margin to achieve stability.

Gain Margin of Nonminimum-Phase Systems

Care must be taken when attempting to extend gain margin as a measure of relative stability to systems with nonminimum-phase loop transfer functions. For such systems a system may be unstable even when the phase-crossover point is to the right of $(-1, j0)$, and thus a positive gain margin may still correspond to an unstable system. Nevertheless, the closeness of the phase crossover to the $(-1, j0)$ point still gives an indication of relative stability.

9-14-2 Phase Margin

The gain margin is only a one-dimensional representation of the relative stability of a closed-loop system. As the name implies, gain margin indicates system stability with respect to the variation in loop gain only. In principle, one would believe that a system with a large gain margin should always be relatively more stable than one with a smaller gain margin. Unfortunately, gain margin alone is inadequate to indicate relative stability when system parameters other than the loop gain are subject to variation. For instance, the two systems represented by the $L(j\omega)$ plots in Fig. 9-56 apparently have the same gain margin. However, locus A actually corresponds to a more stable system than locus B, since with any change in the system parameters that affect the phase of $L(j\omega)$, locus B may easily be altered to enclose the $(-1, j0)$ point. Furthermore, we can show that system B actually has a larger M_r than system A.

To include the effect of phase shift on stability, we introduce the **phase margin (PM),** which requires that we first make the following definitions:

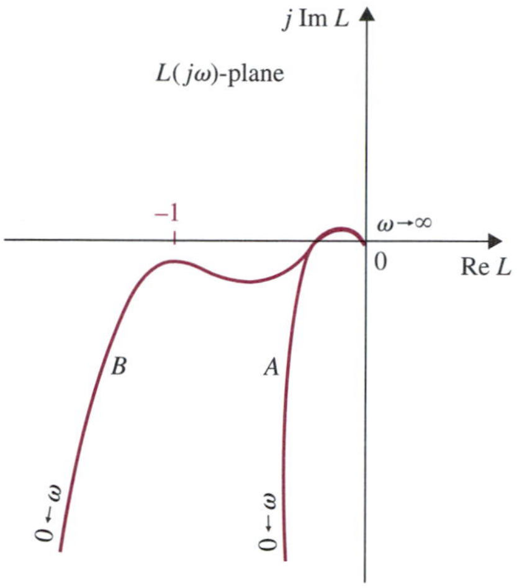

Figure 9-56 Nyquist plots showing systems with the same gain margin but different degrees of relative stability.

Gain Crossover. The gain crossover is a point on the $L(j\omega)$ plot at which the magnitude of $L(j\omega)$ is equal to 1.

Gain-Crossover Frequency. The gain-crossover frequency, ω_g is the frequency of $L(j\omega)$ at the gain crossover, or where

$$|L(j\omega_g)| = 1 \tag{9-153}$$

▲ The definition of phase margin given here is for a system with a minimum-phase loop transfer function.

▲ Phase margin is measured at the gain crossover.

The definition of phase margin is stated as:

Phase margin (PM) is defined as the angle in degrees through which the $L(j\omega)$ plot must be rotated about the origin so that the gain crossover passes through the $(-1, j0)$ point.

Figure 9-57 shows the Nyquist plot of a typical minimum-phase $L(j\omega)$ plot, and the phase margin is shown as the angle between the line that passes through the gain crossover and the origin and the negative real axis of the $L(j\omega)$-plane. In contrast to the gain margin, which is determined by loop gain, phase margin indicates the effect on system stability due to changes in system parameter, which theoretically alter the phase of $L(j\omega)$ by an equal amount at all frequencies. Phase margin is the amount of pure phase delay that can be added to the loop before the closed-loop system becomes unstable.

When the system is of minimum-phase type, the analytical expression of the phase margin, as seen from Fig. 9-57, can be expressed as

$$\text{phase margin (PM)} = \angle L(j\omega_g) - 180^\circ \tag{9-154}$$

where ω_g is the gain-crossover frequency.

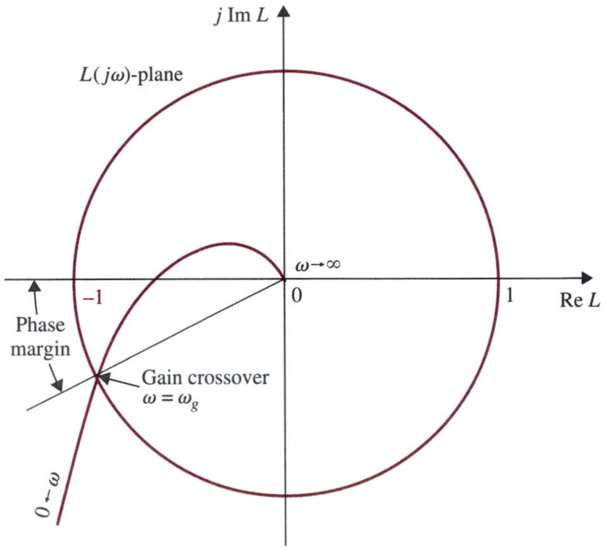

Figure 9-57 Phase margin defined in the $L(j\omega)$-plane.

Phase Margin of Nonminimum-Phase Systems

Care should be taken when interpreting the phase margin from the Nyquist plot of a nonminimum-phase transfer function. When the loop transfer function is of the nonminimum-phase type, the gain crossover can occur in any quadrant of the $L(j\omega)$-plane, and the definition of phase margin given in Eq. (9-154) is no longer valid.

Example 9-14

As an illustrative example on gain and phase margins, consider that the loop transfer function of a control system is

$$L(s) = \frac{2500}{s(s+5)(s+50)} \tag{9-155}$$

The Nyquist plot of $L(j\omega)$ is shown in Fig. 9-58. The following results are obtained from the Nyquist plot:

$$\text{Gain crossover } \omega_g = 6.22 \text{ rad/sec}$$
$$\text{Phase crossover } \omega_p = 15.88 \text{ rad/sec}$$

▲ Phase margin is the amount of pure phase lag that can be added before the system becomes unstable.

The gain margin is measured at the phase crossover. The magnitude of $L(j\omega_p)$ is 0.182. Thus the gain margin is obtained from Eq. (9-148),

$$\text{GM} = 20\log_{10}\frac{1}{|L(j\omega_p)|} = 20\log_{10}\frac{1}{0.182} = 14.82 \text{ dB} \tag{9-156}$$

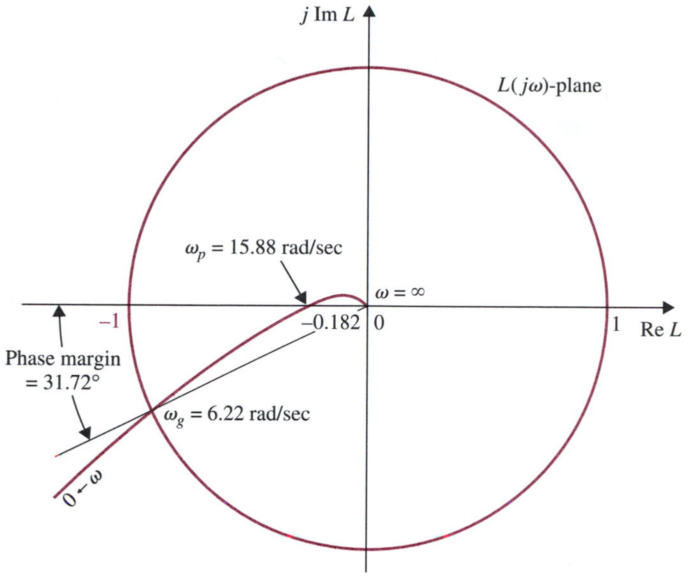

Figure 9-58 Nyquist plot of $L(s) = \dfrac{2500}{s(s+5)(s+50)}$.

The phase margin is measured at the gain crossover. The phase of $L(j\omega_g)$ is 211.72°. Thus the phase margin is obtained from Eq. (9-154),

$$PM = \angle L(j\omega_g) - 180^\circ = 211.72^\circ - 180^\circ = 31.72^\circ \tag{9-157}$$

▲

9-15 Stability Analysis with the Bode Plot

The Bode plot of a transfer function described in Appendix A is a very useful graphical tool for the analysis and design of linear control systems. Before the inception of computers, Bode plots were often called "asymptotic plots" because the magnitude and phase curves can be sketched from their asymptotic properties without detailed plotting. Modern applications of the Bode plot for control systems should be identified with the following advantages and disadvantages:

ADVANTAGES OF THE BODE PLOT

1. In the absence of a computer, a Bode diagram can be sketched by approximating the magnitude and phase with straight-line segments.

2. Gain crossover, phase crossover, gain margin, and phase margin are more easily determined on the Bode plot than from the Nyquist plot.
3. For design purposes, the affects of adding controllers and their parameters are more easily visualized on the Bode plot than on the Nyquist plot.

DISADVANTAGE OF THE BODE PLOT

Absolute and relative stability of only minimum-phase systems can be determined from the Bode plot. For instance, there is no way of telling what the angle Φ_{11} for the Nyquist criterion is on the Bode plot.

With reference to the definitions of gain margin and phase margin given in Figs. 9-55 and 9-57, respectively, the interpretation of these parameters from the Bode diagram is illustrated in Fig. 9-59 for a typical minimum-phase loop transfer function. The

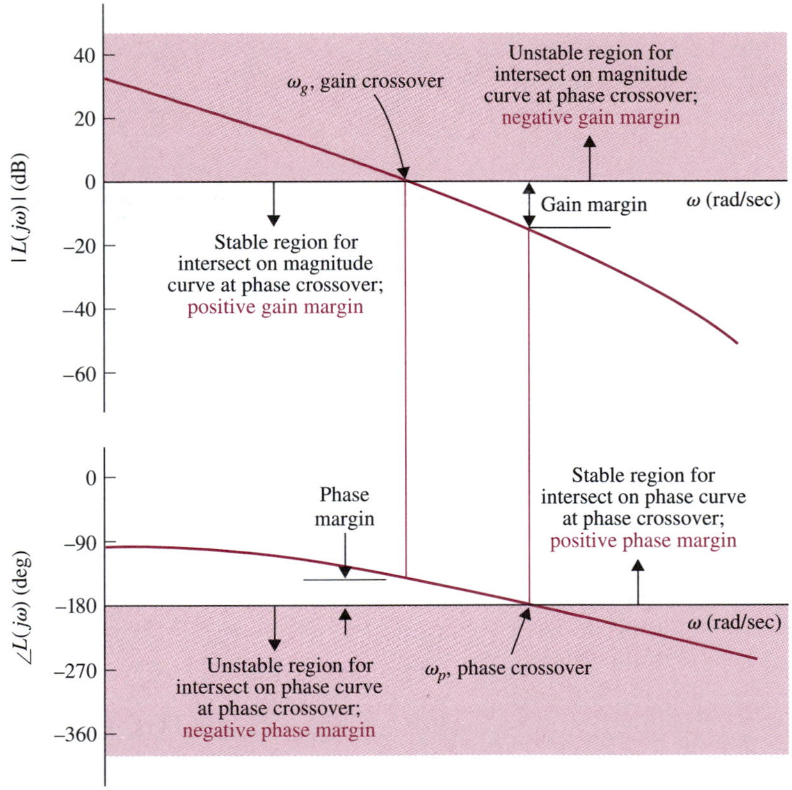

Figure 9-59 Determination of gain margin and phase margin on the Bode plot.

following observations can be made on system stability with respect to the properties of the Bode plot:

▲ Bode plots are useful only for stability studies of systems with minimum-phase loop transfer functions.

1. The gain margin is positive and the system is stable if the magnitude of $L(j\omega)$ at the phase crossover is negative in dB. That is, the gain margin is measured below the 0-dB axis. If the gain margin is measured above the 0-dB axis, the gain margin is negative and the system is unstable.
2. The phase margin is positive and the system is stable if the phase of $L(j\omega)$ is greater than $-180°$ at the gain crossover. That is, the phase margin is measured above the $-180°$ axis. If the phase margin is measured below the $-180°$ axis, the phase margin is negative, and the system is unstable.

Example 9-15

Consider the loop transfer function given in Eq. (9-155); the Bode plot of the function is drawn as shown in Fig. 9-60. The following results are observed easily from the magnitude and phase plots.

The gain crossover is the point where the magnitude curve intersects the 0-dB axis. The gain-crossover frequency ω_g is 6.22 rad/sec. The phase margin is measured at the gain crossover. The phase margin is measured from the $-180°$ axis and is 31.72°. Since the phase margin is measured above the $-180°$ axis, the phase margin is positive and the system is stable.

The phase crossover is the point where the phase curve intersects the $-180°$ axis. The phase-crossover frequency is ω_p = 15.88 rad/sec. The gain margin is measured at the phase crossover and is 14.82 dB. Since the gain margin is measured below the 0-dB axis, the gain margin is positive and the system is stable. The reader should compare the Nyquist plot of Fig. 9-58 with the Bode plot of Fig. 9-60, and the interpretation of ω_g, ω_p, GM, and PM on these plots. ▲

9-15-1 Bode Plots of Systems with Pure Time Delays

The stability analysis of a closed-loop system with a pure time delay in the loop can be conducted easily with the Bode plot. The following example illustrates the standard procedure.

Example 9-16

Consider that the loop transfer function of a closed-loop system is

$$L(s) = \frac{Ke^{T_d s}}{s(s+1)(s+2)} \tag{9-158}$$

Figure 9-61 shows the Bode plot of $L(j\omega)$ with $K = 1$ and $T_d = 0$. The following results are obtained:

Gain-crossover frequency = 0.446 rad/sec
Phase margin = 53.4°

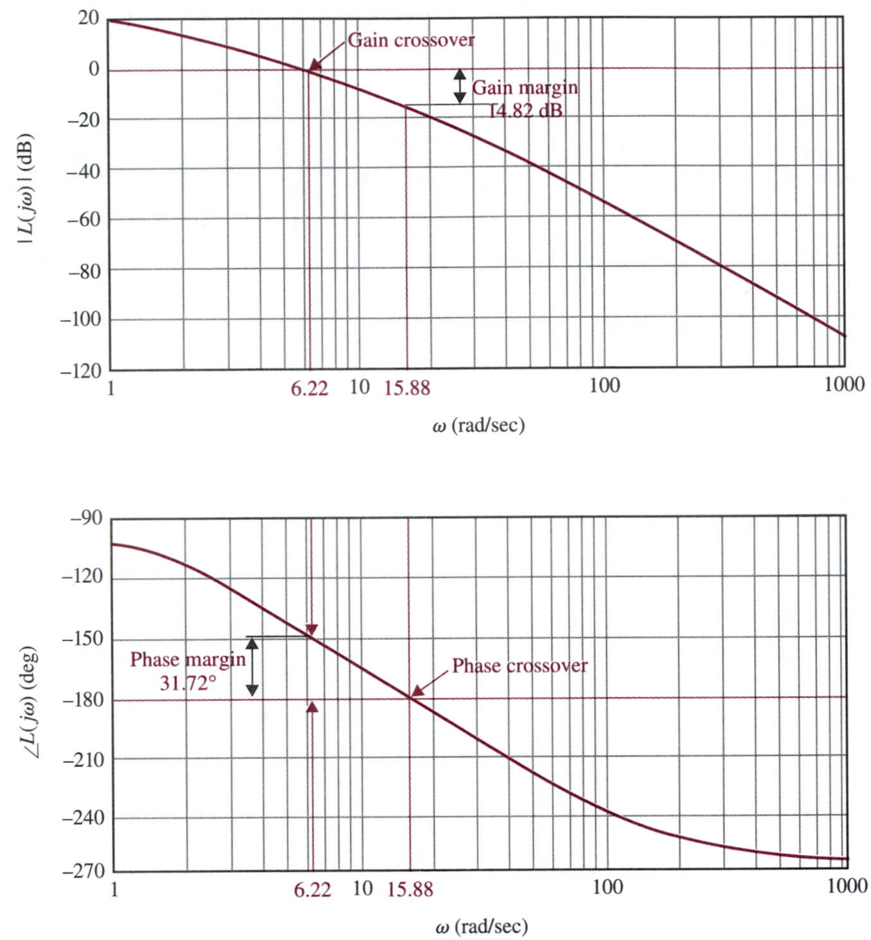

Figure 9-60 Bode plot of $L(s) = \dfrac{2500}{s(s+5)(s+50)}$.

Phase-crossover frequency = 1.416 rad/sec
Gain margin = 15.57 dB

Thus the system without time delay is stable.

The effect of the pure time delay is to add a phase of $-T_d\omega$ radians to the phase curve while not affecting the magnitude curve. The adverse effect of the time delay on stability is apparent, since the negative phase shift caused by the time delay increases rapidly with the increase in ω. To find the critical value of the time delay for stability we set

$$T_d\omega_g = 53.4^\circ \frac{\pi}{180^\circ} = 0.932 \text{ rad} \tag{9-159}$$

Solving for T_d from Eq. (9-159), we get the critical value of T_d to be 2.09 sec.

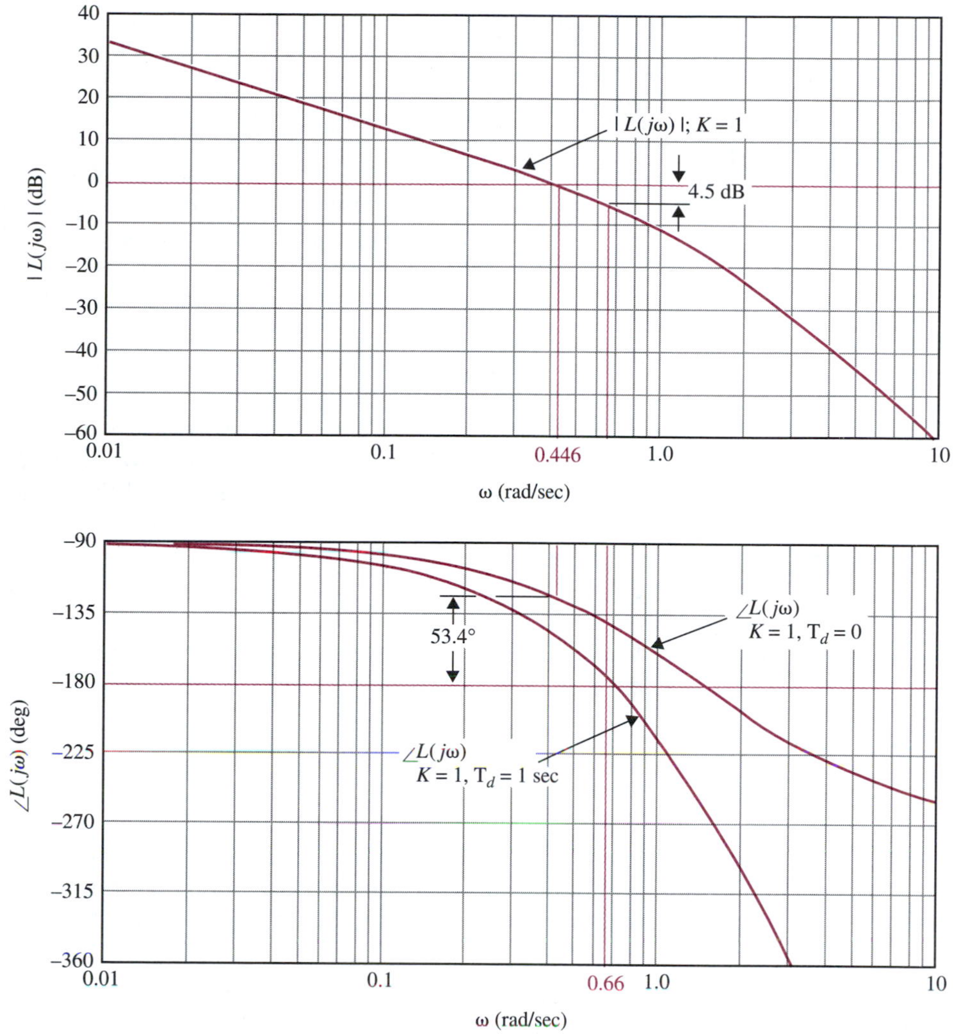

Figure 9-61 Bode plot of $L(s) = Ke^{-T_d s}/[s(s+1)(s+2)]$.

Continuing with the example, we set T_d arbitrarily at 1 sec and find the critical value of K for stability. Figure 9-61 shows the Bode plot of $L(j\omega)$ with this new time delay. With K still equal to 1, the magnitude curve is unchanged. The phase curve droops with the increase in ω, and the following results are obtained:

$$\text{Phase-crossover frequency} = 0.66 \text{ rad/sec} \qquad \text{Gain margin} = 4.5 \text{ dB}$$

Thus, using the definition of gain margin of Eq. (9-148), the critical value of K for stability is $10^{4.5/20} = 1.68$. ▲

9-16 Relative Stability Related to the Slope of the Magnitude Curve of the Bode Plot

In addition to GM, PM, and M_r as relative stability measures, the slope of the magnitude curve of the Bode plot of the loop transfer function at the gain crossover also gives qualitative indication on the relative stability of a closed-loop system. For example, in Fig. 9-60, if the loop gain of the system is decreased from the nominal value, the magnitude curve is shifted downward, while the phase curve is unchanged. This causes the gain-crossover frequency to be lower, and the slope of the magnitude curve at this frequency is less negative; the corresponding phase margin is increased. On the other hand, if the loop gain is increased, the gain-crossover frequency is increased and the slope of the magnitude curve is more negative. This corresponds to a smaller phase margin and the system is less stable. The reason behind these stability evaluations is quite simple. For a minimum-phase transfer function, the relation between its magnitude and phase is unique. Since the negative slope of the magnitude curve is a result of having more poles than zeros in the transfer function, the corresponding phase is also negative. In general, the steeper the slope of the magnitude curve, the more negative the phase. Thus if the gain crossover is at a point where the slope of the magnitude curve is steep, it is likely that the phase margin will be small or negative.

9-16-1 Conditionally Stable System

The illustrative examples given thus far are uncomplicated in the sense that the slopes of the magnitude and phase curves are monotonically decreasing as ω increases. The following example illustrates a **conditionally stable system** that is capable of going through stable/unstable conditions as the loop gain varies.

Example 9-17

Consider that the loop transfer function of a closed-loop system is

$$L(s) = \frac{100K(s+5)(s+40)}{s^3(s+100)(s+200)} \tag{9-160}$$

The Bode plot of $L(j\omega)$ is shown in Fig. 9-62 for $K = 1$. The following results on the system stability are obtained:

$$\text{Gain-crossover frequency} = 1 \text{ rad/sec} \qquad \text{Phase margin} = -78°$$

There are two phase crossovers: one at 25.8 rad/sec and the other at 77.7 rad/sec. The phase characteristics between these two frequencies indicate that if the gain crossover lies in this range, the system would be stable. From the magnitude curve, the range of K for stable operation is found to be between 69 and 85.5 dB. For values of K above and below this range, the phase of

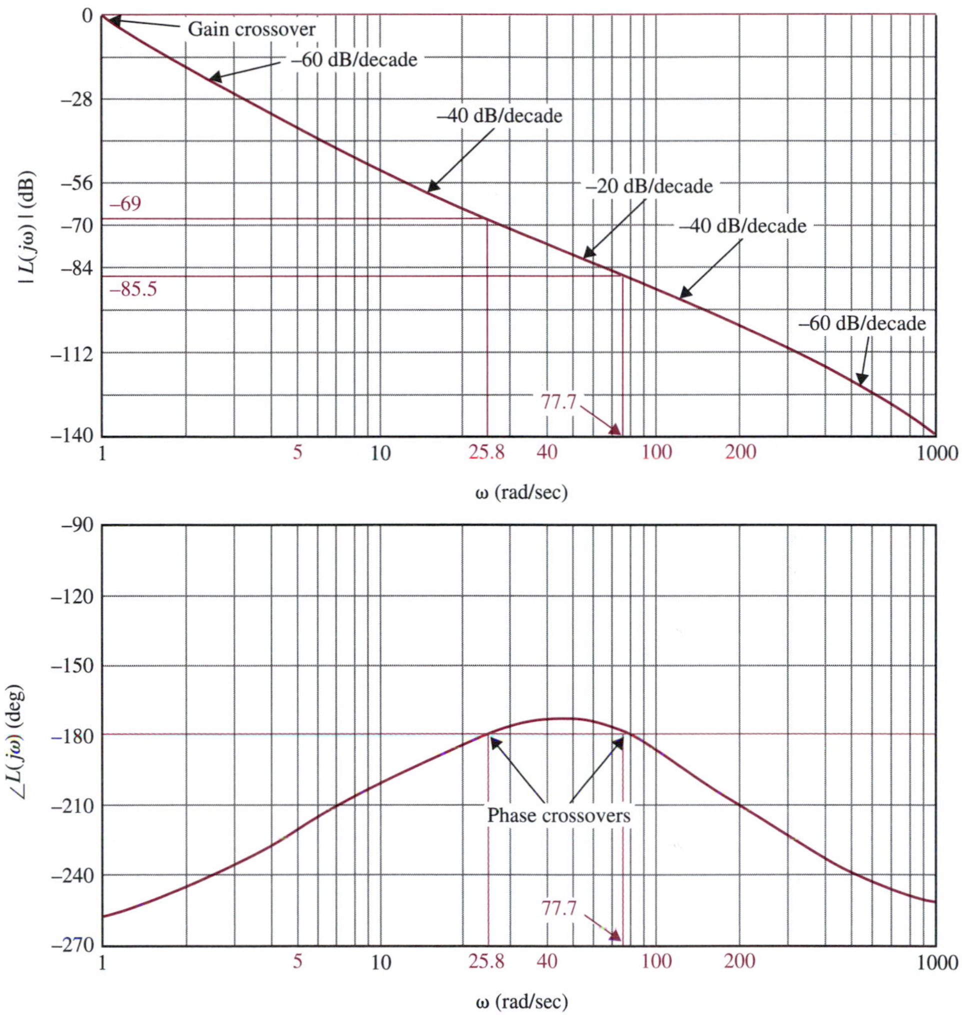

Figure 9-62 Bode plot of

$$L(s) = \frac{100K(s+5)(s+40)}{s^3(s+100)(s+200)} \qquad K = 1$$

$L(j\omega)$ is less than $-180°$, and the system is unstable. This example serves as a good example of the relation between relative stability and the slope of the magnitude curve at the gain crossover. As observed from Fig. 9-62, at both very low and very high frequencies, the slope of the magnitude curve is -60 dB/decade; if the gain crossover falls in either one of these two regions, the phase margin is negative and the system is unstable. In the two sections of the magnitude

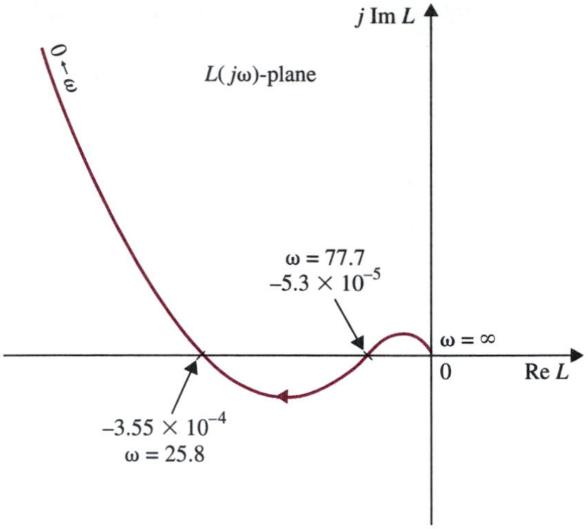

Figure 9-63 Nyquist plot of

$$L(s) = \frac{100K(s+5)(s+40)}{s^3(s+100)(s+200)} \qquad K = 1$$

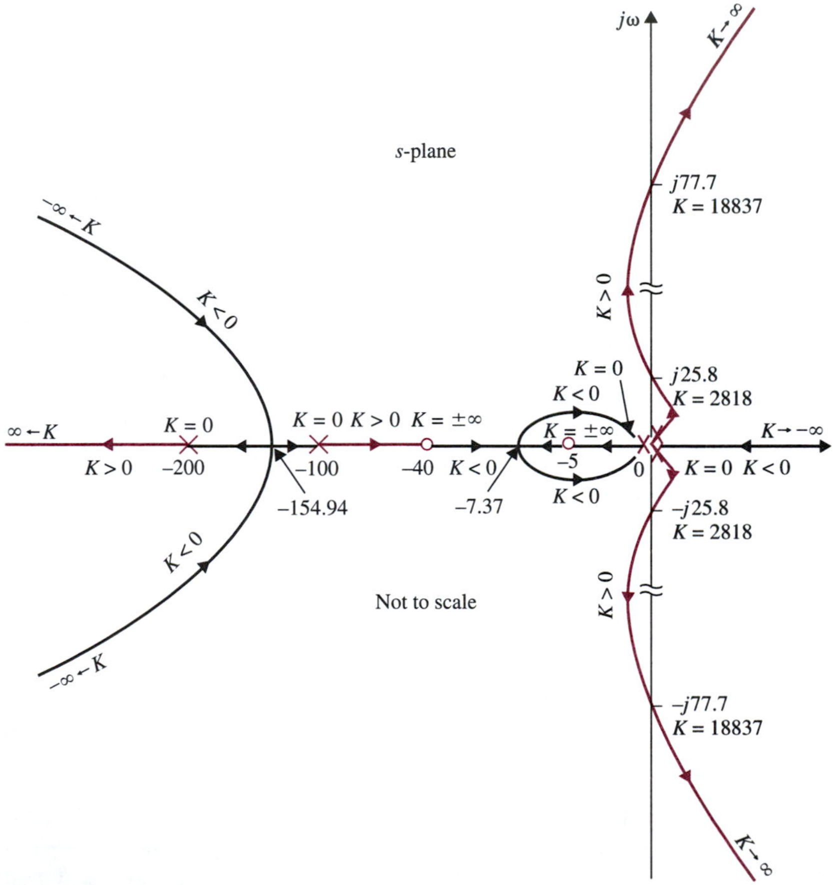

Figure 9-64 Root loci of $G(s) = [100K(s+5)(s+40)]/[s^3(s+100)(s+200)]$.

curve that have a slope of -40 dB/decade, the system is stable only if the gain crossover falls in about half of these regions, but even then the phase margin is small. If the gain crossover falls in the region in which the magnitude curve has a slope of -20 dB/decade, the system is stable.

Figure 9-63 shows the Nyquist plot of $L(j\omega)$. It is of interest to compare the results on stability derived from the Bode plot and the Nyquist plot. The root-locus diagram of the system is shown in Fig. 9-64. The root loci give a clear picture on the stability condition of the system with respect to K. The number of crossings of the root loci on the $j\omega$-axis of the s-plane equals the number of crossings of the phase curve of $L(j\omega)$ of the $-180°$ axis of the Bode plot, and the number of crossings of the Nyquist plot of $L(j\omega)$ with the negative real axis. The reader should check the gain margins obtained from the Bode plot and the coordinates of the crossover points on the negative real axis of the Nyquist plot with the values of K at the $j\omega$-axis crossings on the root loci. ▲

9-17 Stability Analysis with the Magnitude–Phase Plot

The magnitude–phase plot described in Section A-3 of Appendix A is another form of the frequency-domain plot, which has certain advantages for analysis and design in the frequency domain. The magnitude–phase plot of a transfer function $L(j\omega)$ is done in $|L(j\omega)|$ (dB) versus $\angle L(j\omega)$ (degrees). The magnitude–phase plot of the transfer function in Eq. (9-155) is constructed in Fig. 9-65 by use of the data from the Bode plot of Fig. 9-60. The gain and phase crossovers and the gain and phase margins are clearly indicated on the magnitude–phase plot of $L(j\omega)$.

- ▲ The critical point is the intersect of the 0-dB axis and the $-180°$ axis.
- ▲ The phase crossover is where the locus intersects the $-180°$ axis.
- ▲ The gain crossover is where the locus intersects the 0-dB axis.
- ▲ The gain margin is the vertical distance in dB measured from the phase crossover to the critical point.
- ▲ The phase margin is the horizontal distance in degrees measured from the gain crossover to the critical point.

The regions in which the gain and phase crossovers should be located for stability are also indicated. Since the vertical axis for $|L(j\omega)|$ is in dB, when the loop gain of $L(j\omega)$ changes, the locus is simply shifted up and down along the vertical axis. Similarly, when a constant phase is added to $L(j\omega)$, the locus is shifted horizontally without distortion to the curve. If $L(j\omega)$ contains a pure time delay T_d, the effect of the time delay is to add a phase equal to $-\omega T_d \times 180°/\pi$ along the curve.

Another advantage of using the magnitude–phase plot is that *for unity-feedback systems,* closed-loop system parameters such as M_r, ω_r, and BW can all be determined

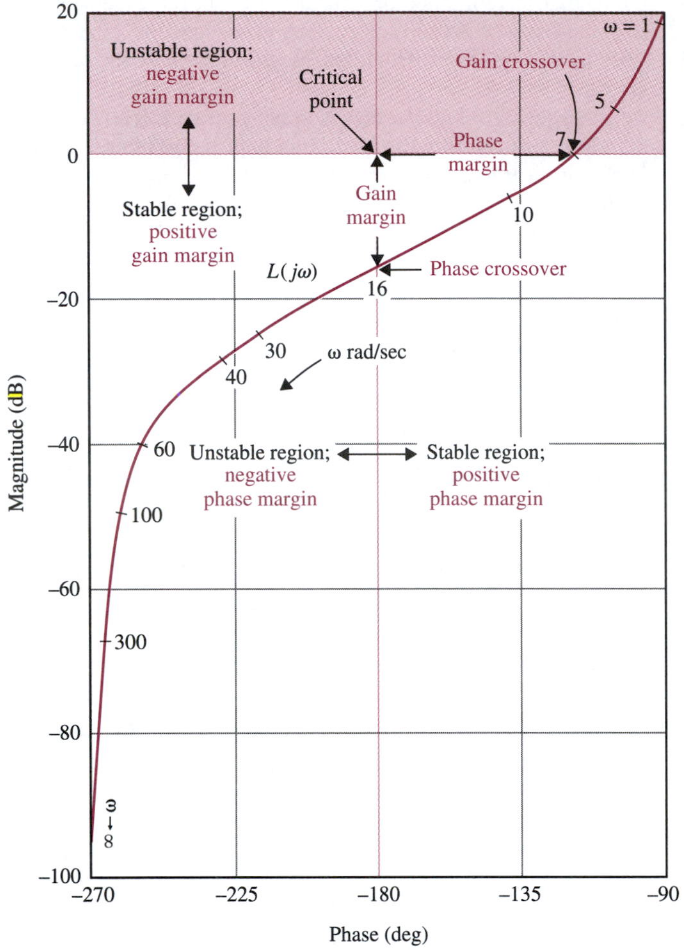

Figure 9-65 Gain–phase plot of $L(s) = 10/[s(1 + 0.2s)(1 + 0.02s)]$.

from the plot with the help of the constant-M loci. These closed-loop performance parameters are not represented on the Bode plot of the forward-path transfer function of a unity-feedback system.

9-18 Constant-M Loci in the $G(j\omega)$-Plane

It was pointed out earlier that analytically, the resonant peak M_r and bandwidth BW are difficult to obtain for high-order systems, and a Bode plot provides information on the closed-loop system only in the form of gain margin and phase margin. It is necessary to develop a graphical method for the determination of M_r, ω_r, and BW using the forward-path transfer function $G(j\omega)$. As we shall see in the following development, the method is directly applicable only to unity-feedback systems, although with some modification it can also be applied to nonunity-feedback systems.

Consider that $G(s)$ is the forward-path transfer function of a unity-feedback system. The closed-loop transfer function is

$$M(s) = \frac{G(s)}{1 + G(s)} \tag{9-161}$$

For sinusoidal steady state, we replace s by $j\omega$; $G(s)$ becomes

$$\begin{aligned} G(j\omega) &= \text{Re}G(j\omega) + j\text{Im}G(j\omega) \\ &= x + jy \end{aligned} \tag{9-162}$$

where for simplicity, x denotes $\text{Re}G(j\omega)$ and y denotes $\text{Im}G(j\omega)$. The magnitude of the closed-loop transfer function is written

$$|M(j\omega)| = \left|\frac{G(j\omega)}{1 + G(j\omega)}\right| = \frac{\sqrt{x^2 + y^2}}{\sqrt{(1 + x)^2 + y^2}} \tag{9-163}$$

For simplicity of notation, let M denote $|M(j\omega)|$; then Eq. (9-163) leads to

$$M\sqrt{(1 + x)^2 + y^2} = \sqrt{x^2 + y^2} \tag{9-164}$$

Squaring both sides of Eq. (9-164) gives

$$M^2[(1 + x)^2 + y^2] = x^2 + y^2 \tag{9-165}$$

Rearranging the last equation yields

$$(1 - M^2)x^2 + (1 - M^2)y^2 - 2M^2x = M^2 \tag{9-166}$$

This equation is conditioned by dividing through by $(1 - M^2)$ and adding the term $[M^2/(1 - M^2)]^2$ on both sides. We have

$$x^2 + y^2 - \frac{2M^2}{1 - M^2}x + \left(\frac{M^2}{1 - M^2}\right)^2 = \frac{M^2}{1 - M^2} + \left(\frac{M^2}{1 - M^2}\right)^2 \tag{9-167}$$

which is finally simplified to

$$\left(x - \frac{M^2}{1 - M^2}\right)^2 + y^2 = \left(\frac{M}{1 - M^2}\right)^2 \qquad M \neq 1 \tag{9-168}$$

For a given value of M, Eq. (9-168) represents a circle with the center at

$$x = \text{Re}\, G(j\omega) = \frac{M^2}{1 - M^2} \qquad y = 0 \tag{9-169}$$

The radius of the circle is

$$r = \left| \frac{M}{1 - M^2} \right| \tag{9-170}$$

Equation (9-168) is invalid for $M = 1$. For $M = 1$, Eq. (9-165) gives

$$x = \text{Re}\, G(j\omega) = -\tfrac{1}{2} \tag{9-171}$$

which is the equation of a straight line parallel to the j Im $G(j\omega)$-axis and passing through the $(-\frac{1}{2}, j0)$ point in the $G(j\omega)$-plane.

When M takes on different values, Eq. (9-168) describes in the $G(j\omega)$-plane a family of circles that are called the **constant-*M* loci,** or the **constant-*M* circles.** Figure 9-66 illustrates a typical set of constant-M circles in the $G(j\omega)$-plane. These circles are symmetrical with respect to the $M = 1$ line and the real axis. The circles to the left of the $M = 1$ locus correspond to values of M greater than 1, and those to the right of the $M = 1$ line are for M less than 1. Equations (9-168) and (9-169) show that when M becomes infinite, the circle degenerates into a point at $(-1, j0)$.

Graphically, the intersection of the $G(j\omega)$ curve and the constant-M circle gives the value of M at the corresponding frequency on the $G(j\omega)$ curve. If we want to keep the value of M_r less than a certain value, the $G(j\omega)$ curve must not intersect the corresponding M circle at any point, and at the same time must not enclose the $(-1, j0)$ point. The constant-M circle with the smallest radius that is tangent to the $G(j\omega)$ curve gives the value of M_r, and the resonant frequency ω_r is read off at the tangent point on the $G(j\omega)$ curve.

Figure 9-67(a) illustrates the Nyquist plot of $G(j\omega)$ for a unity-feedback control system, together with several constant-M loci. For a given loop gain $K = K_1$, the intersects between the $G(j\omega)$ curve and the constant-M loci gives the points on the $|M(j\omega)|$ versus ω curve. The resonant peak M_r is found by locating the smallest circle that is tangent to the $G(j\omega)$ curve. The resonant frequency is found at the point of tangency and is designated as ω_{r1}. If the loop gain is increased to K_2 and if the system is still stable, a constant-M circle with a smaller radius that corresponds to a larger M

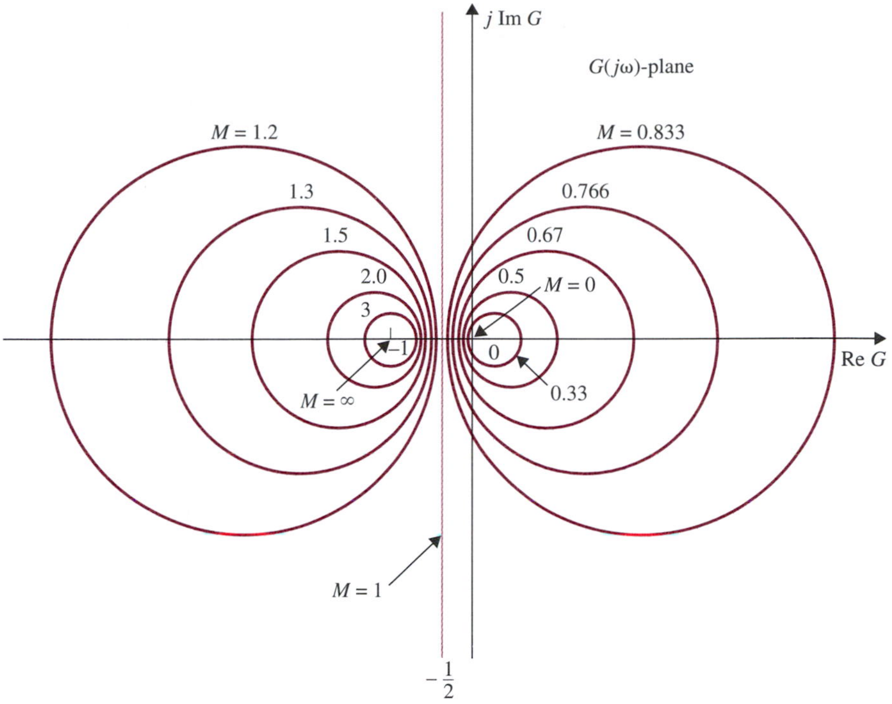

Figure 9-66 Constant-M circles in the polar coordinates.

is found tangent to the $G(j\omega)$ curve, and thus the resonant peak will be larger. The resonant frequency is shown to be ω_{r2}, which is closer to the phase-crossover frequency ω_p than to ω_{r1}. When K is increased to K_3, so that the $G(j\omega)$ curve now passes through the $(-1, j0)$ point, the system is marginally stable and M_r is infinite; ω_{p3} is now the same at the resonant frequency ω_r. When enough points of intersections between the $G(j\omega)$ curve and the constant-M loci are obtained, the magnification curves of $|M(j\omega)|$ versus ω are plotted as shown in Fig. 9-67(b).

▲ When the system is unstable, the constant-M loci and M_r no longer have any meaning.

The bandwidth of the closed-loop system is found at the intersect of the $G(j\omega)$ curve and the $M = 0.707$ locus. For values of K beyond K_3, the system is unstable, and the constant-M loci and M_r no longer have any meaning.

9-19 Constant-Phase Loci in the $G(j\omega)$-Plane

The loci of constant phase of the closed-loop system may be plotted in the $G(j\omega)$-plane by a method similar to that used to secure the constant-M loci. In general, the phase information of the closed-loop system is seldom used in analysis and design, since information on M_r, ω_r, and BW are all derived from the magnitude curve. Figure

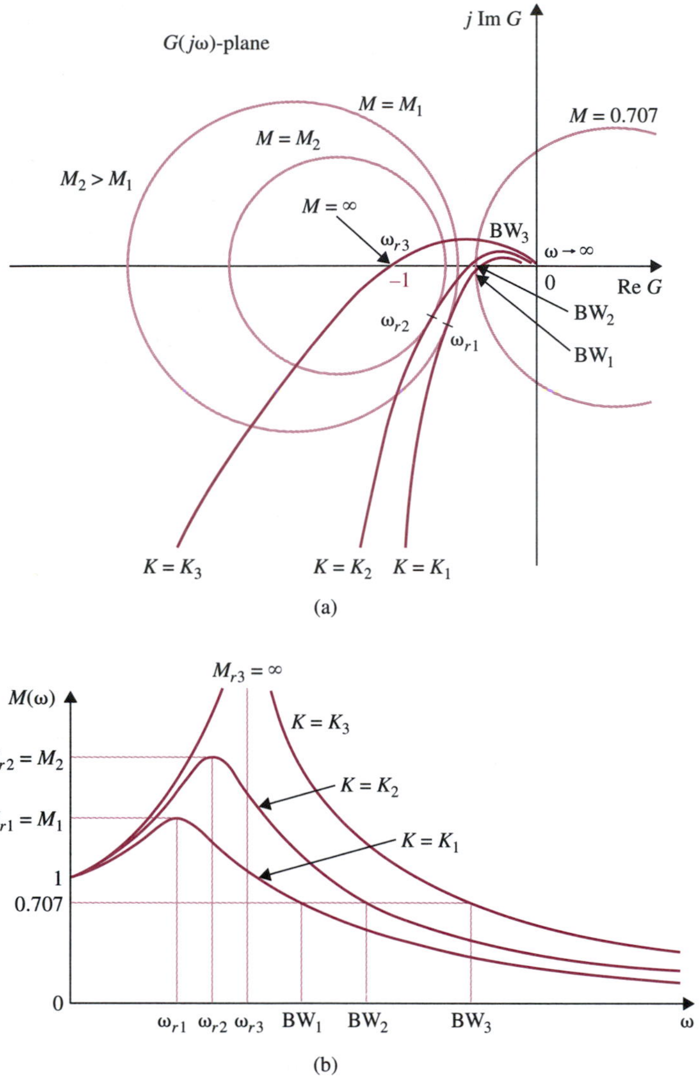

Figure 9-67 (a) Polar plots of $G(s)$ and constant-M loci. (b) Corresponding magnification curves.

9-68 shows a set of constant-phase loci that are called the **constant-N circles** in the $G(j\omega)$-plane. These circles are described by the equation

$$\left(x + \frac{1}{2}\right)^2 + \left(y - \frac{1}{2N}\right)^2 = \frac{1}{4} + \frac{1}{4N^2} \tag{9-172}$$

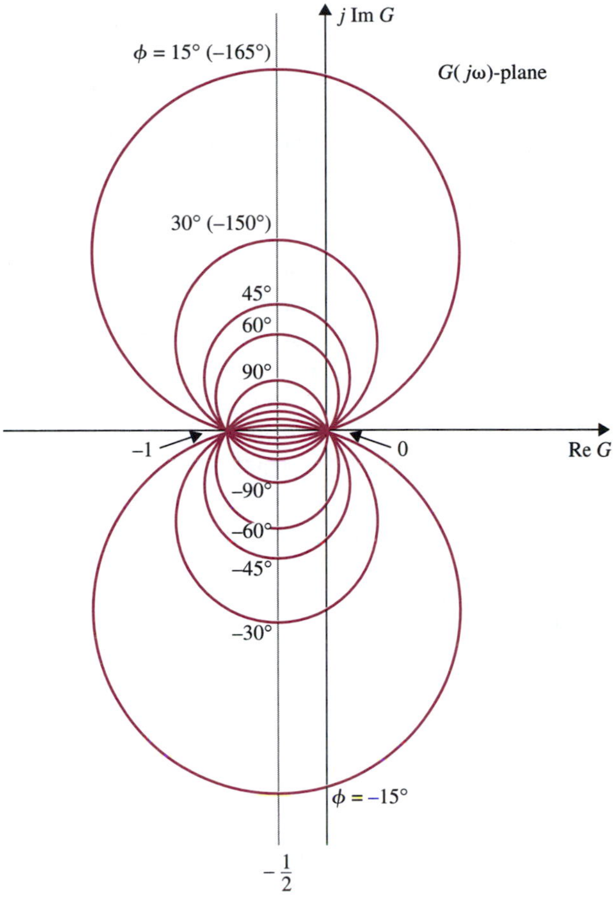

Figure 9-68 Constant-*N* circles in the polar coordinates.

where N is the phase angle in degrees. The centers of these circles are at $(x, y) = (-1/2, 1/2N)$. The radii are given by

$$r = \left(\frac{N^2 + 1}{4N^2}\right)^{1/2} \tag{9-173}$$

9-20 Constant-*M* Loci in the Magnitude–Phase Plane: The Nichols Chart

A major disadvantage in working in the polar coordinates of the Nyquist plot of $G(j\omega)$ is that the curve no longer retains its original shape when a simple modification such as the change of the loop gain is made to the system. Frequently, in design situations, not only must the loop gain be altered, but a series controller may be added to the system. This requires complete reconstruction of the Nyquist plot of the modified $G(j\omega)$. For design work involving M_r and BW as specifications, it is more convenient to work with the magnitude–phase plot of $G(j\omega)$, since when the loop gain is altered, the entire $G(j\omega)$ curve is shifted up or down vertically without distortion. When the phase properties of $G(j\omega)$ are changed independently, without affecting the gain, the magnitude–phase plot is affected only in the horizontal direction.

For the reason noted above, the constant-M loci in the polar coordinates are transferred to the magnitude–phase coordinates, and the resulting loci form a **Nichols chart.** A typical Nichols chart of selected constant-M loci is shown in Fig. 9-69. The constant-N loci can be transferred in a similar manner if needed.

▲ BW is the frequency where the $G(j\omega)$ curve intersects the $M = 0.707$ locus of the constant-M loci.

Once the $G(j\omega)$ curve of the system is constructed in the Nichols chart, the intersects between the constant-M loci and the $G(j\omega)$ trajectory give the value of M at the corresponding frequencies of $G(j\omega)$. The resonant peak M_r is found by locating the smallest of the constant-M locus ($M \geq 1$) that is tangent to the $G(j\omega)$ curve from above. The resonant frequency is the frequency of $G(j\omega)$ at the point of tangency. *The bandwidth of the closed-loop system is the frequency at which the* $G(j\omega)$ *curve intersects the* $M = 0.707$ (-3-dB) *locus.*

The following example illustrates the relationship among the analysis methods using the Bode plot, the magnitude–phase plot, and the Nichols chart.

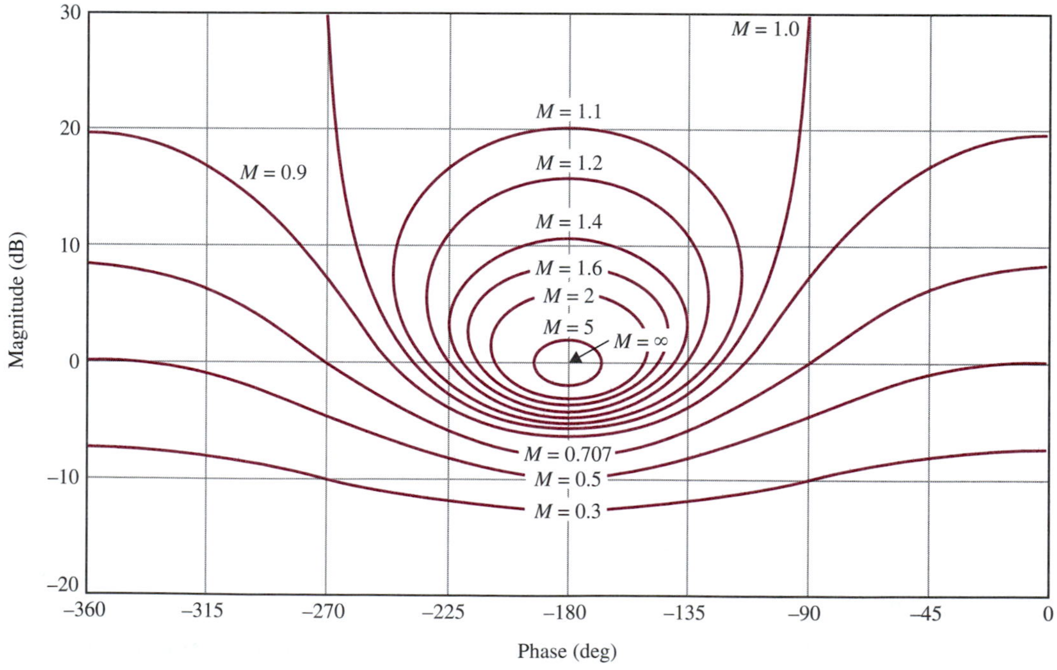

Figure 9-69 Nichols chart.

Example 9-18

Consider the position-control system of the control surfaces of the airship analyzed in Section 7-6. The forward-path transfer function of the unity-feedback system is given by Eq. (7-137) and is repeated here.

$$G(s) = \frac{1.5 \times 10^7 K}{s(s + 400.26)(s + 3008)} \tag{9-174}$$

The Bode plots for $G(j\omega)$ are shown in Fig. 9-70 for $K = 7.248$, 14.5, 181.2, and 273.57. The gain and phase margins of the closed-loop system for these values of K are determined and

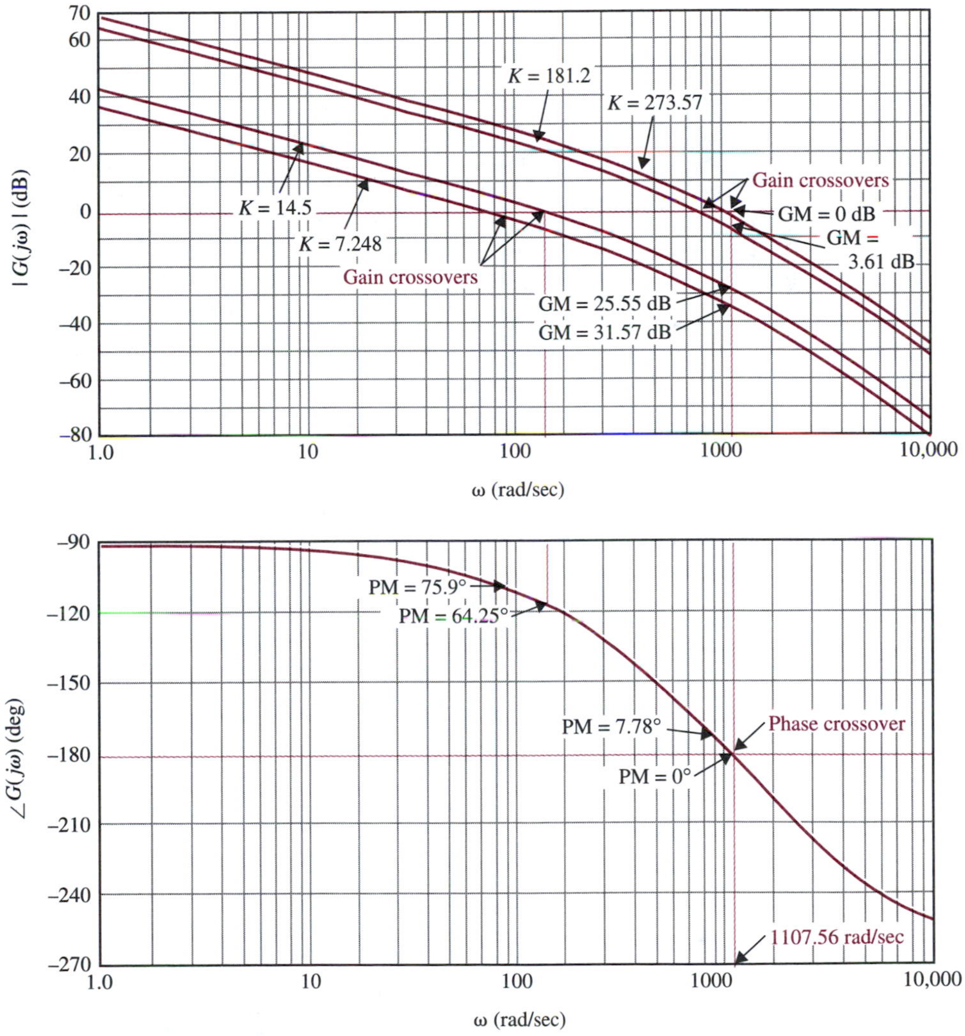

Figure 9-70 Bode diagrams of the system in Example 9-18.

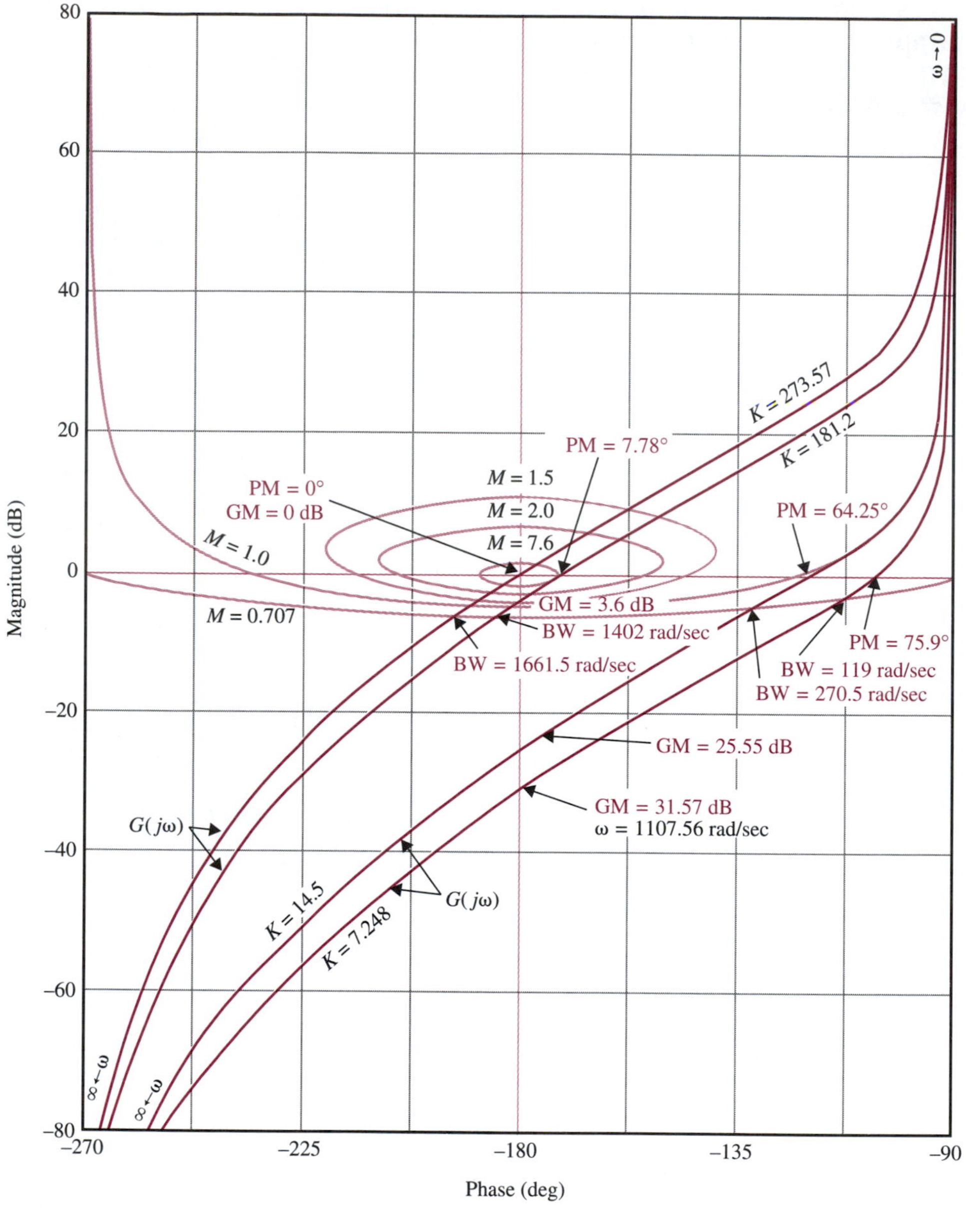

Figure 9-71 Gain–phase plots of the system in Example 9-18.

shown on the Bode plots. The magnitude–phase plots of $G(j\omega)$ corresponding to the Bode plots are shown in Fig. 9-71. These magnitude-phase plots, together with the Nichols chart, give information on the resonant peak M_r, resonant frequency ω_r, and the bandwidth BW. The gain and phase margins are also clearly marked on the magnitude–phase plots. Figure 9-72 shows

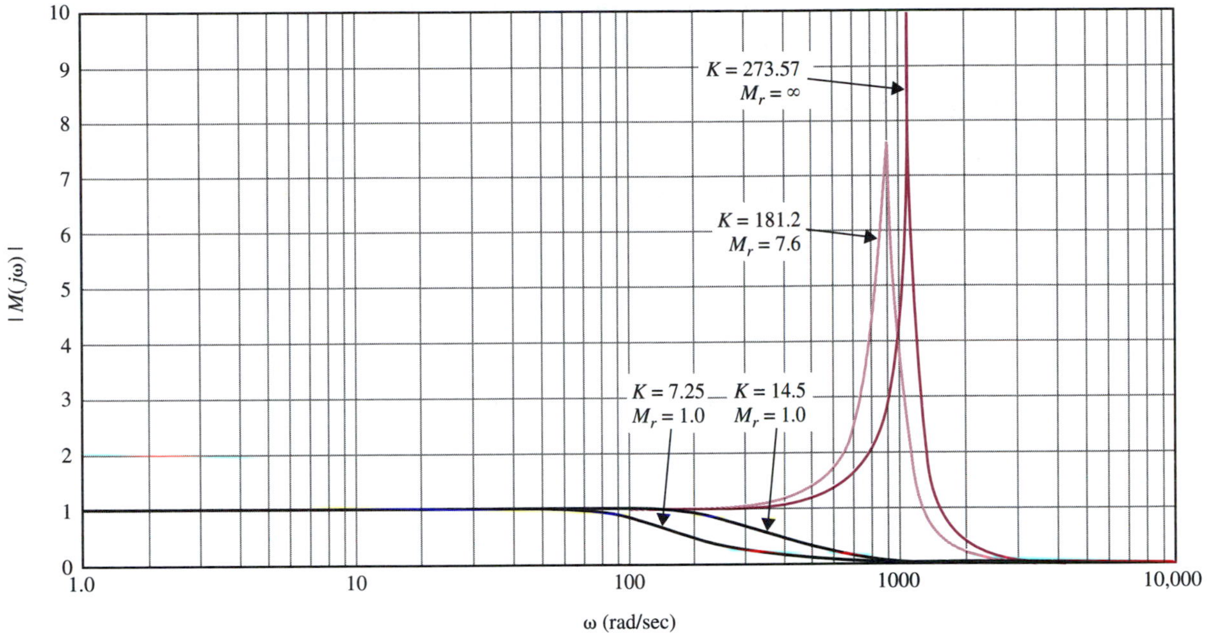

Figure 9-72 Closed-loop frequency responses of the system in Example 9-18.

the closed-loop frequency responses. Table 9-2 summarizes the results of the frequency-domain analysis for the four different values of K together with the time-domain maximum overshoots determined in Section 7-6. ▲

9-21 Computer Solutions

The Bode plot, Nyquist plot, and magnitude–phase plot with Nichols chart analysis can all be conducted with computer programs readily available. The following illustration

Table 9-2 Summary of Frequency-Domain Analysis

K	Maximum Overshoot (%)	M_r	ω_r (rad/sec)	Gain Margin (dB)	Phase Margin (deg)	BW (rad/sec)
7.25	0	1.0	1.0	31.57	75.9	119.0
14.5	4.3	1.0	43.33	25.55	64.25	270.5
181.2	15.2	7.6	900.00	3.61	7.78	1402.0
273.57	100.0	∞	1000.00	0	0	1661.5

is done with the functions in the **CSAD** toolbox. The system described by Eq. (9-174) in Example 9-18 with $K = 181.2$ is used.

Bplot

The function **bplot** of the **CSAD** toolbox computes the frequency response in magnitude (dB) and phase (degrees), and performs the Bode plot. Entering **bplot** <CR> at the **CSAD** prompt >>, the program responds

```
Enter Numerator polynomial vector>271.8e7<CR>

Enter Denominator polynomial vector>pmult([1 0],[1 400.26],[1 3008])<CR>

Enter START frequency w=10^k1, k1 = [-1]>0 <CR>
```
(Choose 1 rad/sec)

```
Enter END frequency w=10^k2, k2 = [1]>4 <CR>
```
(Choose 10,000 rad/sec)

```
            - BODEPLOT OPTIONS -

            Amplitude          Freq range
            Phase              Time delay
            Both               New TF
            Zoom in            Display TF
            Set axes           Margins
            Grid               Roots
            Hold               View Data
            Label              Quit
```

```
Option?>b<CR>
```
(Select to plot both amplitude and phase plots)

The amplitude and phase plots of $G(j\omega)$ are plotted on the monitor screen as shown in Fig. 9-73. The amplitude and phase plots can be made separately by choosing ''Amplitude'' and ''Phase,'' respectively. By selecting ''Margins,'' the performance data are calculated and displayed.

```
Gain margin, dB                 3.581
Phase crossover, rad/sec        1098
Phase margin, degrees           7.779
Gain crossover, rad/sec         888.9
Peak resonance, dB              17.4
Peak resonance, |Mr|            7.412
3 dB bandwidth, rad/sec         1.383
```

The program has a provision for adding a pure time delay to $G(s)$.

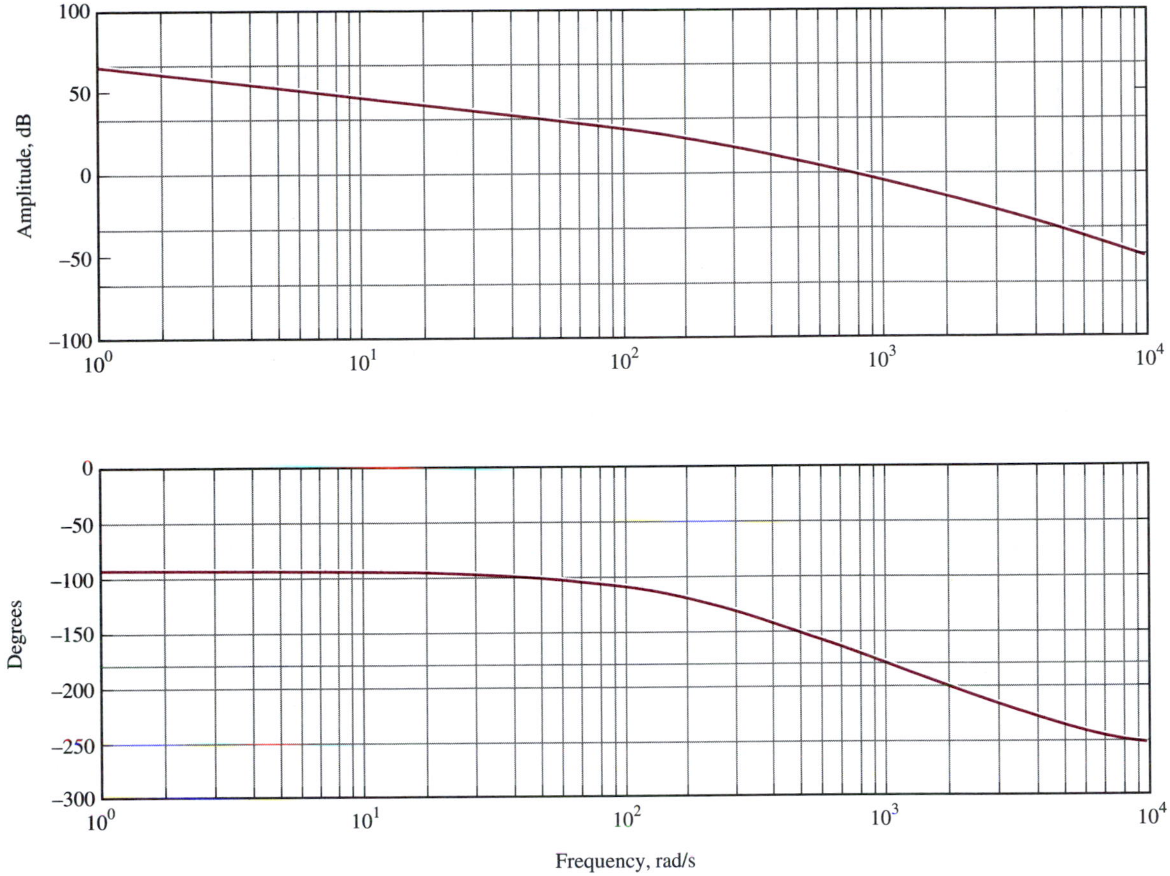

Figure 9-73 Bode plot of the transfer function in Eq. (9-174) with $K =$ 181.2 by **Bode plot** of **CSAD.**

Plrplot

The function **plrplot** of the **CSAD** toolbox computes the frequency response of $G(j\omega)$ and plots the data in the polar coordinates. The program returns the same ''margins'' as given above. The Bode plot of the function in Eq. (9-174) with $K = 181.2$ is plotted in Fig. 9-74 after ''zooming'' the plot to select the significant portion of the graph.

Mvpplot

The function **mvpplot** computes the frequency response of $G(j\omega)$ and plots the data in the amplitude–phase coordinates. Figure 9-75 shows the amplitude–phase plot of the transfer function in Eq. (9-174) with $K = 181.2$, together with the Nichols chart. The program computes the same margins as in **bplot** and **plrplot.**

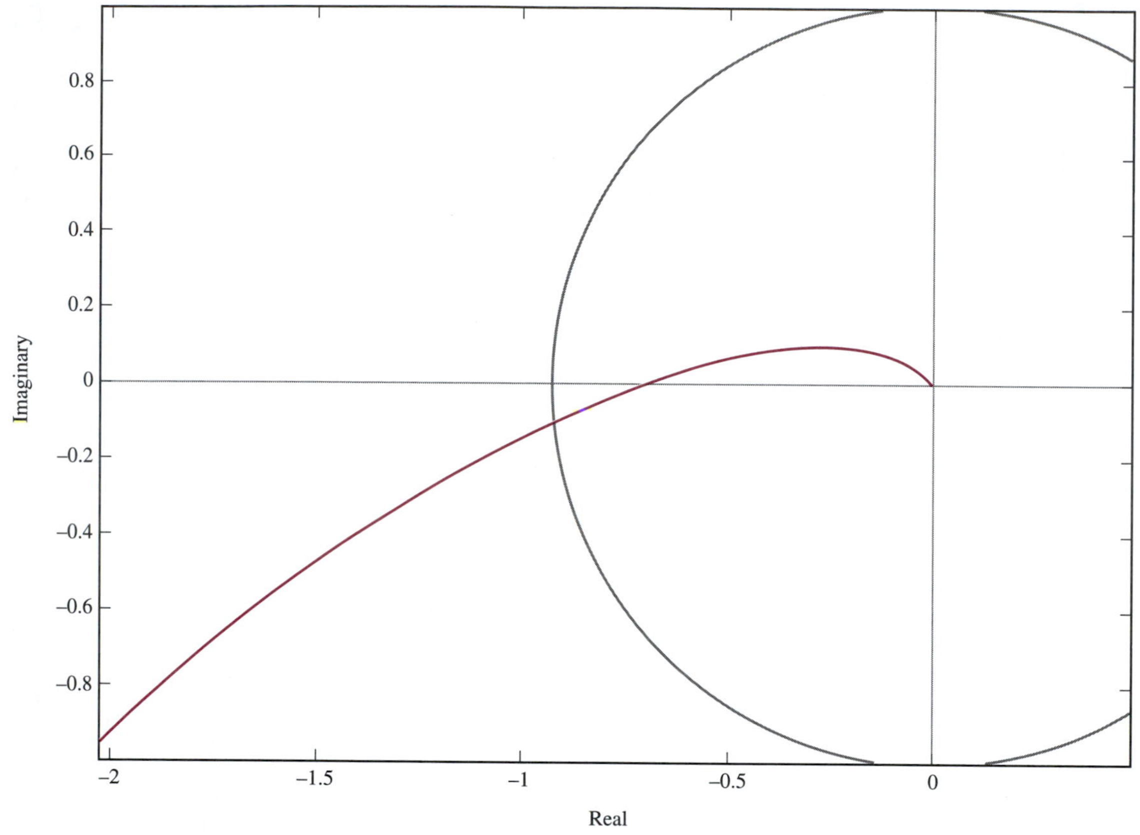

Figure 9-74 Nyquist plot of the transfer function in Eq. (9-174) with $K = 181.2$ by **plrplot** of **CSAD.**

The function **fddesign** of the **CSAD** toolbox can also be used for analysis in the frequency domain. The program is prepared for designing lead and lag controllers; by setting the controller transfer function to unity, it can be used also for analysis purposes.

9-22 Nichols Chart Applied to Nonunity-Feedback Systems

The constant-M loci and the Nichols chart presented in the preceding sections are limited to closed-loop systems with unity feedback whose transfer function is given by Eq. (9-161). When a system has nonunity feedback, the closed-loop transfer function of

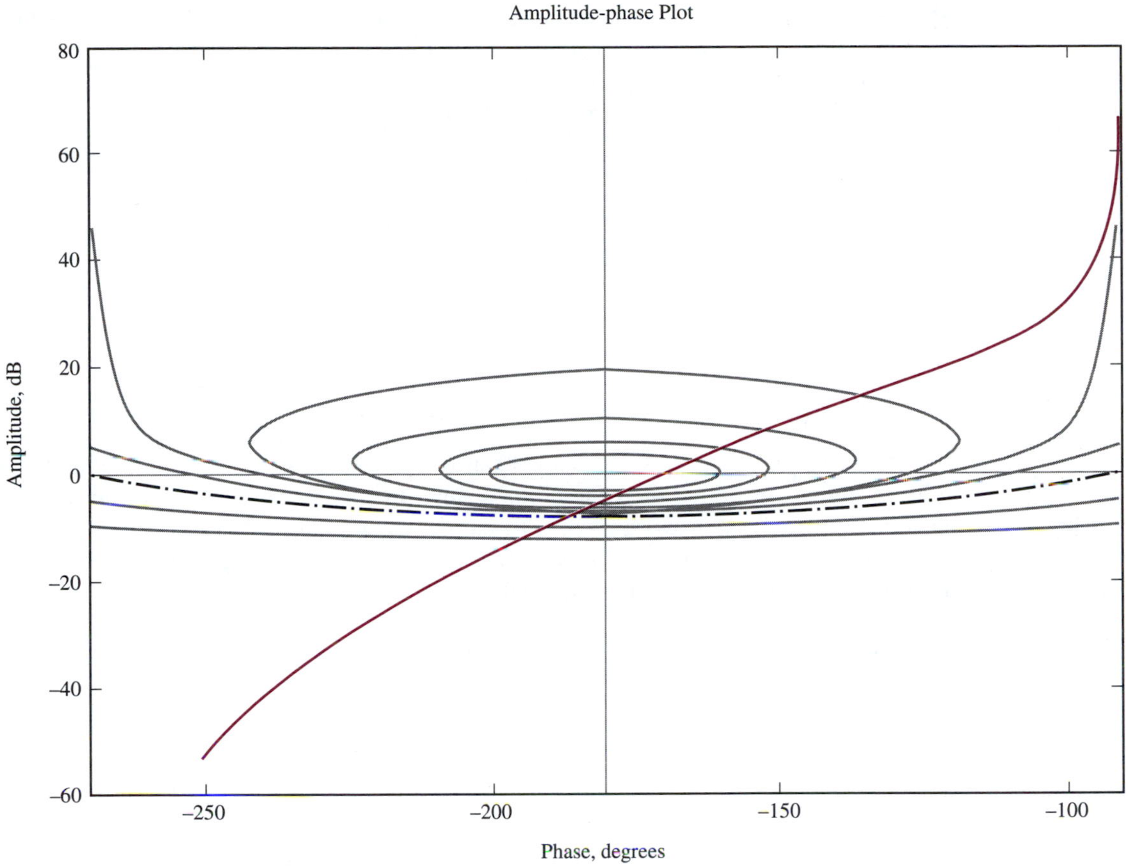

Figure 9-75 Magnitude–phase plot with Nichols chart of the transfer function in Eq. (9-174) for $K = 181.2$ by **mvpplot** of **CSAD.**

the system is expressed as

$$M(s) = \frac{G(s)}{1 + G(s)H(s)} \tag{9-175}$$

where $H(s) \neq 1$. The constant-M loci and the Nichols chart cannot be applied directly to obtain the closed-loop frequency response by plotting $G(j\omega)H(j\omega)$, since the numerator of $M(s)$ does not contain $H(j\omega)$.

By proper modification, the constant-M loci and Nichols chart can still be applied to a nonunity-feedback system. Let us consider the function

$$P(s) = H(s)M(s) = \frac{G(s)H(s)}{1 + G(s)H(s)} \tag{9-176}$$

Apparently, Eq. (9-176) is of the same form as Eq. (9-175). The frequency response of $P(j\omega)$ can be determined by plotting the function $G(j\omega)H(j\omega)$ in the amplitude–phase coordinates along with the Nichols chart. Once this is done, the frequency-response information for $M(j\omega)$ is obtained as follows:

$$|M(j\omega)| = \frac{|P(j\omega)|}{|H(j\omega)|} \tag{9-177}$$

$$\phi_m(j\omega) = \angle M(j\omega) = \angle P(j\omega) - \angle H(j\omega) \tag{9-178}$$

9-23 Sensitivity Studies in the Frequency Domain

▲ Sensitivity study is easily carried out in the frequency domain.

The frequency-domain study of linear control systems has the advantage that higher-order systems can be handled more easily than in the time domain. Furthermore, the sensitivity of the system with respect to parameter variations can easily be interpreted using frequency-domain plots. We shall show how the Nyquist plot and the Nichols chart can be utilized for the analysis and design of control systems based on sensitivity considerations.

Consider a linear control system with unity feedback described by the transfer function

$$M(s) = \frac{G(s)}{1 + G(s)} \tag{9-179}$$

The sensitivity of $M(s)$ with respect to the loop gain K, which is a multiplying factor in $G(s)$, is defined as

$$S_G^M(s) = \frac{dM(s)/M(s)}{dG(s)/G(s)} = \frac{dM(s)}{dG(s)}\frac{G(s)}{M(s)} \tag{9-180}$$

Substituting Eq. (9-179) into Eq. (9-180) and simplifying, we have

$$S_G^M(s) = \frac{1}{1 + G(s)} = \frac{1/G(s)}{1 + 1/G(s)} \tag{9-181}$$

Clearly, the sensitivity function $S_G^M(s)$ is a function of the complex variable s. Figure 9-76 shows the magnitude plot of $S_G^M(s)$ when $G(s)$ is the transfer function given in Eq. (9-155) with $K = 10$. It is interesting to note that the sensitivity of the closed-loop

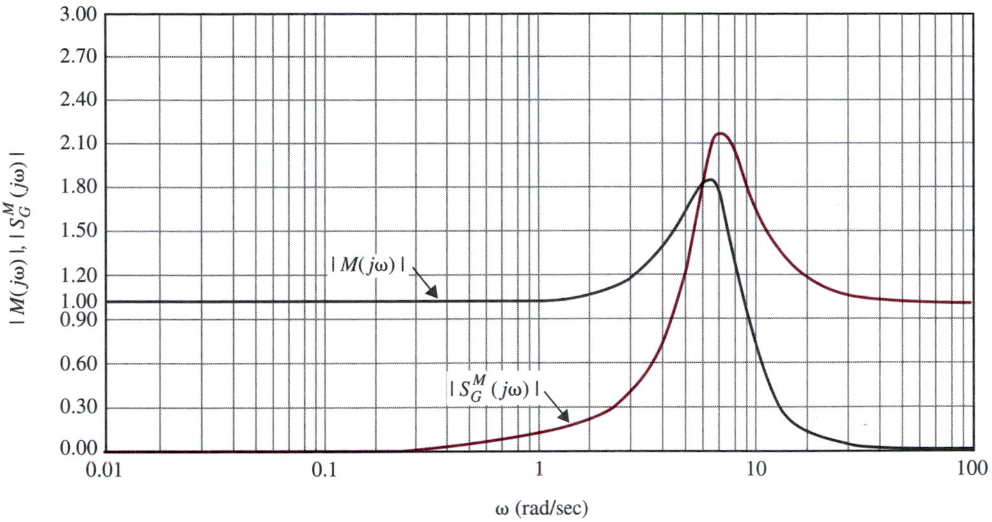

Figure 9-76 $|M(j\omega)|$ and $|S_G^M(j\omega)|$ versus ω for

$$G(s) = \frac{2500}{s(s+5)(s+50)}$$

system is inferior at frequencies greater than 4.8 rad/sec to the open-loop system whose sensitivity to the variation of K is always unity. In general, it is desirable to formulate a design criterion on sensitivity in the following manner:

$$|S_G^M(j\omega)| = \frac{1}{|1 + G(j\omega)|} = \frac{|1/G(j\omega)|}{|1 + 1/G(j\omega)|} \leq k \tag{9-182}$$

where k is a positive real number. Their sensitivity criterion is in addition to the regular performance criteria on the steady-state error and the relative stability.

Equation (9-182) is analogous to the magnitude of the closed-loop transfer function, $|M(j\omega)|$, given in Eq. (9-163), with $G(j\omega)$ replaced by $1/G(j\omega)$. Thus the sensitivity function of Eq. (9-182) can be determined by plotting $1/G(j\omega)$ in the polar coordinates along with the constant-M circles, or by plotting $1/G(j\omega)$ in magnitude–phase coordinates with the Nichols chart. Figure 9-77 shows the polar plots of $G(j\omega)$ of Eq. (9-155) and $1/G(j\omega)$ with $K = 10$. Notice that $G(j\omega)$ is tangent to the $M = 1.84$ circle from below, which means that M_r of the closed-loop system is 1.84. The $1/G(j\omega)$ curve is tangent to the $M = 2.18$ curve from above, and according to Fig. 9-76 is the maximum value of $|S_G^M(s)|$.

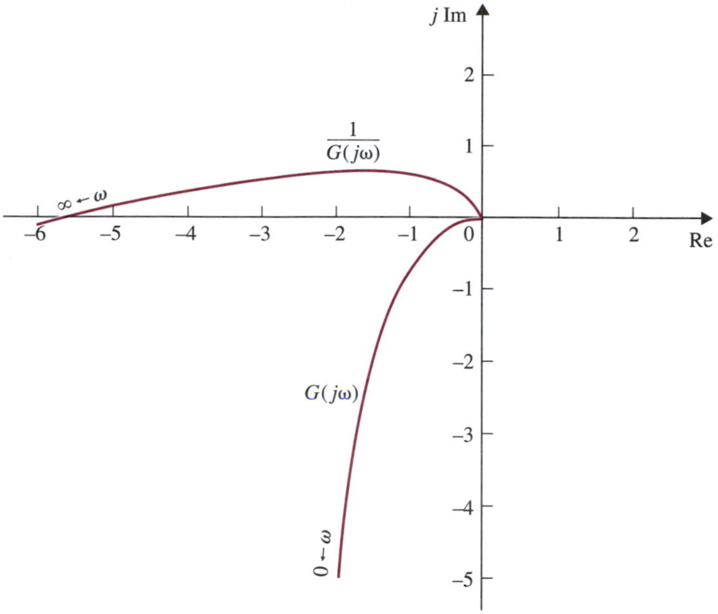

Figure 9-77 Polar plots of $G(j\omega)$ and $1/G(j\omega)$ of

$$G(s) = \frac{2500}{s(s+5)(s+50)}$$

Equation (9-182) shows that for low sensitivity, the loop gain of $G(j\omega)$ must be high, but it is known that in general, high gain could cause instability. Thus the designer is again challenged by the task of designing a system with both high degree of stability and low sensitivity.

Figure 9-78 shows the magnitude–phase plots of $G(j\omega)$ and $1/G(j\omega)$ with the Nichols chart. These results again substantiate those determined from the polar plots. The Bode plots of $G(j\omega)$ and $1/G(j\omega)$ are shown in Fig. 9-79. These are useful for compensation design, since the vertical coordinate is in decibels. The design of robust control systems (low sensitivity) with the frequency-domain methods is discussed in Chapter 10.

9-24 Frequency-Domain Analysis of Discrete-Data Control Systems

All the frequency-domain methods discussed in the preceding sections can be extended to the analysis of discrete-data systems. Consider the discrete-data system shown in

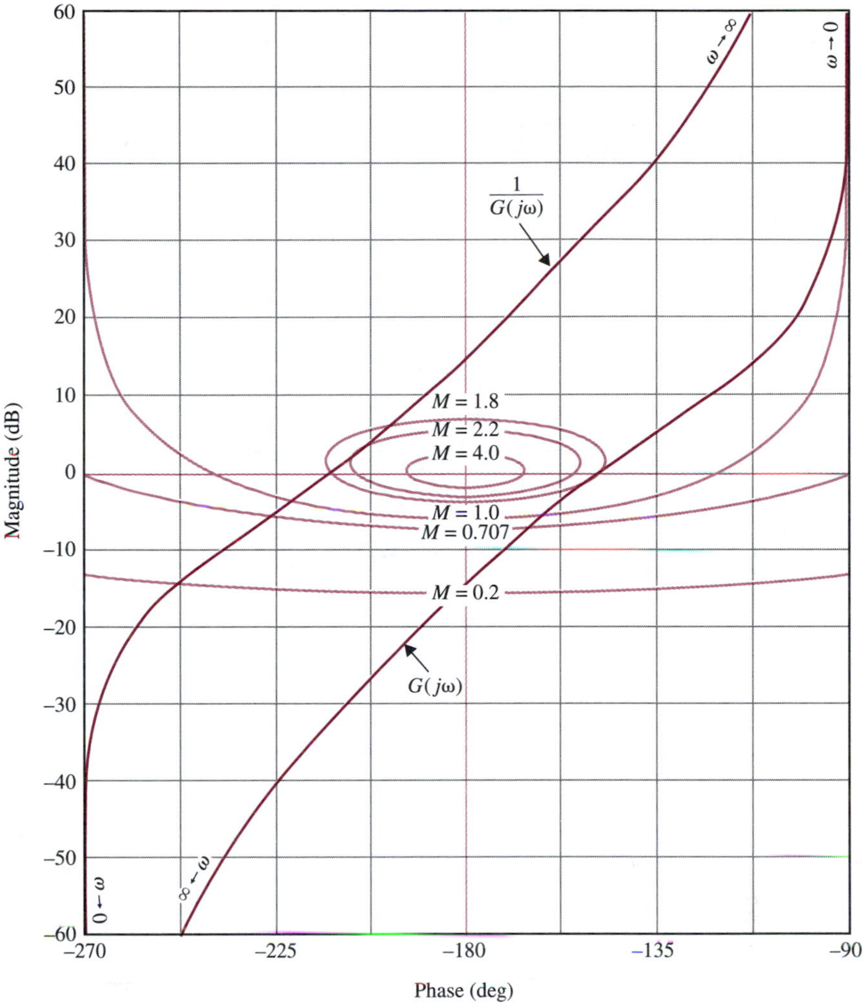

Figure 9-78 Gain–phase plots of $G(j\omega)$ and $1/G(j\omega)$ for

$$G(s) = \frac{2500}{s(s+5)(s+50)}$$

Fig. 9-80. The closed-loop transfer function of the system is

$$\frac{Y(z)}{R(z)} = \frac{G_{h0}G(z)}{1 + G_{h0}G(z)} \tag{9-183}$$

where $G_{h0}G(z)$ is the z-transform of $G_{h0}(s)G(s)$. Just as in the case of continuous-data systems, the absolute and relative stability conditions of the closed-loop discrete-data

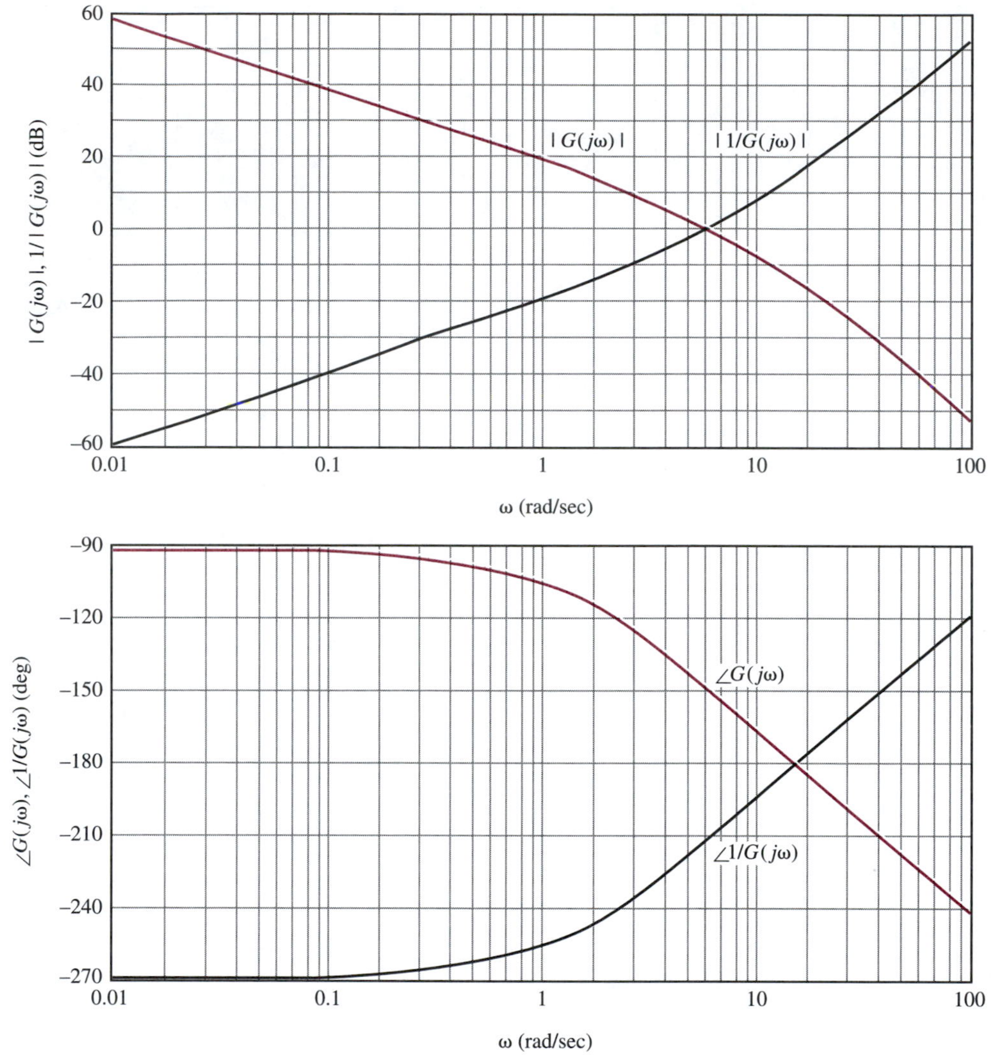

Figure 9-79 Bode plots of $G(j\omega)$ and $1/G(j\omega)$ of

$$G(s) = \frac{2500}{s(s+5)(s+50)}$$

system can be investigated by making the frequency-domain plots of $G_{h0}G(z)$. Since the positive $j\omega$-axis of the s-plane corresponds to real frequency, the frequency-domain plots of $G_{h0}G(z)$ are obtained by setting $z = e^{j\omega T}$ and then letting ω vary from 0 to ∞. This is also equivalent to mapping the points on the unit circle, $|z| = 1$, in the z-plane onto the $G_{h0}G(e^{j\omega T})$-plane. Since the unit circle repeats for every sampling frequency ω_s $(= 2\pi/T)$, as shown in Fig. 9-81, when ω is varied along the $j\omega$-axis, the frequency-

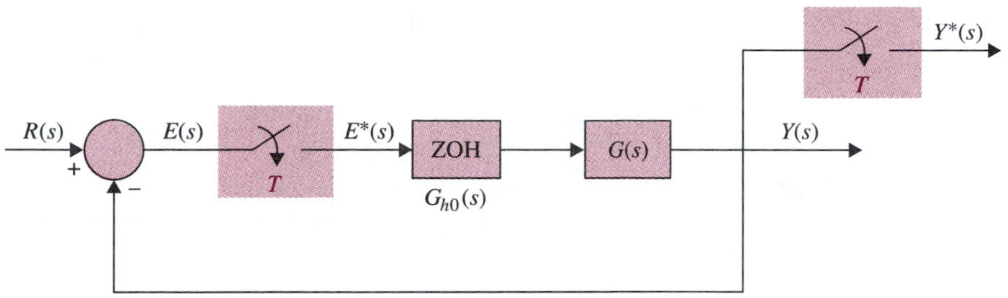

Figure 9-80 Closed-loop discrete-data control system.

domain plot of $G(e^{j\omega T})$ repeats for $\omega = n\omega_s$ to $(n + 1)\omega_s$, $n = 0, 1, 2, \ldots$. Thus it is necessary to plot $G_{h0}G(e^{j\omega T})$ only for the range of $\omega = 0$ to $\omega = \omega_s$. In fact, since the unit circle in the z-plane is symmetrical about the real axis, the plot of $G_{h0}G(e^{j\omega T})$ in the polar coordinates for $\omega = 0$ to $\omega_s/2$ needs to be plotted.

9-24-1 Bode Plot with the *w*-Transformation

The w-transformation introduced in Section 6-7-1, Eq. (6-47), can be used for frequency-domain analysis and design of discrete-data control systems. The transformation is [Eq. (6-46)],

$$z = \frac{(2/T) + w}{(2/T) - w} \tag{9-184}$$

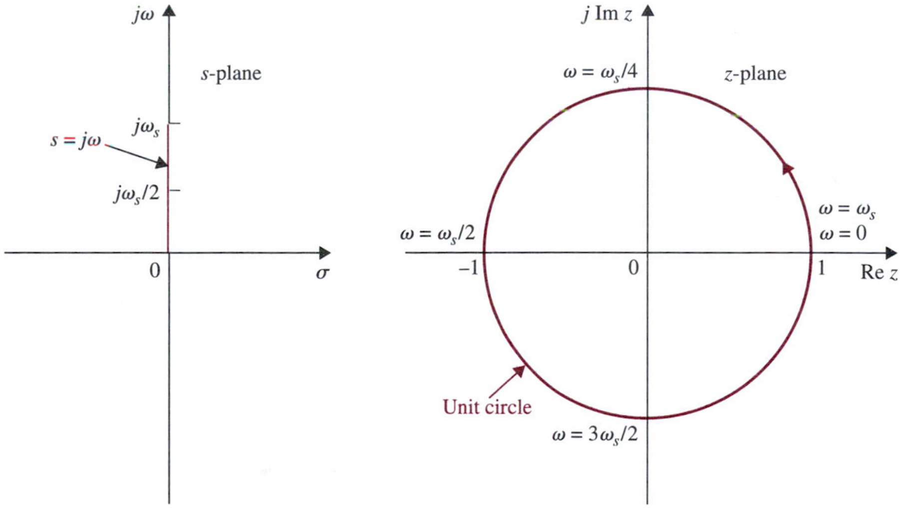

Figure 9-81 Relation between the $j\omega$-axis in the s-plane and the unit circle in the z-plane.

In the frequency domain, we set [Eq. (6-50)],

$$w = j\omega_w = j\frac{2}{T}\tan\frac{\omega T}{2} \tag{9-185}$$

For frequency-domain analysis of a discrete-data system, we substitute Eqs. (9-184) and (9-185) in $G(z)$ to get $G(j\omega_w)$; the latter can be used to form the Bode plot or the polar plot of the system.

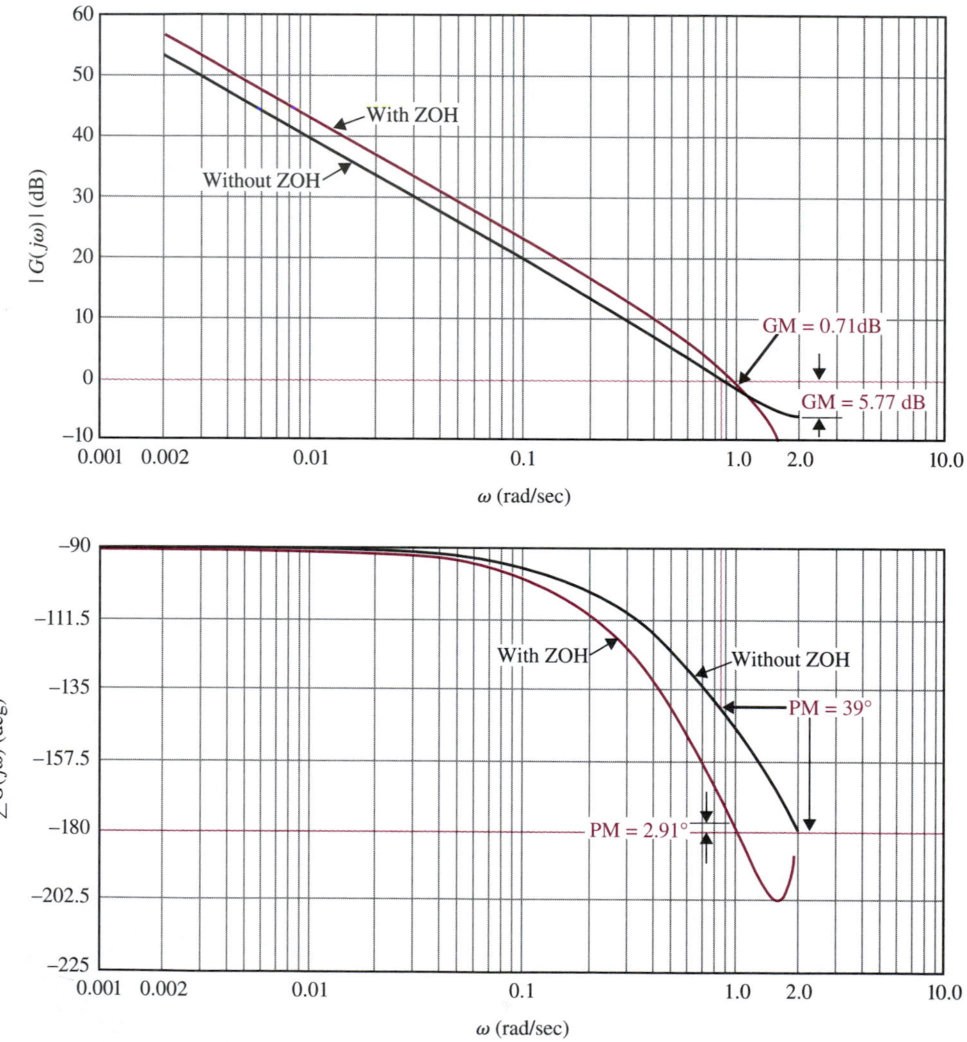

Figure 9-82 Bode plot of $G_{h0}G(z)$ of the system in Fig. 9-80, with $G(s) = 1.57/[s(s+1)]$. $T = 1.57$ sec, and with and without ZOH.

Example 9-19

As an illustrative example on frequency-domain plots of discrete-data control systems, let the transfer function of the process in the system in Fig. 9-80 be

$$G(s) = \frac{1.57}{s(s+1)} \tag{9-186}$$

and the sampling frequency is 4 rad/sec. Let us first consider that the system does not have a zero-order hold, so that

$$G_{h0}G(z) = G(z) = \frac{1.243z}{(z-1)(z-0.208)} \tag{9-187}$$

The frequency response of $G_{h0}G(z)$ is obtained by substituting $z = e^{j\omega T}$ in Eq. (9-186). The polar plot of $G_{h0}G(e^{j\omega T})$ for $\omega = 0$ to $\omega_s/2$ is shown in Fig. 9-82. The mirror image of the locus shown, with the mirror placed on the real axis, represents the plot for $\omega = \omega_s/2$ to ω_s.

The Bode plot of $G_{h0}G(e^{j\omega T})$ consists of the graphs of $|G_{h0}G(e^{j\omega T})|$ in dB versus ω, and $\angle G_{h0}G(e^{j\omega T})$ in degrees versus ω, as shown in Fig. 9-83 for three decades of frequency with the plots ended at $\omega = \omega_s/2 = 2$ rad/sec.

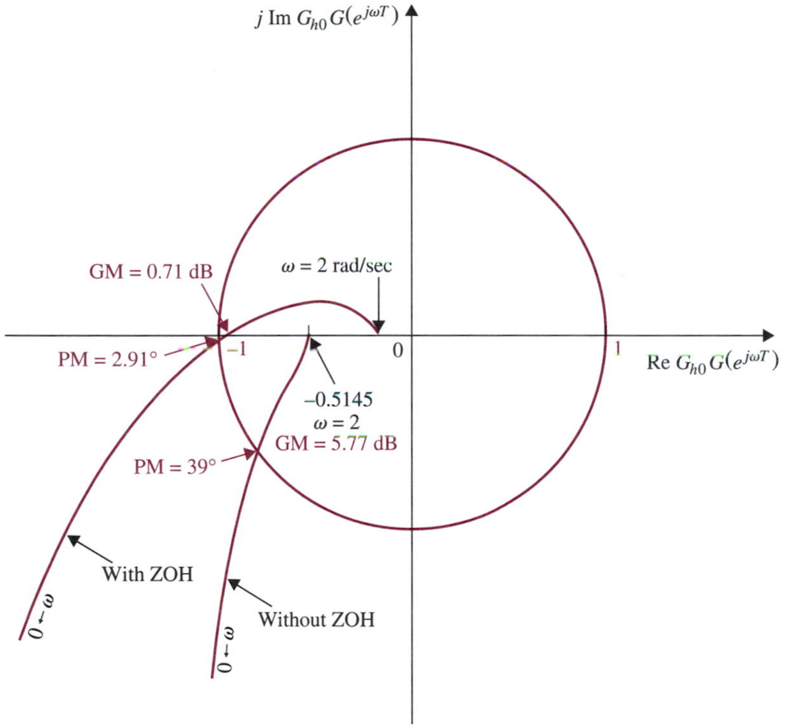

Figure 9-83 Frequency-domain plot of $G_{h0}G(z)$ of the system in Fig. 9-80, with $G(s) = 1.57/[s(s+1)]$, $T = 1.57$ sec, and with and without ZOH.

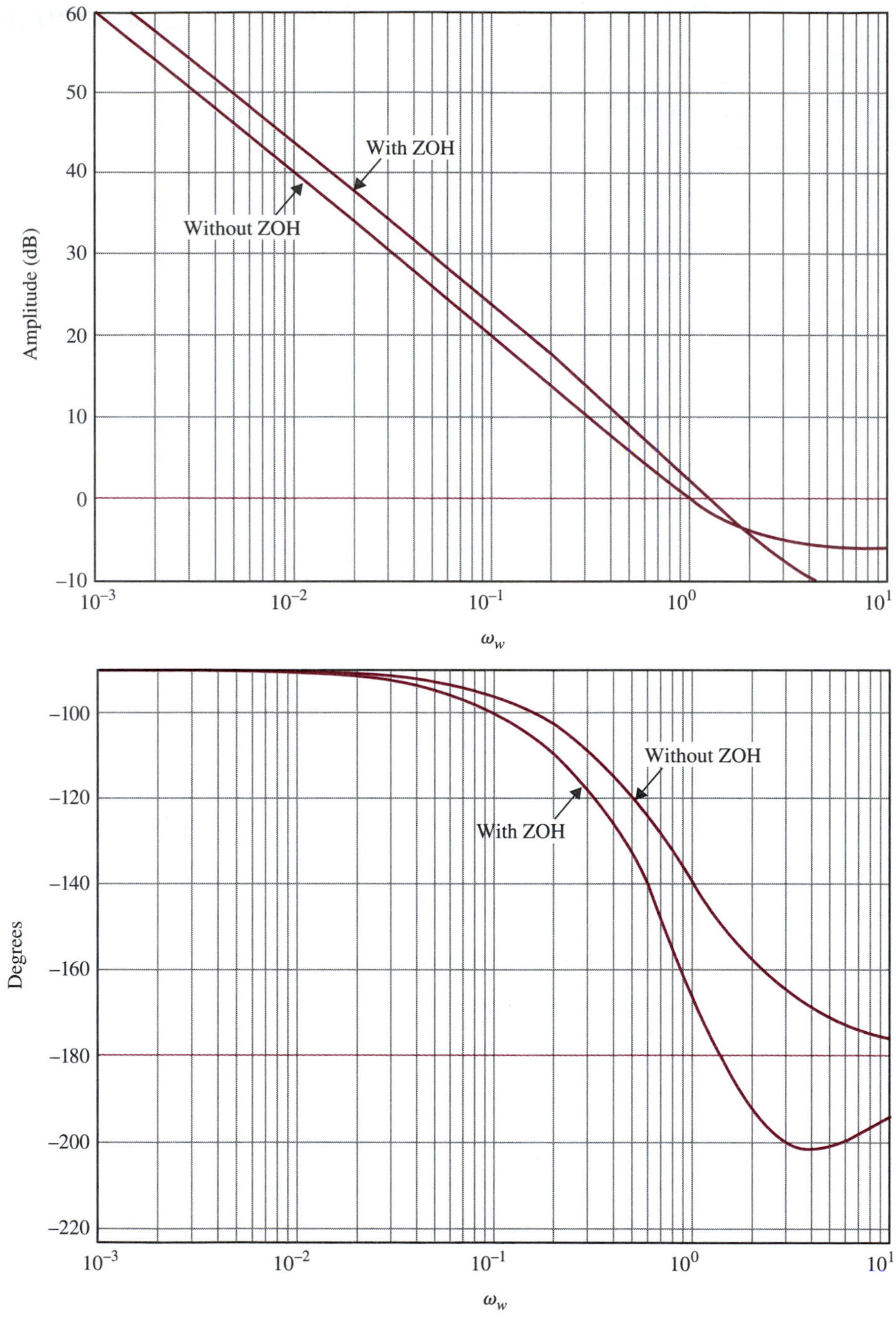

Figure 9-84 Bode plot of $G_{h0}G(z)$ of the system in Fig. 9-80 with $G(s) = 1.57/[s(s+1)]$, $T = 1.57$ sec with and without ZOH. The plots are done with the w-transformation, $w = j\omega_w$.

For the sake of comparison, the forward-path transfer function of the system with a zero-order hold is obtained:

$$G_{h0}G(z) = \frac{1.2215z + 0.7306}{(z-1)(z-0.208)} \tag{9-188}$$

The polar plot and the Bode plot of Eq. (9-188) are shown in Figs. 9-82 and 9-83, respectively. Notice that the polar plot of the system with the zero-order hold intersects the negative real axis at a point that is closer to the $(-1, j0)$ point than that of the system without the zero-order hold. Thus the system with the zero-order hold is less stable. Similarly, the phase of the Bode plot of the system with the zero-order hold is more negative than that of the system without the zero-order hold. The gain margin, phase margin, and peak resonance of the two systems are summarized as follows:

	Gain Margin (dB)	Phase Margin (deg)	M_r
Without ZOH	5.77	39.0	1.58
With ZOH	0.71	2.91	22.64

As an alternative, the Bode plot and polar plot of the forward-path transfer function can be made using the w-transformation of Eqs. (9-184). For the system with ZOH, the forward-path transfer function in the w-domain is

$$G_{h0}G(w) = \frac{1.57(1+0.504w)(1-1.0913w)}{w(1+1.197w)} \tag{9-189}$$

For the system without ZOH,

$$G_{h0}G(jw) = \frac{1-0.6163w^2}{w(1+1.978w)} \tag{9-190}$$

Substituting $w = j\omega_w$ into Eq. (9-190), the Bode plots are made as shown in Fig. 9-84. Notice that the frequency coordinates in Fig. 9-84 are ω_w, whereas those in Fig. 9-82 are the real frequency ω. The two frequencies are related through Eq. (9-185).

The conclusion from this illustrative example is that once z is replaced by $e^{j\omega T}$ in the z-domain transfer function, or using the w-transform, all the frequency-domain analysis techniques available for continuous-data systems can be applied to discrete-data systems. ▲

9-25 Summary

This chapter is devoted to the frequency-domain analysis of linear control systems. The chapter begins by describing typical relationships between the open-loop and closed-loop frequency responses of linear systems. Performance specifications such as the resonance peak M_r, resonant frequency ω_r, and bandwidth BW are defined in the frequency domain. The relationships between these parameters of a second-order prototype system are derived analytically. The effects of adding simple poles and zeros to the loop transfer function on M_r and BW are discussed.

The Nyquist criterion for stability analysis of linear control systems is developed thoroughly. The stability of single-loop control system can be investigated by studying the behavior of the Nyquist plot of the loop transfer function $G(s)H(s)$ for $\omega = 0$ to $\omega = \infty$ with respect to the critical point. The general Nyquist criterion is given in terms of the angular condition given in Eq. (9-82). If $G(s)H(s)$ is a minimum-phase transfer function, the condition of stability is simplified so that the Nyquist plot must not enclose the critical point. Application of the Nyquist criterion to multiloop systems is also outlined.

The relationship between the root loci and the Nyquist plot is pointed out in Section 9-7. The discussion should add more perspective to the understanding of both subjects.

Relative stability is defined in terms of gain margin and phase margin. These quantities are defined in the polar coordinates as well as on the Bode diagram. The gain–phase plot allows the Nichols chart to be constructed for closed-loop analysis. The values of M_r and BW can be found easily by plotting the $G(j\omega)$ locus on the Nichols chart.

The sensitivity function $S_G^M(j\omega)$ is defined as a measure of the variation of $M(j\omega)$ due to variations in $G(j\omega)$. It is shown that the frequency-response plots of $G(j\omega)$ and $G(j\omega)^{-1}$ can readily be used for sensitivity studies.

Finally, the frequency response of discrete-data control systems is investigated. Practically all the frequency-domain specifications defined for analog systems can be extended to discrete-data systems. The frequency-domain plots of discrete-data systems can be made by replacing the z-transform variable z by $e^{j\omega T}$, where T is the sampling period. The w-transform is also applied to the plotting of Bode plots of discrete-data systems.

Review Questions

1. Explain why it is important to conduct frequency-domain analysis of linear control systems.
2. Define resonance peak M_r of a closed-loop control system.
3. Define bandwidth BW of a closed-loop control system.
4. The following quantities are defined:

   Z = number of zeros of $L(s)$ that are in the right-half s-plane
   P = number of poles of $L(s)$ that are in the right-half s-plane
   P_ω = number of poles of $L(s)$ that are on the $j\omega$-axis

   Give the conditions on these parameters for the system to be **(a)** open-loop stable, and **(b)** closed-loop stable.
5. Consider that the Nyquist plot of $L(j\omega)$ is drawn for $\omega = 0$ to $\omega = \infty$. Let Φ_{11} be the total angle traversed by the vector drawn from the $(-1, j0)$ point to the Nyquist plot as ω varies from ∞ to 0. Give the condition between Φ_{11} and the parameters listed in Review Question 4 for closed-loop stability.
6. What condition must be satisfied by the function $L(j\omega)$ so that the Nyquist criterion is simplified to investigating whether or not the $(-1, j0)$ is enclosed by the Nyquist plot?
7. Give all the properties of a minimum-phase transfer function.
8. Give the definitions of gain margin and phase margin.

9. By applying a sinusoidal signal of frequency ω_o to a linear system, the steady-state output of the system will be of the same frequency. (T) (F)
10. For a prototype second-order system, the value of M_r depends solely on the damping ratio ζ. (T) (F)
11. Adding a zero to the loop transfer function will always increase the bandwidth of the closed-loop system. (T) (F)
12. The general effect of adding a pole to the loop transfer function is to make the closed-loop system less stable while decreasing the bandwidth. (T) (F)
13. For a minimum-phase loop transfer function $L(j\omega)$, if the phase margin is negative, the closed-loop system is always unstable. (T) (F)
14. Phase-crossover frequency is the frequency at which the phase of $L(j\omega)$ is 0°. (T) (F)
15. Gain-crossover frequency is the frequency at which the gain of $L(j\omega)$ is 0 dB. (T) (F)
16. Gain margin is measured at the phase-crossover frequency. (T) (F)
17. Phase margin is measured at the gain-crossover frequency. (T) (F)
18. A closed-loop system with a pure time delay in the loop is usually less stable than one without a time delay. (T) (F)
19. The slope of the magnitude curve of the Bode plot of $L(j\omega)$ at the gain crossover usually gives an indication of the relative stability of the closed-loop system. (T) (F)
20. A Nichols chart can be used to find BW and M_r information for a closed-loop system. (T) (F)
21. A Bode plot can be used for stability analysis for minimum- as well as nonminimum-phase transfer functions. (T) (F)

Answers to the true-and-false questions are given following the Problems section.

References

Nyquist Criterion of Continuous-Data Systems

1. **H. NYQUIST,** "Regeneration Theory," *Bell Syst. Tech. J.,* Vol. 11, pp. 126–147, Jan. 1932.
2. **R. W. BROCKETT and J. L. WILLEMS,** "Frequency Domain Stability Criteria—Part I," *IEEE Trans. Automatic Control,* Vol. AC-10, pp. 255–261, July 1965.
3. **R. W. BROCKETT and J. L. WILLEMS,** "Frequency Domain Stability Criteria—Part II," *IEEE Trans. Automatic Control,* Vol. AC-10, pp. 407–413, Oct. 1965.
4. **T. R. NATESAN,** "A Supplement to the Note on the Generalized Nyquist Criterion," *IEEE Trans. Automatic Control,* Vol. AC-12, pp. 215–216, Apr. 1967.
5. **K. S. YEUNG,** "A Reformulation of Nyquist's Criterion," *IEEE Trans. Educ.,* Vol. E-28, pp. 58–60, Feb. 1985.

Sensitivity Function

6. **A. GELB,** "Graphical Evaluation of the Sensitivity Function Using the Nichols Chart," *IRE Trans. Automatic Control,* Vol. AC-7, pp. 57–58, July 1962.

Nyquist Criterion of Discrete-Data Systems

7. **K. S. YEUNG and H. M. LAI,** "A Reformulation of Nyquist Criterion for Discrete Systems," *IEEE Trans. Educ.*, Vol. 31, No. 1, Feb. 1988.

8. **B. C. KUO,** *Digital Control Systems,* 2nd ed., Saunders College Publishing, Philadelphia, 1992.

Problems

The following problems can all be solved using a computer program. Some of the simpler problems should be solved analytically for a better understanding of the fundamental principles.

▲ Analytical solutions of M_r, ω_r, and BW

9-1. The forward-path transfer function of a unity-feedback control system is

$$G(s) = \frac{K}{s(s + 6.54)}$$

Analytically, find the resonance peak M_r, resonant frequency ω_r, and bandwidth BW of the closed-loop system for the following values of K: **(a)** $K = 5$; **(b)** $K = 21.39$; **(c)** $K = 100$. Use the formulas for the second-order prototype system given in the book.

▲ Computer solutions of M_r, ω_r, and BW

9-2. Use a computer program such as **freqrp** of the **ACSP** software package or **bplot** of the **csad** toolbox to solve the following problems. Do not attempt to obtain the solutions analytically. The forward-path transfer functions of unity-feedback control systems are given in the following. Find the resonance peak, M_r, resonant frequency, ω_r, and bandwidth BW of the closed-loop systems. (*Reminder:* Make certain that the system is stable.)

(a) $G(s) = \dfrac{5}{s(1 + 0.5s)(1 + 0.1s)}$ **(b)** $G(s) = \dfrac{10}{s(1 + 0.5s)(1 + 0.1s)}$

(c) $G(s) = \dfrac{500}{(s + 1.2)(s + 4)(s + 10)}$ **(d)** $G(s) = \dfrac{10(s + 1)}{s(s + 2)(s + 10)}$

(e) $G(s) = \dfrac{0.5}{s(s^2 + s + 1)}$ **(f)** $G(s) = \dfrac{100e^{-s}}{s(s^2 + 10s + 50)}$

(g) $G(s) = \dfrac{100e^{-s}}{s(s^2 + 10s + 100)}$ **(h)** $G(s) = \dfrac{10(s + 5)}{s(s^2 + 5s + 5)}$

▲ Correlation between max. overshoot, rise time and M_r and BW

9-3. The specifications on a second-order unity-feedback control system with the closed-loop transfer function

$$M(s) = \frac{Y(s)}{R(s)} = \frac{\omega_n^2}{s^2 + 2\zeta\omega_n s + \omega_n^2}$$

are that the maximum overshoot must not exceed 10 percent and the rise time must be less than 0.1 sec. Find the corresponding limiting values of M_r and BW analytically.

9-4. Repeat Problem 9-3 for maximum overshoot ≤ 20 percent and $t_r \leq 0.2$ sec.

9-5. Repeat Problem 9-3 for maximum overshoot ≤ 30 percent and $t_r \leq 0.2$ sec.

▲ Correlation between frequency and time responses

9-6. The closed-loop frequency response $|M(j\omega)|$ versus frequency of a second-order prototype system is shown in Fig. 9P-6. Sketch the corresponding unit-step response of the system; indicate the values of the maximum overshoot, peak time, and the steady-state error due to a unit-step input.

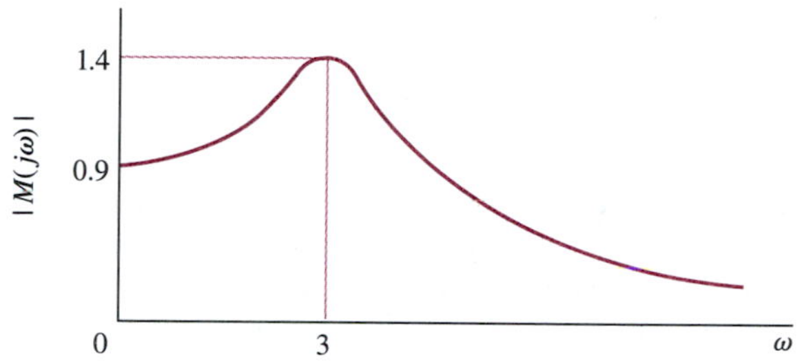

Figure 9P-6

▲ Find M_r and BW

9-7. The forward-path transfer function of a unity-feedback control system is

$$G(s) = \frac{1 + Ts}{2s(s^2 + s + 1)}$$

Find the values of BW and M_r of the closed-loop system for $T = 0.05$, 1, 2, 3, 4, and 5. Use a computer program for solutions.

▲ Find M_r and BW

9-8. The forward-path transfer function of a unity-feedback control system is

$$G(s) = \frac{1}{2s(s^2 + s + 1)(1 + Ts)}$$

Find the values of BW and M_r of the closed-loop system for $T = 0$, 0.5, 1, 2, 3, 4, and 5. Use a computer program for solutions.

▲ Stability analysis with Nyquist criterion

9-9. The loop transfer function $L(s)$ of single-feedback-loop systems are given below. Sketch the Nyquist plot of $L(j\omega)$ for $\omega = 0$ to $\omega = \infty$. Determine the stability of the closed-loop system. If the system is unstable, find the number of poles of the closed-loop transfer function that are in the right-half s-plane. Solve for the intersect of $L(j\omega)$ on the negative real axis of the $L(j\omega)$-plane analytically. You may construct the Nyquist plot of $L(j\omega)$ using any computer program.

(a) $L(s) = \dfrac{20}{s(1 + 0.1s)(1 + 0.5s)}$

(b) $L(s) = \dfrac{10}{s(1 + 0.1s)(1 + 0.5s)}$

(c) $L(s) = \dfrac{100(1+s)}{s(1+0.1s)(1+0.2s)(1+0.5s)}$ **(d)** $L(s) = \dfrac{10}{s^2(1+0.2s)(1+0.5s)}$

(e) $L(s) = \dfrac{5(s-2)}{s(s+1)(s-1)}$ **(f)** $L(s) = \dfrac{50}{s(s+5)(s-1)}$

(g) $L(s) = \dfrac{3(s+2)}{s(s^3+3s+1)}$ **(h)** $L(s) = \dfrac{0.1}{s(s+1)(s^2+s+1)}$

(i) $L(s) = \dfrac{100}{s(s+1)(s^2+2)}$ **(j)** $L(s) = \dfrac{s^2-5s+2}{s(s^3+2s^2+2s+10)}$

(k) $L(s) = \dfrac{-0.1(s^2-1)(s+2)}{s(s^2+s+1)}$ **(l)** $L(s) = \dfrac{10(s+10)}{s(s+1)(s+100)}$

(m) $L(s) = \dfrac{e^{-0.1s}}{s(s+1)(s+10)}$ **(n)** $L(s) = \dfrac{(s+2)e^{-0.2s}}{s(s^2+s+1)}$

▲ Stability analysis with Nyquist criterion

9-10. The loop transfer functions of single-feedback-loop control systems are given in the following. Apply the Nyquist criterion and determine the values of K for the system to be stable. Sketch the Nyquist plot of $L(j\omega)$ with $K = 1$ for $\omega = 0$ to $\omega = \infty$. You may use a computer program to plot the Nyquist plots.

(a) $L(s) = \dfrac{K}{s(s+2)(s+10)}$ **(b)** $L(s) = \dfrac{K(s+1)}{s(s+2)(s+5)(s+15)}$

(c) $L(s) = \dfrac{K}{s^2(s+2)(s+10)}$ **(d)** $L(s) = \dfrac{K(s-2)}{s(s^2-1)}$

(e) $L(s) = \dfrac{K}{s(s+10)(s-2)}$ **(f)** $L(s) = \dfrac{K(s+1)}{s(s^3+3s+1)}$

(g) $L(s) = \dfrac{K(s^2-5s+2)}{s(s^3+2s^2+2s+10)}$ **(h)** $L(s) = \dfrac{K(s^2-1)(s+2)}{s(s^2+s+1)}$

(i) $L(s) = \dfrac{K(s^2-5s+1)}{s(s+1)(s^2+4)}$ **(j)** $L(s) = \dfrac{K(s+1)e^{-s}}{s(s^2+s+1)}$

▲ Application of Nyquist criterion

9-11. The forward-path transfer function of a unity-feedback control system is

$$G(s) = \frac{K}{(s+5)^n}$$

Determine by means of the Nyquist criterion the range of K ($-\infty < K < \infty$) for the closed-loop system to be stable. Sketch the Nyquist plot of $G(j\omega)$ for $\omega = 0$ to $\omega = \infty$ and, **(a)** $n = 2$, **(b)** $n = 3$, and **(c)** $n = 4$.

▲ Application of Nyquist criterion

9-12. Figure 9P-12 shows the Nyquist plots of the loop transfer function $L(j\omega)$ for $\omega = 0$ to $\omega = \infty$ for single-feedback-loop control systems. The number of poles of $L(j\omega)$ that are on the $j\omega$-axis, P_ω, and in the right-half s-plane, P, are indicated for each case. Determine the stability of the closed-loop system by applying the Nyquist criterion. For the unstable systems, give the number of zeros of $1 + L(s)$ that are in the right-half s-plane.

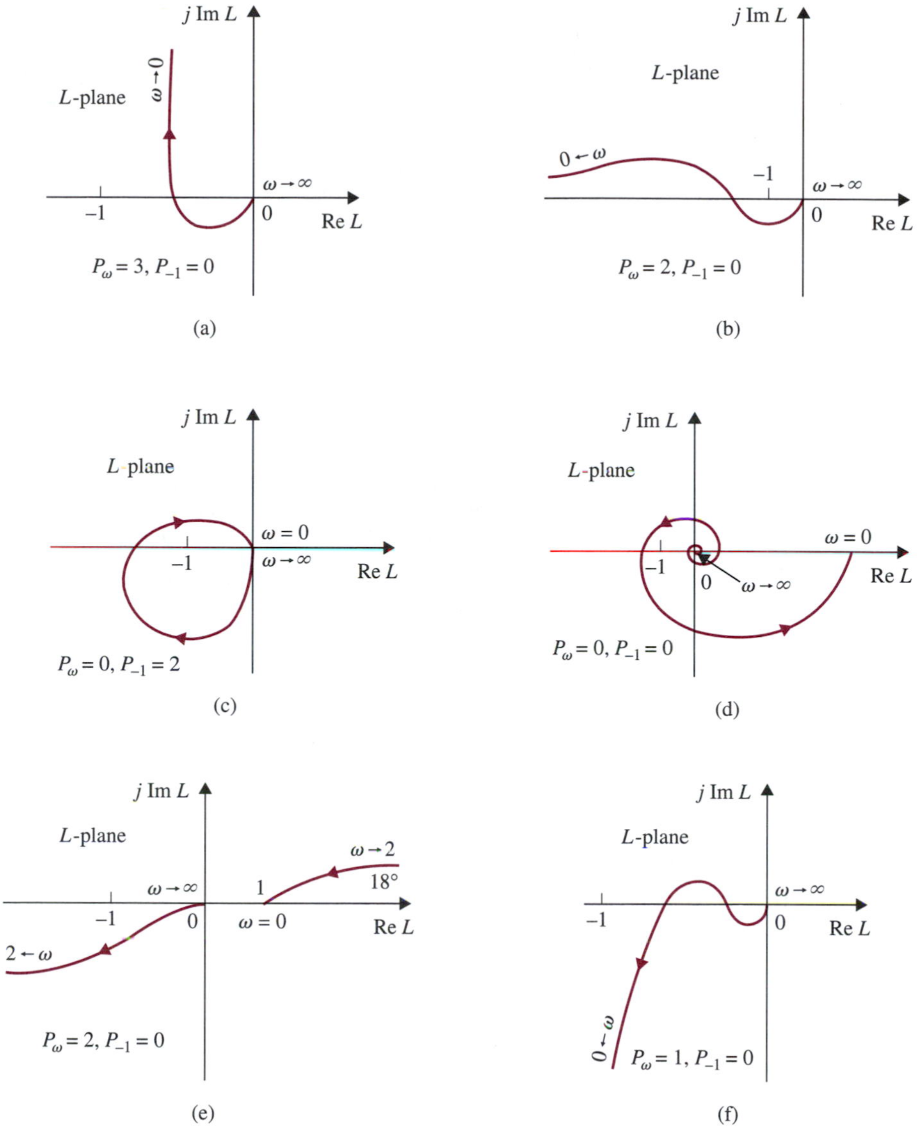

Figure 9P-12

▲ Application of Nyquist criterion to improper functions

9-13. It is mentioned in the book that when the function $L(j\omega)$ has more zeros than poles, it is necessary to plot the Nyquist plot of $1/L(j\omega)$ in order to apply the simplified Nyquist criterion. Determine the stability of the systems described by the function $1/L(j\omega)$ shown in Fig. 9P-13. For each case the values of P_ω and P for the function $1/L(j\omega)$ are given, where P_ω refers to the number of poles of $1/L(j\omega)$ that are on the $j\omega$-axis, and P refers to the number of poles of $1 + 1/L(j\omega)$ that are in the right-half s-plane.

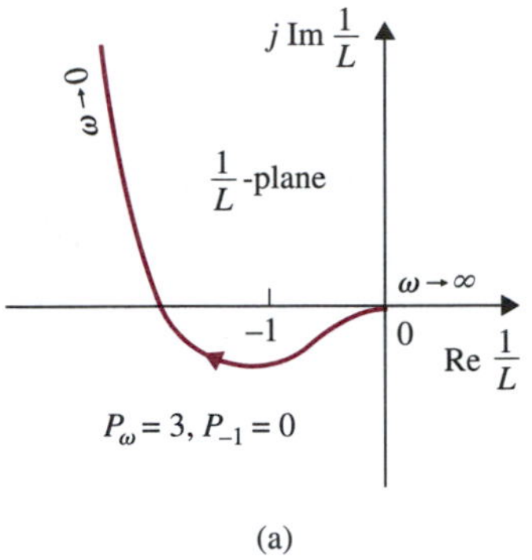

Figure 9P-13

▲ Application of Nyquist criterion

9-14. Figure 9P-14 shows the Nyquist plots of loop transfer function $L(j\omega)$ for $\omega = 0$ to $\omega = \infty$ for single-feedback-loop control systems. The gain K appears as a multiplying factor in $L(s)$. The number of poles of $L(j\omega)$ that are on the $j\omega$-axis and in the right-

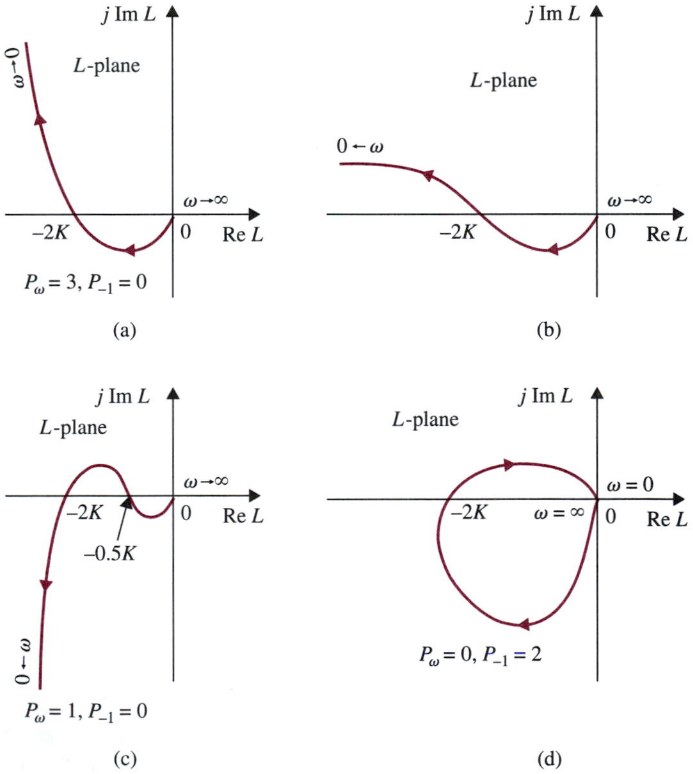

Figure 9P-14

half s-plane are indicated in each case. Determine the range(s) of K for closed-loop system stability.

▲ Application of Nyquist and Routh–Hurwitz criteria

9-15. The characteristic equations of linear control systems are given below. Apply the Nyquist criterion to determine the values of K for system stability. Check the answers by means of the Routh–Hurwitz criterion.

(a) $s^3 + 4Ks^2 + (K + 5)s + 10 = 0$
(b) $s^3 + K(s^3 + 2s^2 + 1) = 0$
(c) $s(s + 1)(s^2 + 4) + K(s^2 + 1) = 0$
(d) $s^3 + 2s^2 + 20s + 10K = 0$
(e) $s(s^3 + 3s + 3) + K(s + 2) = 0$
(f) $s(s^3 + 2s^2 + s + 1) + K(s^2 + s + 1) = 0$

▲ Stability of system with PD controller

9-16. The forward-path transfer function of a unity-feedback control system with a PD (proportional-derivative) controller is

$$G(s) = \frac{10(K_P + K_D s)}{s^2}$$

Select the value of K_P so that the parabolic-error constant K_a is 100. Find the equivalent forward-path transfer function $G_{eq}(s)$ for $\omega = 0$ to $\omega = \infty$. Determine the range of K_D for stability by the Nyquist criterion.

▲ Application of Nyquist and Routh–Hurwitz criteria

9-17. The block diagram of a feedback control system is shown in Fig. 9P-17.

(a) Apply the Nyquist criterion to determine the range of K for stability.
(b) Check the answer obtained in part (a) with the Routh–Hurwitz criterion.

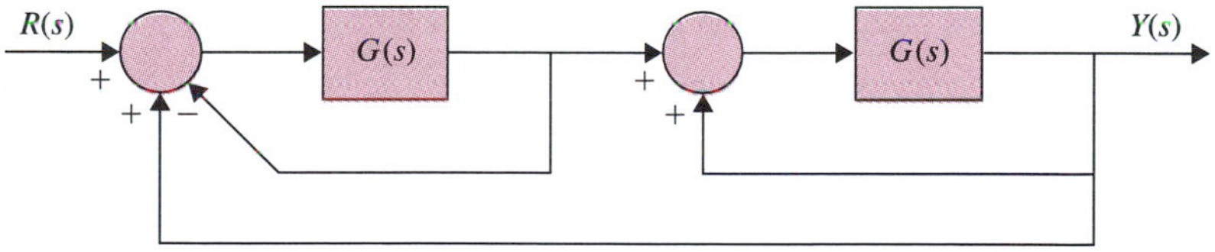

$$G(s) = \frac{K}{(s+4)(s+5)}$$

Figure 9P-17

▲ Stability study of liquid-level control system by Nyquist criterion

9-18. The forward-path transfer function of the liquid-level control system shown in Fig. 6P-13 is

$$G(s) = \frac{K_a K_i n K_I N}{s(R_a J s + K_i K_b)(A s + K_o)}$$

The following system parameters are given: $K_a = 50$, $K_i = 10$, $K_I = 50$, $J = 0.006$, $K_b = 0.0706$, $n = 0.01$, and $R_a = 10$. The values of A, N, and K_o are variable.

(a) For $A = 50$ and $K_o = 100$, sketch the Nyquist plot of $G(j\omega)$ for $\omega = 0$ to ∞ with N as a variable parameter. Find the maximum integer value of N so that the closed-loop system is stable.

(b) Let $N = 10$ and $K_o = 100$. Sketch the Nyquist plot of an equivalent transfer function $G_{eq}(j\omega)$ that has A as a multiplying factor. Find the critical value of K_o for stability.

(c) For $A = 50$ and $N = 10$, sketch the Nyquist plot of an equivalent transfer function $G_{eq}(j\omega)$ that has K_o as a multiplying factor. Find the critical value of K_o for stability.

▲ Stability study of dc-motor control system

9-19. The block diagram of a dc-motor control system is shown in Fig. 9P-19. Determine the range of K for stability using the Nyquist criterion when K_t has the following values: **(a)** $K_t = 0$, **(b)** $K_t = 0.01$, and **(c)** $K_t = 0.1$.

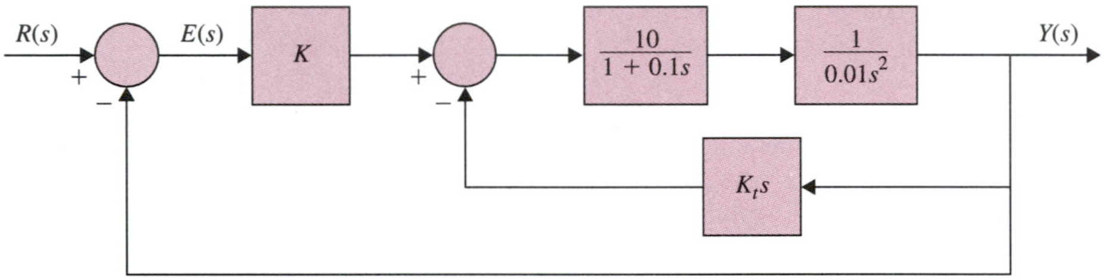

Figure 9P-19

▲ Stability study of dc-motor control system

9-20. For the system shown in Fig. 9P-19, let $K = 10$. Find the range of K_t for stability with the Nyquist criterion.

▲ Stability study of steel-rolling control system

9-21. The steel-rolling control system shown in Fig. 4P-26 has the forward-path transfer function

$$G(s) = \frac{100Ke^{-T_d s}}{s(s^2 + 10s + 100)}$$

(a) When $K = 1$, determine the maximum time delay T_d in seconds for the closed-loop system to be stable.

(b) When the time delay T_d is 1 sec, find the maximum value of K for system stability.

▲ Stability study of steel-rolling control system

9-22. Repeat Problem 9-21 with the following conditions.

(a) When $K = 0.1$, determine the maximum time delay T_d in seconds for the closed-loop system to be stable.

(b) When the time delay T_d is 0.1 sec, find the maximum value of K for system stability.

▲ Stability study of chemical-solution concentration control system

9-23. The system schematic shown in Fig. 9P-23 is devised to control the concentration of a chemical solution by mixing water and concentrated solution in appropriate proportions. The transfer function of the system components between the amplifier output e_a (V) and the valve position x (in.) is

$$\frac{X(s)}{E_a(s)} = \frac{K}{s^2 + 10s + 100}$$

When the sensor is viewing pure water, the amplifier output voltage e_a is zero; when it is viewing concentrated solution, $e_a = 10$ V; 0.1 in. of the valve motion changes the output concentration from zero to maximum. The valve ports can be assumed to be shaped so that the output concentration varies linearly with the valve position. The output tube has a cross-sectional area of 0.1 in.^{2}, and the rate of flow is 10^3 in./sec regardless of the valve position. To make sure that the sensor views a homogeneous solution, is it desirable to place it at some distance D in. from the valve.

(a) Derive the loop transfer function of the system.

(b) When $K = 10$, find the maximum distance D (in.) so that the system is stable. Use the Nyquist stability criterion.

(c) Let $D = 10$ in. Find the maximum value of K for system stability.

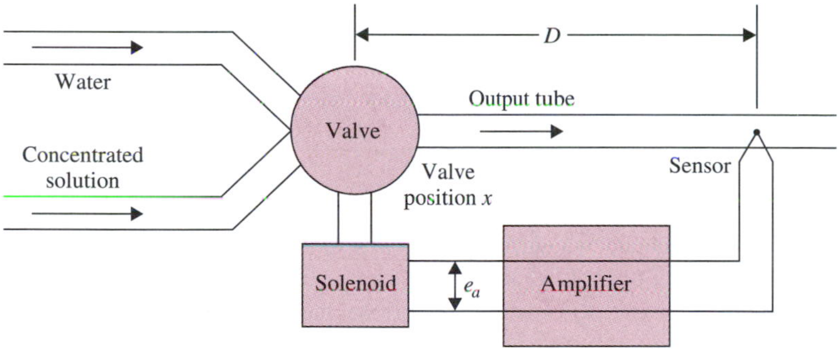

Figure 9P-23

▲ Stability study of chemical-solution concentration

9-24. For the mixing system described in Problem 9-23, the following system parameters are given: When the sensor is viewing pure water, the amplifier output voltage $e_s = 0$ V; when it is viewing concentrated solution, $e_a = 1$ V; 0.1 in. of the valve motion changes the output concentration from zero to maximum. The rest of the system characteristics are the same as in Problem 9-23. Repeat the three parts of Problem 9-23.

▲ Frequency-domain analysis of low-order system approximation

9-25. The approximation of high-order systems by low-order systems discussed in Section 7-9 is based on a frequency-domain criterion. The transfer functions of high-order forward-path transfer functions and the corresponding low-order approximating transfer functions determined in the illustrative examples in Chapter 7 are given below. Compute and compare the values of M_r, ω_r, and BW of the high-order systems with those of the low-order systems, so that the accuracy of the approximations can be judged. Use a computer program for the solutions.

(a) $G_H(s) = \dfrac{8}{s(s^2 + 6s + 12)}$ $\quad G_L(s) = \dfrac{2.31}{s(s + 2.936)}$ (Example 7-8)

(b) $G_H(s) = \dfrac{0.909}{s(1 + 0.5455s + 0.0455s^2)}$ $\quad G_L(s) = \dfrac{0.995}{s(1 + 0.4975s)}$

(Example 7-9, $T = 0.1$)

(c) $G_H(s) = \dfrac{0.5}{s(1 + 0.75s + 0.25s^2)}$ $\quad G_L(s) = \dfrac{0.707}{s(1 + 0.3536s)}$

(Example 7-9, $T = 1.0$)

(d) $G_H(s) = \dfrac{90.3}{s(1 + 0.00283s + 8.3056 \times 10^7 s^2)}$ $\quad G_L(s) = \dfrac{92.943}{s(1 + 0.002594s)}$

(Example 7-10, $K = 7.248$)

(e) $G_H(s) = \dfrac{180.6}{s(1 + 0.00283s + 8.3056 \times 10^7 s^2)}$ $\quad G_L(s) = \dfrac{189.54}{s(1 + 0.002644s)}$

(Example 7-10, $K = 14.5$)

(f) $G_H(s) = \dfrac{1245.52}{s(1 + 0.00283s + 8.3056 \times 10^7 s^2)}$ $\quad G_L(s) = \dfrac{2617.56}{s(1 + 0.0053s)}$

(Example 7-10, $K = 100$)

▲ M_r, ω_r and BW of third-order and second-order systems

9-26. The forward-path transfer function of a unity-feedback control system is

$$G(s) = \frac{1000}{s(s^2 + 105s + 600)}$$

(a) Find the values of M_r, ω_r, and BW of the closed-loop system.

(b) Find the parameters of the second-order system with the open-loop transfer function

$$G_L(s) = \frac{\omega_n^2}{s(s + 2\zeta\omega_n)}$$

that will give the same values for M_r and ω_r as the third-order system. Compare the values of BW of the two systems.

▲ M_r, ω_r and BW of third-order and second-order systems

9-27. Repeat Problem 9-26 for the transfer function given in Problem 9-25(f). (Although the approach used here may render results that are better or at least as good as those obtained by the method outlined in Section 7-9, the match here is restricted only to a prototype second-order system as the approximating low-order system.)

▲ Gain and phase margins from Bode plots

9-28. Sketch or plot the Bode diagrams of the forward-path transfer functions given in Problem 9-2. Find the gain margin, gain-crossover frequency, phase margin, and the phase-crossover frequency for each system.

▲ Gain and phase margins from Bode plots

9-29. The forward-path transfer functions of unity-feedback control systems are given in the following. Plot the Bode diagram of $G(j\omega)/K$ and do the following: **(i)** find the value of K so that the gain margin of the system is 20 dB; **(ii)** find the value of K so that the phase margin of the system is 45°.

(a) $G(s) = \dfrac{K}{s(1 + 0.1s)(1 + 0.5s)}$ **(b)** $G(s) = \dfrac{K(s + 1)}{s(1 + 0.1s)(1 + 0.2s)(1 + 0.5s)}$

(c) $G(s) = \dfrac{K}{(s + 3)^3}$ **(d)** $G(s) = \dfrac{K}{(s + 3)^4}$

(e) $G(s) = \dfrac{Ke^{-s}}{s(1 + 0.1s + 0.01s^2)}$ **(f)** $G(s) = \dfrac{K(1 + 0.5s)}{s(s^2 + s + 1)}$

▲ Application of Nichols chart and gain–phase plots

9-30. The forward-path transfer functions of unity-feedback control systems are given below. Plot $G(j\omega)/K$ in the gain–phase coordinates of the Nichols chart, and do the following: **(i)** find the value of K so that the gain margin of the system is 10 dB; **(ii)** find the value of K so that the phase margin of the system is 45°; **(iii)** find the value of K so that $M_r = 1.2$.

(a) $G(s) = \dfrac{10K}{s(1 + 0.1s)(1 + 0.5s)}$ **(b)** $G(s) = \dfrac{5K(s + 1)}{s(1 + 0.1s)(1 + 0.2s)(1 + 0.5s)}$

(c) $G(s) = \dfrac{10K}{s(1 + 0.1s + 0.01s^2)}$ **(d)** $G(s) = \dfrac{Ke^{-s}}{s(1 + 0.1s + 0.01s^2)}$

▲ Applications of Bode plot

9-31. The Bode diagram of the forward-path transfer function of a unity-feedback control system is obtained experimentally and is shown in Fig. 9P-31 when the forward gain K is set at its nominal value.

(a) Find the gain and phase margins of the system from the diagram as best you can read. Find the gain and phase-crossover frequencies.

(b) Repeat part (a) if the gain is doubled from its nominal value.

(c) Repeat part (a) if the gain is 10 times its nominal value.

(d) Find out how much the gain must be changed from its nominal value if the gain margin is to be 40 dB.

(e) Find out how much the loop gain must be changed from its nominal value if the phase margin is to be 45°.

(f) Find the steady-state error of the system if the reference input to the system is a unit-step function.

(g) The forward path now has a pure time delay of T_d seconds, so that the forward-path transfer function is multiplied by $e^{-T_d s}$. Find the gain margin and the phase margin for $T_d = 0.05$ sec. The gain is set at nominal.

(h) With the gain set at nominal, find the maximum time delay T_d the system can tolerate without going into instability.

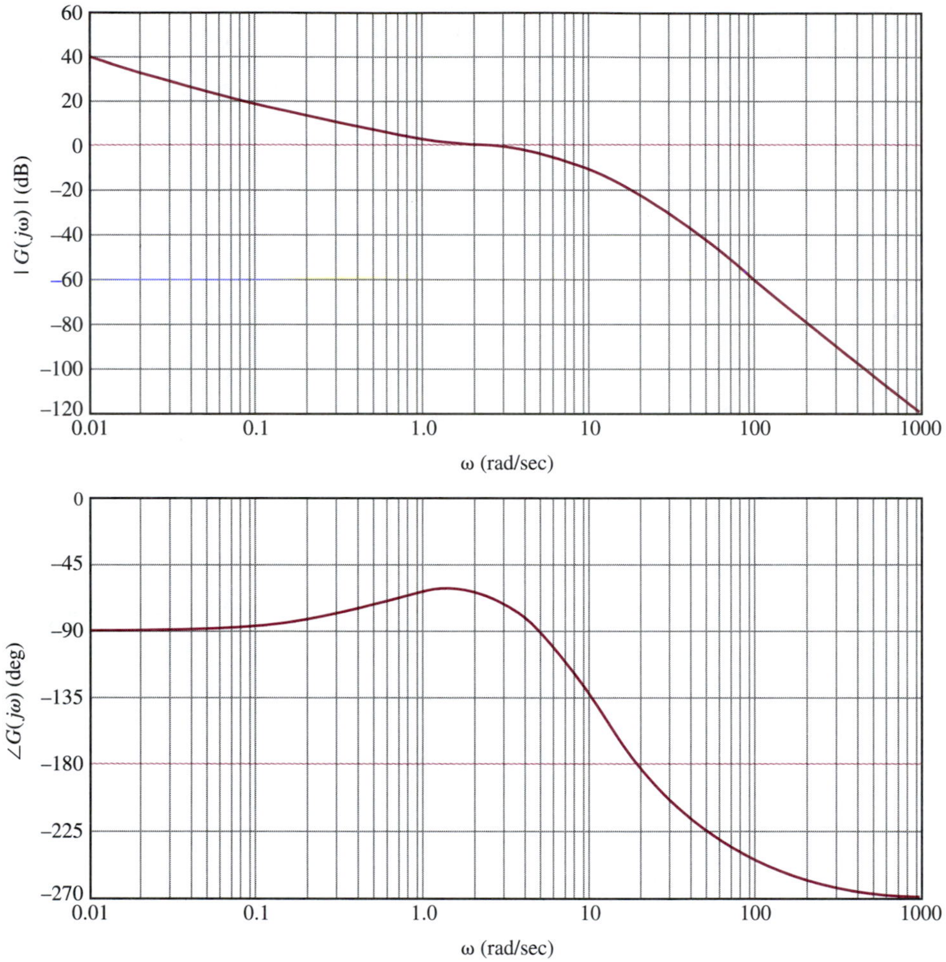

Figure 9P-31

▲ Applications of Bode plots

9-32. Repeat Problem 9-31 using Fig. 9P-31 for the following parts.

(a) Find the gain and phase margins if the gain is four times its nominal value. Find the gain- and phase-crossover frequencies.

(b) Find out how much the gain must be changed from its nominal value if the gain margin is to be 20 dB.

(c) Find the marginal value of the forward-path gain for system stability.

(d) Find out how much the gain must be changed from its nominal value if the phase margin is to be 60°.

(e) Find the steady-state error of the system if the reference input is a unit-step function and the gain is twice its nominal value.

(f) Find the steady-state error of the system if the reference input is a unit-step function and the gain is 20 times its nominal value.

(g) The system now has a pure time delay so that the forward-path transfer function is multiplied $e^{-T_d s}$. Find the gain and phase margins when $T_d = 0.1$ sec. The gain is set at its nominal value.

(h) With the gain set at 10 times its nominal, find the maximum time delay T_d the system can tolerate without going into instability.

▲ Relationships between Bode, Nyquist plot, and root loci

9-33. The forward-path transfer function of a unity-feedback control system is

$$G(s) = \frac{K(1 + 0.2s)(1 + 0.1s)}{s^2(1 + s)(1 + 0.01s)^2}$$

(a) Construct the Bode and Nyquist plots of $G(j\omega)/K$ and determine the range of K for system stability.

(b) Construct the root loci of the system for $K \geq 0$. Determine the values of K and ω at the points where the root loci cross the $j\omega$-axis, using the information found from the Bode plot.

▲ Relations between Bode, Nyquist plot, and root loci

9-34. Repeat Problem 9-33 for the following transfer function:

$$G(s) = \frac{K(s + 1)(s + 2)}{s^2(s^2 + 2s + 2)}$$

▲ Bode plot and relative stability of dc-motor control system

9-35. The forward-path transfer function of the dc-motor control system described in Fig. 4P-17 is

$$G(s) = \frac{6.087 \times 10^8 K}{s(s^3 + 423.42s^2 + 2.6667 \times 10^6 s + 4.2342 \times 10^8)}$$

Plot the Bode diagram of $G(j\omega)$ with $K = 1$, and determine the gain margin and phase margin of the system. Find the critical value of K for stability.

▲ Relative stability of robot-arm control system

9-36. The transfer function between the output position $\Theta_L(s)$ and the motor current $I_a(s)$ of the robot arm modeled in Fig. 4P-21 is

$$G_p(s) = \frac{\Theta_L(s)}{I_a(s)} = \frac{K_i(Bs + K)}{\Delta_o}$$

where

$$\Delta_o(s) = s\{J_L J_m s^3 + [J_L(B_m + B) + J_m(B_L + B)]s^2 + [B_L B_m + (B_L + B_m)B + (J_m + J_L)K]s + K(B_L + B_m)\}$$

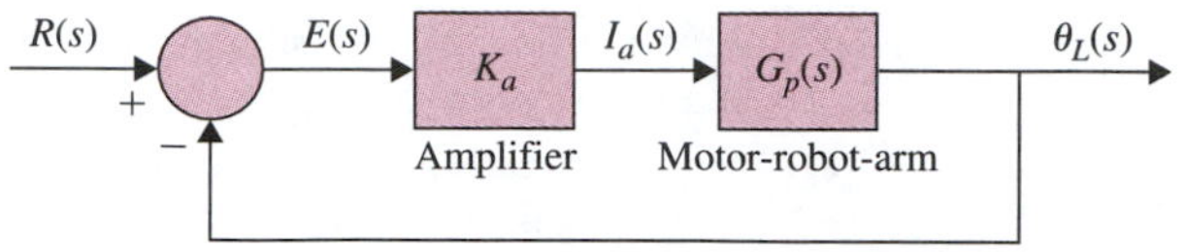

Figure 9P-36

The arm is controlled by a closed-loop system, as shown in Fig. 9P-36. The system parameters are $K_a = 65$, $K = 100$, $K_i = 0.4$, $B = 0.2$, $J_m = 0.2$, $B_L = 0.01$, $J_L = 0.6$, and $B_m = 0.25$.

(a) Derive the forward-path transfer function $G(s) = \Theta_L(s)/E(s)$.

(b) Draw the Bode diagram of $G(j\omega)$. Find the gain and phase margins of the system.

(c) Draw $|M(j\omega)|$ versus ω, where $M(s)$ is the closed-loop transfer function. Find M_r, ω_r, and BW.

▲ Relative stability analysis from gain–phase plot

9-37. The gain–phase plot of the forward-path transfer function of $G(j\omega)/K$ of a unity-feedback control system is shown in Fig. 9P-37. Find the following performance characteristics of the system.

(a) Gain-crossover frequency (rad/sec) when $K = 1$.

(b) Phase-crossover frequency (rad/sec) when $K = 1$.

(c) Gain margin (dB) when $K = 1$.

(d) Phase margin (deg) when $K = 1$.

(e) Resonance peak M_r when $K = 1$.

(f) Resonant frequency ω_r (rad/sec) when $K = 1$.

(g) BW of the closed-loop system when $K = 1$.

(h) The value of K so that the gain margin is 20 dB.

(i) The value of K so that the system is marginally stable. Find the frequency of sustained oscillation in rad/sec.

(j) Steady-state error when the reference input is a unit-step function.

▲ Similar to Problem 9-37

9-38. Repeat parts (a) through (g) of Problem 9-37 when $K = 5$ dB. Repeat part (h) for gain margin = 40 dB.

▲ Relative stability of system with time delay

9-39. For the gain–phase plot of $G(j\omega)/K$ shown in Fig. 9P-37, the system now has a pure time delay of T_d in the forward path, so that the forward-path transfer function becomes $G(s)e^{-T_d s}$.

(a) With $K = 1$, find T_d so that the phase margin is 40°.

(b) With $K = 1$, find the maximum value of T_d so that the system will remain stable.

9-40. Repeat Problem 9-39 with $K = 10$.

9-41. Repeat Problem 9-39(a) so that the gain margin is 5 dB when $K = 1$.

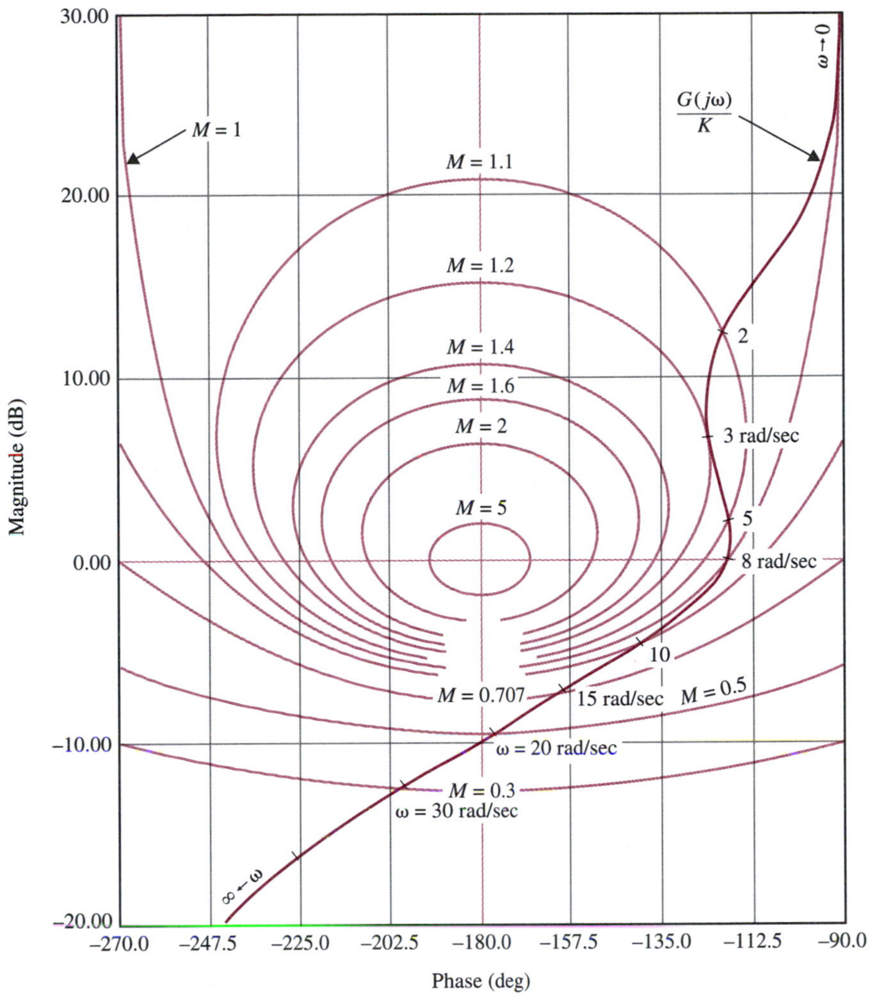

Figure 9P-37

▲ Relative stability of furnace-control system

9-42. The block diagram of a furnace-control system is shown in Fig. 9P-42. The transfer function of the process is

$$G_p(s) = \frac{1}{(1 + 10s)(1 + 25s)}$$

The time delay T_d is 2 sec.

(a) Plot the Bode diagram of $G(s) = Y(s)/E(s)$ and find the gain-crossover and phase-crossover frequencies. Find the gain margin and the phase margin.

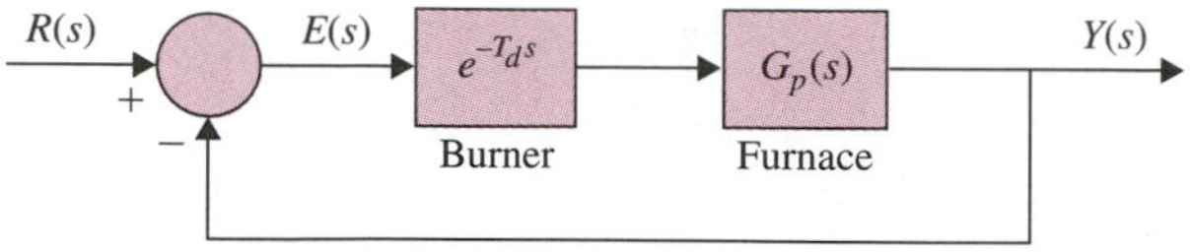

Figure 9P-42

(b) Approximate the time delay by [Eq. (4-170)]

$$e^{-T_d s} \cong \frac{1}{1 + T_d s + T_d s^2/2}$$

and repeat part (a). Comment on the accuracy of the approximation. What is the maximum frequency below which the polynomial approximation is accurate?

(c) Repeat part (b) for approximating the time-delay term by [Eq. (4-171)]

$$e^{-T_d s} \cong \frac{1 - T_d s/2}{1 + T_d s/2}$$

9-43. Repeat Problem 9-42 with $T_d = 1$ sec.

▲ Sensitivity problem

9-44. Plot the $|S_G^M(j\omega)|$ versus ω plot for the system described in Problem 9-36 for $K = 1$. Find the frequency at which the sensitivity is maximum and the value of the maximum sensitivity.

▲ Frequency-domain analysis of discrete-data system

9-45. The forward-path transfer function of a unity-feedback discrete-data control system with sample-and-hold is

$$G_{h0}G(z) = \frac{0.0952z}{(z-1)(z-0.905)}$$

The sampling period is $T = 0.1$ sec.

(a) Plot the Bode plot of $G_{h0}G(z)$ and determine the stability of the closed-loop system.

(b) Apply the w-transformation of Eq. (9-184) to $G_{h0}G(z)$, and plot the Bode plot of $G_{h0}G(w)$. Find the gain and phase margins of the system.

▲ Frequency-domain analysis of liquid-level control system with sampled data

9-46. Consider that the liquid-level control system described in Problem 6-13 is now subject to sample-and-hold operation. The forward-path transfer function of the system is

$$G_{h0}G(z) = \frac{1 - e^{-Ts}}{s}\mathcal{Z}\left(\frac{16.67N}{s(s+1)(s+12.5)}\right)$$

The sampling period is 0.05 sec. The parameter N represents the number of inlet valves. Construct the Bode plot of $G_{h0}G(w)$ and determine the limiting value of N (integer) for the closed-loop system to be stable.

Additional Computer Problems

9-47. Use any computer program to analyze the frequency response of the following unity-feedback control systems. Plot the Bode diagrams, polar plots, and gain–phase plots, and compute the phase margin, gain margin, M_r, and BW.

(a) $G(s) = \dfrac{1 + 0.1s}{s(s+1)(1+0.01s)}$ (b) $G(s) = \dfrac{0.5(s+1)}{s(1+0.2s)(1+s+0.5s^2)}$

(c) $G(s) = \dfrac{(s+1)}{s(1+0.2s)(1+0.5s)}$ (d) $G(s) = \dfrac{1}{s(1+s)(1+0.5s)}$

(e) $G(s) = \dfrac{50}{s(s+1)(1+0.5s^2)}$ (f) $G(s) = \dfrac{0.2(1-s^2)(1+0.5s)}{s(s^2+s+1)}$

(g) $G(s) = \dfrac{(1+0.1s)e^{-0.1s}}{s(s+1)(1+0.01s)}$ (h) $G(s) = \dfrac{10e^{-0.1s}}{s^2+2s+2}$

Answers to True-and-False Review Questions

9.	(T)	10.	(T)	11.	(T)	12.	(T)	13.	(T)	14.	(F)	15.	(T)
16.	(T)	17.	(T)	18.	(T)	19.	(T)	20.	(T)	21.	(F)		

10 Design of Control Systems

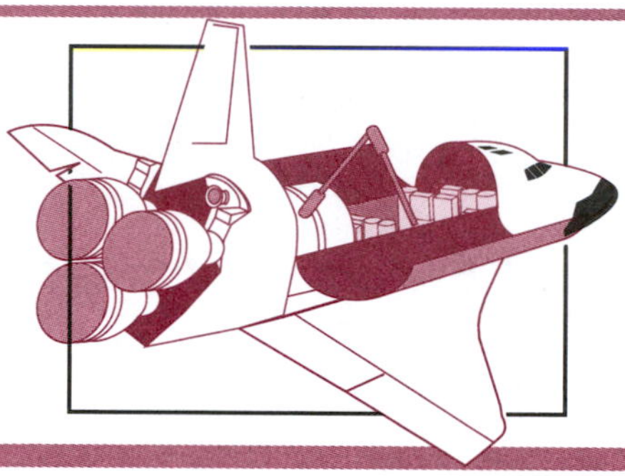

KEYWORDS AND TOPICS

- ▲ Introduction
- ▲ PD Controller
- ▲ PI Controller
- ▲ PID Controller
- ▲ Phase-Lead Controller
- ▲ Phase-Lag Controller
- ▲ Lead-Lag Controller
- ▲ Pole–Zero Cancellation
- ▲ Robust Control
- ▲ Minor-Loop Control
- ▲ Pole-Placement Design
- ▲ Bridged-T Controller
- ▲ State Feedback
- ▲ State Feedback with Integral Control

10-1 Introduction

All the foundations we have laid for analysis in the preceding chapters lead to the ultimate goal of design of control systems. Starting with a controlled process such as that shown by the block diagram in Fig. 10-1, control system design involves the following three steps:

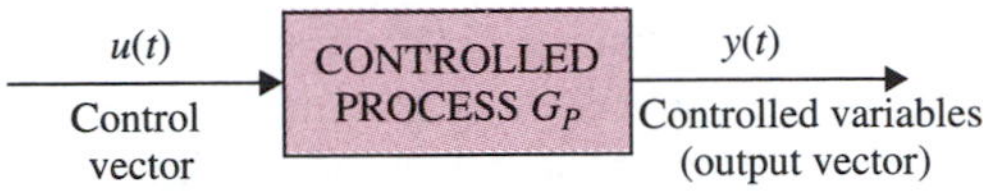

Figure 10-1 Controlled process.

1. Determine what the system should do and how to do it (design specifications).
2. Determine the controller or compensator configuration relative to how it is connected to the controlled process.
3. Determine the parameter values of the controller to achieve the design goals.

These design tasks are explored further in the following sections.

10-1-1 Design Specifications

We often use design specifications to describe what the system should do and how it is done. These specifications are unique to each individual application and often include specifications about **relative stability, steady-state accuracy (error), transient-response,** and **frequency-response characteristics.** In some applications there may be additional specifications on **sensitivity to parameter variations** (i.e., **robustness,** or **disturbance rejection**).

The design of linear control systems can be carried out in either the time or the frequency domain. For instance, **steady-state accuracy** is often specified with respect to a step input, a ramp input, or a parabolic input, and the design to meet a certain requirement is more conveniently carried out in the time domain. Other specifications, such as **maximum overshoot, rise time,** and **settling time,** are all defined for a unit-step input, and therefore are used specifically for time-domain design. We have learned that relative stability is also measured in terms of **gain margin, phase margin,** and **M_r.** These are typical frequency-domain specifications and should be used in conjunction with such tools as Bode plot, polar plot, gain–phase plot, and Nichols chart.

We have shown that for a second-order prototype system, there are simple analytical relationships between some of these time- and frequency-domain specifications. However, for higher-order systems, correlation between time- and frequency-domain specifications are difficult to establish. As pointed out earlier, the analysis and design of control systems is pretty much an exercise of selecting from several alternative methods for solving the same problem. Thus the choice of whether the design should be conducted in the time or frequency domain often depends on the preference of the designer. We should be quick to point out, however, that in most cases, time-domain specifications such as maximum overshoot, rise time, and settling time are usually used as the final measure of system performance. To an inexperienced designer, it is difficult to comprehend the physical connection between frequency-domain specifications such as gain and phase margins, resonance peak, to actual system performance. For instance, does a gain margin of 20 dB guarantee a maximum overshoot of less than 10 percent? To a designer it makes more sense to specify, for example, that the maximum overshoot should be less than 5 percent and a settling time less than 0.01 sec. It is less obvious what, for example, a phase margin of 60° and an M_r of less than 1.1 may bring in system performance. The following outline will hopefully clarify and explain the choices and reasons of using time- versus frequency-domain specifications.

1. Historically, the design of linear control systems was developed with a wealth of graphical tools such as Bode plot, Nyquist plot, gain–phase plot, and Nichols chart, which are all carried out in the frequency domain. The advantage of these tools is that they can all be sketched by following approximation methods without detailed plotting. Therefore, the designer can carry out designs using frequency-domain specifications such as **gain margin, phase margin, M_r,** and so on. High-order systems generally do not pose any particular problem. For certain types of controller, design procedures in the frequency are available which reduce the trial-and-error effort to a minimum.
2. Design in the time domain using such performance specifications as **rise time, delay time, settling time, maximum overshoot,** and so on, is feasible *analytically* only for second-order systems, or systems that can be approximated by second-order systems. General design procedures using time-domain specifications are difficult to establish for systems with order higher than the second.

▲ Historical advantages of frequency-domain designs are diminished by the availability of computer software tools

The development and availability of high-powered and user-friendly computer software is rapidly changing the practice of control system design, which until recently has been dictated by historical development. With modern computer software tools, the designer can go through a large number of design runs using the time-domain specifications within a matter of minutes. This considerably diminishes the historical edge of the frequency-domain design, which is based on the convenience of performing graphical design manually. In addition, it is generally difficult, except for the experienced, to select a meaningful set of frequency-domain specifications that will correspond to the desired time-domain performance requirements. For example, specifying a phase margin of 60° would be meaningless unless we know that it corresponds to a certain maximum overshoot. As it turns out, in general, to control maximum overshoot, one has to specify at least phase margin and M_r. Eventually, establishing an intelligent set of frequency-domain specifications becomes a trial-and-error process that precedes the actual design, which often is also a trial-and-error effort. However, frequency-domain methods are still valuable in interpreting noise rejection, and sensitivity properties of the system, most of all, offer another perspective to the design process. Therefore, in this chapter, the design techniques in the time and frequency domains are treated side by side, so that a comparison and cross-reference can be made between the alternative methods.

10-1-2 Controller Configurations

In general, the dynamics of a linear controlled process can be represented by the block diagram shown in Fig. 10-1. The design objective is to have the controlled variables, represented by the output vector $\mathbf{y}(t)$, behave in certain desirable ways. The problem essentially involves the determination of the control signal $\mathbf{u}(t)$ over the prescribed time interval so that the design objectives are all satisfied.

Most of the conventional design methods in control systems rely on the **fixed-configuration design,** in that at the outset the designer decides the basic configuration of the overall designed system and the place where the controller is to be positioned relative to the controlled process. The problem then involves the design of the elements of the controller. Because most control efforts involve the modification or compensation of the system-performance characteristics, the general design using fixed configuration is also called **compensation.**

Figure 10-2 illustrates several commonly used system configurations with controller compensation. These are described briefly as follows.

- ▲ **Series (cascade) compensation.** Figure 10-2(a) shows the most commonly used system configuration with the controller placed in series with the controlled process, and the configuration is referred to as **series** or **cascade compensation.**
- ▲ **Feedback compensation.** In Fig. 10-2(b) the controller is placed in the minor feedback path and the scheme is called **feedback compensation.**
- ▲ **State feedback compensation.** Figure 10-2(c) shows a system that generates the control signal by feeding back the state variables through constant real gains, and the scheme is known as **state feedback.** The problem with state-feedback control is that for high-order systems, the large number of state variables involved would require a large number of tranducers to sense the state variables for feedback. Thus actual implementation of the state-feedback control scheme may be costly or impractical. Even for low-order systems, often not all the state variables are directly accessible, and an **observer** or **estimator** may be necessary to create the estimated state variables from measurements of the output variables.

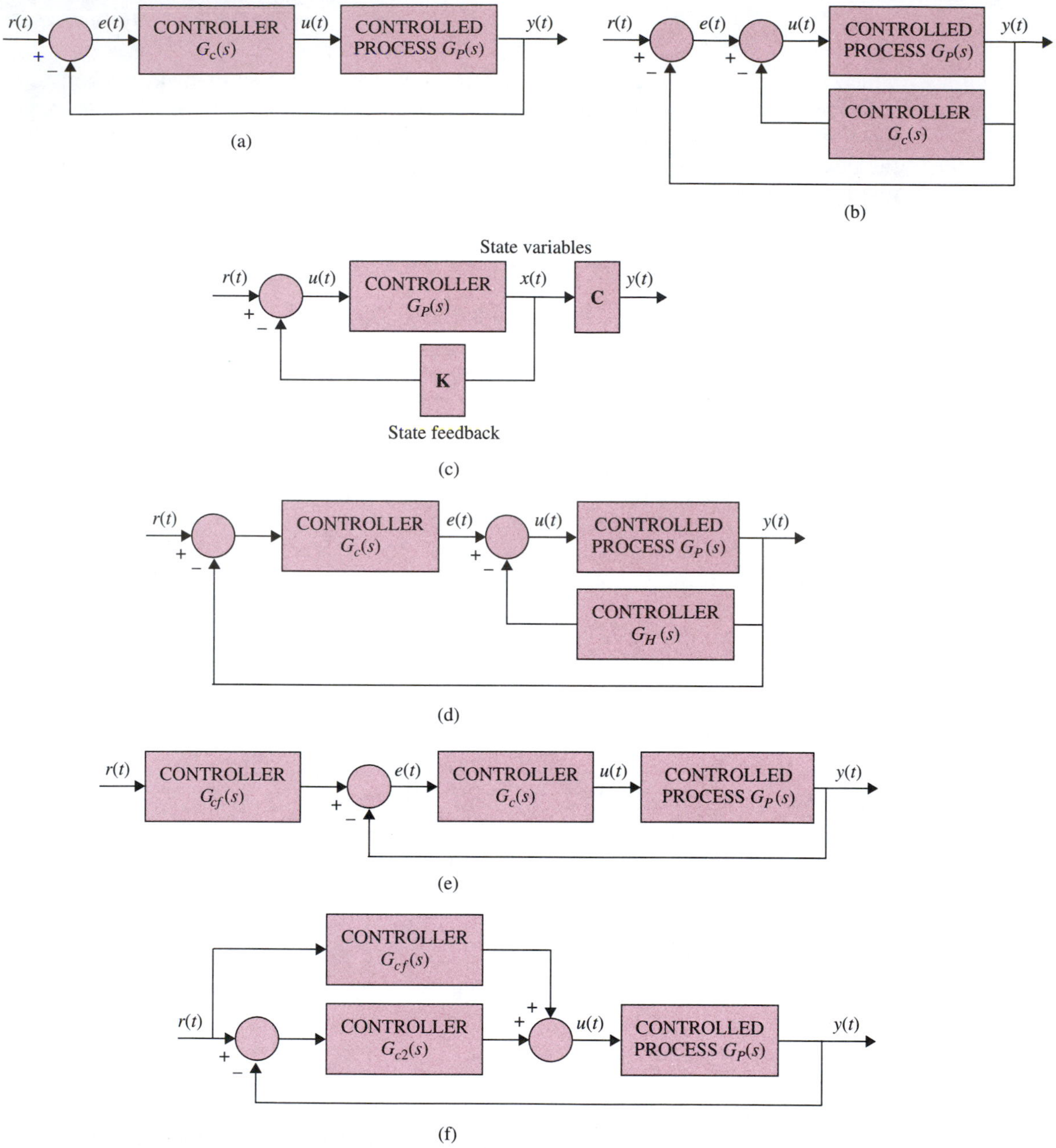

Figure 10-2 Various controller configurations in control system compensation. (a) Series or cascade compensation. (b) Feedback compensation. (c) State-feedback control. (d) Series-feedback compensation (two degrees of freedom). (e) Forward compensation with series compensation (two degrees of freedom). (f) Feedforward compensation (two degrees of freedom).

The compensation schemes shown in Fig. 10-2(a), (b), and (c) all have one degree of freedom in that there is only one controller in each system, even though the controller may have more than one parameter that can be varied. The disadvantage with a one-degree-of-freedom controller is that the performance criteria that can be realized are limited. For example, if a system is to be designed to achieve a certain amount of relative stability, it may have poor sensitivity to parameter variations. Or if the roots of the characteristic equation are selected to provide a certain amount of relative damping, the maximum overshoot of the step response may still be excessive, owing to the zeros of the closed-loop transfer function. The compensation schemes shown in Fig. 10-2(d), (e), and (f) all have two degrees of freedom.

- ▲ **Series-feedback compensation.** Figure 10-2(d) shows the series-feedback compensation for which a series controller and a feedback controller are used.
- ▲ **Feedforward compensation.** Figures 10-2(e) and (f) show the so-called **feedforward compensation.** In Fig. 10-2(e), the feedforward controller $G_{cf}(s)$ is placed in series with the closed-loop system, which has a controller $G_c(s)$ in the forward path. In Fig. 10-2(f) the feedforward controller $G_{cf}(s)$ is placed in parallel with the forward path. The key to the feedforward compensation is that the controller $G_{cf}(s)$ is not in the loop of the system, so that it does not affect the roots of the characteristic equation of the original system. The poles and zeros of $G_{cf}(s)$ may be selected to add or cancel the poles and zeros of the closed-loop system transfer function.

▲ All control system designs will benefit from both the time- and frequency-domain viewpoints.

One of the commonly used controllers in these above-mentioned compensation schemes is a PID controller, which applies a signal to the process that is some combination of a proportional, integral, and derivative of the actuating signal. Since these signal components are easily realized and visualized in the time domain, PID controllers are commonly designed using time-domain methods. In addition to PID-type controllers, lead, lag, lead–lag, and notch controllers are also frequently used. The names of these controllers come from properties of their respective frequency-domain characteristics. As a result, these controllers are often designed using frequency-domain concepts. Despite these design tendencies, however, all control system designs will benefit by viewing the resulting design from both the time- and frequency-domain viewpoints. Thus both methods are used extensively in this chapter.

It should be pointed out that the compensation schemes listed above are by no means exhaustive. Details of these compensation schemes are discussed in ensuing sections. Although the systems illustrated in Fig. 10-2 are all for continuous-data control, the same configurations can be applied to discrete-data control, in which case the controllers are all digital, with the necessary interfacings and signal converters.

10-1-3 Fundamental Principles of Design

After a controller configuration is chosen, the designer must choose a controller type that with proper selection of its element values will satisfy all the design specifications. The types of controllers available for control system design are bounded only by one's imagination. Engineering practice usually dictates that one choose the simplest controller that meets all the design specifications. In most cases, the more complex a controller is, the more it costs, the less reliable it is, and more difficult it is to design. Choosing a specific controller for a specific application is often based on the designer's past experience, and sometimes intuition, and entails as much *art* as it does *science*. As a result, you as a novice may find it difficult initially to make intelligent choices of controllers with confidence. By understanding that confidence comes only through experience, this chapter provides guided experiences that illustrate the basic elements of control system designs.

After a controller is chosen, the next task is to choose controller parameter values. These parameter values are typically the coefficients of one or more transfer functions making up the controller. The basic design approach is to use the analysis tools discussed in previous chapters to determine how individual parameter values influence the design specifications, and finally, system performance. Based on this information, controller parameters are selected so that all design specifications are met. While this process is sometimes straightforward, more often than not it involves many design iterations since controller parameters usually interact with each other and influence design specifications in conflicting ways. For example, a particular parameter value may be chosen so that the maximum overshoot is satisfied, but in the process of varying another parameter value in an attempt to meet the rise-time requirement, the maximum overshoot specification may no longer be met! Clearly, the more design specifications there are and the more controller parameters there are, the more complicated the design process becomes.

In carrying out the design in either the time or the frequency domain, it is important to establish some basic guidelines or design rules. Keep in mind that time-domain design usually relies heavily on the s-plane and the root loci. Frequency-domain design is based on manipulating the gain and phase of the loop transfer function so that the specifications are met.

In general, it is useful to summarize the time- and frequency-domain characteristics so that they can be used as guidelines for design purposes.

1. Complex-conjugate poles of the closed-loop transfer function lead to a step response that is underdamped. If all system poles are real, the step response is overdamped. However, zeros of the closed-loop transfer function may cause overshoot even if the system is overdamped.
2. The response of a system is dominated by those poles closest to the origin in the s-plane. Transients due to those poles farther to the left decay faster.

3. The farther to the left in the s-plane the system's dominant poles are, the faster the system will respond and the greater its bandwidth will be.
4. The farther to the left in the s-plane the system's dominant poles are, the more expensive it will be and the larger its internal signals will be. Although this can be justified analytically, it is obvious that striking a nail harder with a hammer drives the nail in faster but requires more energy per strike. Similarly, a sports car can accelerate faster, but it uses more fuel than an average car.
5. When a pole and zero of a system transfer function nearly cancel each other, the portion of the system response associated with the pole will have a small magnitude.
6. Time- and frequency-domain specifications are loosely associated with each other. Rise time and bandwidth are inversely proportional. Phase margin, gain margin, M_r, and damping are inversely proportional.

10-2 Design with the PD Controller

In all the control systems examples we have discussed thus far, the controller has been typically a simple amplifier with a constant gain K. This type of control action is formally known as **proportional control,** since the control signal at the output of the controller is simply related to the input of the controller by a proportional constant. Intuitively, one should also be able to use the derivative or integral of the input signal, in addition to the proportional operation. Therefore, we can consider a more general continuous-data controller to be one that contains such components as adders (addition or subtraction), amplifiers, attenuators, differentiators, and integrators. The designer's task is to determine which of these components should be used, in what proportion, and how they are connected. For example, one of the best-known controllers used in practice is the PID controller, mentioned in Section 4-9, where the letters stand for **proportional, integral,** and **derivative.** The integral and derivative components of the PID controller have individual performance implications, and their applications require an understanding of the basics of these elements. To gain an understanding of this controller, we consider just the PD portion of the controller first.

Figure 10-3 shows the block diagram of a feedback control system that arbitrarily has a second-order prototype process with the transfer function

$$G_p(s) = \frac{\omega_n^2}{s(s + 2\zeta\omega_n)} \tag{10-1}$$

The series controller is a proportional-derivative (PD) type with the transfer function

$$G_c(s) = K_P + K_D s \tag{10-2}$$

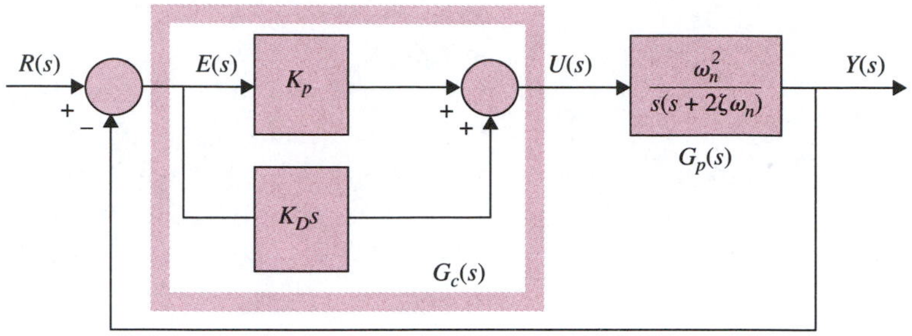

Figure 10-3 Control system with PD control.

Thus the control signal applied to the process is

$$u(t) = K_P e(t) + K_D \frac{de(t)}{dt} \tag{10-3}$$

where K_P and K_D are the proportional and derivative constants, respectively. Using the components given in Table 4-1, two electronic-circuit realizations of the PD controller are shown in Fig. 10-4. The transfer function of the circuit in Fig. 10-4(a) is

$$G_c(s) = \frac{E_o(s)}{E_{in}(s)} = \frac{R_2}{R_1} + R_2 C_1 s \tag{10-4}$$

Comparing Eq. (10-2) with Eq. (10-4), we have

$$K_P = R_2/R_1 \qquad K_D = R_2 C_1 \tag{10-5}$$

The transfer function of the circuit in Fig. 10-4(b) is

$$G_c(s) = \frac{E_o(s)}{E_{in}(s)} = \frac{R_2}{R_1} + R_d C_d s \tag{10-6}$$

Comparing Eq. (10-2) with Eq. (10-6), we have

$$K_P = R_2/R_1 \qquad K_D = R_d C_d \tag{10-7}$$

The advantage with the circuit in Fig. 10-4(a) is that only two op-amps are used. However, the circuit does not allow independent selection of K_P and K_D, as they are commonly dependent on R_2. An important concern of the PD controller is that if the value of K_D is large, a large capacitor C_1 would be required. The circuit in Fig. 10-4(b) allows K_P and K_D to be controlled independently. A large K_D can be compensated by

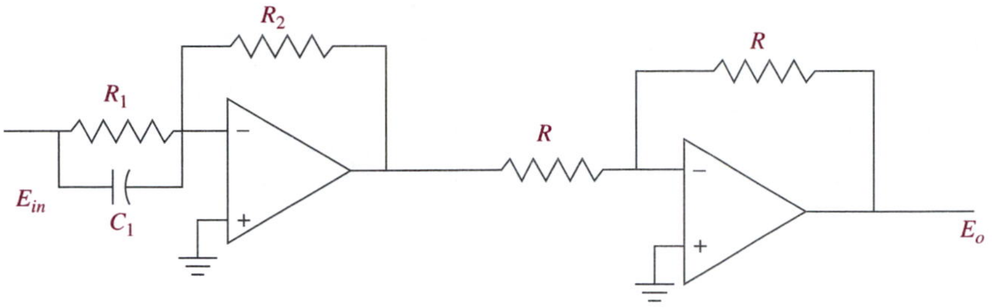

(a)

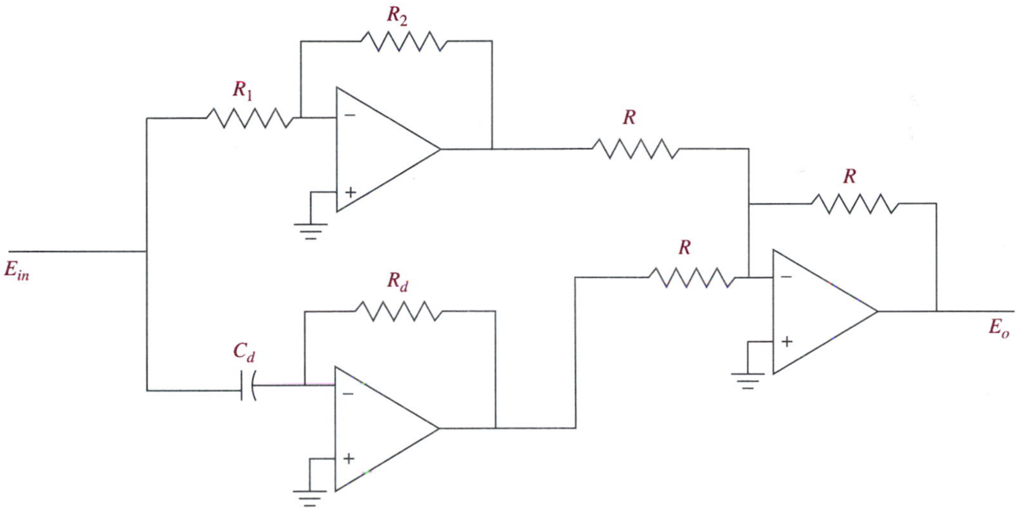

(b)

Figure 10-4 Op-amp circuit realization of the PD controller, $G_c(s) = K_P + K_D s$. (a) Two-op-amp circuit. (b) Three-op-amp circuit.

choosing a large value for R_d, thus resulting in a realistic value for C_d. Although the scope of this book does not include all the practical issues involved in controller transfer function implementation, these issues are of utmost practical importance in practice.

▲ PD control adds a simple zero at $s = -K_P/K_D$ to the forward-path transfer function.

The forward-path transfer function of the compensated system is

$$G(s) = \frac{Y(s)}{E(s)} = G_c(s)G_p(s) = \frac{\omega_n^2(K_P + K_D s)}{s(s + 2\zeta\omega_n)} \tag{10-8}$$

which shows that the PD control is equivalent to adding a simple zero at $s = -K_P/K_D$ to the forward-path transfer function.

10-2-1 Time-Domain Interpretation of PD Control

The effect of the PD control on the transient response of a control system can be investigated by referring to the time responses shown in Fig. 10-5. Let us assume that the unit-step response of a stable system with only proportional control is as shown in Fig. 10-5(a), which has a relatively high maximum overshoot and is rather oscillatory. The corresponding error signal, which is the difference between the unit-step input and the output $y(t)$, and its time derivative $de(t)/dt$ are shown in Fig. 10-5(b) and (c), respectively. The overshoot and oscillation characteristics are also reflected in $e(t)$ and $de(t)/dt$. For the sake of illustration we assume that the system contains a motor of

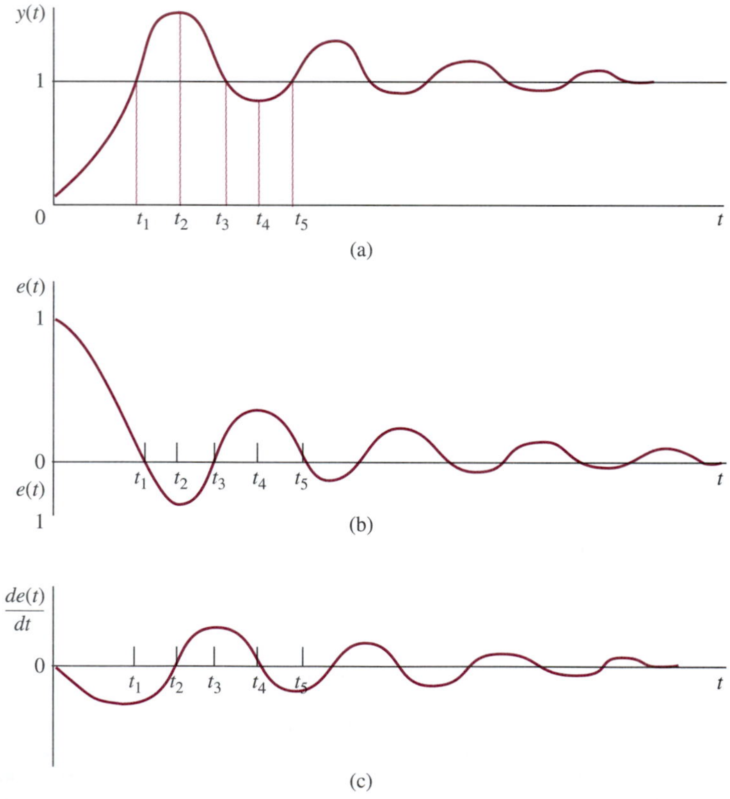

Figure 10-5 Waveforms of $y(t)$, $e(t)$, and $de(t)/dt$, showing the effect of derivative control. (a) Unit-step response. (b) Error signal. (c) Time rate of change of the error signal.

some kind with its torque proportional to $e(t)$. The performance of the system with proportional control is analyzed as follows.

1. During the time interval, $0 < t < t_1$. The error signal $e(t)$ is positive. The motor torque is positive and rising rapidly. The large overshoot and subsequent oscillations in the output $y(t)$ are due to the excessive amount of torque developed by the motor and the lack of damping during this time interval.
2. During the time interval, $t_1 < t < t_3$. The error signal $e(t)$ is negative, and the corresponding motor torque is negative. This negative torque tends to slow down the output acceleration and eventually causes the direction of the output $y(t)$ to reverse and undershoot.
3. During the time interval, $t_3 < t < t_5$. The motor torque is again positive, thus tending to reduce the undershoot in the response caused by the negative torque in the previous time interval. Since the system is assumed to be stable, the error amplitude is reduced with each oscillation, and the output eventually settles to its final value.

Considering the analysis above of the system time response, we can say that the contributing factors to the high overshoot are:

1. The positive correcting torque in the interval $0 < t < t_1$ is too large.
2. The retarding torque in the time interval $t_1 < t < t_2$ is inadequate.

Therefore, to reduce the overshoot in the step response, without significantly increasing the rise time, a logical approach would be to:

1. Decrease the amount of positive correcting torque during $0 < t < t_1$.
2. Increase the retarding torque during $t_1 < t < t_2$.

Similarly, during the time interval, $t_2 < t < t_4$, the negative corrective torque in $t_2 < t < t_3$ should be reduced, and the retarding torque during $t_3 < t < t_4$, which is now in the positive direction, should be increased to improve the undershoot of $y(t)$.

The PD control described by Eq. (10-2) gives precisely the compensation effect required. Since the control signal of the PD control is given by Eq. (10-3), Fig. 10-5(c) shows the following effects provided by the PD controller:

1. For $0 < t < t_1$, $de(t)/dt$ is negative; this will reduce the original torque developed due to $e(t)$ alone.
2. For $t_1 < t < t_2$, both $e(t)$ and $de(t)/dt$ are negative, which means that the negative retarding torque developed will be greater than that with only proportional control.
3. For $t_2 < t < t_3$, $e(t)$ and $de(t)/dt$ have opposite signs. Thus the negative torque that originally contributes to the undershoot is reduced also.

Therefore, all these effects will result in smaller overshoots and undershoots in $y(t)$.

▲ PD is essentially an anticipatory control.

Another way of looking at the derivative control is that since $de(t)/dt$ represents the slope of $e(t)$, the PD control is essentially an *anticipatory* control. That is, by knowing the slope, the controller can anticipate direction of the error and use it to better control the process. Normally, in linear systems, if the slope of $e(t)$ or $y(t)$ due to a step input is large, a high overshoot will subsequently occur. The derivative control measures the instantaneous slope of $e(t)$, predicts the large overshoot ahead of time, and makes a proper corrective effort before the excessive overshoot actually occurs.

▲ Derivative or PD control will have effect on a steady-state error only if the error varies with time.

Intuitively, derivative control affects the steady-state error of a system only if the steady-state error varies with time. If the steady-state error of a system is constant with respect to time, the time derivative of this error is zero, and the derivative portion of the controller provides no input to the process. But if the steady-state error increases with time, a torque is again developed in proportion to $de(t)/dt$, which reduces the magnitude of the error. Equation (10-8) also clearly shows that the PD control does not alter the system type that governs the steady-state error of a unity-feedback system.

10-2-2 Frequency-Domain Interpretation of PD Control

For frequency-domain design, the transfer function of the PD controller is written

$$G_c(s) = K_P + K_D s = K_P\left(1 + \frac{K_D}{K_P}s\right) \tag{10-9}$$

▲ The PD controller is a high-pass filter.

so that it is more easily interpreted on the Bode plot. The Bode plot of Eq. (10-9) is shown in Fig. 10-6 with $K_P = 1$. In general, the proportional-control gain K_P can be combined with a series gain of the system, so that the zero-frequency gain of the PD controller can be regarded as unity. The high-pass filter characteristics of the PD

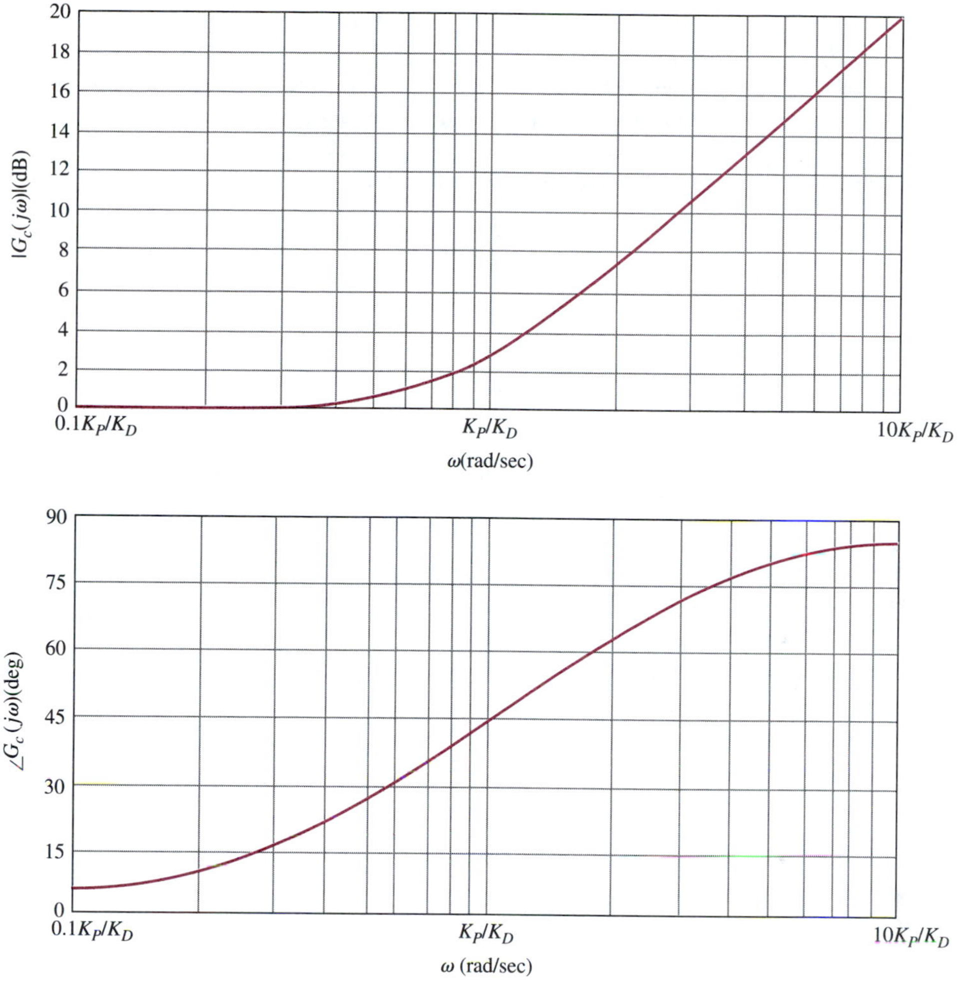

Figure 10-6 Bode diagram of $1 + K_D s/K_P$. $K_p = 1$.

▲ The PD controller has the disadvantage that it accentuates high-frequency noise.

controller are clearly shown by the Bode plot in Fig. 10-6. The phase-lead property may be utilized to improve the phase margin of a control system. Unfortunately, the magnitude characteristics of the PD controller push the gain-crossover frequency to a higher value. Thus *the design principle of the PD controller involves the placing of the corner frequency of the controller, $\omega = K_P/K_D$, such that an effective improvement of the phase margin is realized at the new gain-crossover frequency.* For a given system, there is a range of values of K_P/K_D that is optimal for improving the damping of the system. Another practical consideration in selecting the values of K_P and K_D is in the physical implementation of the PD controller. Other apparent effects of the PD control

▲ The PD controller generally will increase the BW and reduce the rise time of the step response.

in the frequency domain is that due to its high-pass characteristics, in most cases it will increase the BW of the system and reduce the rise time of the step response. The practical disadvantage of the PD controller is that the differentiator portion is a high-pass filter, which usually accentuates any high-frequency noise that enters at the input.

10-2-3 Summary of Effects of PD Control

A properly designed PD controller will affect the performance of a control system in the following ways:

1. Improves damping and reduces maximum overshoot.
2. Reduces rise time and settling time.
3. Increases BW.
4. Improves GM, PM, and M_r.
5. May accentuate noise at higher frequencies.
6. Not effective for lightly damped or initially unstable systems.
7. May require a relatively large capacitor in circuit implementation.

The following example illustrates the effects of the PD controller on the time- and frequency-domain responses of a second-order system.

Example 10-1

Consider the second-order model of the aircraft attitude control system shown in Fig. 7-24. The forward-path transfer function of the system is given in Eq. (7-116), and is repeated:

$$G(s) = \frac{4500K}{s(s + 361.2)} \tag{10-10}$$

Let us set the performance specifications as follows:

Steady-state error due to unit-ramp input ≤ 0.000443
Maximum overshoot ≤ 5 percent
Rise time $t_r \leq 0.005$ sec
Settling time $t_s \leq 0.005$ sec

To satisfy the maximum value of the specified steady-state error requirement, K should be set at 181.17. However, with this value of K, the damping ratio of the system is 0.2 and the maximum overshoot is 52.7 percent, as shown by the unit-step response in Fig. 7-25 and again in Fig. 10-9. Let us consider inserting a PD controller in the forward path of the system, so that the damping and the maximum overshoot of the system are improved while maintaining the steady-state error due to the unit-ramp input at 0.000443.

Time-Domain Design

With the PD controller of Eq. (10-9) and $K = 181.17$, the forward-path transfer function of the system becomes

$$G(s) = \frac{\Theta_y(s)}{\Theta_e(s)} = \frac{815{,}265(K_P + K_D s)}{s(s + 361.2)} \tag{10-11}$$

The closed-loop transfer function is

$$\frac{\Theta_y(s)}{\Theta_r(s)} = \frac{815{,}265(K_P + K_D s)}{s^2 + (361.2 + 815{,}265K_D)s + 815{,}265K_P} \tag{10-12}$$

The ramp-error constant is

$$K_v = \lim_{s \to 0} sG(s) = \frac{815{,}265K_P}{361.2} = 2257.1K_P \tag{10-13}$$

The steady-state error due to a unit-ramp input is $e_{ss} = 1/K_v = 0.000443/K_P$.

Equation (10-12) shows that the effects of the PD controller are to:

1. Add a zero at $s = -K_P/K_D$ to the closed-loop transfer function.
2. Increase the "damping term," which is the coefficient of the s term in the denominator, from 361.2 to $361.2 + 815{,}265K_D$.

The characteristic equation is written

$$s^2 + (361.2 + 815{,}265K_D)s + 815{,}265K_P = 0 \tag{10-14}$$

We can arbitrarily set $K_P = 1$ which is acceptable from the steady-state error requirement. The damping ratio of the system is

$$\zeta = \frac{361.2 + 815{,}265K_D}{1805.84} = 0.2 + 451.46K_D \tag{10-15}$$

which clearly shows the positive effect of K_D on damping. If we wish to have critical damping, $\zeta = 1$, Eq. (10-15) gives $K_D = 0.001772$. We should quickly point out that Eq. (10-11) no longer represents a prototype second-order system; the transient response is also affected by the zero of the transfer function at $s = -K_P/K_D$. It turns out that for this second-order system, as the value of K_D increases, the zero will move very close to the origin and effectively cancels the pole of $G(s)$ at $s = 0$. Thus, as K_D increases, the transfer function in Eq. (10-11) approaches that of a first-order system with the pole at $s = -361.2$, and the closed-loop system will not have any overshoot. In general, for higher-order systems, however, the zero at $s = -K_P/K_D$ may increase the overshoot when K_D becomes very large.

We can apply the root-contour method to the characteristic equation in Eq. (10-14) to examine the effect of varying K_P and K_D. First, by setting K_D to zero, Eq. (10-14) becomes

$$s^2 + 361.2s + 815{,}265K_P = 0 \tag{10-16}$$

The root loci of Eq. (10-16) as K_P varies between 0 and ∞ are shown in Fig. 10-7. When $K_D \neq 0$, the characteristic equation in Eq. (10-14) is conditioned as

$$1 + G_{eq}(s) = 1 + \frac{815{,}265K_D s}{s^2 + 361.2s + 815{,}265K_P} = 0 \tag{10-17}$$

The root contours of Eq. (10-14) with K_P = constant and K_D varies are constructed based on the pole–zero configuration of $G_{eq}(s)$ and are shown in Fig. 10-8 for $K_P = 0.25$ and $K_P = 1$. We see that when $K_P = 1$ and $K_D = 0$, the characteristic equation roots are at $-180.6 + j884.67$

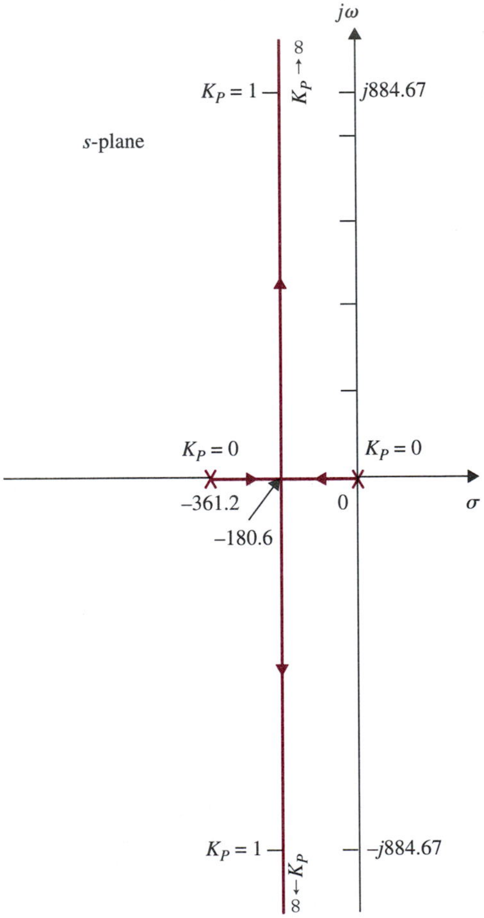

Figure 10-7 Root loci of Eq. (10-16).

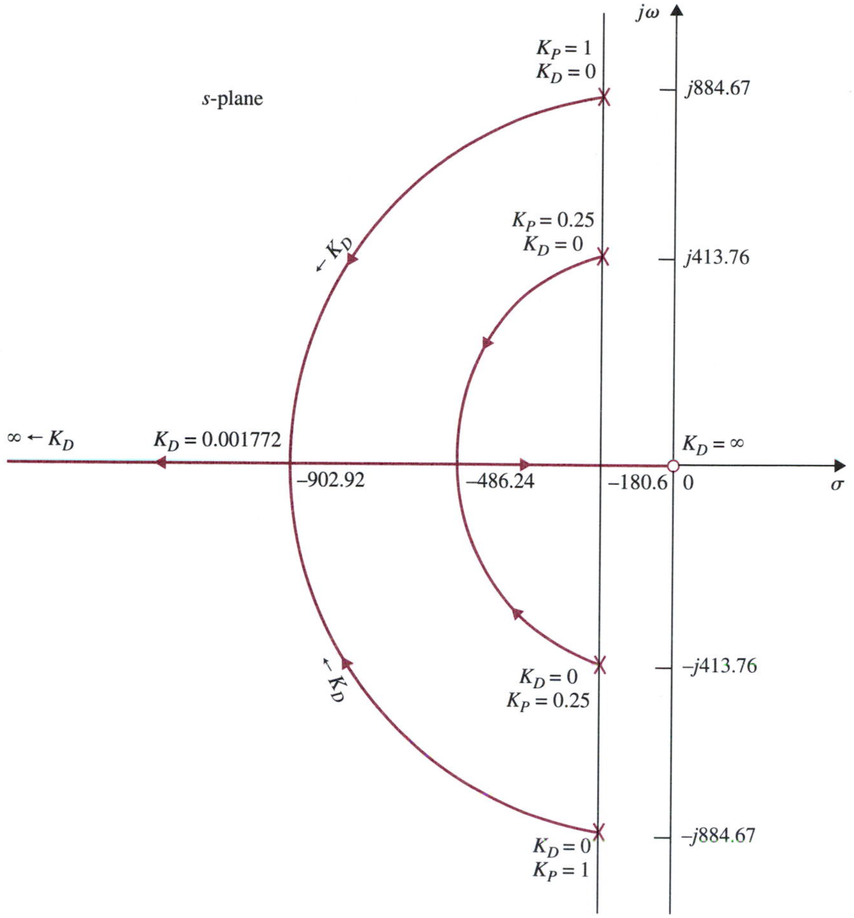

Figure 10-8 Root contours of Eq. (10-14) when $K_P = 0.25$ and 1.0; K_D varies.

and $-180.6 - j884.67$, and the damping ratio of the closed-loop system is 0.2. When the value of K_D is increased, the two characteristic equation roots move toward the real axis along a circular arc. When K_D is increased beyond 0.00177, the roots are real and equal, at -902.92, and the damping is critical. When K_D is increased beyond 0.00177, the two roots become real and unequal, and the system is overdamped. When K_P is 0.25 and $K_D = 0$, the two characteristic equation roots are at $-180.6 + j413.76$ and $-180.6 - j413.76$. As K_D increases in value, the root contours again show the improved damping due to the PD controller. Figure 10-9 shows the unit-step responses of the closed-loop system without PD control and with $K_P = 1$ and $K_D = 0.00177$. With the PD control, the maximum overshoot is 4.2 percent. In the present case, although K_D is chosen for critical damping, the overshoot is due to the zero at $s = -K_P/K_D$ of the closed-loop transfer function. Table 10-1 gives the results on maximum overshoot, rise time,

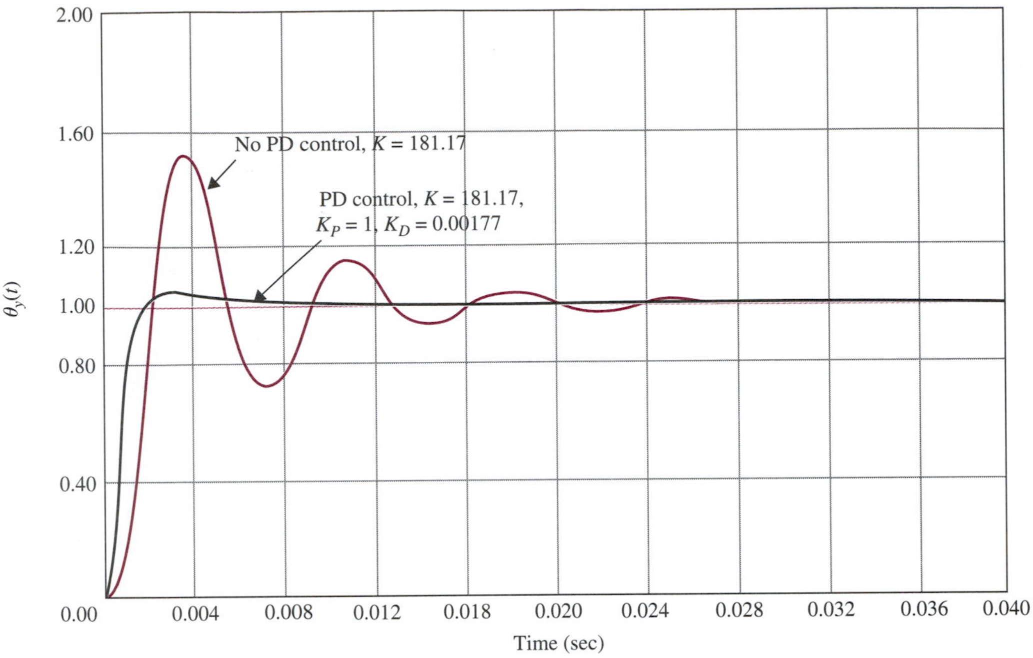

Figure 10-9 Unit-step responses of the attitude control system in Fig. 7-29 with and without PD control.

▲ The PD controller decreases maximum overshoot, rise time, and settling time.

and settling time for $K_P = 1$, $K_D = 0$, 0.0005, 0.00177, and 0.0025. The results in Table 10-1 show that the performance requirements are all satisfied with $K_D \geq 0.00177$. It should be kept in mind that K_D should only be large enough to satisfy the performance requirements. Large K_D corresponds to large BW, which may cause high-frequency noise problems, and there is also the concern of the capacitor value in the op-amp-circuit implementation. The general conclusion is that the PD controller decreases the maximum overshoot, the rise time, and the settling time.

Table 10-1 Attributes of the Unit-Step Responses of the System in Example 10-1 with PD Control

K_D	t_r (sec)	t_s (sec)	Maximum Overshoot (%)
0	0.00125	0.0151	52.2
0.0005	0.0076	0.0076	25.7
0.00177	0.00119	0.0049	4.2
0.0025	0.00103	0.0013	0.7

Another analytic way of studying the effects of the parameters K_P and K_D is to evaluate the performance characteristics in the parameter plane of K_P and K_D. From the characteristic equation of Eq. (10-14), we have

$$\zeta = \frac{0.2 + 451.46K_D}{\sqrt{K_P}} \tag{10-18}$$

Applying the stability requirement to Eq. (10-14), we find that for system stability,

$$K_P > 0 \quad \text{and} \quad K_D > -0.000443$$

The boundaries of stability in the K_P versus K_D parameter plane are shown in Fig. 10-10. The constant-damping-ratio trajectory is described by Eq. (10-18) and is a parabola. Figure 10-10 illustrates the constant-ζ trajectories for $\zeta = 0.5$, 0.707, and 1.0. The ramp-error constant K_v is given by Eq. (10-13), which describes a horizontal line in the parameter plane, as shown in Fig. 10-10. The figure gives a clear picture as to how the values of K_P and K_D affect the various performance criteria of the system. For instance, if K_v is set at 2257.1, which corresponds to

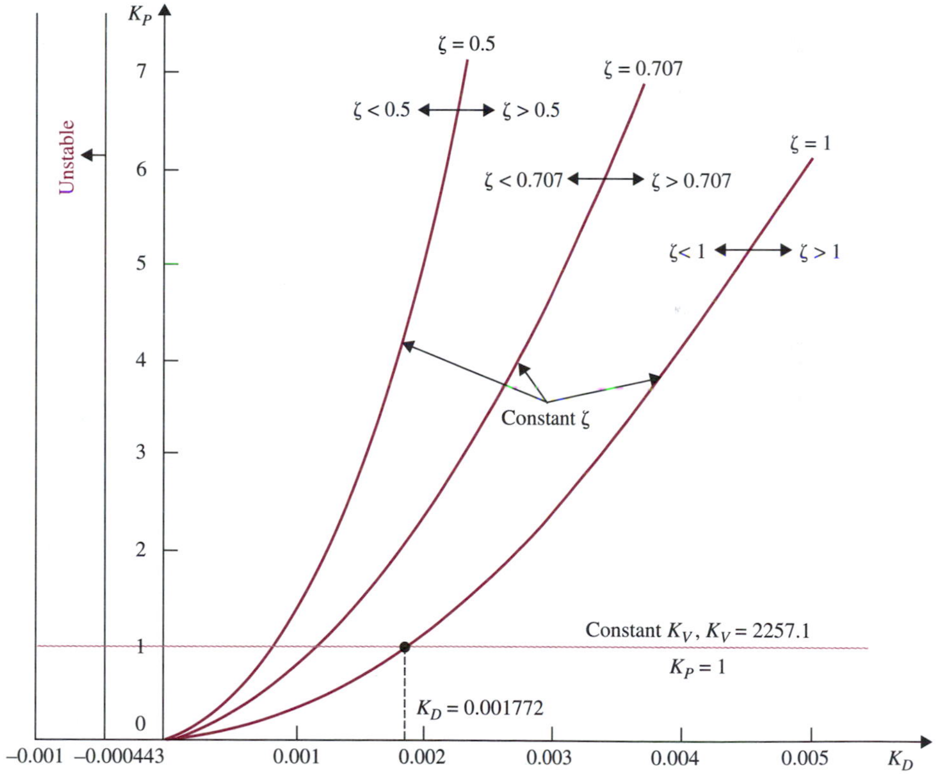

Figure 10-10 K_P *versus* K_D parameter plane for the attitude control system with a PD controller.

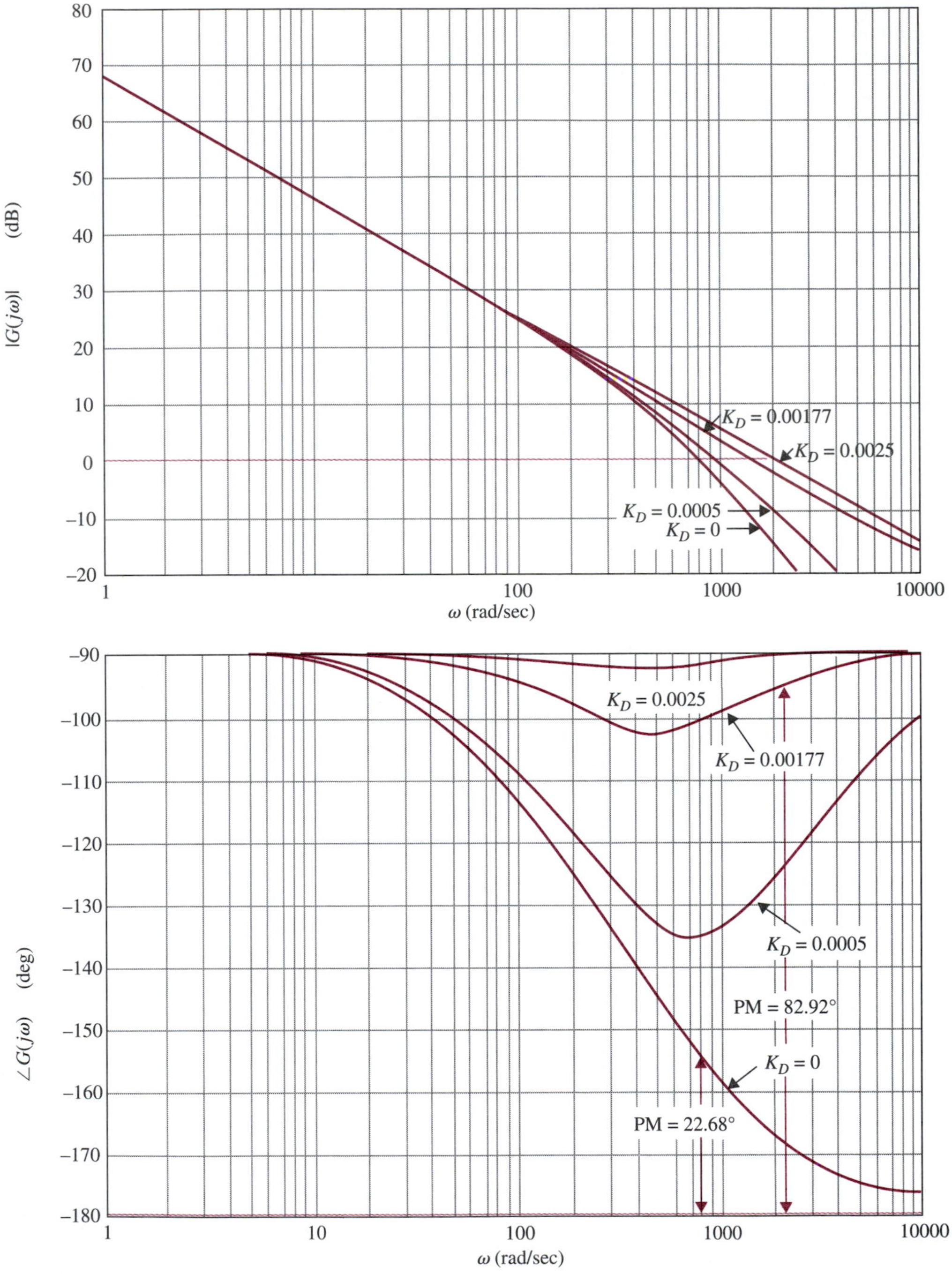

Figure 10-11 Bode plot of $G(s) = \dfrac{815{,}265(1 + K_D s)}{s(s + 361.2)}$.

Table 10-2 Frequency-Domain Characteristics of the System in Example 10-1 with PD Control

K_D	GM (dB)	PM (deg)	Gain CO (rad/sec)	BW (rad/sec)	M_r	t_r (sec)	t_s (sec)	Maximum Overshoot (%)
0	∞	22.68	868	1370	2.522	0.00125	0.0151	52.2
0.0005	∞	46.2	913.5	1326	1.381	0.0076	0.0076	25.7
0.00177	∞	82.92	1502	1669	1.025	0.00119	0.0049	4.2
0.0025	∞	88.95	2046	2083	1.000	0.00103	0.0013	0.7

$K_P = 1$, the constant-ζ loci show that the damping is increased monotonically with the increase in K_D. The intersection between the constant-K_v locus and the constant-ζ locus gives the value of K_D for the desired K_v and ζ.

Frequency-Domain Design

Now let us carry out the design of the PD controller in the frequency domain. Figure 10-11 shows the Bode plot of $G(s)$ in Eq. (10-11) with $K_P = 1$ and $K_D = 0$. The phase margin of the uncompensated system is 22.68°, and the resonant peak M_r is 2.522. These values correspond to a lightly damped system. Let us give the following performance criteria:

Steady-state error due to a unit-ramp input ≤ 0.00443
Phase margin $\geq 80°$
Resonant peak $M_r \leq 1.05$
BW $\leq$ 2000 rad/sec

The Bode plots of $G(s)$ for $K_P = 1$, $K_D = 0$, 0.005, 0.00177, and 0.0025 are shown in Fig. 10-11. The performance measures in the frequency domain for the compensated system with these controller parameters are tabulated in Table 10-2, along with the time-domain attributes for comparison. The Bode plots as well as the performance data were easily generated by using either the **bplot** of **CSAD** or **freqrp** of **ACSP.**

The results in Table 10-2 show that the gain margin is always infinite, and thus the relative stability is measured by the phase margin. This is one example where the gain margin is not an effective measure of the relative stability of the system. When $K_D = 0.00177$, which corresponds to critical damping, the phase margin is 82.92°, the resonant peak M_r is 1.025, and BW is 1669 rad/sec. The performance requirements in the frequency domain are all satisfied. Other effects of the PD control are that the BW and the gain-crossover frequeny are increased. The phase crossover frequency is always infinite in this case. ▲

Example 10-2

Consider the third-order aircraft attitude control system discussed in Chapter 7 with the forward-path transfer function given in Eq. (7-137),

$$G(s) = \frac{1.5 \times 10^7 K}{s(s^2 + 3408.3s + 1{,}204{,}000)} \tag{10-19}$$

The same set of time-domain specifications given in Example 10-1 is to be used. It was shown in Chapter 7 that when $K = 181.17$, the maximum overshoot of the system is 78.88 percent.

Let us attempt to meet the transient performance requirements by use of a PD controller with the transfer function given in Eq. (10-2). The forward-path transfer function of the system with the PD controller and $K = 181.17$ is

$$G(s) = \frac{2.718 \times 10^9 (K_P + K_D s)}{s(s^2 + 3408.3s + 1{,}204{,}000)} \qquad (10\text{-}20)$$

Time-Domain Design

Setting $K_P = 1$ arbitrarily, the characteristic equation of the closed-loop system is written

$$s^3 + 3408.3s^2 + (1{,}204{,}000 + 2.718 \times 10^9 K_D)s + 2.718 \times 10^9 = 0 \qquad (10\text{-}21)$$

To apply the root-contour method, we condition the last equation as

$$1 + G_{eq}(s) = 1 + \frac{2.718 \times 10^9 K_D s}{s^3 + 3408.3s^2 + 1{,}204{,}000s + 2.718 \times 10^9} = 0 \qquad (10\text{-}22)$$

where

$$G_{eq}(s) = \frac{2.718 \times 10^9 K_D s}{(s + 3293.3)(s + 57.49 + j906.6)(s + 57.49 - j906.6)} \qquad (10\text{-}23)$$

The root contours of Eq. (10-21) are plotted as shown in Fig. 10-12, based on the pole–zero configuration of $G_{eq}(s)$. The root contours of Fig. 10-12 reveals the effectiveness of the PD controller for the improvement on the relative stability of the system. Notice that as the value of K_D increases, one root of the characteristic equation moves from -3293.3 toward the origin, while that two complex roots start out toward the left and eventually approach the vertical asymptotes that intersect at $s = -1704$. The immediate assessment of the situation is that if the value of K_D is too large, *the two complex roots will actually have reduced damping, while increasing the natural frequency of the system.* It appears that the ideal location for the two complex characteristic equation roots, from the standpoint of relative stability, is near the bend of the root contour, where the relative damping ratio is approximately 0.707. The root contours of Fig. 10-12 clearly show that if the original system is initially with low damping, or is unstable, the zero introduced by the PD controller may not be able to add sufficient damping, or even stabilize the system.

▲ If a system has very low damping or is unstable, the PD control may not be effective in improving the stability of the system.

Table 10-3 gives the results of maximum overshoot, rise time, settling time, and the roots of the characteristic equation as functions of the parameter K_D. The following conclusions are drawn on the effects of the PD controller on the third-order system.

1. The minimum value of the maximum overshoot, 11.37 percent, occurs when K_D is approximately 0.002.
2. Rise time is improved (reduced) with the increase of K_D.

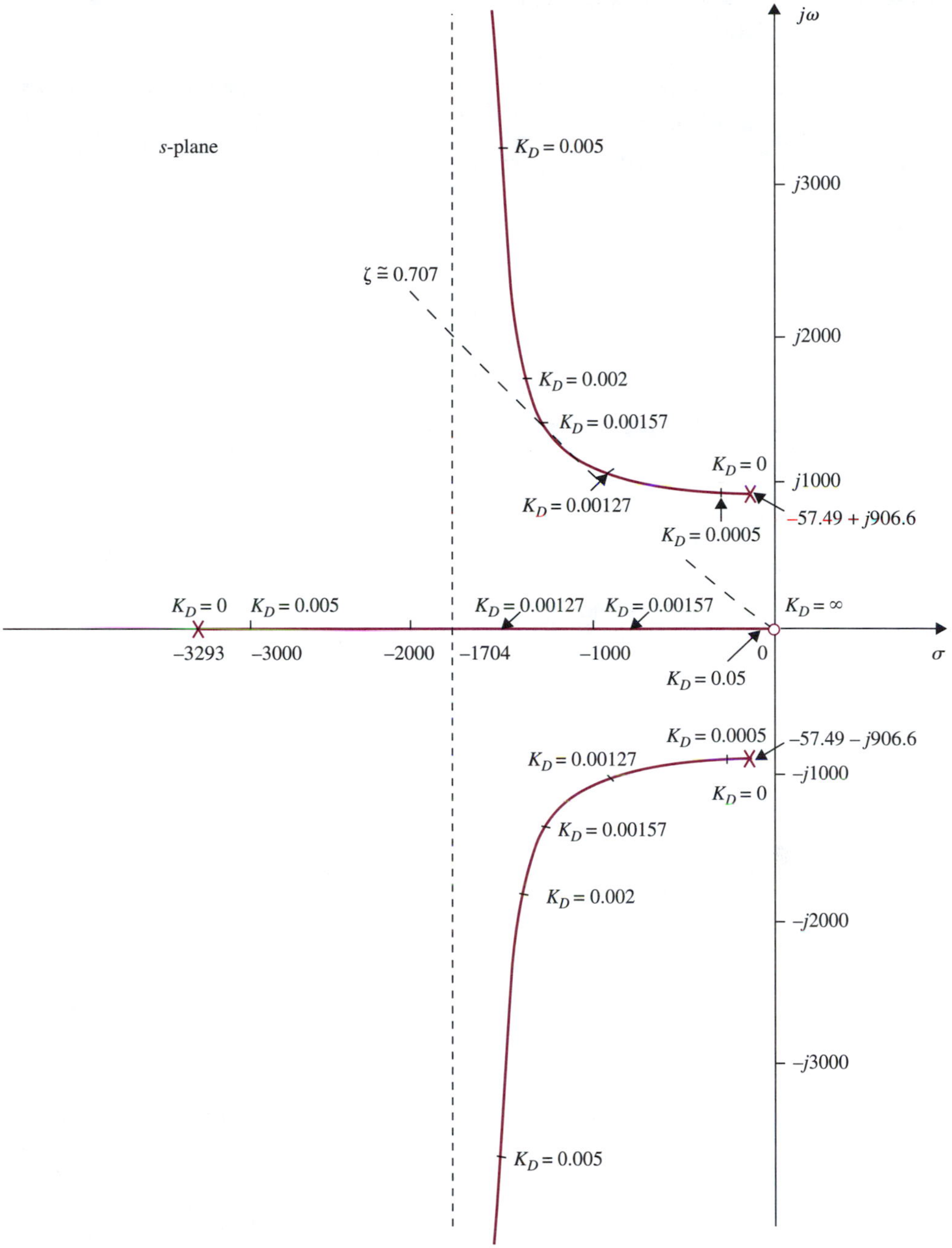

Figure 10-12 Root contours of $s^3 + 3408.3s^2 + (1{,}204{,}000 + 2.718 \times 10^9 K_D)s + 2.718 \times 10^9 = 0$.

Table 10-3 Time-Domain Attributes of the Third-Order System in Example 10-2 with PD Control

K_D	Maximum Overshoot (%)	t_r (sec)	t_s (sec)	Roots of Characteristic Equation	
0	78.88	0.00125	0.0495	−3293.3,	$-57.49 \pm j906.6$
0.0005	41.31	0.00120	0.0106	−2843.07,	$-282.62 \pm j936.02$
0.00127	17.97	0.00100	0.00398	−1523.11,	$-942.60 \pm j946.58$
0.00157	14.05	0.00091	0.00337	−805.33,	$-1301.48 \pm j1296.59$
0.00200	11.37	0.00080	0.00255	−531.89,	$-1438.20 \pm j1744.00$
0.00500	17.97	0.00042	0.00130	−191.71,	$-1608.29 \pm j3404.52$
0.01000	31.14	0.00026	0.00093	−96.85,	$-1655.72 \pm j5032$
0.05000	61.80	0.00010	0.00144	−19.83,	$-1694.30 \pm j11583$

3. Too high a value of K_D will actually increase the maximum overshoot and the settling time substantially. The latter is due to the lowering of damping as K_D is increased indefinitely.

Figure 10-13 shows the unit-step responses of the system with the PD controller for several values of K_D. The conclusion is that while the PD control does improve the damping of the system, it does not meet the maximum-overshoot requirement.

Frequency-Domain Design

The Bode plot of Eq. (10-20) is used to conduct the frequency-domain design of the PD controller. Figure 10-14 shows the Bode plot for $K_P = 1$ and $K_D = 0$. The following performance data are obtained for the uncompensated system:

Gain margin = 3.6 dB	Bandwidth BW = 1408.83 rad/sec
Phase margin = 7.77°	Gain crossover (GCO) = 888.94 rad/sec
Resonant peak M_r = 7.62	Phase crossover (PCO) = 1103.69 rad/sec

▲ For PD control, we should first examine the amount of phase needed to realize the desired PM.

Let us use the same set of frequency-domain performance requirements listed in Example 10-1. The logical way to approach this problem is first to examine how much additional phase is needed to realize a phase margin of 80°. Since the uncompensated system with the gain set to meet the steady-state requirement is only 7.77°, the PD controller must provide an additional phase of 72.23°. This additional phase must be placed at the gain crossover of the compensated system in order to realize a PM of 80°. Referring to the Bode plot of the PD controller in Fig. 10-6, we see that the additional phase is always accompanied by a gain in the magnitude curve. As a result, the gain crossover of the compensated system will be pushed to a higher frequency at which the phase of the uncompensated system would correspond to even a smaller PM. Thus we may run into the problem of diminishing returns. This symptom is parallel to the situation

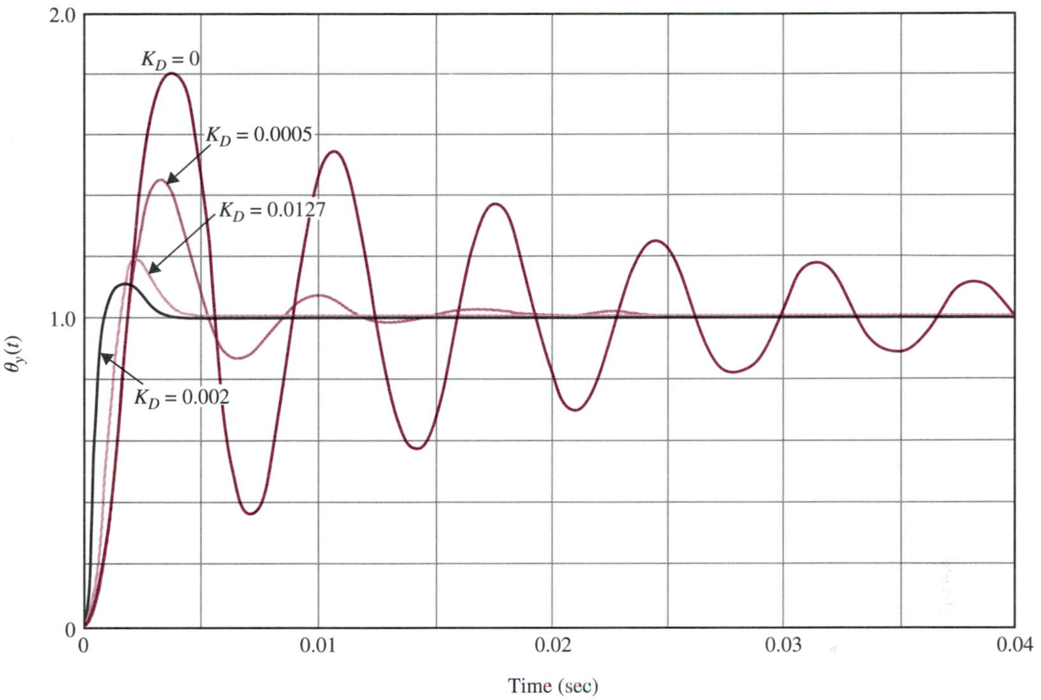

Figure 10-13 Unit-step responses of the system in Example 10-2 with PD controller.

illustrated by the root contour plot in Fig. 10-12, in which the larger K_D would simply push the roots to a higher frequency, and the damping is actually decreased.

The frequency-domain performance data of the compensated system with the values of K_D used in Table 10-3 are obtained from the Bode plots for each case, and the results are shown in Table 10-4. The Bode plots of some of these cases are shown in Fig. 10-14. Notice that the gain margin becomes infinite when the PD controller is added, and the phase margin becomes the dominant measure of relative stability. This is because the phase curve of the PD-compensated system stays above the $-180°$ axis, and the phase crossover is at infinity. When $K_D = 0.002$, the phase margin is at a maximum of 58.42°, and M_r is also minimum at 1.07, which happens to agree with the optimal value obtained in the time-domain design summarized in Table 10-3. When the value of K_D is increased beyond 0.002, the phase margin decreases, which agrees with the findings from the time-domain design that large values of K_D actually decrease damping. However, the BW and the gain crossover increase continuously with the increase in K_D. The frequency-domain design again shows that the PD control falls short in meeting the performance requirements imposed on the system. Just as in the time-domain design, we have demonstrated that if the original system has very low damping, or is unstable, PD control may not be effective in improving the stability of the system. Another situation under which PD control may be ineffective is if the slope of the phase curve near the gain crossover frequency is steep; then the rapid decrease of the phase margin due to the increase of the gain crossover from the added gain of the PD controller may render the additional phase ineffective. ▲

▲ If the slope of the phase curve near the gain crossover is very steep, PD control may not be effective.

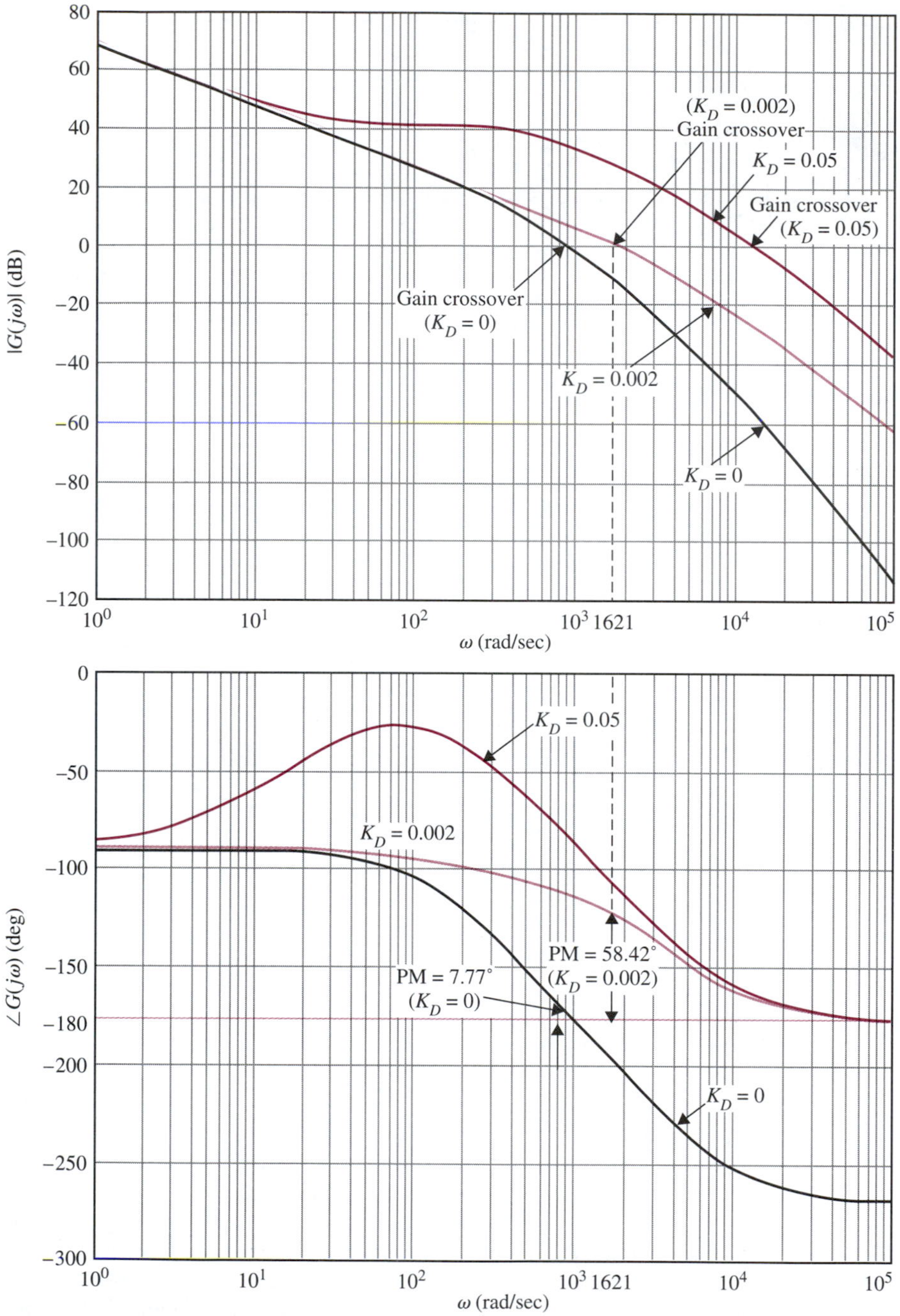

Figure 10-14 Bode diagram of $G(s)$ of system in Example 10-2 with PD control.

Table 10-4 Frequency-Domain Characteristics in the Third-Order System in Example 10-2 with PD Controller

K_D	GM (dB)	PM (deg)	M_r	BW (rad/sec)	Gain CO (rad/sec)	Phase CO (rad/sec)
0	3.6	7.77	7.62	1,408.83	888.94	1103.69
0.0005	∞	30.94	1.89	1,485.98	935.91	∞
0.00127	∞	53.32	1.19	1,939.21	1,210.74	∞
0.00157	∞	56.83	1.12	2,198.83	1,372.30	∞
0.00200	∞	58.42	1.07	2,604.99	1,620.75	∞
0.00500	∞	47.62	1.24	4,980.34	3,118.83	∞
0.01000	∞	35.71	1.63	7,565.89	4,789.42	∞
0.0500	∞	16.69	3.34	17,989.03	11,521.00	∞

10-3 Design with the PI Controller

We see from Section 10-2 that the PD controller can improve the damping and rise time of a control system at the expense of higher bandwidth and resonant frequency, and the steady-state error is not affected unless it varies with time, which is typically not the case for step-function inputs. Thus the PD controller may not fulfill the compensation objectives in many situations.

▲ Given K_P, the value of K_I should not be too small, or the capacitor of the op-amp-circuit implementation would be too large.

The integral part of the PID controller produces a signal that is proportional to the time integral of the input of the controller. Figure 10-15 illustrates a prototype second-order system with a series PI controller. The transfer function of the PI controller is

$$G_c(s) = K_P + \frac{K_I}{s} \tag{10-24}$$

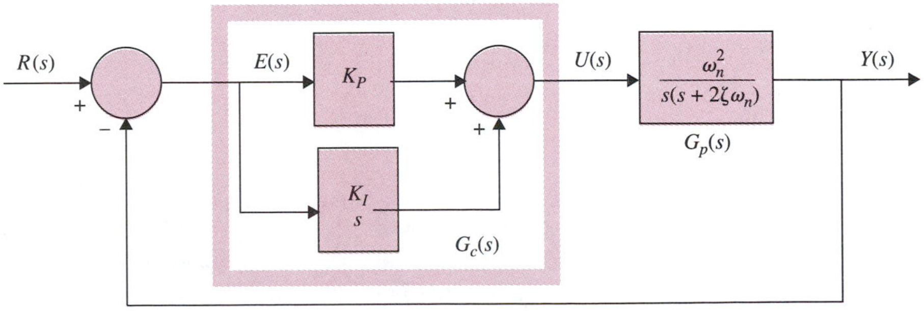

Figure 10-15 Control system with PI control.

Using the circuit elements given in Table 4-1, two op-amp-circuit realizations of Eq. (10-24) are shown in Fig. 10-16. The transfer function of the two-op-amp circuit in Fig. 10-16(a) is

$$G_c(s) = \frac{E_o(s)}{E_{in}(s)} = \frac{R_2}{R_1} + \frac{R_2}{R_1 C_2 s} \tag{10-25}$$

Comparing Eq. (10-24) with Eq. (10-25), we have

$$K_P = \frac{R_2}{R_1} \qquad K_I = \frac{R_2}{R_1 C_2} \tag{10-26}$$

The transfer function of the three-op-amp circuit in Fig. 10-16(b) is

$$G_c(s) = \frac{E_o(s)}{E_{in}(s)} = \frac{R_2}{R_1} + \frac{1}{R_i C_i s} \tag{10-27}$$

Thus the parameters of the PI controller are related to the circuit parameters as

$$K_P = \frac{R_2}{R_1} \qquad K_I = \frac{1}{R_i C_i} \tag{10-28}$$

▲ The value of the capacitor in the PI op-amp circuit is inversely proportional to K_I.

The advantage with the circuit in Fig. 10-16(b) is that the values of K_P and K_I are independently related to the circuit parameters. However, in either circuit, K_I is inversely proportional to the value of the capacitor. Unfortunately, effective PI-control designs usually result in small values of K_I, and thus we must again watch out for unrealistically large capacitor values.

The forward-path transfer function of the compensated system is

$$G(s) = G_c(s)G_p(s) = \frac{\omega_n^2(K_P s + K_I)}{s^2(s + 2\zeta\omega_n)} \tag{10-29}$$

▲ The PI control increases the system type by one, thus improves the steady-state error by one order.

Clearly, the immediate effects of the PI controller are:

1. Adds a zero at $s = -K_I/K_P$ to the forward-path transfer function.
2. Adds a pole at $s = 0$ to the forward-path transfer function. This means that the system type is increased by one to a type-2 system. Thus the steady-

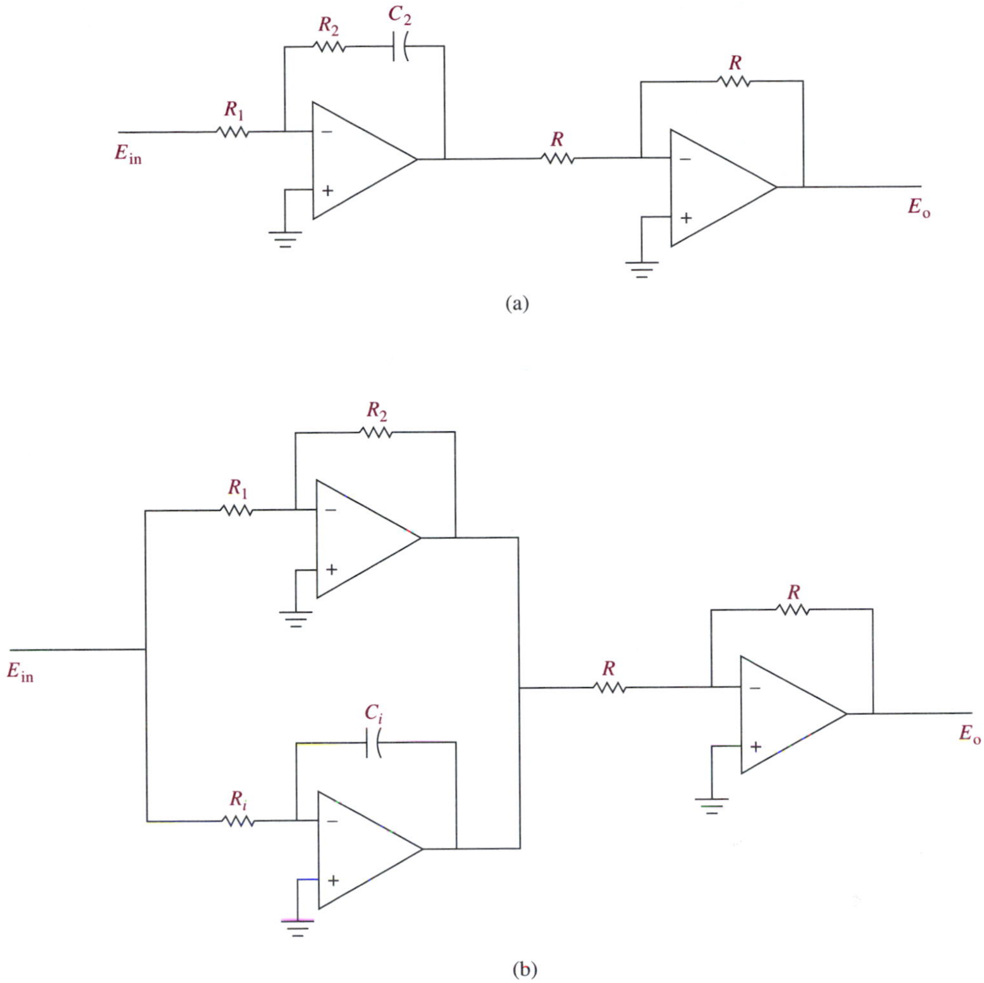

Figure 10-16 Op-amp circuit realization of the PI controller, $G_c(s) = K_p + K_I/s$. (a) Two-op-amp circuit. (b) Three-op-amp circuit.

state error of the original system is improved by one order; that is, if the steady-state error to a given input is constant, the PI control reduces it to zero (provided that the compensated system remains stable).

For the system in Fig. 10-15 with the forward-path transfer function in Eq. (10-29), the system will now have a zero steady-state error when the reference input is a ramp function. However, because the system is now of the third order, it *may be less stable* than the original second-order system or even become *unstable* if the parameters K_P and K_I are not properly chosen.

In the case of a type 1 system with a PD control, the value of K_P is important because the ramp-error constant K_v is directly proportional to K_P, and thus the magnitude of the steady-state error in inversely proportional to K_P when the input is a ramp. On the other hand, if K_P is too large, the system may become unstable. Similarly, for a type 0 system, the steady-state error due to a step input will be inversely proportional to K_P.

▲ The PI controller is essentially a low-pass filter.

When a type-1 system is converted to type 2 by the PI controller, K_P no longer affects the steady-state error, and the latter is always zero for a stable system with a ramp input. The problem is then to choose the proper combination of K_P and K_I so that the transient response is satisfactory.

▲ By proper design PI control can improve the transient and the steady-state performances.

10-3-1 Time-Domain Interpretation and Design of PI Control

The pole–zero configuration of the PI controller in Eq. (10-24) is shown in Fig. 10-17. At first glance it may seem that the PI control will improve the steady-state error at the expense of stability. However, we shall show that if the location of the zero of $G_c(s)$ is selected properly, both the damping and steady-state error can be improved.

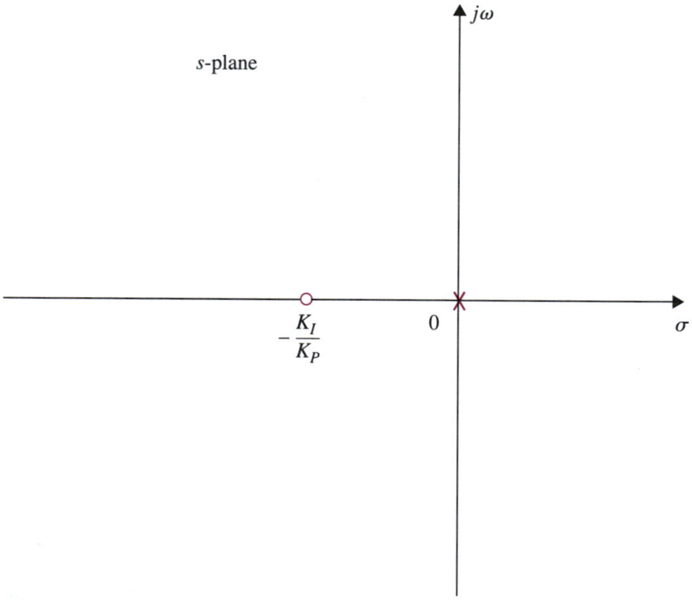

Figure 10-17 Pole–zero configuration of a PI controller.

Since the PI controller is essentially a low-pass filter, the compensated system usually will have a slower rise time and longer settling time. *A feasible method of designing the PI control is to select the zero at* $s = -K_I/K_P$ *so that it is relatively close to the origin and away from the most significant poles of the process, and the values of* K_P *and* K_I *should both be relatively small.*

10-3-2 Frequency-Domain Interpretation and Design of PI Control

For frequency-domain design the transfer function of the PI controller is written

$$G_c(s) = K_P + \frac{K_I}{s} = \frac{K_I[1 + (K_P/K_I)s]}{s} \tag{10-30}$$

The Bode plot of $G_c(j\omega)$ is shown in Fig. 10-18. Notice that the magnitude of $G_c(j\omega)$ at $\omega = \infty$ is $20 \log_{10} K_P$ dB, which represents an attenuation if the value of K_P is less than 1. This attenuation may be utilized to improve the stability of the system. The phase of $G_c(j\omega)$ is always negative, which is detrimental to stability. Thus we should place the corner frequency of the controller, $\omega = K_I/K_P$, as far to the left as the bandwidth requirement allows, so that the phase-lag properties of $G_c(j\omega)$ do not degrade the achieved phase margin of the system.

The frequency-domain design procedure for the PI control to realize a given phase margin is outlined as follows:

1. The Bode plot of the forward-path transfer function $G_p(s)$ of the uncompensated system is made with the loop gain set according to the steady-state performance requirement.
2. The phase and gain margins of the uncompensated system are determined from the Bode plot. For a certain specified phase-margin requirement, the new gain-crossover frequency ω_g' corresponding to this phase margin is found on the Bode plot. The magnitude plot of the compensated transfer function must pass through the 0-dB axis at this new gain-crossover frequency in order to realize the desired phase margin.
3. To bring the magnitude curve of the uncompensated transfer function down to 0 dB at the new gain-crossover frequency ω_g', the PI controller must provide the amount of attenuation equal to the gain of the magnitude curve at the new gain-crossover frequency. In other words, set

$$|G_p(j\omega_g')|_{\text{dB}} = -20\log_{10}K_P \text{ dB} \qquad K_P < 1 \tag{10-31}$$

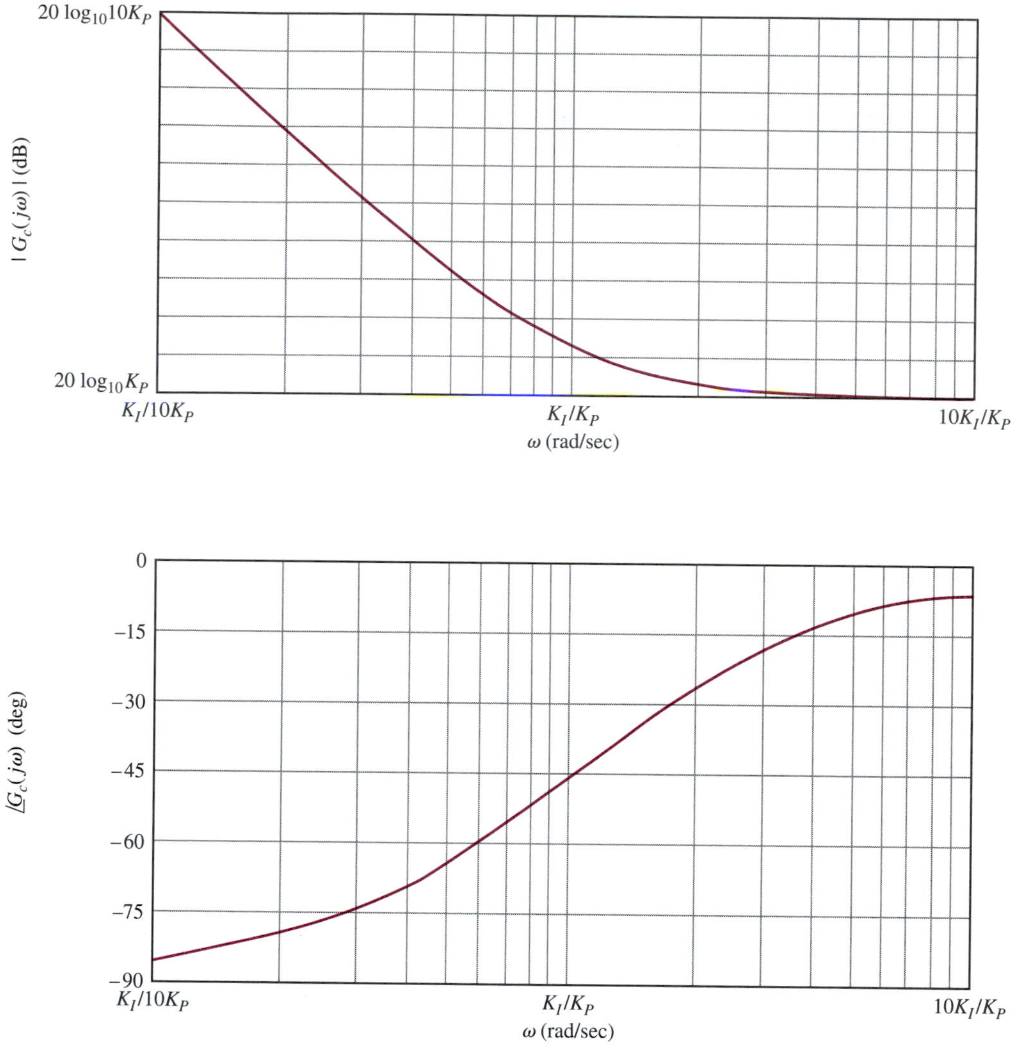

Figure 10-18 Bode diagram of the PI controller. $G_c(s) = K_P + K_I/s$.

from which we have

$$K_P = 10^{-|G_p(j\omega_g')|_{\mathrm{dB}}/20} \qquad K_P < 1 \tag{10-32}$$

Once the value of K_P is determined, it is necessary only to select the proper value of K_I to complete the design. Up to this point, we have assumed that although the gain-crossover frequency is altered by attenuating the magnitude of $G_c(j\omega)$ at ω_g', the original phase is not affected by the PI controller. This is not possible, however, since, as shown in Fig. 10-18, the attenuation property of the PI controller is accompanied with a phase lag that is detrimental to the phase margin. It is apparent that if the corner frequency $\omega = K_I/K_P$ is placed far below ω_g', the phase lag of the PI controller will have a negligible effect on the phase of the compensated system near ω_g'. On the other hand, the value of K_I/K_P should not be too small or the bandwidth of the system will be too low, causing the rise time and settling time to be too long. As a general guideline, K_I/K_P should correspond to a frequency that is at least one decade, sometimes as much as two decades, below ω_g'. That is, we set

$$\frac{K_I}{K_P} = \frac{\omega_g'}{10} \quad \text{rad/sec} \tag{10-33}$$

Within the general guideline, the selection of the value of K_I/K_P is pretty much at the discretion of the designer, mindful of its effect on BW and practical implementation by an op-amp circuit.

4. The Bode plot of the compensated system is investigated to see if the performance specifications are all met.
5. The values of K_I and K_P are substituted in Eq. (10-30) to give the desired transfer function of the PI controller.

If the controlled process $G_P(s)$ is type 0, the value of K_I may be selected based on the ramp-error-constant requirement, and then there would only be one parameter, K_P to determine. By computing the phase margin, gain margin, M_r, and BW of the closed-loop system with a range of values of K_P, the best value for K_P can easily be selected.

Based on the discussions above, we can summarize the advantages and disadvantages of a properly designed PI controller as:

1. Improves damping and reduces maximum overshoot.
2. Increases rise time.
3. Decreases BW.
4. Improves gain margin, phase margin, and M_r.
5. Filters out high-frequency noise.
6. The problem of selecting a proper combination of K_I and K_P so that the capacitor in the circuit implementation of the controller is not excessively large is more acute than in the case of the PD controller.

The following examples will illustrate how the PI control is designed and what its effects are.

Example 10-3

Consider the second-order attitude control system discussed in Example 10-1. Applying the PI controller of Eq. (10-24), the forward-path transfer function of the system becomes

$$G(s) = G_c(s)G_p(s) = \frac{4500KK_P(s + K_I/K_P)}{s^2(s + 361.2)} \tag{10-34}$$

Time-Domain Design

Let the time-domain performance requirements be:

Steady-state error due to parabolic input $t^2u_s(t)/2 \leq 0.2$
Maximum overshoot ≤ 5 percent
Rise time $t_r \leq 0.01$ sec
Settling time ≤ 0.02 sec

We have to relax the rise-time and settling-time requirements from those of Example 10-1 so that we will have a meaningful design for this system. The significance of the requirement on the steady-state error due to a parabolic input is that it indirectly places a minimum requirement on the speed of the transient response.

The parabolic-error constant is

$$K_a = \lim_{s\to 0} s^2G(s) = \lim_{s\to 0} s^2 \frac{4500KK_P(s + K_I/K_P)}{s^2(s + 361.2)}$$
$$= \frac{4500KK_I}{361.2} = 12.46KK_I \tag{10-35}$$

The steady-state error due to the parabolic input $t^2u_s(t)/2$ is

$$e_{ss} = \frac{1}{K_a} = \frac{0.08026}{KK_I} (\leq 0.2) \tag{10-36}$$

Let us set $K = 181.17$, simply because this was the value used in Example 10-1. Apparently, to satisfy a given steady-state error requirement for a parabolic input, the larger the K, the smaller K_I can be. Substituting $K = 181.17$ in Eq. (10-36) and solving K_I for the minimum steady-state error requirement of 0.2, we get the minimum value of K_I to be 0.002215. If necessary, the value of K can be adjusted later.

With $K = 181.17$, the characteristic equation of the closed-loop system is

$$s^3 + 361.2s^2 + 815{,}265K_Ps + 815{,}265K_I = 0 \tag{10-37}$$

Applying Routh's test to Eq. (10-37) yields the result that the system is stable is $0 < K_I/K_P < 361.2$. This means that the zero of $G(s)$ at $s = -K_I/K_P$ cannot be placed too far to the left in the left-half s-plane, or the system would be unstable. Let us place the zero at $-K_I/K_P$ relatively close to the origin. For the present case, the most significant pole of $G_P(s)$, beside the pole at

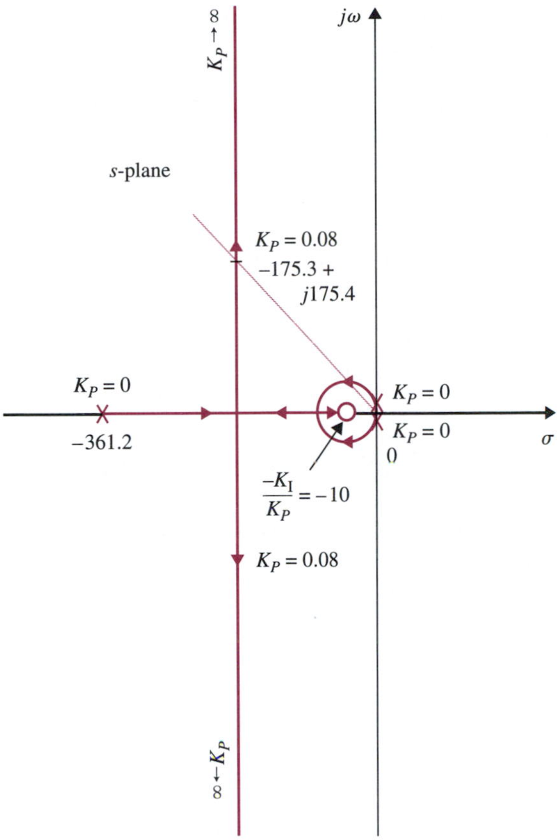

Figure 10-19 Root loci of Eq. (10-37) with $K_I/K_P = 10$; K_P varies.

$s = 0$, is at -361.2. Thus K_I/K_P should be chosen so that the following condition is satisfied:

$$\frac{K_I}{K_P} << 361.2 \tag{10-38}$$

The root loci of Eq. (10-37) with $K_I/K_P = 10$ are shown in Fig. 10-19. Notice that other than the small loop around the zero at $s = -10$, these root loci for the most part are very similar to those shown in Fig. 10-7, which are for Eq. (10-16). With the condition in Eq. (10-38) satisfied, Eq. (10-34) can be approximated by

$$G(s) \cong \frac{815{,}265 K_P}{s(s + 361.2)} \tag{10-39}$$

Where the term K_I/K_P in the numerator is neglected compared with the magnitude of s, which takes on values along the operating points on the complex portion of the root loci that correspond to, say, a relative damping ratio in the range of $0.7 < \zeta < 1.0$. Let us assume that we wish to have a relative damping ratio of 0.707. From Eq. (10-39), the required value of K_P for this damping ratio is 0.08. This should also be true for the third-order system with the PI controller if the value of K_I/K_P satisfies Eq. (10-38). Thus with $K_P = 0.08$, $K_I = 0.8$; the root loci in Fig. 10-19 show that the relative damping ratio of the two complex roots is approximately 0.707. In fact, the three characteristic equation roots are at

$$s = -10.605 \qquad -175.3 + j175.4 \quad \text{and} \quad -175.3 - j175.4$$

The reason for this is that when we "stand" at the root at $-175.3 + j175.4$ and "look" toward the neighborhood near the origin, we see that the zero at $s = -10$ is relatively close to the origin and thus practically cancels one of the poles at $s = 0$. In fact, we can show that as long as $K_P = 0.08$ and the value of K_I is chosen such that Eq. (10-38) is satisfied, the relative damping ratio of the complex roots will be very close to 0.707. For example, let us select $K_I/K_P = 5$; the three characteristic equation roots are at

$$s = -5.145 \qquad -178.03 + j178.03 \quad \text{and} \quad -178.03 - j178.03$$

and the relative damping ratio is still 0.707. Although the real pole of the closed-loop transfer function is moved, it is very close to the zero at $s = -K_I/K_P$, so that the transient due to the real pole is negligible. For example, when $K_P = 0.08$ and $K_I = 0.4$, the closed-loop transfer function of the compensated system is

$$\frac{\Theta_y(s)}{\Theta_r(s)} = \frac{65{,}221.2(s+5)}{(s+5.145)(s+178.03+j178.03)(s+178.03-j178.03)} \tag{10-40}$$

Since the pole at $s = -5.145$ is very close to the zero at $s = -5$, the transient response due to this pole is negligible, and the system dynamics are essentially dominated by the two complex poles.

Table 10-5 gives the attributes of the unit-step responses of the system with PI control for various values of K_I/K_P, with $K_P = 0.08$, which corresponds to a relative damping ratio of 0.707.

The results in Table 10-5 verify the fact that PI control reduces the overshoot but at the expense of longer rise time. For $K_I \leq 1$, the settling times in Table 10-5 actually show a sharp reduction, which is misleading. This is because the settling times for these cases are measured at the points where the response enters the band between 0.95 and 1.00, since the maximum overshoots are less than 5 percent.

The maximum overshoot of the system can be reduced still further than those shown in Table 10-5 by using smaller values of K_P than 0.08. However, the rise time and settling time will be excessive. For example, with $K_P = 0.04$ and $K_I = 0.04$, the maximum overshoot is 1.1 percent, but the rise time is increased to 0.0182 sec and the settling time is 0.024 sec.

For the system considered, improvement on the maximum overshoot slows down for K_I less than 0.08, unless K_P is also reduced. As mentioned earlier, the value of the capacitor of the op-amp circuit is inversely proportional to K_I. Thus, for practical reasons, there is a lower limit on the value of K_I.

Figure 10-20 shows the unit-step responses of the attitude-control system with PI control, with $K_P = 0.08$ and several values of K_I. The unit-step response of the same system with the PD controller designed in Example 10-1, $K_P = 1$ and $K_D = 0.00177$, is plotted in the same figure as a comparison.

Table 10-5 Attributes of the Unit-Step Responses of the System in Example 10-3 with PI Control

K_I/K_P	K_I	K_P	Maximum Overshoot (%)	t_r (sec)	t_s (sec)
0	0	1.00	52.7	0.00135	0.015
20	1.60	0.08	15.16	0.0074	0.049
10	0.80	0.08	9.93	0.0078	0.0294
5	0.40	0.08	7.17	0.0080	0.023
2	0.16	0.08	5.47	0.0083	0.0194
1	0.08	0.08	4.89	0.0084	0.0114
0.5	0.04	0.08	4.61	0.0084	0.0114
0.1	0.008	0.08	4.38	0.0084	0.0115

Frequency-Domain Design

The forward-path transfer function of the uncompensated system is obtained by setting $K_P = 1$ and $K_I = 0$ in the $G(s)$ in Eq. (10-34), and the Bode plot is shown in Fig. 10-21. The phase margin is 22.68° and the gain-crossover frequency is 868 rad/sec.

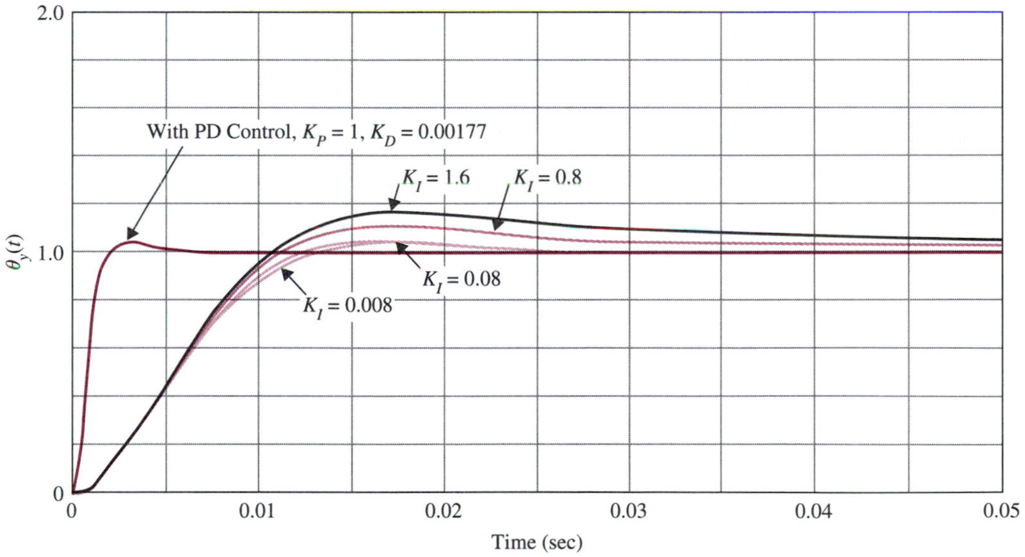

Figure 10-20 Unit-step responses of the system in Example 10-3 with PI control. Also, unit-step response of the system in Example 10-1 with PD control.

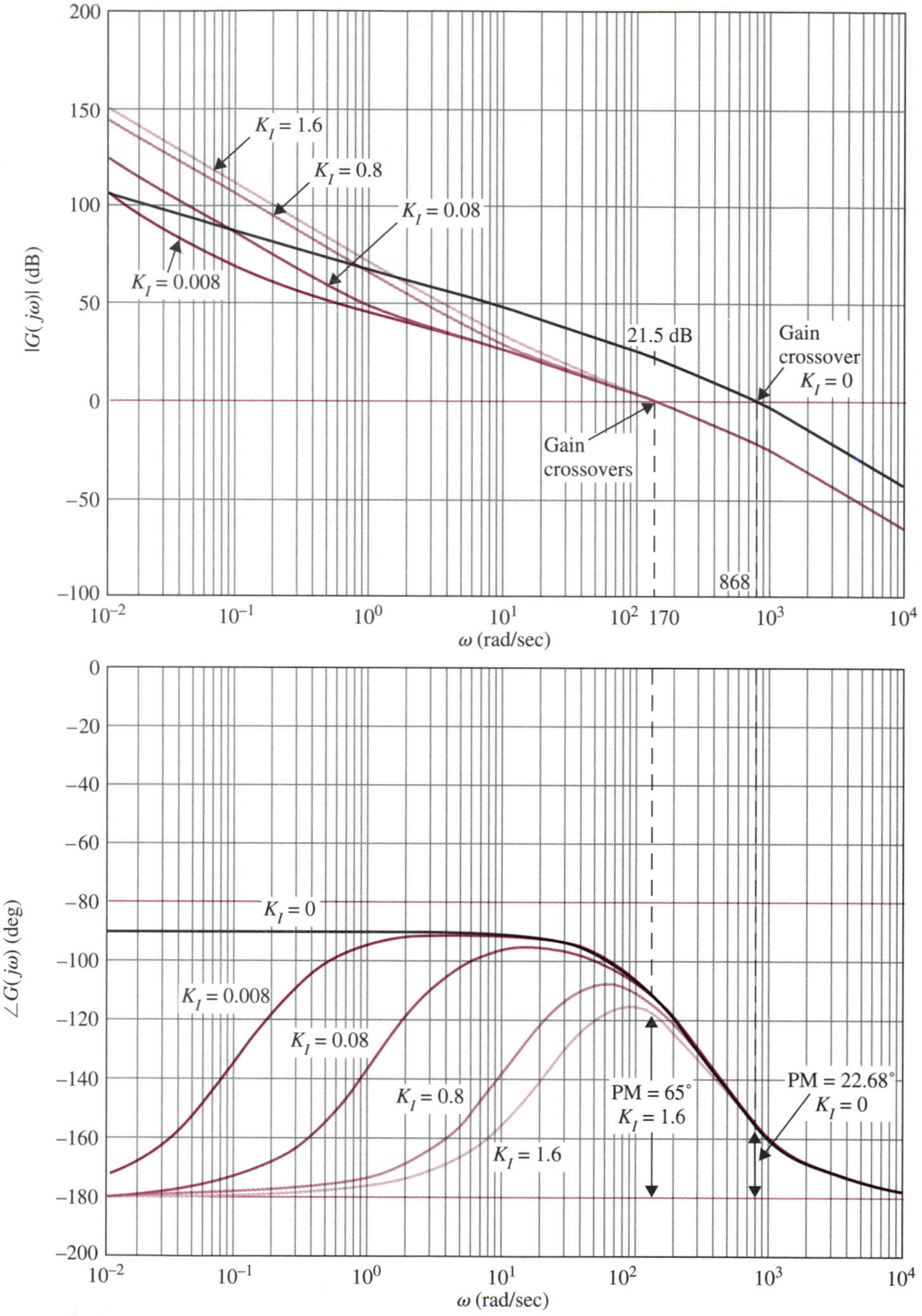

Figure 10-21 Bode plots of control system in Example 10-3 with PI control.

$$G(s) = \frac{815{,}265 K_P(s + K_I/K_P)}{s^2(s + 361.2)} \qquad K_P = 0.08$$

Let us specify that the required phase margin should be at least 65° and that this is to be achieved with the PI controller of Eq. (10-30). Following the procedure outlined in Eqs. (10-31) through (10-33) on the design of the PI controller, we conduct the following steps:

1. Look for the new gain-crossover frequency ω_g' at which the phase margin of 65° is realized. From Fig. 10-21, ω_g' is found to be 170 rad/sec. The magnitude of $G(j\omega)$ at this frequency is 21.5 dB. Thus the PI controller should provide an attenuation of -21.5 dB at $\omega_g' = 170$ rad/sec. Substituting $|G(j\omega_g')| = 21.5$ dB into Eq. (10-32) and solving for K_P, we get

$$K_P = 10^{-|G(j\omega_g')|_{dB}/20} = 10^{-21.5/20} = 0.084 \tag{10-41}$$

   Notice that in the time-domain design conducted earlier, K_P was selected to be 0.08, so that the relative damping ratio of the complex characteristic equation roots will be approximately 0.707. (Perhaps we have cheated a little by selecting the desired phase margin to be 65°. This could not be just a coincidence. Can you believe that we have had no prior knowledge that in this case, $\zeta = 0.707$ corresponds to PM = 65°?)

2. Let us choose $K_P = 0.08$, so that we can compare the design results of the frequency-domain with those of the time-domain design obtained earlier. Equation (10-33) gives the general guideline of finding K_I once K_P is determined. Thus

$$K_I = \frac{\omega_g' K_P}{10} = \frac{170 \times 0.08}{10} = 1.36 \tag{10-42}$$

   As pointed earlier, the value of K_I is not rigid as long as the ratio K_I/K_P is sufficiently smaller than the magnitude of the pole of $G(s)$ at -361.2. As it turns out the value of K_I given by Eq. (10-42) is not sufficiently small for this system.

The Bode plots of the forward-path transfer function with $K_P = 0.08$ and $K_I = 0$, 0.008, 0.08, 0.8, and 1.6 are shown in Fig. 10-21. Table 10-6 shows the frequency-domain properties of the uncompensated system and the compensated system with various values of K_I. Notice that for values of K_I/K_P that are sufficiently small, the phase margin, M_r, BW, and gain-crossover frequency all vary little.

Table 10-6 Frequency-Domain Performance Data of the System in Example 10-3 with PI Control

K_I/K_P	K_I	K_P	GM (dB)	PM (deg)	M_r	BW (rad/sec)	Gain CO (rad/sec)	Phase CO (rad/sec)
0	0	1.00	∞	22.6	2.55	1390.87	868	∞
20	1.6	0.08	∞	58.45	1.12	268.92	165.73	∞
10	0.8	0.08	∞	61.98	1.06	262.38	164.96	∞
5	0.4	0.08	∞	63.75	1.03	258.95	164.77	∞
1	0.08	0.08	∞	65.15	1.01	256.13	164.71	∞
0.1	0.008	0.08	∞	65.47	1.00	255.49	164.70	∞

It should be noted that the phase margin of the system can be improved further by reducing the value of K_P below 0.08. However, the bandwidth of the system will be reduced further. For example, for $K_P = 0.04$ and $K_I = 0.04$, the phase margin is increased to 75.7° and $M_r = 1.01$, but BW is reduced to 117.3 rad/sec. ▲

Example 10-4

Now let us consider using the PI control for the third-order attitude control system described by Eq. (10-19). First, the time-domain design is carried out as follows.

Time-Domain Design

Let the time-domain specifications be:

Steady-state error due to the parabolic input $t^2u_s(t)/2 \leq 0.2$
Maximum overshoot ≤ 5 percent
Rise time $t_r \leq 0.01$ sec
Settling time $t_s \leq 0.02$ sec

These are identical to the specifications given for the second-order system in Example 10-3.

Applying the PI controller of Eq. (10-24), the forward-path transfer function of the system becomes

$$G(s) = G_c(s)G_p(s) = \frac{1.5 \times 10^9 KK_P(s + K_I/K_P)}{s^2(s^2 + 3408.3s + 1{,}204{,}000)} = \frac{1.5 \times 10^9 KK_P(s + K_I/K_P)}{s^2(s + 400.26)(s + 3008)} \tag{10-43}$$

We can show that the steady-state error of the system due to the parabolic input is again given by Eq. (10-36), and arbitrarily setting $K = 181.17$, the minimum value of K_I is 0.002215.

The characteristic equation of the closed-loop system with $K = 181.17$ is

$$s^4 + 3408.3s^3 + 1{,}204{,}000s^2 + 2.718 \times 10^9 K_P s + 2.718 \times 10^9 K_I = 0 \tag{10-44}$$

The Routh's tabulation of the last equation is performed as follows:

s^4	1	$1{,}204{,}000$	$2.718 \times 10^9 K_I$
s^3	3408.3	$2.718 \times 10^9 K_P$	0
s^2	$1{,}204{,}000 - 797{,}465K_P$	$2.718 \times 10^9 K_I$	0
s^1	$\dfrac{1{,}204{,}000K_P - 797{,}465K_P^2 - 3408.3K_I}{1{,}204{,}000 - 797{,}465K_P}$	0	
s^0	$2.718 \times 10^9 K_I$	0	

The stability requirements are

$$\begin{aligned} &K_I > 0 \\ &K_P < 1.5098 \\ &K_I < 353.255K_P - 233.98K_P^2 \end{aligned} \tag{10-45}$$

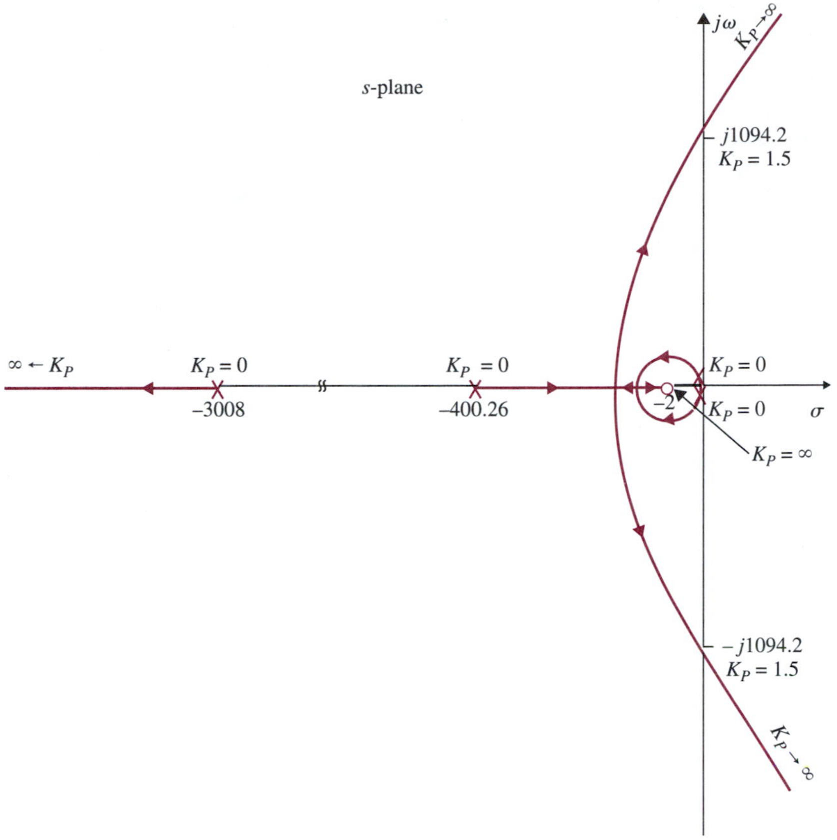

Figure 10-22 (a) Root loci of control system in Example 10-4 with PI control $K_I/K_P = 2$, $0 \le K_P < \infty$.

The design of the PI controller calls for the selection of a small value for K_I/K_P, relative to the nearest pole of $G(s)$ to the origin, which is at -400.26. The root loci of Eq. (10-44) are plotted using the pole–zero configuration of Eq. (10-43). Figure 10-22(a) shows the root loci as K_P varies for $K_I/K_P = 2$. The root loci near the origin due to the pole and zero of the PI controller again form a small loop, and the root loci at a distance away from the origin will be very similar to those of the uncompensated system, which are shown in Fig. 7-28. By selecting the value of K_P properly along the root loci, it may be possible to satisfy the performance specifications given above. To minimize the rise time and settling time, we should select K_P so that the dominant roots are complex conjugate. Table 10-7 gives the performance attributes of several combinations of K_I/K_P and K_P. Notice that although several combinations of these parameters correspond to systems that satisfy the performance specifications, the one with $K_P = 0.075$ and $K_I = 0.15$ gives the best rise and settling times among those shown.

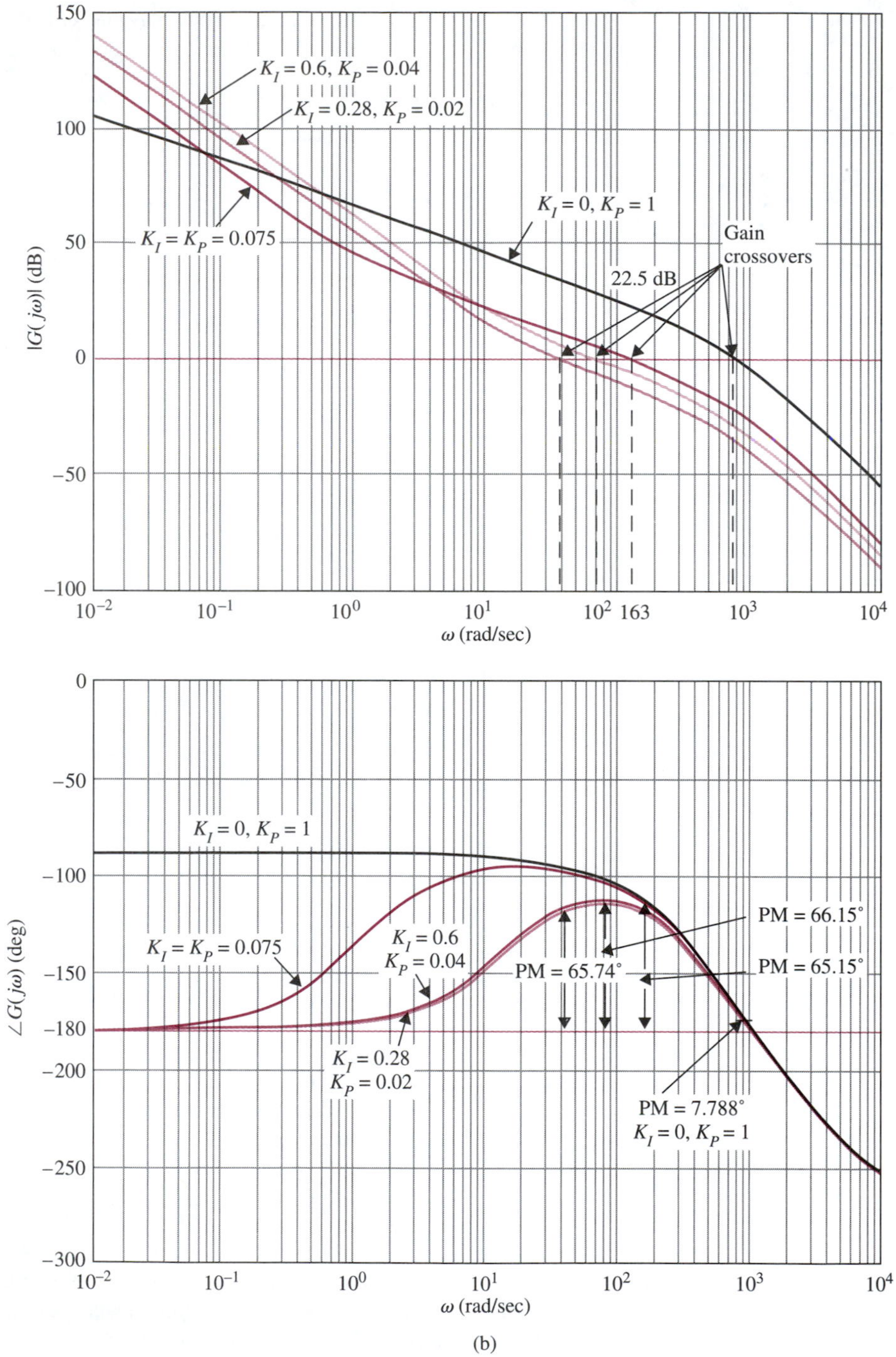

Figure 10-22 (b) Bode plots of control system in Example 10-4 with PI control.

$$G(s) = \frac{2.718 \times 10^9 K_P(s + K_I/K_P)}{s^2(s + 400.26)(s + 3008)}$$

Frequency-Domain Design

The Bode plot of Eq. (10-43) for $K = 181.17$, $K_P = 1$, and $K_I = 0$ is shown in Fig. 10-22(b). The performance data of the uncompensated system are:

Gain margin = 3.578 dB	$M_r = 6.572$
Phase margin = 7.788°	BW = 1378 rad/sec

Let us require that the compensated system has a phase margin of at least 65°, and this is to be achieved with the PI controller of Eq. (10-30). Following the procedure outlined in Eqs. (10-31) through (10-33) on the design of the PI controller, we carry out the following steps:

1. Look for the new gain-crossover frequency ω_g' at which the phase margin of 65° is realized. From Fig. 10-20, ω_g' is found to be 163 rad/sec, and the magnitude of $G(j\omega)$ at this frequency is 22.5 dB. Thus the PI controller should provide an attenuation of −22.5 dB at $\omega_g' = 163$ rad/sec. Substituting $|G(j\omega_g')| = 22.5$ dB into Eq. (10-32), and solving for K_P, we get

$$K_P = 10^{-|G(j\omega_g')|_{dB}/20} = 10^{-22.5/20} = 0.075 \tag{10-46}$$

   This is exactly the result that was selected for the time-domain design that resulted in a system with a maximum overshoot of 4.9 percent when $K_I = 0.15$ or $K_I/K_P = 2$.

2. The suggested value of K_I is found from Eq. (10-33),

$$K_I = \frac{\omega_g' K_P}{10} = \frac{163 \times 0.075}{10} = 1.222 \tag{10-47}$$

Table 10-7 Attributes of the Unit-Step Responses of the System in Example 10-4 with PI Control

K_I/K_P	K_I	K_P	Maximum Overshoot (%)	t_r (sec)	t_s (sec)	Roots of Characteristic Equation
0	0	1	76.2	0.00158	0.0487	−3293, 3, −57.5 ± j906.6
20	1.6	0.08	15.6	0.0077	0.0471	−3035, −22.7, −175.3 ± j180.3
20	0.8	0.04	15.7	0.0134	0.0881	−3021.6, −259, −99, −28
5	0.4	0.08	6.3	0.00883	0.0202	−3035, −5.1, −184 ± j189.2
2	0.08	0.04	2.1	0.02202	0.01515	−3021.7, −234.6, −149.9, −2
5	0.2	0.04	4.8	0.01796	0.0202	−3021.7, −240, −141.2, −5.3
2	0.16	0.08	5.8	0.00787	0.01818	−3035.2, −185.5 ± j190.8, −2
1	0.08	0.08	5.2	0.00792	0.01616	−3035.2, −186 ± j191.4, −1
2	0.15	0.075	4.9	0.0085	0.0101	−3033.5, −187.2 ± j178, −1
2	0.14	0.070	4.0	0.00917	0.01212	−3031.8, −187.2 ± j164, −1

Thus $K_I/K_P = 16.3$. However, the phase margin of the system with these design parameters is only 59.52.

To realize the desired PM of 65°, we can reduce the value of K_P or K_I. Table 10-8 gives the results of several designs with various combinations of K_P and K_I. Notice that the last three designs in the table all satisfy the PM requirements. However, the design ramifications show that:

Reducing K_P would reduce BW and increase M_r.
Reducing K_I would increase the capacitor value in the implementing circuit.

In fact, only the $K_I = K_P = 0.075$ case gives the best all-around performance in both the frequency and time domains. In attempting to increase K_I, the maximum overshoot becomes excessive. This is one example that shows the inadequacy of specifying phase margin only. The purpose of this example is to bring out the properties of the PI controller and the important considerations in its design. No details are explored further.

Figure 10-23 shows the unit-step responses of the uncompensated system and several systems with PI control. ▲

10-4 Design with the PID Controller

From the preceding discussions we see that the PD controller could add damping to a system, but the steady-state response is not affected. The PI controller could improve the relative stability and improve the steady-state error at the same time, but the rise time is increased. This leads to the motivation of using a PID controller so that the

Table 10-8 Performance Summary of System in Example 10-4 with PI Control

K_I/K_P	K_I	K_P	GM (dB)	PM (deg)	M_r	BW (rad/sec)	Maximum Overshoot (%)	t_r (sec)	t_s (sec)
0	0	1	3.578	7.788	6.572	1378	77.2	0.0015	0.0490
16.3	1.222	0.075	25.67	59.52	1.098	264.4	13.1	0.0086	0.0478
1	0.075	0.075	26.06	65.15	1.006	253.4	4.3	0.0085	0.0116
15	0.600	0.040	31.16	66.15	1.133	134.6	12.4	0.0142	0.0970
14	0.280	0.020	37.20	65.74	1.209	66.34	17.4	0.0268	0.1616

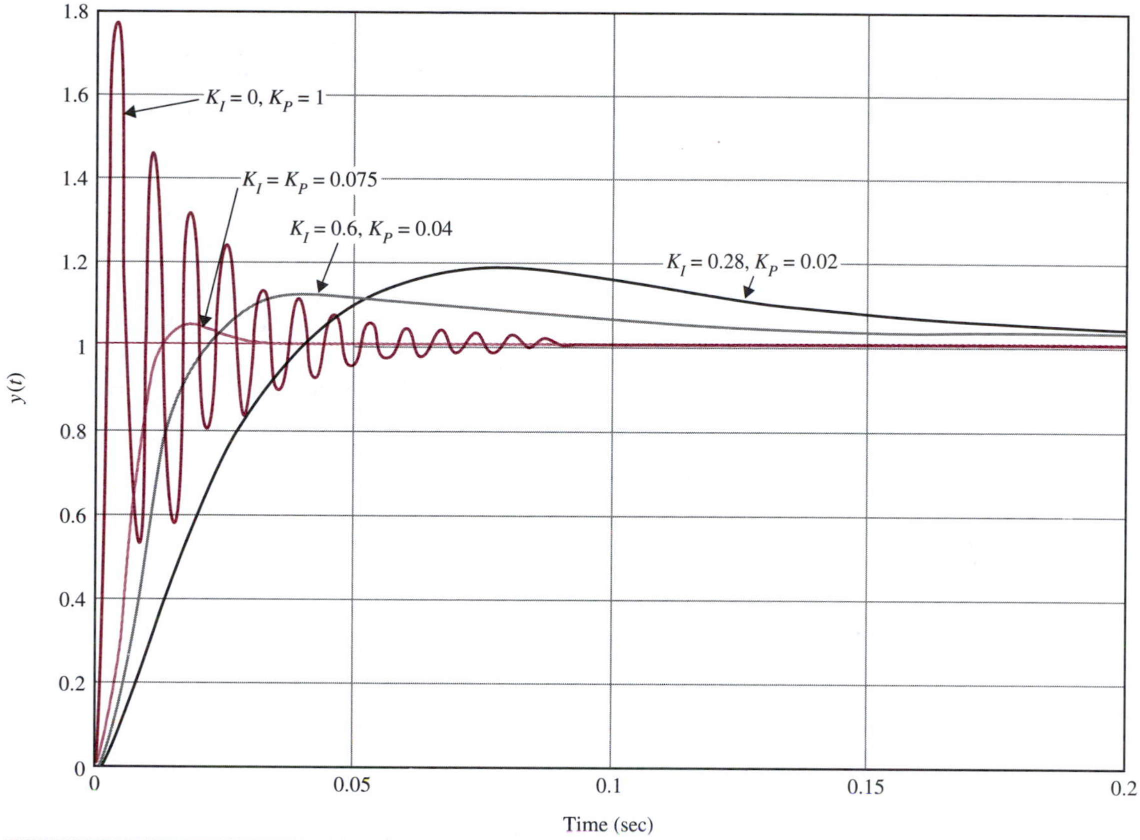

Figure 10-23 Unit-step responses of system with PI control in Example 10-4.

$$G(s) = \frac{2.718 \times 10^9 K_P(s + K_I/K_P)}{s^2(s + 400.26)(s + 3008)}$$

best features of each of the PI and PD controllers are utilized. We can outline the following procedure for the design of the PID controller.

1. Consider that the PID controller consists of a PI portion connected in cascade with a PD portion. The transfer function of the PID controller is written as

$$G_c(s) = K_P + K_D s + \frac{K_I}{s} = (1 + K_{D1}s)\left(K_{P2} + \frac{K_{I2}}{s}\right) \qquad (10\text{-}48)$$

The proportional constant of the PD portion is set to unity, since we need only three parameters in the PID controller. Equating both sides of Eq. (10-48), we have

$$K_P = K_{P2} + K_{D1}K_{I2} \tag{10-49}$$
$$K_D = K_{D1}K_{P2} \tag{10-50}$$
$$K_I = K_{I2} \tag{10-51}$$

2. Consider that the PD portion is in effect only. Select the value of K_{D1} so that a portion of the desired relative stability is achieved. In the time domain, this relative stability may be measured by the maximum overshoot, and in the frequency domain it is the phase margin.
3. Select the parameters K_{I2} and K_{P2} so that the total requirement on relative stability is satisfied.

As an alternative, the PI portion of the controller can be designed first for a portion of the requirement on relative stability, and finally, the PD portion is designed.

The following example illustrates how the PID controller is designed in the time and frequency domains.

Example 10-5

Consider the third-order attitude control system represented by the forward-path transfer function given in Eq. (10-19). With $K = 181.17$, the transfer function is

$$G_p(s) = \frac{2.718 \times 10^9}{s(s + 400.26)(s + 3008)} \tag{10-52}$$

Time-Domain Design

Let the time-domain performance specifications be:

Steady-state error due to a ramp input $t^2u_s(t)/2 \leq 0.2$	rise time $t_r \leq 0.005$ sec
Maximum overshoot ≤ 5 percent	settling time $t_s \leq 0.005$ sec

We realize from Examples 10-2 and 10-4 that these requirements cannot be fulfilled by either the PI control or the PD control acting alone. Let us apply the PD control with the transfer function $(1 + K_{D1}s)$. The forward-path transfer function becomes

$$G(s) = \frac{2.718 \times 10^9(1 + K_{D1}s)}{s(s + 400.26)(s + 3008)} \tag{10-53}$$

Table 10-3 shows that the best PD controller that can be obtained from the maximum overshoot standpoint is with $K_{D1} = 0.002$, and the maximum overshoot is 11.37 percent. The rise and settling times are well within the required values. Next, we add the PI controller, and the forward-path transfer function becomes

$$G(s) = \frac{5.436 \times 10^6 K_{P2}(s + 500)(s + K_{I2}/K_{P2})}{s^2(s + 400.26)(s + 3008)} \tag{10-54}$$

Following the guideline of choosing a relatively small value for K_{I2}/K_{P2}, we let $K_{I2}/K_{P2} = 15$. Equation (10-54) becomes

$$G(s) = \frac{5.436 \times 10^6 K_{P2}(s + 500)(s + 15)}{s^2(s + 400.26)(s + 3008)} \tag{10-55}$$

Table 10-9 gives the time-domain performance characteristics along with the roots of the characteristic equation for various values of K_{P2}. Apparently, the optimal value of K_{P2} is in the neighborhood of 0.2 to 0.4.

Selecting $K_{P2} = 0.3$, and with $K_{D1} = 0.002$ and $K_{I2} = 15K_{P2} = 4.5$, the following results are obtained for the parameters of the PID controller using Eqs. (10-49) through (10-51):

$$\begin{aligned} K_I &= K_{I2} = 4.5 \\ K_P &= K_{P2} + K_{D1}K_{I2} = 0.3 + 0.002 \times 4.5 = 0.309 \\ K_D &= K_{D1}K_{P2} = 0.002 \times 0.3 = 0.0006 \end{aligned} \tag{10-56}$$

Notice that the PID design resulted in a smaller K_D and a larger K_I, which correspond to smaller capacitors in the implementing circuit.

Figure 10-24 shows the unit-step responses of the system with the PID controller, as well as those with PD and PI controls designed in Examples 10-2 and 10-4, respectively. Notice that the PID control, when designed properly, captures the advantages of both the PD and the PI controls.

Table 10-9 Time-Domain Performance Characteristics of Third-Order Attitude Control System with PID Control Designed in Example 10-5

K_{P2}	Maximum Overshoot (%)	t_r (sec)	t_s (sec)	Roots of Characteristic Equation
1.0	11.1	0.00088	0.0025	−15.1, −533.2, −1430 ± j1717.5
0.9	10.8	0.00111	0.00202	−15.1, −538.7, −1427 ± j1571.8
0.8	9.3	0.00127	0.00303	−15.1, −546.5, −1423 ± j1385.6
0.7	8.2	0.00130	0.00303	−15.1, −558.4, −1417 ± j1168.7
0.6	6.9	0.00155	0.00303	−15.2, −579.3, −1406 ± j897.1
0.5	5.6	0.00172	0.00404	−15.2, −629, −1382 ± j470.9
0.4	5.1	0.00214	0.00505	−15.3, −1993, −700 ± j 215.4
0.3	4.8	0.00271	0.00303	−15.3, −2355, −519 ± j263.1
0.2	4.5	0.00400	0.00404	−15.5, −2613, −390 ± j221.3
0.1	5.6	0.00747	0.00747	−16.1, −284, −284 ± j94.2
0.08	6.5	0.00895	0.04545	−16.5, −286.3, −266 ± j4.1

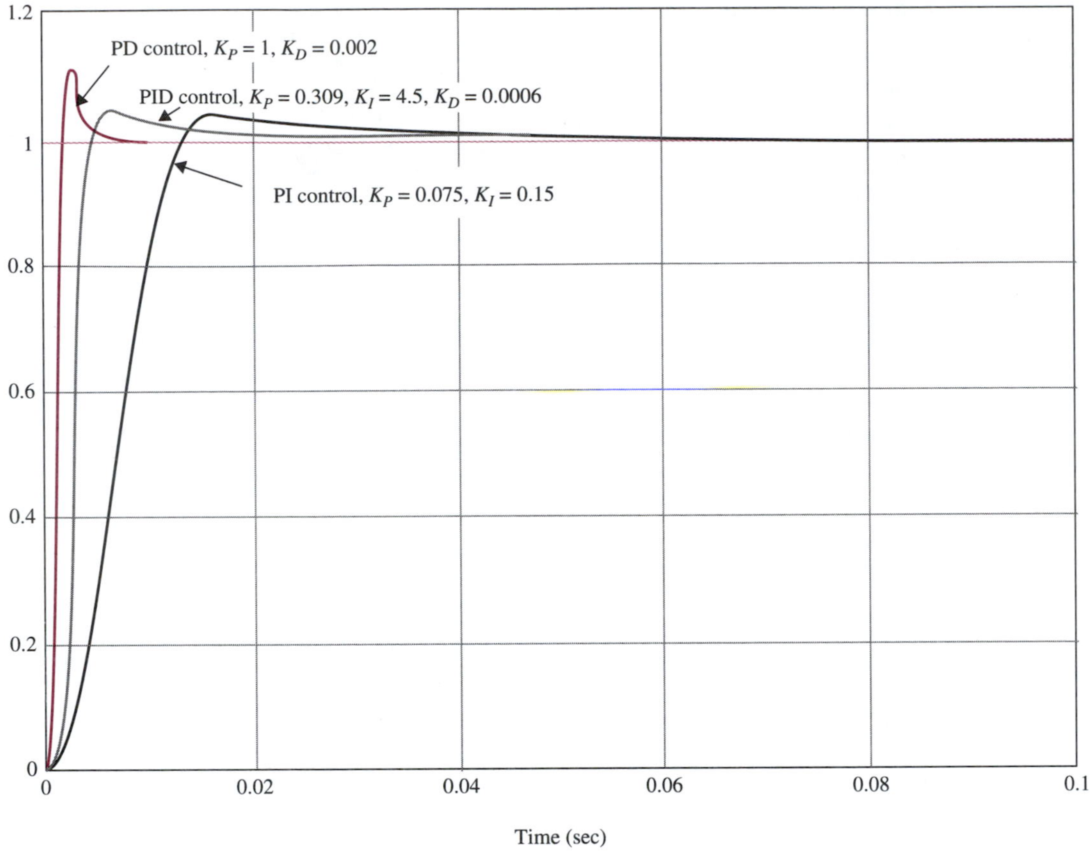

Figure 10-24 Step responses of system in Example 10-5 with PD, PI, and PID controls.

Frequency-Domain Design

The PD control of the third-order attitude control systems was already carried out in Example 10-2, and the results were tabulated in Table 10-3. When $K_P = 1$ and $K_D = 0.002$, the maximum overshoot is 11.37 percent, but this is the best that the PD control could offer. Using this PD controller, the forward-path transfer function of the system is

$$G(s) = \frac{2.718 \times 10^9(1 + 0.002s)}{s(s + 400.26)(s + 3008)} \tag{10-57}$$

and its Bode plot is shown in Fig. 10-25. Let us estimate that the following set of frequency-domain criteria corresponds to the time-domain specifications given in this problem.

Phase margin $\geq 70°$ BW ≥ 1000 rad/sec
$M_r \leq 1.1$

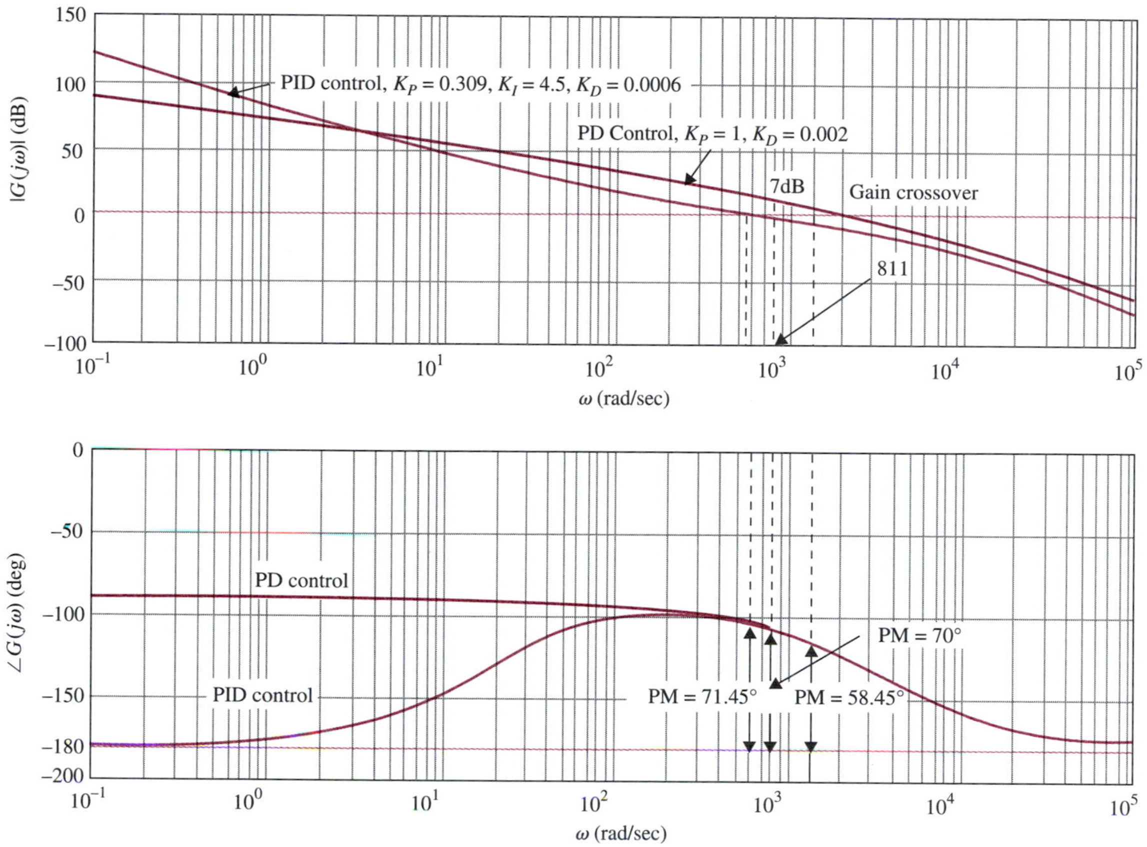

Figure 10-25 Bode plots of system in Example 10-5 with PD and PID controls.

From the Bode diagram in Fig. 10-25 we see that to achieve a phase margin of 70°, the new phase-crossover frequency should be $\omega_g' = 811$ rad/sec, at which the magnitude of $G(j\omega)$ is 7 dB. Thus, using Eq. (10-32), the value of K_{P2} is calculated to be

$$K_{P2} = 10^{-7/20} = 0.45 \tag{10-58}$$

Notice that the desirable range of K_{P2} found from the time-domain design with $K_{I2}/K_{P2} = 15$ is from 0.2 to 0.4. The result given in Eq. (10-58) is slightly out of the range. Table 10-10 shows the frequency-domain performance results with $K_D = 0.002$, $K_{I2}/K_{P2} = 15$, and several values of K_{P2} starting with 0.45. It is interesting to note that as K_{P2} continues to decrease, the phase margin increases monotonically, but below $K_{P2} = 0.2$, the maximum overshoot actually increases. In this case, the phase margin results are misleading, but the resonant peak M_r is a more accurate indication of this. ▲

Table 10-10 Frequency-Domain Performance of System in Example 10-5 with PID Control

K_{P2}	K_{I2}	GM (dB)	PM (deg)	M_r	BW (rad/sec)	t_r (sec)	t_s (sec)	Maximum Overshoot (%)
1.00	0	∞	58.45	1.07	2607	0.0008	0.00255	11.37
0.45	6.75	∞	68.5	1.03	1180	0.0019	0.0040	5.6
0.40	6.00	∞	69.3	1.027	1061	0.0021	0.0050	5.0
0.30	4.50	∞	71.45	1.024	1024	0.0027	0.00303	4.8
0.20	3.00	∞	73.88	1.031	528.8	0.0040	0.00404	4.5
0.10	1.5	∞	76.91	1.054	269.5	0.0076	0.0303	5.6
0.08	1.2	∞	77.44	1.065	216.9	0.0092	0.00469	6.5

10-5 Design with Phase-Lead Controller

The PID controller and its components in the form of PD and PI controls represent simple forms of controllers that utilize derivative and integration operations in the compensation of control systems. In general, we can regard the design of controllers of control systems as a filter design problem; then there are a large number of possible schemes. From the filtering standpoint, the PD controller is a high-pass filter, the PI controller is a low-pass filter, and the PID controller is a bandpass or band-attenuate filter, depending on the values of the controller parameters. The high-pass filter is often referred to as a **phase-lead controller** since positive phase is introduced to the system over some frequency range. The low-pass filter is also known as **phase-lag controller,** since the corresponding phase introduced is negative. These ideas related to filtering and phase shifts are useful if designs are carried out in the frequency domain.

The transfer function of a simple lead or lag controller is expressed as

$$G_c(s) = K_c \frac{s + z_1}{s + p_1} \tag{10-59}$$

where the controller is high-pass or phase-lead if $p_1 > z_1$, and low-pass or phase-lag if $p_1 < z_1$.

The op-amp circuit implementation of Eq. (10-59) is given in Table 4-1(g) and is repeated in Fig. 10-26 with an inverting amplifier. The transfer function of the circuit is

$$G_c(s) = \frac{E_o(s)}{E_{in}(s)} = \frac{C_1}{C_2} \frac{s + 1/R_1C_1}{s + 1/R_2C_2} \tag{10-60}$$

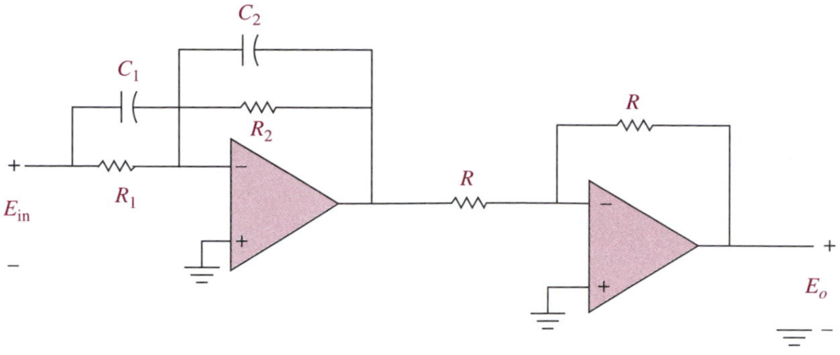

Figure 10-26 Op-amp circuit implementation of $G(s) = K_c \frac{s + z_1}{s + p_1}$.

Comparing the last two equations, we have

$$K_c = C_1/C_2$$
$$z_1 = 1/R_1C_1$$
$$p_1 = 1/R_2C_2 \qquad (10\text{-}61)$$

We can reduce the number of design parameters from four to three by setting $C = C_1 = C_2$. Then Eq. (10-60) is written as

$$G_c(s) = \frac{R_2}{R_1}\left(\frac{1 + R_1Cs}{1 + R_2Cs}\right) = \frac{1}{a}\left(\frac{1 + aTs}{1 + Ts}\right) \qquad (10\text{-}62)$$

where

$$a = \frac{R_1}{R_2} \qquad (10\text{-}63)$$

$$T = R_2C \qquad (10\text{-}64)$$

10-5-1 Time-Domain Interpretation and Design of Phase-Lead Control

In this section we first consider that Eqs. (10-60) and (10-62) represent a phase-lead controller ($z_1 < p_1$ or $a > 1$). In order that the phase-lead controller not degrade the steady-state error, the factor a in Eq. (10-62) should be absorbed by the forward-path gain K. Then for design purposes, $G_c(s)$ can be written as

$$G_c(s) = \frac{1 + aTs}{1 + Ts} \qquad (a > 1) \tag{10-65}$$

The pole-zero configuration of Eq. (10-65) is shown in Fig. 10-27. Based on the discussions given in Chapter 8 on the effects of adding a pole–zero pair, with the zero closer to the origin, to the forward-path transfer function, we realize that phase-lead controller can improve the stability of the closed-loop system if its parameters are chosen properly. The design of phase-lead control is essentially that of placing the pole and zero of $G_c(s)$ so that the design specifications are satisfied. The root-contour method can be used to indicate the proper ranges of the parameters. Computer programs such as **PIDESIGN** and **LLDESIGN** of **CSAD/MATLAB** can be used to speed up the cut-and-try procedure considerably. The following guidelines can be made with regard to selection of the parameters a and T.

1. Moving the zero $-1/aT$ toward the origin should improve rise and settling times. If the zero is moved too close to the origin, the maximum overshoot may again increase since $-1/aT$ also appears as a zero of the closed-loop transfer function.
2. Moving the pole at $-1/T$ farther away from the zero and the origin should reduce the maximum overshoot, but if the value of T is too small, rise and settling times will again increase.

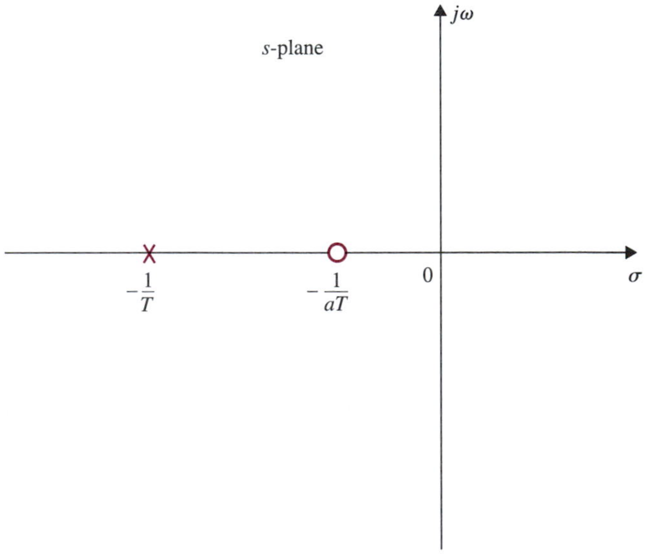

Figure 10-27 Pole–zero configuration of the phase-lead controller.

We can make the following general statements with respect to the effects of phase-lead control on the time-domain performance of a control system.

1. When used properly, it can increase damping of the system.
2. Improves rise and settling times.
3. In the form of Eq. (10-65), phase-lead control does not affect the steady-state error, since $G_c(0) = 1$.

10-5-2 Frequency-Domain Interpretation and Design of Phase-Lead Control

The Bode plot of the phase-lead controller of Eq. (10-65) is shown in Fig. 10-28. The two corner frequencies are at $\omega = 1/aT$ and $\omega = 1/T$. The maximum value of the phase, ϕ_m, and the frequency at which it occurs, ω_m, are derived as follows. Since ω_m is the geometric mean of the two corner frequencies, we write

$$\log_{10}\omega_m = \frac{1}{2}\left(\log_{10}\frac{1}{aT} + \log_{10}\frac{1}{T}\right) \tag{10-66}$$

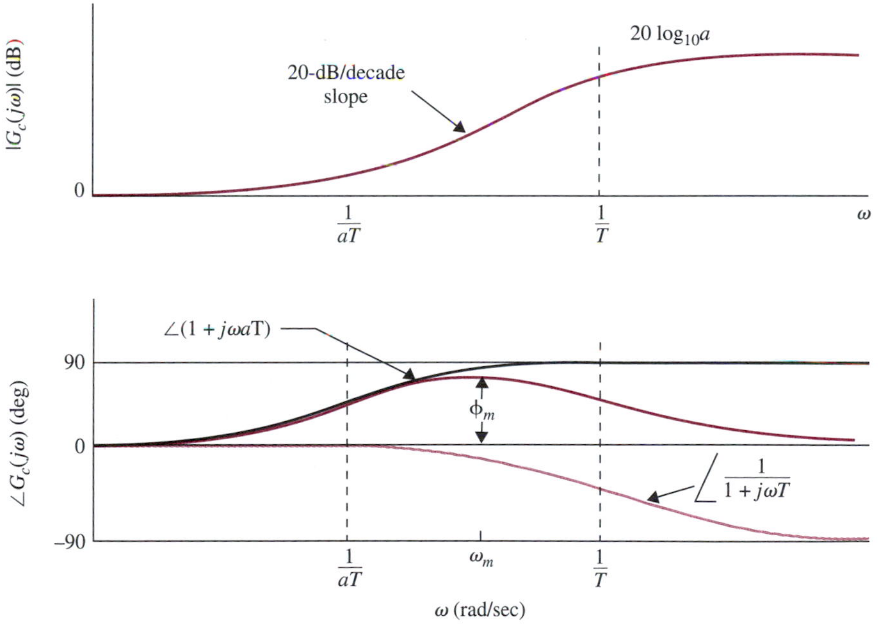

Figure 10-28 Bode plot of phase-lead controller

$$G_c(s) = a\frac{s + 1/aT}{s + 1/T} \qquad a > 1.$$

Thus

$$\omega_m = \frac{1}{\sqrt{a}\,T} \tag{10-67}$$

To determine the maximum phase ϕ_m, the phase of $G_c(j\omega)$ is written

$$\angle G_c(j\omega) = \phi(j\omega) = \tan^{-1}\omega a T - \tan^{-1}\omega T \tag{10-68}$$

from which we get

$$\tan\phi(j\omega) = \frac{\omega a T - \omega T}{1 + (\omega a T)(\omega T)} \tag{10-69}$$

Substituting Eq. (10-67) into Eq. (10-69) for ω, we have

$$\tan\phi_m = \frac{a-1}{2\sqrt{a}} \tag{10-70}$$

or

$$\sin\phi_m = \frac{a-1}{a+1} \tag{10-71}$$

Thus, by knowing ϕ_m, the value of a is determined from

$$a = \frac{1 + \sin\phi_m}{1 - \sin\phi_m} \tag{10-72}$$

The relationship between the phase ϕ_m and a and the general properties of the Bode plot of the phase-lead controller provide an advantage of designing in the frequency domain. The difficulty is, of course, in the correlation between the time- and frequency-domain specifications. The general outline of phase-lead controller design in the frequency domain is given as follows. It is assumed that the design specifications simply include steady-state error and phase-margin requirements.

1. The Bode diagram of the uncompensated process $G_p(j\omega)$ is constructed with the gain constant K set according to the steady-state-error requirement. The value of K has to be adjusted upward once the value of a is determined.

2. The phase margin and the gain margin of the uncompensated system are determined, and the additional amount of phase lead needed to realize the phase margin is determined. From the additional phase lead required, the desired value of ϕ_m is estimated accordingly, and the value of a is calculated from Eq. (10-72).
3. Once a is determined, it is necessary only to determine the value of T, and the design is in principle completed. This is accomplished by placing the corner frequencies of the phase-lead controller, $1/aT$ and $1/T$, such that ϕ_m is located at the new gain-crossover frequency ω_g', so that the phase margin of the compensated system is benefited by ϕ_m. It is known that the high-frequency gain of the phase-lead controller is 20 $\log_{10} a$ dB. Thus, to have the new gain crossover at ω_m, which is the geometric mean of $1/aT$ and $1/T$, we need to place ω_m at thc frequency where the magnitude of the uncompensated $G_p(j\omega)$ is $-10 \log_{10} a$ dB, so that adding the controller gain of $10 \log_{10} a$ dB to this makes the magnitude curve go through 0 dB at ω_m.
4. The Bode diagram of the forward-path transfer function of the compensated system is investigated to check that all performance specifications are met; if not, a new value of ϕ_m must be chosen and the steps repeated.
5. If the design specifications are all satisfied, the transfer function of the phase-lead controller is established from the values of a and T.

If the design specifications also include M_r and/or BW, these must be checked using either the Nichols chart or the output data from a computer program.

We use the following example to illustrate the design of the phase-lead controller in the time and frequency domains.

Example 10-6

The block diagram of the sun-seeker control system described in Section 4-7 is shown again in Fig. 10-29. The system may be mounted on a space vehicle so that it will track the sun with high accuracy. The variable θ_r represents the reference angle of the solar ray, and θ_o denotes the vehicle axis. The objective of the sun-seeker system is to maintain the error between θ_r and θ_o, α, near zero. The parameters of the system are as follows:

- ▲ $R_F = 10{,}000\ \Omega$
- ▲ $K_b = 0.0125$ V/rad/sec
- ▲ $K_i = 0.0125$ N-m/A
- ▲ $R_a = 6.25\ \Omega$
- ▲ $J = 10^{-6}$ kg-m^2
- ▲ $K_s = 0.1$ A/rad
- ▲ K = to be determined
- ▲ $B = 0$
- ▲ $n = 800$

The forward-path transfer function of the uncompensated system is

$$G_p(s) = \frac{\Theta_o(s)}{\mathrm{A}(s)} = \frac{K_s R_F K K_i / n}{R_a J s^2 + K_i K_b s} \tag{10-73}$$

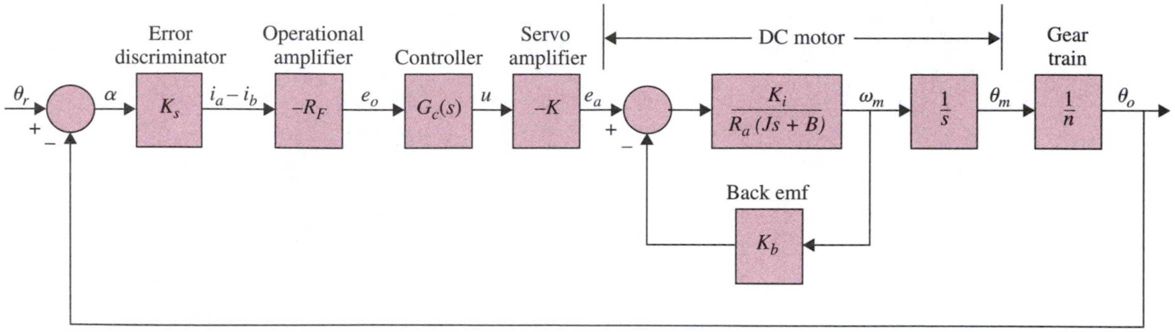

Figure 10-29 Block diagram of sun-seeker control system.

where $\Theta_o(s)$ and $A(s)$ are the Laplace transforms of $\theta_o(t)$ and $\alpha(t)$, respectively. Substituting the numerical values of the system parameters in Eq. (10-73), we get

$$G_p(s) = \frac{\Theta_o(s)}{A(s)} = \frac{2500K}{s(s+25)} \tag{10-74}$$

Time-Domain Design

The time-domain specifications of the system are as follows:

1. The steady-state error of $\alpha(t)$ due to a unit-ramp function input for $\theta_r(t)$ should be ≤ 0.01 rad per rad/sec of the final steady-state output velocity. In other words, the steady-state error due to a ramp input should be ≤ 1 percent.
2. The maximum overshoot of the step response should be less than 5 percent, or as small as possible.
3. Rise time $t_r \leq 0.02$ sec.
4. Settling time $t_s \leq 0.02$ sec.

The minimum value of the amplifier gain, K, is determined initially from the steady-state requirement. Applying the final-value theorem to $\alpha(t)$, we have

$$\lim_{t\to\infty} \alpha(t) = \lim_{s\to 0} sA(s) = \lim_{s\to 0} \frac{s\Theta_r(s)}{1 + G_p(s)} \tag{10-75}$$

For a unit-ramp input, $\Theta_r(s) = 1/s^2$. By using Eq. (10-74), Eq. (10-75) leads to

$$\lim_{t\to\infty} \alpha(t) = \frac{0.01}{K} \tag{10-76}$$

Thus for the steady-state value of $\alpha(t)$ to be ≤ 0.01, K must be ≥ 1. Let us set $K = 1$, the worst case from the steady-state error standpoint, the characteristic equation of the uncompensated system is

$$s^2 + 25s + 2500 = 0 \tag{10-77}$$

We can show that the damping ratio of the uncompensated system with $K = 1$ is only 0.25, which corresponds to a maximum overshoot of 44.4 percent. Figure 10-30 shows the unit-step response of the system with $K = 1$.

A space has been reserved in the forward path of the block diagram of Fig. 10-29 for a controller with transfer function $G_c(s)$. Let us consider using the phase-lead controller of Eq. (10-62), although in the present case, a PD controller or one of the other types of phase-lead controllers may also be effective in satisfying the performance criteria given.

The forward-path transfer function of the compensated system is written

$$G(s) = \frac{2500K(1 + aTs)}{as(s + 25)(1 + Ts)} \tag{10-78}$$

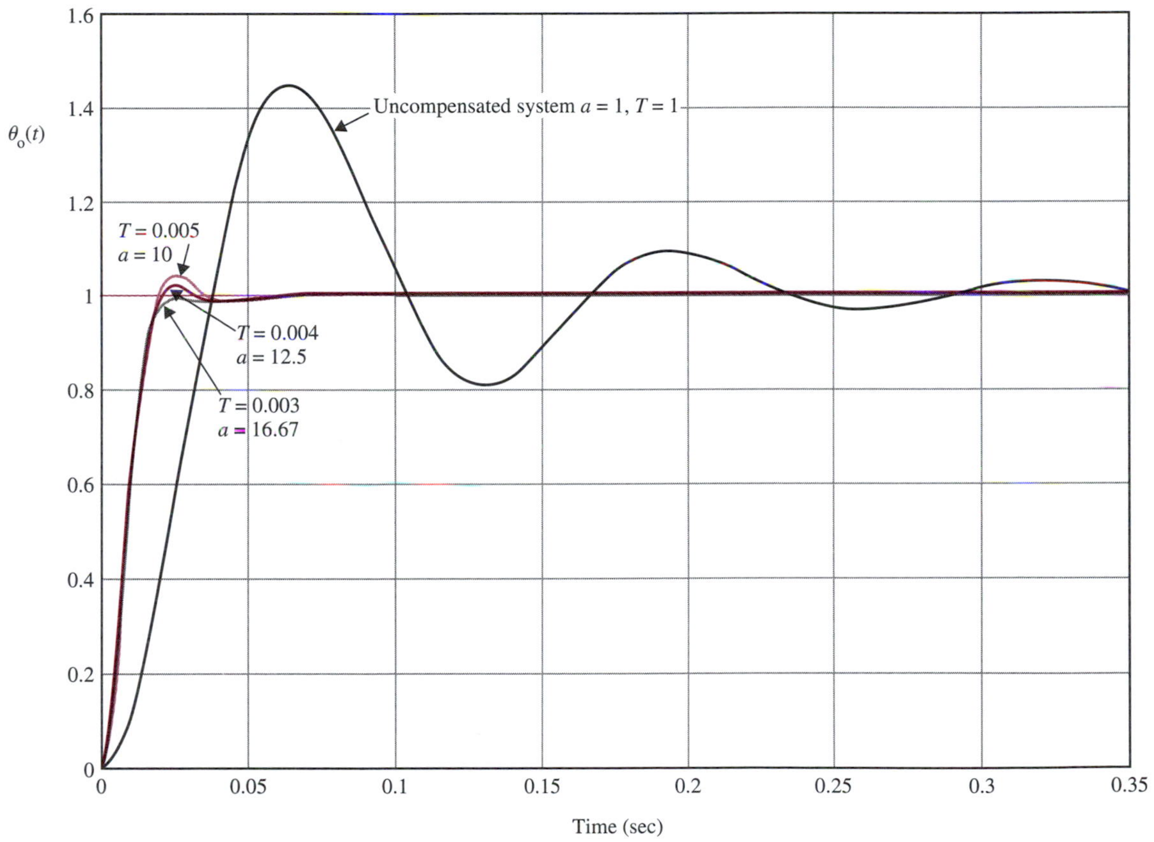

Figure 10-30 Unit-step responses of sun-seeker system in Example 10-6.

For the compensated system to satisfy the steady-state error requirement, K must satisfy

$$K \geq a \tag{10-79}$$

Let us set $K = a$. The characteristic equation of the system is

$$(s^2 + 25s + 2500) + Ts^2(s + 25) + 2500aTs = 0 \tag{10-80}$$

We can use the root-contour method to show the effects of varying a and T of the phase-lead controller. Let us first set $a = 0$. The characteristic equation of Eq. (10-80) becomes

$$s^2 + 25s + 2500 + Ts^2(s + 25) = 0 \tag{10-81}$$

Dividing both sides of Eq. (10-81) by the terms that do not contain T, we get

$$1 + G_{eq1}(s) = 1 + \frac{Ts^2(s + 25)}{s^2 + 25s + 2500} = 0 \tag{10-82}$$

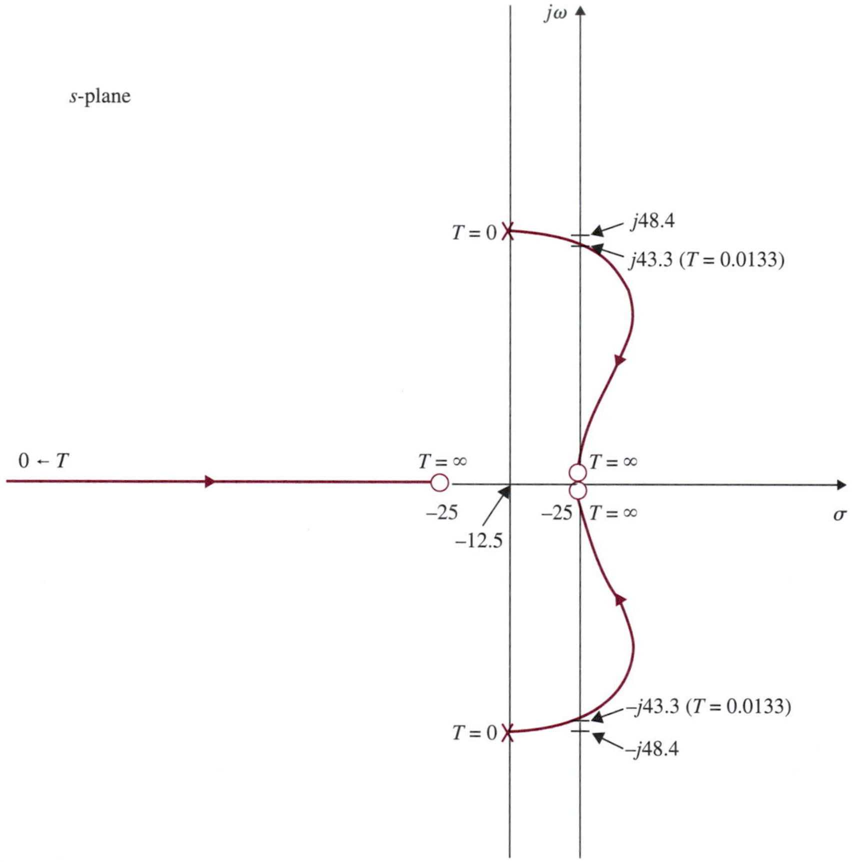

Figure 10-31 Root contours of the sun-seeker system with $a = 0$, and T varies from 0 to ∞.

Thus the root contours of Eq. (10-81) when T varies are determined using the pole–zero configuration of $G_{eq1}(s)$ in Eq. (10-82). These root contours are drawn as shown in Fig. 10-31. Notice that the poles of $G_{eq1}(s)$ are the roots of the characteristic equation when $a = 0$ and $T = 0$. The root contours in Fig. 10-31 clearly show that adding the factor $(1 + Ts)$ to the denominator of Eq. (10-74) alone would not improve the system performance, since the characteristic equation roots are pushed toward the right-half plane. In fact, the system becomes unstable when T is greater than 0.0133. To achieve the full effect of the phase-lead controller, we must restore the value of a in Eq. (10-80). To prepare for the root contours with a as the variable parameter, we divide both sides of Eq. (10-80) by the terms that do not contain a, and the following equation results:

$$1 + aG_{eq2}(s) = 1 + \frac{2500aTs}{s^2 + 25s + 2500 + Ts^2(s + 25)} = 0 \tag{10-83}$$

For a given T, the root contours of Eq. (10-80) when a varies are obtained based on the poles and zeros of $G_{eq2}(s)$. Notice that the poles of $G_{eq2}(s)$ are the same as the roots of Eq. (10-81). Thus for a given T, the root contours of Eq. (10-80) when a varies must start ($a = 0$) at points on the root contours of Fig. 10-31. These root contours end ($a = \infty$) at $s = 0, \infty, \infty$, which are the zeros of $G_{eq2}(s)$. The complete root contours of Eq. (10-80) are now shown in Fig. 10-32 for several values of T, and a varies from 0 to ∞.

▲ For effective phase-lead control, the value of T should be small.

From the root contours of Fig. 10-32, we see that for effective phase-lead control, the value of T should be small. For large values of T, the natural frequency of the system increases rapidly as a increases, and very little improvement is made on the damping of the system.

Let us choose $T = 0.01$ arbitrarily. Table 10-11 shows the attributes of the unit-step response when the value of aT is varied from 0.02 to 0.1. The program **lldesign** of **CSAD/MATLAB** was used for calculations of the time responses. The results show that the smallest maximum overshoot is obtained when $aT = 0.05$, although the rise and settling times decrease continuously as aT increases. However, the smallest value of the maximum overshoot is 16.2 percent, which exceeds the design specification. Next, we set $aT = 0.05$ and vary T from 0.01 to 0.001. Table 10-12 shows the attributes of the unit-step responses. As the value of T decreases, the maximum overshoot decreases, but the rise and settling times increase. The cases that satisfy the design requirements are indicated in the table. Figure 10-30 shows the unit-step responses of the phase-lead-compensated system with three sets of controller parameters.

Choosing $T = 0.004$, $a = 12.5$, the transfer function of the phase-lead controller is

$$G_c(s) = a\frac{s + 1/aT}{s + 1/T} = 12.5\frac{s + 20}{s + 250} \tag{10-84}$$

The transfer function of the compensated system is

$$G(s) = G_c(s)G_p(s) = \frac{31{,}250(s + 20)}{s(s + 25)(s + 250)} \tag{10-85}$$

To find the op-amp-circuit realization of the phase-lead controller, we arbitrarily set $C = 0.1\ \mu\text{F}$. The resistors of the circuit are found using Eqs. (10-63) and (10-64) as, $R_1 = 500{,}000\ \Omega$ and $R_2 = 40{,}000\ \Omega$.

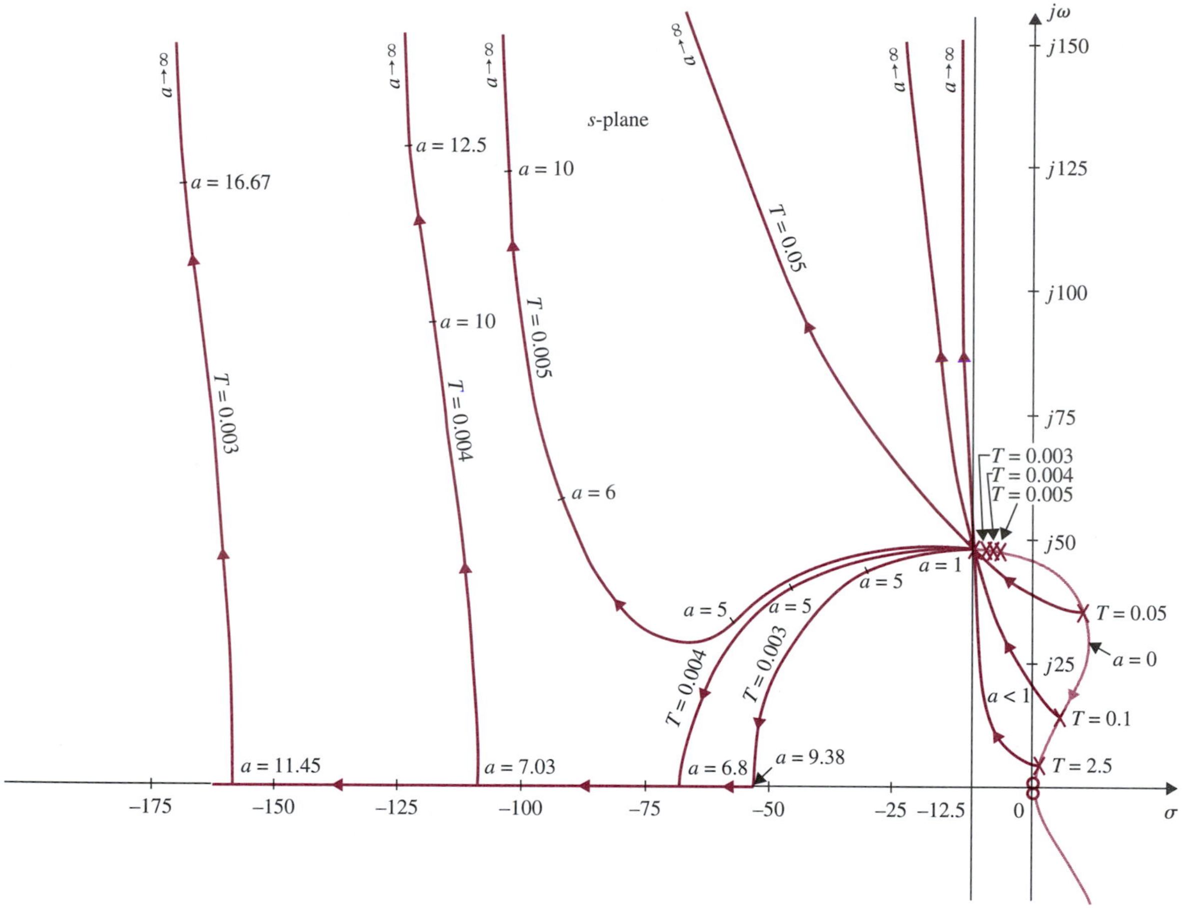

Figure 10-32 Root contours of the sun-seeker system with a phase-lead controller.

Frequency-Domain Design

Let us specify that the steady-state error requirement is the same as that given above. For frequency-domain design, the phase margin is to be greater than 45°. The following design steps are taken:

1. The Bode diagram of Eq. (10-74) with $K = 1$ is plotted as shown in Fig. 10-33.
2. The phase margin of the uncompensated system, read at the gain-crossover frequency, $\omega_g = 47$ rad/sec, is 32°. Since the minimum desired phase margin is 45°, at least 13 more degrees of phase lead should be added to the loop at the gain-crossover frequency.

Table 10-11 Attributes of Unit-Step Response of System with Phase-Lead Controller in Example 10-6: $T = 0.01$

aT	a	Maximum Overshoot (%)	t_r (sec)	t_s (sec)
0.02	2	26.6	0.0222	0.0830
0.03	3	18.9	0.0191	0.0665
0.04	4	16.3	0.0164	0.0520
0.05	5	16.2	0.0146	0.0415
0.06	6	17.3	0.0129	0.0606
0.08	8	20.5	0.0112	0.0566
0.10	10	23.9	0.0097	0.0485

Table 10-12 Attributes of Unit-Step Responses of System with Phase-Lead Control in Example 10-6: $aT = 0.05$

T	a	Maximum Overshoot (%)	t_r (sec)	t_s (sec)
0.01	5.0	16.2	0.0146	0.0415
0.005	10.0	4.1	0.0133	0.0174
0.004	12.5	1.1	0.0135	0.0174
0.003	16.67	0	0.0141	0.0174
0.002	25.0	0	0.0154	0.0209
0.001	50.0	0	0.0179	0.0244

3. The phase-lead controller of Eq. (10-65) must provide the additional 13° at the gain-crossover frequency of the compensated system. However, by applying the phase-lead controller the magnitude curve of the Bode plot is also affected in such a way that the gain-crossover frequency is shifted to a higher frequency. Although it is a simple matter to adjust the corner frquencies, $1/aT$ and $1/T$, of the controller so that the maximum phase of the controller, ϕ_m, falls exactly at the new gain-crossover frequency, the original phase curve at this point is no longer 32°, and could be considerably less, as the phase of most control processes decreases with the increase in frequency. In fact, if the phase of the uncompensated process decreases rapidly with increasing frequency near the gain-crossover frequency, the single-stage phase-lead controller is no longer effective.

In view of the difficulty estimating the necessary amount of phase lead, it is essential to include some safety margin to account for the inevitable phase drop-off.

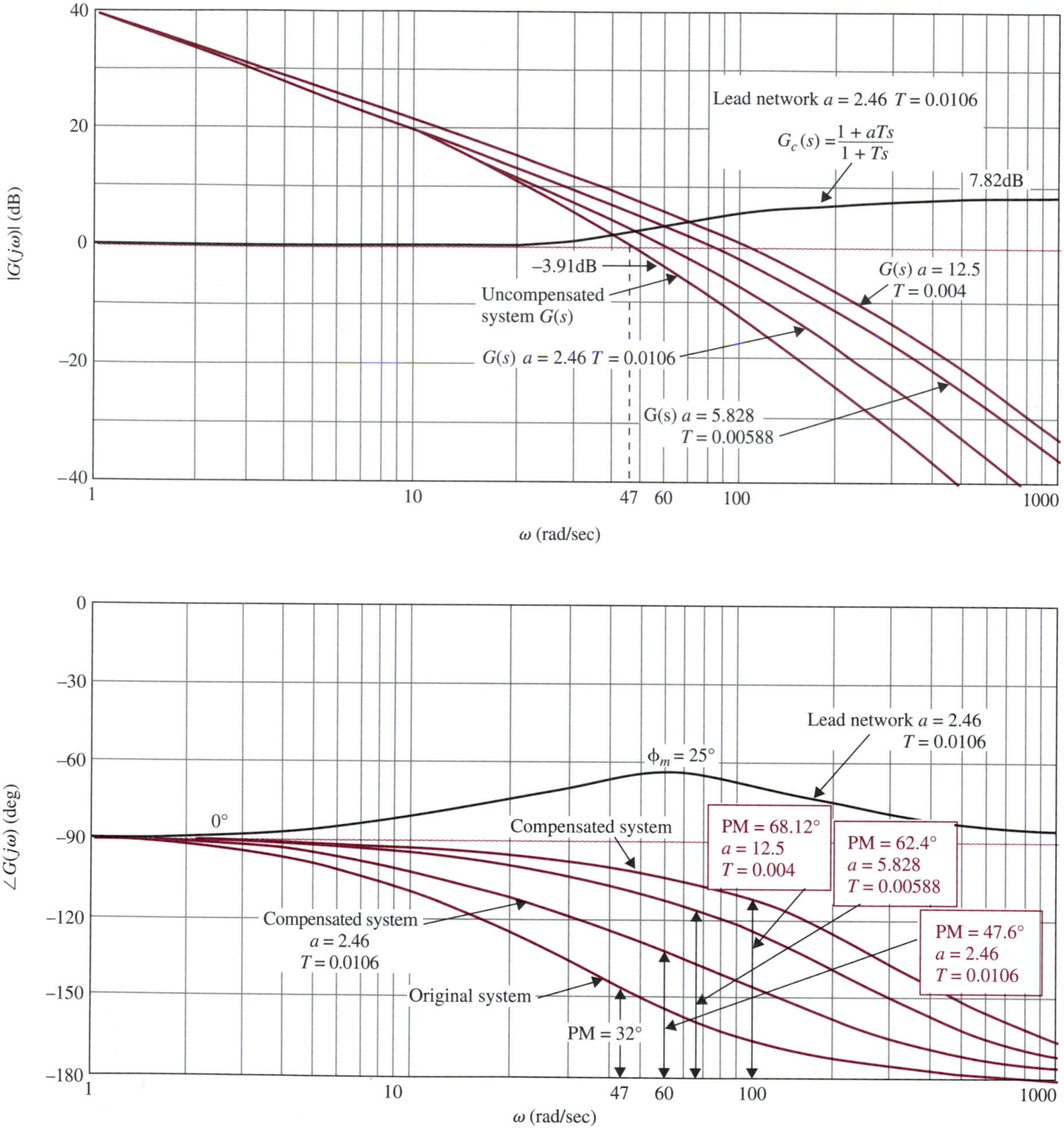

Figure 10-33 Bode diagrams of phase-lead compensated and uncompensated systems in Example 10-6.

$$G(s) = \frac{2500(1 + aTs)}{s(s + 25)(1 + Ts)}$$

Therefore, in the present case, instead of selecting a ϕ_m of a mere 13°, we let ϕ_m be 25°. Using Eq. (10-72), we have

$$a = \frac{1 + \sin 25°}{1 - \sin 25°} = 2.46 \tag{10-86}$$

4. To determine the proper location of the two corner frequencies, $1/aT$ and $1/T$, of the controller, it is known from Eq. (10-67) that the maximum phase lead ϕ_m occurs at the geometric mean of the two corner frequencies. To achieve the maximum phase margin with the value of a determined, ϕ_m should occur at the new gain-crossover frequency ω_g', which is not known. The following steps are taken to ensure that ϕ_m occurs at ω_g'.

(a) The high-frequency gain of the phase-lead controller of Eq. (10-65) is

$$20 \log_{10} a = 20 \log_{10} 2.46 = 7.82 \text{ dB} \tag{10-87}$$

(b) The geometric mean ω_m of the two corner frequencies $1/aT$ and $1/T$ should be located at the frequency at which the magnitude of the uncompensated process transfer function $G_p(j\omega)$ in dB is equal to the negative value in dB of one-half of this gain. This way, the magnitude curve of the compensated transfer function will pass through the 0-dB axis at $\omega = \omega_m$. Thus ω_m should be located at the frequency where

$$|G_p(j\omega)|_{\text{dB}} = -10 \log_{10} 2.46 = -3.91 \text{ dB} \tag{10-88}$$

From Fig. 10-33, this frequency is found to be $\omega_m = 60$ rad/sec. Now using Eq. (10-67), we have

$$\frac{1}{T} = \sqrt{a}\omega_m = \sqrt{2.46} \times 60 = 94.1 \text{ rad/sec} \tag{10-89}$$

Then $1/aT = 94.1/2.46 = 38.21$ rad/sec. The transfer function of the phase-lead controller is

$$G_c(s) = a\frac{s + 1/aT}{s + 1/T} = 2.46\frac{s + 38.21}{s + 94.1} \tag{10-90}$$

The forward-path transfer function of the compensated system is

$$G(s) = G_c(s)G_p(s) = \frac{6150(s + 38.21)}{s(s + 25)(s + 94.1)} \tag{10-91}$$

Figure 10-33 shows that the phase margin of the compensated system is actually 47.6°.

Table 10-13 Attributes of System with Phase-Lead Control in Example 10-6

a	T	PM (deg)	M_r	Gain CO (rad/sec)	BW (rad/sec)	Maximum Overshoot (%)	t_r (sec)	t_s (sec)
1	1	28.03	2.06	47.0	74.3	44.4	0.0255	0.2133
2.46	0.0106	47.53	1.26	60.2	98.2	22.3	0.0204	0.0744
5.828	0.00588	62.36	1.03	79.1	124.7	7.7	0.0169	0.0474
12.5	0.0040	68.12	1.00	113.1	172.5	1.1	0.0135	0.0174

In Fig. 10-34, the magnitude and phase of the original and compensated systems are plotted on the Nichols chart for display only. These plots can be made by taking the data directly from the Bode plots of Fig. 10-33. The values of M_r, ω_r, and BW can all be determined from the Nichols chart. However, the performance data are more easily obtained with a computer program such as **bplot** or **lldesign** of **CSAD.** Checking the time-domain performance of the compensated system, we have the following results:

$$\text{maximum overshoot} = 22.3\% \qquad t_r = 0.02045 \text{ sec} \qquad t_s = 0.07439 \text{ sec}$$

which fall short of the time-domain specifications listed earlier. Figure 10-33 also shows the Bode plot of the system compensated with a phase-lead controller with $a = 5.828$ and $T = 0.00588$. The phase margin is improved to 62.4°. Using Eq. (10-71), we can show that the result of $a = 12.5$ obtained in the time-domain design actually corresponds to $\phi_m = 58.41$. Adding this to the original phase of 28°, the corresponding phase margin would be 86.41°. The time- and frequency-domain attributes of the system with the three phase-lead controllers are summarized in Table 10-13. The results show that with $a = 12.5$ and $T = 0.004$, even the projected phase margin is 86.41°. The actual value is 68.12, due to the fall-off of the phase curve at the new gain crossover. ▲

Example 10-7

In this example we illustrate the application of a phase-lead controller to a third-order system with relatively high loop gain. Let us consider that the inductance of the dc motor of the sun-seeker system described in Fig. 10-29 is not zero. The following set of system parameters are given:

▲ $R_F = 10{,}000\ \Omega$	▲ $K_b = 0.0125$ V/rad/sec
▲ $K_i = 0.0125$ N-m/A	▲ $R_a = 0.625\ \Omega$
▲ $J = 10^{-6}$ kg-m^2	▲ $K_s = 0.3$ A/rad
▲ K = to be determined	▲ $B = 0$
▲ $n = 800$	▲ $L_a = 10^{-3}$ H

The transfer function of the dc motor is written

$$\frac{\Omega_m(s)}{E_a(s)} = \frac{K_i}{s(L_a J s^2 + J R_a s + K_i K_b)} \tag{10-92}$$

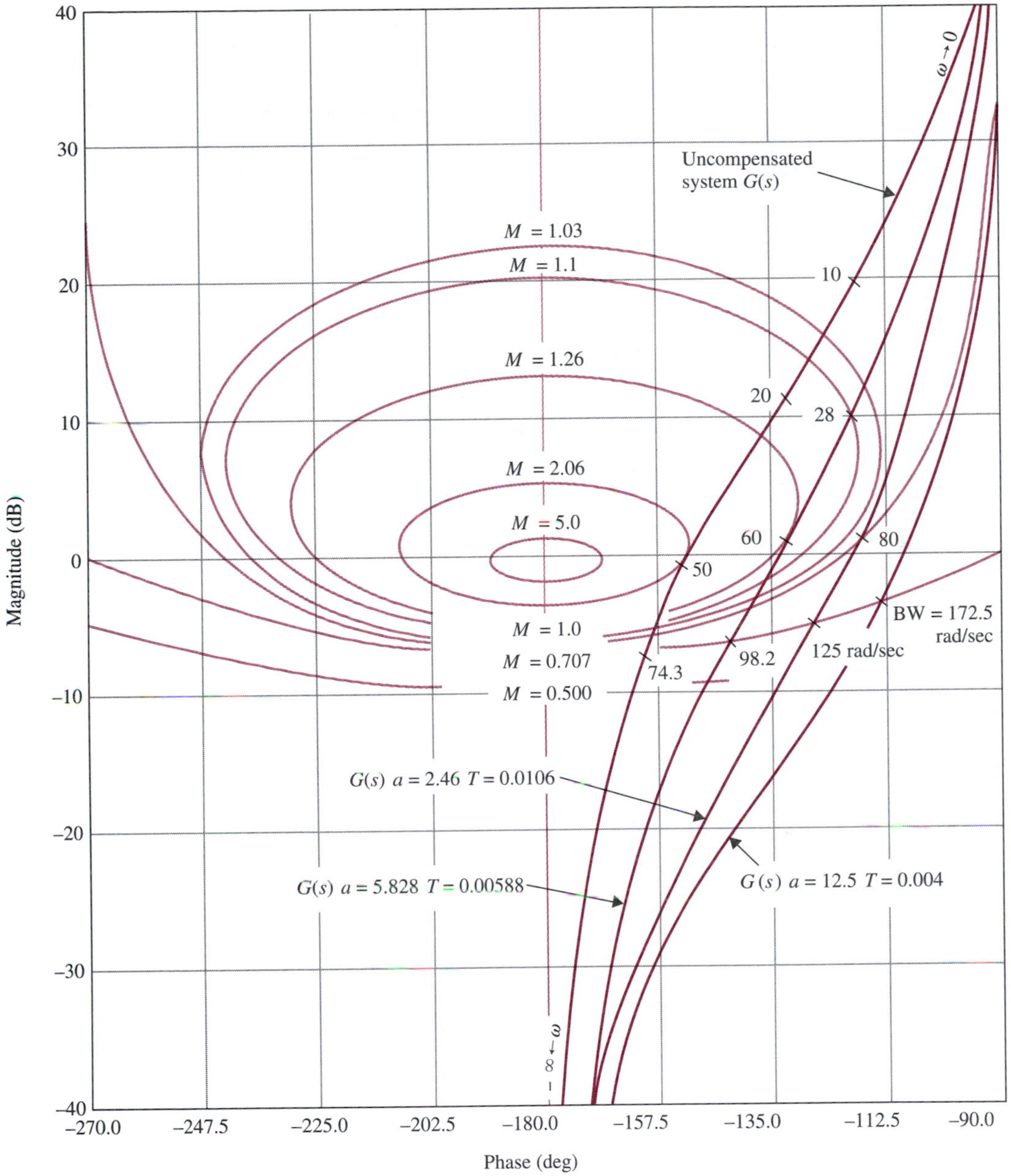

Figure 10-34 Plots of $G(s)$ in the Nichols chart for the system in Example 10-6.

$$G(s) = \frac{2500(1 + aTs)}{s(s + 25)(1 + Ts)}$$

The forward-path transfer function of the system is

$$G_p(s) = \frac{\Theta_o(s)}{A(s)} = \frac{K_s R_F K K_i}{s(L_a J s^2 + J R_a s + K_i K_b)} \tag{10-93}$$

Substituting the values of the system parameters in Eq. (10-93), we get

$$G_p(s) = \frac{\Theta_o(s)}{A(s)} = \frac{4.6875 \times 10^7 K}{s(s^2 + 625s + 156{,}250)} \tag{10-94}$$

Time-Domain Design

The time-domain specifications of the system are given as follows:

1. The steady-state error of $\alpha(t)$ due to a unit-ramp function input for $\theta_r(t)$ should be $\leq$ 1/300 rad/rad/sec of the final steady-state output velocity.
2. The maximum overshoot of the step response should be less than 5 percent, or as small as possible.
3. Rise time $t_r \leq 0.004$ sec.
4. Settling time $t_s \leq 0.02$ sec.

The minimum value of the amplifier gain K is determined initially from the steady-state requirement. Applying the final-value theorem to $\alpha(t)$, we get

$$\lim_{t\to\infty} \alpha(t) = \lim_{s\to 0} sA(s) = \lim_{s\to 0} \frac{s\Theta_r(s)}{1 + G_p(s)} \tag{10-95}$$

Substituting Eq. (10-94) into Eq. (10-95) and $\Theta_r(s) = 1/s^2$, we have

$$\lim_{t\to\infty} \alpha(t) = \frac{1}{300K} \tag{10-96}$$

Thus for the steady-state value of $\alpha(t)$ to be $\leq$ 1/300, K must be ≥ 1. Let us set $K = 1$; the forward-path transfer function in Eq. (10-94) becomes

$$G_p(s) = \frac{4.6875 \times 10^7}{s(s^2 + 625s + 156{,}250)} \tag{10-97}$$

We can show that the closed-loop sun-seeker system with $K = 1$ has the following attributes for the unit-step response.

Maximum overshoot = 43% Rise time $t_r = 0.004797$ sec Settling time $t_s = 0.04587$ sec

To improve the system response, let us select the phase-lead controller described by Eq. (10-62). The forward-path transfer function of the compensated system is

$$G(s) = G_c(s)G_p(s) = \frac{4.6875 \times 10^7 K(1 + aTs)}{as(s^2 + 625s + 156{,}250)(1 + Ts)} \tag{10-98}$$

Now to satisfy the steady-state requirement, K must be readjusted so that $K \geq a$. Let us set $K = a$. The characteristic equation of the phase-lead compensated system becomes

$$(s^3 + 625s^2 + 156{,}250s + 4.6875 \times 10^7) + Ts^2(s^2 + 625s + 156{,}250) + 4.6875 \times 10^7\, aTs = 0 \tag{10-99}$$

We can use the root-contour method to examine the effects of varying a and T of the phase-lead controller. Let us first set a to zero. The characteristic equation of Eq. (10-99) becomes

$$(s^3 + 625s^2 + 156{,}250s + 4.6875 \times 10^7) + Ts^2(s^2 + 625s + 156{,}250) = 0 \tag{10-100}$$

Dividing both sides of Eq. (10-100) by the terms that do not contain T, we get

$$1 + G_{eq1}(s) = 1 + \frac{Ts^2(s^2 + 625s + 156{,}250)}{s^3 + 625s^2 + 156{,}250s + 4.6875 \times 10^7} = 0 \tag{10-101}$$

The root contours of Eq. (10-100) when T varies are determined from the pole–zero configuration of $G_{eq1}(s)$ in Eq. (10-101), and are drawn as shown in Fig. 10-35. When a varies from 0 to ∞, we divide both sides of Eq. (10-99) by the terms that do not contain a, and we have

$$1 + G_{eq2}(s) = 1 + \frac{4.6875 \times 10^7\, aTs}{s^3 + 625s^2 + 156{,}250s + 4.6875 \times 10^7 + Ts^2(s^2 + 625s + 156{,}250)} = 0 \tag{10-102}$$

For a given T, the root contours of Eq. (10-99) when a varies are obtained based on the poles and zeros of $G_{eq2}(s)$. The poles of $G_{eq2}(s)$ are the same as the roots of Eq. (10-100). Thus the root contours when a varies start ($a = 0$) at the root contours for variable T. Figure 10-35 shows the dominant portions of the root contours when a varies for $T = 0.01, 0.0045, 0.001, 0.0005, 0.0001$, and 0.00001. Notice that due to the fact that the uncompensated system is lightly damped, for the phase-lead controller to be effective, the value of T should be very small. Even for very small values of T, there is only a small range of a that could bring increased damping, but the natural frequency of the system increases with the increase in a. The root contours in Fig. 10-35 show the approximate locations of the dominant characteristic equation roots where maximum damping occur. Table 10-14 gives the roots of the characteristic equation and the unit-step-response attributes for the cases that correspond to near-smallest maximum overhsoot for the T selected. Figure 10-36 shows the unit-step response when $a = 500$ and $T = 0.00001$. Although the maximum overshoot is only 3.8 percent, the undershoot in this case is greater than the overshoot.

Frequency-Domain Design

The Bode plot of $G_p(s)$ in Eq. (10-97) is shown in Fig. 10-37. The performance attributes of the uncompensated system are:

$$\text{PM} = 29.74°$$
$$M_r = 2.156$$
$$\text{BW} = 426.5 \text{ rad/sec}$$

We would like to show that the frequency-domain design procedure outlined earlier does not work effectively here, since the phase curve of $G_p(j\omega)$ shown in Fig. 10-37 has a very steep slope near

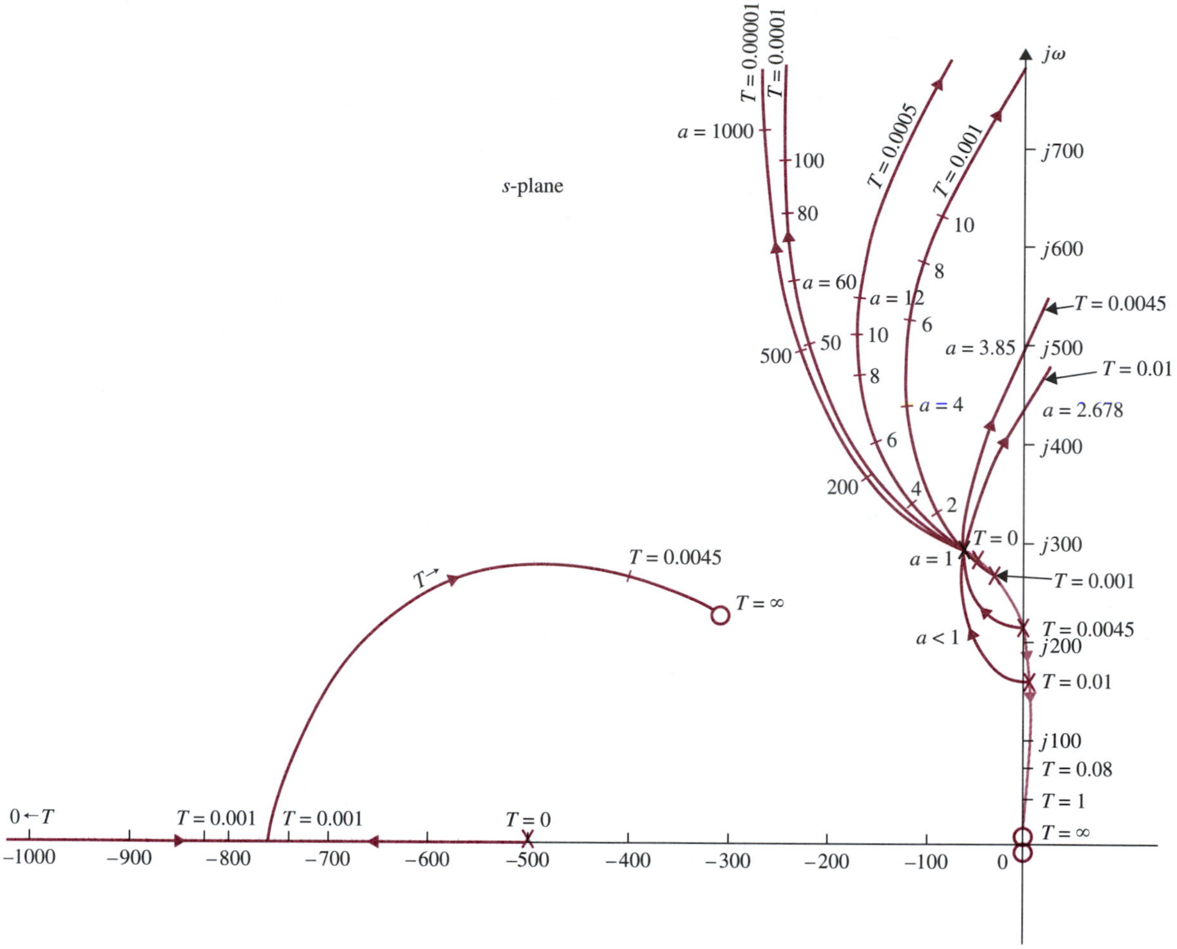

Figure 10-35 Root contours of sun-seeker system in Example 10-7 with phase-lead control.

$$G_c(s) = \frac{1 + aTs}{1 + Ts}$$

the gain crossover. For example, if we wish to realize a phase margin of 65°, we need at least 65 − 29.74 = 35.26° of phase lead, or $\phi_m = 35.26°$. Using Eq. (10-72), the value of a is calculated to be

$$a = \frac{1 + \sin \phi_m}{1 - \sin \phi_m} = \frac{1 + \sin 35.26°}{1 - \sin 35.26°} = 3.732 \tag{10-103}$$

Let us choose $a = 4$. Theoretically, to maximize the utilization of ϕ_m, ω_m should be placed at the new gain crossover, which is located at the frequency where the magnitude of $G_p(j\omega)$ is

Table 10-14 Roots of Characteristic Equation and Time Response Attributes of System with Phase-Lead Control in Example 10-7

T	a	Roots of Characteristic Equation	Maximum Overshoot (%)	t_r (sec)	t_s (sec)
0.001	4	$-189.6,\ -1181.6,\ -126.9 \pm j439.5$	21.7	0.0037	0.0184
0.0005	9	$-164.6,\ -2114.2,\ -173.1 \pm j489.3$	13.2	0.00345	0.0162
0.0001	50	$-147,\ -10{,}024,\ -227 \pm j517$	5.4	0.00348	0.0150
0.00005	100	$-147,\ -20{,}012,\ -233 \pm j515$	4.5	0.00353	0.0150
0.00001	500	$-146.3,\ -10^5,\ -238 \pm j513.55$	3.8	0.00357	0.0146

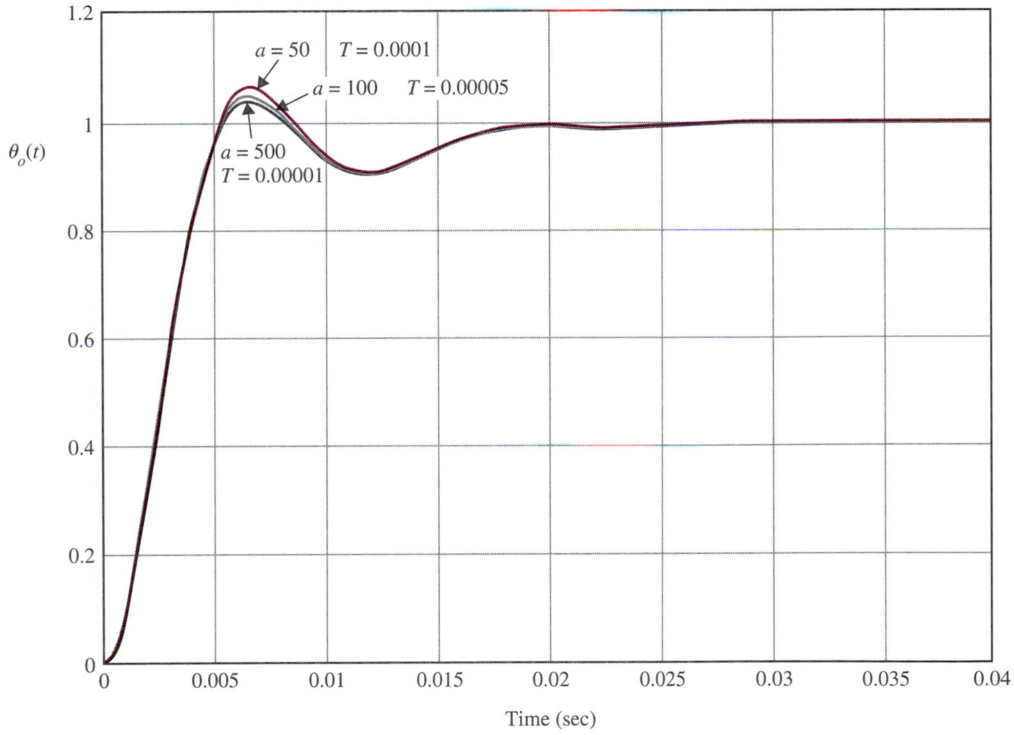

Figure 10-36 Unit-step responses of sun-seeker system in Example 10-7 with phase-lead controller.

$$G_c(s) = \frac{1 + aTs}{1 + Ts}$$

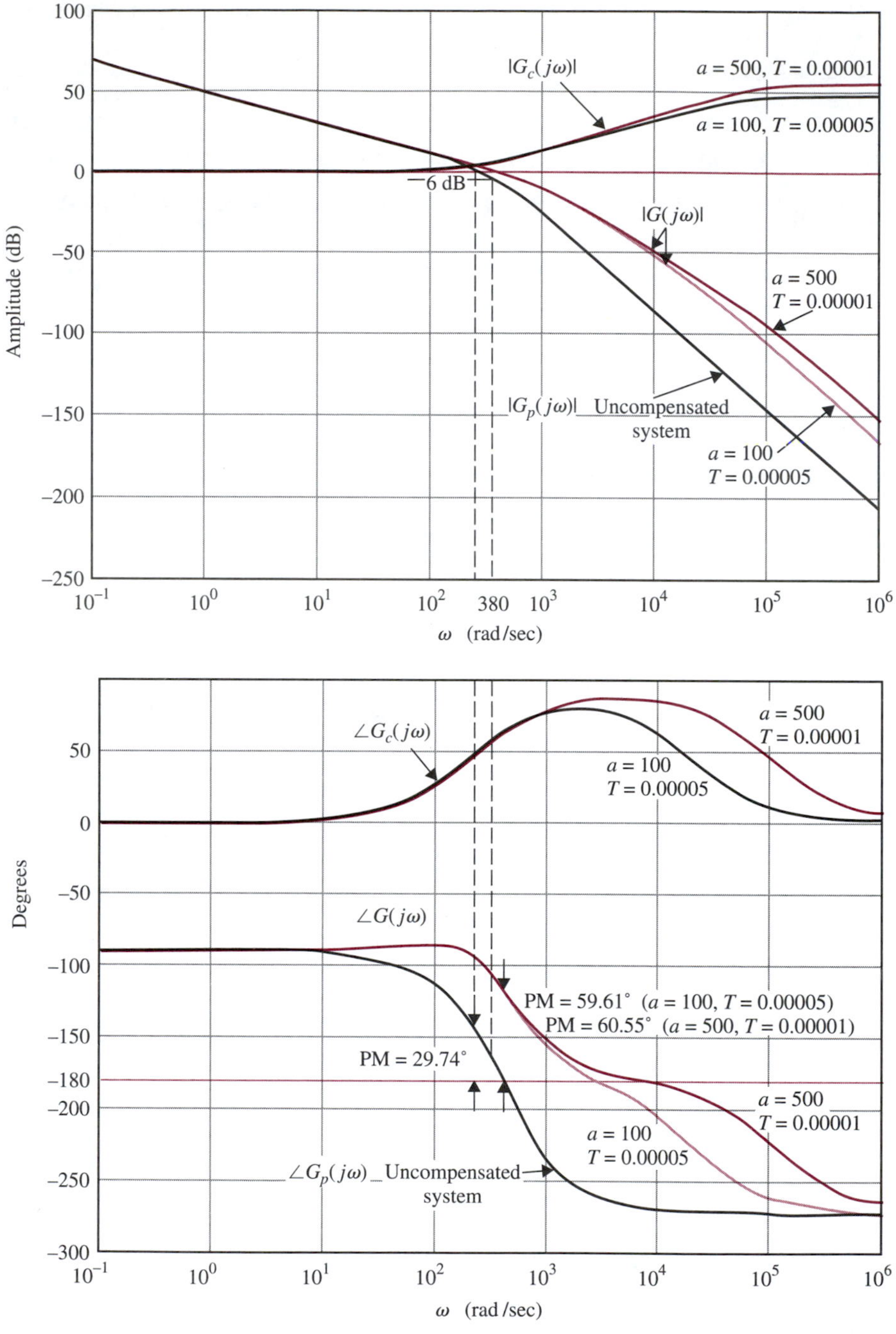

Figure 10-37 Bode plots of phase-lead controller and forward-path transfer functions of sun-seeker system in Example 10-7 with phase-lead control.

$$G_c(s) = \frac{1 + aTs}{1 + Ts}$$

$-10 \log_{10} a$ dB $= -10 \log_{10} 4 = -6$ dB. From the Bode plot in Fig. 10-37 this frequency is found to be 380 rad/sec. Thus we let $\omega_m = 380$ rad/sec. The value of T is found by using Eq. (10-67).

$$T = \frac{1}{\omega_m \sqrt{a}} = \frac{1}{380\sqrt{4}} = 0.0013 \tag{10-104}$$

However, checking the frequency response of the phase-lead compensated system with $a = 4$ and $T = 0.0013$, we found that the phase margin is only improved to 38.27°, and $M_r = 1.69$. The reason is due to the steep negative slope of the phase curve of $G_p(j\omega)$. The fact is that at the new gain-crossover frequency of 380 rad/sec, the phase of $G_p(j\omega)$ is $-170°$, as against $-150.26°$ at the original gain crossover; a drop of almost 20°! From the time-domain design, the first line in Table 10-14 shows that when $a = 4$ and $T = 0.001$, the maximum overshoot is 21.7 percent.

Checking the frequency response of the phase-lead compensated system with $a = 500$ and $T = 0.00001$, the following performance data are obtained:

$$\text{PM} = 60.55° \qquad M_r = 1 \qquad \text{BW} = 664.2 \text{ rad/sec}$$

This shows that the value of a has to be increased substantially just to overcome the steep drop of the phase characteristics when the gain crossover is moved upward.

Figure 10-37 shows the Bode plots of the phase-lead controller and the forward-path transfer functions of the compensated system with $a = 100$, $T = 0.0005$, and $a = 500$, $T = 0.00001$. A summary of performance data is given in Table 10-15.

Selecting $a = 100$ and $T = 0.00005$, the phase-lead controller is described by the transfer function

$$G_c(s) = \frac{1}{a}\frac{1 + aTs}{1 + Ts} = \frac{1}{100}\frac{1 + 0.005s}{1 + 0.00005s} \tag{10-105}$$

Using Eqs. (10-63) and (10-64) and letting C = 0.01 μF, the circuit parameters of the phase-lead controller are found to be

$$R_2 = \frac{T}{C} = \frac{5 \times 10^{-5}}{10^{-8}} = 5000\ \Omega \tag{10-106}$$

$$R_1 = aR_2 = 500{,}000\ \Omega \tag{10-107}$$

Table 10-15 Attributes of System with Phase-Lead Controller in Example 10-7

T	a	PM (deg)	GM (dB)	M_r	BW (rad/s)	Maximum Overshoot (%)	t_r (sec)	t_s (sec)
1	1	29.74	6.39	2.16	430.4	43	0.00478	0.0459
0.00005	100	59.61	31.41	1.009	670.6	4.5	0.00353	0.015
0.00001	500	60.55	45.21	1.000	664.2	3.8	0.00357	0.0146

The forward-path transfer function of the compensated system is

$$\frac{\Theta_o(s)}{A(s)} = \frac{4.6875 \times 10^7(1 + 0.005s)}{s(s^2 + 625s + 156{,}250)(1 + 0.00005s)} \tag{10-108}$$

where the amplifier gain K has been set to 100, to satisfy the steady-state requirement.

From the results of the last two illustrative examples, we can summarize the effects and limitations of the single-stage phase-lead controller as follows. ▲

10-5-3 Effects of Phase-Lead Compensation

1. The phase-lead controller adds a zero and a pole, with the zero to the right of the pole, to the forward-path transfer function. The general effect is to add more damping to the closed-loop system. The rise and settling times are reduced in general.
2. The phase of the forward-path transfer function in the vicinity of the gain-crossover frequency is increased. This improves the phase margin of the closed-loop system.
3. The slope of the magnitude curve of the Bode plot of the forward-path transfer function is reduced at the gain-crossover frequency. This usually corresponds to an improvement in the relative stability of the system in the form of improved gain and phase margins.
4. The bandwidth of the closed-loop system is increased. This corresponds to faster time response.
5. The steady-state error of the system is not affected.

10-5-4 Limitations of Single-Stage Phase-Lead Control

In general, phase-lead control is not suitable for all systems. Successful application of single-stage phase-lead compensation to improve the stability of a control system is hinged on the following conditions:

1. Bandwidth considerations: If the original system is unstable or has a low stability margin, the additional phase lead required to realize a certain desired phase margin may be excessive. This may require a relatively large value of a for the controller, which, as a result, will give rise to a large

bandwidth for the compensated system, and the transmission of high-frequency noise entering the system at the input may become objectionable. However, if the noise enters the system near the output, the increased bandwidth may be beneficial to noise rejection. The larger bandwidth also has the advantage of robustness (i.e., the system is insensitive to parameter variations and noise rejection as described before).

2. If the original system is unstable or has a low stability margin, the phase curve of the Bode plot of the forward-path transfer function has a steep negative slope near the gain-crossover frequency. Under this condition, the single-stage phase-lead controller may not be effective because the additional phase lead at the new gain crossover is added to a much smaller phase angle than that at the old gain crossover. The desired phase margin can be realized only by using a very large value of a for the controller. The amplifier gain K must be set to compensate a, so a large value for a requires a high-gain amplifier, which could be costly.

   As shown in Example 10-7, the compensated system may have a larger undershoot than overshoot. Often, a portion of the phase curve may still dip below the 180° axis, resulting in a **conditionally stable** system, even though the desired phase margin is satisfied.

3. The maximum phase lead available from a single-stage phase-lead controller is less than 90°. Thus if a phase lead of more than 90° is required, a multistage controller should be used.

10-5-5 Multistage Phase-Lead Controller

When the design with a phase-lead controller requires an additional phase of more than 90°, a multistage controller should be used. Figure 10-38 shows an op-amp-circuit realization of a two-stage phase-lead controller. The input–output transfer function of

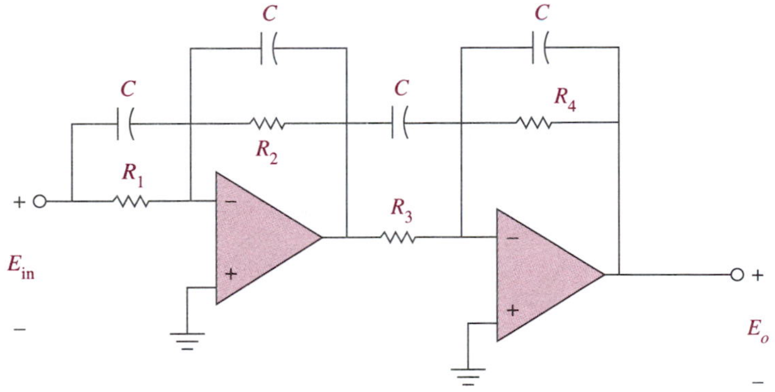

Figure 10-38 Two-stage phase-lead (phase-lag) controller.

the circuit is

$$G_c(s) = \frac{E_o(s)}{E_{in}(s)} = \left(\frac{s + 1/R_1C}{s + 1/R_2C}\right)\left(\frac{s + 1/R_3C}{s + 1/R_4C}\right)$$
$$= \frac{R_2R_4}{R_1R_3}\left(\frac{1 + R_1Cs}{1 + R_2Cs}\right)\left(\frac{1 + R_3Cs}{1 + R_4Cs}\right) \tag{10-109}$$

or

$$G_c(s) = \frac{1}{a_1a_2}\left(\frac{1 + a_1T_1s}{1 + T_1s}\right)\left(\frac{1 + a_2T_2s}{1 + T_2s}\right) \tag{10-110}$$

where $a_1 = R_1/R_2$, $a_2 = R_3/R_4$, $T_1 = R_2C$, and $T_2 = R_4C$.

The design of a multistage phase-lead controller in the time domain becomes more cumbersome, since now there are more poles and zeros to be placed. The root-contour method also becomes unattractive, since there are more variable parameters. The frequency-domain design in this case does represent a better choice of the design method. For example, for a two-stage controller, we can choose the parameters of the first stage of a two-stage controller so that a portion of the phase margin requirement is satisfied, and then the second stage fulfills the remaining requirement. In general, there is no reason why the two stages cannot be identical. The program **freqcad** of **ACSP** has a feature of selecting multistage controllers. The following example illustrates the design of a system with a two-stage phase-lead controller.

Example 10-8

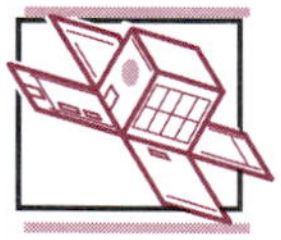

For the sun-seeker system designed in Example 10-7, let us alter the rise-time and settling-time requirements to be:

Rise time $t_r \leq 0.001$ sec
Settling time $t_s \leq 0.005$ sec

The other requirements are not altered. One way to meet faster rise-time and settling-time requirements is to increase the forward-path gain of the system. Let us consider that the forward-path transfer function is

$$G_p(s) = \frac{\Theta_o(s)}{A(s)} = \frac{156{,}250{,}000}{s(s^2 + 625s + 156{,}250)} \tag{10-111}$$

Another way of interpreting the change in the forward-path gain is that the ramp-error constant is increased to 1000 (up from 300 in Example 10-6). The Bode plot of $G_p(s)$ is shown in Fig. 10-39. The closed-loop system is unstable, with a phase margin of $-15.43°$.

To satisfy the specified relative stability requirement, let us set the desired phase margin at 80°. Apparently, this phase margin cannot be realized with a single-stage phase-lead controller. It appears that a two-stage controller would be adequate. If the program **freqcad** of **ACSP** is available, the design procedure would simply be by entering the transfer function of $G_p(s)$ of Eq. (10-111) in the program, select the two-stage phase-lead controller option, and set the desired

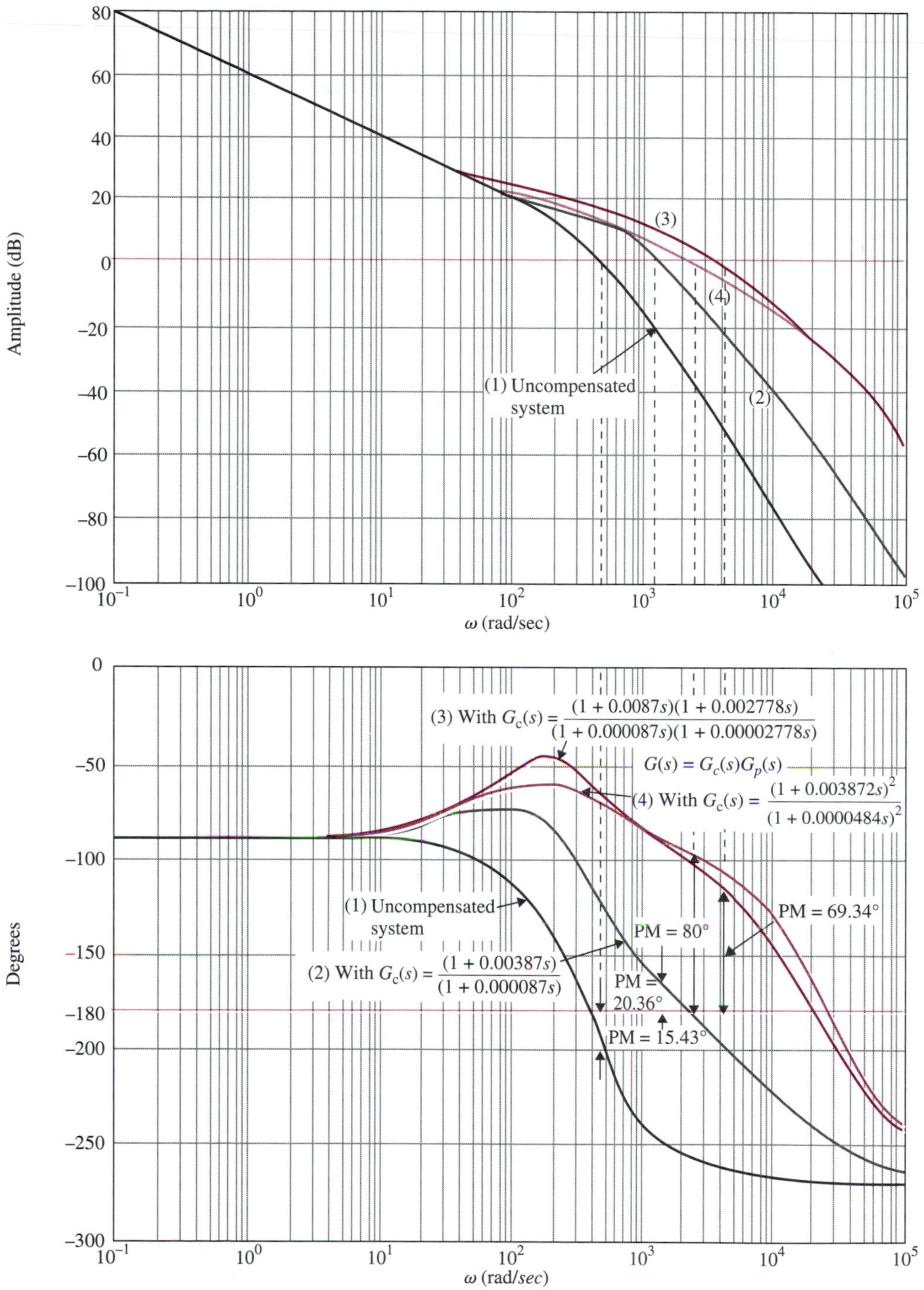

Figure 10-39 Bode plots of uncompensated and compensated sun-seeker system in Example 10-7 with two-stage phase-lead controller.

$$G_p(s) = \frac{156{,}250{,}000}{s(s^2 + 625s + 156{,}250)}$$

phase margin to 80°. The program returns the parameters of the identical two-stage controller as

$$a_1 = a_2 = 80 \qquad T_1 = T_2 = 0.0000484$$

Without the program **freqcad,** one has to go through some trial-and-error steps in arriving at a satisfied controller. As it turns out, since we have two stages of controllers at our disposal, the design has a multitude of flexibility. In an attempt to be objective, we ignore the results given by the **freqcad** program and set out by letting $a_1 = 100$ for the first stage of the phase-lead controller. The phase lead provided by the controller is obtained from Eq. (10-71),

$$\phi_m = \sin^{-1}\left(\frac{a_1 - 1}{a_1 + 1}\right) = \sin^{-1}\left(\frac{99}{101}\right) = 78.58° \tag{10-112}$$

To maximize the effect of ϕ_m, the new gain crossover should be at

$$-10\log_{10} a_1 = -10\log_{10} 100 = -20 \text{ dB} \tag{10-113}$$

From Fig. 10-39 the frequency that corresponds to this gain on the amplitude curve is approximately 1150 rad/sec. Substituting $\omega_{m1} = 1150$ rad/sec and $a_1 = 100$ in Eq. (10-67), we get

$$T_1 = \frac{1}{\omega_{m1}\sqrt{a_1}} = \frac{1}{1150\sqrt{100}} = 0.000087 \tag{10-114}$$

The forward-path transfer function with the one-stage phase-lead controller is

$$G(s) = \frac{156{,}250{,}000(1 + 0.0087s)}{s(s^2 + 625s + 156{,}250)(1 + 0.000087s)} \tag{10-115}$$

The Bode plot of the last equation is drawn as shown in Fig. 10-39. We see that the phase margin of the interim design is only 20.36°. Next, we arbitrarily set the value of a_2 of the second stage at 100. From the Bode plot of the transfer function of Eq. (10-115) in Fig. 10-39, we find that the frequency at which the magnitude of $G(j\omega)$ is -20 dB is approximately 3600 rad/sec. Thus

$$T_2 = \frac{1}{\omega_{m2}\sqrt{a_2}} = \frac{1}{3600\sqrt{100}} = 0.00002778 \tag{10-116}$$

The forward-path transfer function of the sun-seeker system with the two-stage phase-lead controller is

$$G(s) = \frac{156{,}250{,}000(1 + 0.0087s)(1 + 0.002778s)}{s(s^2 + 625s + 156{,}250)(1 + 0.000087s)(1 + 0.00002778s)} \tag{10-117}$$

Figure 10-39 shows the Bode plots of the sun-seeker system with the two-stage phase-lead controller designed above, and that obtained with the **freqcad** program, $a_1 = a_2 = 80$ and $T_1 = T_2 = 0.0000484$. As seen from Fig. 10-39, the phase margin of the system with $G(s)$ given in Eq. (10-117) is only 69.34°. However, the system satisfies all the time-domain specifications.

Table 10-16 Attributes of Sun-Seeker System in Example 10-8 with Two-Stage Phase-Lead Controller

$a_1 = a_2$	T_1	T_2	PM (deg)	M_r	BW (rad/sec)	Maximum Overshoot (%)	t_r (sec)	t_s (sec)
80	0.0000484	0.0000484	80	1	5686	0	0.00095	0.00475
100	0.000087	0.0000278	69.34	1	5686	0	0.000597	0.00404
70	0.0001117	0.000039	66.13	1	5198	0	0.00063	0.00404

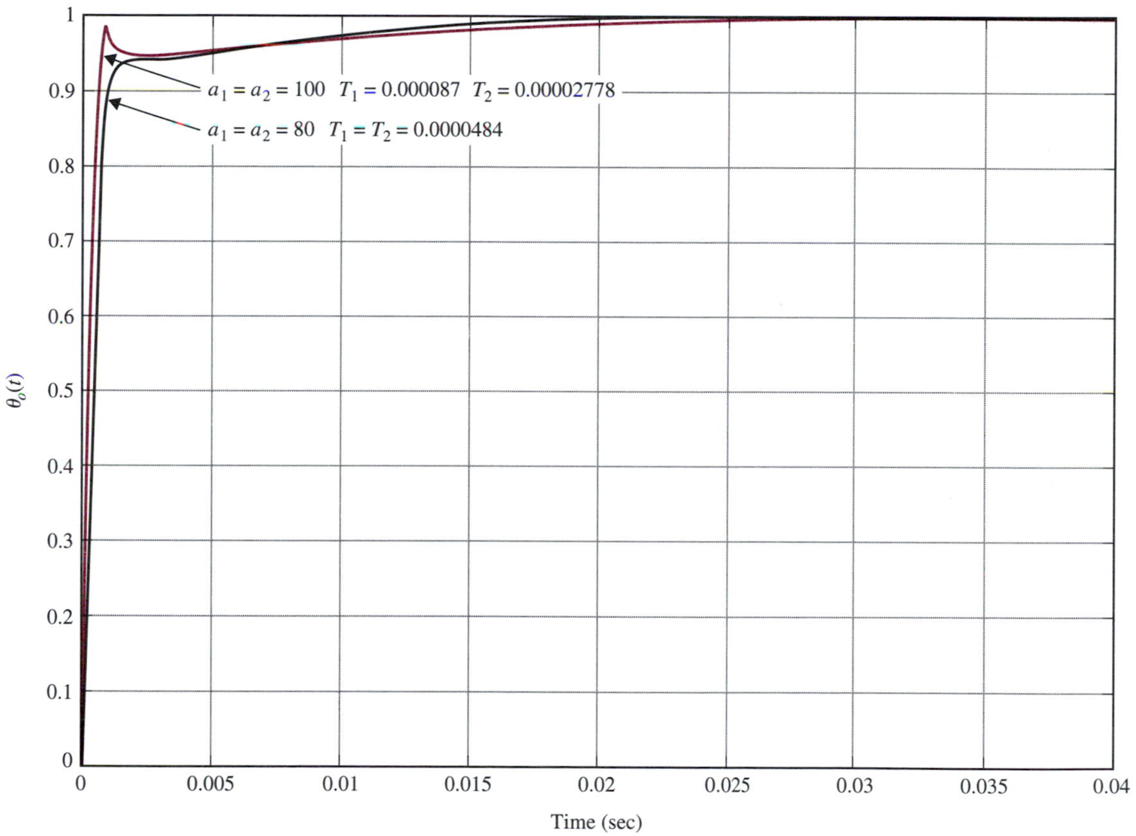

Figure 10-40 Unit-step responses of sun-seeker system in Example 10-7 with two-stage phase-lead controller.

$$G_c(s) = \frac{1 + a_1 T_1 s}{1 + T_1 s}\frac{1 + a_2 T_2 s}{1 + T_s s} \qquad G_p(s) = \frac{156{,}250{,}000}{s(s^2 + 625s + 156{,}250)}.$$

Thus the estimated phase margin of 80° was overly stringent. To show that the design is not critical, we can select $a_1 = a_2 = 70$. Following similar design steps as above, we arrived at $T_1 = 0.0001117$ and $T_2 = 0.000039$. The final phase margin is only 66.13°, but the time-domain requirements are still satisfied. Table 10-16 summarizes all the attributes of the system performance with these three controllers. The unit-step responses of the system with the two-stage phase-lead controller for $a_1 = a_2 = 80$ and 100 are shown in Fig. 10-40. ▲

10-5-6 Sensitivity Considerations

The sensitivity function defined in Section 9-23, Eq. (9-181) can be used as a design specification to indicate the robustness of the system. In Eq. (9-181), the sensitivity of the closed-loop transfer function with respect to the variations of the forward-path transfer function is defined as

$$S_G^M(s) = \frac{\partial M(s)/M(s)}{\partial G(s)/G(s)} = \frac{G^{-1}(s)}{1 + G^{-1}(s)} = \frac{1}{1 + G(s)} \tag{10-118}$$

The plot of $|S_G^M(j\omega)|$ versus frequency gives indication on the sensitivity of the system as a function of frequency. The ideal robust situation is for $|S_G^M(j\omega)|$ to assume a small value (<<1) over a wide range of frequencies. As an example, the sensitivity function of the sun-seeker system designed in Example 10-7 with the one-stage phase-lead controller with $a = 100$ and $T = 0.00005$ is plotted as shown in Fig. 10-41. Note that the sensitivity function is low at low frequencies, and is less than unity for $\omega < 400$ rad/sec. Although the sun-seeker system in Example 10-7 does not need a multistage phase-lead controller, we shall show that if a two-stage phase-lead controller is used, not only the value of a will be substantially reduced, resulting in lower gains for the op-amps, but the system will be more robust. Following the design procedure outlined in Example 10-8, a two-stage phase-lead controller is designed for the sun-seeker system with the process transfer function described by Eq. (10-97). The parameters of the controller are: $a_1 = a_2 = 5.83$ and $T_1 = T_2 = 0.000673$. The forward-path transfer

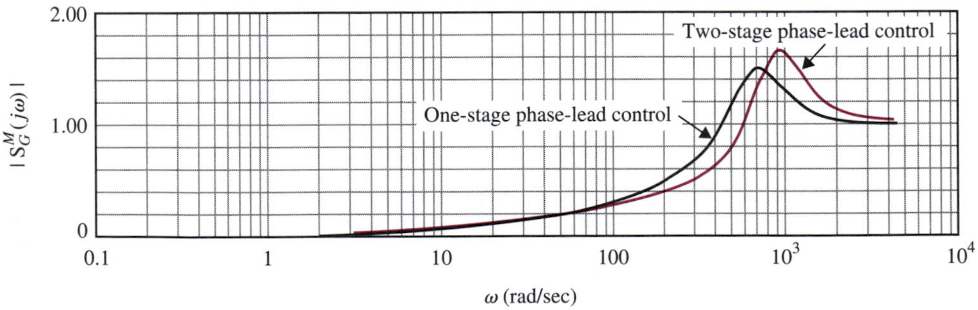

Figure 10-41 Sensitivity functions of sun-seeker system in Example 10-7.

▲ Systems with phase-lead control will generally be more robust due to the increase in bandwidth.

function of the compensated system is

$$G(s) = \frac{4.6875 \times 10^7 (1 + 0.0039236s)^2}{s(s^2 + 625s + 156{,}250)(1 + 0.000673s)^2} \tag{10-119}$$

Figure 10-41 shows that the sensitivity function of the system with the two-stage phase-lead controller is less than unity for $\omega < 600$ rad/sec. Thus the system with the two-stage phase-lead controller is more robust than the system with the single-stage controller. The reason for this is that the more robust system has a higher bandwidth. In general, systems with phase-lead control will be more robust due to the higher bandwidth. However, Fig. 10-41 shows that the system with the two-stage phase-lead controller has a higher sensitivity at high frequencies.

10-6 Design with Phase-Lag Controller

The transfer function in Eq. (10-62) represents a phase-lag controller or low-pass filter when $a < 1$. The transfer function is repeated as follows:

$$G_c(s) = \frac{1}{a}\left(\frac{1 + aTs}{1 + Ts}\right) \qquad a < 1 \tag{10-120}$$

10-6-1 Time-Domain Interpretation and Design of Phase-Lag Control

The pole–zero configuration of $G_c(s)$ is shown in Fig. 10-42. Unlike the PI controller, which provides a pole at $s = 0$, the phase-lag controller affects the steady-state error only in the sense that the zero-frequency gain of $G_c(s)$ is greater than unity. Thus any

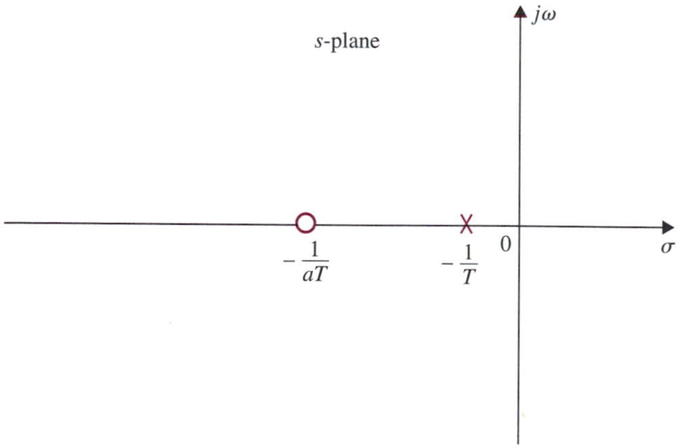

Figure 10-42 Pole-zero configuration of phase-lag controller.

$$G_c(s) = \frac{1}{a}\frac{1 + aTs}{1 + Ts} \qquad a < 1$$

error constant that is finite and nonzero will be increased by the factor $1/a$ from the phase-lag controller.

Since the pole at $s = -1/T$ is to the right of the zero at $-1/aT$, effective use of the phase-lag controller to improve damping would have to follow the same design principle of the PI control presented in Section 10-3. Thus *the proper way of applying the phase-lag control is to place the pole and zero close together. For type 0 and type 1 systems the combination should be located near the origin in the s-plane.* Figure 10-43 illustrates the design strategies in the s-plane for type 0 and type 1 systems. Phase-lag control should not be applied to a type 2 system.

The design principle described above can be explained by considering that the controlled process of a type 0 control system is

$$G_p(s) = \frac{K}{(s + p_1)(s + \bar{p}_1)(s + p_3)} \tag{10-121}$$

where p_1 and $\bar{p}_1$ are complex-conjugate poles, such as the situation shown in Fig. 10-43.

Just as in the case of the phase-lead controller, we can drop the gain factor $1/a$ in Eq. (10-120), since whatever the value of a is, the value of K can be adjusted to compensate. Applying the phase-lag controller of Eq. (10-120), without the factor $1/a$, to the system, the forward-path transfer function becomes

$$G(s) = G_c(s)G_p(s) = \frac{K(1 + aTs)}{(s + p_1)(s + \bar{p}_1)(s + p_3)(1 + Ts)} \qquad a < 1 \tag{10-122}$$

Let us assume that the value of K is set to meet the steady-state-error requirement. Also, assume that with the selected value of K, the system has low damping or is even unstable. Now let $1/T \cong 1/aT$, and place the pole–zero pair near the pole at $-1/p_3$, as shown in Fig. 10-43. Figure 10-44 shows the root loci of the system with and without the phase-lag controller. Since the pole–zero combination of the controller is very close to the pole at $-p_3$, the shape of the loci of the dominant roots with and without the phase-lag control will be very similar. This is easily explained by writing Eq. (10-122) as

$$\begin{aligned} G(s) &= \frac{Ka(s + 1/aT)}{(s + p_1)(s + \bar{p}_1)(s + p_3)(s + 1/T)} \\ &\cong \frac{Ka}{(s + p_1)(s + \bar{p}_1)(s + p_3)} \end{aligned} \tag{10-123}$$

Since a is less than 1, the application of phase-lag control is equivalent to reducing the forward-path gain from K to Ka, *while not affecting the steady-state performance of the system.* Figure 10-44 shows that the value of a can be chosen so that the damping of the compensated system is satisfactory. Apparently, the amount of damping that can be added is limited if the poles $-p_1$ and $-\bar{p}_1$ are very close to the imaginary axis. Thus

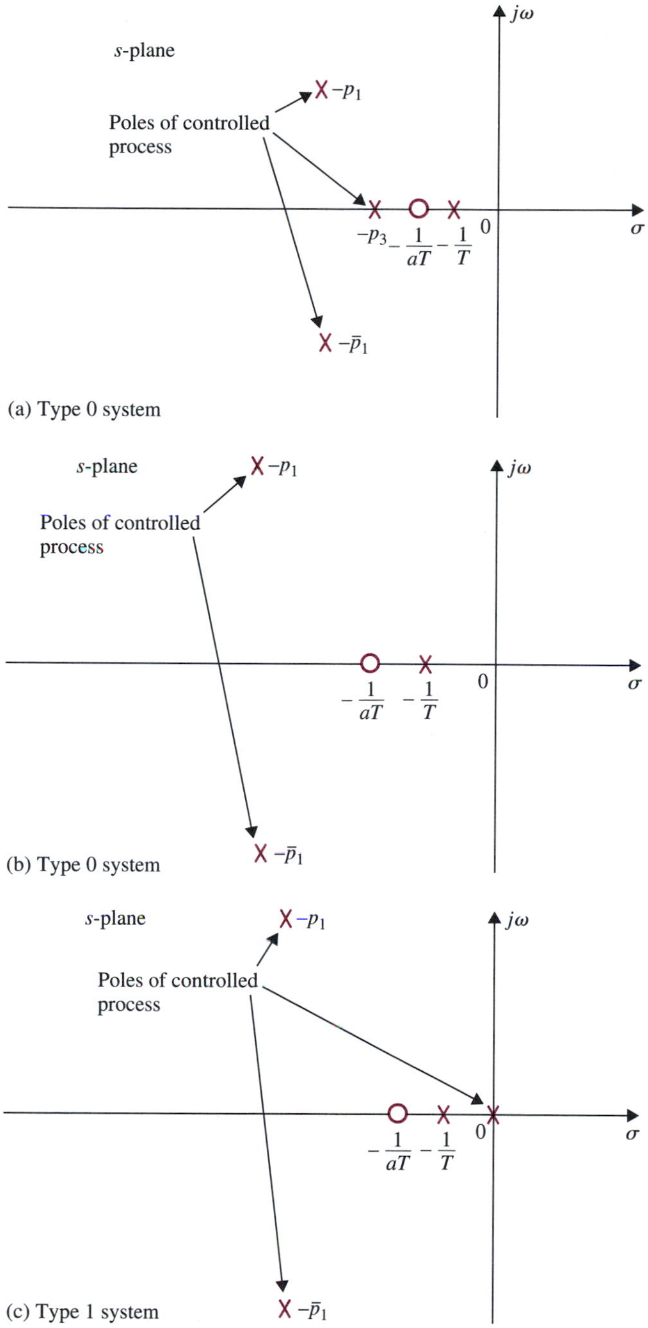

Figure 10-43 Design strategies for phase-lag control for type 0 and type 1 systems.

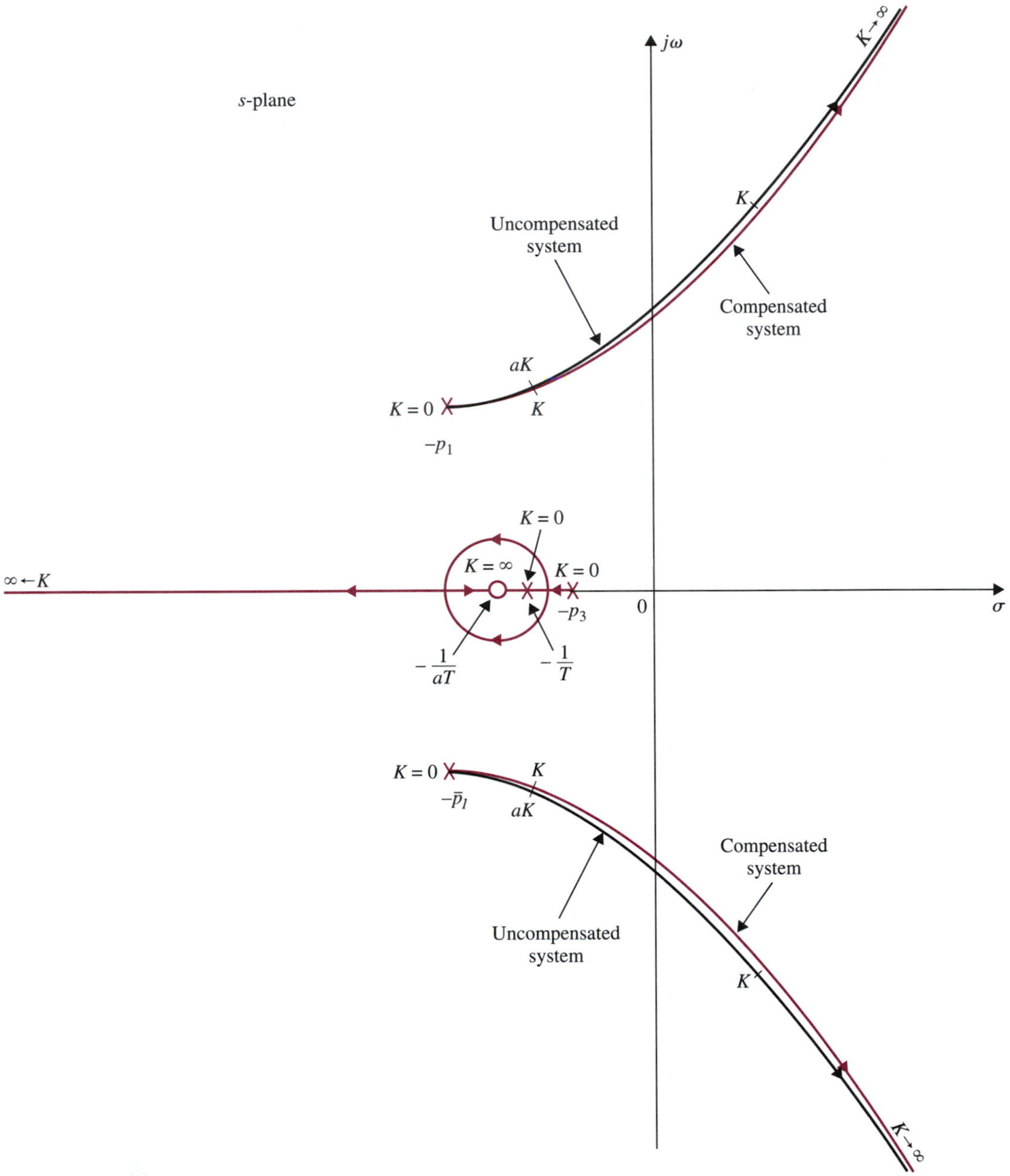

Figure 10-44 Root loci of uncompensated and phase-lag-compensated system.

▲ Phase-lag control generally will increase rise time and settling time.

we can select a using the following equation:

$$a = \frac{K \text{ to realize the desired damping}}{K \text{ to realize the steady-state performance}} \tag{10-124}$$

The value of T should be so chosen that the pole and zero of the controller are very close together, and close to $-p_3$. In the time domain, phase-lag control generally has the effect of increasing the rise time and settling time.

10-6-2 Frequency-Domain Interpretation and Design of Phase-Lag Control

The transfer function of the phase-lag controller can again be written as

$$G_c(s) = \frac{1 + aTs}{1 + Ts} \qquad a < 1 \tag{10-125}$$

by assuming that the gain factor, $1/a$ is eventually absorbed by the forward gain K. The Bode diagram of Eq. (10-125) is shown in Fig. 10-45. The magnitude curve has

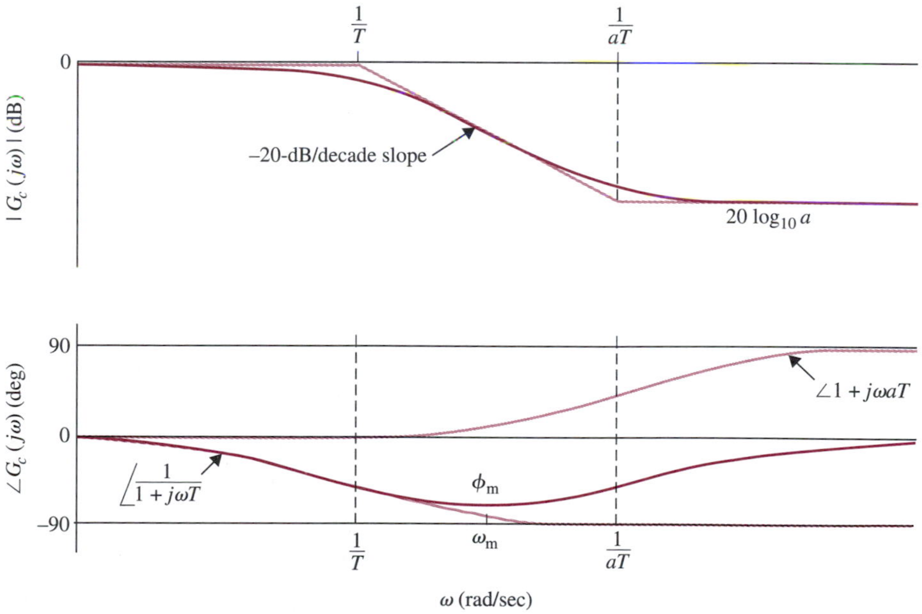

Figure 10-45 Bode diagram of the phase-lag controller, $G_c(s) = (1 + aTs)/(1 + Ts)$; $a < 1$.

▲ The objective of phase-lag control in the frequency domain is to move the gain crossover to the frequency where the desired phase margin is realized, while keeping the phase curve relatively unchanged at the new gain crossover.

corner frequencies at $\omega = 1/aT$ and $1/T$. Since the transfer functions of the phase-lead and phase-lag controllers are identical in form, except for the value of a, the maximum phase lag ϕ_m of the phase curve of Fig. 10-45 is given by

$$\phi_m = \sin^{-1}\left(\frac{a-1}{a+1}\right) \qquad a < 1 \tag{10-126}$$

Figure 10-45 shows that the phase-lag controller essentially provides an attenuation of $20 \log_{10} a$ at high frequencies. Thus, unlike the phase-lead control that utilizes the maximum phase lead of the controller, *phase-lag control utilizes the attenuation of the controller at high frequencies.* This is parallel to the situation of introducing an attenuation of a to the forward-path gain in the root-locus design. For phase-lead control, the objective of the controller is to increase the phase of the open-loop system in the vicinity of the gain crossover while attempting to locate the maximum phase lead at the new gain crossover. In phase-lag control, the objective is to move the gain crossover to a lower frequency, where the desired phase margin is realized, while keeping the phase curve of the Bode plot relatively unchanged at the new gain crossover.

The design procedure for phase-lag control using the Bode plot is outlined as follows:

1. The Bode plot of the forward-path transfer function of the uncompensated system is drawn. The forward-path gain K is set according to the steady-state performance requirement.
2. The phase and gain margins of the uncompensated system are determined from the Bode plot.
3. Assuming that the phase margin is to be increased, the frequency at which the desired phase margin is obtained is located on the Bode plot. This frequency is also the new gain-crossover frequency ω'_g, where the compensated magnitude curve crosses the 0-dB axis.
4. To bring the magnitude curve down to 0 dB at the new gain-crossover frequency ω'_g, the phase-lag controller must provide the amount of attenuation equal to the value of the magnitude curve at ω'_g. In other words,

$$|G_p(j\omega'_g)| = -20\log_{10} a \quad \text{dB} \qquad a < 1 \tag{10-127}$$

Solving for a from the last equation, we get

$$a = 10^{-|G_p(j\omega'_g)|/20} \qquad a < 1 \tag{10-128}$$

Once the value of a is determined, it is necessary only to select the proper value of T to complete the design. From the phase characteristics shown in

Fig. 10-45, it is observed that if the corner frequency $1/aT$ is placed far below the new gain-crossover frequency ω_g' the phase lag of the controller will not appreciably affect the phase of the compensated system near ω_g'. On the other hand, the value of $1/aT$ should not be too small because the bandwidth of the system will be too low, causing the system to be too sluggish and less robust. Usually, as a general guideline, the frequency $1/aT$ should be approximately one decade below ω_g'; that is,

$$\frac{1}{aT} = \frac{\omega_g'}{10} \quad \text{rad/sec} \tag{10-129}$$

Then

$$\frac{1}{T} = \frac{a\omega_g'}{10} \quad \text{rad/sec} \tag{10-130}$$

5. The Bode plot of the compensated system is investigated to see if the phase margin requirement is met; if not, the values of a and T are readjusted, and the procedure is repeated. If design specifications involve gain margin, or M_r, or BW, these should be checked and satisfied.

Since the phase-lag control brings in more attenuation to a system, if the design is proper, the stability margins will be improved, but at the expense of lower bandwidth. The only benefit of lower bandwidth is reduced sensitivity to high-frequency noise and disturbances.

The following example illustrates the design of the phase-lag controller and all its ramifications.

Example 10-9

In this example we use the second-order sun-seeker system described in Example 10-6 to illustrate the principle of design of phase-lag control. The forward-path transfer function of the uncompensated system is

$$G_p(s) = \frac{\Theta_o(s)}{A(s)} = \frac{2500K}{s(s+25)} \tag{10-131}$$

Time-Domain Design

The time-domain specifications of the system are:

1. The steady-state error of $\alpha(t)$ due to a unit-ramp function input for $\theta_r(t)$ should be ≤ 1 percent.
2. The maximum overshoot of the step response should be less than 5 percent, or as small as possible.

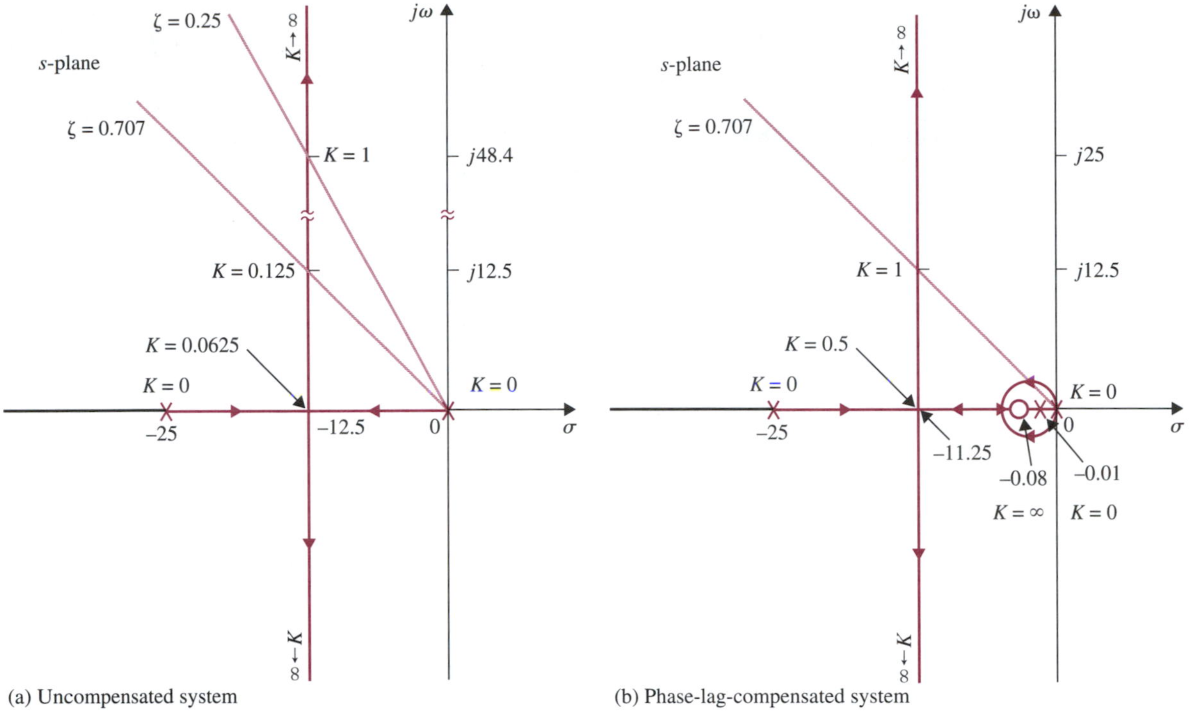

Figure 10-46 Root loci of sun-seeker system in Example 10-9.

$$G_p(s) = \frac{2500K}{s(s+25)} \qquad G_c(s) + \frac{1+aTs}{1+Ts} \qquad a = 0.125, \quad T = 100$$

3. Rise time $t_r \le 0.5$ sec.
4. Settling time $t_s \le 0.5$ sec.
5. Due to noise problems, the bandwidth of the system must be <50 rad/sec.

Notice that the rise- and settling-time requirements have been relaxed considerably from the phase-lead design in Example 10-6. The root loci of the uncompensated system are shown in Fig. 10-46(a).

As in Example 10-6, we set $K = 1$ initially. The damping ratio of the uncompensated system is 0.25, and the maximum overshoot is 44.4 percent. Figure 10-47 shows the unit-step response of the system with $K = 1$.

Let us select the phase-lag controller with the transfer function given in Eq. (10-120). The forward-path transfer function of the compensated system is

$$G(s) = G_c(s)G_p(s) = \frac{2500K(s + 1/aT)}{s(s+25)(s+1/T)} \tag{10-132}$$

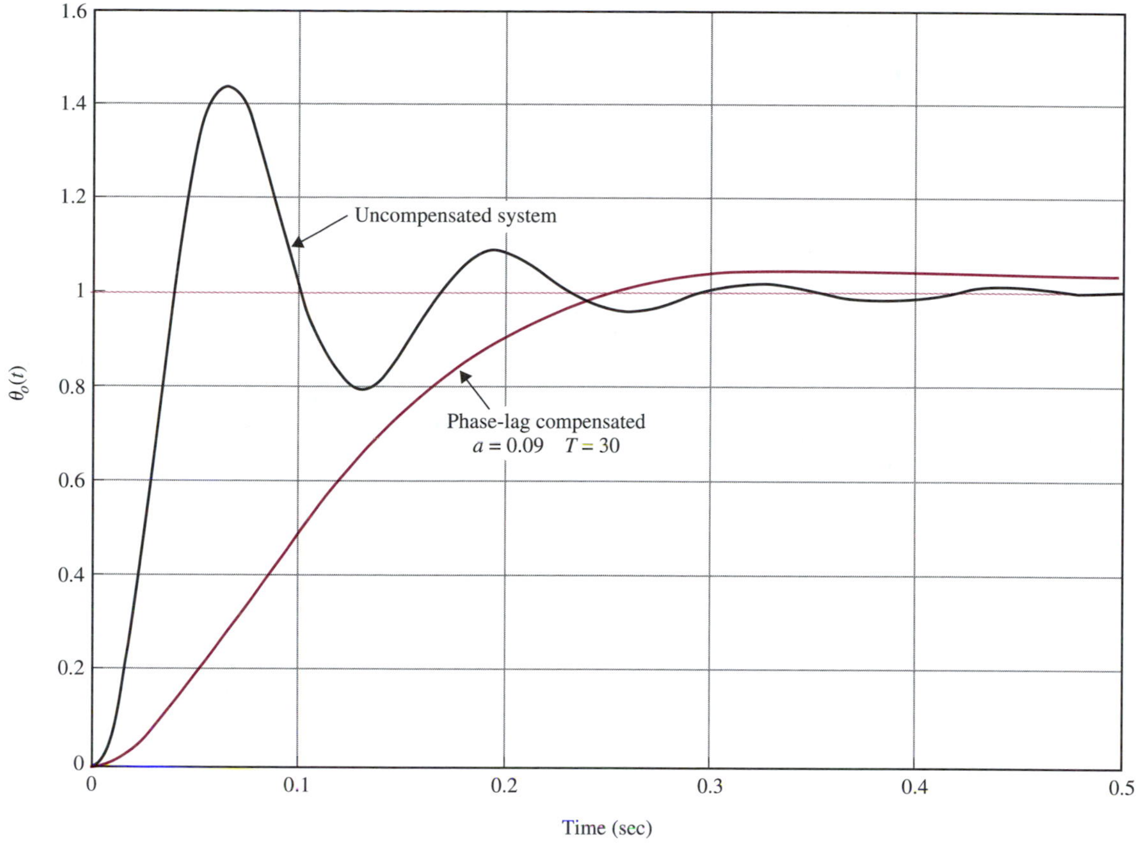

Figure 10-47 Unit-step responses of uncompensated and compensated sun-seeker system with phase-lag control in Example 10-9.

$$G_p(s) = \frac{2500}{s(s+25)} \qquad G_c(s) + \frac{1 + aTs}{1 + Ts} \qquad a = 0.09, \quad T = 30$$

If the value of K is maintained at 1, the steady-state error will be a percent, which is better than that of the uncompensated system, since $a < 1$. *For effective phase-lag control, the pole and zero of the controller transfer function should be placed close together, and then for the type 1 system, the combination should be located relatively close to the origin of the s-plane.* From the root loci of the uncompensated system in Fig. 10-46(a), we see that if K could be set to 0.125, the damping ratio would be 0.707 and the maximum overshoot of the system would be 4.32 percent. By setting the pole and zero of the controller close to $s = 0$, the shape of the loci of the dominant roots of the compensated system will be very similar to those of the uncompensated system. We can find the value of a using Eq. (10-124); that is,

$$a = \frac{K \text{ to realize the desired damping}}{K \text{ to realize the steady-state performance}} = \frac{0.125}{1} = 0.125 \qquad (10\text{-}133)$$

Thus if the value of T is sufficiently large when $K = 1$, the dominant roots of the characteristic equation will correspond to a damping ratio of approximately 0.707. Let us arbitrarily select $T = 100$. The root loci of the compensated system are shown in Fig. 10-46(b). The roots of thecharacteristic equation when $K = 1$, $a = 0.125$, and $T = 100$ are

$$s = -0.0805, \quad -12.465 + j12.465 \quad \text{and} \quad -12.465 - j12.465$$

which corresponds to a damping ratio of exactly 0.707. If we had chosen a smaller value for T, the damping ratio would be slightly off 0.707. From a practical standpoint, the value of T cannot be too large, since from Eq. (10-64), $T = R_2C$, a large T would correspond to either a large capacitor or an unrealistically large resistor. To reduce the value of T, and simultaneously, satisfying the maximum overshoot requirement, a should also be reduced. However, a cannot be reduced indefinitely, or the zero of the controller at $-1/aT$ would be too far to the left on the real axis. Table 10-17 gives the attributes of the time-domain performance of the phase-lag-compensated sun-seeker system with various a and T. The ramifications of the various design parameters are clearly displayed. Thus a suitable set of controller parameters would be $a = 0.09$ and $T = 30$. With $T = 30$, selecting $C = 1\ \mu$F would require R_2 to be 30 MΩ. A smaller value for T can be realized by using a two-stage phase-lag controller. The unit-step response of the compensated system with $a = 0.09$ and $T = 30$ is shown in Fig. 10-47. Notice that the maximum overshoot is reduced at the expense of rise time and settling time. Although the settling time of the compensated system is shorter than that of the uncompensated system, actually, it takes much longer time for the phase-lag-compensated system to reach steady state.

It would be enlightening to explain the design of the phase-lag controller by means of the root contours. The root-contour design conducted earlier in Example 10-6 using Eqs. (10-80) through (10-83) for phase-lead control and Figs. 10-31 and 10-32 is still valid for phase-lag control, except that in the present case, $a < 1$. Thus, in Fig. 10-32 only the portions of the root contours that correspond to $a < 1$ are applicable for phase-lag control. These root contours clearly show that for effective phase-lag control, the value of T should be relatively large. In Fig. 10-48 we illustrate further that the complex poles of the closed-loop transfer function are rather insensitive to the value of T when the latter is relatively large.

Table 10-17 Attributes of Performance of Sun-Seeker System in Example 10-9 with Phase-Lag Control

a	T	Maximum Overshoot (%)	t_r (sec)	t_s (sec)	BW (rad/sec)	Roots of Characteristic Equation
1.000	1	44.4	0.0255	0.2133	75.00	$-12.500 \pm j48.412$
0.125	100	4.9	0.1302	0.1515	17.67	$-0.0805,\ -12.465 \pm j12.465$
0.100	100	2.5	0.1517	0.2020	13.97	$-0.1009,\ -12.455 \pm j9.624$
0.100	50	3.4	0.1618	0.2020	14.06	$-0.2037,\ -12.408 \pm j9.565$
0.100	30	4.5	0.1594	0.1515	14.19	$-0.3439,\ -12.345 \pm j9.484$
0.100	20	5.9	0.1565	0.4040	14.33	$-0.5244,\ -12.263 \pm j9.382$
0.090	50	3.0	0.1746	0.2020	12.53	$-0.2274,\ -12.396 \pm j8.136$
0.090	30	4.4	0.1719	0.2020	12.68	$-0.3852,\ -12.324 \pm j8.029$
0.090	20	6.1	0.1686	0.5560	12.84	$-0.5901,\ -12.230 \pm j7.890$

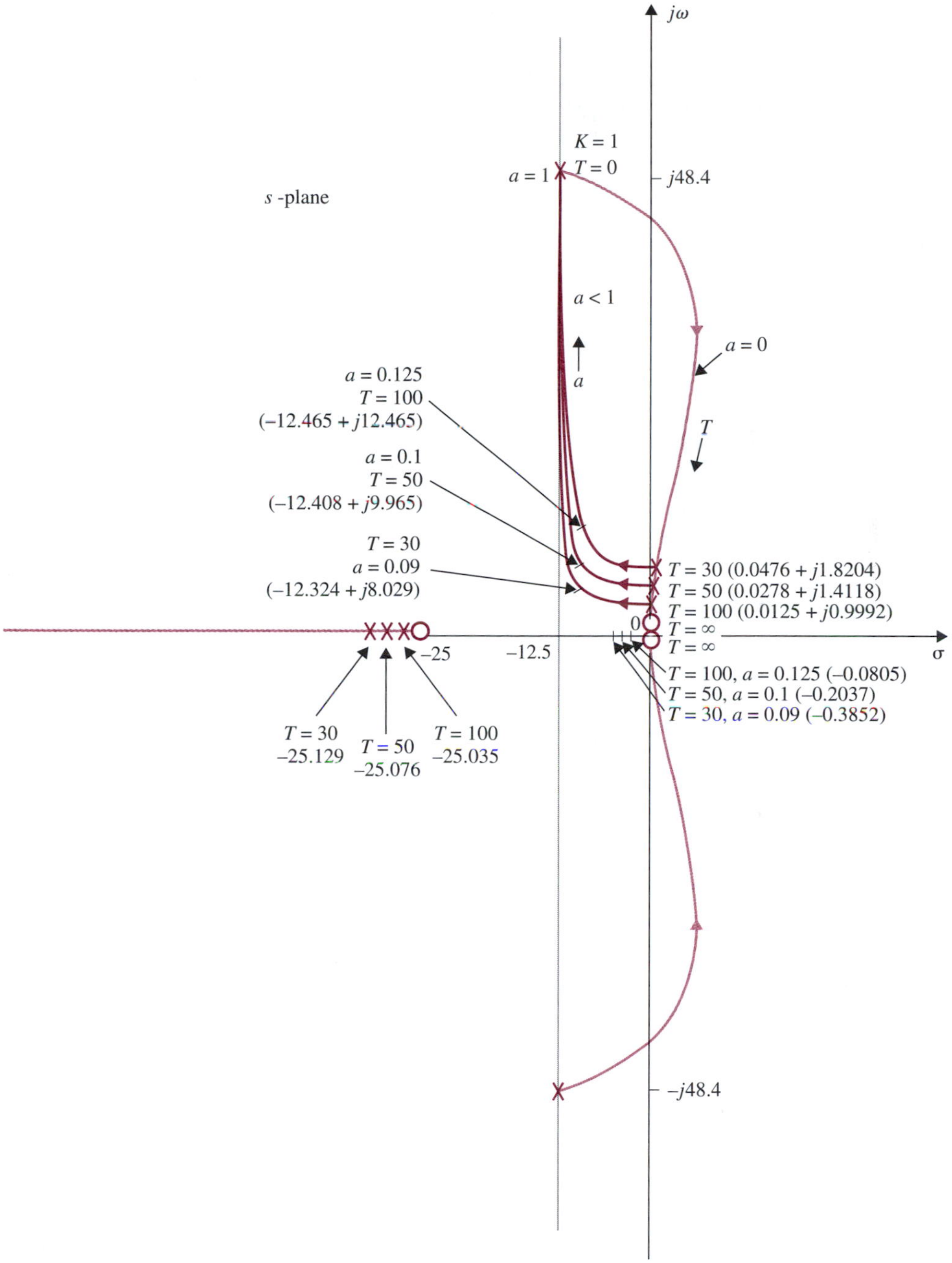

Figure 10-48 Root contours of sun-seeker system in Example 10-9 with phase-lag control.

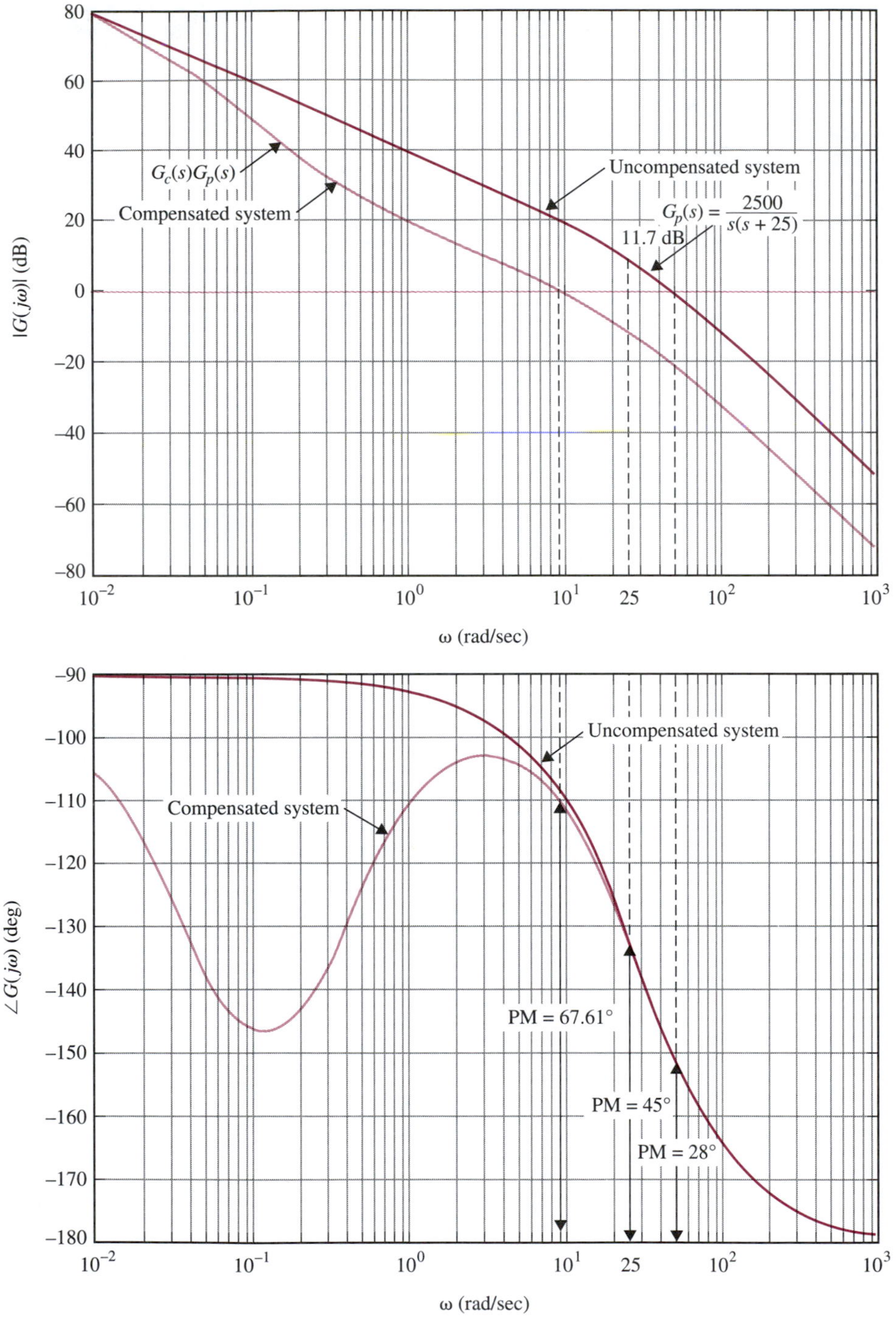

Figure 10-49 Bode plot of uncompensated and compensated systems with phase-lag control in Example 10-9.

$$G_c(s) = \frac{1 + 3s}{1 + 30s} \qquad G_p(s) = \frac{2500}{s(s + 25)}$$

Frequency-Domain Design

The Bode plot of $G_p(j\omega)$ of Eq. (10-131) is shown in Fig. 10-49 for $K = 1$. The Bode plot shows that the phase margin of the uncompensated system is only 28°. Not knowing what phase margin will correspond to a maximum overshoot of less than 5 percent, we conduct the following series of designs using the Bode plot in Fig. 10-49. Starting with a phase margin of 45°, we observe that this phase margin can be realized if the gain-crossover frequency ω_g' is at 25 rad/sec. This means that the phase-lag controller must reduce the magnitude curve of $G_p(j\omega)$ to 0 dB at $\omega = 25$ rad/sec while it does not appreciably affect the phase curve near this frequency. Since the phase-lag controller still contributes a small negative phase when the corner frequency $1/aT$ is placed at $\frac{1}{10}$ of the value of ω_g', it is a safe measure to choose ω_g' at somewhat less than 25 rad/sec, say, 20 rad/sec.

From the Bode plot, the value of $|G_p(j\omega_g')|_{\text{dB}}$ at $\omega_g' = 20$ rad/sec is 11.7 dB. Thus, using Eq. (10-128), we have

$$a = 10^{-|G_p(j\omega_g')|/20} = 10^{-11.7/20} = 0.26 \tag{10-134}$$

The value of $1/aT$ is chosen to be at 1/10 the value of $\omega_g' = 20$ rad/sec. Thus

$$\frac{1}{aT} = \frac{\omega_g'}{10} = \frac{20}{10} = 2 \text{ rad/sec} \tag{10-135}$$

and

$$T = \frac{1}{2a} = \frac{1}{0.52} = 1.923 \tag{10-136}$$

Checking out the unit-step response of the system with the designed phase-lag control, we found that the maximum overshoot is 24.5 percent. The next step is to try aiming at a higher phase margin. Table 10-18 gives the various design results by using various desired phase margin up to 80°.

Examining the results in Table 10-18, we see that none of the cases satisfies the maximum overshoot requirement of ≤5 percent. The $a = 0.044$ and $T = 52.5$ case yields the best maximum overshoot, but the value of T is too large to be practical. Thus we single out the case with $a =$

Table 10-18 Performance Attributes of Sun-Seeker System in Example 10-9 with Phase-Lag Control

Desired PM (deg)	a	T	Actual PM (deg)	M_r	BW (rad/sec)	Maximum Overshoot (%)	t_r (sec)	t_s (sec)
45	0.26	1.923	46.78	1.27	33.37	24.5	0.0605	0.2222
60	0.178	3.75	54.0	1.19	25.07	17.5	0.0823	0.303
70	0.1	10	63.87	1.08	14.72	10.0	0.1369	0.7778
80	0.044	52.5	74.68	1.07	5.7	7.1	0.3635	1.933

0.1 and $T = 10$ and refine the design by increasing the value of T. As shown in Table 10-17, when $a = 0.1$ and $T = 30$, the maximum overshoot is reduced to 4.5 percent. The Bode plot of the compensated system is shown in Fig. 10-49. The phase margin is 67.61°.

The unit-step response of the phase-lag-compensated system shown in Fig. 10-47 points out a major disadvantage of the phase-lag control. Since the phase-lag controller is essentially a low-pass filter, the rise time and settling time of the compensated system are usually increased. However, we shall show by the following example that phase-lag control can be more versatile and has a wider range of effectiveness in improving stability than the single-stage phase-lead controller, especially if the system has low or negative damping. ▲

Example 10-10

Consider the sun-seeker system designed in Example 10-8, with the forward-path transfer function given in Eq. (10-111). Let us restore the gain K so that a root-locus plot can be made for the system. Then Eq. (10-111) is written

$$G_p(s) = \frac{156{,}250{,}000K}{s(s^2 + 625s + 156{,}250)} \tag{10-137}$$

The root loci of the closed-loop system are shown in Fig. 10-50. When $K = 1$, the system is unstable, and the characteristic equation roots are at -713.14, $44.07 + j466.01$, and $44.07 - j466.01$.

Example 10-8 shows that the performance specification on stability cannot be achieved with a single-stage phase-lead controller. Let the performance criteria be:

Maximum overshoot $\leq$ 5 percent
Rise time $t_r \leq 0.02$ sec
Settling time $t_s \leq 0.02$ sec

Let us assume that the desired relative damping ratio is 0.707. Figure 10-50 shows that when $K = 0.10675$, the dominant characteristic equation roots of the uncompensated system are at $-172.77 \pm j172.73$, which corresponds to a damping ratio of 0.707. Thus the value of a is determined from Eq. (10-124),

$$a = \frac{K \text{ to realize the desired damping}}{K \text{ to realize the steady-state performance}} = \frac{0.10675}{1} = 0.10675 \tag{10-138}$$

Table 10-19 Performance Attributes of Sun-Seeker System in Example 10-10 with Phase-Lag Control

a	T	BW (rad/sec)	PM (deg)	Maximum Overshoot (%)	t_r (sec)	t_s (sec)
0.1	20	173.5	66.94	1.2	0.01273	0.01616
0.1	10	174	66.68	1.6	0.01262	0.01616
0.1	5	174.8	66.15	2.5	0.01241	0.01616
0.1	2	177.2	64.56	4.9	0.01601	0.0101

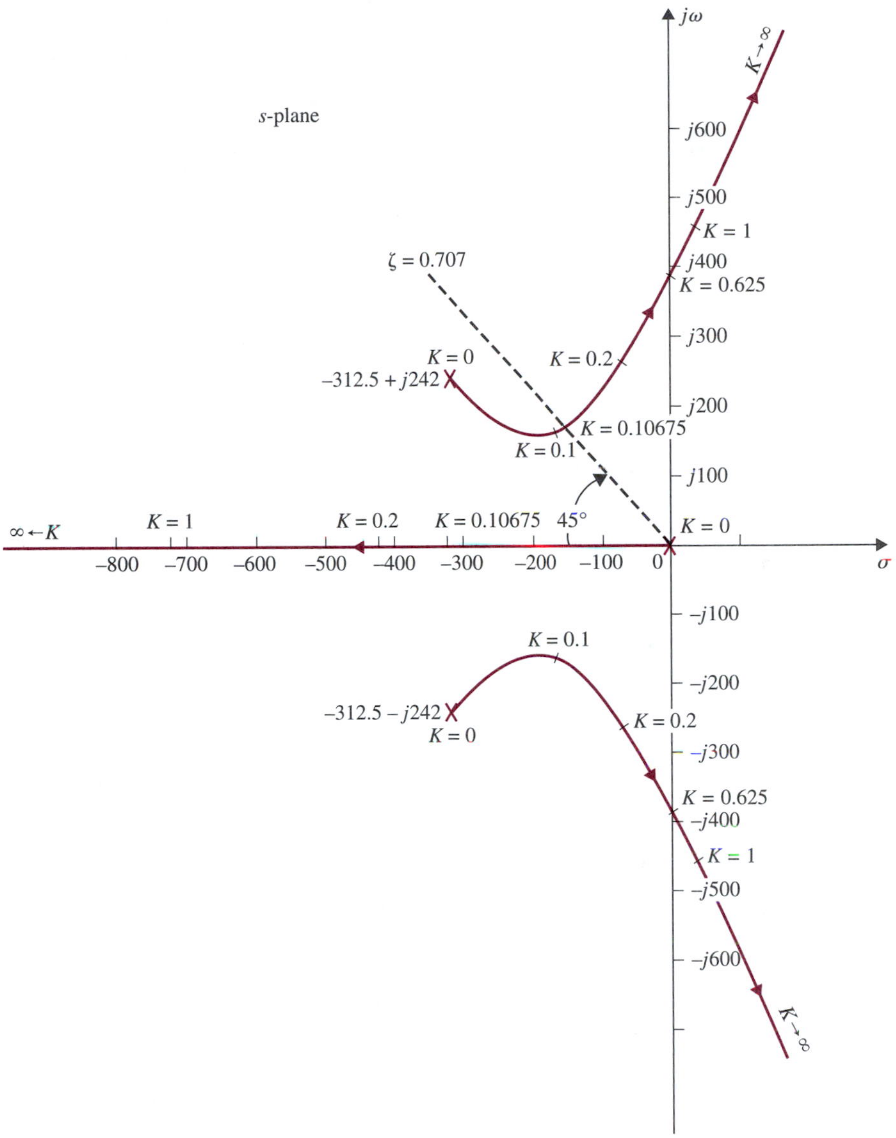

Figure 10-50 Root loci of uncompensated system in Example 10-10.

$$G_p(s) = \frac{156{,}250{,}000K}{s(s^2 + 625s + 156{,}250)}$$

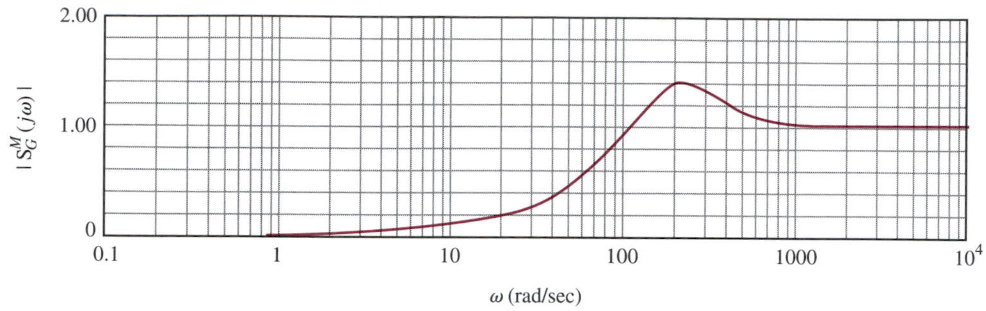

Figure 10-51 Sensitivity function of phase-lag-compensated system in Example 10-10.

Let $a = 0.1$. Since the loci of the dominant roots are far away from the origin in the s-plane, the value of T has a wide range of flexibility. Table 10-19 shows the performance results when $a = 0.1$ for various values of T.

Therefore, the conclusion is that only one stage of the phase-lag controller is needed to satisfy the stability requirement, whereas two stages of the phase-lead controller are needed, as shown in Example 10-8.

Sensitivity Function

The sensitivity function $|S_G^M(j\omega)|$ of the phase-lag-compensated system with $a = 0.1$ and $T = 20$ is shown in Fig. 10-51. Notice that the sensitivity function is less than unity for frequencies up to only 102 rad/sec. This is due to the low bandwidth of the system as a result of phase-lag control. ▲

10-6-3 Effects and Limitations of Phase-Lag Control

From the results of the preceding illustrative examples, the effects and limitations of phase-lag control on the performance of linear control systems can be summarized as follows:

1. For a given forward-path gain K, the magnitude of the forward-path transfer function is attenuated near and above the gain-crossover frequency, thus allowing improvement of the relative stability of the system.
2. The gain-crossover frequency is decreased, and thus the bandwidth of the system is reduced.
3. The rise and settling times of the system are usually longer, since the bandwidth is usually decreased.

4. The system is more sensitivity to parameter variations in the sense that the sensitivity function is greater than unity for all frequencies approximately greater than the bandwidth of the system.

10-7 Design with Lead–Lag Controller

We have learned from preceding sections that phase-lead control generally improves rise time and damping but increases the natural frequency of the closed-loop system. On the other hand, when applied properly, phase-lag control improves damping but usually results in a longer rise and settling times. Therefore, each of these control schemes has its advantages, disadvantages, and limitations, and there are many systems that cannot be compensated satisfactorily by either scheme acting alone. It is natural, therefore, whenever necessary, to consider using a combination of the lead and lag controllers, so that the advantages of both schemes are utilized.

The transfer function of a simple lag-lead (or lead-lag) controller can be written

$$G_c(s) = G_{c1}(s)G_{c2}(s) = \underbrace{\left(\frac{1 + a_1T_1s}{1 + T_1s}\right)}_{\text{lead}}\underbrace{\left(\frac{1 + a_2T_2s}{1 + T_2s}\right)}_{\text{lag}} \qquad a_1 > 1, \quad a_2 < 1 \qquad (10\text{-}139)$$

The gain factors of the lead and lag controllers are not included, since as discussed earlier, these gain and attenuation are eventually compensated by the adjustment of the forward gain K.

Since the lead–lag controller transfer function in Eq. (10-139) now has four unknown parameters, its design is not as straightforward as the single-stage phase-lead or phase-lag controller. *In general, the phase-lead portion of the controller is used primarily for the purpose of achieving a shorter rise time and higher bandwidth, and the phase-lag portion is brought in to provide major damping of the system.* Either phase-lead or phase-lag control can be designed first. We shall use the following example to illustrate the design steps.

Example 10-11

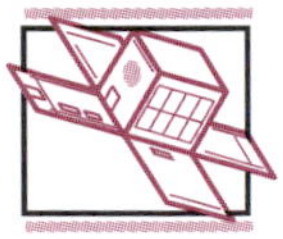

As an illustrative example of designing a lead–lag controller, let us consider the sun-seeker system of Example 10-8. The uncompensated system with $K = 1$ is shown to be unstable. A two-stage phase-lead controller was designed in Example 10-9, and a single-stage phase-lag controller was designed in Example 10-10.

Based on the design in Example 10-8, we can first select a phase-lead control with $a = 70$ and $T_1 = 0.00004$. The remaining phase-lag control can be designed using either the root-locus method or the Bode plot method. Table 10-20 gives the results by letting $T_2 = 2$, which is an insensitive parameter, and various values of a. The results in Table 10-20 show that the optimal value of a_2, from the standpoint of minimizing the maximum overshoot, for $a_1 = 70$

Table 10-20 Performance Attributes of Sun-Seeker System in Example 10-11 with Lead–Lag Control: $a_1 = 70$, $T_1 = 0.00004$

a_2	T_2	PM (deg)	M_r	BW (rad/sec)	Maximum Overshoot (%)	t_r (sec)	t_s (sec)
0.1	20	81.81	1.004	122.2	0.4	0.01843	0.02626
0.15	20	76.62	1.002	225.5	0.2	0.00985	0.01515
0.20	20	70.39	1.001	351.4	0.1	0.00668	0.00909
0.25	20	63.87	1.001	443.0	4.9	0.00530	0.00707

and $T_2 = 0.00004$, is approximately 0.2. Compared with the single-stage phase-lag control designed in Example 10-9, the BW is increased to 351.4 rad/sec from 66.94 rad/sec, and the rise time is reduced to 0.00668 sec from 0.01273 sec. The system with the lead–lag controller should be more robust, since the magnitude of the sensitivity function should not increase to unity until near the BW of 351.4 rad/sec. As a comparison, the unit-step responses of the system with the two-stage phase-lead control, the single-stage phase-lag-control and the lead-lag control are shown in Fig. 10-52.

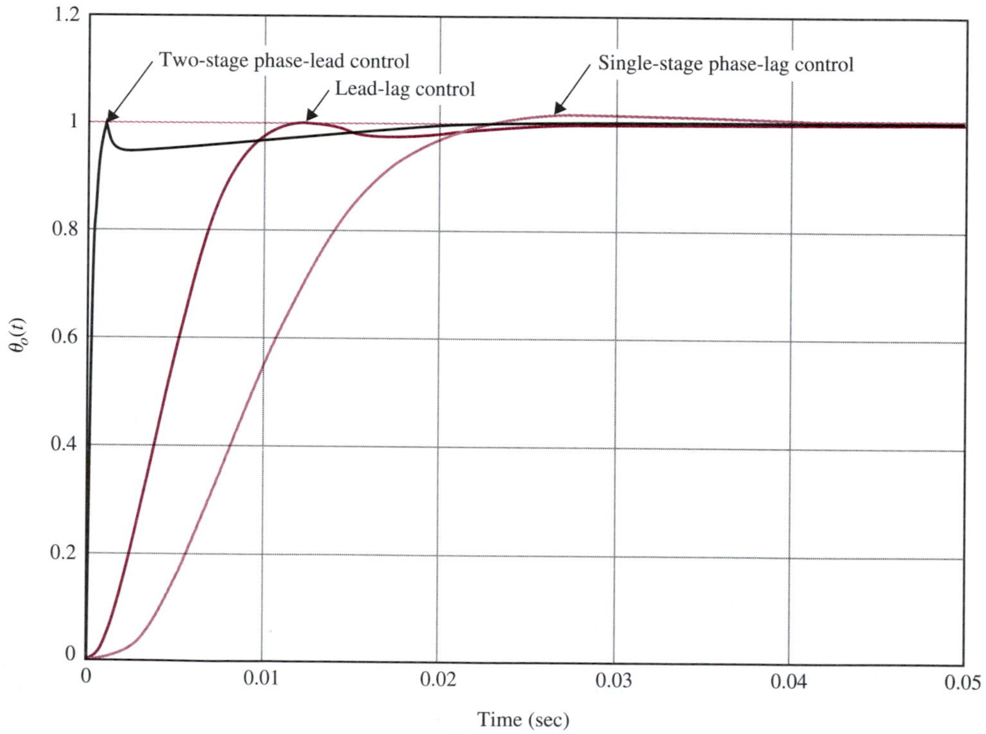

Figure 10-52 Sun-seeker system in Example 10-11 with single-stage phase-lag control, lead–lag control, and two-stage phase-lead control.

It should be noted that the bandwidth and rise time of the sun-seeker system can be further increased and reduced, respectively, by using a larger value of a_1 for the phase-lead portion of the controller. However, the resulting step response will have a large undershoot, although the maximum overshoot can be kept small. ▲

10-8 Pole–Zero-Cancellation Design: Notch Filter

The transfer functions of many controlled processes contain one or more pairs of complex-conjugate poles that are very close to the imaginary axis of the s-plane. These complex poles usually cause the closed-loop system to be lightly damped or unstable. One immediate thought is to use a controller that has a transfer function with zeros selected such as to cancel the undesirable poles of the controlled process, and the poles of the controller are placed at more desirable locations in the s-plane to achieve the desired dynamic performance. For example, if the transfer function of a process is

$$G_p(s) = \frac{K}{s(s^2 + s + 10)} \tag{10-140}$$

in which the complex-conjugate poles may cause stability problems in the closed-loop system when the value of K is large, the suggested series controller may be of the form

$$G_c(s) = \frac{s^2 + s + 10}{s^2 + as + b} \tag{10-141}$$

The constants a and b may be selected according to the performance specifications of the closed-loop system.

There are practical difficulties with the pole–zero-cancellation design scheme that should be mentioned to prevent the method from being used indiscriminantly. The problem is that in practice *exact* cancellation of poles and zeros of transfer functions is rarely possible. One reality is that the transfer function of the process, $G_p(s)$, is usually determined through testing and physical modeling; linearization of a nonlinear process and approximation of a complex process are unavoidable. Thus the "true" poles and zeros of the transfer function of the process may not be modeled accurately. In fact, the true order of the system may be even higher than that represented by the transfer function used for modeling purposes. Another difficulty is that the dynamic properties of the process may vary, even very slowly, due to aging of the system components or changes in the operating environment, so that the poles and zeros of the transfer function may move during operation of the system. The parameters of the controller are constrained by the actual physical components available, and cannot be assigned arbitrarily. For these and other reasons, even if we could design the poles and zeros of the transfer function of the controller precisely, exact pole–zero cancellation is almost never possible in practice. We show in the following that in most cases, exact cancellation is *not* really necessary to effectively negate the influence of the undesirable poles using pole–zero-cancellation compensation schemes.

▲ In the real world, exact cancellation of poles and zeros is almost never possible.

Let us assume that a controlled process is represented by

$$G_p(s) = \frac{K}{s(s+p_1)(s+\overline{p}_1)} \tag{10-142}$$

where p_1 and $\overline{p}_1$ are the two complex-conjugate poles that are to be canceled. Let the transfer function of the series controller be

$$G_c(s) = \frac{(s+p_1+\varepsilon_1)(s+\overline{p}_1+\overline{\varepsilon}_1)}{s^2+as+b} \tag{10-143}$$

where ε_1 is a complex number whose magnitude is very small and $\overline{\varepsilon}_1$ is its complex conjugate. The open-loop transfer function of the compensated system is

$$G(s) = G_c(s)G_p(s) = \frac{K(s+p_1+\varepsilon_1)(s+\overline{p}_1+\overline{\varepsilon}_1)}{s(s+p_1)(s+\overline{p}_1)(s^2+as+b)} \tag{10-144}$$

Because of inexact cancellation, we cannot discard the terms $(s+p_1)(s+\overline{p}_1)$ in the denominator of Eq. (10-144). The closed-loop transfer function is

$$\frac{Y(s)}{R(s)} = \frac{K(s+p_1+\varepsilon_1)(s+\overline{p}_1+\overline{\varepsilon}_1)}{s(s+p_1)(s+\overline{p}_1)(s^2+as+b)+K(s+p_1+\varepsilon_1)(s+\overline{p}_1+\overline{\varepsilon}_1)} \tag{10-145}$$

The root-locus diagram in Fig. 10-53 explains the effect of inexact pole–zero cancellation. Notice that the two closed-loop poles as a result of inexact cancellation lie between the pairs of poles and zeros at $s = -p_1, -\overline{p}_1$, and $-p_1-\varepsilon_1, -\overline{p}_1-\overline{\varepsilon}_1$, respectively. Thus these closed-loop poles are very close to the open-loop poles and zeros that are meant to be canceled. Equation (10-145) can be approximated as

$$\frac{Y(s)}{R(s)} \cong \frac{K(s+p_1+\varepsilon_1)(s+\overline{p}_1+\overline{\varepsilon}_2)}{(s+p_1+\delta_1)(s+\overline{p}_1+\overline{\delta}_1)(s^3+as^2+bs+K)} \tag{10-146}$$

where δ_1 and $\overline{\varepsilon}_1$ are a very small complex-conjugate numbers that depend on ε_1, $\overline{\varepsilon}_1$, and all the other parameters. The partial-fraction expansion of Eq. (10-146) is

$$\frac{Y(s)}{R(s)} \cong \frac{K_1}{s+p_1+\delta_1} + \frac{K_2}{s+\overline{p}_1+\overline{\delta}_1} + \text{terms due to the remaining poles} \tag{10-147}$$

We can show that K_1 is proportional to $\varepsilon_1 - \delta_1$, which is a very small number. Similarly, K_2 is also very small. *This exercise simply shows that although the poles at $-p_1$ and $-\overline{p}_1$ cannot be canceled precisely, the resulting transient-response terms due to inexact cancellation will have insignificant amplitudes, so that unless the controller zeros*

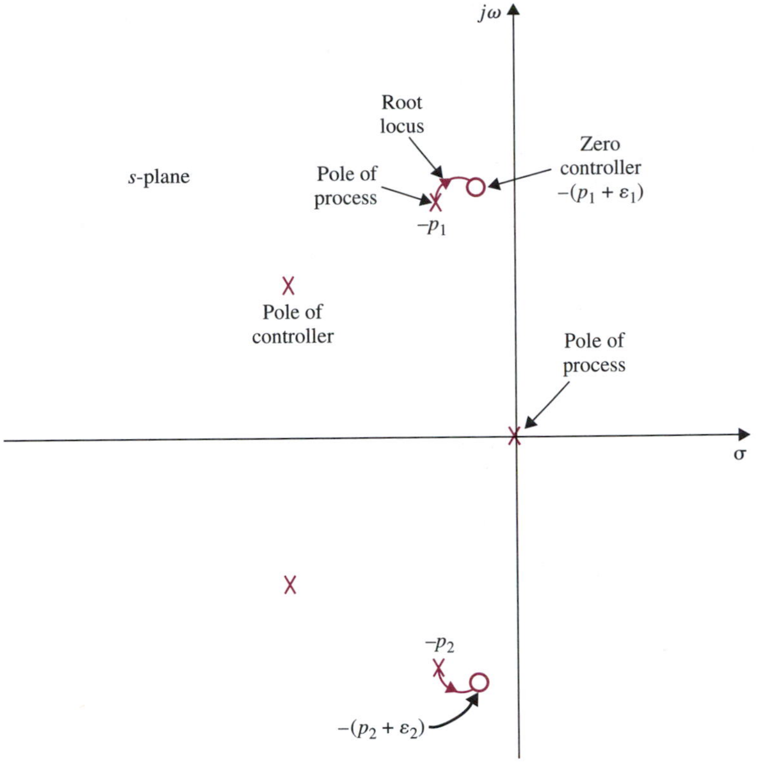

Figure 10-53 Pole–zero configuration and root loci of inexact cancellation.

earmarked for cancellation are too far off target, the effect can be neglected for all practical purposes. Another way of viewing this problem is that the zeros of $G(s)$ are retained as the zeros of closed-loop transfer function $Y(s)/R(s)$, so that from Eq. (10-146), we see that the two pairs of poles and zeros are close enough to be canceled from the transient-response standpoint.

▲ Never attempt to cancel open-loop poles that are in the right-half s-plane.

Keep in mind that we should never attempt to cancel poles that are in the right-half s-plane, since any inexact cancellation will result in an unstable system. Inexact cancellation of poles could cause difficulties if the unwanted poles of the process transfer function are very close to or right on the imaginary axis of the s-plane. In this case, inexact cancellation may also result in an unstable system. Figure 10-54(a) illustrates a situation in which the relative positions of the poles and zeros intended for cancellation result in a stable system, whereas in Fig. 10-54(b), inexact cancellation is unacceptable. Although the relative distance between the poles and zeros intended for cancellation is small, which results in terms in the time response that have very small amplitudes, these responses will grow without bound as time increases.

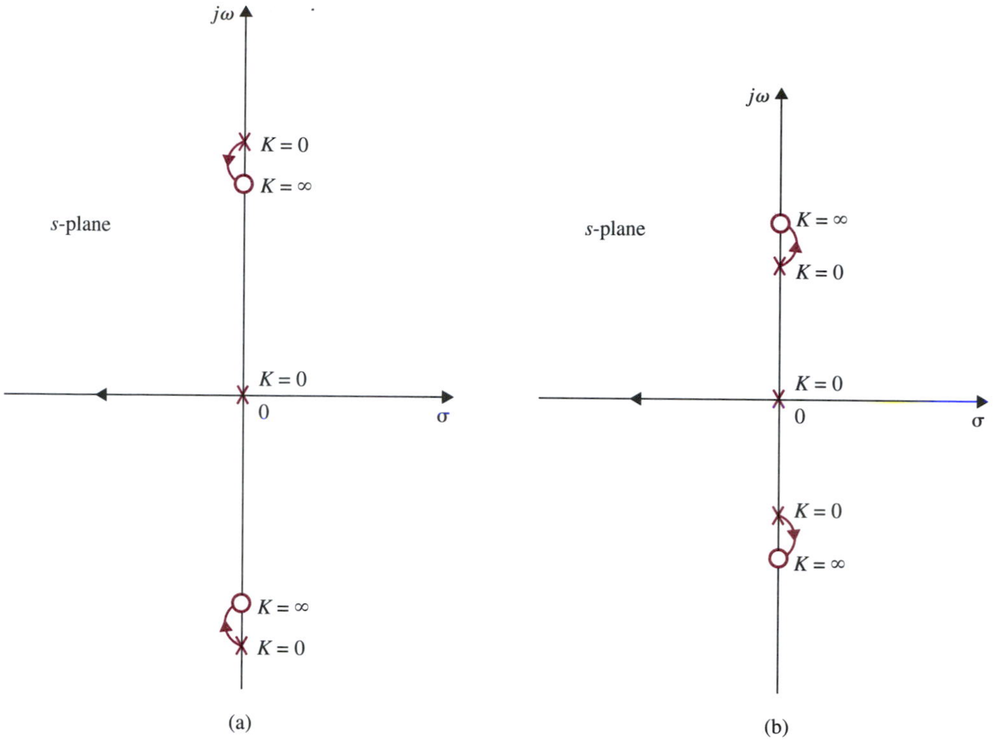

Figure 10-54 Root loci showing the effects of inexact, pole–zero cancellations.

10-8-1 Second-Order Active Filter

Transfer functions with complex poles and/or zeros can be realized by electric circuits with op-amps. Consider the transfer function

$$G_c(s) = \frac{E_2(s)}{E_1(s)} = K\frac{s^2 + b_1 s + b_2}{s^2 + a_1 s + a_2} \tag{10-148}$$

where a_1, a_2, b_1, and b_2 are real constants. The active-filter realization of Eq. (10-148) can be accomplished by using the direct decomposition scheme of state variables discussed in Section 5-9. A typical op-amp circuit is shown in Fig. 10-55. The parameters of the transfer function in Eq. (10-148) are related to the circuit parameters as follows:

$$K = -\frac{R_6}{R_7} \tag{10-149}$$

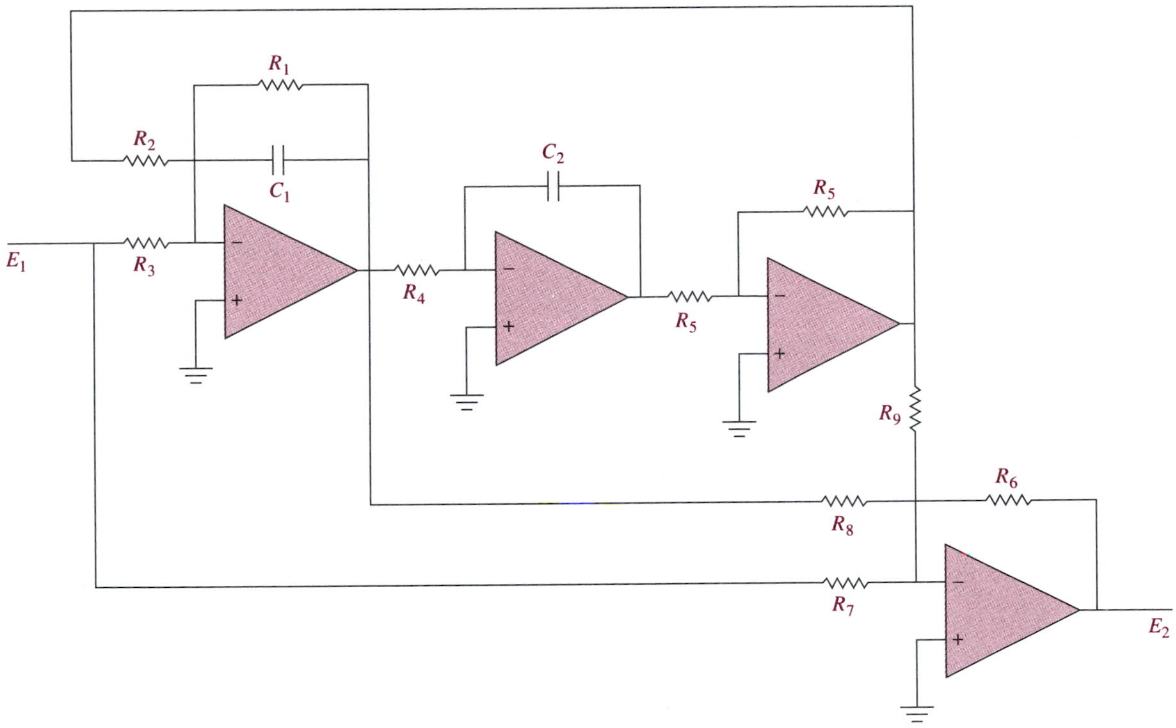

Figure 10-55 Op-amp circuit realization of the second-order transfer function

$$\frac{E_2(s)}{E_1(s)} = K\frac{s^2 + b_1 s + b_2}{s^2 + a_1 s + a_2}$$

$$a_1 = \frac{1}{R_1 C_1} \tag{10-150}$$

$$a_2 = \frac{1}{R_2 R_4 C_1 C_2} \tag{10-151}$$

$$b_1 = \left(1 - \frac{R_1 R_7}{R_3 R_8}\right) a_1 \qquad (b_1 < a_1) \tag{10-152}$$

$$b_2 = \left(1 - \frac{R_2 R_7}{R_3 R_9}\right) a_2 \qquad (b_2 < a_2) \tag{10-153}$$

Since $b_1 < a_1$, the zeros of $G_c(s)$ in Eq. (10-148) are less damped and are closer to the origin in the s-plane than the poles. By setting various combinations of R_7 and R_8, and R_9 to infinity, a variety of second-order transfer functions can be realized. Note that all the parameters can be adjusted independently of one another. For example, R_1 can be adjusted to set a_1; R_4 can be adjusted to set a_2; b_1 and b_2 are set by adjusting R_8 and R_9, respectively. The gain factor K is controlled independently by R_6.

10-8-2 Frequency-Domain Interpretation and Design

While it is simple to grasp the idea of pole–zero-cancellation design in the s-domain, the frequency domain provides added perspective to the design principles. Figure 10-56

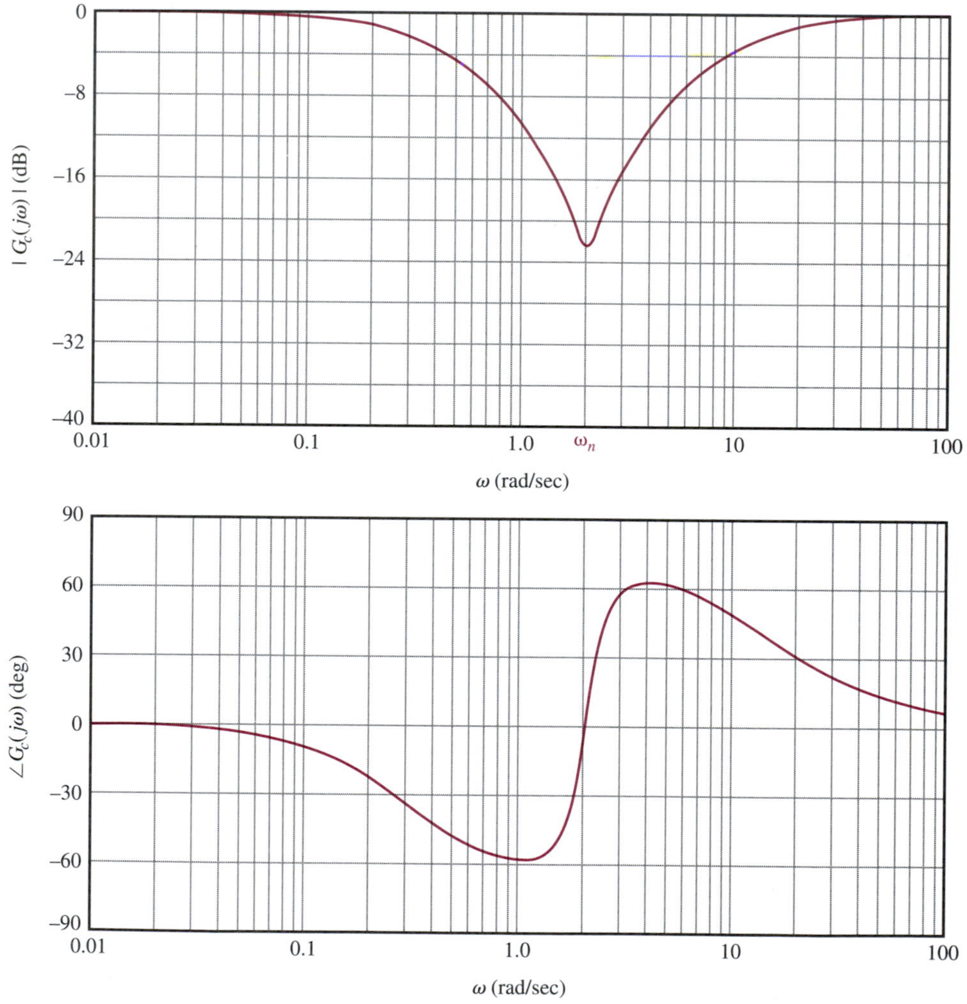

Figure 10-56 Bode plot of a notch controller with the transfer function $G(s) = (s^2 + 0.8s + 4)/(s + 0.384)(s + 10.42)$.

illustrates the Bode plot of the transfer function of a typical second-order controller with complex zeros. The magnitude plot of the controller typically has a ''notch'' at the resonant frequency ω_n. The phase plot is negative below and positive above the resonant frequency, while passing through zero degrees at the resonant frequency. The attenuation of the magnitude curve and the positive-phase characteristics can be used effectively to improve the stability of a linear system. Because of the ''notch'' characteristic in the magnitude curve, the controller is also referred to in the industry as a **notch filter** or **notch controller.**

From the frequency-domain standpoint, the notch controller has advantages over the phase-lead and phase-lag controllers in certain design conditions, since the magnitude and phase characteristics do not affect the high- and low-frequency properties of the system. Without using the pole–zero-cancellation principle, the design of the notch controller for compensation in the frequency domain involves determination of the amount of attenuation required and the resonant frequency of the controller.

Let us express the transfer function of the notch controller in Eq. (10-148) as

$$G_c(s) = \frac{s^2 + 2\zeta_z \omega_n s + \omega_n^2}{s^2 + 2\zeta_p \omega_n s + \omega_n^2} \qquad (10\text{-}154)$$

where we have made the simplification by assuming that $a_2 = b_2$. The attenuation provided by the magnitude of $G_c(j\omega)$ at the resonant frequency ω_n is

$$|G_c(j\omega_n)| = \frac{\zeta_z}{\zeta_p} \qquad (10\text{-}155)$$

Thus, knowing the maximum attention required at ω_n, the ratio of ζ_z/ζ_p is known.

The following example illustrates the design of the notch controller based on pole–zero cancellation and required attenuation at the resonant frequency.

Example 10-12

Complex-conjugate poles in system transfer functions are often due to compliances in the coupling between mechanical elements. For instance, if the shaft between the motor and load is nonrigid, the shaft is modeled as a torsional spring, which could lead to complex-conjugate poles in the process transfer function. Figure 10-57 shows a speed-control system in which the coupling between the motor and the load is modeled as a torsional spring. The system equations are

$$T_m(t) = J_m \frac{d\omega_m(t)}{dt} + B_m\omega_m(t) + J_L \frac{d\omega_L(t)}{dt} \qquad (10\text{-}156)$$

$$K_L[\theta_m(t) - \theta_L(t)] + B_L[\omega_m(t) - \omega_L(t)] = J_L \frac{d\omega_L(t)}{dt} \qquad (10\text{-}157)$$

$$T_m(t) = K\omega_e(t) \qquad (10\text{-}158)$$

$$\omega_e(t) = \omega_r(t) - \omega_L(t) \qquad (10\text{-}159)$$

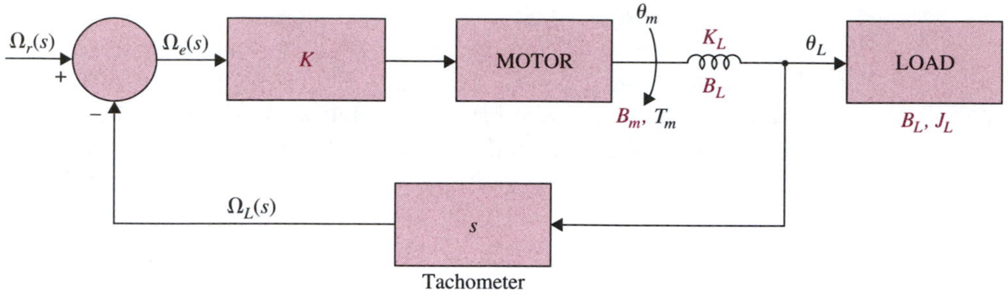

Figure 10-57 Block diagram of speed-control system in Example 10-12.

where

$$
\begin{aligned}
T_m(t) &= \text{motor torque} \\
\omega_m(t) &= \text{motor angular velocity} \\
\omega_L(t) &= \text{load angular velocity} \\
\theta_L(t) &= \text{load angular displacement} \\
\theta_m(t) &= \text{motor angular displacement} \\
J_m &= \text{motor inertia} = 0.0001 \text{ oz-in.-sec}^2 \\
J_L &= \text{load inertia} = 0.0005 \text{ oz-in.-sec}^2 \\
B_m &= \text{viscous-friction coefficient of motor} = 0.01 \text{ oz-in.-sec} \\
B_L &= \text{viscous-friction coefficient of shaft} = 0.001 \text{ oz-in.-sec} \\
K_L &= \text{spring constant of shaft} = 100 \text{ oz-in./rad} \\
K &= \text{amplifier gain} = 1
\end{aligned}
$$

The loop transfer function of the system is

$$
\begin{aligned}
G_p(s) &= \frac{\Omega_L(s)}{\Omega_e(s)} \\
&= \frac{B_L s + K_L}{J_m J_L s^3 + (B_m J_L + B_L J_m + B_L J_L)s^2 + (K_L J_L + B_m B_L + K_L J_m)s + B_m K_L}
\end{aligned}
\tag{10-160}
$$

By substituting the system parameters in the last equation, $G_p(s)$ becomes

$$
\begin{aligned}
G_p(s) &= \frac{20{,}000(s + 100{,}000)}{s^3 + 112s^2 + 1{,}200{,}200s + 20{,}000{,}000} \\
&= \frac{20{,}000(s + 100{,}000)}{(s + 16.69)(s + 47.66 + j1094)(s + 47.66 - j1094)}
\end{aligned}
\tag{10-161}
$$

Thus the shaft compliance between the motor and the load creates two complex-conjugate poles in $G_p(s)$ that are lightly damped. The resonant frequency is approximately 1095 rad/sec, and the closed-loop system is unstable. The complex poles of $G_p(s)$ will cause the speed response to oscillate even if the system were stable.

Pole-Zero Cancellation Design with Notch Controller

The performance specifications of the system are:

1. The steady-state speed of the load due to a unit-step input should have an error of not more than 1 percent.
2. Maximum overshoot of output speed ≤ 5 percent
3. Rise time $t_r < 0.5$ sec
4. Settling time $t_s < 0.5$ sec

To compensate the system, we need to get rid, or, perhaps more realistically, minimize the effect, of the complex poles of $G_p(s)$ at $s = -47.66 + j1094$ and $-47.66 - j1094$. Let us select a notch controller with the transfer function given in Eq. (10-154) to improve the performance of the system. The complex-conjugate zeros of the controller should be so placed that they will cancel the undesirable poles of the process. Therefore, the transfer function of the notch controller should be

$$G_c(s) = \frac{s^2 + 95.3s + 1{,}198{,}606.6}{s^2 + 2\zeta_p\omega_n s + \omega_n^2} \tag{10-162}$$

The forward-path transfer function of the compensated system is

$$G(s) = G_c(s)G_p(s) = \frac{20{,}000(s + 100{,}000)}{(s + 16.69)(s^2 + 2\zeta_p\omega_n s + \omega_n^2)} \tag{10-163}$$

Since the system is type 0, the step-error constant is

$$K_P = \lim_{s\to 0} G(s) = \frac{2 \times 10^9}{16.69 \times \omega_n^2} = \frac{1.198 \times 10^8}{\omega_n^2} \tag{10-164}$$

For a unit-step input, the steady-state error of the system is written

$$e_{ss} = \lim_{t\to\infty} \omega_e(t) = \lim_{s\to 0} s\Omega_e(s) = \frac{1}{1 + K_p} \tag{10-165}$$

Thus for the steady-state error to be less than or equal to 1 percent, $K_P \geq 99$. The corresponding requirement on ω_n is found from Eq. (10-164),

$$\omega_n \leq 1210 \tag{10-166}$$

We can show that from the stability standpoint, it is better to select a large value for ω_n. Thus let $\omega_n = 1200$ rad/sec, which is at the high end of the allowable range from the steady-state error standpoint. However, the design specifications given above can only be achieved by using a very large value for ζ_p. For example, when $\zeta_p = 15{,}000$, the time response has the following performance attributes.

Maximum overshoot = 3.7 percent
Rising time t_r = 0.1897 sec
Settling time t_s = 0.256 sec

Although the performance requirements are satisfied, the solution is unrealistic, since the extremely large value for ζ_p cannot be realized by physically available controller components.

Let us choose $\zeta_p = 10$ and $\omega_n = 1000$ rad/sec. The forward-path transfer function of the system with the notch controller is

$$G(s) = G_c(s)G_p(s) = \frac{20{,}000(s + 100{,}000)}{(s + 16.69)(s + 50)(s + 19{,}950)} \tag{10-167}$$

We can show that the system is stable, but the maximum overshoot is 71.6 percent. Now we can regard the transfer function in Eq. (10-167) represents a new design problem. There are a number of possible solutions to the problem of meeting the design specifications given above. We can introduce a phase-lag controller or a PI controller, among other possibilities.

Second-Stage Phase-Lag Controller Design

Let us design a phase-lag controller as the second-stage controller for the system. The roots of the characteristic equation of the system with the notch controller are at $s = -19{,}954$, $-31.328 + j316.36$, and $-31.328 - j316.36$. The transfer function of the phase-lag controller is

$$G_{c1}(s) = \frac{1 + aTs}{1 + Ts} \qquad a < 1 \tag{10-168}$$

where for design purpose we have omitted the gain factor $1/a$ in Eq. (10-168).

Let us select $T = 10$ for the phase-lag controller. Table 10-21 gives time-domain performance attributes for various values of a. The best value of a from the overall performance

Table 10-21 Time-Domain Performance Attributes of System in Example 10-12 with Notch-Phase-Lag Controller

a	T	aT	Maximum Overshoot (%)	t_r (sec)	t_s (sec)
0.001	10	0.01	14.8	0.1244	0.3836
0.002	10	0.02	10.0	0.1290	0.3655
0.004	10	0.04	3.2	0.1348	0.1785
0.005	10	0.05	1.0	0.1375	0.1818
0.0055	10	0.055	0.3	0.1386	0.1889
0.006	10	0.06	0	0.1400	0.1948

standpoint appears to be 0.005. Thus the transfer function of the phase-lag controller is

$$G_{c1}(s) = \frac{1 + aTs}{1 + Ts} = \frac{1 + 0.05s}{1 + 10s} \tag{10-169}$$

The forward-path transfer function of the compensated system with the notch-phase-lag controller is

$$G(s) = G_c(s)G_{c1}(s)G_p(s) = \frac{20{,}000(s + 100{,}000)(1 + 0.05s)}{(s + 16.69)(s + 50)(s + 19{,}950)(1 + 10s)} \tag{10-170}$$

The unit-step response of the system is shown in Fig. 10-58. Since the step-error constant is 120.13, the steady-state speed error due to a step input is 1/120.13, or 0.83 percent.

Second-Stage PI Controller Design

A PI controller can be applied to the system to improve the steady-state error and the stability simultaneously. The transfer function of the PI control is written

$$G_{c2}(s) = K_P + \frac{K_I}{s} = K_P\left(\frac{s + K_I/K_P}{s}\right) \tag{10-171}$$

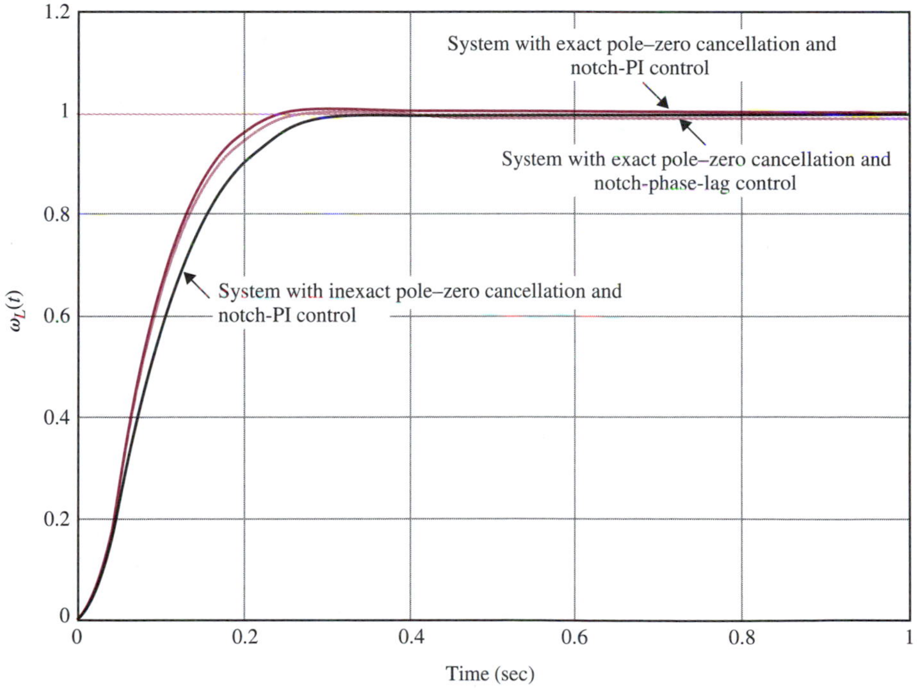

Figure 10-58 Unit-step responses of speed-control system in Example 10-12.

We can design the PI controller based on the phase-lag controller by writing Eq. (10-169) as

$$G_{c1}(s) = 0.005\left(\frac{s+20}{s+0.1}\right) \tag{10-172}$$

Thus we can set $K_P = 0.005$ and $K_I/K_P = 20$. Then $K_I = 0.1$. Figure 10-58 shows the unit-step response of the system with the notch-PI controller. The attributes of the step response are:

Maximum overshoot = 1 percent
Rise time $t_r = 0.1380$ sec
Settling time $t_s = 0.1818$ sec

which are extremely closed to those with the notch-phase-lag controller, except that in the present case the steady-state velocity error is zero when the input is a step function.

Sensitivity Due to Imperfect Pole–Zero Cancellation

As mentioned earlier, exact cancellation of poles and zeros is almost never possible in real life. Let us consider that the numerator polynomial of the notch controller in Eq. (10-162) cannot be exactly realized by physical resistor and capacitor components. Rather, the transfer function of the notch controller is more realistically chosen as

$$G_c(s) = \frac{s^2 + 100s + 1{,}000{,}000}{s^2 + 20{,}000s + 1{,}000{,}000} \tag{10-173}$$

Figure 10-58 shows the unit-step response of the system with the notch controller in Eq. (10-173). The attributes of the unit-step response are:

Maximum overshoot = 0.4 percent
Rise time $t_r = 0.17$ sec
Settling time $t_s = 0.2323$ sec

Frequency-Domain Design

To carry out the design of the notch controller, we refer to the Bode plot of Eq. (10-161) shown in Fig. 10-59. Due to the complex conjugate poles of $G_p(s)$, the magnitude plot has a peak of 24.86 dB at 1095 rad/sec. From the Bode plot in Fig. 10-59 we see that we may want to bring the magnitude plot down to −20 dB at the resonant frequency of 1095 rad/sec so that the resonance is smoothed out. This requires an attenuation of −44.86 dB. Thus, from Eq. (10-155),

$$|G_c(j\omega_n)| = -44.86 \text{ dB} = \frac{\zeta_z}{\zeta_p} = \frac{0.0435}{\zeta_p} \tag{10-174}$$

where ζ_z is found from Eq. (10-162). Solving for ζ_p from Eq. (10-174), we get $\zeta_p = 7.612$. The attenuation should be placed at the resonant frequency of 1095 rad/sec; thus $\omega_n = 1095$ rad/sec. The notch controller of Eq. (10-162) becomes

$$G_c(s) = \frac{s^2 + 95.3s + 1{,}198{,}606.6}{s^2 + 16{,}670.28s + 1{,}199{,}025} \tag{10-175}$$

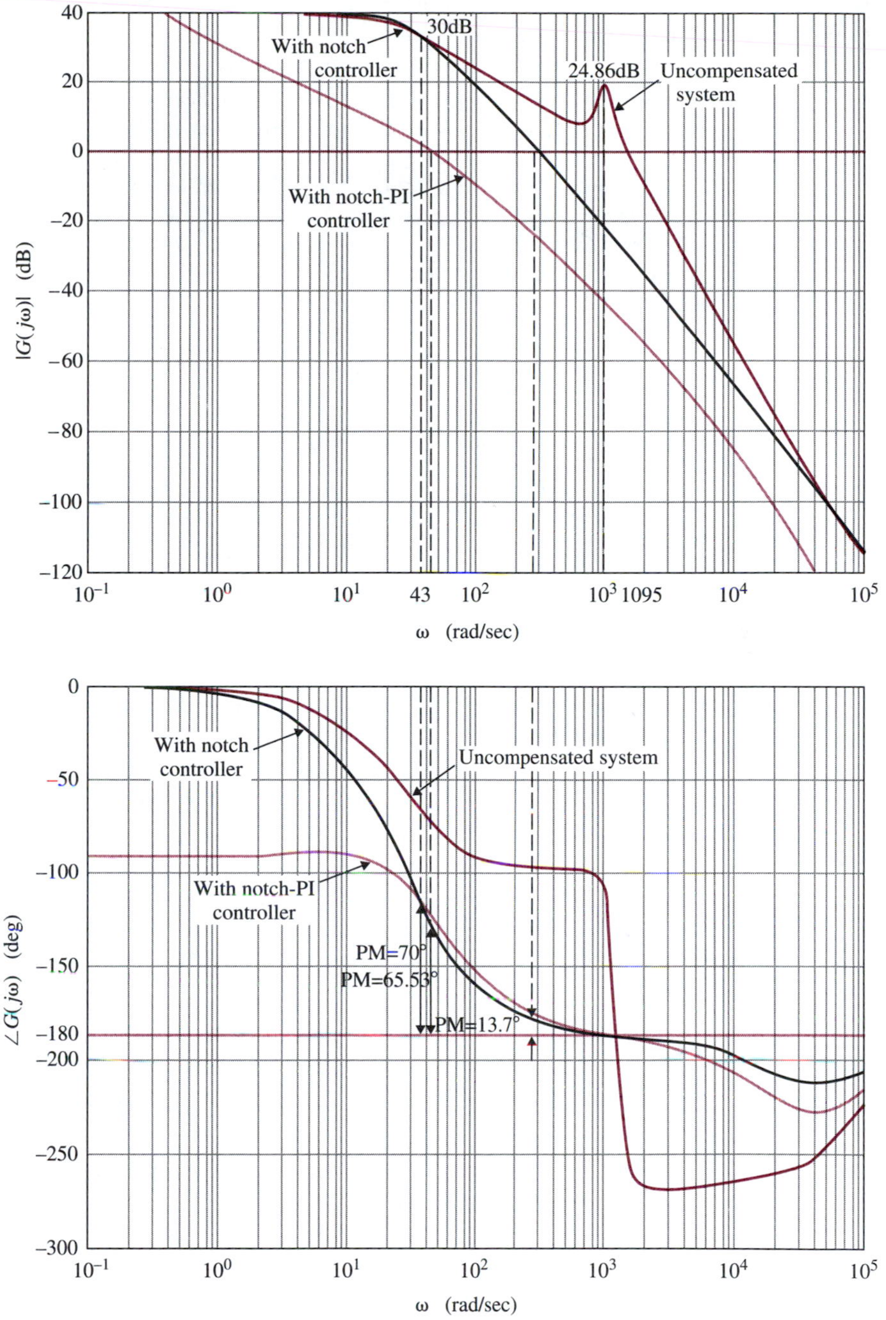

Figure 10-59 Bode plots of uncompensated speed-control system in Example 10-12, and with notch controller, and with notch-PI controller.

$$G_c(s) = \frac{s^2 + 95.3s + 1{,}198{,}606.6}{s^2 + 16{,}670.28s + 1{,}199{,}025} \qquad G_{c2}(s) = 0.0316 + \frac{0.35}{s}$$

The Bode plot of the system with the notch controller in Eq. (10-175) is shown in Fig. 10-59. We can see that the system with the notch controller has a phase margin of only 13.7° and M_r is 3.92.

To complete the design, we can use a PI controller as a second-stage controller. Following the guideline given in Section 10-3 on the design of a PI controller, we assume that the desired phase margin is 70°. From the Bode plot in Fig. 10-59, we see that to realize a phase margin of 70°, the new gain-crossover frequency should be $\omega_g' = 43$ rad/sec, and the magnitude of $G(j\omega_g')$ is 30 dB. Thus, from Eq. (10-32),

$$K_P = 10^{-|G(j\omega_g')|_{dB}/20} = 10^{-30/20} = 0.0316 \qquad (10\text{-}176)$$

The value of K_I is determined using the guideline given by Eq. (10-33),

$$K_I = \frac{\omega_g' K_P}{10} = \frac{43 \times 0.0316}{10} = 0.135 \qquad (10\text{-}177)$$

Since the original system is type 0, the final design needs to be refined by adjusting the value of K_I. Table 10-22 gives the performance attributes when $K_P = 0.0316$ and K_I is varied from 0.135. From the best maximum-overshoot, rise-time, and settling-time measures, the best value of K_I appears to be 0.35. The forward-path transfer function of the compensated system with the notch-PI controller is

$$G(s) = \frac{20{,}000(s + 100{,}000)(0.0316s + 0.35)}{s(s + 16.69)(s^2 + 16{,}670.28s + 1{,}199{,}025)} \qquad (10\text{-}178)$$

Figure 10-59 shows the Bode plot of the system with the notch-PI controller, with $K_P = 0.0316$ and $K_I = 0.35$. The unit-step responses of the compensated system with $K_P = 0.0316$ and $K_I = 0.135$, 0.35, and 0.40 are shown in Fig. 10-60. ▲

Table 10-22 Performance Attributes of System in Example 10-12 with Notch-PI Controller Designed in Frequency Domain

K_P	K_I	PM (deg)	M_r	Maximum Overshoot (%)	t_r (sec)	t_s (sec)
0.0316	0.1	76.71	1.00	0	0.2986	0.5758
0.0316	0.135	75.15	1.00	0	0.2036	0.4061
0.0316	0.200	72.22	1.00	0	0.0430	0.2403
0.0316	0.300	67.74	1.00	0	0.0350	0.1361
0.0316	0.350	65.53	1.00	1.6	0.0337	0.0401
0.0316	0.400	63.36	1.00	4.3	0.0323	0.0398

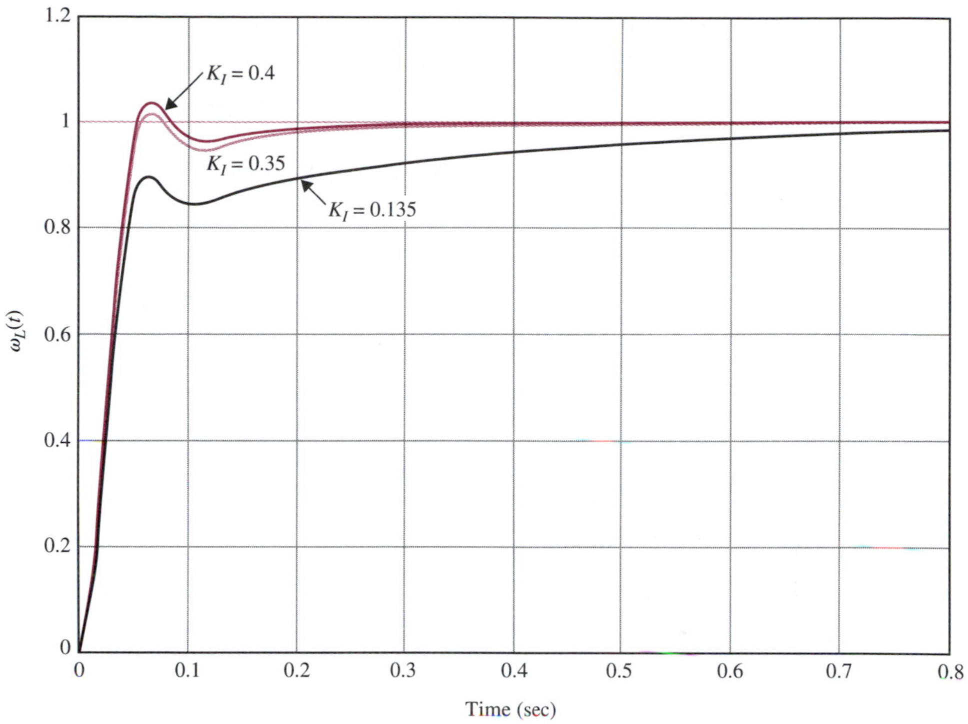

Figure 10-60 Unit-step responses of speed-control system in Example 10-12 with notch-PI controller.

$$G_c(s) = \frac{s^2 + 95.3s + 1{,}198{,}606.6}{s^2 + 16{,}670.28s + 1{,}199{,}025} \qquad G_{c2}(s) = 0.0316 + \frac{K_I}{s}$$

10-9 Forward and Feedforward Controllers

The compensation schemes discussed in the preceding sections all have one degree of freedom in that there is essentially one controller in the system, although the controller can contain several stages connected in series or in parallel. The limitations of a one-degree-of-freedom controller were discussed in Section 10-1. The two-degree-of-freedom compensation scheme shown in Fig. 10-2(d) through (f) offer design flexibility when a multiple number of design criteria have to be satisfied simultaneously.

From Fig. 10-2(e), the closed-loop transfer function of the system is

$$\frac{Y(s)}{R(s)} = \frac{G_{cf}(s)G_c(s)G_p(s)}{1 + G_c(s)G_p(s)} \tag{10-179}$$

and the error transfer function is

$$\frac{E(s)}{R(s)} = \frac{1}{1 + G_c(s)G_p(s)} \tag{10-180}$$

Thus the controller $G_c(s)$ can be designed so that the error transfer function will have certain desirable characteristics, and the controller $G_{cf}(s)$ can be selected to satisfy performance requirements with reference to the input–output relationship. Another way of describing the flexibility of a two-degree-of-freedom design is that the controller $G_c(s)$ is usually designed to provide a certain degree of system stability and performance, but since the zeros of $G_c(s)$ always become the zeros of the closed-loop transfer function, unless some of the zeros are canceled by the poles of the process transfer function, $G_p(s)$, these zeros may cause a large overshoot in the system output even when the relative damping as determined by the characteristic equation is satisfactory. In this case and for other reasons, the transfer function $G_{cf}(s)$ may be used for the control or cancellation of the undesirable zeros of the closed-loop transfer function, while keeping the characteristic equation intact. Of course, we can also introduce zeros in $G_{cf}(s)$ to cancel some of the undesirable poles of the closed-loop transfer function that could not be otherwise affected by the controller $G_c(s)$. The feedforward compensation scheme shown in Fig. 10-2(f) serves the same purpose as the forward compensation, and the difference between the two configurations depends on system and hardware implementation considerations.

It should be kept in mind that while the forward and feedforward compensations may sound powerful, in that they can be used directly for the addition or deletion of poles and zeros of the closed-loop transfer function, there is a fundamental question involving the basic characteristics of feedback. If the forward or feedforward controller is so powerful, why do we need feedback at all? Since the $G_{cf}(s)$ in the systems of Fig. 10-2(e) and (f) are outside the feedback loop, the system is susceptible to parameter variations in $G_{cf}(s)$. Therefore, in reality, these types of compensation cannot be satisfactorily applied to all situations.

Example 10-13

As an illustration of the design of the forward and feedforward compensators, consider the second-order sun-seeker system with phase-lag control designed in Example 10-9. One of the disadvantages of the phase-lag compensation is that the rise time is usually quite long. Let us consider that the phase-lag-compensated sun-seeker system has the forward-path transfer function,

$$G(s) = G_c(s)G_p(s) = \frac{2500(1 + 10s)}{s(s + 25)(1 + 100s)} \tag{10-181}$$

The time-response attributes are:

$$\text{maximum overshoot} = 2.5\% \qquad t_r = 0.1637 \text{ sec} \qquad t_s = 0.2020 \text{ sec}$$

We can improve the rise time and the settling time while not appreciably increasing the overshoot by adding a PD controller $G_{cf}(s)$ to the system, as shown in Fig. 10-61(a). This effectively adds

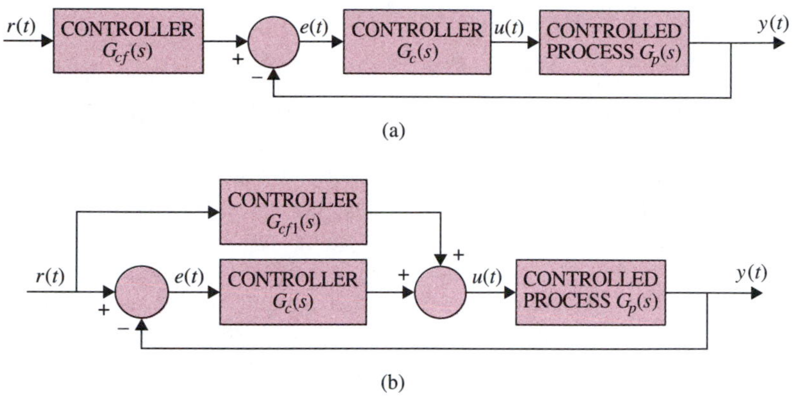

Figure 10-61 (a) Forward compensation with series compensation. (b) Feedforward compensation with series compensations.

a zero to the closed-loop transfer function, while not affecting the characteristic equation. Selecting the PD controller as

$$G_{cf}(s) = 1 + 0.05s \tag{10-182}$$

the time-domain performance attributes are:

maximum overshoot = 4.3% $\quad t_r = 0.1069$ sec $\quad t_s = 0.1313$ sec

If, instead, the feedforward configuration of Fig. 10-61(b) is chosen, the transfer function of $G_{cf1}(s)$ is directly related to $G_{cf}(s)$; that is, equating the closed-loop transfer functions of the two systems in Fig. 10-62(a) and (b), we have

$$\frac{[G_{cf1}(s) + G_c(s)]G_p(s)}{1 + G_c(s)G_p(s)} = \frac{G_{cf}G_c(s)G_p(s)}{1 + G_c(s)G_p(s)} \tag{10-183}$$

Solving for $G_{cf1}(s)$ from the last equation yields

$$G_{cf1}(s) = [G_{cf}(s) - 1]G_c(s) \tag{10-184}$$

Thus, with $G_{cf}(s)$ as given in Eq. (10-182), we have the transfer function of the feedforward controller,

$$G_{cf1}(s) = 0.05s\left(\frac{1 + 10s}{1 + 100s}\right) \tag{10-185}$$

▲

10-10 Design of Robust Control Systems

In many control system applications, the system designed must not only satisfy the damping and accuracy specifications, but the control must also yield performance that is **robust** (insensitive) to external disturbance and parameter variations. We have shown that feedback in conventional control systems has the inherent ability of reducing the effects of external disturbance and parameter variations. Unfortunately, robustness with the conventional feedback configuration is achieved only with a high loop gain, which is normally detrimental to stability. Let us consider the control system shown in Fig. 10-62. The external disturbance is denoted by the signal $d(t)$, and we assume that the amplifier gain K is subject to variation during operation. The input–output transfer function of the system when $d(t) = 0$ is

$$M(s) = \frac{Y(s)}{R(s)} = \frac{KG_{cf}(s)G_c(s)G_p(s)}{1 + KG_c(s)G_p(s)} \tag{10-186}$$

and the disturbance–output transfer function when $r(t) = 0$ is

$$T(s) = \frac{Y(s)}{D(s)} = \frac{1}{1 + KG_c(s)G_p(s)} \tag{10-187}$$

In general, *the design strategy is to select the controller $G_c(s)$ so that the output $y(t)$ is insensitive to the disturbance over the frequency range in which the latter is dominant, and the feedforward controller $G_{cf}(s)$ is designed to achieve the desired transfer function between the input $r(t)$ and the output $y(t)$.*

Let us define the sensitivity of $M(s)$ due to the variation of K as

$$S_K^M = \frac{\text{percent change in } M(s)}{\text{percent change in } K} = \frac{dM(s)/M(s)}{dK/K} \tag{10-188}$$

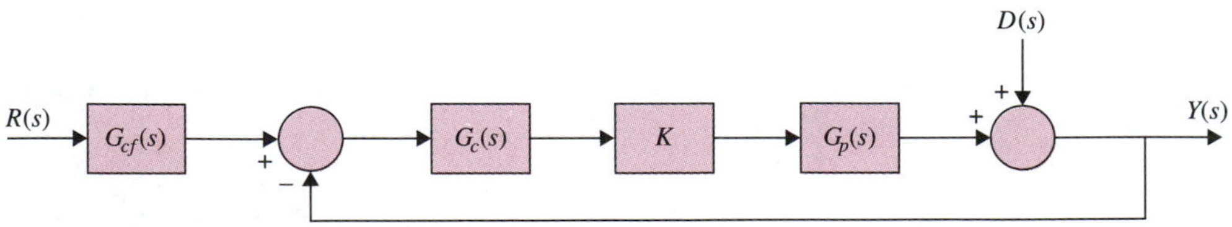

Figure 10-62 Control system with disturbance.

Then, for the system in Fig. 10-62,

$$S_K^M = \frac{1}{1 + KG_c(s)G_p(s)} \tag{10-189}$$

▲ Disturbance suppression and robustness can often be designed with the same control schemes.

which is identical to Eq. (10-187). Thus the sensitivity function and the disturbance–output transfer function are identical, which means that disturbance suppression and robustness with respect to variations of K can be designed with the same control schemes.

The following example shows how the two-degree-of-freedom control system of Fig. 10-62 can be used to achieve a high-gain system that will satisfy both the performance and robustness requirements, as well as for noise rejection.

Example 10-14

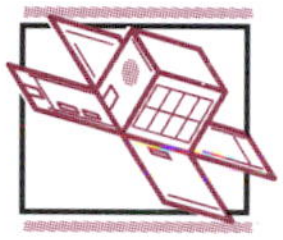

Let us consider the second-order sun-seeker system in Example 10-9, which is compensated with phase-lag control. The forward-path transfer function is

$$G_p(s) = \frac{2500K}{s(s+25)} \tag{10-190}$$

where $K = 1$. The forward-path transfer function of the phase-lag-compensated system with $a = 0.1$ and $T = 100$ is

$$G(s) = G_c(s)G_p(s) = \frac{2500K(1+10s)}{s(s+25)(1+100s)} \qquad (K = 1) \tag{10-191}$$

Since the phase-lag controller is a low-pass filter, the sensitivity of the closed-loop transfer function $M(s)$ with respect to K is poor. The bandwidth of the system is only 13.97 rad/sec; it is expected that $|S_K^M(j\omega)|$ will be greater than unity at frequencies beyond 13.97 rad/sec. Figure 10-63 shows the unit-step responses of the system when $K = 1$, the nominal value, and $K = 0.5$ and 2.0. Notice that if for some reason, the forward gain K is changed from its nominal value, the system response of the phase-lag-compensated system varies substantially. The attributes of the step responses and the characteristic equation roots are shown in Table 10-23 for the three values of K. Figure 10-64 shows the root loci of the system with the phase-lag control. The two complex roots of the characteristic equation vary substantially as K varies from 0.5 to 2.0.

▲ One design strategy for robustness is to place the zeros of the controller near the desired closed-loop poles.

The design strategy of the robust controller is to place two zeros of the controller near the desired closed-loop poles which according to the phase-lag-compensated system are at $s = -12.455 \pm j9.624$. Thus we let the controller transfer function be

$$G_c(s) = \frac{(s+13+j10)(s+13-j10)}{269} = \frac{s^2+26s+269}{269} \tag{10-192}$$

The forward-path transfer function of the system with the robust controller is

$$G(s) = \frac{9.2937K(s^2+26s+269)}{s(s+25)} \tag{10-193}$$

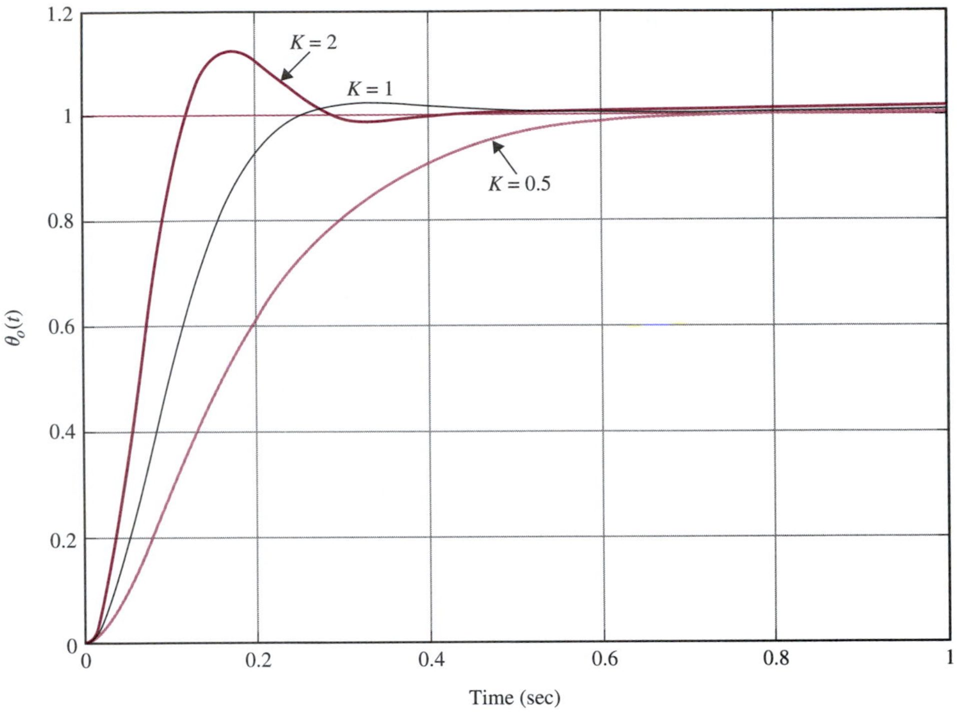

Figure 10-63 Unit-step responses of second-order sun-seeker system with phase-lag control.

$$G(s) = \frac{2500K(1 + 10s)}{s(s + 25)(1 + 100s)}$$

Table 10-23 Attributes of Unit-Step Response of Second-Order Sun-Seeker System with Phase-Lag Control in Example 10-14

K	Maximum Overshoot (%)	t_r (sec)	t_s (sec)	Roots of Characteristic Equation
2.0	12.6	0.07854	0.2323	-0.1005, $-12.4548 \pm j18.51$
1.0	2.6	0.1519	0.2020	-0.1009, $-12.4545 \pm j9.624$
0.5	1.5	0.3383	0.4646	-0.1019, -6.7628, -18.1454

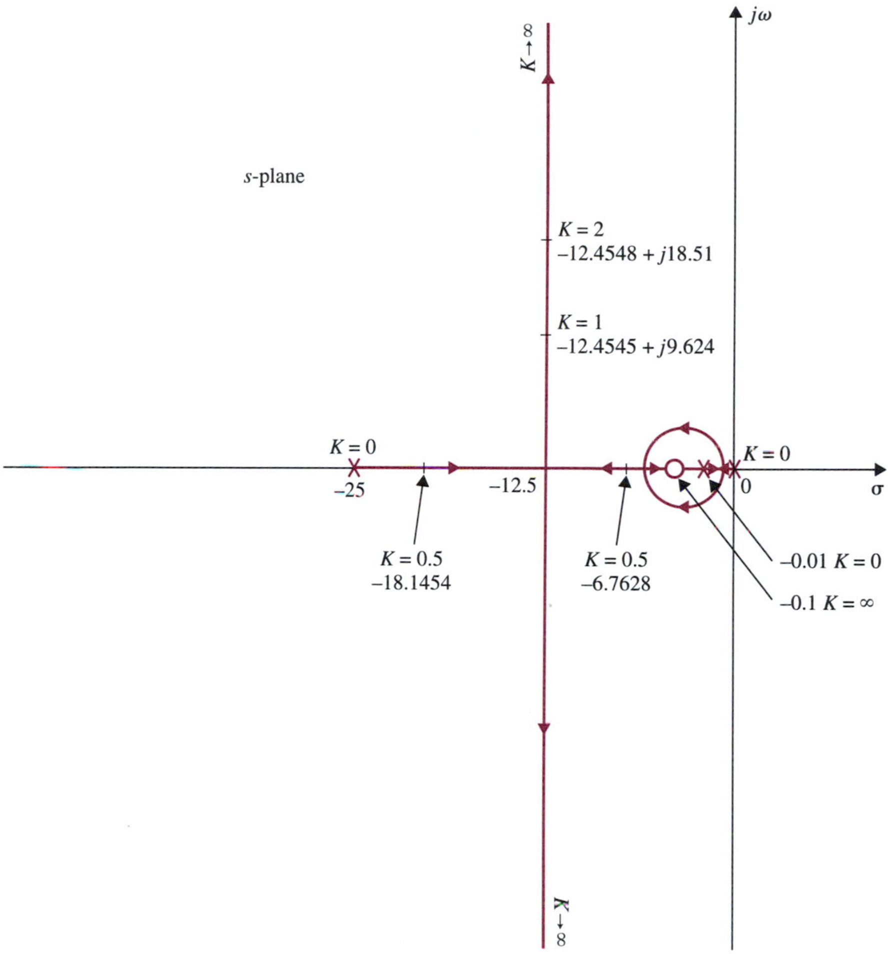

Figure 10-64 Root loci of second-order sun-seeker system with phase-lag control.

$$G(s) = \frac{2500K(1 + 10s)}{s(s + 25)(1 + 100s)}$$

Figure 10-65 shows the root loci of the system with the robust controller. By placing the two zeros of $G_c(s)$ near the desired characteristic equation roots, the sensitivity of the system is greatly improved. In fact, the root sensitivity near the two complex zeros at which the root loci terminate is very low. Figure 10-65 shows that when K approaches infinity, the two characteristic equation roots approach $-13 \pm j10$.

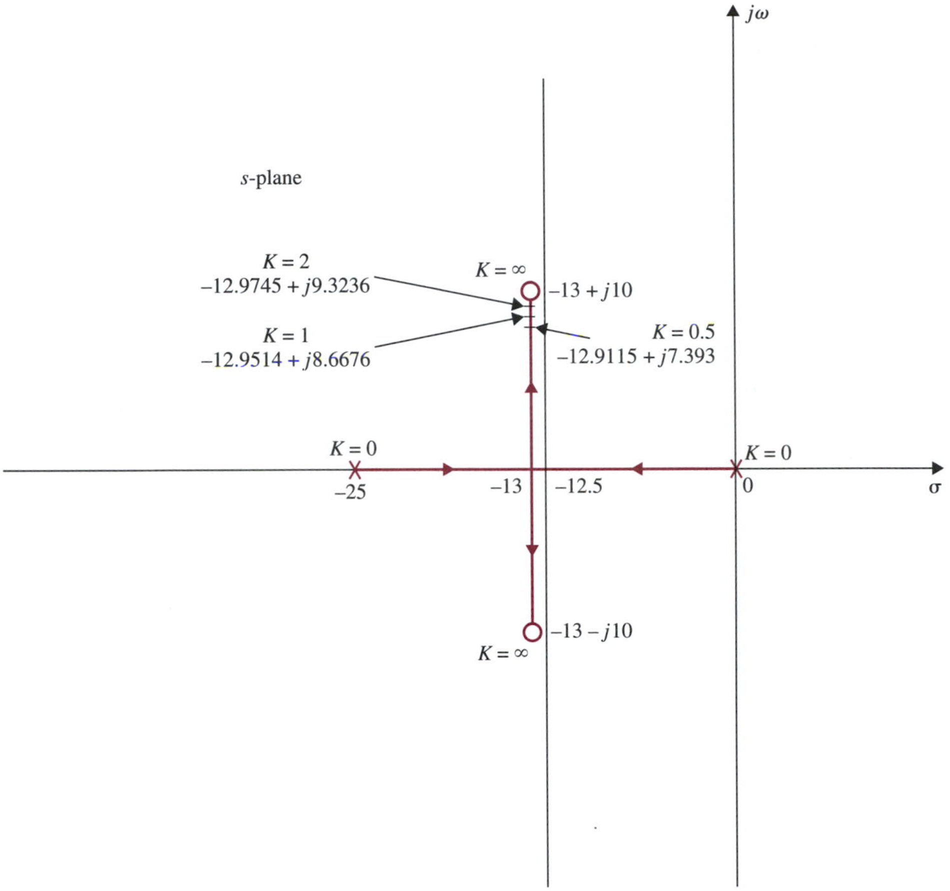

Figure 10-65 Root loci of second-order sun-seeker system with robust controller.

$$G(s) = \frac{9.2937K(s^2 + 26s + 269)}{s(s + 25)}$$

Since the zeros of the forward-path transfer function are identical to the zeros of the closed-loop transfer function, the design is not complete by using only the series controller $G_c(s)$, since the closed-loop zeros will essentially cancel the closed-loop poles. This means that we must add the forward controller, as shown in Fig. 10-62, where $G_{cf}(s)$ should contain poles to cancel the zeros of $s^2 + 26s + 269$ of the closed-loop transfer function. Thus the transfer function of the forward controller is

$$G_{cf}(s) = \frac{269}{s^2 + 26s + 269} \tag{10-194}$$

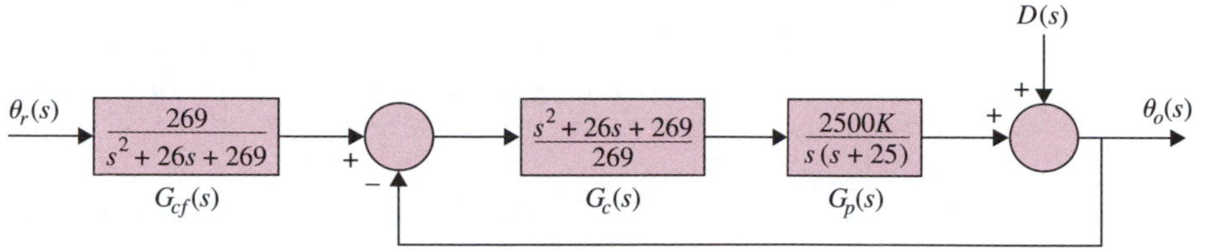

Figure 10-66 Second-order sun-seeker system with robust controller and forward controller.

The block diagram of the overall system is shown in Fig. 10-66. The closed-loop transfer function of the compensated system with $K = 1$ is

$$\frac{\Theta_o(s)}{\Theta_r(s)} = \frac{242.88}{s^2 + 25.903s + 242.88} \tag{10-195}$$

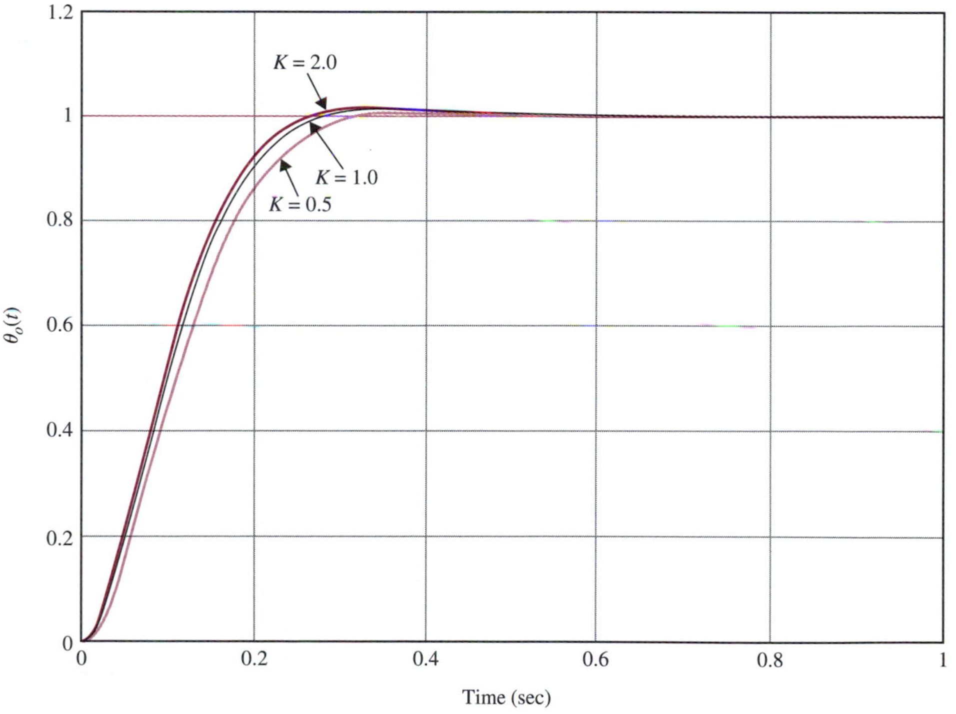

Figure 10-67 Unit-step responses of second-order sun-seeker system with robust controller and forward controller.

Table 10-24 Attributes of Unit-Step Response of Second-Order Sun-Seeker System with Robust Control in Example 10-14

K	Maximum Overshoot (%)	t_r (sec)	t_s (sec)	Roots of Characteristic Equation
2.0	1.3	0.1576	0.2121	$-12.9745 \pm j9.3236$
1.0	0.9	0.1664	0.2222	$-12.9514 \pm j8.6676$
0.5	0.5	0.1846	0.2525	$-12.9115 \pm j7.3930$

The unit-step responses of the system for $K = 0.5$, 1.0, and 2.0 are shown in Fig. 10-67 (on page 783), and their attributes are given in Table 10-24. As shown, the system is now very insensitive to the variation of K.

Since the system in Fig. 10-66 is now more robust, it is expected that the disturbance effect would be reduced. However, we cannot evaluate the effect of the controllers in the system of Fig. 10-66 by applying a unit-step function as $d(t)$. The true improvement on the noise rejection

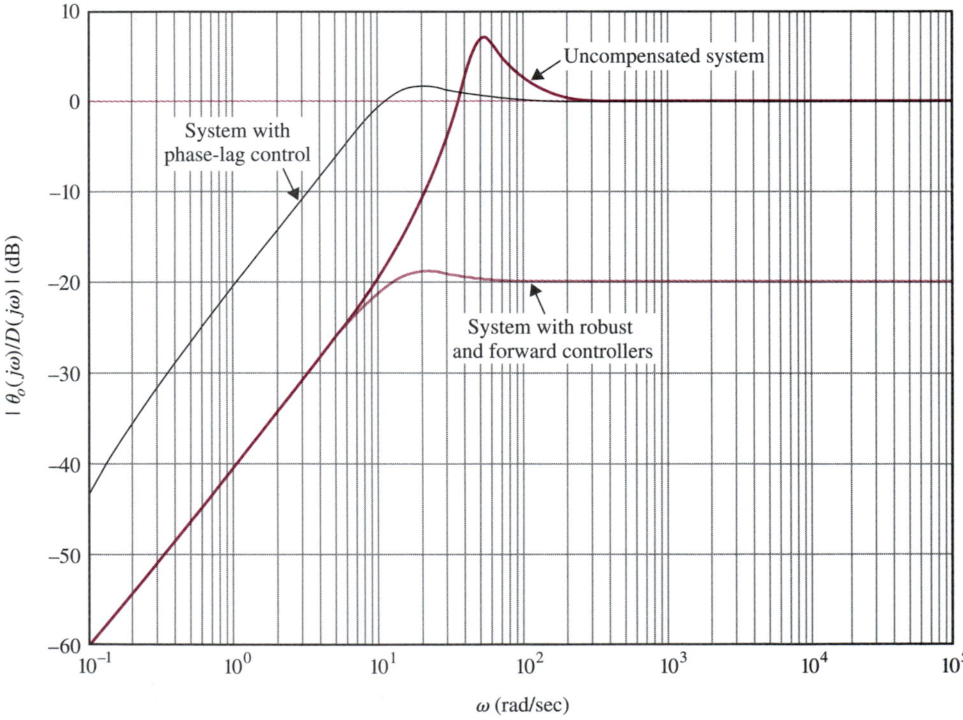

Figure 10-68 Frequency amplitude responses of output due to noise of second-order sun-seeker system.

properties is more appropriately analyzed by investigating the frequency response of $\Theta_o(s)/D(s)$. The noise-to-output transfer function is written from Fig. 10-66,

$$\frac{\Theta_o(s)}{D(s)} = \frac{1}{1 + G_c(s)G_p(s)} = \frac{s(s+25)}{10.2937s^2 + 266.636s + 2500} \tag{10-196}$$

The amplitude Bode plot of Eq. (10-196) is shown in Fig. 10-68 (page 784), along with those of the uncompensated system and the system with phase-lag control. Notice that the magnitude of the frequency response between $D(s)$ and $\Theta_o(s)$ is much smaller than those of the system without compensation and with phase-lag control. The phase-lag control also accentuates the noise for frequencies up to approximately 40 rad/sec, although adding more stability to the system. ▲

Example 10-15

In this example a robust controller with forward compensation is designed for the third-order sun-seeker system which is designed in Example 10-10 with phase-lag control. The forward-path transfer function of the uncompensated system is

$$G_p(s) = \frac{156{,}250{,}000K}{s(s^2 + 625s + 156{,}250)} \tag{10-197}$$

where $K = 1$. The root loci of the closed-loop system are shown in Fig. 10-50, which lead to the phase-lag control with results shown in Table 10-19. Let us select the parameters of the phase-lag controller as $a = 0.1$ and $T = 20$. The dominant roots of the characteristic equation are $s = -187.73 \pm j164.93$.

Let us place the two zeros of the second-order robust controller at $-180 \pm j166.13$, so that the controller transfer function is

$$G_c(s) = \frac{s^2 + 360s + 60{,}000}{60{,}000} \tag{10-198}$$

To ease the high-frequency realization problem of the controller, we may add two nondominant poles to $G_c(s)$. The following analysis is carried out with $G_c(s)$ given in Eq. (10-198), however. The root loci of the compensated system are shown in Fig. 10-69. Thus, by placing the zeros of the controller very close to the desired dominant roots, the system is very insensitive to

Table 10-25 Attributes of Unit-Step Response and Characteristic Equation Roots of Third-Order Sun-Seeker System with Robust and Forward Controls in Example 10-15

K	Maximum Overshoot (%)	t_r (sec)	t_s (sec)	Roots of Characteristic Equation
0.5	1.0	0.01115	0.01616	$-1558.1,\quad -184.5 \pm j126.9$
1.0	2.1	0.01023	0.01414	$-2866.6,\quad -181.3 \pm j147.1$
2.0	2.7	0.00966	0.01313	$-5472.6,\quad -180.4 \pm j156.8$
10.0	3.2	0.00924	0.01263	$-26{,}307,\quad -180.0 \pm j164.0$

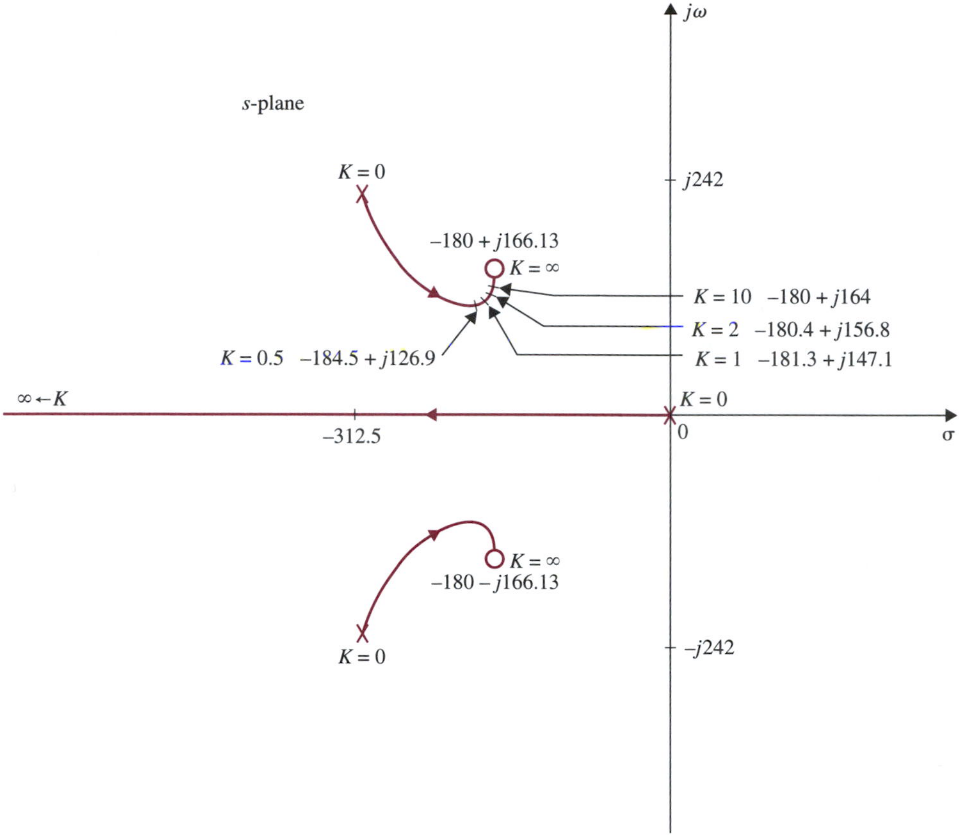

Figure 10-69 Root loci of third-order sun-seeker system with robust and forward controllers.

changing value of K near and beyond the nominal value of K. The forward controller has the transfer function

$$G_{cf}(s) = \frac{60{,}000}{s^2 + 360s + 60{,}000} \tag{10-199}$$

The attributes of the unit-step response for $K = 0.5$, 1.0, 2.0, and 10.0 and the corresponding characteristic equation roots are given in Table 10-25. ▲

Example 10-16

In this example we consider the design of a position-control system that has a variable load inertia. This type of situation is quite common in control systems. For example, the load inertia seen by the motor in an electronic printer will change when different printwheels are used. The system should have satisfactory performance for all the printwheels intended to be used with the system.

To illustrate the design of a robust system that is insensitive to the variation of load inertia, consider that the forward-path transfer function of a unity-feedback control system is

$$G_p(s) = \frac{KK_i}{s[(Js + B)(Ls + B) + K_iK_b]} \tag{10-200}$$

The system parameters are:

K_i = motor torque constant = 1 N-m/A
K_b = motor back-emf constant = 1 V/rad/sec
R = motor resistance = 1 Ω
L = motor inductance = 0.01 H
B = motor and load viscous-friction coefficient ≅ 0
J = motor and load inertia, varies between 0.01 and 0.02 N-m/rad/sec^2
K = amplifier gain

Substituting these system parameters into Eq. (10-200), we get

$$J = 0.01: \quad G_p(s) = \frac{10{,}000K}{s(s^2 + 100s + 10{,}000)} \tag{10-201}$$

$$J = 0.02: \quad G_p(s) = \frac{5000K}{s(s^2 + 100s + 5000)} \tag{10-202}$$

The performance specifications are:

Ramp-error constant $K_v \geq 200$
Maximum overshoot ≤ 5 percent or as small as possible
Rise time $t_r \leq 0.05$ sec
Settling time $t_s \leq 0.05$ sec

These specifications are to be maintained for $0.01 \leq J \leq 0.02$.

To satisfy the ramp-error constant requirement, the value of K must be at least 200. Figure 10-70 shows the root loci of the uncompensated system for $J = 0.01$ and $J = 0.02$. We see that regardless of the value of J, the uncompensated system is unstable for $K > 100$.

To achieve robust control, let us choose the system configuration of Fig. 10-61(a). We introduce a second-order series controller with the zeros placed near the desired dominant characteristic equation of the compensated system. The zeros should be so placed that the dominant characteristic equation roots should be insensitive to the variation in J. This is done by placing the two zeros at $-55 \pm j45$, although the exact location is unimportant. By choosing the two controller zeros as designated, the root loci of the compensated system show that the two complex roots of the characteristic equation will be very close to these zeros for various J, especially when the value of K is large. The transfer function of the robust controller is

$$G_c(s) = \frac{s^2 + 110s + 5050}{5050} \tag{10-203}$$

As in Example 10-15, we may add two nondominant poles to $G_c(s)$ to ease the high-frequency realization problem of the controller. The analysis is carried out with $G_c(s)$ given in Eq. (10-203).

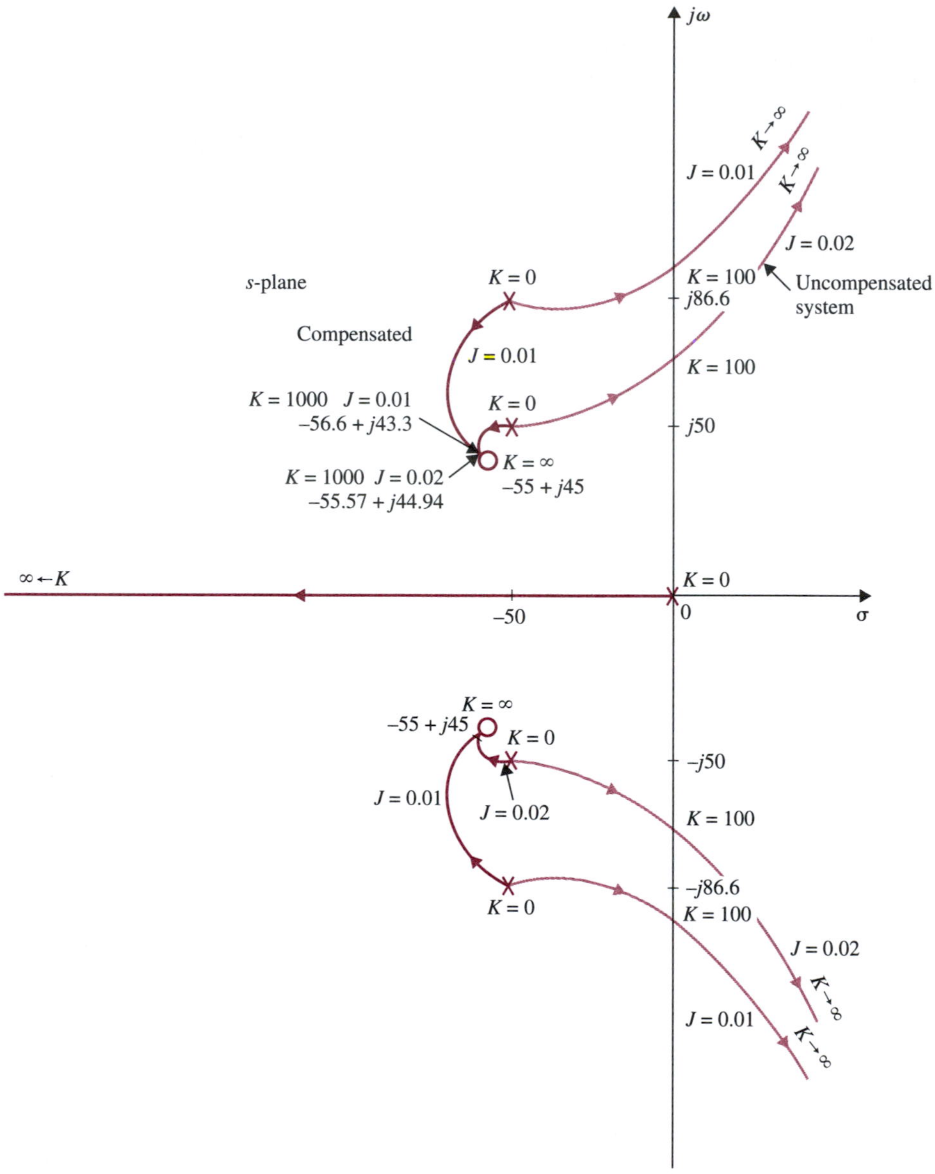

Figure 10-70 Root loci of position-control system in Example 10-16 with robust and forward controllers.

Let $K = 1000$, although 200 would have been adequate to satisfy the K_v requirement. Then, for $J = 0.01$, the forward-path transfer function of the compensated system is

$$G(s) = G_c(s)G_p(s) = \frac{1980.198(s^2 + 110s + 5050)}{s(s^2 + 100s + 10{,}000)} \tag{10-204}$$

Table 10-26 Attributes of Unit-Step Response and Characteristic Equation Roots of System with Robust and Forward Controllers in Example 10-16

J (N-m/rad/sec²)	Maximum Overshoot (%)	t_r (sec)	t_s (sec)	Roots of Characteristic Equation
0.01	1.6	0.03453	0.04444	$-1967, \quad -56.60 \pm j43.3$
0.02	2.0	0.03357	0.04444	$-978.96, \quad -55.57 \pm j44.94$

and for $J = 0.02$,

$$G(s) = \frac{990.99(s^2 + 110s + 5050)}{s(s^2 + 100s + 5000)} \tag{10-205}$$

To cancel the two zeros of the closed-loop transfer function, the transfer function of the forward controller is

$$G_{cf}(s) = \frac{5050}{s^2 + 110s + 5050} \tag{10-206}$$

The attributes of the unit-step response and the characteristic equation roots of the compensated system with $K = 1000$, $J = 0.01$, and $J = 0.02$ are given in Table 10-26. ▲

10-11 Minor-Loop Feedback Control

▲ A tachometer can be used in a minor feedback loop for improving stability.

The control schemes discussed in the preceding sections have all utilized series controllers in the forward path of the main loop or feedforward path of the control system. Although series controllers are the most common because of their simplicity in implementation, depending on the nature of the system, sometimes there are advantages in placing the controller in a minor feedback loop, as shown in Fig. 10-2(b). For example, a tachometer may be coupled directly to a dc motor not only for the purpose of speed indication, but more often for improving the stability of the closed-loop system by feeding back the output signal of the tachometer. The motor speed can also be generated by processing the back emf of the motor electronically. In principle, the PID controller or phase-lead and phase-lag controllers can all, with varying degree of effectiveness, be applied as minor-loop feedback controllers. Under certain conditions, minor-loop control can yield systems that are more robust, that is, less sensitive to external disturbance or internal parameter variations.

10-11-1 Rate-Feedback or Tachometer-Feedback Control

The principle of using the derivative of the actuating signal to improve the damping of a closed-loop system can be applied to the output signal to achieve a similar effect. In other words, the derivative of the output signal is fed back and added algebraically to the actuating signal of the system. In practice, if the output variable is mechanical displacement, a tachometer may be used to convert mechanical displacement into an electrical signal that is proportional to the derivative of the displacement. Figure 10-71 shows the block diagram of a control system with a secondary path that feeds back the derivative of the output. The transfer function of the tachometer is denoted by $K_t s$, where K_t is the tachometer constant, usually expressed in volts/radian per second for analytical purposes. Commercially, K_t is given in the data sheet of the tachometer typically in volts per 1000 rpm. The effects of rate or tachometer feedback can be illustrated by applying it to a second-order prototype system. Consider that the controlled process of the system shown in Fig. 10-71 has the transfer function

$$G_p(s) = \frac{\omega_n^2}{s(s + 2\zeta\omega_n)} \qquad (10\text{-}207)$$

The closed-loop transfer function of the system is

$$\frac{Y(s)}{R(s)} = \frac{\omega_n^2}{s^2 + (2\zeta\omega_n + K_t\omega_n^2)s + \omega_n^2} \qquad (10\text{-}208)$$

and the characteristic equation is

$$s^2 + (2\zeta\omega_n + K_t\omega_n^2)s + \omega_n^2 = 0 \qquad (10\text{-}209)$$

From the characteristic equation, *it is apparent that the effect of the tachometer feedback is the increase of the damping of the system, since K_t appears in the same term as the damping ratio ζ.* In this respect, tachometer-feedback control has exactly the same

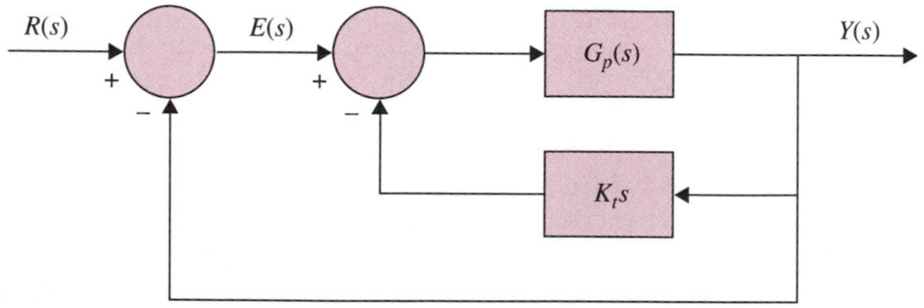

Figure 10-71 Control system with tachometer feedback.

effect as the PD control. However, the closed-loop transfer function of the system with PD control in Fig. 10-3 is

$$\frac{Y(s)}{R(s)} = \frac{\omega_n^2(K_P + K_D s)}{s^2 + (2\zeta\omega_n + K_D\omega_n^2)s + \omega_n^2 K_P} \qquad (10\text{-}210)$$

Comparing the two transfer functions in Eqs. (10-208) and (10-210), we see that the two characteristic equations are identical if $K_P = 1$ and $K_D = K_t$. However, Eq. (10-210) has a zero at $s = -K_P/K_D$, whereas Eq. (10-208) does not. Thus the response of the system with tachometer feedback is uniquely defined by the characteristic equation, whereas the response of the system with the PD control also depends on the zero at $s = -K_P/K_D$, which could have a significant effect on the overshoot of the step response.

With reference to the steady-state analysis, the forward-path transfer function of the system with tachometer feedback is

$$\frac{Y(s)}{E(s)} = \frac{\omega_n^2}{s(s + 2\zeta\omega_n + K_t\omega_n^2)} \qquad (10\text{-}211)$$

▲ For a type-1 system, tachometer feedback decreases the ramp-error constant, but does not affect the step-error constant.

Since the system is still type 1, the basic characteristics of the steady-state error are not altered by the tachometer feedback; that is, when the input is a step function, the steady-state error is zero. However, for a unit-ramp function input, the steady-state error of the system is $(2\zeta + K_t\omega_n)/\omega_n$, whereas that of the system with PD control in Fig. 10-3 is $2\zeta/\omega_n$. Thus *for a type 1 system, tachometer feedback decreases the ramp-error constant* K_v while not affecting the step-error constant K_P.

10-11-2 Minor-Loop Feedback Control with Active Filter

Instead of using a tachometer, to reduce cost, and for space saving, an active filter with *RC* elements and op-amps can be used in the minor feedback loop for compensation. We illustrate this approach with the following example.

Example 10-17

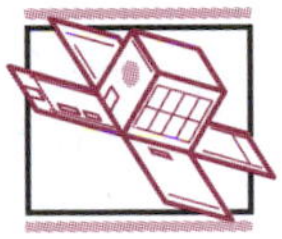

Consider that for the second-order sun-seeker system in Example 10-9, instead of using a series controller in the forward path, we adopt the minor-loop feedback control, as shown in Fig. 10-72(a), with

$$G_p(s) = \frac{2500}{s(s + 25)} \qquad (10\text{-}212)$$

and

$$H(s) = \frac{K_t s}{1 + Ts} \qquad (10\text{-}213)$$

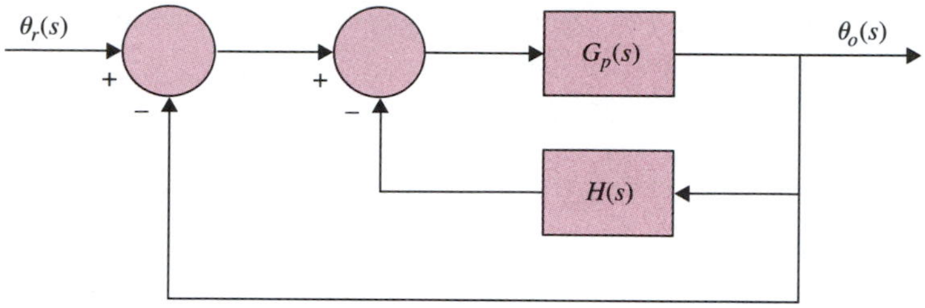

(a)

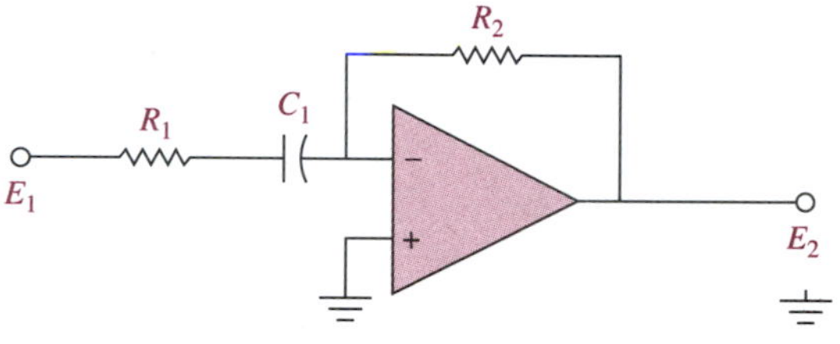

(b)

Figure 10-72 (a) Sun-seeker control system with minor-loop control. (b) Op-amp circuit realization of $K_t s/(1 + Ts)$.

To maintain the system as type 1, it is necessary that $H(s)$ contains a zero at $s = 0$. Equation (10-213) can be realized by the op-amp circuit shown in Fig. 10-72(b). This circuit cannot be applied as a series controller in the forward path, since it acts as an open circuit in the steady state when the frequency is zero. As a minor-loop controller, the zero-transmission property to dc signals does not pose any problem.

The forward-path transfer function of the system in Fig. 10-72(a) is

$$\frac{\Theta_o(s)}{\Theta_e(s)} = G(s) = \frac{G_p(s)}{1 + G_p(s)H(s)} = \frac{2500(1 + Ts)}{s[(s + 25)(1 + Ts) + 2500K_t]} \tag{10-214}$$

The characteristic equation of the system is

$$Ts^3 + (25T + 1)s^2 + (25 + 2500T + 2500K_t)s + 2500 = 0 \tag{10-215}$$

To show the effects of the parameters K_t and T, we construct the root contours of Eq. (10-215) by first considering that K_t is fixed and T is variable. Dividing both sides of Eq. (10-215) by the terms that do not contain T, we get

$$1 + \frac{Ts(s^2 + 25s + 2500)}{s^2 + (25 + 2500K_t)s + 5000} = 0 \tag{10-216}$$

When the value of K_t is relatively large, the two poles of the last equation are real, with one very close to the origin. It is more effective to choose K_t so that the poles of Eq. (10-216) are complex.

Figure 10-73 shows the root contours of Eq. (10-215) with $K_t = 0.02$, and T varies from 0 to ∞. When $T = 0.006$, the characteristic equation roots are at -56.72, $-67.47 + j52.85$, and $-67.47 - j52.85$. The attributes of the unit-step response are:

maximum overshoot = 0% $\quad t_r = 0.04485$ sec $\quad t_s = 0.06061$ sec $\quad t_{max} = 0.4$ sec

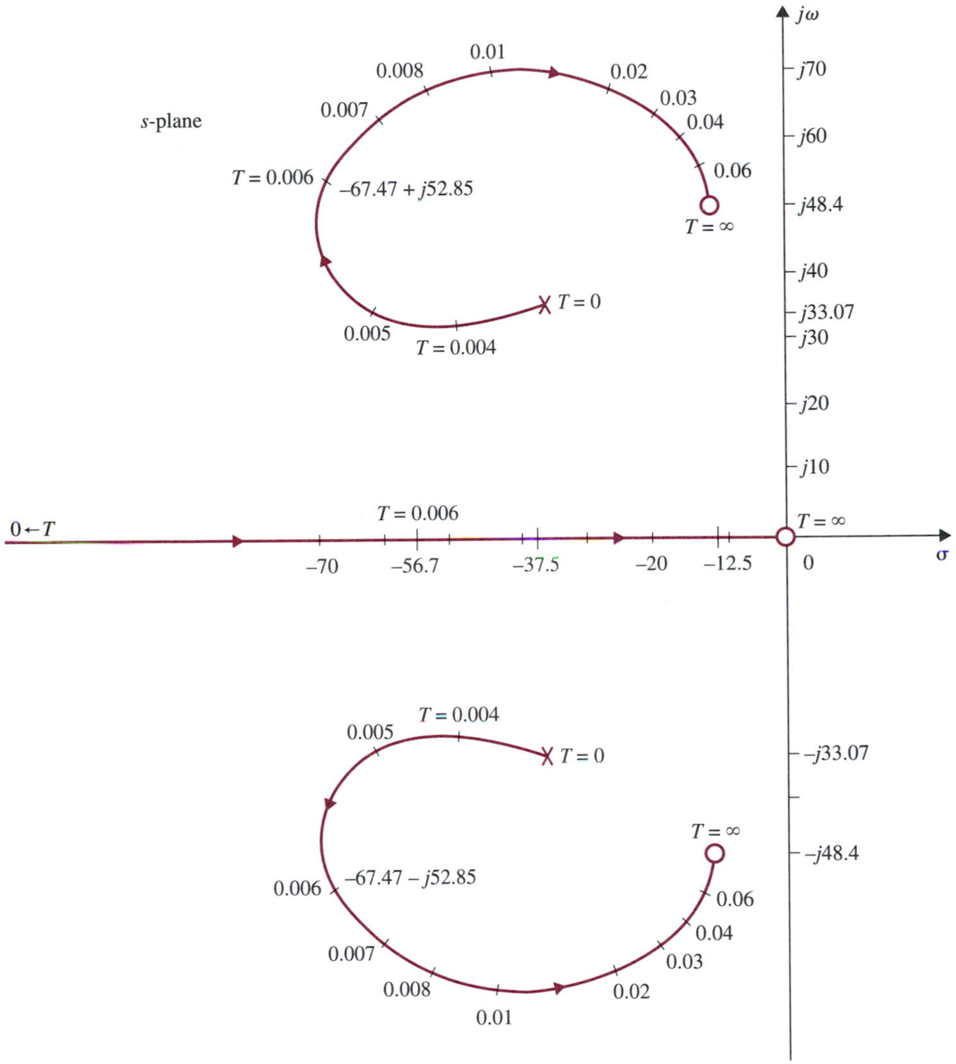

Figure 10-73 Root contours of $Ts^3 + (25T + 1)s^2 + (25 + 2500K_t + 2500T)s + 2500 = 0$. $K_t = 0.02$.

The ramp-error constant of the system is

$$K_v = \lim_{s \to 0} sG(s) = \frac{100}{1 + 100K_t} \tag{10-217}$$

Thus, just as with tachometer feedback, the minor-loop feedback controller of Eq. (10-213) reduces the ramp-error constant K_v, although the system is still type 1. ▲

10-12 State-Feedback Control

A majority of the design techniques in modern control theory is based on the state-feedback configuration. That is, instead of using controllers with fixed configurations in the forward or feedback path, control is achieved by feeding back the state variables through real constant gains. The block diagram of a system with state-feedback control is shown in Fig. 10-2(c). A more detailed block diagram is shown in Fig. 10-74.

We can show that the PID control and the tachometer-feedback control discussed earlier are all special cases of the state-feedback control scheme. In the case of tachometer-feedback control, let us consider the second-order prototype system described in Eq. (10-207). The process is decomposed by direct decomposition and is represented by the state diagram of Fig. 10-75(a). If the states $x_1(t)$ and $x_2(t)$ are physically accessible, these variables may be fed back through constant real gains $-k_1$ and $-k_2$, respectively, to form the control $u(t)$, as shown in Fig. 10-75(b). The transfer function of the system with state feedback is

$$\frac{Y(s)}{R(s)} = \frac{\omega_n^2}{s^2 + (2\zeta\omega_n + k_2)s + k_1} \tag{10-218}$$

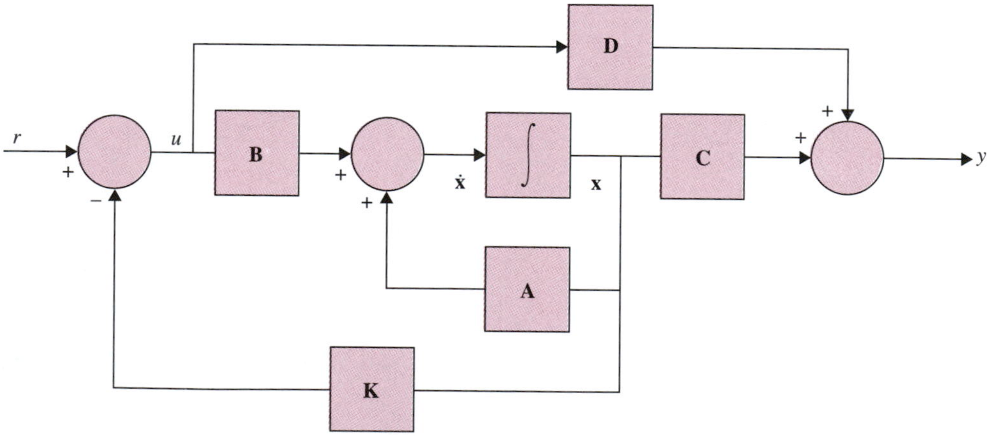

Figure 10-74 Block diagram of control system with state feedback.

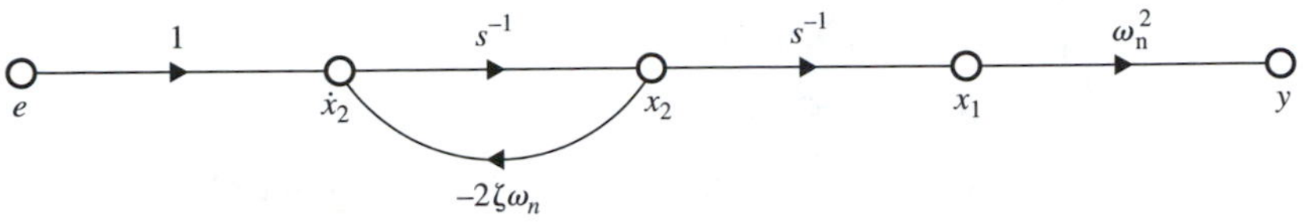

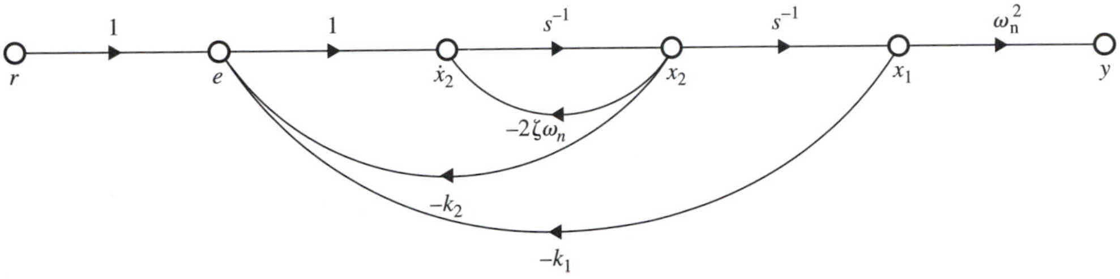

Figure 10-75 Control of a second-order system by state feedback.

For comparison purposes, we display the transfer functions of the systems with tachometer feedback, Eq. (10-208), and with PD control in Eq. (10-210) as follows:

$$\text{Tachometer feedback:}\quad \frac{Y(s)}{R(s)} = \frac{\omega_n^2}{s^2 + (2\zeta\omega_n + K_t\omega_n^2)s + \omega_n^2} \tag{10-219}$$

$$\text{PD control:}\quad \frac{Y(s)}{R(s)} = \frac{\omega_n^2(K_P + K_D s)}{s^2 + (2\zeta\omega_n + K_D\omega_n^2)s + \omega_n^2 K_P} \tag{10-220}$$

Thus tachometer feedback is equivalent to state feedback if $k_1 = \omega_n^2$ and $k_2 = K_t\omega_n^2$. Comparing Eq. (10-218) with Eq. (10-220), we see that the characteristic equation of the system with state feedback would be identical to that of the system with PD control if $k_1 = \omega_n^2 K_P$ and $k_2 = K_D\omega_n^2$. However, the numerators of the two transfer functions are different.

The class of systems with zero reference input, $r(t) = 0$, is commonly known as **regulators.** When $r(t) = 0$, the control objective is to drive any arbitrary initial conditions of the system to zero in some prescribed manner: for example, as quickly as possible. Then a second-order system with PD control is the same as state-feedback control.

It should be emphasized that the comparisons made above are all for second-order systems. For higher-order systems, the PD control and tachometer-feedback control are equivalent to only feeding back the state variables x_1 and x_2, while state-feedback control can feed back all the state variables.

Since PI control increases the order of the system by one, it cannot be made equivalent to state feedback through constant gains. We show in Section 10-14 that if we combine state feedback with integral control we can again realize PI control in the sense of state-feedback control.

10-13 Pole-Placement Design through State Feedback

When root loci are utilized for the design of control systems, the general approach may be described as that of **pole placement;** the poles here refers to that of the closed-loop transfer function, which are also the roots of the characteristic equation. Knowing the relation between the closed-loop poles and the system performance, we can effectively carry out the design by specifying the location of these poles.

The design methods discussed in the preceding sections are all characterized by the property that the poles are selected based on what can be achieved with the fixed-controller configuration and the physical range of the controller parameters. A natural question would be: *Under what condition can the poles be placed arbitrarily?* This is an entirely new design philosophy and freedom that apparently can be achieved only under certain conditions.

When we have a controlled process of the third order or higher, the PD, PI, the single-stage phase-lead or phase-lag controller would not be able to control independently all the poles of the system, since there are only two free parameters in each of these controllers.

To investigate the condition required for arbitrary pole placement in an nth-order system, let us consider that the process is described by the following state equation:

$$\frac{d\mathbf{x}(t)}{dt} = \mathbf{A}\mathbf{x}(t) + \mathbf{B}u(t) \tag{10-221}$$

where $\mathbf{x}(t)$ is an $n \times 1$ state vector, and $u(t)$ is the scalar control. The state-feedback control is

$$u(t) = -\mathbf{K}\mathbf{x}(t) + r(t) \tag{10-222}$$

where $\mathbf{K}$ is the $1 \times n$ feedback matrix with constant-gain elements. By substituting Eq. (10-222) into Eq. (10-221), the closed-loop system is represented by the state equation

$$\frac{d\mathbf{x}(t)}{dt} = (\mathbf{A} - \mathbf{B}\mathbf{K})\mathbf{x}(t) + \mathbf{B}r(t) \tag{10-223}$$

▲ If a system is completely state controllable, its eigenvalues can be arbitrarily assigned through state feedback.

It will be shown in the following that if the pair [**A, B**] is completely controllable, a matrix **K** exists that can give an arbitrary set of eigenvalues of $(\mathbf{A} - \mathbf{B}\mathbf{K})$; that is, the n roots of the characteristic equation

$$|s\mathbf{I} - \mathbf{A} + \mathbf{B}\mathbf{K}| = 0 \tag{10-224}$$

can be arbitrarily placed. To show that this is true, we refer to the findings in Section 5-8 that if a system is completely controllable, it can always be represented in the controllable canonical form (CCF); that is, in Eq. (10-221),

$$\mathbf{A} = \begin{bmatrix} 0 & 1 & 0 & \cdots & 0 \\ 0 & 0 & 1 & \cdots & 0 \\ \vdots & \vdots & \vdots & \ddots & \vdots \\ 0 & 0 & 0 & \cdots & 1 \\ -a_0 & -a_1 & -a_2 & \cdots & -a_{n-1} \end{bmatrix} \qquad \mathbf{B} = \begin{bmatrix} 0 \\ 0 \\ \vdots \\ 0 \\ 1 \end{bmatrix} \tag{10-225}$$

The feedback gain matrix **K** is expressed as

$$\mathbf{K} = [k_1 \quad k_2 \quad \cdots \quad k_n] \tag{10-226}$$

where $k_1, k_2, \ldots, k_n$ are real constants. Then

$$\mathbf{A} - \mathbf{BK} = \begin{bmatrix} 0 & 1 & 0 & \cdots & 0 \\ 0 & 0 & 1 & \cdots & 0 \\ \vdots & \vdots & \vdots & \ddots & \vdots \\ 0 & 0 & 0 & \cdots & 1 \\ -a_0 - k_1 & -a_1 - k_2 & -a_2 - k_3 & \cdots & -a_{n-1} - k_n \end{bmatrix} \tag{10-227}$$

The eigenvalues of **A** − **BK** are then found from the characteristic equation

$$|s\mathbf{I} - (\mathbf{A} - \mathbf{BK})| = s^n + (a_{n-1} + k_n)s^{n-1} + (a_{n-2} + k_{n-1})s^{n-2} + \cdots + (a_0 + k_1) = 0 \tag{10-228}$$

Clearly, the eigenvalues can be arbitrarily assigned, since the feedback gains k_1, k_2, $\ldots$, k_n are isolated in each coefficient of the characteristic equation. Intuitively, it makes sense that a system must be controllable for the poles to be placed arbitrarily. If one or more state variables are uncontrollable, the poles associated with these state variables are also uncontrollable and cannot be moved as desired. The following example illustrates the design of a control system with state feedback.

Example 10-18

Consider the magnetic-ball-suspension system analyzed in Section 5-16. This is a typical regulator system for which the control problem is to maintain the ball at its equilibrium position. It is shown in Section 5-16 that the system without control is unstable.

The linearized state model of the magnetic-ball system is represented by the state equation,

$$\frac{d\,\Delta\mathbf{x}(t)}{dt} = \mathbf{A}^*\,\Delta\mathbf{x}(t) + \mathbf{B}^*\,\Delta v(t) \tag{10-229}$$

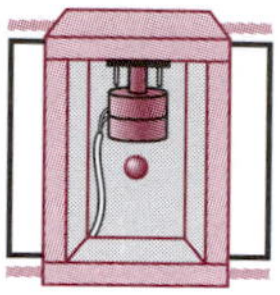

where $\Delta\mathbf{x}(t)$ denotes the linearized state vector and $\Delta v(t)$ the linearized input voltage. The coefficient matrices are

$$\mathbf{A}^* = \begin{bmatrix} 0 & 1 & 0 \\ 64.4 & 0 & -16 \\ 0 & 0 & -100 \end{bmatrix} \qquad \mathbf{B}^* = \begin{bmatrix} 0 \\ 0 \\ 100 \end{bmatrix} \tag{10-230}$$

The eigenvalues of $\mathbf{A}^*$ are $s = -100$, -8.025, and 8.025. Thus the system without feedback control is unstable.

Let us give the following design specifications:

1. The system must be stable.
2. For any initial disturbance on the position of the ball from its equilibrium position, the ball must return to the equilibrium position with zero steady-state error.
3. The time response should settle to within 5 percent of the initial disturbance in not more than 0.5 sec.
4. The control is to be realized by state feedback

$$\Delta v(t) = -\mathbf{K}\,\Delta\mathbf{x}(t) = -[k_1 \quad k_2 \quad k_3]\,\Delta\mathbf{x}(t) \tag{10-231}$$

where k_1, k_2, and k_3 are real constants.

A state diagram of the "open-loop" ball-suspension system is shown in Fig. 10-76(a), and the same of the "closed-loop" system with state feedback is shown in Fig. 10-76(b). We must select the desired location of the eigenvalues of $s\mathbf{I} - \mathbf{A}^* + \mathbf{B}^*\mathbf{K}$ so that requirement 3 above on the time response is satisfied. Without entirely resorting to trial and error, we can start with the following decisions:

1. The system dynamics should be controlled by two dominant roots.
2. To achieve a relatively fast response, the two dominant roots should be complex.
3. The damping that is controlled by the real parts of the complex roots should be adequate, and the imaginary parts should be high enough for the transient to die out sufficiently fast.

After a few trial-and-error runs using the function **svdesign** of **CSAD/MATLAB,** we found that the following characteristic equation roots should satisfy the design requirements:

$$s = -20 \qquad s = -6 + j4.9 \qquad s = -6 - j4.9$$

The corresponding characteristic equation is

$$s^3 + 32s^2 + 300s + 1200 = 0 \tag{10-232}$$

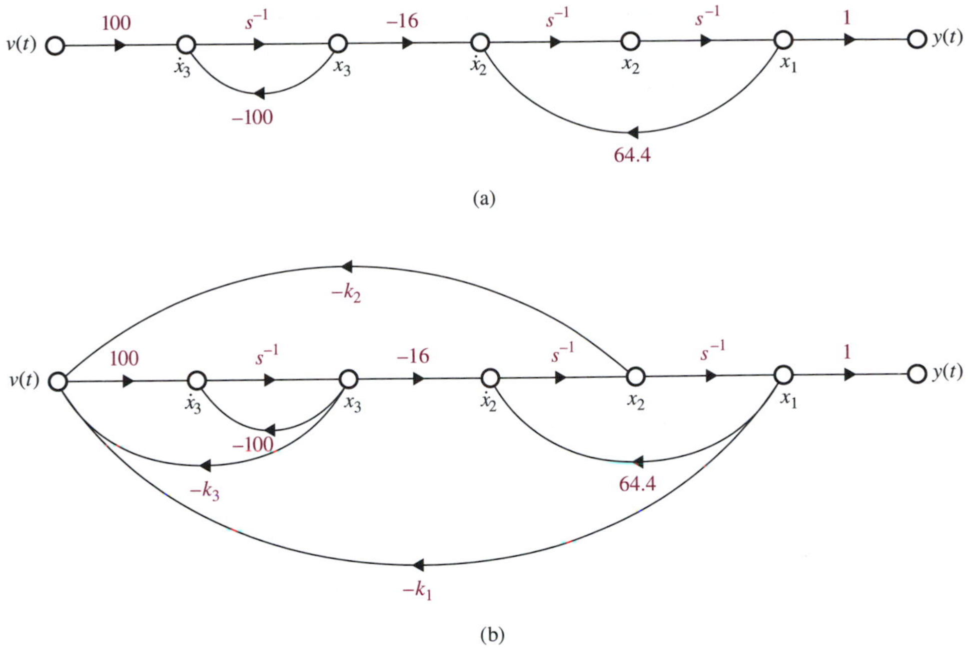

Figure 10-76 (a) State diagram of magnetic-ball-suspension system. (b) State diagram of magnetic-ball-suspension system with state feedback.

The characteristic equation of the closed-loop system with state feedback is written

$$|s\mathbf{I} - \mathbf{A}^* + \mathbf{B}^*\mathbf{K}| = \begin{vmatrix} s & -1 & 0 \\ -64.4 & s & 16 \\ 100k_1 & 100k_2 & s + 100 + 100k_3 \end{vmatrix} \tag{10-233}$$

$$= s^3 + 100(k_3 + 1)s^2 - (64.4 + 1600k_2)s - 1600k_1 - 6440(k_3 + 1) = 0$$

which can also be obtained directly from Fig. 10-76(b) using the SFG gain formula. Equating like coefficients of Eqs. (10-232) and (10-233), we get the following simultaneous equations:

$$\begin{aligned} 100(k_3 + 1) &= 32 \\ -64.4 - 1600k_2 &= 300 \\ -1600k_1 - 6440(k_3 + 1) &= 1200 \end{aligned} \tag{10-234}$$

Solving the last three equations and being assured that the solutions exist and are unique, we get the feedback gain matrix,

$$\mathbf{K} = [k_1 \quad k_2 \quad k_3] = [-2.038 \quad -0.22775 \quad -0.68] \tag{10-235}$$

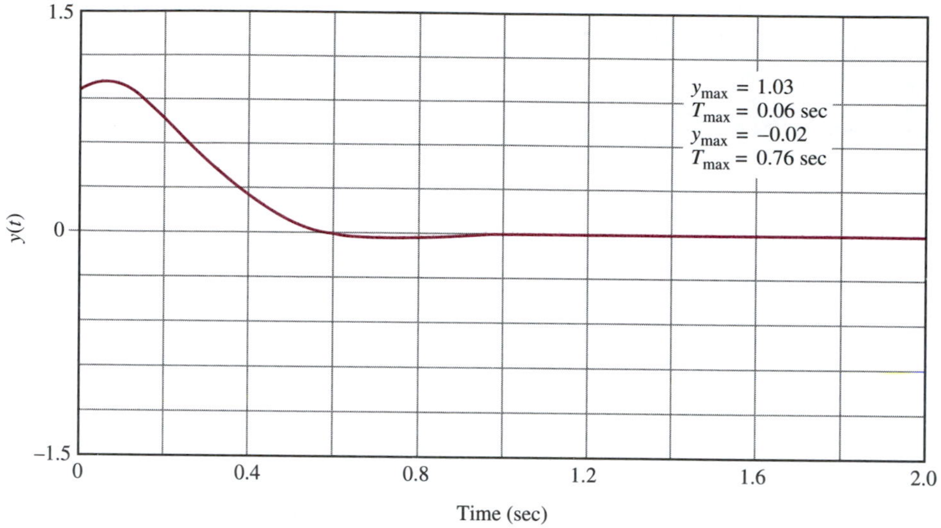

Figure 10-77 Output response of magnetic-ball-suspension system with state feedback, subject to initial condition $y(0) = x_1(0) = 1$.

Figure 10-77 shows the output response $y(t)$ when the system is subject to the initial condition

$$\mathbf{x}(0) = \begin{bmatrix} 1 \\ 0 \\ 0 \end{bmatrix} \tag{10-236}$$

▲

Example 10-19

In this example we design a state-feedback control for the second-order sun-seeker system treated in Example 10-9. The transfer function of the controlled process is given in Eq. (10-131) with $K = 1$. The CCF state diagram of the process is shown in Fig. 10-78(a). The problem involves the design of a state-feedback control with

$$\theta_e(t) = -\mathbf{K}\mathbf{x}(t) = -[k_1 \quad k_2]\mathbf{x}(t) \tag{10-237}$$

The state equations are represented in vector-matrix form as

$$\frac{d\mathbf{x}(t)}{dt} = \mathbf{A}\mathbf{x}(t) + \mathbf{B}\theta_e(t) \tag{10-238}$$

where

$$\mathbf{A} = \begin{bmatrix} 0 & 1 \\ 0 & -25 \end{bmatrix} \qquad \mathbf{B} = \begin{bmatrix} 0 \\ 1 \end{bmatrix} \tag{10-239}$$

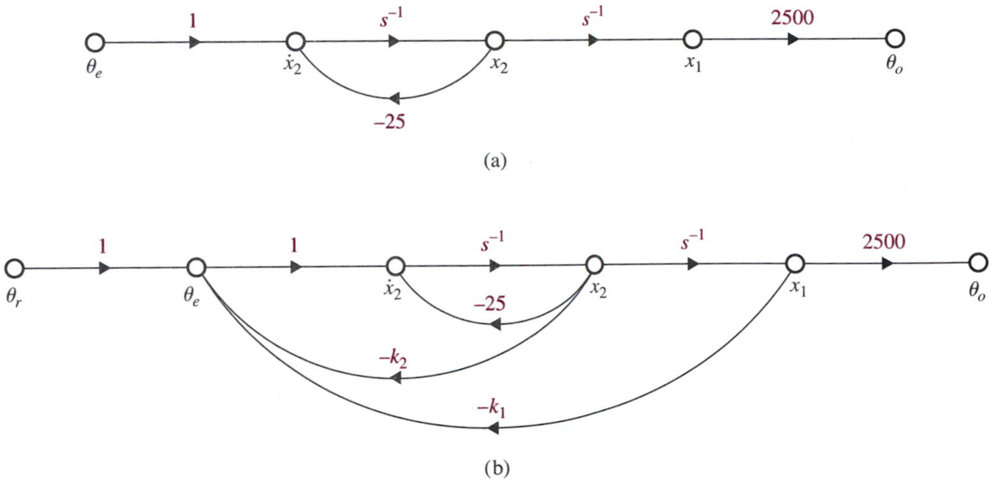

Figure 10-78 (a) State diagram of second-order sun-seeker system.
(b) State diagram of second-order sun-seeker system with state feedback.

The output equation is

$$\theta_o(t) = \mathbf{C}\mathbf{x}(t) \tag{10-240}$$

where

$$\mathbf{C} = [1 \quad 0] \tag{10-241}$$

The design objectives are:

1. The steady-state error due to a step function input should equal 0.
2. With the state-feedback control, the unit-step response should have minimum overshoot, rise time, and settling time.

The transfer function of the system with state feedback is written

$$\frac{\Theta_o(s)}{\Theta_r(s)} = \frac{2500}{s^2 + (25 + k_2)s + k_1} \tag{10-242}$$

Thus for zero steady-state error to a step input, the constant terms in the numerator and denominator must be equal, or $k_1 = 2500$. This means that while the system is completely controllable, we cannot arbitrarily assign the two roots of the characteristic equation, which is now

$$s^2 + (25 + k_2)s + 2500 = 0 \tag{10-243}$$

In other words, only one of the roots of Eq. (10-243) can be assigned arbitrarily. The problem is solved using the function **svdesign** of **CSAD/MATLAB.** After a few trial-and-error runs, we found out that the maximum overshoot, rise time, and settling time are all at a minimum when $k_2 = 75$. The two roots are $s = -50$ and -50. The attributes of the unit-step response are:

$$\text{maximum overshoot} = 0\% \qquad t_r = 0.06717 \text{ sec} \qquad t_s = 0.09467 \text{ sec}$$

The state-feedback gain matrix is

$$\mathbf{K} = [2500 \quad 75] \tag{10-244}$$

▲ State-feedback control produces a type-0 system.

The lesson that we learned from this illustrative example is that state-feedback control generally produces a system that is type 0. For the system to track a step input without steady-state error, which requires a type 1 or higher-type system, the feedback gain k_1 of the system in the CCF state diagram cannot be assigned arbitrarily. This means that for an nth-order system, only $n - 1$ roots of the characteristic equation can be placed arbitrarily. ▲

10-14 State Feedback with Integral Control

The state-feedback control structured in Section 10-13 has one deficiency in that it does not improve the type of the system. As a result, the state-feedback control with constant-gain feedback is generally useful only for regulator systems for which the system does not track inputs if all the roots of the characteristic equation are to be placed at will.

In general, most control systems must track inputs. One solution to this problem is to introduce integral control, just as with PI controller, together with the constant-gain state feedback. The block diagram of a system with constant-gain state feedback and integral control feedback of the output is shown in Fig. 10-79. The system is also

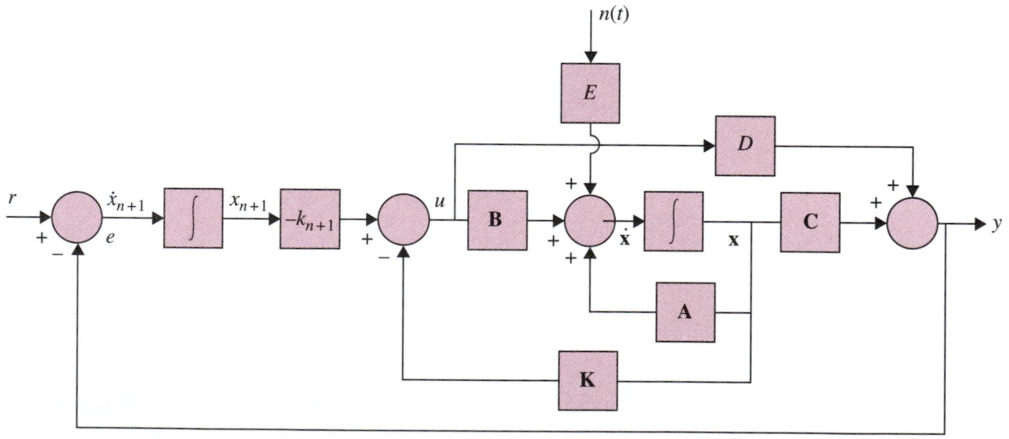

Figure 10-79 Block diagram of control system with state feedback and integral output feedback.

subject to a noise input $n(t)$. For a SISO system, the integral control adds one integrator to the system. As shown in Fig. 10-79, the output of the $(n+1)$st integrator is designated as x_{n+1}. The dynamic equations of the system in Fig. 10-79 are written as

$$\frac{d\mathbf{x}(t)}{dt} = \mathbf{A}\mathbf{x}(t) + \mathbf{B}u(t) + \mathbf{E}n(t) \tag{10-245}$$

$$\frac{dx_{n+1}(t)}{dt} = r(t) - y(t) \tag{10-246}$$

$$y(t) = \mathbf{C}\mathbf{x}(t) + Du(t) \tag{10-247}$$

where $\mathbf{x}(t)$ is the $n \times 1$ state vector; $u(t)$ and $y(t)$ are the scalar actuating signal and output, respectively; $r(t)$ is the scalar reference input; and $n(t)$ is the scalar disturbance input. The coefficient matrices are represented by $\mathbf{A}$, $\mathbf{B}$, $\mathbf{C}$, D, and $\mathbf{E}$, with appropriate dimensions. The actuating signal $u(t)$ is related to the state variables through constant-state and integral feedback,

$$u(t) = -\mathbf{K}\mathbf{x}(t) - k_{n+1}x_{n+1}(t) \tag{10-248}$$

where

$$\mathbf{K} = [k_1 \quad k_2 \quad k_3 \quad \cdots \quad k_n] \tag{10-249}$$

with constant real gain elements, and k_{n+1} is the scalar integral-feedback gain.

Substituting Eq. (10-248) into Eq. (10-245), and combining with Eq. (10-246), the $n+1$ state equations of the overall system with constant-gain and integral feedback are written

$$\frac{d\overline{\mathbf{x}}(t)}{dt} = (\overline{\mathbf{A}} - \overline{\mathbf{B}}\overline{\mathbf{K}})\overline{\mathbf{x}}(t) + \begin{bmatrix} \mathbf{0} \\ 1 \end{bmatrix} r(t) + \overline{\mathbf{E}}n(t) \tag{10-250}$$

where

$$\overline{\mathbf{x}}(t) = \begin{bmatrix} \dfrac{d\mathbf{x}(t)}{dt} \\ \dfrac{dx_{n+1}(t)}{dt} \end{bmatrix} \quad [(n+1) \times 1] \tag{10-251}$$

$$\overline{\mathbf{A}} = \begin{bmatrix} \mathbf{A} & \mathbf{0} \\ -\mathbf{C} & 0 \end{bmatrix} \quad [(n+1) \times (n+1)] \qquad \overline{\mathbf{B}} = \begin{bmatrix} \mathbf{B} \\ D \end{bmatrix} \quad [(n+1) \times 1] \tag{10-252}$$

$$\overline{\mathbf{K}} = [\mathbf{K} \quad k_{n+1}] = [k_1 \quad k_2 \quad \cdots \quad k_n \quad k_{n+1}] \quad [1 \times (n+1)] \tag{10-253}$$

$$\overline{\mathbf{E}} = \begin{bmatrix} \mathbf{E} \\ 0 \end{bmatrix} \quad [(n+1) \times 1] \tag{10-254}$$

Substituting Eq. (10-248) into Eq. (10-247), the output equation of the overall system is written

$$y(t) = \overline{\mathbf{C}}\overline{\mathbf{x}}(t) \tag{10-255}$$

where

$$\overline{\mathbf{C}} = [\mathbf{C} - D\mathbf{K} \quad D\mathbf{K}] \quad [1 + (n+1)] \tag{10-256}$$

The design objectives are:

1. The steady-state value of the output $y(t)$ follows a step-function input with zero error; that is,

$$e_{ss} = \lim_{t \to \infty} e(t) = 0 \tag{10-257}$$

2. The $n + 1$ eigenvalues of $(\overline{\mathbf{A}} - \overline{\mathbf{B}}\overline{\mathbf{K}})$ are placed at desirable locations.

For the last condition to be possible, the pair $[\overline{\mathbf{A}}, \overline{\mathbf{B}}]$ must be completely controllable.

The following example illustrates the applications of state feedback with integral control.

Example 10-20

We have shown in Example 10-19 that with constant-gain-state-feedback control, the second-order sun-seeker system can have only one of its two roots placed at will for the system to track a step input without steady-state error. Now let us consider the same second-order sun-seeker system in Example 10-19 except that an integral control is added to the forward path. The state diagram of the overall system is shown in Fig. 10-80. The coefficient matrices are

$$\mathbf{A} = \begin{bmatrix} 0 & 1 \\ 0 & 25 \end{bmatrix} \quad \mathbf{B} = \begin{bmatrix} 0 \\ 1 \end{bmatrix} \quad \mathbf{C} = [2500 \quad 0] \quad D = 0 \tag{10-258}$$

From Eq. (10-252),

$$\overline{\mathbf{A}} = \begin{bmatrix} \mathbf{A} & \mathbf{0} \\ -\mathbf{C} & 0 \end{bmatrix} = \begin{bmatrix} 0 & 1 & 0 \\ 0 & -25 & 0 \\ -2500 & 0 & 0 \end{bmatrix} \quad \overline{\mathbf{B}} = \begin{bmatrix} \mathbf{B} \\ D \end{bmatrix} = \begin{bmatrix} 0 \\ 1 \\ 0 \end{bmatrix} \tag{10-259}$$

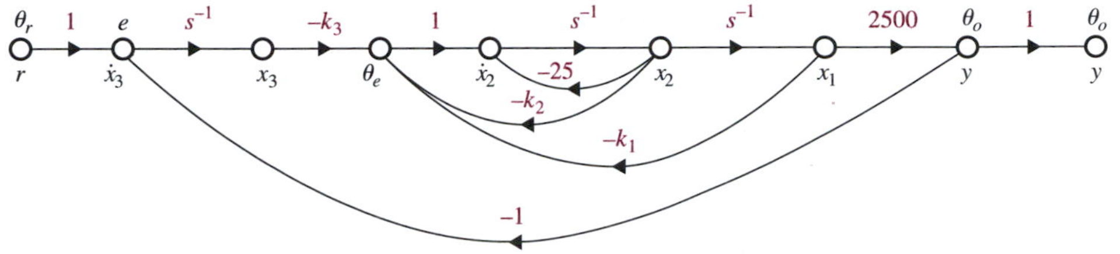

Figure 10-80 Sun-seeker system with state feedback and integral control in Example 10-20.

We can show that the pair $[\overline{\mathbf{A}}, \overline{\mathbf{B}}]$ is completely controllable. Thus the eigenvalues of $(s\mathbf{I} - \overline{\mathbf{A}} + \overline{\mathbf{B}}\overline{\mathbf{K}})$ can be placed arbitrarily. Substitution of $\overline{\mathbf{A}}$, $\overline{\mathbf{B}}$, and $\overline{\mathbf{K}}$ in the characteristic equation of the closed-loop system with state and integral feedback, we have

$$|s\mathbf{I} - \overline{\mathbf{A}} + \overline{\mathbf{B}\mathbf{K}}| = \begin{vmatrix} s & -1 & 0 \\ k_1 & s + 25 + k_2 & k_3 \\ -2500 & 0 & s \end{vmatrix} \tag{10-260}$$
$$= s^3 + (25 + k_2)s^2 + k_1 s + 2500k_3 = 0$$

which can also be found from Fig. 10-80 using the SFG gain formula.

The design objectives are:

1. The steady-state output must follow a step function input with zero error.
2. The rise time and settling time must be less than 0.05 sec.
3. The maximum overshoot of the unit-step response must be less than 5 percent.

Since all three roots of the characteristic equation can be placed arbitrarily, it is not realistic to require minimum rise and settling times, as in Example 10-19.

Again, to realize fast rise and settling times, the roots of the characteristic equation should be placed far to the left in the s-plane, and the natural frequency should be high. Keep in mind that *roots with large magnitudes will lead to high gains for the state-feedback matrix.*

The function **svdesign** of **CSAD/MATLAB** can be used to carry out the design efficiently. After a few trial-and-error runs, the design specifications can be satisfied by placing the roots at

$$s = -200 \qquad -50 + j50 \quad \text{and} \quad -50 - j50$$

The desired characteristic equation is

$$s^3 + 300s^2 + 25{,}000s + 1{,}000{,}000 = 0 \tag{10-261}$$

Equating like coefficients of Eq. (10-260) and (10-261), we get

$$k_1 = 25{,}000 \qquad k_2 = 275 \quad \text{and} \quad k_3 = 400$$

The attributes of the unit-step response are:

$$\text{maximum overshoot} = 4\% \qquad t_r = 0.03247 \text{ sec} \qquad t_s = 0.04667 \text{ sec}$$

Notice that the high feedback gain of k_1, which is due to the large values of the roots selected, may pose physical problems; if so, the design specifications may have to be revised. ▲

Example 10-21

In this example we illustrate the application of state feedback with integral control to a system with a disturbance input.

Consider a dc-motor control system that is described by the following state equations:

$$\frac{d\omega(t)}{dt} = \frac{-B}{J}\omega(t) + \frac{K_i}{J}i_a(t) - \frac{1}{J}T_L u_s(t) \tag{10-262}$$

$$\frac{di_a(t)}{dt} = \frac{-K_b}{L}\omega(t) - \frac{R}{L}i_a(t) + \frac{1}{L}e_a(t) \tag{10-263}$$

where

$i_a(t)$ = armature current, A
$e_a(t)$ = armature applied voltage, V
$\omega(t)$ = motor velocity, rad/sec
B = viscous-friction coefficient of motor and load = 0
J = moment of inertia of motor and load = 0.02 N-m/rad/sec^2
K_i = motor torque constant = 1 N-m/A
K_b = motor back-emf constant = 1 V/rad/sec
T_L = constant load torque (magnitude not known), N-m
L = armature inductance = 0.005 H
R = armature resistance = 1 Ω

The output equation is

$$y(t) = \mathbf{C}\mathbf{x}(t) = [1 \quad 0]\mathbf{x}(t) \tag{10-264}$$

The design problem is to find the control $u(t) = e_a(t)$ through state feedback and integral control such that

1. $\lim_{t\to\infty} i_a(t) = 0$ and $\lim_{t\to\infty} \frac{d\omega(t)}{dt} = 0$ (10-265)
2. $\lim_{t\to\infty} \omega(t) = \text{step input } r(t) = u_s(t)$
3. The eigenvalues of the closed-loop system with state feedback and integral control are at $s = -300$, $-10 + j10$, and $-10 - j10$.

Let the state variables be defined as $x_1(t) = \omega(t)$ and $x_2(t) = i_a(t)$. The state equations in Eqs. (10-262) and (10-263) are written in vector-matrix form.

$$\frac{d\mathbf{x}(t)}{dt} = \mathbf{A}\mathbf{x}(t) + \mathbf{B}u(t) + \mathbf{E}n(t) \tag{10-267}$$

where $n(t) = T_L u_s(t)$.

$$\mathbf{A} = \begin{bmatrix} -\dfrac{B}{J} & \dfrac{K_i}{J} \\ -\dfrac{K_b}{L} & -\dfrac{R}{L} \end{bmatrix} = \begin{bmatrix} 0 & 50 \\ -200 & -200 \end{bmatrix} \quad (10\text{-}268)$$

$$\mathbf{B} = \begin{bmatrix} 0 \\ \dfrac{1}{L} \end{bmatrix} = \begin{bmatrix} 0 \\ 200 \end{bmatrix} \quad (10\text{-}269)$$

$$\mathbf{E} = \begin{bmatrix} -\dfrac{1}{J} \\ 0 \end{bmatrix} = \begin{bmatrix} -50 \\ 0 \end{bmatrix} \quad (10\text{-}270)$$

From Eq. (10-252),

$$\overline{\mathbf{A}} = \begin{bmatrix} \mathbf{A} & \mathbf{0} \\ -\mathbf{C} & 0 \end{bmatrix} = \begin{bmatrix} 0 & 50 & 0 \\ -200 & -200 & 0 \\ -1 & 0 & 0 \end{bmatrix} \qquad \overline{\mathbf{B}} = \begin{bmatrix} \mathbf{B} \\ D \end{bmatrix} = \begin{bmatrix} 0 \\ 200 \\ 0 \end{bmatrix} \quad (10\text{-}271)$$

$$\overline{\mathbf{C}} = [\mathbf{C} - D\mathbf{K} \quad D\mathbf{K}] = [1 \quad 0 \quad 0] \qquad \overline{\mathbf{E}} = \begin{bmatrix} \mathbf{E} \\ 0 \end{bmatrix} = \begin{bmatrix} -50 \\ 0 \\ 0 \end{bmatrix} \quad (10\text{-}272)$$

The control is given by

$$u(t) = -\mathbf{K}\mathbf{x}(t) - k_{n+1}x_{n+1}(t) = \overline{\mathbf{K}\mathbf{x}}(t) \quad (10\text{-}273)$$

where

$$\overline{\mathbf{K}} = [k_1 \quad k_2 \quad k_3] \quad (10\text{-}274)$$

Figure 10-81 shows the state diagram of the overall designed system. The coefficient matrix of the closed-loop system is

$$\overline{\mathbf{A}} - \overline{\mathbf{B}\mathbf{K}} = \begin{bmatrix} 0 & 50 & 0 \\ -200 - 200k_1 & -200 - 200k_2 & -200k_3 \\ -1 & 0 & 0 \end{bmatrix} \quad (10\text{-}275)$$

The characteristic equation is

$$|s\mathbf{I} - \overline{\mathbf{A}} + \overline{\mathbf{B}\mathbf{K}}| = s^3 + 200(1 + k_2)s^2 + 10{,}000(1 + k_1)s - 10{,}000k_3 = 0 \quad (10\text{-}276)$$

which is more easily determined by applying the gain formula of SFG to Fig. 10-81.

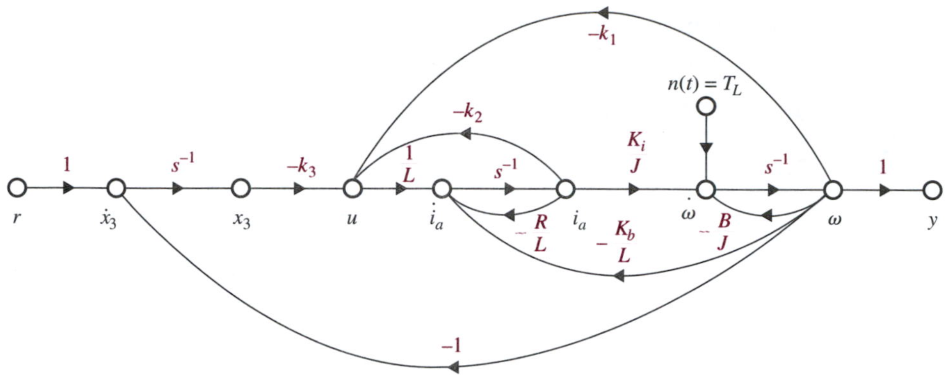

Figure 10-81 Dc motor control system with state feedback and integral control and disturbance torque in Example 10-21.

For the three roots assigned, the last equation must equal

$$s^3 + 320s^2 + 6200s + 60{,}000 = 0 \tag{10-277}$$

Equating the like coefficients of Eqs. (10-276) and (10-277), we get

$$k_1 = -0.38 \qquad k_2 = 0.6 \qquad k_3 = -6.0$$

Applying the SFG gain formula to Fig. 10-81 between the inputs $r(t)$ and $n(t)$ and the states $\omega(t)$ and $i_a(t)$, we have

$$\begin{bmatrix} \Omega(s) \\ I_a(s) \end{bmatrix} = \frac{1}{\Delta_c(s)} \begin{bmatrix} -\frac{1}{J}\left(s^2 + \frac{R}{L}s + \frac{k_2}{L}s\right) & -\frac{k_3 K_i}{JL} \\ -\frac{1}{J}\left(-\frac{K_b}{L}s - \frac{k_1}{L}s + \frac{k_3}{L}\right) & -\frac{k_3}{L}\left(s + \frac{B}{J}\right) \end{bmatrix} \begin{bmatrix} \frac{T_L}{s} \\ \frac{1}{s} \end{bmatrix} \tag{10-278}$$

where $\Delta_c(s)$ is the characteristic polynomial of the closed-loop system.

Applying the final-value theorem to the last equation, the steady-state values of the state variables are found to be

$$\lim_{t\to\infty} \begin{bmatrix} \omega(t) \\ i_a(t) \end{bmatrix} = \lim_{s\to 0} s \begin{bmatrix} \Omega(s) \\ I_a(s) \end{bmatrix} = \begin{bmatrix} 0 & K_i \\ 1 & B \end{bmatrix} \begin{bmatrix} T_L \\ 1 \end{bmatrix} = \begin{bmatrix} 1 \\ T_L \end{bmatrix} \tag{10-279}$$

Thus the motor velocity $\omega(t)$ will approach the constant reference input step function $r(t) = u_s(t)$ as t approaches infinity, independent of the disturbance torque T_L. Substituting the system

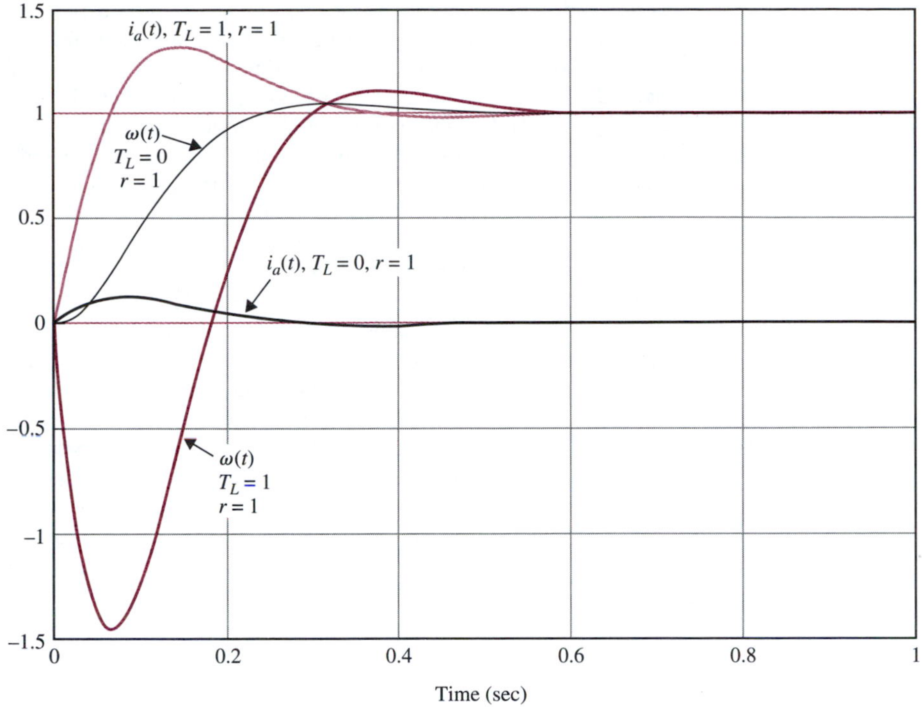

Figure 10-82 Time responses of dc motor control system with state feedback and integral control and disturbance torque in Example 10-21.

parameters into Eq. (10-278), we get

$$\begin{bmatrix} \Omega(s) \\ I_a(s) \end{bmatrix} = \frac{1}{\Delta_c(s)} \begin{bmatrix} -50(s+320)s & 60{,}000 \\ 6200s + 60{,}000 & 1200s \end{bmatrix} \begin{bmatrix} \dfrac{T_L}{s} \\ \dfrac{1}{s} \end{bmatrix} \tag{10-280}$$

where $\Delta_c(s)$ is the left-hand side of Eq. (10-277). Figure 10-82 shows the time responses of $\omega(t)$ and $i_a(t)$ when $T_L = 1$ and $T_L = 0$. The reference input is a unit-step function. ▲

10-15 Summary

This chapter is devoted to the design of linear control systems. The chapter begins by giving some fundamental considerations on system design. The specifications in the time and frequency domains are reviewed. Fixed configurations of compensation schemes used in practice, such as series, forward and feedforward, feedback and minor loop, and state feedback, are illustrated. The types of controllers considered are the PD, PI, PID, phase-lead, phase-lag, lead–lag, pole–zero cancellation, and notch filters. Designs are carried out in the time domain/s-domain as well as the frequency domain. The time-domain design is characterized by specifications such as the relative damping ratio, maximum overshoot, rise time, delay time, settling time, or simply the location of the characteristic-equation roots, keeping in mind that the zeros of the system transfer function also affect the transient response. The performance is generally measured by the step response and the steady-state error. Frequency-domain designs are generally carried out using Bode diagrams, or gain–phase plots. The performance specifications in the frequency domain are the phase margin, gain margin, M_r, BW, and so on.

The effect of feedforward control on noise and disturbance reduction is demonstrated. A section is devoted to the design of robust control systems. State feedback through constant-feedback gains and with dynamic feedback are discussed. The state-feedback design is more versatile than the conventional fixed-configuration controller design, since the characteristic-equation roots are directly controlled. An unstable system that is controllable can always be stabilized by state-feedback control. The disadvantage with the state feedback is that all the states must be sensed and fed back for control, which may not be practical.

Although the design techniques covered in this chapter are outlined with analytical procedures, the material heavily promotes the use of computers. The availability of user-friendly computer programs has totally changed the complexity of control systems design.

Review Questions

1. What is a PD controller? Write its input–output transfer function.
2. A PD controller has the constants K_D and K_P. Give the effects of these constants on the steady-state error of the system. Does the PD control change the type of a system?
3. Give the effects of the PD control on the rise time and settling time of a control system.
4. How does the PD controller affect the bandwidth of a control system?
5. Once the value of K_D of a PD controller is fixed, increasing the value of K_P will increase the phase margin monotonically. **(T) (F)**
6. If a PD controller is so designed that the characteristic-equation roots have better damping than the original system, the maximum overshoot of the system is always reduced. **(T) (F)**
7. What does it mean when a control system is described as being robust?
8. A system compensated with a PD controller is usually more robust than a system compensated with a PI controller. **(T) (F)**
9. What is a PI controller? Write its input–output transfer function.
10. A PI controller has the constants K_P and K_I. Give the effects of the PI controller on the steady-state error of the system. Does the PI control change the system type?
11. Give the effects of the PI control on the rise and settling times of a control system.
12. How does the PI controller affect the bandwidth of a control system?
13. What is a PID controller? Write its input–output transfer function.

14. Give the limitations of the phase-lead controller.
15. How does the phase-lead controller affect the bandwidth of a control system?
16. Give the general effects of the phase-lead controller on rise and settling times.
17. For the phase-lead controller, $G_c(s) = (1 + aTs)/(1 + Ts)$, $a > 1$, what is the effect of the controller on the steady-state performance of the system?
18. The phase-lead controller is generally less effective if the uncompensated system is very unstable to begin with. **(T) (F)**
19. The maximum phase that is available from a single-stage phase-lead controller is 90°. **(T) (F)**
20. The design objective of the phase-lead controller is to place the maximum phase lead at the new gain-crossover frequency. **(T) (F)**
21. The design objective of the phase-lead controller is to place the maximum phase lead at the frequency where the magnitude of the uncompensated $G_p(j\omega)$ is $-10 \log_{10} a$, where a is the gain of the phase-lead controller. **(T) (F)**
22. The phase-lead controller may not be effective if the negative slope of the uncompensated process transfer function is too steep near the gain-crossover frequency. **(T) (F)**
23. For the phase-lag controller, $G_c(s) = (1 + aTs)/(1 + Ts)$, $a < 1$, what is the effect of the controller on the steady-state performance of the system?
24. Give the general effects of the phase-lag controller on rise time and settling time.
25. How does the phase-lag controller affect the bandwidth?
26. For a phase-lag controller, if the value of T is large and the value of a is small, it is equivalent to adding a pure attenuation of a to the original uncompensated system at low frequencies. **(T) (F)**
27. The principle of design of the phase-lag controller is to utilize the zero-frequency attenuation property of the controller. **(T) (F)**
28. The corner frequencies of the phase-lag controller should not be too low or the bandwidth of the system will be too low. **(T) (F)**
29. Give the limitations of the pole–zero-cancellation control scheme.
30. How does the sensitivity function relate to the bandwidth of a system?

Answers to the true-and-false questions are given following the Problems section.

References

1. **J. C. WILLEMS and S. K. MITTER,** "Controllability, Observability, Pole Allocation, and State Reconstruction," *IEEE Trans. Automatic Control,* Vol. AC-16, pp. 582–595, Dec. 1971.
2. **H. W. SMITH and E. J. DAVISON,** "Design of Industrial Regulators," *Proc. IEE (London)*, Vol. 119, pp. 1210–1216, Aug. 1972.
3. **F. N. BAILEY and S. MESHKAT,** "Root Locus Design of a Robust Speed Control," *Proc. Incremental Motion Control Symposium,* pp. 49–54, June 1983.

Problems

Most of the following problems can be solved using a computer program. This is highly recommended if the reader has access to such a program.

▲ Design of series controller to satisfy steady-state-error requirement

10-1. The block diagram of a control system with a series controller is shown in Fig. 10P-1. Find the transfer function of the controller $G_c(s)$ so that the following specifications are satisfied:

The ramp-error constant K_v is 5.
The closed-loop transfer function is of the form

$$M(s) = \frac{Yh(s)}{R(s)} = \frac{K}{(s^2 + 20s + 200)(s + a)}$$

where K and a are real constants. Find the values of K and a.

The design strategy is to place the closed-loop poles at $-10 + j10$ and $-10 - j10$, and then adjust the values of K and a to satisfy the steady-state requirement. Thus the closed-loop transfer is function selected. The value of a is large, so that it will not affect the transient response appreciably. Find the maximum overshoot of the designed system.

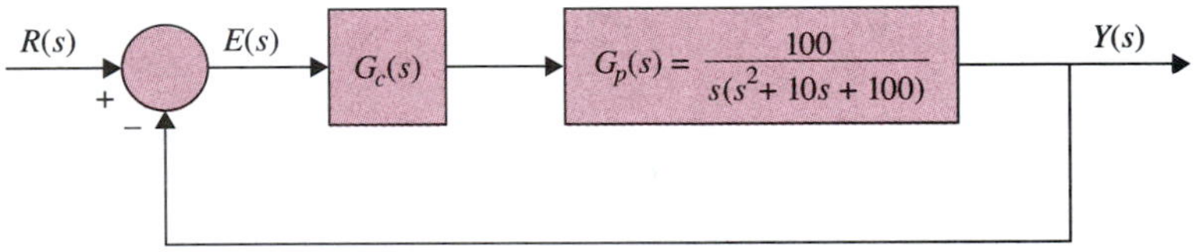

Figure 10P-1

▲ Design of series controller to satisfy steady-state-error requirement

10-2. Repeat Problem 10-1 if the ramp-error constant is to be 9. What is the maximum value of K_v that can be realized? Comment on the difficulties that may arise in attempting to realize a very large K_v.

▲ Design of system with PD controller

10-3. A control system with a PD controller is shown in Fig. 10P-3.

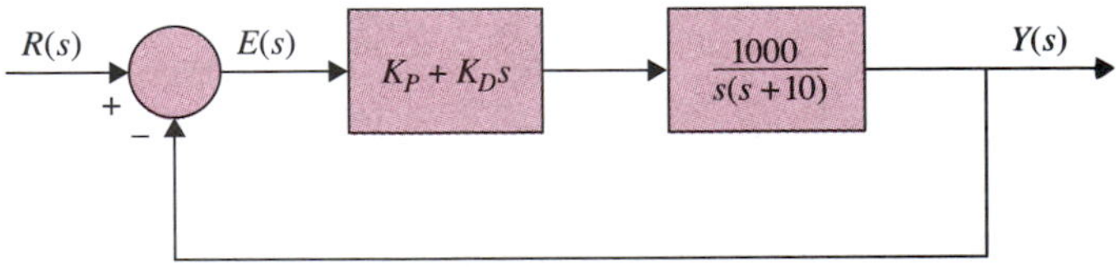

Figure 10P-3

(a) Find the values of K_P and K_D so that the ramp-error constant K_v is 1000 and the damping ratio is 0.5.

(b) Find the values of K_P and K_D so that the ramp-error constant K_v is 1000 and the damping ratio is 0.707.

(c) Find the values of K_P and K_D so that the ramp-error constant K_v is 1000 and the damping ratio is 1.0.

▲ Design of system with PD controller

10-4. For the control system shown in Fig. 10P-3, set the value of K_P so that the ramp-error constant is 10,000. Vary the value of K_D from 0.2 to 1.0 in increments of 0.2, and determine the values of phase margin, gain margin, M_r, BW, and the maximum overshoot of the unit-step response of the system.

▲ Design of aircraft attitude control system with PD controller

10-5. Consider the second-order model of the aircraft attitude control system shown in Fig. 7-24. The transfer function of the process is

$$G_p(s) = \frac{4500K}{s(s + 361.2)}$$

(a) Design a series of PD controller with the transfer function $G_c(s) = K_D + K_P s$ so that the following performance specifications are satisfied:

Steady-state error due to a unit-ramp input ≤ 0.001
Maximum overshoot ≤ 5 percent
Rise time $t_r \le 0.005$ sec
Settling time $t_s \le 0.005$ sec

(b) Repeat part (a) for all the specifications listed, and in addition, the bandwidth of the system must be less than 850 rad/sec.

▲ Design of liquid-level control system with PD controller

10-6. Figure 10P-6 shows the block diagram of the liquid-level control system described in Problems 6-13 and 7-29. The number of inlets is denoted by N. Set $N = 20$. Design the PD controller so that with a unit-step input, the tank is filled to within 5 percent of the reference level in less than 3 sec without overshoot.

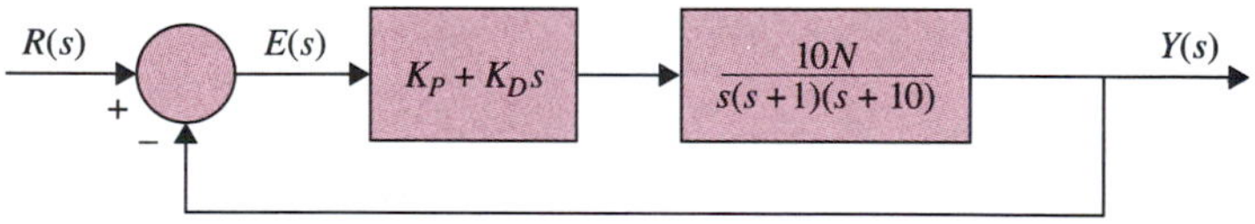

Figure 10P-6

▲ Design of liquid-level control system with PD controller

10-7. For the liquid-level control system described in Problem 10-6,

(a) Set K_P so that the ramp-error constant is 1. Vary K_D from 0 to 0.5 and find the value K_D that gives the maximum phase margin. Record the gain margin, M_r, and BW.

(b) Plot the sensitivity functions $|S_G^M(j\omega)|$ of the uncompensated system and the compensated system with the values of K_D and K_P determined in part (a). How does the PD controller affect the sensitivity?

▲ Design of a system with PI controller

10-8. A control system with a type-0 process $G_p(s)$ and a PI controller is shown in Fig. 10P-8.

(a) Find the value of K_I so that the ramp-error constant K_v is 10.

(b) Find the value of K_P so that the magnitude of the imaginary parts of the complex roots of the characteristic equation of the system is 15 rad/sec. Find the roots of the characteristic equation.

(c) Sketch the root contours of the characteristic equation with the value of K_I as determined in part (a) and for $0 \le K_P < \infty$.

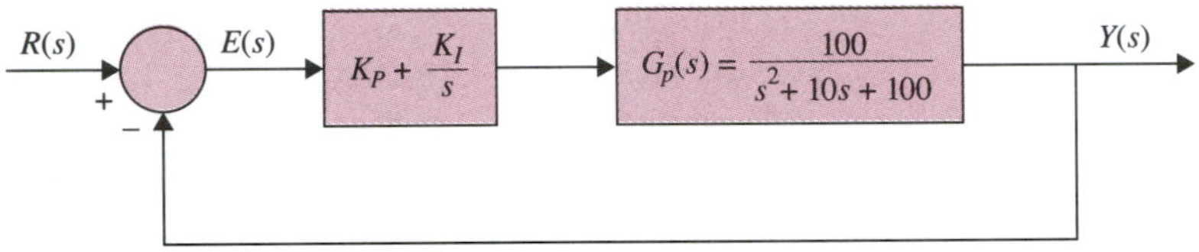

Figure 10P-8

10-9. For the control system described in Problem 10-8:

▲ Design of a system with PI controller

(a) Set K_I so that the ramp-error constant is 10. Find the value of K_P so that the phase margin is maximum. Record the values of the phase margin, gain margin, M_r, and BW. How does this optimal K_P relate to the root contours constructed in Problem 10-8(c)?

(b) Plot the sensitivity functions $|S_G^M(j\omega)|$ of the uncompensated and the compensated systems with the values of K_I and K_P selected in part (a). Comment on the effect of the PI control on sensitivity.

▲ Design of system with PI controller

10-10. For the control system shown in Fig. 10P-8:

(a) Find the value of K_I so that the ramp-error constant K_v is 100.

(b) With the value of K_I found in part (a), find the critical value of K_P so that the system is stable. Sketch the root contours of the characteristic equation for $0 \le K_P < \infty$.

(c) Show that the maximum overshoot is high for both large and small values of K_P. Use the value of K_I found in part (a). Find the value of K_P when the maximum overshoot is a minimum. What is the value of this maximum overshoot?

▲ Design of system with PI controller

10-11. Repeat Problem 10-10 for $K_v = 10$.

▲ Design of system with PID controller

10-12. A control system with a type-0 process and a PID controller is shown in Fig. 10P-12. Design the controller parameters so that the following specifications are satisfied:

Ramp-error constant $K_v = 100$ Maximum overshoot < 2 percent
Rise time $t_r < 0.01$ sec

Plot the unit-step response of the designed system.

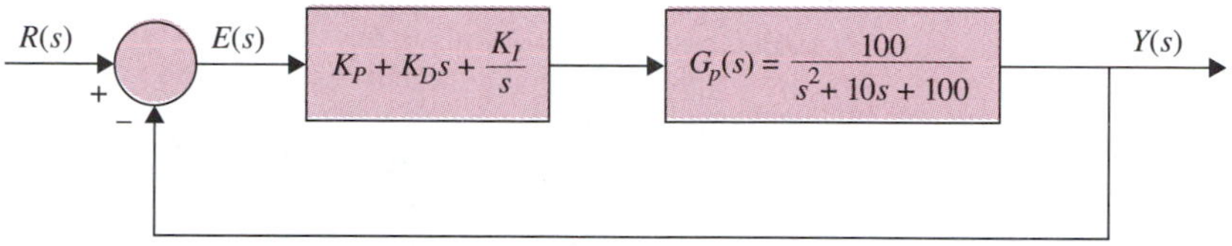

Figure 10P-12

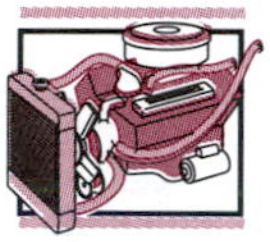

▲ Design of auto engine control

10-13. A considerable amount of effort is being spent by automobile manufacturers to meet the exhaust-emission-performance standards set by the government. Modern automotive-power-plant systems consist of an internal combustion engine that has an internal cleanup device called the catalytic converter. Such a system requires control of the engine air/fuel ratio (A/F), the ignition-spark timing, exhaust-gas recirculation, and injection air. The control system problem considered in this exercise deals with the control of the air/fuel ratio. In general, depending on fuel composition and other factors, a typical stoichiometric A/F is 14.7 : 1, that is, 14.7 grams of air to each gram of fuel. An A/F value greater or less than stoichiometry will cause high hydrocarbons, carbon monoxide, and nitrous oxides in the tailpipe emission. The control system whose block diagram is shown in Fig. 10P-13 is devised to control the air/fuel ratio so that a desired output variable is maintained for a given command signal. Figure 10P-13 shows that the sensor senses the composition of the exhaust-gas mixture entering the catalytic converter. The electronic controller detects the difference or the error between the command and the sensor signals and computes the control signal necessary to achieve the desired exhaust-gas composition. The output variable $y(t)$ denotes the effective air/fuel ratio. The transfer function of the engine is given by

$$\frac{Y(s)}{U(s)} = G_p(s) = \frac{e^{-T_d s}}{1 + \tau s}$$

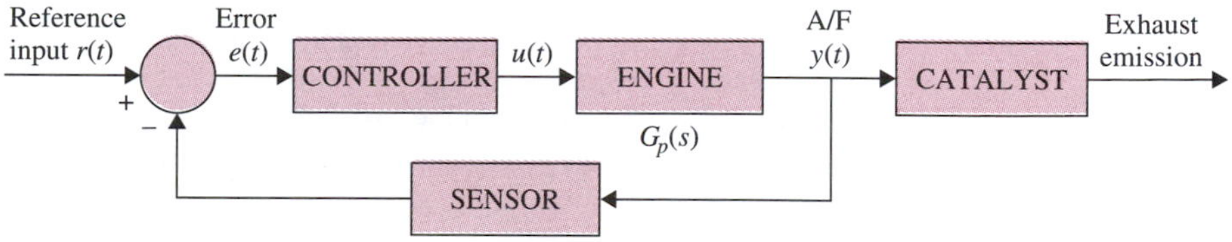

Figure 10P-13

where T_d is the time delay and is 0.2 sec. The time constant τ is 0.25 sec. Approximate the time delay by a power series:

$$e^{-T_d s} \cong \frac{1}{1 + T_d s + T_d^2 s^2/2}$$

(a) Let the controller be a PI controller so that

$$G_c(s) = \frac{U(s)}{E(s)} = K_P + \frac{K_I}{s}$$

Find the value of K_I so that the ramp-error constant K_v is 2. Determine the value of K_P so that the maximum overshoot of the unit-step response is a minimum and the settling time is a minimum. Give the values of the maximum overshoot and the settling time. Plot the unit-step response of $y(t)$. Find the marginal value of K_P for system stability.

(b) Can the system performance be further improved by using a PID controller?

$$G_c(s) = \frac{U(s)}{E(s)} = K_P + K_D s + \frac{K_I}{s}$$

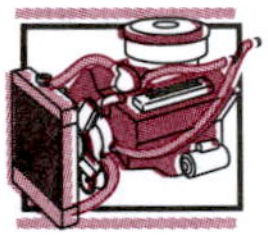

▲ Design of auto engine control system, frequency domain

10-14. One of the advantages of the frequency-domain analysis and design methods is that systems with pure time delays can be treated without approximation. Consider the automobile-engine-control system treated in Problem 10-13. The process has the transfer function

$$G_p(s) = \frac{e^{-0.2s}}{1 + 0.25s}$$

Let the controller be of the PI type so that $G_c(s) = K_P + K_I/s$. Set the value of K_I so that the ramp-error constant K_v is 2. Find the value of K_P so that the phase margin is a maximum. How does this ''optimal'' K_P compare with the value of K_P found in Problem 10-13(a)? Find the critical value of K_P for system stability. How does this value of K_P compare with the critical value of K_P found in Problem 10-13?

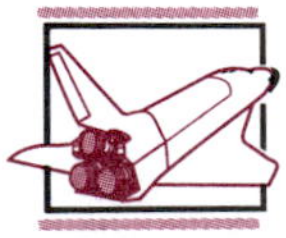

▲ Design of space telescope control system

10-15. The telescope for tracking stars and asteroids on the space shuttle may be modeled as a pure mass M. It is suspended by magnetic bearings so that there is no friction, and its attitude is controlled by magnetic actuators located at the base of the payload. The dynamic model for the control of the z-axis motion is shown in Fig. 10P-15(a). Since there are electrical components on the telescope, electric power must be brought to the telescope through a cable. The spring shown is used to model the wire-cable attachment, which exerts a spring force on the mass. The force produced by the magnetic actuators is denoted by $f(t)$. The force equation of motion in the z direction is

$$f(t) - K_s z(t) = M\frac{d^2 z(t)}{dt^2}$$

where $K_s = 1$ lb/ft, $M = 150$ lb(mass), $f(t)$ is in pounds, and $z(t)$ is measured in feet.

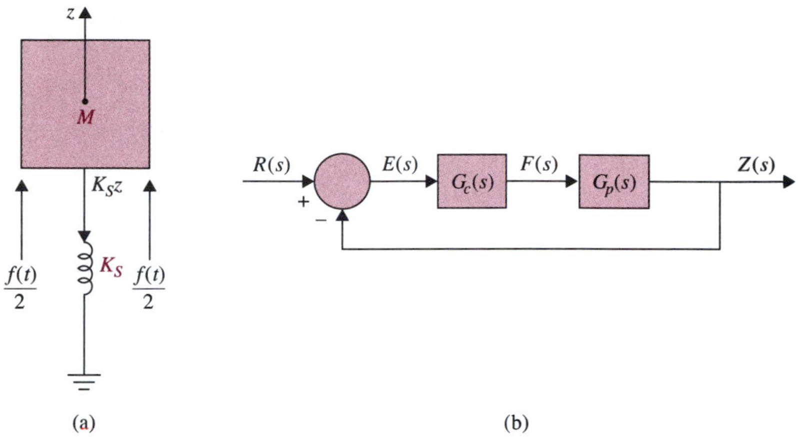

Figure 10P-15

(a) Show that the natural response of the system output $z(t)$ is oscillatory without damping. Find the natural undamped frequency of the open-loop space-shuttle system.

(b) Design the PID controller

$$G_c(s) = K_P + K_D s + \frac{K_T}{s}$$

shown in Fig. 10-15(b) so that the following performance specifications are satisfied:

Ramp-error constant $K_v = 100$
The complex characteristic equation roots correspond to a relative damping ratio of 0.707 and a natural undamped frequency of 1 rad/sec

Compute and plot the unit-step response of the designed system. Find the maximum overshoot. Comment on the design results.

(c) Design the PID controller so that the following specifications are satisfied:

Ramp-error constant $K_v = 100$ Maximum overshoot <5 percent

Compute and plot the unit-step response of the designed system. Find the roots of the characteristic equation of the designed system.

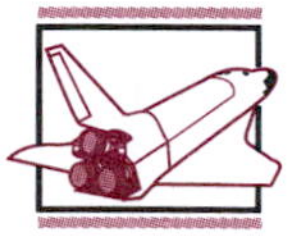

▲ Design of space telescope control system

10-16. Repeat Problem 10-15(b) with the following specifications:

Ramp-error constant $K_v = 100$
The complex characteristic equation roots correspond to a relative damping ratio of 1.0 and a natural undamped frequency of 1 rad/sec

▲ Design of inventory control system

10-17. An inventory control system is modeled by the following state equations:

$$\frac{dx_1(t)}{dt} = -2x_2(t)$$

$$\frac{dx_2(t)}{dt} = -2u(t)$$

where $x_1(t)$ is the level of inventory, $x_2(t)$ the rate of sales of product, and $u(t)$ the production rate. The output equation is $y(t) = x_1(t)$. One unit of time is 1 day. Figure 10P-17 shows the block diagram of the closed-loop inventory control system with a series controller. Let the controller be a PD controller, $G_c(s) = K_P + K_D s$.

(a) Find the parameters of the PD controller, K_P and K_D, so that the roots of the characteristic equation correspond to a relative damping ratio of 0.707 and $\omega_n = 1$ rad/sec. Plot the unit-step response of $y(t)$ and find the maximum overshoot.

(b) Find the values of K_P and K_D so that the overshoot is zero and the rise time is less than 0.06 sec.

(c) Design the PD controller so that $M_r = 1$ and BW ≤ 40 rad/sec.

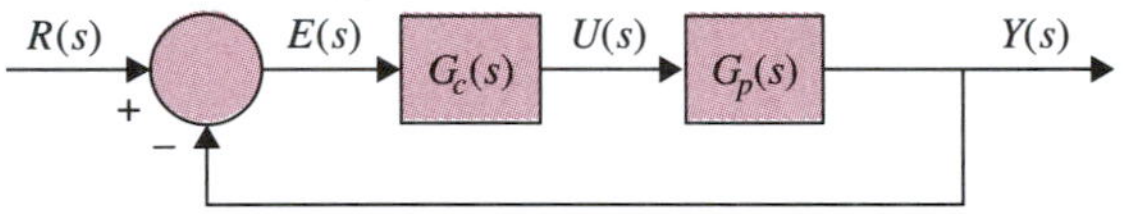

Figure 10P-17

▲ Design of phase-lead controller

10-18. The block diagram of a type 2 control system with a series controller $G_c(s)$ is shown in Fig. 10P-18. The objective is to design a PD controller so that the following specifications are satisfied:

Maximum overshoot <10 percent   Rise time <0.5 sec

(a) Obtain the characteristic equation of the closed-loop system and determine the ranges of the values of K_P and K_D for stability. Show the region of stability in the K_D versus K_P plane.

(b) Construct the root loci of the characteristic equation with $K_D = 0$ and $0 \leq K_P < \infty$. Then construct the root contours for $0 \leq K_D < \infty$ and several fixed values of K_P ranging from 0.001 to 0.01.

(c) Design the PD controller to satisfy the performance specifications given above. Use the information on the root contours to help your design. Plot the unit-step response of $y(t)$.

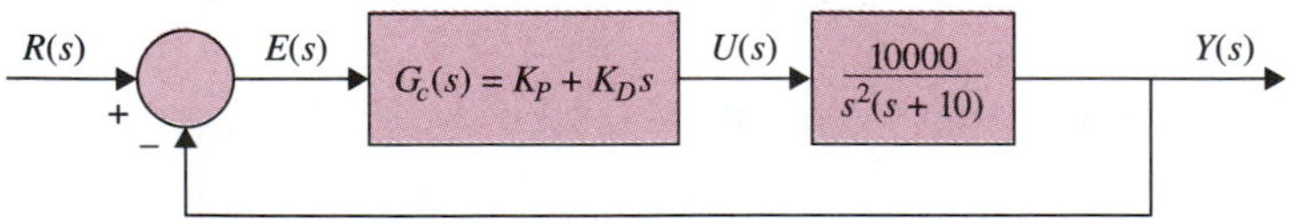

Figure 10P-18

(d) Check the design results obtained in part (c) in the frequency domain. Determine the phase margin, gain margin, M_r, and BW of the designed system.

▲ Design of PD controller for broom-balancing system

10-19. Consider the broom-balancing control system described in Problems 4-23 and 5-41. The **A*** and **B*** matrices are given in Problem 5-41 for the small-signal linearized model.

$$\Delta\dot{\mathbf{x}}(t) = \mathbf{A}^* \,\Delta\mathbf{x}(t) + \mathbf{B}^* \,\Delta r(t)$$
$$\Delta y(t) = \mathbf{C}\,\Delta\mathbf{x}(\mathrm{t})$$
$$\mathbf{D}^* = [0 \quad 0 \quad 1 \quad 0]$$

Figure 10P-19 shows the block diagram of the system with a series PD controller. Determine if the PD controller can stabilize the system; if so, find the values of K_P and K_D. If the PD controller cannot stabilize the system, explain why not.

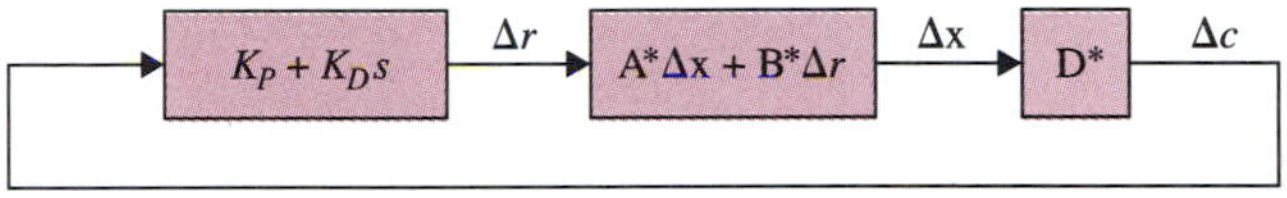

Figure 10P-19

▲ Design of series controller

10-20. The process of a unity-feedback control system has the transfer function

$$G_p(s) = \frac{100}{s^2 + 10s + 100}$$

Design a series controller (PD, PI, or PID) so that the following performance specifications are satisfied:

Steady-state error due to a step input = 0 Rise time < 0.02 sec
Maximum overshoot < 2 percent

Carry out the design in the frequency domain and check the design in the time domain.

▲ Design of phase-lead controller for inventory-control system

10-21. For the inventory control system shown in Fig. 10P-17, let the controller be of the phase-lead type:

$$G_c(s) = \frac{1 + aTs}{1 + Ts} \qquad a > 1$$

Determine the values of a and T so that the following performance specifications are satisfied:

Steady-state error due to a step input = 0 Maximum overshoot <5 percent

(a) Design the controller using the root contours with T and a as variable parameters. Plot the unit-step response of the designed system. Plot the Bode diagram of $G(s) = G_c(s)G_p(s)$, and find the PM, GM, M_r, and BW of the designed system.

(b) Design the phase-lead controller so that the following performance specifications are satisfied:

$$\text{Steady-state error due to a step input} = 0 \qquad M_r < 1.1$$
$$\text{Phase margin} > 75°$$

Construct the Bode diagram of $G(s)$ and carry out the design in the frequency domain. Find the attributes of the time response of the designed system.

▲ Design of phase-lead controller

10-22. Consider that the process of a unity-feedback control system is

$$G_p(s) = \frac{1000}{s(s+10)}$$

Let the series controller be a single-stage phase-lead controller:

$$G_c(s) = \frac{1 + aTs}{1 + Ts} \qquad a > 1$$

(a) Determine the values of a and T so that the zero of $G_c(s)$ cancels the pole of $G_p(s)$ at $s = -10$. The damping ratio of the designed system should be unity. Find the attributes of the unit-step response of the designed system.

(b) Carry out the design in the frequency domain using the Bode plot. The design specifications are:

$$\text{Phase margin} > 75° \qquad M_r < 1.1$$

Find the attributes of the unit-step response of the designed system.

▲ Design of phase-lead controller for liquid-level control system

10-23. Consider that the controller in the liquid-level control system shown in Fig. 10P-6 is a phase-lead controller:

$$G_c(s) = \frac{1 + aTs}{1 + Ts} \qquad a > 1$$

(a) For $N = 10$, select the values of a and T so that the maximum overshoot is minimized. The value of a must not exceed 1000. Find the attributes of the unit-step response of the designed system. Plot the unit-step response.

(b) For $N = 10$, design the phase-lead controller in the frequency domain. Find the values of a and T so that the phase margin is maximized subject to the condition that BW < 100. The value of a must not exceed 1000.

▲ Design of single-stage and two-stage phase-lead controllers

10-24. The transfer function of the process of a unity-feedback control system is

$$G_P(s) = \frac{6}{s(1 + 0.2s)(1 + 0.5s)}$$

(a) Construct the Bode diagram of $G_p(j\omega)$ and determine the PM, GM, M_r, and BW of the system.

(b) Design a series single-stage phase-lead controller with the transfer function

$$G_c(s) = \frac{1 + aTs}{1 + Ts} \qquad a > 1$$

so that the phase margin is maximum. The value of a must not be greater than 1000. Determine PM and M_r of the designed system. Determine the attributes of the unit-step response.

(c) Use the system designed in part (b) as a basis, design a two-stage phase-lead controller so that the phase margin is at least 85°. The transfer function of the two-stage phase-lead controller is

$$G_c(s) = \left(\frac{1 + aT_1 s}{1 + T_1 s}\right)\left(\frac{1 + bT_2 s}{1 + T_2 s}\right) \qquad a > 1, \quad b > 1$$

where a and T_1 are determined in part (b). The value of b should not exceed 1000. Find the values of PM and M_r of the designed system. Find the attributes of the unit-step response.

(d) Plot the unit-step responses of the output in parts (a), (b), and (c).

▲ Design of phase-locked-loop system, phase-lead controller

10-25. A phase-locked-loop dc-motor-speed-control system is described in Problem 4-20. The block diagram of the system is shown in Fig. 10P-25. The system parameters and transfer functions are given as follows:

Reference speed command, f_r = 120 pulses/sec
Phase-detector gain, K_p = 0.06 V/pulse/sec
Amplifier gain, K_a = 20
Encoder gain, K_e = 5.73 pulses/rad

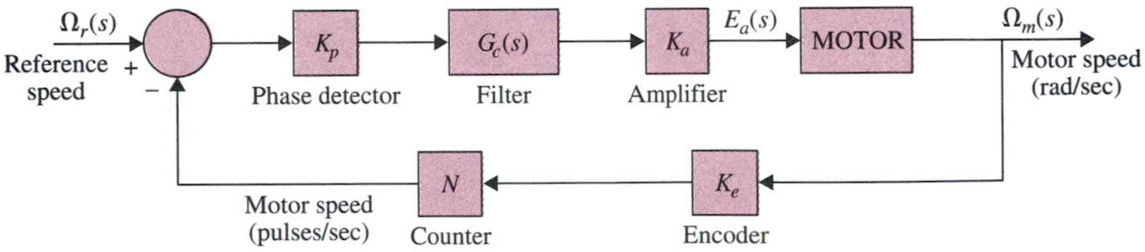

Figure 10P-25

Counter gain, $N = 1$
Motor transfer function,

$$\frac{\Omega_m(s)}{E_a(s)} = \frac{10}{s(1 + 0.05s)}$$

(a) Let the filter (controller) transfer function be of the form

$$G_c(s) = \frac{E_o(s)}{E_i(s)} = \frac{1 + R_2Cs}{R_1Cs}$$

where $R_1 = 2 \times 10^6\ \Omega$ and $C = 1\ \mu$F. Determine the value of R_2 so that the complex roots of the closed-loop characteristic equation have a maximum relative damping ratio. Sketch the root loci of the characteristic equation for $0 \le R_2 < \infty$. Compute and plot the unit-step responses of the motor speed $f_\omega(t)$ (pulses/sec) with the values of R_2 found, when the input is 120 pulses/sec. Convert the speed in pulses/sec to rpm.

(b) Let the filter transfer function be

$$G_c(s) = \frac{1 + aTs}{1 + Ts} \qquad a > 1$$

where $T = 0.01$. Find a so that the complex roots of the characteristic equation have a maximum relative damping ratio. Compute and plot the unit-step response of the motor speed $f_\omega(t)$ (pulses/sec) when the input is 120 pulses/sec.

(c) Design the phase-lead controller in the frequency domain so that the phase margin is at least 60°.

▲ Liquid-level control system design, phase-lag control

10-26. Consider that the controller in the liquid-level control system shown in Fig. 10P-18 is a single-stage phase-lag controller:

$$G_c(s) = \frac{1 + aTs}{1 + Ts} \qquad a < 1$$

(a) For $N = 20$, select the values of a and T so that the two complex roots of the characteristic equation correspond to a relative damping ratio of approximately 0.707. Plot the unit-step response of the output $y(t)$. Find the attributes of the unit-step response. Plot the Bode plot of $G_c(s)G_p(s)$ and determine the phase margin of the designed system.

(b) For $N = 20$, design the phase-lag controller in the frequency domain so that the phase margin is approximately 60°. Plot the unit-step response of the output $y(t)$, and find the attributes of the unit-step response.

▲ Phase-lead and phase-lag controller design

10-27. The controlled process of a unity-feedback control system is

$$G_p(s) = \frac{K}{s(s + 5)^2}$$

The series controller has the transfer function

$$G_c(s) = \frac{1 + aTs}{1 + Ts}$$

(a) Design a phase-lead controller ($a > 1$) so that the following performance specifications are satisfied:

Ramp-error constant $K_v = 10$ Maximum overshoot is near minimum

The value of a must not exceed 1000. Plot the unit-step response and give its attributes.

(b) Design a phase-lead controller in the frequency domain so that the following performance specifications are satisfied:

Ramp-error constant $K_v = 10$ Phase margin is near maximum

The value of a must not exceed 1000.

(c) Design a phase-lag controller ($a < 1$) so that the following performance specifications are satisfied:

Ramp-error constant $K_v = 10$ Rise time $t_r < 2$ sec
Maximum overshoot < 1 percent Settling time $t_s < 2.5$ sec

Find the PM, GM, M_r, and BW of the designed system.

(d) Design the phase-lag controller in the frequency domain so that the following performance specifications are satisfied:

Ramp-error constant $K_v = 10$ Phase margin $\geq 70°$

Check the unit-step response attributes of the designed system and compare with those obtained in part (c).

▲ Design of dc-motor control system with shaft compliance

10-28. The controlled process of a dc-motor control system with unity feedback has the transfer function

$$G_p(s) = \frac{6.087 \times 10^{10}}{s(s^3 + 423.42s^2 + 2.6667 \times 10^6 s + 4.2342 \times 10^8)}$$

Due to the compliance in the motor shaft, the process transfer function contains two lightly damped poles that will cause oscillations in the output response. The following performance criteria are to be satisfied:

Maximum overshoot < 1 percent
Rise time $t_r < 0.15$ sec
Settling time $t_s < 0.15$ sec
Output response should not have oscillations
Ramp-error constant is not affected

(a) Design a series phase-lead controller,

$$G_c(s) = \frac{1 + aTs}{1 + Ts} \qquad a > 1$$

so that all the step-response attributes (except for the oscillations) are satisfied.

(b) To eliminate the oscillations due to the motor shaft compliance, add another stage to the controller with the transfer function

$$G_{c1}(s) = \frac{s^2 + 2\zeta_z \omega_n s + \omega_n^2}{s^2 + 2\zeta_p \omega_n s + \omega_n^2}$$

so that the zeros of $G_{c1}(s)$ will cancel the two complex poles of $G_p(s)$. Set the value of ζ_p so that the two poles of $G_{c1}(s)$ will not have appreciable affect on the system response. Determine the attributes of the unit-step response to see if all the requirements are satisfied. Plot the unit-step responses of the uncompensated system, the compensated system with the phase-lead controller designed in part (b).

▲ Design of computer-tape-drive system, PI controller

10-29. A computer-tape-drive system utilizing a permanent-magnet dc motor is shown in Fig. 10P-29(a). The closed-loop system is modeled by the block diagram in Fig. 10P-29(b). The constant K_L represents the spring constant of the elastic tape, and B_L denotes the

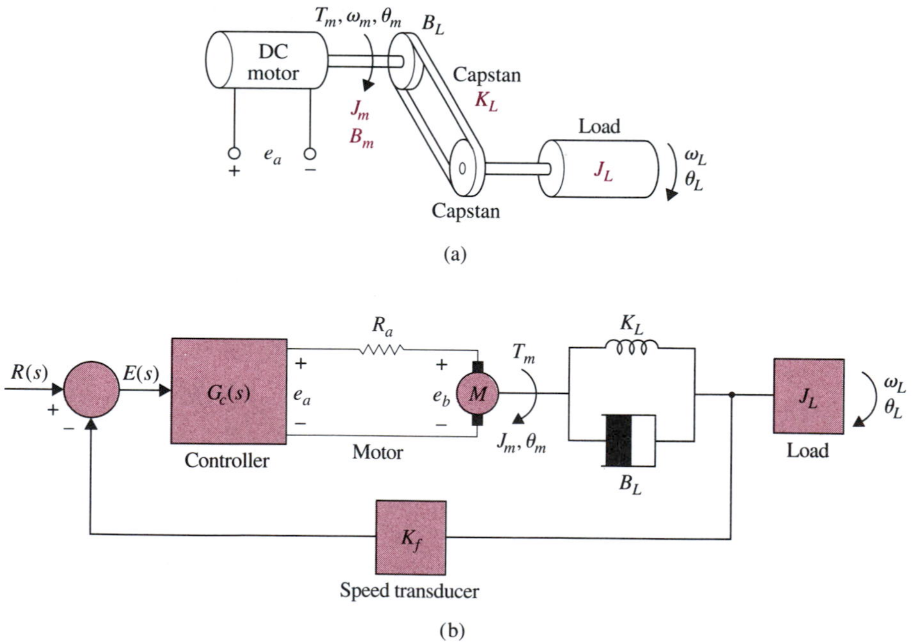

Figure 10P-29

viscous-friction coefficient between the tape and the capstans. The system parameters are as follows:

K_i = motor torque constant = 10 oz-in./A
K_b = motor back-emf constant = 0.0706 V/rad/sec
B_m = motor friction coefficient = 3 oz-in./rad/sec
$R_a = 0.25\ \Omega$ $\quad L_a \cong 0$ H
K_L = 3000 oz-in./rad $\quad B_L$ = 10 oz-in./rad/sec
J_L = 6 oz-in./rad/sec^2 $\quad K_f$ = 1 V/rad/sec
J_m = 0.05 oz-in./rad/sec^2

(a) Write the state equations of the system between e_a and θ_L using θ_L, ω_L, θ_m, and ω_m as state variables and e_a as the input. Draw a state diagram using the state equations. Derive the transfer functions:

$$\frac{\Omega_m(s)}{E_a(s)} \quad \text{and} \quad \frac{\Omega_L(s)}{E_a(s)}$$

(b) The objective of the system is to control the speed of the tape, ω_L, accurately. Consider that PI controller with the transfer function $G_c(s) = K_P + K_I/s$ is to be used. Find the values of K_P and K_I so that the following specifications are satisfied:

Ramp-error constant $K_v = 100$
Rise time < 0.02 sec
Settling time < 0.02 sec
Maximum overshoot <1 percent or at minimum

Plot the unit-step response of $\omega_L(t)$ of the system.

(c) Design the PI controller in the frequency domain. The value of K_I is to be selected as in part (b). Vary the value of K_P and compute the values of PM, GM, M_r, and BW. Find the value of K_P so that PM is maximum. How does this value of K_P compare with the result obtained in part (b)?

▲ Motor-control system with flexible shaft, notch controller

10-30. Figure 10P-30 shows the block diagram of a motor-control system that has a flexible shaft between the motor and the load. The transfer function between the motor torque

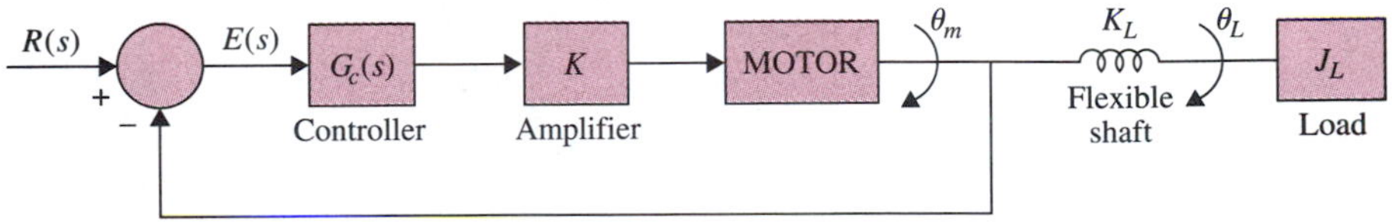

Figure 10P-30

and motor displacement is

$$G_p(s) = \frac{\Theta_m(s)}{T_m(s)} = \frac{J_L s^2 + B_L s + K_L}{s[J_m J_L s^3 + (B_m J_L + B_L J_m)s^2 + (K_L J_m + K_L J_L + B_m B_L)s + B_m K_L]}$$

where $J_L = 0.01$, $B_L = 0.1$, $K_L = 10$, $J_m = 0.01$, $B_m = 0.1$, and $K = 100$.

(a) Compute and plot the unit-step response of $\theta_m(t)$. Find the attributes of the unit-step response.

(b) Design a second-order notch controller with the transfer function

$$G_c(s) = \frac{s^2 + 2\zeta_z \omega_n s + \omega_n^2}{s^2 + 2\zeta_p \omega_n s + \omega_n^2}$$

so that its zeros cancel the complex poles of $G_p(s)$. The two poles of $G_c(s)$ should be selected so that they do not affect the steady-state response of the system and the maximum overshoot is a minimum. Compute the attributes of the unit-step response and plot the response.

(c) Carry out the design of the second-order controller in the frequency domain. Plot the Bode diagram of the uncompensated $G_p(s)$, and find the values of PM, GM, M_r, and BW. Set the two zeros of $G_c(s)$ to cancel the two complex poles of $G_p(s)$. Determine the value of ζ_p by determining the amount of attenuation required from the second-order notch controller, and using Eq. (10-155). Find the PM, GM, M_r, and BW of the compensated system. How does the frequency-domain-design results compare with the results in part (b)?

▲ Design of notch controller

10-31. The transfer function of the process of a unity-feedback control system is

$$G_p(s) = \frac{500(s + 10)}{s(s^2 + 10s + 1000)}$$

(a) Plot the Bode diagram of $G_p(s)$ and determine the PM, GM, M_r, and BW of the uncompensated system. Compute and plot the unit-step response of the system.

(b) Design a series second-order notch controller with the transfer function

$$G_c(s) = \frac{s^2 + 2\zeta_z \omega_n s + \omega_n^2}{s^2 + 2\zeta_p \omega_n s + \omega_n^2}$$

so that its zeros cancel the complex poles of $G_p(s)$. Determine the value of ζ_p using the method outlined in Section 10-8-2. Find the PM, GM, M_r, and BW of the designed system. Compute and plot the unit-step response.

(c) Design the series second-order notch controller so that its zeros cancel the complex poles of $G_p(s)$. Determine the value of ζ_p so that the following specifications are satisfied:

Maximum overshoot < 1 percent   Settling time < 0.5 sec
Rise time < 0.4 sec

▲ Forward controller design

10-32. Design the controllers $G_{cf}(s)$ and $G_c(s)$ for the system shown in Fig. 10P-32 so that the following specifications are satisfied:

Ramp-error constant $K_v = 50$
Dominant roots of the characteristic equation at $-5 \pm j5$ approximately
Rise time < 0.1 sec
Maximum overshoot ≤ 8 percent
System must be robust when K varies ± 20 percent from the nominal value, with the rise time and overshoot staying within specifications

Compute and plot the unit-step responses to check the design.

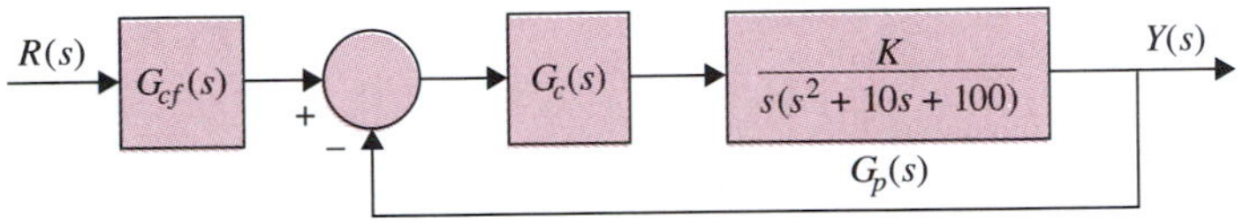

Figure 10P-32

▲ Forward controller design, motor-control system

10-33. Figure 10P-33 shows the block diagram of a motor-control system. The transfer function of the controlled process is

$$G_p(s) = \frac{1000K}{s(s+a)}$$

where K denotes the aggregate of the amplifer gain and motor torque constant, and a is the inverse of the motor time constant. Design the controllers $G_{cf}(s)$ and $G_c(s)$ so that the following performance specifications are satisfied.

Ramp-error constant $K_v = 100$ when $a = 10$
Rise time < 0.3 sec
Maximum overshoot < 8 percent
Dominant characteristic equation roots $= -5 \pm j5$
System must be robust when a varies between 8 and 12

Compute and plot the unit-step responses to verify the design.

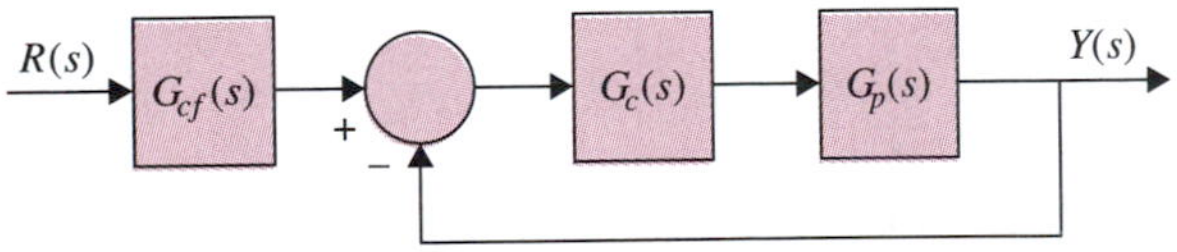

Figure 10P-33

▲ Design of system with tachometer feedback

10-34. Figure 10P-34 shows the block diagram of a dc-motor control system with tachometer feedback. Find the values of K and K_t so that the following specifications are satisfied.

Ramp-error constant $K_v = 1$
Dominant characteristic equation roots correspond to a damping ratio of approximately 0.707; if there are two solutions, select the larger value of K

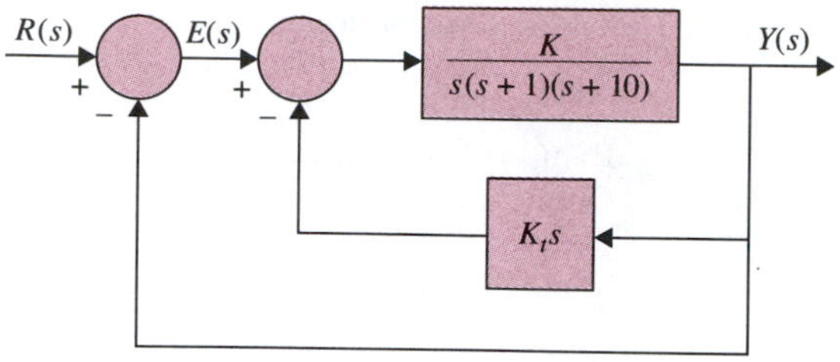

Figure 10P-34

▲ Design of system with tachometer feedback

10-35. Carry out the design with the specifications given in Problem 10-34 for the system shown in Fig. 10P-35.

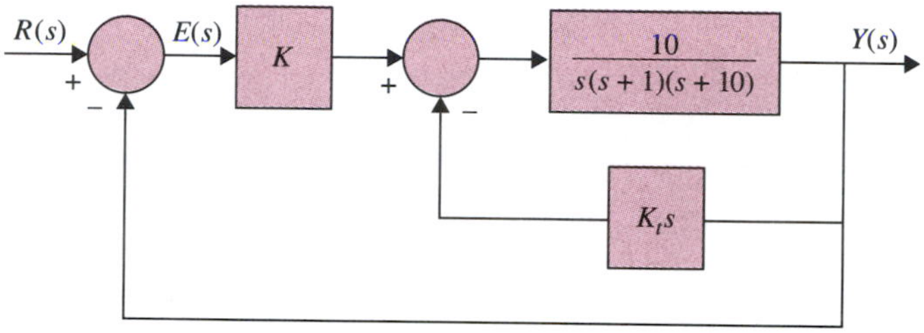

Figure 10P-35

▲ Design of system with tachometer feedback

10-36. The block diagram of a control system with a type 2 process is shown in Fig. 10P-36. The system is to be compensated by tachometer feedback and a series controller. Find the values of a, T, K, and K_t so that the following performance specifications are satisfied:

Ramp-error constant $K_v = 100$
Dominant characteristic equation roots correspond to a damping ratio of 0.707

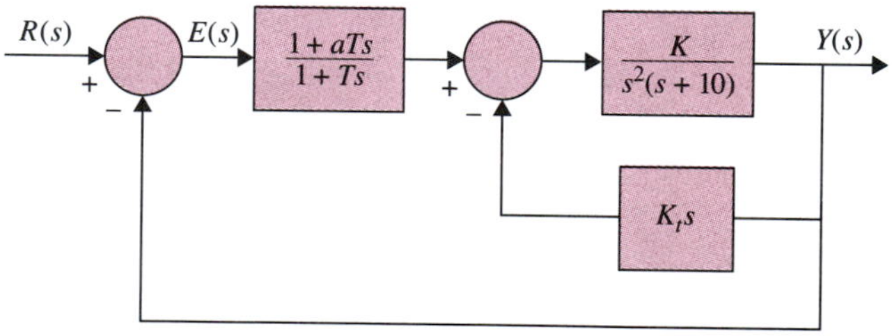

Figure 10P-36

▲ Design of aircraft attitude control system. Tachometer feedback

10-37. The aircraft attitude control system described in Section 7-6 is modeled by the block diagram shown in Fig. 10P-37. The system parameters are:

K = variable	$K_s = 1$	$K_1 = 10$	$K_2 = 0.5$	K_t = variable
$R_a = 5$	$L_a = 0.003$	$K_i = 9.0$	$K_b = 0.0636$	$J_m = 0.0001$
$J_L = 0.01$	$B_m = 0.005$	$B_L = 1.0$	$N = 0.1$	

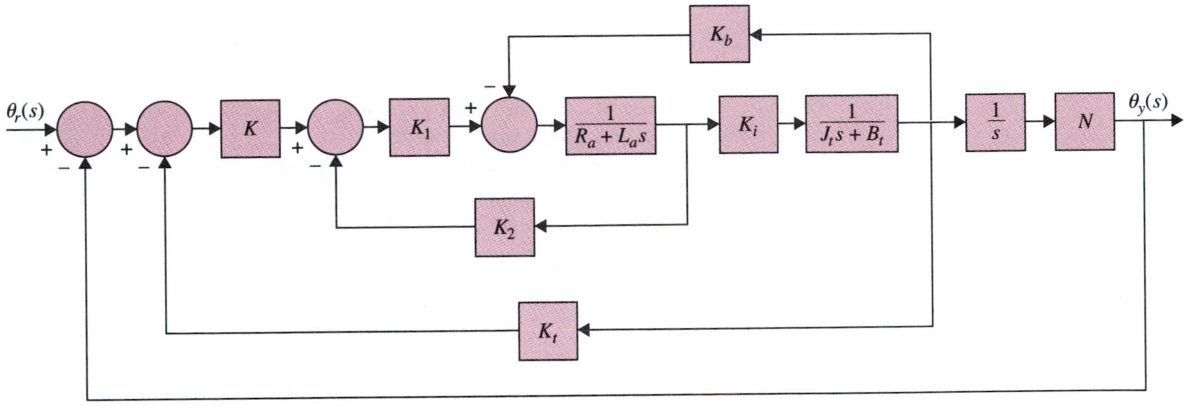

Figure 10P-37

Find the values of K and K_t so that the following specifications are satisfied:

Ramp-error constant $K_v = 100$
Relative damping ratio of the complex roots of the characteristic equation is approximately 0.707

Plot the unit-step response of the designed system.

▲ Design of position-feedback control system, single- and two-stage phase-lead controllers

10-38. Figure 10P-38 shows the block diagram of a position-control system with a series controller $G_c(s)$.

(a) Determine the minimum value of the amplifier gain K so that the steady-state value of the output $y(t)$ due to a unit-step torque disturbance is ≤ 0.01.

(b) Show that the uncompensated system is unstable with the minimum value of K determined in part (a). Construct the Bode diagram for the open-loop transfer function $G(s) = Y(s)/E(s)$, and find the values of PM and GM.

(c) Design a single-stage phase-lead controller with the transfer function

$$G_c(s) = \frac{1 + aTs}{1 + Ts} \qquad a > 1$$

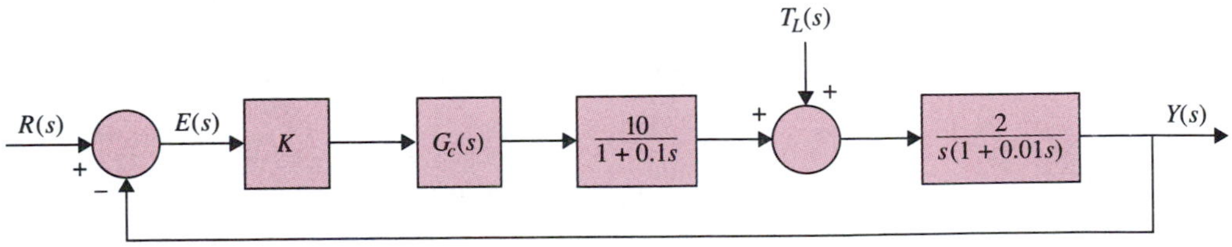

Figure 10P-38

so that the phase margin is 30°. Show that this is nearly the highest phase margin that can be achieved with a single-stage phase-lead controller. Find GM, M_r, and BW of the compensated system.

(d) Design a two-stage phase-lead controller using the system arrived at in part (c) as a basis, so that the phase margin is 55°. Show that this is the best PM that can be obtained for this system with a two-stage phase-lead controller. Find GM, M_r, and BW of the compensated system.

▲ Design of two-stage phase-lead controller, phase-lag controller, lag-lead controller

10-39. The transfer function of the process of a unity-feedback control system is

$$G_p(s) = \frac{60}{s(1 + 0.2s)(1 + 0.5s)}$$

Show that due to the relative high gain, the uncompensated system is unstable.

(a) Design a two-stage phase-lead controller with

$$G_c(s) = \left(\frac{1 + aT_1 s}{1 + T_1 s}\right)\left(\frac{1 + bT_2 s}{1 + T_2 s}\right) \qquad a > 1, \quad b > 1 \tag{1}$$

so that the phase margin is greater than 60°. Conduct the design by first determining the values of a and T_1 to realize a maximum phase margin that can be achieved with a single-stage phase-lead controller. The second stage of the controller is then designed to realize the balance of the 60° phase margin. Determine GM, M_r, and BW of the compensated system. Compute and plot the unit-step response of the compensated system.

(b) Design a single-stage phase-lag controller with

$$G_c(s) = \frac{1 + aTs}{1 + Ts} \qquad a < 1$$

so that the phase margin of the compensated system is greater than 60°. Determine GM, M_r, and BW of the compensated system. Compute and plot the unit-step response of the compensated system.

(c) Design a lag-lead controller with $G_c(s)$ as in Eq. (1) in part (a). Design the phase-lag portion first by setting the phase margin at 40°. The resulting system is then compensated by the phase-lead portion to achieve a total of 60° of phase margin. Determine GM, M_r, and BW of the compensated system. Compute and plot the unit-step response of the compensated system.

▲ Design of steel-rolling control system

10-40. The block diagram of the steel-rolling system described in Problem 4-26 is shown in Fig. 10P-40. The transfer function of the process is

$$G_p(s) = \frac{Y(s)}{E(s)} = \frac{5e^{-0.1s}}{s(1 + 0.1s)(1 + 0.5s)} \qquad K_s = 1$$

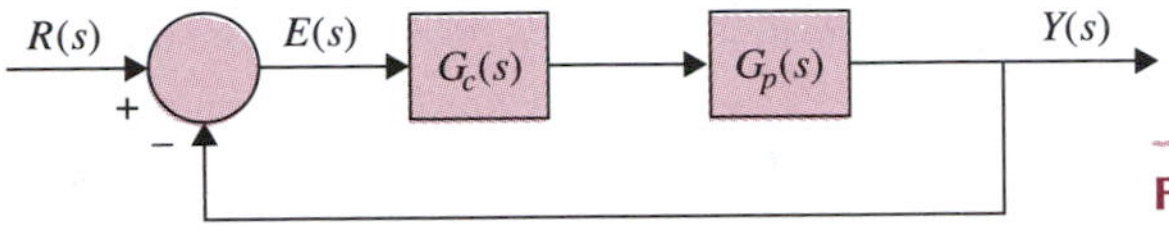

Figure 10P-40

(a) Approximate the time delay by

$$e^{-0.1s} \cong \frac{1 - 0.05s}{1 + 0.05s}$$

Design a series controller of your choice so that the phase margin of the compensated system is at least 60°. Determine GM, M_r, and BW of the compensated system. Compute and plot the unit-step responses of the compensated and the uncompensated systems.

(b) Repeat part (a) without using the approximation of the time delay.

▲ Design of respiratory control system, PI controller

10-41. Human beings breathe in order to provide for gas exchange for the entire body. A respiratory control system is needed to ensure that the body's needs for this gas exchange are adequately met. The criterion of control is adequate ventilation, which ensures satisfactory levels of both oxygen and carbon dioxide in the arterial blood. Respiration is controlled by neural impulses that originate within the lower brain and are transmitted to the chest cavity and diaphragm to govern the rate and tidal volume. One source of signals is the chemoreceptors located near the respiratory center, which are sensitive to carbon dioxide and oxygen concentrations. Figure 10P-41 shows the block diagram of a simplified model of the human respiratory control system. The objective is to control the effective ventilation of the lungs so that a satisfactory balance of concentrations of carbon dioxide and oxygen is maintained in the blood circulated at the chemoreceptor.

(a) Plot the Bode diagram of the transfer function $G(s) = Y(s)/E(s)$ when $G_c(s) = 1$. Find the PM and GM. Determine the stability of the system.

(b) Design a PI controller, $G_c(s) = K_P + K_I/s$, so that the following specifications are satisfied:

Ramp-error constant $K_v = 1$ Phase margin is maximized

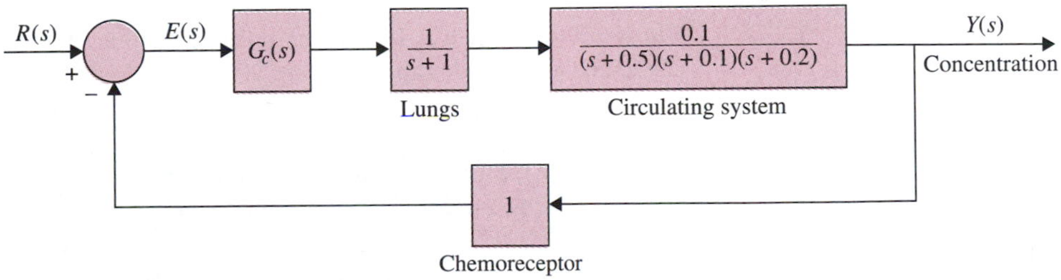

Figure 10P-41

Plot the unit-step response of the system. Find the attributes of the unit-step response.

(c) Design a PI controller so that the following specifications are satisfied:

Ramp-error constant $K_v = 1$ Maximum overshoot is minimized

Plot the unit-step response of the system. Find the attributes of the unit-step response. Compare the design results in parts (b) and (c).

▲ State-feedback control

10-42. The block diagram of a control system with state feedback is shown in Fig. 10P-42. Find the real feedback gains, k_1, k_2, and k_3 so that

The steady-state error $e_{ss}[e(t)$ is the error signal] due to a step input is zero
The complex roots of the characteristic equation are at $-1 + j$ and $-1 - j$

Find the third root. Can all three roots be assigned arbitrarily while still meeting the steady-state requirement?

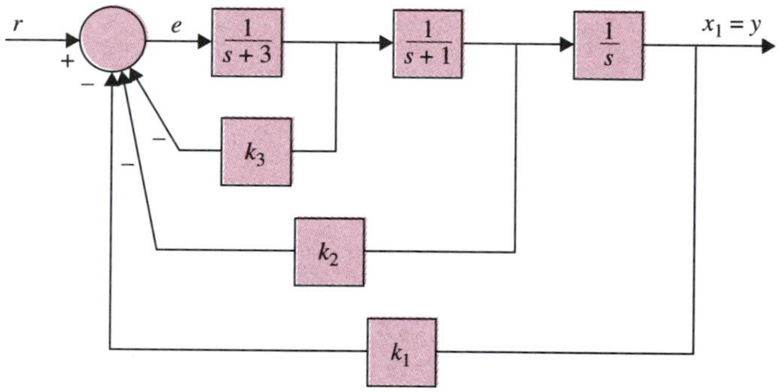

Figure 10P-42

▲ State-feedback control

10-43. The block diagram of a control system with state feedback is shown in Fig. 10P-43(a). The feedback gains k_1, k_2, and k_3 are real constants.

(a) Find the values of the feedback gains so that

The steady-state error $e_{ss}[e(t)$ is the error signal] due to a step input is zero.
The characteristic equation roots are at $-1 + j$, $-1 - j$, and -10.

(b) Instead of using state feedback, a series controller is implemented, as shown in Fig. 10P-43(b). Find the transfer function of the controller $G_c(s)$ in terms of k_1, k_2, and k_3 found in part (a) and the other system parameters.

10-44. Problem 10-19 has revealed that it is impossible to stabilize the broom-balancing control system described in Problems 4-23 and 5-41 with a series PD controller. Consider that

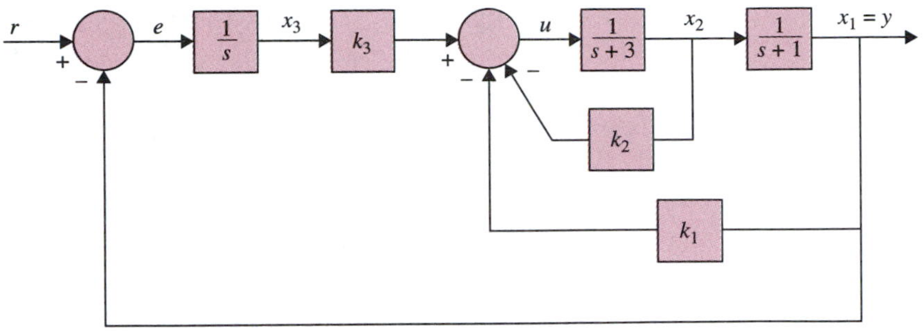

(a)

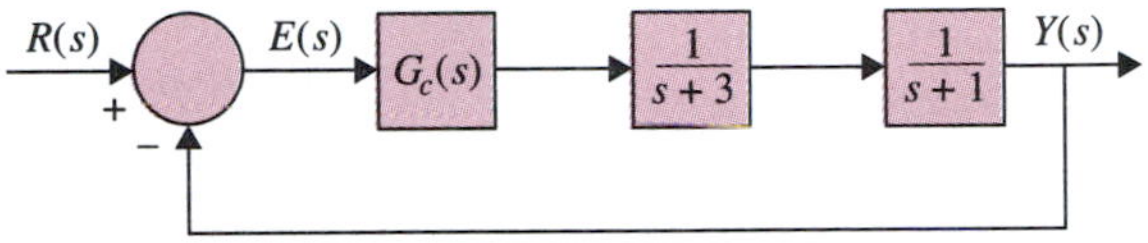

(b)

Figure 10P-43

▲ Broom-balancing system with state-feedback control

the system is now controlled by state feedback with $\Delta r(t) = -\mathbf{K}\mathbf{x}(t)$, where

$$\mathbf{K} = [k_1 \quad k_2 \quad k_3 \quad k_4]$$

(a) Find the feedback gains k_1, k_2, k_3, and k_4 so that the eigenvalues of $\mathbf{A}^* - \mathbf{B}^*\mathbf{K}$ are at $-1 + j$, $-1 - j$, -10, and -10. Compute and plot the responses of $\Delta x_1(t)$, $\Delta x_2(t)$, $\Delta x_3(t)$, and $\Delta x_4(t)$ for the initial condition, $\Delta x_1(0) = 0.1$, $\Delta\theta(0) = 0.1$, and all other initial conditions are zero.

(b) Repeat part (a) for the eigenvalues at $-2 + j2$, $-2 - j2$, -20, and -20. Comment on the difference between the two systems.

▲ Ball-suspension system with state-feedback control

10-45. The linearized state equations of the ball-suspension control system described in Problem 4-25 are expressed as

$$\Delta\dot{\mathbf{x}}(t) = \mathbf{A}^* \, \Delta\mathbf{x}(t) + \mathbf{B}^* \, \Delta i(t)$$

where

$$\mathbf{A}^* = \begin{bmatrix} 0 & 1 & 0 & 0 \\ 115.2 & -0.05 & -18.6 & 0 \\ 0 & 0 & 0 & 1 \\ -37.2 & 0 & 37.2 & -0.1 \end{bmatrix} \qquad \mathbf{B}^* = \begin{bmatrix} 0 \\ -6.55 \\ 0 \\ -6.55 \end{bmatrix}$$

Let the control current $\Delta i(t)$ be derived from the state feedback $\Delta i(t) = -\mathbf{K}\,\Delta\mathbf{x}(t)$, where

$$\mathbf{K} = [k_1 \quad k_2 \quad k_3 \quad k_4]$$

(a) Find the elements of $\mathbf{K}$ so that the eigenvalues of $\mathbf{A}^* - \mathbf{B}^*\mathbf{K}$ are at $-1 + j$, $-1 - j$, -10, and -10.

(b) Plot the responses of $\Delta x_1(t) = \Delta y_1(t)$ (magnet displacement) and $\Delta x_3(t) = \Delta y_2(t)$ (ball displacement) with the initial condition

$$\Delta\mathbf{x}(0) = \begin{bmatrix} 0.1 \\ 0 \\ 0 \\ 0 \end{bmatrix}$$

(c) Repeat part (b) with the initial condition

$$\Delta\mathbf{x}(0) = \begin{bmatrix} 0 \\ 0 \\ 0.1 \\ 0 \end{bmatrix}$$

Comment on the responses of the closed-loop system with the two sets of initial conditions used in (b) and (c).

▲ Temperature control of electric furnace, state feedback, PI control

10-46. The temperature $x(t)$ in the electric furnace shown in Fig. 10P-46 is described by the differential equation

$$\frac{dx(t)}{dt} = -2x(t) + u(t) + n(t)$$

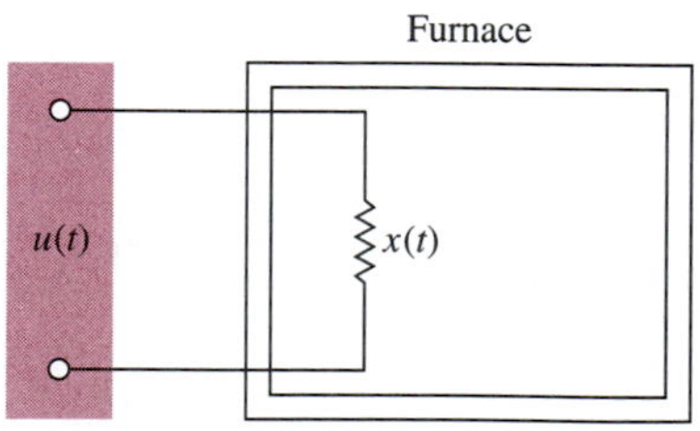

Figure 10P-46

where $u(t)$ is the control signal and $n(t)$ the constant disturbance of unknown magnitude due to heat loss. It is desired that the temperature $x(t)$ follows a reference input r that is a constant.

(a) Design a control system with state and integral control so that the following specifications are satisfied:

$$\lim_{t\to\infty} x(t) = R = \text{constant}$$

The eigenvalues of the closed-loop system are at -10 and -10

Plot the responses of $x(t)$ for $t \geq 0$ with $r = 1$ and $n(t) = -1$, and then with $r = 1$ and $n(t) = 0$, all with $x(0) = 0$.

(b) Design a PI controller so that

$$G_c(s) = \frac{U(s)}{E(s)} = K_P + \frac{K_I}{s}$$

$$E(s) = R(s) - X(s) \qquad \text{where} \quad R(s) = R/s$$

Find K_P and K_I so that the characteristic equation roots are at -10 and -10. Plot the responses of $x(t)$ for $t \geq 0$ with $R = 1$ and $n(t) = -1$, and then with $R = 1$ and $n(t) = 0$, all $x(0) = 0$.

Answers to True-and-False Review Questions

5. (F) **6.** (F) **8.** (T) **18.** (T) **19.** (T) **20.** (T) **21.** (T)
22. (T) **26.** (T) **27.** (T) **28.** (T)

11 Design of Discrete-Data Control Systems

KEYWORDS AND TOPICS

- ▲ Digital Controllers
- ▲ Phase-Lead Control
- ▲ Deadbeat-Response Design
- ▲ Digital PID Controller
- ▲ Phase-Lag Control
- ▲ State-Feedback Design
- ▲ Root-Locus Design

11-1 Introduction

The design of discrete-data control systems is similar in principle to the design of continuous-data control systems. The design objective is basically that of determining the controller so that the system will perform in accordance with specifications. In fact, in most situations, the controlled process is the same, except that in discrete-data systems the controller is designed to process sampled or digital data.

The design of discrete-data control systems treated in this chapter is intended only for introductory purposes. An in-depth coverage of the subject may be found in books dedicated to digital control. In this chapter we deal only with the design of a control

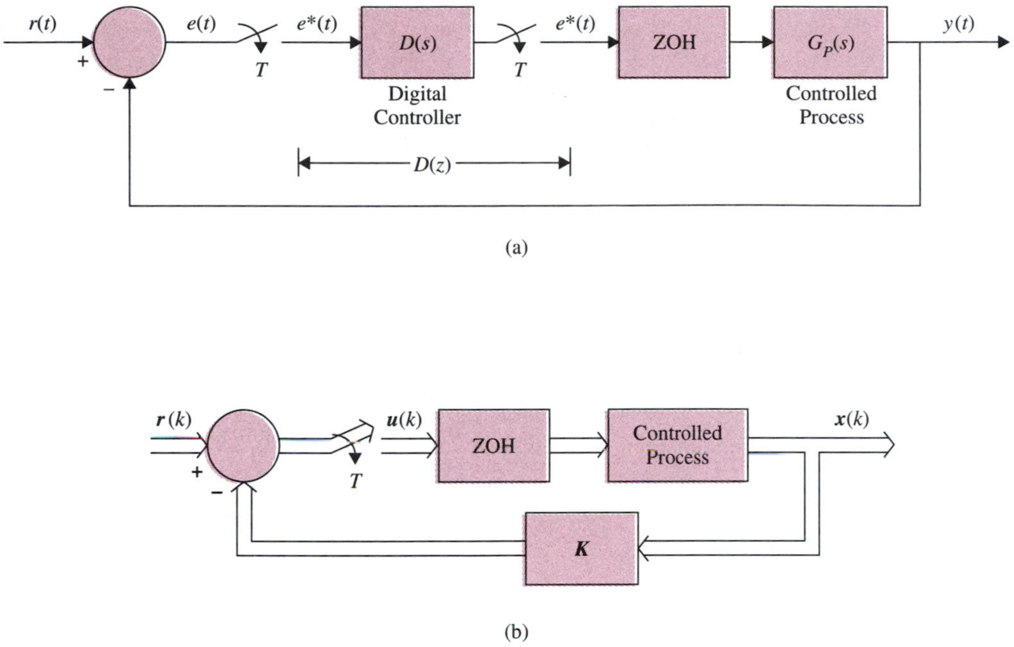

Figure 11-1 (a) Digital control system with cascade digital controller. (b) Digital control system with state feedback.

system with cascade digital controller and a system with digital state feedback. The block diagrams of these systems are shown in Fig. 11-1.

Just as with the design of continuous-data control systems, the design of discrete-data control systems can be carried out in either the frequency domain or the time domain. With the availability of all the computer programs, the design of digital control systems can be carried out with a minimum amount of trial-and-error effort.

11-2 Digital Implementation of Analog Controllers

It seems that most people learn how to design continuous-data systems before digital, the latter, if at all. Therefore, it is not surprising that most engineers prefer to design continuous-data systems over digital control systems. Ideally, if the designer intends to use digital control, the system should be designed so that the dynamics of the controller can be described by a z-transfer function or difference equations. However, there are situations under which the analog controller is already designed, but the availability and advantages of digital control suggest that the controller be implemented by digital elements. Thus the problems discussed in this section are twofold; first, we investigate how continuous-data controllers such as PID, phase-lead or phase-lag controllers, and others can be approximated by digital controllers; second, the problem of implementing digital controllers by digital processors is investigated.

11-2-1 Digital Implementation of the PID Controller

The PID controller in the continuous-data domain is described by

$$G_c(s) = K_P + K_D s + \frac{K_I}{s} \tag{11-1}$$

The proportional component K_P is implemented digitally by a constant gain K_P. Since a digital computer or processor has finite word length, the constant K_P cannot be realized with infinite resolution.

The time derivative of a function $f(t)$ at $t = kT$ can be approximated by the **backward-difference rule,** using the values of $f(t)$ measured at $t = kT$ and $(k - 1)T$; that is,

$$\left.\frac{df(t)}{dt}\right|_{t=kT} = \frac{1}{T}\Big(f(kT) - f[(k-1)T]\Big) \tag{11-2}$$

To find the z-transfer function of the derivative operation described before, we take the z-transform on both sides of Eq. (11-2). We have

$$\mathcal{Z}\left(\left.\frac{df(t)}{dt}\right|_{t=kT}\right) = \frac{1}{T}(1 - z^{-1})F(z) = \frac{z-1}{Tz}F(z) \tag{11-3}$$

Thus the z-transfer function of the digital differentiator is

$$G_D(z) = K_D\frac{z-1}{Tz} \tag{11-4}$$

where K_D is the proportional constant of the derivative controller. Replacing z by e^{Ts} in Eq. (11-4), we can show that as the sampling period T approaches zero, $G_D(z)$ approaches $K_D s$, which is the transfer function of the analog derivative controller. In general, the choice of the sampling period is extremely important. The value of T should be sufficiently small, so that the digital approximation is adequately accurate.

There are a number of numerical integration rules that can be used to digitally approximate the integral controller K_I/s. The three basic methods of approximating the area of a function numerically are **trapezoidal integration, forward-rectangular integration,** and **backward-rectangular integration.** These are described as follows.

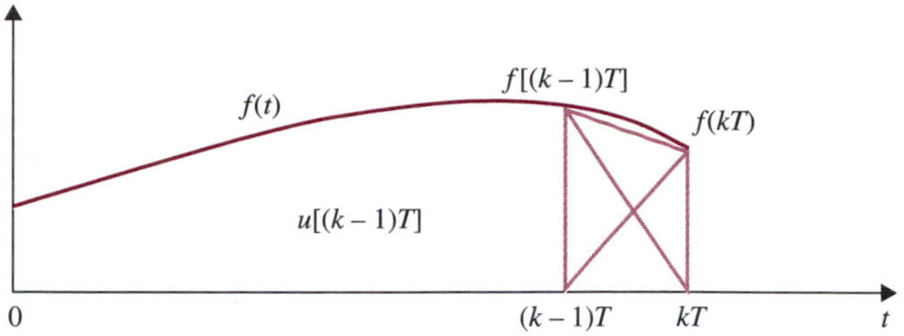

Figure 11-2 Trapezoidal-integration rule.

Trapezoidal Integration

The **trapezoidal-integration rule** approximates the area under the function $f(t)$ by a series of trapezoids, as shown in Fig. 11-2. Let the integral of $f(t)$ evaluated at $t = kT$ be designated as $u(kT)$. Then

$$u(kT) = u[(k-1)T] + \frac{T}{2}\{f(kT) + f[(k-1)T]\} \tag{11-5}$$

where the area under $f(t)$ for $(k-1)T \leq t < kT$ is approximated by the area of the trapezoid in the interval. Taking the z-transform on both sides of Eq. (11-5), we have the transfer function of the digital integrator as

$$G_I(z) = K_I\frac{U(z)}{F(z)} = \frac{K_I T(z+1)}{2(z-1)} \tag{11-6}$$

where K_I is the proportional constant.

Forward-Rectangular Integration

For the forward-rectangular integration, we approximate the area under $f(t)$ by rectangles, as shown in Fig. 11-3. The integral of $f(t)$ at t = kT is approximated by

$$u(kT) = u[(k-1)T] + Tf(kT) \tag{11-7}$$

By taking the z-transform on both sides of Eq. (11-7), the transfer function of the digital integrator using the forward-rectangular rule is

$$G_I(z) = K_I\frac{U(z)}{f(z)} = \frac{K_I Tz}{z-1} \tag{11-8}$$

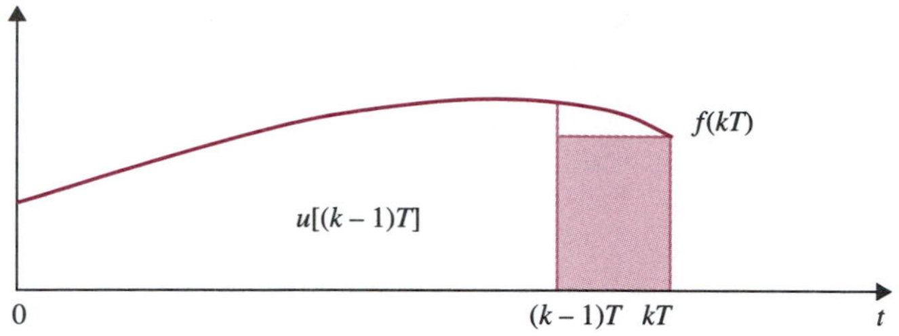

Figure 11-3 Forward-rectangular integration rule.

Backward-Rectangular Integration

For the backward-rectangular integration, the digital approximation rule is illustrated in Fig. 11-4. The integral of $f(t)$ at $t = kT$ is approximated by

$$u(kT) = u[(k-1)T] + Tf[(k-1)T] \tag{11-9}$$

The z-transfer function of the digital integrator using the backward-rectangular integration rule is

$$G_I(z) = K_I \frac{U(z)}{F(z)} = \frac{K_I T}{z-1} \tag{11-10}$$

By combining the proportional, derivative, and integration operations described before, the digital PID controller is modeled by the following transfer functions.

Trapezoidal Integration

$$G_c(z) = \frac{(K_P + TK_I/2 + K_D/T)z^2 + (TK_I/2 - K_P - 2K_D/T)z + K_D/T}{z(z-1)} \tag{11-11}$$

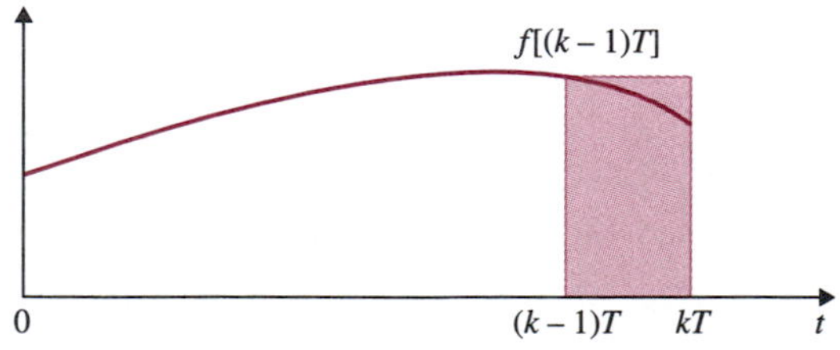

Figure 11-4 Backward-rectangular integration rule.

Forward-Rectangular Integration

$$G_c(z) = \frac{(K_P + K_D/T + TK_I)z^2 - (K_P + 2K_D/T)z + K_D/T}{z(z-1)} \tag{11-12}$$

Backward-Rectangular Integration

$$G_c(z) = \frac{(K_P + K_D/T)z^2 + (TK_I - K_P - 2K_D/T)z + K_D/T}{z(z-1)} \tag{11-13}$$

When $K_I = 0$, the transfer function of the digital PD controller is

$$G_c(z) = \frac{(K_P + K_D/T)z - K_D/T}{z} \tag{11-14}$$

Once the transfer function of a digital controller is determined, the controller can be implemented by a digital processor or computer. The operator z^{-1} is interpreted as

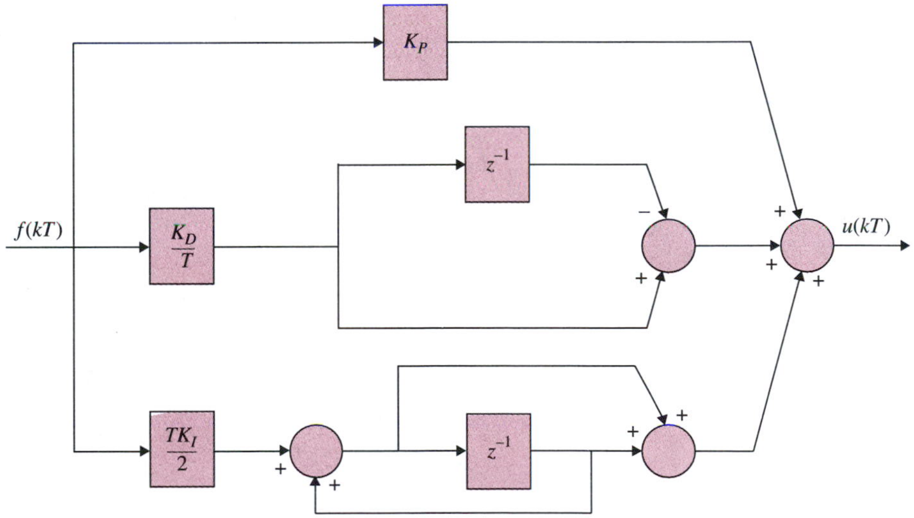

Figure 11-5 Block diagram of a digital-program implementation of the PID controller.

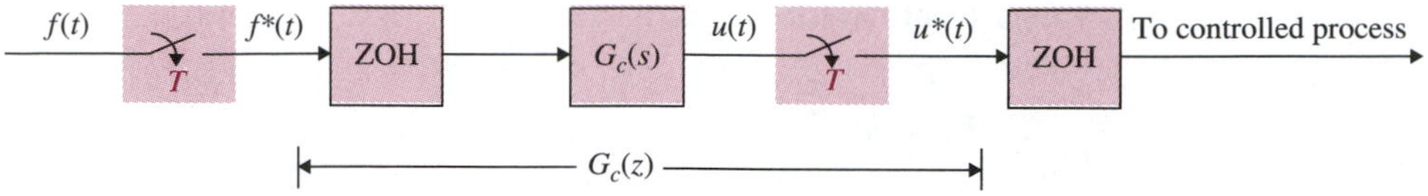

Figure 11-6 Realization of a digital controller by an analog controller with sample-and-hold units.

a time delay of T seconds. In practice, the time delay is implemented by storing a variable in some storage location in the computer and then taking it out after T seconds have elapsed. Figure 11-5 illustrates a block diagram representation of the digital program of the PID controller using the trapezoidal-integration rule.

11-2-2 Digital Implementation of Lead and Lag Controllers

In principle, any continuous-data controller can be made into a digital controller simply by adding sample-and-hold units at the input and the output terminals of the controller, and selecting a sampling frequency as small as practical. Figure 11-6 illustrates the basic scheme, and $G_c(s)$ as the transfer function of the continuous-data controller and $G_c(z)$ the equivalent digital controller. The sampling period T should be sufficiently small so that the dynamic characteristics of the continuous-data controller are not lost through the digitization. The system configuration shown in Fig. 11-6 actually suggests that given the continuous-data controller $G_c(s)$, the equivalent digital controller $G_c(z)$ can be obtained by the arrangement shown. On the other hand, given the digital controller $G_c(z)$, we can realize it by using an analog controller $G_c(s)$ and sample-and-hold units, as shown in Fig. 11-6.

Example 11-1

As an illustrative example, consider that the continuous-data controller in Fig. 11-6 is represented by the transfer function

$$G_c(s) = \frac{s+1}{s+1.61} \tag{11-15}$$

From Fig. 11-6 the transfer function of the digital controller is written

$$G_c(z) = \frac{U(z)}{F(z)} = (1 - z^{-1})\mathscr{Z}\left[\frac{s+1}{s(s+1.61)}\right]$$

$$= \frac{z - (0.62e^{-1.61T} + 0.38)}{z - e^{-1.61T}} \tag{11-16}$$

The digital-program implementation of Eq. (11-16) is shown in Fig. 11-7. ▲

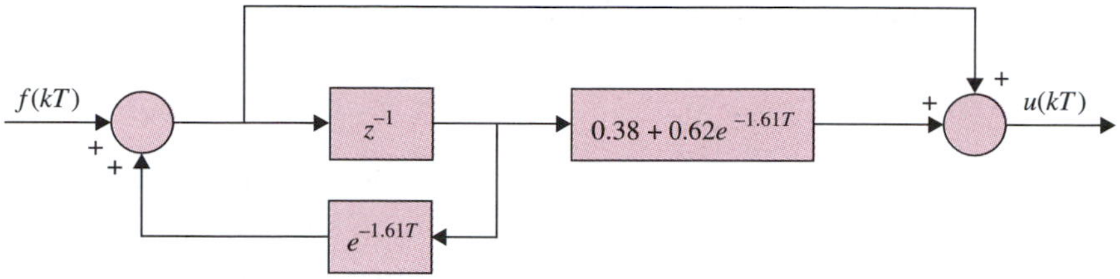

Figure 11-7 Digital-program realization of Eq. (11-16).

11-3 Digital Controllers

▲ One advantage of the digital controller is that its program can easily be altered.

Digital controllers can be realized by digital networks, digital computers, microprocessors, or digital signal processors (DSPs). A distinct advantage of digital controllers implemented by microprocessors or DSPs is that the control algorithm contained in the controller can easily be altered by changing the program. Changing the components of a continuous-data controller is rather difficult once the controller has been built.

11-3-1 Physical Realizability of Digital Controllers

The transfer function of a digital controller can be expressed as

$$G_c(z) = \frac{E_2(z)}{E_1(z)} = \frac{b_0 + b_1 z^{-1} + \cdots + b_m z^{-m}}{a_0 + a_1 z^{-1} + \cdots + a_n z^{-n}} \tag{11-17}$$

where n and m are positive integers. The transfer function $G_c(z)$ is said to be physically realizable if its output does not precede any input. This means that the series expansion of $G_c(z)$ should not have any positive powers in z. In terms of the $G_c(z)$ given in Eq. (11-17), if $b_0 \neq 0$, then $a_0 \neq 0$. If $G_c(z)$ is expressed as

$$G_c(z) = \frac{b_m z^m + b_{m-1} z^{m-1} + \cdots + b_1 z + b_0}{a_n z^n + a_{n-1} z^{n-1} + \cdots + a_1 z + a_0} \tag{11-18}$$

then the physical realizability requirements is $n \geq m$.

The decomposition techniques presented in Chapter 5 can be applied to realize the digital controller transfer function by a digital program. We consider that a digital program is capable of performing arithmetic operations of addition, subtraction, multiplication by a constant, and shifting. The three basic methods of decomposition are discussed in the following for digital programming.

Digital Program by Direct Decomposition

Applying direct decomposition to Eq. (11-17), we have the following equations:

$$E_2(z) = \frac{1}{a_0}(b_0 + b_1 z^{-1} + \cdots + b_m z^{-m})X(z) \tag{11-19}$$

$$X(z) = \frac{1}{a_0}E_1(z) - \frac{1}{a_0}(a_1 z^{-1} + a_2 z^{-2} + \cdots + a_n z^{-n})X(z) \tag{11-20}$$

Figure 11-8 shows the signal-flow graph of a direct digital program of Eq. (11-17) by direct decomposition for $m = 2$ and $n = 3$. The branches with gains of z^{-1} represent time delays or shifts of one sampling period.

Digital Program by Cascade Decomposition

The transfer function $G_c(z)$ can be written as a product of first- or second-order transfer functions each realizable by a simple digital program. The digital program of the overall transfer function is then represented by these simple digital programs connected in cascade. Equation (11-17) is written in factored form as

$$G_c(z) = G_{c1}(z)G_{c2}(z)\cdots G_{cn}(z) \tag{11-21}$$

where the individual factors can be expressed as follows:

Real pole and zero: $$G_{ci}(z) = K_i\frac{1 + c_i z^{-1}}{1 + d_i z^{-1}} \tag{11-22}$$

Complex-conjugate poles (no zeros): $$G_{ci}(z) = \frac{K_i}{1 + d_{i1}z^{-1} + d_{i2}z^{-2}} \tag{11-23}$$

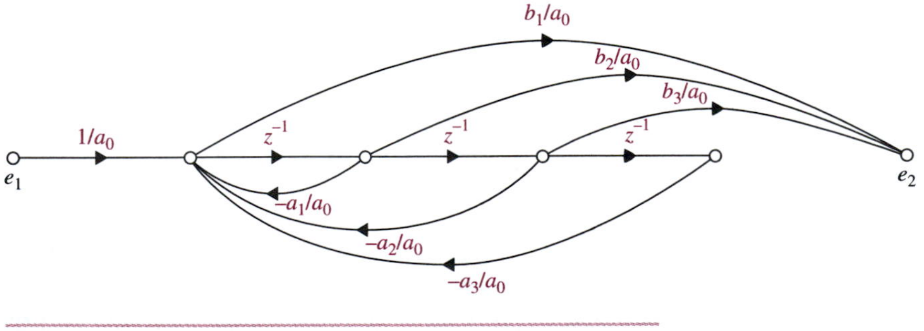

Figure 11-8 Signal-flow graph of digital program by direct decomposition of Eq. (11-17) with $n = 3$ and $m = 2$.

Complex-conjugate poles with one zero: $$G_{ci}(z) = K_i \frac{1 + c_i z^{-1}}{1 + d_{i1} z^{-1} + d_{i2} z^{-2}} \tag{11-24}$$

and several other possible forms up to the second order.

Digital Program by Parallel Decomposition

The transfer function in Eq. (11-17) can be expanded into a sum of simple first- or second-order terms by partial-fraction expansion. These terms are then realized by digital programs connected in parallel.

Example 11-2

Consider the following transfer function of a digital controller:

$$G_c(z) = \frac{E_2(z)}{E_1(z)} = \frac{10(1 + 0.5z^{-1})}{(1 - z^{-1})(1 - 0.2z^{-1})} \tag{11-25}$$

Since the leading coefficients of the numerator and denominator polynomials in z^{-1} are all constants, the transfer function is physically realizable. The transfer function $G_c(z)$ is realized by the three types of digital programs discussed above.

Direct Digital Programming

Equation (11-25) is written

$$G_c(z) = \frac{E_2(z)}{E_1(z)} = \frac{10(1 + 0.5z^{-1})X(z)}{(1 - z^{-1})(1 - 0.2z^{-1})X(z)} \tag{11-26}$$

Expanding the numerator and denominator of the last equation and equating, we have

$$E_2(z) = (10 + 5z^{-1})X(z) \tag{11-27}$$
$$X(z) = E_1(z) + 1.2z^{-1}X(z) - 0.2z^{-2}X(z) \tag{11-28}$$

The last two equations are realized by the digital program shown in Fig. 11-9.

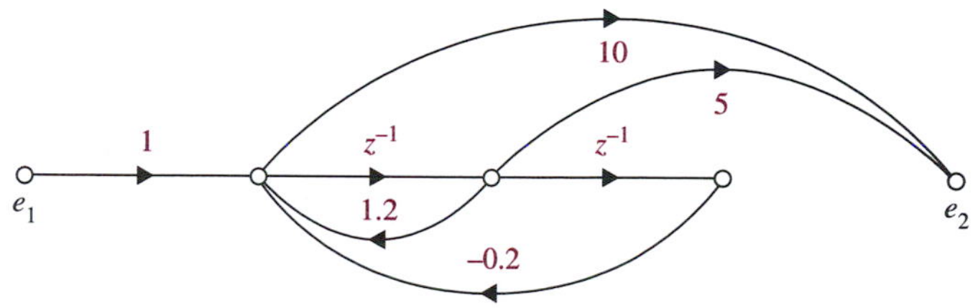

Figure 11-9 Direct digital program of Eq. (11-25).

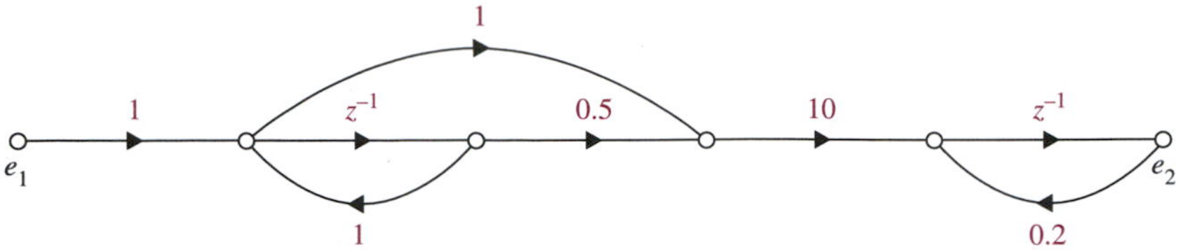

Figure 11-10 Cascade digital program of Eq. (11-25).

Cascade Digital Programming

The right-hand side of Eq. (11-25) is divided into two factors in one of several possible ways.

$$G_c(z) = \frac{E_2(z)}{E_1(z)} = \frac{1 + 0.5z^{-1}}{1 - z^{-1}} \frac{10}{1 - 0.2z^{-1}} \tag{11-29}$$

Figure 11-10 shows the signal-flow graph of the cascade digital program of the controller.

Parallel Digital Programming

The right-hand side of Eq. (11-25) is expanded by partial fraction into two separate terms.

$$G_c(z) = \frac{E_2(z)}{E_1(z)} = \frac{18.75}{1 - z^{-1}} - \frac{8.75}{1 - 0.2z^{-1}} \tag{11-30}$$

Figure 11-11 shows the signal-flow graph of the parallel digital program of the controller. ▲

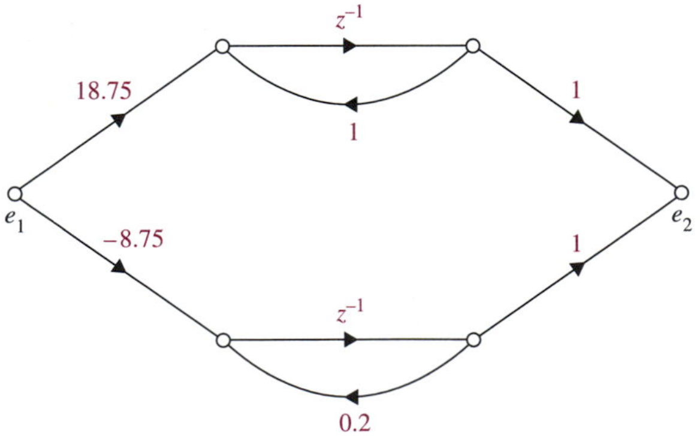

Figure 11-11 Pallel digital program of Eq. (11-25).

11-4 Design of Discrete-Data Control Systems in the Frequency Domain and the z-Plane

The w-transformation introduced in Section 7-10 can be used to carry out the design of discrete-data control systems in the frequency domain. Once the transfer function of the controlled process is transformed into the w-domain, all the design techniques for continuous-data control systems can be applied to the design of discrete-data systems.

11-4-1 Phase-Lead and Phase-Lag Controllers in the w-Domain

Just as in the s-domain, the single-stage phase-lead and phase-lag controllers in the w-domain can be expressed by the transfer function

$$G_c(w) = \frac{1 + a\tau w}{1 + \tau w} \tag{11-31}$$

where $a > 1$ corresponds to phase lead, and $a < 1$ corresponds to phase lag. When w is replaced by $j\omega_w$, the Bode plots of Eq. (11-31) are identical to those of Figs. 10-28 and 10-45 for $a > 1$ and $a < 1$, respectively. Once the controller is designed in the w-domain, the z-domain controller is obtained by substituting the w-transformation relationship in Eq. (6-47); that is,

$$w = \frac{2}{T}\frac{z - 1}{z + 1} \tag{11-32}$$

The following example illustrates the design of a discrete-data control system using the w-transformation in the frequency domain and the z-plane.

Example 11-3

Consider the sun-seeker control system described in Section 4-7 and shown in Fig. 10-29. Now let us assume that the system has discrete data, so that there is a sample-and-hold in the forward path. The sampling period is 0.01 sec. The transfer function of the controlled process is

$$G_p(s) = \frac{2500}{s(s + 25)} \tag{11-33}$$

The z-transfer function of the forward path, including the sample-and-hold, is

$$G_{h0}G_p(z) = (1 - z^{-1})\mathscr{Z}\left(\frac{2500}{s^2(s + 25)}\right) \tag{11-34}$$

Carrying out the z-transform in Eq. (11-34) with $T = 0.01$ sec, we get

$$G_{h0}G_p(z) = \frac{0.1152z + 0.106}{(z-1)(z-0.7788)} \tag{11-35}$$

The closed-loop transfer function of the discrete-data system is

$$\frac{\Theta_o(z)}{\Theta_r(z)} = \frac{G_{h0}G_p(z)}{1 + G_{h0}G_p(z)} = \frac{0.1152z + 0.106}{z^2 - 1.6636z + 0.8848} \tag{11-36}$$

The unit-step response of the uncompensated system is shown in Fig. 11-12. The maximum overshoot is 66 percent.

Let us carry out the design in the frequency domain using the w-transformation of Eq. (6-46),

$$z = \frac{(2/T) + w}{(2/T) - w} \tag{11-37}$$

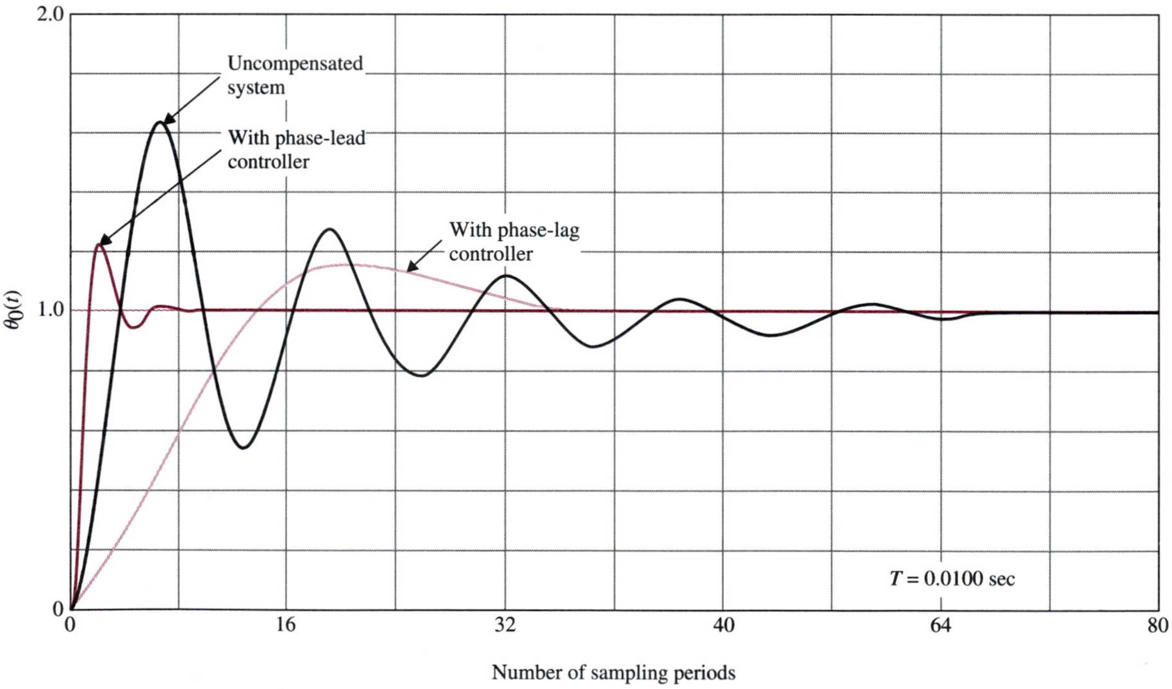

Figure 11-12 Step responses of discrete-data sun-seeker system in Example 11-3.

Substituting Eq. (11-37) into Eq. (11-35), we have

$$G_{h0}G_p(w) = \frac{100(1 - 0.005w)(1 + 0.000208w)}{w(1 + 0.0402w)} \tag{11-38}$$

The Bode plot of the last equation is shown in Fig. 11-13. The gain and phase margins of the uncompensated system are 6.39 dB and 14.77°, respectively.

Phase-Lag Controller Design in the Frequency Domain

Let us first design the system using a phase-lag controller with the transfer function given in Eq. (11-31) with $a < 1$. Let us require that the phase margin of the system be at least 50°. From the Bode plot in Fig. 11-13, a phase margin of 50° can be realized if the gain-crossover point is at $\omega_w = 12.8$ and the gain of the magnitude curve of $G_{h0}G_p(j\omega_w)$is 16.7 dB. Thus we need -16.7 dB of attentuation to bring the magnitude curve down so that it will cross the 0-dB axis at $\omega_w = 12.8$. We set

$$20 \log_{10} a = -16.7 \text{ dB} \tag{11-39}$$

from which we get $a = 0.1462$. Next, we set $1/a\tau$ to be at least one decade below the gain-crossover point at $\omega_w = 12.8$. We set

$$\frac{1}{a\tau} = 1 \tag{11-40}$$

Thus

$$\frac{1}{\tau} = a = 0.1462 \tag{11-41}$$

The phase-lag controller in the w-domain is

$$G_c(w) = \frac{1 + a\tau w}{1 + \tau w} = \frac{1 + w}{1 + 6.84w} \tag{11-42}$$

Substituting the z–w-transform relation, $w = (2/T)(z - 1)/(z + 1)$ in Eq. (11-42), the phase-lag controller in the z-domain is obtained,

$$G_c(z) = 0.1468\frac{z - 0.99}{z - 0.9985} \tag{11-43}$$

The Bode plot of the forward-path transfer function with the phase-lag controller of Eq. (11-42) is shown in Fig. 11-13. The phase margin of the compensated system is improved to 55°. The unit-step response of the phase-lag-compensated system is shown in Fig. 11-12. The maximum overshoot is reduced to 16 percent.

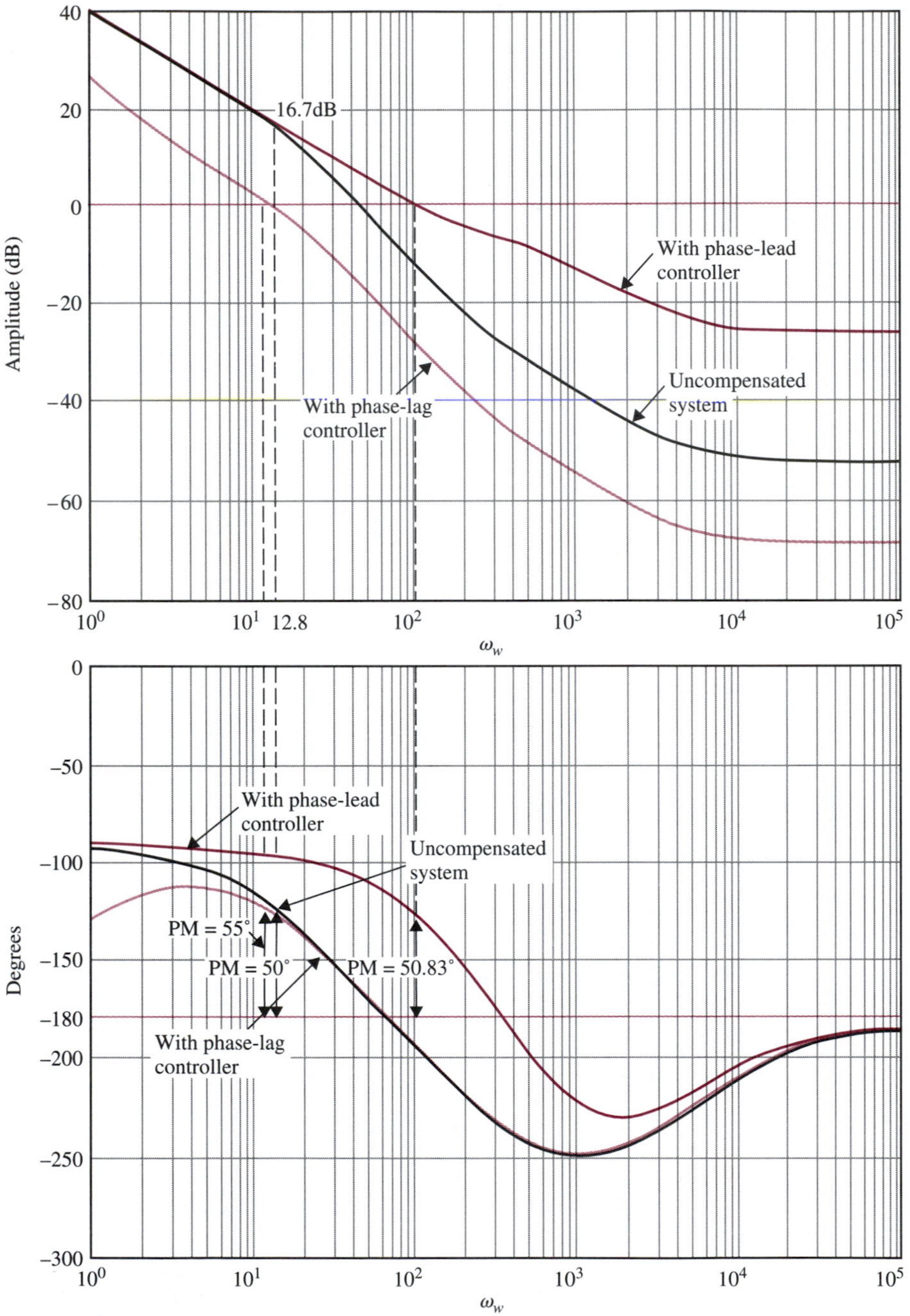

Figure 11-13 Bode plots of discrete-data sun-seeker system in Example 11-3.

Phase-Lead Controller Design in the Frequency Domain

A phase-lead controller is obtained by using $a > 1$ in Eq. (11-31). The same principle for the design of phase-lead controllers of continuous-data systems described in Chapter 10 can be applied here. Since the slope of the phase curve near the gain crossover is rather steep, as shown in Fig. 11-13, it is expected that some difficulty may be encountered in designing a phase-lead controller for the system. Nevertheless, we can assign a relatively large value for a rather than using the amount of phase lead required as a guideline.

Let us set $a = 20$. The gain of the controller at high values of ω_w is

$$20 \log_{10} a = 20 \log_{10} 20 = 26 \text{ dB} \tag{11-44}$$

From the design technique outlined in Chapter 10, the new gain crossover should be located at the point where the magnitude curve is at $-26/2 = -13$ dB. Thus the geometric mean of the two corner frequencies of the phase-lead controller should be at the point where the magnitude of $G_{h0}G_p(j\omega)$ is -13 dB. From Fig. 11-13 this is found to be a $\omega_w = 115$. Thus

$$\frac{1}{\tau} = 115\sqrt{a} = 514 \tag{11-45}$$

The w-domain transfer function of the phase-lead controller is

$$G_c(w) = \frac{1 + a\tau w}{1 + \tau w} = \frac{1 + 0.03888w}{1 + 0.001944w} \tag{11-46}$$

The transfer function of the phase-lead controller in the z-domain is

$$G_c(z) = \frac{8.7776z - 6.7776}{1.3888z + 0.6112} \tag{11-47}$$

The Bode plot of the phase-lead-compensated system is shown in Fig. 11-13. The phase margin of the compensated system is 50.83°. The unit-step response of the phase-lead-compensated system is shown in Fig. 11-12. The maximum overshoot is 27 percent, but the rise time is faster.

Digital PD-Controller Design in the z-Plane

The digital PD controller is described by the transfer function in Eq. (11-14) and is repeated as

$$G_c(z) = \frac{(K_P + K_D/T)z - K_D/T}{z} \tag{11-48}$$

To satisfy the condition that $G_c(1) = 1$, so that $G_c(z)$ does not affect the steady-state error of the system, we set $K_P = 1$. Applying the digital PD controller as a cascade controller to the sun-seeker system, the forward-path transfer function of the compensated system is

$$G_c(z)G_{h0}G_p(z) = \frac{(1 + 100K_D)z - 100\,K_D}{z}\,\frac{0.1152z + 0.106}{(z - 1)(z - 0.7788)} \tag{11-49}$$

We can use the root-contour method to investigate the effects of varying K_D. The characteristic equation of the closed-loop system is

$$z(z^2 - 1.6636z + 0.8848) + 11.52K_D(z - 1)(z + 0.9217) = 0 \tag{11-50}$$

Dividing both sides of the last equation by the terms that do not contain K_D, the equivalent forward-path transfer function with K_D appearing as a multiplying factor is

$$G_{eq}(z) = \frac{11.52K_D(z - 1)(z + 0.9217)}{z(z^2 - 1.6636z + 0.8848)} \tag{11-51}$$

The root contours of the system for $K_D > 0$ are shown in Fig. 11-14. These root contours show that the effectiveness of the digital PD controller is limited for this system, since the contours do not dip low enough toward the real axis. In fact, we can show that the best value of K_D from the standpoint of overshoot is 0.022 and the maximum overshoot is 28 percent.

Digital PI-Controller Design in the z-Plane

The digital PI controller introduced in Section 11-2-1 can be used to improve the steady-state performance by increasing the system type and, at the same time, improve the relative stability by using the "dipole" principle. Let us select the backward-rectangular integration implementation of the PID controller given by Eq. (11-13). With $K_D = 0$, the transfer function of the digital PI controller becomes

$$G_c(z) = \frac{K_P z - (K_P - K_I T)}{z - 1} = K_P \frac{z - (1 - K_I T/K_P)}{z - 1} \tag{11-52}$$

The principle of the "dipole" design of the PI controller is to place the zero of $G_c(z)$ very close to the pole at $z = 1$. The effective gain provided by the controller is essentially equal to K_P.

To create a root-locus problem, the transfer function in Eq. (11-35) is written

$$G_{h0}G_p(z) = \frac{K(z + 0.9217)}{(z - 1)(z - 0.7788)} \tag{11-53}$$

where $K = 0.1152$. The root loci of the system are drawn as shown in Fig. 11-15. The roots of the characteristic equation when $K = 0.1152$ are $0.8318 + j0.4392$ and $0.8318 - j0.4392$. As shown earlier, the maximum overshoot of the system is 66 percent. If K is reduced to 0.01152, the characteristic equation roots are at $0.8836 + j0.0926$ and $0.8836 - j0.0926$. The maximum overshoot is reduced to only 3 percent. We can show that if the gain in the numerator of Eq. (11-33) is reduced to 250, the maximum overshoot of the system would be reduced to 3 percent. This means that to realize a similar improvement on the maximum overshoot, the value of K_P in Eq. (11-52) should be set to 0.1. At the same time, we let the zero of $G_c(z)$ be at 0.995. Thus

$$G_c(z) = 0.1 \frac{z - 0.995}{z - 1} \tag{11-54}$$

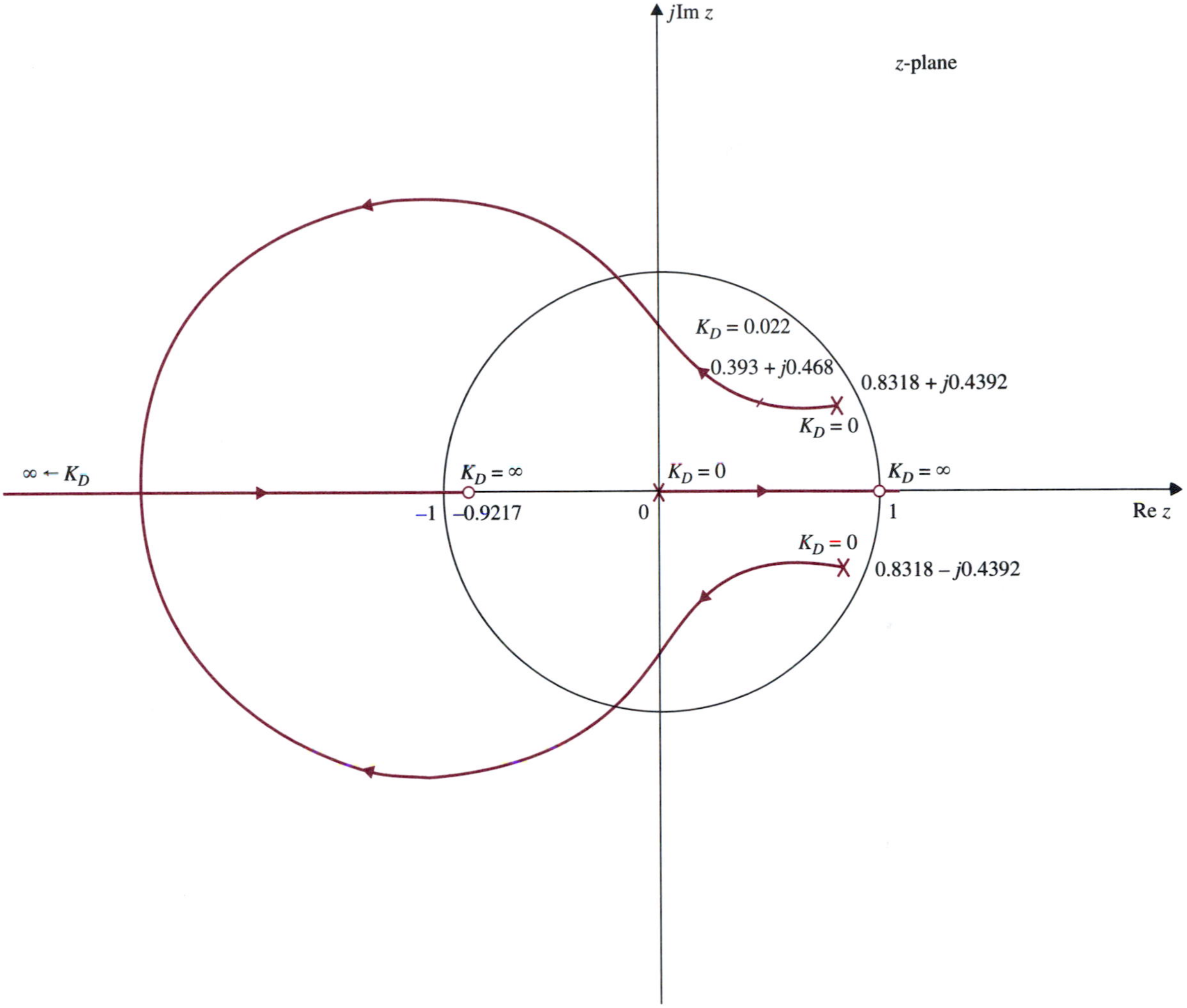

Figure 11-14 Root contours of sun-seeker system in Example 11-3 with digital PD controller. K_D varies.

The corresponding value of K_I is 0.05. The forward-path transfer function of the system with the PI controller becomes

$$G_c(z)G_{h0}G_p(z) = \frac{0.1K(z + 0.995)(z + 0.9217)}{(z - 1)^2(z - 0.7788)} \tag{11-55}$$

where $K = 0.1152$. The root loci of the compensated system are shown in Fig. 11-15. When $K = 0.1152$, the two dominant roots of the characteristic equation are at $0.8863 + j0.0898$ and $0.8863 - j0.0898$. The third root is at 0.9948, which is very close to the pole of $G_c(z)$ at $z = 1$,

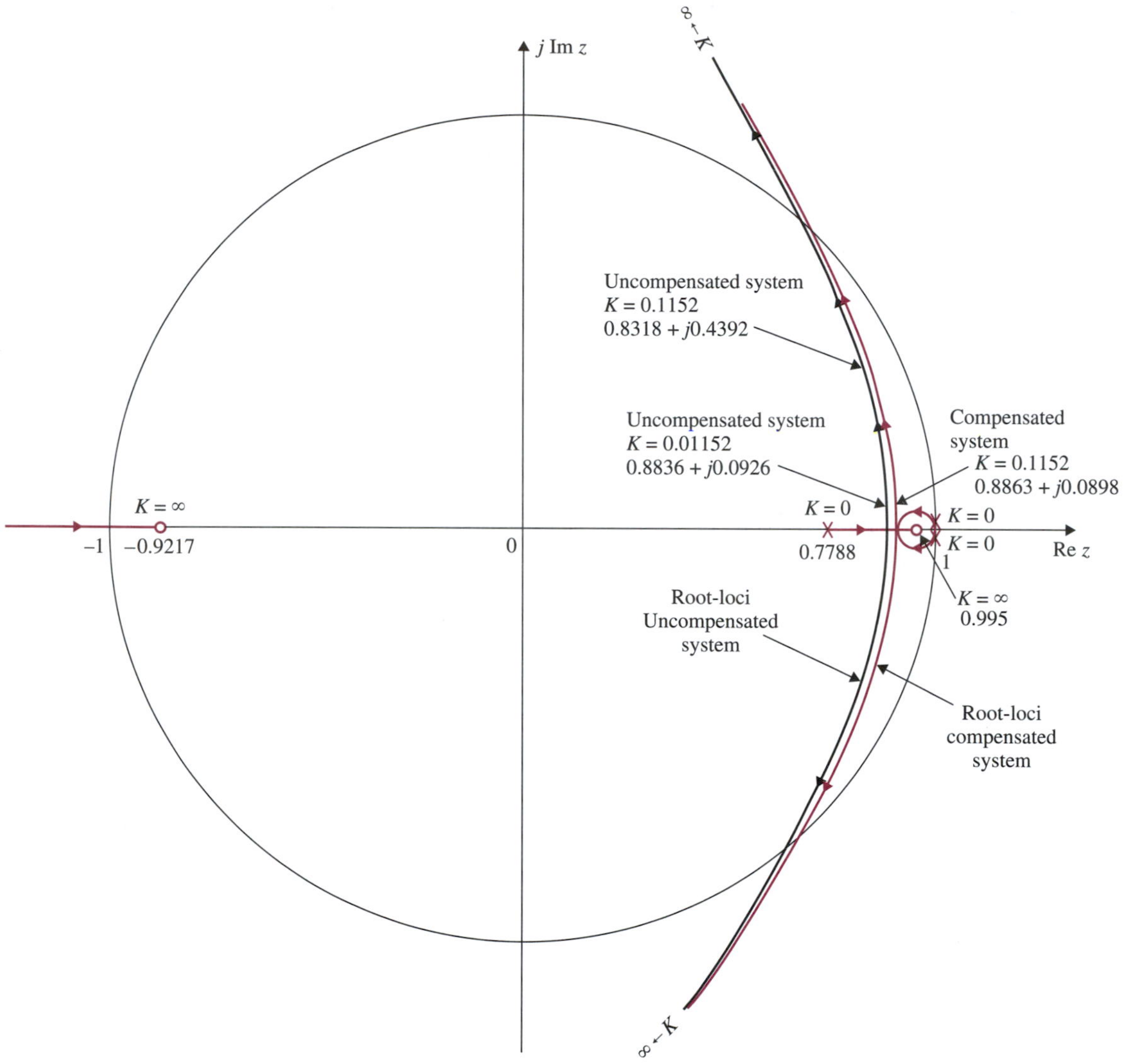

Figure 11-15 Root loci of sun-seeker system in Example 11-3, with and without digital PI controller.

and thus the effect on the transient response is negligible. We can show that the actual maximum overshoot of the system with the forward-path transfer function in Eq. (11-55) is approximately 8 percent.

This design problem simply illustrates the mechanics of designing a phase-lead and a phase-lag controller using the *w*-transformation method in the frequency domain and digital PD and PI controllers in the *z*-plane. No attempt was made in optimizing the system with respect to a set of performance specifications. ▲

11-5 Design of Discrete-Data Control Systems with Deadbeat Response

One difference between a continuous-data control system and a discrete-data control system is that the latter is capable of exhibiting a **deadbeat response.** A deadbeat response is one that reaches the desired reference trajectory in minimum time without error. In contrast, a continuous-data system reaches the final steady-state trajectory or value theoretically only when time reaches infinity. The switching operation of sampling allows the discrete-data systems to be able to have a finite transient period. Figure 11-16 shows a typical deadbeat response of a discrete-data system subject to a unit-step input. The output response reaches the desired steady-state with zero error in a minimum number of sampling periods without intersampling oscillations.

H. R. Sirisena [5] showed that given a discrete-data control system with the controlled process described by

$$G_{h0}G_p(z^{-1}) = \frac{Q(z^{-1})}{P(z^{-1})} \tag{11-56}$$

for the system to have a deadbeat response to a step input, the transfer function of the digital controller is given by

$$G_c(z) = \frac{P(z^{-1})}{Q(1) - Q(z^{-1})} \tag{11-57}$$

where $Q(1)$ is the value of $Q(z^{-1})$ with $z^{-1} = 1$.

The following example illustrates the design of a discrete-data system with deadbeat response using Eq. (11-57).

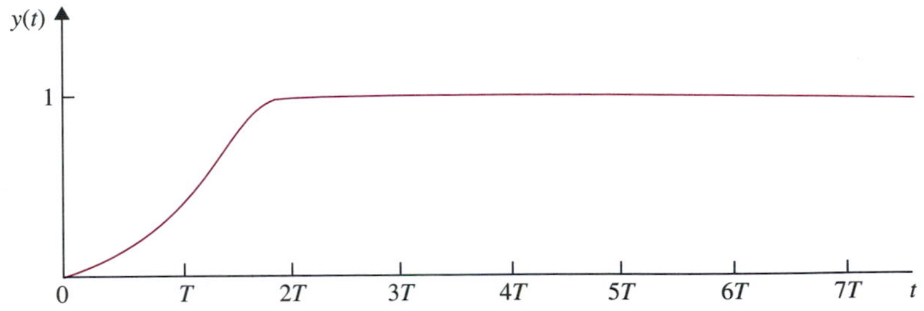

Figure 11-16 A typical deadbeat response to a unit-step input.

Example 11-4

Consider the discrete-data sun-seeker system discussed in Example 11-3. The forward-path transfer function of the uncompensated system is given in Eq. (11-35), and is written as

$$G_{h0}G_p(z^{-1}) = \frac{Q(z^{-1})}{P(z^{-1})} = \frac{0.1152z^{-1}(1 + 0.9217z^{-1})}{(1 - z^{-1})(1 - 0.7788z^{-1})} \tag{11-58}$$

Thus

$$Q(z^{-1}) = 0.1152z^{-1}(1 + 0.9217z^{-1}) \tag{11-59}$$
$$P(z^{-1}) = (1 - z^{-1})(1 - 0.7788z^{-1}) \tag{11-60}$$

and $Q(1) = 0.22138$.

The digital controller for a deadbeat response is obtained by using Eq. (11-57).

$$G_c(z^{-1}) = \frac{P(z^{-1})}{Q(1) - Q(z^{-1})} = \frac{(1 - z^{-1})(1 - 0.7788z^{-1})}{0.22138 - 0.1152z^{-1} - 0.106z^{-2}} \tag{11-61}$$

Thus

$$G_c(z) = \frac{(z - 1)(z - 0.7788)}{0.22138z^2 - 0.1152z - 0.106} \tag{11-62}$$

The forward-path transfer function of the compensated system is

$$G_c(z)G_{h0}G_p(z) = \frac{0.1152(z + 0.9217)}{0.22138z^2 - 0.1152z - 0.106} \tag{11-63}$$

The closed-loop transfer function of the compensated system is

$$\frac{\Theta_o(z)}{\Theta_r(z)} = \frac{G_c(z)G_{h0}G_p(z)}{1 + G_c(z)G_{h0}G_p(z)} = \frac{0.5204(z + 0.9217)}{z^2} \tag{11-64}$$

For a unit-step function input, the output transform is

$$\Theta_o(z) = \frac{0.5204(z + 0.9217)}{z(z - 1)} = 0.5204z^{-1} + z^{-2} + z^{-3} + \cdots \tag{11-65}$$

Thus the output response reaches the unit-step input in two sampling periods.

To show that the output response is without intersampling ripples, we evaluate the output velocity of the system; that is, $\omega_o(t) = d\theta_o(t)/dt$. The z-transfer function of the output velocity is written

$$\frac{\Omega_o(z)}{\Theta_r(z)} = \frac{G_c(z)(1 - z^{-1})\mathcal{Z}[2500/s(s + 25)]}{G_c(z)G_{h0}G_p(z)} = \frac{100(z - 1)}{z^2} \tag{11-66}$$

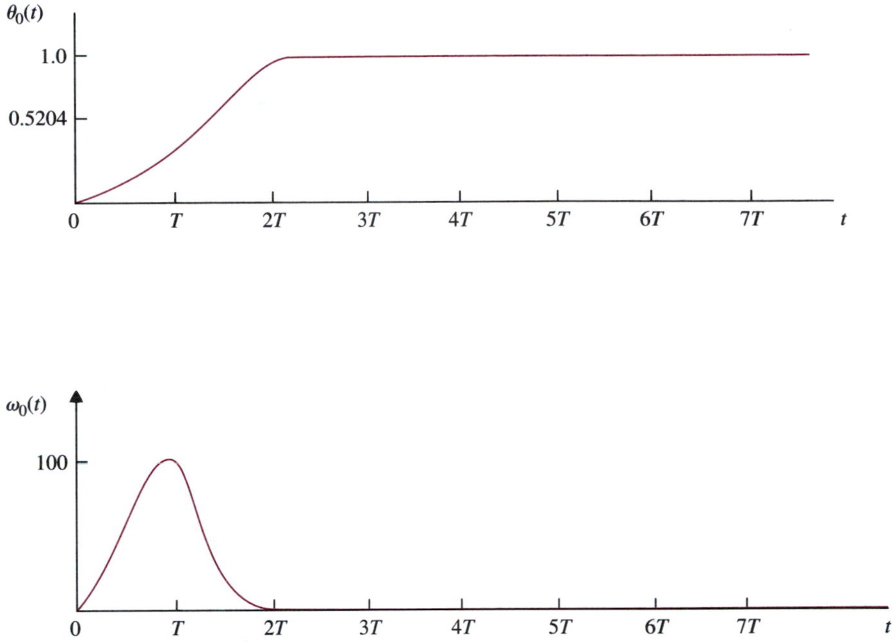

Figure 11-17 Output position and velocity responses of discrete-data sun-seeker system in Example 11-4.

The output velocity response due to a unit-step input is

$$\Omega_o(zx) = \frac{100}{z} = 100z^{-1} \tag{11-67}$$

Thus the output velocity becomes zero after the second sampling period, which proves that the position response is deadbeat without intersampling ripples. The responses of $\theta_o(t)$ and $\omega_o(t)$ are shown in Fig. 11-17.

▲ The root sensitivity of a system with deadbeat response is poor.

The characteristic of a system with deadbeat response is that the poles of the closed-loop transfer function are all at $z = 0$. Since these are multiple-order poles, from the standpoint of root sensitivity discussed in Chapter 8, the root sensitivity of a system with deadbeat response is very high. ▲

11-6 Pole-Placement Design with State Feedback

Just as for continuous-data systems, pole-placement design through state feedback can be applied to discrete-data systems. Let us consider the discrete-data system described by the following state equation:

$$\mathbf{x}[(k+1)T] = \mathbf{A}\mathbf{x}(kT) + \mathbf{B}u(kT) \tag{11-68}$$

where $\mathbf{x}(kT)$ is an $n \times 1$ state vector and $u(kT)$ is the scalar control. The state-feedback control is

$$u(kT) = -\mathbf{Kx}(kT) + r(kT) \tag{11-69}$$

where $\mathbf{K}$ is the $1 \times n$ feedback matrix with constant-gain elements. By substituting Eq. (11-69) into Eq. (11-68), the closed-loop system is represented by the state equation

$$\mathbf{x}[(k+1)T] = (\mathbf{A} - \mathbf{BK})\mathbf{x}(kT) \tag{11-70}$$

Just as in the continuous-data case treated in Section 10-12, we can show that if the pair $[\mathbf{A}, \mathbf{B}]$ is completely controllable, a matrix $\mathbf{K}$ exists that can give an arbitrary set of eigenvalues of $(\mathbf{A} - \mathbf{BK})$; that is, the n roots of the characteristic equation

$$|z\mathbf{I} - \mathbf{A} + \mathbf{BK}| = 0 \tag{11-71}$$

can be placed arbitrarily. The following example illustrates the design of a discrete-data control system with state feedback and pole placement.

Example 11-5

Consider that the sun-seeker system described in Example 10-9 is subject to sampled data so that the state diagram of the system without feedback is shown in Fig. 11-18(a). The sampling period is 0.01 sec. The dynamics of the zero-order hold is represented by the branch with a transfer function of $1/s$. The closed-loop system with state feedback for the time interval $(kT) \leq t \leq (k+1)T$ is portrayed by the state diagram shown in Fig. 11-18(b), where the

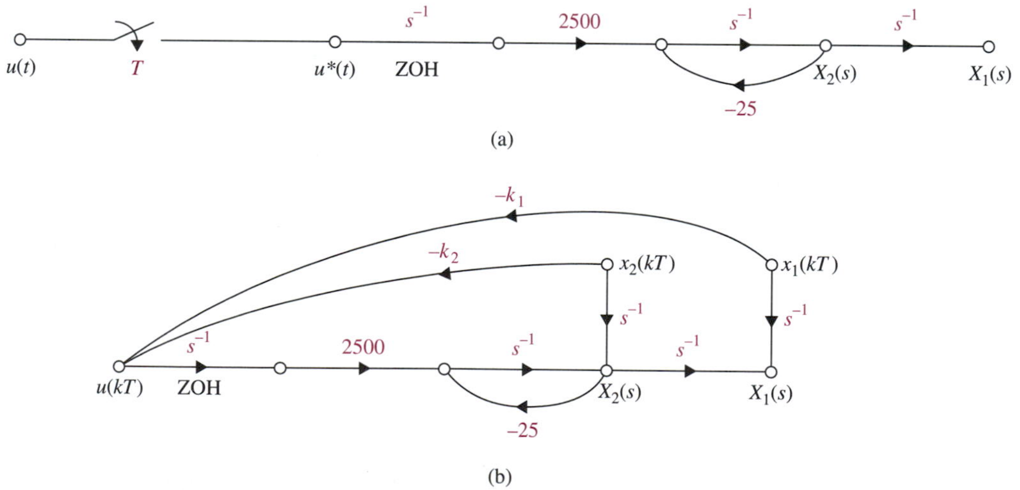

Figure 11-18 (a) Signal-flow graph of open-loop discrete-data sun-seeker system. (b) Signal-flow graph of discrete-data sun-seeker system with state feedback.

feedback gains k_1 and k_2 form the feedback matrix

$$\mathbf{K} = [\mathrm{k}_1 \quad k_2] \tag{11-72}$$

Applying the SFG gain formula to Fig. 11-18(b), with $X_1(s)$ and $X_2(s)$ as outputs and $x_1(kT)$ and $x_2(kT)$ as inputs, we have

$$X_1(s) = \left[\frac{1}{s} - \frac{2500k_1}{s^2(s+25)}\right] x_1(kT) + \left[\frac{1}{s(s+25)} - \frac{2500k_2}{s^2(s+25)}\right] x_2(\mathrm{kT}) \tag{11-73}$$

$$X_2(s) = \frac{-2500k_1}{s(s+25)} x_1(kT) - \frac{2500k_2}{s(s+25)} x_2(kT) + \frac{1}{s+25} x_2(kT) \tag{11-74}$$

Taking the inverse Laplace transform on both sides of Eqs. (11-73) and (11-74) and letting $t = (k+1)T$, we have the discrete-data state equations as

$$\begin{aligned} x_1[(k+1)T] &= (1 - 0.1152k_1)x_1(kT) + (0.2212 - 0.1152k_2)x_2(kT) \\ x_2[(k+1)T] &= -22.12k_1 x_1(kT) + (0.7788 - 22.12k_2)x_2(kT) \end{aligned} \tag{11-75}$$

Thus the coefficient matrix of the closed-loop system with state feedback is

$$\mathbf{A} - \mathbf{BK} = \begin{bmatrix} 1 - 0.1152k_1 & 0.2212 - 0.1152k_2 \\ -22.12k_1 & 0.7788 - 22.12k_2 \end{bmatrix} \tag{11-76}$$

The characteristic equation of the closed-loop system is

$$\begin{aligned} |z\mathbf{I} - \mathbf{A} + \mathbf{BK}| &= \begin{vmatrix} z - 1 + 0.1152k_1 & -0.2212 + 0.1152k_2 \\ 22.12k_1 & z - 0.7788 + 22.12k_2 \end{vmatrix} \\ &= z^2 + (-1.7788 + 0.1152k_1 + 22.12k_2)z + 0.7788 + 4.8032k_1 - 22.12k_2 = 0 \end{aligned} \tag{11-77}$$

Let the desired location of the characteristic equation roots be at $z = 0, 0$. The conditions on k_1 and k_2 are

$$-1.7788 + 0.1152k_1 + 22.12k_2 = 0 \tag{11-78}$$

$$0.7788 + 4.8032k_1 - 22.12k_2 = 0 \tag{11-79}$$

Solving for k_1 and k_2 from the preceding two equations, we get

$$k_1 = 0.2033 \quad \text{and} \quad k_2 = -0.07936 \tag{11-80}$$

▲

11-7 Summary

In this chapter the design of linear discrete-data control system is discussed. Many practical design of discrete-data control systems are carried out by using a digitized version of the continuous-data controller. Digital implementation of the PID controller is first presented by using three different integration rules for the integration portion: trapezoidal integration, forward-rectangular integration, and backward-integration integration. Digital implementation of continuous-data controllers can also be carried out by adding sample-and-hold units at the input and the output terminals of the controller.

Given the transfer function of a digital controller in the z-domain, the controller can be realized by three basic types of decomposition or programming: direct digital programming, parallel decomposition, and cascade decomposition. By using the w-transformation of Section 7-10, most of the design techniques in the frequency domain for continuous-data systems can be applied to the design of discrete-data systems. Examples on how to design phase-lead and phase-lag controllers for discrete-data control systems are given. One of the distinctive features of a discrete-data system is that it is capable of achieving a deadbeat response. A method of designing a discrete-data system with deadbeat response, and without intersampling ripples, is described. Finally, pole-placement design of discrete-data systems with state feedback is presented.

The purpose of this chapter is to give an overview of the fundamentals of design of discrete-data control systems. For in-depth treatment of the subject, the reader should refer to books dedicated to discrete-data control systems listed in the reference section of this chapter.

Review Questions

1. Give the discrete-data approximation of the derivative of a time function $f(t)$, $df(t)/dt$, by the backward-difference rule.

2. Describe the three digital integration rules described in this chapter.

3. Give the conditions for the following transfer function to be physically realizable.

$$G_c(z) = \frac{b_m z^m + b_{m-1} z^{m-1} + \cdots + b_1 z + b_0}{a_n z^n + a_{n-1} z^{n-1} + \cdots + a_1 z + a_0}$$

4. Give the advantage of using the w-transformation for the design of discrete-data systems.

5. What is a deadbeat response?

6. What is the characteristic of the poles of the transfer function of a system that has a deadbeat response?

7. What can you say about the root sensitivity of a system with a deadbeat response?

References

1. **G. F. FRANKLIN, J. D. POWELL, and M. L. WORKMAN,** *Digital Control of Dynamic Systems,* 2nd ed., Addison-Wesley Publishing Co., Reading, MA, 1990.

2. **B. C. KUO,** *Digital Control Systems,* 2nd ed., Saunders College Publishing, Philadelphia, 1992.
3. **K. OGATA,** *Discrete-Time Control Systems,* Prentice Hall, Englewood Cliffs, NJ, 1987.
4. **C. L. PHILLIPS and H. T. NAGLE, JR.,** *Digital Control System Analysis and Design,* Prentice Hall, Englewood Cliffs, NJ, 1984.
5. **H. R. SIRISENA,** ''Ripple-Free Deadbeat Control of SISO Discrete Systems,'' IEEE Trans. Automatic Control, Vol. AC-30, pp. 168–170, February 1985.

Problems

▲ Digital integration

11-1. Find the digital equivalents using the following integration rules for the controllers given: backward-rectangular integration rule, forward-rectangular integration rule, and trapezoidal-integration rule. Use the backward-difference rule for derivatives.

(a) $G_c(s) = 2 + \dfrac{200}{s}$ **(b)** $G_c(s) = 10 + 0.1s$ **(c)** $G_c(s) = 1 + 0.2s + \dfrac{5}{s}$

▲ Digital program implementation of digital controllers

11-2. A continuous-data controller with sample-and-hold units is shown in Fig. 11P-2. The sampling period is 0.1 sec. Find the transfer function of the equivalent digital controller. Draw a digital-program implementation diagram for the digital controller. Carry out the analysis for the following continuous-data controllers.

(a) $G_c(s) = \dfrac{10}{s + 12}$ **(b)** $G_c(s) = \dfrac{10(s + 1.5)}{s + 10}$

(c) $G_c(s) = \dfrac{s}{s + 1.55}$ **(d)** $G_c(s) = \dfrac{1 + 0.4s}{1 + 0.01s}$

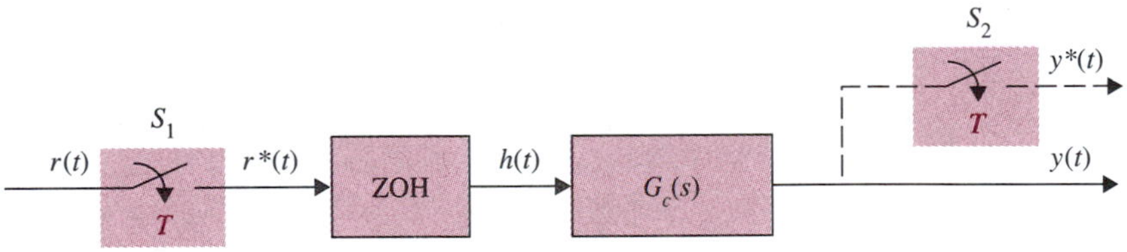

Figure 11P-2

▲ Physical realizability of digital transfer functions

11-3. Determine which of the following digital transfer functions are physically realizable.

(a) $G_c(z) = \dfrac{10(1 + 0.2z^{-1} + 0.5z^{-2})}{z^{-1} + z^{-2} + 1.5z^{-3}}$ **(b)** $G_c(z) = \dfrac{1.5z^{-1} - z^{-2}}{1 + z^{-1} + 2z^{-2}}$

(c) $G_c(z) = \dfrac{z + 1.5}{z^3 + z^2 + z + 1}$ **(d)** $G_c(z) = z^{-1} + 0.5z^{-3}$

(e) $G_c(z) = 0.1z + 1 + z^{-1}$ **(f)** $G_c(z) = \dfrac{z^{-1} + 2z^{-2} + 0.5z^{-3}}{z^{-1} + z^{-2}}$

▲ Inventory-control system with digital PD controller

11-4. The transfer function of the process of the inventory-control system described in Problem 10-17 is

$$G_p(s) = \frac{4}{s^2}$$

The block diagram of the system with a PD controller and sample-and-hold is shown in Fig. 11P-4.

$$G_c(s) = 1 + 10s$$

Find the transfer function of the equivalent digital PD controller using [Eq. (11-48)]

$$G_c(z) = \frac{(K_P + K_D/T)z - K_D/T}{z}$$

Select a sampling period T so that the maximum overshoot of $y(kT)$ will be less than 1 percent.

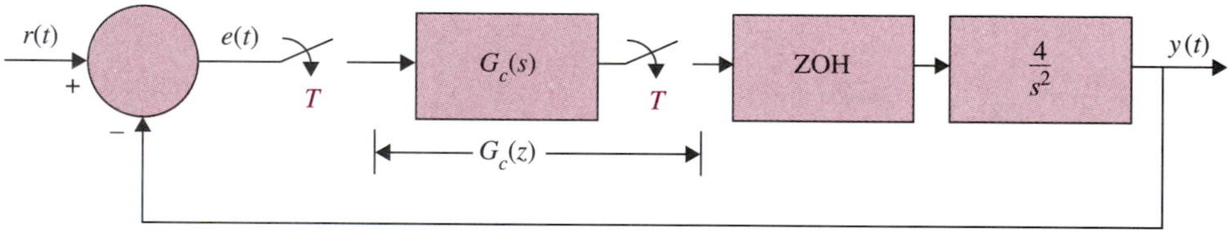

Figure 11P-4

▲ Inventory-control system with digital PD controller

11-5. Figure 11P-4 shows the block diagram of the inventory-control system described in Problem 10-17 with a digital PD controller. The sampling period is 0.01 sec. Consider that the digital PD controller has the transfer function

$$G_c(z) = K_P + \frac{K_D(z - 1)}{Tz}$$

(a) Find the values of K_P and K_D so that two of the three roots of the characteristic equation are at 0.5 and 0.5. Find the third root. Plot the unit-step response $y(kT)$ for $k = 0, 1, 2, \ldots$.

(b) Set $K_P = 1$. Find the value of K_D so that the maximum overshoot of $y(kT)$ is a minimum.

▲ Inventory-control system with digital phase-lead controller

11-6. For the inventory-control system described in Problem 11-5, design a phase-lead controller using the w-transformation so that the phase margin of the system is at least 60°. Can you design a phase-lag controller in the w-domain? If not, explain why not. Use $T = 0.1$ sec.

▲ Aircraft attitude control system with digital controller

11-7. The process transfer function of the second-order aircraft attitude control system described in Problem 10-5 is

$$G_p(s) = \frac{4500K}{s(s + 361.2)}$$

Consider that the system is to be compensated by a series digital controller $G_c(z)$ through a sample-and-hold. $T = 0.001$ sec.

(a) Find the value of K so that the discrete ramp-error constant K_v^* is 1000.

(b) With the value of K found in part (a), plot the unit-step response of the output $y^*(t)$, and find maximum overshoot.

(c) Design the digital controller so that the output is a deadbeat response to a step input. Plot the unit-step response.

▲ Sun-seeker system with digital controller, deadbeat response

11-8. The sun-seeker system described in Example 11-3 is considered to be controlled by a series digital controller with the transfer function $G_c(z)$. The sampling period is 0.05 sec. Design the controller so that the output of the system is a deadbeat response to a unit-step input. Plot the unit-step response of the designed system.

▲ Sun-seeker system with state feedback

11-9. Design the state-feedback control for the sun-seeker system in Example 11-5 so that the characteristic equation roots are at $z = 0.5$ and 0.5.

▲ State-feedback control

11-10. Consider the digital control system

$$\mathbf{x}[(k + 1)T] = \mathbf{A}\mathbf{x}(kT) + \mathbf{B}u(kT)$$

where

$$\mathbf{A} = \begin{bmatrix} 0 & 1 \\ -1 & -1 \end{bmatrix} \qquad \mathbf{B} = \begin{bmatrix} 0 \\ 1 \end{bmatrix}$$

The state-feedback control is described by $u(kT) = -\mathbf{K}\mathbf{x}(kT)$, where $\mathbf{K} = [k_1 \quad k_2]$. Find the values k_1 and k_2 so that the roots of the characteristic equation of the closed-loop system are at 0.5 and 0.7.

A Appendix
Frequency-Domain Plots

KEYWORDS AND TOPICS

- ▲ Polar Plot
- ▲ Bode Plot
- ▲ Magnitude–Phase Plot
- ▲ Gain-Crossover Point
- ▲ Phase-Crossover Point
- ▲ Minimum-Phase and Nonminimum-Phase Functions

Let $G(s)$ be the forward-path transfer function of a linear control system with unity feedback. The frequency-domain analysis of the closed-loop system can be conducted from the frequency-domain plots of $G(s)$ with s replaced by $j\omega$. The function $G(j\omega)$ is generally a complex function of the frequency ω, and can be written as

$$G(j\omega) = |G(j\omega)|\angle G(j\omega) \tag{A-1}$$

where $|G(j\omega)|$ denotes the magnitude of $G(j\omega)$ and $\angle G(j\omega)$ is the phase of $G(j\omega)$.

The following frequency-domain plots of $G(j\omega)$ versus ω are often used in the analysis and design of linear control systems in the frequency domain.

1. *Polar plot.* A plot of the magnitude versus phase in the polar coordinates as ω is varied from zero to infinity.
2. *Bode plot.* A plot of the magnitude in decibels versus ω (or $\log_{10}\omega$) in semilog (or rectangular) coordinates.
3. *Magnitude–phase plot.* A plot of the magnitude in decibels versus the phase on rectangular coordinates with ω as a variable parameter on the curve.

Computer-Aided Construction of the Frequency-Domain Plots

The data for plotting of the frequency-domain plots are usually quite time consuming to generate if the computation is to be carried out manually, especially if the function is of high order. In practice, a digital computer should be used to do the computation as well as plotting of the graph. Many software packages are available commercially and can be used for the construction of the frequency-domain plots. The program **FREQRP** of the **ACSP** software package and **CSAD/MATLAB** functions **bplot** and **plrplot** can all be used for this purpose. From an analytical standpoint, the analyst and designer should still be familiar with the properties of the frequency-domain plots, so that proper interpretations can be made on these plots made by the computer.

A-1 Polar Plots

The polar plot of a function of the complex variable s, $G(s)$, is a plot of the magnitude of $G(j\omega)$ versus the phase of $G(j\omega)$ on polar coordinates as ω is varied from zero to infinity. From the mathematical viewpoint, the process can be regarded as the mapping

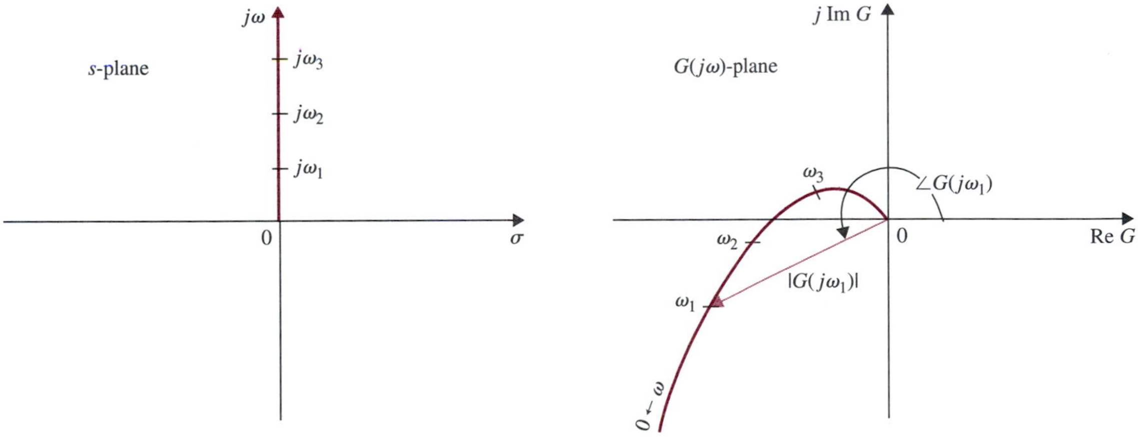

Figure A-1 Polar plot shown as a mapping of the positive half of the $j\omega$-axis in the s-plane onto the $G(j\omega)$-plane.

of the positive half of the imaginary axis of the s-plane onto the $G(j\omega)$-plane. A simple example of this mapping is shown in Fig. A-1. For any frequency $\omega = \omega_1$, the magnitude and phase of $G(j\omega_1)$ are represented by a vector in the $G(j\omega)$-plane. In measuring the phase, counterclockwise is referred to as positive, and clockwise is negative.

Example A-1

To illustrate the construction of the polar plot of a function $G(s)$, consider the function

$$G(s) = \frac{1}{1 + Ts} \tag{A-2}$$

where T is a positive constant. Setting $s = j\omega$, we have

$$G(j\omega) = \frac{1}{1 + j\omega T} \tag{A-3}$$

In terms of magnitude and phase, Eq. (A-3) is written

$$G(j\omega) = \frac{1}{\sqrt{1 + \omega^2 T^2}} \angle -\tan^{-1} \omega T \tag{A-4}$$

When ω is zero, the magnitude of $G(j\omega)$ is unity, and the phase of $G(j\omega)$ is at 0°. Thus, at $\omega = 0$, $G(j\omega)$ is represented by a vector of unit length directed in the 0° direction. As ω increases, the magnitude of $G(j\omega)$ decreases and the phase becomes more negative. As ω increases, the length of the vector in the polar coodinates decreases and the vector rotates in the clockwise (negative) direction. When ω approaches infinity, the magnitude of $G(j\omega)$ becomes zero and the phase reaches $-90°$. This is presented by a vector with an infinitesimally small length directed along the $-90°$-axis in the $G(j\omega)$-plane. By substituting other finite values of ω into Eq. (A-4), the exact plot of $G(j\omega)$ turns out to be a semicircle, as shown in Fig. A-2. ▲

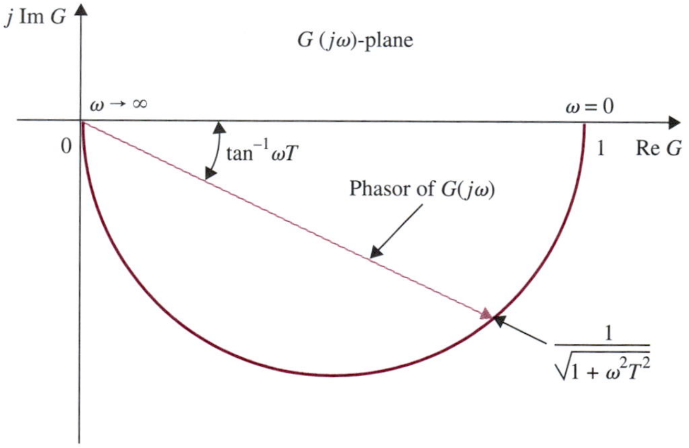

Figure A-2 Polar plot of $G(j\omega) = 1/(1 + j\omega T)$.

Example A-2

As a second illustrative example, consider the function

$$G(j\omega) = \frac{1 + j\omega T_2}{1 + j\omega T_1} \tag{A-5}$$

where T_1 and T_2 are positive real constants. Equation (A-5) is written

$$G(j\omega) = \sqrt{\frac{1 + \omega^2 T_2^2}{1 + \omega^2 T_1^2}} \angle(\tan^{-1}\omega T_2 - \tan^{-1}\omega T_1) \tag{A-6}$$

The polar plot of $G(j\omega)$, in this case, depends on the relative magnitudes of T_1 and T_2. If T_2 is greater than T_1, the magnitude of $G(j\omega)$ is always greater than unity as ω is varied from zero to infinity, and the phase of $G(j\omega)$ is always positive. If T_2 is less than T_1, the magnitude of $G(j\omega)$ is always less than unity and the phase is always negative. The polar plots of $G(j\omega)$ of Eq. (A-6) that correspond to the two above-mentioned conditions are shown in Fig. A-3. ▲

In many control system applications, such as the Nyquist stability criterion, an exact plot of the frequency response is not essential. Often, a rough sketch of the polar plot of the transfer function $G(j\omega)H(j\omega)$ is quite adequate for stability anaysis in the frequency domain. The general shape of the polar plot of a function $G(j\omega)$ can be determined from the following information:

1. The behavior of the magnitude and phase of $G(j\omega)$ at $\omega = 0$ and $\omega = \infty$
2. The intersections of the polar plot with the real and imaginary axes, and the values of ω at these intersections

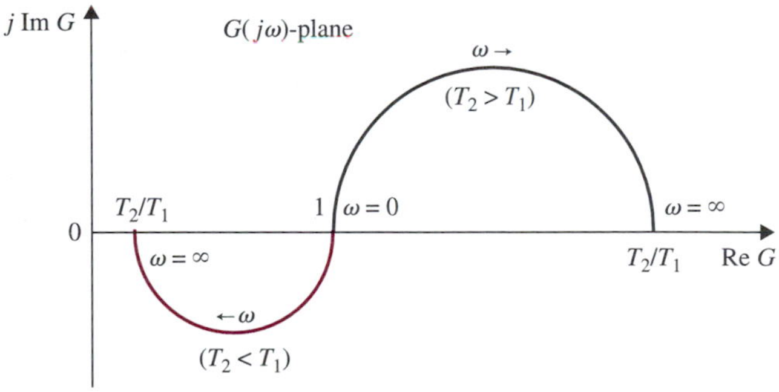

Figure A-3 Polar plots of $G(j\omega) = (1 + j\omega T_2)/(1 + j\omega T_1)$.

Example A-3

In frequency-domain analysis of control systems, often we have to determine the basic properties of a polar plot. Consider the following transfer function:

$$G(s) = \frac{10}{s(s+1)} \tag{A-7}$$

By substituting $s = j\omega$ in Eq. (A-7), the magnitude and phase of $G(j\omega)$ at $\omega = 0$ and $\omega = \infty$ are computed as follows:

$$\lim_{\omega \to 0} |G(j\omega)| = \lim_{\omega \to 0} \frac{10}{\omega} = \infty \tag{A-8}$$

$$\lim_{\omega \to 0} \angle G(j\omega) = \lim_{\omega \to 0} \angle 10/j\omega = -90^\circ \tag{A-9}$$

$$\lim_{\omega \to \infty} |G(j\omega)| = \lim_{\omega \to \infty} \frac{10}{\omega^2} = 0 \tag{A-10}$$

$$\lim_{\omega \to \infty} \angle G(j\omega) = \lim_{\omega \to \infty} \angle 10/(j\omega^2) = -180^\circ \tag{A-11}$$

Thus the properties of the polar plot of $G(j\omega)$ at $\omega = 0$ and $\omega = \infty$ are ascertained. Next, we determine the intersections, if any, of the polar plot with the two axes of the $G(j\omega)$-plane. If the polar plot of $G(j\omega)$ intersects the real axis, at the point of intersection, the imaginary part of $G(j\omega)$ is zero; that is,

$$\mathrm{Im}[G(j\omega)] = 0 \tag{A-12}$$

To express $G(j\omega)$ as the sum of its real and imaginary parts, we must rationalize $G(j\omega)$ by multiplying its numerator and denominator by the complex conjugate of its denominator. There-

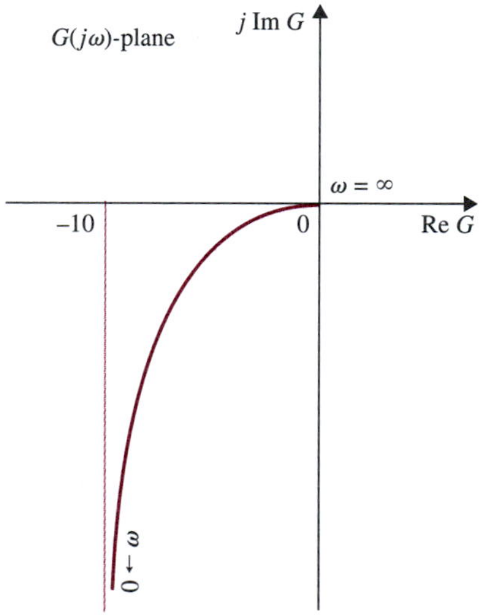

Figure A-4 Polar plot of $G(s) = 10/[s(s+1)]$.

fore, $G(j\omega)$ is written

$$G(j\omega) = \frac{10(-j\omega)(-j\omega + 1)}{j\omega(j\omega + 1)(-j\omega)(-j\omega + 1)} = \frac{-10\omega^2}{\omega^4 + \omega^2} - j\frac{10\omega}{\omega^4 + \omega^2} = \text{Re}[G(j\omega)] + j\text{Im}[G(j\omega)] \tag{A-13}$$

When we set $\text{Im}[G(j\omega)]$ to zero, we get $\omega = \infty$, meaning that the $G(j\omega)$ plot intersects only with the real axis of the $G(j\omega)$-plane at the origin.

Similarly, the intersection of $G(j\omega)$ with the imaginary axis is found by setting $\text{Re}[G(j\omega)]$ of Eq. (A-13) to zero. The only real solution for ω is also $\omega = \infty$, which corresponds to the origin of the $G(j\omega)$-plane. The conclusion is that the polar plot of $G(j\omega)$ *does not* intersect any one of the axes at any finite nonzero frequency. Under certain conditions we are interested in the properties of the $G(j\omega)$ at infinity, which corresponds to $\omega = 0$ in this case. From Eq. (A-13) we see that $\text{Im}[G(j\omega)] = \infty$ and $\text{Re}[G(j\omega)] = -10$ at $\omega = 0$. Based on this information, as well as knowledge of the angles of $G(j\omega)$ at $\omega = 0$ and $\omega = \infty$, the polar plot of $G(j\omega)$ is easily sketched without actual plotting, as shown in Fig. A-4. ▲

Example A-4

Given the transfer function

$$G(s) = \frac{10}{s(s + 1)(s + 2)} \tag{A-14}$$

We want to make a rough sketch of the polar plot of $G(j\omega)$. The following calculations are made for the properties of the magnitude and phase of $G(j\omega)$ at $\omega = 0$ and $\omega = \infty$:

$$\lim_{\omega \to 0} |G(j\omega)| = \lim_{\omega \to 0} \frac{5}{\omega} = \infty \tag{A-15}$$

$$\lim_{\omega \to 0} \angle G(j\omega) = \lim_{\omega \to 0} \angle 5/j\omega = -90^\circ \tag{A-16}$$

$$\lim_{\omega \to \infty} |G(j\omega)| = \lim_{\omega \to \infty} \frac{10}{\omega^3} = 0 \tag{A-17}$$

$$\lim_{\omega \to \infty} \angle G(j\omega) = \lim_{\omega \to \infty} \angle 10/(j\omega^3) = -270^\circ \tag{A-18}$$

To find the intersections of the $G(j\omega)$ plot on the real and imaginary axes of the $G(j\omega)$-plane, we rationalize $G(j\omega)$ to give

$$G(j\omega) = \frac{10(-j\omega)(-j\omega + 1)(-j\omega + 2)}{j\omega(j\omega + 1)(j\omega + 2)(-j\omega)(-j\omega + 1)(-j\omega + 2)} \tag{A-19}$$

After simplification, Eq. (A-19) is written

$$G(j\omega) = \text{Re}[G(j\omega)] + j\text{Im}[G(j\omega)] = \frac{-30}{9\omega^2 + (2 - \omega^2)^2} - \frac{j10(2 - \omega^2)}{9\omega^3 + \omega(2 - \omega^2)^2} \tag{A-20}$$

Setting $\text{Re}[G(j\omega)]$ to zero, we have $\omega = \infty$ and $G(j\infty) = 0$, which means that the $G(j\omega)$ plot intersects the imaginary axis only at the origin. Setting $\text{Im}[G(j\omega)]$ to zero, we have $\omega =$

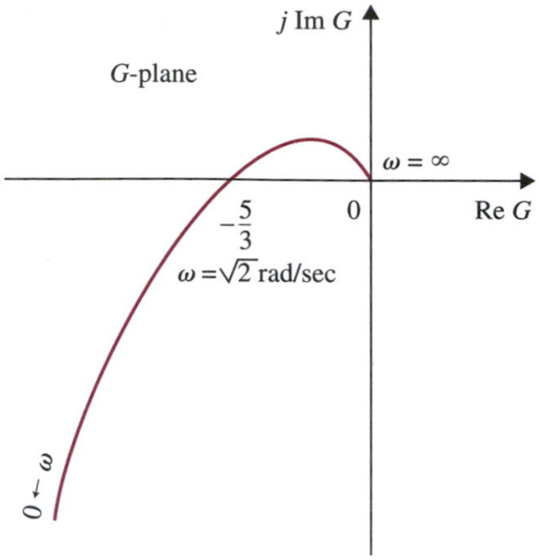

Figure A-5 Polar plot of $G(s) = 10/[s(s+1)(s+2)]$.

$\pm\sqrt{2}$ rad/sec. This gives the point of intersection on the real axis at

$$G(\pm j\sqrt{2}) = -\tfrac{5}{3} \tag{A-21}$$

The result of $\omega = -\sqrt{2}$ rad/sec has no physical meaning, as the frequency is negative; it simply represents a mapping point on the negative $j\omega$-axis of the s-plane. In general, if $G(s)$ is a rational function of s (a quotient of two polynomials of s), the polar plot of $G(j\omega)$ for negative values of ω is the mirror image of that for positive ω, with the mirror placed on the real axis of the $G(j\omega)$-plane. From Eq. (A-20) we also see that $\text{Re}[G(j0)] = \infty$ and $\text{Im}[G(j0)] = \infty$. With this information it is now possible to make a sketch of the polar plot for the transfer function in Eq. (A-14), as shown in Fig. A-5. ▲

Although the method of obtaining the rough sketch of the polar plot of a transfer function as described is quite straightforward, in general, for complicated transfer functions that may have multiple crossings on the real and imaginary axes of the transfer-function plane, the algebraic manipulation may again be quite involved. Furthermore, the polar plot is basically a tool for analysis; it is somewhat awkward for design purposes. We shall show in the next section that approximate information on the polar plot can always be obtained from the Bode plot, which can be sketched without any calculations. Thus for more complicated transfer functions, other than using the digital computer, sketches of the polar plots, if necessary, are preferably obtained with the help of Bode plots.

A-2 Bode Plot (Corner Plot or Asymptotic Plot)

The Bode plot of the function $G(j\omega)$ is composed of two plots, one with the amplitude of $G(j\omega)$ in decibels (dB) versus $\log_{10}\omega$ or ω, and the other with the phase of $G(j\omega)$ in degrees as a function of $\log_{10}\omega$ or ω. The Bode plot is also known as the **corner**

▲ A Bode plot is also known as corner plot or asymptotic plot.

plot or the **asymptotic plot** of $G(j\omega)$. These names stem from the fact that the Bode plot can be constructed by using straight-line approximations that are asymptotic to the actual plot.

In simple terms, the Bode plot has the following features:

1. Since the magnitude of $G(j\omega)$ in the Bode plot is expressed in dB, product and division factors in $G(j\omega)$ became additions and subtractions, respectively. The phase relations are also added and subtracted from each other algebraically.
2. The magnitude plot of the Bode plot of $G(j\omega)$ can be approximated by straight-line segments, which allow the simple sketching of the plot without detailed computation.

Since the straight-line approximation of the Bode plot is relatively easy to construct, the data necessary for the other frequency-domain plots, such as the polar plot and the magnitude versus phase plot, can be easily generated from the Bode plot.

Consider the function

$$G(s) = \frac{K(s + z_1)(s + z_2) \cdots (s + z_m)}{s^j(s + p_1)(s + p_2) \cdots (s + p_n)} e^{-T_d s} \tag{A-22}$$

where K and T_d are real constants, and the z's and the p's may be real or complex (in conjugate pairs) numbers. In Chapter 8, Eq. (A-22) is the preferred form for root-locus construction, since the poles and zeros of $G(s)$ are easily identified. For constructing the Bode plot manually, $G(s)$ is preferably written in the following form:

$$G(s) = \frac{K_1(1 + T_1 s)(1 + T_2 s) \cdots (1 + T_m s)}{s^j(1 + T_a s)(1 + T_b s) \cdots (1 + T_n s)} e^{-T_d s} \tag{A-23}$$

where K_1 is a real constant, the T's may be real or complex (in conjugate pairs) numbers, and T_d is the real time delay. If the Bode plot is to be constructed with a computer program, then either forms of Eq. (A-22) or eq. (A-23) can be used.

Since practically all the terms in Eq. (A-23) are of the same form, without loss of generality, we can use the following transfer function to illustrate construction of the Bode diagram.

$$G(s) = \frac{K(1 + T_1 s)(1 + T_2 s)}{s(1 + T_a s)(1 + 2\zeta s/\omega_n + s^2/\omega_n^2)} e^{-T_d s} \tag{A-24}$$

where K, T_d, T_1, T_2, T_a, ζ, and ω_n are real constants. It is assumed that the second-order polynomial in the denominator has complex-conjugate zeros.

The magnitude of G in dB is obtained by multiplying the logarithm to the base 10 of $|G(j\omega)|$ by 20; we have

$$
\begin{aligned}
|G(j\omega)|_{\text{dB}} = 20\log_{10}|G(j\omega)| &= 20\log_{10}|K| + 20\log_{10}|1 + j\omega T_1| + 20\log_{10}|1 + j\omega T_2| \\
&\quad -20\log_{10}|j\omega| - 20\log_{10}|1 + j\omega T_a| - 20\log_{10}|1 + j2\zeta\omega - \omega^2/\omega_n^2|
\end{aligned}
\tag{A-25}
$$

The phase of $G(j\omega)$ is

$$
\begin{aligned}
\angle G(j\omega) &= \angle K + \angle(1 + j\omega T_1) + \angle(1 + j\omega T_2) - \angle j\omega - \angle(1 + j\omega T_a) \\
&\quad - \angle(1 + 2\zeta\omega/\omega_n - \omega^2/\omega_n^2) - \omega T_d \qquad \text{rad}
\end{aligned}
\tag{A-26}
$$

In general, the function $G(j\omega)$ may be of higher order than that of Eq. (A-24) and have many more factored terms. However, Eqs. (A-25) and (A-26) indicate that additional terms in $G(j\omega)$ would simply produce more similar terms in the magnitude and phase expressions, so that the basic method of construction of the Bode plot would be the same. We have also indicated that, in general, $G(j\omega)$ can contain just five simple types of factors:

1. Constant factor: K
2. Poles or zeros at the origin of order p: $(j\omega)^{\pm p}$
3. Poles or zeros at $s = -1/T$ of order q: $(1 + j\omega T)^{\pm q}$
4. Complex poles and zeros of order r: $(1 + j2\zeta\omega/\omega_n - \omega^2/\omega_n^2)^{\pm r}$
5. Pure time delay $e^{-j\omega T_d}$, where T_d, p, q, and r are positive integers.

Equations (A-25) and (A-26) verify one of the unique characteristics of the Bode plot in that each of the five types of factors listed can be considered as a separate plot; the individual plots are then added or subtracted accordingly to yield the total magnitude in dB and the phase plot of $G(j\omega)$. The curves can be plotted on semilog graph paper or linear rectangular-coordinate graph paper, depending on whether ω or $\log_{10}\omega$ is used as the abscissa.

We shall now investigate sketching the Bode plot of different types of factors.

Real Constant K

Since

$$K_{\text{dB}} = 20\log_{10} K = \text{constant} \tag{A-27}$$

and

$$\angle K = \begin{cases} 0^\circ & K > 0 \\ 180^\circ & K < 0 \end{cases} \tag{A-28}$$

The Bode plot of the real constant K is shown in Fig. A-6 in semilog coordinates.

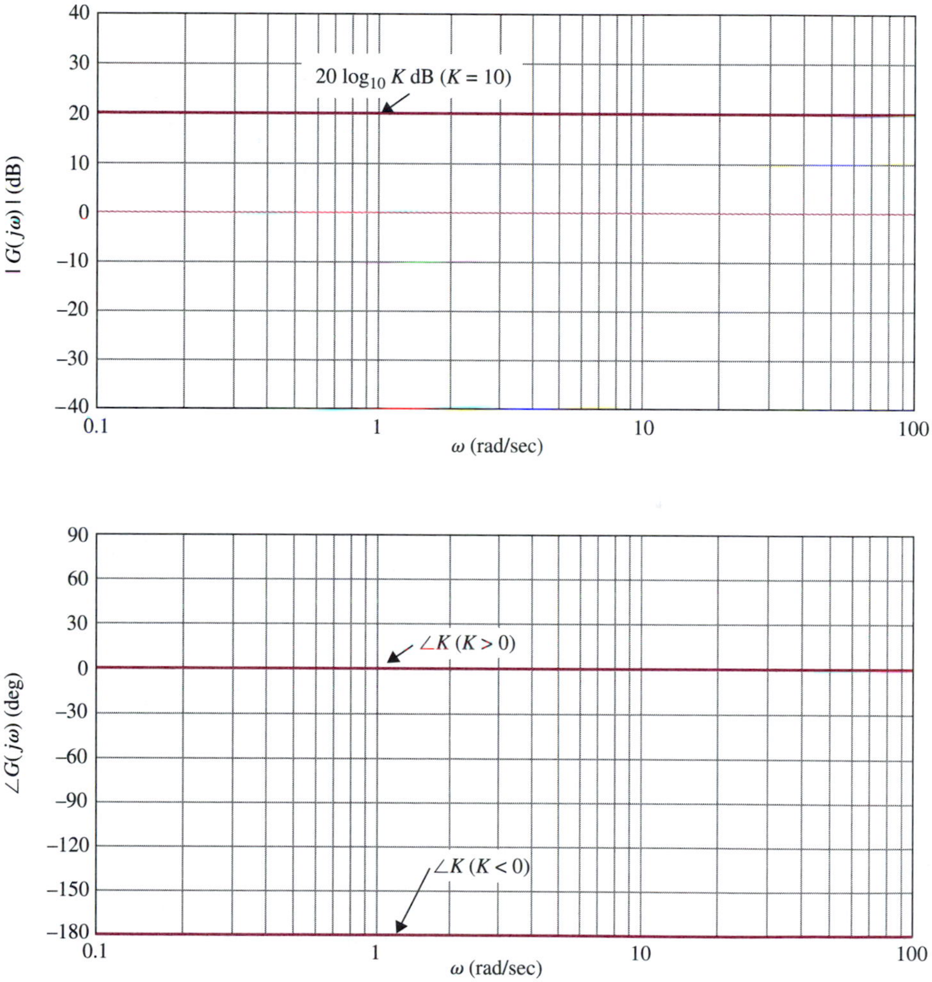

Figure A-6 Bode plot of constant K.

Poles and Zeros at the Origin, $(j\omega)^{\pm p}$

The magnitude of $(j\omega)^{\pm p}$ in dB is given by

$$20 \log_{10}|(j\omega)^{\pm p}| = \pm 20p \log_{10} \omega \qquad \text{dB} \tag{A-29}$$

for $\omega \geq 0$. The last expression for a given p represents a straight line in either semilog or rectangular coordinates. The slopes of these lines are determined by taking the derivative of Eq. (A-29) with respect to $\log_{10} \omega$; that is,

$$\frac{d}{d \log_{10} \omega}(\pm 20p \log_{10} \omega) = \pm 20p \qquad \text{dB/decade} \tag{A-30}$$

These lines pass through the 0-dB axis at $\omega = 1$. Thus a unit change in $\log_{10} \omega$ corresponds to a change of $\pm 20p$ dB in magnitude. Furthermore, a unit change in $\log_{10} \omega$ in the rectangular coordinates is equivalent to one **decade** of variation in ω, that is, from 1 to 10, 10 to 100, and so on, in the semilog coordinates. Thus the slopes of the straight lines described by Eq. (A-29) are said to be $\pm 20p$ dB/decade of frequency.

Instead of using decades, sometimes **octaves** are used to represent the separation of two frequencies. The frequencies ω_1 and ω_2 are separated by one octave if $\omega_2/\omega_1 = 2$. The number of decades between any two frequencies ω_1 and ω_2 is given by

$$\text{number of decades} = \frac{\log_{10}(\omega_2/\omega_1)}{\log_{10} 10} = \log_{10}\left(\frac{\omega_2}{\omega_1}\right) \tag{A-31}$$

Similarly, the number of octaves between ω_2 and ω_1 is

$$\text{number of octaves} = \frac{\log_{10}(\omega_2/\omega_1)}{\log_{10} 2} = \frac{1}{0.301} \log_{10}\left(\frac{\omega_2}{\omega_1}\right) \tag{A-32}$$

Thus the relation between octaves and decades is

$$\text{number of octaves} = 1/0.301 \text{ decades} = 3.32 \text{ decades} \tag{A-33}$$

Substituting Eq. (A-33) into Eq. (A-30), we have

$$\pm 20p \text{ dB/decade} = \pm 20p \times 0.301 \cong 6p \qquad \text{dB/octave} \tag{A-34}$$

For the function $G(s) = 1/s$, which has a simple pole at $s = 0$, the magnitude of $G(j\omega)$ is a straight line with a slope of -20 dB/decade passing through the 0-dB axis at $\omega = 1$ rad/sec.

The phase of $(j\omega)^{\pm p}$ is written

$$\angle(j\omega)^{\pm p} = \pm p \times 90° \tag{A-35}$$

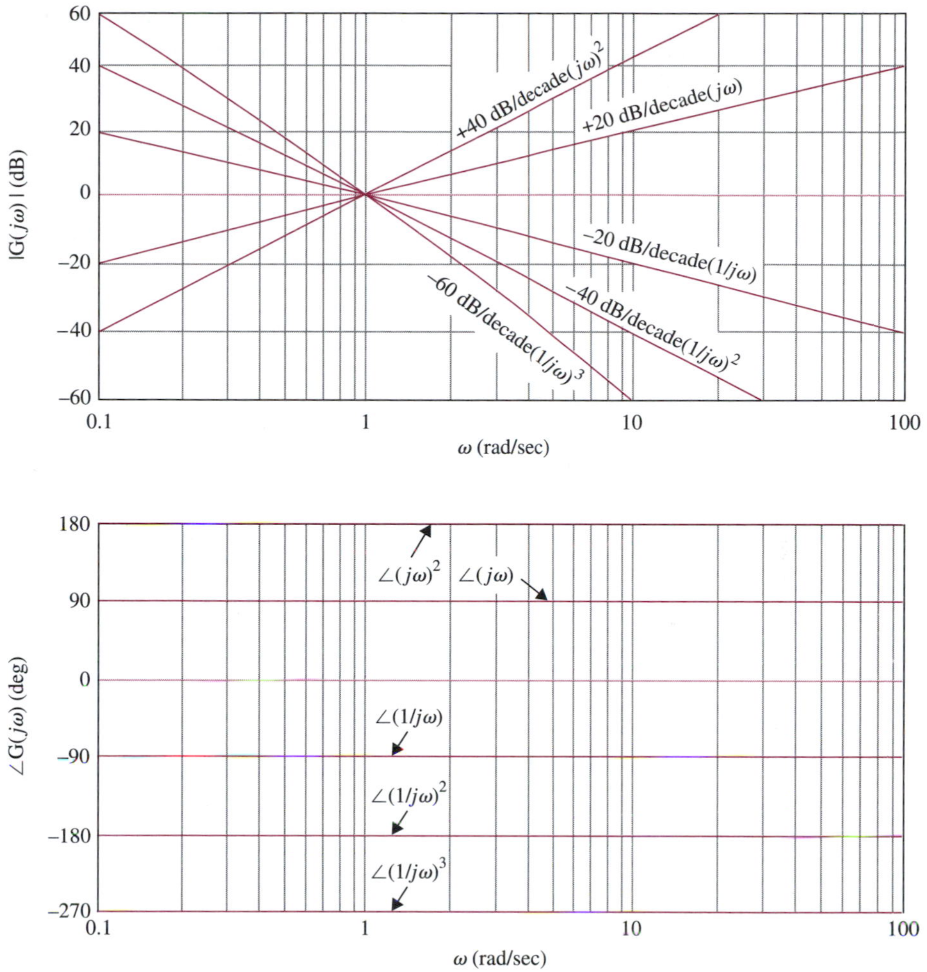

Figure A-7 Bode plots of $(j\omega)^p$.

The magnitude and phase curves of the function $(j\omega)^{\pm p}$ are shown in Fig. A-7 for several values of P.

Simple Zero, $1 + j\omega T$

Consider the function

$$G(j\omega) = 1 + j\omega T \tag{A-36}$$

where T is a positive real constant. The magnitude of $G(j\omega)$ in dB is

$$|G(j\omega)|_{\text{dB}} = 20 \log_{10}|G(j\omega)| = 20 \log_{10}\sqrt{1 + \omega^2 T^2} \tag{A-37}$$

To obtain asymptotic approximations of $|G(j\omega)|_{\text{dB}}$, we consider both very large and very small values of ω. At very low frequencies, $\omega T << 1$, Eq. (A-37) is approximated by

$$|G(j\omega)|_{\text{dB}} \cong 20\log_{10} 1 = 0 \quad \text{dB} \tag{A-38}$$

since ω^2T^2 is neglected when compared with 1.

At very high frequencies, $\omega T >> 1$, we can approximate $1 = \omega^2T^2$ by ω^2T^2; then Eq. (A-37) becomes

$$|G(j\omega)|_{\text{dB}} \cong 20\log_{10}\sqrt{\omega^2T^2} = 20\log_{10}\omega T \tag{A-39}$$

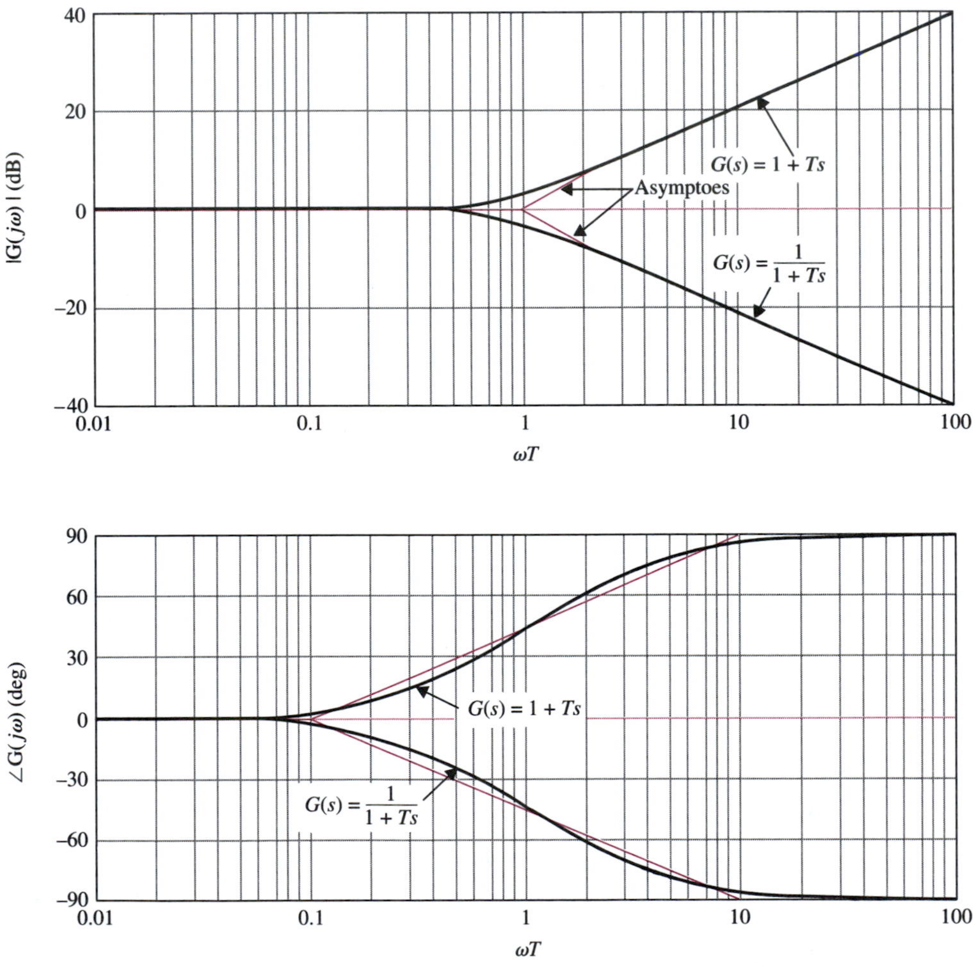

Figure A-8 Bode plots of $G(s) = 1 + Ts$ and $G(s) = 1/(1 + Ts)$.

Equation (A-38) represents a straight line with a slope of 20 dB/decade of frequency. The intersect of these two lines is found by equating Eq. (A-38) to Eq. (A-39), which gives

$$\omega = 1/T \tag{A-40}$$

This frequency is also the intersect of the high-frequency approximate plot and the low-frequency approximate plot, which is the 0-dB axis. The frequency given in Eq. (A-40) is also known as the **corner frequency** of the Bode plot of Eq. (A-36), since the asymptotic plot forms the shape of a corner at this frequency, as shown in Fig. A-8. The actual $|G(j\omega)|_{dB}$ plot of Eq. (A-36) is a smooth curve, and deviates only slightly from the straight-line approximation. The actual values and the straight-line approximation of $|1 + j\omega T|_{dB}$ as functions of ωT are tabulated in Table A-1. The error between the actual magnitude curve and the straight-line asymptotes is symmetrical with respect to the corner frequency $\omega = 1/T$. It is useful to remember that the error is 3 dB at the corner frequency, and 1 dB at 1 octave above ($\omega = 2/T$) and 1 octave below ($\omega = 1/2T$) the corner frequency. At 1 decade above and below the corner frequency, the error is dropped to approximately 0.3 dB. Based on these facts, the procedure of making a sketch of $|1 + j\omega T|_{dB}$ is as follows:

1. Locate the corner frequency $\omega = 1/T$ on the frequency axis.
2. Draw the 20-dB/decade (or 6-dB/octave) line and the horizontal line at 0 dB, with the two lines intersecting at $\omega = 1/T$.
3. If necessary, the actual magnitude curve is obtained by adding the errors to the asymptotic plot at the strategic frequencies. Usually, a smooth curve can be sketched simply by locating the 3-dB point at the corner frequency and the 1-dB points at one octave above and below the corner frequency.

Table A-1

ωT	$\log_{10}\omega T$	$\|1 + j\omega T\|$	$\|1 + j\omega T\|_{dB}$	Straight-Line Approximation $\|1 + j\omega T\|_{dB}$	Error (dB)	$\angle(1 + j\omega T)$ (deg)
0.01	−2	1.0	0.000043	0	0.00043	0.5
0.10	−1	1.04	0.043	0	0.043	5.7
0.50	−0.3	1.12	1	0	1	26.6
0.76	−0.12	1.26	2	0	2	37.4
1.00	0	1.41	3	0	3	45.0
1.31	0.117	1.65	4.3	2.3	2	52.7
2.00	0.3	2.23	7.0	6.0	1	63.4
10.00	1.0	10.4	20.043	20.0	0.043	84.3
100.0	2.0	100.005	40.00043	40.0	0.00043	89.4

The phase of $G(j\omega) = 1 + j\omega T$ is

$$\angle G(j\omega) = \tan^{-1} \omega T \tag{A-41}$$

Similar to the magnitude curve, a straight-line approximation can be made for the phase curve. Since the phase of $G(j\omega)$ varies from 0° to 90°, we can draw a line from 0° at one decade below the corner frequency to 90° at one decade above the corner frequency. As shown in Fig. A-8, the maximum deviation between the straight-line approximation and the actual curve is less than 6°. Table A-1 gives the values of $\angle(1 + j\omega T)$ versus ωT.

Simple Pole, $1/(1 + j\omega T)$

For the function

$$G(j\omega) = \frac{1}{1 + j\omega T} \tag{A-42}$$

the magnitude, $|G(j\omega)|$ in dB, is given by the negative of the right side of Eq. (A-37), and the phase $\angle G(j\omega)$ is the negative of the angle in Eq. (A-41). Therefore, it is simple to extend all the analysis for the case of the simple zero to the Bode plot of Eq. (A-42). The asymptotic approximations of $|G(j\omega)|_{\mathrm{dB}}$ at low and high frequencies are

$$\omega T << 1: \qquad |G(j\omega)|_{\mathrm{dB}} \cong 0 \text{ dB} \tag{A-43}$$

$$\omega T >> 1: \qquad |G(j\omega)|_{\mathrm{dB}} \cong -20 \log_{10} \omega T \tag{A-44}$$

Thus the corner frequency of the Bode plot of Eq. (A-42) is still at $\omega = 1/T$, except that at high frequencies, the slope of the straight-line approximation is -20 dB/decade. The phase of $G(j\omega)$ is 0° at $\omega = 0$ and $-90°$ when $\omega = \infty$. The magnitude in dB and phase of the Bode plot of Eq. (A-42) are shown in Fig. A-8. The data in Table A-1 are still useful for the simple-pole case if appropriate sign changes are made to the numbers. For instance, the numbers in the $|1 + j\omega T|_{\mathrm{dB}}$, the straight-line approximation of $|1 + j\omega T]_{\mathrm{dB}}$, the error (**dB**), and the $\angle(1 + j\omega T)$ columns should all be negative. At the corner frequency, the error between the straight-line approximation and the actual magnitude curve is -3 dB.

Quadratic Poles and Zeros

Now consider the second-order transfer function

$$G(s) = \frac{\omega_n^2}{s^2 + 2\zeta\omega_n s + \omega_n^2} = \frac{1}{1 + (2\zeta/\omega_n)s + (1/\omega_n^2)s^2} \tag{A-45}$$

We are interested only in the case when $\zeta \leq 1$, since otherwise $G(s)$ would have two unequal real poles and the Bode plot can be obtained by considering $G(s)$ as the product of two transfer functions with simple poles. By letting $s = j\omega$, Eq. (A-45) becomes

$$G(j\omega) = \frac{1}{[1 - (\omega/\omega_n)^2] + j2\zeta(\omega/\omega_n)} \tag{A-46}$$

The magnitude of $G(j\omega)$ in dB is

$$20 \log_{10}|G(j\omega)| = -20 \log_{10}\sqrt{[1 - (\omega/\omega_n)^2]^2 + 4\zeta^2(\omega/\omega_n)^2} \tag{A-47}$$

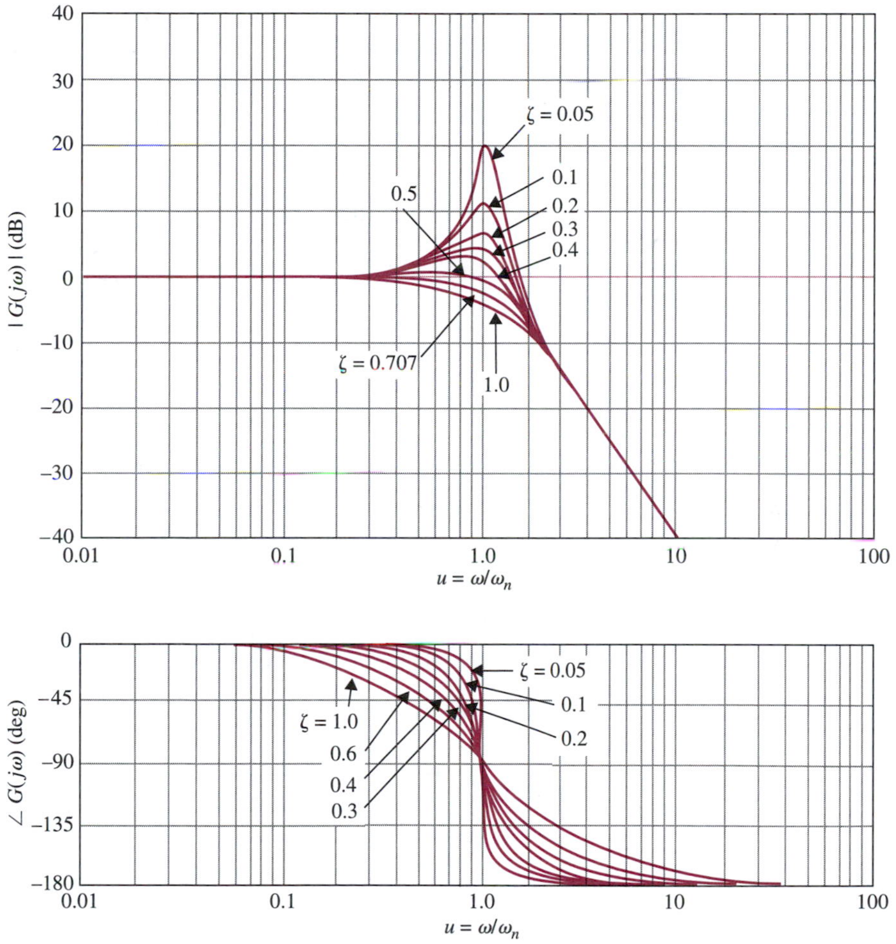

Figure A-9 Bode plots of $G(s) = 1/[1 + 2\zeta(s/\omega_n) + (s/\omega_n)^2]$.

At very low frequencies, $\omega/\omega_n << 1$; Eq. (A-47) can be approximated as

$$|G(j\omega)|_{dB} = 20\log_{10}|G(j\omega)| \cong -20\log_{10}1 = 0\text{dB} \qquad \text{(A-48)}$$

Thus the low-frequency asymptote of the magnitude plot of Eq. (A-45) is a straight line that lies on the 0-dB axis. At very high frequencies, $\omega/\omega_n >> 1$; the magnitude in dB of $G(j\omega)$ in Eq. (A-45) becomes

$$|G(j\omega)|_{dB} \cong -20\log_{10}\sqrt{(\omega/\omega_n)^4} = -40\log_{10}(\omega/\omega_n) \quad \text{dB} \qquad \text{(A-49)}$$

This equation represents a straight line with a slope of -40 dB/decade in the Bode plot coordinates. The intersection of the two asymptotes is found by equating Eq. (A-48) to Eq. (A-49), yielding the corner frequency at $\omega = \omega_n$. The actual magnitude curve of $G(j\omega)$ in this case may differ strikingly from the asymptotic curve. The reason for this is that the amplitude and phase curves of the second-order $G(j\omega)$ depend not only on the corner frequency ω_n, but also on the damping ratio ζ, which does not enter the asymptotic curve. The actual and asymptotic curves of $|G(j\omega)|_{dB}$ are shown in Fig. A-9 for several values of ζ. The errors between the two sets of curves are shown in Fig. A-10 for the same set of values of ζ. The standard procedure of constructing the second-order $|G(j\omega)|_{dB}$ is first to locate the corner frequency ω_n and the -40-dB/decade line to the right of ω_n. The actual curve is obtained by making corrections to the asymptotes by using either the data from the error curves of Fig. A-10 or the curves in Fig. A-9 for the corresponding ζ.

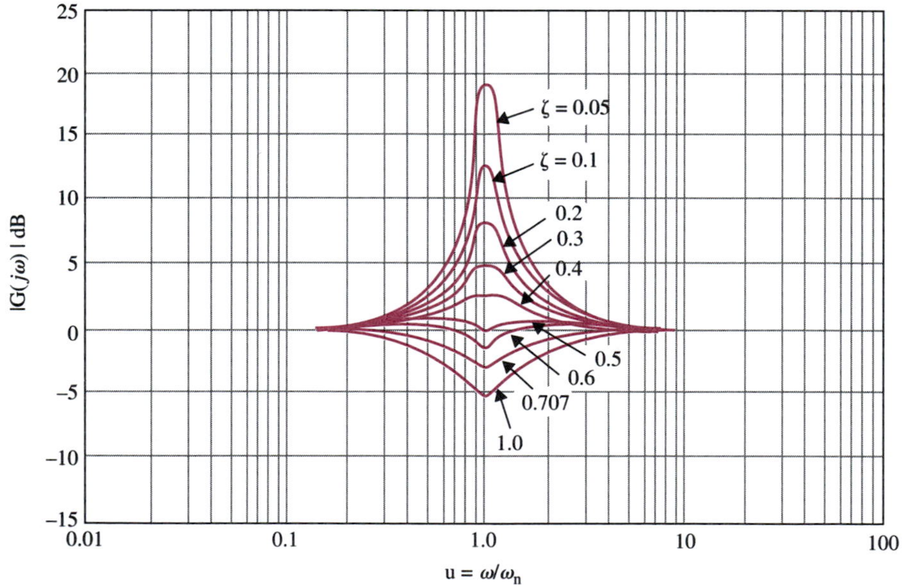

Figure A-10 Errors in magnitude curves of Bode plots of $G(s) = 1/[1 + 2\zeta(s/\omega_n) + (s/\omega_n)^2]$.

The phase of $G(j\omega)$ is given by

$$\angle G(j\omega) = -\tan^{-1}\left\{\frac{2\zeta\omega}{\omega_n}\left[1 - \left(\frac{\omega}{\omega_n}\right)^2\right]\right\} \quad \text{(A-50)}$$

and is plotted as shown in Fig. A-9 for various values of ζ.

The analysis of the Bode plot of the second-order transfer function of Eq. (A-45) can be applied to the second-order transfer function with two complex zeros. For

$$G(s) = 1 + \frac{2\zeta}{\omega_n}s + \frac{1}{\omega_n^2}s^2 \quad \text{(A-51)}$$

The magnitude and phase curves are obtained by inverting those in Fig. A-9. The errors between the actual and the asymptotic curves in Fig. A-10 are also inverted.

Pure Time Delay, $e^{-j\omega T_d}$

▲ The magnitude of the pure time delay term is unity for all ω.

The magnitude of the pure time delay term is equal to unity for all values of ω. The phase of the pure time-delay term is

$$\angle e^{-j\omega T_d} = -\omega T_d \quad \text{(A-52)}$$

which decreases linearly as a function of ω. Thus for the transfer function

$$G(j\omega) = G_1(j\omega)e^{-j\omega T_d} \quad \text{(A-53)}$$

the magnitude plot $|G(j\omega)|_{\text{dB}}$ is identical to that of $|G_l(j\omega)|_{\text{dB}}$. The phase plot $\angle G(j\omega)$ is obtained by subtracting ωT_d radians from the phase curve of $G_1(j\omega)$ at various ω.

Example A-5

As an illustrative example on the construction of Bode plot manually, consider the function

$$G(s) = \frac{10(s+10)}{s(s+2)(s+5)} \quad \text{(A-54)}$$

The first step is to express $G(s)$ in the form of Eq. (A-23) and set $s = j\omega$, keeping in mind that for computer plotting, this step is unnecessary; we have

$$G(j\omega) = \frac{10(1 + j0.1\omega)}{j\omega(1 + j0.5\omega)(1 + j0.2\omega)} \quad \text{(A-55)}$$

Equation (A-54) shows that $G(j\omega)$ has corner frequencies at $\omega = 2$, 5, and 10 rad/sec. The pole at $s = 0$ gives a magnitude curve that is a straight line with slope of -20 dB/decade, passing through the $\omega = 1$ rad/sec point on the 0-dB axis. The complete Bode plot of the magnitude and phase of $G(j\omega)$ is obtained by adding the component curves together, point by point, as shown in Fig. A-11. The actual curves can be obtained by a computer program and are shown in Fig. A-11. ▲

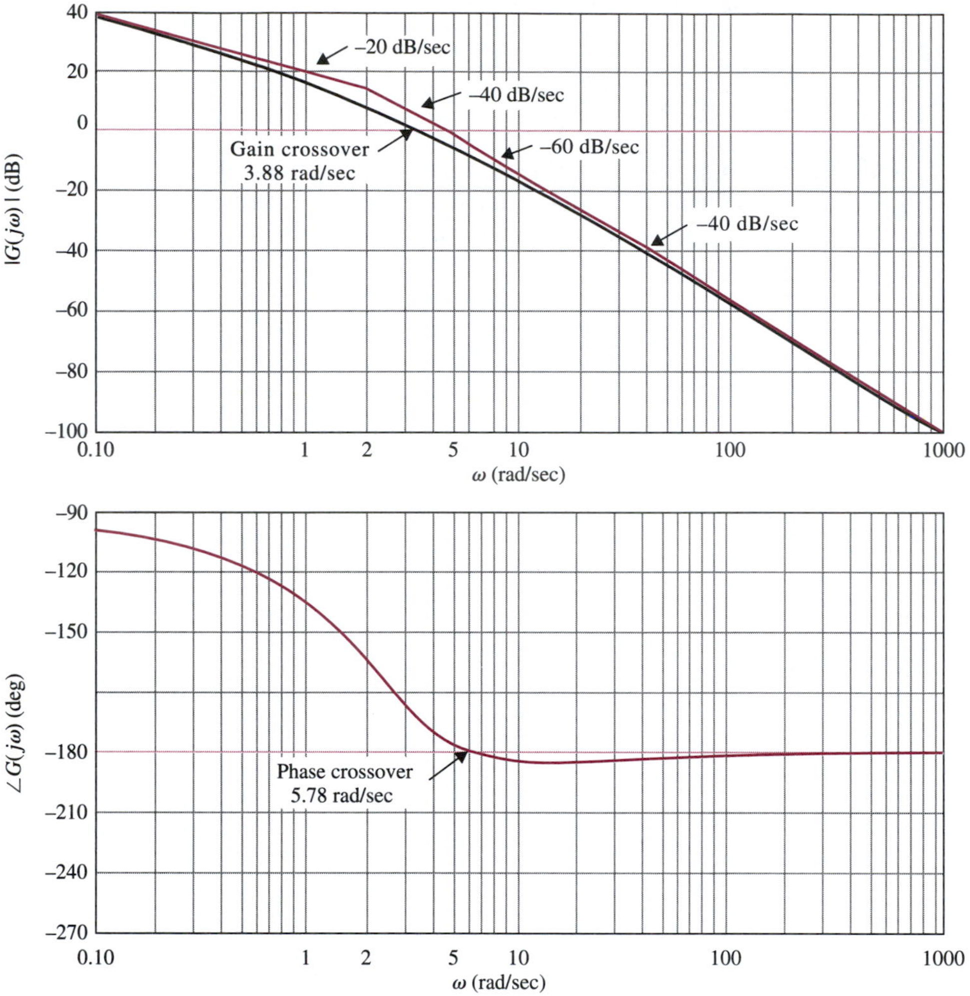

Figure A-11 Bode plot of $G(s) = 10(s + 10)/[s(s + 2)(s + 5)]$.

A-3 Magnitude–Phase Plot

The magnitude–phase plot of $G(j\omega)$ is a plot of the magnitude of $G(j\omega)$ in dB versus its phase in degrees, with ω as a parameter on the curve. One of the most important applications of this type of plot is that when $G(j\omega)$ is the forward-path transfer function of a unity-feedback control system, the plot can be superposed on the Nichols chart (see Chapter 9) to give information on the relative stability and frequency response of the system. When the gain factor K of the transfer function varies, the plot is simply raised or lowered vertically according to the value of K in dB. However, in the construction of the plot, the property of adding the curves of the individual components of the transfer function in the Bode plot does not carry over to this case. Thus it is best to make the magnitude–phase plot by computer or transfer the data from the Bode plot.

Example A-6

As an illustrative example, the polar plot and the magnitude–phase plot of Eq. (A-54) are shown in Figs. A-12 and A-13, respectively. The Bode plot of the function is shown in Fig. A-11. The relationships among these three plots are easily identified by comparing the curves in Figs. A-11, A-12, and A-13. ▲

A-4 Gain- and Phase-Crossover Points

Gain- and phase-crossover points on the frequency-domain plots are important for analysis and design of control systems. These are defined as follows.

Gain-Crossover Point. The gain-crossover point on the frequency-domain plot of $G(j\omega)$ is a point at which $|G(j\omega)| = 1$ or $|G(j\omega)|_{dB} = 0$ dB. The frequency at the gain-crossover point is called the **gain-crossover frequency ω_g.**

Phase-Crossover Point. The phase-crossover point on the frequency-domain plot of $G(j\omega)$ is a point at which $\angle G(j\omega) = 180°$. The frequency at the phase-crossover point is called the **phase-crossover frequency ω_p.**

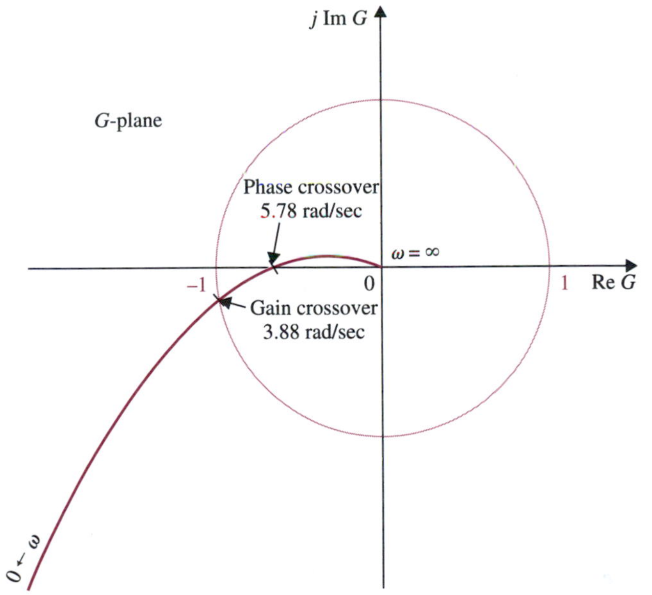

Figure A-12 Polar plot of $G(s) = 10(s + 10)/[s(s + 2)(s + 5)]$.

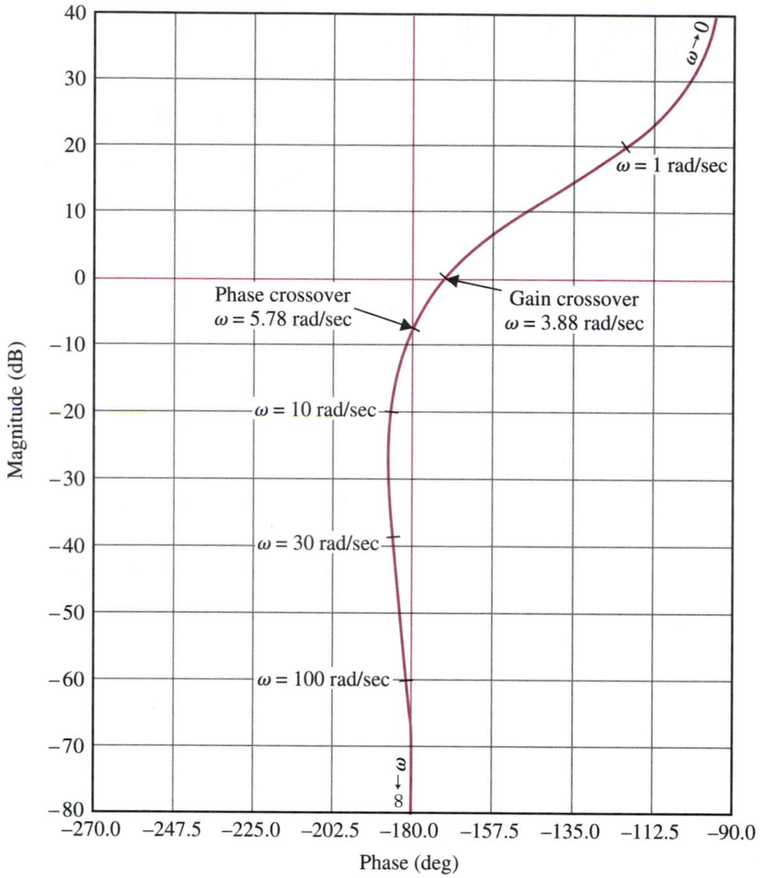

Figure A-13 Magnitude-versus-phase plot of $G(s) = 10(s + 10)/[s(s + 2)(s + 5)]$.

The gain and phase crossovers are interpreted with respect to three types of plots.

Polar Plot. The gain-crossover point (or points) is where $|G(j\omega)| = 1$. The phase-crossover point (or points) is where $\angle G(j\omega) = 180°$ (see Fig. A-12).

Bode Plot. The gain-crossover point (or points) is where the magnitude curve $|G(j\omega)|_{dB}$ crosses the 0-dB axis. The phase-crossover point (or points) is where the phase curve crosses the 180° axis (see Fig. A-11).

Magnitude–Phase Plot. The gain-crossover point (or points) is where the $G(j\omega)$ curve crosses the 0-dB axis. The phase-crossover point (or points) is where the $G(j\omega)$ curve crosses the 180° axis (see Fig. A-13).

A-5 Minimum-Phase and Nonminimum-Phase Functions

▲ The magnitude and phase characteristics of a minimum-phase function are uniquely related.

A majority of the process transfer functions encountered in linear control systems do not have poles or zeros in the right-half s-plane. This class of transfer functions is called the **minimum-phase transfer functions.** When a transfer function has either a pole or a zero in the right-half s-plane, it is called a **nonminimum-phase transfer function.**

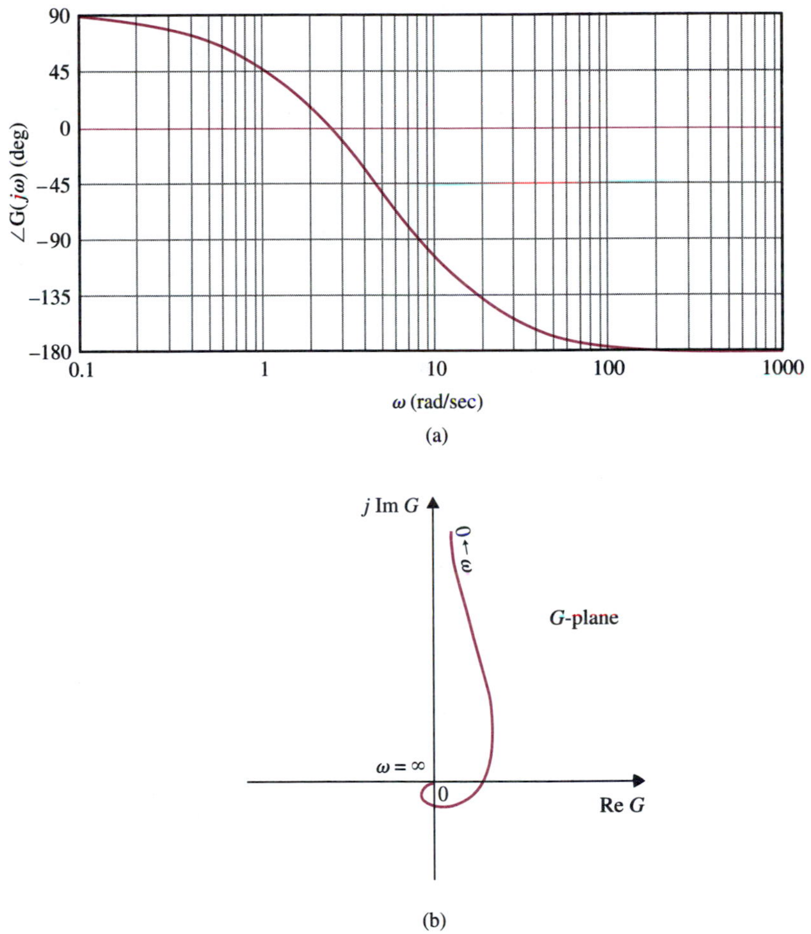

Figure A-14 (a) Phase curve of the Bode plot. (b) Polar plot. $G(s) = 10(s - 10)/[s(s + 2)(s + 5)]$.

Minimum-phase transfer functions have an important property in that their magnitude and phase characteristics are uniquely related. In other words, given a minimum-phase function $G(s)$, knowing its magnitude characteristics, $|G(j\omega)|$, completely defines the phase characteristics, $\angle G(j\omega)$. Conversely, given $\angle G(j\omega)$, $|G(j\omega)|$ is completely defined.

Nonminimum-phase transfer functions do not have the unique magnitude–phase relationships. For instance, given the function

$$G(j\omega) = \frac{1}{1 - j\omega T} \tag{A-56}$$

the magnitude of $G(j\omega)$ is the same whether T is positive (nonminimum phase) or negative (minimum phase). However, the phase of $G(j\omega)$ is different for positive and negative T.

Additional properties of the minimum-phase transfer functions are as follows:

1. For a minimum-phase transfer function $G(s)$ with m zeros and n poles, excluding the poles at $s = 0$, if any, when $s = j\omega$, and as ω varies from ∞ to 0, the total phase variation of $G(j\omega)$ is $(n - m)\pi/2$.
2. The value of a minimum-phase transfer function cannot become zero or infinity at any finite nonzero frequency.
3. A nonminimum-phase transfer function will always have a more positive phase shift as ω is varied from ∞ 0.

Example A-7

As an illustrative example on the properties of the nonminimum-phase transfer function, consider that the zero of the transfer function of Eq. (A-54) is in the right-half s-plane; that is,

$$G(s) = \frac{10(s - 10)}{s(s + 2)(s + 5)} \tag{A-57}$$

▲ Do not use the Bode plot and the gain–phase plot of a nonminimum-phase transfer function for stability studies.

The magnitude plot of the Bode diagram of $G(j\omega)$ is identical to that of the minimum-phase transfer function in Eq. (A-54) and is shown in Fig. A-11. The phase curve of the Bode plot of $G(j\omega)$ of Eq. (A-57) is shown in Fig. A-14(a), and the polar plot is shown in Fig. A-14(b). Notice that the nonminimum-phase function has a net phase shift of 270° (from −180° to +90°) as ω varies from ∞ to 0, whereas the minimum-phase transfer function if Eq. (A-54) has a net phase change of only 90° (from −180° to −90°) over the same frequency range.

Care should be taken when using the Bode diagram for the analysis and design of systems with nonminimum-phase transfer functions. For stability studies, the polar plot, when used along with the Nyquist criterion discussed in Chapter 9, is more convenient for nonminimum-phase systems. Bode diagrams of nonminimum-phase forward-path transfer functions *should not* be used for stability analysis of closed-loop control systems. The same is true for the magnitude–phase plot. ▲

Appendix B

Laplace Transform Table

Laplace Transform $F(s)$	Time Function $f(t)$
1	Unit-impulse function $\delta(t)$
$\frac{1}{s}$	Unit-step function $u_s(t)$
$\frac{1}{s^2}$	Unit-ramp function t
$\frac{n!}{s^{n+1}}$	t^n (n = positive integer)
$\frac{1}{s+\alpha}$	$e^{-\alpha t}$
$\frac{1}{(s+\alpha)^2}$	$te^{-\alpha t}$
$\frac{n!}{(s+\alpha)^{n+1}}$	$t^n e^{-\alpha t}$ (n = positive integer)
$\frac{1}{(s+\alpha)(s+\beta)}$	$\frac{1}{\beta-\alpha}(e^{-\alpha t}-e^{-\beta t})$ $(\alpha \neq \beta)$
$\frac{s}{(s+\alpha)(s+\beta)}$	$\frac{1}{\beta-\alpha}(\beta e^{-\beta t}-\alpha e^{-\alpha t})$ $(\alpha \neq \beta)$
$\frac{1}{s(s+\alpha)}$	$\frac{1}{\alpha}(1-e^{-\alpha t})$
$\frac{1}{s(s+\alpha)^2}$	$\frac{1}{\alpha^2}(1-e^{-\alpha t}-\alpha t e^{-\alpha t})$
$\frac{1}{s^2(s+\alpha)}$	$\frac{1}{\alpha^2}(\alpha t-1+e^{-\alpha t})$
$\frac{1}{s^2(s+\alpha)^2}$	$\frac{1}{\alpha^2}\left[t-\frac{1}{\alpha}+\left(t+\frac{2}{\alpha}\right)e^{-\alpha t}\right]$

Laplace Transform Table (continued)

Laplace Transform $F(s)$	Time Function $f(t)$
$\dfrac{s}{(s+\alpha)^2}$	$(1-\alpha t)e^{-\alpha t}$
$\dfrac{\omega_n^2}{s^2+\omega_n^2}$	$\sin \omega_n t$
$\dfrac{s}{s^2+\omega_n^2}$	$\cos \omega_n t$
$\dfrac{\omega_n^2}{s(s^2+\omega_n^2)}$	$1-\cos \omega_n t$
$\dfrac{\omega_n^2(s+\alpha)}{s^2+\omega_n^2}$	$\omega_n\sqrt{\alpha^2+\omega_n^2}\sin(\omega_n t+\theta)$ where $\theta=\tan^{-1}(\omega_n/\alpha)$
$\dfrac{\omega_n}{(s+\alpha)(s^2+\omega_n^2)}$	$\dfrac{\omega_n}{\alpha^2+\omega_n^2}e^{-\alpha t}+\dfrac{1}{\sqrt{\alpha^2+\omega_n^2}}\sin(\omega_n t-\theta)$ where $\theta=\tan^{-1}(\omega_n/\alpha)$
$\dfrac{\omega_n^2}{s^2+2\zeta\omega_n s+\omega_n^2}$	$\dfrac{\omega_n}{\sqrt{1-\zeta^2}}e^{-\zeta\omega_n t}\sin \omega_n\sqrt{1-\zeta^2}\,t \quad (\zeta<1)$
$\dfrac{\omega_n^2}{s(s^2+2\zeta\omega_n s+\omega_n^2)}$	$1-\dfrac{1}{\sqrt{1-\zeta^2}}e^{-\zeta\omega_n t}\sin(\omega_n\sqrt{1-\zeta^2}\,t+\theta)$ where $\theta=\cos^{-1}\zeta \quad (\zeta<1)$
$\dfrac{s\omega_n^2}{s^2+2\zeta\omega_n s+\omega_n^2}$	$\dfrac{-\omega_n^2}{\sqrt{1-\zeta^2}}e^{-\zeta\omega_n t}\sin(\omega_n\sqrt{1-\zeta^2}\,t-\theta)$ where $\theta=\cos^{-1}\zeta \quad (\zeta<1)$
$\dfrac{\omega_n^2(s+\alpha)}{s^2+2\zeta\omega_n s+\omega_n^2}$	$\omega_n\sqrt{\dfrac{\alpha^2-2\alpha\zeta\omega_n+\omega_n^2}{1-\zeta^2}}\,e^{-\zeta\omega_n t}\sin(\omega_n\sqrt{1-\zeta^2}\,t+\theta)$ where $\theta=\tan^{-1}\dfrac{\omega_n\sqrt{1-\zeta^2}}{\alpha-\zeta\omega_n} \quad (\zeta<1)$
$\dfrac{\omega_n^2}{s^2(s^2+2\zeta\omega_n s+\omega_n^2)}$	$t-\dfrac{2\zeta}{\omega_n}+\dfrac{1}{\omega_n^2\sqrt{1-\zeta^2}}e^{-\zeta\omega_n t}\sin(\omega_n\sqrt{1-\zeta^2}\,t+\theta)$ where $\theta=\cos^{-1}(2\zeta^2-1) \quad (\zeta<1)$

Appendix C

z-Transform Table

Laplace Transform	Time Function	z-Transform
1	Unit impulse $\delta(t)$	1
$\frac{1}{s}$	Unit step $u_s(t)$	$\frac{z}{z-1}$
$\frac{1}{1-e^{-Ts}}$	$\delta_T(t) = \sum_{n=0}^{\infty} \delta(t-nT)$	$\frac{z}{z-1}$
$\frac{1}{s^2}$	t	$\frac{Tz}{(z-1)^2}$
$\frac{1}{s^3}$	$\frac{t^2}{2}$	$\frac{T^2 z(z+1)}{2(z-1)^3}$
$\frac{1}{s^{n+1}}$	$\frac{t^n}{n!}$	$\lim_{\alpha \to 0} \frac{(-1)^n}{n!} \frac{\partial^n}{\partial \alpha^n} \left[\frac{z}{z-e^{-\alpha T}} \right]$
$\frac{1}{s+\alpha}$	$e^{-\alpha t}$	$\frac{z}{z-e^{-\alpha T}}$
$\frac{1}{(s+\alpha)^2}$	$te^{-\alpha t}$	$\frac{Tze^{-\alpha T}}{(z-e^{-\alpha T})^2}$
$\frac{\alpha}{s(s+\alpha)}$	$1 - e^{-\alpha t}$	$\frac{(1-e^{-\alpha T})z}{(z-1)(z-e^{-\alpha T})}$
$\frac{\omega}{s^2+\omega^2}$	$\sin \omega t$	$\frac{z \sin \omega T}{z^2 - 2z\cos \omega T + 1}$
$\frac{\omega}{(s+\alpha)^2+\omega^2}$	$e^{-\alpha t} \sin \omega t$	$\frac{ze^{-\alpha T} \sin \omega T}{z^2 - 2ze^{-\alpha T} \cos \omega T + e^{-2\alpha T}}$
$\frac{s}{s^2+\omega^2}$	$\cos \omega t$	$\frac{z(z-\cos \omega T)}{z^2 - 2z \cos \omega T + 1}$
$\frac{s+\alpha}{(s+\alpha)^2+\omega^2}$	$e^{-\alpha t} \cos \omega t$	$\frac{z^2 - ze^{-\alpha T} \cos \omega T}{z^2 - 2ze^{-\alpha T} \cos \omega T + e^{-2\alpha T}}$

Answers for Selected Problems

The answers given in the following are for certain selected problems. These answers are given for checking purposes, and thus are often for only certain parts of a problem.

Chapter 2

2-1 **(a)** Poles: $s = 0, 0, -1, -10$; zeros: $s = -2, \infty, \infty, \infty$.

(c) Poles: $s = 0, -1, -1, +j, -1, -j$; zero: $s = -2$.

2-3 **(a)** $G(s) = \dfrac{1 - e^{-s}}{s(1 + e^{-s})}$ $\quad G_T(s) = \dfrac{1}{s}(1 - e^{-s})^2$

(b) $G(s) = \dfrac{2(1 - e^{-0.5s})}{s^2(1 + e^{-0.5s})}$

2-6 **(a)** $g(t) = \dfrac{1}{6} - \dfrac{1}{2}e^{-2t} + \dfrac{1}{3}e^{-3t} \quad t \geq 0$

(e) $g(t) = 0.5t^2e^{-t} \quad t \geq 0$

2-11 **(d)**

$$\mathbf{A}^{-1} = \begin{bmatrix} 1.3 & 0.5 & -0.3 \\ 1 & 0 & 0 \\ -0.2 & 0 & 0.2 \end{bmatrix}$$

2-13 **(a)**

$$F(z) = \frac{ze^{-3}}{(z - e^{-3})^2}$$

(d)

$$F(z) = \frac{ze^{-2}(z + e^{-2})}{(z - e^{-2})^3}$$

2-16 **(c)** $F(z) = \dfrac{0.541z}{z - 1} - \dfrac{0.541z}{z + 0.85}$ $\quad f(k) = 0.541[1 - (-0.85)^k] \quad k = 0, 1, 2, \cdots$

2-18 **(b)** $x(k) = \cos\left(\dfrac{k\pi}{2}\right) \quad k = 0, 1, 2, \cdots$

2-19 **(d)** $395.15

Chapter 3

3-1 **(a)** $\frac{Y(s)}{R(s)} = \frac{3s+1}{s^3+2s^2+5s+6}$ **(d)** $\frac{Y(s)}{R(s)} = \frac{1+2e^{-s}}{2s^2+s+5}$

3-7 $\left.\frac{E(s)}{R(s)}\right|_{N=0} = \frac{1+G_3(s)H_1(s)-G_1(s)G_3(s)}{\Delta}$ $\Delta = 1 + G_2(s)G_3(s) + G_3(s)H_1(s)$

3-15 **(b)** $\left.\frac{Y(s)}{E(s)}\right|_{N=0} = \frac{10(s+4)}{s^2+6s-20}$

3-17 $G_d(s) = \frac{s(s+10)}{10}$

3-20 **(b)** $\Delta = 1 - 2[G(s)]^2$

3-21 **(b)** $y(t) = (20 - 25e^{-t} + 5e^{-5t})u_s(t)$

3-22 **(b)** Characteristic equation: $s^2 + 7s + 25 = 0$

3-23 **(c)** Characteristic equation: $s^3 + 5s^2 + 6s + 10 = 0$

(d) Transfer function: $\frac{Y(s)}{R(s)} = \frac{1}{s^3+5s^2+6s+10}$

3-25 **(c)** Characteristic equation: $s^3 + 21.01s^2 + 30.198s + 10 = 0$

3-27 **(d)** $\frac{Y(z)}{R(z)} = \frac{0.46z+0.33}{(z-1)(z-0.368)}$

3-28 **(c)** $Y(z) = \frac{T}{z-1}$

Chapter 4

4-1 **(b)** Force equations: $\frac{d^2y_1}{dt^2} = -\frac{(B_1+B_2)}{M}\frac{dy_1}{dt} + \frac{B_2}{M}\frac{dy_2}{dt} + \frac{1}{M}f$ $\frac{dy_2}{dt} = \frac{dy_1}{dt} - \frac{K}{B_2}y_2$

4-3 **(b)** Torque equations: $\frac{d^2\theta_1}{dt^2} = -\frac{K}{J}(\theta_1-\theta_2) + \frac{1}{J}T$ $K(\theta_1-\theta_2) = B\frac{d\theta_2}{dt}$

4-7 **(b)** Optimal gear ratio is $n^* = \sqrt{J_m/J_L}$

4-8 **(b)** Transfer function: $\frac{Y(s)}{T_m(s)} = \frac{r}{s[(J_m+Mr^2)s+B_m]}$

4-12 (a) $\dfrac{H_i(s)}{H_e(s)} = -(R_a + L_a s)$ (b) $\left.\dfrac{\Omega_m(s)}{\Omega_r(s)}\right|_{T_L=0} \cong \dfrac{1}{K_b H_e(s)}$

4-14 (d) Forward-path transfer function:

$$\frac{\Theta_o(s)}{\Theta_e(s)} = \frac{KK_sK_in}{s[J_TL_as^2 + (R_aJ_T + B_mL_a)s + R_aB_m + K_iK_b]}$$

4-17 (c) Forward-path transfer function:

$$G(s) = \frac{608.7 \times 10^6 K}{s(s^3 - 423.42s^2 + 2.6667 \times 10^6 s + 4.2342 \times 10^8)}$$

4-19 (c) Characteristic equation: $Js^2 + (JK_L + B)s + K_2B + K_3K_4e^{-T_Ds} = 0$

4-22 (b) State equations: $i_a(t)$ as input. $\dfrac{dx(t)}{dt} = v(t)$ $\dfrac{dv(t)}{dt} = -Bv(t) + K_iK_fi_a^2(t)$

4-26 (b) Forward-path transfer function:

$$\frac{Y(s)}{E(s)} = \frac{KK_inG_c(s)e^{-T_Ds}}{s\{(R_a + L_as)[(J_m + J_L)s + B_m] + K_bK_i\}}$$

4-27 (c)

$$\frac{E_o(s)}{E(s)} = \frac{-R_FR_2\left(s + \dfrac{1}{R_2C}\right)}{(R_1 + R_2)\left(s + \dfrac{1}{(R_1 + R_2)C}\right)}$$

Chapter 5

5-3 (a) Eigenvalues: $s = -0.5 + j1.323, -0.5 - j1.323$

(c) State transition matrix: $\phi(t) = \begin{bmatrix} e^{3t} & 0 \\ 0 & e^{-3t} \end{bmatrix}$

(g) Characteristic equation: $\Delta(s) = s^3 + 15s^2 + 75s + 125 = 0$

5-6 (a) Eigenvalues of **A**: 2.325, $-0.3376 + j0.5623$, $-0.3376 - j0.5623$

(b) Output transfer function: $\dfrac{Y(s)}{U(s)} = \dfrac{s+2}{(s+1)^2}$

5-8 (c) $\mathbf{S} = [\mathbf{B} \quad \mathbf{AB} \quad \mathbf{AB}^2] = \begin{bmatrix} 1 & -1 & 0 \\ 1 & -2 & 4 \\ 1 & -6 & 23 \end{bmatrix}$ $\mathbf{P} = \mathbf{SM} = \begin{bmatrix} 9 & 6 & 1 \\ 6 & 5 & 1 \\ -3 & 1 & 1 \end{bmatrix}$

5-11 (c) $\mathbf{S} = \begin{bmatrix} 2 & 1 + 2\sqrt{2} \\ \sqrt{2} & 2 + \sqrt{2} \end{bmatrix}$ **S** is singular.

5-13 (a) $\phi(t) = \begin{bmatrix} \cos t & \sin t \\ -\sin t & \cos t \end{bmatrix}$

5-16 (a) Forward-path transfer function: $G(s) = \dfrac{Y(s)}{E(s)} = \dfrac{5(K_1 + K_2 s)}{s[s(s+4)(s+5)+10]}$

Closed-loop transfer function: $M(s) = \dfrac{5(K_1 + K_2 s)}{s^4 + 9s^3 + 20s^2 + (10 + 5K_2)s + 5K_1}$

5-19 (b) State transition matrix: $\phi(t) = \begin{bmatrix} e^{-t} & -te^{-t} \\ 0 & e^{-t} \end{bmatrix}$

5-28 (b) State equations:

$$\begin{bmatrix} \dot{x}_1 \\ \dot{x}_2 \\ \dot{x}_3 \\ \dot{x}_4 \end{bmatrix} = \begin{bmatrix} -2 & 20 & -1 & 0 \\ 0 & -10 & 1 & 0 \\ -0.1 & 0 & -20 & 1 \\ 0 & 0 & 0 & -5 \end{bmatrix} \begin{bmatrix} x_1 \\ x_2 \\ x_3 \\ x_4 \end{bmatrix} + \begin{bmatrix} 0 & -1 \\ 0 & 0 \\ 0 & 0 \\ 30 & 0 \end{bmatrix} \begin{bmatrix} u \\ T_D \end{bmatrix}$$

5-32 (b) Characteristic equation: $s^3 + 22s^2 + 170s + 600 = 0$

5-42 The system is controllable.

5-43 (a) Transfer function: $\dfrac{\Theta_v(s)}{R(s)} = \dfrac{K_I H}{J_v s^2 (J_G s^2 + K_P s + K_I + K_N)}$

5-45 (b) For controllability, $k_2 \neq -\dfrac{11}{2}$ For observability, $|\mathbf{V}| = -1 + 3k_1 - 3k_2 \neq 0$

5-52 Forward-path transfer function: $G(z) = \dfrac{0.632}{z - 0.368}$

Discrete-data state equation: $x[(k+1)T] = 0.264x(kT) + 0.632r(kT)$

Chapter 6

6-1 (b) Poles are at $s = -5, -j\sqrt{2}, j\sqrt{2}$ Two poles on imaginary axis. Marginally stable.

(d) Poles are at $s = -5, -1 + j, -1 - j$ All poles in the left-half s-plane. Stable.

6-2 **(b)** No roots in RHP. **(f)** Two roots in RHP

6-3 **(e)** Conditions for stability: $K > 2$ and $K < -2.9055$

6-5 **(b)** Condition for stability: $K > 4.6667$

6-9 Stability requirement: $K_t > 0.081$

6-12 **(a)** There is one root in the region to the right of $s = -1$ in the s-plane.

6-17 **(b)** $F(z)$ has two roots outside the unit circle.

6-20 **(c)** Stability requirement: $0 < K < 3.9535$

Chapter 7

7-3 **(c)** $K_p = \infty \quad K_v = K \quad K_a = 0$

7-5 **(a)** Unit-step Input: $e_{ss} = \dfrac{1}{K_H}\left(1 - \dfrac{b_0 K_H}{a_0}\right) = \dfrac{2}{3}$ Unit-ramp Input: $e_{ss} = \infty$

(d) Unit-step Input: $e_{ss} = 0$ Unit-ramp Input: $e_{ss} = 0.05$
Unit-parabolic Input: $e_{ss} = \infty$

7-6 **(c)** Unit-step Input: $e_{ss} = 2.4$ Unit-ramp and Unit-parabolic Inputs: $e_{ss} = \infty$

7-10 **(a)** Stability requirement: $K_t > 0.02$ and $K > 0$

7-13 Rise time $t_r = 0.2$ sec $\quad \omega_n = 10.82$ rad/sec $\quad K = 4.68 \quad K_t = 0.0206$

7-15 System transfer function: $\dfrac{Y(s)}{R(s)} = \dfrac{802.59}{s^2 + 25.84s + 802.59}$ $\quad y_{\max} = 1.2$ (20% overshoot)

7-24 **(b)** $k_2^2 = 59 + 10k_1$ **(c)** $k_2 = 13.14$

7-27 **(a)** Stability requirement: $0 < K < 14001.9$

(c) Stability requirement: $0 < K < \infty$

7-29 **(f)** $\alpha = 5 \quad \left.\dfrac{Y(s)}{D(s)}\right|_{r=0} = \dfrac{100s(s+2)}{s^3 + 100s^2 + 699s + 1000}$

7-32 **(a)** $G_L(s) = \dfrac{0.995}{s(s + 0.895)}$ **(e)** $G_{dL}(s) = \dfrac{4.975}{s(s + 0.2225)}$

7-35 **(a)** $G_L(s) = \dfrac{0.2222}{s(s + 0.7888)}$

7-36 **(a)** $\dfrac{Y(z)}{R(z)} = \dfrac{0.0234z + 0.021904}{z^2 - 1.7953z + 0.8406}$

7-38 **(b)** Characteristic equation: $z^3 + 1.6805z^2 + 2.424z - 0.331 = 0$

Chapter 8

8-2 **(a)** $K > 0$: $\theta_1 = 135°$ $K < 0$: $\theta_1 = -45°$ **(e)** $\theta_1 = -108.435°$

8-4 **(a)** Breakaway-point Equation: $2s^5 + 20s^4 + 74s^3 + 110s^2 + 48s = 0$
Breakaway Points: -0.7272, -2.3887

8-5 **(h)** Intersect of Asymptotes: $\sigma_1 = -4$ Breakaway Points: 0, -4, -8

(i) Breakaway Points: -2.07, 2.07, $-j1.47$, $j1.47$

8-6 **(d)** $\zeta = 0.707$, $K = 8.4$

8-16 **(a)** $\sigma_1 = -1.5$, breakaway points: ($K > 0$) 0, -3.851

8-21 **(a)** Breakaway-point Equation: $2s^2 + 3(1 + \alpha)s + 6\alpha = 0$
For no breakaway point other than at $s = 0$, $0.333 < \alpha < 3$.

8-25 **(b)** The system is marginally stable for $-3.333 < K < 23.333$; unstable for all other values of K.

Chapter 9

9-1 **(c)** For $K = 100$, $\omega_n = 10$ rad/sec, $\zeta = 0.327$, $M_r = 1.618$, $\omega_r = 9.45$ rad/sec

9-2 **(b)** $M_r = 15.34$, $\omega_r = 4$ rad/sec, BW $= 6.223$ rad/sec

(e) $M_r = 1.57$, $\omega_r = 0$ rad/sec, BW $= 1.12$ rad/sec

9-5 Maximum $M_r = 1.496$, Minimum BW $= 1.4106$ rad/sec

9-9 **(a)** $\Phi_{11} = 270°$, $Z = 2$ **(b)** $\Phi_{11} = (Z - 1)180° = 180°$, $Z = 2$

(m) $\Phi_{11} = (Z - 0.5)180° = -90°$, $Z = 0$

9-10 **(h)** The system is stable for $-1/3 < K < 0$.

9-11 **(a)** The system is stable for $-25 < K < \infty$.

9-12 **(b)** $\Phi_{11} = (Z - 1)180° = -180°$, $Z = 0$.

9-14 **(a)** The system is stable for $K > 0.5$.

9-15 **(a)** The system is stable for $K > 0.5$.

(e) The system is unstable for all values of K.

9-18 **(a)** For stability, $N < 3$.

9-21 **(a)** For stability, $T_d = 1.47$ sec.

9-24 **(b)** Maximum $D = 15.7$ in.

9-28 **(e)** GM = 6.82 dB, PM = 50.27 deg.

(h) GM = infinite, PM = 13.4 deg.

9-29 **(c)** K can be increased by 28.71 dB.

9-30 **(d)** K must be decreased by -2.92 dB.

9-36 **(b)** PM = 2.65°, GM = 10.51 dB, $M_r = 17.72$, $\omega_r = 5.75$ rad/sec, BW = 9.53 rad/sec

9-38 **(g)** BW = 30 rad/sec

9-39 **(b)** $T_d = 0.1244$ sec

9-43 **(b)** GM = 30.72 dB, PM = infinite

9-46 The maximum number for N for stability is 7.

Chapter 10

10-3 **(a)** $K_P = 10$, $K_D = 0.09$

10-5 **(b)** $K_D \cong 0.00175$

10-7 **(a)** $K_D = 0.09$, PM = 93.8°, GM = ∞, $M_r = 1$, $BW = 1.41$ rad/sec

10-9 **(a)** $K_P = 1.1$, $K_I = 10$, $K_v = 10$, PM = 31.51°

10-13 **(a)** t_s = minimum (0.851 sec) when $K_P = 0.6$. $K_P < 4.9718$ for stability.

10-16 **(b)** $K_D = 399.85$, $K_P = 348.93$

10-20 $K_P = 50$, $K_D = 6$, $K_I = 100$

10-22 **(a)** Set $1/aT$ so that the pole of $G(s)$ at $s = -10$ is cancelled.

(b) Set the additional phase lead at 67 degrees. The new gain crossover frequency = 70 rad/sec.

10-25 **(a)** $R_2 = 2.65 \times 10^5$ gives a relative damping ratio of 0.65.

10-27 **(a)** $a = 1000$, $T = 0.0004$ **(c)** $a = 0.074$

10-30 **(b)** $\zeta_p = 1.222$, $\omega_n = 45$ rad/sec

10-34 $K = 59.3$

10-38 **(c)** $G_c(s) = \dfrac{1 + 0.0487s}{1 + 0.00348s}$

10-40 **(a)** $G_c(s) = \dfrac{1 + 15.63s}{1 + 117.13s}$

10-43 **(a)** $k_1 = 11$

10-45 **(a)** $k_1 = -64.84$, $k_3 = 20.34$

10-46 **(b)** $K_P = 2$, $K_I = 200$

Chapter 11

11-1 **(b)** $G_c(z) = (10T + 0.1) - 0.1z^{-1}$

11-3 **(d)** Physically realizable.

11-4 $T = 0.01$ sec, $G(z) = \dfrac{0.02(z + 1)(10.01z - 10)}{z(z - 1)^2}$

11-6 $G_c(z) = \dfrac{21.21(z - 0.9222)}{(z + 0.4344)}$

11-9 $k_1 = 0.058$, $k_2 = 0.035$

Index

D

Z